Human
Development

Human Development

Seventh Edition

James W. Vander Zanden
The Ohio State University

Revised by Thomas L. Crandell
and Corinne Haines Crandell
Broome Community College

McGraw-Hill

Boston   Burr Ridge, IL   Dubuque, IA   Madison, WI   New York   San Francisco   St. Louis
Bangkok   Bogotá   Caracas   Lisbon   London   Madrid
Mexico City   Milan   New Delhi   Seoul   Singapore   Sydney   Taipei   Toronto

McGraw-Hill Higher Education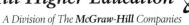

*A Division of The **McGraw-Hill** Companies*

HUMAN DEVELOPMENT, SEVENTH EDITION

Copyright © 2000, 1997, 1993, 1989, 1985, 1981, 1978 by The McGraw-Hill Companies, Inc. All rights reserved. Printed in the United States of America. Except as permitted under the United States Copyright Act of 1976, no part of this publication may be reproduced or distributed in any form or by any means, or stored in a data base or retrieval system, without the prior written permission of the publisher.

This book is printed on acid-free paper.

2 3 4 5 6 7 8 9 0 VNH/VNH 0 9 8 7 6 5 4 3 2 1 0

ISBN 0–07–229345–4

Editorial director: *Jane E. Vaicunas*
Executive editor: *Mickey Cox*
Editorial coordinator: *Stephanie Cappiello*
Senior marketing manager: *James Rozsa*
Senior project manager: *Gloria G. Schiesl*
Senior production supervisor: *Sandra Hahn*
Design manager: *Stuart D. Paterson*
Photo research coordinator: *John C. Leland*
Senior supplement coordinator: *Audrey A. Reiter*
Compositor: *ElectraGraphics, Inc.*
Typeface: *10/12 Stempel Garamond*
Printer: *Von Hoffmann Press, Inc.*

Cover/interior design: *Elise Lansdon*
Cover images: *Background photo Stock Boston, center image FPG International*
Photo research: *Connie Gardner Picture Research*

The credits section for this book begins on page 683 and is considered an extension of the copyright page.

Library of Congress Cataloging-in-Publication Data

Vander Zanden, James Wilfrid.
    Human development / James W. Vander Zanden. — 7th ed. / revised
  by Thomas L. Crandell and Corinne Haines Crandell.
      p.   cm.
    Includes bibliographical references and indexes.
    ISBN 0–07–229345–4
    1. Developmental psychology.   I. Crandell, Thomas L.
  II. Crandell, Corinne Haines.   III. Title.
BF713.V36   2000
155—dc21   99–22649
                              CIP

www.mhhe.com

Contents in Brief

Contents

Part Five

Middle Childhood: 7 to 12   261

Preface

Preface

The beginning of the new millennium is an exciting time for the field of human development. Researchers are increasingly moving toward a contextual perspective that is consistent with the complexity of development. While not losing sight of individual variation, developmentalists are expanding their horizons to encompass the environmental (social and ecological) context in which behavior occurs. Simultaneously, they are placing greater reliance on time-extended research designs and enlarging the breadth of their research objectives.

More developmental psychologists are reaching out to other disciplines and embracing a multidisciplinary, collaborative approach that draws on concepts and contributions from sociology, social psychology, anthropology, women's studies, biology, medicine, and social history. Moreover, researchers are studying ethnic populations that had been largely ignored a decade ago. Our cross-cultural knowledge base of human development is expanding, with the Internet providing nearly instant access to published empirical findings from research conducted around the world. The result is that the field has much to offer society in its efforts to cope with serious social problems and an ever-growing aging and diverse population. Individual students also benefit by seeing that the study of human development is contingent on a diverse body of knowledge that incorporates a variety of views and theoretical approaches. In sum, human development is emerging as a truly dynamic and relevant field for the twenty-first century.

We hope that students who read this textbook will find answers to their questions about their own lives, much as we have done in our research and writing of this book. It is our earnest desire that courses in human development and developmental psychology help people move toward Abraham Maslow's ideal and become self-actualized men and women. From such courses, they should acquire a new vision of the human experience and a sharpening of their observational and analytical skills, which can help them lead fuller, richer, and more fruitful lives.

Most of us share the belief that education is not the sum of 8, 12, 16, or even 20 years of schooling. Instead, it is a lifelong habit, a striving for growth and wise living. Education is something we retain when we have stored or sold our texts, recycled our lecture notes, and forgotten the minutiae we learned for an exam. Therefore, textbooks must present controversy and unanswered questions. Otherwise students will believe that facts are the stuff of education, and they will derive a false sense of security from cramming their heads full of information rather than refining their minds with thoughtful analysis. The stuff of human development is ultimately about real people living their lives in a real world, and many of the boxes in this edition of Human Development offer students an opportunity to think critically about social issues and how these issues relate to their personal lives and world.

James Vander Zanden's Personal Note from His Preface to the Sixth Edition

As I have noted in the prefaces to previous editions, I deem myself fortunate to be associated with the field of human development. It has afforded me a great many insights that have benefited my own life and that of my family. Several years prior to beginning the first edition, my wife became ill and subsequently died. Our younger son was then an infant and his older brother a toddler. Consequently, except for teaching part-time at Ohio State University, I dropped out of academic life for about five years and functioned more or less as a full-time parent. I found that researching and writing *Human Development* (and subsequently seeing it through later editions) offered profound help in the rearing of my sons. In the early editions, the boys were frequently about, playing in the yard or the living room while I worked on the book in the adjoining den. As is characteristic of youngsters, they were periodically in and out of the den on one matter or another. From time to time, I would take breaks and visit or play with them and their friends. Time has marched on. My sons are now young adults, have received their Ph.D. degrees in computer science (one from Cornell University at age 24 and the other from the University of Illinois at age 25), and are living happy, productive, and rewarding lives. Even though they are no longer home, we continue to share a warm, caring, and rich relationship.

As a male who reared his youngsters in a single-parent home, I find truth in the argument that equal opportunity for women in public spheres is severely impaired by a gender-role differentiation in which women are assigned primary responsibility for raising children. The child-rearing years are also the years that are typically most critical in the development of a career. During this period of the life span, professors secure tenure at good universities; lawyers and accountants become partners in top firms; business managers make it onto the fast track; and blue-collar workers find positions that generate high earnings and seniority. Rearing youngsters is time consuming and disruptive of the activities that commonly make for an orderly and successful career. It is a tragic commentary on our society that those individuals most immediately charged with caring for and raising children are penalized for doing so in countless social and economic ways, particularly in the workplace. I, too, encountered these difficulties. Yet, in hindsight, I would not exchange the rewards and satisfactions I found in parenthood for all the laurels offered by the academic community.

Seeing this textbook through its various editions has been a highly personal and satisfying experience for another reason. As a youngster, I experienced considerable abuse. Indeed, at 2½ years of age I underwent surgery to repair severe internal damage and bleeding inflicted by my father. Due to this and continuing abuse, it is hardly surprising that I had a troubled childhood and adolescence. Nor is it accidental that, in adulthood, I became intrigued by the study of human behavior and made it my career. *Human Development* is a testimony to my own search for answers and my dedication to the betterment of the human condition.

Organization and Focus

This textbook views human physical, cognitive, emotional, and social growth as blending in an unending, dynamic process. In terms of its approach to the study of the life span, *Human Development* emphasizes development in context. This approach focuses on the development of people within families and the larger ecological context implied by this theme. By examining the groundbreaking work of developmentalists such as Tiffany Field, Urie Bronfenbrenner, Lev Vygotsky, K. Warner Schaie, Daniel Levinson, and Paul and Margret Baltes, students will fully understand the complex network of developmental tasks that shape us as we move through the life span.

Much like the course of human life, this new edition reflects both continuity and change. Like previous editions, the seventh edition of Human Development features a chronological approach to studying the life span, and consists of 19 chapters. The first two chapters orient the student to the central research methods and the diversity of theories utilized in the study of human development. Chapter 3 examines beginnings: reproduction, heredity and genetics, and the prenatal period. Chapter 4 presents birth and the first two years of infant growth. Chapters 5 and 6 include infant cognitive, language, emotional, and social development. From Chapter 7, "Early Childhood," to Chapter 18, "Late Adulthood," each stage of the life span has been organized into two chapters: Physical, cognitive, and moral development are examined in the opening chapter, and emotional and social development follow in the subsequent chapter. Chapter 19 deals with dying and death.

Thinking Critically

As we have said, a course on human development should do more than provide students with a body of scientific findings. Rote memorization of definitions and facts does not do justice to the dynamic nature of this subject matter. We must encourage students to think critically and creatively about their own development and how it is shaped by the world around them. This text will provide students with a deeper understanding of the human experience and the variety of factors that directly or indirectly mold our life course.

These new abilities will not be limited only to the classroom, however. The challenging, real-life topics we discuss include assisted reproductive technologies for infertile couples, and genetic counseling and testing (Chapter 3); early intervention services for infants born at risk (Chapter 4); differing theories of language acquisition (Chapter 5); choosing a child-care provider (Chapter 6); nutrition and health issues for young children, early signs of developmental delays in children, and moral development in childhood (Chapter 7); the special needs of gifted and talented children (Chapter 8); typical health issues in middle childhood, assessment of children's intelligence (including emotional intelligence), and school programs for children at risk (Chapter 9); nurturing a healthy self-concept, the impact of divorce on children and the impact of living in a single-parent family, and after-school care and supervision (Chapter 10); adolescent career development and choice (Chapter 12); determining whether an adolescent has an alcohol or drug problem (Chapter 12); the cohabitation lifestyle, and Levinson's stages of a man's life and stages of a woman's life (Chapter 14); physical and health changes in middle adulthood, sexual functioning in middle adulthood and the increasing risk of

AIDS and STDs, maximizing cognitive abilities in middle adulthood, and a schedule of physical checkups at midlife (Chapter 15); living in a stepfamily (Chapter 16); exercise and longevity (Chapter 17); faith and aging (Chapter 18); and preparing a living will, a guide for professionals and families coping with a terminally ill person who requests physician-assisted suicide, warning signs of suicide at all life stages, and the controversial issue of near-death experience (Chapter 19).

Commitment to Diversity

Past editions of *Human Development* have been lauded for their sensitivity and coverage of issues of race, class, gender, and ethnicity. The seventh edition continues this legacy by integrating information on cross-cultural, minority, and gender differences wherever possible. The seventh edition of *Human Development* utilizes an integrative approach to demonstrate our commitment to diversity, as well as addresses some issues in boxes entitled "Human Diversity." This edition, in particular, presents recent findings of developmental research that shed light on important issues for growing populations of Hispanic Americans and Asian Americans. The special needs and mandated services for children born at risk are also presented, and the realities of an inclusive lifestyle for those who are differently abled are covered across the life course. Also, this edition addresses the development of lesbian and gay individuals from adolescence into late adulthood.

Attention to both classic and emerging issues in human development is an imperative component of our task as teachers and authors. Specific examples of this approach include discussions of developing an emotional bond with children born at risk (Chapter 6); changing demographics and implications for health of minority children (Chapter 7); preparing for children with differences in early childhood settings (Chapter 8); individual differences in children's cognitive development, bilingualism and ESL instruction, and the diversity of family structures and their impact on child development (Chapter 9); adoption of children from other cultures (Chapter 10); cultural practices of female circumcision, international comparison of adolescent intellectual performance, and cultural aspects of adolescent identity formation (Chapter 11); adolescent sexual orientation and behaviors (Chapter 12); examination of young adult health across cultures, young adult gay/lesbian/bisexual attitudes and behaviors, and traditional and nontraditional college students coping with stress (Chapter 13); diversity in adult lifestyle options, lesbian and gay parenting, cultural dislocation of first-generation immigrants, and arranged marriages across cultures (Chapter 14); strategies for success for middle-age college students

(Chapter 15); life without a middle age (Chapter 16); theories of biological aging (Chapter 17); ethnic diversity and the aging population, and lesbian and gay elderly adjustment to aging (Chapter 18); faith and facing death, and cross-cultural perspectives on dying, death, and grief (Chapter 19).

New to the Seventh Edition

This edition has been reorganized to make it even easier for instructors and students to use. We had four main goals in revising this edition:

1. To reorganize the chapters to achieve greater continuity about development across the life span. Concepts, issues, theories, and research findings are grouped more precisely into physical, cognitive, moral, emotional, or social development for each stage of the life span.
2. To present the textual material and illustrations in a more readable format, supported by additional applications to real-world situations. The use of critical thinking questions as advance organizers, the placement of headings, and the section summary questions make the information more easily assimilated and memorable.
3. To present students with the most up-to-date research in the many domains of study across the human life span. More specifically, we have incorporated recent research findings across the adult stages of development and relevant cross-cultural studies across the life span whenever pertinent.
4. To introduce students to new information and strategies for managing many experiences and challenges that will face them across the life course.

Expanded Section Coverage on Crucial Issues in Life Span Development

In addition to including coverage of such topics as early intervention services for children born at risk (Chapter 7) and diversity in adult lifestyle options (Chapter 14), the seventh edition of *Human Development* is unrivaled in its detailed coverage of numerous critical issues. Each chapter features new research findings and updates coverage of classic studies in human development. This unique quality manifests our commitment to students' learning and overall breadth of knowledge. We begin by addressing the changing conception of age and aging (Chapter 1). Lev Vygotsky's sociocultural theory has been added (Chapter 2). Information on genetics and assisted reproductive technologies has been updated (Chapter 3). The research findings have been updated on the increasing numbers of babies being born at risk due to substance abuse and the significant impact of poverty (Chapter 4). Findings about the impact of an increasing number of children growing up in fatherless homes are presented (Chapter 4). There is expanded coverage of information pertinent to raising a child with developmental delays, as well as understanding the needs of a child who is determined to be intellectually gifted or talented (Chapters 8 and 9). More recently there is a research focus specific to emotional health and its relationship to cognitive growth and job satisfaction, healthy social relationships and family life, and overall life satisfaction (Chapters 6, 8, 16 and 18); the increasing number of Hispanic American and Asian American children who need extensive services from the educational system (Chapters 7 and 9); single parenthood and its association with poverty (Chapter 14); changing trends in teenage substance abuse and changing trends in teenage sexual behaviors (Chapter 11); strategies for middle-aged college students (Chapter 15); adult diversity of lifestyle options and delay of marriage (Chapter 13); the increasing number of Americans reaching middle adulthood and late adulthood (Chapters 15 and 17); the challenges of the "'sandwich" generation at midlife (Chapter 16); the increasing longevity of life (Chapter 17); grandparents parenting grandchildren (Chapter 18); the hospice movement (Chapter 19); physician-assisted suicide (Chapter 19); and the worldwide AIDS epidemic (Chapters 11, 13, 15, and 19).

Positive Approach to Adulthood and Aging

The text features an extensive, honest discussion of the aging process, from young adulthood through late adulthood. Topics examined include the latest research and theory on biological aging, Alzheimer's disease, memory and cognitive functioning, Elderhostel programs for the aging, theories of adjustment, sexuality in late adulthood, institutional and adult day care, psychosocial aging, faith and aging, and the bereavement of widows and widowers. Many of the issues on aging are presented with cross-cultural views.

Pedagogy and Design of This Text

We have incorporated a number of in-text learning aids throughout this edition, including chapter previews, critical thinking questions, in-text review questions, key terms, chapter topical summary statements, and a glossary of terms at the end of the book. The design of the text has been updated with a carefully planned color and background schema that enhances student learning.

The *chapter preview* serves as a cognitive bridge, or advance organizer, between the concepts learned in the previous chapter and the new concepts to be learned in the current chapter. The *critical thinking questions* were carefully and creatively devised to encourage students to challenge their own beliefs about critical issues of human development relevant to that chapter. *In-text review questions* were provided at critical intervals to provide students with an opportunity to assess and review what they just read and to serve as a positive reinforcer. *Updated illustrations* were carefully selected by content and strategically placed to serve as a visual schema and to maintain interest for the learner. The *segue* at the end of each chapter serves as a post organizer and helps the readers associate and relate information learned and prepares the students for new information to follow in the next chapter. *Topical summary statements* provide an organizational framework to help students understand and integrate the material that has been learned. These statements can also be used by students who choose to look at the chapter in a holistic fashion, prior to reading the specific content of the chapter. The *key terms* provide students with the basic vocabulary to be learned in each chapter and are associated with the most important concepts. The lists of *Internet web sites* were carefully selected to provide the reader with resources to follow up on issues of interest or discussion. The *end-of-the-book glossary* provides definitions of the key terms in the text for easy referral.

Practical and Informative Boxed Material

In an effort to highlight the most current issues in a comprehensive and accessible manner, three different kinds of boxes are carefully woven into the chapter narrative. The "Information You Can Use" boxes provide practical information that can help students make better-informed decisions as they encounter real-life situations. The "Human Diversity" boxes further examine special topics related to issues of race, class, gender, ethnicity, and culture. "Further Developments" boxes take an in-depth look at specific issues across the life span.

The Most Current Research and Theory

The seventh edition of *Human Development* includes comprehensive discussions of the ground broken by inspirational researchers and theorists such as Bruner, Vygotsky, Field, Maccoby, Elkind, Gilligan, Ainsworth, Sroufe, Kübler-Ross, Kagan, Belsky, Baumrind, Izard, and many others. By featuring the most current findings in research and theory, we may truly see evidence of our increased understanding of development over the life span.

New Photo Program

In thumbing through this new edition, one will undoubtedly note the beauty and creativity of our new photo program. The photos and illustrations in *Human Development* display our continued commitment to clarification of concepts and issues of diversity. Sensitivity to race, class, gender, ethnicity, and ability (or disability) is of tantamount importance, and this is reflected in the photos we have chosen for this edition. We are especially pleased to have been granted permission to use an official photo of the McCaughey septuplets.

New References

The seventh edition is both a useful teaching tool and a thoroughly updated resource for students and instructors. The references in each chapter have been streamlined to allow for easier reading of the text. Hundreds of new references have been added to the seventh edition of *Human Development* and are integrated throughout the text. Additionally, at the conclusion of each chapter, the reader will be able to explore up-to-date research findings and relevant professional organizations by connecting to web sites on the Internet. These web sites can be hot-linked through the text web site at http://www.mhhe.com/crandell7.

Supplements

The supplements listed here may accompany the seventh edition of *Human Development.* Please contact your local McGraw-Hill representative for details concerning policies, prices, and availability as some restrictions may apply. The supplement package to accompany the seventh edition of *Human Development* completes the necessary resources for teaching and learning from this text. The *Student Study Guide* has been revised by Tom Crandell and Craig Vivian, a Ph.D. candidate at Cornell University. Students are given the resources to better understand the material in each chapter through outlines, learning objectives, matching questions, factual multiple-choice questions, conceptual multiple-choice questions, and essay questions. Internet resources are also available for an additional place to look for further information or paper topics.

The *Instructor's Manual* and the *Test Bank* are now available as separate supplements, both written by Jada Kearns of Valencia Community College. The *Instructor's Manual* includes resources to benefit any classroom, such as learning objectives, chapter summaries, lecture topics, classroom activities, student projects, updated and expanded video suggestions, and a list of Internet sites. The *Test Bank* includes over 1,500 factual, conceptual, and applied multiple-choice and essay questions. The *Test Bank* is also available electronically to aid with the creation of exams. Our web site at http://www.mhhe.com/developmental provides more information on this title as well as text specific resources and web links.

Acknowledgments

In truth, authors have but a small part in the production of textbooks. Consider the thousands upon thousands of researchers who have dedicated themselves to the scholarly investigation of human behavior and life-span development. Consider the labors of countless journal editors and reviewers who assist them in fashioning intelligible reports out of their research findings. And consider the enormous effort expended by the personnel of research-grant agencies and reviewers who seek to funnel scarce resources to the most promising studies. Indeed, a vast number of scholars across the generations have contributed to our contemporary reservoir of knowledge regarding human development. Textbook authors simply seek to assemble the research in a coherent and meaningful manner. More specifically, a number of reviewers helped us shape and guide the manuscript into its final form. They appraised the clarity of expression, technical accuracy, and completeness of coverage. Their help was invaluable, and we are deeply indebted to them. For the seventh edition, we extend thanks to:

Stephen Burgess
Southwestern Oklahoma State University

Deborah M. Cox
Madisonville Community College

Rhoda Cummings
University of Nevada, Reno

Robert J. Griffore
Michigan State University

Patricia E. Guth
Westmoreland County Community College

Harry W. Hoemann
Bowling Green State University

Russell A. Isabella
University of Utah

Jada D. Kearns
Valencia Community College

Joyce Splann Krothe
Indiana University

Timothy Lehmann
Valencia Community College

Pamela A. Meinert
Kent State University

Linda W. Morse
Mississippi State University

Joyce Munsch
Texas Tech University

Ana Maria Myers
Polk Community College

Gail Overbey
Southeast Missouri State University

Lisa Pescara-Kovach
University of Toledo

Robert F. Schultz
Fulton Montgomery Community College

Elliot M. Sharpe
Maryville University

Laurence Simon
Kingsborough Community College

Lynda Szymanski
College of St. Catherine

Robert S. Weisskirch
California State University Fullerton

Peggy Williams-Petersen
Germanna Community College

For the sixth edition, our thanks go to:

Janet Coy
St. Mary's College of Maryland

Linda Hilliard
Midlands Technical College

Harry Hoemann
Bowling Green State University

Carol Laman
Houston Community College

Timothy Lehmann
Valencia Community College

Jennifer Trapp Myers
University of North Carolina at Chapel Hill.

We also want to express gratitude to Cecil B. Nichols, Miami-Dade Community College, for his input and kind remarks. For the fifth edition, thanks go to:

Dana H. Davidson
University of Hawaii

Timothy Lehmann
Valencia Community College

Bonnie R. Seegmiller
Hunter College

Paul A. Susen
Mount Wachusett Community College

Dennis Thompson
Georgia State University

Joseph M. Tinnin
Richland College

Alvin Y. Wang
University of Central Florida.

A special note of thanks to Craig Vivian, from Cornell University, a talented and creative writer and researcher who has made meaningful contributions to this seventh edition by reviewing the sections on cognitive and moral development and some of the diversity issues. Our thanks also go to Lesa Carter, from Cornell University, who reviewed much of the text on diversity issues relevant to African Americans. Thank you to Bev Frost, for providing us with the pleasant office space for several months of solitude for writing. Our dedicated research assistants and information resource specialists, Karen Pitcher and Larry Jenkins, were invaluable in providing us with current empirical research findings on topics across the entire life span and reminded us that attention to detail is paramount. We are extremely grateful for the courteous and meticulous service provided by Gary Hitchcock and Howard Nickerson, our college's Copy Center professionals, who contributed to our ability to meet many deadlines (and reminded us that laughter is the best medicine). We also appreciate the sunny disposition of our mailroom staff, Ethel Buchanan and Joan Drew, who packaged and prepared documentation for delivery to our various editors across the country and without whom we would not have been able to meet our deliverable dates to get this text ultimately into the hands of the reader.

We are indebted to everyone at McGraw-Hill who helped to produce this book and give special thanks to the following professionals in the domain of publishing: to Jane Vaicunas, editorial director, for supporting our work over the past several years; to our executive editor, Mickey Cox, for encouraging us through these long months and for being willing to make the tough decisions; to Stephanie Cappiello, our developmental editor, for guiding us through the whitewaters of producing a textbook and maintaining poise and tactful guidance throughout the project; to Gloria Schiesl, for overseeing the project through the production process and keeping us on schedule; to Connie Gardner, for bringing to visual life the concepts we knew would enhance the understanding of the material in this text; to our copy editor, Wendy Nelson, and to ElectraGraphics for improving the communication value of this text and helping us achieve the overall educational goals we set, and to Diane Kraut, for searching for the necessary permissions from a variety of authors and sources. This project has been a total team undertaking at all times. We sincerely appreciate the encouragement and enthusiasm each person brought to this undertaking and the professional competence each one exhibited in bringing this new edition to fruition.

Finally, we wish to acknowledge our children, Jim, Colleen, Becky, and Patrick, who taught us the real joys of life, and to whom we dedicate this book.

Thomas L. Crandell

Corinne H. Crandell

About the Authors

Authors

We bring an uncommon blend of academic, professional, and personal experiences to this text on your behalf.

Corinne Haines Crandell Having earned a B.S. degree from the University Center at Albany and an M.S. degree from the State University of New York at Oneonta in psychology and counseling, Corinne has had a variety of instructional experiences at the community college level teaching psychology classes for many years, has been a college counselor, has co-authored developmental psychology study guides, instructor's manuals, and computerized study guides for the past 10 years, and developed the first distance learning course in developmental psychology for the SUNY Learning Network, offered over the Internet since 1997. Recently she supervised student interns in Broome Community College's human services program at nearly 40 social service agencies. For 5 years she taught in a middle school and worked with children with learning disabilities in grades 4 through 8. Additionally, she was the coordinator of the gifted and talented program for a private school district. Corinne has coached and judged in the regional Odyssey of the Mind program, has been a board member for 5 years at our local Association for Retarded Citizens, and currently teaches religious education to high school students.

Thomas L. Crandell After earning an M.A. in counseling psychology from Scranton University, Tom taught a variety of undergraduate psychology courses at Broome Community College and worked as a college counselor for several years. At age 34, he continued his formal education at Cornell University in pursuit of a Ph.D. in psychology and education. While at Cornell, Tom received a research assistantship sponsored by the Office of Naval Research, and he subsequently helped to initiate and develop one of the most productive reading research programs in the country. His experimental findings on cognitive styles and reader comprehension have been adopted by researchers and practitioners worldwide. He first won international recognition when his doctoral dissertation was selected as one of the top five in the country by the International Reading Association. In addition to being a professor, he has been a consultant in educational, business, and legal settings for the past 20 years and has authored numerous articles on the design of instructional materials for ease of reading. Many of these design strategies have been incorporated into the seventh edition of this text. In 1996 he earned the "Distinguished Article of the Year" award in the Frank R. Smith Competition by the *Journal for the Society of Technical Communications.* He is a long-standing member of the American Psychological Association and has become an expert and consultant in the design of instructional materials, reading research, and evaluation. Tom has also coached youth basketball and soccer and has taught adult religious education courses and marriage preparation classes through his church. He has maintained a healthy lifestyle with a passion for basketball and golf throughout his years of professional growth and development.

We have been teaching students from the community college to the graduate level in a variety of professional capacities for nearly 30 years. During this time, we have seen our student population become more diverse, composed of a blend of traditional and nontraditional learners from rural, urban, and distant cultures. As our student population began to include students with learning disabilities and those without high school diplomas, we prepared ourselves to understand the individual learning needs of our students. Tom has become an educational consultant and expert on the role of cognitive styles in classroom instruction, and Corinne has completed additional graduate studies in reading, special education, and learning disabilities.

As a parent and a stepparent of a blended family, we bring to this text a wealth of knowledge and understanding about the issues facing single parents, stepfamilies, and the emotional needs of children (and parents) living in a blended family. Becky, nearly 30 and our daughter/stepdaughter with Down syndrome, has enriched our lives in countless ways. Her brother Jim has become a special education teacher, her sister Colleen has become a caring physician, and her younger brother Patrick is a high school student who has learned compassion for people with differences. Tom has developed and taught a course on intelligence and mental retardation to over 2,000 college students who have become special education teachers, psychologists, sociologists, social workers, clergy, nurses, physical therapists, speech therapists, recreational therapists, and informed parents.

Part One

The Study of Human Development

Knowledge about human development provides us with the opportunity to improve the human condition and to live life more satisfactorily. We begin in Chapter 1 with an overview of how a diversity of social science and life science researchers (collectively known as developmentalists) approach the monumental task of studying humans over the course of the life span. Our discussion includes the goals of the scientific community, the recognized framework for studying the life span, what aspects of development warrant extensive examination, and what scientific methods are used to conduct research with human subjects. Chapter 2 discusses the major developmental theories over the past century. Over the past 100 years, social scientists, biologists, and chemists focused on studying discrete aspects of human development. Earlier introspective methods about subconscious experience and contemporary measurable evidence about microscopic genetic codes, neurons, and hormones all contribute to our understanding of the human condition. Many contemporary researchers are focusing on how to integrate theory and scientific findings from across cultures into a more meaningful whole about human development.

Chapter 1 Introduction

Critical Thinking Questions

1. Developmental change takes place in three fundamental domains: the physical, the cognitive, and the emotional-social. Which domain has been most important for your becoming who you are? Will any one of the domains become more important as you get older?

2. Make a list of three aspects of yourself that have changed over the last ten years and three that have remained constant. How do you feel about both the "dynamic" and the "static" aspects of yourself?

3. If someone had researched your personal development over time, where would they have noticed the most change? the least change? If they continued their research, where would they probably see the most and least amount of change over the next ten years, in your opinion?

4. If we could answer most of the important questions about human development by continuously studying ten individuals who interact with each other from birth until death, would the knowledge gained from the study justify keeping them isolated from the rest of the world for their entire lives?

Human development can be described as a process of becoming someone different while remaining in some respects the same person over an extended period of time. The goals of scientists who study development are to describe, explain, predict, and have the ability to control developmental changes. Development takes place in three essential areas—physical, cognitive, and emotional-social. In other words, you develop when your body, mind, and social relationships change. These changes occur through growth, maturation, and learning. But many factors— institutions, society, and family—affect individuals and can help or hinder their personal development. The ecological approach to development examines the mutual accommodations between the developing person and four levels of environmental influence: the microsystem, the mesosystem, the exosystem, and the macrosystem.

Age is one of the most important indicators of what an individual should be doing. Over the last century, research on development has focused on several areas, including emotion, cognition, self, behavior, thought, and nature. Research is essential to understanding human development, and there are diverse methods for obtaining analyzable information. Valid research findings often help improve the quality of life over the life span.

The Major Concerns of Science

The most incomprehensible thing about the world is that it is comprehensible.

Albert Einstein

A sign posted in a Western cowboy bar says: "I ain't what I ought to be. I ain't what I'm going to be, but I ain't what I was."

This thought captures the sentiment that lies behind much contemporary interest in the study of human development. It is hoped that with knowledge, we will be able to lead fuller, richer, and more fruitful lives. Knowledge offers us the opportunity to improve the human condition by helping us to achieve self-identity, freedom, and self-fulfillment.

Continuity and Change in Development

The motto in the bar directs our attention to still another fact—that to live is to *change.* Indeed, life is never static but always in flux. Nature has no fixed entities, only transition and transformation. According to modern physics—particularly quantum mechanics—the objects you normally see and feel consist of nothing more than patterns of energy that are forever moving and altering. From electrons to galaxies, from amoebas to humans, from families to societies, every phenomenon exists in a state of continual "becoming." The fertilized egg you developed from was smaller than the period at the end of this sentence. All of us undergo dramatic changes as we pass from the embryonic and fetal stages through infancy, childhood, adolescence, adulthood, and old age. We start small, grow up, and grow old, just as countless generations of our forebears have done.

Change occurs across many dimensions—the biological, the psychological, and the social-emotional. Life-span perspectives on human development focus on long-term sequences and patterns of change in human behavior. This perspective is unique in tracing the ways people develop and change across the life span.

Contradictory as it may seem, life also entails *continuity.* At age 70 we are in many ways the same persons we were at 5 or 25. Many aspects of our biological organism, our gender roles, and our thought processes carry across different life periods. Features of life that are relatively lasting and uninterrupted give us a sense of identity and stability over time. As a consequence of such continuities, most of us experience ourselves not as just so many disjointed bits and pieces but rather as wholes—larger, independent entities that possess a basic oneness—and much of the change in our lives is not accidental or haphazard.

Development: A Definition

Scientists refer to the elements of change and constancy over the life span as *development.* **Development** is defined as the orderly and sequential changes that occur with the passage of time as an organism moves from conception to death. Development occurs through processes that are biologically programmed within the organism and processes of interaction with the environment that transform the organism.

What we have been saying adds up to this: Human development over the life span is a process of becoming something different while remaining in some respects the same. Perhaps what is uniquely human is that we remain in an unending state of development. Life is always an unfinished business, and death is its only cessation.

Traditionally life-span development has primarily been the province of psychologists. Most commonly the field is called *developmental psychology* or, if focused primarily on children, *child development* or *child psychology.* Psychology itself is often defined as the scientific study of behavior and mental processes. **Developmental psychology** is the branch of psychology that deals with how individuals change with time while remaining in some respects the same. Child psychology is the branch of psychology that studies the development of children.

The field of life-span development has expanded to include not only psychology but biology, women's studies, medicine, sociology, gerontology, and anthropology. A multidisciplinary approach stimulates fresh perspectives and advances in knowledge. Researchers increasingly welcome aid and collaboration from any qualified person, whether or not that person is in the same discipline.

What Are the Goals of Developmental Psychologists?

Scientists who study human development focus on four major goals:

1. To *describe* the changes that typically occur across the life span. When, for instance, do children generally begin to speak? What is the nature of this first speech? Does speech alter with time? In what sequence does the average child link sounds to form words or sentences?
2. To *explain* these changes—to specify the determinants of developmental change. What behaviors, for instance, underlie the child's first use of words? What part does biological "pretuning" or "prewiring" play in the process? What is the role of learning in language acquisition? Can the process

Human Diversity

Rethinking Women's Biology

Over the past fifteen years we have seen the emergence of "identity politics." Marginalized groups are staking out claims of equality based on the differences between groups and not on a minority's similarities to dominant social groups (Bem, 1998). These ideas have also surfaced in academe as researchers are challenged to rethink their approaches to describing the world objectively. As you read the discussion below, think about how your own development might have been different if you had not been "pushed" quite as hard by society to become "male" or "female." Then apply some of these ideas to race and class to see the implications of this statement: "It takes a whole village to *develop* a child."

Women's biology is a social construct and a political concept, not just a scientific matter, and this is meant in at least three ways. The first can be summed up in Simone de Beauvoir's dictum "One isn't born a woman, one becomes a woman." This does not mean that the environment shapes us; rather, it means that the concepts "woman" and "man" are social constructs that little girls and boys try to fit as they grow up. Some of us are better at it than others, but we all try. Our efforts have biological as well as social consequences—in fact, this is a false dichotomy because our biological and social attributes are interrelated. How active we are, what clothes we wear, what games we play, what and how much we eat, what kinds of schools we go to, what work we do—all of these affect our biological and social being in ways we cannot sort out. In this sense, one isn't born a woman or man, one becomes one.

The concept of women's biology is socially constructed, and political, in a second way: Women have not been the dominant definers of their biology. Women's biology has been described mainly by physicians and scientists who, for historical reasons, have been mostly economically privileged, university-educated men; and these men have had strong personal and political interests in describing women in ways that make it appear "natural" for women to play roles that are important for men's well-being, personally and as a group. Men's self-serving, ideological descriptions of women's biology date back at least to Aristotle. Present-day scientists and their theories continue to characterize women as weak, emotional, and at the mercy of raging hormones; they "construct" the female as being centered around the functions of her reproductive organs. No one has suggested that men are valuable only for their reproductive capacity, but throughout history women have been looked on as though they were walking ovaries and wombs. Some cultures of the world today continue to value—or devalue—a woman for her ability to produce sons as the father's heirs,

Mary Whiton Calkins, 1863–1930, Early Developmentalist and First Female President of the APA. Mary Calkins, from Buffalo, New York, attended Smith College in Massachusetts in 1880. She then trained at Harvard under the direction of William James (though she was not allowed to register as a student) and set up an experimental lab and taught the first experimental psychology course at Wellesley College (which hired only female instructors). Though she wrote a scholarly thesis in 1896 on memory of numerals with color and sat for the Ph.D. exam at Harvard and performed brilliantly, she was denied the degree. She was an early pioneer in human development and is known for her valuable research on both memory and the psychology of the self. She also published a text in introductory psychology, was elected by her colleagues in 1905 as the first female president of the American Psychological Association, and in 1918 was elected the first woman president of the American Philosophical Association.

and in some cultures women are still considered the "property" of the husband.

At the turn of the twentieth century, when American women tried to enroll in colleges and universities, scientists originally claimed that women could not be educated because their brains were too small. When that claim became untenable, scientists granted that women could be educated the same as men but questioned whether women should

Continued on page 6

Continued from page 5

be—whether it was "good" for women. They based their concerns on the claim that girls needed to devote much energy to establishing the proper functioning of their ovaries and womb; and if they divert this energy to their brains by studying, their reproductive organs would shrivel, they would become sterile, and the human species would die out. This view was held by most men and women, and women who attempted to develop their intellectual capacities endured obstacles and ridicule while paving the way for other women who followed. Now, a century later, women earn approximately half of all doctorates in psychology.

The "scientific" logic that barred women from intellectual and occupational pursuits was steeped in gender and class prejudice. The notion that women's reproductive organs need careful nurturing was used to exclude upper-class girls and young women from higher education. But it was not used to spare the working-class, poor, or black women who were laboring in the factories and homes of the upper class. If anything, these women were said to have too many children. In fact, the poor woman's ability to have many children despite working so hard was taken as evidence that she was less highly evolved than upper-class women.

Our concepts of "female" and "male" are said to be socially and politically constructed because any society's interpretation of what is and is not normal naturally affects what its females and males do and become. Thus norms—standards of behavior evident in the majority—not merely describe how we are but prescribe how we *should be*.

be accelerated? What factors produce language and learning difficulties?

3. To *predict* developmental changes. What are the language capabilities of a 6-month-old infant likely to be at 14 months of age? Or what are the expected consequences for language development if a child suffers from an inherited disorder?

4. To be able to use their knowledge to intervene in the course of events in order to *control* them (Miller, 1995; Kipnis, 1994). For example, researchers have found that if infants are put on a special diet after birth, intellectual impairment from one inherited disorder (PKU) can often be minimized (Welsch et al., 1990).

But even as scientists strive for knowledge and control, they must continually remind themselves of the dangers described by eminent physicist J. Robert Oppenheimer (1955): "The acquisition of knowledge opens up the terrifying prospects of controlling what people do and how they feel." We return to the matter of ethical standards in scientific research later in this chapter; *Human Diversity: "Rethinking Women's Biology"* looks at the negative implications for females of the fact that, historically, scientists have mainly been males. The four scientific goals—*describing, explaining, predicting,* and having the ability to *control* developmental changes—should be kept in mind as you examine the different domains and theories of human development in this book.

Questions

How is development defined? What are the four main goals of the study of human development?

A Framework for Studying Development

Life is never a material, a substance to be molded. If you want to know, life is the principle of self-renewal; it is constantly renewing and remaking and changing and transfiguring itself.

Boris Pasternak, *Doctor Zhivago*

If we are to organize information about human development from a variety of perspectives, we need some sort of framework that is both meaningful and manageable. Studying human development involves considering many details simultaneously. A framework provides us with *categories* for bringing together bits of information that we believe are related to one another. Categories let us simplify and generalize large quantities of information by clustering certain components. A framework, then, helps us find our way in an enormously complex and diverse field. One way to organize information about development is in terms of four basic categories:

- The major domains of development
- The processes of development
- The context of development
- The timing of developmental events

Let's look at each of these categories to see how they fit within a given framework.

Major Domains of Development

Developmental change takes place in three fundamental domains: the physical, the cognitive, and the emotional-social. Think how much you have changed in the years since you first entered school. Your body, the way you think, and how you interact with others are aspects of "you" that have undergone transformations and will continue to do so.

Physical development involves the changes that occur in a person's body, including changes in weight and height; in the brain, heart, and other organ structures and processes; and in skeletal, muscular, and neurological features that affect motor skills. Consider, for instance, the physical changes that take place at adolescence, which together are called *puberty.* At puberty young people undergo revolutionary changes in growth and development. Adolescents suddenly catch up with adults in size and strength. Accompanying these changes is the rapid development of the reproductive system and the attainment of reproductive capability—the ability to conceive children.

Cognitive development involves those changes that occur in mental activity, including changes in sensation, perception, memory, thought, reasoning, and language. Again consider adolescence. Young people gradually acquire several substantial intellectual capacities. Compared with children, for instance, adolescents are much better able to think about abstract concepts such as democracy, justice, and morality. Moreover, young people become capable of dealing with hypothetical situations, and they achieve the ability to monitor and control their own mental experiences and thought processes.

Emotional-social development (also referred to as psychosocial development) includes changes in an individual's personality, emotions, and relationships with others. All societies distinguish between individuals viewed as children and individuals viewed as adults, and our relationships with children usually are qualitatively different from the relationships we have with adults. Adolescence is a period of social redefinition in which young people undergo changes in their social roles and status. Our society distinguishes between people who are "underage," or minors, and those who have reached the age of majority, or adults. Adults are permitted to drive cars, to drink alcohol, to serve in the military, and to vote. How each of us becomes a unique adult can be seen as the result of interaction between the personal "self" and the social environment we find ourselves occupying. As we will see in Chapter 11, some societies recognize adolescence or entry into adulthood through a special initiation ceremony—a rite of passage.

Even though we are differentiating these domains of development, we do not want to lose sight of the unitary nature of the individual human. Physical, cognitive, and emotional-social factors are intertwined in every aspect of development. Indeed, scientists are increasingly aware that what happens in any one domain depends largely on what happens in the others.

Processes of Development

Development meets us at every turn. Infants are born. The jacket the the 2-year-old wears in the spring is outgrown by winter. At puberty, youth exhibit a marked spurt in size and acquire various secondary sexual characteristics. Individuals commonly leave their parents' homes and set out on careers, establish families of their own, see their own children leave home, retire, and so

Initiation Ceremonies and Religious Rites of Passage Some societies induct their youth into adult status by means of initiation ceremonies. The youth know that if they accomplish the goals set forth for them, they will acquire adult status. This youth is accomplishing his Bar Mitzvah.

on. The concepts of *growth, maturation,* and *learning* are important to our understanding of these events.

Growth takes place through metabolic processes from within. One of the most noticeable features of early development is the *increase in size* that occurs with age. The organism takes in a variety of substances, breaks them down into their chemical components, and then reassembles them into new materials. Most organisms get larger as they become older. For some organisms, including human beings, growth levels off as they approach sexual maturity. Others—many plant and fish forms—continue the growth process until they die.

Maturation is another aspect of development. It concerns the more or less automatic unfolding of biological potential in a set, irreversible sequence. Both growth and maturation involve biological change. But whereas *growth* refers to the increase in the number of an individual's cells, maturation concerns the development of the individual's organs and limbs in relation to their ability to function. In other words, maturation reflects the unfolding of genetically prescribed, or "preprogrammed," patterns of behavior. Such changes are relatively independent of environmental events, as long as environmental conditions remain normal. As we will see in Chapter 4, an infant's motor development after birth follows a regular sequence—grasping, sitting, crawling, standing, walking. Similarly, at about 10 to 14 years of age, puberty brings many changes, including ovulation in girls and the production of live sperm in boys, providing the potential for reproduction.

Learning is still another component of development. It is the more or less permanent modification in behavior that results from the individual's experience in the environment. Learning occurs across the entire life span—in the family, among peers, at school, on the job, and in many other spheres. Learning differs from maturation in that maturation typically occurs without any specific experience or practice. Learning, however, depends on both growth and maturation, which underlie a person's *readiness* for certain kinds of activity, physical and mental. The ability to learn is clearly critical, for it allows each of us to adapt to changing environmental conditions. Hence, learning provides the important element of flexibility in behavior.

Questions

Can you determine what should be considered a normal relationship among growth, maturation, and learning? For example, if the learning component develops quickly relative to the growth and maturation components, we would call the child "precocious." What might happen if a child matures too quickly?

As we will emphasize in Chapter 2, the biological forces of growth and maturation should not be contrasted with the environmental forces of learning. Too often the *nature-nurture controversy* is presented as a dichotomy—nature *or* nurture. Rather, it is the *interaction* between heredity and environment that gives an individual her or his unique characteristics. We find that as we interact with the world about us—as we act upon, transform, and modify the world—we in turn are shaped and altered by the consequences of our actions (Kegan, 1988; Vygotsky, 1978; Piaget, 1963). *We literally change ourselves through our actions.* For instance, as we pass through life, our biological organism is altered by dietary practices, activity level, alcoholic and drug intake, smoking habits, illness, exposure to X rays and radiation, and so on. Furthermore, as many of us enter school, finish school, seek a job, marry, settle on a career, have children, become grandparents, and retire, we arrive at new self-conceptions and identities. In these and many other ways, we are engaged in a lifetime process in which we are forged and shaped as we interact with our environment (Tudge & Winterhoff, 1993). In brief, development occurs *throughout* our lifetime—the prenatal period, infancy, childhood, adolescence, adulthood, and old age.

The Context of Development

I have made a ceaseless effort not to ridicule, not to bewail, nor to scorn human actions, but to understand them.

Spinoza

If we are to understand human development, we must consider the environmental context in which it occurs. In his **ecological approach** to development, Urie Bronfenbrenner (1997, 1986, 1979) proposes that the study of developmental influences must include the person's interaction with the environment, the person's changing physical and social settings, the relationship among those settings, and how the entire process is affected by the society in which the settings are embedded. Bronfenbrenner examines the *mutual* accommodations between the developing person and these changing contexts in terms of four levels of environmental influence: the *microsystem,* the *mesosystem,* the *exosystem,* and the *macrosystem* (see Figure 1-1).

The **microsystem** consists of the network of social relationships and the physical settings in which a person is involved each day. Consider, for instance, Maria and Sumi. Both are seventh-graders who live in a large U.S. city. In many ways their lives and surroundings seem similar. Yet in truth, they live in rather different worlds. See if you can spot important differences:

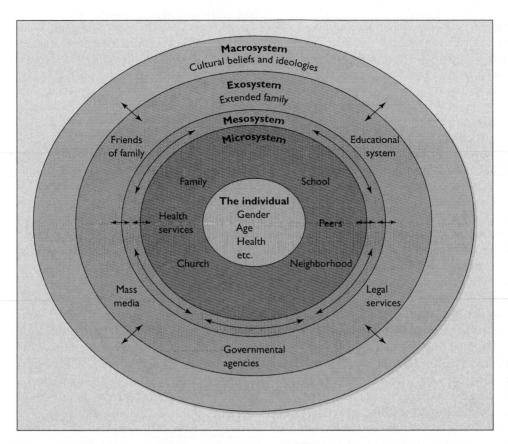

Figure 1-1 Bronfenbrenner's Ecological Theory of Development This shows the four levels of environmental influence: the microsystem, the mesosystem, the exosystem, and the macrosystem.

Maria Maria is the oldest of three children. Her family immigrated to the United States when she was an infant. Both of her parents work outside the home at full-time jobs, but they are usually able to arrange their schedules so that one parent is home when the children return from school. Should the parents be delayed, the children know to go to a neighbor—a grandmotherly figure—to spend the afternoon. Maria often helps her mother or father prepare a dinner "just like we used to eat in Nicaragua." The family members who do not cook on a given evening are the ones who later clean up. Homework is taken seriously by Maria and her parents. The children are allowed to watch television each night, but only after they have completed their homework. Her parents encourage the children to speak Spanish at home but insist that they speak English outside the home. Maria is enthusiastic about her butterfly collection, and family members help her hunt butterflies on family outings. She is somewhat of a loner but has one very close friend.

Sumi Sumi is 12 and lives with her parents and a younger brother. Both of her parents have full-time jobs that require them to commute more than an hour each way. Pandemonium occurs on weekday mornings as the family members prepare to leave for school and work. Sumi is on her own until her parents return

home in the evening after they first pick up her brother at a day-care center. Sumi's parents have demanding work schedules, and one of them is usually working on the weekends. Her mother assumes responsibility for the evening meal, which typically had been kim-chee, bok-choi, rice, fish, or soup, but fast food is starting to replace traditional meals on a regular basis. Sumi's father does not do housework; when he is not working, he can be found with friends at a local bar. Although Sumi realizes that getting a good education is important, she has difficulty concentrating in school. She spends a good deal of time with her friends, all of whom enjoy riding the bus downtown to go to the movies. On these occasions they "hang" and occasionally shoplift or smoke a little dope. Her parents disapprove of her friends, so Sumi keeps her friends and her parents apart.

The **mesosystem** consists of the *interrelationships* among the various settings in which the developing person is immersed. Consider again the cases of Maria and Sumi. Both come from two-parent families in which both parents work. Yet the events in their home environments are having substantially different effects on their schooling. Maria's family setting is supportive of academic achievement. Without necessarily being aware of it, Maria's parents are employing a principle of the Russian educator A. S. Makarenko (1967), who was quite successful in working with wayward adolescents in the 1920s: "The maximum support with the maximum of challenge." Although Sumi's parents also stress the importance of doing well at school, Sumi is not experiencing the same gentle but firm shove that encourages Maria to move on and develop into a capable young adult. Sumi's family has dispensed with the amenities of family self-discipline in favor of whatever is easiest. Moreover, Sumi has become heavily dependent on peers, one of the strongest predictors of problem behavior in adolescence (Pulkkinen, 1982).

An environment that is "external" to the developing person is called an **exosystem.** The exosystem consists of social structures that directly or indirectly affect a person's life: school, the world of work, mass media, government agencies, and various social networks. The development of children like Maria and Sumi is influenced not only by what happens in the environments in which they spend time but also by what occurs in the settings in which their parents live. Stress in the workplace often carries over to the home, where it has consequences for the parents' marriage. Children who feel rootless or caught in conflict at home find it difficult to pay attention in school. Like Sumi, they often look to a group of peers with similar histories, who, having no welcoming place to go and little to do that challenges them, seek excitement on the streets. Despite encountering job stresses somewhat similar to those of Sumi's parents, Maria's parents have made a deliberate effort to create arrangements that work against Maria's alienation.

The **macrosystem** consists of the overarching *cultural patterns* of a society that are expressed in family, educational, economic, political, and religious institutions. We have seen how the world of work contributes to alienation in Sumi's family. When we look to the broader societal context, we note that the United States is beginning to catch up with other industrialized nations in providing child-care services and other benefits designed to promote the well-being of families (see Chapters 6 and 8). But only some American parents enjoy such benefits as maternity and paternity leaves, flex time, job-sharing arrangements, and personal leave to care for sick children. Along with most U.S. families today, the families of Maria and Sumi are experiencing the unraveling of extended family, neighborhood, and other institutional support systems that in the past were central to the health and well-being of children and their parents.

The ecological approach allows us to view the developing person's environment as a nested arrangement of structures, each contained within the next. The most immediate structure is the setting in which the person presently carries out his or her daily activities; each ensuing structure is progressively more encompassing, until we reach the most inclusive or societal level. These dynamic interlocking structures challenge us to consider the risks and opportunities for development at each level. For instance, such problems as homelessness, child abuse and neglect, school violence, and psychopathology can be insightfully viewed as products of contextual factors that interact with individual and institutional vulnerabilities, particularly the family (Toro et al., 1997).

The ecological approach allows us to see people not in some contrived experimental vacuum but actively immersed in a real world of everyday life. Imagine how much more extensive the information gathered would be if a researcher were allowed to record your day-to-day experiences as opposed to interviewing you in a clinical setting. However, this seeming advantage is also the ecological approach's major disadvantage: We usually have enormous difficulty studying people in contexts where a great many factors are operating simultaneously. Because so many factors bear on a person, we find it impractical, indeed impossible, to take them all into account. Only as we control a large array of factors can we secure a "fix" on any one of them. We will have more to say on these matters later in the chapter when we consider the nature of developmental research.

Question

Can you give examples of how each of the ecological systems have affected your own development?

The Timing of Developmental Events

Time plays an important role in development. Traditionally, the passage of time has been treated as synonymous with chronological age. There has been an emphasis on the changes that occur within individuals as they grow older. More recently, social and behavioral scientists have broadened their focus. They take into account changes that occur over time, not only within the person but also in the environment, and examine the dynamic relation between these two processes. Paul B. Baltes has contributed to our understanding of these changes by identifying three sets of influences that, mediated through the individual, act and interact to produce development (Baltes & Baltes, 1998; Schulz & Heckhausen, 1996):

1. **Normative age-graded influences.** This first set of influences has a strong relation to chronological age. Among youth in early adolescence, like Maria and Sumi, these influences include the physical, cognitive, and psychosocial changes that we discussed earlier. Maria and Sumi are entering puberty, a condition associated with biological maturation. But they have also encountered *age-graded* social influences, such as the abrupt

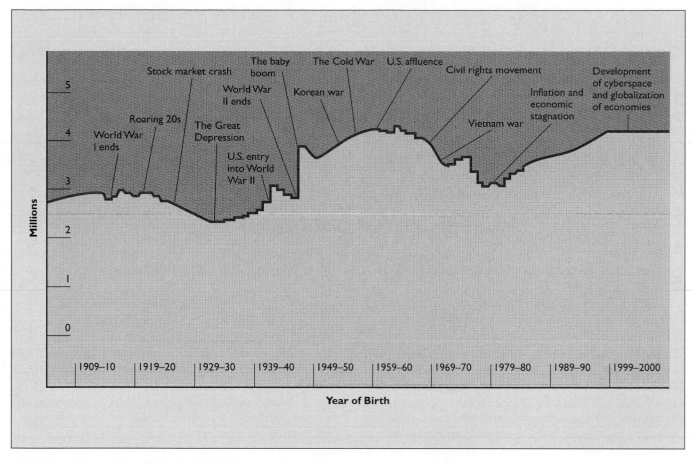

Figure 1-2 The Relative Size of Various Age Cohorts in the U.S.
Source: U.S. Census Bureau.

transition from a highly structured elementary school setting to a less structured and more complex middle school or junior high environment.

2. **Normative history-graded influences.** This second set of influences concerns historical factors. Although there is considerable cultural similarity among the members of a society, each age cohort is unique because it is exposed to a unique segment of history. An **age cohort** (also called a *birth cohort*) is a group of persons born in the same time interval (see Figure 1-2). Because society changes over time, the members of different age-cohorts age in different ways. The members of each new generation enter and leave childhood, adolescence, adulthood, and old age at a similar point in time, and so each generation's members experience certain decisive economic, social, political, and military events at similar junctures in their lives. As a consequence of the unique events of the era in which they live out their lives—for instance, the Great Depression of the 1930s, World War II, the Korean War, the prosperity of the 1950s, the Vietnam War—each generation tends to fashion a somewhat unique style of thought and life.

3. **Nonnormative life events.** This third set of influences involves unique turning points at which people change some direction in their lives. A person might suffer severe injury in an accident, win millions in a lottery, undergo a religious conversion, give birth to multiples of children at one time, secure a divorce, or set out on a new career at midlife or later. Nonnormative influences do not impinge on everyone, nor do they necessarily occur in easily discernible sequences or patterns. Although these determinants have significance for individual life histories, the determinants are not closely associated with either age or history.

Questions

Who is in your age cohort? What are some normative history-graded influences in your life? in your parents' lives? Have you experienced any nonnormative events?

Normative History-Graded Influences Each age cohort is exposed to a unique segment of history. Even when generations attempt to repeat a great event, it can never be experienced exactly the same, as evident in the original 1969 Woodstock gathering and Woodstock 1994.

Not surprisingly, each age cohort of U.S. youth over the past 80 years has acquired a somewhat different popular image: the youth of the roaring twenties, the political radicals of the 1930s, the wild kids of the World War II years, the silent generation of the 1950s, the love generation of the 1960s, the "Me" generation of the 1970s, the materialistic generation of the 1980s, and Generation X in the 1990s. Each of these generations confronted an environment different from that faced by earlier generations (see the discussion of generations in Chapter 11).

These influences do not operate only in one direction. Consider age cohorts. They are not simply acted on by social and historical forces. Because people of different cohorts age in distinct ways, they contribute to changes in society and alter the course of history. As society moves through time, statuses and roles are altered. Older occupants of social positions are replaced by younger entrants influenced by more recent life experiences. The flow of new generations results in some loss to the cultural inventory, a reevaluation of its components, and the introduction of new elements.

In particular, although parental generations play a crucial part in predisposing their offspring to specific values and behaviors, new generations are not necessarily bound to replicate the views and perspectives of their elders. These observations call our attention to the important part that cultural and historical factors play in development. What is true in the United States and other Western societies might not be true in other parts of the world. And what is true for the first decade of the 2000s might not have been true in the 1960s or the 1770s. Accordingly, if social and behavioral scientists wish to determine whether their findings hold in general for human behavior, they have to look to other societies and historical periods to test their ideas. Examining behavior from a cross-cultural perspective is a more common approach

in psychological research today. Since the advent of telecommunications, we can access relevant documents and publish findings easily in reputable online journals, which commonly feature text, pictures, charts, audio, and video clips. As we enter into the twenty-first century, technological developments should aid researchers as they continue to explore human development from a worldwide perspective.

Questions

Look at the pictures of the two Woodstock festivals above, and imagine what the members of that age cohort thought about the times they were living in. What might they think about that era now, when they reflect back on it? What do you think of that era from witnessing it through media such as movies, music, and literature? Is it difficult to understand the experiences of another age cohort?

Partitioning the Life Span: Cultural and Historical Perspectives

The Age-Old Question: Who Am I ?

Because nature confronts everyone with a biological cycle beginning with conception and continuing through old age and death, all societies must deal with the life cycle. Age is a major dimension of social organization.

For instance, all societies use age to allow or disallow benefits, activities, and endeavors. People are assigned roles independently of their unique abilities or qualities. Like one's sex, age is a master status, which governs entry to many other statuses and makes its own distinct imprint on them. Within the United States, for

Nonnormative Life Events Some people experience an event in life that creates a unique turning point or challenge in their lives.

instance, age operates *directly* as a criterion for driving a car without supervision (recently many states have raised this age from 16 to 18), voting (age 18), becoming president (age 35), and receiving Social Security retirement benefits (age 62). Age also operates *indirectly* as a criterion for certain roles through its linkage with other factors. For example, age linked with reproductive capacity limits entry into the parental role. Age linked with 12 years of elementary and secondary school usually permits entry into college. Figure 1-3 portrays the age distribution of the U.S. population.

Because age is a *master status,* a change in chronological age accompanies most changes in role over a person's life span—entering school, completing school, getting one's first job, marrying, having children, being promoted at work, seeing one's youngest child marry, becoming a grandparent, retiring, and so on. Age is a critical dimension by which individuals locate themselves within society and in turn are located by others (Settersten, 1997).

Consider, for example, which activities might or might not attract a nontraditional student at a university. Would a middle-aged mother of two want to join a sorority? Would she be welcome if she did want to join?

Age functions as a reference point that allows people to orient themselves in terms of *what* or *where* they are within various social networks—such as the family, the school, the church, and the world of work. It is one ingredient that provides people with the answer to the question "Who am I?" In brief, it helps people establish their identities.

Cultural Variability

The part that *social definitions* play in dividing the life cycle is highlighted when one compares the cultural practices of different societies. **Culture** refers to the social heritage of a people—those learned patterns of thinking, feeling, and acting that are transmitted from one generation to the next. Upon the organic age grid, societies weave varying social arrangements. A 14-year-old girl might be expecting to be a junior high school student in one culture, a mother of two children in another; a 45-year-old man might be at the peak of a business career, still moving up in a political career, or retired from a career in major league baseball—or dead and worshipped as an ancestor in some other society. All societies divide biological time into socially relevant units; and although birth, puberty, and death are biological facts of life, it is society that gives each its distinctive meaning and assigns each its social consequences.

Some cultures even extend their concepts of age to include two additional periods: the unborn and the deceased. Australian aborigines think of the unborn as the spirits of departed ancestors. These spirits restlessly seek to enter the womb of a passing woman and be reborn as a human child (Murdock, 1934). Similarly, Hindus regard the unborn as the spirits of persons or animals who lived in former incarnations (Davis, 1949). Some societies consider the dead to be continuing members of the community. Anthropologist Ralph Linton found that when a Tanala of Madagascar died, the person was viewed as merely surrendering one set of rights and duties for another:

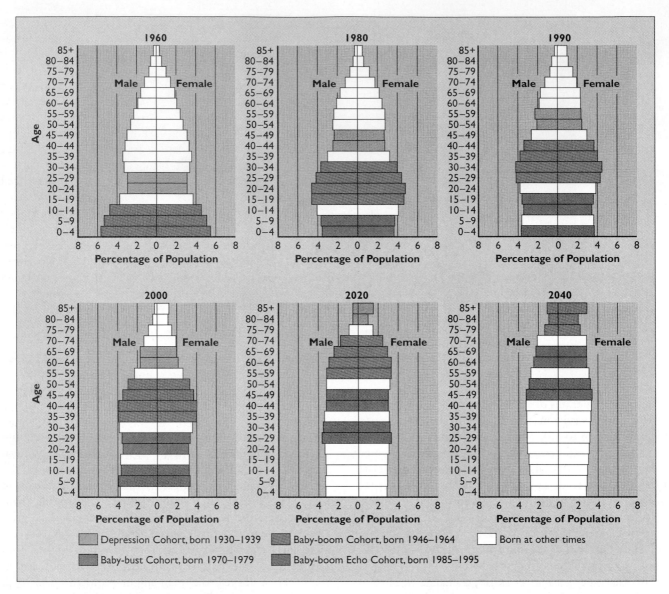

Figure 1-3 U.S. Population Pyramids, 1960–2040 The figure portrays population as a pyramid, the "tree of ages." Age groups are placed on a vertical scale, with the youngest at the bottom and the oldest at the top. On the horizontal axis are plotted the percentage of people in each age group. By 2020 the traditional pyramid will resemble a rectangle, with about the same percentage of people in each age group from birth through age 65. The pyramids trace the demographic impact of the Great Depression cohort (born 1930–1939), the baby-boom cohort (born 1946–1964), the baby-bust cohort (born 1970–1979), and the baby-boom echo cohort (born 1985–1995).

From: Leon F. Bouvier and Carol J. De Vita, "The Baby Boom—Entering Midlife," *Population Bulletin* 46, 3. Copyright 1991. Reprinted by permission of the Population Reference Bureau.

Thus a Tanala clan has two sections which are equally real to its members, the living and the dead. In spite of rather half-hearted attempts by the living to explain to the dead that they are dead and to discourage their return, they remain an integral part of the clan. They must be informed of all important events, invited to all clan ceremonies, and remembered at every meal. In return they allow themselves to be consulted, take an active and helpful interest in the affairs of the community, and act as highly efficient guardians of the group's mores [rules]. (Linton, 1936, pp. 121–122)

Viewed this way, all societies are divided into **age strata**—social layers that are based on time periods in life. Age strata organize people in society in much the same way that the earth's crust is organized by stratified geological layers. Grouping by age strata has certain similarities to class stratification. Both involve the differentiation and ranking of people as superior or inferior, higher or lower. But unlike movement up or down the class ladder, the mobility of individuals through the age strata is not dependent on motivational and recruitment

Varying Cultural Practices Societies differ culturally in a good many ways. Consequently, youngsters grow up with different social definitions of the behavior that is and is not appropriate for them as members of particular age groups. Pictured above is a family practicing a cultural tradition.

factors. In large measure, mobility from one age stratum to the next is biologically determined and irreversible.

Societies show considerable differences in the prestige they accord various age positions, such as old age. Countries such as Japan and China provide more positive attention and care for the elderly than countries in the West (Dein, 1997). In many rural societies, such as imperial China, elders have enjoyed a prominent and authoritative position (Lang, 1946). Among the agricultural Palaung of North Burma, long life was considered a great privilege that befell those who had lived virtuously in a previous incarnation. People showed their respect by taking great care not to step on the shadow of an older person, and young women sought to appear older than their actual age, because women acquired honor and privilege in proportion to their years (Milne, 1924). In sharp contrast, the elderly in our society today have a more restricted functional position and considerably less prestige. Our favored age stratum is youth. Americans commonly define adulthood as a period of responsibility. In contrast, the youth subculture is portrayed as irresponsible and carefree, dominated by the theme of "having a good time." In short, each culture shapes the processes of development in its own image, defining the stages it recognizes as significant.

People's behavior within various age strata is regulated by **social norms** or expectations that specify what constitutes appropriate and inappropriate behavior for individuals at various periods in the life span. In some cases, an informal consensus provides the standards by which people judge each other's behavior. Hence, the notion that you ought to "act your age" pervades many spheres of life. Within the United States, for instance, it is thought that a child of 6 is "too young" to baby-sit for other youngsters. By the same token, a man of 60 is thought to be "too old" to "party." In other cases, laws set floors and ceilings in various institutional spheres. For instance, there are laws regarding marriage without parental consent, entry into the labor force, and eligibility for Social Security and Medicare benefits. We need only think of such terms as *childish, juvenile, youth culture, adolescence, senior citizen,* and *the generation gap* to be aware of the potency of age in determining expectations about behavior in our own society. Indeed, we even find apartment dwellings exclusively for a particular age group, such as young singles, and cities designed for a particular age group, such as retired people.

Changing Conceptions of Age

In the United States we commonly think of the life span in terms of prenatal development, infancy, childhood, adolescence, adulthood, middle age, and old age. Yet the French historian Philippe Ariès (1962) said that in the Middle Ages the concept of childhood was not defined as we know it today. Children were regarded as small adults. Child rearing meant little more than allowing children to participate in adult affairs.

The world we think proper for children—the world of fairy tales, games, toys, and books—is of comparatively recent origin. J. H. Plumb (1972) observes that our words for young males—*boy, garçon, Knabe*—were, until the seventeenth century, used to mean a male in an independent position: a man of 30, 40, or 50. No special word existed for a young male between the ages of 7 and 16. The word *child* expressed kinship, not an age stage. The arts and documents of the medieval world portray adults and children mingling, wearing the same clothes, and engaging in many of the same activities. Only around the year 1600 did a new concept of childhood emerge. The notion of *adolescence* is even more recent, dating from the nineteenth and early twentieth centuries in the United States due to compulsory school legislation, child labor laws, and special legal procedures for "juveniles" that made a social fact of adolescence.

Now our society appears to be evolving an additional new stage between adolescence and adulthood: youth—men and women of traditional college age. Recent developments—rising prosperity, the increase in educational level, and the enormously high educational demands of a postindustrial society—have prolonged the transition to adulthood (Modell, 1989; Keniston, 1970).

The notion of "old age" has also undergone change in the Western world. Literary evidence indicates that

Further Developments

Childhood: Two Hundred Years of Profound Change

Children's lives have been profoundly transformed during the past 200 years by revolutionary changes in American society, particularly in the family and the economy (Calvert, 1992; Hernandez, 1993a, 1993b). For hundreds of years, agriculture and the two-parent farm family constituted the primary forms of economic production and family organization in Western nations. In 1830, a large majority of colonial American youngsters—nearly 70 percent—lived in two-parent farm families. As Figure 1-4 reveals, by 1930 the proportion of children living in two-parent farm families had dropped to a minority of less than 30 percent. During these same one hundred years, children living in nonfarm families with breadwinner fathers and homemaker mothers rose sharply from only 15 percent to a majority of 55 percent. This shift fundamentally altered the nature of childhood. In two-parent farm families, family members typically lived and worked side-by-side to sustain themselves in small community settings. But the shift to urban life transformed these patterns. Fathers spent their workdays away from home earning the family's economic livelihood while mothers remained in the home to care for the children and carry out multiple domestic responsibilities. A more recent and continuing shift in children's lives has been associated with dramatic increases in women's participation in the labor force and in the incidence of mother-child families in which no father is present in the home.

Childhood for Native American children differed greatly from that of colonial children. Some tribes reared children exclusively with mothers and grandmothers; in other tribes males were the mentors and disciplinarians. Native American children learned to imitate their parents' work at an earlier age—with boys hunting, fishing, and gathering fruits and nuts, and girls sewing and farming.

African American children were raised much more collectively than colonial children due to the harsh conditions of slavery, which meant that mothers and fathers were physically separated from the children during the day. At the day's end, children would join parents at home and help prepare supper. Even though one former slave recounts "my stay upon the farm had been pleasant. I played among the wild flowers and wandered in high glee over hill and hollow, . . . and knew not that I was a slave and the son of a slave," it was a more common experience to live in squalid housing, to see families broken up due to one or more members being sold, and to be forced to call the white woman on the plantation "mother"—a practice intended to undermine the authority of the biological mother (Horn, 1993).

The fundamental shift from farming to an urban life has been accompanied by a dramatic decline in large families. Among white adolescents born at the conclusion of the Civil War in 1865, some 82 percent lived in families with five or more children. By 1930 this figure had dramatically fallen to only 30 percent. During these same sixty-five years, families with one to four children more than tripled, from 18 percent to 70 percent. Consequently, the median number of siblings in families with teenagers dropped from 7.3 siblings to 2.6 siblings per family. This shift dramatically changed the level of competition for resources that youngsters experienced at home. Indeed, by 1930, nearly 60 percent of American children were either only children or had merely one or two siblings (Hernandez, 1993a, 1993b).

The past two hundred years have witnessed still another revolutionary change in children's lives. School enrollments have risen sharply (Horan & Hargis, 1991). In the seventy years between 1870 and 1940, school enrollments rose from about 50 percent to about 95 percent for children age 7 to 13

during the Renaissance men were already considered "old" in their forties. Currently, another division is emerging, one between "young-old" and "old-old." Young-old signifies a postretirement period in which there is physical vigor, new leisure time, and new opportunities for community service and self-fulfillment. In the past old-old characterized a minority of elderly persons who are in need of special care and support (Neugarten, 1982a, 1982b). Recently, questions have been raised about our "knowledge" of aging in persons over 70. Most of the scientific studies looking at aging have used subjects aged 60 to 70. We might need to differenti-

ate between young-old and old-old when researching those labeled "old" (Baltes & Baltes, 1998).

These emerging distinctions have tended to blur many of our assumptions regarding the rights and responsibilities of people with respect to social age. All across adulthood, age has increasingly become a poor predictor of the timing of major life events, such as changes in health, work status, family status, interests, and needs. As Bernice L. Neugarten and Dail A. Neugarten (1987) observe: "We have conflicting images rather than stereotypes of age: the 70-year-old in a wheelchair, but also the 70-year-old on the tennis court;

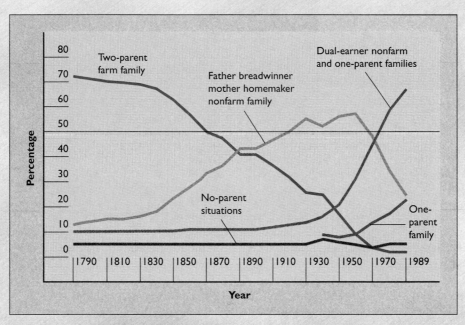

Figure 1-4 Types of Families with Children to Age 17, 1790–1989. Notice the recent dramatic rise in dual-earner nonfarm and one-parent families.
Source: U.S. Census Bureau

and to about 79 percent for children age 14 to 17. Simultaneously, the number of days students spent in classrooms doubled (from 21 percent of the total days in the year in 1870 to 42 percent of the days in 1940). As children have come to spend larger portions of the year in formal educational settings, they have less time at home with their parents and siblings. Additionally, as increasing numbers of mothers have entered the paid labor force, growing numbers of younger children are spending substantial portions of their nonweekend days in day-care centers and preschool settings. The motivating force underlying these three major social changes has been the desire of parents to alter their family's relative social and economic status (Hareven, 1982; Hernandez, 1993a, 1993b).

We will have more to say on these matters in the chapters that follow. For now, one caveat is in order: Although most historians agree that recent decades have witnessed a progressive improvement in the lives and treatment of children, we should not overlook the stark circumstances of one-fourth of U.S. youngsters under age 6 who live their lives in poverty, the dysfunctional nature of many of the nation's large school systems, and a mounting body of evidence that family violence, child abuse, and incest are much more common than most Americans have suspected.

the 18-year-old who is married and supporting a family, but also the 18-year-old college student who brings his laundry home to his mother each week" (pp. 30–32).

However, even though some timetables are losing their significance, others are becoming more compelling. Many young people might feel they are a failure if they have not "made it" in corporate life by the time they are 35. A young woman who has delayed marriage because of her career might feel under enormous pressure to marry and bear a child upon approaching her mid to late thirties. Historical definitions of social age, then, influence the standards a person uses in giving meaning to the

life course, in accommodating to others, and in contemplating the time that is past and the time that remains.

Connecting Historical "Areas of Concern"

We have chosen three historical points in time—the early years of developmental psychology, the 1950s through the 1960s, and the present era—to show how beliefs about "what research is currently needed" have shifted over the years, while keeping in mind that developmental issues of the past have served to define the field of human development today.

What were the critical questions being asked about human development by early pioneers? Are there any concerns that characterize the field at any specific point in historical time? Have there been any major shifts in the how these concerns have been approached by researchers over the last century?

What Were the Concerns of Early Developmentalists?
In the early 1900s, five major areas of research in human development stood out, and these issues are still actively researched today:

- Emotional development
- Biology and behavior issues
- Cognitive development
- Conscious and unconscious thought
- Role of the self

The Turn of the 20th Century   Researchers of human development at the turn of the twentieth century were concerned with discovering and explaining aspects of human development scientifically. Remember, science and the scientific method were still relatively new ways of examining phenomenon but were optimistically thought to be the keys that could unlock the secrets of nature and humanity.

Regarding *emotional development*, Charles Darwin wanted to show that emotional expressions were innate and not learned behaviors. He also wanted to demonstrate that the same biological processes existed in various individuals in diverse cultures. Therefore he collected pictures of individuals from different cultures displaying emotions such as anger, sadness, joy, and surprise, in order to show the similarities across cultures. A second area that interested researchers was the *biological basis for behavior.* Two issues were hotly debated: the relationship between evolution and development, and the relationship between nature and nurture and their effects on development. Both of these concerns are evident in the work of Darwin and Freud (Cairns, 1983).

A third area in the field's early history was *cognitive development.* As we will see in Chapter 2, Jean Piaget is considered to be the founder of the cognitive development movement. There were, however, other early contributors. One figure who foreshadowed many of Piaget's notions about cognitive development was Alfred Binet. Binet, known as the developer of the first intelligence (IQ test) also recognized that adult's thinking differed qualitatively from children's thinking. And it was Binet who fashioned ways of testing these differences.

A fourth area was the idea that *conscious* and *unconscious* thought can be differentiated. Both Binet and Freud dealt with this issue. Before Freud articulated his theory, people did not imagine that there were "hidden" areas of the mind guiding their behavior (Cairns, 1983).

A fifth area was the *role of the self* in development. Many of our field's early researchers were preoccupied with the notion of self. None were more focused on this issue than J. Mark Baldwin. His two seminal works, *Mental Development in the Child and the Race* (1895) and *Social and Ethical Interpretations of Mental Development* (1897) were extremely provocative to developmental psychologists at the time. The latter book was the first work by an American psychologist on social-cognitive development in childhood. These themes fell in and out of favor throughout the fifty years that followed. By the middle of the twentieth century, new themes surfaced.

The 1950s and 1960s   In mid century, behavioral theory dominated psychology and in turn shaped the concerns of developmental psychology. Much effort at the time was spent on testing the assumptions of psychoanalytic theory but cast in the language of behavioral theory. Some research was conducted to understand personality and social development across the life span. Another major interest was in modifying children's behavior. B. F. Skinner, a behaviorist, studied the basic principles of learning in children. The 1950s and 1960s also witnessed the emergence of an intensive investigation of how infants perceive the world.

Contemporary Issues   Today there appears to be renewed interest in several of the themes that characterized developmental psychology at the turn of the twentieth century. Researchers today are particularly interested in the biological bases of behavior, emotional development, the emergence of social relationships, and empirical study of children's cognitive capacities.

Today more attention is also being given to the *biological bases of behavior,* partly due to advances in the field of behavioral genetics. Developmentalists are interested in the role that genetics (biology) plays in the unfolding of behavior across the life span. For example, what genes might be responsible for certain traits in people, such as introversion or extroversion? How do hormones affect behavior, especially during infancy and adolescence? Biological themes are reminiscent of Darwin's early efforts to apply evolutionary principles to human behavior.

In the past thirty to forty years, there has also been more research on the timing of *emotional development* and the role of emotions in social interaction. Recent research on emotions has included studies on self-conscious emotions such as guilt, shame, pride, empathy, and envy—emotions that Freud considered but that received little attention until recently.

There is a rigorous effort today to explain social interactive processes, particularly the nature of *social relationships,* as demonstrated in parent-infant attachment (bonding) studies (see Chapter 6). The definition of family has been expanded, focusing not only on the mother–infant relationship but also on the roles of fathers, siblings, and grandparents in the child's development.

And researchers are returning to issues of the roles played by consciousness, reflection, and motivation in development. Investigators have developed methods that permit examination of the impact of unconscious processes on *cognition*. These methods are being applied systemically to a range of current topics, from eyewitness testimony to adoption issues.

Since the 1960s researchers have also demonstrated children's remarkable *precocity*—their ability to understand physical and perceptual phenomena at younger ages than formerly believed.

Back to the Future   As history demonstrates time and time again, issues often disappear only to return in a slightly different form at a later date. Social acceptance or rejection of new ideas is often based on timing: creative thinkers are usually ahead of their time, and often their genius is not recognized until society catches up with their progressive notions.

So, too, in the field of human development. There appears to be much consistency between the perspectives of the major developmental themes of our pioneers a century ago and current interests. Why are we returning to the concerns of the past? First, our forebears were far-sighted and innovative thinkers. They raised enlightened and enduring questions. Second, as developmentalists established the discipline, they improved it through the use of technologically advanced methods and statistical techniques that enable us to re-pose the earlier questions and address them using modern technology and more sophisticated research methods. In the next section, we will learn more about contemporary methods and techniques used in the study of human development.

Questions

What types of developmental issues that were being studied a century ago are still under investigation today? What types of human development issues are of more recent concern to social scientists?

The Nature of Developmental Research

To him who devotes his life to science, nothing can give more happiness than increasing the number of his discoveries, but his cup of joy is full when the results of his studies immediately find practical application.

Louis Pasteur

The task of science is to make the world intelligible to us. Albert Einstein once observed that "the whole of science is nothing more than a refinement of everyday thinking." So we do scientific research in much the same way

that we ask questions and come to conclusions in our everyday lives. We make guesses and mistakes; we argue our conclusions with one another; we try out our ideas to see what fits, and we get rid of what doesn't. There is one important way scientific inquiry differs from ordinary inquiry: It specifies a systematic and formal process for gathering facts and searching for a logical explanation of them. This process, called the **scientific method,** is a series of steps we follow that allow us and others to be clear about what we studied, how we studied it, and what were our conclusions. Sufficient detail must be given to allow others to replicate our research and verify our conclusions. These are the steps of the scientific method that provide a framework for objective inquiry: (1) select a researchable problem; (2) formulate a **hypothesis**—a tentative proposition that can be tested; (3) test the hypothesis; (4) draw conclusions about the hypothesis; and (5) make the findings of the study available to the scientific community.

How do we use this method to help us understand and explain human development? First, let's consider some questions we might ask about development: Are there certain characteristics measurable in childhood that are predictive of success in different areas of adult life? Which children are more prone to violence, and what can be done to teach these children conflict resolution techniques? Who is likely to be affected by anorexia, and what steps can we take to save this person's life? Does everyone's memory decline with age? What aspects of personality and social stress are related to longevity? Are there peaks in the frequency of sexual activity for men and women, or does it generally increase or decrease with age? It is easy to choose any one of these questions and come up with a researchable problem. Perhaps through your reading and life experience you could suggest a hypothesis—a proposition that can be tested scientifically—that might answer one of these questions. Next you need to test your hypothesis. To do that requires choosing a research design that will provide valid (accurate) and reliable (consistent) information to support or reject your hypothesis.

Research Design   In developmental psychology, our research focuses on change that occurs over time or with age. Three basic kinds of methods are used: (1) longitudinal, (2) cross-sectional, and (3) sequential. Experimental methods, although powerful, are seldom used in developmental studies because usually it is not possible to exercise the control necessary for experimental design; interesting variables such as spatial ability, memory, and physical characteristics cannot be assigned to groups of individuals or manipulated in terms of quantity or quality presented. Other methods used in developmental research include case studies, observational methods, surveys and cross-cultural studies. These are briefly discussed in the next sections.

The Longitudinal Method

The **longitudinal method** is used when we study the same individuals at different points in their lives. We can then compare the group at these regular intervals and describe their behavior and characteristics of interest. This method allows us to look at change sequentially and offers insight into why people turn out similarly or differently in adulthood.

The Terman Life-Cycle Study, a classic longitudinal study—indeed, the grandparent of life-course research—was begun by psychologist Lewis Terman in 1921–1922 (Friedman & Brownell, 1995; Cravens, 1992). Terman followed 1,528 gifted boys and girls from California public schools—who later nicknamed themselves "Termites"—and a control group of children of average intelligence from preadolescence through adulthood. These subjects were studied at 5- to 10-year intervals ever since. The 856 boys and 672 girls were selected on the basis of their intelligence quotients, or IQs (between 135 and 200 on the Stanford-Binet scale), which were said to represent the top 1 percent of the population. He found that the gifted ones were generally taller, heavier, and stronger than youngsters with average IQs. Moreover, they tended to be more active socially and to mature faster than average children (Terman & Merrill, 1937). One of the effects of the study has been to dispel the belief that the acceleration of bright children in school is harmful.

After Terman's death, other psychologists continued the project (Friedman & Brownell, 1995; Shneidman, 1989; Holahan, 1988; Rafferty, 1984; Sears, 1977), and their research has provided longitudinal data on religion and politics, health, marriage, emotional development, family history and careers, longevity, and cause of death. One notable recent finding is that Termites whose parents divorced had a greater risk of early death (the average age of death for men was 76, compared with age 80 for those whose parents remained married; for women the corresponding ages of death were 82 and 86 years). Researchers speculate that the stress and anxiety associated with their parents' strife took its toll in earlier mortality (Friedman & Brownell, 1995).

Limitations of the Longitudinal Method   Although the longitudinal method allows us to study development over time, the approach does have a number of disadvantages. One major problem is selective attrition and dropout. Subjects drop out because they become ill or die, because they move away and are difficult to locate, or simply because they become disinterested in continuing the study. Selective attrition means simply that the subjects who drop out tend to be different from the subjects who remain in the study. For example, those who remain might come from the most cooperative and stable families or be more intelligent or successful. These changes can bias the sample of subjects as it becomes smaller over time (surprisingly enough, only 10 percent

of the "termites" were unaccounted for in 1995). Other problems include testing and tester consistency over the length of the study. It is impossible to test every person at every scheduled testing on every test item. People get sick or go on vacation. They become upset, so that part of the test must be omitted. They might refuse to comply on some items. And children or their parents occasionally forget appointments (Willett, Singer, & Martin, 1998; Schaie, Campbell, & Rawlings, 1988; Bayley, 1965). Comparable data might not be collected from every subject at every time interval. Likewise changes, such as turnover or burnout, in the staff that tests or observes the subjects can result in inconsistencies in the measurements taken.

More importantly, longitudinal studies cannot control for unusual events during this group's life span. The effects of such economic and social events can make it difficult to generalize the findings from one age cohort to another age cohort born 10 or 20 years later. Such events can distort the amount or direction of the change reported:

> War, depression, changing cultures, and technological advances all make considerable impacts. What are the differential effects on two-year-olds of depression-caused worries and insecurities, of TV or no TV, of the shifting climate of the baby-experts' advice from strict-diet, let-him-cry, no-pampering schedules to permissive, cuddling, "enriching" loving care? (Bayley, 1965, p. 189)

The length of time and amount of money required to complete a long-term study can also be prohibitive (Brooks-Gunn, Phelps, & Elder, 1991). For example, eighteen national agencies collaborated to fund research on the well-being of America's children ("America's Children," 1998). It would be appropriate to consider a cost-benefit analysis prior to choosing this methodology. Finally, there is the problem of finding out tomorrow what relevant factors should have been taken into account yesterday. Once set in motion the project is difficult to alter even when newer techniques might improve the overall design. For example, computerized testing or a survey on a web page could make it easier for subjects to record their data and would be less expensive than bringing subjects into a research lab, but would the data be comparable to the data collected earlier? What would be the effect of having no tester-subject interaction? Would the responses by the subject be affected by her or his computer experience or comfort level working electronically?

Despite these limitations, longitudinal studies can provide us with important information. For example, if our research interest is the effect of aging on spatial ability, we might design the following study: Select a sample of 20-year-olds and measure their *spatial ability*. (Spatial ability deals with visual form relationships, such as drawing a specific triangle from memory, arranging blocks to recreate a pattern, or reading a graph or map [see Figure 1-5]. Tests of spatial ability are used as a means of determining vocational skill and to predict success with specific school-related subjects such as geometry and mechanical drawing.) We would bring this same group (cohort) back to our lab every 10 years and repeat this measurement. Figure 1-5 describes our hypothetical results on the spatial ability tests, the number of subjects tested each time, and their age at the time of testing.

Question

Based on the "results" shown in Figure 1-5, what can you say about age and subjects' spatial ability over time?

The Cross-Sectional Method

The hallmark of the longitudinal method is taking successive measurements of the same individuals. In contrast, the **cross-sectional method** investigates development by simultaneously comparing *different age groups.* Unlike our longitudinal research example described above, this time we would investigate spatial ability and age by selecting a group of 20-year-olds, a group of 30-year-olds, a group of 40-year-olds, etc., through our last group, 70-year-olds. We would test the spatial ability of all six groups at the same time. What savings in time and money! You don't have to wait 50 years for the data to be complete, nor do you have to worry about locating your subjects and bringing them back for retesting. Staff turnover is not an issue nor in most cases are there problems with subject cooperation and testing inconsistencies. Figure 1-6 summarizes the hypothetical results from the cross-sectional study of spatial ability and age. These findings seem to fit the hypothesis that spatial ability declines with age. But can we really say that? Could there be other differences (besides age) among these selected groups that are affecting spatial ability?

Limitations of the Cross-Sectional Method The *confounding* of age and cohort is the major disadvantage in cross-sectional research. "Confounding" in research means the elements are mingled so they cannot be distinguished or separated. We can never be sure that the reported age-related differences between subjects are not the product of other differences between the groups. For instance, the groups might differ in social environment, intelligence, or diet. So the comparability of the groups can be substantiated only through careful sampling and measurement techniques. For example, if you were to investigate in your own family how many years of formal schooling your grandparents received, versus your parents, versus yourself, you are most likely to determine that your generation has the resources to earn a

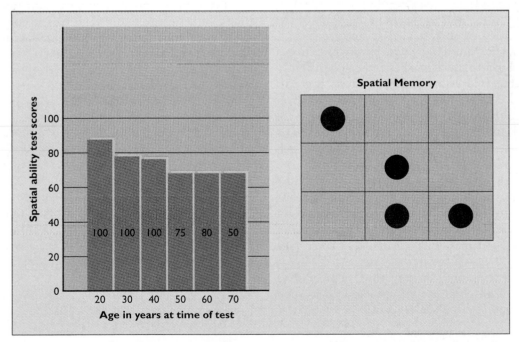

Figure 1-5 Longitudinal Research Study Showing the Effects of Age on Spatial Memory Ability In this study, the same group of adults was tested at 10-year intervals. Spatial memory is a measure of short-term visual memory for abstract material. The performance task is to view a 3 X 3 grid for 5 seconds and then recreate that same pattern from memory by placing red and black chips on a blank grid. The scores shown here are the average test scores at each age. The number of adults located and tested is indicated inside each bar. What if the adults not tested at ages 50, 60, and 70 were the ones who scored the lowest on spatial ability in the earlier time period?

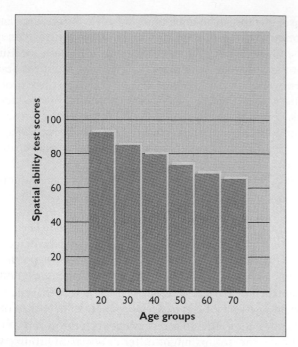

Figure 1-6 Cross-Sectional Study Showing the Effects of Age on Spatial Memory Ability In this study, adults of various ages were tested on spatial ability. The scores shown here are the average test scores for each age group.

higher level of education, delaying full-time work status and most likely delaying childbearing. We know that the life experiences of a typical 20-year-old today are much different from those of a 20-year-old in the 1930s or the 1950s.

These problems are highlighted by cross-sectional studies of intelligence. Such studies rather consistently show that average scores on intelligence tests begin to decline around 20 years of age and continue to drop throughout adulthood. But as we will see in Chapter 15, cross-sectional studies do not make allowance for *cohort differences* in performance on intelligence tests. Each successive generation of Americans has received more schooling than the preceding generation. Consequently, the overall performance of each generation of Americans on intelligence tests improves. Improvement caused by increasing education creates the erroneous conclusion that intelligence declines with chronological age (see Chapter 15).

Robert Kastenbaum shows how comparisons among people of different age groups can lead to such faulty conclusions:

Occasionally I have the opportunity to chat with elderly people who live in the communities nearby Cushing Hospital. I cannot help but observe that many of these people speak with an Italian accent. I also chat with young adults who live in these same communities. They do *not* speak with an Italian accent. As a student of human behavior and development, I am interested in this discrepancy. I indulge in some deep thinking and

come up with the following conclusion: as people grow older they develop Italian accents. This must surely be one of the prime manifestations of aging on the psychological level. (Quoted in Botwinick, 1978, p. 364)

Sequential Methods

All **sequential methods** involve measuring more than one cohort over time. This combination of collecting data over time as well as across groups overcomes the age/cohort confounding found in cross-sectional studies as well as the effect of unique events found in longitudinal designs. If we consider our example of age and spatial ability, this time we could select a sample of 25-year-olds and a sample of 35-year-olds, measure their spatial ability and then bring each cohort back for successive measurements at specific time intervals. Figure 1-7 provides some actual data on spatial ability using a sequential design. Spatial ability was measured in adults born in 1930 for the ages 35, 45, 55, and 65. For adults born in 1940, measurements were taken at ages 25, 35, 45, and 55. Each spatial ability score reported in the table is the group *average* for a particular time interval. First, study the scores of the 1930 group over time and then look at the scores for the 1940 group. Finally, compare the scores of the two groups at a particular age—for example, when both groups were measured at age 35.

Figure 1-8 shows the relationship of time and age sampling for a design that is longitudinal, cross-sectional, and sequential. This is a complex figure but worth spending a little time to understand it. The columns represent the year of birth of the subjects, and the rows indicate the year that the measurement was taken. The cell entries indicate the age of the subjects, given their birth year and year of measurement. For example, subjects born in 1910 who were tested in 1925 were 15 years old when the first

	Year of birth			
Year of spatial ability measurement		1930		1940
1965	70	(50)	75	(50)
1975	65	(49)	70	(48)
1985	60	(45)	65	(45)
1995	55	(50)	60	(40)

Figure 1-7 A Sequential Design Study to Test the Effects of Age on Spatial Memory Ability Spatial ability in adults born in 1930 and 1940 was measured four times at 10-year intervals. The numbers reported in the box are the average spatial ability scores for that group at that time interval. The number of subjects measured is reported in parentheses.

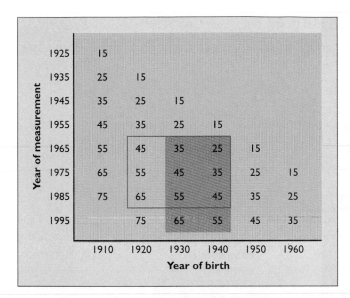

Figure 1-8 A Cross-Sequential Study Showing the Effects of Age on Spatial Memory Ability Longitudinal studies follow one group of subjects over time. Any column would represent one group (same birth year) with multiple measurements (year of measurement) taken at 10-year intervals. Any row would represent cross-sectional research with multiple age groups (different birth years) measured at one point in time (year of measurement). An example of a sequential study is the box outlined in the middle. The shaded area represents the cross-sequential study of spatial ability in adults born in 1930 and 1940, with each age group measured in 1965, 1975, 1985, and 1995.

measurement was taken. Longitudinal studies would be represented by the scores in any part of one particular column, and cross-sectional studies would be represented by the scores in any portion of one particular row. Sequential studies would be represented by a box including more than one column and more than one row. See the box outlined in Figure 1-8. The sequential design described above is represented by the shaded area in the same figure.

Questions

How does age affect spatial ability? Does this same relationship hold true for the two groups in our example? How would you compare and explain the differences in scores for the two groups of 35-year-olds? Finally, how does having two groups help control for selective attrition?

Limitations of Sequential Methods Sequential methods can be complex and difficult to analyze if the groups measured longitudinally (over time) are found to be very different in the variable under study. For example, if 25-year-olds in 1935 have significantly lower spatial ability scores than 25-year-olds in 1965, it is difficult to combine these scores for an overall measurement. Doing so

might distort the sequential changes in spatial ability throughout the study. The issues of time and money continue to be a limitation as when any group is followed over a longer period of time.

Questions

Which three research designs are frequently used to describe developmental change across the life span? For each design, can you describe situations where it would be appropriate for researchers to use it?

The Experimental Method

The **experimental method** is one of the most rigorously objective techniques available to science. An **experiment** is a study in which the investigator manipulates one or more variables and measures the resulting changes in the other variables to attempt to determine the cause of a specific behavior. Experiments are "questions put to nature." They are the only effective technique for establishing a cause-and-effect relationship. This is a relationship in which a particular characteristic or occurrence (X) is one of the factors that causes another characteristic or occurrence (Y). Scientists design an experimental study so that it is possible to determine whether X does or does not cause Y. To say that X *causes* Y is simply to indicate that whenever X occurs, Y will follow at some later time.

In an experiment, researchers try to find out whether a causal relationship exists between two variables, X and Y. They systematically vary the first variable (X) and observe the effects on the second variable (Y). Factor X, the factor that is under study and is manipulated in an experiment, is the **independent variable.** It is independent of what the subject or subjects do. The independent variable is assumed to be the *causal factor* in the behavior being studied. Researchers must also attempt to control for **extraneous variables,** factors that could confound the outcome of the study; these could include the age and gender of the subjects, the time of day the study is conducted, the educational attainment of the subjects, and so on (see Figure 1-9).

The study is planned such that the subjects in the **experimental group** are administered the independent variable (some refer this as the "treatment"). In comparison, the **control group** of subjects should be identical to the experimental group except they will not be administered the independent variable while they perform the same task as the experimental group.

We need to determine if the independent variable has made any difference in the performance of the experimental group of subjects. We call the end result of the experiment—the factor that is affected—the **dependent variable,** which is some measure of the subjects' behavior.

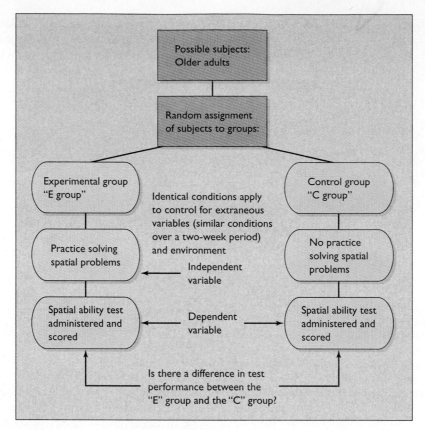

Figure 1-9 Sample Experimental Design Elements of a psychological experiment to assess the effects of practicing spatial ability problems on older adults' performance on spatial memory ability.

For instance, dependent variables are often administered in the form of paper-and-pencil tests or performance tests, for the researcher must quantify data in some measurable way. The researcher then performs various statistical analyses to be able to compare results and to look for any significant differences (e.g., How did the performance of the experimental group vary from the performance of the control group?) (see Figure 1-9).

Let's think back to the relationship between age and spatial ability. In a true experiment we would have to vary age systematically and measure its effect on spatial ability. The problem is that we can't manipulate age—we cannot assign the same subject to different age groups. But what if, after reviewing the literature and the results from our longitudinal and cross-sectional studies, we hypothesize that the decrease in spatial ability in older participants is due, in part, to a lack of recent experience in spatial ability tasks? If we wanted to test that hypothesis, we could design an experiment such as the one presented in Figure 1-9. In this spatial ability experiment, we would develop a 2-week training and practice course that would give older participants a chance to practice special spatial ability problems. We would randomly assign half of the older subjects to the practice course (the experimental group who will receive the training) while

the other half of the subjects (the control group) would receive a 2-week period of time together to share information and have social conversation. In this example, the training in spatial relations would be the independent variable (practice with spatial relations versus no practice). At the end of the 2-week period, we would most likely use some type of paper-and-pencil or performance test to measure the dependent variable (spatial ability). If the average spatial ability scores for the group who trained and practiced were significantly higher than the average for the group who did not practice, our hypothesis would be supported.

Experiments must be replicated by other researchers with different groups of subjects to see if there is consistency in results before any major theory can be significantly substantiated.

Limitations of the Experimental Method in Developmental Psychology   It is difficult to use an experimental approach in developmental psychology, for several reasons. The first, as indicated above, is the inability to assign subjects to the variable of interest. Developmental psychologists cannot manipulate many of the variables they study—such as age, gender, abusive family background, or ethnicity. These variables come with the individual along with many other variables that can confuse us when interpreting their effects on the dependent variable. Second, many of the questions we ask involve the effects of stressful or dangerous experiences, such as tobacco or alcohol use, medical procedures, or the withholding of treatments thought to be beneficial. Manipulations of these variables would be unethical, if not impossible. Third, some argue that how people behave or perform in an experimental lab setting is not how they actually behave in a "real-world" setting. Fourth, planning, designing, conducting, and evaluating a true experimental design is very time-consuming and costly, as you can imagine from reflecting on the actual small-group lab experiment illustrated in Figure 1-10.

Questions

Why is the experimental method considered to be the only one that can determine the cause of a specific behavior? Why is experimental design seldom used in most developmental studies?

The Case-Study Method

The **case-study method** is a special type of longitudinal study that focuses on a single individual rather than a group of subjects. Its aim is the same as that of other longitudinal approaches—the accumulation of developmental information. An early form of the case study method was the "baby biography." Over the past two centuries, a small number of parents have kept detailed observational diaries of their children's behavior. Charles Darwin, for example, wrote a biographical account of his infant son.

A good deal of the early work of Jean Piaget, an influential Swiss developmental psychologist, was based on the *case-study approach* (Wallace, Franklin, & Keegan, 1994; Gratch & Schatz, 1988). Piaget (1952) carefully observed the behavior of his three children—Lucienne, Laurent, and Jacqueline—and used this information to formulate hypotheses about cognitive development. One of his best-known case studies involved his unique technique for studying the mathematical concept of *conservation*. Piaget would begin with two identical

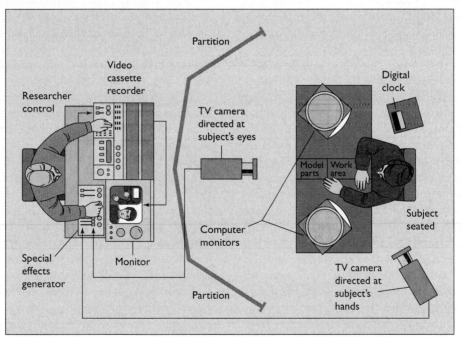

Figure 1-10 Experimental Research in a Small-Group Lab Many college campuses provide researchers with special facilities and equipment for conducting experiments and observing subjects. This small-group lab consists of two areas separated by a room divider. Researchers at the control console can observe subjects in the experimental area without themselves being observed (and thus do not interfere with the spontaneous behavior of the subjects). This small-group laboratory also contains video equipment so the experimenter can view on a television screen the subject's behavior in the experimental area. This format is devised so individual subjects do not observe and influence one another's perceptions and interpretations of the task to be performed. Moreover, the videotape equipment permits researchers to record behavior. Later the experimenter can analyze the behavior more closely (in this case a second-by-second analysis of the subject's hand actions in an assembly task, an analysis of when the subject looked at instructions in picture format or instructions by text and pictures, and analysis of whether the assembly task was performed correctly).

balls of clay, show them to the child, and let the child hold and manipulate the balls until the child agreed that the two contained the same amount of clay. Then, in view of the child, Piaget would flatten one ball of clay into a pancake shape. He would then ask the child if both pieces still had same amount of clay or if the pancake or the ball had more. Between the ages of 4 and 5, children consistently said the ball had more. When tested at the age of 6 or 7, children consistently indicated that the two pieces still contained the same amount of clay, an indication of cognitive maturation (Piaget, 1952).

Case studies have also had a prominent place in the clinical treatment of maladjusted and emotionally disturbed individuals. Sigmund Freud and his followers have stressed the part that early experience plays in mental illness. According to this view, the task of the therapist is to help patients reconstruct their own histories so that, in the process, they can resolve their inner conflicts. An example of a classic case study, published in the late 1950s, is *The Three Faces of Eve,* about a woman with multiple-personality disorder. More recently, the clinical approach has been extended to the study of healthy individuals. Case studies are often used by researchers who study individuals who exhibit behavior that is an exception from the norm, such as a child who is a genius or a person who has committed serial murders.

Limitations of the Case-Study Method   The case-study method has a number of drawbacks. The data are recorded on only one individual, and it is difficult to generalize from one case to the whole population of interest. Of course, if the kind of case study is repeated many times, on many individuals, as was the research that Piaget did with children, the findings are more valuable. A second problem is the extended interaction between the observer or experimenter and the subject under study. Because case studies normally involve frequent contact between researcher and subject over a long period of time, researchers and subjects become familiar with each other, and the objectivity of the results may be in question. The experimenter might even become a part of the subject's treatment, and the same results would not necessarily be found by a different researcher.

The Social Survey Method

Researchers are often interested in studying the incidence of specific behaviors or attitudes in a large population of people. Suppose researchers want to discover the prevalence and characteristics of people who are homeschooling their own children, or the frequency and type of drug use among teenagers, or the impact of a public campaign promoting senior day care. Using the **social survey method,** researchers ask questions to a sample of individuals who are representative of the population of

individuals likely to be affected. These questions can be asked through personal interviews, by phone, or by mail in questionnaire form. When surveys are mailed, the researchers must rely on the people selected to answer the questions and return the document for analysis. For example, in the past several decades in the United States, census takers came to our homes and interviewed us while filling out a very long questionnaire. The national census in 2000 will not be conducted in quite the same way, because that has become far too costly and labor intensive, and many people are working during the daytime hours when the census taker would work. Census researchers also must try to interview homeless and institutionalized Americans, as well as immigrants. Social scientists at the Census Bureau have a tremendous logistical task in such a large-scale survey, and it will take years to analyze the data collected.

For survey research to produce useful results, the sample of subjects must be representative, and the questions must be well designed, easy to answer, and clear. For example, open-ended questions that require long answers might have a lower return rate than a simple check off list. However, open-ended questions might give the researcher more details about a subject's attitude and practice. A survey that is too lengthy or detailed—no matter how significant or timely the questions might be—is likely to get a very poor response rate (see Figure 1-11).

The representativeness of the sample is based on **random sampling.** There are different methods of random sampling, but the basic premise is that each member of the population sampled has an equally likely probability of being chosen. This allows the researcher to generalize her or his findings from the sample to the population of interest. In other words, if we want to be able to talk about the societal issue of teenage pregnancy, we cannot simply interview 500 students from one high school and report the results as representing the national population of teenagers (for instance, students in an urban high school probably are not representative of students in suburban or rural schools). If we interview 100 students from Tampa, 100 from New York City, 100 from Topeka, 100 Kokomo, and 100 from Ithaca, perhaps then we can talk about this issue on a national scale. The design and clarity of the questions are particularly important when a document is to be mailed and respondents will not be able to ask for assistance in understanding the questions. Social scientists interested in conducting surveys usually take advanced statistics courses to learn how to design and analyze a reliable survey.

Limitations of the Social Survey Method   The greatest concerns in survey research are response rate and bias. Are the subjects who chose to respond (less than a

"Next question: I believe that life is a constant striving for balance, requiring frequent tradeoffs between morality and necessity, within a cyclic pattern of joy and sadness, forging a trail of bittersweet memories until one slips, inevitably, into the jaws of death. Agree or disagree?"

Figure 1-11 Survey Questions Should Be Simple and Short to Be Comprehensible

Source: The New Yorker Collection 1989 George Price from cartoonbank.com. All rights reserved.

50 percent return is common) different from the subjects who chose not to fill out the survey or agree to be interviewed? This is similar to the problem of *selective attrition* in longitudinal research. In addition, many people who are surveyed give answers they think the researcher expects or that they think will make them seem mature or "good," while other respondents might exaggerate when responding. Many adults are sensitive to questions that touch on matters they consider private (such as sexual practices, income, and political or religious beliefs) and are unable or unwilling to give accurate answers on these subjects. Finally, the survey method has limited use with children and cannot be used at all with infants.

The Naturalistic Observation Method

In **naturalistic observation,** researchers intensively watch behavior as it occurs and record it by means of notepad, videotape, or other method. Observers must be careful not to disturb or affect the events under investigation. This method produces more detail and greater depth of insight than the social survey method (Cahill,

1990; Willems & Alexander, 1982), but it is effective only for a smaller range of subjects. For example, naturalistic observation has been used extensively to observe young children's interactions in a nursery or preschool setting, but it would be much harder to use this method to study adults at work.

An advantage of naturalistic observation is that it is independent of the subject's ability or willingness to report on given matters. Many people lack sufficient self-insight to tell the researcher about certain aspects of their behavior. Or if their behavior is illegal, socially unacceptable, or deviant, they might be reluctant to talk about it. The following account of an interview with a mother about her child-rearing practices is a good example (Maccoby & Maccoby, 1954, p. 484):

> During the interview she held her small son on her lap. The child began to play with his genitals. The mother, without looking directly at the child, moved his hand away and held it securely for a while. . . . [Later] in the interview the mother was asked what she ordinarily did when the child played with himself. She replied that he never did this—he is a very "good" boy. She was evidently unconscious of what had transpired in the very presence of the interviewer.

Using more systematic techniques through observation can provide greater objectivity in collecting and analyzing the data. One technique, **time sampling,** involves counting the occurrence of some specific behavior over a period of different time intervals. Another technique would be to record actions observed at particular time intervals, such as every 30 seconds. Researchers who do not want to lose the sequential flow of events focus on a class of behaviors, such as fighting on a playground, and record the time lapse for each episode; this approach is termed **event sampling.** Still other researchers use precoded behavior categories. They determine beforehand what behaviors they will observe and then record these behaviors using code symbols. The use of videotaping in observational studies has allowed researchers much greater reliability in coding as well as greater flexibility in choosing events or behaviors observed (see *Information You Can Use:* "Tips for Observing Children").

Limitations of Naturalistic Observation Naturalistic observation can provide a rich source of ideas for more extensive future study. But it is not a particularly strong technique for testing hypotheses. The researcher lacks control over the behavior of the subjects being observed. Furthermore, no independent variable is "manipulated." Consequently, the theorizing associated with naturalistic observation (such as trying to figure out why a behavior occurs) tends to be highly speculative. The observer might be biased, have certain expectations, and look for those behaviors to record. Still another problem with this method is that the observer's presence can alter the behavior he or

Information You Can Use

Tips for Observing Children

One of the best ways to learn about children is to observe them. To provide access to the full drama, color, and richness of the world of children, many instructors have their students watch children in the laboratory or in the field. Here are a number of tips that may prove helpful for observing children:

- The minimal aids you will need for observation generally include paper, pen, a timepiece, and a writing board.
- Record the date, the time interval, the location, the situation, and the age and sex of the subject or subjects.
- Most observations take place in nursery school settings. Add diversity to your report by observing children in parks, streets, stores, vacant lots, homes, and swimming pools.
- Have the purpose of your research firmly in mind. You should explicitly define and limit in advance the range of situations and behaviors you will observe. Will you watch the entire playground, giving a running account of events? Will you concentrate upon one or two individuals? Will you record the activities of an entire group? Or will you focus only on certain types of behavior, such as aggression?
- Once the target behavior is identified, describe both the behavior and the social context in which it occurs. Include not only what a child says and does but also what others say and do to the child. Report spoken words, cries, screams, startle responses, jumping, running away, and related behaviors.
- Describe the relevant body language—the nonverbal communication of meaning through physical movements and gestures. Body language includes smiles, frowns, scowls, menacing gestures, twisting, and other acts that illuminate the intensity and affect of behavior.
- Give descriptions of behavior, not interpretations that generalize about behavior.

- Make notes in improvised shorthand. Immediately after an observation session, transcribe your notes into a full report. The longer the interval between your full recording of observations and the events themselves, the less accurate, the less detailed, and the more biased your report will be.
- Limit your periods of observation to half an hour, which is about as long as a researcher can remain alert enough to perceive and remember the multitude of simultaneous and sequential occurrences.
- At times, children will notice your observing them. If they ask what you are doing, be truthful. Explain it openly and frankly. According to Wright and Barker (1950), children under the age of 9 generally display little self-consciousness when being observed.
- Keep in mind that one of the greatest sources of unreliability in observation is the researcher's selective perceptions influenced by his or her own needs and values. For example, observers who sharply disapprove of aggressive behaviors tend to overrecord these behaviors. Remember at all times that objectivity is your goal.
- Use time sampling for some observations. Time your field notes at intervals of a minute or even 30 seconds. You may wish to tally the children's behavior in terms of helping, resistance, submission, giving, and other responses.
- Use event sampling of behavioral sequences or episodes for some observations. Helen Dawe's 1934 study of the quarrels of preschool children provides a good model. Dawe made "running notes" on prepared forms that gave space for recording (a) the name, age, and sex of every subject, (b) the duration of the quarrel, (c) what the children were doing at the onset of the quarrel, (d) the reason for the quarrel, (e) the role of each subject, (f) specific motor and verbal behavior, (g) the outcome, and (h) the aftereffects. The advantage of event sampling is that it allows you to structure the field of observation into natural units of behavior.

she is observing—all of us tend to act differently when we know we are under close scrutiny. In spite of these shortcomings, there is support for observing behavior as it takes place spontaneously within its natural context. Indeed, some researchers argue that observation of subjects in their natural setting affords greater justice to the rich, genuine, and dynamic quality of human life.

Question

What are the advantages and limitations of each of the following research methods: experimental, case study, social survey, and naturalistic observation?

Cross-Cultural Studies

No animal lives under more diverse conditions than man, and no species exhibits more behavioral variation from one population to another.

Robert A. Levine

Scientists often are interested in discovering which theories hold for all societies, which hold for only certain types of societies, and which hold for only one particular society. The **cross-cultural method** is used to answer these kinds of questions. When researchers can compare data from two or more societies and cultures, then culture, rather than individuals, is the subject of analysis. Cross-cultural studies might focus upon a single issue, such as child-care practices, puberty rites at adolescence, or living conditions of the elderly, or a wide variety of behaviors and customs (Harkness, 1992; Nugent, Lester, & Brazelton, 1991).

Studies dealing with grandparenthood provide a good example of cross-cultural research. According to anthropologist A. R. Radcliffe-Brown (1940), tensions between parents and children tend to draw grandparent and grandchild together. To test this hypothesis, a number of researchers examined cross-cultural data (Apple, 1956; Nadel, 1951). They found close and warm relationships between children and their grandparents *only* in cultures where grandparents do not serve as disciplinarians. Where grandparents have a disciplinary role,

grandparents and grandchildren do not have easy, friendly, playful relations. Other investigators have found strong ethnic differences in styles of grandparenting. For example, Hispanic American grandparents are more likely to have compassionate relationships with their grandchildren and see them more often, providing more help than Anglo grandparents (Bengtson, 1985).

Limitations of Cross-Cultural Studies Like other research approaches, the cross-cultural study has limitations. First, the quality of the data varies from casual, unprofessional accounts by explorers and missionaries to the most sophisticated fieldwork by trained anthropologists. Second, data for some research problems are lacking for many cultures. Third, the data tend to focus on the typical behaviors and practices of a people but seldom provide information on individual differences among them. Nonetheless, as distinguished anthropologist George Peter Murdock has written, cross-cultural research has demonstrated that it is "unsafe" for the scientist "to generalize his knowledge of Euro-American societies, however profound, to mankind [humankind] in general" (Murdock, 1957, p. 251). For example, let us say that a social scientist wants to study the prevalence of the personality trait of "shyness" in young children both in the United States and in Japan. How we define "shyness" in children in the United States (embarrassment at being called upon, or being very quiet or meek, and so on) might be a prevalent personality trait of most Japanese children, where modesty, self-discipline, and

Cross-Cultural Research By comparing data from two or more societies, scientists are able to generalize about human development under varying cultural circumstances. One aspect of development that has interested social scientists recently is the part the father plays in rearing children. Here a Balinese father of Indonesia relaxes with his family.

respect for parents, teachers, and adults is still the norm. For a child to be singled out for the highest grades or the best performance at a task might be an embarrassment or shameful to the Japanese child who has been taught the Confucian philosophy of unity within the group. American society, on the other hand, tends to reward the outspoken, more assertive child, who we say will get ahead in life. Cross-cultural researchers must be careful not to impose their own cultural views upon the behavior under study.

> **Question**
>
> *What are some of the advantages and limitations of using cross-cultural studies?*

Research Analysis

After the research design has been implemented and the subjects chosen and measured, the data are ready to be analyzed statistically. In studies of development we use two broad categories of analysis. First, we can compare different age groups on the variable of interest by simply calculating an average score for each group. We did this informally when we looked at the data on spatial ability using longitudinal and cross-sectional methodologies. We can report sample means and measures of variability and perform various statistical tests to determine how probable it is that the observed differences could have occurred by chance. Second, we can look at the relationship between two variables using correlational analysis. This type of analysis allows us to quantify the association or relationship in terms of strength and direction.

Correlational Analysis

Sometimes social scientists and medical researchers want to know the degree to which two or more behaviors are associated with each other. Correlational analysis does not prove causation, but it can be used for predictive purposes (Aronson, Brewer, & Carlsmith, 1985). For example, in American society over the past decade or so, we have heard through the media that there is an association or relationship between eating a high-fat diet and poor health, such as increased incidence of obesity, higher risk of heart attack, and other overall health risks. Likewise, we also have heard that eating fruits and vegetables daily typically can reduce one's level of bad cholesterol, promoting better health. These are examples of relationships. Having a low cholesterol reading by itself is not the cause of excellent health—but it is one of the many factors predictive of better health.

If two conditions occur and rise or fall in value together, then there is some measure of a positive relationship. For example, eating chocolate bars daily is likely to be associated with a higher cholesterol reading. That is, as the first condition (amount of chocolate eaten) increases, the second condition (cholesterol level) also increases. This is an example of a positive correlation. On the other hand, two conditions that tend to occur in opposition to each other—for instance, when people eat more fruits and vegetables, their cholesterol scores tend to decrease—those conditions have a negative correlation. Medical researchers and social scientists are always searching for these types of associations in order to improve our health.

Plotting the data on a graph or through the use of a mathematical formula can help us determine the extent and direction of these relationships. A **correlation coefficient** (r) is the numerical expression of the degree or extent of relationship between two variables or conditions (note that the word explains itself: *co-relation* (meaning "with relation"). A correlation coefficient can range from -1.00 to $+1.00$. If it is $+1.00$, we say there is a perfect positive relationship between two variables (as one variable increases, the other increases). If it is -1.00, there is a perfect negative relationship between two variables (as one variable increases, the other decreases). If a correlation is near .00, then there is no relationship between the variables. For example, an educated guess about the relationship between how many chocolate bars we eat daily and our intelligence quotient should lead us to the conclusion that there is most likely no relationship whatsoever (r = .00) (see Figure 1-12).

In social and developmental research, we seldom find perfect correlations, but moderately strong correlations can be found in either a positive or a negative direction and can be helpful in explaining certain relationships. For example, one of the strongest positive correlations examined is that of IQ scores of identical twins. The IQ score of one identical twin is strongly predictive of the IQ score of the other twin. A well-publicized example of a negative correlation is the relationship between the number of hours children watch television and their grades in school: the more hours of TV children watch per day, the lower their grades.

> **Questions**
>
> *How would you begin to examine the hypothesis that students who study more get higher grades? What conditions would you need to examine? What type of correlation is this?*

Ethical Standards for Human Development Research

*H*urt not others in ways that you yourself would find hurtful.

A verse from Buddhist Scripture

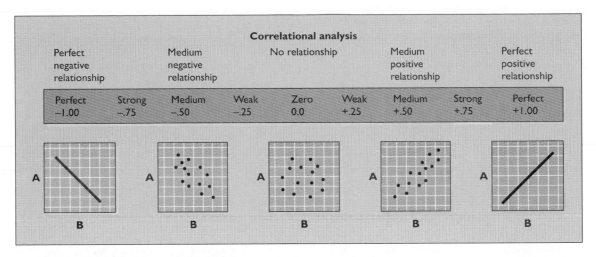

Figure 1-12 Degrees of Relationship Using Correlational Analysis The graphs depict a possible range of relationships between two variables. By plotting our data on a scatter diagram, we can actually begin to see something about the strength and direction (+ or –) of the correlation. The graph on the far left is an example of a perfect negative relationship (unlikely or rare in the real world). As one variable increases, the other would have to decrease in the same measure. Notice the direction of the dots—close together and going up to the left. This illustrates a stronger relationship closer to –1.00. Eating fruits and vegetables to lower bad cholesterol is an example of a negative correlation. With no relationship, the points scatter all over the graph, which tells us the relationship is close to .00. For example, if one plotted the weather and IQ, the chances are great that there is no relationship whatsoever. In a perfect positive relationship, to the far right, as one variable goes up in value, the other variable goes up the same. This would be a +1.00 correlation. An example of a perfect positive relationship occurs when you get paid for the hours you work: if you work 1 hour at $6 per hour, you will make $6; if you work 2 hours, you will make $12; if you work 5 hours, you will make $30. Perfect relationships are rare in real life (and in reality your take-home paycheck is lower than the full amount: the more you make, the more is taken out).

Any research on human development involves some risks and raises some ethical questions, yet research on humans is essential to make progress in understanding the developmental process. How can we learn about how people interact, raise their children, and make decisions about marriage and work, while at the same time safeguarding their privacy? Anytime we study human subjects, we must balance the need to know with the need to protect subjects' personal rights and privacy. The following guidelines and principles established by the American Psychological Association (1992) must be followed any time humans are the subject of our research.

Informed consent: First, the researcher, must obtain written consent to participate from each subject. This consent must be voluntary. For example, research often takes place on college campuses, and it is easy to imagine participation being linked to grades, extra credit, or even less tangible rewards such as staying in a professor's good graces or a recommendation for graduate school or a job. At any time before, during, or after the research, an individual can withhold her or his consent. Second, the researchers must tell each subject the purpose of the research and the risks and benefits of participation, emphasize the voluntary nature of participation, and provide them with a way to communicate with a key person involved in the research. Finally, the participant must be able to understand what he or she is being asked to do, as well as its purpose, risks, benefits, and results. Subjects with cognitive disabilities, subjects diagnosed with a mental illness, and children and adolescents should have special safeguards to ensure that their rights are protected.

Right to privacy: Participants must be assured that the information they share and all research records of their behaviors will be kept confidential. Any data collected must be coded and reported in such a way that the participant cannot be identified. Even if names are not used, the data should never be reported individually—unless specific permission to do so is given by the participant.

The Society for Research in Child Development has also issued a set of guidelines. Research with children raises particularly complex and sensitive issues, because the legal and moral legitimacy of experimentation on human beings depends fundamentally on the consent of the subject (Stanley & Sieber, 1992). Children have very limited capacity to give informed consent; very young children have none. Even so, research suggests that many children over 9 years of age are able to make sensible decisions about whether to take part in research (Thompson, 1990; Fields, 1981). Even though parents and legal guardians are empowered to decide whether minors under their control will be used as research subjects, children should not be viewed simply as pawns that can be manipulated at will by their elders. Finally, both the

APA and the SRCD state that the experimenter must assume responsibility for detecting and correcting any undesirable results that might follow from an individual's participation in the research.

Question

How do we protect human subjects from unethical studies?

Segue

As we have seen, the study of human development is dynamic, and researchers from many disciplines contribute to our understanding of the range and depth of human behavior. The growth, learning, and maturation of individuals and groups are examined across cultures, because each society's interpretation of what is and is not normal and natural affects the lives of the individuals within each culture. Developmentalists study the domains, processes, context, and timing of events to understand both the changes and the continuities of human behavior.

All societies divide the life span in terms of age, defining stages that range from the moment of conception to the moment of death. Societies differ in the prestige they accord various age groupings, particularly with respect to young children and the elderly. Also, societies specify expectations for appropriate behavior within each age grouping. The most profound changes that have occurred in families over the past two hundred years in Western societies include a shift from rural to urban and suburban life, smaller families, more time spent in formal schooling, more women entering the paid labor force, parents and children spending less time together, and a significant increase in the numbers of senior citizens who are living much longer. What developmentalists began studying a century ago is still of prime interest today: the role of biology and behavior, how a person's sense of self develops, how and in what ways a person's environment shapes emotional and social development, maturation in cognitive development, and improvement in the quality of each person's life. Several scientific research methods are used to examine many developmental issues of individuals and groups over the course of the life span.

When we do research with human subjects, we assume major responsibilities. We must study problems and human behaviors in a rigorous and disciplined way, collect and analyze data objectively, and report our results honestly. At the same time, we have the responsibility to protect the rights and privacy of the subjects we are studying and to protect subjects from harm.

In Chapter 2, we provide an overview of the strengths and weaknesses of major theories of human cognitive, moral, emotional, and social development over a lifetime.

Summary

The Major Concerns of Science

1. The study of human development involves the exploration of both change and continuity. Social scientists refer to the elements of change and constancy over the life span as development.

2. The field of human development has four major goals: (a) to identify and describe the changes that occur across the human life span, (b) to explain these changes, (c) to predict occurrences in human development, and (d) to intervene in the course of events in order to control them.

A Framework for Studying Development

3. Developmental change takes place in three fundamental domains: the physical, the cognitive, and the emotional-social.

4. The concepts of growth, maturation, and learning are important to our understanding of human development. We must not contrast the biological forces of growth and maturation with the environmental forces of learning—the interaction between heredity and environment gives an organism its unique characteristics.

5. Urie Bronfenbrenner has proposed an ecological approach to development in which he examines the mutual accommodations between the developing person and four levels of environmental influence: the microsystem, mesosystem, exosystem, and macrosystem.

6. Time plays an important role in development. Traditionally, the passage of time has been treated as synonymous with chronological age. But more recently, social and behavioral scientists have broadened their focus to take into account the changes that occur over time in the environment—and the dynamic relation between change in the

person and change in the environment. One way of looking at these changes is to consider normative age-graded influences, normative history-graded influences, and nonnormative life events.

Partitioning the Life Span: Cultural and Historical Perspectives

7. All societies must deal in one fashion or another with the life cycle. They divide this cycle into strata that reflect social definitions. Such definitions often vary from one culture to another and from one historical period to another.

8. Age is a master status, so most changes in roles over a person's life span are accompanied by a change in chronological age.

9. In the United States, we commonly think of the life span in terms of prenatal development, infancy, childhood, adolescence, adulthood, and old age. Yet this was not always so, and it seems to be changing even now.

10. In the early 1900s, the areas of human development researchers investigated included emotional development, biology and behavior issues, cognitive development, conscious and unconscious thought, and the role of the self. These are still topics of research today.

The Nature of Developmental Research

11. In developmental psychology, research focuses on change that occurs over time or with age. Three basic methods are used: (1) the longitudinal method, (2) the cross-sectional method, and (3) the sequential method.

12. The longitudinal method for studying human development measures the same individuals at regular intervals between birth and death. It allows researchers to describe change sequentially.

13. The longitudinal method undertakes successive measurements on the *same* individuals, while the cross-sectional method compares different groups of people of different ages at the same time.

14. The sequential method involves measuring more than one age group over time. This combination of collecting data over time as well as across groups

overcomes the age/cohort problem found in cross-sectional studies as well as the effect of unique events found in longitudinal designs.

15. The experimental method is one of the most rigorously objective techniques available to science. It offers the only effective technique for establishing cause-and-effect relationships.

16. The case-study method describes one individual over time. This method provides rich detail and description, but its findings cannot easily be generalized to other individuals, other settings, and other time periods. Case studies are often used to examine people with exceptionalities.

17. The social survey method uses questionnaires and interviews to measure attitudes and behaviors of a large number of people.

18. Naturalistic observation enables a researcher to study people independently of their ability or willingness to report on themselves. The techniques of naturalistic observation range from reports on the most casual uncontrolled experiences to videotape records taken in a laboratory setting.

19. The cross-cultural method allows scientists to specify which theories in human development hold true for all societies, which hold for only certain types of societies, and which hold for only a particular society.

Research Analysis

20. Data collected in human developmental studies can be analyzed descriptively (by group or variable of interest). Correlational analysis quantifies the relationship between two variables in terms of strength and direction.

Ethical Standards for Human Development Research

21. Having made the decision to conduct research, a scientist must carry out the investigation with respect for the people who participate as subjects and must protect their privacy and welfare. The American Psychological Association and the Society for Research in Child Development have developed strict guidelines that must be followed any time humans are the subjects of research.

Key Terms

age cohort *(11)*

age strata *(14)*

case-study method *(25)*

cognitive development *(7)*

control group *(23)*

correlation coefficient *(30)*

cross-cultural method *(29)*

cross-sectional method *(21)*

culture *(13)*

dependent variable *(23)*

development *(4)*

developmental psychology *(4)*

ecological approach *(8)*

emotional-social development *(7)*

event sampling *(27)*

exosystem *(10)*

experiment *(23)*

experimental group *(23)*

experimental method *(23)*

extraneous variables *(23)*

growth *(8)*

hypothesis *(19)*
independent variable *(23)*
informed consent *(31)*
learning *(8)*
longitudinal method *(20)*
macrosystem *(10)*
maturation *(8)*
mesosystem *(9)*

microsystem *(8)*
naturalistic observation *(27)*
nonnormative life events *(11)*
normative age-graded influences *(10)*
normative history-graded
 influences *(11)*
physical development *(7)*
random sampling *(26)*

right to privacy *(31)*
scientific method *(19)*
sequential methods *(22)*
social norms *(15)*
social survey method *(26)*
time sampling *(27)*

Following Up on the Internet

Web sites for this chapter focus on professional organizations in the field of human development. Please access the text web site at http://www.mhhe.com/crandell7 for up-to-date hot-linked Internet addresses for the following organizations and resources.

The American Psychological Association (APA)

APA Division 7: Developmental Psychology

The American Psychological Society

International Association for Cross Cultural Psychology

Galaxy: Social Sciences

Chapter 2 Developmental Theories

Critical Thinking Questions

1. Suppose someone comes up with a new theory of development called the "Food Theory," which states that human development can be explained in terms of the foods we eat. Because no two people eat exactly the same foods, it follows that no two people develop in exactly the same way. Why would you accept or reject this theory?

2. If genetic scientists took one of your cells and cloned you—and then gave the cloned infant to the same caregivers you had—do you think the clone would turn out to be just like you?

3. Think for a moment about the individuals who come up with theories about human development. Do you think factors in their own personal development play a role in the formulation of their theories?

4. In what ways is a theory that tries to explain how humans develop similar to a theory that attempts to explain how the universe developed?

Theories allow us to see the world coherently and to act on the world in a rational way. Many theories have evolved over the past century in Western cultures that attempt to explain how human personality develops, why we behave as we do, what environmental conditions motivate us to act certain ways, and how these factors are interrelated. Some of these theories base their explanations on critical physical and social-emotional circumstances in our earliest years of life; some on the impact of environmental influences of our family, community, and culture; some on our distinct learning and thought processes; some on successful completion of specific developmental "tasks" at each stage over the life span; and some on how a healthy—or unhealthy—sense of self shapes our personality and behaviors. Over the past decade or so, the universal applicability of traditional theoretical models of development has been challenged. Many of the long-standing theories presented in this chapter were formulated by Western white males about Western white males. Some newer theories seek to explain the development of women, nonwhites, and people in non-Western cultures.

Cross-cultural social scientists are putting the older theoretical models to the test on a broader scale in scholarly debates in university settings, in international conferences, and in chat rooms and online discussion groups. This is leading to newer perspectives on individual development in all domains. Recently, the American Psychological Association established a new division, International Psychology, and there are emerging many more cross-cultural associations that encourage professionals from all disciplines to collaborate and examine human development on a grander scale.

Theory: A Definition

Nothing is so practical as a good theory.

Kurt Lewin

Many Americans hold theory in low regard. The word *theory* suggests to them a detached, ivory-tower irrelevance to everyday life. College students often complain, "Why do we have to bother with all these theories? Why not just let the facts speak for themselves!" Unfortunately, facts do not "speak for themselves"; facts are silent. Before facts can speak to us, we have to find relationships among them. For example, you might baby-sit, care for younger brothers or sisters, have children of your own, or anticipate having children. What do you do when they misbehave? Do you scold them, threaten them, spank them, forbid them to engage in a favorite activity, reason with them, ignore them, or demonstrate the behavior you expect? What you do is based on your theory—whether explicit or not—about how children learn. Perhaps the theory is embedded in a proverb or maxim, such as Spare the Rod and Spoil the Child, You Got to Toughen Kids Up for Life, Just Give Them Loads of Love, or Spanking Children Causes Emotional Problems.

By formulating a theory, we attempt to make sense of our experiences. We must somehow "catch" fleeting events and find a way to describe and explain them. Only then can we predict and influence the world around us. Theory is the "fabric" we weave to accomplish these ends. A fine garment is crafted out of a variety of pieces of fabric and thread, carefully sewn together, and worn for a particular purpose. Similarly, a **theory** is a set of interrelated statements that provides an explanation for a class of events. It is "a way of binding together a multitude of facts so that one may comprehend them all at once" (G. A. Kelly, 1955, p. 18). The value of the knowledge yielded by the application of theory lies in the control it gives us over our experience. Theory serves as a guide to action.

More specifically, a theory performs a number of functions. First, it allows us to organize our observations and to deal meaningfully with information that would otherwise be chaotic and useless. As French mathematician Jules-Henri Poincaré (1854–1912) observed: "Science is built up with facts, as a house is with stones, but a collection of facts is no more a science than a heap of stones is a house." Second, theory allows us to see relationships among facts and uncover implications that would not otherwise be evident in isolated bits of data. Third, it stimulates inquiry as we search for knowledge about many different and often puzzling aspects of behavior. A theory, then, inspires research that can be used to verify, disprove, or modify that theory. So research continually challenges us to craft new and better theories (see Figure 2-1). In human development, as in other social and behavioral sciences, it is often difficult to determine how *conclusively* the evidence supports a theory, let alone to choose among competing theories. It is a considerably easier task to decide whether the evidence is *harmonious* with a theory (Lieberson, 1992). The various functions of theory will become more evident as we examine some major types of theories of human development.

Psychoanalytic Theories

The history of psychology—like the history of the twentieth century—could not be written without discussing the contributions of Sigmund Freud. Both supporters

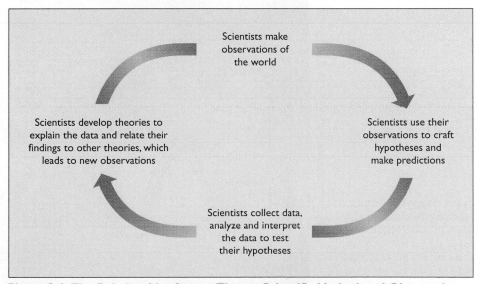

Figure 2-1 The Relationships Among Theory, Scientific Method, and Observations of the World

and critics of his theory of personality regard it as a revolutionary milestone in the history of human thought (Robinson, 1993; Macmillan, 1991). His notions of how behavior is motivated have been embedded in the work done by a multitude of philosophers, social scientists, and mental health practitioners. And characters in countless plays and novels have been built on Freud's view of the individual. Central to **psychoanalytic theory** is the view that personality is fashioned progressively as the individual passes through various **psychosexual stages:** oral, anal, phallic, latency, and genital. He also proposed that people operate from three states of being: the *id,* which seeks self-gratification; the *superego,* which seeks to do what is morally proper; and the *ego,* the rational mediator between the id and superego. Let us turn, then, to a consideration of psychoanalytic theory.

Sigmund Freud: Psychosexual Stages of Development

Freud was born in 1856 and lived most of his life in Vienna. As a child, he was identified as a gifted student and scholar. Though his interests were in research, he was encouraged to become a physician because that was considered the most prestigious profession at that time. Initially, in his medical practice, he distinguished himself through his research on the human nervous system. Early in his career, Freud used hypnosis in treating patients with nervous (which he called *neurotic*) disorders. But he soon became disenchanted with this method. Some of his patients exhibited nervous disorders that could not be attributed to anything physical, per se. Freud hypothesized that something else caused his patients such distress—something the patient was unaware of. He began experimenting with free association of ideas and with dream analysis to tap patients' "unconscious" thoughts, and from this he developed his famous psychoanalytic approach. Many American and European psychologists and psychiatrists were directly or indirectly influenced by Freud's teachings. Freud continued to work in Vienna until 1938, when the Nazis invaded Austria. A Jew, Freud escaped to England. He died in London a year later of cancer. There is a museum in London that is dedicated to Freud's life and his work on the role of the unconscious and psychosexual stages.

The Role of the Unconscious   Freud stressed the role in our behavior of *unconscious* motivation—stemming from impulses buried below the level of awareness. According to Freud, human behavior arises out of a

Sigmund Freud with his daughter Anna in 1939   Although Freud reached the pinnacle of his fame in the period between 1919 and his death in 1939, he formulated most of the essentials of his psychoanalytic theory in the decade between 1893 and 1903. Anna Freud, building upon her father's work, pioneered in the psychoanalytic study and treatment of children.

Information You Can Use

Freudian Analysis of *Hansel and Gretel*

Literature should enrich a child's life. To do so, it must entertain and arouse curiosity, clarify emotions, be attuned to the child's difficulties, but also suggest solutions to those difficulties. Fairy tales offer deep insights into the human condition, and interpreting such tales in terms of the psychoanalytic model of development can shed light on the important messages that are carried to the conscious and unconscious mind of the child. The following is a paraphrase of Bruno Bettelheim's analysis of a classic fairy tale. See if you agree with Bettelheim's view that this theory explains the universal appeal of fairy tales as well as the psychological benefits they have for children.

The Analysis

Hansel and Gretel begins realistically. The parents are poor and they worry about how they will be able to take care of their children. Together at night they discuss their predicament and how they can deal with it. Even on this surface level, the fairy tale conveys an important, although unpleasant, truth: Poverty and deprivation do not improve one's character but rather make one more selfish, less sensitive to the sufferings of others, and thus prone to embark on evil deeds. The fairy tale expresses in words and actions the things that go on in the children's minds. In terms of the child's dominant anxiety, Hansel and Gretel believe that their parents are talking about a plot to desert them. A small child, awakening hungry in the darkness of the night, feels threatened by complete rejection and desertion, which they experience in the form of fear of starvation.

The mother represents the source of all food to the children, so it is she who now is characterized as abandoning them, as if in a wilderness. The child's anxiety and deep disappointment when Mother is no longer willing to meet all oral demands leads the child to believe that Mother has become unloving, selfish, rejecting. Because the children know they need their parents desperately, they attempt to return home after being deserted. In fact, Hansel succeeds in finding their way back from the forest the first time they are abandoned. Before a child has the courage to embark on a voyage of finding herself or himself, of becoming an independent person through meeting the world, initiative can only be developed by trying to return to passivity, to secure eternally dependent gratification. Hansel and Gretel find that this will not work in the long run, and they are frustrated by their inability to find a solution to their problem. Their reliance on food for safety has failed them (the bread crumbs they dropped to mark the path home were eaten by the birds). So now they give in completely to their oral regression. The gingerbread house represents an existence based on primitive satisfactions.

Carried away by their uncontrolled craving, the children destroy the house that could have given them shelter and safety. The gingerbread house is an image nobody forgets: The picture is incredibly appealing and tempting, showing what a terrible risk one runs by giving in to temptation. The house stands for the attractiveness of giving in to oral greediness. A gingerbread house that one can eat up symbolizes the mother, who in fact nurses the infant from her body, and so Hansel and Gretel are eating the good mother, who offers her body as a source of nourishment—the original all-giving mother, whom every child hopes to find again later out in the world when the child's own mother begins to make demands and impose restrictions. But as the story tells, such unrestrained giving in to gluttony threatens destruction. Regression to the earlier state of being, when at her breast one lived symbiotically off the mother, does away with all independence. The witch, who is the personification of the destructive aspects of orality, is as bent on eating up the children as they are on demolishing her gingerbread house. When the children give in to untamed id impulses, they risk being destroyed. Turning the tables on the witch is justified because children who have little experience and are still learning self-control are not to be measured by the same yardstick as adults.

The witch's evil designs finally force the children to recognize the dangers of unrestrained oral greed and dependence. To survive, they must develop initiative and realize that their only recourse lies in intelligent planning and acting. They must exchange subservience to the pressures of the id for acting in accordance with the ego. In the story, the ego makes use of intelligent assessment of the situation and engages in shrewd goal-directed behavior, such as substituting the bone for a finger and tricking the witch into climbing into the oven. Only when the dangers of remaining fixated at orality are recognized does the way for further development open up. It turns out that the good, giving mother was hidden in the bad, destructive mother, because there are treasures to be gained—the children get the witch's jewels. With jewels in hand, they can once again find the good parent. This suggests that as children transcend their oral anxiety and free themselves from reliance on oral satisfaction for security, they can also free themselves from the image of the threatening mother (the witch) and rediscover the good parents, whose greater wisdom (the shared jewels) then benefits all.

The children's experience at the witch's house has purged them of their oral fixations; they arrive home as more mature children, ready to rely on their own intelligence and initiative to solve life's problems. As dependent children, they had been burdens on their parents, but on their return they have become the family's support, as they bring home the treasure.

Continued on page 40

Continued from page 39

These treasures are the children's newly won independence in thought and action, a new self-reliance that is the opposite of the passive dependence that characterized them at the beginning of the tale. *Hansel and Gretel* ends with the heroes returning to the home from which they had started out and now finding happiness there. This is psychologically correct, because a young child, driven into an adventure by oral or oedipal problems, cannot hope to find happiness outside the home. If all is to go well with the child's development, the problems must be worked out while the child is still dependent on the parents. Only with good relations with parents can a child successfully mature into adolescence.

From: THE USES OF ENCHANTMENT by Bruno Bettelheim. Copyright © 1975, 1976 by Bruno Bettelheim. Reprinted by permission of Alfred A. Knopf, Inc.

struggle between societal prohibitions and the instinctual drives associated with sex and aggression. As a consequence of certain behaviors' being forbidden and punished, many instinctual impulses are driven out of our awareness early in our lives. Nonetheless, they still affect our behavior. They find new expression in slips of the tongue ("Freudian slips"), dreams, bizarre symptoms of mental disorder, religion, the arts, literature, and myth. For Freud, the early years of childhood have critical importance; what happens to an individual later in life is merely a ripple on the surface of a personality structure that is fashioned firmly during the child's first five to six years (see *Information You Can Use:* "Freudian Analysis of *Hansel and Gretel*").

Psychosexual Stages   Freud said that all human beings, starting in infancy, pass through a series of **psychosexual stages.** Each stage is dominated by the development of sensitivity in a particular erogenous, or pleasure-giving, zone of the body. Furthermore, each stage poses for individuals a unique conflict that they must resolve before they pass on to the next stage. Should individuals be unsuccessful in resolving the conflict, the resulting frustration becomes chronic and remains a central feature of their psychological makeup. Alternatively, individuals might become so addicted to the pleasures of a given stage that they are unwilling to move on to later stages. As a result of either frustration or overindulgence, individuals experience fixation, or a complex, at a particular stage of development (see Figure 2-2). **Fixation** is the tendency to stay at a particular stage: The individual is troubled by the conflict characteristic of the stage and seeks to reduce tension by means of the behavior characteristic of that stage.

Figure 2-2 Freud: The Oedipus Complex in Boys

© The New Yorker Collection 1991 Lee Lorenz from cartoonbank.com. All rights reserved.

"During the next stage of my development, Dad, I'll be drawing closer to my mother—I'll get back to you in my teens."

The characteristics of Freud's three key psychosexual stages of development—the *oral, anal,* and *phallic* stages—are described in detail in Table 2-1. Freud also identified two later stages, the *latency* period and the *genital* period. He considered these stages less important to the development of the basic personality structure than the stages from birth to age 7. The latency period corresponds to the middle childhood years. During this phase, Freud thought children suppress most of their sexual feelings and become interested in games, sports, and friendships—boys associate with boys, girls with girls. Sexual reawakening occurs at puberty, launching the genital period. In this stage the equilibrium of the latency period is upset. Young people begin experiencing romantic infatuations, emotional upheavals, and the desire to have a satisfactory sexual relationship.

Appraisal of Freud's Work   For decades Freud's ideas dominated much clinical therapy. To many people Freud

seemed to open an entirely new psychological world. His emphasis on environment, not biology or heredity, as the primary factor in mental health and illness was particularly hopeful. In fact, people were so fascinated with the novelty of Freud's insights that few questioned their truth. Nonetheless, scientists have come to recognize that Freudian theory is difficult to evaluate. It makes few predictions that can be tested by accepted scientific procedures (Roazen, 1990; Colby & Stoller, 1988). Freudians say that only a personal psychoanalysis can reveal the truth of the theory's assertions. Unconscious motivation is, by definition, not in the conscious mind. Consequently, scientists lack the means to observe and study such motivation objectively.

Freud constructed his developmental stages almost entirely on the basis of inferences from adult patients. Recent historical research has depicted Freud as having occasionally claimed cures when there were none and as having suppressed or distorted the facts of cases to prove

Table 2-1   Freud's Key Psychosexual Stages

Characteristic	Oral	Anal	Phallic
Time period	Birth to approximately 18 months	Approximately 18 months to 3 years	Approximately the third to seventh year
Pleasurable body zones	Mouth, lips, and tongue	Anus, rectum, and bladder	The genitals
Most pleasurable activity	Sucking during the early phase; biting during the later phase	In the early phase, expelling feces and urine; in the later phase, retaining feces and urine	Masturbation
Sources of conflict	Terminating breast-feeding	Toilet training	In boys, the Oedipal complex: boys feel sexual love for the mother and hostile rivalry toward the father, leading them to fear punishment through castration by the father. In girls, the Electra complex: girls feel sexual love for the father and hostile rivalry toward the mother, leading them to conclude that they have been castrated (because they lack a penis). Their sense of castration gives girls a feeling of inferiority that finds expression in "penis envy."
Common problems associated with fixation	An immature, dependent personality with overwhelming and insatiable demands for mothering; a verbally abusive and demanding personality; or a personality characterized by excessive "oral" behaviors, such as alcoholism, smoking, compulsive eating, and nail biting.	A hostile, defiant personality that has difficulty relating to people in positions of authority; a superconformist personality characterized by preoccupation with rules, regulations, rigid routines, compulsive neatness, and orderliness; or a stingy, miserly personality.	Sexual problems in adulthood—impotence or frigidity; homosexuality; inability to handle competitive relationships.
Social relationships	Infants cannot differentiate between self and nonself. Consequently, they are self-centered and preoccupied with their own needs.	Since parents interfere with elimination pleasures, the child develops ambivalent attitudes toward the parents. As children resolve the conflict between their needs for parental love and for instinctual gratification, they evolve lifelong attitudes toward cleanliness, orderliness, punctuality, submissiveness, and defiance.	A successful resolution of phallic conflict leads the child to identify with the parent of the same sex. In this fashion the child achieves a sense of masculinity or femininity and gives up the incestuous desire for the parent of the opposite sex.

his theoretical points (Raymond, 1991b; Goleman, 1990a). In addition, despite stressing the importance of the early years, Freud rarely worked with children. However, other child psychoanalysts, such as his daughter Anna, did apply his theories to the treatment of children.

Feminist scholars find the psychoanalytic hypothesis of female "penis envy" highly problematic, sexist, and based on the biases of the male-dominated culture of the Victorian era of the late 1800s (Slipp, 1993) (see *Human Diversity:* "Rethinking Women's Biology" in Chapter 1). "Beginning with Freud, and ever since, the psychoanalytic theory of human development has been conceived in terms of male development, with female development either ignored, treated as an afterthought, or forced into parallel lines of reasoning" (Josselson, 1988). Although one of Freud's most famous followers, Erik Erikson, was sympathetic to women's struggles and cognizant of the culturally imposed limitations they faced, he also did not use female exemplars in formulating his theory (Archer, 1992). The unique ways women experience the world might mean that life stages are different for women than they are for men. "Just as women do not define themselves by their work, they also do not define themselves as mother to someone or wife of someone. The hallmark of identity achievement [for women] is the balancing among work, relationships, and interests . . . relationships are primary in their lives" (Josselson, 1988).

Freudian theorists tended not only to ignore women's experience but to blame them for others' psychological difficulties. For example, as recently as the 1950s, Bruno Bettleheim, a Freudian psychoanalyst, claimed that autism resulted from children being raised in a family devoid of warmth and love, or by what he termed "refrigerator mothers." This put a heavy, unnecessary burden on mothers at that time who were attempting to understand a child with such a complex disorder.

Psychiatrist Jean Baker Miller posed her own *relational theory* in contrast to Freud's and Erikson's views. She states that personality growth occurs within relationships and that infants respond to the emotions of caregivers. The goal is to continue to form intimate relationships, not to strive for autonomy and individuation (Miller, 1991). Finally, critics charge that Freudian theory is a poor guide to *healthy* personality development because his patients were suffering from emotional difficulties (Torrey, 1992).

Over the past 35 years, interest in the duration of breast-feeding, severity of weaning, age of toilet training, and other psychoanalytic variables has gradually waned, and the cures expected from psychoanalysis have proven elusive. It is not unusual to encounter someone who has been in psychoanalysis for years. By the early 1970s, a new generation of U.S. psychiatrists was turning to psychobiology, considering defects of nature, not nurture, to be the primary factors in mental illness. These psychi-

Jean Baker Miller and Women's Development Psychiatrist Miller expanded on psychoanalytic theory and posed the relational theory of women's development.

atrists claimed that neurochemical factors, not childhood traumas, best explained mental illness and addictions—hence they looked to genes and the biochemistry of the brain, not to bad parenting, to explain how mental illness is transmitted from one generation to another. The shift away from Freudian theory in no way detracts from the revolutionary significance of Freud's work. Perhaps, more than anything else, Freud deserves considerable credit for directing attention to the importance of early social experience in human development and how those experiences impact the later stages of life.

Questions

What are the distinct features of Freud's psychoanalytic theory? What are the strengths and weaknesses of this theory?

Erik Erikson: Psychosocial Stages of Development

One of Freud's major contributions was to stimulate the work of other theorists and researchers. One of the most talented and imaginative of these theorists was Erik Erikson. A neo-Freudian psychoanalyst of Danish extraction, Erikson (1902–1994) came to the United States in 1933. While acknowledging Freud's genius and monumental contributions, Erikson moved away from the fatalism implicit in Freudian theory, challenging Freud's notion that the personality is primarily established during the first five to six years of life. He observed that if

Erik H. Erikson Although Erikson did not publish his first book until 1950, when he was 48 years old, he then became a leading figure in the psychoanalytic study of human growth and development. He continued writing until his death in his early 90s. His later writings have drawn upon field work with the Oglala Sioux of South Dakota and the Yurok of Northern California, and the clinical treatment of disturbed children and adolescents. Erikson also wrote biographies of Martin Luther and Mohandas Gandhi.

everything goes back to early childhood, then everything becomes someone else's fault, and this undermines trust in one's own capabilities. Erikson concluded that the personality continues to develop over the entire life span. His more optimistic view emphasizes success, greatness, and the flowering of human potential. As his work progressed, Erikson also departed from Freud in another respect. He wove the external landscapes provided by culture, society, and history into Freudian notions of the internal dimensions of the mind.

The Nature of Psychosocial Development Erikson's chief concern is with **psychosocial development,** or development of the person within a social context. In contrast, Freud focused chiefly on the tension occurring as sexual energy sought release, or psycho*sexual* development. Erikson formulated eight major stages of development (see Table 2-2). Each stage poses a unique developmental task and simultaneously confronts individuals with a crisis (Erikson prefers the term *opportunity*) that they must struggle through. As employed by Erikson (1968a, p. 286), a crisis is not "a threat of catastrophe but a turning point, a crucial period of increased vulnerability and heightened potential." He sees great

people of history, such as Martin Luther and Mohandas Gandhi, as achieving greatness by virtue of the fit between their personal crises and the crises of their times. Their solutions—as expressed in their ideas—become cultural solutions to broader social problems.

According to Erikson (1982, 1959), individuals develop a "healthy personality" by mastering "life's outer and inner dangers." Development follows the **epigenetic principle,** a term he borrowed from biology—"anything that grows has a ground plan, and . . . out of this ground plan the parts arise, each having its time of special ascendancy, until all parts have arisen to form a functioning whole" (Erikson, 1968b, p. 92). Hence, according to Erikson, each part of the personality has a particular time in the life span when it must develop if it is going to develop at all—much like the development of the fetus in the womb, whereby each part of the body must develop when its time approaches. Should a capacity not develop on schedule, the rest of the individual's personality development is unfavorably altered. The individual is then hindered in dealing effectively with reality (see *Further Developments: "Theories of Emotions"*). However, Erikson did insist that there must be a healthy balance between both sides of each crisis that we go through. For instance, a healthy mastery of Stage 1 culminates in a preponderance of *trust,* but also produces a healthy dose of *mistrust:* You cannot trust every person you meet on the street and avoid mishap—you must develop a bit of mistrust to get along in this world. But in the end you should lean toward trust, and not mistrust, in your psychosocial development.

Erikson's Eight Stages Erikson was the first theorist to offer a model of development that extended over the entire life cycle. His epigenetic diagram (Table 2-2) depicts the eight stages, beginning with "Trust vs. Mistrust" and ending with "Integrity vs. Despair." By following the diagonal progression, you can see how individuals develop through different crises that must be experienced and worked through. Note that in Table 2-2 the items directly beneath "Identity vs. Identity Confusion" are all manifestations of the previous stages of development that will contribute to the successful formation of identity. For example, if in Stage 1 trust is attained as the predominant outcome, then at the time of identity construction in Stage 5, the individual will be more inclined to mutual recognition of self and others. Then it will be easier to differentiate the "me" from everyone else and positively state "This is who I am." Alternatively, if mistrust is the predominant outcome, autistic isolation will prevent the interpersonal interactions necessary for the constructions of a healthy identity.

Again, using Stage 1 as an example, we can see that the horizontal item beginning with "Temporal Perspective" is really another aspect of *trust* at the stage of development called "Identity vs. Identity Confusion."

Table 2-2   Erikson's Eight Stages of Psychosocial Development

	Stage 1 Birth to 1 year	2 2–3	3 4–5	4 6–puberty	5 Adolescence	6 Young Adulthood	7 Adulthood	8 Late Adulthood
VIII								**Integrity vs. despair**
VII							**Generativity vs. stagnation**	
VI						**Intimacy vs. isolation**		
V	Temporal perspective vs. time confusion	Self-certainty vs. self-consciousness	Role experimentation vs. role fixation	Apprenticeship vs. work paralysis	**Identity vs. identity confusion**	Sexual polarization vs. bisexual confusion	Leader- and followership vs. authority confusion	Ideological commitment vs. confusion of values
IV				**Industry vs. inferiority**	Task identification vs. sense of futility			
III			**Initiative vs. guilt**		Anticipation of roles vs. role inhibition			
II		**Autonomy vs. shame and doubt**			Will to be oneself vs. self-doubt			
I	**Trust vs. mistrust**				Mutual recognition vs. autistic isolation			

From: *Youth and Crisis,* by Erik H. Erikson. Copyright © 1968 by W. W. Norton & Company, Inc. Reprinted by permission of W. W. Norton & Company, Inc.

Erikson's model poses that if trust is established early on in Stage 1, then during adolescence the individual will be able to correctly interpret temporal relations between people and events—and will not, for instance, become angry or distraught if friends show up late to a social gathering. If people resolve each stage satisfactorily, they continue to develop in a healthy manner. Let's briefly examine each crisis, or opportunity, and the preferable outcome:

1. *Trust vs. mistrust* (birth to 1 year): Children come to trust or mistrust themselves and others, depending on their early experiences, usually with caretakers. Favored outcome: Trust in self, parents, and the world.
2. *Autonomy vs. shame and doubt* (2 to 3 years): As children become mobile, they must decide whether to assert their wills. Favored outcome: A sense of self-control without a loss of self-esteem.
3. *Initiative vs. guilt* (4 to 5 years): As children become curious, they begin to explore and manipulate objects. They can be encouraged or made to feel guilty. Favored outcome: The child learns to acquire direction and purpose in activities.
4. *Industry vs. inferiority* (6 to 12 years): Children become concerned with how things work and how they are made. Favored outcome: A sense of mastery and competence.
5. *Identity vs. identity confusion* (adolescence, 13 to 24 years): Children must answer the question "Who am I?" Favored outcome: Ego-identity, a coherent sense of self.
6. *Intimacy vs. isolation* (young adulthood): Being able to reach out and make contact with others. Favored outcome: Working toward a career and becoming intimate with someone.
7. *Generativity vs. stagnation* (adulthood): Looking beyond oneself and embracing society and future generations. Favored outcome: Beginning a family or being concerned with those outside immediate family.
8. *Integrity vs. despair* (old age): In the final years, taking stock of the past. Favored outcome: A sense of satisfaction when looking back over one's life.

Further Developments
Theories of Emotions

Although theories of emotion have a long history, only recently have researchers examined emotive powers positively and with the same interest as they've show in cognition. One reason emotions have traditionally been given "second status" in academia is that emotions are very hard to quantify and measure. Emotions have also historically been linked to abnormal or irrational behaviors.

Aristotle, for example, endorsed the theory that a balance of bodily fluids determines the individual's temperament. He associated anger with overheated blood and argued that the desire to retaliate for a personal injury or offense will sustain anger indefinitely. Descartes believed that ideas are innate and that the body and mind are distinct entities; he attempted to locate emotions in the nervous system. Spinoza regarded emotions as excessive impulses and promoted rational self-control as the means of freeing the self from "emotional bondage." Rousseau insisted that the infant is born with noble emotions, which society adulterates. Kant suggested that innate dispositions are neither good nor bad, and that people need to be guided through life experience and free self-expression in order to control the emotions that are produced. Theodule Ribot, a French psychologist, claimed that emotions are rooted in the individual's innate physical structure, that attitudes and feelings are inborn: some people are more sensitive or more emotional than others because they are born that way.

Darwin thought that strong emotions are important for the survival of species; for instance, a strong emotion like fear in the face of danger enables one to run away and live to face another day. G. S. Hall noted that emotions such as joy, sadness, fear, and anger tend to be expressed more frequently and intensely in childhood and youth. In adolescence, social forces start to thwart or redirect the expression of emotions, leading to other manifestations such as violence.

Freud, influenced by two men, J. Breuer and F. Mesmer, was intrigued with the possibilities of using hypnosis to deal with emotional conflicts in patients. He also attributed fear and anxiety to birth trauma. Later he rejected this view and decided that emotional disturbances were not much different from other neuroses, being much more a matter of degree than difference. William James argued that emotion consists of the feeling or perception of changes occurring in bodily organs; for example, if one sees a dangerous object, one begins to *tremble and run* and *then* experience fear, so that the emotion follows the physical movement.

In response to James's theory, several researchers at Harvard in the early twentieth century countered that emotions depend on neural activity in the brain cortex. They removed part of the hypothalamus from a cat and reported that they had eliminated all angry reactions from the cat. John Dewey thought that the brain and all other bodily structures function in harmonious relation to each other, creating a series of feelings depending on the environment. Watson concluded from his observations and experiments that fear, rage, and love are inherited or developed shortly after birth, and that all other emotions are learned later through classical conditioning. More recently, the study of emotions has included the effects of certain drugs on brain chemistry, as well as the influence of genetic and environmental factors.

This overview is not meant to be comprehensive, but it is hoped that you will get a clearer idea of how theory can change over time to explain complex behaviors.

Source: Adapted from Samuel Smith, *Ideas of the Great Psychologists,* 1983.

Appraisal of Erikson's Work   Erikson's work provides a welcome balance to traditional Freudian theory. Although not neglecting the powerful effects of childhood experience, Erikson draws our attention to the continual process of personality development that takes place *throughout* the life span. His view is a more optimistic view than Freud's. Whereas Freud was primarily concerned with pathological outcomes, Erikson holds open the prospect of healthy and positive resolutions of our identity crises. Erikson's portrait of the life cycle allows "second chances" for opportunities missed and paths not taken. It has always been a general tenet of U.S. individualism that people can improve themselves and continually refashion their fate by changing their social situation, so Erikson's perspective has captured the imagination of the U.S. public. The language Erikson provided—"identity," "identity crisis," "the life cycle"—plays a major role in the U.S. way of thinking about adolescence and, beyond this, about the widest range of adult trials and tribulations (Turkle, 1987).

One legitimate criticism of Erikson's work is that all of the subjects of his psychobiographies and most of his case samples were males (Josselson, 1988). However, since the early 1970s, identity development in women has been looked at more closely using Erikson's identity statuses as a base (Marcia, 1991). Josselson (1988) studied women's identity statuses and found that "a woman's identity at the close of adolescence forms the template for her adulthood." The issues most important to her female subjects were social and religious, not occupational or political. Josselson's findings agree with Jean Baker Miller's relational theory: "Women's sense of self becomes very much organized around being able to make and then to maintain affiliations and relationships" (Josselson, 1988). Gilligan's (1982a) theory also views female identity as rooted in connections to others and in relationships: "Women conceptualize and experience the world in a different voice, and men and women operate with different internal models" (Gilligan, 1982a, p. 7). A comprehensive concept of identity must incorporate both female and male ways of developing (Pescitelli, 1998).

Questions

How does Erikson's theory differ from Freud's theory? What crisis characterizes each of Erikson's eight psychosocial stages?

Behavioral Theory

Psychoanalytic theory focuses on the mental and emotional processes that shape human personality. The data it uses come largely from the self-observations provided by *introspection*. Behavioral theory contrasts sharply with this approach. As its name suggests, **behavioral theory** is concerned with the observable behavior of people—what they actually do and say. Behavioral psychologists believe that if psychology is to be a science, its data must be directly observable and measurable.

Behavioral theorists have traditionally divided *behavior* into units called **responses** and divided the *environment* into units called **stimuli**. Behaviorists are especially interested in how people *learn* to behave in par-

ticular ways, and hence the approach is also termed *learning theory*. Historically, behaviorism has emphasized two types of learning: (1) *classical*, or respondent, conditioning and (2) *operant*, or *instrumental*, conditioning (see Figure 2-3).

Classical conditioning is based upon the work of Ivan Pavlov (1849–1936), a Russian physiologist. Pavlov gained international renown and a Nobel Prize for his early research dealing with the role of gastric juices in digestion in dogs. Subsequently, Pavlov pursued an observation he made while conducting his gastric experiments with dogs. He noted that a dog would initially salivate only when food was placed in its mouth. With the passage of time, however, the dog's mouth would water *before* it tasted the food. Indeed, the mere sight of the food or even the sound of the experimenter's footsteps would cause salivation.

Pavlov was intrigued by the anticipatory flow of saliva in the dogs, a phenomenon he termed "psychic secretion." He saw the study of "psychic secretions" as a powerful and objective means for investigating the mechanisms by which organisms adapt to their environment. So Pavlov devised a series of experiments in which he rang a bell immediately before feeding a dog. After doing this a number of times, the dog's mouth would water at the sound of the bell even though food did not follow.

In his experiments, Pavlov dealt with a behavior that is biologically preprogrammed within a dog through genetic inheritance—the salivation reflex. The *reflex* is an involuntary and unlearned response that is automatically activated by a given stimulus, the presence of food in the animal's mouth. By pairing the sound of the bell with food, Pavlov established a new relationship or connection between a stimulus (the sound of the bell) and a response (salivation) that previously had not existed. This

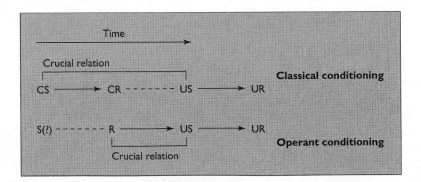

Figure 2-3 Comparison of Classical (Pavlovian) and Operant Conditioning In both processes, the sequence of events is nearly the same. In classical (Pavlovian) conditioning, it is the relation between the CS (conditioned stimulus) and the US (unconditioned stimulus) which is responsible for the CR (conditioned response). After conditioning, the CS (conditioned stimulus) *causes* the CR (conditioned response) (see arrow). In operant conditioning, it is the relationship between the response (R) and the unconditioned stimulus (US), which is crucial (see arrow). Training dolphins to perform acrobatic acts is an example of operant conditioning.

phenomenon is called **classical conditioning**—a process of stimulus substitution in which a new, previously neutral stimulus is substituted for the stimulus that naturally elicits a response. Two illustrations might be helpful: Consider a bright student who develops intense nausea associated with fear when confronted with a test situation. As a child, this student had a teacher who denied recess to youngsters who did poorly on tests and assigned them extra work. Or consider the case of a boy of small size and slender build who develops anxiety about physical education classes after being compelled to compete against bigger and stronger children.

Classical conditioning depends on the prior existence of a reflex that can occur in the service of a new stimulus; in other words, you already have some reflex you can work with. But the conditioning of reflexes does not take us very far, because we usually lack some preexisting unconditioned stimulus with which we can link a new stimulus. So psychologists have searched for alternative mechanisms. One of these mechanisms is probably familiar to you if you have seen animals perform tricks. When dolphins perform acrobatic jumps, they are always rewarded with food immediately afterward. In this procedure the dolphin is made to *enact* the behavior and then is rewarded with fish; the food *follows* the response, or the trick, and reinforces that particular behavior. When teaching a dolphin to do tricks, trainers employ **operant conditioning**—a type of learning in which the consequences of a behavior alter the strength of that behavior. *Operants* are behaviors that are susceptible to control by changing the effects that follow them; they are responses that "operate" or act upon the environment and generate consequences. So when a dolphin engages in behavior that produces food, the behavior is strengthened by this consequence and therefore is more likely to recur in the future (in contrast to classical conditioning, where the food produces the behavior) (see Figure 2-3).

To summarize, classical conditioning derives from preexisting reflexes; operant conditioning does not. In classical conditioning a stimulus is said to *elicit* the response, whereas in operant conditioning the response is *emitted.* In classical conditioning antecedents determine the response probability; in operant conditioning it is determined by *consequences.*

We owe much to the earlier work of behaviorist John Watson (1878–1958), who said that people do not go through distinct stages but do go through a continuous process of behavior changes due to responses to environmental influences (external stimuli). Later, B. F. Skinner (1904–1990), at Harvard University, promoted our understanding of operant conditioning, especially the role of rewards and punishments. During the 1950s and 1960s no U.S. psychologist enjoyed greater prominence or commanded greater influence than did Skinner. Among the concepts popularized by Skinner is that of **reinforcement**—the process whereby one event

B. F. Skinner   In the twenty years following World War II, B. F. Skinner (above left) was the dominant figure in American psychology. His experimental work with pigeons pioneered many facets of behavioral theory. As a strict behaviorist, Skinner did not concern himself with what goes on inside the organism. Instead, he stressed the part that learning processes (environment) play in an organism's acquisition of various behaviors.

strengthens the probability of another event's occurring. Skinner showed that much of life is structured by arranging reinforcing consequences, or "payoffs." For instance, businesses reward appropriate employee work behaviors with wages, commissions, and bonuses; and a doctor must make certain that a patient feels he or she has benefited from an office visit in order to induce the patient to return for further treatment.

Many of the principles of learning have found a use in **behavior modification.** This approach applies learning theory and experimental psychology to the problem of altering maladaptive behavior. According to behaviorists, pathological behavior is acquired just as normal behavior is acquired—through the process of learning. They claim that the simplest technique for eliminating an *unwanted* behavior is usually to stop reinforcing it. Interestingly enough, by attending to a child's inappropriate behavior (e.g., by scolding), we can be reinforcing exactly what we want to diminish. The next time you are in a grocery store, observe how a parent reacts when his or her child wants a package of candy at the checkout counter. The parent might start out by saying no, then give in to the child's demand in order to avoid a commotion. Would you care to guess what is going to happen the next time that same child comes to the grocery store checkout line? (Remember, the candy is the reinforcer.)

But behavior modification can also involve more deliberate intervention in the form of rewards or punishments. Rewards, as reinforcers, normally are selected by the individual whose behavior is to be changed. Behavior modification has helped obese people lose weight and has helped people overcome *phobias,* such as fear of high places, of taking tests, of sexual inadequacy, of closed-in spaces, and of speaking before an audience.

Our understanding of conditioning has undergone major transformations over the past three decades (Rosales-Ruiz & Baer, 1997; Chiesa, 1992). Psychologists no longer view conditioning as a simple, mechanical process involving the association of two events that happen to occur rather closely in time. Organisms do not pair events in a vacuum. The environmental context—the overshadowing of some stimuli, the blocking of others, and the highlighting of still others—is critically important. As seen from a cognitive learning perspective, organisms learn only when events violate their expectations (Williams, LoLordo, & Overmier, 1992). Over time, organisms build an image of the external world and continually compare this image with reality, selectively associating the most informative or predictive stimuli with certain events.

For conditioning to take place, a stimulus must tell the organism something useful about events in the world that the organism does not already know. For example, suppose you retrieve a baseball from a bed of poison ivy. A few hours later, your skin becomes red and itchy, and tiny blisters develop. You are unlikely to link the two events. But should a physician, friend, or coworker point out to you that you are allergic to poison ivy, you grasp the relationship between the blisters and the offending plant. You then take care to avoid contact with poison ivy in the future. You have learned!

Question
Consider the growing problem of the eating disorder anorexia nervosa in younger Americans. How would a behavioral psychologist explain the development of this condition?

Humanistic Theory

*T*o be what we are, and to become what we are capable of becoming, is the only end of life.

Robert Louis Stevenson

In the past 40 years or so, a "third force" in psychology has arisen in reaction to the established traditions of psychoanalysis and behaviorism. Commonly termed **humanistic psychology,** it maintains that humans are different from all other organisms in that they actively intervene in the course of events to control their destinies and shape the world around them. Humanistic psychologists, such as Abraham Maslow (1970, 1968) and Carl R. Rogers (1970), share a common concern with maximizing the human potential for self-direction and freedom of choice. They take a **holistic approach,** one that views the human condition in its totality and each person as more than a collection of physical, social, and psychological components.

One of the key concepts advanced by Maslow is the **hierarchy of needs,** depicted in Figure 2-4. Maslow felt that human beings have certain basic needs that they must meet before they can go on to fulfill their other developmental needs. At the bottom of Maslow's pyramid are fundamental requirements to satisfy physiological needs (including needs for food, water, and sex) and safety needs. Next, Maslow identified a set of psychological needs centering on belongingness (love) needs and self-esteem needs. At the top of the pyramid, he placed the need to realize one's unique potential to the fullest, in a process he termed **self-actualization.**

To Maslow, such people as Abraham Lincoln, Albert Einstein, Walt Whitman, Eleanor Roosevelt, and Martin Luther King are good examples of self-actualizers. From their lives he constructed what he believed to be a composite picture of self-actualized persons (Maslow, 1970):

- They have a firm perception of reality.
- They accept themselves, others, and the world for what they are.
- They often are spontaneous in thought and behavior.
- They are problem-centered rather than self-centered.
- They have an air of detachment and a need for privacy.
- They are autonomous and independent.
- They resist mechanical and stereotyped social behaviors, although they are not deliberately or flamboyantly unconventional.
- They are sympathetic to the condition of other human beings and seek to promote the common welfare.
- They establish deep and meaningful relationships with a few people rather than superficial bonds with a great many people.
- They have a democratic world perspective.
- They transcend their environment rather than merely cope with it.
- They have a considerable fund of creativeness.
- They are susceptible to *peak experiences* marked by rapturous feelings of excitement, insight, and happiness.

Maslow and other humanistic psychologists believe that scientific inquiry should be directed toward helping people achieve freedom, hope, self-fulfillment, and strong identities. The goal of humanistic therapy is to help a person become more self-actualized; that is, to guide the client through self-directed change, building self-esteem along the way (in contrast to psychoanalysis and behavior modification, which are directed more by the thera-

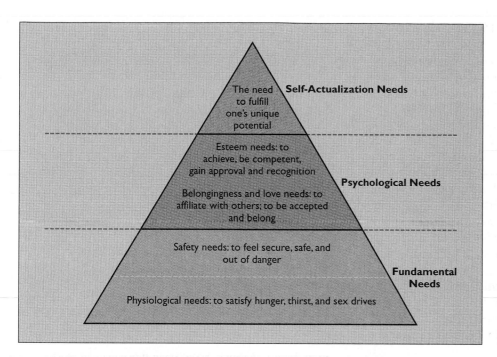

Figure 2-4 Maslow's Hierarchy of Human Needs According to the humanistic psychologist Abraham Maslow, fundamental needs must be satisfied before an individual is free to progress to psychological needs, which in turn must be met before the person can realize self-actualization needs.

From: MOTIVATION AND PERSONALITY 3rd ed. By Abraham H. Maslow. Copyright © 1954, 1987 by Harper & Row Publishers, Inc. Copyright 1970 by Abraham H. Maslow. Reprinted by permission of Addison-Wesley Educational Publishers, Inc.

pist). However, many other psychologists are skeptical about their humanistic colleagues. Indeed, important differences characterize their intellectual style (Kimble, 1984). Psychoanalytic and behavioral psychologists see their primary task as one of increasing the storehouse of scientific knowledge, whereas humanists primarily focus on improving the human condition. Moreover, the former view behavior as determined by underlying laws that can be revealed by using the scientific method. Many humanistic psychologists assert that there is nothing lawful about human behavior except perhaps at the level of statistical averages; they investigate behavior by relying on intuition and insight. Additionally, critics charge that humanistic psychology turns people inward, encouraging an intense concern with the self, breeding a narcissistic outlook, one that says that if each of us works on becoming more fully human ourselves, then social ills such as racism, homelessness, hunger, and militarism will vanish.

Questions

Can you characterize a "self-actualized" individual, as described by Maslow? Do you know such an individual in your own life?

Cognitive Theory

In its early formulations, behaviorism regarded human life as if it were a "black box." Behaviorism's proponents viewed input or stimuli as entering the "box" at one end and coming out the other end as output or responses. They had little concern about what was inside. But over the past 40 years, psychologists have become increasingly interested in what goes on inside the box. They term these internal factors **cognition**—acts or processes of knowing. Like humanistic theory, **cognitive theory** takes issue with a number of behaviorist tenets. Cognition involves how we go about representing, organizing, treating, and transforming information as we devise our behavior. It encompasses such phenomena as sensation, perception, imagery, retention, recall, problem solving, reasoning, and thinking. Cognitive psychologists are especially interested in the cognitive structures and processes that allow a person to mentally represent events that transpire in the environment. The initial impetus to the study of cognition in the United States came from the work of a Swiss developmental psychologist, Jean Piaget (1896–1980).

Jean Piaget: Cognitive Stages in Development

Intelligence is an adaptation. . . . Life is a continuous creation of increasingly complex forms and a progressive balancing of these forms with the environment.

Jean Piaget, *The Origins of Intelligence in Children*, 1952

Like Freud, Piaget has come to be recognized as a giant of twentieth-century psychology (Beilin, 1992). Anyone who studies Freud and Piaget will never again see children in quite the same way. Whereas Freud was primarily concerned with *personality development,* Piaget concentrated on changes that occur in the child's *mode of thought.* Central to Piaget's work are the **cognitive stages** in development—sequential periods in the growth or maturing of an individual's ability to think—to gain knowledge, self-awareness, and awareness of the environment.

Adjustment as Process   When Piaget began to work with children in the early 1920s, little was known about the process by which thinking develops. To the extent that they considered the matter at all, most psychologists assumed that children reason in essentially the same way as adults. Piaget soon challenged this view. He insisted that the thought of infants and children is not a miniature version of adult thought; it is qualitatively distinct. As children grow up, the form of their thought changes. When they say that their shadow follows them about when they go for a walk or that dreams come through the window, they are not being illogical. Rather, they are operating from a mental framework that is different from that of adults.

Piaget depicted children as engaged in a continual interaction with their environment. They act upon, transform, and modify the world in which they live. In turn, they are shaped and altered by the consequences of their own actions. New experiences interact with an existing structure or mode of thought, thereby altering this structure and making it more adequate. This modified structure in turn influences the child's new perceptions. These new perceptions are then incorporated into a more complex structure. In this fashion, experience modifies structure and structure modifies experience. Hence, Piaget viewed the individual and the environment as engaged in continuing interaction. This interaction leads to new perceptions of the world and new organizations of knowledge (Brown, 1996; Beilin, 1992, 1990).

Basically, Piaget saw development as *adaptation.* Beginning with the simple reflexes they have at birth, children gradually modify their repertoire of behaviors to meet environmental demands. By interacting with their environment during play and other activities, children construct a series of *schemas*—concepts or models—for coping with their world. In Piaget's theory, **schemas** are cognitive structures that people evolve for dealing with specific kinds of situations in their environment. Thus, as portrayed by Piaget, children's thoughts reflect not so much the bits of information that they acquire but the schemas or mental frameworks by which they interpret information from the environment.

As Piaget viewed adaptation, it involves two processes: assimilation and accommodation. **Assimilation** is the process of taking in new information and interpreting it so that it conforms to a currently held model

Jean Piaget at Work   Piaget spent more than 50 years observing children in informal settings. His work convinced him that a child's mind is not a miniature model of the adult's. We often overlook this fact when attempting to teach children by using adult logic.

of the world. Piaget said that children typically stretch a schema as far as possible to fit new observations. But life periodically confronts them with the inescapable fact that some of their observations simply do not fit their current schemas. Then, disequilibrium or imbalance occurs. As a result, children are compelled to reorganize their view of the world to fit new experience. In effect, they are required to invent increasingly better schemas or theories about the world as they grow up. **Accommodation** is the process of changing a schema to make it better match the world of reality. Unlike assimilation, in which new experiences are fit into existing conceptions of the world, accommodation involves changing a conception in order to make better sense of the world.

A balance between the processes of accommodation and assimilation is **equilibrium.** When in equilibrium, the child assimilates new experiences in terms of the models she or he arrived at through accommodation. But equilibrium eventually gives way again to the process of accommodation and the creation of new schemes or models. Thus, as viewed by Piaget, cognitive development is marked by alternating states of *equilibrium* and *disequilibrium.* Each stage consists of particular sets of schemes that are in a relative stage of equilibrium at some point in a child's development.

In one study, Piaget and other researchers asked children whether they had ever had a bad dream. One 4-year-old said that she had dreamed about a giant and explained, "Yes, I was scared, my tummy was shaking and

I cried and told my mommy about the giant." When asked, "Was it a real giant or was it just pretend?" she responded, "It was really there but it left when I woke up. I saw its footprint on the floor" (Kohlberg & Gilligan, 1971, p. 1057).

According to Piaget, this child's response is not to be dismissed as the product of wild imagination. Viewed from the perspective of her current schema, the happenings in dreams are real. As the child matures, she will have new experiences that will cause her to question the schema. She might observe, for instance, that there is not really a footprint on the floor. This assimilation of new information will result in disequilibrium. Through accommodation, she will then change her schema to make it a better fit with reality. She will recognize that dreams are not real events. She will then formulate a new schema that will establish a new equilibrium. In this new schema, a child her age will typically depict dreams as imaginary happenings. She will still believe, however, that her dreams can be seen by other people.

The process of accommodation continues through additional steps of the same sort. Soon after realizing that dreams are not real, the child comes to recognize that they cannot be seen by others. In the next step the child conceives of dreams as internal, but nevertheless material, events. Finally, somewhere between 6 and 8 years of age, the child becomes aware that dreams are thoughts that take place within her mind.

Characteristics of Piaget's Cognitive Stages Piaget contended that biological growth combines with children's interaction with their environment to take them through a series of separate, age-related *stages*. The stage concept implies that the course of development is divided into steplike levels. Clear-cut changes in behavior occur as children advance up the developmental staircase, with no skipping of stages allowed. Although teaching and experience can speed up or slow down development, Piaget believed that neither can change the basic order of the stages (Piaget, 1970). Piaget distinguished four stages in the development of cognition or intelligence. They are summarized in Table 2-3 and will be treated in greater detail in later chapters concerned with cognitive growth.

Table 2-3 Piaget's Stages of Cognitive Development

Developmental Stage	Major Cognitive Capabilities	Example
Sensorimotor stage (birth to 2 years)	Infants discover the relationships between sensations and motor behavior.	They learn that their hands are part of themselves whereas a ball is not.
	Children master the *principle of object permanence*.	Piaget observed that when a baby of 4 or 5 months is playing with a ball and the ball rolls out of sight behind another toy, the child does not look for it even though it remains within reach. Piaget contended that infants do not realize that objects have an independent existence. Around the age of 8 months, the child grasps the fact of object constancy and will search for toys that disappear from view.
Preoperational stage (2 to 7 years)	Children develop the capacity to employ *symbols*, particularly language.	Children use symbols to portray the external world internally—for instance, to talk about a ball and form a mental image of it.
	Egocentrism prevails.	Children of 4 and 5 years consider their own point of view to be the only possible one. They are not yet capable of putting themselves in another's place. A 5-year-old who is asked why it snows will answer by saying, "So children can play in it."
Stage of concrete operations (7 to 11 years)	Children show the beginning of rational activity. They are able to "conserve" mass, weight, number, length, area, and volume.	Youngsters come to master various logical operations, including arithmetic, class and set relationships, measurement, and conceptions of hierarchical structures. Before this stage children do not appreciate that a ball of clay can change to a sausage shape and still be the same amount of clay.
	Children gain the ability to "conserve" quantity.	Before this stage, children cannot understand that when water is poured out of a full glass into a wider glass that the water fills only halfway, the amount of water remains unchanged. Instead, children "concentrate" on only one aspect of reality at a time. They see that the second glass is half empty and conclude that there is less water in it. Now children come to understand that the quantity of water remains the same.
Stage of formal operations (11 years and older)	Youths acquire a greater ability to deal with abstractions.	When younger children are confronted with the problem, "If coal is white, snow is ———," they insist that coal is black. Adolescents, however, respond that snow is black.
	Youths can engage in scientific thought.	At this stage, youths can discuss Newtonian principles about the behavior of spherical objects.

Appraisal of Piaget's Work   U.S. scientists largely ignored Piaget's discoveries until about 1960. Today, however, the study of cognitive factors in development is of central interest to U.S. psychologists. For the most part, psychologists credit Piaget with drawing their attention to the possibility that an unsuspected order might underlie some aspects of children's intellectual development (Levin & Druyan, 1993). Nonetheless, many early American followers of Piaget, such as John H. Flavell, have become disenchanted with the Piagetian model. Flavell (1992) says that the notion of stages implies long periods of stability, followed by abrupt change. But he believes that development does not happen this way. The most important changes happen gradually, over months and even years. In short, human cognitive development is too varied in its mechanisms, routes, and rates to be accurately portrayed by an inflexible stage theory. According to Flavell, then, growing up is much less predictable than Piaget thought.

A mounting body of evidence also suggests that Piaget underestimated the cognitive capabilities of infants and young children. For instance, the kinds of memory Piaget found in 18-month-old babies researchers now find in babies at 6 months of age. Of course, Piaget did not have many of the methods that are now available to scholars, including equipment and procedures to measure the electrical activity of the brain. Today's methods reveal that by 6 months of age the brain's electrical activity already falls into telltale patterns, showing that babies can distinguish between familiar and unfamiliar pictures and documenting that they have developed mental representations allowing them to compare old and new information (Raymond, 1991a). The operational thinking capabilities of children from 2 to 7 years of age also are considerably greater than Piaget recognized (Novak & Gowin, 1989).

Furthermore, research on other cultures has revealed both striking similarities and marked differences in the performance of children on various cognitive tasks. Certain aspects of cognitive development among children in these cultures seem to conflict with particular assumptions of Piagetian theory (Chapman, 1988; Fischer & Silvern, 1985). We should remember that no theory—particularly one that offers such a comprehensive explanation of development—can be expected to withstand the tests of further investigation without undergoing some criticism (Brown, 1996; Beilin, 1990).

It is too soon to determine the ultimate impact Piaget's theory will have on our understanding of cognitive development. Yet we must recognize that we would not know as much as we do about children's intellectual development without the monumental contributions of Piaget. He noted many ways in which children seem to differ from adults, and he shed light on how adults acquire fundamental concepts such as the concepts of space, time, morality, and causality (Sugarman, 1987). Contemporary researchers have attempted to integrate aspects of Piaget's theory into cognitive learning and information-processing theories, which are discussed below (Brown, 1996; Demtrious, 1988).

Cognitive Learning and Information Processing

Piaget's work gave a major impetus to cognitive psychology and to research into the part played by inner, mental activity in human behavior (Sperry, 1993). Cognitive psychologists view the contents of conscious experience and their subjective qualities as dynamic, emergent properties of *brain activity* (inseparably interfused with and tied to the brain's cellular and biochemical properties and processes). Reversing classical behavioral notions, cognitive theorists affirm that the world we live in is driven not solely by mindless physical forces but also by subjective human attitudes, values, and aims.

These psychologists are finding, for instance, that mental schemes—often called "scripts" or "frames"—function as selective mechanisms that influence the information individuals attend to, how they structure it, how much importance they attach to it, and what they then do with it (Vander Zanden, 1987; Markus, 1977). And as we noted earlier in the chapter, psychologists are also finding that learning consists of more than merely bringing two events together. People are not simply acted upon by external stimuli. They actively engage their environment, evaluate different stimuli, and devise their actions accordingly.

Classic behavioral theory also fails to explain many changes in our behavior that result from interactions with people in a social context. Indeed, if we learned solely by direct experience—by the reward or punishment for our actions—most of us would not survive to adulthood. If, for example, we depended on direct experience to learn how to cross the street, most of us would already be traffic fatalities. Similarly, we probably could not develop skill in playing baseball, driving a car, solving mathematical problems, cooking meals, or even brushing our teeth if we were restricted to learning through direct reinforcement. We can avoid tedious, costly, trial-and-error experimentation by *imitating* the behavior of socially competent models. By watching other people, we learn new responses without first having had the opportunity to make the responses ourselves. This process is termed **cognitive learning.** (It is also termed *observational learning, social learning,* and *social modeling.*) The approach is represented by the work of theorists such as Albert Bandura (1989a, 1986, 1977), Walter Mischel (1973), and Ted L. Rosenthal and Barry J. Zimmerman (1978).

The cognitive learning theory of Bandura (1989a, 1989b) relies heavily on notions of information-processing theory, which holds that individuals perform a series of discrete mental operations on incoming information

Social Modeling Often Influences Child Behavior How little we may be aware that children imitate our behaviors.

they actively seek out all sorts of environments (Grusec, 1992).

Cognitive learning theorists say that our capacity to use symbols gives us a powerful way to comprehend and deal with our environment. Language and imagery allow us to represent events, analyze our conscious experience, communicate with others, plan, create, imagine, and engage in foresightful action. Symbols are the foundation of reflective thought and enable us to solve problems without first having to enact all the various solutions. Indeed, stimuli and reinforcements exert little impact on our behavior unless we first represent them mentally (Rosenthal & Zimmerman, 1978; Bandura, 1977).

Cognitive theorists reject the portrayal of children as "blank slates" who passively and unselectively imitate whatever the environment presents to them. Rather, they portray children as active, constructive thinkers and learners. Children's cognitive structures and processing strategies lead them to select meaningful information from an array of sensory input and to mentally represent and transform this information. Children actively seek knowledge, develop their own theories about the world around them, and continually subject these theories to knowledge-extending and knowledge-refining tests. So, to a considerable degree, children manufacture their own development as they interact with the environment (Flavell, 1992).

Social learning theories and *cognitive learning theories* have been criticized for their lack of attention to significant developmental changes that can impact behavior. Bandura attempted to respond to this matter in his later theoretical writings, but he and his associates undertook little accompanying research that specifically addressed developmental issues. Consequently, approaches that emphasize more clearly age-related changes in development have moved to the forefront of interest for many developmental psychologists (Grusec, 1992).

and then mentally store the conclusions drawn from the process (Mayer, 1996). Bandura's theory emphasizes how children and adults mentally operate on their social experiences and how these mental operations in turn influence their behavior. People abstract and integrate information that they encounter in the course of their social experiences, including their exposure to models, verbal discussions, and encounters with discipline.

By means of this abstraction and integration, individuals mentally represent their environments and themselves, particularly in terms of the expectations they hold for the outcomes of their behavior and the perceptions they evolve of the actual effectiveness of their actions. Bandura portrayed people not as weather vanes who constantly shift their behavior in accordance with momentary influences but rather as stewards of values, social standards, and commitments. That is, individuals judge and regulate their own behavior. They evolve beliefs about their own specific abilities and characteristics (what Bandura called "self-efficacy") and then use these beliefs in fashioning what they say and do. Moreover, children and adults not only respond to environments,

> **Question**
>
> *Freud and Piaget are often considered to be the great "stage" theorists of developmental psychology. In what ways are their theories similar, and in what ways are they dissimilar, in their views of human development?*

Evolutionary Adaptation Theory

Historically, psychologists have focused primarily on the environmental influences that come to bear on organisms in fashioning and directing their behavior. But over the past decade or so they have paid increasing attention to the part that biological patterns play in enhancing an organism's adaptation to a particular environment. Adaptation involves changes in species rather than in individual organisms and approaches based on adaptation view human behavior as the unique product of our evolutionary history—we behave as we do because our species has evolved in human ways.

Adaptation theories rest on the theory of evolution advanced in 1859 by the English naturalist Charles Darwin (Scarr, 1993; Charlesworth, 1992). Darwin proposed a specific mechanism by which adaptation occurs—**natural selection.** The idea behind natural selection is remarkably simple: The different organisms produce more offspring than the available resources can support. Because a habitat can support only a limited number of organisms, each organism must compete for a "place." Those that are best adapted to the environment survive and pass on their genetic characteristics to their offspring. The others perish, with few or no offspring. So later generations resemble their better-adapted ancestors. The result is evolutionary change. *Ethology* is a major outgrowth of Darwinian theory.

Ethology is the study of the behavior patterns of organisms from a biological point of view. Behavior is seen as part of the adaptational package of an organism and as being as necessary for its survival as its heartbeat or its skeletal structure. Recent research suggests that adaptation is also important in the development of language, religion, warfare, reading, writing, and so forth (Buss et al., 1998). Ethologists say that natural selection molds a species' behavior in accordance with environmental costs and benefits (see Scarr, 1993).

It follows that organisms are *genetically prepared* for some responses (Eibl-Eibesfeldt, 1989). For instance, much learning in many insects and higher animals is guided by information inherent in the genetic makeup of the organism (MacDonald, 1992). The organism is pre-programmed to learn particular things and to learn them in particular ways. As we will see in Chapter 5, Noam Chomsky says that the basic structure of human language is biologically channeled by an inborn language-generating mechanism. Such a mechanism helps to explain why we learn speech so much more easily than we learn inherently simpler tasks such as addition and subtraction.

Ethologists also hold that human babies are biologically preadapted with behavior systems like crying, smiling, and cooing that elicit caring by adults (Zebrowitz, Olson, & Hoffman, 1993). Similarly, babies' being "cute"—with large heads, small bodies, and distinctive facial features—induces others to want to pick them up and cuddle them. Ethologists call these behaviors and features **releasing stimuli.** They function as especially potent activators of parenting.

A number of psychologists, among whom John Bowlby (1969) is perhaps the most prominent, compare the development of strong bonds of attachment between human caretakers and their offspring to the process of imprinting encountered among some bird and animal species. **Imprinting** is a process of attachment that occurs only during a relatively short period and is so resistant to change that the behavior appears to be innate. Konrad Lorenz (1935), the Nobel Prize–winning ethologist, has shown that there is a short period of time early in the lives of goslings and ducklings when they slavishly follow the first moving object they see—their mother, a human being, even a rubber ball. Once this imprinting has occurred, it is irreversible. The object becomes "Mother" to the birds, so that thereafter they prefer it to all others and in fact will follow no other. Imprinting (Lorenz uses his native German word *Prägung*, which literally means "stamping in") differs from other forms of learning in at least two ways. First, imprinting can take place only during a relatively short period, termed a **critical period.** (For example, the peak period for the imprinting effect among domestic chickens occurs about 17 hours after hatching and declines rapidly thereafter.)

Konrad Lorenz Here, young goslings follow the eminent Austrian ethologist rather than their mother. Because he was the first moving object that they saw during the critical imprinting period, they came to prefer him to all other objects.

Human Diversity

Jerome Kagan on the Early Years

Most Americans—both psychologists and the general public—believe that early experience etches an indelible mark upon the mind. A long intellectual tradition in Western culture has portrayed the adult as locked in a core personality fashioned before the age of 6. It is thus hardly surprising that a good many American parents experience considerable anxiety and guilt over the adequacy of their child rearing.

Jerome Kagan (1984), a Harvard University developmental psychologist, challenges the prevailing belief (also see Rowe, 1994; Scarr, 1993). Since 1971 he and his associates have done research among children in Guatemala that has led Kagan to conclude that children are considerably more resilient than we think (Kagan, 1984).

Among the children Kagan studied is a group in San Marcos, an isolated Indian village on Lake Atitlán. During the first year of life, these infants spend most of their time in the small, dark interiors of windowless huts. Their mothers are busy with domestic chores and rarely speak to their babies or play with them. The Indians believe that babies can be harmed by the outside sun, air, and dust and rarely permit them to crawl on the hut floors or to venture beyond the doorway. Judged by U.S. standards, the infants appear to be severely retarded—undernourished, listless, apathetic, fearful, dour, and extraordinarily quiet. Indeed, to Kagan the children have a ghostlike quality. Observations and tests suggest that they are 3 to 12 months behind U.S. children in acquiring psychological and cognitive skills ranging from the simple ability to pay attention all the way up to the development of meaningful speech.

Nonetheless, Kagan finds that by age 11 the children show no traces of their early "retardation." On tests of perceptual inference, perceptual analysis, recall, and recognition, they perform in a manner comparable to U.S. middle-class children. According to Kagan, they begin to overcome their early retardation when they become mobile at around 15 months. They leave the dark huts and begin to play with other children, and in the ensuing years they experience challenges that demand and foster intellectual growth and development.

Kagan suggests that a child's experiences can slow down or speed up the emergence of basic abilities by several months or even years. Thus, Kagan reaches the highly controversial conclusion that children are biologically preprogrammed with basic mental competencies—an inherent blueprint that equips them with the essentials for perceptual and intellectual functioning—and that the content of this functioning differs from culture to culture (Kagan & Klein, 1973, p. 960):

> The San Marcos child knows much less than the American about planes, computers, cars, and the many hundreds of other phenomena that are familiar to the Western youngster, and he is a little slower in developing some of the basic

Continued on page 56

Guatemalan Indian Children By American standards, Guatemalan Indian children's early experiences are devoid of surroundings that stimulate their psychological and intellectual development. Even so, youngsters like the 11-year-olds shown here appear normally adjusted and healthy.

Continued from page 55

cognitive competencies of our species. But neither appreciation of these events nor the earlier cognitive maturation is necessary for a successful journey to adulthood in San Marcos. The American child knows far less about how to make canoes, rope, tortillas, or how to burn an old *milpa* in preparation for June planting. Each knows what is necessary, each assimilates the cognitive conflicts that are presented to him, and each seems to have the potential to display more talent than his environment demands of him.

Kagan says that his research has implications for the United States' educational problems. He suggests that the poor test performance of economically impoverished and minority-group children should not be taken as evidence of permanent or irreversible defects in intellectual ability. To class children arbitrarily at age 7 as competent or incompetent makes as much sense, Kagan insists, as classifying children as reproductively fertile or sterile depending on whether they have reached physiological puberty by age 13.

Second, as already mentioned, imprinting is irreversible; it is highly resistant to change, so that the behavior appears to be innate.

Some developmental psychologists have applied ethological notions to human development. However, many prefer the term **sensitive period** to "critical period," for it implies greater flexibility in the time dimension and greater reversibility in the later structure. According to this concept, particular kinds of experience affect the development of an organism during certain times of life more than they do at other times (Bornstein, 1989). As we saw in our earlier discussion of Freud, the notion of sensitive periods is central to psychoanalytic thought. Freud's view that infancy is the crucial period in molding an individual's personality was the basis of his famous aphorism "No adult neurosis without an infantile neurosis."

Theorists who adhere to the supposition that there are sensitive periods generally insist that it is almost impossible to make up a deficit in a person's development at some later period (White, 1975; Bower, 1974). However, most life-span developmentalists reject the idea that the first five years of a child's life are all-important. More recent research suggests that the long-term effects of short, traumatic incidents are generally negligible in young children (Werner, 1989). Jerome Kagan comes to a somewhat similar conclusion on the basis of studies that he and his associates have conducted in Guatemala since 1971 (see *Human Diversity:* "Jerome Kagan on the Early Years").

Many individuals have the capacity to heal and arrive at adulthood without substantial psychological maladjustment. Thus, although all authorities agree that development is influenced by the child's early experiences, they disagree considerably regarding how important these experiences are. Chapter 6 returns to the matter of critical periods in relation to attachment, sensory deprivation, and maternal deprivation in early childhood.

Ecological Theory

As mentioned in Chapter 1, Urie Bronfenbrenner, from Cornell, proposes an **ecological theory** that centers on the relationship between the developing individual and the changing environment. These interactions cannot be captured entirely in the laboratory, for, as Bronfenbrenner (1979) points out, "Development never takes place in a vacuum; it is always embedded and expressed through behavior in a particular environment" (Bronfenbrenner, p. 27). One cannot grasp human development by simply observing and measuring individuals' behavior in clinical settings that are divorced from their relevant social, physical, and cultural environments.

Bronfenbrenner's ideas have been influenced by Freud, Piaget, Vygotsky, and, most importantly, Kurt Lewin. According to Lewin's field theory, the "dialogue" between the person and the environment can be expressed in the formula $B = f(PE)$: *B*ehavior is determined by the interaction between the *P*erson and the *E*nvironment. Bronfenbrenner modified the formula to reflect the distinction between behavior and development so that his formula reads $D = f(PE)$: *D*evelopment is the result of the interaction between the *P*erson and the *E*nvironment. By substituting development for behavior in the equation, he highlights the importance of time and, with that, change and the significance of the longitudinal study as essential to understanding the human condition.

In proposing the ecological model as a research tool, Bronfenbrenner wants to move away from the traditional focus that sees either the environment (*E*) or the person (*P*)—instead of the relationship between them—as the most important aspect of development. Furthermore, he wants to focus on the process of development rather than concentrate on isolated variables at a single point in time. Think of someone you know who either dropped out of school or considered dropping out. Bronfenbrenner suggests that an approach focusing solely on factors such as the yearly income of the family, intellectual ability, or ethnicity in order to explain the student's disengagement from school will miss most of the information relevant to this particular student's situation. Instead of trying to match categories or labels with certain outcomes, researchers must look at the relationships among variables in different environments. If you were to read an article in a research journal that sought to explain your friend's "dropping out" primarily in terms of distinct categories to which she or he belonged, you would probably be dissatisfied with the explanation, knowing that the reasons were much more complex or historical than those offered by the researcher.

Finally, Bronfenbrenner's theory is important as a way of capturing how people make sense of their circumstances and how their understanding, in turn, influences their behavior. You have probably been in a situation where a number of people reacted differently to the same experience. How each person defined that situation—based on his or her personal history, expectations, feelings, and so forth—determined how he or she behaved. It is important to keep in mind while studying development not only that different people see things differently, but that the same person, as she or he develops cognitively, physically, and psychosocially, will see the same phenomenon differently throughout the life span. For example, a person will probably have very different reactions to a film about war if the same film is seen both before and after the person has fought in a real war.

Question

Every morning when she rises, a woman from the Birikati Islands in the Pacific pulls one hair from her head, places it in a container, and then goes out to check her fish trap. You need to come up with a reason that explains why she does this. You have two possible ways to collect data and arrive at an understanding of why she does this. You do not speak the language but you can use an interpreter for one day, or you can observe her for one week without being able to talk to her. Which of the previous theories, in your opinion, comes closest to explaining her behavior? Explain.

Sociocultural Theory

Lev Vygotsky, a Russian psychologist who is credited with creating one of the outstanding schools of Soviet psychology, is known for his **sociocultural theory** of psychological development. The major theses of his work are as follows:

- The development of individuals occurs during the early formative years and has a specifically

historical character, content, and form; in other words, development will be different depending on when and where you grow up.
- Development takes place during changes in a person's social situation or during changes in what activities the person undertakes.
- Activities are usually done in groups during social interaction.
- Individuals observe an activity and then internalize the basic form of that activity.
- It is essential to being able to internalize activities that systems of signs and symbols (like language) be available.
- A person assimilates the values of a particular culture by interacting with other people in that culture.

It is important to note that Vygotsky assumes that the development of the individual is determined by the *activity of groups*. The child will interact with another person, assimilate the social aspects of the activity, and take that information and internalize it. In this way social values become personal values (Vygotsky, 1978).

Consequently, according to Vygotsky, in order to understand the mind we must first understand how the functions of the mind are shaped by psychological processes, especially language. Vygotsky's theory provides a developmental perspective on how such mental functions as thinking, reasoning, and remembering are facilitated through language and how such functions are anchored in the child's interpersonal relationships (Tappan, 1997). The child, according to Vygotsky, will observe something happening between others and then will be able to take that observation and mentally incorporate it. One of Vygotsky's examples is the way children use language: First a child will be told "Say please and thank you" by its parents. The child will also see people saying "Please" and "Thank you" to each other. Then the child will begin to say these words aloud. By saying "Please" and "Thank you" aloud, the child is internalizing the words and the concepts they stand for in a social setting. Only after assimilating the words' meanings can the child individually start to act in a polite manner. It follows that development is always a social process for Vygotsky, and child–adult interaction plays an important role. So it should come as no surprise that for Vygotsky, the way to understand development is to observe the individual in a social activity.

Controversies

A clash of doctrines is not a disaster—it is an opportunity.
Alfred North Whitehead

Each theory has its proponents and its critics. Yet the theories are not mutually exclusive; we need not accept one and reject the others. As we pointed out at the beginning of this chapter, theories are simply tools—mental constructs that allow us to visualize (that is, to describe and analyze) something. Any theory limits the viewer's experience, presenting a tunnel perspective. But a good theory also extends the horizon of what is seen, functioning like a pair of binoculars. It provides rules of inference through which new relationships can be discovered and suggestions as to how the scope of a theory can be expanded.

Furthermore, different tasks call for different theories. For instance, behavioral theory helps us understand why U.S. children typically learn English and Russian children learn Russian. At the same time ethological theory, one of the evolutionary adaptation theories, directs our attention to ways in which the human organism is neurally prewired for certain activities, so that, in interaction with an appropriate environment, young children typically find that their acquisition of language comes rather "naturally"—a type of *easy learning*. Simultaneously, psychoanalytic theory alerts us to personality differences and to differing child-rearing practices that influence a child's learning to talk. Cognitive theory encourages us to consider the stages of development and the mental processes that are involved in the acquisition of language. Finally, humanistic theory reminds us that people are not passive beings mechanically buffeted about by environmental and biological forces but are themselves creative beings who actively pursue language competence. The distinction between mechanistic and organismic models helps to clarify some of these theories.

Mechanistic and Organismic Models

Some psychologists attempt to classify developmental theories in terms of two basic categories: a mechanistic worldview and an organismic worldview (see Table 2-4).

The **mechanistic model** represents the universe as a machine composed of elementary particles in motion. All phenomena, no matter how complex, are viewed as ultimately reducible to these fundamental units and their relationships. Each human being is regarded as a physical object, a kind of elaborate machine. Like other parts of the universal machine, the organism is inherently at rest. It is inherently passive and responds only when an external power source is applied. This view is the *reactive organism model*. In keeping with this worldview, human development is portrayed as a gradual, uninterrupted, chainlike sequence of events. Indeed, one can question whether a machine can be said to "develop"; it changes only when some external agent adds, subtracts, or alters the machine's parts (Sameroff & Cavanagh, 1979). Change, then, cannot occur without influence from the environment. Individual differences are the central focus of mechanistic approaches. Behavioral learning theories fall within this tradition.

In contrast, the **organismic model** focuses not on elementary particles but on the whole. The distinctive interrelations among the lower-level components are seen as imparting to the whole characteristics not found in the components alone. Hence, the whole differs in kind from its parts. The human being is seen as an organized configuration. The organism is inherently active—it is the source of its own acts rather than being activated by external forces. This view is the *active organism model*. Viewed from this perspective, human development is characterized by discrete, steplike levels or states. Human beings are portrayed as developing by constantly restructuring themselves. The new structures that will be formed are determined by the *interaction* between the environment and the organism (Gottlieb, 1991). The stage theories of Freud, Erikson, and Piaget fall within the organismic tradition.

However, most psychologists prefer an **eclectic approach.** This perspective allows them to select and choose from the various theories and models those aspects that provide the best fit for the descriptive and analytical task at hand. Perhaps we can gain a better understanding and appreciation of these controversies by considering an illustration, continuity and discontinuity in development.

Continuity and Discontinuity in Development

Most psychologists agree that development involves orderly sequences of change that depend on growth

Table 2-4   Mechanistic and Organismic Paradigms

Characteristic	Mechanistic Paradigm	Organismic Paradigm
Metaphor	The machine	The organism
Focus	The parts	The whole
Source of motivation	Intrinsically passive	Intrinsically active
Nature of development	Gradual, uninterrupted adding, subtracting, or altering of parts (continuity)	Discrete, steplike levels or states (discontinuity)

and maturation as individuals interact with their environment. Other psychologists emphasize discontinuity in sequences of change. Those who support continuity say that development produces smooth, gradual, and incremental change. Those who stress continuity typically fall within the mechanistic camp. Those who accentuate discontinuity usually fall within the organismic camp.

The two different models of development can be clarified by considering two analogies. According to the *continuity model,* human development is analogous to the growth of a leaf. After a leaf sprouts from a seed, it grows by simply becoming larger. The change is gradual and uninterrupted. Psychologists who emphasize the part that learning plays in behavior tend to take this point of view. They see the learning process as lacking sharp developmental states between infancy and adulthood. Learning is cumulative, building on itself.

According to the *discontinuity model,* human development is analogous to the developmental changes that produce a butterfly. Once a caterpillar hatches from an egg, it feeds on vegetation. After a time it fastens itself to a twig and spins a cocoon within which the pupa develops. One day the pupal covering splits open and the butterfly emerges. Psychologists who adopt the discontinuity model see human development as similar to the process of insect metamorphosis. Each individual is seen as passing through a set sequence of stages in which change constitutes a difference of kind rather than merely of degree. Each stage is characterized by a distinct and unique state in ego formation, identity, or thought. The theories of Sigmund Freud, Erik Erikson, and Jean Piaget are of this sort.

How we view development depends in part on our vantage point. To return to our analogies, when we first observe a caterpillar and then a butterfly, we are struck by the dramatic qualitative change. But when we observe the developmental changes that occur within the cocoon, we have a different impression. We see that butterfly-like characteristics are gradually acquired, and consequently we are more likely to describe the process as continuous (Lewis & Starr, 1979). However, if we look at a seed and then a tree, we are impressed by the magnitude of the change that has occurred.

Increasingly, psychologists are less inclined to divide themselves into sharply opposing camps on the issue of continuity and discontinuity in development. They, too, recognize that much depends on one's vantage point and hence see both continuities and discontinuities across the life span (Colombo, 1993; Lewis, 1993). In sum, then, social and behavioral scientists increasingly have come to see development as residing in a relation between organism and environment—in a transaction or collaboration: People work with and affect their environment and, in turn, it works with and affects them.

Nature Versus Nurture

*T*he way a question is asked limits and disposes the ways in which any answer to it—right or wrong—may be given.

Susanne K. Langer, *Philosophy in a New Key*

Time and again it has been officially claimed that heredity-environment questions are dead, that they have been definitely answered for all time. Yet in one fashion or another, each generation resurrects them, threshes them out once more, and then presumes once again to set them to permanent rest. For example, a prevailing question in contemporary U.S. society is why some of our children and adolescents are so violent. Is the child's tendency to be violent due to an inherited genetic flaw, or due to the type of home environment, or due to a combination of both factors? Some of the difficulties associated with the nature-nurture controversy stem from the fact that investigators often operate from different assumptions. Various schools of thought ask different questions and hence come up with different answers. How we phrase our questions structures the alternatives by which the questions are answered.

Scientists began by asking *which* factor, heredity or environment, is responsible for a given trait, such as a mental disorder or a person's level of intelligence. Later, they sought to establish *how much* of the observed differences among people are due to differences in heredity and *how much* to differences in environment. And recently, some scientists have insisted that a more fruitful question is *how* specific hereditary and environmental factors *interact* to influence various characteristics (Plomin & McClearn, 1993; Anastasi, 1958). Each of these questions leads to its own theories, interpretations, and methods of inquiry.

The "Which" Question   Most students can recall debating in a class the question "Which is more important, heredity or environment?" Yet most scientists today reject this formulation. They believe that phrasing the issue in terms of *heredity versus environment* has caused the scientific community, and society at large, untold difficulties. Counterposing heredity to environment is similar in some respects to debating whether sodium or chlorine is more important in ordinary table salt. The point is that we would not have salt if we did not have both sodium and chlorine (see Figure 2-5).

The "How Much" Question   As scientists came to recognize the inappropriateness of the "which" question, some of them reformulated the issue. Granting that both heredity and environment are essential for the emergence of any characteristic, they asked, *"How much* of each is

Figure 2-5 Gene-Environment Interaction A person who has a gene for "fatness" might actually weigh less than one with a gene for "leanness," if the former lives on a scanty diet and the latter on an abundant diet.

From: T. Dobzhansky. MANKIND EVOLVING. Copyright 1962, Yale University Press. Reprinted by permission.

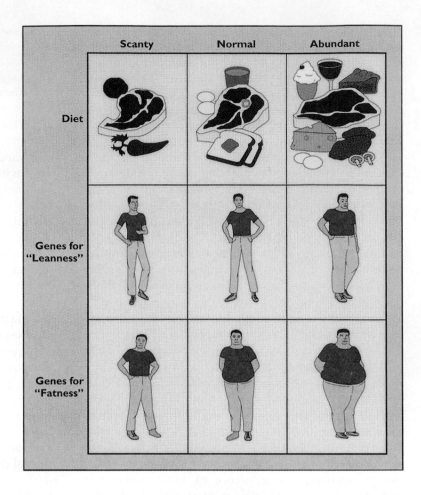

required to produce a given trait?" For example, they asked, "What percentage of a person's level of intelligence is attributable to heredity, and how much depends on environment?" The same question could be asked of a given mental disorder.

Scientists have traditionally sought answers to the "how much" question by measuring the resemblance among family members with respect to a particular trait (Segal, 1993). Botanists use similar procedures to discover the separate contributions of heredity and environment by taking cuttings from a single plant and then replanting the parts in different environments: one at sea level, another at an intermediate elevation, and still another in the alpine zone of a mountain range. Each cutting develops into a new plant under different environmental conditions. Because the cuttings are genetically identical, any observed *differences* in vigor, size, leaves, stems, and roots are directly traceable to differences in environment (Dobzhansky, 1962).

Such deliberate experimentation is not possible with humans. Nonetheless, nature occasionally provides us with the makings of a natural experiment. From time to time a fertilized egg, by some accident, gets split into two parts termed **identical** or **monozygotic twins.** Genetically, each is essentially a carbon copy of the other. The

study of identical twins reared under different environmental conditions is the closest approach possible to the experiments with plant cuttings (see "The Minnesota Twin Project" later in this chapter).

In contrast with identical twins, **fraternal** or **dizygotic twins** come from two eggs fertilized by two different spermatozoa. They are simply siblings who happen to develop separately in the womb at the same time and are (usually) born at the same time. Important evidence can be obtained and comparisons can be made between identical twins reared apart and fraternal twins reared together. Many scientists believe that such comparisons reveal valuable information about the relative contributions that heredity and environment make to a particular trait or behavior.

By studying children who were adopted at birth and reared by foster parents, one can compare some characteristic of the adopted children, such as IQ score or the presence of a particular mental disorder, with that of their biological parents and their foster parents. In this fashion researchers attempt to weigh the relative influences of the genetic factor and the home environment.

The "How" Question   A number of scientists, such as the psychologist Anne Anastasi (1958), believe that the task of science is to discover *how* hereditary and environmental factors work together to produce behavior. They argue that the "how much" question, like the "which" question, is unproductive. The "how much" question assumes that nature and nurture are related to each other in such a way that the contribution of the one is *added* to the contribution of the other to produce a particular behavior.

Anastasi, among others (Thelen, 1995; Lykken et al., 1992), disputes this view. She argues that, as applied to human life, neither heredity nor environment exists separately. They are always interconnected, continually interacting. Consequently, Anastasi says, it is a hopeless task to identify "which" of the two factors produces a particular behavior or to determine "how much" each contributes. However, Anastasi recognizes that the role played by hereditary factors is more central in some aspects of development than in others. She thus sets forth the notion of the **continuum of indirectness.** At one end of the continuum are the contributions of heredity that are most direct, such as physical characteristics like eye color and chromosomal disorders like Down syndrome. At the other end of the continuum are the contributions of heredity that are quite indirect, such as the social stereotypes that the members of a given society attach to various categories of skin color and hair texture.

Medawar (1977, p. 14) also notes that we might not be able to attach exact percentages to the contributions of heredity and environment. Because heredity and environment interact in a relationship of varying dependence, what appears to be a hereditary contribution in one context can be seen as an environmental contribution in another. An example is provided by phenylketonuria (PKU), a severe form of mental retardation that is transmitted genetically. PKU results from the inability of the body to metabolize phenylalanine, a common ingredient of our diet. But if a child who has inherited a susceptibility to the disease is given a diet free of phenylalanine, there is no buildup of toxic materials, and the child's development is essentially normal. Hence, PKU can be viewed as entirely environmental in origin, because PKU shows up in the presence of phenylalanine but not in its absence.

Heredity and environment interact in complex ways. Genes influence the kinds of environment we seek, what we attend to, and how much we learn (Plomin & Daniels, 1987). For instance, psychologists Sandra Scarr and Kathleen McCartney (1983) contend that each stage in a child's psychological development is ushered in by an increment in the child's biological maturation. Only after the child is genetically receptive is the environment able to have any significant effect on her or his behavioral development. Scarr and McCartney believe that children's genetic predispositions tailor their environment in three ways—passively, evocatively, and actively:

- *Passive relationship:* Parents give their children both genes and an environment that are favorable (or unfavorable) to the development of a particular capability. For example, parents who are gifted in social skills are likely to provide their children with an enriched social environment.
- *Evocative relationship:* A child evokes particular responses from others because of the child's genetically influenced behavior. For instance, children who are socially engaging typically elicit from other people more social interaction than passive, sober children do.
- *Active relationship:* Children seek out environments that they find compatible with their temperament and genetic propensities. For example, sociable children search for playmates and even create imaginary playmates if real ones are not at hand.

In short, what children experience in any given environment is a function of genetic individuality and developmental status (Scarr, 1985b). Scientists, then, are increasingly able to apply rigorous measurements to some aspects of the old nature-nurture controversy. In particular, new and valuable insights are coming from a rapidly growing field of study that undertakes to embrace, forge, and integrate insights from both psychology and genetics—*behavioral genetics.*

Behavioral Genetics

Over the past 50 years or so, the scientific community has been taking a more tolerant attitude toward heredity's contribution to the variation we observe in people's behavior (Plomin, 1994; Plomin & McClearn, 1993). Interest in the hereditary aspects of behavior had been subdued for nearly half a century by both behaviorism and psychoanalytic theory. The renewed interest in biological factors is due partly to exciting new discoveries in microbiology and genetics (advancing technologies let us examine cell structures both microscopically and chemically) and partly to the failure of social scientists to document a consistently strong relationship between measures of environmental experience and behavioral outcome. The pendulum seems to be swinging away from the environmentalists toward the side of the biologists. Indeed, some scholars worry that the pendulum is moving too rapidly toward a biological determinism that is as extreme as the earlier loyalty of some social and behavioral scientists to an environmental explanation of behavior (Kagan, 1994; McDonald, 1994).

Jerome Kagan: Timidity Studies One area of recent investigation is extreme timidity ("shyness"). Jerome Kagan and his associates (Kagan & Snidman, 1991; Kagan, 1989) followed 41 children in longitudinal

Shyness How does this child's behavior appear to you? Timidity, or shyness, can be demonstrated in several ways, and cultures vary in the value they place on this trait.

research for eight years, studying "behavioral inhibition." The researchers found that 10 to 15 percent of those studied seem to be born with a biological predisposition that makes them unusually fearful of unfamiliar people, events, or even objects like toys. These youngsters have intense physical responses to mental stress: Their dilated pupils, faster and more stable heart rates, and higher levels of salivary cortisol (a hormone found in saliva) indicate that their nervous system is accelerated by even mildly stressful conditions.

Other researchers have found that shy biological parents tend to have shy children—even when the youngsters are adopted by socially outgoing parents (Daniels & Plomin, 1985). In addition, shy boys are more likely than their peers to delay entry into marriage, parenthood, and stable careers; to attain less occupational achievement and stability; and—when late in establishing stable careers—to experience marital instability. Shy girls are more likely than their peers to follow a conventional pattern of marriage, childbearing, and homemaking (Caspi, Elder, & Bem, 1988). Although some children are inherently inhibited and "uptight," they can be helped, by good parenting, to cope with their shyness. In other

words, a predisposition like that for timidity can be enhanced or reduced, but not eliminated, by nurturing child-rearing experiences. Kagan offers this additional bit of advice: "Look at whether the child is happy. Some shy kids are. And they often end up doing well in school . . . they become computer scientists, historians. We need these people, too" (quoted by Elias, 1989, p. 1D).

The Minnesota Twin Project The results of an ongoing project at the University of Minnesota similarly suggest that genetic makeup has a marked impact on personality (Bouchard et al., 1990, Goleman, 1986). Researchers put 348 pairs of identical twins, including 44 pairs who were reared apart, through six days of extensive testing that included analysis of their blood, brain waves, intelligence, and allergies. All the twins took several personality tests, answering more than 15,000 questions on subjects ranging from personal interests and values to aggressiveness, aesthetic judgment, and television and reading habits.

Of 11 key personality traits or clusters of traits analyzed in the study, 7 revealed a stronger influence for hereditary factors than for child-rearing factors. The Minnesota researchers found that the cluster that rated highest for heritability was "social potency" (a tendency toward leadership or dominance); "social closeness" (the need for intimacy, comfort, and help) was rated lowest. Although they had not expected traditionalism (obedience to authority and strict discipline) to be more an inherited than an acquired trait, it is one of the traits with a strong genetic influence.

The Minnesota researchers do not believe that a single gene is responsible for any one of the traits. Instead, each trait seems to be determined by a large number of genes in combination, so that the pattern of inheritance is complex—what is called **polygenic inheritance** (see Chapter 3). Such findings do not mean that environmental factors are unimportant. It is not full-blown personality traits that are inherited but rather tendencies or predilections. Such family factors as extreme deprivation, incest, or abuse would have a larger impact—though a negative one—than the Minnesota research reveals.

The message for parents is not that it matters little how they care for and rear their children but that it is a mistake to treat all children the same. Children can—and often do—experience the same events differently, and this uniqueness nudges their personalities down different roads. In studies of thousands of children in Colorado, Sweden, and England, researchers found that siblings often respond to the same event (a parent's absence, a burglarized home), and interpret the same behavior (a mother's social preening), in quite different ways (Dunn & Plomin, 1990). Birth order, school experiences, friends, and chance events often add up to very different childhoods for siblings (McGuire et al., 1994).

Identical Twins Separated at Birth Frequently Reveal Startling Similarities
Although they were separated as infants, identical twins Mark Newman and Jerry Levey grew up to share a firefighting avocation. They are among the twins studied by psychologist Thomas Bouchard, Jr., and his associates at the University of Minnesota. Upon being reunited, the men discovered that they both drank only Budweiser and that they both held the bottle awkwardly with their little finger stretched beneath the bottom of the bottle. Moreover, both men were bachelors and compulsive flirts, and both engaged in raucous good humor. Indeed, the twins said it was "spooky" that they tended to make the same remarks at the same time and to use similar gestures.

Youngsters perceive events through unique filters, each of which is skewed by how earlier experiences affected them. Because each child carries about her or his own customized version of the environment, it seems that growing up in the same family actually works to make siblings different. Even youngsters as young as 14 months of age are acutely aware of the minute-by-minute differences in parental attention and affection doled out to their brothers and sisters, as evidenced by the skill they display in yanking back the spotlight. Parents who try to be evenhanded are foiled by their own consistency because they cannot control the way children perceive these efforts. So in guiding and shaping children, parents should respect their individuality, adapt to it, and cultivate those qualities that will help each child cope with life. For a child who is timid, good parenting would involve providing experiences in which success will encourage the child to take more risks. If another child is fearless, good parenting will involve cultivating qualities that temper risk taking with intelligent caution. Remember, though, that cultures differ in the value they place on such personality traits as risk taking or timidity, so that good parenting will differ from culture to culture.

Some scholars fear that the results of the Minnesota research will be used to blame the poor and downtrodden for their misfortunes. Political liberals have long believed that crime and poverty are primarily the by-products of unhealthy social environments. So they are distrustful of biological or genetic explanations of behavior. Other scholars point out that the research holds promise for preventive medicine. If researchers can find a genetic predisposition for various disorders, we can then work on changing the environment with diet, medication, or other interventions. For example, if offspring of alcoholics are found to have genes that render them susceptible to alcoholism, they could be taught from childhood to avoid alcohol. Scientists can also develop new treatments. For instance, if they find a gene that increases a person's risk for schizophrenia or bipolar disorder (also known as manic-depressive illness), researchers can find the protein that the gene codes for and better understand the basic mechanism of the disease. Once they understand the basic mechanism, they can search for new ways to treat the disease. In sum, the potential dangers of genetic research are large, but so are its potential benefits.

Segue

In Chapter 2 we have considered several major types of theory dealing with human development:

- Psychoanalytic theories draw our attention to the importance of early experience in the fashioning of personality and to the role of unconscious motivation.

- Behavioral theories emphasize the part that learning plays in prompting people to act in the ways that they do.

- Humanistic theory reminds us that individuals are capable of intervening in the course of life's events to influence and shape their own beings.
- Cognitive theories highlight the importance of various mental capabilities and problem-solving skills that arm human beings with a powerful potential to adapt and cope.
- Evolutionary adaptation theories allow us to bring into focus various biological patterns that predispose and prepare human beings for particular kinds of behavior.
- Ecological theory focuses on the process of development and stress the importance of the relationship between the developing individual and the changing environment.
- Sociocultural theories focus on the interaction between the individual and others in a social activity and how individuals assimilate and internalize cultural meaning.

With Chapter 3, we will begin to take you on a journey through all of the stages of life, from conception through birth, infancy, early childhood, middle childhood, late childhood, adolescence, early adulthood, middle adulthood, late adulthood, and dying and death. In each of these life stages, you will need to understand the developmental theories presented in Chapter 2.

We have also written this text to help you broaden your understanding of human development from several other perspectives. Throughout the following chapters you will encounter a prudent blending of research findings and theories from the hard sciences of biology, chemistry, and genetics as well as from the social sciences of psychology, sociology, anthropology, history, and political science. In addition, we have incorporated some findings from cross-cultural research. You will come to realize that contemporary developmentalists live and work around the world, conduct research and collaborate on a global scale, and disseminate findings such that the "newest" theories are more easily accessible than in the past. We encourage you to use your *critical thinking skills* to evaluate the diversity of theories you will undoubtedly encounter, both in our text, in your classroom, and in the online world.

Summary

Theory: A Definition

1. Theory is a tool that allows us to organize a large array of facts so that we can understand them. If we understand how nature works, we have the prospect of gaining some control over our destiny.

Psychoanalytic Theories

2. Sigmund Freud postulated that human development involves a series of psychosexual stages: oral, anal, phallic, latency, and genital. Each stage poses a unique conflict that the individual must resolve before passing on to the next stage.

3. Critics complain that Freudian theory is difficult to evaluate because it makes few predictions that can be tested by accepted scientific procedures. Freud's work is also criticized for neglecting women's development. Nonetheless, Freud's work is generally regarded as a revolutionary milestone in the history of human thought.

4. Erik Erikson, a theorist in the psychoanalytic tradition, identifies eight psychosocial stages, each of which confronts the individual with a major conflict that the individual must successfully resolve in order to achieve healthy development. Erikson's theory draws our attention to the continual process of personality development that takes place throughout a person's life span.

Behavioral Theory

5. Behavioral theory contrasts sharply with psychoanalytic theory. Its proponents believe that if psychology is to be a science, it must look to data that are directly observable and measurable, and not rely on introspection.

6. Behaviorists are especially interested in how people learn to behave in particular ways. They deem learning to be a process whereby individuals, as a result of their experience, establish an association or linkage between two events, a process called "conditioning." Behaviorists divide the environment into units called "stimuli" and the behavior elicited by stimuli into units called "responses."

Humanistic Theory

7. Humanistic psychology, which is sometimes called the "third force" in psychology, arose in reaction to psychoanalysis and behaviorism. It maintains that

human beings are different from all other organisms in that they actively intervene in the course of events to control their destinies and to shape the world around them. Its proponents seek to maximize the human potential for self-direction and freedom of choice.

Cognitive Theory

8. Jean Piaget has come to be recognized as a giant of twentieth-century psychology. For Piaget the critical question in the study of growing children is how they adjust to the world they live in. *Schema, assimilation, accommodation,* and *equilibrium* are key concepts underlying the four Piagetian stages of cognitive development: the sensorimotor stage, the preoperational stage, the stage of concrete operations, and the stage of formal operations.

9. Cognitive learning theorists say that our capacity to use symbols affords us a powerful means for comprehending and dealing with our environment. Verbal and imagined symbols allow us to represent events; to analyze our conscious experience; to communicate with others; to plan, to create, to imagine; and to engage in foresightful action.

Evolutionary Adaptation Theory

10. Evolutionary adaptation theory studies the behavior patterns of organisms from a biological point of view. Its proponents rely heavily on the evolutionary theory of Charles Darwin. Ethologists say that evolution applies not only to anatomy and physiology but also to predispositions for certain types of behavior. Organisms are said to be genetically prepared for some responses.

Ecological Theory

11. Urie Bronfenbrenner has devised an ecological theory that centers on the relationship between the developing individual and the changing environment. Bronfenbrenner's theory is important as a way of capturing how people make sense of their circumstances and how their understanding, in turn, influences their behavior.

Sociocultural Theory

12. Sociocultural theories focus on the interaction between the individual and others in a social activity, and how individuals assimilate and internalize cultural meanings.

Controversies

13. Each of the major types of theory we considered has its proponents and critics. Yet the theories are not mutually exclusive; we need not accept one and reject the others. Different tasks simply call for different theories. Thus, most psychologists prefer an eclectic approach to development.

14. When scientists came to recognize the inappropriateness of the "which" question, some of them took a somewhat different approach. They sought to establish *how much* of the observed differences among people are due to heredity and *how much* to differences in environment. Recently, a number of scientists have argued that the "how much" question, like the "which" question, leads to no productive end. They have insisted that a more fruitful approach is to be found in the question of *how* specific hereditary and environmental factors work together to influence various characteristics.

15. New and valuable insights are coming from a rapidly growing field of study that undertakes to embrace, forge, and integrate insights from both psychology and genetics—behavioral genetics. Jerome Kagan and his associates have shown the part genetic factors play in extreme timidity. Researchers at the University of Minnesota have similarly examined how genetic makeup impacts personality.

Key Terms

accommodation *(50)*	critical period *(54)*	humanistic psychology *(48)*
assimilation *(50)*	eclectic approach *(58)*	identical (monozygotic) twins *(60)*
behavior modification *(47)*	ecological theory *(56)*	imprinting *(54)*
behavioral theory *(46)*	epigenetic principle *(43)*	mechanistic model *(58)*
classical conditioning *(47)*	equilibrium *(50)*	natural selection *(54)*
cognition *(49)*	ethology *(54)*	operant conditioning *(47)*
cognitive learning *(52)*	fixation *(40)*	organismic model *(58)*
cognitive stages *(50)*	fraternal (dizygotic) twins *(60)*	polygenic inheritance *(62)*
cognitive theory *(49)*	hierarchy of needs *(48)*	psychoanalytic theory *(38)*
continuum of indirectness *(61)*	holistic approach *(48)*	psychosexual stages *(38, 40)*

psychosocial development *(43)*   schemas *(50)*   stimuli *(46)*
reinforcement *(47)*   self-actualization *(48)*   theory *(37)*
releasing stimuli *(54)*   sensitive period *(56)*
response *(46)*   sociocultural theory *(57)*

Following Up on the Internet

Web sites for this chapter focus on the historical study of human development and major theories of various aspects of development. Please access the text web site at http://www.mhhe.com/crandell7 for up-to-date hot-linked Internet addresses for the following topics.

Classics in the History of Psychology

Erikson's Eight Stages of Psychosocial Development

Theorists in Psychology

Cross-Cultural Psychology

Twin Studies

PartTWO

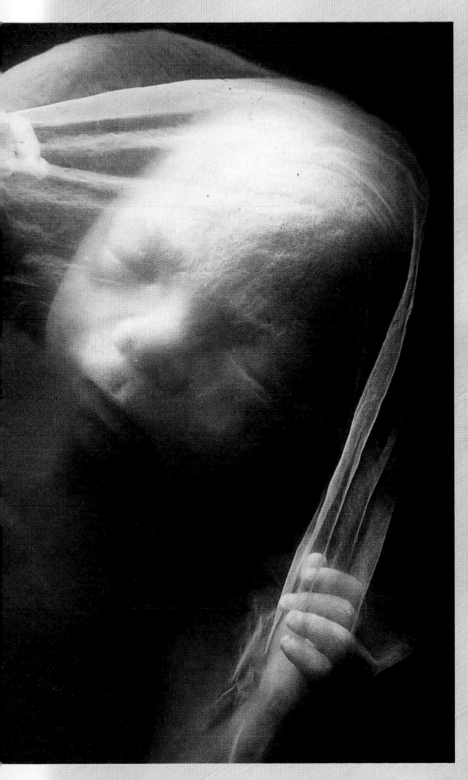

Beginnings

The biological foundations of heredity, reproduction, and prenatal development are discussed in Chapter 3. Today we think we know much more about the beginnings of life, yet medical research continues to astound us. The Human Genome Project, a worldwide collaborative study to map all human genes, will lead to even greater understanding of, and perhaps treatments for, hereditary defects. Chapter 3 explains sophisticated genetic testing and assisted reproductive technologies that have led to advances in fertility. More recent technological advances have led to the possibility of becoming pregnant after menopause. The stages of prenatal development leading up to birth have remained the same, though sophisticated imaging techniques allow us to observe the tiniest humans readying themselves for living outside of the womb.

Chapter 3

Reproduction,
Heredity and
Genetics, and
Prenatal Development

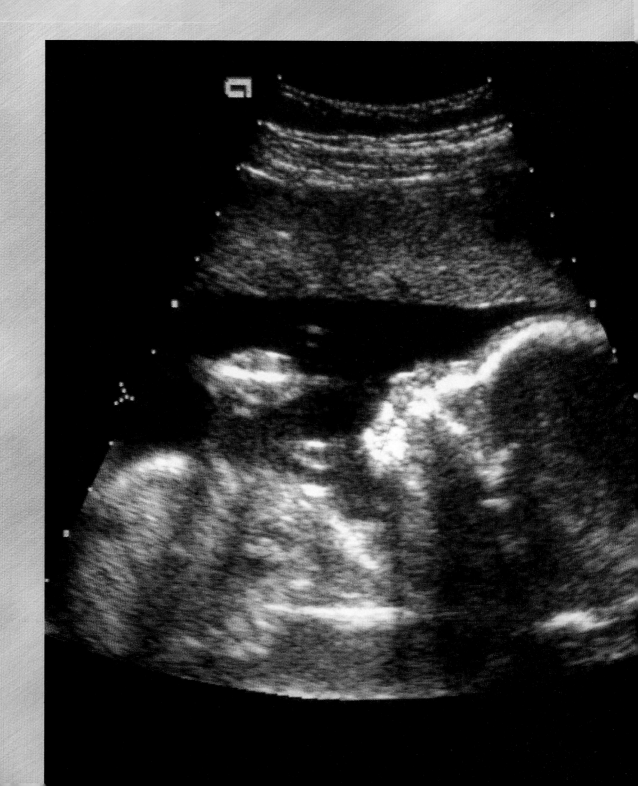

Critical Thinking Questions

1. The media report that women are having babies both younger and older than ever before. Is this actually happening? Do you know of anyone who is 12 and has had a baby? Do you know of anyone who gave birth in her late forties or early fifties?

2. Is there an optimum time during a menstrual cycle to conceive or avoid conception?

3. Is there a specific trait, such as red hair or green eyes or a special talent, that you have inherited that you have traced through your own family history?

4. What are some of the biochemical agents that can compromise the health of a developing baby and mother?

Like all other living things, the majority of human beings are capable of producing new individuals and thus ensuring the survival of the species. With the use of assisted reproductive technologies (such as artificial fertilization, human egg and sperm donations, cryogenic preservation, implantation techniques) and birth alternatives (such as selected surrogacy, intrauterine surgery, grandmothers bearing their own grandchildren), many humans who were previously deemed infertile can choose to reproduce. Infertile couples and singles—both women and men—can now choose to have their own biological offspring, instead of adopting a child or remaining childless. In another technological miracle, Japanese researchers unveiled the first "womb tank" in 1998, a technological wonder that could potentially revolutionize the bearing of children and childbirth. And the idea of human cloning, once merely a futuristic idea in science-fiction novels, is a current—though ethically questionable—possibility.

It's almost as if the idea of woman + man = child *is the old-fashioned way to recreate the species. What once was a private experience has now become both public Internet entertainment and big business. For people with the resources, there certainly is a kaleidoscope of opportunities to procreate.*

Reproduction

Reproduction is the term biologists use for the process by which organisms create more organisms of their own kind. Biologists depict reproduction as the most important of all life processes.

Two kinds of mature sex cells, or **gametes,** are involved in human reproduction: the male gamete, or **sperm,** and the female gamete, or **ovum** (egg). In the process called **fusion** (fertilization), a male sperm enters and unites with a female ovum to form a **zygote** (fertilized egg). Sperm cells are not visible to the naked eye, being only six-hundredths of a millimeter long (.00024 inch). A sperm cell consists of an oval head, a whiplike tail, and a connecting middle piece, or collar, and moves by lashing its tail. A normal adult man's testes might produce as many as 300 million or more mature sperm each day, each of a unique genetic composition.

Ova, on the other hand, are not self-propelled and are moved along by small cilia structures in the woman's reproductive tract (see Table 3-1). During a woman's fertile years between puberty and menopause, normally one ovum matures and is released each month. Each ovum, too, has a unique composition of genetic material and is about the size of the period at the end of this sentence, which can barely be seen by the naked eye. At most, only some 400 to 500 of the immature ova ultimately reach maturity. The rest degenerate and are absorbed by the body (Nilsson & Hamberger, 1990).

The Male Reproductive System

The primary male reproductive organs are a pair of **testes** normally lying outside the body in a pouchlike structure, the scrotum (see Figure 3-1). Sperm are produced and stay viable at a temperature a little lower than normal body temperature (about 96 degrees Fahrenheit). The scrotum holds and protects the testes and keeps them from being held too close to the man's warmer body. The testes produce sperm and the male sex hormones, called *androgens.* The principal androgens are testosterone and androsterone. The androgens are responsible for producing masculine secondary sexual characteristics, including facial and body hair, increased muscle mass, and a deeper voice.

Sperm are produced in winding tubules, within each testis. They are then emptied into the *epididymis,* a long, slender, twisted tube, where they are stored. During sexual arousal and ejaculation, the sperm pass from the epididymis along muscular ducts into the *urethra.* On the way they are mixed with secretions (which will nourish the sperm on their journey out of the man's body and into the woman's body) from the *seminal vesicles* and the *prostate gland.* The mixture of the sperm and secretions is termed *semen,* which will be ejaculated through the male's urethra—a tube that also connects with the bladder—and is surrounded by the man's external reproductive organ, the **penis.**

Sperm production and viability are influenced by many factors, including the man's own physical health,

---

Table 3-1   The Sperm and the Ovum

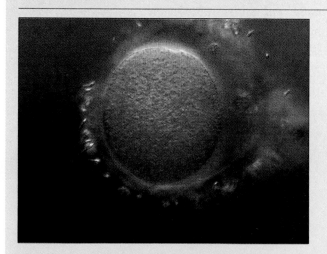

The Sperm and the Ovum The larger ovum is at the right, and many sperm are at the left of the ovum, each trying to penetrate the cell wall and deposit its genetic material.

Description	Vulnerabilities
Sperm are unusually small cells with very little cytoplasm. Once sperm are ejaculated with semen into the female's vagina, they make their way through the cervix and the uterus, and then fewer of them move into the fallopian tubes (oviducts). Of the millions of sperm that enter the vagina, only a few hundred complete the journey. After one sperm succeeds in penetrating and entering the ovum, a biochemical change occurs in the ovum's cell wall and prevents penetration by any other sperm. The ovum (egg cell) is the largest human cell, between 0.1 and 0.2 millimeter in diameter.	Gametes (sex cells) from the male and female carry their own unique genetic material. Their genes carry dominant and recessive traits. Trisomy 21 (Down syndrome) is an example of an inherited disorder that occurs at the moment of conception when the sperm and egg begin pairing up genetic material. The man's and woman's sex cells are vulnerable to disease, substance abuse, and biochemical hazards in the workplace and home environment both prior to sexual activity and after fertilization. Infertility can be caused by these factors.

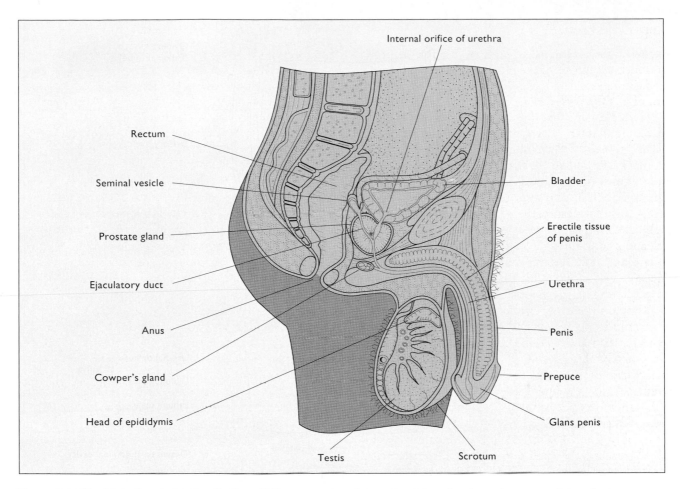

Figure 3-1 The Male Reproductive System   This cross section of the male pelvic region shows the organs of reproduction.

work, and recreational environment—even tight clothing can affect the temperature of the scrotum and harm sperm development. We learned a great deal about the male contribution to fertility and birth defects from Americans who returned from the Vietnam War after being exposed to Agent Orange, a potent chemical defoliant that has been linked to an increase in birth defects by these servicemen and Vietnamese citizens even years later. Americans who served during Desert Storm and are suspected of having been exposed to biological warfare toxins are also being studied for long-term health and reproductive effects. Contrary to what was thought 30 years ago, we now know that smoking, drinking alcohol, ingesting psychoactive drugs, and unprotected sexual activity can affect the health of a man's reproductive organs and developing sperm.

The Female Reproductive System

A woman's reproductive system is composed of the organs that produce ova (eggs), are involved in sexual in-

tercourse, allow fertilization of the ovum, nourish and protect the fertilized ovum until it is developed, and are involved in giving birth. The primary female reproductive organs are a pair of ovaries, almond-shaped structures that lie in the pelvis (see Figure 3-2). While still in her mother's womb, a female embryo's developing ovaries produce more than 400,000 immature ova. After puberty, the **ovaries** produce mature ova and the female sex hormones, *estrogen* and *progesterone*. These hormones are responsible for the development of female secondary characteristics, including breast (mammary gland) development, body hair, and hip development.

Typically, once a female reaches puberty, an ovum is expelled from one of her ovaries on a monthly schedule. For most women this is approximately every 28 days, though for some women the cycle varies, particularly during the first few, and last few, years of menstruation. The ovum is moved through the **fallopian tube,** or *oviduct,* where it may be fertilized if sperm are present. The fallopian tube is lined with tiny, hairlike projections called *cilia* that propel the ovum along its course through

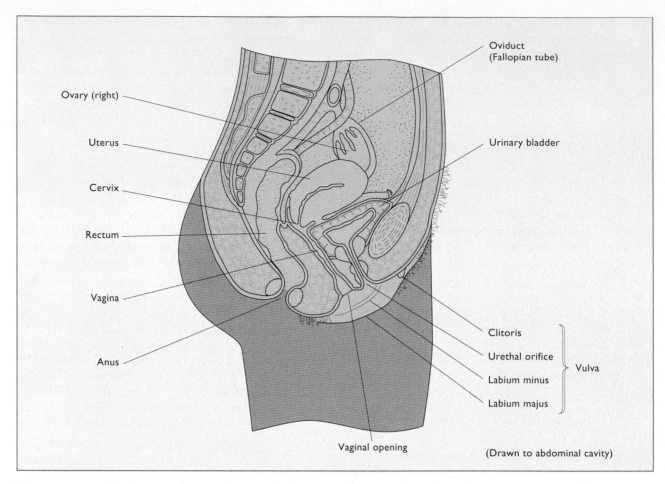

Figure 3-2 The Female Reproductive System This cross section of the female pelvic region shows the organs of reproduction.

the fallopian tube into the womb, or uterus. This short progression takes a few days; the fallopian tube is about 6 inches long and has the circumference of a human hair (Nilsson & Hamberger, 1990).

The pear-shaped **uterus,** a hollow, thick-walled, muscular organ, will house and nourish the developing embryo. The uterus prepares itself each month for potential conception with a blood-rich lining, which each month it sheds for four to six days (in menstruation) if conception (fertilization) does not occur. The unfertilized ovum is expelled from the body through the narrow lower end of the uterus, called the *cervix,* which projects into the vagina. The **vagina** is a muscular passageway that is capable of considerable dilation. The penis is inserted into the vagina during sexual intercourse, and the infant passes through the vagina at birth. Surrounding the external opening of the vagina are the external genitalia, collectively termed the *vulva.* The vulva contain the fleshy folds, known as the *labia,* and the *clitoris* (a small erectile structure comparable in some ways to the penis).

A woman's home and working environment, nutritional habits, health care, and sexual behaviors have a great impact on the health of her reproductive system, as

well as the health of a potential fetus, both before and after conception (see *Information You Can Use:* "Major Drug and Chemical Teratogens" on p. 94).

How and When Fertilization Occurs

The Menstrual Cycle The series of changes associated with a woman's **menstrual cycle** begins with menstruation, the maturing of an ovum and ovulation, and the eventual expulsion of an unfertilized ovum from the body through the vagina. A healthy woman's ovaries produce a mature ovum, or egg cell, every 25 to 32 days, the average being every 28 days (Planned Parenthood of Central and Northern Arizona, 1998). A woman's health (illness, disease, stress or exercise) can affect her menstrual cycle. Variations among women in the length of the ovarian cycle are common and normal. Young women who have just begun menstruating and older women who are approaching or in their forties are quite likely to have irregular cycles or to skip cycles. Day 1 of a cycle is the first day of menstruation. Toward the middle of each menstrual cycle (around days 13 to 15), usually one ovum reaches maturity in a follicle of an ovary and passes into one of the fallopian tubes (oviducts). This is the most optimum time for conception to occur, because a mature ovum is viable for approximately 24 hours. If there is no fusion with a sperm in the fallopian tube, the ovum begins to degenerate after 24 hours and will be expelled from the body during menstruation.

Ovulation   An ovary contains many follicles, but normally only one undergoes full development in each ovarian cycle. Initially a *follicle* in an ovary consists of a single layer of cells; but as it grows, the cells proliferate, producing a fluid-filled sac that surrounds the primitive ovum, which contains the mother's genetic contribution. Most women's ovaries seem to alternate releasing an ovum every other month, though it has been found that when one ovary is diseased or has been removed, the other ovary will ovulate each month. Through the influence of the hypothalamus in the brain instructing the pituitary gland to release hormones, the maturing follicle in the ovary ruptures, and the ovum is discharged. This discharge of the ovum from the follicle in the ovary is called **ovulation.** When the mature follicle ruptures, releasing its ovum, it undergoes rapid change. Still a part of the ovary, the follicle transforms itself into the *corpus luteum,* a small growth recognizable by its golden pigment. The corpus luteum secretes *progesterone* (a female hormone), which enters the bloodstream and causes the mucous lining along the inner wall of the uterus to pre-

pare itself for the potential implantation of the newly fertilized egg. If conception and implantation do not occur, the corpus luteum degenerates and eventually disappears. If pregnancy occurs, the corpus luteum continues to develop and produces progesterone until the placenta takes over the same function. The corpus luteum then becomes superfluous and disappears (Nilsson & Hamberger, 1990).

Fertilization   The union (or fusion) of a sperm and an ovum is called **fertilization.** This normally occurs in the upper end of the fallopian tube and results in a new structure called the **zygote** (see Table 3-2). At the time of sexual intercourse, a man customarily ejaculates 100 to 500 million sperm into the woman's vagina. Sperm can ascend the cervical canal only during those few crucial days when the woman's cervix is open and produces strands of mucous that allow some of the sperm to enter the uterus and fallopian tubes. Sperm have a high mortality rate within the female tract because of its high acidity and other factors related to the health of the sperm. Some sperm are viable up to 48 hours in the female's reproductive tract. The one sperm that fuses with the ovum has won against gigantic odds, several hundred million to one. Even then, however, the newly fertilized egg is extremely vulnerable. For various reasons, about one-third of all zygotes die shortly after fertilization.

Table 3-2   Conception

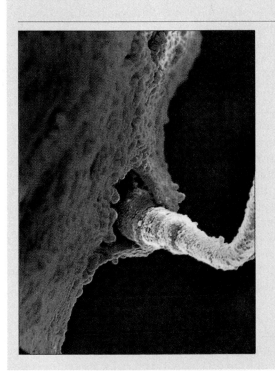

Description	Vulnerabilities
Fertilization occurs when the two gametes, ovum (egg) and sperm, have fused. Their DNA has joined, creating a new structure called a zygote. Additional sperm that did not penetrate the cell wall of the egg (normally nearly 100 or so) continue to attempt to penetrate the cell wall of the ovum. This action, plus the movement of the cilia of the tubal lining, promotes a counterclockwise motion of the zygote. The zygote will proceed down the fallopian tube, a journey of approximately 6 inches that will take 3 to 4 days.	The fallopian tube has about the same circumference as a human hair. If the fallopian tube is scarred or blocked, the zygote will be unable to proceed. Damage or obstruction to the fallopian tube can occur because of such factors as pelvic inflammatory disease (PID), sexually transmitted disease (STDs), and endometreosis, to name a few. There is potential for incomplete fusion If the sperm is defective or if the biochemistry of the cell wall of the ovum is not functioning properly.

Conception   The photo shows a sperm, its tail thrashing, burrowing into an ovum.

If fertilization fails to take place, the decreasing levels of ovarian hormones (estrogen and progesterone) lead to **menstruation** about 14 days after ovulation. The thickened layers of tissue lining the uterus are not needed for the support of a zygote, so they deteriorate and are shed over a three- to seven-day period, when the debris from the wall of the uterus, a small amount of blood, and other fluids are discharged from the vagina. Before the end of a menstrual flow, the pituitary gland secretes hormones into the bloodstream that direct one of the ovarian follicles to begin rapid growth. As a consequence, the cycle starts anew.

Multiple Conception   If more than one ovum matures and is released, the woman might conceive multiple, nonidentical siblings (*dizygotic* or fraternal twins). Identical twins (*monozygotic* twins) result from one fertilized egg splitting into two identical parts after conception.

For couples selecting *in vitro* fertilization (see p. 75), multiple conception also occurs in a petri dish in a medical laboratory environment. The resultant embryos (usually several) may be transplanted into a woman's uterus in hopes that at least one will implant itself into the uterine wall and continue to develop. Since 1980, the rate of twin births has risen 37 percent; and for triplets and larger sets, the rate has increased 312 percent. These dramatic increases are due to the success of fertility treatments. The National Center for Health Statistics reports that in 1996 nearly 6,000 babies were born in triplets, quadruplets, or even larger sets of multiple births, the largest number ever and a trend that shows no sign of abating (Vobejda, 1998a; National Center for Health Statistics, 1998).

> **Question**
>
> *Can you identify the primary sex organs in males and females, and can you explain their functions in the process of human reproduction?*

Conceiving or Avoiding Conception

Do you and your partner want to conceive a child in the near future? Or do you and your partner not want to conceive a child right now? Do you know when a woman is most likely to be "fertile"? As mentioned above, at midcycle an ovum in the fallopian tube is viable for approximately 24 hours. Sperm are viable for fertilization for approximately 48 hours once they are introduced into the vagina. Women who have difficulty conceiving must become aware of their optimal time of conception and their daily body temperature, especially during their days of ovulation.

Recent investigations have discovered optimum times of the calendar year as well. In some regions of the world,

people mating during the optimal fertility season have twice the chance of conceiving than they have at other times. The optimal period for conception seems to be when the sun shines for about 12 hours a day, and the temperature hovers between 50 and 70 degrees Fahrenheit. Although high temperatures are known to interfere with sperm production, heat alone is not the explanation, because researchers find no relevant difference between men who work outside in the heat and men who work in air-conditioned settings. Most likely, an internal biological clock, fine-tuned by the length of daylight, contributes to the seasonal differences (Sperling, 1990). Fertility peaks are currently less notable in Western nations than they were prior to the era of industrialization; this difference is associated with indoor lighting and temperature control. But infertility rates in industrialized countries have been rising for the past three decades. It is suspected that this is due to women delaying childbirth (an older woman's ova are less likely to be fertilized than a younger woman's ova), an increase in pelvic inflammatory diseases, an increase in uterine cells growing outside the uterus (endometriosis), and lower sperm counts in men.

Infertility and Assisted Reproductive Technology   In the United States alone, from 1988 to 1995, the number of women who were affected by fertility problems rose from 4.9 million to 6.1 million, a 25 percent increase (Lemonick, 1997). Many of these women seek out **assisted reproductive technologies (ARTs)** to increase their chances of becoming pregnant. The Center for Disease Control and Prevention (CDC) reports that 59,142 assisted reproduction cycles were carried out in the United States in 1996 (CDC, 1999). Since 1978, when the world's first test-tube baby was born in England, medical researchers have created many fertility drugs and microscopic procedures that have dramatically transformed infertility treatments. As of December 1997, there were 315 fertility clinics in the United States and hundreds more around the world. The goal of these infertility clinics is to offer hope to childless couples, single women, and those who postpone childbearing because of illness, disease, career, or late marriage or remarriage (Lemonick, 1997; Centers for Disease Control, 1999).

Swan and colleagues (1997) analyzed the data from 61 studies published between 1938 and 1990 on male reproductive dysfunction, with special attention to declining semen quality. They concluded that in Denmark, England, and the United States, sperm counts have fallen significantly. Research studies are ongoing worldwide in this important area of reproduction.

Assisted Reproductive Technologies (ARTs)   There are several assisted reproductive technology options because there are multiple reasons why conception might not occur through sexual intercourse. A man's sperm count might be too low, he might have sustained injury or disease of the testicles, or his sperm might be unhealthy or have slow motility. A woman's fallopian tubes might be blocked, scarred, or missing from disease or injury. The follicles in her ovaries might not be producing healthy ova. A woman might have undergone chemotherapy that destroyed her ova. The endometrial lining of the uterus might not be able to host a developing embryo. In some instances there seems to be no physiological reason why a couple cannot conceive. The scientific search to help couples conceive has been a worldwide research effort. Louise Brown, the world's first in vitro baby, was born in England. Australia claims the first baby born from a frozen embryo. And it was researchers in a Belgian lab that found a way to inject sperm directly into an egg cell (Lemonick, 1997).

In vitro fertilization (IVF) is fertilization outside the body in a petri dish in a medical lab environment. In an attempt to accomplish pregnancy, these steps are involved: (1) Stimulate the ovaries to produce several viable eggs, (2) retrieve several eggs from the ovaries, (3) fertilize several eggs with sperm of spouse or donor in the IVF lab, (4) let the embryos develop for a few days in a special culture in a petri dish, and (5) place the embryos into the uterus for implantation (embryo transfer) and wait to see if implantation takes on the wall of the uterus. *GIFT* (gamete intrafallopian transfer) utilizes many of the above steps, but after the woman's eggs are retrieved from her ovaries, the eggs and sperm are placed into the woman's fallopian tubes to facilitate fertilization in the body. *ZIFT* (zygote intrafallopian transfer) is similar to the *GIFT* procedure, except laboratory-fertilized embryos (zygotes) are placed in a healthy fallopian tube. *ICSI* (intracytoplasmic sperm injection) is injection of a sperm directly into an egg using a microscopically fine needle. In vitro fertilization procedures are now available for a woman past menopause. These IVF procedures can use the woman's own eggs and her partner's sperm, or donor eggs or donor sperm (see Figure 3-3). The McCaughey (pronounced "McCoy") septuplets born November 19, 1997, were conceived using some form of assisted reproductive technology; and multiple births, as mentioned above, are on the rise in the United States because of these procedures (Greene, 1998).

Future Assisted Reproductive Technologies   Recently, techniques have been developed to retrieve and save viable eggs and sperm. The science of *cryopreservation* (a technique for preserving cells or embryos through freezing) is making it possible for people to store eggs, sperm, and embryos for an indefinite period for potential future use ("Internet Health," 1996). For a person who is diagnosed at an early age with uterine, ovarian, or testicular cancer and requires chemotherapy, radiation, or surgical removal of the reproductive organs, egg or sperm re-

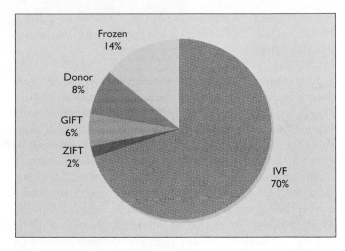

Figure 3-3 Types of ART Procedures—United States, 1995
Source: Centers for Disease Control and Prevention.

trieval and storage can be done prior to other treatments. A woman's eggs are more fragile than a man's sperm, but in October 1997 a woman gave birth to twin boys conceived from donor eggs that had been frozen for two years (Lemonick, 1997). Doctors are also removing ovarian tissue and testicular tissue for later reimplantation, developing methods for longer embryo growth outside the womb (larger embryos would mean fewer would need to be implanted, resulting in fewer multiple births), and microscopically transferring chromosomes and/or nuclei from older eggs to younger eggs (Lemonick, 1997). A newer method, *cytoplasm transfer,* is in its early stages. Cytoplasm, the nonnuclear part of a cell, can be removed from the eggs of younger women and injected into the eggs of older women. As of December 1997, one birth had been recorded after using this procedure.

Richard Seed, a physicist trained at Harvard University, announced on January 6, 1998, on National Public Radio's "All Things Considered" that he and his colleagues are setting up a clinic to clone human babies and predicted that as many as 200,000 human clones a year would be produced, once his process is perfected. Fertility researchers in Scotland, who cloned the sheep Dolly using mammary cells, are also working on a human cloning technique. On January 10, 1998, President Clinton called for a ban on human cloning in the United States (NPR, 1998). He then reconvened the National Bioethics Advisory Commission, a committee of bioethicists and clergy established in 1995, to examine this issue (Palca, 1996). In January 1998, nineteen European nations signed an agreement to prohibit the cloning of human beings (to "defend human dignity against the abuse of scientific techniques") while permitting the cloning of cells for research purposes (Lowrie, 1998). Britain and Germany did not sign the protocol—German law prohibits all research on human embryos, whereas British scientists are at the forefront of cloning research. (See "Following Up on the Internet" at the end of this chapter.)

Ethical Dilemmas of Baby Making   Most people have a very strong urge to reproduce. Yet assisted reproduction is one of the least regulated medical specialties in the United States. France and Australia have passed legislation determining how long an embryo is allowed to grow in a lab before it is disposed of. Most European countries require fertility clinics to be licensed. Likewise, in these European countries national health insurance covers the cost of these procedures, whereas in the United States, high-quality assisted reproduction is mainly for a more affluent clientele (the average cost of such procedures is $8,000). The lack of legislation and regulation regarding this growing industry in our country leaves the U.S. consumer in a more vulnerable situation (Lemonick, 1997).

But social scientists, medical professionals, bioethicists, and laypeople raise many questions about ARTs. How long should we allow an embryo to develop in a lab? Who do these embryos belong to if the recipients change their minds? What type of research is being conducted with gene transfer and gene therapies? If we condemn those mothers who abuse their fetuses with drug and alcohol consumption, what should we think about researchers who have free license to destroy developing embryos? Will we also be growing human fetal tissue for transplantation? Is it right for a surviving partner to take the gametes from a deceased partner? How will children conceived through donor technologies feel when they find out they were created from anonymous donors and have no access to records regarding their medical and biological inheritance? Congress has passed no legislation governing assisted reproductive technologies to protect American citizens from potential abuse of these technologies or to set a standard for the medical research community.

The researchers and fertility specialists who support assisted reproductive technologies assert they are not subverting any moral standard. Scientists like Richard Seed say they are bringing humankind closer to God by helping infertile couples fulfill their dream to reproduce. Those who support eventual human cloning assert that a clone could never be an exact duplicate of the original person because environment plays such a big role in producing an individual's personality, motivation, and life outcomes.

Birth Control Methods   The recent reduction in rates of teenage pregnancy has been attributed to public education campaigns directed toward adolescents who are sexually active. Adolescents and adults who are sexually active are now better informed about the risks of sexual activity, including HIV and AIDS, sexually transmitted diseases, and socioeconomic consequences. They also are more aware of the variety of birth control options to prevent pregnancy and have legal access to abortion to terminate pregnancy.

Contraception   Methods of birth control include male and female condoms, the cervical cap, the diaphragm, birth control pills, vaginal contraceptives, Depo Provera (an injection of hormones given every 12 weeks to inhibit ovulation), intrauterine devices, Norplant subcutaneous inserts to inhibit ovulation, the sponge, vaginal contraceptive chemical barriers, the rhythm method, and male or female sterilization ("Planned Parenthood," 1998). In September 1998, a "morning after" contraceptive kit called PREVEN came on the market in the United States. These pills prevent or delay ovulation or prevent a fertilized egg from implanting in the uterine wall and are reported to be 75 percent effective (Hajela, 1998). RU-486, the French abortion pill that ends a pregnancy several weeks after it has begun, is not available in the United States at the time of this writing. Some women are reported to use repeat abortions as a method of birth control.

Abortion   Abortion is the spontaneous or induced expulsion of the fetus prior to the time of viability, and it occurs most often during the first 20 weeks of the human gestation period. Since 1973, when the U.S. Supreme Court legalized abortion in the *Roe v. Wade* decision, an increasing number of women have been obtaining abortions earlier in pregnancy when health risks are at their lowest ("Planned Parenthood," 1998). An especially volatile issue is partial-birth abortion, performed toward the latest stage of pregnancy when a fetus could be viable outside the womb. With an estimated 35 million abortions performed in the United States from 1973 to 1996 ("Planned Parenthood," 1998), the abortion issue has spawned a highly charged moral battle between two large factions of citizens, the pro-life and the pro-choice advocates (Abortion Surveillance, 1996; Koonin et al., 1998).

The pro-life advocates point to the extensive loss of human life, the reported lasting psychological harm to the women who have selected this choice, and the moral decay of a society that allows this to happen. The National Right-to-Life Committee (1998) has identified post-abortion syndrome as a cluster of symptoms some women experience after the abrupt termination of a pregnancy by elective abortion: depression, a sense of worthlessness, personal relationship disorders, sexual dysfunction, damaged self-esteem, a sense of victimization, and, for a small number of women, suicide. The woman who initiated the *Roe v. Wade* lawsuit to secure a safe, legal abortion in 1973 is now an outspoken pro-life advocate.

The pro-choice advocates state that abortion provides many benefits to women and couples who choose not to have a baby and that every woman deserves the right to control what happens to her body. Illegal abortions performed prior to 1973 put women at a much greater health risk than those performed in medical settings today. Couples with a high risk of producing a fetus with genetic defects have a choice to continue or terminate the pregnancy. Pro-choice advocates further state that the abortion procedure, performed during early stages of pregnancy, is statistically safer to the mother than childbirth. The serious complication rate of abortion is less than 1 percent. Of the 2.6 million unintended pregnancies each year in the United States, approximately 1.4 million are terminated by medically safe, legal abortions ("Planned Parenthood," 1998).

Both sides of this emotional issue are gathering greater force, affecting the outcomes of election campaigns at state and national levels, and might impact the future of our society.

The Expanding Reproductive Years

Of great interest to medical professionals is the fact that American and European girls are starting ovulation earlier than they did 25 to 30 years ago. Some girls today begin menstruating at 8 or 9 years old; the average age is 11 to 12. Earlier in this century women did not begin to menstruate until 14 years old, on the average. "When puberty begins earlier than expected, it is natural for both the parent and the child to feel overwhelmed. . . . In most Western cultures early maturity for boys is equated with power but for girls, it becomes sexualized" (Kato, 1996). For some girls, along with precocious puberty comes earlier sexual activity, and many of these girls become "babies having babies." This is becoming a critical societal issue because these young girls do not generally have the resources to support themselves and the children they are bearing. The health care system, the social welfare system, the educational system, governing agencies, and American business and industry are collaborating to educate American youth about the risks of teenage (or younger) pregnancy.

Some nutritionists suggest that better nutrition is promoting more body fat, which triggers early puberty (Nilsson & Hamberger, 1990). Some developmentalists suggest that the family stressors of divorce and interparental conflict are associated with early onset of menstruation (Wierson, Long, & Forehand, 1993). Other nutritional theorists suggest that a major source of the protein in our diet is contaminated with steroids that promote quick development of the animals (chickens, pigs, cattle) and then promote faster development of children. At this time there does not seem to be any scientific consensus regarding the cause of the increasing incidence of earlier puberty. The incidence of early teen pregnancies had reached an all-time high through the 1980s and early 1990s. Even though it was recently reported that the rate at which American teenagers are becoming pregnant has fallen to its lowest point since 1975, dropping 14 percent since 1990, U.S. teenage birth rates remain relatively high in comparison to those of other countries (Vobejda, 1998b).

Throughout history, a woman's reproductive years have typically started during puberty (today, at about age 12), and ceased after her last menstrual cycle (today, at about age 50), called menopause. With the reproductive technology methods available today, some women are choosing to have a child after menopause. A woman past menopause has no viable ova of her own, so a donated ovum is sought. The older woman is treated with a regimen of female hormones to stimulate her uterus to become a healthy environment for in vitro (artificial) fertilization. Sperm are provided from the woman's spouse or from a donor. Fertilization occurs in a medical lab environment instead of by means of natural conception, and the embryo is grown in vitro to a size at which it can be implanted into the woman's womb. Responsible, reproductive clinics screen this older mother-to-be to ensure that she has good health and the potential to endure the months of pregnancy and the birth process, whether natural or cesarean. Interestingly enough, the sensationalized instances of women in their fifties and early sixties having children have created public clamor against performing ART procedures with women over a certain age (see Figure 3-4). Yet the father's age does not come under the same close scrutiny or criticism. A man's reproductive years can extend well into his eighties or older. One man was reported to be 103 when he fathered a child (Broder, 1996).

Additionally, wholesale commercialization of egg donation is gaining momentum. Although the medical community has agreed for ethical reasons that a donor should not be paid for her eggs, she can be compensated for her time and the inconvenience of surgical extraction ("Eggs for Sale," 1998). In the February 25, 1998, *New York Times,* St. Barnabus Medical Center in New Jersey advertised that it would pay egg donors $5,000, an increase of $2,000 over the standard rate (Kolata, G., 1998). One of the major concerns with marketing reproductive cells is the issue of what genetic traits are being selected by those with the purchasing power. Are couples selecting to have more boys than girls? What heritable traits do the majority of consumers hope to have in a child? Heredity and genetics make a significant contribution to what we look like, how we might behave, our overall health, and our life's potential.

Question

What are some of the assisted reproductive technologies, and why are some methods considered to be so controversial?

Figure 3-4 Birth After Menopause For the first time in history, assisted reproductive technologies have made it possible for some postmenopausal women to be implanted with an embryo and carry a baby to term. This cartoon reflects some of the controversy with this issue. One American woman was 63 when she gave birth in 1997 (after concealing her true age from her physician). MIKE SMITH reprinted by permission of United Features Syndicate, Inc.

Heredity and Genetics

Perhaps at the heart of the debate about assisted reproductive technology and the expanding range of reproductive years is our understanding that each person's genetic makeup is very complex, with a range of flaws likely. Do we really want to parent a younger physical replica of ourselves (though scientists say even identical twins aren't totally genetically alike)? How many embryos may we create and destroy before accepting the "perfect" one? With normal sexual reproduction, many things have the potential to malfunction. What will happen when scientists have free license to recreate the human species—and all its genetic combinations?

As previously mentioned, another debate in the fields of psychology, psychobiology, and sociobiology involves what contribution hereditary or environmental factors make to our unfolding physical, intellectual, social, and emotional development. This is called the *nature (biology) versus nurture (environment) debate.* Sociobiologists work to define general laws of the evolution and biology of social behavior in many species. Psychologists tend to focus on the nature of psychological mechanisms and adaptability to one's environment. Healthy debate among professionals in each of these fields contributes to our growing knowledge about the significant role of biological inheritance.

We now take a look at our biological inheritance, referred to as **heredity.** Each of us has inherited a specific genetic code from our biological parents, and fertilization is the major event determining our biological inheritance. We begin life as a single fertilized cell, or zygote, that contains all the hereditary material passed on to us from our parents and their ancestors. Precisely blueprinted in this original cell are the 200 billion or so cells that we possess nine months later at birth. **Genetics** is the scientific study of biological inheritance.

What Are Chromosomes and Genes?

Around the early 1900s, the use of microscopes to study cellular tissue led to the discovery of chromosomes.

Chromosomes are long, threadlike structures made of protein and nucleic acid that contain the hereditary materials found in the nuclei of any cell. Upon fertilization, for human beings, the 23 chromosomes of the ovum are combined with the 23 chromosomes of the sperm, for a total of 46 chromosomes, usually referred to as 23 pairs (see Figure 3-5). Exceptions to this normal pairing will be discussed later in this section.

Each chromosome contains a linear arrangement of thousands of smaller units that divide it into regions called **genes.** Genes are like beads on a string, and each gene has its own specific location on the chromosome. Genes transmit inherited characteristics passed from biological parents to children. Each human cell contains approximately 100,000 genes and 3 billion letters of chemical code, which are composed of the chemical **DNA (deoxyribonucleic acid).** DNA is the active biochemical substance in genes that programs the cells to manufacture vital protein substances, including enzymes, hormones, antibodies, and other structural proteins (National Human Genome Research Institute [NHGRI], 1998).

This DNA code of life is carried in a large molecule shaped like a double helix or twisted rope ladder (see Figure 3-6). The **human genome** is a map of the genetic makeup of all the genes on their appropriate chromosomes and is being studied extensively by hundreds of researchers at university settings worldwide through the *Human Genome Project.* This project was launched in 1990 by the U.S. National Institutes of Health and the Department of Energy. Its goal is to understand the human hereditary instructions (the 100,000 or so human genes) and to read the entire genetic script (3 billion bits of information) by the year 2005. (See "Following Up on the Internet" at the end of this chapter.) Not only do our genes influence what we look like, but molecular errors in our genes are responsible for an estimated 3,000 to 4,000 clearly hereditary diseases (National Human Genome Research Institute [NHGRI], 1998b).

Geneticists predict many breakthroughs over the next 15 to 20 years in the early detection and treatment of many diseases. The goal of the new science of *gene therapy* is to correct or replace the altered gene. The gene for *cystic fibrosis,* the most common lethal hereditary disease among Caucasians, was discovered in 1989, and the first human gene therapy efforts are under way in federally approved clinical trials. Scientists can test for these diseases directly as well as *prenatally.* About 5 percent of the project budget is spent trying to resolve the ethical, legal, and social issues likely to arise from this research (NHGRI, 1998a).

Nearly all cells in the human body are formed through a kind of cell division called **mitosis,** during which every chromosome in the cell splits lengthwise to form a new pair. Through this process of nuclear division, the cell replicates itself by dividing into two "daughter" cells with the same hereditary information. Unlike other cells in the human body, the *gametes*—ova and sperm—have only 23 chromosomes each, not the usual 23 pairs. Gametes are formed by a more complex kind of cell division called meiosis. **Meiosis** involves two cell divisions during which the chromosomes are reduced to half their original number. Each gamete receives only one chromosome from each pair in every parental cell. This is half the usual number, allowing each parent to contribute half the total number of chromosomes and genetic material at fertilization. Thus, upon fertilization, the newly formed zygote contains 23 pairs of chromosomes (see Figure 3-7).

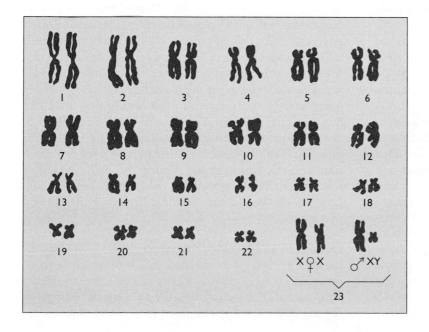

Figure 3-5 Human Chromosomes Every cell nucleus contains twenty-three pairs of chromosomes—two of each type. Each parent provides one member of the pair. Chromosomes differ in size and shape. For convenience in talking about them, scientists arrange the twenty-three pairs in descending order by size, and number them accordingly. The members of each chromosome pair look alike, with the exception of the twenty-third pair in males. As the drawing shows, the twenty-third chromosome is the sex-determining chromosome. An XX combination of chromosomes in this pair produces a female; an XY combination, a male.

Figure 3-6 A Model of the DNA Molecule
The double-chained structure of a DNA molecule is coiled in a helix. During cell division the two chains pull apart, or "unzip." Each half is now free to assemble a new complementary half. At the right side of this illustration, we see a larger depiction of genes, which transmit inherited characteristics. DNA in genes programs cells to manufacture vital protein substances. Scientists around the world are involved in the Human Genome Project and are attempting to understand the location and function of every gene by 2005.

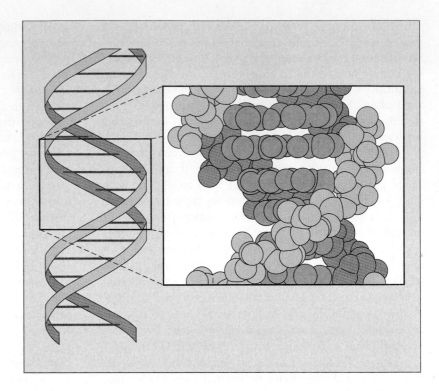

Determination of an Embryo's Sex

In many societies of the world, even today, a male child is expected to inherit the family name or carry on the family business or provisional role in a culture. Consequently, in many cultures the birth of a male child is celebrated but the birth of a female child might not be. An extreme example is China. To control population growth, China has family-planning laws that allow only one child per family; some couples abort female children because they want a male child instead. Social scientists are studying the consequence for a society of having more male children than female (Albright & Kunstel, 1999).

Historical literature records many incidents of women being faulted, divorced, or worse for not producing a male child. However, through the study of genetics we know that it is the male's sperm that carries the chromosome that determines the sex of a child. Of the 46 chromosomes (23 pairs) that each human normally possesses, 22 pairs are similar in size and shape in both men and women. These pairs are called **autosomes.** The 23rd pair, the **sex chromosomes**—one from the mother and one from the father—determine the baby's sex. Each of the mother's ova has an X chromosome. However, a sperm can contain either an X chromosome or a Y chromosome. If an ovum with X is fertilized by a sperm with X, then the zygote will be a female (XX). If an ovum with an X is fertilized by a sperm with Y, then the zygote will be a male (XY). The Y chromosome determines that a child will be male. Approximately six to eight weeks into embryonic development, the male embryo starts producing the male hormone testosterone, which promotes the development of male characteristics.

Principles of Genetics

Have you ever wondered how your own heredity has influenced your characteristics and development? Much of our original understanding of genes and the science of genetics came from studies conducted by an Austrian monk, Gregor Johann Mendel (1822–1884). By crossing varieties of peas (short, tall; red flowers, white flowers) in his small monastery garden, Mendel was able to formulate the basic principles of heredity. Mendel hypothesized that independent units he called *factors* determined inherited characteristics. Today we call these units *genes.* Mendel reasoned that the genes that control a single hereditary characteristic must exist in pairs. Advances in microbiology and genetics have confirmed Mendel's hunch. Genes occur in pairs, one on a maternal chromosome and the other on the corresponding paternal chromosome. The two genes in a pair occupy a specific position on each chromosome. Currently, each and every human gene is being mapped for both its location and function by the Human Genome Project, as mentioned previously.

Dominant and Recessive Characteristics Each member of a pair of genes is called an allele. An **allele** (pronounced "al eel"), then, is a gene at a given location on a chromosome. There can be only two alleles per person for any characteristic, one from each parent (one on the

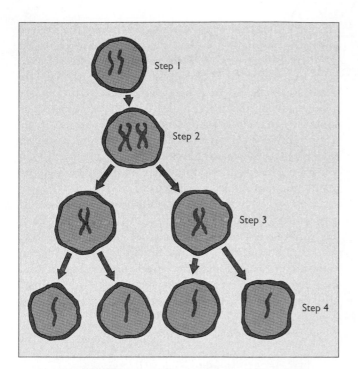

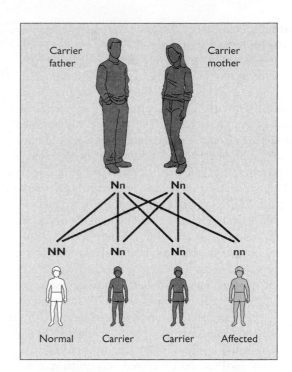

Figure 3-7   Meiosis   This simplified diagram illustrates how gametes (ova and sperm) are formed through meiosis. In step 1 each chromosome teams up with its partner. During step 2 the first meiotic division occurs; both chromosomes of each pair are duplicated. (For simplicity, the figure portrays only one pair, although human beings normally have twenty-three pairs.) Each original chromosome and its exact copy, termed a *chromatid,* are joined at the center. Step 3 involves a second meiotic division. Each of the two original chromosomes and its copy become a part of an intermediary cell. In step 4 each chromatid—the original and its copy—segregates into a separate ovum or sperm gamete. Thus, the intermediary cells undergo cell division without chromosome duplication. This process produces four gametes.

Figure 3-8   Transmission of Recessive, Single-Gene Defects   Both parents, usually themselves unaffected by the faulty gene, carry a normal gene *(N)* that dominates its faulty recessive counterpart *(n).* The odds for each child are (1) a 25 percent chance of being *NN,* normal: inheriting two *N*s and accordingly being free of the faulty recessive gene; (2) a 50 percent chance of being *Nn* and therefore a carrier like both parents; and (3) a 25 percent risk of being *nn,* affected: inheriting a "double dose" of *n* genes, which may cause a serious genetic disease.

maternal chromosome and one on the paternal chromosome). Mendel demonstrated that one allele (the **dominant character**) completely masks or hides the other allele (the **recessive character**). Mendel used a capital letter of the alphabet to signify the dominant allele *(A)* and a lowercase letter to signify the recessive allele *(a).* When both alleles from the parents are the same, this is referred to as a **homozygous** characteristic *(AA)* or *(aa).* When the alleles are different, this is called a **heterozygous** characteristic *(Aa).* The characteristic of the dominant allele *(A)* will be expressed (see Figure 3-8).

Not all traits are transmitted as simply. Some inherited traits, or defects, are the result of the complex interaction of many genes, which is called **polygenic inheritance.** Personality, intelligence, aptitudes, and abilities are examples of polygenic inheritance.

Phenotypes and Genotypes   By cross-pollinating red-flowered and white-flowered pea plants, Mendel demonstrated the distinction between the **genotype,** the actual genetic makeup of an organism, and the **phenotype,** the observable (or expressed) characteristics of the organism. In human beings, the phenotype includes physical, physiological, and behavioral traits. Humans also possess dominant and recessive genes, as mentioned earlier.

As an illustration, consider your eye color. Brown is the dominant eye color in humans worldwide. Blue and green are recessive eyes colors. When looking at your own eyes in a mirror (without colored contact lenses), you are observing the *phenotype.* Potentially, the underlying *genotype* could be one of three possibilities for eye color, with *B* representing brown: *BB, Bb,* or *bb.* If you are a brown-eyed person, you have either the *BB* or the *Bb* genotype. If you are a blue-eyed or green-eyed person, you have inherited the two recessive alleles for eye color and have the *bb* genotype. Do you have natural dark hair or light colored hair (blonde or red)? The observed color of your natural hair is your *phenotype.* Again, using *B* for brown or black, there are three possible *genotypes: BB, Bb,* or *bb.* If you have natural dark hair color, your *genotype* is dominant with either *BB* or *Bb.* If you have naturally light hair color, your *genotype* is recessive and is *bb.* To modify what a TV comic Flip Wilson used to say: *What you see is not necessarily what you get.*

Multifactorial Transmission If we consider that heredity and genetics provide us with our basic biochemical structure and an unfolding plan over the course of development, what role does our environment (or ecological systems, as Bronfenbrenner would say) play in our development? Are you who you are because of this genetic blueprint that has been passed along to you, or does your environment promote or detract from that blueprint? Do you, yourself, have any influence on your own life's development? The recognition that environmental factors interact with genetic factors to produce traits is termed **multifactorial transmission.** For example, consider Leann Rimes, who was born with a predisposition toward musical talent. From an early age she loved singing and sang publicly. With her parents' encouragement and support (praise, lessons, time, financial sacrifice, and management), she developed her ability throughout her childhood. By her early teen years, just a few years ago, she had received recognition and praise from top professionals in the country music field. Her songs have been "topping the charts." Her self-motivation *and* the encouragement from others promoted the development of her natural abilities. Had she not had the encouragement and support of these people in her *environment* through her early years, she might not have developed her singing talent.

Additionally, some physical characteristics are the result of multifactorial transmission. For example, the age for the onset of puberty and the age of menopause are believed to be preprogrammed genetically, but nutrition, physical fitness, stress, illness, and disease can advance or delay these preprogrammed events.

Sex-Linked Inherited Characteristics Genes are inherited independently only if they are on different chromosomes. Genes that are linked, or appear on the same chromosome, are inherited together. Good examples of linked genes are **sex-linked traits.** The X chromosome, for instance, contains many genes that are not otherwise related to sexual traits. *Hemophilia,* a hereditary defect that interferes with the normal clotting process of blood, is a sex-linked characteristic carried by the X chromosome. There are about 150 other known sex-linked disorders, including red-green color blindness, a type of muscular dystrophy, certain forms of night blindness, Hunter's disease (a severe form of mental retardation), juvenile glaucoma (hardening of the fluids within the eyeball), and male-pattern baldness.

The vast majority of sex-linked genetic defects occur in men, because men have only one X chromosome. In women, the harmful action of a gene on one X chromosome is usually suppressed by a dominant gene on the other chromosome. Thus, though women themselves normally are unaffected by a given sex-linked disorder, they can be carriers. A man is affected if he receives from his mother an X chromosome bearing the genetic defect (see Figure 3-9). A man cannot receive the abnormal gene from his father. Males transmit an X chromosome only to their daughters, never to their sons, who always receive a father's Y chromosome. A common example of a sex-linked trait is male-pattern baldness (which begins with thinning hair at the crown of the man's head and can lead to extensive baldness by the late twenties and early thirties). The mother inherits this trait from her father, but she herself is not affected. However, her sons have a 50 percent chance of being affected. Another illustration of a sex-linked trait is red-green color blindness. The majority affected by this are males, who often do not realize they see things "differently" until they start driving and have difficulty recognizing

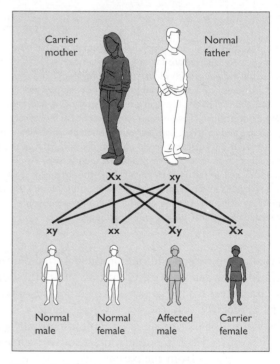

Figure 3-9 Transmission of Sex-linked Genetic Defects In most sex-linked genetic disorders, the female sex chromosome of an unaffected mother (a woman who does not herself show the disorder) carries one faulty chromosome (X) and one normal chromosome (x). The father carries normal x and y chromosomes. The statistical odds for each male child are (1) a 50 percent risk of inheriting the faulty X chromosome and hence the disorder and (2) a 50 percent chance of inheriting normal x and y chromosomes. For each female child the statistical odds are (1) a 50 percent risk of inheriting one faulty X chromosome and hence becoming a carrier like her mother and (2) a 50 percent chance of inheriting no faulty gene.

the red and green on traffic lights. To those affected, these two colors look more like a brownish color. Newer traffic lights have a shutter-like clear covering that opens up and closes quickly to emit a pulsing bright light on the red and green so those affected with red-green color blindness can more accurately distinguish the command. Also, the position of the lights (red on top and green on bottom) is another cue. Preschool and kindergarten teachers teaching color recognition are likely to be the first to spot this in those children who are affected.

In rare instances a female can inherit these disorders if the X chromosome she receives from her mother and the X chromosome she receives from her father *both* bear a gene for a given disease or disorder.

Question

How are traits passed from parents to offspring? Give an example for each of the following: dominant trait, recessive trait, and sex-linked trait.

Genetic Counseling and Testing

Over the past 20 years or so, our increased knowledge of genetics has given rise to the field of **genetic counseling.** People who are deciding whether to create a child might have concerns about inherited diseases in their family backgrounds and might seek genetic counseling in order to find out the risk of passing along this particular disorder or disease. In some genetic disorders, the disease shows up during the first year of life; in others, the symptoms do not become apparent until much later. Some disorders are much more serious than others. A variety of diagnostic tests are now available to parents with substantiated histories of genetic disease and to couples who, for whatever reason, wish to determine whether a fetus has defects. Some genetic defects are more prevalent in people of certain racial or ethnic backgrounds. For example, sickle cell anemia occurs among people of African descent, and Tay Sachs disease occurs in Ashkenazic Jews (see *Human Diversity:* "Genetic Counseling and Testing").

Though genetic counseling can be quite beneficial in many circumstances, ethicists also debate the pros and cons of genetic counseling. How is a decision made that a fetus should live or die? Is our society moving into a disguised eugenics movement? Is genetic information kept private? Is genetic information a reason to excuse certain illegal behaviors in a court of law? What do we do after we test a 2-year-old for a disease and find out the child has inherited a disorder? How much does it cost to raise a child with severe disabilities? What are the hidden emotional and social costs to the family and to society? There are many sides to this issue, and no conclusive answers.

Genetic and Chromosomal Abnormalities

Some disorders are associated with the presence of too few or too many chromosomes (rather than the normal 23 pairs, in humans). One common disorder is *Down syndrome,* a disorder that occurs in 1 out of every 800 live births (National Down Syndrome Society, 1998). In the United States, nearly 350,000 families have a child with Down syndrome. Approximately 5,000 children with Down syndrome are born each year. As the mortality rate associated with Down syndrome is decreasing, the prevalence of individuals with Down syndrome in American society will increase. Some experts project that the number of people with Down syndrome will double in the next 10 years because many couples today are postponing parenting until later in life.

There are three causes of Down syndrome, but in about 95 percent of cases there are three copies of the 21st chromosome, a condition called *trisomy 21.* In these individuals the total number of chromosomes is 47 instead of the normal 46. The extra chromosome alters the course of development and causes the characteristics associated with Down syndrome: flat facial profile, upward slanted eyes, protruding lower jaw, poor Moro reflex, hyperflexibility of joints, excess skin on neck, protruding underlip, enlarged tongue, short neck, very short fifth finger on each hand, one long crease across the palm

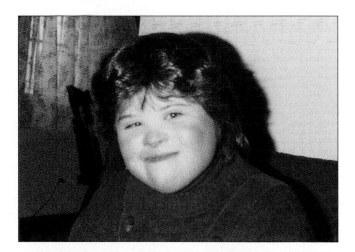

A Young Woman with Down Syndrome Becky has been an active participant in our community over the past 28 years. During her formal schooling in childhood and adolescence, she swam on a regular basis at the YMCA, attended religious education classes and was confirmed at age 16, toured with her school's Bell Choir as a fund-raiser, participated in Special Olympics, acted in school plays, and "danced up a storm" on the party nights. She works in paid employment at our local ARC workshop, visits her sister in New York City, goes to the movies regularly, and awaits placement in a group home setting. Today young children with Down syndrome are mainstreamed in regular classrooms and experience more social acceptance by peers.

Human Diversity

Genetic Counseling and Testing

Prenatal diagnosis employs a variety of techniques to determine the health and condition of an unborn fetus. Without knowledge gained by prenatal diagnosis, there could be an untoward outcome for the fetus or the mother or both. Congenital anomalies (existing at birth) account for 20 to 25 percent of perinatal deaths. Prenatal diagnosis is helpful for (1) managing the remaining weeks of the pregnancy, (2) determining the outcome of the pregnancy, (3) planning for possible complications with the birth process or health risks in the newborn, (4) deciding whether to continue a pregnancy, and (5) finding conditions that could affect future pregnancies (MedLib Utah, 1998). With greater frequency genetic testing has been employed to detect life-threatening conditions and to treat the fetus while still in the uterus. For example, medical scientists are currently attempting to transplant healthy genes directly into the affected fetus to alter the infant's genetic blueprint. The aim is to treat the condition by editing a defective gene out of the infant's hereditary code.

In recent years both invasive and noninvasive techniques have been developed to identify many genetic and chromosomal problems during prenatal development. A variety of sensitive tests are now available to parents with substantiated histories of genetic disease and to couples who, for whatever reason, wish to determine whether the fetus is normal. **Amniocentesis** is one commonly used invasive procedure conducted normally between the 14th to 20th week of gestation. A physician inserts a long, hollow needle through a woman's abdomen into the uterus, drawing out a small amount of the amniotic fluid that surrounds the fetus. A fetal ultrasound is done prior to and during this procedure to view the positions of fetal organs and extremities (see Figure 3-10).

The amniotic fluid contains cells of the unborn child. These are grown in a laboratory culture for a few weeks and analyzed for the presence of various genetic abnormalities. Genetic and chromosomal defects, including those associated with Down syndrome and at least seventy other inherited

biochemical disorders, can be detected in this manner. Some doctors recommend that amniocentesis be used routinely for women who become pregnant after the age of 35. After that age the risk of giving birth to a baby with genetic problems (1 in roughly 192 pregnancies) is greater than the risk of miscarriage due to the procedure (approximately 1 in 200). Risks with amniocentesis are uncommon but, besides miscarriage, can include maternal Rh sensitization, which can be treated

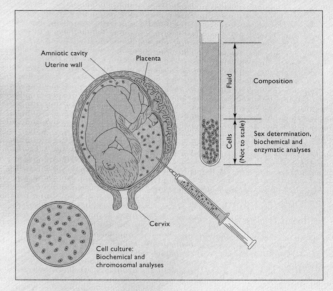

Figure 3-10 Amniocentesis Amniocentesis is a procedure for detecting hereditary defects in the fetus. A sterile needle is inserted into the amniotic cavity, and a small amount of amniotic fluid is withdrawn. This sample is centrifuged to separate fetal cells from the fluid. The cells are then grown in a laboratory culture and analyzed for chromosomal and genetic abnormalities.

From: SEX AND HUMAN LIFE by Pengelley. Copyright 1974 by Addison-Wesley. Reprinted by permission of Addison Wesley Longman.

of each hand, borderline to moderate IQ, global developmental delays, and increased incidence of respiratory, cardiovascular, and other manifestations (National Down Syndrome Society ([NDSS]), 1998).

Today, early intervention services help children with Down syndrome develop to their full potential. Those who receive good medical care and are included in school and community activities can be expected to adapt successfully, develop social skills, find work, participate in decisions that affect them, and make a positive contribution to society. Some adults with Down syndrome are marrying and forming their own families

(children of trisomy-21 parents have a 50 percent chance of having normal intelligence) (NDSS, 1998).

Can genetic counseling predict who is likely to have a child with Down syndrome? A few relevant factors are known. Down syndrome affects people of all races and economic levels. The additional chromosome that causes Down syndrome is more likely to originate from the mother (95 percent chance) than from the father (5 percent chance). Women 35 and older have a significantly increased risk of having a child with Down syndrome. A 35-year-old woman has a 1 in 400 chance, a 40-year-old woman has a 1 in 110 chance, and at age 45 the incidence

with RhoGam. Genetic counselors determine the risk factors involved for each case. Some women think the risk of amniocentesis is unacceptably high (MedLib Utah, 1998).

Ultrasound scanning, or **ultrasonography,** allows physicians to determine the size and position of the fetus, the size and position of the placenta, the amount of amniotic fluid, and the appearance of fetal anatomy. This noninvasive procedure uses sonar to bounce sound waves off the fetus. The developing embryo can be visualized by week 6 of prenatal development. The result is a picture that is safer for the mother and fetus than that afforded by standard X rays (MedLib Utah, 1998). Still another procedure, **fetoscopy,** allows a physician to examine the fetus directly through a lens after inserting a very narrow tube into the uterus (a blood sample or skin biopsy can also be taken from the fetus). However, this procedure is more risky than amniocentesis.

Another invasive method is **chorionic villus biopsy (CVS),** which can be performed between 9.5 and 12.5 weeks of gestation. The chorionic villi are hairlike projections of the membrane that surrounds the embryo during early pregnancy. The physician, guided by ultrasound, inserts a thin catheter through the vagina and cervix and into the uterus and, employing suction, removes a small plug of villous tissue. Although the chorion is not an anatomical part of the embryo itself, it is embryonic rather than maternal in origin. The most common test employed on cells obtained by CVS is chromosome analysis to determine the karyotype of the fetus. CVS has a small but significant risk of causing miscarriage (0.5 to 1 percent higher than the risk from amniocentesis), and might cause maternal Rh sensitization.

Fortunately, there are a number of blood tests that might eventually render amniocentesis and chorionic villus sampling unnecessary. With **maternal blood sampling** for fetal blood cells, researchers have detected Down syndrome and a number of other common birth defects by analyzing fetal cells shed into the mother's bloodstream. However, it is difficult to get many fetal blood cells using this technique (MedLib Utah, 1998).

Maternal serum alpha-fetoprotein (MSAFP) is a newer diagnostic method that analyzes two major blood proteins—albumin and alpha fetoprotein (AFP). When the fetus has neural tube defects such as anencephaly or spina bifida, more AFP crosses the placenta and reaches the mother's blood. The incidence of such defects is 1 to 2 births per 1,000 in the United States. MSAFP detects defects in the fetal abdominal wall and can also be useful in screening for Down syndrome and other trisomies. MSAFP has the greatest sensitivity at 16 to 18 weeks of gestation.

The measurement of maternal serum estriol in the third trimester will give a general indication of the well-being of a fetus. The level of estriol in the mother's urine is measured (the substrate for estriol begins as DHEA made by the fetal adrenal glands). The fetal placenta metabolizes the DHEA into estriol. This estriol crosses to the maternal circulation and is excreted by maternal urine or by maternal bile from the liver. Estriol is lower with Down syndrome and with anencephaly (MedLib Utah, 1998).

Results of genetic testing procedures can bring much relief or produce anguish and sorrow. Parents who learn that they are carriers of genetic diseases often feel ashamed and guilty, mortified to be stigmatized as "genetic defectives." They are also faced with the difficult decision of whether they should proceed with an abortion if the test results indicate the presence of defects in the fetus. Parents who learn that the child they are expecting is abnormal can experience psychological problems including denial, severe guilt, depression, termination of sexual relations, marital discord, and divorce. Clearly, some parents need psychological counseling in conjunction with genetic counseling.

Genetic screening also creates new opportunities for discrimination (Faden & Kass, 1993). Your genetic profile could be used to determine who you could marry, what jobs you could apply for, and whether insurance companies would consider you a good risk. Another dilemma is whether you would want to know if you had inherited a harmful gene.

becomes approximately 1 in 35 (NDSS, 1998). However, younger women also give birth to children with Down syndrome. Some genetic researchers suggest that the current statistics are misleading because the total number of older women (over 35) having babies is typically far fewer than the total number of younger women who are having babies. A number of other disorders are linked to sex chromosome abnormalities.

For those who believe they may have a significant chance of bearing a child at risk of a genetic disorder, genetic counseling and testing can determine with a high degree of accuracy whether the fetus has this defect, can

help couples prepare for a child with a disability, or can lead to termination of a pregnancy. (See "Following Up on the Internet" at the end of this chapter.)

Questions

Why is the entire human genetic code being mapped by scientists in research labs around the world? If you were going to have a baby, would you seek out genetic testing? Why or why not?

Prenatal Development

Never again in his life, in so brief a period will this human being grow so rapidly or so much, or develop in so many directions, as he does between conception and birth. . . . During this critical period, the development of the human body exhibits the most perfect timing and the most elaborate correlations that we ever display in our entire lives. The building and launching of a satellite, involving thousands of people and hundreds of electronic devices, is not nearly so complex an operation as the building and launching of a human being. (Physical anthropologist Ashley Montagu, 1964, pp. 20–21)

Whether conception occurs naturally or as a result of an assisted reproductive technique, between conception and birth the human being grows from a single cell, barely visible to the naked eye, to a mass of about 7 pounds containing some 200 billion cells.

The **prenatal period** is the period between conception and birth. It normally averages about 266 days, or 280 days from the last menstrual period. Embryologists divide prenatal development into three stages: The **germinal period** extends from conception to the end of the second week; the second, the **embryonic period** extends from the end of the second week to the end of the eighth week; and the third, the **fetal period** extends from the end of the eighth week until birth. Developmental biology is progressing rapidly, giving us a better understanding of how the body and its specialized organs and tissues are formed (Visible Embryo Project, 1998; Barinaga, 1994).

The Germinal Period

The germinal period is characterized by (1) the growth of the zygote and (2) the establishment of a linkage between the zygote and the support system of the mother. After fertilization, the zygote begins a three- to four-day journey down the fallopian tube toward the uterus (see Figure 3-11). The zygote is moved along by the action of the cilia and the active contraction of the walls of the oviduct. Within a few hours of fertilization, growth begins with the initiation of mitosis. In *mitosis* the zygote divides, forming 2 cells identical in makeup to the first cell. In turn, each of these cells divides, making 4 cells. The 4 cells then divide into 8, 8 into 16, 16 into 32, and so on.

The early mitotic cell divisions in development are called *cleavage* and occur very slowly. The first cleavage takes about 24 hours; each subsequent cleavage takes 10 to 12 hours. These cell divisions soon convert the zygote into a hollow fluid-filled ball of cells termed a **blastocyst** (see Table 3-3). The blastocyst should continue to develop and travel into the uterus. When a blastocyst remains trapped in the fallopian tube, the pregnancy is *ectopic* or *tubal.* This will cause the mother a great deal of pain, and the blastocyst will have to be surgically removed or the fallopian tube will burst, causing hemorrhaging.

Once the blastocyst enters the uterine cavity, it floats freely for 2 or 3 days. When it is about 6 to 7 days old and composed of some 100 cells, the blastocyst makes contact with the *endometrium,* the wall of the uterus. The endometrium in turn becomes vascular, glandular,

Figure 3-11 Early Human Development: The Course of the Ovum and Embryo The drawing depicts the female reproductive system, the fertilization of the ovum, and the early growth of the blastocyst, which will soon become an embryo.

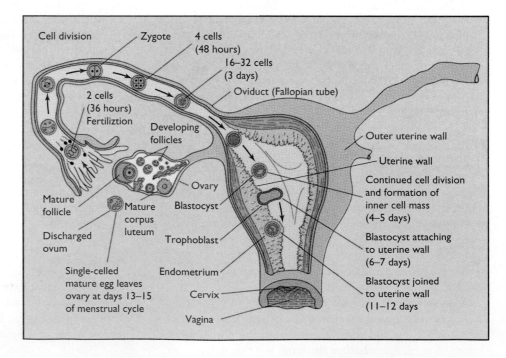

Table 3-3 Blastocyst, Cleavage, and Implantation

	Description	Vulnerabilities
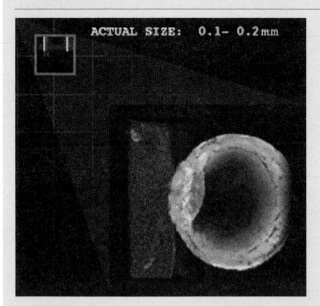	The early process of *mitotic cell division* reduces cell size within the same original structure. During this process, called cleavage, the mass of the embryo remains constant. At the 16-cell stage, cells begin to adhere to each other. At the 100-cell stage, the cell cluster is filled with gel-like fluids, has entered the upper end of the uterus, and is now called a *blastocyst*. The outer layer of cells are the trophoblast, and the inner cluster is called the inner cell mass. By the 9th to 11th day of development, the blastocyst buries itself in the nutrient-rich lining of the mother's uterus (implantation).	During implantation, a variety of chemicals (proteins and hormones) enter the mother's bloodstream from the blastocyst, and a complex interchange of "information" begins. The mother's uterus might reject the blastocyst as a foreign cell (it has its own unique genetic makeup—different from the mother's). If anything goes wrong during this stage, the mother's own immune system might destroy the blastocyst (Nilsson & Hamberger, 1990). If the mother's uterus is diseased from pelvic inflammatory disease or sexually transmitted diseases, the blastocyst might not be able to implant itself into the uterine wall and will not survive.

Implantation of the Blastocyst Around the 9th to 11th day, the blastocyst implants itself in the nutrient-rich lining of the uterus.

and thick. The blastocyst "digests" its way into the endometrium through the action of enzymes and gradually becomes completely buried in it. As a result, the embryo develops within the wall of the uterus and not in its cavity. This invasion of the uterus by the blastocyst creates a small pool of maternal blood. During the germinal period, the organism derives its nourishment from the eroded tissue and maternal blood that flow through spaces in the outer layer of cells of the blastocyst.

By the eleventh day, the blastocyst has completely buried itself in the wall of the uterus, in a process called **implantation** (see Table 3-3). The hormone progesterone from the ovary prepares the uterine lining for implantation. This increase in progesterone is also a signal to the brain that the woman is pregnant, and most pregnant women cease to menstruate (in rare instances, a woman might menstruate and not realize she is pregnant). At this stage in development, the organism is about the size of a pinhead. As of yet, the mother is seldom aware of any symptoms of pregnancy. The blastocyst, now made up of hundreds of cells, is busy surrounding itself with chemicals to prevent the uterine immune system from destroying it, and, the cervix has been sealed by a plug of mucus (Nilsson & Hamberger, 1990).

During the implantation process, the blastocyst begins separating into two layers. The outer layer of cells, called the *trophoblast*, is responsible for embedding the embryo in the uterine wall. The inner surface of the trophoblast becomes the nonmaternal portions of the placenta, the *amnion* and *chorion*. The amnion forms a closed sac around the embryo and is filled with a watery amniotic fluid to keep the embryo moist and protect it against shock or adhesions. The chorion is a membrane that surrounds the amnion and links the embryo to the placenta. The internal disc or cluster of cells that compose the blastocyst, called the *inner cell mass*, produces the embryo. The entire process is controlled by genes. Some genes turn on rapidly as the embryo develops; others turn on slowly; and still others operate throughout the prenatal period and beyond. The patterns of gene activity are complex and involve many different genes (Nilsson & Hamberger, 1990; Marx, 1984).

Toward the end of the second week, mitotic cell division proceeds more rapidly. The embryonic portion of the inner cell mass begins to separate into three layers: the *ectoderm* (the outer layer), which is the source of future cells forming the nervous system, the sensory organs, the skin, and the lower part of the rectum; the *mesoderm* (the middle layer), which gives rise to the skeletal, muscular, and circulatory systems and the kidneys; and the *endoderm* (the inner layer), which develops into the digestive tract (including the liver, the pancreas, and the gall bladder), the respiratory system, the bladder, and portions of the reproductive organs (Nilsson & Hamberger, 1990).

The Embryonic Period

The **embryonic period** lasts from the end of the second week to the eighth week. It spans that period of pregnancy from the time the blastocyst completely implants itself in the uterine wall to the time the developing organism becomes a recognizable human fetus. During this period the developing organism is called an **embryo** and normally experiences (1) rapid growth, (2) the establishment of a placental relationship with the mother, (3) the early structural appearance of all the chief organs, and (4) the development, in form at least, of a recognizably human body. All of the major organs are developing now, except the sex organs, which will begin to develop within several weeks; at that point male embryos begin to produce the hormone testosterone, and male sex organs begin to differentiate from female organs.

The embryo becomes attached to the wall of the uterus by means of the placenta. The **placenta** forms from uterine tissue and the trophoblast of the blastocyst and functions as an exchange terminal that permits entry of food materials, oxygen, and hormones into the embryo from the mother's bloodstream and the exit of carbon dioxide and metabolic wastes from the embryo into the mother's bloodstream. The placenta is a partially permeable membrane that does not permit the passage of blood cells between the two organisms. This feature provides a safeguard against the mingling of the blood of the mother with that of the embryo. Were the mother's and embryo's blood to intermix, the mother's body would reject the embryo as foreign material.

The transfer between the placenta and the embryo occurs across a web of fingerlike projections, the *villi*, that extend into blood spaces in the maternal uterus. The villi begin developing during the second week, growing outward from the chorion. When the placenta is fully developed at about the seventh month of pregnancy, it is shaped like a pancake or disc, 1 inch thick and 7 inches in diameter. From the beginning, the **umbilical cord** links the embryo to the placenta and is a conduit carrying two arteries and one vein. This connecting structure, or lifeline, is attached to the middle of the fetal abdomen.

Development that commences with the brain and head areas and then works its way down the body is called **cephalocaudal development.** This direction of development ensures an adequate nervous system to support the proper functioning of other systems. During the early part of the third week, the developing embryo begins to take the shape of a pear, the broad, knobby end of which becomes the head. The cells in the central portion of the embryo also thicken and form a slight ridge

that is referred to as the primitive streak. The *primitive streak* divides the developing embryo into right and left halves and eventually becomes the spinal cord. The tissues grow in opposite directions away from the axis of the primitive streak, a process termed **proximodistal development.** Cephalocaudal development and proximodistal development are illustrated in Chapter 4 on pages 127 and 128.

By the twenty-eighth day, the head region takes up roughly one-third of the embryo's length. Also about this time, a brain and a primitive spinal cord become evident. As development progresses during the second month, the head elevates, the neck emerges, and rudiments of the nose, eyes, mouth, and tongue appear. Another critical system—the circulatory system—also develops early. By the end of the third week, the heart tube has already begun to beat in a halting manner.

Within four weeks of conception, the embryo is about ⅕ inch long—nearly 10,000 times larger than the fertilized egg. About this time the mother usually becomes suspicious that she is pregnant. Her menstrual period is generally two weeks overdue. She might feel a heaviness, fullness, and tingling in her breasts; simultaneously, the nipples and surrounding areolas may enlarge and darken. Also at this time, about one-half to

Embryo with Primitive Streak Barely 6 weeks old and measuring 15 mm (just over ½ inch), the embryo has its transparent back turned toward us. The primitive streak, visible through the thin skin, will become its spinal cord. The embryo is encircled by its amniotic sac, with the ragged chorionic villi and the umbilical cord to the right. The yolk sac hovers to the left.

two-thirds of all pregnant women experience a morning queasiness or nauseous feeling. The condition, called "morning sickness," can persist for several weeks or months and varies in intensity from woman to woman.

The developing embryo is particularly sensitive to the invasion of drugs, diseases, and environmental toxins in the mother's body because so many of its major body systems are developing. The mother's excessive use of alcohol, nicotine, or caffeine, and her use of other more potent chemical agents such as crack cocaine, heroin, and strong prescription medications can certainly harm the development of this embryo's organs and structures. Each organ and structure has a **critical period** during which it is most vulnerable to damaging influences. *Information You Can Use:* "Major Drug and Chemical Teratogens" on page 94 discusses a number of important diseases and chemical agents that can harm the developing embryo.

The Fetal Period

The final stage in prenatal life—the **fetal period**—begins with the ninth week and ends with birth. During this time the organism is called a **fetus,** and its major organ systems continue to develop and assume their specialized functions. By the end of the eighth week, the organism definitely resembles a human being. It is complete with face, arms, legs, fingers, toes, basic trunk and head muscles, and internal organs. The fetus now builds on this basic form (see *Further Developments:* "Highlights of Embryo and Fetus Maturation").

Development during the fetal period is less dramatic than that during the embryonic period. Even so, signifi-

cant changes occur. By the tenth week the fetal face acquires a truly human appearance. During the third month the fetus develops skeletal and neurological structures that lay the foundation for spontaneous movements of the arms, legs, and fingers. By the fourth month, stimulation of the infant's body surfaces activates a variety of reflex responses. About the beginning of the fifth month, the mother generally begins to feel the spontaneous movements of the fetus (called *quickening,* a sensation like a moving butterfly in the abdominal region). Also during the fifth month, a fine, downy, woolly fuzz (*lanugo hair*) begins to cover the fetal body.

At six months the eyebrows and lashes are well defined; the body is lean but strikingly human in proportions; the skin is wrinkled. At seven months the fetus (now weighing about 2½ pounds and measuring about 15 inches in length) gives the appearance of a dried-up, aged person, with red, wrinkled skin covered by a waxy coating (*vernix*). The fetus is now a viable organism and can cry weakly. At eight months, fat is being deposited around the body, the fetus gains an additional two pounds, and its neuromuscular activity increases.

At nine months the dull redness of the skin fades to pink, the limbs become rounded, and the fingernails and toenails are well formed. At full term (38 weeks) the body is plump; the skin has lost most of its lanugo hair, although the body is still covered with vernix; and all the organs necessary to carry on independent life are functioning. The fetus is now ready for birth.

Recent research with sheep suggests that hormones released by the fetal pituitary and adrenal glands are involved in starting the process of labor and birth. In sheep, immediately prior to birth, the production of

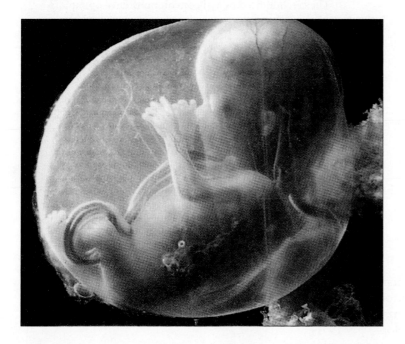

The Fetus at Three Months At 3 months the fetus is a little more than 3 inches long and weighs nearly an ounce. Its head is disproportionally large and appears increasingly like that of a human. By now the external ears and eyelids have formed. The umbilical cord increases in size to accommodate the growing organism. The fetus is developing the skeletal and neurological structures that provide the foundation for moving in its capsule, where it floats weightlessly much in the manner of an astronaut in space.

Further Developments

Highlights of Embryo and Fetus Maturation

Following are the highlights of growth and development during the embryonic, fetal, and postnatal stages.

First Month

Fertilization occurs.

Blastocyst implants itself in the lining of the uterus.

The blastocyst differentiates into the trophoblast and the inner cell mass.

The inner cell mass differentiates into three layers: the ectoderm, the mesoderm, and the endoderm.

Nervous and circulatory systems begin to develop.

Second Month

The head develops rapidly, at first accounting for about half of the embryo's size.

The face and neck begin to form.

The eyes rapidly begin converging toward the center of the face.

Mouth and jaws form and become well differentiated.

Major organs of the digestive system become differentiated.

Buds for the limbs form and grow.

The circulatory system between the embryo and the placenta is completed, and heartbeats begin.

Third Month

The digestive organs begin functioning.

Centers of ossification (bone formation) appear.

Buds for all temporary teeth form.

Sex organs develop rapidly, especially in males.

Arms, legs, and fingers make spontaneous movements.

Fourth Month

The face acquires a human appearance.

The lower parts of the body show rapid growth, with the body outgrowing the head.

Reflex movement becomes brisker.

Most bones are distinctly indicated throughout the body.

In males the testes are in position for later descent into the scrotum; in females the uterus and vagina are recognizable.

Fifth Month

Lanugo hair (a fine, woolly fuzz) covers the whole body.

Vernix (a waxy coating) collects.

The nose and ears begin ossification.

Fingernails and toenails begin to appear.

Sixth Month

Eyebrows and lashes are well defined.

The body is lean but strikingly human in its head and body proportions.

The cerebral cortex becomes layered.

The eyes are completely formed.

Seventh Month

The fetus is capable of living outside the uterus.

The fetus looks like a dried-up, aged person with red, wrinkled skin.

The cerebral fissures and convolutions develop rapidly.

Eighth and Ninth Months

Subcutaneous fat is deposited for later use.

The fingernails reach beyond the fingertips.

A good deal of the lanugo hair is shed.

Initial myelination of the brain takes place. (Myelin is a fatty substance that forms a sheath around the axons of nerve cells.)

The activity of the chief organs is stepped up.

Vernix is present over the entire body.

From: W. J. Robbins, *Growth.* Copyright © 1928. Reprinted by permission of Yale University Press.

adrenocorticotropic hormone (ACTH) by the pituitary increases and is then followed by a rise in cortisol from the adrenal gland. The increased cortisol alters the enzymatic balance in the mother's uterus, and this development seems to trigger the beginning of labor. Scientists are continuing with this research to discover whether this model for sheep is accurate for human beings (Palca, 1991).

Loss by Miscarriage

An estimated 14 to 18 percent of all known pregnancies end in miscarriage, and one-third of all women experience a miscarriage at some time in their reproductive years (DeFrain, Millspaugh, & Xiaolin, 1996). A **miscarriage** occurs when the zygote, embryo, or fetus is naturally expelled from the uterus before it can survive outside the mother's womb. The medical term for miscarriage is *spontaneous abortion,* which is most likely preceded by cramping or bleeding. All women who miscarry want to know the cause of their loss.

Approximately 75 percent of miscarriages occur before 12 weeks of development, which normally indicates a problem with the implantation of the embryo into the uterine wall or some form of genetic mutation causing abnormality of the embryo (DeFrain, Millspaugh, & Xiaolin, 1996). Later miscarriages are commonly the result of some structural problem with the uterus, problems with implantation, or a cervix that does not stay closed but begins to open (DeFrain, Millspaugh, & Xiaolin, 1996). In some women the cervix is temporarily "sewn" to prevent premature birth. The pregnant woman might also have sustained a serious accident, trauma, or illness that triggers a miscarriage. Additionally, we are just beginning to understand the significant impact of environmental factors on prenatal development or loss as well (see *Information You Can Use:* "Major Drug and Chemical Teratogens" on p. 94). Sometimes miscarriage is both unpredicted and inexplicable (Puddifoot & Johnson, 1997).

Substantial research findings document the mother's deep emotional distress after a miscarriage, yet Western cultures have no established ritual to mourn the loss of a child by miscarriage. Over the past 20 years, some researchers have conducted empirical studies on the psychosocial impact of this loss on mothers, fathers, and other family members. We know the mother is likely to experience guilt because she might mistakenly perceive this loss as a personal failure, a form of punishment for her habits, personal practices, or ambitions (Dunn et al., 1991). The mother and father are greatly in need of social and family support at this time. Despite this, women consistently report unhelpful responses from family, friends, physicians, and other health care professionals following a miscarriage (Conway, 1995). Typically people overtly sympathize with her distress, while attempting to comfort her by saying "these things happen," or "you already have children," or "you can have another one." Established support groups and visits by the clergy are normally very helpful to the family.

The partner's emotional reactions at the loss of a baby have largely been overlooked. Most people assume the man's role is one of "support" and that he is more "peripheral" to the situation (Smart, 1992). One recent study in the United Kingdom focused on 20 men selected randomly from a larger pool of subjects whose partners had miscarried. The immediate focus of the males' attention was with their partner's condition, but the men also expressed considerable grief for the lost child during the interviews. Results of this study demonstrated that men's scores on measures of grief and stress were not dissimilar to those reported by female cohorts (Johnson & Puddifoot, 1997). Nowadays more fathers are present during ultrasound imaging of the fetus and acknowledge the scan is real evidence of the existence of the baby. Their emotional reactions to loss by miscarriage range from statements of no emotional response to much stronger ones such as "It's hard to explain . . . it's like having your guts ripped out." The fathers also suggested that it would take little change for health authorities to inform both the mother and father in a more sensitive manner and to keep them better informed (Puddifoot & Johnson, 1997).

Like for any other death of a loved one, grieving can take a long time—even a lifetime. Most parents resolve the issue after a lengthy period of shock, pain, disorganization, and redefinition (DeFrain, Millspaugh, & Xiaolin, 1996). Both mothers and fathers report having disturbing nightmares about crying babies, flashbacks over many years, feelings of "going crazy," feelings of guilt and blame, and thoughts of suicide (DeFrain, Millspaugh, & Xiaolin, 1996). Over the past 20 years, DeFrain and colleagues (1996) have developed and field-tested a psychometric instrument that measures the extensive psychosocial effects on family members of infant loss, including miscarriage and stillbirth (see Chapter 19). The loss of a baby can have profound effects on parents and surviving siblings. There are sites on the Internet devoted to remembering those special little ones, and many communities offer group support.

Prenatal Environmental Influences

To most of us, the concept of "environment" refers to a human being's surroundings *after* birth. In truth, environmental influences are operating from the moment of conception, if not actually before with the mother and father's own health. (The mother's health during pregnancy is discussed further in Chapter 13.) The fertilized ovum undertakes a hazardous week-long journey down the fallopian tube and around the uterus, encountering throughout a highly variable and chemically active medium. We generally think of the uterus as providing a sheltered, warm, and protected environment for prenatal development. But even after implanting itself in the uterus, the embryo is vulnerable to maternal disease, malnutrition, use of cigarettes, and licit and illicit drugs, and biochemical malfunctioning.

Most pregnancies end with the birth of normal, healthy babies. Nonetheless, as stated above, 14 to 18 percent of all conceptions result in spontaneous abortion or stillbirths (the baby dies before birth). Nearly 25 percent of these are the product of chromosomal and gene abnormalities; the remainder result from problems associated with the prenatal environment (Mueller, 1983). Another 6 to 7 percent of conceptions lead to babies born with birth defects. Some of these defects arise from environmental factors. Scientists term any agent that contributes to birth defects or anomalies a **teratogen.** The field of study concerned with birth defects is called **teratology.**

Drugs and Other Chemical Agents   The thalidomide tragedy in the early 1960s awakened the medical profession and the public alike to the potential dangers of drugs for pregnant women. Thousands of German, English, and Canadian women who had been prescribed thalidomide (a tranquilizer) for nausea or "morning sickness" during the embryonic stage of pregnancy gave birth to infants with malformed or missing limbs. The use of thalidomide is on the rise again in various countries around the world, because it reduces nausea in AIDS patients. We now know that many drugs and chemical agents cross the placenta and affect the embryonic and fetal systems (Hutchings, 1989). Quinine (used in the treatment of malaria) can cause congenital deafness. Barbiturates (sedating drugs) can affect the oxygen supply to the fetus and result in brain damage. Antihistamines can increase the mother's susceptibility to spontaneous abortion. And even aspirin and a number of antibiotics have been tentatively linked in several cases to heart, hearing, and other problems in fetal development (Barr et al., 1990).

According to current medical opinion, pregnant women should not take drugs except for conditions that seriously threaten their health—and then *only* under the supervision of a physician. Some authorities advise against extra vitamins (particularly vitamin A) unless prescribed by a physician. Likewise, miscellaneous reports of increased rates of spontaneous abortion and birth defects among coffee drinkers prompted the federal Food and Drug Administration (FDA) to recommend in 1980 that women stop or reduce their consumption of caffeinated coffee, tea, and cola drinks during pregnancy. Concerning the increasing maternal use of illicit drugs, Chavkin and Breitbart (1997) remind us of America's dualistic societal response: anger and blame directed at women who use alcohol and other drugs, and neglect and a consequent lack of appropriate treatment. Often the focus is on the addicted pregnant woman, and the debate posits a woman's right to autonomy and privacy in opposition to the future child's right to be born free from harm. *Information You Can Use:* "Major Drug and Chemical Teratogens" on page 94 provides information on major drug and chemical teratogens and fetal effects.

Maternal Infectious and Noninfectious Diseases
Under some circumstances infections that cause illness in the mother can harm the fetus (Kalter & Warkany, 1983a, 1983b). When the mother is directly infected, viruses, bacteria, or malarial parasites might cross the placenta and infect the child. In other cases the fetus can be indirectly affected by a high fever in the mother or by toxins produced by bacteria in the mother's body. The exact time during the fetus's development at which an infection occurs in the mother has an important bearing. As described earlier, the infant's organs and structures emerge according to a fixed sequence and timetable. Each organ and structure has a *critical period* during which it is most vulnerable to damaging influences.

Rubella and Other Agents   If the mother contracts *rubella* (German measles) in the first three months of pregnancy, there is a substantial risk of blindness, deafness, brain damage, and heart disease in the offspring. In 10 to 20 percent of the pregnancies complicated by rubella, spontaneous abortion or stillbirth ensues. However, should the mother contract the disease in the last three months of pregnancy, there is usually no major damage. As a preventative measure, a woman must be inoculated with a preventive vaccine at least three months before the beginning of pregnancy. Various other viral, bacterial, and protozoan agents are suspected either of being transmitted to the fetus or of otherwise interfering with normal development. These agents include hepatitis, influenza, poliomyelitis, malaria, typhoid, typhus, mumps, smallpox, scarlet fever, gonorrhea, and cytomegalovirus infection.

Syphilis   Syphilis continues to pose a serious health problem to both mother and child, especially for those women who are trading sex for drugs or have multiple sex partners (Lewin, 1992c). Congenital syphilis causes fetal or perinatal death in 40 percent of the infants infected (Centers for Disease Control and Prevention, [CDC], 1998c; Mobley, McKeown, Jackson, & Sy, 1998). If not detected at birth, the disease takes a gradual toll in a deterioration of thought, speech, and motor abilities before the child finally dies. More than three-fourths of known syphilitic pregnant women do not show clinical evidence of the disease and are not taking penicillin treatment. Consequently, the Wasserman test for syphilis should be administered to all pregnant women. Although congenital syphilis nearly vanished as a scourge a decade or so ago, it reappeared with a vengeance in the late 1980s.

Genital Herpes   Genital herpes infects roughly 22 percent of adults in the United States and between 7 and 28 percent of Europeans, depending on the country (Seppa, 1998). Infants delivered through an infected birth canal stand a risk of contracting the disease. Infants whose mothers have had recurrent episodes of herpes infection apparently acquire neutralizing antibodies that lower their risk of contracting the disease. But since 33 percent of infected babies will die and another 25 percent will suffer permanent brain damage, some obstetricians advise delivery by Cesarean section to minimize the chances of infection. Pregnant women with genital herpes also have a considerably higher risk of miscarriage than women without herpes. The antiviral drug famciclovir is currently under study for its potential protection against genital herpes (Seppa, 1998).

HIV/AIDS   On the average, one in four babies born to mothers with HIV develops AIDs. The likelihood of infecting an offspring apparently increases with the amount of the virus in a pregnant woman's blood (Altman, 1994b). Government health officials also report that the drug AZT dramatically reduces transmission of the HIV virus from infected women to their newborns (Altman, 1994a). The disease weakens the immune systems of its victims, leaving them prey to unusual infections, forms of cancer, and early death. Fighting the spread of the disease is especially difficult because most infants with HIV are born to mothers with no outward signs of disease. Many of the mothers are intravenous drug users and are at high risk of contracting HIV from shared needles (American Academy of Family Physicians, 1998).

Half to three-quarters of infants infected with HIV have distinct head and face abnormalities, including small head circumference; a prominent, boxlike appearance of the forehead; a "scooped out" profile; slanting eyes; blue in the whites of the eyes; and a short, flattened nose. Because craniofacial structures typically are developed during the first trimester of pregnancy, the abnormal features suggest that the virus attacks the fetus early in development. Most prenatally infected youngsters develop symptoms of the disease within the first 12 months of life; these symptoms include recurrent bacterial infections, swollen lymph glands, failure to thrive, neurological impairments, and delayed development. Many of these mothers are from low socioeconomic environments, lack medical insurance, and require financial assistance and social services (Task Force on Pediatric AIDS, 1989; CDC, 1998b; "Pregnancy and HIV," 1998).

Diabetes   Diabetes is a metabolic disease characterized by a deficiency of insulin and an excess of sugar in the blood and urine. If medical measures are not taken to control the mother's diet and to administer insulin artificially, there is a 50 percent probability that the fetus will be aborted or stillborn. In babies born to mothers with diabetes, as many as one in four have deformities such as unformed spines, displaced hips, misplaced hearts, and extra ear skin (CDC, 1986). Only approximately 0.3 percent of all U.S. pregnancies occur in women with established diabetes, but the estimated 10,000 to 14,000 infants born annually to women with established diabetes are at high risk for mortality, prematurity, and congenital defects. Risks of maternal complications are also associated with diabetes during pregnancy.

Maternal Sensitization: The Rh Factor   The Rh factor is a condition of incompatibility of a factor in the mother's and child's blood that under some circumstances produces a serious and often fatal form of anemia and jaundice in the fetus or newborn—a disorder termed *erythroblastosis fetalis*. About 85 percent of all whites have the Rh factor; they are called Rh-positive. About 15 percent do not have it; they are Rh-negative. Among blacks only about 7 percent are Rh-negative, and among Asians the figure is less than 1 percent. Rh-positive blood and Rh-negative blood are incompatible. Each blood factor is transmitted genetically in accordance with Mendelian rules (Rh-positive is dominant). For the most part, the maternal and fetal blood supplies are separated by the placenta. On occasion, however, a capillary in the placenta ruptures and a small amount of maternal and fetal blood mixes. Likewise, some admixture usually occurs during the "afterbirth," when the placenta separates from the uterine wall.

An incompatibility results between the mother's and the infant's blood when an Rh-negative mother has a baby with Rh-positive blood. Under these conditions, which occur in about 1 of every 200 pregnancies among whites, the mother's body produces antibodies that cross the placenta and attack the baby's blood cells. *Erythroblastosis fetalis* can now be prevented if an Rh-negative mother is given anti-Rh antibodies (Rhogam) shortly after the birth of her first child. If an Rh-negative mother has already been sensitized to Rh-positive blood by several pregnancies in the absence of Rhogam therapy, her infant can be given an interuterine transfusion.

Maternal Stress   The effect of maternal emotions on the unborn infant has long been a subject of folklore. Most of us are well aware that being frightened by a snake, a mouse, a bat, or some other creature will not cause a pregnant woman to give birth to a child with a distinctive personality or birthmark. Medical science does suggest, however, that severe, prolonged anxiety in an expectant mother can have a harmful effect on her child (Glover, 1997). Of course, no direct neural linkage exists between the mother and the fetus. When the mother is anxious or under stress, various hormones such as epinephrine (adrenaline) and acetylcholine are released into her bloodstream. These hormones can pass through the placenta and enter the blood of the fetus. Should a pregnant woman feel that she is experiencing prolonged and unusual stress, she would be well advised to consult a physician, a trained therapist, or someone in the clergy.

Information You Can use
Major Drug and Chemical Teratogens

Smoking

The nicotine found in tobacco is a mild stimulant drug. When an expectant mother smokes, her bloodstream absorbs nicotine. The drug is then transmitted through the placenta to the embryo, and this in turn increases fetal activity. This poses a serious hazard, because the mortality rate for premature infants (low birthweight) is higher than that for full-term infants. Furthermore, major congenital abnormalities are more prevalent among the infants of women who smoke (Zimmer & Zimmer, 1998). Problems associated with smoking and pregnancy continue to present special challenges to public health in virtually all societies. The consensus for intervention is to target older and less educated females, as well as blacks, and those who are unemployed during pregnancy and to inform all of the importance of early prenatal care (Zimmer & Zimmer, 1998).

Marijuana

Marijuana, a psychoactive drug, is the most frequently used illicit drug in America (Calhoun & Alforque, 1996). An accumulating body of evidence suggests that marijuana use by expectant mothers has detrimental effects on fetal development and neonatal behavior, including infants' neurological development (Lester & Dreher, 1989). Infants exposed to marijuana prenatally are sometimes identified at birth and during the neonatal period by their low birthweight and size, respiratory problems, slow weight gain, and increased risk of sudden infant death syndrome (SIDS) (Calhoun & Alforque, 1996). The severity of the effects of marijuana use are difficult to determine, given the confounding effects of tobacco, alcohol, and other hard drugs that these mothers are likely to also be using (Sluder, Kinnison, & Cates, 1997).

Hard Drugs

Heroin, methadone, cocaine and its derivative "crack" are capable of producing a wide range of birth deformities (Zimmer & Zimmer, 1998; Calhoun & Alforque, 1996). Newborn infants exposed *in utero* to heroin are identified by premature birth size and weight, excessive tremulous behavior, profuse sweating, excessive sneezing, excessive yawning, poor sleep patterns, poor swallowing ability, poor sucking or eating ability, and increased risk of SIDS (Calhoun & Alforque, 1996). Cocaine is the second most frequently used illicit drug in the United States (Calhoun & Alforque, 1996). "Crack babies" are identified at birth by premature birth size and weight, tremulous behavior, nasal stuffiness, prolonged high-pitch crying, high temperature, poor sucking or feeding ability, respiratory problems, regurgitation problems, excessive hyperactivity, and rigidity (Lester, 1997; 1995). These newborns go through the same type of withdrawal symptoms as addicted adults. These infants are often exposed prenatally to narcotics, tobacco smoke, limited nutrition, or sexually transmitted diseases their mothers contracted during sexual behavior engaged in to support a drug habit. The dimension of hard drug use is staggering in long-term health effects and the cost to society. According to the National Association for Perinatal Addiction Research and Education, about 1 out of every 10 newborns in the United States—about 375,000 each year—is exposed in the womb to one or more illicit drugs (Calhoun & Alforque, 1996).

Alcohol

Alcohol is now recognized as *the* leading teratogen to which the fetus is likely to be exposed. *In utero* alcohol exposure is thought to be the most common cause of mental retardation (MedicineNet, 1997). We were forewarned long ago. In the Old Testament, Samson's mother is admonished by an angel: "Thou shalt conceive, and bear a son. Now therefore beware, I pray thee, and drink not wine nor strong drink." *Fetal alcohol syndrome* (FAS) was formally reported in a French pediatric journal in 1968 by P. Lemoine and

---

Maternal stress and anxiety are linked with complications of pregnancy, mainly prematurity and low birthweight for gestational age (Glover, 1997; Copper et al., 1996). Lou and colleagues (1994) followed 3,021 women through pregnancy and obtained results about their stress using the questionnaire method. They found that maternal stress and smoking contributed independently and significantly to a lower gestational age, lower birthweight, smaller head circumference, and poorer scores on the neonatal neurological exam (Lou et al., 1994). Some stress and anxiety are an inescapable feature of expectant motherhood, and moderate amounts of stress probably have no harmful effect on the fetus (Istvan, 1986).

Maternal Age   Recent medical evidence suggests that babies born to teenagers are especially likely to be premature and undersized. Because young mothers are often poor and ill-educated, many experts assumed their living conditions explained their pregnancy problems. But new evidence shows that middle-class teenagers are almost twice as likely as older women to deliver premature babies (Leary, 1995). Also, researchers generally agree that adolescent mothers seem to provide lower-

colleagues from Nantes, France, and again in 1973 in the British medical journal *The Lancet* by U.S. researchers K. L. Jones and colleagues from Seattle. FAS is a cluster of severe physical and mental defects caused by alcohol damage to the developing fetus. No amount of good nutrition and postnatal care erase the growth retardation, head and facial abnormalities, skeletal, heart, and brain damage (MedicineNet, 1997; Calhoun & Alforque, 1996). Recent findings suggest that even as few as two drinks a week during pregnancy can do damage.

Oral Contraceptives

Exposure to oral contraceptives has also been linked to birth defects. An estimated 10 percent of pregnant women in the United States use oral contraceptives early in their pregnancies. The birth defects involve the heart, limbs, anus, esophagus, vertebral column, or central nervous system (Nora & Nora, 1975).

Toxins in the Environment

Expectant women often encounter potentially toxic agents in everyday substances, including hair spray, cosmetics, insecticides, household cleansers, food preservatives, and polluted air and water. The specific risk associated with these agents remains to be determined. It is known, however, that contact with chemical defoliants should be avoided. The National Cancer Institute has linked chemical sprays used to destroy jungle and forest areas in Vietnam with a substantial increase in the number of malformed Vietnamese infants. In 1979 the federal Environmental Protection Agency ordered an emergency ban on herbicides containing dioxin (including 2, 4, 5-T and Silvex) because of "significant evidence" linking the spraying of dioxin herbicides with an "alarming" rate of miscarriages among women in Oregon. Miscarriages and birth defects also occur at two to three times average rates in areas of California where water is contaminated by chemicals used in high-tech electronics manufacturing (Miller, 1985).

Toxins in the Workplace

In recent years medical authorities have become increasingly concerned over the hazards to the reproductive organs and processes that are found in the places where people work (Stone, 1994). Studies reveal, for instance, that continuous exposure to a variety of gaseous anesthetic agents used in hospital settings is associated with an increased number of spontaneous abortions among women workers *and* among the wives of exposed male workers. The children of these people also have a higher incidence of congenital malformations. A California study disclosed that 29.7 percent of pregnant nurses working in operating rooms had spontaneous miscarriages. The figure was only 8.8 percent among pregnant general-duty nurses (Bronson, 1977). The University of Massachusetts School of Public Health has found that women working in so-called clean rooms of semiconductor makers—where computer chips are etched with acids and gases—have a miscarriage rate nearly twice the national average (Meier, 1987). Studies are also under way to determine the risks posed by video display terminals (VDTs) following reports of high miscarriage rates among some VDT users in several industries. VDTs emit unique wavelengths of ionizing radiation.

Sperm are no less susceptible to damage from environmental causes than are ova. This fact has often been overlooked because most research studies the reproductive systems of women, not of men. Nonetheless, a number of studies suggest a link between male exposure to chemical agents and reproduction. Mercury, some solvents, and various pesticides and herbicides can contribute to male infertility, spontaneous abortions, and birth defects. In sum, it makes sense to clean up the workplace not only for mothers but for fathers as well (Blakeslee, 1991b).

quality parenting than adult mothers do. Pregnancy and motherhood are stressful to the adolescent, who is dealing with the demands of parenting concurrent with establishing her own identity and confronting the developmental tasks of adolescence. Adolescent mothers are more likely to experience educational and economic limitations and complicating family problems (Field et al., 1998; Bass & Jackson, 1997). The current consensus tends to be that a healthy woman in her thirties, forties, or older enjoys a good prospect of giving birth to a healthy infant and remaining well herself, provided that she is under supervised medical attention. This conclusion is especially good news for women who are postponing childbirth, often to allow for completing an education or establishing a career (Angier, 1991). Women who elect to have children after menopause (using assisted reproductive technologies) will be under careful study to determine to what extent, if any, their health is compromised by pregnancy in later midlife.

Maternal Nutrition and Prenatal Care The unborn infant's nourishment comes from the maternal bloodstream through the placenta. Consequently, nutritional deficiencies in the mother, particularly severe ones, are reflected in fetal development. Babies of poorly nourished mothers are more likely to be underweight at birth, to die in infancy, to suffer rickets, and to have physical and neurological defects, low vitality, and certain forms of mental retardation. Poor maternal nutrition—associated with war, famine, poverty, drug addiction, and poor dietary practice—has long-term insidious effects on brain growth and intelligence.

Proper early prenatal care is significantly correlated with babies born with adequate birthweight and fewer birth complications. Yet in the United States, many women either wait until after their fourth month of pregnancy to secure care, or they make fewer than half the number of recommended prenatal visits to physicians. One-third of single mothers receive inadequate care, a rate *three times higher* than that for married women. Minority women and their children are most affected: 30 percent of Hispanic mothers, 27 percent of blacks, and 16 percent of whites do not secure adequate prenatal care (Singh, Forrest, & Torres, 1990). Even though the public health community stresses the need for early prenatal care, the civil and criminal justice systems are moving toward punishing substance-abusing women who are pregnant (Chavkin & Breitbart, 1997). The rates of drug addiction and poor prenatal care are highest for single, young, minority mothers, who might decide not to seek prenatal care for fear of losing existing children or being prosecuted. Research studies over the past 30 years overwhelming confirm that good prenatal care pays off, both in significantly improved health for newborns and their mothers and in lower costs to society.

Segue

In this chapter we have given you an introduction to the marvelous intricacies and complexities of the female and male reproductive systems. Newer assisted reproductive technologies have brought hope for those who had previously been classified as infertile. The microscopic hereditary code transferred from both parents to the zygote at conception provides a blueprint for our physical makeup and a timing device for various changes in our body's makeup during our lifetime. It is truly a "miracle of life" to realize that each of us started out with such a complex code, and that it was embedded in a fertilized egg smaller than the period at the end of this sentence. There are many structures in the mother's womb that must function properly to support the embryo/fetus during its course of development and birth. However, for optimum health of both mother and fetus, the mother-to-be must get proper nutrition and sleep, abstain from smoking cigarettes or using alcohol and other drugs, avoid toxins in the home and work environments, exercise moderately to prepare for labor and delivery, and try to keep stress to a minimum whenever possible.

In the next chapter, we discuss the variety of approaches employed to prepare for the coming birth of a child, methods used to ease mothers and the fetus through labor and delivery, and the exciting first two years of infant development.

Summary

Reproduction

1. Like all other living things, human beings are capable of producing new individuals of their own kind and thus ensuring the survival of the species. This process is called reproduction. Two kinds of cells are involved in human reproduction: the female gamete, or ovum, and the male gamete, or sperm. A sperm fuses with an ovum to form a zygote (fertilized egg) in a process called fertilization or conception. Each ovum and sperm contributes its own unique genetic material to the zygote during fertilization.

2. The male's primary reproductive organs are a pair of testes located in a pouchlike structure called the scrotum. Healthy sperm are produced daily in the testes at a temperature a little lower than normal body temperature. The testes produce the principal male sex hormone, called androgens (testosterone and androsterone), which produce male secondary sex characteristics.

3. During sexual arousal and ejaculation, mature sperm are released from the tubules, emptied into the epididymis, and mixed with nurturant fluids from the prostate gland and seminal vessicles. This mixture is called semen. Semen is ejaculated out of a man's body through the urethra in the man's penis.

4. A male's sperm production (sperm count and

health) can be compromised by illness, sexually transmitted diseases, and exposure to biochemical substances in the home or work environment.

5. The primary female reproductive organs are a pair of ovaries that lie deep in her pelvic area. Each ovary produces mature eggs (ova) and the female sex hormones estrogen and progesterone. During ovulation a mature egg normally is moved into the fallopian tube (oviduct), then into the pear-shaped uterus (womb), through the cervix, into the vagina, and out of the body through the folds of skin outside of the vagina, called the vulva.

6. The series of regular hormonal and physical changes associated with producing a mature ovum is called the menstrual cycle. The average length of the menstrual cycle is about 28 days (and can vary from 25 to 32 days). The beginning of the menstrual cycle is the first day the woman menstruates. A mature ovum is produced toward the middle of each monthly ovarian cycle (normally between days 13 to 15) when an ovum reaches maturity and passes into one of the fallopian tubes. If sperm are present, fertilization may take place in the upper end of the fallopian tube. Secretions of progesterone from the corpus luteum prepare the uterus for implantation of the zygote in the wall of the uterus.

7. If fertilization of the ovum by the sperm does not take place, decreasing levels of female hormones lead to menstruation (shedding of the thickened uterine lining) about 12 to 14 days later. For most women, this menstrual cycle continues for 30 to 40 years, unless pregnancy, disease, illness, stress, or surgical intervention occurs.

8. A woman might conceive multiple, nonidentical children (fraternal twins, triplets, etc.) if two ova (or more) are released during the same menstrual cycle. Identical twins are the result of one fertilized egg splitting into two identical parts after conception. Multiple conceptions are occurring more frequently due to in vitro fertilization procedures. Such conceptions can put the mother and the babies at risk.

9. In general, many women in the United States are delaying childbirth until their thirties or forties. Some women, particularly those over 35, might be infertile. Medical researchers have created several fertility drugs and assisted reproductive technologies over the last 20 years to aid in conception. A common method is in vitro fertilization (IVF), where fertilization occurs outside the womb in a dish in a medical lab. The resultant embryos are then implanted back into the woman's uterus for potential implantation and maturation. There is ethical debate regarding the use of ART methods to produce human embryos for reproduction or experimentation.

10. Public education campaigns about several birth control methods have been targeted toward younger adolescents, women at risk of HIV and STDs, and those with multiple sex partners. Recently, a "morning after" kit of pills became available to the public. Elected abortion is legal in the United States, and pro-life and pro-choice advocates are contesting each other with powerful political efforts. Partial-birth abortion is highly controversial.

11. More recently the ages of conception have expanded. Both younger adolescents and a few postmenopausal women are having babies. The teenage pregnancy rate was recently reported to be dropping in comparison to the rates of the late 1980s and early 1990s. Many teenage pregnancies put a great financial strain on families and on society.

Heredity and Genetics

12. Our biological inheritance is called heredity. A major debate by those in the field of human development is the nature-nurture controversy. To what extent does our genetic "blueprint" in our chromosomes and genes make us who we are? How much does parenting and environment contribute to influencing our personality, motivation, skills, and traits?

13. Chromosomes are long threadlike structures made of protein and nucleic acid that are located in the nucleus of a cell. Each human chromosome is shaped like a rope ladder (called the double helix) and contains nearly 100,000 hereditary markers called genes (like the beads on a string). Genes are composed of DNA (deoxyribonucleic acid), which actively programs cells to manufacture vital substances for life. Upon normal fertilization, the 23 chromosomes of the ovum and the 23 chromosomes of the sperm combine to create a genetically unique zygote with 46 chromosomes (23 pair). The Human Genome Project is a worldwide research effort to map all of the nearly 3 billion letters of chemical code by the early 2000s. Geneticists predict many breakthroughs over the next 15 to 20 years in early detection and treatment of inherited diseases and disorders.

14. Mitosis is the type of cell division through which nearly all cells of the human body (except the sex cells) replicate themselves. In the cell nucleus, each single chromosome splits lengthwise to form a new pair, called "daughter" cells, which have the same hereditary information as the original. Meiosis is a

replication process that takes place in the gametes (the sperm and ovum). Meiosis involves two cell divisions during which the chromosomes are reduced to half their original number. Each gamete receives only 23 chromosomes, not 23 pair as in other cells. Half the usual number allows the sperm and egg to each contribute one-half of the genetic material at fertilization.

15. The male's sperm carries the chromosome that determines the sex of a child. Of the 23 pair of chromosomes that most humans possess, 22 are similar in size and shape in both men and women. These are the autosomes. The 23rd pair, the sex chromosomes (one from the mother and one from the father), determines the baby's sex. The mother's ovum contributes an X chromosome; the sperm contributes either an X (female) or Y (male) chromosome. An ovum (X) fertilized by a sperm with X will produce a female child (XX). An ovum (X) fertilized by a sperm with Y will produce a male (XY). The Y chromosome from the father determines the child will be a male (XY). Approximately 6 to 8 weeks into embryonic development, the male embryo produces the hormone testosterone, which promotes masculinization of the fetus.

16. Each member of a pair of genes at a specific location on the chromosome is called an allele. Mendel demonstrated that one allele can be dominant and hide the traits of the other allele, which is then considered a recessive character. A dominant allele is labeled with a capital letter, such as *(A)*, and the recessive allele is signified by a lowercase letter *(a)*. When both inherited alleles from parents are the same, this is called homozygous and represented as either *(AA)* or *(aa)*. When inherited alleles from parents are different, this is called heterozygous and represented as *(Aa)*. Not all traits are inherited from a single gene. Some traits, such as personality and intelligence, are the result of the complex interaction of many genes, which is called polygenic inheritance.

17. An organism's genotype is its actual genetic makeup, whereas the phenotype is its observable characteristics. In humans, the phenotype includes physical, physiological, and behavioral traits. Humans also possess dominant and recessive traits. For example, brown hair is a dominant characteristic, whereas red hair is recessive. Brown hair, however, can be either *BB* or *Bb,* whereas red hair must be *bb*. In humans, environmental factors interact with genetic factors to produce traits, and this is called multifactorial transmission.

18. Genes can also be linked together during inheritance. The X chromosome can carry a sex-linked trait such as hemophilia, red-green color blindness, and approximately 150 other traits and disorders. The vast majority of sex-linked genetic defects occur in males. Though a woman can be a carrier of a sex-linked disorder, she herself rarely exhibits that characteristic.

19. The field of genetic counseling and testing has arisen out of the past 20 years of genetic research to apply knowledge about the impact of genetics. Couples might be concerned about a family history of inherited disease, particularly diseases that might be specific to their ethnic/racial background. A variety of diagnostic tests are now available and are used to counsel those who are most likely to be affected. A number of known disorders are associated with the presence of too few or too many chromosomes. Prenatal diagnosis employs a variety of techniques to determine the health and condition of an unborn fetus.

Prenatal Development

20. The prenatal period normally lasts an average of 266 days. Embryologists divide it into three stages: the germinal period, the embryonic period, and the fetal period.

21. The germinal period is characterized by the growth of the zygote (the fertilized egg) and the establishment of an initial linkage between the zygote and the support system of the mother through implantation in the uterine wall. The fertilized egg divides through mitosis. In about six or seven days the more developed structure called the blastocyst begins the process of differentiation into the trophoblast and the inner cell mass. Toward the end of the second week, the embryonic portion of the inner cell mass begins to differentiate into three layers: ectoderm, mesoderm, and endoderm.

22. The embryonic period lasts from the end of the second week to the end of the eighth week. It is characterized by rapid growth, the establishment of a placental relationship with the mother, the differentiation in early structural form of the chief organs, and the appearance of recognizable features of a human body.

23. The fetal period begins with the ninth week and ends with birth. During this period the differentiation of the major organ systems continues, and the organs themselves become competent to assume their specialized functions. The fetus becomes a sensory aware and active being inside the mother's womb. It can move around,

suck its thumb, kick its feet, and hear sounds outside the mother's womb.

24. Environmental influences in the home and workplace can affect the developing organism from the moment of conception and throughout the prenatal period. The pregnant mother is highly advised to avoid cigarettes, caffeine, alcohol, and other psychoactive substances, and to protect herself against sexually transmitted diseases. She must get proper nutrition, get plenty of rest, avoid excessive stress, and exercise moderately to prepare herself for a healthy pregnancy and labor and delivery in nine months.

Key Terms

abortion (76)
allele (80)
amniocentesis (84)
assisted reproductive technology (ART) (74)
autosomes (80)
blastocyst (86)
cephalocaudal development (88)
chorionic villus biopsy (CVS) (85)
chromosomes (79)
critical period (89)
deoxyribonucleic acid (DNA) (79)
dominant character (81)
embryo (88)
embryonic period (86, 88)
fallopian tubes (71)
fertilization/fusion (73)
fetal period (86, 89)
fetoscopy (85)
fetus (89)
gametes (70)

genes (79)
genetic counseling (83)
genetics (78)
genotype (81)
germinal period (86)
heredity (78)
heterozygous (81)
homozygous (81)
human genome (79)
implantation (87)
in vitro fertilization (IVF) (75)
maternal blood sampling (85)
meiosis (79)
menstrual cycle (72)
menstruation (74)
miscarriage (91)
mitosis (79)
multifactorial transmission (82)
ovaries (71)
ovulation (73)
ovum (70)

penis (70)
phenotype (81)
placenta (88)
polygenic inheritance (81)
prenatal diagnosis (84)
prenatal period (86)
proximodistal development (88)
recessive character (81)
reproduction (70)
sex chromosomes (80)
sex-linked traits (82)
sperm (70)
teratogen (92)
teratology (92)
testes (70)
ultrasonography (85)
umbilical cord (88)
uterus (72)
vagina (72)
zygote (70, 73)

Following Up on the Internet

Web sites for this chapter focus on heredity, genetics, and human reproduction. Please access the text web site at http://www.mhhe.com/crandell7 for up-to-date hot-linked Internet addresses for the following organizations and topics.

American Society for Reproductive Medicine

Sexually Transmitted Disease Center from the Journal of the American Medical Association

Family Planning, Infertility, Miscarriage, Stillbirth, and other Complications in Prenatal Development

The Human Genome Project from the National Institutes of Health

The Visible Embryo Project

The Merck Manual: Heredity & Genetics Information on genetic and chromosomal defects

Part Three

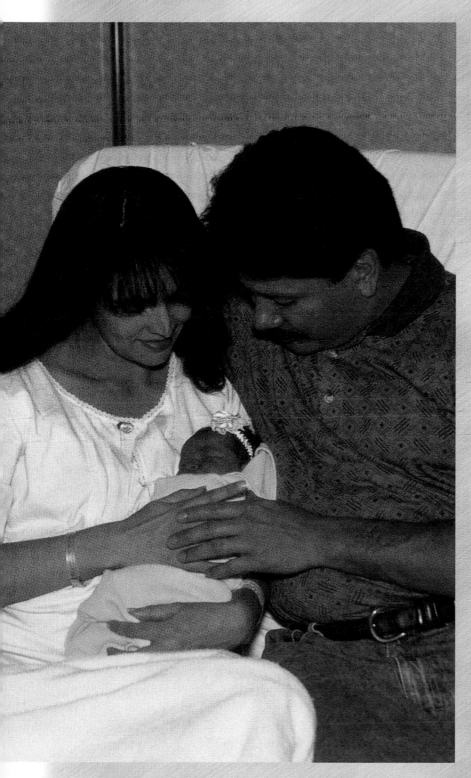

Birth and Infancy
The First Two Years

Chapter 4 describes the preparation of the mother for labor, delivery, and birth of the neonate. We will discuss several birth and delivery methods, including natural, or prepared, childbirth. Potential birth complications and birth defects are also examined. Then we will turn to the infant's first two years of physical, sensory, and motor development. In Chapter 5, we examine the processes of cognitive and language development, which allow infants to take greater command of their environment. Chapter 6 describes the emotional and social development of infants, including the significant influences of attachment, temperament, and parenting practices.

Chapter 4

Birth and Physical Development
The First Two Years

Critical Thinking Questions

1. If you were to design the "perfect" birth experience, where would you want to be and who would you want to assist you? If you have an older child, would you want to include this child in the experience? Why or why not?

2. Most births result in healthy newborns. Hypothetically, though, if your newborn has a disorder, how do you think you might react, and who would you turn to to cope?

3. Recognizing that parenting styles vary, do you think you might be a "responsive" parent, attending to a crying child's every need, or would you think that approach "spoils the child"?

4. If you were planning to become a day-care provider for young toddlers in your home, what would you plan to do to provide them with physical activity and sensory stimulation?

Extensive research in prenatal development over the past 30 years has revealed that the tiny human demonstrates physical, cognitive, and emotional behaviors. Routine use of noninvasive diagnostic and imaging technologies have made it possible for us to view the developing embryo and fetus in its preparation for birth. The moment of first seeing a fetal ultrasound or hearing the fetal heartbeat is a peak moment in life for many expectant parents.

Touch, the first sense and the cornerstone of human experience and communication, begins in the womb (Montagu, 1978). The first dramatic movement that symbolizes life itself is the first heartbeat at three weeks after conception. Hand-to-head, hand-to-face, hand-to-mouth movements and mouth opening, closing, and swallowing are present at 10 weeks of development (Tajani & Ianniruberto, 1990). The fetus lives in a stimulating environment of sound, vibration, and motion, and when a mother laughs or coughs, her fetus moves within seconds (Chamberlain, 1998). Voices reach the womb, and patterns of pitch and rhythm, as well as music, reach the fetus. A mother's voice is particularly powerful (Shahidullah & Hepper, 1992). The fetus reacts to amniocentesis (usually done between weeks 14 and 16) by shrinking away from the needle, a reaction easily observed on the ultrasound. Rapid eye movement sleep, a manifestation of dreaming, is observed as early as 23 weeks of gestation (although the eyelids remain closed) (Birnholz, 1981).

Remarkable as it is, then, most full-term infants arrive at birth with all sensory systems functioning. Newborns are real, separate persons—not the "blank slates" they were thought to be only a few decades ago. Tests made at birth reveal exquisite taste and odor discrimination and definite preferences, and visual tests demonstrate how remarkably a newborn can imitate a variety of facial expressions. When newborns are awake, their eyes constantly seek out the environment (Slater et al., 1991). Throughout the newborn's first two years of life, extraordinary emotional and social development accompany the maturation of its physical and cognitive systems.

Birth

Preparing for Childbirth

The births of all things are weak and tender, and therefore, our eyes should be intent on beginnings.

Montaigne

In the 1940s the English obstetrician Grantly Dick-Read (1944) began popularizing the view that pain in childbirth could be greatly reduced if women understood the birth process and learned to relax properly. Childbirth, he argued, is essentially a normal and natural process. He trained prospective mothers to relax, to breathe correctly, to understand their anatomy and the process of labor, and to develop muscular control of their labor through special exercises. He also advocated training the father as an active participant in both prenatal preparation and delivery. During the same time, Russian doctors began to apply Pavlov's theories of the conditioned reflex to delivery practices and reasoned that society conditioned women to be tense and fearful during labor. If pain was a response conditioned by society, it could be replaced by a different, more positive response. Accordingly, the **psychoprophylactic method** evolved, which encouraged women to relax and concentrate on the manner in which they breathed when a contraction occurred.

In 1951 Fernand Lamaze (1958), a French obstetrician, visited maternity clinics in the Soviet Union. When he returned to France, he introduced the fundamentals of the psychoprophylactic method. Lamaze emphasized the active participation of the mother in every phase of labor. He developed a precise and controlled breathing drill in which women in labor respond to a series of verbal cues by panting, pushing, and blowing. The Lamaze method has proved popular with U.S. physicians and prospective parents who prefer natural childbirth, and nearly every hospital and many private medical organizations offers Lamaze childbirth preparation classes.

For many Americans, the term **natural childbirth** has come to be equated with a variety of approaches that stress the preparation of the mother and the father for childbirth and their active involvement in the process. But the term actually refers to an awake, aware, and unmedicated mother-to-be. A woman proceeding through a Lamaze delivery might use a number of cognitive techniques that distract her from the activities of the labor room and provide an additional source of support. These techniques include using visual focus and sucking on hard candies or ice chips (Wideman & Singer, 1984).

Natural childbirth offers a number of advantages. Childbirth education classes can do much to relieve the mother's anxiety and fear; along with natural childbirth classes, prospective parents get to see a birthing room. Many couples find their joint participation in labor and delivery a joyous, rewarding occasion. And the mother takes no medication or is given it only sparingly in the final phase of delivery, at her request. There is conflicting research on the effects of these analgesic and sedative drugs. Some studies show that infants whose mothers received medication tend to perform less well on standard behavioral tests at one month to one year of age than do infants whose mothers were not medicated during labor (Sanders-Phillips, Strauss, & Gutberlet, 1988). Other studies show no discernible long-term effects of drugs administered to the mother during the birth process (Kolata, 1979). Safe obstetric practice seems to suggest caution in the administration of these drugs, because what passes into the mother's system might affect the baby during labor and delivery.

Medical authorities are increasingly concluding that no mother should ever labor and deliver alone (Collins et al., 1993). Evidence suggests that women who have a friendly companion with them during childbirth have faster, simpler deliveries, have fewer complications, and are more affectionate toward their babies. This insight has led to the reemergence of *doula* services (*doula* is a Greek word meaning "someone who nurtures and cares for new mothers"). *Doulas* and *midwives,* as acknowledged members of the maternity care team, provide emotional care and physical comfort and are usually licensed and often affiliated with **obstetricians** (physicians who specialize in conception, prenatal development, birth, and the woman's postbirth care). For centuries in European cultures, doulas or midwives, attended women in labor and delivery until approximately the 1600s or 1700s. From that time until the late 1960s, physicians in Western societies claimed that labor and delivery came exclusively under their medical domain.

Although natural childbirth clearly offers advantages for many couples, it is more suitable for some couples and some births than others. In some cases the pain becomes so severe that wise and humane practice calls for medication. Although the average intensity of labor pain is quite high, women substantially differ in their experience of it. And despite their pain and discomfort, most women say that childbirth is one of the greatest, if not *the* greatest, experience of their lives (Picard, 1993). However, both proponents and critics of natural, or prepared, childbirth agree that women who are psychologically or physically unprepared for it should not consider themselves inadequate or irresponsible if they resort to conventional practice. Indeed, many practitioners view prepared childbirth training and pain-relief remedies as compatible, complementary procedures.

Birthing Accommodations

With smaller families and more childbirth classes, most couples are seeking an obstetrician and a hospital that view uncomplicated pregnancies as a normal process rather than as an illness. And they are rebelling against regimented and impersonal hospital routines. They do not want the delivery of their babies to be a surgical procedure unless such surgery is necessary or planned. They also note that infections can be spread in hospitals, sometimes by physicians and staff. Consequently, more couples are choosing to have babies at home (Belkin, 1992). The medical establishment has reacted with alarm to this trend, insisting that home births pose risks to both the mother and the child.

Dissatisfaction with the maternity care options provided by physicians has contributed to the revival of **midwifery** (pronounced "mid-wiff'-ery"). All fifty states have legalized the provision of prenatal care and delivery by midwives so long as the practitioners are registered nurses. Midwives who are not nurses are also seeking legal status. Midwives were fairly common in the United States until the late 1930s. By then, medical groups succeeded in suppressing midwifery by portraying it as dangerous and dirty. The demand for midwives is now being fueled by middle-class and affluent professional women who shun the impersonalness of high-tech hospital care and by poor and rural women who lack access to or cannot afford the high costs of traditional obstetrical and gynecological care (Brody, 1993a).

In response to the home-birth movement, many hospitals have introduced **birthing rooms.** Such rooms have a homelike atmosphere complete with wallpapered walls, window drapes, potted plants, color television, a queen-size bed, and other comforts. Medical equipment is nor-

mally out of view. The woman can give birth assisted by a nurse-midwife or an obstetrician and her husband or other partner. Other relatives, friends, or even the baby's brothers and sisters might be present. Should complications arise, the woman can be quickly moved to a regular delivery room. This arrangement allows for a homelike birth with proximity to hospital life-saving equipment. The mother and child might then return home about 6 to 24 hours after an uncomplicated delivery.

Other hospitals, while retaining more traditional childbirth procedures, have introduced family-centered hospital care, in which birth is made a family experience. The plan is usually coupled with **rooming in,** an arrangement in which the infant stays in a bassinet beside the mother's bed. This practice runs counter to the U.S. hospital tradition of segregating infants in a sterile nursery. Rooming in allows the mother to get acquainted with her child and integrates the father early into the child-care process. Under the supervision of the nursing staff, parents gain skill in nursing/feeding, bathing, diapering, and caring for their infant. Women who desire to breast-feed their babies can begin the process with the sympathetic help and support of the trained hospital staff.

Birthing centers are opening in many urban areas. These primary care facilities are used only for low-risk deliveries because they lack high-tech equipment. Should complications arise, patients are transferred to nearby hospitals (DeWitt, 1993). Another recent trend has shortened hospital stays for new mothers and their newborns. In today's cost-conscious climate of managed health care, maternity stays have declined from the weeklong sojourn common in the 1950s, and still prevalent in many European nations, to a national average of about two-and-a-half days. Three days is now typical for cesarean births. Not surprisingly, many health professionals criticize the fact that most health insurance allows only such abbreviated stays, saying that mothers need more hospital time to rest and recover and to acquire basic child-care skills. Although many problems with newborns surface early, some conditions, including jaundice and heart murmurs, tend to appear only after the first six hours (Lord, 1994), and the mother herself might experience medical complications hours after delivery.

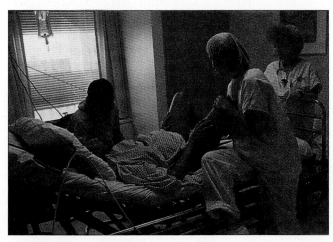

Birthing Room The scene above reflects a family-centered childbirth experience in a birthing room with rooming-in for the newborn. How different is this from childbirth up through the 1950s in the United States?

Questions

What are the differences among giving birth in a birthing room in a hospital, in a medically supervised birthing center, or in a home delivery with a midwife or doula attending? What are some factors to consider in making this important decision?

Stages of the Birth Process

Birth is the transition between dependent existence in the uterus and life as a separate organism. In less than a day, a radical change occurs. The fetus is catapulted from its warm, fluid, sheltered environment in the womb into the larger world. The infant is compelled to depend exclusively on its own biological systems. Birth, then, is a bridge between two stages of life. Normally around 266 days of prenatal development, some factor, not yet identified but suspected to be a hormonal signal from the brain to the blood, prompts the beginnings of birth. In this section we will explain the stages of the birth process, labor, delivery, and crucial neonatal assessment at birth.

A few weeks before birth, the head of the infant generally turns downward, which ensures that it will be born head first. (A small percentage of babies are born buttocks or feet first, in the *breech position.* This most often requires a surgical delivery.) The uterus simultaneously sinks downward and forward. These changes are termed **lightening.** They "lighten" the mother's discomfort, and she now breathes more easily, because the pressure on her diaphragm and lungs is reduced (see Figure 4-1). At about the same time the mother might begin experiencing mild "tuning-up" contractions (*Braxton-Hicks contractions*), which are a prelude to the more vigorous contractions of labor.

The birth process consists of three stages: labor, delivery, and afterbirth. Either at the beginning of labor or sometime during it, the *amniotic sac* that surrounds and cushions the fetus ruptures, releasing the amniotic fluid, which should then flow as a clear liquid from the vagina. This "water breaking" is usually the first signal to the mother-to-be that labor is impending. It is imperative that she call her obstetrician, doula, or midwife at this time. Again, a woman should not attempt to deliver alone. The duration of this first stage of labor varies considerably depending on several factors: the age of the mother, her number of prior pregnancies, and potential complications of the pregnancy. During **labor** the strong muscle fibers of the uterus rhythmically contract, pushing the infant downward toward the birth canal (the *vagina*). Simultaneously, the muscular tissue that forms the thick lower opening of the uterus (the *cervix*) relaxes, becoming both shortened and widened, thus permitting the infant's passage (see Figure 4-1).

Normally labor averages about 14 hours for women having their first babies. Women who have already had at least one baby average about 8 hours. Initially, the uterine contractions are spaced about 15 to 20 minutes apart and last for about 25 seconds to 30 seconds. As the intervals shorten to 3 to 5 minutes, the contractions become stronger and last for about 45 seconds or longer. As the mother's uterine contractions increase in intensity and occur more frequently, her cervix opens wider (dilates) (see Figure 4-1). Eventually it will expand enough to allow the baby's head and body to pass through.

Delivery begins once the infant's head passes through the cervix (the neck of the uterus) and ends when the baby has completed its passage through the birth canal. This stage generally requires 20 to 80 minutes, but can be shorter in deliveries of subsequent children. During delivery, contractions last for 60 to 65 seconds and come at 2- to 3-minute intervals. The mother aids each contraction by "bearing down" (pushing) with her abdominal muscles at recommended times. With each contraction, the baby's head and body emerges more. **Crowning** occurs when the widest diameter of the baby's head is at the mother's vulva (the outer entrance to the vagina). Sometimes an incision called an *episiotomy* is made between the vagina and the rectum if the opening of the vagina does not stretch enough to allow passage of the baby's head. This type of surgical intervention has come under much criticism but might be necessary to prevent complications of delivery in some births. Once the head has passed through the birth canal, the rest of the body quickly follows. The second stage of labor is now over, unless there are multiple births involved (see Figure 4-1). The doctor or health professional will quickly suction mucus from the baby's throat with a hand-operated suctioning device. The newborn, called the **neonate,** is still connected to the mother by the umbilical cord, which will be attended to within a few minutes. The newborn will be quickly assessed for its level of alertness and health and might be placed on the mother's warm body or quickly cleaned of the *vernix caseosa* (a white, waxy substance that covers its body) and placed into the father's or mother's awaiting arms.

After the baby's birth, the uterus commonly stops its contractions for a few minutes. The contractions then resume, and the placenta and the remaining umbilical cord are expelled from the uterus through the vagina. This process, expelling the **afterbirth,** may last for about 20 minutes. During this process, the father of the baby (should he choose to do so) may assist with the clamping and "cutting the cord" to separate the newborn baby from mother. In just the last few years, a technique has been used to collect and preserve the blood from the placenta and umbilical cord using cryopreservation for later transplantation. The "progenitor" cells normally found in bone marrow are vital in treating such life-threatening diseases as leukemia, types of cancer, or some immune or genetic disorders. In some hospitals parents have the option of storing the blood from their baby's cord and placenta. After the safe birth of the baby, a five-minute procedure is used to collect the blood contained in the placenta and umbilical cord. Some physicians are promoting the development of placental-and-cord-blood banks (Silberstein & Jefferies, 1996; McCowage, 1995).

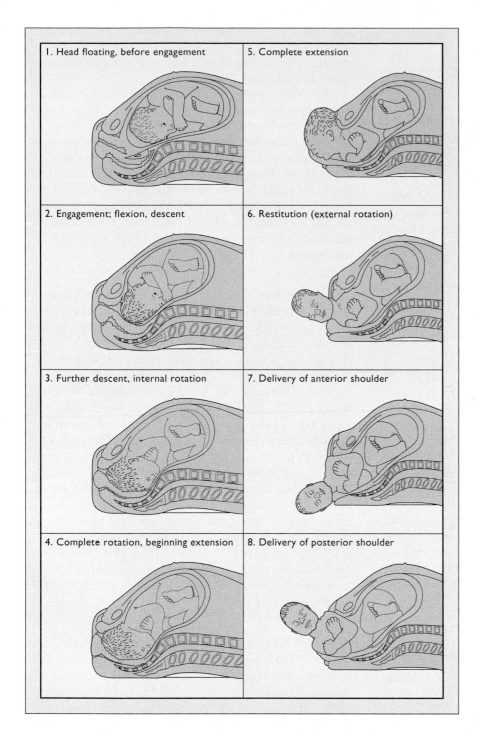

1. Head floating, before engagement
2. Engagement; flexion, descent
3. Further descent, internal rotation
4. Complete rotation, beginning extension
5. Complete extension
6. Restitution (external rotation)
7. Delivery of anterior shoulder
8. Delivery of posterior shoulder

Figure 4-1 Normal Birth The principal movements in the mechanism of labor and delivery.

F. Gary Cunningham, MD, Paul C. MacDonald, MD, Norman F. Gant, MD, Kenneth J. Levano, MD, Larry C. Gilstrap, III, MD, Gary D.V. Hankins, MD, and Steven L. Clark, MD. WILLIAMS OBSTETRICS, 20th EDITION. Copyright Appleton and Lange. Reprinted by permission.

Most parents cherish the miracle of birth as a peak moment of their lives. However, personal and cultural attitudes can temper reactions to the birth of a baby. Was this baby planned for and wanted? Has the mother's health been compromised in some way? Is the baby's father present at the birth? Is it already known the child has a birth defect? Is the family already overburdened with many children? Is this a young teenager's first pregnancy? Is this baby the product of rape or incest? Is this a surrogate pregnancy? Is there a planned adoption?

These types of circumstances certainly impact the mother's and father's level of acceptance and emotional reaction at the first sight of their newborn child.

Questions

What is usually the first indication to a woman that she is about to give birth? What are the events that occur in the progressive stages during delivery culminating with birth?

The Baby's Birth Experience

In 1975 Frederick Leboyer, a French obstetrician, captured popular attention with his best-selling book *Birth Without Violence* (1975). Birth for the baby, says Leboyer, is an exceedingly traumatic experience. Leboyer calls for a more gentle entry into the world via lowered sound and light levels in the delivery room, the immediate soothing of the infant through massaging and stroking, and a mild, warm bath for the newborn.

However, claims like Leboyer's are exceedingly controversial. Canadian researchers have found that Leboyer's method offers no special clinical or behavioral advantages to the infant or mother that are not offered by a gentle, conventional delivery (Nelson et al., 1980). Other researchers report that despite surface appearances, the stresses of a normal delivery are usually not harmful. The fetus produces unusually high levels of the stress hormones adrenaline and noradrenaline that equip it to withstand the stress of birth. This surge in hormones protects the infant from asphyxia during delivery and prepares the infant to survive outside the womb. It clears the lungs and changes physiological properties to promote normal breathing while simultaneously ensuring that a rich supply of blood goes to the heart and brain (Lagercrantz & Slotkin, 1986). We should keep in mind that it is the shock of birth that triggers babies' gasping efforts to breathe for themselves. Some babies breathe on their own prior to the traditional slap on the bottom. Parents can take comfort from knowing that from the baby's standpoint the stress of labor during normal birth is likely to be less unhappy and more beneficial than common sense might suggest, for the neonate's blood flow must reverse itself, a specific valve in the heart must close, and the baby's lungs must begin to function on their own. By the same token, however, Leboyer fostered a more humane view of childbirth management.

Electronic Fetal Monitoring Normally during the process of labor and delivery in a hospital setting, a velcro-type strap connected to a monitor is placed around the mother's abdomen and back. The fetal heartbeat is monitored continuously and registered on a strip of paper. The baby's pulse slows down during strong contractions, but it regains its original rate in between. Its heartbeat is likely to be twice as fast as the mother's. Even though normally the baby's body is well prepared to withstand the stress of delivery, surgical intervention can be necessary if the heart-rate monitor indicates that the fetus is in distress. Use of this monitor and newer computer devices can be crucial to the survival of some babies.

Newborn Appearance At the moment of birth, infants are covered with *vernix,* a thick, white, waxy substance. Some newborns still have their lanugo, the fetus's fine, woolly facial and body hair, which disappears by age 4 months. The matting of their hair with vernix gives newborns an odd, pasty look.

On the average, a full-term newborn is 19 to 22 inches long and weighs 5½ to 9½ pounds. Their heads are often misshapen and elongated as a product of molding. In *molding* the soft skull "bones" become temporarily distorted to accommodate passage through the birth canal. Babies born by C-section do not have this same elongated look. In most infants the chin recedes and the lower jaw is underdeveloped. Bowleggedness is the rule, and the feet might be pigeon-toed. Even more crucial is the neonate's behavioral status, to which we now turn our attention.

The Apgar Test The average birthweight for babies is about 7 pounds 8 ounces (3.4 kilograms) (Leach, 1998). However, weight is only one factor in assessing the health of a neonate. The normalcy of the baby's condition at birth is usually appraised by the physician or attending nurse in terms of the **Apgar scoring system,** a method developed by an anesthesiologist, Virginia Apgar (1953). The infant is assessed at one minute and again five minutes after birth on the basis of five conditions: heart rate, respiratory effort, muscle tone, reflex irritability (the infant's response to a catheter placed in its nostril), and body color. Each of the conditions is rated 0, 1, or 2 (see Figure 4-2). The ratings of the five conditions are then summed (the highest possible score is 10). At 60 seconds after birth, about 6 percent of all infants receive scores of 0 to 2, 24 percent have scores of 3 to 7, and 70 percent have scores of 8 to 10. A score of less than 5 indicates the need for prompt diagnosis and medical intervention. Infants with the lowest Apgar scores have the highest mortality rate.

Brazelton Neonatal Behavioral Assessment Scale Dr. T. Berry Brazelton, a noted pediatrician, author, and television and Internet physician, devised the Neonatal Behavioral Assessment Scale (NBAS) to be used several hours after birth or during the week after birth. Additionally, it is used by many researchers studying infant development. An examiner uses the 27 subtests of the NBAS to assess four categories of development: physiological, motor, states, and interaction with people (Brazelton, Nugent, & Lester, 1987). A low score might indicate potential cognitive impairment in a neonate or a need for more stimulation as provided by early intervention methods. An excellent, extensive discussion of neonatal appraisal can be found at the appraisal web site listed in "Following Up on the Internet" at the end of this chapter.

Questions

How is a newborn's health assessed at birth? Suppose that Stephen, a neonate, is born with arms and legs fixed, that his pulse is below 100 beats per minute, and that he sneezes, has a pale appearance all over, and has slow, irregular respiration. Referring to the Apgar scale, what is his health status? Does he need any medical intervention, or is his health status within the normal range?

Parent-Infant Bonding

During most of human history, babies have been placed immediately on their mothers' bodies after birth. According to the studies conducted by anthropologist Meredith Small (1998), in most cultures around the world babies are still placed on their mothers this way. The practice of separating babies from their mothers has arisen only in the past 100 years and only in Western cultures (Small, 1998). In 1896 when Martin Cooney invented one of the first incubators to aid premature infants, he advocated separation of mother and child for health benefits (this was back when microorganisms had first been discovered to exist). Cooney's methods saved more than 5,000 babies, and by the 1940s incubators for most newborns became standard practice in hospitals (Small, 1998). Nurses monitored the newborns in these incubators, infant and mother survival rates went up, and more mothers chose to give birth in hospitals rather than at home. New mothers, often heavily sedated, were given a glimpse of their babies: mothers were whisked to maternity wards to recuperate while all infants were whisked to the nursery (Small, 1998).

By the late 1960s, a reversal of birthing practices began in the Western world. John Bowlby had proposed his attachment theory, and Harry Harlow had conducted and published results from his infant monkey experiments (Harlow, 1971). In 1976, two obstetricians, Marshall Klaus and John Kennell, theorized that there is a critical early period for human mother-infant bonding. By 1978 the American Medical Association proclaimed as its official policy the promotion of bonding between newborns and their mothers (Small, 1998). Today, evolutionary biologists agree that it is natural (adaptive) for mothers and their newborns to bond. This requires close proximity, constant interaction, and emotional attachment. Bonding is especially necessary in our species, because human infants are dependent beings, who need much care, protection, and teaching (Small, 1998). But psychologists and medical professionals now recognize that not spending the first few minutes or hours together will not leave a permanent gap in the relationship. **Parent-infant bonding** is considered by most as a *process* of interaction and mutual attention that occurs over time and builds an emotional bond. Still, some scholars find the whole bonding doctrine to be "without scientific merit" and "believe it fosters unwarranted social stereotypes that portray motherhood as the 'font of emotional support'" (Eyer, 1992; Sluckin, Herbert, & Sluckin, 1983).

Proponents of natural childbirth argue that it facilitates emotional bonding between parents and their child. This is a time for intimacy which is only the beginning of parent-infant bonding. It is a time of gentle touching and looking at each other, and some mothers may select to breastfeed their newborn.

Mothers who have cesarean deliveries or parents who adopt children should not conclude that they have missed out on something fundamental to a healthy child-parent relationship. A growing body of research suggests that parents who do not have contact with their infant immediately after delivery typically can bond as strongly with the youngster as parents who do have such contact (Eyer, 1992).

	Sign	0 Points	1 Point	2 Points
A	Activity (muscle tone)	Absent	Arms and legs flexed	Active movement
P	Pulse	Absent	Below 100 bpm	Above 100 bpm
G	Grimace (reflex irritability)	No response	Grimace	Sneeze, cough, pulls away
A	Appearance (skin color)	Blue-gray, pale all over	Normal, except for extremities	Normal over entire body
R	Respiration	Absent	Slow, irregular	Good, crying

Figure 4-2   Apgar Scoring for Newborns
A score is given for each sign at 1 minute and 5 minutes after the birth. If there are problems with the baby, an additional score is given at 10 minutes. A score of 7–10 is considered normal, while 4–7 might require some resuscitative measures, and a baby with apgars of 3 and below requires immediate resuscitation.

Further Developments

Effects of Fatherlessness on Children

Fatherlessness is our most urgent social problem.

Wade Horn, Ph.D., U.S. Commissioner for Children, Youth and Families, Bush Administration

How Many American Children Are Living Without Fathers?

In the United States, extensive research is being conducted about the effects of absentee fathers on the lives of the many children being born to an increasing number of unwed mothers. Between 1980 and 1994, the birth rate for unmarried women increased from 29 to 47 per thousand, concurrent with the increase in nonmarital cohabitation (Wallman, 1998). In 1980, 77 percent of American children lived with two parents; in 1997, 68 percent lived with two parents (Wallman, 1998). In 1997, almost a quarter (24 percent) of children lived with only their mothers, 4 percent lived with their fathers, and 4 percent lived with neither parent (Wallman, 1998) (see Figure 4-3). Nearly 23 million American children do not live with their biological fathers. And 40 percent of the children of divorced parents haven't seen their fathers in the past year, states Dr. Horn. "Over the last three decades we have engaged in a great social experiment to determine what will happen if large numbers of children are reared without their fathers, and the conclusion is that the children will suffer greatly" (Horn, 1998).

The percentage of children living with two parents has been declining among all racial and ethnic groups. In 1997, 75 percent of white children lived with two parents, compared to 35 percent of black children and 64 percent of children of Hispanic origin (Wallman, 1998).

What Are the Effects of Fatherlessness on Children?

Unless you come in contact with children in day care, in school, in health care, or in other child-care settings, you may be unaware of the societal impact of this change in the family structure in the United States. Numerous studies confirm that children of unmarried mothers are at a higher risk of having adverse birth outcomes, such as low birthweight and infant mortality, and are more likely to live in poverty than children of married mothers (Wallman, 1998). Children living in poverty are more likely to suffer from poor general health, to have high levels of blood lead, and to have no usual source of health care. They are also more likely to experience housing problems and hunger, less likely to be enrolled in early childhood education, and less likely to have a parent working full-time all year (Wallman, 1998). These children are more likely to have psychological problems, abuse drugs and alcohol, and fail in school. Seventy percent of the kids in state reform institutions grew up without the presence of a father (Horn, 1998).

Why Fathers Count

Fathers parent differently than mothers do. Mothers tend to be more verbal with children, while fathers tend to be more physical. Fathers often engage in rough-and-tumble play with their sons, which we are discovering serves as practice for boys to develop control over their aggression. The combination of the mother's nurturing and the father's tendency to encourage achievement both contribute to the childhood experience. Fathers usually provide positive role models for daughters as well. "What we are saying," says Dr. Horn, "is that children really need both a mother and a father" (Horn, 1998). Fathers in intact homes also normally provide eco-

Bonding with several caretakers is required in some societies where there is a style of communal living. Child development researcher Edward Tronick has shown that Efé Pygmy infants are cared for by a number of adults in the village, and the infant-adult relationship can be even more communal. An Efé infant might spend 50 percent of its day with other caretakers and might be nursed by any of several women who are lactating. Yet the baby clearly knows who its mother and father are (Belsky & Rovine, 1993; Tronick, Morelli, & Ivey, 1992).

Developmentalists also recognize that some mothers and fathers have difficulty forming this attachment. Attachment can be difficult for mothers who have had a

particularly complicated labor and birth or whose infants are premature, malformed, or initially unwanted. Because some births are high risk, not every parent-child relationship begins calmly.

Paternal Bonding   In many cultures, expectant fathers might experience *couvade syndrome*—complaints of uncomfortable physical symptoms, dietary changes, and weight gain because of their partner's pregnancy (Small, 1998). A study of 147 expectant fathers in the Milwaukee area found that about 90 percent of them experienced "pregnancy" symptoms similar to those of their wives. For instance, the men had nausea in the first trimester

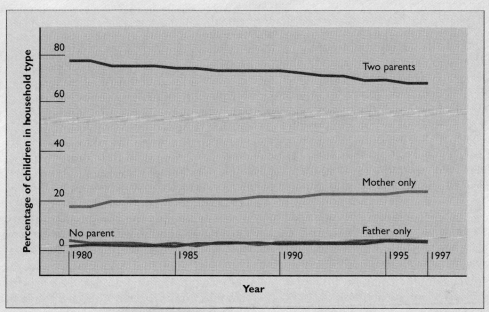

Figure 4-3 Percentage Distribution of Children Under Age 18 by Presence of Parents in Household, 1980–1997 The number of parents living with a child is generally linked to the amount and quality of human and economic resources available to that child. Children who live in a household with one parent are substantially more likely to have family incomes below the poverty line than are children who grow up in a household with two parents.

Source: U.S. Bureau of the Census, *March Current Population Survey,* 1998.

nomic security that is necessary to raising children in American society today (Parke, 1996).

How Can We Promote Father Involvement in Families?

First, we must recognize the importance of fathers. We have to understand that fathers provide a different style of caring, discipline, and parenting; these traits also promote both physically and psychologically healthier children who feel cared about. Second, we may have to rethink the institution of marriage as a stabilizing force that provides a sense of permanence for children. Studies have shown that the presence and involvement of fathers in the nurturing and development of their children confers benefits that are irreplaceable by any father substitute, whether the substitute is the state, a grandparent, a male friend, or a stepparent (Martin, 1998).

and backaches in the last trimester. The majority of the men reported weight gains ranging from 2 to 15 pounds, and they all lost weight in the first four weeks after the babies were born. Couvade could be one way fathers express a bond of sympathy with the expectant mother and a change in social roles (Lewis, 1985). In addition, expectant fathers commonly become more concerned about their ability to provide for and protect an expanding family (Kutner, 1990).

Anthropologists have found that biological fathers have a very important parenting role in societies where family life is strong, women contribute to subsistence, the family is an integrated unit of parents and offspring, and men are not preoccupied with being warriors. Although the degree of fathering across cultures varies, the potential for human males to contribute to infant care is great. There is also documented evidence that fathers can be intimately connected to their children (Small, 1998). To their benefit, many American fathers in the 1990s are attending childbirth preparation classes, participating in the birth of their children, and taking an active role in parenting. Developmental psychologists studying attachment theory also find that increased contact between fathers and their newborn infants creates a bond, making them better fathers (Keller, Hildebrandt, & Richards, 1985). After the birth, new fathers and mothers report similar emotions when first viewing the newborn. When given a chance, fathers explore a new baby's body in the same pattern as mothers do: fingers first, then palms of hands, arms, legs, and then trunk (Small, 1998). New fathers, like mothers, also instinctively raise the pitch and cadence of their speech (speaking what is called "motherese") with their newborn. It is clear that the more fathers interact with their babies, the more mutual attachment occurs (Small, 1998).

Unfortunately, though, some American men (from all racial/ethnic and socioeconomic statuses) have made little or no effort to bond with or support the children they are producing, and some men father many children by several women without regard to the future welfare of the mothers or babies. There are far-reaching consequences for these neglected children and American society (see *Further Developments:* "Effects of Fatherlessness on Children").

Question

The issue of absentee fathers is a highly charged one in American society today. Some single-parent moms are doing the best they can and do not wish the father to be involved. Others are angry that biological fathers are allowed by the government and courts to become "deadbeat" dads. What is your analysis of the impact of fatherlessness?

Complications of Pregnancy and Birth

Although most pregnancies and births proceed without complications, there are exceptions. Currently in the United States slightly more than one in seven women experience complications during labor and delivery that are due to conditions existing prior to pregnancy (including diabetes, pelvic abnormalities, hypertension, and infectious diseases) (National Library of Medicine, 1998). During pregnancy or in the labor or birth process, complications can arise that require surgical intervention. The purpose of good prenatal care and diagnostics under medical supervision is to minimize complications. But if complications develop, much can be done through medical intervention to help the mother and save the child. For example, about 1 percent of babies are born with **anoxia,** oxygen deprivation caused by the umbilical cord's having become squeezed or wrapped around the baby's neck during delivery. As mentioned in the previous chapter, in some pregnancies the mother and baby have incompatible Rh factors in their blood, and medical procedures can prevent serious complications. Efforts used to be concentrated on saving the mother if there were complications; today saving both the mother and the infant is a high priority.

In 1996 the United States had the highest rate of low-birthweight births in two decades, but technological innovations have decreased the infant mortality rates (Wallman, 1998). Larger urban hospitals are likely to have a neonatology unit, staffed with *perinatologists* and *neonatologists* who specialize in managing complicated, high-risk pregnancies, birth, and postbirth experience. The next sections discuss some complications of labor and birth, including cesarean delivery and infants born at risk.

Cesarean Section Delivery   Approximately 40 percent of women who experience complications in labor or delivery will have a surgical delivery called the **cesarean section** ("C-section"). In the United States, nearly 25 percent of all babies are born by cesarean section (National Library of Medicine, 1998). In this surgical technique, the physician enters the uterus through an abdominal incision and removes the infant. In 1970 the C-section rate was about 5 percent; by 1988, it had peaked at 24.7 percent, and by 1990 the rate decreased slightly to 23.5 percent. The dramatic rise in cesarean deliveries in the United States can be attributed to many factors: older mothers giving birth, bigger babies, improved technology to save high-risk infants, women's choice to schedule a surgical delivery, obstetrician choice to select this procedure, and the rise in malpractice suits.

Sometimes the C-section is a planned surgery if there are known risks to the mother or fetus (as with a mother who is diabetic, has high blood pressure, is HIV positive, or has *placenta praevia*—the placenta is lower in the uterus than the fetus's head and would be expelled first). Or it might be discovered during labor and delivery that an emergency cesarean section is necessary—for instance, when the mother's pelvis is too small to allow passage of the infant's head, or when the baby is positioned abnormally, as in *breech* presentation (buttocks or feet first rather than head first) or *transverse* presentation (a sideways or vertical position). Around 4 to 5 percent of all cesareans are done to deliver a breech baby. An obstetrician might attempt a maneuver called "external vision" at about 37 weeks if the baby hasn't turned so it is head first. In skilled hands, 60 to 70 percent of the babies turn—but the procedure is not without risk and could trigger premature labor (Sears & Sears, 1994).

A cesarean is major surgery and entails some risks, especially to the mother. Cesarean delivery can provoke anxiety, especially in couples not prepared for it. However, with the use of a *spinal block,* an injection that blocks pain from the waist down, the mother can be awake during delivery, the father can be in the delivery room, and both can share in the moment of birth. When a woman chooses a general anesthetic, she will be unaware of the birth, and the father is usually not allowed in the delivery room (though he is allowed to be just outside and is likely to hold the infant almost immediately after birth). Some women who have had cesareans feel "cheated" out of the experience of a natural delivery, but most are grateful for the option to deliver a healthy infant. Women who have had cesareans also typically experience more discomfort and the temporary incapacitation that accompanies recuperation from surgery. As a response to these problems, childbirth classes now usually include units on cesarean birth options, and hospital media materials promote the theme, "Having a Section Is Having a Baby." Additionally, more women now are giving birth vaginally after having had a cesarean delivery in an earlier pregnancy; this is referred to as VBAC, vaginal birth after cesarean.

Researchers have found that when a trained woman companion or *doula* provides constant support during labor and delivery, the need for cesarean sections, forceps deliveries, and other such measures is significantly reduced. However, every mother-to-be should plan for the unexpected in scheduling her place and method of delivery.

At-Risk Infants Development of at-risk infants is a topic of increasing importance. Advances in medical technology are saving many newborns who previously would not have survived. Simultaneously, the number of babies born unusually small is rising in the United States.

On an Average Day in the United States*

10,860	Babies are born
1,424	Babies are born to teen mothers
790	Babies are born low birthweight
467	Babies are born to mothers who receive late or no prenatal care
412	Are born with a birth defect
145	Babies are born very low birthweight
90	Babies die before reaching first birthday
20	Babies die as a result of a birth defect

Premature Infants Prematurity seems to be the principal culprit. In 1994, 11 percent of all births were preterm, with an 18.1 percent prematurity rate for black infants and 9.6 percent for white infants (March of Dimes, 1998). A **premature infant** has been traditionally defined as a baby weighing less than 5 pounds 8 ounces at birth or having a gestational age of less than 37 weeks. Although low birthweights are associated with prematurity, apparently it is developmental immaturity rather than low birthweight per se that is the primary source of the difficulties. More than half the low-birthweight babies born in the United States are not preterm; **small-for-term infants** typically do well (Brody, 1995b), as often happens with twins and other multiple births. Another factor is that multiple births are on the rise because of in

vitro fertilization procedures, and with multiple births there is a higher risk of low birthweight, complications, and premature birth. Between 1984 and 1994, the multiple-birth rate increased nearly 27 percent (March of Dimes, 1998). For preterm infants, however, survival rate correlates closely with birthweight, with better survival for the larger and more mature infants. Nonetheless, in a number of the nation's better hospitals, physicians are saving 80 to 85 percent of the infants weighing 2.2 to 3.2 pounds and, even more remarkably, 50 to 60 percent of those weighing 1.6 to 2.2 pounds. However, treating premature babies is hardly routine and frequently costs thousands of dollars a day. Significantly, the March of Dimes (1998) reports that low-birthweight infants are more than 20 times more likely than normal-weight babies to die before their first birthday.

Intensive-care nurseries (neonatology units) for preterm babies are quite foreign to parents who are unprepared to encounter their infant in a see-through incubator (Kolata, 1991b). The child might be receiving oxygen through plastic tubes inserted in the nose or windpipe. Banks of blipping lights, blinking numbers, and beeping alarms of electronic equipment and computerized devices monitor the baby's vital signs.

A leading cause of death in premature infants is a condition called *respiratory distress syndrome (RDS)*. Some 8,000 to 10,000 infant deaths each year are linked with RDS; another 40,000 newborns suffer from it. One difficulty is that premature infants lack a substance known as *surfactant*, a lubricant found in the amniotic fluid surrounding a fetus in the womb. Surfactant helps inflate the air sacs in the lungs after birth and prevents the lungs from collapsing or sticking together after each breath. The fetus normally does not develop surfactant until about week 35. Recently, researchers have found that by providing premature infants with the substance, many otherwise fatal complications can be avoided and the babies can be saved.

Prematurity affects development in a variety of ways (Liaw & Brooks-Gunn, 1993; Duffy, Als, & McAnulty, 1990). First, the relative immaturity of the premature infant makes it a less viable organism in coping with the stresses of birth and postnatal life and more susceptible to infections. Second, the developmental difficulties shown by the premature infant might be associated with the same prenatal disorders that caused the baby to be born early (maternal malnutrition, drug use, poverty, and maternal diabetes). Some research findings suggest that the long-term status of preterm infants is more likely to be related to the socioeconomic status, education, and supportive home environment than to the preterm status itself. Third, once delivered, the premature infant is often placed in an incubator (isolette) and connected to many tubes and monitoring devices.

Research over the past 20 years has demonstrated that preemies are more likely to survive if they have

*Source: National Center for Health Statistics, 1993 and 1994. Final data prepared by the March of Dimes, 1996. Numbers are approximations.

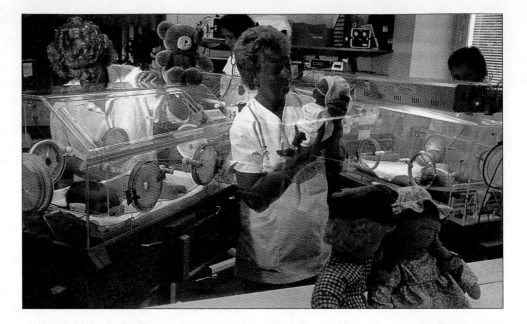

Neonatal Intensive Care The incubator, combined with many other technical advances, has helped to save the lives of many premature babies, enabling many to develop normally. Parents may be a little overwhelmed with the technology at first, but the neonatal professional staff trains the family to help in the care of the newborn (Neonatology on the Web, 1998).

normal skin contact and other stimulation, especially from the mother and father. New parents who want to be involved as caregivers for their preterm infants are encouraged to gently massage, cuddle, talk to, sing to, and rock their child (Scafidi et al., 1990). Finally, although premature infants constitute fewer than 10 percent of all babies, they constitute 23 to 40 percent of battered children, depending on the study quoted. Some researchers attribute this result to the likelihood that the prolonged separation between premature infants and their parents impairs the parent-child bonding process (Kennell, Voos, & Klaus, 1979). In addition, the characteristics of premature infants—their high-pitched cry, their fragility and smallness, their greater irritability, their lower levels of visual alertness, and their shriveled appearance—make them less cute and responsive, even somewhat repelling beings, in the eyes of their parents and caretakers (Landry et al., 1990; Zarling, Hirsch, & Landry, 1988).

Most premature infants show no abnormalities or mental retardation. Winston Churchill, who was born prematurely, lived to be 91 and led an active, productive life. Recent advances in the monitoring of premature babies have allowed physicians to anticipate and, in many cases, prevent or minimize some problems through therapeutic interventions. The result has been a reduction in the overall complications and mortality associated with premature birth. However, about 16 to 18 percent of babies weighing less than 1,500 grams (3 pounds, 4 ounces) have lasting neurological impairment; for babies under 750 grams (1 pound, 10 ounces), 25 to 35 percent are so affected (Brody, 1991). Overall, medical science continues to make important and exciting strides in helping preterm babies (Bendersky & Lewis, 1994; Bradley et al., 1994; Brooks-Gunn et al., 1993).

A recent large-scale study carried out over four years at sites across the United States involving 985 premature babies, including many inner-city children, has confirmed the benefits to be realized through developmental interventions. By the time the youngsters were 3 years old, those who had received the developmental interventions achieved substantially higher scores on intelligence tests and tests of social functioning. The IQs of the babies weighing from 4.4 to 5.5 pounds at birth were 13.2 points higher than those of babies of comparable weight in the control group. The IQs of babies weighing less than 4.4 pounds at birth were 6.6 points higher than those of comparable-weight control-group youngsters. Moreover, control-group babies were nearly three times more likely to be mentally retarded (Kolata, 1990). Additionally, much work is being done—and still more needs to be done—to prevent premature births (Radetsky, 1994).

Questions

What are some of the causes of premature birth? How does the medical community today save many of these babies born at high risk?

Postmature Infants   A baby that is delivered more than 2 weeks after the usual 40 weeks of gestation in the womb is classified as a **postmature infant.** Most postmature babies are healthy, but they must be watched carefully for a few days. Some babies are heavier because their mothers are diabetic or pre-diabetic, or an extra amount of sugar has crossed the placenta. Such babies might have metabolic problems for the first few days

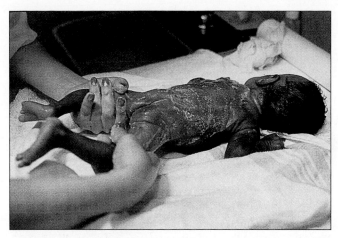

Drug-Exposed Infants Bathing is one of the few ways nurses have found to comfort drug-exposed newborns. Soap and warm water soothe the frantic babies. The bath also removes the sweat that envelops them as they go through withdrawal.

after birth and require closer medical scrutiny. Postmature infants are likely to be larger, posing more complications for both mother and infant during delivery. A mother's options include induced labor and cesarean delivery.

Babies Born with Crack Addiction The majority of babies born are not addicted to any substances. However, it has been conservatively estimated that about 11 percent of babies—particularly babies born to mothers who live in poverty, are of minority status, are drug users, and reside in large urban areas—are at a higher risk of being born addicted to crack (a derivative of cocaine); such babies are called "crack" babies (Kobre, 1998a, 1998b). This is actually a misnomer, or a misguidingly simple label, for a complex problem. There are many variables interacting. As crack addicts themselves, the mothers are highly likely to be using other drugs, such as alcohol, marijuana, amphetamines, or heroin. Many have sexually transmitted diseases contracted during sexual activity they engaged in to promote their drug addiction. Many have most likely resisted prenatal care and have deprived the fetus of proper nutrition, blood, and oxygen for normal growth and development.

"Crack" babies are likely to be premature and to have low birthweight, and many have smaller-than-normal head size. These neonates experience severe pain and withdrawal symptoms. They cry more frequently than unexposed babies. Typically, when held, these babies tend to arch their backs, pull away, and cry until they exhaust themselves. Cocaine-exposed infants are more jittery, have more muscle tension, and are hard to move because they are stiff (Turner, 1996). Nurses have noted that these babies don't like to be touched and have a high, shrill cry different from other babies. These be-

haviors can further influence how the parents treat the baby.

Babies Born with HIV HIV is the insidious virus that causes AIDS. One of the most important decisions a pregnant woman with HIV will make is whether to take medicine for her infection during pregnancy; there are reasons both for and against taking HIV treatment during pregnancy. The best recommendation for a pregnant woman with HIV is to see her health care provider early and often during the pregnancy to maintain her own health and the health of her fetus. This suggestion is likely to be ignored by women from cultures that rely solely on alternative medical practices or who lack access to standard medical care; for instance, in Zimbabwe nearly 35 percent of women are infected with HIV (James, 1998). A woman who has HIV or AIDS should potentially plan a cesarean delivery so her blood doesn't mix with the neonate's as it would during a normal delivery ("Pregnancy and HIV—What Women and Doctors Need to Know," 1998). After delivery, a mother with HIV or AIDS should not breast-feed her baby.

A baby can contract HIV during pregnancy, during labor and delivery, or during breast-feeding—though the virus is most likely to be transmitted during labor and delivery (National Institute of Allergy and Infectious Disease, 1997). With aggressive medical treatment of the mother during pregnancy and of the infant for six weeks after birth, only about 10 percent of infants born to HIV-infected women become infected with the virus (Engle, 1998). This results in an estimated 1,300 to 2,000 children born with HIV annually in the United States. Worldwide, according to Dr. John Saba, a clinical research specialist with the United Nations Program on HIV/AIDS, over 500,000 infants are infected with HIV from their mothers (James, 1998). The good news, today, is that about 90 percent of babies born to HIV-positive mothers do not get HIV. All babies of women with HIV will test positive for the HIV antibodies at first, but this doesn't mean the baby is infected. A baby who is not infected will lose the mother's antibodies between 6 and 18 months of age and start to test negative for HIV. The baby who is infected with HIV will continue to test positive for HIV (National Institute of Allergy and Infectious Disease, 1997). Babies born infected with HIV appear normal at birth, but 10 to 20 percent develop AIDS and die by the age of 2. Most of the rest develop AIDS before the age of 6 (National Institute of Allergy and Infectious Disease, 1997).

A recent report from a large European registry of HIV-infected children gave the encouraging news that nearly 50 percent of the children with perinatally acquired HIV were alive at age 9 (National Institute of Allergy and Infectious Disease, 1997). The prognosis for pediatric HIV has improved overall with early diagnosis and early treatment (Cadman, 1997). In 1998, the International Association for Physicians in AIDS Care launched a project in several states to create pediatric resource centers for HIV-positive children. They also organized and held the first meeting of drug companies, clinicians, and government officials to promote more extensive pediatric HIV treatment research (Centers for Disease Control, 1998b).

After the baby with HIV is born, the parents and caretakers must face not only the emotional strains of caring for a sick baby, but also the financial strains. As of 1996, the estimated annual cost for health care for an HIV-infected child was $9,400; for those children with AIDS, the estimated health care costs were $38,000 (Gorsky, Farnham, & Straus, 1996). The National Pediatric and Family HIV Resource Center (NPHRC) offers state-of-the-art information to families and professionals caring for infants and children with HIV/AIDS.

Babies with Fetal Alcohol Syndrome (FAS) Research has shown that when the mother-to-be drinks during her pregnancy, not only does the placenta "soak it up like a sponge," but the alcohol remains in the amniotic fluid longer than it does in the mother's system. The toxic effect of repeated exposure to this common teratogen has known consequences for fetal development. According to the National Organization on Fetal Alcohol Syndrome (NOFAS), FAS is a group of physical and mental birth defects resulting from a woman's drinking, or binge drinking, during her pregnancy (National Organization on Fetal Alcohol Syndrome, 1998). Symptoms can include growth deficiencies both before and after birth, central nervous system dysfunction resulting in lowered IQ and learning disabilities, physical malformations of the face and cranial areas, and other organ dysfunctions (see Chapter 3). A child does not outgrow FAS. Some children are diagnosed with FAE, fetal alcohol effects, which is a manifestation of fewer of these effects. FAS is the leading known cause of mental retardation, and it appears in every race, social class, and culture (Streissguth & Kanter, 1997).

It is now estimated that one in five women consumes alcohol and other drugs during pregnancy. And annual costs related to the care of these children and adults is estimated at $7.5 billion to $9.7 billion (National Organization on Fetal Alcohol Syndrome, 1998). Over the past 25 years, the Centers for Disease Control and Prevention have reported a large increase in the number of babies diagnosed with FAS per every 10,000 live births: 1 case in 1979; 6.7 cases in 1993; 19.5 cases by 1997 (National Organization on Fetal Alcohol Syndrome, 1998). In July 1998, South Dakota became the first state to enforce treatment programs for alcoholic pregnant women. Friends and relatives can commit a pregnant alcoholic woman to an emergency detoxification center, and judges can order pregnant alcoholic women into treatment facilities (Zeller, 1998). At this time, FAS is the only known birth defect that is 100 percent preventable. Like other children born with birth defects, these children are eligible from birth for early intervention services. Parents, caretakers, and teachers should realize that these children exhibit such characteristics as short attention span, inability to follow directions, difficulty communicating, impulsivity, and failure to recognize the consequences of actions (Streissguth & Kanter, 1997).

Babies with Prenatal Exposure to Chemical Toxicants
According to Landrigan (1997) over 160,000 of the 697,000 Gulf War veterans have reported to the Gulf War Registry. Many of these veterans have experienced their own arrays of physical symptoms that defy any known classification (Landrigan, 1997). Other unofficial sources classify these symptoms as Gulf War illness (GWI) or Gulf War syndrome (GWS) (McAlvaney, 1996). These veterans were exposed to 21 different "reproductive toxicants," and the anthrax vaccination and antibotulism medicine they had to take might have played a part (McAlvaney, 1996). Sixty-five percent of the children born to these veterans after the war are sick or were born with congenital birth defects (multiple deformities that are statistically unlikely to have occurred from chance alone). Affected families have emotional strain, financial drain, and disbelief that they have been unrecognized by the U.S. government, military, and insurance systems. However, healthcare providers, educators, and mental health professionals need to be aware of the special needs of this population of families who live with chronic illness. In the early 1960s, American society experienced a similar high incidence of miscarriage and limb deformities with the widespread use of the prescription drug *Thalidomide* for pregnant women's "morning sickness." Veterans exposed to Agent Orange during the Vietnam War in the 1960s and 1970s produced many children with birth defects. The high incidence of miscarriages and birth deformities in the 1970s in Love Canal, near Buffalo, New York, were proven to be directly related to air and water contamination. We must not forget how vulnerable a developing embryo or fetus is to what we are exposed to and ingest.

Support for Babies with Disorders Parents who give birth to an infant diagnosed with any disorder at birth might want to get connected immediately to local professionals and support groups that focus on that disorder. Knowledge can alleviate much fear and anxiety at this early stage and give parents the hope they need to parent this child as normally as possible. There are also thousands of support groups with Web sites on the Internet for those families that live in more remote locations or have a child with a rare disorder (see "Following Up on

Information You Can Use

Early Intervention Services

Over the past 15 years, our federal, state, and local governments as well as private organizations have invested heavily in funding and research to provide better early-intervention services and medical care for our growing number of at-risk infants and children. Programs such as Head Start, Women, Infants, & Children (WIC), community day care, free health clinics, public preschool, Even Start, early special education, and so forth, employ professionals in the fields listed below. Some positions require a minimum of a master's degree, others require a Ph.D., Psy.D., or M.D. Caring for our children who have been born at risk has become big business in the United States. Longitudinal studies are being conducted to evaluate the effectiveness of a variety of early-intervention strategies to help children develop their competencies.

If you are interested in pursuing a career in any of the following areas, you can search further in your campus career development center or career placement office. Those offices and your campus library will have a variety of resources, such as the *Dictionary of Occupational Titles, Selected Characteristics of Occupations Defined in the Dictionary of Occupational Titles, The Worker Traits Data Book,* and the *Handbook for Analyzing Jobs.* Internet Web sites for professional organizations can provide a wealth of specific career information.

Early Intervention Professional Occupations

Art therapy	Education, special	Pediatrics, behavioral	Psychoanalysis
Behavioral therapy	Family practice	Pediatrics, developmental	Psychology, clinical
Child development, generalist	Music therapy	Pediatrics, general	Psychology, developmental
Child life	Neurology	Physical therapy	Recreational therapy
Education, elementary	Nursing	Psychiatry, adult	Social work
Education, preschool	Occupational therapy	Psychiatry, child	Speech and language therapy

the Internet" at the end of this chapter). Although pediatricians and specialists are learning more about the physiological aspects of many disorders, their medical training often does not include learning about the social and emotional consequences of raising a child with a difficulty.

Public Law 99-457: Early Intervention Services for Infants Born at Risk   Public Law 99-457, enacted by the U.S. Congress in October 1986, was designed to provide early intervention services (free education, training, therapy) to families with children from birth to 5 years that have disabilities and special needs. There is an infant component for those from birth to age 2, and there is a preschool component for those aged 3 to 5. But no one foresaw the onset of the crack epidemic about 15 years ago. Between 375,000 and 700,000 "crack" babies have been born so far in the United States, and treatment for these children has used up much of the money from this federally sponsored program. However, several studies have shown that with early intervention, these children can be saved. Crack-exposed children who went to an infant development program improved significantly in physical coordination, language, and problem-solving skills. A comparison group of crack-exposed children whose families chose not to participate did not show improvement at the time of follow-up testing (Kobre, 1998c). Early intervention that provides interaction with other children, individualized work on children's problem areas (including mood swings and poor stress management), and parent counseling can help at-risk children improve significantly (Kobre, 1998c). (See *Information You Can Use:* "Early Intervention Services.")

Infant Mortality   Babies are not supposed to die; it feels contrary to the natural order of life. Consequently, the death of an infant is a traumatic experience—some say it is the worst experience in life. Parents, family, and friends will go through a period of grief and mourning, just as when an older loved one dies. For more than 20 years, birth defects have been the leading cause of infant mortality. **Infant mortality** is the death of an infant within the first year of life. **Stillbirth** is the rare occurrence of fetal death in the womb or death that occurs during labor and delivery. In 1995, birth defects accounted for 6,554 infant deaths in the United States. Prematurity or low birthweight was the second leading

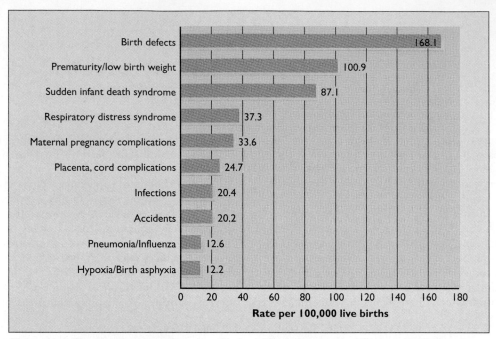

Figure 4-4 The Ten Leading Causes of Infant Mortality, United States, 1995 For more than 20 years, birth defects have been the leading cause of infant mortality in the United States. In 1995, birth defects accounted for 6,554 infant deaths, reflecting a rate of 168.1 per 100,000 live births. Prematurity/low birthweight (LBW) was the second leading cause of infant mortality (100.9), followed by sudden infant death syndrome (SIDS—87.7), respiratory distress syndrome (RDS—37.3), and maternal pregnancy complications (33.6). Together, these five causes accounted for more than half of all U.S. infant deaths in 1995.

Source: National Center for Health Statistics; prepared by March of Dimes Perinatal Data Center, 1998.

cause of infant mortality, followed by sudden infant death syndrome (SIDS), respiratory distress, and maternal pregnancy complications (March of Dimes, 1998) (see Figure 4-4). Following are some of the basic statistics on infant mortality in the United States:

- Every hour four babies die.
- Birth defects are the leading cause of mortality.
- One in five infant deaths is due to birth defects.
- Black infants are more than twice as likely to die before their first birthday as white infants.
- The U.S. infant mortality rate is higher than for 24 other countries.

Postpartum Experience for Mom (and Dad)

Whether a newborn is born healthy or at risk, every new mother needs time to adjust and adapt, both physically and emotionally, after delivery of her child. Normally the *postpartum period* lasts several weeks. However, some women need several months to adapt. Each woman experiences a range of hormone fluctuations after giving birth. Her progesterone levels were elevated during the course of her pregnancy (and some women say they feel their best when they are pregnant); now her highly active endocrine system is attempting to bring her body back to some type of pre-pregnancy balance, and this can pro-

mote mood swings. Most women, no matter what their discomfort, want their babies with them as often as possible during their hospital stay. However, the many postpartum variables can cause a woman to feel quite fatigued and apprehensive about meeting her responsibilities as new mother, wife or partner, and possibly mother to other children. "How can I cope with all of this?" might be on her mind (see Figure 4-5).

Some women choose to leave the hospital with their newborn a few hours after giving birth; others might need to stay several days to recuperate from surgery or complications. The postpartum period is a time of adjustment for the father, too. He is most likely trying to manage at home and work while beginning to adapt to his role as "Dad" to the new baby. The mother is going to need his help more than ever, especially if she is recuperating from surgery or if the baby was born at risk. Some women (and men) seem to need much more time to adapt, both physically and emotionally. These women experience what is called **postpartum depression (PPD).** The following thoughts and behaviors are signs of postpartum depression: crying spells, depression, not sleeping enough or wanting to sleep too much, changes in appetite, anxiety, feelings of not being able to cope, thoughts of not wanting to take care of the baby, or thoughts of wanting to harm the baby. These are indica-

Figure 4-5 The post-partum period is a time of adjustment for the new mother and father

© Lynn Johnston Productions Inc./Dist. By United Features Syndicate, Inc.

tions that the mother needs to seek professional counseling from her gynecologist, medical doctor, and/or a therapist familiar with PPD (most cities have some form of Women's Health Connection through a local hospital). Newborns demand a great deal of time and attention, both night and day, and might spend quite a bit of time crying—*I am hungry. I need to be changed. I am tired. I am cold. I need comfort. I am sick. I want attention* (Harnish, Dodge, & Valenta, 1995). Most of us have never had anyone else be so *dependent* on us as a newborn!

If the new mother's depression goes untreated, it can linger, affecting her family—especially the children, who are at higher risk of developing depression themselves (DeAngelis, 1997; Abrams et al., 1995). For the past 20 years, Dr. Tiffany Field has been researching the effects of maternal depression on newborns, infants, and children. Her findings reveal that depressed mothers produce depressed newborns. "The newborns have elevated stress hormones, brain activity suggestive of depression, show little facial expression, and have other depressive symptoms as well" (Field, 1998). Depressed infants are slower to learn to walk, weigh less, and are less responsive than other babies. Dr. Field's intervention strategy with these mothers and infants includes coaching the mothers into interacting with their babies and massaging the full body of the infant for 15 minutes per day. In addition to physical care, we know that infants have cognitive, emotional, and social needs to be met. To develop normally, they need eye contact, need to be spoken to, need gentle massages, and need to be played with ("Mirror Images," 1998).

Questions

What is an early-intervention program, and which babies and families are eligible for these services?
What are some signs of postpartum depression?

Development of Basic Competencies

To the watchful observer, newborns continue to communicate their perceptions and abilities. Newborns tell us what they hear, see, and feel in the same ways other organisms do—through systematic responses to stimulating events. In brief, newborns are active human beings who are eager to understand and engage their physical and social world. The hallmark of the first two years of life—the period we call **infancy**—is the enormous amount of energy children spend exploring, learning about, and mastering their world. Few characteristics of infants are more striking than their relentless and persistent pursuit of competence. They continually initiate activities by which they can interact effectively with the environment. Healthy children are active creatures. They seek stimulation from the world around them. In turn, they act on their world, chiefly on caretakers, to achieve the satisfaction of their needs.

According to sleep researcher James McKenna (quoted in Small, 1998, p. 35), "There is no such thing as a baby: there is a baby and someone." McKenna and other infant researchers have discovered that infant biology is intimately connected to the biology of the adults who are responsible for their care. This symbiotic relationship, called **entrainment**, is "a kind of biological feedback system across two organisms, in which the movement of one influences the other . . . the physiology of the two individuals is so entwined that, in a biological sense, where one goes, the other follows, and vice versa" (Small, 1998). Entrainment is first and foremost a physical relationship (touching, nursing, cleaning, massaging, etc.). The connection is also visual and auditory: a newborn recognizes its mother's voice (and usually its father's voice, too) and prefers it over other sounds.

Child development expert Edward Tronick points out that a baby's most powerful, adaptive, and necessary skill is the ability to engage an adult on a social level to

Table 4-1   Infant States

Regular Sleep: Infants are at full rest; little or no motor activity occurs; facial muscles are relaxed; spontaneous eye movement is absent; respirations are regular and even.

Irregular Sleep: Infants engage in spurts of gentle limb movements and more general stirring, squirming, and twisting; eye movement is occasional and rapid; facial grimaces (smiling, sneering, frowning, puckering, and pouting) are frequent; the rhythm of respiration is irregular and faster than in regular sleep.

Drowsiness: Infants are relatively inactive; on occasion they squirm and twist their bodies; they open and close their eyes intermittently; respiratory patterns are regular but faster than in regular sleep.

Alert Inactivity: Although infants are inactive, their eyes are open and have a bright, shining quality; respirations are regular but faster than during regular sleep.

Waking Activity: Infants may be silent or moan, grunt, or whimper, spurts of diffuse motor activity are frequent; their faces may be relaxed or pinched, as when crying; their rate of respiration is irregular.

Crying: Vocalizations are strong and intense; motor activities are vigorous; the babies' faces are contorted; their bodies are flushed bright red. In some infants, tears can be observed as early as 24 hours after birth.

meet its needs (Small, 1998). Some infant researchers refer to entrainment as a type of synchronicity, a physical reaction, between a baby and its caretakers. Child expert Dr. T. Berry Brazelton (1998) says that this synchronicity of movements and physical reactions between parents and infant is vital to infant development. He further suggests that infants who fail to thrive lack this physical engagement with their mother (as can happen when the baby is institutionalized, or the mother is severely depressed or addicted to drugs and doesn't attend to the infant's needs). An infant with **failure to thrive** does not take nourishment and therefore is severely underweight for its age and gender. Child experts suggest that the child's physical and cognitive development will not proceed normally without physical engagement with the mother, but parent-baby engagement can be taught or learned (Small, 1998).

Newborn States

Interest in neonate sleeping patterns has been closely linked with interest in newborn states. The term **states,** according to Peter H. Wolff (1966), refers to a continuum of alertness ranging from regular sleep to vigorous activity (see Table 4-1). The noted pediatrician T. Berry Brazelton (1978) says that states are the infant's first line of defense. By means of changing state, infants can shut out certain stimuli and thereby inhibit their responses. A change in state is also the way infants set the stage for actively responding (Blass & Ciaramitaro, 1994). Consequently, the newborn's use of various states reflects a high order of nervous system control (Korner et al., 1988). The *Brazelton Neonatal Behavioral Assessment Scale* evaluates a neonate's early behavior by assessing how the baby moves from sleep states to alert states of consciousness (Brazelton, Nugent, & Lester, 1987).

Reflexes   The newborn comes equipped with a number of behavioral systems, or reflexes, that are ready to be activated. A **reflex** is a relatively simple, involuntary, and unlearned response to a stimulus. In other words, it is a response that is triggered automatically through built-in circuits. Some reflexes, such as coughing, blinking, and yawning last throughout life. Others disappear over the first weeks and reappear as learned voluntary behaviors as the infant's brain and body develop. Reflexes are the evolutionary remains of actions seen in animals lower in the phylogenic scale (Cratty, 1970). Reflexes are good indicators of neurological development in infants. Researchers estimate that the human is born with at least 27 reflexes. Table 4-2 illustrates two of these.

Sleeping   The major "activity" of newborns is sleeping. Newborns normally sleep 16 or more hours per day, in seven or eight naps. Sleep and wakefulness alternate roughly in 4-hour cycles—3 hours in sleep and 1 hour awake. Unless ill or uncomfortable, neonates will sleep wherever they are (a crib, a stroller, or cradled in Mom or Dad's arms). By six weeks the naps become longer, with infants taking only two to four naps during the day. Around this age, many begin to sleep through most of the night, though others will not sleep through the night for many months yet. "Lack of sleep, and more especially broken sleep, is the very worst part of parenting for many people," states British psychologist Penelope Leach (1998) in her book *Your Baby and Child.* As the infant matures into a 1- to 2-year-old toddler, sleeptime is usually reduced to one naptime during the day and an extended sleep at night. See *Human Diversity:* "Co-Sleeping, a Cross-Cultural View."

Sudden Infant Death Syndrome (SIDS)   Each year, some 4,000 U.S. families experience the devastating tragedy and agony of **sudden infant death syndrome (SIDS),** or crib death (Peth-Pierce, 1997). SIDS occurs most often during the third or fourth month of life, although it can happen up to a year old. Parents put their seemingly healthy baby down for a nap or to bed at night and return to find that the infant has died. There is

Table 4-2 Reflex Behaviors

Stepping Reflex		Sucking Reflex	
When the infant is held upright with the soles of the feet touching a firm surface, the infant will take deliberate "steps" as if walking. This behavior disappears after the first week, then reappears in several months as a learned, voluntary behavior.		When a newborn's mouth or lips are touched, it automatically sucks on the object in its mouth.	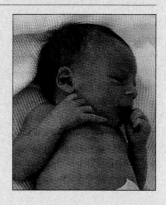

typically no warning. SIDS is one of the leading causes of infant death (after birth defects and accidents). It is one of medicine's unsolved mysteries, and considerable money is being spent researching the cause. Over the years several theories have been proposed, including parental neglect or abuse, birth defects, a seasonal rise in infectious diseases in the winter months, immature brain development in the brain stem, a mother who smoked before the infant's birth, caretakers who smoke after the infant's birth, inborn metabolic disorders, immune system disorders, kidney defects, abnormal pineal gland control over circadian rhythms, and immature brain-stem cell functioning (Blakeslee, 1989). Most SIDS experts agree that the two most important things parents can do to reduce their babies' risk of SIDS are to place babies on their backs during sleep and refrain from smoking (Bock & Dubois, 1998; McKenna, 1996).

Crying Crying in the newborn is an unlearned, involuntary, highly adaptive response that incites the parents to caretaking activities. Humans find few sounds more disconcerting and unnerving than the infant's cry. Physiological studies reveal that the sound of a baby's cry triggers an increase in parents' blood pressure and heart rate (Donate-Bartfield & Passman, 1985). Some parents feel rejected because of the crying and reject the child in turn. But simply because caretakers have difficulty getting babies to stop crying does not mean they are doing a bad job. Crying is the chief way that babies communicate. Different cries—each with distinctive pitch, rhythm, and duration—convey different messages.

The Language of Crying Most parents typically learn the "language" of crying rather quickly (Bisping et al., 1990). Babies have one cry that means hunger, one that means discomfort, one that means it needs attention, one

that means frustration, and others for such problems as pain or illness. Children's cries become more complex over time. Around the second month, the irregular or fussy cry appears (Fogel & Thelen, 1987). At about 9 months, the child's cry becomes less persistent and more punctuated by pauses while the youngster checks how the cry is affecting a caregiver (Bruner, 1983). Babies exposed prenatally to cocaine and other drugs by their mothers present special problems. These newborns commonly experience withdrawal symptoms consisting of irritability and incessant shrill crying, inability to sleep, restlessness, hyperactive reflexes, tremors, and occasionally convulsions. Some of these symptoms subside after the infant has gone through withdrawal (Hawley & Disney, 1992). In non-Westernized cultures if babies are wrapped tightly next to the mother's front or back during the day, they exhibit less crying and are more subdued than American babies tend to be (Small, 1998).

Shaken Baby Syndrome If they cannot get the baby to stop crying, some parents or caretakers wind up feeling distraught and helpless and act out angrily toward the baby. Until more recently, **shaken baby syndrome (SBS)** was a medical diagnosis that few of us knew about. This syndrome occurs when a baby's head is violently shaken back and forth or strikes something, resulting in bruising or bleeding of the brain, spinal cord injury, and eye damage. Health-care and day-care professionals look for a glassy-eyed look or lethargy in an infant if they suspect the baby has been shaken or abused. Bruises and vomiting can also indicate SBS (Duhaime et al., 1998). Infants less than 6 months old are particularly vulnerable to SBS. Of those diagnosed each year, one-third die, one-third suffer brain damage, and one-third recover. Researchers have found that men (fathers or boyfriends)

Human Diversity

Co-Sleeping, a Cross-Cultural View

Cultural Attitudes Toward Infant Sleep Arrangements

Ethnopediatricians (social scientists influenced by Vygotsky's view that child development is inseparable from society and culture) now focus on the importance of the sleep environment because they believe that the sleep environment is crucial to infant health and development. Moreover, anthropologists studying cultures around the world have discovered that for most of human history, babies and children have slept with their mothers, or perhaps with both parents, because of the physical nature of their huts and one-room dwellings. Dwellings with more than one room used to be mainly for the affluent. Today, the majority of people around the world still live and sleep in one-room dwellings (Small, 1998). Culture, customs, and traditions handed down through generations influence how we sleep, with whom we sleep, and where we sleep. One study of 186 nonindustrial societies revealed that in 67 percent of the cultures, children sleep in the company of others. More significantly, in all 186 societies, infants slept with a parent or parents until at least 1 year old (Small, 1998). "The United States consistently stands out as the only society in which babies are routinely placed in their own beds and in their own rooms" (Small, 1998). Babies in other cultures sleep in various environments—in a swaddling cloth on their mother's back, in a hanging basket, in a hammock made of skin or fiber, on a futon, on a mattress made of bamboo, and so on.

Anthropologist Gilda Morelli and colleagues (1992) studied the sleeping arrangements of parents and babies in the United States and in a group of Mayan Indians in Guatemala. The Mayan babies always slept with their mothers for the first year and sometimes the second. Mayan mothers reported no sleep difficulties because they turned and nursed their babies whenever the babies cried with hunger. Mayan mothers also view the mother and child as one unit. In the American sample, no babies slept with their parents. Seventeen of the 18 mothers reported having to wake up and get up for nighttime feedings. Mayan mothers expressed shock and disapproval

when they discovered how American babies were put to bed. It did not matter to them that babies squirm or there was no privacy: They saw closeness at night between mother and baby as what all parents should do for their children (Morelli et al., 1992). Americans in the study reported that co-sleeping was worrisome and somehow emotionally and psychologically unhealthy. Typically, American mothers are advised by pediatricians and child-care experts that sleeping alone is safer for the baby, so the mothers follow this advice.

In a study of Hispanic Americans in New York City, 80 percent of the Hispanic children slept in the same room as their parents (Schachter et al., 1989). In a comparative study of whites and African Americans, 55 percent of the white parents and 70 percent of the African American parents responded that they co-slept with their babies (Lozoff, Wolf, & Davis, 1984).

In Japan, there appears to be a collective attitude toward the relationship needed between a mother and infant. "A baby is a pure spirit, essentially good by design and in need of being incorporated into the maternal self" (Wolf et al., 1996). Japanese mothers are given pamphlets that tell them they should be "responsive and gentle and communicate frequently with their babies, to entwine the infant to its mother and bring the baby into the family fold." Unlike American mothers, Japanese mothers are not interested in making their babies become independent, but rather in making sure they become part of the mother, a connected social being (Small, 1998). Japanese babies and children are placed on futons in the parental bedroom, for the Japanese concept of family includes sharing the night.

The non-western view of co-sleeping appears to be to promote attachment with the infant, whereas Western cultures value independence and self-sufficiency in their children. The ideology of privacy when sleeping appeared in the United States during the 1800s, when housing expanded. American parents have been taught that it is morally correct for infants to sleep alone (Small, 1998). A few other industrialized nations have set sleep expectations for children, too.

are more likely to inflict this injury, followed by female babysitters, then mothers (Duhaime et al., 1998). Boy babies are more likely to be the victims of this abuse than girl babies (Duhaime et al., 1998). SBS usually results from an impulsive, angry response to an infant's crying. Parents should take extra precaution to select babysitters carefully and to never leave a child with a stranger. Younger babies may accept a new sitter more easily; however, most infants by 7 to 8 months develop a fear of

strangers called "stranger anxiety" and are more apt to cry for long periods of time when left with a strange caretaker.

What if a parent feels frustrated by a baby's incessant crying? First, check the infant to make sure nothing is wrong (see the next section, "Soothing the Infant"). If the child's physical and comfort needs have been met, the parent who fears hurting the child should leave the room, shut the door, go to another room, and calm

Customs and Baby Sleep Arrangements Culture, customs, and traditions influence whether a baby sleeps swaddled on a mother's back, in a woven basket, in a hammock, on a futon, on a mattress made of bamboo, in a crib, or co-sleeps with parents.

Dutch parents believe children should be regulated in sleep and all other matters. Babies and children are put to sleep at the same time every night, and if they wake up, they are expected to entertain themselves. A regular routine for the infant is a must in the Dutch family (Small, 1998).

Helping the Baby Sleep Through the Night

American parents often struggle to get their infant to sleep through the night. Some of their strategies to induce sleep include placing devices or stuffed animals in the crib that play a mother's recorded heartbeat, taking car rides, using automatic infant swings, and playing "white noise" machines or quiet music to mask other noises in the home. In American culture, sleeping patterns have become a marker of infant maturity and development: *Is the infant sleeping through the night yet?* Typically, by 3 or 4 months of age, the infant brain has matured enough to develop a circadian rhythm—brain recognition of day and night that it did not experience while

in the womb. James McKenna (1996), sleep researcher, has discovered that babies, like adults, sleep different amounts, and each culture helps determine how much that sleep should progress.

McKenna has also conducted co-sleeping experiments in a sleep lab environment with mothers and infants. He discovered that co-sleepers are physiologically entwined and react to each other's movements and breathing. Because babies are born neurologically immature, they have episodes of breathing lapses. Co-sleeping babies sleep differently than those that sleep alone. Co-sleeping infants respond more to the patterns and rhythms of the mother's breathing, and McKenna suggests that this is a way of teaching infants how to breathe. Co-sleeping mothers pay much more attention to their infants (kissing, touching, repositioning). To McKenna (1996), "it is no coincidence that management of breathing ability comes developmentally at three to four months of age—just at the same period when babies are most vulnerable to SIDS."

down. Calling a trusted family member or friend to come help might provide some relief. The infant is trying to communicate its discomfort, and crying is the only way she can communicate with you. If this happens regularly, the infant needs to be examined carefully by a pediatrician to check for a condition called colic—**colic** is a condition of discomfort, of unknown cause, in which the baby cries for an hour or more, typically every day at about the same time, for up to several weeks.

Soothing the Infant Meeting physical and emotional needs is likely to soothe a crying baby. If a baby is crying because it is hungry, or sleepy, it is likely that it will briefly soothe itself by sucking its own fingers. A baby who contents himself this way might not need a pacifier. There are pros and cons to giving a baby a pacifier. Infants who can learn to comfort themselves might be learning to meet their own needs, compared to those that learn to howl until someone puts a pacifier in their mouth. The first order of business is to try to determine why the baby is crying in the first place. Has the baby eaten enough? Is he warm enough? Is her diaper dry? Has the caretaker given comforting touches and words recently? Many older infants like to have a comfort object, such as a specific blanket or stuffed animal, at naptimes or at particularly stressful times. Some infants like rhythmical behaviors, such as rocking in a chair or riding in a stroller. In many cultures of the world, babies spend most of their time carried in a sling on the back or side of an adult, which is warm as well as physically soothing. Most importantly, parents should try to remain calm and not show anger and stress around babies, for the infant's behavior often mirrors the behavior of its caretakers (recall the "synchronicity" concept between child and caretaker mentioned earlier in this chapter).

Feeding   The first few weeks of feeding a newborn can be worrisome, because a baby will feed only when its internal state signals that it needs nutrition. After a few weeks, the caretaker will be more aware of this particular baby's pattern of feeding. When fully awake, neonates spend a good deal of time feeding. Indeed, their hunger and sleep patterns are closely linked. Newborns might feed 8 to 14 times during the day. Some infants prefer to feed at short intervals, perhaps every 90 minutes, all day. Others have intervals of 3 to 4 hours or longer. Fortunately, infants come to require fewer daily feedings as they grow older. Most seem to eat three to five meals a day by the time they are 1 year old. As they grow into toddlers, all babies begin to vary in how much they will eat at one time and what they will eat. A child who earlier ate very little might begin to eat a lot more, or one that ate well might now eat very little. Child health experts suggest that this is normal and that young children experience "growth spurts," occasional periods of time when they need considerably more energy to accommodate their bodies' growth.

On-Demand Feeding   Several decades ago, doctors recommended strict feeding schedules for infants. But pediatricians have come to recognize that babies differ markedly, and they now encourage parents to feed their baby when it is hungry—let the infant pick its own times in the 24-hour cycle to feed, in on-demand feeding. Whatever schedule parents and caretakers follow, they must decide whether the baby will be breast-fed or bottle-fed. Before 1900 the vast majority of mothers breast-fed their babies or employed a "wet nurse" to do this (a wet nurse is a lactating woman employed to nurse others' babies). But in the ensuing years, bottle-feeding with infant formulas became increasingly popular, so that by 1946 only 38 percent of women left the hospital with a nursing baby. This figure dropped to 21 percent by 1956. Since then, breast-feeding has gained in popularity. In

1993, some 56 percent of new American mothers breast-fed their babies (Painter, 1994b).

Breast-Feeding   There is a large body of evidence that breast-feeding the infant for the first several months of life is best (Slusser & Powers, 1997). Breast-feeding offers a number of advantages. Mother's milk for the first three to five days provides *colostrum,* a substance that provides antibodies that build up the newborn's immune system. The newborn can normally digest breast milk easily, because it more watery than formula-based milk. Also, a baby's stools will be of a more liquid consistency, and elimination might be easier than if formula fed. Today, many American women and those from non-Westernized cultures believe that the intimate contact afforded by breast-feeding creates a sense of security and well-being in the infant, and that this favorably influences its later personality. Breast-feeding is also less costly and less time-consuming than purchasing and preparing formula. The mother's milk is always ready and at the proper temperature. Other benefits attributed to breast-feeding include a reduced chance of developing allergies or asthma and fewer ear infections. Mothers are likely to return to their pre-pregnancy weight more rapidly.

The chief drawback of breast-feeding is that it can limit the physical freedom of the mother. She must be available to the infant every few hours, night and day, unless she expresses her milk and stores it in bottles so that the father and other caretakers can feed the child. Today there are manual and mechanical breast pumps that allow the mother to release her milk and refrigerate it for later feedings. Another disadvantage of breast-feeding might be that the mother does not know how much milk the baby is getting; if she were feeding the baby formula from precisely measured bottles, she would know the baby's exact intake. However, a breast-feeding baby who is gaining weight and eliminating several times a day is getting adequate nutrition. Another drawback to breast-feeding is that the mother may find she needs to limit her intake of caffeine (for instance, in coffee, tea, and sodas), which is a mild stimulant. Other foods such as broccoli, cauliflower, cabbage, and spicy foods can also affect the baby's developing gastrointestinal system and cause crying, crankiness, or irritability.

Formula (Bottle) Feeding   The advantage of bottle (formula) feeding is that it gives mothers physical freedom and easily lets fathers and other caretakers become involved in feeding the infant. Also, mothers who are taking medications (e.g., antidepressants, anticonvulsants, insulin, AZT) can still feed their infants. Commercial formulas tend to fill babies up more, so they can go longer between feedings. Mothers who bottle-feed can continue to provide nurturant contact with the baby.

One drawback is that formula-fed babies tend to pass bulkier stools and are more likely to experience the discomfort of constipation than a breast-fed baby.

Additional Cautions Regarding Infant Nutrition
Breast-feeding infants should be monitored regularly by a health care professional if the mother is taking medications or drugs. As stated earlier, a mother with HIV should not breast-feed her baby, because the virus might be transmitted to the baby in the breast milk. However, in some developing countries breast milk might be the only available food for an infant. Also, the Committee on Nutrition of the American Academy of Pediatrics recommends that breast-fed infants be given certain supplements such as vitamin D, iron, and fluoride. Mothers in developing countries, who are now targeted by suppliers of infant formula, might inadvertently prepare formula with contaminated water—putting the baby's health at risk (Spotlight on the Baby Milk Industry, 1998). Pediatricians have more recently discovered evidence that cow enzymes and antibodies are likely factors in infant colic, because babies' immature digestive and excretory systems are unable to process these enzymes and antibodies. Formulas with a soy base (vegetable base) do not contain such agents. Some infants cannot digest milk-based or soy-based formulas and need a special pediatric formula to sustain their nutritional needs for development.

Graduating to "Regular" Foods   Over the course of the first two years of development, a child will gradually begin to eat "regular" table foods and beverages, such as breads and cereals, mashed vegetables and fruits, and eventually small portions of meats. An adequate, balanced diet is extremely important for continued health and brain growth. Two developmental milestones occur when a child can pick up food using the forefinger and thumb and hold a cup and drink from it unaided by adults. It is highly recommended that parents use moderation in giving the child sweetened foods and beverages, since the child's baby teeth could begin to decay. All children exhibit "likes" and "dislikes" when it comes to the flavors and textures of foods, but it is a good idea to introduce a variety of tastes into the child's diet during this time. How much a child can eat at one time varies from child to child and from age to age in development.

Mastering Toilet Training   By about 1½ to 2 years of age, or later, most young children show an interest in toilet training. This is an especially important developmental milestone in American families, because many of our youngsters are taken into public settings, such as day care, for long periods of time each day. Along with other muscles in the body that are developing and allowing the child to walk, climb, or run, the muscles in the toddler's

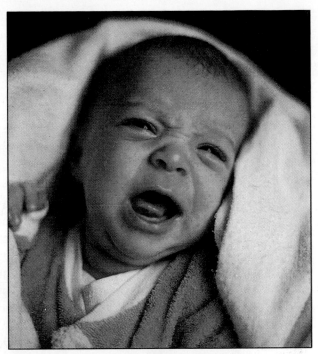

Newborn States A baby's major job is to regulate its internal states, which include sleeping, feeding, eliminating, and communicating with caregivers.

anal and urinary tract are developing. When these are strong enough, the child will let a caretaker know that he or she is ready to be toilet trained. No young child should be forced to sit on a toilet for long periods of time nor left alone on the toilet. Freud said in his psychoanalytic theory that attitudes toward one's sexuality form during toilet-training times. Words used to identify body parts and elimination of waste vary from culture to culture, within cultures, and within families.

Questions

What are several states that babies are learning to regulate during the first few years of life? Who do you think is in charge of managing these, the parents or the infant?

Brain Growth and Development

Growth is the only evidence of life.

John Henry

Infants do grow at a surprising rate and change in wonderful ways. Their development is especially dramatic during the first two years of life. Indeed, the change from

Table 4-3	Growth Norms: Height and Weight Averages for Children			
Age	Average Height for Girls (inches)	Average Weight for Girls (pounds)	Average Height for Boys (inches)	Average Weight for Boys (pounds)
3–6 months	24½	14½	25½	16
6–9 months	26¾	18½	27½	19¾
9–12 months	28¾	20½	29¾	22½
1–1½ years	30¾	23½	31	24¼
1½–2 years	32	24¾	33¼	27
2 years	35	29	36	30½
3 years	38½	33¼	39	34¾
4 years	41¾	38¾	42	39¾
5 years	44	42½	44	44½
6 years	46	47½	46¾	48½
7 years	48	53½	49	54½
8 years	50¾	60¾	51	61¼
9 years	53¼	69	53¼	69
10 years	55½	77	55¼	74½
11 years	58½	87½	57¼	85
12 years	60½	94	59	89
13 years	60¼	103	61	99

© Baby Bag Online, 1996. Http://www.babybag.com.

the dependent newborn to the walking, talking, socially functioning child whom we meet hardly 600 days later is awesome. These maturational changes take place because of growth in key systems of the body. The *pituitary gland,* in conjunction with the *hypothalamus* (a structure at the base of the brain composed of a tightly packed cluster of nerve cells), secretes hormones that play a critical part in regulating children's growth (Guillemin, 1982). Too little of the growth hormones creates a dwarf, and too much creates a giant. Predictable changes occur at various age levels. Many investigators (Meredith, 1973; Gesell, 1928) have analyzed the developmental sequence of various characteristics and skills. From these studies psychologists have evolved standards, called **norms,** for evaluating a child's developmental progress relative to the average of the child's age group. Although children differ considerably in their individual *rate* of maturation, they show broad similarities in the *sequence* of developmental change. Among infants, length and weight are the indices most strongly correlated with behavioral development and performance (Lasky et al., 1981).

Charts in pediatricians' offices that are based on norms show relatively smooth continuous curves of growth, suggesting that youngsters grow in a steady, slow fashion (see Table 4-3). In contrast, parents often comment that their growing child "shot up overnight." Intriguing findings by Michelle Lampl and her col-leagues seemingly show that babies grow in fits and starts, with long intervals between growth spikes (Kolata, 1992a; Lampl et al., 1995). They find that babies remain the same size for 2 to 63 days and then spurt up by from ⅕ to 1 full inch in less than 24 hours. Additionally, it seems that a few days prior to growing, youngsters often become hungry, out of sorts, fussy, agitated, and sleepy. This research suggests that growth operates in an on/off fashion, much like a light switch. However, other researchers challenge the findings, insisting that human growth occurs continuously (Heinrichs et al., 1995).

The Rate of Growth of Key Systems and the Brain

Not all parts of the body grow at the same rate. The growth curve for lymphoid tissue—the thymus and lymph nodes—is quite different from that for tissue in the rest of the body. At 12 years of age lymphoid tissue is more than double the level it will reach in adulthood; after age 12 it declines until maturity. In contrast, the reproductive system grows very slowly until adolescence, at which point its growth accelerates. The internal organs, including the kidneys, liver, spleen, lungs, and stomach, keep pace with the growth in the skeletal system, and these systems therefore show the same two growth spurts in infancy and adolescence.

The Newborn and Brain Development   The nervous system develops more rapidly than other systems. At birth the brain already weighs about 350 grams; at one year it is about 1,400 grams; by 7 years of age, the brain is almost adult in weight and size (Restak, 1984). The circumference of the baby's skull is measured to verify continued cerebral cortex growth with ultrasound imaging while in the womb, at birth, and at every scheduled medical checkup during childhood. Those parts of the hindbrain that control basic processes such as circulation, respiration, and consciousness are operative at birth. Most neonatal reflexes, like sucking, rooting, and grasping, are organized at the subcortical level (the part of the brain that guides basic biological functioning, including sleeping, heart rate, hunger, and digestion). The parts that control processes less critical to immediate survival, including physical mobility and language, mature after birth. The rapid growth of the brain during the first two years of life is associated with the development of neural pathways and connections among nerve cells, particularly in the cerebral cortex (the part of the brain responsible for learning, thinking, reading, and problem solving). More efficient brains are characterized by a complexity of neuronal interactions and a richness of synaptic connections (see Figure 4-6). The rapid development of the cortex during the first 12 months provides the foundation for children's less stereotyped and more flexible behavior (Chugani & Phelps, 1986).

Diagnostics and Imaging   PET (positron emission tomography) scanners are providing evidence that the biological or metabolic activity of the brain undergoes substantial change between birth and adulthood. Neuroscientists using PET scans have found that the metabolic rate of the baby's brain is about two-thirds that of the adult's. By the age of 2 the rate approximates that of the adult's, with rapid increases occurring in the activity of the cerebral cortex. By age 3 or 4 the metabolic rate of the child's brain is about twice that of the adult's. The

brain stays supercharged until the age of 10 or 11. Metabolic rate then tapers off, reaching the adult rate at about age 13 or 14 (Blakeslee, 1986). Thanks to new scanning and imaging technologies, including powerful brain scans, scientists are now able to form a much clearer picture of the brain's inner workings. This has allowed greater insight into early development.

Environmental Stimulation Factor   Researchers now confirm that the way parents interact with children in the early years and the experiences they provide have a big impact on an infant's emotional development, learning abilities, and later functioning ("Brain Facts," 1999; Yarrow et al., 1984). A baby is born with an unfinished brain, one that lets the child develop neural pathways in direct response to its world. From birth, the baby's brain is rapidly creating neural connections. Researchers are finding that the quality of caregiving has an even greater effect on brain development than most people previously suspected. Of course, heredity also plays a role. Recent research suggests that the expression of parental love affects the way the brain forms its complex connections: Looking into the baby's eyes and holding and stroking the baby stimulates the brain to release hormones that promote growth; singing and talking to a baby stimulates the sense of hearing ("Brain Facts," 1999). Some researchers refer to this as "emotional intelligence." If an infant's brain is not exposed to visual or auditory experiences, the child will have difficulty mastering language and will have difficulty with visual and auditory tasks ("Brain Facts," 1999). Every potential caregiver, as well as every family member, is a source of love, learning, comfort, and stimulation.

Principles of Development   As mentioned earlier, human development proceeds according to two major principles. Development according to the *cephalocaudal principle* proceeds from the head to the feet. Improvements in structure and function come first in the head

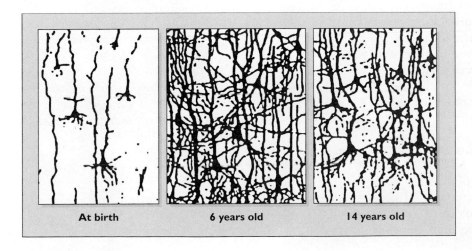

| At birth | 6 years old | 14 years old |

Figure 4-6 Synaptic Density in the Human Brain A single neuron can connect with as many as 15,000 other neurons. The incredibly complex network of connections that results is often referred to as the brain's "wiring" or "circuitry." Experience and environmental stimulation shape the way circuits are made in the brain.

region, then in the trunk, and finally in the leg region. At birth the head is disproportionately large. In adults the head makes up only about one-tenth to one-twelfth of the body, but in newborns it is about one-fourth of the body. In contrast, the arms and legs of newborns are disproportionately short. From birth to adulthood the head doubles in size, the trunk trebles, the arms and hands quadruple in length, and the legs and feet grow fivefold (Bayley, 1956). Motor development likewise follows the cephalocaudal principle. Infants first learn to control the muscles of the head and neck. Then, they learn to control the arms, the abdomen, and, last, the legs. Thus, when they begin to crawl, they use the upper body to propel themselves, dragging the legs passively behind. Only later do they begin to use the legs as an aid in crawling. Similarly, babies learn to hold their heads up before they acquire the ability to sit, and they learn to sit before they learn to walk (Bayley, 1936; 1935).

The other major pattern of human development follows the *proximodistal principle:* development from near to far, outward from the central axis of the body toward the extremities. Early in infancy, babies must move their head and trunk in order to orient their hands when grasping an object. Only later can they use their arms and legs independently, and it is still longer before they can make refined movements with their wrists and fingers. On the whole, control over movement travels down the arm as children become able to perform increasingly precise and sophisticated manual and grasping operations. Another way of expressing the same principle is to say that, in general, large-muscle control precedes fine-muscle control. Thus, the child's ability to jump, climb, and run (activities involving the use of large muscles) develops ahead of the ability to draw or write (activities involving smaller muscles).

Motor Development

Reaching, grasping, crawling, and walking—behaviors that infants become able to perform with considerable proficiency—have proven to be highly complex and problematic tasks to engineers who design computers and robots that can perform these tasks. And in laboratory tests professional athletes have become exhausted from mimicking baby movements. Not surprisingly, much of the early psychological research devoted to motor development was primarily descriptive, much as we might describe the mechanical activities of a computer or robot (Halverson, 1931; McGraw, 1935). To crawl, walk, climb, and grasp objects with precision, babies must have reached certain levels of skeletal and muscular development. As their heads become smaller relative to their bodies, their balance improves (imagine how difficult it must be to move around with a head that is one-fourth of one's total size). As children's legs become stronger and longer, they can master various locomotive activities. As their shoulders widen and their arms lengthen, their manual and mechanical capacities increase. Motor development occurs in accordance with maturational processes that are built into the human organism and are activated by a child's interaction with the environment (Thelen, 1995, 1986, 1981).

Rhythmic Behaviors   Probably the most interesting motor behavior displayed by young infants involves bursts of rapid, repeated movements of the limbs, torso, or head (Thelen, 1995, 1981). Infants kick, rock, bounce, bang, rub, thrust, and twist. They seem to follow the dictum "If you can move it at all, move it rhythmically." Such behaviors are closely related to motor development and provide the foundation for the more skilled behav-

The Cephalocaudal principle
Physical development and motor development come first in the head and neck region, then in the trunk and upper body, and finally in the legs and feet. Thus, when infants begin to crawl, they use their upper body to propel themselves, dragging the legs behind. Then they get up on all fours, using the legs to aid in crawling.

iors that will come later. Hence, rhythmical patterns that involve the legs, like kicking, begin gradually, increase at about 1 month, peak immediately prior to a child's initiation of crawling at about 6 months, and then taper off. Likewise, rhythmical hand and arm movements appear before complex manual skills. Thus, bouts of rhythmic movement seem to be transitional behaviors between uncoordinated activity and complex voluntary motor control. They represent a state in motor maturation that is more complex than that found in simple reflexes yet less variable and flexible than that found in later, cortically controlled behavior.

Locomotion The infant's ability to walk, which typically evolves among U.S. youngsters between 11 and 15 months of age, is the climax of a long series of developments (Thelen, 1995, 1986). These developments progress in a sequence that follows the cephalocaudal principle. First, children gain the ability to lift up the head and, later, the chest. Next, they achieve command of the trunk region, which enables them to sit up. Finally, they achieve mastery of their legs as they learn to stand and to walk. For most infants the seventh month brings a surge in motor development. Usually, children begin by *crawling*—moving with the abdomen in contact with the floor. They maneuver by twisting the body and pulling and tugging with the arms. Next, they may progress to *creeping*—moving on hands and knees while the body is parallel with the floor. Some children also employ *hitching*—sitting and sliding along the floor by "digging in" and pushing themselves backward with the heels. In this form of **locomotion** they often use the arms to aid in propulsion. Indeed, an occasional infant varies the procedure by sitting and then, employing each arm as a pendulum, bouncing across the floor on its buttocks.

By 7 or 8 months children resemble perpetual-motion machines, as they relentlessly tackle new tasks. At 8 months they pull themselves to a standing position but usually have difficulty getting back down again. Often they fall over backward but, undaunted, keep practicing. The urge to master new motor skills is so powerful in infants at this age that bumps, spills, falls, and other obstacles only momentarily discourage them. Before age 1, many infants are "cruising" (standing and walking while holding on to furniture).

New sights, sounds, and experiences challenge cognitive structures that in turn lead the child to develop new motor skills. For example, the first time children discover they can crawl up the stairs, they will not forget, and they will continually go back to the stairs to repeat the process over and over; however, it will most likely be weeks before they will know how to get back down the stairs, and they will need parental supervision with such developmental tasks.

The stages and the timing of motor development are based largely on studies of infants from Western cultures. But the possibility that there are considerable differences among cultures in the timing of motor development has been raised by a number of studies of African infants (Keefer et al., 1982; Ainsworth, 1967). Marcelle Geber and R. F. Dean (1957a, 1957b) tested nearly 300 infants living in an urban area of Uganda. They found that these babies were clearly accelerated in motor development relative to American white infants. The Ugandan infants' precocity is greatest during the first 6 months of life, after which the gap between the two groups tends to decrease. It closes by the end of the second year. Other cross-cultural differences exist. Hopi Indian children of the American Southwest, for instance, begin walking about a month or so later than Anglo

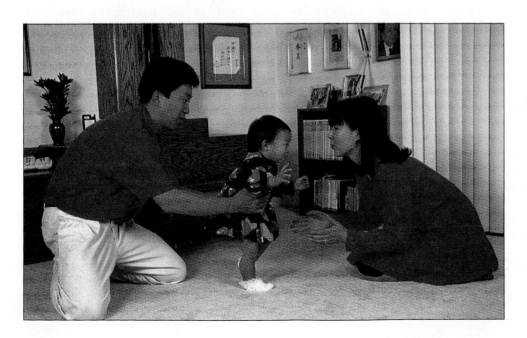

Walking Unaided Walking is one of the most exciting developmental milestones (for the child and parents) occurring around the toddler's first birthday or later. This culmination of physical and motor development brings with it greater independence and mobility in her environment.

children do (Dennis & Dennis, 1940). In sum, cross-cultural research reveals that variations occur across populations in the course and timing of motor development (LeVine, 1970).

Manual Skills The child's development of manual skills proceeds through a series of orderly stages in accordance with the *proximodistal principle*—from the center of the body toward the periphery. At 2 months of age infants merely make swiping movements toward objects with the upper body and arms; they do not attempt to grasp objects. At 3 months of age their reaching consists of clumsy shoulder and elbow movements. Their aim is poor, and their hands are fisted. After about 16 weeks children approach an object with hands open. Around this same time, infants spend a considerable amount of time looking at their own hands. Around 20 weeks children become capable of touching an object in one quick, direct motion of the hand; occasionally, some of them succeed in grasping it in an awkward manner.

Infants of 24 weeks employ a corralling and scooping approach with the palm and fingers. At 28 weeks they begin to oppose the thumb to the palm and other fingers. At 36 weeks they coordinate their grasp with the tips of the thumb and forefinger. Caretakers must take extra precautions once infants can pick up small objects, because infants seem to put all objects in their mouth for the next several months. By about 52 weeks infants master a more sophisticated forefinger grasp (Ausubel & Sullivan, 1970; Halverson, 1931). By the age of 24 months, most children can hold and use such items as eating utensils, a crayon or paintbrush, a ball, or a toothbrush.

The Senses

What is the world like to the infant? Increasingly, sophisticated monitoring equipment is permitting us to pinpoint what the infant sees, hears, smells, tastes, and feels. Social and behavioral scientists now recognize that infants are capable of doing much more, and doing it much earlier, than was believed possible even 20 years ago. Indeed, these new techniques and the insights they have given us have been hailed by some psychologists as "a scientific revolution."

During the first 6 months of life, there is a considerable discrepancy between infants' vast sensory capabilities and their relatively sluggish motor development. Their sensory apparatus yields perceptual input far beyond their capacity to use it. As a result of maturation, experience, and practice, they have already acquired the ability to extract information from the environment at a phenomenal rate. When these perceptual abilities become linked with the big spurt in motor development that begins around the seventh month, the child surges ahead in an awesome fashion. Hence, 10 to 11 months

later, at 18 months of age, the child is an accomplished social being. Let us consider the processes of sensation and perception. *Sensation* refers to the reception of information by our sense organs. *Perception* concerns the interpretation or meaning that we assign to sensation.

Vision

*T*he Eye altering alters all.
                William Blake, *The Mental Traveller*, 1800–1810

A full-term newborn is equipped at birth with a functional and intact visual apparatus. However, the eyes are immature. The retina and the optic nerve, for instance, are not fully developed (Abramov et al., 1982). Neonates also seem to lack visual accommodation. The muscles that control the lenses are not fully developed. As a result, the lenses are fixed in focus for about a month. Thus, only objects that are about 7 to 9 inches from the neonate's eye are in focus (Bronson, 1994). In part this fixed focus might be a blessing, for it limits the amount of stimulus input infants must cope with. As one would expect, infants' visual scanning capabilities become progressively more sophisticated with the passage of time (Bronson, 1997, 1994; Granrud, 1993).

Research by Robert L. Fantz and colleagues (1975) confirms that neonates are able to distinguish among various visual patterns. In one experiment Fantz (1963) found that infants from 10 hours to 5 days old looked longer at a schematic black-and-white face than at a patch of newsprint; longer at newsprint than at a black-and-white bull's-eye; and longer at a bull's-eye than at a plain red, white, or yellow circle. Fantz also demonstrated that neonates prefer pattern over color differences, "complex" over "simple" figures, oval over plain shapes, and curved over straight contours (Fantz, 1966; Fantz & Miranda, 1975). Other researchers find that neonates can follow slowly moving objects with their eyes (Haith & Goodman, 1982). Of course, infants show considerable variation in the duration of their visual fixations and in the speed with which they process information from their surroundings (Freeseman, Colombo, & Coldren, 1993; Jacobson et al., 1992).

Let us take a closer look at the developing visual capabilities of youngsters by examining a number of specific components: visual constancy, depth perception, perception of form, and perception of the human face.

Visual Constancy Perception is oriented toward things, not toward their sensory features. We can perceive features such as blueness, squareness, or softness, but we generally experience them as qualities of objects. We are aware of a blue car, a square block, or a soft pillow, not of "blueness," "squareness," or "softness" as distinct entities. We fashion and build our world in terms of things, objects that endure and that we encounter again and again. It seems that 3- to 5-month-old infants are able to recognize object boundaries and object unity by detecting surface separations or contours (Spelke, van Hofsten, & Kestenbaum, 1989).

One of the most intriguing aspects of perceptual experience is *visual constancy*—the tendency for objects to look the same to us despite fluctuations in sensory input. We perceive the colors, sizes, and shapes of objects as relatively unchanging regardless of changes in the colors, sizes, and shapes in their images projected on the retina of our eye (Dannemiller, 1989). For instance, we see the color of coal as black even though the amount of light reflected from it changes. This phenomenon is called *color constancy.* Likewise, an object does not appear to shrink as we move father away from it, even though the size of the image on our retina becomes smaller. This phenomenon is *size constancy.* Finally, we see a door as a rectangle even though its shape becomes a trapezoid as the door opens, with the edge that is toward us appearing wider than the hinged edge. This phenomenon is *shape constancy.*

Since the earliest days of their discipline, psychologists have been interested in how the visual constancies arise in children. Among the indicators that investigators have used are changes in the infant's eye orientation, sucking rate, body movements, skin conductance, heart rate, and conditioned responses. Most researchers believe that color, size, and shape constancies become apparent in infants at about 4 to 6 months of age (Catherwood, Crassini, & Freiberg, 1989). However, T. G. R.

Bower (1976) claims to find visual constancy among infants at 6 weeks of age, a capability he attributes to an innate genetic mechanism. Perhaps for the present it is best if we keep an open mind about visual constancy and await further research.

Depth Perception   Eleanor Gibson and her husband James Gibson have both had distinguished careers as psychologists (Pick, 1992; Reed, 1988). When their daughter was about 2 years old, the Gibsons visited the Grand Canyon in Arizona. When their child ventured close to the rim of the canyon, James Gibson, an authority on perception, assured his wife that a child their daughter's age was in little danger. He told her that 2-year-olds can recognize the depth of a drop-off as well as an adult can. Eleanor Gibson was not impressed by this scholarly observation and made the child move well back from the rim. Some years later, the memory of this episode led Eleanor Gibson to undertake a visual cliff experiment with the assistance of one her students, Richard D. Walk (Gibson & Walk, 1960). In this technique an infant is placed on a center board between two glass surfaces upon which the child can crawl (see Figure 4-7). The shallow side is covered on the underside with a checkered material. At the deep side an illusion of a cliff is created with a checkered material placed several feet below the glass. The infant's mother stands alternately at the shallow and deep sides and coaxes the infant to crawl toward her. If infants can perceive depth, they should be willing to cross the shallow side but not the cliff side, because the cliff side looks like a chasm.

Gibson and Walk tested 36 infants between 6½ months and 14 months of age. Twenty-seven of the

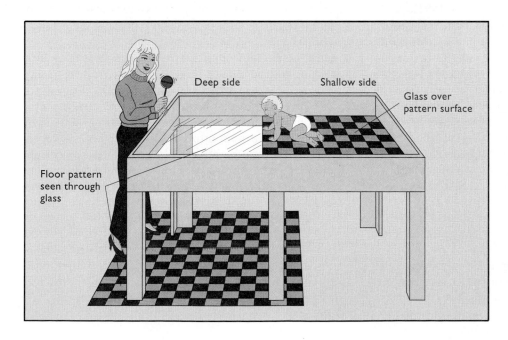

Deep side   Shallow side

Glass over pattern surface

Floor pattern seen through glass

Figure 4-7 The Visual Cliff Experiment In the visual cliff experiment, the child is placed on a center board that has a sheet of glass extending outward on either side. A checkered material is placed on one side about 40 inches below the glass, thus providing the illusion of depth. Despite its mother's coaxing and the presence of a safe glass surface, a 6-month-old infant generally will not crawl across the "chasm." The infant will, however, venture across the shallow side of the apparatus to reach its mother.

Source: Adapted from E. J. Gibson and R. D. Walk, "The 'Visual Cliff,'" *Scientific American,* Vol. 202 (1960), p. 65.

infants ventured off the center board and crawled across the shallow side toward their mothers. Only 3, however, could be enticed to cross the cliff side. A number of infants actually crawled away from their mothers when beckoned from the deep side; others cried, presumably because they could not reach their mothers without crossing the chasm. Some patted the glass on the deep side, ascertaining that it was solid, but nonetheless backed away. Apparently, they were more dependent upon the visual evidence than on the evidence provided by their sense of touch. This research suggests that the vast majority of babies can perceive a dropoff and avoid it by the time they become capable of creeping.

Infants' *binocular vision*—the ability to tell the distances of various objects and to experience the world three-dimensionally—undergoes a sudden burst between 3 and 5 months of age (Yonas, Granrud, & Pettersen, 1985). The fact that the ability arises quite suddenly and rapidly suggests to some psychologists that it represents a change in the visual cortex, the portion of the brain responsible for vision. During the postnatal period, neural connections undergo substantial growth and elaboration. Apparently, these developmental changes result in the two eyes working in concert and allow the brain to extract reliable three-dimensional information from perceptual processes.

Perception of Form   Over the first two years of life, the way infants focus on and organize visual events changes. During the first two months, babies attend to stimuli that move and to those that contain a high degree of contrast. From birth, infants actively engage in visual scanning. As children get older, their scanning patterns become more exhaustive and less redundant. Consequently, the information they collect is more directly relevant to the task (Bronson, 1994; Granrud, 1993).

Researchers study infants' visual scanning by means of corneal photography. They train a movie camera and lights on an infant's eye and give the baby something to look at. The lights that they beam at the baby's eye are filtered so that the infant does not see them, but the movie film picks up their reflections. Later, the researchers develop the film, project it, and measure where the light reflections fall on the infant's eyeball. This process allows them to make a fairly precise map of the parts of an object that are capturing an infant's interest and the course of the baby's scanning activity (Maurer & Maurer, 1976).

Corneal photography reveals that newborn infants, when viewing an object such as a triangle, tend to focus on a relatively limited portion of the figure. Furthermore, if they are shown a black triangle on a white field, the infants' eyes hover on the corners of the triangle, where the contrast between black and white is strongest. Because infants do not usually scan the sides of the tri-

angle, it is doubtful that they perceive the figure in its entirety (Salapatek, 1975; Salapatek & Kessen, 1966). Although infants of 2 to 4 months of age perceive the parts of a figure, they do not bring them together in a figure arrangement until they are about 4 to 5 months of age. If they are shown a cross inside a circle, younger infants are more likely to see it as a cross and a circle rather than as a cross within a circle. But as they mature, they increasingly respond to the whole rather than to the individual parts (Linn et al., 1982).

Perception of the Human Face   One of the infant's visual preferences is for the human face. This interest is highly adaptive, because an adult face is a critical element in the infant's natural environment (Toda & Fogel, 1993). Robert L. Fantz (1963) found, for instance, that infants from 10 hours to 5 days old looked longer at a facelike stimulus than they did at newsprint, a bull's-eye, or different color disks. There is some question, however, whether the face is interesting to infants during the first 10 weeks of life because it is a face or because it is a complex object (Maurer & Salapatek, 1976). However, newborns ranging from 12 to 36 hours of age produce significantly more sucking responses in order to see an image of their mothers' faces than they do to see an image of strangers' faces (Walton, Bower, & Bower, 1992).

Daphine Maurer and Philip Salapatek (1976) found, in a study employing corneal photography, that 2-month-old infants tend to inspect the external contour of the face, usually devoting long periods of time to a particular area, such as the hairline, chin, or ear. One-month-old babies apparently can discriminate the faces of their mothers from the faces of strangers, probably by differences in the hairline or chin.

At about 5 to 7 weeks of age a dramatic change occurs in face looking. Infants invariably inspect one or more internal features of the face, especially the eyes. Indeed, talking to a child increases its scanning in the eye area. It is very likely that this activity carries special social meaning for the infant's caretakers and enhances parent-infant bonding (Haith, Bergman, & Moore, 1977).

Current research suggests that infants can detect the horizontally paired eyes in the upper part of the head by the third to fourth month. The mouth becomes differentiated by the fifth month and the broader facial configuration by the fifth or sixth month. And by the sixth to seventh month, infants come to recognize individual faces (Caron et al., 1973; Gibson, 1969). And they prefer faces that we as adults typically define as being "attractive" (Langlois et al., 1991). Such findings are consistent with Eleanor Gibson's (1969) view that object perception in infancy begins first with the differentiation of an object's parts and later progresses to the larger structure in which the parts are embedded. In sum, infants have already developed rather sophisticated perceptual capabilities by the time they reach 6 or 7 months of age (Younger, 1992).

Hearing At the time of birth, the hearing apparatus of the neonate is remarkably well developed. Indeed, the human fetus can hear noises three months before birth (Shahidullah, Scott, & Hepper, 1993; Birnholz & Benacerraf, 1983). However, for several hours or even days after delivery, the neonate's hearing might be somewhat impaired. Vernix and amniotic fluid frequently stop up the external ear passage, while mucus clogs the middle ear. These mechanical blockages disappear rapidly after birth.

Educators have long recognized that hearing plays an important part in the process by which children acquire language. But research by William S. Condon and Louis W. Sander (1974a, 1974b) purporting to show that newborns are attuned to the fine elements of adult speech took the scientific community by surprise. The researchers made videotapes of interactions between neonates and adults and minutely analyzed them frame by frame. To ordinary viewers, the hands, feet, and head of an infant appear uncoordinated, clumsily flexing, twitching, and moving about in all directions. But Condon and Sander say closer examination reveals that the infant's movements are synchronized with the sound patterns of the adult's speech. For example, if an infant is squirming about when an adult begins to talk, the infant coordinates the movements of brows, eyes, limbs, elbows, hips, and mouth to start, stop, and change with the boundaries of the adult's speech segments (phonemes, syllables, or words). The newborns, who were from 12 hours to 2 days old, were equally capable of synchronizing their movements with Chinese or English.

Condon and Sander conclude that if infants, from birth, move in precise, shared rhythm with the speech patterns of their culture, then they participate in millions of repetitions of linguistic forms long before they employ them in communication. By the time children begin to speak, they have already laid down within themselves the form and structure of their people's language system. However, other researchers have not been able to replicate the Condon-Sander findings. Indeed, John M. Dowd and Edward Z. Tronick (1986) conclude that speech-movement synchrony requires reaction times inconsistent with an infant's limited motor abilities and that the methodology used by Condon and Sander is flawed in many ways. Admittedly, it is exceedingly difficult to determine the level of infant auditory capability, and the available evidence is ambiguous (Werner et al., 1992). So again we encounter considerable controversy surrounding the question "How much do infants know, and when do they know it?"

Taste and Smell Both taste (gustation) and smell (olfaction) are present at birth. Infant taste preferences

can be determined by measuring sucking behavior (Blass & Ciaramitaro, 1994). Young infants relax and suck contentedly when provided with sweet solutions, although they prefer sucrose over glucose (Engen, Lipsitt, & Peck, 1974). Infants react to sour and bitter solutions by grimacing and breathing irregularly (Rosenstein & Oster, 1988; Jensen, 1932). The findings for salt perception is less clear; some researchers find that newborns do not discriminate a salty solution from nonsalty water, whereas other investigators find that salt is a negative experience for newborns (Bernstein, 1990). Charles K. Crook and Lewis P. Lipsitt (1976) found that newborns decrease their sucking speed when receiving sweet fluid, which suggests that they savor the liquid for the pleasurable taste. This would indicate that the hedonistic aspects of tasting are present at birth (Acredolo & Hake, 1982).

The olfactory system is unique both in what it responds to and in how it responds. As an environmental monitor, it seems to operate rather cautiously and prefers the familiar to the novel. Much of the time the olfactory system monitors the environment without the organism being aware of the process. The familiar simply fades into the background. But should novel odors come into sensory range, the system promptly brings them to conscious awareness. It is this feature—alerting the organism to potential danger and so increasing the organism's chances for survival—that accounts for the system's evolutionary value (Engen, 1991).

Infants respond to different odors, and the vigor of the response corresponds to the intensity and quality of the stimulant. Trygg Engen, Lewis P. Lipsitt, and Herbert Kaye (1963) tested olfaction in 2-day-old infants. At regular intervals they held a cotton swab saturated with anise oil (which has a licorice smell) or asafetida (which smells like boiling onions) under an infant's nose. A polygraph recorded the babies' bodily movements, respiration, and heart rate. When they first detected an odor, infants moved their limbs, their breathing quickened, and their heart rate increased. With repeated exposure, however, infants gradually came to disregard the stimulant. The olfactory thresholds decreased drastically over the first few days of life, meaning that the neonates became increasingly sensitive to nasal stimulants. Other researchers have confirmed that neonates possess well-developed olfactory abilities (Rieser, Yonas, & Wikner, 1976; Self, Horowitz, & Paden, 1972).

Cutaneous Senses Heat, cold, pressure, and pain—the four major cutaneous sensations—are present in neonates (Humphrey, 1978). Kai Jensen (1932) found that a bottle of hot or cold milk (above 124 degrees Fahrenheit or below 72 degrees Fahrenheit) caused an irregular sucking rhythm in neonates. On the whole, however, neonates are relatively insensitive to small differences in thermal stimuli. Neonates also respond to body pressure. Touching activates many of the reflexes discussed earlier in the chapter. Finally, we infer from infants' responses that they experience sensations of pain. For instance, observation of neonate and infant behavior suggests that gastrointestinal upsets are a major source of discomfort. As infants receive required vaccination injections during the first few years of life, it is quite obvious that they sense pain and discomfort (Izard et al., 1987). And male infants increase their crying during circumcision, providing additional evidence that neonates are sensitive to pain.

Circumcision   Each year within a few days of birth, more than one million U.S. male newborns undergo an operation that many medical professionals now believe is unnecessary. Circumcision is the surgical removal of the foreskin (prepuce) that covers the tip (glans) of the penis. Through the years this procedure has been a religious rite for Jews and Muslims. Among some African and South Pacific peoples, circumcision is performed at puberty to mark the passage of a youth to adulthood. In contrast, circumcision has never been common in Europe. Until about 20 years ago, U.S. physicians promoted circumcision as a health measure and a protection against cancer of the penis (and, in female sexual partners, cancer of the cervix). The procedure was also viewed as a means to prevent venereal disease and urinary tract infections. Because physicians now maintain that there are few valid medical indications for routine circumcision of the newborn, the practice has been on the decline.

Interconnections Among the Senses   Our sensory systems commonly operate in concert with one another. We expect to see things we hear, feel things we see, and smell things we taste. We often employ information we gain from one sensory system to "inform" our other systems (Acredolo & Hake, 1982). For instance, even newborns move both their head and their eyes in efforts to locate the source of sounds, especially when the sounds are patterned and sustained.

Developmental psychologists, psychobiologists, and comparative psychologists have advanced two opposing theories about how the interconnections among systems evolve. Take the development of sensory and motor coordination in infants. One viewpoint holds that infants only gradually achieve an integration of eye-hand activities as they interact with their environment. In the process of adapting to the larger world, infants are seen as progressively forging a closer and sharper coordination between their sensory and motor systems. According to Jean Piaget, infants initially lack cognitive structures for knowing the external world. Consequently, they must actively construct mental schemes that will allow them to structure their experience.

The opposing theory holds that eye-hand coordination is biologically prewired in the infant's nervous system at birth and emerges according to a maturational schedule. This interpretation is favored by T. G. R. Bower (1976). Bower finds that newborn infants engage in visually initiated reaching. Apparently, when neonates look at an object and reach out for it, both the looking and the reaching are part of the same response by which the infants orient themselves toward the object (von Hofsten, 1982).

In ensuing months, however, an increase in visual guidance occurs during the approach phase of reaching (Ashmead et al.,1993). Perhaps a cautious conclusion to be drawn from this research is that the eye-hand coordination of newborns and very young infants is biologically prewired. But visual guidance in eye-hand coordination becomes more important as older babies monitor and progressively reduce the "gap" between the seen target and the seen hand (McKenzie et al., 1993). So for this later sort of reaching, the youngster must attend to its hand. In sum, eye-hand coordination changes early in life from using the felt hand to using the seen hand (Bushnell, 1985). These findings suggest that some behavior and intellectual development do not occur in a strictly cumulative and incremental manner. Development is often characterized by patterns of skill acquisition, loss, and reacquisition on new foundations and levels.

Segue

One fact stands out in any consideration of infants' developing motor and sensory abilities: Infants actively search out and respond to their environment (von Hofsten, 1993). Their natural endowment predisposes them to begin learning how the world about them operates. As they mature, they refine their ability to take information from one sense and transfer it to another. All the senses, including seeing, hearing, and touch, create a system that is a whole. Information gained from multiple systems is often more important than that gained from one sense, precisely because it is interactive (Stein & Meredith, 1993). A growing body of contemporary research points to the multicausal, fluid, contextual, and self-organizing nature of developmental change, the unity of motor behavior and perception, and the role of exploration and selection in the emergence of new behavior (Thelen, 1995). In Chapter 5 we will turn our attention to cognition and maturing intellectual abilities, such as use of language, which grows from the roots in motor behavior, sensation, and perception that we have discussed in this chapter.

Summary

Birth

1. The period of gestation for a human baby from conception to birth is around 266 days, nearly 9 months, from the first day of the last menstrual cycle.

2. There are a variety of approaches that help a mother prepare for delivery. Lamaze classes help the mother-to-be to prepare for an unmedicated delivery. Other women know they have to schedule a cesarean delivery because of their own preexisting health condition, such as diabetes.

3. Some women choose to deliver in hospital birthing rooms, in birthing centers, or in home delivery with an experienced midwife or doula attending. The majority of American women are attended by obstetricians who specialize in prenatal care, birth, and postbirth care.

4. Most American hospitals offer family-centered hospital care in which childbirth can be a family experience. Natural (prepared) childbirth and rooming in are common features of this program.

5. Frederick Leboyer captured popular attention with his opinion that infant birth needs to be a gentler entry into the world, advocating lower sound and light levels, a warmer delivery room, newborn massage, and a warm bath. However, most authorities believe that ordinary stresses of birth do not exceed the infant's physical or neurological capacity.

6. The birth process consists of three stages: labor, delivery, and afterbirth. At the beginning of labor, the amniotic sac ruptures, releasing the amniotic fluid that has cushioned the fetus. During the several hours of labor, the strong muscle fibers of the mother's uterus contract rhythmically, pushing the baby toward the birth canal. The first periodic contractions normally are about 15 to 20 minutes apart and come more quickly and intensely as delivery nears. Delivery begins when the baby's head passes through the cervix and ends with the passage of the baby through the birth canal. Birth concludes when the mother's body expels the remaining umbilical cord and placenta, termed afterbirth.

7. In a small minority of cases, complications arise during pregnancy and childbirth. The purpose of good prenatal care under medical supervision is to minimize complications. If complications should develop, however, much can be done through medical intervention and technology to help the mother and infant. Among the possible complications are births that require cesarean delivery and premature births. Larger hospitals today have neonatology units to help these premature, very-low-birthweight, or at-risk infants survive.

8. The average birthweight for babies is around 7 pounds 8 ounces (3.4 kilograms). However, other health factors are assessed at 1 minute and 5 minutes after birth using the Apgar scale: heart rate, respiratory effort, muscle tone, alertness, and skin tone. Another scale, the Brazelton Neonatal Behavioral Assessment Scale (NBAS), might be used during the first week to assess four categories of development. The newborn is now called a neonate, and the first several weeks of life are called the neonatal period.

9. Parent-infant bonding is now considered to be a process of interaction and mutual attention between parents and their child that occurs over a period of time, whether the child is delivered by natural birth, delivered by cesarean, or adopted.

10. Because the birth rate for unmarried women has risen dramatically since the early 1980s, there is extensive research being conducted on the effects of fatherlessness on children. Children of single parents are more likely to live in poverty and to have no regular source of health care. Normally, both mothers and fathers contribute to the childhood experience.

11. Although American infants are born without complications, a very small percentage are born stillborn or die shortly after birth. A small percentage are born with drug addictions and experience withdrawal; these include "crack babies" and babies with fetal alcohol syndrome (FAS). A smaller percentage of babies are born with HIV. Others are born with genetic disorders or experience birth complications. Infants who are determined to be at risk are eligible for early-intervention services.

12. After birth, some mothers experience postpartum depression (PPD), which might include crying spells, depression, sleep changes, appetite changes, anxiety, and thoughts of not being able to cope with taking care of the baby. Maternal depression has a significant impact on the infant.

Development of Basic Competencies

13. Sleeping, crying, and feeding are the chief behaviors of the newborn. An infant's responses at any given time are related to its state. The following states have been identified in the neonate: regular sleep, irregular sleep, drowsiness, alert inactivity, waking

activity, and crying. Pediatricians and social scientists are examining cultural attitudes toward co-sleeping with infants; their findings might be useful for preventing sudden infant death syndrome (SIDS). The main recommendations offered in the United States for preventing SIDS include placing the infant on its back to sleep and avoiding smoking around a baby.

14. The infant's hunger and sleep patterns are closely linked. Newborns spend much of their waking time in feeding. Both breast-feeding and bottle-feeding offer advantages and disadvantages. The newborn is equipped at birth with a number of reflexes— behavioral systems that are readily activated.

15. Not all body systems grow at the same rate: (a) The nervous system grows more rapidly than other systems. Environmental stimulation and emotional comfort stimulate brain growth. (b) At 12 years of age, a child's lymphoid tissue is more than double the level it will reach in adulthood; after 12 it declines until maturity. (c) The reproductive system grows very slowly until adolescence, at which point its growth accelerates. (d) The skeletal and internal organ systems show two growth spurts, one in early infancy and the other at adolescence.

16. Development follows two patterns: the cephalocaudal principle and proximodistal principle. The sequence of motor development proceeds in accordance with the cephalocaudal principle. Children gain mastery first over the head muscles, then the trunk muscles, and finally the leg muscles.

17. The child's development of manual skills proceeds through a series of orderly stages in accordance with the proximodistal principle—from the center of the body toward the periphery. On the whole, large-muscle control precedes fine-muscle control.

18. Young infants display bursts of rapid, repeated, rhythmic movements of the limbs, torso, and head. The behaviors are closely related to motor development and provide the foundation for later, more skilled actions.

19. The development of locomotion proceeds in a specific sequence, but children vary in their rate of development. First they lift the head and then the chest; then they sit, crawl, and creep; then stand and cruise around furniture; and eventually they walk unaided, typically between 11 months to 15 months.

20. Perception is oriented toward things, not toward their sensory features. This fact is demonstrated by visual constancies: color constancy, size constancy, and shape constancy.

21. The visual cliff experiment reveals that children possess depth perception at a very early age. Over the first two years of life, infants typically undergo a patterned sequence of changes in their method of focusing on and organizing visual events.

22. The eye-hand coordination of newborns and very young infants apparently is biologically prewired. The eye-hand coordination of somewhat older babies, however, is accomplished by monitoring and progressively reducing the "gap" between the seen target and the seen hand. Eye-hand coordination, then, changes early in life from using the felt hand to using the seen hand. Taste and smell are present at birth, as are the cutaneous senses of heat, cold, pressure, and pain.

23. Infants are actively searching out and responding to their environment. Information gained from their senses is interactive.

Key Terms

afterbirth *(106)*

anoxia *(112)*

Apgar scoring system *(108)*

birth *(106)*

birthing centers *(105)*

birthing rooms *(105)*

cesarean section (C-section) *(112)*

colic *(123)*

crowning *(106)*

delivery *(106)*

entrainment *(119)*

failure to thrive *(120)*

infancy *(119)*

infant mortality *(117)*

labor *(106)*

lightening *(106)*

locomotion *(129)*

midwifery *(105)*

natural childbirth *(104)*

neonate *(106)*

norms *(126)*

obstetrician *(104)*

parent-infant bonding *(109)*

postmature infant *(114)*

postpartum depression (PPD) *(118)*

premature infant *(113)*

psychoprophylactic method *(104)*

reflex *(120)*

rooming in *(105)*

shaken-baby syndrome (SBS) *(121)*

small-for-term infant *(113)*

states *(120)*

stillbirth *(117)*

sudden infant death syndrome (SIDS) *(120)*

Following Up on the Internet

Web sites for this chapter focus on preparation for birth and delivery, neonate assessment, infant development, and parent-infant attachment. Please access the text web site at http://www.mhhe.com/crandell7 for up-to-date hot-linked Internet addresses for the following organizations and resources.

March of Dimes Perinatal Center

ParenthoodWeb

Childbirth.Org

Major Components of Appraisal by Ed Hammer, Ph.D. of the Behavioral Pediatrics Clinic, Texas Tech School of Medicine

Neonatology on the Web

SIDS Network

Chapter 5

Infancy
Cognitive and Language Development

Critical Thinking Questions

1. Learning is defined in terms of three criteria: There must be some change in behavior, this change must be relatively stable, and the change must result from experience. Can learning take place if only two of the three conditions are present? Why or why not?

2. It has been said that even if animals could speak, we would not understand them. Do you agree? Why or why not?

3. Most U.S. children learn to comprehend more than 14,000 words between the ages of 2 and 6; they average 6 to 9 new words per day. Why do they learn at that rate, and why don't we continue to learn at that rate when we are older?

4. Speech develops through a series of stages, ending in the construction of 2- and 3-word sentences for the beginning language learner. Do you think you could learn a second language easier and faster today if you went through a similar process? In other words, could you learn French if you were restricted to 2- and 3-word sentences for the first two years of learning the language at college? At what age do you think children should start learning a second language?

Our cognitive and language abilities are probably our most distinctive features as human beings. Cognitive skills enable us to gain knowledge of our social and physical environment. Language enables us to communicate with one another. Without either, human social organization would be impossible. Even if we lacked these abilities, we might still have families, for the family organization is not peculiar to human beings—it appears elsewhere in the animal kingdom. But without cognitive and language abilities, our families would probably not have the structure we recognize as typically human. We would lack rules about incest, marriage, divorce, inheritance, and adoption. We would have no political, religious, economic, or military organizations; no codes of morality; no science, theology, art, or literature. We would have virtually no tools. In sum, we would be without culture, and we would not be human (White, 1949). This chapter surveys the processes by which cognition and language develop during the early years of infancy, from birth through age 2.

We do not live to think, but, on the contrary, we think in order that we may succeed in surviving.

José Ortega y Gasset

Cognitive Development

Man is obviously made to think. It is his whole dignity and his whole merit.

Pascal, *Pensées*

As discussed in Chapter 2, cognition refers to the process of knowing and encompasses such phenomena as sensation, perception, imagery, retention, recall, problem solving, reasoning, and thinking. We receive raw sensory information and transform, elaborate, store, recover, and use this information in our daily activities (Neisser, 1967).

Making Connections

Mental activity allows us to "make something" out of our perceptions. We do so by relating some happening to other events or objects in our experience. We use information from our environment and our memories to make decisions about what we say and do. Because these decisions are based on information available to us and on our ability to process the information intelligently, we view them as rational (Anderson, 1990). This capacity allows us to intervene in the course of events with conscious deliberation.

For instance, if we show youngsters aged 13 to 24 months the simple steps involved in "making spaghetti" with clay, a garlic press, and a plastic knife and then allow them to undertake the task for themselves, they are able to recall the sequence of events and repeat them—sometimes eight months later. Clearly, these youngsters are obtaining knowledge from their senses, imitating the actions of others, and remembering the information—all evidence of higher cognitive functioning. Indeed, a mounting body of evidence suggests that 16- and 20-month-olds are capable of organizing their recall of novel events around causal relations—they know that "what happens" occurs in such a way that one event ordinarily follows another event and that this same sequence of events will again unfold in the same manner in the future (Oakes, 1994; Bauer & Mandler, 1989).

Psychologists are increasingly coming to view infants as very complex creatures who are capable of experiencing, thinking about, and processing enormous amounts of information (Perris, Myers, & Clifton, 1990). Building on the developing competencies detailed in the previous chapter, infants begin to form associations be-

tween their own behavior and events in the external world in the early months of their lives. As they do so, they progressively gain a conception of the world as an environment that possesses stable, recurrent, and reliable components and patterns. Such conceptions allow them to begin functioning as effective beings who cause events to happen in the world about them and who evoke social responses from others (Bakeman et al., 1990). Let us begin our exploration of these matters with a consideration of infant learning.

Learning: A Definition

Learning is a fundamental human process. It permits us to adapt to our environment by building on previous experience. Psychologists have traditionally defined learning in terms of three criteria:

- There must be some change in behavior.
- This change must be relatively stable.
- The change must result from experience.

Learning, then, involves a relatively permanent change in a capability or behavior that results from experience. As we discussed in Chapter 2, theories of learning fall into three broad categories:

1. *Behavioral theories* emphasize that people can be conditioned.
2. *Cognitive theories* focus on how to fashion the cognitive structures by which individuals think about their environment.
3. *Social learning theories* stress the need to provide models for people to imitate.

These three kinds of theories highlight some of the common influences on individuals that facilitate learning.

How Soon Do Infants Start Learning?

With the use of more sophisticated visual diagnostics during prenatal development and other sophisticated research conducted immediately after birth and during the first few days as a neonate, researchers are beginning to see that the fetus exhibits some early learning.

Learning in the Womb Psychologist Anthony De-Casper and his colleagues (1980) have investigated whether learning can occur among babies still in the womb. They believe they have found evidence that some kind of learning is occurring, although they do not know its exact mechanism. The researchers devised a nipple apparatus that activates a tape recorder. By sucking in one pattern, newborns would hear their own mother's voice; by sucking in another pattern, they would hear another woman's voice. The babies (some just hours old) tended to suck in a way that would allow them to hear their

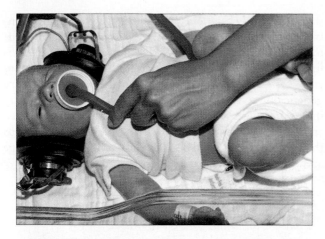

Can a Fetus Learn? This baby is participating in Anthony DeCasper's research on learning in fetuses and newborns. The infant sucks to hear a tape recording of his mother reading a story that she read aloud, on a regular basis, while pregnant. The rate of sucking is much higher for the familiar story than for another one.

mother's voice. The researchers concluded that the infants' preferences were affected by their auditory experiences before birth.

In other tests, 16 pregnant women read Dr. Seuss's *The Cat in the Hat* to their unborn children twice a day for the last six weeks of gestation—for a total of about five hours. After they were born, the infants were allowed to choose, by means of their sucking behavior, to hear either a recording of their mother reading *The Cat in the Hat* or a recording of their mother reading stories by other authors having a different meter. The infants chose to hear *The Cat in the Hat.* Since this initial research, other studies have indicated that in the last few months of pregnancy the fetus can hear and distinguish sounds (Kisilevsky, 1995; Lecanuet, Granier-Deferre, & Busnell, 1995).

Some parents have interpreted these findings to mean that they can give their infants a developmental head start by teaching them things before they are born. For instance, they might try to create a musical prodigy by playing classical music to the fetus. An example is the recent marketing of the Tummy Tutor: this device plays music and speech in various languages, and the pregnant woman wears it on her abdomen. Although 20 years ago we gave fetuses and newborns little credit for having cognitive and learning capabilities, today the pendulum might be swinging optimistically too far in the opposite direction. We need to remind ourselves that much of the work in this area remains conjecture and that it is usually wise to resist the urge to apply preliminary research findings (Cohen et al., 1988; Weiss, Zelazo, & Swain, 1988). Many of the responses that babies show are adaptations to specific stimuli for which newborns appear to

be biologically prepared (Sameroff & Cavanagh, 1979), so in a moment we will take another look at Piaget's work, particularly his notions regarding the sensorimotor period.

Newborn Learning Developmental psychologists have long been interested in knowing whether newborns can learn—or, more particularly, whether they can adjust their behavior according to whether it succeeds or fails. Arnold J. Sameroff (1968) conducted a study on infant learning involving neonatal sucking techniques and tentatively suggests that the answer is yes. It is generally recognized that two nursing methods are available to newborns—*expression* involves pressing the nipple against the roof of the mouth with the tongue and squeezing milk out of it, and *suction* involves creating a partial vacuum by reducing the pressure inside the mouth and thus pulling the milk from the nipple. Sameroff devised an experimental nipple that permitted him to regulate the supply of milk an infant received. He provided one group of babies with milk only when they used the expressive method (squeezing the nipple); he gave the second group milk only when they used the suction method.

He found that the infants adapted their responses according to which technique was reinforced. For instance, the group that was given milk when they used the expressive method diminished their suction responses—indeed, in many cases they abandoned the suction method during the training period. In a second experiment Sameroff (1968) was able to induce the babies, again through reinforcement, to express milk at one of two different pressure levels. These results suggest that learning can occur among 2- to 5-day-old, full-term infants. Various other researchers have likewise demonstrated learning in newborns (Cantor, Fischel, & Kaye, 1983; DeCasper & Carstens, 1981).

Question

In what ways have researchers demonstrated that a fetus and a neonate can learn?

Piaget: The Sensorimotor Period

Man's mind stretched to a new idea never goes back to its original dimensions.

Oliver Wendell Holmes

As we saw in Chapter 2, the Swiss developmental psychologist Jean Piaget contributed a great deal to our understanding of how children think, reason, and solve problems. Perhaps more than any other person, Piaget was responsible for the rapid growth of interest in cognitive development over the past four decades. In many respects the breadth, imagination, and originality of his work overshadowed other research in the field.

Piaget charted a developmental sequence of stages during which the child constructs increasingly complex notions of the world, and he described how the child acts at each level and how this activity leads to the next level. His most detailed analysis was of the first two years of life, which he calls the "sensorimotor period." In Piaget's terminology, **sensorimotor** refers to the coordination of motor activities with sensory inputs (perceptions), which is the major tasks of the sensorimotor period. In this period of development, babies develop the capacity to look at what they are listening to and learn to guide their grasping and walking by visual, auditory, or tactile cues. In sum, the infant comes to *integrate* the motor and perceptual systems. This integration lays the foundation for the development of new adaptive behaviors.

A second characteristic of the sensorimotor period is that babies develop the capacity to view the external world as a permanent place. Infants fashion a notion of **object permanence**—they come to view a thing as having a reality of its own that extends beyond their immediate perception of it. As adults, we take this notion for granted. However, infants do not necessarily do so during the first six to nine months in the sensorimotor period. However, sometime after six to nine months, a baby becomes capable of searching for an object that an adult has hidden under a cloth. The child will search for it on the basis of information about where the object went. In so doing, the infant understands that the object exists even when it cannot be seen. This developing ability provides a fixed point for constructing conceptions of space, time, and cause.

According to Piaget, a third characteristic of the sensorimotor period is the inability of infants to represent the world to themselves internally. They are limited to the immediate here and now. Because they cannot fashion symbolic mental representations of the world, they "know" the world only through their own perceptions and their own actions upon it. For example, children in the sensorimotor stage know food only as something they can eat and manipulate with their fingers, and they cannot conceive of it apart from these activities. Infants have a mental picture of food only insofar as actual sensory input reveals the food's existence. This mental picture disappears when the sensory input ceases. According to Piaget, infants are unable to form a static mental image of food "in their heads" in the absence of the actual visual display. "Out of sight, out of mind" is an appropriate description of how the infant perceives the external world during the sensorimotor stage. In sum, in the sensorimotor period infants coordinate the ways they interact with their environment, give the environment permanence, and begin to "know" the environment, although their knowledge of the environment is limited to their sensual interactions with it. The infant enters the sensorimotor period with more than twenty genetically given reflexes (for instance, give any healthy infant an object and the infant will grab it, and it often ends up in the child's mouth). The child then enters into the next developmental period ready to develop language and other symbolic ways of representing the world.

> **Question**
>
> *What are an infant's major cognitive accomplishments during the sensorimotor period, and at approximately what months do these typically occur?*

Neo- and Post-Piagetian Research

Piaget's work has stimulated other psychologists to investigate children's cognitive development. They have been intrigued by the idea that infants do not think about objects and events in the same ways adults do, and they have particularly studied object permanence in infants (Demetriou, Efklides, & Platsidou, 1993; Bjorklund & Green, 1992). This ongoing work is revising and refining Piaget's insights. For example, researchers have found that infants possess a set of object search skills more sophisticated than Piaget had imagined (Rochat & Striano, 1998). Many of the errors youngsters make in searching for items do not reflect an absence of basic concepts of objects and space—even by 4 months of age they might understand that an object continues to exist when their view of it is blocked, but they might not yet be capable of coordinating their movements to search for it (Baillargeon & DeVos, 1991; Yates & Bremner, 1988).

Playing Is Learning Moreover, developmental psychologists find that children do not develop an interest in objects and object skills in a social vacuum. Rather, caretakers can set the stage for youngsters by "playing" with them, giving babies clues as to what they should do and when they should do it (Bornstein & O'Reilly, 1993; Bruner, 1991). Additionally, by playing with infants, parents provide experiences that youngsters cannot generate by themselves (Vygotsky, 1978).

In the course of such activities, infants acquire and refine their capacities for intersubjectivity, so that by the end of the first year they share attention, emotional feelings, and intentions with others. All the while infants gain a sense of their society's culture and gain some of the skills essential to living in that culture. Caregivers—

the curators of culture—transmit the knowledge, attitudes, values, and behaviors essential for effective participation in society, and they help to gradually transform infants into genuine social beings capable of manipulating objects and acting in concert with others. In playing with their youngsters, caretakers provide sociocultural guidance for the children's later cognitive and language performance (Bakeman et al., 1990). However, this is often not the case with youngsters whose mothers suffer from clinical depression.

Consequences of Maternal Depression Clinically depressed mothers might have such debilitating symptoms that they are virtually incapable of fulfilling their children's needs. Clinical depression is an emotional disorder characterized by a mood drop that can last for months, even years. As depression deepens, it commonly involves insomnia, disinterest in work, low energy, loss of appetite, reduced sexual desire, persistent sadness, hopeless feelings, and profound overall emotional despair; even routine tasks become difficult to perform. Additionally, many depressed people report difficulty concentrating, remembering things, and getting their thoughts together. Some also suffer considerable anxiety as part of their depression. Many factors have been implicated in clinical depression. In some women, depression is complicated by drug abuse, and some 10 to 12 percent of new mothers experience prolonged postpartum depression ("Mirror Images," 1998).

There is a direct relationship between the severity of a mother's depression and the quality of the care she provides for her youngsters (Seifer & Sameroff, 1987). Health professionals report that depressed mothers often appear sad, are given to frequent sighs, fail to interact playfully with their youngsters, seem insensitive to their babies' needs, and focus their gaze downward. Clinical depression also affects the quality and organization of the home environment (Abrams et al., 1995; Egeland, Jacobvitz, & Sroufe, 1988).

Because depressed mothers have a reduced capacity for caregiving and nurturing, their youngsters tend to lag behind in their emotional, language, and social development. The babies of depressed women are more withdrawn, unresponsive, and inattentive than other youngsters. They may cry and fuss a good deal, appear apathetic and listless, have problems sleeping and feeding, and fail to grow normally (failure to thrive). Matters can be complicated by a mother's disciplinary ineptness. For example, depressed mothers might alternately ignore their children and lash out at them with strict prohibitions. Such behavior is baffling to youngsters, and they might respond by being negative, challenging limits, resenting punishment, and becoming unusually argumentative. A vicious cycle is likely to ensue wherein this difficult behavior reinforces the mother's depression and

sense of parental inadequacy. School-age children of depressed women might exhibit negative self-concepts, poorer peer skills, hyperactivity, academic problems, and childhood psychopathology (Gelfand & Teti, 1990).

Clinical depression is usually a treatable disorder that responds to antidepressant medication and other psychiatric interventions. However, before depression can be treated, it must first be recognized and brought to the attention of appropriate medical personnel and therapists.

Bruner on Modes of Cognitive Representation

One of the first U.S. psychologists to appreciate the importance of Piaget's work was Jerome S. Bruner. Bruner is a distinguished psychologist in his own right who has served as president of the American Psychological Association. Many of his research papers show a strong Piagetian influence, especially in the way he treats the stages of cognitive development.

Through the years, however, Bruner and Piaget developed differences of opinion about the roots and nature of intellectual growth. Most particularly, the two disagreed over Bruner's (1970) view that "the foundations of any subject may be taught to anybody at any age in some form." Piaget, in contrast, held to a rigorous stage approach, in which knowledge of certain subjects can be gained only when all the components of that knowledge are present and properly developed.

One of Bruner's primary contributions to our understanding of cognitive development concerns the changes that occur in children's favored modes for representing the world as they grow older (Bruner, Oliver, & Greenfield, 1966). According to Bruner, at first (during the time period that Piaget called the sensorimotor period) the representative process is *enactive*: Children represent the world through their motor acts. In the preschool and kindergarten years, *ikonic* representation prevails: Children use mental images or pictures that are closely linked to perception. In the middle school years, the emphasis shifts to *symbolic* representation: Children use arbitrary and socially standardized representations of things; this enables them to internally manipulate the symbols that are characteristic of abstract and logical thought. Thus, according to Bruner, we "know" something in three ways:

- Through doing it (enactive)
- Through a picture or image of it (ikonic)
- Through some symbolic means such as language (symbolic)

Take, for instance, our "knowing" a knot. We can know the knot by tying it; we can have a mental image of the knot as an object on the order of either a pretzel or "bunny ears" (or a mental "motion picture" of the knot being formed); and we can represent a knot linguistically by combining four alphabetical letters, *k-n-o-t* (or by linking utterances in sentences to describe the process of tying string). Through these three such general means, human beings increase their ability to achieve and use knowledge.

Continuity in Cognitive Development from Infancy

Psychologists have long been interested in knowing whether mental competence and intelligence later in life can be predicted from cognitive performance in infancy. Until relatively recently, psychologists believed that there was little continuity between early and later capabilities. But now they are increasingly concluding that individual differences in mental performance in infancy are, to a moderate extent, developmentally continuous across childhood and perhaps beyond (Haith & Benson, 1998; Colombo, 1993).

Decrement and Recovery in Attentiveness   *Information-processing models* of intelligence have contributed to this reassessment. For people to mentally represent and process information concerning the world about them, they must first pay attention to various aspects of their environment. Two components of attention seem most indicative of intelligence in youngsters:

- *Decrement of attention*—losing interest in watching an object or event that is unchanging
- *Recovery of attention*—regaining interest when something new happens

Youngsters who tire more quickly when looking at or hearing the same thing are more efficient processors of information. So are those who prefer the novel over the familiar. These infants typically like more complex tasks, show advanced sensorimotor development, explore their environment rapidly, play in relatively sophisticated ways, and solve problems rapidly. Similarly, the rapidness with which adults learn something is associated with measures of their intelligence. Given equivalent opportunities, more intelligent people learn more than less intelligent people in the same amount of time. Not surprisingly, then, psychologists are finding that decrement and recovery of attention seem to predict childhood cognitive competence more accurately than do more traditional tests of infant development.

The patterns children reveal in attending to information reflect their cognitive capabilities and, more particularly, their ability to construct workable schemas (à la Piaget) of what they see and hear. As Harriet L. Rheingold (1985) points out, mental development proceeds through transformations of novelty into familiarity. Each thing in the environment begins as something new,

and so development progresses as infants turn something new into something known. In turn, once you know something, it provides a context for recognizing what is new, and so the known provides the foundation for further mental development. Both the familiar and the novel, then, compel attraction and are reciprocal processes central to lifetime adaptation (see *Information You Can Use:* "Reducing Retardation Rates and Boosting Babies' Brain Power").

Language and Thought

Language is the armory of the human mind, and at once contains the trophies of its past and the weapons of its future conquests.

Samuel Taylor Coleridge

Human beings are set apart from other animals by their possession of a highly developed system of language communication. This system allows them to acquire and transmit the knowledge and ideas provided by the culture in which they live. To be sure, a number of scientists claim that skills characteristic of the use of language have been developed in a dozen or so chimpanzees (Savage-Rumbaugh et al., 1993). But although the skills exhibited by the chimps are clearly related to human skills, they are hardly equivalent to our intricate and subtle language capacity. And the methods by which chimps typically must be trained are quite different from the spontaneous ways in which children learn a language. Apes learn to deal with signs sluggishly and often only after being plied with bananas, cola, and M&Ms (Gould, 1983). **Language** is a structured system of sound patterns (words and sentences) that have socially standardized meanings. Language provides a set of symbols that rather thoroughly catalog the objects, events, and processes in the human environment (DeVito, 1970). Humans process and interpret language in the left cerebral cortex (see Figure 5-2).

The Functional Importance of Language

Language is by its very nature a communal thing; that is, it expresses never the exact thing but a compromise—that which is common to you, me and everybody.

Thomas Ernest Hulme

Language makes two vital contributions to human life: It enables us to communicate with one another (interindividual communication), and it facilitates individual

Information You Can Use

Reducing Retardation Rates and Boosting Babies' Brain Power

At one time it was thought that the brain is hardwired and that its wiring cannot be changed. However, a mounting body of evidence now suggests that enriched environments can produce physical changes in the developing brain. Imaging studies using techniques like PET scans reveal a positive correlation between positive environmental changes and an increase in synaptic connections among brain cells. "Pushing certain buttons in the brain" through enrichment experiences—good nutrition, toys, playmates, learning opportunities, and parental counseling—can prevent a substantial amount of mental retardation and developmental disability: *Early intervention* can make the future brighter for many youngsters whose development otherwise would be stunted (Shonkoff et al., 1992).

These findings have led many parents to wonder if they too can boost their baby's intellectual development. Psychologists have long noted that good parenting can have profound and positive consequences for youngsters (Hart & Risley, 1992). The emotional quality of the parent-infant relationship certainly plays a key part in children's early cognitive and language competence. Parental behaviors affect infants' competence in a number of ways (Olson, Bates, & Bayles, 1984). First, children's learning is directly enhanced if parents provide them with immediate positive feedback when they say or do novel, creative, or adaptive things. Second, children's developmental competence is encouraged when parents provide a nonrestrictive environment that allows them to engage in exploratory behavior. And third, children who are securely attached to their caretakers are more apt than others to undertake competent exploration of their surroundings. Effective parents are aware of their children's developmental needs and guide their own behavior to meet these needs.

These findings have had both positive and negative influences. On the positive side, the research results encourage interventions with infants who are at risk for delayed cognitive development, particularly those from impoverished homes (Seitz, Rosenbaum, & Apfel, 1985). The key interventions involve encouraging parents to increase their levels of positive verbal and object-centered communication with their offspring (Olson, Bates, & Bayles, 1984). For instance, regarding a crying or defiant infant, parents can be helped to determine why the behaviors are occurring, to infer the needs, motivations, and limitations in the youngster that likely underlie the behaviors, and to select the most appropriate responses (Dix, Ruble, & Zambarano, 1989).

Continued on page 146

Assisting Youngsters Who Have Developmental Delays
A new and exciting body of scientific evidence (including that provided by brain scanning, such as PET scans) suggests that the connection between the "brain" and the "mind" is a two-way street. For instance, scientists have found that behavioral cognitive therapy techniques not only help some participants with their psychological problems but also change their brains' physical structures. In brief, participants who learn via a series of behavioral techniques to resist various destructive urges end up altering their brains. Such evidence has encouraged others to pursue new research and avenues for assisting children with developmental delays. Modeling behavior helps this child learn appropriate mouth movements for better expressive language skills.

Continued from page 145

On the negative side, the research findings have caused some parents to indulge in what educational psychologists call "hot-housing" or trying to "jump-start" youngsters toward success. The image of a toddler calling out "Five!" when peering at five red dots on a white flash card or reading aloud from *The Cat in the Hat* brings joy to the hearts of many parents. Yet too many parents are pushing very young children too hard to gain academically oriented skills (see Figure 5-1). Thus far, the only proven beneficiaries of preschool programs have been culturally deprived youngsters. Many children who are pressured to learn through inappropriate methods begin to dislike learning. Young children learn best from their own experience—from self-directed activity, exploring real objects, talking to people, and solving real-life problems, such as how to balance a stack of blocks. And they seem to benefit from having stories read to them on a regular basis. When caretakers intrude in children's self-directed learning and insist on their own priorities for their learning, such as mathematics or reading, they interfere with children's own impulses and initiative. Parents and other caretakers, then, must take into consideration the style of learning appropriate for the very young (Elkind, 1987).

"Oh, yes, indeed. We all keep a sharp eye out for those little clues that seem to whisper 'law' or 'medicine.'"

Figure 5-1 Pushing Children Too Hard to Gain Academic Skills Some parents want their children to get a head start on language or math skills through early enrichment and stimulation.

thinking (intraindividual communication). The first contribution, called **communication,** is the process by which people transmit information, ideas, attitudes, and emotions to one another. This feature of language allows human beings to coordinate complex group activities. They fit their developing lines of activity to the developing actions of others on the basis of the "messages" they provide one another. Thus, language provides the foundation for family and for economic, political, religious, and educational institutions—indeed, for society itself.

Language has enabled human beings, alone of all animals, to transcend biological evolution. Evolutionary processes took millions of years to fashion amphibians—creatures that can live on land or in water. In contrast, a second kind of "amphibians"—astronauts who can live in the earth's atmosphere or in the space outside it—have "evolved" in a comparatively short time (Brown & Herrnstein, 1975). But in the second case, human anatomy did not alter so that it could survive in space—rather, human beings increased their knowledge to the point where they could employ it to complement and supplement their anatomy; in this manner they made themselves spaceworthy.

The second contribution of language is that it facilitates *thought* and other cognitive processes. Language enables us to encode our experiences by assigning names to them. It provides us with concepts by which we dissect the world around us and categorize new information. Thus, language helps us to partition the environment into manageable areas that are relevant to our concerns. Language also allows us to deal with past experiences and to anticipate future experiences through reference. It enlarges the scope of our environment and experience. It is this second function, the relation of language to thought, that has been the subject of intense debate. According to one view, language is merely *the con-*

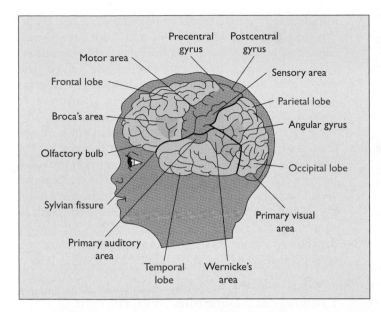

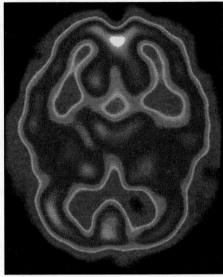

Red in PET scan shows brain's most active areas.

Figure 5-2 Where and How Babies Process Sound and Language The drawing at the left shows the left hemisphere of the cerebral cortex. The temporal lobes of each hemisphere interpret the sounds heard by each ear. Often one side is more "dominant" in hearing than the other, though both sides hear a given sound. For most individuals speech-language are located in Broca's and Wernicke's areas. The child is also learning language by observing a speaker's mouth and facial expressions and is also using the visual/occipital cortex. It really becomes a whole brain effort! Note the full brain involvement in language on the PET scan to the right. An infant's experience with hearing the sounds of a language actively shapes the physical structure of the brain. After a few years of experience hearing one language, a child is able to distinguish the sounds typically heard and loses the ability to distinguish those sounds not used often. In Asian cultures, the sounds for L and R are likely to give older youth and adults difficulty in distinguishing those sounds. PET scans have shown that such sounds are decoded in distinctly separate parts of the brains of those of us who speak English but in the same part of the brain for those from Asian cultures.

tainer of thought. The contrary view holds that language is the *determinant of thought.* Let us examine each of these positions more closely.

Language as the Container of Thought

Language is the dress of thought.

Samuel Johnson

Those who hold that language is the container of thought say that thought takes place independently of language. They believe that words are necessary not for thought but only for conveying it to others. For instance, some types of thought are primarily nonverbal visual images and "feelings." Probably you are most aware of language as a vehicle for conveying thought when someone asks you to describe something—your mother, the view from your room, the main street in your hometown. You seek to translate a mental picture into words. But often you find that the task of verbally describing pictures is complex and difficult.

Piaget (1962, 1952) took the view that structured language presupposes the prior development of other kinds of mental representation. On the basis of his studies,

Piaget concluded that language has only a limited role in young children's mental activity. According to Piaget, children form *mental images* of objects (water, food, a ball) and events (drinking, sucking, holding), and these images are based on mental reproduction or imitation, not on word labels. Thus, the children's task in acquiring words is to map language onto their preexisting concepts.

In our discussion in the pages ahead, we will see that Piaget vastly oversimplifies matters. In some areas, representation does precede language. For instance, William Zachry (1978) finds that solid progress in mental representation is necessary for some forms of language production. He suggests that children gain the ability to represent motor schemes (certain generalized activities) internally as images. Thus, the various actions associated with bottles would come to be represented by such mental pictures as holding a bottle, sucking a bottle, pouring from a bottle, and so on. Later, the child comes to represent the "bottle activities" by the word *bottle.* The word *bottle* then becomes a semantic "marker" that represents the qualities associated with a bottle—holdable, suckable, pourable, and so on. In this matter, Zachry says, words come to function as semantic markers for mental pictures.

Researchers also find that children approach the task of "word learning" equipped with preexisting cognitive

Acquiring Language Although maturational processes may "ready" children for language use, they do not guarantee that children will acquire the capacity to use language. Youngsters also require social interaction that encourages them to develop and refine their language capabilities.

biases that lead them to prefer some possible meanings over others (Reber, 1993; Hall, 1991). More specifically, children seem predisposed to "whole object" meanings for nouns—they assume that a new noun refers to an entire object rather than to one of its parts. By way of illustration, consider the noun *dog*. Youngsters must learn that *dog* can refer both to a specific object (for instance, Fido) as well as to the category Dog, yet the word *dog* does not apply to individual aspects of the object (for instance, its nose or tail); relationships between the object and other objects (for example, between a dog and its toy), or the object's behavior (for instance, the dog's eating, barking, or sleeping)

Were children to weigh these and countless other possible meanings before arriving at the correct mapping of the word *dog*, they would be overwhelmed by an unmanageable sea of dog-related inputs. Youngsters need not follow such a laborious route. They bring a bias in thinking with them, one that allows "fast mapping"—the capturing of the basic elements of an experience to the exclusion of other experiences. In addition, children are cognitively biased toward an assumption that words refer to mutually exclusive categories (Taylor & Gelman, 1989). So we see that *some* aspects of linguistic development are linked to a preexisting level of conceptual development.

It also seems that infants as young as 4 months of age possess the ability to partition the color spectrum into four basic hues—blue, green, yellow, and red. For example, infants respond differently to two wavelengths selected from adjacent adult hue categories, such as "blue" at 480 millimicrons and "green" at 510 millimicrons. However, infants do not respond differently to two wavelengths selected from the same adult hue category

although separated by a similar physical distance (30 millimicrons), such as "blue" at 450 and "blue" at 480 millimicrons. It seems, then, that the mental representations of infants are organized into blue, green, yellow, and red—rather than as exact wavelength codes, which, for adults, makes up the color spectrum (Bornstein & Marks, 1982). Only later do the children come to name these categories. Such findings suggest that color organization precedes and is not a product of the categories (the verbal labels *blue, green, yellow,* and *red*) provided by language and culture (Soja, 1994). More recent research confirms that infants spontaneously form categories during the prelinguistic period (Roberts, 1988). In some respects, then, children's knowledge of language depends on a prior mastery of concepts about the world to which words will refer (Coldren & Colombo, 1994).

Language as a Determinant of Thought

The second viewpoint is that language develops parallel with, or even prior to, thought. According to this viewpoint, language shapes thought. This theory follows in the traditions of George Herbert Mead (1934), Benjamin L. Whorf (1956), and Lev Vygotsky (1962).

This perspective emphasizes the part that concepts play in our partitioning stimuli into manageable units and into areas that are relevant to our concerns. Through **conceptualization**—grouping perceptions into *classes* or *categories* on the basis of certain similarities—children and adults alike can identify and classify informational input. Without our ability to categorize, our lives would be chaotic. By virtue of categories, objects need not be treated as unique. In using categories, adults and infants "tune out" certain stimuli and "tune in" others (Needham & Baillargeon, 1998). Consequently, they are able to view the same object as being the same despite the fact that the object varies from perspective to perspective and from moment to moment. And people are able to treat two different but similar objects as equivalent—as being the same kind of thing. In enabling human beings to make the mental leap from the specific to the general, categories give us the basis for more advanced cognitive thinking. Significantly, researchers find that language increases the time infants look at objects beyond the time that the actual verbal labeling occurs, suggesting that infants are biased to look at objects in the presence of language (Baldwin & Markman, 1989).

Concepts also perform a second service. They enable individuals to go beyond the immediate information provided to them. People can mentally manipulate concepts and imaginatively link them to fashion new adaptations. This attribute of concepts allows human beings to make additional inferences about the unobserved properties of objects and events (Bruner, Goodnow, & Austin, 1956). Human beings have an advantage over other animals in that we can use words in the conceptualization process. Some social and behavioral scientists claim that the activity of naming, or verbally labeling, offers three advantages:

- Facilitating thought by producing linguistic symbols for integrating ideas
- Expediting memory storage and retrieval via a linguistic code
- Influencing perception by sensitizing people to some stimuli and desensitizing them to others

Critics contend, however, that it is easy to oversimplify and overstate the relationship between language and various cognitive processes. Eric H. Lenneberg (1967) and Katherine Nelson (1972) note that often children's first words are names of preexisting cognitive categories. As pointed out in the previous section, color organization in infants precedes learned categories provided by language. The suggestion here is that language is not the sole source for the internal representation on which thought depends. Nor is language the sole source for the representation of information in memory (Perlmutter & Myers, 1976). And language has at best only a minor impact upon perception.

The language-as-the-container-of-thought viewpoint argues that thought shapes language, while the language-as-a-determinant-of-thought viewpoint asserts that language shapes thought. Even though these perspectives stand in sharp contrast, many linguists and psychologists believe that the two views are not mutually exclusive—that language and thought reciprocally influence and shape each other, operating not as well-bounded separate spheres but in tandem (Pinker, 1991; Acredolo & Goodwyn, 1988). Indeed, the interaction

The Role of Naming Naming—verbally labeling—facilitates the process by which children come to put a "handle" on their experiences and render them meaningful. Accordingly, language expedites much mental activity and functioning. Notice this child's ability to use the pointing gesture (nonverbal language).

between the two is exceedingly complex, because thought (cognition) and language are each composed of many separate, underlying skills and mechanisms (Bloom, 1993; Yamada, 1991). The problem in some respects parallels the learning and nativist controversy surrounding the acquisition of language, a matter to which we now turn our attention.

Question

Historically, there have been two opposing views in scientific circles regarding the relationship between language and thought. The first view is that thought takes place independently of language and that language is merely the container of already established thought. The second is that language shapes thought by providing the concepts or categories into which individuals mentally sort their perceptual stimuli. Is someone who is both blind and deaf without thoughts—or even without language?

Theories of Language Acquisition

But I gotta use words when I talk to you. But here's what I was going to say.

T. S. Eliot, *Sweeney Agonistes*

How are we to explain the development of speech in children? Is language acquired through learning processes? Or is the human organism biologically "preprogrammed" for language usage? These questions expose a nerve in the long-standing nature-versus-nurture controversy, with environmentalists and nativists (hereditarians) vigorously and heatedly disagreeing on their answers.

Learning and Interactionist Theories

Growing numbers of psychologists are exploring the language environment in which infants and children are reared (Hoff-Ginsberg, 1986; Wexler & Culicover, 1981). Some have followed in the tradition of B. F. Skinner (1957), who argues that language is acquired in the same manner as any other behavior, namely, through learning processes of reinforcement (Hayes & Hayes, 1992). Others have studied the interaction between caretakers and youngsters that contributes to the acquisition of language (Baumwell, Tamis-LeMonda, & Bornstein, 1997). Indeed, language use might begin quite early.

As we noted earlier in the chapter, DeCasper's research suggests that babies might have a sensitivity to speech that starts even before birth. While they are in the uterus, we believe, they hear "the melody of language." After birth this sensitivity provides them with clues about which sounds belong together. The ability of neonates to discriminate between speech samples spoken in their mother's native language and in an unfamiliar language could derive from the unique melodic qualities found in the linguistic signals to which they were exposed prenatally (Fernald, 1990). Other researchers suggest that babies intensively tune in to the subtleties of their native language for the first six to eight months before refining their listening practices and that they eventually come to ignore sounds that do not exist in their native language (Werker & Stager, 1997).

Caretaker Speech   Much recent research has focused on caretaker speech. In caretaker speech mothers and fathers systematically modify the language that they employ with adults when addressing infants and young people. **Caretaker speech** differs from everyday speech in its simplified vocabulary, higher pitch, exaggerated intonation, short simple sentences, and high proportion of questions and imperatives. Parents use caretaker speech with preverbal infants in numerous European languages, Japanese, and Mandarin Chinese (Fernald & Morikawa, 1993; Papousek, Papousek & Symmes,1991). For their part, young infants show a listening preference for caretaker speech with its higher overall pitch, wider pitch excursions, more distinctive pitch contours, slower tempo, longer pauses, and increased emphatic focus (Cooper & Aslin, 1990; Fernald, 1985).

Speech characterized by the first two characteristics of caretaker speech—simplified vocabulary and higher pitch—is termed *baby talk.* Baby talk has been documented in numerous languages, from Gilyak and Comanche (languages of small, isolated, preliterate Old World and New World communities) to Arabic and Marathi (languages spoken by people with literary traditions). Furthermore, adults phonologically simplify vocabulary for children—"wa-wa" for *water,* "choo-choo" for *train,* "tummy" for *stomach,* and so on. Baby talk also serves the psychological function of marking speech as affectionate (Moskowitz, 1978).

The Interactional Nature of Caretaker Speech   The interactional nature of caretaker speech actually begins with birth (Rheingold & Adams, 1980). Hospital staff, both men and women, speak to the newborns in their care. The speech focuses primarily on the baby's behavior and characteristics and on an adult's own caretaking activities. Moreover, the caretakers speak as though the infants understand them. Their words reveal that they view the newborns as persons with feelings, wants, wishes, and preferences. Similarly, a burp, smile, yawn, cough, or sneeze typically elicits a comment to the child from the caretaker (Snow, 1977). Often the utterances

are in the form of questions, which the caretakers then answer as they imagine the children might respond. If a baby smiles, a parent might say, "You're happy, aren't you?" Or if the child burps, the caretaker might say, "Excuse me!" Indeed, caretakers impute intention and meaning to infants' earliest behavior, making the babies appear more adept than they in fact are. These imputations facilitate children's language acquisition much in the manner of self-fulfilling prophecies. Infants with depressed mothers are handicapped in this respect because their mothers are less likely to use the exaggerated intonation contours of "motherese" (see below) and because their mothers are slower to respond to their early attempts at vocalization (Bettes, 1988).

Motherese   When infants are still in their babbling phase, adults often address long, complex sentences to them. But when infants begin responding to adults' speech, especially when they start uttering meaningful, identifiable words (at around 12 to 14 months), caretakers invariably speak what is called **motherese**—a simplified, redundant, and highly grammatical sort of language.

When speaking motherese, parents tend to restrict their utterances to the present tense, to concrete nouns, and to comments on what the child is doing or experiencing. And they typically focus on what objects are named ("That's a doggie!" or "Johnnie, what's this?"), the color of objects ("Bring me the yellow ball. The yellow ball. No, the yellow ball. That's it. The yellow ball!"), and where objects are located ("Hey, Lisa! Lisa! Where's the kitty? Where's the kitty? See. On the steps. See over there on the steps!"). The pitch of the caretaker's voice is correlated with the child's age: The younger the child, the higher the pitch of speech. In ad-

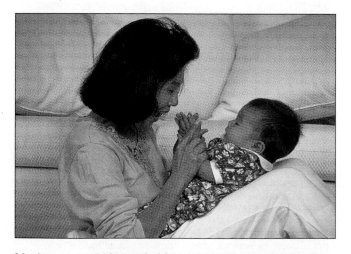

Motherese and Nonverbal Language Caretakers typically speak to infants in a simplified, redundant, and highly grammatical sort of language. Babies communicate to caretakers through their own nonverbal cues before they can speak any words.

dition, the intonation of infant-directed motherese—the melody inherent in mother speech—offers more reliable cues of a speaker's communicative intent than does speech directed by adults to other adults (Sokolov, 1993; Fernald, 1990). Motherese seems to derive less from parents' intent to provide brief language lessons than from their efforts to communicate to their youngsters. And as we will see later in this chapter, infants also use intonation effectively to express desires and intentions before they master conventional phonetic forms (Lewis, 1936/1951). Briefly restated, caretaker speech is simple, high-pitched, and used to talk to preverbal infants, while motherese is employed when the caretaker assumes that the infant can begin to interact with the environment.

Innateness Theory

Research by Noam Chomsky (1980, 1968, 1965, 1957), Eric H. Lenneberg (1969, 1967), Peter D. Eimas (1985), and Steven Pinker (1994) focuses not on the part that learning plays in language acquisition, but rather on the biological endowments that human beings bring to the environmental context. Such views are held by *nativists*, who view human beings as having evolved in ways that make some kinds of behavior, like language acquisition, easier and more natural than others.

Youngsters are said to begin life with the underpinnings of later speech perception and comprehension, as they begin life with the specialized anatomy of the vocal tract and the speech centers in the brain. Nativists contend that human beings are "prewired" by their brain circuitry for language use—that the potential for the language acquisition has been "built into" human beings by genes and only needs to be elicited by an appropriate "triggering mechanism" in the same way that nutrition triggers growth. Even deaf children manifest a natural inclination to develop a structured means of communication (Kolata, 1992c)—see *Human Diversity:* "Deaf Children Invent Their Own Sign Language."

Limitations of Learning and Interactionist Theories

In many respects the nativist position has been shaped, sharpened, and clarified as its proponents have done battle against learning and interactionist theories. According to nativists, the major inadequacies of learning and interactionist theories include these:

- *Children Acquire Language with Little Difficulty.* Even very young children master an incredibly complex and abstract set of rules for transforming strings of sounds into meanings. By way of illustration, consider that there are 3,628,800 ways to rearrange the 10 words in the following sentence:

Try to rearrange any ordinary sentence consisting of ten words.

However, only one arrangement of the words is grammatically meaningful and correct. Nativists say that a youngster's ability to distinguish the one correct sentence from the 3,628,799 incorrect possibilities cannot arise through experience alone (Allman, 1991). Likewise, consider how formidable a foreign language such as Japanese or Arabic seems to you.

- *Adult Speech Is Inconsistent, Garbled, and Sloppy.* Reflect for a moment on how a conversation carried on in an unfamiliar language sounds to you; probably more like one giant word than neat packages of words. Or listen to a conversation between two adults; it is full of false starts, "ums," and many "filler phrases" such as "you know." Indeed, linguists have experimentally shown that even in our own language we often cannot make out a word correctly if it is taken out of context. From recorded conversations linguists splice out individual words, play them back to people, and ask the people to identify the words. Listeners can generally understand only about half the words, although the same words were perfectly intelligible to them in the original conversation (Cole, 1979).
- *Children's Speech Is Not a Mechanical Playback of Adult Speech.* Children combine words in unique ways and also make up words. Expressions such as "I buyed," "foots," "gooder," "Jimmy hurt hisself," and the like, reveal that children do not imitate adult speech in a strict fashion. Rather, according to nativists, children are fitting their speech to underlying language systems with which they are born and so exceptions are not initially mastered.

Chomsky's Theory of Language Development

Noam Chomsky, a linguist at the Massachusetts Institute of Technology, has provided a nativist theory of language development that has had a major impact on education and psychology over the past 40 years (Chomsky, 1975, 1968, 1965, 1957). Supporters and critics alike acknowledge that Chomsky's theoretical formulations have provided many new directions in the study of linguistics.

Central to Chomsky's position is the observation that mature speakers of a language can understand and produce an infinite set of sentences, even sentences they have never before heard, read, or uttered and therefore could not have learned. The explanation for this, argues Chomsky, is that human beings possess an inborn language-generating mechanism, which he terms the **language acquisition device (LAD)**. Chomsky sees the human brain as wired to simplify the chaos of the auditory world by sorting through incoming frequencies and

Human Diversity

Deaf Children Invent Their Own Sign Language

Oral language (speech) is part of the environment for most infants, but not for deaf children. Nonetheless, deaf infants manage to create a sign language of their own—stereotyped gestures that refer to objects around them. Susan Goldin-Meadow and Heidi Feldman (1977) studied six deaf children ranging in age from 17 to 49 months. The children's parents had normal hearing. Despite their children's deafness, the parents wanted the children to depend on oral communication. Consequently, they did not expose the children to a manual sign language. The researchers observed the children in their homes at periodic intervals. At the time they were studied, the children had learned only a few spoken words. In contrast, each child had individually developed a language-like system of communication that included properties found in the language of hearing children.

The children would indicate first the object to be acted on, next the action itself, and finally the recipient of the action (should there be one). For instance, one child pointed at a shoe and then pointed at a table to request that the shoe (the object) be put (the act) on the table (the recipient). On another occasion the child pointed at a jar (the object) and then produced a twisting motion (the act) to comment on how his mother had opened the jar. Interestingly, even when the children were playing alone, they employed signs to "talk" to themselves, as hearing children would.

Once the researchers had determined that the children had acquired a sign language, their next task was to discover who had first elaborated the signs, the children or their parents. The researchers concluded that most of each child's communication system originated with the child and not invented by the parents. Some of the children used complex combinations of words before their parents did with them. Moreover, although the parents produced as many different characterizing signs as the children, only about a quarter of the signs were common to both parties.

Goldin-Meadow and Feldman (1977, p. 403) were especially impressed by the deaf children's achievements, in comparison with the ability of Lucy, Washoe, and other chimpanzees to employ sign language:

> While chimpanzees seem to learn from manual language training, they have never been shown to spontaneously develop a language-like communication system without such training—even when the chimp is lovingly raised at a human mother's knee. On the other hand, even under difficult circumstances, the human child reveals a natural inclination to develop a structured communication system.

Additional research by Goldin-Meadow with deaf children has confirmed that children have a strong bias to communicate in language-like ways (Goldin-Meadow & Mylander, 1984). Other researchers find that deaf babies of deaf parents babble with their hands in the same rhythmic, repetitive fashion as do hearing babies who babble with their voices (Petitto & Marentette, 1991). Sounds such as "goo-goo" and

shunting speech sounds into 40 or so intelligible *phonemes* (the smallest units of language). In the process of language acquisition, children merely need to learn the peculiarities of their society's language, not the basic structure of language. Although Chomsky's theory has attracted a good deal of attention as well as controversy, it is difficult to test by established scientific procedures and hence remains neither verified nor disproved.

In support of his view, Chomsky points out that the world's languages differ in *surface structure*—for example, in the words they use. But they have basic similarities in their composition, which he calls "deep structure." The most universal features of *deep structure* include having nouns and verbs and the possibility of posing questions, giving commands, and expressing negatives. Chomsky suggests that through preverbal, intuitive rules—*transformational grammar*—individuals turn deep structure into surface structure, and vice versa.

He says such transformational grammar is biologically built into the functioning of the human organism. Chomsky does not claim that a child is genetically endowed with a specific language (English, French, Chinese); he simply maintains that children possess an inborn capacity for generating productive rules (grammar) for language and that this capacity is what goes to work when a young child is first learning a language.

Questions

Environmentalists argue that language is learned. Nativists insist that human beings possess an inborn language-generating mechanism. Can you think of another domain regarding which these two schools of thought might have the same argument? Is language unique in that it is uniquely human?

Learning Sign Language Deaf children pass through the same maturing stages that are observed in the vocalizations of hearing children. In addition, research suggests that babbling, whether manual or vocal, is an inherent feature of the maturing brain as it acquires language.

"da-da-da" that hearing babies make arise at around the same time as the babbling signs and motions arise among deaf youngsters. Most of the hand motions of the deaf infants are actual elements of American Sign Language—gestures that do not in themselves mean anything but that have the potential to indicate something when pieced together with other gestures.

Significantly, deaf children also pass through the same stages that are observed in the vocal babbling of hearing children: They string together signs and motions in much the same way that hearing youngsters string together sounds. Such gestures appear to have the same functional significance as the babble noises of hearing babies, for they are far more systematic and deliberate than the random finger flutters and fist clenches of hearing youngsters. Such findings suggest that language is distinct from speech and that speech is only one of the signal systems available to us for communicating with one another. In addition, the research suggests that babbling, whether manual or vocal, is an inherent feature of the maturing brain as it acquires the structure of language.

A Resolution of Divergent Theories

Most psychologists agree that there is a biological basis for language, but they continue to disagree over how much the input from parents and other caretakers matters. The most satisfactory approach seems to be one that looks to the strengths of each theory and focuses on the complex and many-sided aspects of the development of language capabilities. Indeed, language acquisition cannot be understood by examining learning or genetic factors in isolation. No aspect by itself can produce a language-using human. Instead of asking which factor is most important, we need to study the ongoing process by which the factors dynamically come together.

In conclusion, infants are biologically adapted to acquire language. They possess a genetically determined plan that leads them toward language usage. Their attentional and perceptual apparatus seems biologically pre-

tuned to make phonetic distinctions. But simply because human beings possess a biological predisposition for the development of language does not mean that environmental factors play no part in language acquisition. Indeed, language is acquired only in a social context (Huttenlocker et al., 1991). Youngsters' earliest vocalizations, even their cries, are interpreted by caregivers, who in turn use these interpretations to determine how they will respond to the youngsters.

Language Development

What is involved in learning to talk? This question has fascinated people for centuries. The ancient Greek historian Herodotus reports on the research of Psammetichus, ruler of Egypt in the seventh century B.C.—the first attempt at a controlled psychological experiment in recorded history. The king's research was based on the notion that vocabulary is transmitted genetically and that children's babbling sounds are words from the world's first language:

> Psammetichus . . . took at random, from an ordinary family, two newly born infants and gave them to a shepherd to be brought up amongst his flocks, under strict orders that no one should utter a word in their presence. They were to be kept by themselves in a lonely cottage, and the shepherd was to bring in goats from time to time, to see that the babies had enough milk to drink, and to look after them in any other way that was necessary. All these arrangements were made by Psammetichus because he wished to find out what word the children would first utter. . . .The plan succeeded; two years later the shepherd, who during that time had done everything he had been told to do, happened one day to open the door of the cottage . . . [and both children ran up to him and] pronounced the word "becos." (Herodotus, 1964, pp. 102–103)

When the king learned that the children had said "becos," he undertook to discover the language to which the word belonged. From the information produced by his inquiries, he concluded that "becos" was the Phrygian word for "bread." As a consequence, the Egyptians reluctantly yielded their claim to being the most ancient people and admitted that the Phrygians surpassed them in antiquity.

Communication Processes

Becoming competent in a language is not simply a matter of employing a system of rules for linking sounds and meaning (Ellis, 1992; Feyereisen & de Lannoy, 1991). Language also involves the ability to use such a system for communication and furthermore to keep such systems separate when one has access to two or more (see *Further Developments:* "Bilingualism").

Nonverbal Communication or Body Language   The essence of language is the ability to talk to one another. Yet spoken language is only one channel or form of message transmission. We also communicate by body language (also termed "kinesics"), which is the nonverbal communication of meaning through physical movements and gestures. For instance, we wink an eye to demonstrate intimacy; we lift an eyebrow in disbelief; we tap our fingers to show impatience; we rub our noses or scratch our heads in puzzlement.

We also communicate by gaze: We look at the eyes and face of another person and make eye contact. One way we use gazing is in the sequencing and coordination of speech. Typically, speakers look away from a listener as they begin to talk, shutting out stimulation and planning what they will say. At the end of the utterance, they look at the listener to signal that they have finished and

are yielding the floor; and in between they give the listener brief looks to derive feedback information.

By the end of the second year, most children seem to pattern their eye contact in the way adults in their environment do: looking up when they are done speaking to signal that they are finished and looking up when the other person is through speaking to confirm that the floor is about to be offered back to them. Even so, children show considerable differences in the consistency and frequency of these behaviors (Rutter & Kurkin, 1987). And some groups actually avoid eye contact, such as African Americans, who view staring as aggressive, and American Indians, who view staring as rude.

Another nonverbal behavior is pointing. Pointing can be observed in infants as young as 2 months old, though pointing at such a young age is not an intentional act (Trevarthen, 1977). In contrast, pointing at the end of the first year is an intentional act (Fogel & Thelen, 1987). Pointing is a *nonverbal* precursor of language. Mothers commonly employ pointing when talking to their youngsters. Children use the gesture to mark out features of a book or to call attention to an activity.

Another form of communication is *paralanguage*—the stress, pitch, and volume of vocalizations by which we communicate expressive meaning. Paralanguage involves *how* something is said, not *what* is said. Tone of voice, pacing of speech, and extralinguistic sounds (such as sighs) are examples of paralanguage. By the late babbling period, infants already control the intonation, or pitch modulation, of their utterances (Moskowitz, 1978). By about 9 months, new social cognitive competencies emerge in joint attention, social referencing, and communicative gestures (Tomasello, 1995).

Most of the research on language development has focused on language production, the ability of children to string together sounds so as to communicate a message in a meaningful fashion. Until recently, little research dealt with language *reception*, the quality of receiving or taking in messages. Yet children's receptive capacities tend to outdistance their productive capabilities. For instance, even very young babies are able to make subtle linguistic discriminations—as between the sounds *p* and *b* (Eimas, 1985).

Older children similarly make finer distinctions in comprehension than they reveal in their own language productions (Bates, Bretherton, & Snyder, 1988). Consider the now somewhat classic conversation the linguist Roger Brown had with a young child (Moskowitz, 1978): The child made reference to "fis," and Brown repeated "fis." The child was dissatisfied with Brown's pronunciation of the word. After a number of exchanges between Brown and the child, Brown tried "fish," and the child, finally satisfied, replied, "Yes, fis." Although the child was as yet unable to pronounce the distinction between *s* and *sh*, he knew that such a sound difference did exist.

Further Developments
Bilingualism

In overall cognitive capabilities, children are less proficient than adults (Bjorklund & Green, 1992). For instance, adults are less egocentric and process information more rapidly than children do. But the picture is quite different for language acquisition. Newborns are language universalists. It seems that normal, healthy youngsters can learn any sound in any language and distinguish among the vocal sounds that human beings utter. In contrast, adults are language specialists. They have considerable difficulty perceiving speech sounds that are not in their native tongue. For instance, Japanese infants can distinguish between the English sounds *la* and *ra*, but Japanese adults cannot because their language does not contrast these sounds (Kuhl et al., 1992).

Some linguists contend there is a *critical period* of language acquisition (see Chapter 2) and that language learning occurs primarily in childhood. Some aspects of the nervous system seemingly lose their plasticity with age, so that by the onset of puberty the organization of the brain is basically fixed, making the learning of a new language difficult (Lenneberg, 1967).

Given this state of affairs, it is hardly surprising that proficiency in a second language is related to the age at which training in the language begins (Johnson & Newport, 1989). Put another way, adults who learned a second language early in childhood are more proficient with the language than

adults who learn a second language later in life. Similar results are found for people with hearing deficiencies who learn American Sign Language as their first language (Newport, 1990).

Evidence suggests that, regarding the ability to acquire a second language, there is a gradual decline across childhood rather than a sudden discontinuity at puberty. Indeed, the decline begins very early in life. Patricia K. Kuhl and her colleagues (1992) report that experience alters sound perception by 6 months of age. For instance, experiments with 6-month-old babies in the United States and Sweden reveal that American youngsters routinely ignore the different pronunciations of the *i* sound because in the United States they hear the same sound. But American infants can distinguish slight variations in *y* sounds. The reverse is true of Swedish babies—they ignore variations in *y* sounds but notice variations in the *i* sound.

These findings have substantial implications for educators. The best time to learn a new language is early in life. The cognitive structures of young children seem especially suited for learning both a first and a second language. This ability is gradually lost across the childhood years. Although adults are able to acquire a second language, they rarely attain the same proficiency as individuals who acquire the language in childhood (Bjorklund & Green, 1992). The sooner bilingual education begins, the better.

As we shall see in later chapters on childhood, the large influx of immigrant children into the United States over the past decade has been cause for a flurry of research on bilingualism. A major controversy in many school districts across the country has been whether or not young children should be taught in their native language while learning the English language over a period of years—or whether non-English-speaking children should be immersed in the English language from the first days in preschool or kindergarten. The state of California recently changed its previous language-support policy in its schools and legislated that English must be the primary language of instruction in the schools.

Second-Language Acquisition in Very Young Children
This mother, hoping to take advantage of the ease with which young children learn a second language, has enrolled her child in language classes at a very young age.

Questions

You are recent immigrants into a new country. Your children are entering the school system and do not speak the language. How would you want your child to be taught, and what research findings would you use to support this approach? How do you think you would fare learning the new language yourself? Would you speak your native language at home with your children, or would you attempt to speak only the new language at home?

The Sequence of Language Development

Until a couple of decades ago, linguists assumed that children merely spoke an imperfect version of adult language, one that reflected a child's handicaps of limited attention, limited memory span, and other cognitive deficits. However, linguists now generally accept that children speak their own language—a language with characteristic patterns that develop through a series of stages (Tomasello, 1992).

Children reveal tremendous individual variation in the rate and form of language development (Reznick & Goldfield, 1992). Indeed, some children don't begin to talk until well into their third year, whereas others are producing long sentences at this point. Such variations do not appear to have implications for adult language skill, provided that the child is otherwise normal. Table 5-1 summarizes the typical milestones in language development of the "average" child.

From Vocalization to Babbling   Crying is the most noticeable sound uttered by the newborn. As we saw in Chapter 4, variations on the basic rhythm include the "mad" and "pain" cries. Although it serves as the infant's primary means of communication, crying cannot be considered true language. Young infants also produce a number of other sounds, including yawns, sighs, coughs, sneezes, and belches. Between the sixth and eighth week, infants diversify their vocalizations and, when playing alone, employ new noises, including "Bronx cheers," gurgling, and tongue noise games. Around their third month, infants begin making cooing sounds and squealing-gurgling noises, which they sustain for 15 to 20 seconds.

Babbling   In babbling, around the sixth month, infants produce sequences of alternating vowels and consonants that resemble one-syllable utterances, such as "da-da-da." Babies appear to engage in cooing and babbling for their own sake. Indeed, infants seem to play with sounds, enjoying the process and exploring their capabilities. Frequently babbled sounds consist of *n, m, y, w, d, t,* or *b,* followed by a vowel such as the eh sound in *bet.* It is probably no coincidence that in many languages the words for mother and father begin with these sounds (e.g., *mama, nana, papa, dada, baba*). Consonants such as *l, r, f,* and *v* and consonant clusters such as *st* are rare.

Deaf infants also go through the cooing and babbling phase, even though they might have never heard any spoken sounds. They babble in much the same fash-

Table 5-1   Milestones in Language Development

Age	Characteristic Sounds
1 month	Cries; makes small throaty noises.
2 months	Begins producing vowel-like cooing noises, but the sounds are unlike those of adults.
3 months	Cries less, coos, gurgles at the back of the throat, squeals, and occasionally chuckles.
4 months	Cooing becomes pitch-modulated; vowellike sounds begin to be interspersed with consonantal sounds; smiles and coos when talked to.
6 months	Vowel sounds are interspersed with more consonantal sounds (*f, v, th, s, sh, z, sz,* and *n* are common), which produce babbling (one-syllable utterances); displays pleasure with squeals, gurgles, and giggles, and displeasure with growls and grunts.
8 months	Displays adults' intonation in babbling; often uses two-syllable utterances such as "mama" or "baba"; imitates sounds.
10 months	Understands some words and associated gestures (may say "no" and shake head); may pronounce "dada" or "mama" and use holophrases (words with many different meanings).
12 months	Employs more holophrases, such as "baby," "bye-bye," and "hi"; many imitate sounds of objects, such as "bow-wow"; has greater control over intonation patterns; gives signs of understanding some words and simple commands (such as "Show me your nose").
18 months	Possesses a repertoire of 3 to 50 words; may begin using two-word utterances; still babbles, but employs several syllables with intricate intonation pattern.
24 months	Has repertoire of more than 50 words; uses two-word utterances more frequently; displays increasing interest in verbal communication.
30 months	Rapid acceleration in learning new words; speech consists of two or three words and even five words; sentences have characteristic child grammar and rarely are verbatim imitations of adult speech; intelligibility of the speech is poor, although children differ in this regard.
36 months	Has a vocabulary of some 1,000 words; about 80 percent of speech is intelligible, even to strangers; grammatical complexity is roughly comparable to colloquial adult language.
48 months	Language well established; deviations from adult speech are more in style than in grammar.

Adapted from F. Caplan, *The First Twelve Months of Life* (New York: Putnam Publishing Group, Inc., 1967), pp. 128–130. Copyright © 1967. By permission of copyright owner.

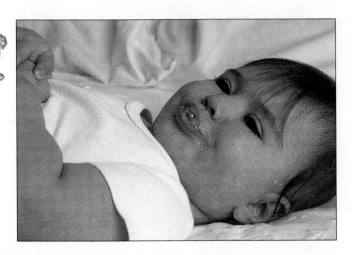

Early Communication: Cooing Around 3 months of age, infants enjoy making "cooing" sounds. These are usually simple vowel sounds with the mouth open, such as "ooooh," "eeeeeh," and "ahhhh." Typically before an infant begins to express harder consonant sounds by around 6 months, he or she will enjoy making "bubbles" with the saliva from the mouth, with simple spitting. The infant is learning to move the tongue and lips in readiness for language.

ion as normal infants, despite the fact that they cannot hear themselves (Petitto & Marentette, 1991; Lenneberg, 1967). This behavior suggests that a hereditary mechanism underlies the early cooing and babbling process. However, later on, deaf babies' babbling sounds have a somewhat more limited range than hearing children's. Furthermore, unless congenitally deaf children are given special training, their language development is retarded (Folven & Bonvillian, 1991).

These observations suggest that although vocal behavior emerges spontaneously, it flourishes only in the presence of adequate environmental stimulation. And deaf babies, of course, are incapable of "talking back," the process of vocal contagion and model imitation noted by Piaget. However, children do not seem to learn language simply by hearing it spoken. A boy with normal hearing but with deaf parents who communicated by the American Sign Language was daily exposed to television so that he might learn English (because the child suffered from asthma, he was confined to his home). Consequently, he interacted only with people at home who, as family members or visitors, communicated in sign language. By the time he reached 3 years of age, he was fluent in sign language but he neither understood nor spoke English. This observation suggests that in order to learn a language, children must be able to interact with people in that language (Hoff-Ginsberg & Shatz, 1982).

Receptive Vocabulary Between 6 to 9 months, caretakers will notice that the child understands some words. This is a favorite time for parents, who begin to ask the baby questions: "Where is Mommy's nose?" and the baby points to mommy's nose. Or Daddy might say, "Wave bye-bye to Daddy" as he leaves the house, and the child demonstrates understanding by waving bye-bye. A child who is not yet speaking can respond appropriately when asked to go get things, like his or her toys or the pots and pans in the kitchen. All of these actions demonstrate that children have developed a **receptive vocabulary** long before they speak their first word—that is, long before they have an *expressive vocabulary*.

Holophrases Most developmental psychologists agree that most children speak their first word at about 10 to 13 months of age. However, the precise age at which a child arrives at this milestone is often difficult to determine. The child's first word is so eagerly anticipated by many parents that they read meaning into the infant's babbling—for instance, they note "mama" and "dada" but ignore "tete" and "roro." Hence, one observer might credit a child with a "first word" where another observer would not. Behavioral theorists suggest that at this time parents reinforce, or reward, the infant with their smiles and encouragement. In turn, the child repeats the same expression over and over, such as "da," which quickly becomes "dada" (or what we interpret to be *daddy*). The *d* sound is easier for the infant to say; the *m* sound (as in "mama") requires the child to purse the lips together, so it might not be heard for another few months.

Children's first truly linguistic utterances are **holophrases**—single words that convey different meanings depending on the context in which they are used. Using a holophrase, a child can imply a complete thought. G. De Laguna (1929) first noted the characteristics of holophrases more than 60 years ago:

> It is precisely because the words of the child are so indefinite in meaning, that they can serve such a variety of uses. . . . A child's word does not . . . designate an object or a property or an act; rather it signifies loosely and vaguely the object together with its interesting properties and the acts with which it is commonly associated in the life of the child. . . . Just because the terms of the child's language are themselves so indefinite, it is left to the particular setting and context to determine the specific meaning for each occasion. In order to understand what the baby is saying, you must see what the baby is doing.

The utterance "mama," not uncommon in the early repertoire of English-language youngsters, is a good illustration of a holophrase. In one situation it may communicate "I want a cookie"; in another, "Let me out of my crib"; and in another, "Don't take my toy away from me." A holophrase is most often a noun, an adjective, or a self-invented word. Only gradually do the factual and emotional components of the infant's early words become clearer and more precise.

Nelson and colleagues (1978) found that children typically pass through three phases in their early learning of language. About 10 to 13 months of age, they become capable of matching a number of words used by adults to already existing concepts, or mental images, such as the concept "bottle" discussed earlier in the chapter. One study reveals that the average child of 13 months understands about 50 words; in contrast, the average child does not speak 50 words until six months later (Benedict, 1976). It is also interesting to note that caretakers have successfully taught preverbal children signs for words such as *more, cat,* and *hungry,* which they use in place of spoken language.

In the second phase, usually occurring between 11 and 15 months of age, children themselves begin to speak a small number of words. These words are closely bonded to a particular context or action.

Overextension   In the third phase—from 16 to 20 months—children produce many words, but they tend to *extend* or *overgeneralize* a word beyond its core sense. For instance, one child, Hildegard, first applied the word *tick-tock* to her father's watch, but then she broadened the meaning of the word, first to include all clocks, then all watches, then a gas meter, then a firehose wound on a spool, and then a bathroom scale with a round dial (Moskowitz, 1978). In general, children overextend meanings on the basis of similarities of movement, texture, size, and shape. Overgeneralization apparently derives from discrepancies between comprehension and production. For example, one child, Rachel, overextended *car* in her own verbal productions to include a wide range of vehicles. But she could pick out a motorcycle, a bicycle, a truck, a plane, and a helicopter in response to their correct names. Once her vocabulary expanded—once she acquired the productive labels for these concepts—the various vehicles began to emerge from the *car* cluster (Rescorla, 1976).

Children tend to first acquire words that relate to their own actions or to events in which they are participants (Shore, 1986). Nelson (1973) noted that children begin by naming objects whose most salient property is change—the objects do things like roll (ball), run (dog, cat, horse), growl (tiger), continually move (clock), go on and off (light), and drive away (car, truck). The most obvious omissions in children's early vocabulary are immobile objects (sofas, tables, chests, sidewalks, trees, grass).

Children also typically produce a holophrase when they are engaged in activities to which the holophrase is related. Marilyn H. Edmonds (1976, p. 188) observed that her subjects

> named the objects they were acting on, saying "ball" as they struggled to remove a ball from a shoe; they named where they placed objects, saying "bed" as they put their dolls to bed; they named their own actions, saying "fall" when they fell, they asserted possession, saying "mine" as they recovered objects appropriated by siblings; they denied the actions of their toys, yelling "no" when a toy cow fell over; and so forth.

Very often, a child's single-word utterances are so closely linked with action that the action and speech appear fused: In a Piagetian sense, a word becomes "assimilated" to an existing sensorimotor scheme—the word is fitted or incorporated into the child's existing behavioral or conceptual organization (see Chapter 2). It is as if the child has to produce the word in concert with the action. Edmonds (1976, p. 188) cites the case of a child at 21 months of age who said "car" 41 times in 30 minutes as he played with a toy car.

Two-Word Sentences   When they are about 18 to 22 months olds, most children begin to use two-word sentences. Examples include "Allgone sticky," said after washing hands; "More page," a request to an adult to continue reading aloud; and "Allgone outside," said after a door is closed behind the child ("allgone" is treated as one word, because "all" and "gone" do not appear separately in these children's speech). Most of the two-word sentences are not acceptable adult English sentences, and most are not imitations of parental speech. Typical constructions are "More wet," "No down," "Not fix," "Me drink," "Allgone lettuce," and "Other fix" (Clark, Gelman, & Lane, 1985; Braine, 1963). Two-word sentences represent attempts by children to express themselves in their own way through their own unique linguistic system.

As with holophrases, one often must interpret children's two-word sentences in terms of the context. Lois Bloom (1970), for instance, observed that one of her young subjects, Kathryn, employed the utterance "Mommy sock" in two different contexts with two different meanings. "Mommy sock" could mean that Mommy was in the act of putting a sock on Kathryn, or it could mean that Kathryn had just found a sock that belonged to Mommy.

Children's actual utterances are simpler than the linguistic structures that underlie them (Brown, 1973). Dan I. Slobin (1972, p. 73) observes that even with a two-word horizon, children can convey a host of meanings:

Identification: "See doggie."
Location: "Book there."
Repetition: "More milk."
Nonexistence: "Allgone thing."
Negation: "Not wolf."
Possession: "My candy."
Attribution: "Big car."
Agent-action: "Mama walk."
Agent-object: "Mama book" (meaning, "Mama read book").

Action-location: "Sit chair."
Action-direct object: "Hit you."
Action-indirect object: "Give papa."
Action-instrument: "Cut knife."
Question: "Where ball?"

Children also use intonation to distinguish meanings, as when a child says, "Baby chair" to indicate possession and "Baby chair" to indicate location.

Telegraphic Speech Children who begin to use short, precise words in two- or three-word combinations are demonstrating **telegraphic speech.** The third word frequently fills in the part that was implied in the two-word statement (Slobin, 1972). "Want that" becomes "Jerry wants that" or "Mommy milk" becomes "Mommy drink milk."

Psycholinguist Roger Brown (1973) characterizes the language of 2-year-old children as telegraphic speech. Brown observes that words in a telegram cost money, so that we have good reason to be brief. Take the message "My car has broken down and I have lost my wallet; send money to me at the American Express in Paris." We would word the telegram, "Car broken down; wallet lost; send money American Express Paris." In this manner we omit eleven words: *my, has, and, I, have, my, to, me, at, the, in.* The omitted words are pronouns, prepositions, articles, conjunctions, and auxiliary verbs. We retain the nouns and verbs:

> The adult user of English when he writes a telegram operates under a constraint on length and the child when he first begins to make sentences also operates under some kind of constraint that limits length. The curious fact is that the sentences the child makes are like adult telegrams in that they are largely made up of nouns and verbs (with a few adjectives and adverbs) and in that they generally do not use prepositions, conjunctions, articles, or auxiliary verbs (Brown, 1973).

Between 12 to 26 months of age, a child is most likely to be using nouns and verbs, with some adjectives and adverbs, such as "Daddy go bye-bye," "Me go out," "Me want drink" (Brown, 1973).

By 27 to 30 months, a child is starting to form plurals: "Annie want cookies," "Mommy get shoes." The use of articles are now evident in speech: "The cat goes Meow," "Annie wants a cookie." Some prepositions (indicating placement) are also used: "Jimmy in bed now."

Questions

What are the several steps in infant language development from vocalizations, such as crying, to speaking an actual sentence? Is there an average age range for this developmental milestone?

The Significance of Language Development

Parents anxiously await normal language development in their young children, for in American culture language expression and understanding are considered indicators of intellectual ability or intellectual delay. Children vary in their timing of their expressive language. Parents and grandparents generally speak constantly to a first-born child, and first borns are the most likely to speak early or within the expected range of development. Later-born children with siblings sometimes do not have to speak to get what they want; if older siblings take care of the child and anticipate the child's needs, then it is likely that this child will simply delay expressing language. Some children speak very little, and then suddenly they surprise everyone by speaking in short sentences! However, if caretakers notice that a child does not follow simple instructions or does not speak simple words according to a normal timetable, they are advised to schedule a checkup with the child's pediatrician. Some children with delayed speech have hearing impairments; others may be eligible for speech therapy to promote normal speech development. We do know that children who have language delays or cannot be understood when they speak are at a risk for later social isolation upon entering group activities, such as day care, preschool, or kindergarten.

Some American parents are enrolling their toddlers in nursery or preschool programs where instruction in a second language is the norm, whereas parents who recently immigrated into this country are struggling to immerse their children in the English-speaking culture. Some children have bilingual parents and seem to learn both languages with ease at home.

Segue

The issue of the increasing numbers of non-English-speaking young children in the United States has prompted much debate on when language learning should begin, what teaching methods are most effective in the earliest years, what language or languages should be spoken in the child's home, and how all of these programs will be funded. People who work in such areas as health care, teaching, and early child care are particularly aware of how difficult it is to communicate without a common language. (We encourage readers of this text to learn a second language, for you will find it extremely beneficial when you seek employment.)

In Chapter 6, we discuss the influential role of the home environment and caretakers on the young child's emotional development and the expanding social contexts within which the child's developing sense of self evolves.

Summary

Cognitive Development

1. Learning, which permits us to adapt to our environment by building on previous experience, is defined in terms of three criteria: There must be some change in behavior, this change must be relatively stable, and the change must result from experience. To facilitate learning, we can condition people, provide a model for them to imitate, and shape and fashion the cognitive structures by which individuals think about their environment.

2. Jean Piaget characterized children's cognitive development during the first two years of life as "sensorimotor." The child's major task during this period is to integrate the perceptual and motor systems in order to arrive at progressively more adaptive behavior.

3. Another hallmark of the sensorimotor period is the child's progressive refinement of the notion of object permanence.

4. Piaget's work has stimulated other psychologists to investigate children's cognitive development. This ongoing work is revising and refining Piaget's insights. Researchers find that infants possess a set of object skills more sophisticated than Piaget had imagined. Moreover, psychologists find that, in order to become interested in objects and object skills, youngsters must be provided with experiences that they cannot generate by themselves, such as playing with others.

5. Individual differences in mental performance in infancy are, to a moderate extent, developmentally continuous across childhood and perhaps beyond. Children's patterns of attending to information reflect their cognitive capabilities, particularly their ability to construct workable schemes of the visual and auditory worlds.

Language and Thought

6. Human beings are set apart from other animals by their highly developed system of language communication. Language enables people to communicate with one another (interindividual communication) and facilitates thought (intraindividual communication).

7. Historically, there have been two opposing scientific views on the relationship between language and thought: (1) that thought takes place independently of language, and language is merely the container of already established thought; and (2) that language shapes thought by providing the concepts or categories into which individuals mentally sort their perceptual stimuli.

Theories of Language Acquisition

8. Environmentalists and nativists disagree strongly about the determinants of language. Environmentalists argue that language is learned. Nativists insist that human beings possess an inborn language-generating mechanism.

9. Many scientists have concluded that the acquisition of language cannot be understood by examining either learning or genetic factors in isolation from one another. Complex interactions take place among biochemical processes, maturational factors, learning strategies, and the social environment.

Language Development

10. The essence of language is the ability to transmit messages. Yet language is only one form of message transmission. We also communicate through body language (the nonverbal communication of meaning through physical movements and gestures) and paralanguage (the stress, pitch, and volume of vocalizations by which we communicate expressive meaning).

11. Children speak their own language, one that has its own characteristic patterns. Speech develops through a series of stages: early vocalizations (primarily crying), cooing and babbling, holophrastic speech, two-word sentences, and three-word sentences.

12. Children vary in the timing of using language. While some speak early, others show signs of delay and might need speech therapy.

Key Terms

caretaker speech (150)

communication (146)

conceptualization (148)

holophrase (157)

language (144)

language acquisition device (LAD) (151)

motherese (150)

object permanence (142)

receptive vocabulary (157)

sensorimotor (142)

telegraphic speech (159)

Following Up on the Internet

Web sites for this chapter focus on cognitive and language development in infancy. Please access the text web site at http://www.mhhe.com/crandell7 for up-to-date hot-linked Internet addresses for the following organizations, topics, and resources.

Child Development Journal

Infant Studies

Jean Piaget Society

Normal Speech Development

National Institute for Deafness & Other Communication Disorders

International Society of Infant Studies

Chapter 6

Infancy
Emotional and Social Development

Critical Thinking Questions

1. How would you feel if your baby consistently preferred to be held by other people instead of you? Would you try to change your baby's preferences?

2. Why do we sometimes experience two emotional states simultaneously? For example, a child will laugh and cry at the same time. Have you ever loved and hated someone at the same time? Is it really possible?

3. Babies are said to have different temperaments: Some are difficult, some are slow to warm up, and some are easy going. If you and your baby end up at opposite ends of the temperament continuum and one of you needed to change, who would have more difficulty changing—you or the baby? Why?

4. If you could not raise your child in a traditional home environment (i.e., two parents), what would you opt for—an extended family where many adults take care of the child, a single-parent household, or a day-care center? What factors did you consider in arriving at your decision?

In Chapters 4 and 5, we focused on the infant as a growing and cognitively developing being. With those areas of development as a foundation, we now examine multiple ways young children are socialized into the larger human group. In America many families have moved away from the three-generational model of child care in the home evident in many other countries. For example, in Japan, China, and India, it is the norm for grandparents to live with their children and help care for their grandchildren. In the United States, grandparents and others in the extended family might live nearby, but they are just as likely to live hundreds or thousands of miles away.

As we enter the 2000s, there is an increasing diversity of family structures and fewer two-parent families. Adding to this mix of societal concerns is the changing demographic nature of the United States (see Figure 6-1). And with more mothers employed outside the home than ever, a growing number of families need assistance from the larger community to provide high-quality care and supervision for growing children.

Young children in the United States are supervised, and therefore socialized, by a larger array of people than ever before. A diversity of programs have evolved over the past decade to assist in the important task of child care. Additionally, the television in the home has come to be a powerful socialization influence. This is of great concern to social scientists who study attachment, because many psychologists believe that children's emotional ties in their early years are extremely significant and serve as models for their later relationships. Perhaps most importantly, the child's own personality and temperament are underlying factors in the successful development of emotional and social bonds.

In the orphanage, children become sad and many of them die of sadness.

From the diary of a Spanish bishop, 1760

Above all else, people are social beings and human infants need to become socialized into their own group. As Harriet L. Rheingold (1969b, p. 781), a developmental psychologist, puts it, "The human infant is born into a social environment; he can remain alive only in a social environment; and from birth he takes his place in that environment." Humanness, then, is a social product (Candland, 1993; Tronick, Morelli, & Ivey, 1992). This fact is starkly reflected in two separate cases involving children who were reared in extreme isolation. The cases of Anna and Isabelle are similar in a number of respects. Both girls were born illegitimate, and their mothers, out of shame, had hidden them in secluded rooms over a period of years. Anna and Isabelle received only enough care to keep them alive. When they were discovered by local au-

thorities, both were extremely retarded, showing few, if any, human capabilities or responses. In Anna's case

[the child] could not talk, walk, or do anything that showed intelligence. She was in an extremely emaciated and undernourished condition . . . completely apathetic, lying in a limp, supine position and remaining immobile, expressionless, and indifferent to everything. Anna was placed in an institution for retarded children, where she died of hemorrhagic jaundice at 10 years of age (Davis, 1949, pp. 204–205).

In contrast with Anna, after Isabelle was discovered, she received special training from members of the staff at Ohio State University. Within a week after training was begun, she attempted her first vocalization. Isabelle rapidly progressed through the stages of social and cultural learning that are considered typical of U.S. children. She finished the sixth grade at age 14 and was judged by her teachers to be a competent and well-adjusted student. Isabelle is reported to have completed high school, married, and had her own normal family. A report on the two cases by sociologist Kingsley Davis (1949, pp. 207–208) concluded:

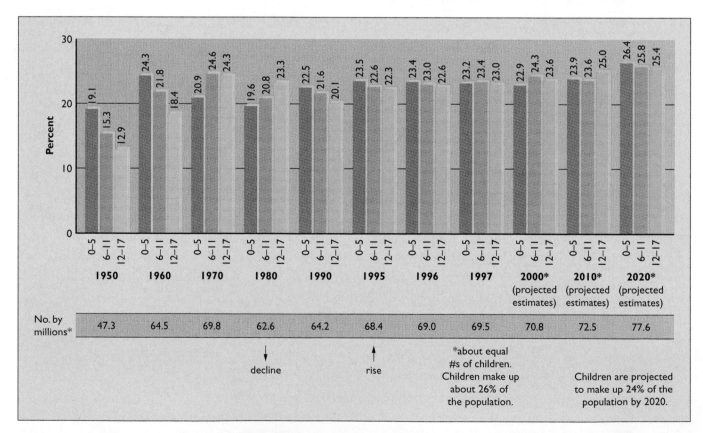

Figure 6-1   The Changing Demographics of Childhood: Number of Children in the U.S. Under 18, Selected Years 1950–1997 and projected 2000–2020 The number of children in the U.S. has increased significantly since 1950. Currently, children make up about 26 percent of the U.S. population. The Bureau of the Census projects a marked increase in young children aged 0 to 5 over the next 20 years.

Source: U.S. Bureau of the Census, Current Population Reports. Retrieved July 2, 1998 from the World Wide Web: http://www.childstats.gov/ac1998/pop1.htm

Isolation up to the age of six, with failure to acquire any form of speech and hence missing the whole world of cultural meaning, does not preclude the subsequent acquisition of these. . . . Most of the human behavior we regard as somehow given in the species does not occur apart from training and example by others. Most of the mental traits we think of as constituting the human mind are not present unless put there by communicative contact with others.

In the 1990s, many psychologists who are interested in the emotional and cognitive development of deprived children have focused on Romanian infants raised in orphanages or adopted by American and European families during the late 1980s and early 1990s—reported to be over 100,000 infants and children. When the Romanian dictatorship was overthrown in 1989, it was discovered that thousands of Romanian children had been confined to orphanages by the dictatorship. With the collapse of the Romanian economy, thousands more children were given up by their parents, who could no longer afford to provide for them. Some infants were institutionalized; older children were abandoned to the streets. The infants raised in crowded, stark institutions, deprived of normal stimulation and nurturance by any significant caregiver, were considered to be the "lucky" ones because they at least had food and shelter. But only their basic physical needs were met due to overcrowding and understaffing (Kaler & Freeman, 1997).

What were the results of this mass institutionalization and abandonment to the streets of tens of thousands of young children? Certainly some of those who have survived, so severely deprived of early nurturant care, will be studied throughout their lifetimes. Many of these adopted infants, though normal in appearance, display psychopathological behaviors that their adoptive families might not be prepared to handle. Some of these adoptive parents have organized support groups to deal with the severe effects of their children's earlier physical neglect (not having been held or touched), cognitive deprivation (lack of sensory stimulation other than self-abusive behaviors and lack of language skills), emotional abuse (lack of anyone who demonstrated love and caring), and social deprivation (having had only other infants in the orphanage to interact with) (Kaler & Freeman, 1997).

Over the past century, researchers have demonstrated that physical contact and sensory stimulation improve the sensorimotor functioning of children in institutionalized settings (Saltz, 1973; White, 1969). Indeed, even a small amount of extra handling highlights the value of "enrichment," at least over the short term. As pointed out in Chapter 2, children show greater resilience than child psychologists believed was the case only a few decades ago. But note that, as in most other matters, children differ. Some show greater vulnerability to deprivation experiences than others do (Langmeier & Matějček, 1974;

Rutter, 1974). Overall, the research on institutionalized infants supported the trend toward earlier adoption and the use of foster homes rather than institutions.

The Development of Emotional and Social Bonds

We are molded and remolded by those who have loved us; and though the love may pass, we are nevertheless their work, for good or ill.

François Mauriac

Theories of Personality Development

Early in the 1900s, psychoanalytic and psychosocial psychologists began to focus on the developmental impact of healthy emotional development in infancy and early childhood. Later, behavioral and cognitive theories emerged. And more recently, ecological theories have been proposed (Thomas, Chess, & Birch, 1970).

The Psychoanalytic View   In the late 1800s and early 1900s, Sigmund Freud revolutionized Western view of infancy by stressing the part early infant and childhood experience plays in fashioning the adult personality (see Chapter 2). Central to Freud's thinking was the idea that adult neurosis has its roots in childhood conflicts associated with the meeting of instinctual needs, such as sucking, expelling urine and feces, assertion of self, and pleasure (Freud, 1930/1961). Over the past 70 years, Freud's views have had an important influence on child-rearing practices in the United States. According to Freudians, the systems of infant care that produce emotionally healthy personalities include breast-feeding, a prolonged period of nursing, gradual weaning, an on-demand nursing schedule, delayed and patient bowel and bladder training, and freedom from punishment.

Many pediatricians, clinical psychologists, and family counselors have accepted major tenets of Freudian theory, especially as popularized by Dr. Benjamin Spock through his best-selling book *Baby and Child Care*, first published in 1946. Through multiple revisions and editions, it is said to have sold more copies than any other book worldwide excepting the Bible. Freud stated that infants become "fixated" in the oral stage if they are not allowed to continue to nurse at the breast or suck from a bottle until they are physically ready and motivated to drink from a cup. Dr. Spock, for example, promoted an on-demand feeding schedule during infancy in contrast to a rigid schedule of feeding every two hours as determined by pediatricians.

Many psychologists, especially those influenced by the Freudian tradition, hold that children's relationships in their early years are extremely significant and serve as prototypes for their later relationships with people. Viewed from this perspective, the flavor, maturity, and stability of a person's relationships derive from her or his early ties. The goal of psychoanalytic therapy is to use such methods as dream analysis, free association, and hypnosis during therapy sessions to discover any traumatic events from early childhood concealed in the patient's subconscious that might be causing adult personality disturbances.

Psychoanalytic research, however, has produced few empirical findings, relying more on case studies and observational recordings. In an extensive study of child-rearing practices, Sewell and Mussen (1952) found no tie between the type of feeding schoolchildren had received as infants and their oral symptoms such as nail biting, thumb sucking, and stuttering. Likewise, a longitudinal study carried out by Heinstein (1963) revealed no significant differences in the later behavior of bottle-fed and breast-fed babies. Research has also demonstrated that other variables emphasized in psychoanalytic literature, including those associated with bladder and bowel training, are not related to later personality characteristics (Schaffer & Emerson, 1964; Behrens, 1954; Sewell, & Mussen, 1952).

Psychologists now generally concede that such practices in and of themselves have few demonstrable effects on later development. Many psychologists today believe that children are not the psychologically fragile and vulnerable beings depicted by Freud. They are considerably more resilient and less easily damaged by traumatic events and emotional stress than was once thought to be true (Werner, 1990). Nor can parents expect to inoculate their children with love against future misfortune, misery, and psychopathology. And contrary to Freudian expectation, many people who had nurturant and devoted parents during their early years feel unloved as young adults. Psychologists today stress that development occurs across the entire life span and that no one period is more critical than another. They do not deny that development is influenced by early experiences, but they portray human beings as considerably more resilient than Freud and his followers imagined them to be.

Erikson's Psychosocial View   As we noted in Chapter 2, Erikson maintained that the essential task of infancy is the development of a basic trust in others. He believed that during infancy children learn whether the world is a good and satisfying place where one's needs are met by others or a source of discomfort, frustration, and misery. If the child's basic needs are met with genuine and sensitive care, the child develops a "basic trust" in people and a foundation of self-trust (a sense of being "all right" and a complete self). In Erikson's view, a baby's first social achievement is the willingness to let its mother move out

Learning Basic Trust   According to Erik Erikson, an infant's first social achievement is coming to terms with the fact that his mother periodically moves out of sight. As the child develops confidence that she will return, he is less likely to cry and rage when she departs.

of sight without undue anxiety or rage, because "she has become an inner certainty as well as an outer predictability" (Erikson, 1963, p. 247). The psychosocial psychologists highlighted the importance of resolving progressive conflicts for healthy emotional and social development throughout life.

Behavioral Psychology Theory   The strict behaviorism (also called learning theory) of John Watson in the early 1900s and later B. F. Skinner had essentially eliminated the study of emotions from the curriculum of behavioral science during the 1940s and 1950s. Watson's claim that he could take 12 healthy infants and form them into anything he wanted them to be is famous. Behaviorists recognize that infants are endowed with innate emotions (including fear, rage, and love) yet are unconcerned about the child's subconscious or inner feelings. They are more concerned with the outward display of emotions through observable behaviors, then rewarding "appropriate" behaviors or extinguishing "inappropriate" behaviors. Through keen observation and a system of rewards or punishments, children's behaviors can be shaped or controlled. Reinforcement schedules, time-outs, and other behavioral techniques are used to create desired patterns of behavior and expression of emotions. Most early childhood education programs that follow principles of traditional behavioral theory do not give priority to emotional development (Hyson, 1994, p. 35). By mastering specific academic and self-regulatory skills, children are supposed to gain positive feelings and self-confidence.

The Cognitive Perspective

During the 1960s and 1970s, developmental psychologists centered their attention on components and stages of cognitive growth, a legacy of Jean Piaget's body of work on cognitive development in children. Legions of child psychologists devoted their attention to how children reasoned and solved problems, starting with the infant's experience of sensory stimulation. They saw emotion as peripheral, of interest mostly when it interfered with rational thought or found deviant expression in the form of mental illness.

However, over the past decade or so, a renewed interest in the study of emotions is correcting the predominantly behavioral or cognitive view of human development. In the process, psychologists are rejecting the image of humankind as simply a "stimulus-response black box" or a "thinking machine" (Kagan, 1993; Goode, 1991).

The Ecological View

Urie Bronfenbrenner's ecological theory (1997) posits that a variety of environmental influences, ranging from the child's family through global economic forces, contribute to the emotional and social development of children. Naturally the nuclear family, siblings, single parent, grandparent, or main caretaker is going to be the major influence, at least initially. As we know, many children are also in child-care or preschool environments, which means teachers and caretakers exert their influence several hours per day. The availability of high-quality child care and low-cost nutrition programs in a community directly affects each and every child's development. Opportunities for employment—or lack of employment—in a community can make or break a family. The policies established by state legislators also impact the services offered at a local level to families, especially for education and health care. Federal decisions to allocate funds to preserve large corporations generally mean that there will be significantly less funding for programs such as parental leaves of absence for child care. Federal laws can either keep families together or, through welfare incentives or judicial decrees, reward families to keep the father out of the home or involved at only a minimal level. On a cultural level, a society's views can fluctuate regarding the value of the nuclear family or the health and welfare of its youngest citizens. Some European countries have national parental leave policies with compensation for early child care; the United States does not. A dramatic example of governmental impact on families regards abortion and family planning: Since the U.S. Supreme Court ruled that women have the right to abortion, there have been at least 35 million abortions performed in the United States, according to Planned Parenthood (1998) estimates.

Question

What is the central focus of the psychoanalytic, psychosocial, behavioral, cognitive, and ecological theories of personality development during the early years of life?

Attachment

Attachment is an affectional bond that one individual forms for another and that endures across time and space (Ainsworth, 1995, 1993, 1992; Ainsworth & Bell, 1974). An attachment is expressed in behaviors that promote proximity and contact. Among infants these behaviors include approaching, following, clinging, and signaling (smiling, crying, and calling). Through these activities a child demonstrates that specific people are important, satisfying, and rewarding. Some writers call this constellation of socially oriented reactions "dependency," whereas laypeople refer to it simply as "love."

What Is the Course of Attachment? H. Rudolph Schaffer and Peggy E. Emerson (1964) studied the development of attachment in 60 Scottish infants over their first 18 months of life. They identified three stages in the development of infant social responsiveness:

1. *First stage:* During the first two months of life, infants are aroused by all parts of their environment. They seek arousal equally from human and nonhuman aspects.
2. *Second stage:* Infants display indiscriminate attachment. During this stage, which occurs around the third month, infants become responsive to human beings as a general class of stimuli. They protest the withdrawal of any person's attention, whether the person is familiar or strange.
3. *Third stage:* When they are about 7 months old, babies show signs of specific attachment. They begin displaying a preference for a particular person and, over the next three to four months, make progressively more effort to be near this attachment object.

Children differ greatly in the age at which specific attachment occurs. Among the 60 babies in the Schaffer and Emerson study, 1 showed specific attachment at 22 weeks, whereas 2 did not display it until after their first birthdays. Cross-cultural differences also play a part in this development (Van IJzendoorn, 1990; Oppenheim, Sagi, & Lamb, 1988).

Mary Ainsworth found that infants in Uganda show specific attachment at about 6 months of age—a month or so earlier than the Scottish infants studied by Schaffer and Emerson. Similarly, it was found that separation protest occurred earlier among infants in Guatemala than among

Fostering Secure Infant-Parent Attachment Close proximity and contact promote affectional bonds that individuals form for one another. Cross-cultural differences play a part in the development of attachment.

those in the United States (Lester et al., 1974). Researchers attribute the precocity of the Ugandan and Guatemalan infants to cultural factors. Ugandan infants spend most of their time in close physical contact with their mothers (they are carried about on their mother's back), and they are rarely separated from her. In the United States, infants are placed in their own rooms shortly after birth. Such separation is virtually unknown in Guatemala, where most rural families live in a one-room rancho. It might be that separation is a more noticeable event to Guatemalan children, leading them to respond to it earlier than children reared in the United States.

Schaffer (1996, 1971) suggests that the onset of separation protest is directly related to a child's level of object permanence. Social attachment depends on the ability of infants to differentiate between their mother and strangers and on their ability to recognize that their mother continues to exist even when she is not visible. In terms of Piaget's cognitive theory, outlined in Chapter 5, these abilities do not appear until late in the sensorimotor stage. Indeed, Silvia M. Bell (1970) finds that in some instances the concept of person permanence—the notion that an individual exists independently of immediate visibility—might appear in a child before the concept of ob-

ject permanence. Studies by other researchers also confirm that protests over parental departures are related to a child's level of cognitive development (Kagan, 1997; Kagan, Kearsley, & Zelazo, 1978).

How Do Attachments Form?   Psychologists have advanced two explanations of the origins or determinants of attachment, one based on an *ethological* perspective and the other on a *learning* perspective. Psychoanalytically oriented ethologist John Bowlby (1988, 1969) said that attachment behaviors have biological underpinnings that can be best understood from a Darwinian evolutionary perspective. For the human species to survive despite an extended period of infant immaturity and vulnerability, both mothers and infants are endowed with innate tendencies to be close to each other. This reciprocal bonding functioned to protect the infant from predators when humans lived in small nomadic groups. (Bowlby, 1988, 1969)

According to Bowlby, human infants are biologically preadapted with a number of behavioral systems ready to be activated by appropriate "elicitors" or "releasers" within the environment. For instance, close physical contact—especially holding, caressing, and rocking—serves to soothe and quiet a distressed, fussing infant. Indeed, an infant's crying literally compels attention from a caretaker, and smiles accomplish much the same end (Spangler & Grossmann, 1993). Rheingold observes:

> As aversive as the cry is to hear, just so rewarding is the smile to behold. It has a gentling and relaxing effect on the beholder that causes him to smile in turn. Its effect upon the caretaker cannot be exaggerated. Parents universally report that with the smile the baby now becomes "human." With the smile, too, he begins to count as a person, to take his place as an individual in the family group, and to acquire a personality in their eyes. Furthermore, mothers spontaneously confide that the smile of the baby makes his care worthwhile. In short, the infant learns to use the coin of the social realm. As he grows older and becomes more competent and more discriminating, the smile of recognition appears; reserved for the caretaker, it is a gleeful response, accompanied by vocalizations and embraces. (Rheingold, 1969a, p. 784)

Sucking, clinging, calling, approaching, and following are other kinds of behavior that promote both contact and proximity. Viewed from the evolutionary perspective, a child is genetically programmed to a social world and "in this sense is social from the beginning" (Ainsworth, Bell, & Stayton, 1974). Parents, in turn, are said to be genetically predisposed to respond with behaviors that complement the infant's behaviors (Ainsworth, 1993; Ainsworth et al., 1979). The baby's small size, distinctive body proportions, and infantile head shape apparently elicit parental caregiving (Alley, 1983).

Programmed to Elicit Parenting? The noted ethologist Konrad Lorenz believed that human beings are genetically programmed for parenting behavior. It seems that caretaking tendencies are aroused by "cuteness." When Lorenz compared human infants with other young animals, including kittens and puppies, he noticed that they all seem to display a similar set of sign stimuli that arouse parental response. Apparently, short faces, prominent foreheads, round eyes, and plump cheeks all stir up parental feelings.

Whereas Bowlby believes that inborn mechanisms account for attachment behaviors, learning theorists attribute attachment to socialization processes. According to psychologists such as Robert R. Sears (1963, 1972), Jacob L. Gerwitz (1972), and Sidney W. Bijou and Donald M. Baer (1965), the mother is initially a neutral stimulus for her child. She comes to take on rewarding properties, however, as she feeds, warms, dries, and snuggles her baby and otherwise reduces the infant's pain and discomfort. Because the mother is associated with the satisfaction of the infant's needs, she acquires secondary reinforcing properties—her mere physical presence (her talking, smiling, and gestures of affection) becomes valued in its own right. Thus, according to learning theorists, the child comes to acquire a need for the presence of the mother. In brief, attachment develops.

Learning theorists stress that the attachment process is a two-way street. The mother also finds gratification in her ability to terminate the child's piercing cries and with it to allay her own discomfort that is associated with the nerve-wracking sound. Furthermore, infants reward their caretakers with smiles and coos. Thus, as viewed by learning theorists, the socialization process is reciprocal and derives from a mutually satisfying and reinforcing relationship (Adamson, 1996).

Who Are the Objects of Attachment? In their study of Scottish infants, Schaffer and Emerson found that the mother was most commonly (in 65 percent of the cases) the first object of specific attachment. However, in 5 percent of cases the first attachment was to the father or a grandparent. And in 30 percent of cases, initial attachments occurred simultaneously to the mother and another person. Furthermore, the number of a child's attachments increased rapidly. By 18 months only 13 percent displayed attachment to only one person, and almost one-third of the babies had five or more attachment persons. Indeed, the concept of attachment as originally formulated was too narrow (Bronfenbrenner, 1979). Because infants have ongoing relationships with their fathers, grandparents, and siblings, these psychologists suggest the focus of theory and research should fall on the social network, an encompassing web of ties to significant others (Stern, 1985).

What Are the Functions of Attachment? Ethologically oriented psychologists point out that attachment has adaptive value in keeping infants alive. It promotes proximity between helpless, dependent infants and protective caretakers. But attachment also fosters social and cognitive skills. According to this view, four complementary systems coordinate the behavior of the child and the environment (Lamb & Bornstein, 1987):

- The *attachment behavioral* system leads to the development and maintenance of proximity and contact with adults.
- The *fear-wariness behavioral* system encourages youngsters to avoid people, objects, or situations that might be a source of danger to them; this system is often called "stranger wariness" and will be discussed later in the chapter.
- The *affiliative behavioral* system encourages infants, once the wariness response diminishes, to enter into social relationships with other members of the human species.
- The *exploratory behavioral* system provides babies with feelings of security that permit them to

explore their environment knowing that they are safe in the company of trusted and reliable adults (Jones, 1985).

Do Early Attachment Patterns Predict Later Relationships? Freud (1940, p. 188) saw the child's relationship to the mother as having lifelong consequences, calling it "unique, without parallel, established unalterably for a whole lifetime as the first and strongest love-object and as the prototype of all later love relations—for both sexes." This Freudian premise has underlain much research on attachment behaviors, drawing attention to the mother's influence on her developing child.

Mary Ainsworth and her colleagues devised a procedure called the **strange situation** to capture the quality of attachment in the infant-parent relationship (Ainsworth, 1983; Ainsworth & Wittig, 1969). In the strange situation, a mother and her infant enter an unfamiliar playroom where they find interesting toys as well as a stranger. After a few minutes, the mother leaves and the youngster is given an opportunity to explore the toys and interact with the unfamiliar adult. When the mother returns, the infant's behavior is observed, and then the procedure is repeated for a total of eight times with slightly different variations.

In the course of her research, Ainsworth became intrigued by differences in the behavior of children, especially the way they react upon reunion with their mothers. About 60 percent of the infants used the mother as a secure base from which to explore the unfamiliar environment and as a source of comfort following separation. When their mother returned to the room, the infants would greet her warmly, show little anger, or indicate they wanted to be picked up and comforted. These children Ainsworth called *securely attached* infants (pattern B attachments). Other infants, about 20 percent, snubbed or avoided the mother on her return. These children were called *avoidant* infants (pattern A attachments). Still others, about 10 to 15 percent, were reluctant to explore the new setting when they entered the playroom and would cling to the mother and hide from the stranger. However, when the mother returned after her brief absence, these children would mix active contact seeking with squirming, continued crying, and pushing the mother away. These youngsters were called *resistant* infants (pattern C attachments). More recently, a number of researchers have added another category, *disorganized/disoriented* infants (pattern D attachments). Whereas infants in the A, B, and C categories possess a coherent strategy for dealing with the stress of separation and reunion, D-pattern youngsters seem to lack coherent coping mechanisms (Jacobsen, Edelstein, & Hofmann, 1994). Moreover, D-pattern youngsters apparently are at greater risk for social maladaptation in childhood (Lyons-Ruth, Alpern, & Repacholi, 1993).

Ainsworth contended that the A, B, and C patterns of attachment behavior in the strange situation reflect the quality of maternal caregiving children receive during their first 12 months of life. She traced the origins of the A and C patterns to a disturbed parent-child relationship, one in which the mother was rejecting, interfering, or inconsistent in caring for the child. The mothers would often over- or understimulate their baby, fail to match their behavior to that of the child, be cold, irritable, and insensitive, and afford perfunctory care. In contrast, the securely attached infants (pattern B attachment) seemed to have received consistent, sensitive, and responsive mothering (Ainsworth, Bell, & Stayton, 1974). D patterns are thought to be linked to parents who suffered from abuse or deep emotional loss and have not yet resolved these issues. And even though mothers are influenced by the temperaments of their youngsters (for instance, whether they are irritable and difficult), this factor does not seem to be critical in determining a mother's responsiveness to her child's signals and needs (Belsky & Rovine, 1987). On the whole, other researchers have confirmed the Ainsworth findings, although the relationship is not as strong as Ainsworth initially suggested (Cassidy & Berlin, 1994). Moreover, the quality of a relationship, once formed, is not necessarily permanent (Belsky, 1996a).

Early attachment behaviors are predictive of children's functioning in other areas. *Mastery motivation* seems to be associated with secure attachment (Yarrow et al., 1984). And some studies report a relationship between attachment security and cognitive development, although hardly ever a relationship with language acquisition. The effects of early attachments similarly carry over to later social relationships. For instance, classifications derived from the strange situation predict social functioning in the preschool with teachers and peers. Youngsters judged to have secure attachments to their mothers are socially more competent in preschool, sharing more and showing a greater capacity to initiate and sustain interaction. Such children are also more accepting of their mothers' showing attention to their older brothers and sisters, and secure older siblings are more likely to assist and care for their younger brothers and sisters than are insecure older siblings (Teti & Ablard, 1989). And the B-pattern children seem more resilient and robust when placed in stressful or challenging circumstances (Stevenson-Hinde & Shouldice, 1995).

These findings are consistent with the speculations of attachment theorists that young children who enjoy a secure attachment to their parents develop internal "representational models" of their parents as loving and responsive and of themselves as worthy of nurturance, love, and support. By contrast, youngsters with insecure attachments develop "representational models" of the caregiver as unresponsive and unloving and of themselves as unworthy of nurturance, love, and support (Bowlby, 1988, 1969; Denham, Renwick, & Holt, 1991).

Others, however, disagree with attachment theory, especially as it relates to subsequent human development. They argue that the critical factor in becoming a well-adjusted or maladjusted adult lies in how we adapt to life's experiences and not to how securely we are attached to a caregiver at the end of our first year of life (Lewis, 1998). Some evidence suggests that the various patterns of attachment might be transmitted across generations via parents' states of mind, their internal working models of attachment relationships that they subtly communicate to their children (Benoit & Parker, 1994; Bowlby, 1969).

Cultural Differences in Attachment Behaviors Not surprisingly, not only do child-rearing practices differ from one society to another, but attachment patterns differ also (Harwood, 1992). A-patterns are relatively more prevalent in western European nations and C-patterns in Israel and Japan (Van IJzendoorn & Kroonenberg, 1988). Moreover, two separate studies that examined attachment theory have proclaimed contradictory findings: De Wolff and IJzendoorn (1997) conclude that the mother's sensitivity is an important condition if attachment security is to occur, whereas Goldsmith and Alansky (1987) claim that there is at most a weak association between a mother's sensitivity and attachment. The ongoing debate between these two views has led to a call for a new definition of sensitivity that conceptualizes it as a characteristic that pertains to the interactions within the parent/child dyad and hence is meaningless when it refers only to the mother or the child (Vanden Boom, 1997).

Stranger Anxiety and Separation Anxiety Wariness of strangers, an expression of the fear-wariness behavior system, usually emerges about a month or so after specific attachment begins. **Stranger anxiety,** a wariness of unknown people, seems to be rather common among 8-month-old infants, seems to peak at 13 to 15 months, and decreases thereafter (see Figure 6-2). When encountering a strange person, particularly when a trusted caretaker is absent, many youngsters frown, whimper, fuss, look away, and even cry (Morgan & Ricciuti, 1969; Waters, Matas, & Sroufe, 1975). Even at 3 and 4 months of age some babies stare fixedly at a strange person, and occasionally this prolonged inspection leads to crying (Bronson, 1972).

Another common behavior 8-month-old infants display is **separation anxiety,** distress shown when a familiar caregiver leaves. Parents would do well to introduce grandparents and baby-sitters to an infant during the earlier stages of infancy so there is a caretaker that both the parents and the baby trust when it comes time to go out for an evening together. Sometimes the baby's distress at being left alone with a stranger is so intense that the baby cries the whole time the parents are gone. Most likely those of you who have earned money by baby-sitting can relate to this experience. This can be an extremely volatile situation, because the baby-sitter might not be able to tolerate the noise of lengthy crying and might attempt unhealthy measures to quiet the child. Infant abuse and infant homicide have occurred under these circumstances (see "Shaken Baby Syndrome" in Chapter 4). All local YWCAs provide baby-sitter training, and some high schools offer parenting classes.

Thirty years ago psychologists assumed that "fear of the stranger" was a developmental milestone that occurred in normal children. But then some researchers found that youngsters do not invariably fear unfamiliar people (Rheingold & Eckerman, 1973). In fact, a more common reaction is to be accepting and make friendly overtures. Furthermore, mothers show significantly

Cultural Differences in Attachment An Indonesian mother facilitates physical and social contact with her child by a culturally evolved carrying contrivance. A solid attachment between child and caretaker promotes the youngster's emotional and social development; it eases and forestalls episodes of fussing and crying. Notice, though, that this child appears a little unsure of the photographer.

Figure 6-2 Stranger Anxiety Strangers often approach infants in public places. By 7 to 8 months, most American children demonstrate "stranger anxiety" by turning away, crying, or clinging to parents.

"She's a little traumatized—this is her first Wal-Mart."

more wariness than their infants do. Parents', subtle negative behaviors on the approach of an unfamiliar person are cues that are communicated to their infants, and an infant's growing wariness toward strangers might simply reflect the fact that it is becoming an increasingly sophisticated participant in human interaction (Mangelsdorf, 1992).

Questions

How do parents and infants become emotionally "attached" to each other? How does their behavior show that they are becoming attached?

The Role of Emotion

It is terribly amusing how many different climates of feeling one can go through in a day.

Anne Morrow Lindbergh, *Bring Me a Unicorn*

Emotions play a critical part in our daily existence. Indeed, if we lacked the ability to experience love, joy, grief, and anger, we would not recognize ourselves as human. Emotions set the tone for much of our lives, and at times they even override our most basic needs: Fear can preempt appetite, anxiety can wreak havoc on a student's performance on an exam, despair can lead a person to a fatal flirtation with a pistol.

Most of us have a gut-level feeling of what we mean by the term *emotion,* yet we have difficulty putting the feeling into words. Psychologists have similar problems. Indeed, they have characterized emotion in different ways. Some have viewed it as a reflection of physiological changes that occur in our bodies, including rapid heartbeat and breathing, muscle tension, perspiration, and a "sinking feeling" in the stomach. Others have portrayed it as the subjective feelings that we experience—the "label" we assign to a state of arousal. Still others have depicted it in terms of the expressive behavior that we display, including crying, moaning, laughing, smiling, and frowning. Yet emotion is not one thing but many, and it is best characterized as a combination of all these components. We will say that **emotion** is the physiological changes, subjective experiences, and expressive behaviors that are involved in such feelings as love, joy, grief, and anger.

Emotions, then, are not simply "feelings" but rather processes by which individuals establish, maintain, and terminate relations between themselves and their environment (Campos et al., 1993). For instance, people who are happily involved in a conversation are likely to continue conversing, and their facial expressions and behaviors signal the others in the conversation that they too should keep up their interaction; sad people tend to feel that they cannot successfully attain some goal, and their sadness signals others that they need help.

Functions of Emotions   Charles Darwin was intrigued by the expression of emotions and proposed an evolutionary theory for them. In *The Expression of Emotions in Man and Animals,* published in 1872, Darwin contended that many of the ways in which we express emotions are inherited patterns that have survival value. He observed, for instance, that dogs, tigers, monkeys, and human beings all bare their teeth in the same way during rage. In so doing, they communicate to members of their own and other species important messages regarding their inner dispositions.

Contemporary psychologists have followed up on Darwin's leads, noting that emotions perform a number of functions:

- Emotions help humans survive and adapt to their environment. For instance, fear of the dark, fear of being alone, and fear of sudden happenings are adaptive because there is an association between these feared things and potential danger.
- Emotions serve to guide and motivate human behavior. For example, our emotions influence whether we categorize events as dangerous or beneficial, and they provide the motivation for patterning our subsequent behavior.
- Emotions support communication with others. For instance, by reading other people's facial, gestural, postural, and vocal cues, we gain indirect access to their emotional states. Knowing that a friend is afraid or sad allows us to more accurately predict the friend's behavior and to respond to it appropriately.

Being able to "read" another person's emotional reactions also permits **social referencing,** the practice whereby an inexperienced person relies on a more experienced person's interpretation of an event to regulate his or her subsequent behavior. Before they are 1 year old, most infants engage in social referencing. They typically look at their parents when confronted with new or unusual events. They then base their behavior on the emotional and informational messages their parents communicate (Rosen, Adamson, & Bakeman, 1992). Social referencing varies with age; it probably begins at about 6 months of age and increases thereafter. Apparently youngsters care little whether it is their fathers or their mothers who are doing the signaling (Hirshberg & Svejda, 1990).

Youngsters 10 months of age use others' emotional expressions to appraise events like those encountered in the visual cliff experiment conducted by Gibson and Walk (1960) (see Chapter 4). When they approach the il-

lusory "dropoff," they look to their mothers' expressions and modify their own behavior accordingly. When their mothers present an angry or fearful face, most youngsters will not cross the "precipice." But when their mothers present a joyful face, they will cross it. Infants show similar responses to their mothers' vocalizations that convey fear or joy. This research reveals that infants actively seek out information from others to supplement their own information and that they are capable of using this information to override their own perceptions and evaluations of an event (Campos et al., 1983).

Emotional Development in Infants   Psychologists influenced by ethological theory have played a central role in the recent burst of studies on the emotional life of children (see Chapter 2). They have been influenced by Darwin's ideas and the more recent work of Paul Ekman (1994, 1980, 1972). Ekman and his colleagues have shown subjects from widely different cultures photographs of faces that people from Western societies would judge to display six basic emotions: happiness, sadness, anger, surprise, disgust, and fear. They find that subjects in the United States, Brazil, Argentina, Chile, Japan, and New Guinea label the same faces with the same emotion. Apparently, people associate specific emotions with specific facial expressions.

Ekman takes these findings as evidence that the human central nervous system is genetically prewired for the facial expression of emotion: The face provides a window by which other people can gain access to our inner emotional life and by which we gain similar access to their inner life. But the window is not entirely open. Early in life we learn to disguise or inhibit our emotions. We might smile even though we are depressed, look calm when we are irate, and put on a confident face in the presence of danger.

One thing is clear: Infants have emotions (Weinberg & Tronick, 1994). Psychologist Carroll E. Izard has been a central figure in the study of children's emotional development and has introduced his own "differential emotions theory" (Izard & Ackerman, 1995). Like Ekman, Izard contends that each emotion has its own distinctive facial pattern. Izard says that a person's facial expression colors what the thinking brain "feels." For example, the muscular responses associated with smiling make you aware that you are joyful. And when you experience rage, a specific pattern of muscle firings that is physiologically linked with anger "informs" your brain that you feel rage and not anguish or humiliation. In sum, according to Izard the feedback from sensations generated by your facial and related neuromuscular responses yield the distinctive subjective experiences that you recognize as different types of feeling. (Smile or frown while you read the next paragraph, and see whether it affects how you feel.)

Carroll Izard: Infant Facial Archetypes of Emotion. This photo from Izard's initial study depicts the facial expression of joy, one that Izard believes is a basic, universal emotion. When the mouth forms a smile, the cheeks are lifted, and the eyes twinkle, we interpret the expression as joy.

Izard finds that babies have intense feelings from the moment of birth. But at first their inner feelings are limited to distress, disgust, and interest. In the course of their maturation, new emotions—one or two feelings at a time—develop in an orderly fashion. Izard says that emotions are preprogrammed on a biological clock: Infants gain the social smile (joy) at around 4 to 6 weeks of age; anger, surprise, and sadness at about 3 to 4 months; fear at 5 to 7 months; shame, shyness, and self-awareness at about 6 to 8 months; and contempt and guilt during the second year.

Psychologist Joseph Campos disagrees with Izard (Campos et al., 1993). He believes that all the basic emotions are in place at birth, depending on a prewired process for which neither experience nor social input is required. Campos says that many of an infant's emotions do not become apparent to observers until later, so the first experience of an emotion does not necessarily coincide with its first expression. Further complicating matters, the early signs of an emotion might differ from later ones. For instance, when 4-month-olds are frustrated, they direct their anger at the immediate cause (a person's hand), whereas at 7 months they direct their anger at the person (Stenberg, Campos, & Emde, 1983). And although infants typically do not show fear until they are 5 to 7 months old, infants as young as 3 months who have been abused by a male show fear when a male stranger approaches them. Abused youngsters similarly display sadness much earlier than usual.

Whether emotions emerge only in the course of maturation or are present at birth, we know that infants' emotional expressions become more graded, subtle, and complex beyond the first year of life (Denham, Zoller, & Couchoud, 1994). Moreover, infants show an increasing ability to discriminate among vocal clues and facial expressions, particularly happy, sad, and angry ones, in the first five months of life (Soken & Pick, 1992). Even more significantly, infants progressively come to modify their displays of emotion and their behaviors on the basis of their appreciation of their mother's displays of emotion and behaviors (Cohen & Tronick, 1987).

> **Question**
>
> *Does a smile signal emotion, or should emotion be defined solely in terms of physiological states such as rapid heartbeat, sweaty palms, or trembling?*

"What do you call a nerd fifteen years from now?" Answer: "Boss"

A children's joke from Goleman (1995)
Emotional Intelligence: Why It Can Matter More Than IQ

Emotional Intelligence   As the human brain evolved, the limbic system, which is the center of emotions, developed over the brain stem. The brain stem controls basic life processes such as blood pressure, the sleep/wake cycle, and respiration. Continued evolution produced the cerebral cortex above and around the limbic system. The cerebral cortex allows us to think, register sensations, analyze, problem solve, and plan ahead. The limbic system is a "go-between" for the brain stem and the cerebral cortex. Suppose that you are tired as you read this paragraph. "Bottom-up," the brain stem registers that the body is fatigued and requires sleep and forwards those signals to the limbic system. The limbic system communicates "feelings" of weariness or irritability to the cortex, which is the decision maker in the brain. The cortex senses these "feelings" and decides either to continue with this task or stop to sleep. Or suppose you are driving your car with the radio on. "Top-down," your cerebral cortex "hears" a familiar song of upbeat tempo on your car radio. The cortex in turn triggers the emotion of happiness and excitement in the limbic system, which in turn sends a message to the lower brain stem to increase your pulse and respiration as you sing along. In each example, the limbic system is the intermediary between the brain stem and the cerebral cortex. Consequently, some scientists argue that emotions, via the limbic system, play a much more significant role in rational thinking than ever before realized.

New scanning, imaging, and diagnostic methods that can map *process* as well as *function* are daily revealing new information about neural pathways. We used to believe that sensory systems sent information directly to specific lobes of the cortex, then to the limbic system for processing for emotional interpretation and reaction: We sense, then think, then feel, then react. However, the newest discoveries in neurobiology and neuropsychology tell us that we sense, then feel, then nearly simultaneously react and think about what we are experiencing.

This is demonstrated time and again in life-threatening situations: People simply react to save a life, then think about their actions later on. Our rational mind, if given enough time to think about an action, probably would prevent us from putting our body over those of children exposed to random gunfire. Normal neural circuitry from the limbic system to the frontal lobes (which interpret and mediate emotional signals) is crucial to effective thought. Signals of strong emotion from the amygdala in the limbic system to the frontal cortex (as can occur with continual fear and dread in child abuse) can create deficits in a child's intellectual abilities, crippling the capacity to learn (Goleman, 1995, p. 27). Adults are familiar with this effect. When we are emotionally upset (e.g., at receiving an unexpected pink slip at work) or emotionally overjoyed (e.g., at winning a lottery), we are likely to say, "I just can't think straight."

Such intellectual deficits due to strong emotion are likely to show up in a child's continual agitation and impulsivity. One study gave neuropsychological tests to primary school boys with above-average IQ scores who were doing poorly in school and found that the boys had impaired frontal cortex functioning (Goleman, 1995). They were impulsive, anxious, and disruptive in class, all of which suggested faulty frontal-cortical control of their limbic system urges. These are the children at highest risk for later problems such as academic failure, alcoholism, and criminality, because their emotional life is impaired. "These emotional circuits are sculpted by experience throughout childhood—and we leave those experiences utterly to chance at our peril" (Goleman, 1995, p. 27). Neurologist and brain researcher Dr. Antonio Damasio posits that feelings are typically indispensable for rational decisions (Goleman, 1995, p. 28).

"In a sense, we have two brains or two minds—and two different kinds of intelligence: rational [assessed by standard IQ measures] and emotional. How we do in life is determined by both—it is not just IQ, but emotional intelligence that matters. Indeed, intellect cannot work at its best without emotional intelligence" (Goleman, 1995). **Emotional intelligence** is a new concept and includes such abilities as being able to motivate oneself, persisting in the face of frustrations, controlling impulses and delaying gratification, empathizing, hoping, and regulating one's moods to keep distress from overwhelming one's ability to think (Goleman, 1995). Howard Gardner (1993b, 1983), a psychologist at the Harvard School of Education, in *Frames of Mind* posited a theory of *multiple intelligences*, including emotional intelligence:

> *Inter*personal intelligence is the ability to understand other people: what motivates them, how they work, how to work cooperatively with them. Successful salespeople, politicians, teachers, clinicians, and religious leaders are all likely to be individuals with high degrees of interpersonal intelligence. *Intra*personal intelligence . . . is a correlative ability, turned inward. It

is a capacity to form an accurate, veridical model of oneself and to be able to use that model to operate effectively in life. (quoted in Goleman, 1995, p. 39)

There is much evidence that "people who are emotionally adept—who know and manage their feelings well, and who read and deal effectively with other people's feelings—are at an advantage in any domain of life" (Goleman, 1995). Researchers who support the concept of emotional intelligence say that these crucial competencies should be learned and improved upon by children, especially if we want to reduce the tide of youth and adult violence in American society.

Stages in Childrens' Emotional Development   Closely related to recent findings about emotional intelligence is the Greenspan model. Stanley and Nancy Greenspan (1985) chart the emotional progress of the typical healthy child from birth through age 4, outlining six stages:

Stage 1. Self-regulation and interest in the world (0–3 months). Infants learn to calm themselves, and they develop a multisensory interest in the world.

Stage 2. "Falling in love" (2–7 months). Infants develop a joyful interest in the human world and engage in cooing, smiling, and hugging.

Stage 3. Developing intentional communications (3–10 months). Infants develop a human dialogue with the important people of their lives (for instance, they lift their arms to a caretaker, give and take a toy offered to them, gurgle in response to a caretaker's speech, and enjoy peekaboo games).

Stage 4. Emergence of an organized sense of self (9–18 months). Toddlers learn how to integrate their behavior with their emotions, and they begin to acquire an organized sense of self (for example, they will run to greet a parent returning home or lead a caretaker to the refrigerator in order to show hunger, instead of simply crying for dinner).

Stage 5. Creating emotional ideas (18–36 months). Toddlers begin to acquire an ability to create their own mental images of the world and to use ideas to express emotions and regulate their moods.

Stage 6. Emotional thinking—the basis for fantasy, reality, and self-esteem (30–48 months). Young children expand these capacities and develop "representational differentiation," or emotional thinking; they distinguish among feelings and understand how they are related; and they learn to tell fantasy from reality.

According to the Greenspans' model, even in infancy children are actively constructing and regulating their environments. Early attachment relationships lead to purposeful communication, and then to the toddler's creation of a coherent, positive sense of self. These early accomplishments lay the foundation for the young child's use of language, pretend play, and engagement in "emotional" thinking (Hyson, 1994, p. 58). According to the Greenspans, the more negative factors interfere with a child's mastering of emotional milestones, the more likely it is that the child's intellectual and emotional development will be compromised later on. For example, those of us who have taught in elementary school can see that a child will have difficulty mastering the multiplication tables if Mom and Dad have just separated. In "current events" discussions, elementary-age children will express their fears about moving to a new neighborhood, their anxiety about Mom or Dad remarrying and new step-siblings moving into the home, or Dad's going to jail for selling drugs. Their main concerns are of a personal, emotional nature, and many children are just "bursting" to let them out.

Stability of Emotional Expression   Izard and others have also turned up evidence of continuity, or stability, of emotional expression in children (Izard et al., 1993). The amount of sadness a child shows during a brief separation from its mother seems to predict the amount of sadness the same child shows six months later. And the amount of anger a child shows when receiving a painful inoculation at 2 to 7 months of age predicts the amount of anger the same child displays at 19 months of age (Izard & Malatesta, 1987). Additionally, emotional expressions and behaviors in infancy are related to personality characteristics at 5 years of age.

Emotions and Cultural Norms   Izard and his colleagues (1993) measured the fixation time, heart rate, and facial expressions of infants as they looked at a human face, a store mannequin face, and an object with scrambled facial features. They then examined the same youngsters five years later. The infants who had looked longest at the faces were the most sociable; those who looked the least showed the greatest shyness and withdrawal behavior. Izard concludes that emotional expressions have a strong biological component and that behaviors in infancy tell us something about the personality or enduring characteristics of a person. And if infants have distinctive personalities, then caretakers should treat them in ways that recognize their different characteristics and needs. Once again, it appears that parenting practices need to be tailored to the individual child's capabilities (Izard, 1991).

Izard does not deny that, in some measure, learning conditions and experience modify a child's personality. For instance, a mother's mood affects how her infant feels and acts (Termine & Izard, 1988). When mothers of 9-month-old youngsters display a sad face, their children often display the same facial expression and engage in less

vigorous play than when their mothers appear happy. Izard believes that "the interactional model of emotional development is probably correct. Biology provides some thresholds, some limits, but within these limits, the infant is certainly affected by the mother's moods and emotions" (quoted by Trotter, 1987, p. 44). Indeed, much parental socialization is directed toward teaching children how to modulate their feelings and expressive behavior to conform to cultural norms. Some studies have found that infants and toddlers have much difficulty with emotional self-regulation (Eisenberg et al., 1998); other studies suggest that children as young as 4 years are able to regulate their emotions, depending on the parent or caregiver they are interacting with and the expected outcomes of those emotional expressions (Zeman et al., 1997). Issues of continuity and discontinuity in children's emotional expressions invariably lead investigators to the matter of infant temperament and, more particularly, to differences in temperament among youngsters (Izard et al., 1993; Kohnstamm, Bates, & Rothbart, 1989).

Temperament

Research in infant temperament has become quite active in recent years. **Temperament** refers to the relatively consistent, basic dispositions that underlie and modulate much of a person's behavior. The temperamental qualities studied most often by developmental psychologists are those that are obvious to parents. They include irritability, a happy mood, ease of being soothed, motor activity, sociability, attentiveness, adaptability, intensity of arousal, regulation of arousal states, and timidity (Goldsmith, 1997; Chess & Thomas, 1996; Kagan, 1993).

For example, as we noted in Chapter 2, Jerome Kagan (1997) finds that some children are born with a tendency, or "vulnerability," toward extreme timidity when faced with an unfamiliar person or situation and that other children are not. This difference persists as they grow up and has profound social consequences. On the first day at school, timid youngsters tend to remain quietly watchful at the outskirts of activity, whereas uninhibited youngsters are all smiles and eager to approach other children. In our culture, uninhibited people are more popular than inhibited ones, and many timid youngsters are pressured by parents and others to be more outgoing. Kagan believes this bias is regrettable. Instead, he says, we should provide our children with an environment that is, within reason, respectful of individual differences. Although uninhibited children might grow up to be popular adults, inhibited children might invest more energy in schoolwork and become intellectuals if they are afforded settings that value academic achievement. Kagan notes that although we push our inhibited children toward the uninhibited end of the scale, other cultures, such as China, tend to view uninhibited behavior as disrespectful and unseemly (Guillen, 1984).

Individuality in Temperament   Alexander Thomas and his associates (Thomas & Chess, 1987; Thomas, et al., 1963) have come to conclusions quite similar to those of Kagan from their studies of more than 200 children. They found that babies show a distinct individuality in temperament during the first weeks of life and that this is independent of their parents' handling or personality styles. Thomas views temperament as the stylistic component of behavior—the *how* of behavior, as opposed to the why of behavior (motivation) or the what of behavior (its content). Thomas names three most common types of babies:

1. *Difficult babies.* These babies wail and cry a great deal, have violent tantrums, spit out new foods, scream and twist when their faces are washed, eat and sleep in irregular patterns, and are not easy to pacify (10 percent of all infants are difficult babies).
2. *Slow-to-warm-up babies.* These infants have low activity levels, adapt very slowly, tend to be withdrawn, seem somewhat negative in mood, and show wariness in new situations (15 percent of babies are slow to warm up).
3. *Easy babies.* These infants generally have sunny, cheerful dispositions and adapt quickly to new routines, foods, and people (40 percent of all infants are easy babies).

The remaining 35 percent of infants show mixtures of traits that do not readily fit into these categories. Thomas and colleagues also found that all infants possess nine components of temperament that emerge quickly after birth and continue relatively unchanged into adulthood:

1. *Activity level:* The proportion of active versus inactive periods
2. *Rhythmicity:* The regularity of hunger, sleep, bowel movements
3. *Distractibility:* The degree to which extraneous stimuli alter behavior
4. *Approach withdrawal:* The response to a new person or object
5. *Adaptability:* The ease with which a child adapts to changes
6. *Attention span and persistence:* The amount of time a child pursues an activity, and whether they are easily distracted
7. *Intensity of reaction:* The energy of response
8. *Threshold of responsiveness:* The degree of stimulation needed to evoke a response
9. *Quality of mood:* The amount of pleasant, friendly behavior versus unpleasant and unfriendly behavior

Goodness of Fit   Thomas and Chess (1987) have introduced the notion of "goodness of fit" to refer to the match between the characteristics of infants and their families. In a good match, the opportunities, expectations, and demands of the environment are in accord with the child's temperament. A good match fosters optimal development. Conversely, a poor fit makes for a stormy household and contributes to distorted development and maladaptive functioning. Thomas emphasizes that parents need to take their baby's unique temperament into account in their child-rearing practices.

Children do not react in the same ways to the same developmental influences. Domineering, authoritarian parental behavior can make one child anxious and submissive and lead another to be defiant and antagonistic. As a consequence, Thomas and his colleagues (1963, p. 85) conclude, "There can be no universally valid set of rules that will work equally well for all children everywhere."

Parents with difficult babies often feel considerable anxiety and guilt. "What are we doing wrong?" they ask. Thomas has a reassuring answer for such parents:

> The knowledge that certain characteristics of their child's development are not primarily due to parental malfunctioning has proven helpful to many parents. Mothers of problem children often develop guilt feelings because they assume that they are solely responsible for their children's emotional difficulties. This feeling of guilt may be accompanied by anxiety, defensiveness, increased pressures on the children, and even hostility toward them for "exposing" the mother's inadequacy by their disturbed behavior. When parents learn that their role in the shaping of their child is not an omnipotent one, guilt feelings may lessen, hostility and pressures may tend to disappear, and positive restructuring of the parent-child interaction can become possible. (Thomas et al., 1963, p. 94)

All this research highlights the importance of adjusting child-rearing practices to the individual infant. A given environment does not have identical functional consequences for all children; if you have siblings, you know this to be the case. Much depends on the child's own temperamental makeup. Babies are individualists from the moment they draw breath.

Children, then, are not simply passive beings, merely acted on by environmental forces that condition and reinforce them in one fashion or another (Rickman & Davidson, 1994). They are active agents in the socialization process; they are influenced by, but also influence, their caretakers. For instance, even very young infants seek to control their mothers' actions. The infant and the mother might be looking at one another. Should the baby look away and upon turning back find the mother's gaze turned away, the baby will start fussing and whimpering. When the mother again gazes at the infant, she or he stops the commotion. Infants quickly learn elaborate means for securing and maintaining their caretaker's attention (Lewis, 1998, 1995; Heimann, 1989). Indeed, as Harriet L. Rheingold (1968, p. 283) observes: "Of men and women [the infant] makes fathers and mothers." Thus, to a surprising degree, parents are the product of the children born to them. They themselves are molded by the very children they are trying to rear. Of particular importance in these complex processes are the nature and quality of the emotional bonds or ties that evolve between youngsters and their caretakers—in brief, attachment (Belsky, 1996a, 1996b).

Questions

How well do you rate yourself in the realm of emotional intelligence? How would you describe your own temperament in respect to the three basic concepts proposed by Thomas and colleagues?

Early Parenting and Caregiving Practices

Parent-Infant Interaction

In keeping with how strongly the psychoanalytic tradition has influenced U.S. life, researchers, clinical psychologists, and the court system have focused almost exclusively on the mother-child tie. The historic neglect of the father's role in child rearing has been linked with the notion that women are somehow more inclined toward child care and parenting than men are. One speculation is that biological predeterminers contribute to a more nurturant, "maternal" disposition in women and a more instrumental, "paternal" disposition in men (Rossi, 1977; Harlow, 1971; Erikson, 1964). Others suggest that the differences between men and women are products of the distinct, socially defined roles assumed by mothers and fathers (Parsons, 1955). The "sole caretaker" role is a heavy responsibility for a mother. If her child ever exhibits severe psychological or sociopathic disturbances (e.g., commits murder, rape, theft), the mother is immediately suspected of poor parenting, child neglect, or outright abuse.

Up until the 1990s, the American court system also favored the psychoanalytic view that mothers were the best parent and awarded child custody to mothers in the majority of cases. Only more recently has research by developmental psychologists examined the significant role that both mothers and fathers play in children's lives, above and beyond financial support.

When men abandon the upbringing of their children to their wives, a loss is suffered by everyone, but perhaps most of all by themselves. For what they lose is the possibility of growth in themselves for being human, which the stimulation of bringing up one's children gives.

Ashley Montagu

Social definitions of fatherhood have alternated over the past century between two poles: fathers as providers and fathers as nurturers (Atkinson & Blackwelder, 1993; Griswold, 1993). At the same time, the role of the father was largely ignored. Over the past 15 years, however, we have seen something of a revolution in the thinking of child developmentalists about the importance of fathers in the emotional, cognitive, and social development of young children (Biller, 1993). Increasingly, researchers are concluding that men have at least the potential to be as good caretakers of children as women are. Parke (1979) observed the behavior of both middle-class and lower-class parents of newborns on hospital maternity wards. Fathers were just as responsive as mothers to their infants' vocalizations and movements. Fathers touched, looked at, talked to, rocked, and kissed their babies in much the same fashion as mothers do. However, in response to their infants' vocalizations, fathers were more likely than mothers to increase their vocalization rate; mothers, in contrast, were more likely to react with touching. And fathers, when alone with their infants, are as protective, giving, and stimulating as mothers. In fact, fathers are more likely than mothers to hold their babies and to look at them. Fathers also play more physical games with their children. They tend to touch their infants in rhythmic tapping patterns. And they also

Fatherhood Social definitions of fatherhood have changed over the past century. Today, many fathers are taking a more active role in parenting. Each parent gives the child somewhat different kinds of experiences.

play more rough-and-tumble games, such as tossing the baby in the air (Parke, 1998, 1996; Power & Parke, 1983).

Moreover, being a father often makes a major contribution to a man's self-concept, personality functioning, and overall satisfaction with life. Men are increasingly coming to recognize that a closer relationship with children is beneficial to both child and adult. During the 1990s more men have taken advantage of an opportunity to be "stay-at-home Dads," doing much of the child care, while the mother has taken on the role of major breadwinner. In some American families, during the day the mother works and the father stays home with the children; then in the evening the father works and the mother takes over child care (or vice versa). Additionally, 4 percent of fathers today are raising children as single parents (U.S. Bureau of the Census, 1997).

Significantly, infants make a considerable contribution to the development of fatherliness through eliciting, evoking, provoking, promoting, and nudging fathering from men (Pruett, 1987). Not surprisingly, John Snarey (1993), drawing from a four-generation, four-decade study of men, finds that fathers who participate actively in child rearing are more likely to become "societally generative" at midlife. Viewed from the framework provided by Erik Erikson (see Chapter 2), such men in their middle years display "a clear capacity for establishing, guiding, or caring for the next generation through sustained responsibility for the growth, well-being, or leadership of younger adults or of the larger society" (Snarey, 1993, p. 98). In sum, good fathering really does matter both for the child and the father.

Fathers are important in still another respect. Studies show that a mother performs better in the parenting role when the father provides her with emotional support and encouragement (Belsky, 1996a, 1996b, 1990; Dickstein & Parke, 1988). The man who gives warmth, love, and ego gratification to his wife helps her feel good about herself, and she is then more likely to pass on these feelings to their child.

Good, Better, Best?   One should not conclude that either mothering or fathering is superior to the other. They are simply different. Each parent affords the child somewhat different kinds of experiences. However, many gender differences in parenting are not universal across cultures. For instance, many gender differences in American play styles are not found among Swedish fathers and mothers (Lamb et al., 1982). The ability to nurture is not the property of one sex or the other, and societies can differ considerably in their definitions of the parenting roles. For instance, in a survey of 141 societies, fathers in 45 societies (32 percent) maintained a "regular, close" or "frequent, close" proximity with the infant. At the other extreme, in 33 societies (23 percent), fathers rarely or never were in close proximity to the infant (Crano, 1998; Crano & Aronoff, 1978).

All of this is not to say that mothers and fathers are interchangeable; they make their own contributions to the care and development of children. Research suggests that the mother-child and father-child relationships might well be qualitatively different and might have different influences on a child's development (Biller, 1993; Harris & Morgan, 1991). Lamb (1977) finds, for instance, that mothers most often hold babies to perform caretaking functions, whereas fathers most often hold babies to play with them. Indeed, according to some studies, fathers spend four to five times as much time playing with their infants as in diapering them, feeding them, washing them, and the like. And Biller (1976, 1974) notes that whereas mothers are more likely to inhibit their children's exploration of the environment, fathers tend to encourage curiosity and challenge their children to attempt new cognitive and motor activities.

Rather than look for critical developmental phases in the lives of children, as did Freud and his followers, today's psychologists are increasingly examining children's everyday interactions with both their parents. They find that infants learn a good many lessons from the continuities that take place in these early relationships. It seems to be that small moments—not the dramatic episodes or traumas—give rise to most of the expectations that children evolve and bring to their later relationships. By way of illustration, consider infants' small acts of assertion that are associated with their development of autonomy. At 4 months of age children assert themselves by averting their eyes from the caretaker's eyes; by 12 months they can crawl or walk away; and at 18 months they can say no. Such acts of will allow children to evolve a sense of themselves as independent beings. How parents and other caretakers react to these acts can affect the infant's ability to achieve normal independence. Patterns of lifelong social relationships begin with such simple encounters, but, contrary to the Freudian view, the psychological imprints of the early years are not irrevocably set for the entire life span. Relationships throughout life continually shape and reshape our working schemes or models of social relationships. Viewed in this fashion, no single event in childhood is the source of a psychological difficulty in adulthood. Rather, the problems arise from an accretion of small episodes (Parke & Kellam, 1994; Luster & Okagaki, 1993).

Single Parenting and Absentee Fathers   Regarding fathering, there seems to be a number of crosscurrents at work in the United States. On the one hand, increasing numbers of children are living in homes where their fathers are not present (see *Further Developments:* "Effects of Fatherlessness on Children" in Chapter 4). The media has focused particular attention upon unwed fathers: Sensational anecdotes portray them as shiftless and irresponsible "deadbeats" who provide neither financial nor psychological support for their children. However, evidence is slowly emerging that challenges such generalizations and shows the diversity of this population (Jacobsen & Edmondson, 1993; Lerman & Ooms, 1993).

On the other hand, in homes where the father is present, fathers are taking a more active role in caring for their children (even though mothers still provide most of the child care and perform most of the household tasks). As more women have entered the paid workforce, the pressure on fathers to share child care has grown. For instance, 1991 Census Bureau data revealed that fathers with working wives were the primary caretakers for 23 percent of the nation's children under age 5 (7 percent of unmarried fathers cared for their children) (Chira, 1993a).

These trends have the utmost social significance, because research shows that fathering is not an unnecessary or superfluous matter. It seems that boys are more affected by the father's absence than girls are (Cooksey & Craig, 1998). In comparison with other boys, boys from fatherless homes exhibit less well internalized standards of moral judgment. They tend to evaluate the seriousness of misbehavior according to the probability of detection or punishment rather than in terms of interpersonal relations and social responsibility (Hoffman, 1971). Research data also reveal that the absence of a positive father-son relationship impairs a boy's overall academic achievement and his IQ test performance (Biller, 1993; Nugent, 1991). The extent of the impairment is greater the younger the boy was when he lost his father and the longer the father has been absent (Blanchard & Biller, 1971; Carlsmith, 1964). Moreover, one of the most consistently reported effects of the father's absence on a son is deterioration in school performance (Lamb, 1981). However, the mother's remarriage, especially if it occurs early in the child's life, appears to be associated with an improvement in intellectual performance (Santrock, 1972; Lessing, Zagorin, & Nelson, 1970).

Note, however, that the presence of the father is no guarantee of adequate fathering (Belsky, et al., 1991). The quality of the father-child relationship seems to be more important than the father's mere physical presence in the home (Lamb, 1981). Perhaps in the final analysis, the matter is not so much one of fathering or mothering as one of parenting.

Sibling-Infant Interaction

It is well known that the birth of a new baby has a significant impact on older siblings. All siblings are going to have to adjust and accept less attention from their parents, while still getting their needs met within the family situation. According to research by Dunn (1993), some older siblings become "initiators" with parents to get their needs met, while others become more withdrawn about losing attention to the newcomer in the family and

appear to resent the baby more. Wise parents prepare other children in age-appropriate ways for the arrival of a new baby in the family. Some families decide to have older children attend the birth of the new sibling in a birthing room or birth at home; others decide to introduce the new baby to siblings when they bring the baby home from the hospital.

From the wealth of research on infant cognitive, social, and emotional development over the past 30 years, we know that babies need and appreciate sensory and emotional stimulation, and an older sibling is just the perfect person to assist. In many cultures around the world, older siblings are also required to assist with the physical responsibilities of taking care of the baby (Dunn & Munn, 1985). Some older children have been known to cry "Take the baby back"; other siblings treat the new arrival as if it is their own beloved baby. Most siblings display a strong degree of caring, attachment, and protectiveness toward a new child in the family. Older siblings typically serve as the models for younger

Sibling Interactions and Culture Americans are often shocked to see children care for younger siblings in other cultures, but people from other cultures would be shocked to see that American babies spend so much time alone in infant carriers, cribs, and strollers.

siblings, and younger siblings want to "tag along" with older children. This can create tension and disagreement at times, but siblings normally learn to get along with each other and to share with each other. As we shall see in Chapter 18, we normally have our longest lasting relationships with siblings.

Question

If you have siblings, how important are your relationships with them now?

Grandparents and Extended Family Interaction

Today, nearly a million families in the United States are made up of grandparents raising one or more grandchildren. According to the U.S. Census Bureau, 5 percent of all U.S. children and 12.3 percent of African American children live with grandparents who are the primary caregivers (Casper & Bryson, 1998). Some of these grandparents are foster parents, while others have legal custody of their grandchildren. And some have no legal status, per se. In a 1996 survey, the American Association of Retired Persons (AARP) found the following to be the most common reasons why American grandparents are raising their grandchildren:

- Drug abuse (45 percent)
- Child abuse (21.5 percent)
- Abandonment (6 percent)
- Teenage pregnancy (6 percent)
- Parent unable (5 percent)
- Death of parent (5 percent)
- Other (institutionalization, or incarceration in jail, etc.) (11 percent)

Raising one's grandchildren has not always been expected to be part of the normal developmental sequence for the majority of American elderly. Across the country, grandparents are establishing support groups and gathering to share ideas, resources, and comfort. It is natural for grandparents to be angry and upset with their own child for being incapable of parenting the grandchildren. However, in the grandchildren's best interest, experts recommend that grandparents attempt to discuss the positive contributions of the parents with grandchildren. Many grandparents are capable of providing infants and young children with the love, care, stimulation, and security they need for normal emotional and social development.

Some grandparents love their grandchildren unconditionally and might tend to overindulge them, but others feel overburdened by child care at a later age in life. Raising grandchildren requires strength and energy, as well as additional financial resources (U.S. Department of Health & Human Services, 1997d).

How Important Are Early Child-Care Practices?

In American society today, child care is likely to include nonrelatives and day-care arrangements (Clarke-Stewart, Gruber, & Fitzgerald, 1994).

Child Day-Care Centers   Two important social forces have given impetus to the child day-care movement in the United States. First, more than half of U.S. mothers of preschool children are now in the workforce, compared with 14 percent in 1950 and 29 percent in 1970. Second, there has been a substantial increase in single-parent families. According to the Gallup Organization, most Americans (62 percent) believe that care in a child's own home is the best arrangement for infants, with child-care centers rating a distant second (16 percent) (Crispell, 1994). When it comes to toddlers, 44 percent still think that home is best, but 31 percent prefer child-care centers. However, most infants with working mothers spend the day with relatives in either their own home or another home. Yet home-based day care, even provided by relatives, is often of poor quality (Chira, 1994). Additionally, most parents strive for continuity in care, but about one-half of employed mothers who pay for child care change their arrangements each year (Shellenbarger, 1991) (see *Information You Can Use:* "Choosing a Child-Care Provider").

Many critics of day care say that children require continuity, stability, and predictability in their care. Followers of the psychoanalytic tradition emphasize that a child's emotional breadth and depth and capacity to love are derived from the experience of love in the early years. But in a center children must share the attention of a day-care worker with other youngsters. Add to this vacations and job turnover, and a child ends up with no one special person to be close to. Complicating matters, problem children are rarely the favorites of overburdened workers, yet they are usually the children who need love the most (Charlesworth & Hartup, 1967).

More than two decades ago Jerome Kagan, a noted Harvard developmental psychologist, also thought day care for children was a poor idea. But he has since reversed his opinion on the basis of additional research (Kagan, Kearsley, & Zelazo, 1978; Kagan, 1997). Kagan and his associates set up a day-care facility for 33 children from middle- and working-class homes in the South End neighborhood of Boston. The children entered the center at around 4 months of age and left it at 29 months. The caretakers were carefully selected on the basis of their nurturing qualities, and the center afforded a ratio of one worker for each three or four children.

Like other investigators of day care, Kagan and his associates compared the infants placed in day-care facilities with home-reared children of the same age, social

Information You Can Use

Choosing a Child-Care Provider

When you are looking for a child-care provider, remember the following five steps:

- Interview caregivers.
- Visit the center.
- Check references.
- Choose quality care.
- Stay involved.

Interview Caregivers

When interviewing caregivers, call around and ask them all these questions:

- What is the cost, and is there financial assistance available or a sliding scale based on income?
- How many children and what age groups do your work with?
- Do you provide transportation and/or meals?
- Do you provide hot meals or snacks?
- How long have you been in business?
- What is the philosophy of the center?
- Are you licensed, accredited, or certified?
- Can I make unannounced visits?

Visit the Center (More Than Once Is Recommended)

- Look for responsive, warm, and caring interactions between the adults and the children.
- See if the children are happy and comfortable with the caregivers.
- Examine both indoors and outdoors to see if the facility is clean, safe, and functional, especially places where children nap, eat, and use the bathroom.

- Make sure that a variety of toys, books, puzzles, and climbing equipment are available, because these provide a stimulating and interesting environment that will help nurture your child's physical and cognitive development.
- Make sure the children are getting individual attention.
- Watch how the caregivers resolve conflicts between children.

Ask Questions of the Center

- Can I visit anytime?
- How do you discipline the children?
- How do you handle a sick child? An emergency?
- What training have the staff received, e.g., CPR or First Aid certified?
- Are all children and employees required to be immunized?
- May I see a copy of your license and certification?
- Where and how long do the children nap?
- May I have a list of current and former parents who have used your center?

Check References (What to Ask Other Parents)

- Would you recommend this caregiver without reservations?
- Was the caregiver reliable?
- How did the caregiver discipline your child?
- Did you enjoy this caregiver and the center?
- Was the caregiver respectful of your values and culture?
- If you no longer use this center, why did you leave?

---

class, and ethnic background. The researchers found that the day-care children were not much different intellectually, emotionally, or socially from their home-raised counterparts. And the day-care children were neither more nor less attached to their mothers than the other children. However, at 29 months of age the home-raised youngsters were somewhat more socially advanced than the day-care children.

Much day-care research has been undertaken in centers associated with universities. Such centers have high staff–child ratios and well-designed programs directed at fostering the children's cognitive, emotional, and social development. Many of the child-care workers are highly motivated and dedicated students preparing for careers as

teachers. Yet most of the day care currently available to North American parents is not of this type and quality. Unfortunately, in most day-care centers, group size is large, the ratio of caretakers to children is low, and the staff is untrained or poorly supervised, all of which compromises a child's well-being (Belsky, 1993; Hofferth & Phillips, 1991). An additional problem associated with low-quality day-care facilities is that they commonly function as networks for spreading a variety of diseases, particularly hepatitis A, diarrhea, dysentery, and other intestinal illnesses. Because children enter and leave many of the centers in an erratic pattern, there is considerable opportunity for the mixing of infected and susceptible children (Cottle, 1998; "Early Child Care," 1998).

Good-Quality Day Care Although some day care in the United States is of poor quality, evidence suggests that high quality day care can have beneficial consequences for the social, intellectual, and emotional development of young children.

Check the Licensing and/or Referral Program

- What regulations must child-care providers meet?
- Are there any complaints on record against the provider I am considering?

Choose Quality Care

After examining your options, ask yourself:

- Which child-care provider will be best for my child?
- Which provider can meet my child's special needs?
- Does the provider share values similar to my own, or am I uncomfortable with the differences?
- Is the provider affordable and available when I need its services?
- Would I enjoy spending the day here?
- Do I feel good about my decision?

Stay Involved

What you can do after choosing a child-care provider:

- Try to arrange your schedule so you can talk with the caregiver and ask questions about your child's day.
- Talk to your child about the center and the caregiver.
- Visit your child at the center and observe activities.
- Anticipate problems that might occur and talk to the caregiver.
- Find out from various sources how your child is interacting; ask caregiver, other parents, other children.
- Network with other parents.
- Promote good working conditions for your child-care provider.

Most researchers recognize that we still have much to learn about the impact of day care on children, parents, the family, and social institutions in general (Roggman et al., 1994; Belsky, 1990). For now, perhaps the safest conclusion we can draw, based on research, is this: High-quality day care is an acceptable alternative child-care arrangement with possible benefits for both cognitive development and parent-child relationships (NICHD, 1999). Children display remarkable resilience. Infants around the world are raised under a great variety of conditions; the day-care arrangement is just one of them. The effects of day care depend to some extent on the amount of time a child spends at a center and on the quality of parent-child interaction when the family is to-

gether. And as we have pointed out in previous chapters, home care, in and of itself, does not guarantee secure attachments or healthy social and emotional development.

Note that the United States has not come to terms with the requirements and ramifications of rearing a generation of children outside the home while their parents work. Most states are lax in their regulation of day-care centers, and there are few federal policies and guidelines. Moreover, the vast majority of parents cannot afford to buy quality day care at full market prices. Yet, simultaneously, child-care workers are poorly paid: They earn 30 percent to 60 percent less per hour than kindergarten or elementary school teachers, and they earn 40 percent less than other workers with the same levels of education

(Cottle, 1998). It is even worse for the children of low-income households, where poor-quality child-care arrangements can compromise development (Sherman, 1997). In contrast, West European nations such as Sweden have well-developed and competent national child day-care systems, and parents are financially compensated for staying at home to raise children.

Multiple Mothering Traditionally in the United States, the preferred arrangement for raising children has been the nuclear family, which consists of two parents and their children. The view that mothering should be provided by one figure has been celebrated and extolled by many professionals as the key to good mental health. Yet this view is a culture-bound perspective, for children throughout the world are successfully reared in situations of **multiple mothering**—an arrangement in which responsibility for a child's care is dispersed among several people.

In some cases one major mother figure shares mothering with a variety of mother surrogates, including aunts, grandmothers, older cousins, non-kin neighbors, or co-wives. For instance, within the United States, Jacquelyne Faye Jackson (1993) shows that a multiple caregiver arrangement—based on shared caregiving by a number of parent figures irrespective of maternal marital status—is normative for African American infants. Another example of diffused nurturance is that found among the Ifaluk of Micronesia:

> For the Westerner, the amount of handling the infant receives is almost fantastic. The infant, particularly after it can crawl, is never allowed to remain in the arms of one person. In the course of a half-hour conversation, the baby might change hands ten times, being passed from one person to another. . . . The adults, as well as the older children, love to fondle the babies and to play with them, with the result that the infant does not stay with one person very long. . . . Should an infant cry, it is immediately picked up in an adult's arms, cuddled, consoled or fed. . . . There is little distinction between one's own relatives and "strangers." If he needs something, anyone will try to satisfy his need. Every house is open to him and he never has to learn that some houses are different from others. (Spiro, 1947, pp. 89–97)

Another fascinating approach to child care is found in the collective form of social and economic life in Israeli agricultural settlements (kibbutzim). From early infancy children are reared in a nursery with other children by two or three professional caretakers. Their own mothers visit them regularly and are primarily responsible for meeting their children's affectional needs. The burden of discipline and punishment falls primarily upon the professional caretakers (Devereux et al., 1974). Despite this arrangement of "concomitant mothering," systematic observation, testing, and clinical assessment have demonstrated that kibbutz children are within the normal range in intelligence, motor development, mental health, and social adjustment (Aviezer et al., 1994; Butler & Ruzany, 1993; Lamb, Sternberg, & Prodromidis, 1992).

Infants At Risk Attachment plays an important role in children's early development. But what are the consequences for children who lack caring parents? In other words, how important is "mother love" or "father love"? During the 1930s and 1940s studies by Margaret A. Ribble (1943), René Spitz (1946), and William Goldfarb (1945) did much to draw both scientific and humanitarian attention to the problems of homeless and neglected children. This work popularized the concept of maternal deprivation—the view that the absence of normal mothering can result in psychological damage and physical deterioration in children. In the intervening decades clinical and behavioral scientists have come to recognize that *maternal deprivation* is actually a catchall term. It encompasses many conditions, including insufficient sensory stimulation, the failure to form attachment bonds, the disruption of attachment bonds, unstable or rejecting mothering, inadequate intellectual stimulation, and even malnutrition. Because the term has become so highly charged and synonymous with all that is destructive in child care, most researchers use terms that refer to specific conditions, such as *social deprivation* and *sensory deprivation*. *Further Developments:* "Children Living in Poverty" discusses risks associated with childhood poverty.

If you must beat a child, use a string.

Talmud: Baba Bathra

Child Neglect and Abuse Children are dependent on parents and caregivers to take care of them during childhood, and the majority of American children are well cared for by their parents. However, according to the Center on Child Abuse Prevention Research, during 1997 a case of child neglect or abuse was reported every 10 seconds in the United States—a total of nearly 2 million reports and 1 million confirmed victims in one year (Wang, 1998). **Neglect** is defined as the absence of adequate social, emotional, and physical care (see Figure 6-3). Neglect cases make up a sizable proportion of cases in the child protection system (National Committee to Prevent Child Abuse, 1998). **Child abuse** is defined as nonaccidental physical attack on or injury to children by individuals caring for them (we will consider sexual abuse in Chapter 8). Much of the past research in this area of study has focused on physical abuse and not on emotional and social neglect.

Further Developments

Children Living in Poverty

Currently there are 14.5 million children living below the poverty line in America (Sherman, 1997). These children live in environments with an increased risk of lead poisoning, limited learning opportunities, and severe emotional distress due to family disintegration caused by economic strains. Their families cannot afford adequate housing, adequate nutritious food, or proper child care. This means that more than one in five U.S. children suffer in the following ways:

- *Health:* An increased risk of stunted growth, anemia, and less chance that the child will survive to her or his first birthday.
- *Education:* More repeated school years, lower test scores, and less education due to dropping out or expulsion.
- *Work:* Lower wages and lower overall lifetime earnings. In economic terms, the costs of keeping poverty-stricken children in schools longer, as well as having to supply them with free breakfasts and lunches, special education services and tutoring, coupled with large medical expenses due to initial poor health, means that every year that we decide not to address issues of child poverty translates into a monetary loss of as much as $130 billion. To get an idea of how children in poverty live, imagine that your family must survive on $11 a day, which according to the Census Bureau is the amount you would receive if your family income were at the poverty threshold ($12,516 for a family of three). Out of that $11 per day, your family needs to eat, pay rent, and be clothed—plus meet any other expenses incurred on a day-to-day basis. For most people living in poverty, over 30 percent of their income is spent on housing, so in your hypothetical case, you immediately take our $3.50 to keep a roof over your head, leaving you with $7.50 for the day.

Thought Experiment

You have $52.50 to live on this week ($7.50 × 7 days). See how quickly that amount disappears when you buy just the food you plan on eating over the next 7 days. Then you realize that it takes more than one full day's amount of money to put a week's worth of gasoline in the car. In other words, if you put $10 of gas in the car, you do not have enough for food for the week. A pair of jeans could cost a few days' or even the whole week's budget. You can forget buying CDs, going to the movies, or playing any sport that requires equipment, such as tennis, golf, or skiing. Even going to the doctor's office can be so expensive that you might have to decide to avoid getting health care you need. Children raised in poverty are:

- Twice as likely to die from accidents. Why do you think this is true? Give an explanation.
- Three times more likely to die. Why do you think this is true? Give an explanation.
- Four times more likely to die in fires. Why do you think this is true? Give an explanation.
- Five times more likely to die from infectious diseases. Why do you think this is true? Give an explanation.
- Six times more likely to die from other diseases. Why do you think this is true? Give an explanation.

Seen from a developmental perspective, children's health, emotional and cognitive growth, and social interactions are all adversely affected by the harsh environments caused by poverty. In short, poverty steals the promise from a child's future, and in the end, it impacts everyone's development.

This information was adapted from *Poverty Matters* (p. 29) by A. Sherman (1997) from the Children's Defense Fund. For further information, visit CDF on the web, *www.childrensdefense.org*, or write CDF, Family Income Division, 25 E Street NW, Washington, DC 20001.

To generalize about parents who abuse children is difficult. Multiple factors are usually involved, and these vary for different individuals, times, and social environments. Increasingly, however, researchers are looking at child maltreatment from an ecological perspective and examining the complex social context and web in which the behavior is embedded (Baumrind, 1994; Belsky, 1993). Research suggests that child abuse is more prevalent among families living in poverty (Sherman, 1997), but child abuse is not confined to lower-socioeconomic-status households; it is found across the class spectrum. Child abuse is also related to social stress in families. For instance, high levels of marital conflict, interspousal physical violence, and job loss are associated with a higher incidence of child maltreatment (Dodge, Bates, & Pettit, 1990). In addition, child abuse is more common among parents suffering from mental illness and substance addiction (Walker, Downey, & Bergman, 1989). And families that are socially isolated and outside neighborhood support networks are more likely to abuse children than are families with rich social ties (Trickett & Susman, 1988).

Psychiatrists Brandt G. Steele and Carl B. Pollock (1968) made intensive studies of 60 families in which

Figure 6-3 Social and Emotional Deprivation Occurs Across Socioeconomic Class Lines

"Quality time . . . quality time . . ."

significant child abuse had occurred. The parents came from all segments of the population: from all socioeconomic strata, all levels of intelligence and education, and most religious and ethnic groups. Steele and Pollock found a number of elements common to many child abusers. On the whole, these parents demanded a great deal from their infants, far more than the babies could understand or respond to:

> Henry J., in speaking of his 16-month-old son, Johnny, said, "He knows what I mean and understands it when I say 'come here.' If he doesn't come immediately, I go and give him a gentle tug on the ear to remind him of what he's supposed to do." In the hospital it was found that Johnny's ear was lacerated and partially torn away from his head. (Steele & Pollock, 1968, p. 110)

The parents also felt insecure and unsure of being loved, and they looked to the child as a source of reassurance, comfort, and affection. A parent, Kathy, made this poignant statement:

> "I have never felt really loved all my life. When the baby was born, I thought he would love me; but when he cried all the time, it meant he didn't love me, so I hit him." Kenny, age three weeks, was hospitalized with bilateral subdural hematomas [multiple bruises]. (Steele & Pollock, 1968, p. 110)

The Intergenerational Cycle of Violence Steele and Pollock found that all 60 child abusers studied had been raised in the same authoritarian style that they were recreating with their own children. Indeed, abusive parents tend to be less satisfied with their youngsters and to

experience child rearing as more difficult and less enjoyable than do nonabusive parents (Trickett & Susman, 1988). Other researchers have confirmed that abusive parents are themselves likely to have been abused or to have witnessed domestic violence when they were children (Simons et al., 1991). Indeed, evidence suggests that the pattern is unwittingly transmitted from parent to child, generation after generation—what researchers and professionals have called a "cycle of violence" and the "intergenerational transmission of violence."

Proponents of social learning theory say that violent, aggressive children have learned that behavior from their parents, who are powerful models for children (Tomison, 1996). Some researchers contend there is a biological or genetic component to aggressive behavior, and that aggressiveness is an individual characteristic based on the child's own temperament (Muller, Hunter, & Stollack, 1995)—that is, the child's inherited disposition perpetuates the cycle of maltreatment. The third explanation for intergenerational transmission of violence is the interaction of environmental (social learning) and biological/genetic factors. Kaufman and Zigler (1993, quoted in Tomison, 1996) suggest that a genetic component for the expression of antisocial behavior puts the individual at risk for expressing violent behavior, and the interaction of both genetic and environmental factors produces the greatest risk for acting violently. It is generally acknowledged that no single factor can explain how maltreatment is transmitted generationally.

Even so, having been abused does not always lead to being abusive: Among adults who experienced abusive childhoods, between one-fifth and one-third abuse their own youngsters (Kaufman & Zigler, 1987). However, the greater the frequency of violence experienced in childhood, the greater the chance that the victim will grow up to be a violent parent (Straus, Gelles, & Steinmetz, 1980). Milner and colleagues (1990) state that merely witnessing physical abuse as a child has been associated with an increased abuse potential as an adult. Corby (1993) found that the vast majority of studies (with the exception of incest research) have focused investigations on mothers' behaviors, despite the fact that men account for more than half of all physical abuse.

Psychiatrists estimate that 90 percent of abusing parents are treatable if they receive competent counseling (Helfer & Kempe, 1977). Most parents want to be good parents. Parent education programs—classes that teach parenting skills—often help prevent fathers or mothers who have abused their children from doing so again (Peterson & Brown, 1994).

Abusing parents usually do not abuse all their children; commonly they select one child to be the victim. Some children appear to be more "at risk for abuse" than other children. They include children who were premature infants, were born out of wedlock, possess congenital anomalies or other handicaps, or were "difficult" babies. Overall, a child viewed by an abuse-prone parent as being "strange" or "different" is more at risk than are other children in the family (Brenton, 1977). *Human Diversity:* "Interaction with Infants with Disabilities" suggests ways to provide good parenting to infants born with disabilities.

Signs of Abuse and Maltreatment   As mentioned above, children who are born at-risk are more likely to be mistreated. Maltreated children show a variety of symptoms. Because teachers are the only adults outside the family whom many children see with any consistency, they are often in a position to detect signs of child abuse or neglect and to begin to remedy the situation by reporting it to the proper authorities (Besharov, 1990). In fact, most states require teachers and health care professionals to report cases of child abuse, and the law provides them with legal immunity for erroneous reports made in good faith. Teachers and others in educational settings typically account for the highest percentage of reporting abuse and neglect. The American Humane Association has published a list of the signs teachers should look for as possible tip-offs of child abuse or neglect. They include the following:

- Does the child have bruises, welts, or contusions?
- Does the child complain of beatings or maltreatment?
- Does the child frequently arrive early at school or stay late? (The child may be seeking an escape from home.)

- Is the child frequently absent or late?
- Is the child aggressive, disruptive, destructive, shy, withdrawn, passive, or overly compliant?
- Is the child inadequately dressed for the weather, unkempt, dirty, undernourished, tired, in need of medical attention, or frequently injured?

In addition, being neglected or abused as a child increases an individual's later risk for delinquency, adult criminal behavior, and violent criminal behavior. Even so, the majority of neglected and abused youngsters do not become delinquent, criminal, or violent. A good many other events in children's lives—for instance, their natural abilities, their temperaments, their networks of social support, and their participation in therapy—may mediate the adverse consequences of child abuse and neglect (Widom, 1989a, 1989b; Egeland, Jacobvitz, & Sroufe, 1988).

Neglect puts great burdens on youngsters. Psychologist Byron Egeland's longitudinal study (1993) of children who are at risk because of poor quality of care found that emotionally unresponsive mothers tend to ignore their youngsters when the children are unhappy, uncomfortable, or hurt; they do not share their children's pleasures; and consequently, the children find that they cannot look to their mothers for security and comfort. Both physically abused and emotionally deprived children typically have low self-esteem, poor self-control, and negative feelings about the world. However, whereas physically abused youngsters tend to show high levels of rage, frustration, and aggression, those reared by emotionally unavailable mothers tend to be withdrawn and dependent and to exhibit more severe mental and behavioral damage as they become older. Because they come to view and experience the world in deviant ways, many of them later perpetuate the abusive patterns of their parents and mistreat their own children (Dodge, Bates, & Pettit, 1990).

Breaking the Cycle of Violence   Several researchers find there are several factors that seem to "buffer" children from being abused. These include mothers having strong social support systems, being involved in community activities, having less rigid expectations about what children should be able to do, encountering fewer stressful life events, having a supportive partner/spouse, making a conscious decision not to repeat the history of abuse, having positive school experiences as a child, and having a strong, supportive religious affiliation (Tomison, 1996). Fry (1993) suggests several approaches to breaking the cycle of violence:

- Promoting a cultural attitude that physical force is unnecessary and unacceptable (outlawing corporal punishment)

Human Diversity

Interaction with Infants with Disabilities

Every child comes into a family somewhat like a rock thrown into a pond. The ripples caused by the new arrival affect everyone. Nobody in the family remains exactly the same. Everyone changes.

Perske, 1981

I Am Just Like You

Everybody quietly hopes for an exceptional infant, but few parents get one, at least not exceptional in the "gifted" sense. Many people who eventually become great leaders (such as Winston Churchill, Nelson Rockefeller), scientists (such as Albert Einstein, Stephen Hawking), or artists (such as Stevie Wonder, Whoopie Goldberg) do so out of a personal need to compensate for deficiencies and defects in their lives—often astounding their parents in the process.

Parents who give birth to a child with disabilities often experience a variety of strong emotions, including denial, anguish, pain, guilt, panic, depression, and a deep sense of loss. Some parents of children with handicaps confess experiencing all of these feelings. The best advice is to find professionals, advocates, relatives, and friends who will enter your struggle with you and support you and your "special" child. Professionals in the field of special education and disabilities encourage parents to do the following:

- Love the child exactly as he or she is. Look at your child, touch your child, speak lovingly to your child, and treat your child as normally as possible. Your child is a person first!
- Start early intervention services as soon as possible after birth. You will connect with professionals and advocates knowledgeable about your child's needs. Be aware that we live in a special time in history when there is more professional knowledge than ever before about disorders, birth defects, and children born at-risk (Greenspan, 1991).
- Ask questions and get informed. If support services are not available in your community, find parental and professional support via professional journals or over the Internet. Parents need to know they are not alone. Even parents informed of a rare diagnosis can connect with someone else in the world who will understand and who is experiencing the same challenge.
- Be aware that the family expends extra energy when a youngster has a disability, and over time most parents discover untapped energies and resilience to become people they never knew they could become.

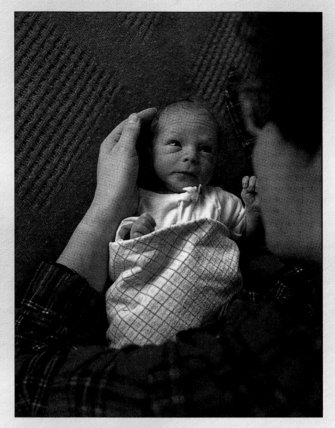

Attachment and Infants with Disabilities   Emily, born with a heart condition, has difficulty eating and requires a feeding tube. Her mother views infant massage as a way to enhance her infant's emotional and physical health. Some pediatricians say massage creates emotional contact and attachment between the caregiver and infant.

- The practice of inclusion allows all children with a handicap to be educated in an age-appropriate regular classroom environment in the public school system. It is likely that an atypical child is eligible for free nursery and preschool programs as well.
- Recognize the main developmental principle that all children, regardless of their abilities, have strengths and skills that will emerge at their own individual rate of growth. Celebrate your child's developmental milestones.
- Recognize that your child can become a positive, contributing member of society in some way.
- Be aware that change can be for the better.

- Training all children in nonviolent conflict resolution and problem solving
- Training parents in healthy child-rearing techniques
- Intervening in abusive situations as soon as possible

Questions

Why do child neglect and abuse occur? Is there an association between these harmful behaviors and socioeconomic status or parenting style?

Seque

This chapter has documented social and behavioral scientists' long-standing interest in how children's personality development is related to their early emotional and social experiences. Initially social scientists focused upon maternal deprivation, believing that it was enough simply to ask about the mother's influence on the child. As time passed, research began to focus on fathers, and then siblings, grandparents, aunts, uncles, and other extended family members. The circle has widened even further in recent years to child-care providers and preschool teachers.

Sharply divergent views exist among Americans regarding the desirability of day-care facilities. The impulse to decide whether day care is "good" or "bad" has at times seemed more a matter of ideology than an issue for science. In any event, modern societies and social scientists are increasingly confronting this question: How are we to manage the successful care and rearing of future generations of children when parents spend a substantial portion of time at work away from home?

As we examine in the next section, children during the preschool and early elementary school years continue to grow, experience remarkable cognitive development, and experience a diversity of social influences.

Summary

The Development of Emotional & Social Bonds

1. Humanness is a social product that arises as children interact with the significant people in their environment.
2. Children reared in institutional settings generally do not develop well compared with children reared in home surroundings. The impairment experienced by many institutionalized children derives from social and sensory deprivation.
3. Freudians stress that the development of emotionally healthy personalities is associated with breast-feeding, a prolonged period of nursing, gradual weaning, a self-demand nursing schedule, delayed and patient bowel and bladder training, and freedom from punishment. However, research has provided little or no support for these psychoanalytic assumptions.
4. Erik Erikson stressed the importance of an infant's early years in fashioning a basic trust in others, an essential foundation for successful social functioning.
5. Behaviorists are concerned with the outward display of emotions through observable behavior, not the thoughts that caused the emotions. They employ a system of rewards and punishments to shape children's behaviors, which in turn, they believe, changes the children's feelings.
6. According to the cognitive view, emotions have a signifcant impact on our thoughts and behaviors.
7. Bronfenbrenner's ecological theory proposes that the emotional and social development of children are influenced by several levels of environment: the home and family, the neighborhood, the school and church, the local community, and national and cultural influences.
8. Psychologists influenced by ethological theory have played a central role in the recent burst of studies on the emotional life of children. Carroll E. Izard finds that babies have intense feelings from the moment of birth. At first their inner feelings are limited to distress, disgust, and interest, but as babies mature, they develop new emotions in a regular, orderly fashion.
9. Attachment behaviors promote proximity between helpless, dependent infants and protective caretakers. But attachment also fosters social cognitive skills. Psychologists commonly use the strange-situation procedure to capture the quality of the attachment in the parent-child relationship. The patterns of attachment behavior revealed by the strange-situation procedure reflect the quality

of maternal caregiving children receive during their first 12 months of life. Attachment behaviors vary across cultures.

10. A good many children display stranger anxiety during the second six months of life. However, infants are both attracted to and wary of novel objects, so the situation does much to determine which reaction will be activated.

11. Emotions seem to have evolved as adaptive processes that enhance survival. They influence the mental operations we use in information processing. And by reading other people's facial, gestural, postural, and vocal cues, we gain indirect access to their emotional states. Children appear to progress through several stages in emotional development.

12. Gardner and Goleman present the idea that emotional intelligence involves emotions useful in successful adaptations in relationships and at work.

13. Emotional expression in youngsters is associated with their differing temperaments. The temperamental qualities that are studied most often by developmental psychologists are typically those that are also obvious to parents. They include irritability, a happy mood, ease of being soothed, motor activity, sociability, attentiveness, adaptability, intensity of arousal, regulation of arousal states, and timidity.

Early Parenting and Caregiving Practices

14. The mother-child and father-child relationships are qualitatively different and apparently have a different impact on a child's development. Historically, in most cultures the mother has been considered the nurturer and the father has been the breadwinner. In recent years, many American mothers have also taken on the role of breadwinner, and some fathers have adopted a more nurturant style of parenting. In the United States there is a growing number of single mothers and absentee fathers.

15. Siblings play an integral part in the emotional and social development of infants and young children. In some cultures, older siblings take on much of the responsibility for caring for younger ones.

16. Nearly 5 percent of grandparents in the United States are the primary caretakers of their grandchildren. This added responsibility requires more energy and financial resources.

17. Infants around the world are raised under a great variety of conditions, including multiple mothering, home care, and child day care. Children display remarkable resilience. Research suggests that multiple mothering and high-quality day care are acceptable child-care arrangements.

18. The impact of child abuse and neglect is far reaching. Studies show that adult abusers were abused as children themselves. Much research and new programs are being conducted to institute societal programs to break this cycle of violence. In the meantime, many professionals affiliated with children have a responsibility to report suspected neglect and abuse.

19. With more babies being born at-risk or prematurely, more newborns have disabilities and special challenges. These infants are people first and need to be loved and cherished as special people, too. Throughout history, many people who started life with some type of disability or handicap have made great contributions to humankind.

Key Terms

attachment *(167)*	multiple mothering *(184)*	strange situation *(170)*
child abuse *(184)*	neglect *(184)*	stranger anxiety *(171)*
emotion *(172)*	separation anxiety *(171)*	temperament *(176)*
emotional intelligence *(175)*	social referencing *(173)*	

Following Up on the Internet

Web sites for this chapter focus on emotional and social development in infancy. Please access the text web site at http://www.mhhe.com/crandell7 for up-to-date hot-linked Internet addresses for the following organizations, topics, and resources.

Child Development Abstracts

Single Parenting

Institute for Research on Poverty

National Clearinghouse on Child Abuse & Neglect

Our Kids: Devoted to Children with Special Needs

Early Childhood News

Part Four

Early Childhood
2 to 6

Chapter 7 is the first of two chapters focusing on early childhood, the developmental stage between 2 and 6 years of age. During this period children acquire autonomy, evolve new ways of relating to other people, and gain a sense of themselves and their effectiveness in the world. Healthy children experience physical growth, coordination of motor skills, and an energetic zest for play. Proper nutrition, good health, and stimulating sensory experiences provide a foundation for continued cognitive growth and language development. Children also begin to learn a sense of right or wrong based upon preoperational thought processes. In Chapter 8, we will examine the young child's growing self-awareness in the domains of emotions and gender. Within the social context of family and friends, child-care settings, and kindergarten, young children acquire a set of guidelines about expressing their emotional needs.

Chapter 7

Early Childhood
Physical and Cognitive Development

Critical Thinking Questions

1. How do we know what a young child needs for growth of both brain and body? Why are injuries the leading cause of death in young children?

2. A young girl 5 years old can hear a piece of music once and then sit at the piano and play it perfectly, but she does not know her name and will never learn to read words or music. Is she intelligent? What does it mean to be intelligent, and how do we measure intelligence?

3. Why is it that an 8-year-old child is unlikely to remember her first trip to the zoo taken at age 3? Can you remember what you ate for breakfast one week ago? Why or why not?

4. Imagine that you are shipwrecked on an island where everyone wears masks so that you cannot see each other's faces. How would you be able to figure out what other people were really thinking? And how do you think your communication would be different?

Between the ages of 2 and 6, children enlarge their repertoire of behaviors. As young children develop physically and cognitively, they become capable beings in their own right. Most are healthy, energetic, and curious about mastering their world. Their growing bodies and increasing strength permit them to climb higher, jump longer, yell louder, and hug harder. For young children, every day is truly a new day. They are expanding their vocabulary, asking questions, and entertaining with their own wit and humor.

As Erikson suggests, young children begin to struggle with their own conflicting needs and rebel against parental controls while acquiring a sense of autonomy or independence. These occasional upheavals are popularly called "the terrible twos," when toddlers begin to assert their own will and have temper tantrums. At the same time, children come to see themselves as individuals who are separate from their parents, though still dependent on them (Crockenberg & Litman, 1990; Erikson, 1963). How does the young child's mind operate to remember what is significant and what is trivial? We will examine several theories of early cognitive development and memory, as well as physical and moral development.

Passing hence from infancy, I came to boyhood or rather it came to me, displacing infancy. Nor did that depart—(for whither went it?)—and yet it was no more.

St. Augustine, *Confessions*

Physical Development

Early childhood lays the cognitive and social foundations for the more complex life of the school years. Underpinning these intellectual skills are continued brain growth, physical development, the perfection of gross and fine motor skills, and maturation of the sensory systems. These developments can take place in a normal progression only if children are healthy and receive the proper nutrition, exercise, and health care to support their developing body systems. Among the most encompassing factors that impede both physical and cognitive development in early childhood are the far-reaching outcomes of living in poverty. A special section of this chapter is devoted to health risk factors for many American children, particularly minority children.

Physical Growth and Motor-Skill Development

As we have noted in earlier chapters, growth is unevenly distributed over the first 20 years of life. From birth to age 5, the rate, or velocity, of growth in height declines sharply. You may have heard it said that "young children sprout like weeds," since about twice as much of this growth occurs between the ages of 1 and 3 as between the ages of 3 and 5. After age 5, the rate of growth in height levels off so that the velocity is practically constant until puberty. Additionally, relative to the growth norms of their age group, broadly built children tend to grow faster than average, and slenderly built children slower than average (Tanner, 1971).

One of the most striking and perhaps most fundamental characteristics of growth is what James M. Tanner (1970, p. 125), a noted authority on the subject, calls its "self-stabilizing" or "target-seeking" quality:

Children, no less than rockets, have their trajectories, governed by the control systems of their genetical constitution and powered by energy absorbed from the natural environment. Deflect the child from its growth trajectory by acute malnutrition or illness, and a restoring force develops so that as soon as the missing food is supplied or the illness terminated, the child catches up toward its original curve. When it gets there, it slows down again to adjust its path onto the old trajectory once more.

Thus, children display a compensatory or remedial property of "making up" for arrested growth when normal conditions are restored (unless the cause of the interruption is severe or prolonged).

During the preschool and early elementary years, children also become better coordinated physically. Walking, climbing, reaching, grasping, and releasing are no longer simply activities in their own right but have become the means for new endeavors. Their developing skills give children new ways to explore the world and to accomplish new things (see Table 7-1). It is common today to see a young child riding a two-wheel bike (without training wheels), to be throwing a Frisbee with parents at the park, and to be agile on the ski slopes.

Gross Motor Skills   Healthy young children ages 2 to 6 definitely are not couch potatoes, as you can see from this chapter's opening photograph of children busy with a colorful parachute. At this age, children run, jump, or hop every chance they get. Their arm and leg muscles are developing, and children in this age group need and benefit from plenty of exercise every day. Four-year-olds are more comfortable with their bodies and push their physical limits by exploring jungle gyms and other play structures. Coordination between upper and lower body develops, and tasks like running are done much more efficiently. It has been well established that young children go through three distinct stages of learning to walk and reach a mature pattern of walking by approximately 4 years of age (Lee & Chen, 1996). Five-year-olds can be daredevils; they swing, jump, and try acrobatics that

Table 7-1   Motor and Skills Development Among Preschoolers

Age 2	Age 3	Age 4	Age 5
Can run	Can stand on one foot	Can do stunts on a tricycle	Can skip
Can kick a large ball	Can hop on both feet	Can descend a ladder, alternating feet	Can hop on one foot for 10 feet
Can jump 12 inches	Can ride a tricycle	Can gallop	Can copy squares
Can navigate stairs alone	Can propel a wagon with one foot	Can cut on a line with a scissors	Can copy letters and numbers
Can construct tower of six to eight blocks	Can copy a circle	Can make crude letters	Can throw a ball well
Can turn pages of a book singly	Can draw a straight line	Can catch a ball with elbows in front of the body	Can fasten buttons that are visible to the eye
Can put on simple clothing	Can pour from a pitcher	Can dress self	Can catch a ball with elbows at the sides
Can hold a glass with one hand	Can catch a ball with arms extended		

cause their parents to hold their breath. It is hard to believe that the 5-year-old who proficiently skates and skips found it difficult to walk very far without falling down just a few years previously. Children with any type of physical disability should still be encouraged to be active in any way they can so that they can develop strength and coordination skills and enjoy the physical act of movement. Recreational therapists can suggest activities that individual children can accomplish.

Fine Motor Skills   Whereas gross motor skills are the capabilities involving larger body parts, fine motor skills involve small body parts. Fine motor skills develop more slowly than gross motor skills, so 3-year-olds who no longer need to concentrate intensely on the task of running will still require much mental energy to stack blocks, construct with Legos, use a paintbrush, sculpt with play dough, maneuver a crayon, or tap keys on a computer keyboard. They still tend to force a puzzle piece into the hole or slide and wiggle it until it pops into place. Five-year-olds normally have their hands, arms, legs, and feet under tight command and are bored with the simple acts of coordination, preferring instead to walk on a balance beam, build high block structures, and begin the task of tying shoelaces. Interestingly enough, today with the use of Velcro on young children's shoes and sneakers, many children are delayed in learning the necessary task of tying shoelaces until they are 8 or 9 years old.

Unfortunately, some 5 percent of youngsters have noticeable difficulties with coordination, and perhaps 50 percent of the children who have these problems at age 5 still have them at age 9. Increasingly, psychologists and teachers are paying attention to these children with poor physical coordination. Assisting less adroit children to become more successful at physical activities can be important. Motor skills form a large part of youngsters' self-concepts and how they perceive others. Researchers find that children who have coordination problems are at greater risk for significant social problems later in elementary school because clumsiness often interferes with youngsters' social relationships. Boys perceived as less adept are most frequently impacted: They tend to have fewer friends than their more coordinated peers. Psychologists and educators are developing ways to help these children improve in areas that initially might seem unrelated (Kutner, 1993).

Children maintain the top-heavy look—the head being large relative to the body—until the end of the preschool years, but they become thinner and lose the baby fat that characterizes the infant and toddler. Children under 2 tend to be chubby, whereas children 2 to 6 years old are, for the most part, slim. International height differences have been attributed to ethnic origin and to nutrition, which is related to income—low-income rural children are shorter than higher-income urban children. Other factors that contribute to height differences in-

Fine Motor Skills Involve More Complex Hand-Eye Coordination   Patrick, 4, is describing his newest creation from Legos, which required extensive use of fine motor skills. Putting together blocks, beads, puzzle pieces, and such are favorite activities of young children.

clude whether the mother smoked during pregnancy and, in the United States, race—blacks are on average taller than whites. So you can see that height and weight variations in children stem from both genetic and environmental factors.

Question

Why would a pediatrician be worried if a child's fine or gross motor skills seemed delayed?

Sensory Development

Visual, Tactile, and Kinesthetic Senses   Once the infant develops the physical coordination to grasp objects he sees, caretakers will notice the child stroking objects of various textures, putting any new objects in the mouth, patting the dog or cat, or pulling the earrings off adults' ears! The tendency to put objects in the mouth persists well into toddlerhood. To develop their visual and tactile senses, young children not only explore objects visually but enjoy touching them, especially interesting materials such as sand, water, food, grass, finger paints, play dough, soap, washcloths, feathers, and wooden floors. They use their mouth, hands, and feet to help them discover all the fascinating differences in the objects in their environment. The mess the child will make is minor compared to the child's happiness and the complex neural connections developing in the child's brain because of these experiences. Eventually the young child will be able to throw a ball, use a fork for feeding, flush the toilet, open the pages of a book, get dressed, and so on, utilizing the maturing visual, tactile, and kinesthetic senses of the muscles for deliberate movements.

Hearing and Language Development

A baby's language capabilities are dependent upon a healthy auditory system as well as the growth and development of muscles in the mouth, tongue, and larynx. The sounds made by babies less than a year old are considered to be "universal" across cultures, and a child can listen to and learn the sounds of more than one language during this time. The auditory sense can be temporarily impaired by colds, ear infections, sinus congestion, sore throats, and allergies (particularly by the phlegm that can build up if the child is allergic to dairy products). One of the most common illnesses at this time is called *otitis media,* a type of inner-ear infection that causes fluid buildup in the child's ear. This is extremely painful. The typical symptom is that the child will be crying or screaming in the middle of the night, inconsolably, after lying down for a period of time has caused the fluid to settle painfully in the ear. Typically medication is needed to reduce the accumulation of fluid. A child who has a series of ear infections over a short period of time, or for whom medical attention is delayed, might have some hearing loss. Chronic hearing problems interfere with learning and using language and can therefore cause serious delays in cognitive development if left undetected.

Olfactory and Gustatory Sensations

Children exhibit a range of responses to smells in the environment and tastes of foods and beverages. Some children are more sensitive to these sensory stimuli than others. Because both smell and taste are involved in eating foods, mealtime is when parents notice a "finicky" eater or a child who eats anything you put on the plate. The combination of the child's developing taste buds and the intensity of new sensations as foods are introduced into the child's diet can make for very interesting mealtimes. Most young children do not readily accept spicy, strong-tasting, bitter, or pungent foods (such as mashed spinach for toddlers!) Consequently, sweeteners are added to

Information You Can Use

Young Children and Lead Poisoning

One of the most serious health hazards for young children is lead poisoning. Lead is often found in old paint on walls of older homes and apartment buildings and in tap water drawn from old lead plumbing. It is estimated that in the United States alone, 2.5 million children under age 12, including 1 in 4 black children younger than age 6, suffers lead poisoning (Brody et al., 1994). Young children who live in old homes and apartments are known to put the peeling paint chips in their mouths and chew them. Unfortunately, lead exposure is linked with attention deficit disorder, hyperactivity, aggression and antisocial behavior, delinquency, and later crime (Brody et al., 1994). Another medical research study associates lead in children's blood to lower IQ scores, poorer attendance rates, reading disabilities, and poorer vocabulary test scores (Needleman et al., 1996). Children under the age of 5 who live in poverty are four times more likely than children in high-income families and three times more likely than children in middle-income families to have dangerous levels of lead in their bloodstream. The fol-

lowing measures would reduce the danger of harm from lead exposure:

- Improve the child's nutrition. Adequate levels of iron, calcium, and other nutrients reduce the absorption of lead into the blood.
- Move into newer or better maintained homes.
- Test both paint and water for lead levels.
- Purchase filters to remove lead from tap water.
- Repaint and clean a home regularly to reduce dust from old layers of lead paint.

Prenatal exposure to lead has been linked with low birthweight, stunted growth, hearing loss, and damage to children's blood production and kidney development. The outcome of this health hazard is associated with infant mortality, high costs of neonatal care, a lifetime of special education services and/or criminal justice penalties, and a reduced earning potential for a lifetime.

Adapted from Arloc Sherman, *Poverty Matters: The Cost of Child Poverty in America* (Washington DC: Children's Defense Fund, 1997), p. 16. Reprinted by permission.

children's liquid medications. Identifying for a child what a smell is (a rose) or what a taste is (strawberry) helps the child learn to recognize and cognitively categorize smells and tastes. One important reminder here, though, is that all dangerous substances, such as household cleansers, should be kept out of the child's reach, as should all potted plants with poisonous leaves. Caretakers should point out to children what *not* to put into their mouths (the dog's food, a marble on the floor, soap or shaving cream, etc.). One benign-appearing thing young children will put in their mouths and swallow is chips of old paint containing lead; swallowing substances containing lead can seriously harm the child's developing brain and nervous system (see *Information You Can Use:* "Young Children and Lead Poisoning").

The Brain and the Nervous System

The brain and central nervous system normally continue rapid development during early childhood. A rich complexity of connections among neurons will continue to develop if the child is stimulated through using her senses and through developing social interactions with others. At the age of 5, the child's brain will weigh about 90 percent of its final, adult weight, whereas the child's body will be only about one-third of its final weight. When you see a 6-year-old wearing an adult's hat, you will see that the differences in head size are not so great. Given that children's brains are quickly increasing in size, it is no surprise that their increasingly complex cognitive abilities are impressive.

As children's physical worlds expand, typically accompanied by new social horizons, children are confronted with new developmental requirements. They actively seek new opportunities for affecting and regulating their environment and, in doing so, achieve a sense of their own effectiveness. These processes underlie and stimulate their cognitive development. Children who are born at risk (for instance, due to premature birth, birth defects, or being born into poverty) are likely to experience slower development and slower maturation of their senses and therefore might not get the opportunity to develop to their full potential. These children are eligible for intensive early intervention services, and an Individual Family Service Plan (IFSP) will be developed through the young child's public school district (see Figure 7-1).

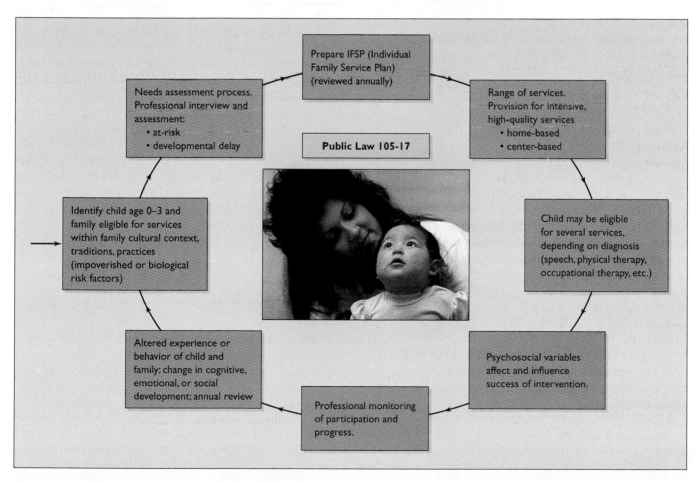

Figure 7-1 The Early Intervention Services Process Many young children born at-risk or living in poverty conditions are considered eligible for an Individual Family Services Plan (IFSP) through their local public school district. Seven-month-old Sabrina's 29-year-old mother drank during her pregnancies. Another 3-year-old child in the family also has been diagnosed with FAS (fetal alcohol syndrome).

Nutrition and Health Issues

By the second year of life, toddlers will be able to eat most family foods at the same time as the rest of the family, as long as they get some nutritious snacks in between. Most freshly prepared foods adults eat can be served to children in smaller portions cut up or mashed. Commercially prepared foods for young children might provide the necessary nutrition, but restaurant and take-out foods tend to be high in sodium and fats. Children generally should not be fed highly spiced, fatty foods with artificial colorings, additives, and preservatives. It is recommended that after the age of 2, children gradually adopt a diet that by age 5 provides no more than 30 percent of energy from fat (Dietary Guidelines Advisory Committee, 1995). The longer a parent or caretaker can wait to introduce nutritionally deficient sugar-coated cereals, pastries, and sugar-concentrated drinks, the better. A glass of cool, clean water is healthy for anyone of any age, and many children tolerate cow's milk very well. Small portions of fresh juices, fruits, and vegetables cut up and attractively presented as finger foods are usually quite appealing to young children, although a recent study found that only 16 percent of children entering school ate the recommended daily five servings of vegetables or fruits (Perry et al., 1998).

Pediatricians normally recommend that young children need a "mixed diet." This means a variety of foods and beverages eaten in different combinations so that over a period of time the child will get the proper nourishment.

Some Myths About Cow's Milk   No single food, or one beverage, is necessary. Any specific food is likely to be cooked in or with other foods. It is the sum total of what the child eats and drinks that counts. For generations, medical practitioners and powerful dairy advertising convinced parents that children's bones would not grow without daily doses of cow's milk, which is a source of calcium, protein, and vitamin D. Perhaps you were one of those children who hated milk and were told that you would not be allowed to leave the table until you had drunk your milk. Nowadays, "cow's milk is recognized as a food that *some* children do better without. . . . The valuable proteins, minerals, and vitamins milk contains are in other foods, too" (Leach, 1998, p. 233). Some children are lactose intolerant or allergic to milk; they can get the necessary nutrients in soy products, tomatoes and green and leafy vegetables. The revision authors' young son developed several symptoms, including severe cramps, a red rash on his face and body, indigestion, rhinitis (runny nose), changed bowel habits, and chronic irritability after cow's milk was introduced into his diet during his second year of life, and his parents, too, were hesitant to remove dairy products from his diet. As their wise pediatrician (from the Philippines) told them, most mammals stop nursing their offspring within the first year of life. Think how large an elephant's bones become, and its diet is mainly water, grasses, and other "woodsy" foodstuffs. The pediatrician told them that their son would still thrive if he ate a variety of fresh vegetables and other sources of calcium. The enzymes in cow's milk are more difficult for the human digestive system to process, compared to human breast milk, though many children have no problems digesting cow's milk. For children who cannot tolerate cow's milk, alternative sources of calcium and protein include fresh goat's milk and soy milk (from the soybean plant). These are now readily available in most supermarkets as well as in health food stores.

Good Health Also Means Sufficient Calories   Children must have an adequate supply of calories every day, to keep their bodily functions running smoothly and to provide the fuel they need for growth and their high-energy activities. Malnourished children have little energy and are lethargic. Staple carbohydrates, eaten in different forms all over the world, include rice, wheat, potatoes, corn, beans, yams, and sweet potatoes. Young children can get an ample supply of protein from meat, fish, poultry, eggs or foods that contain eggs, cheese, yogurt, peanut butter, and similar foods. Children in vegetarian families commonly get their carbohydrates and protein from a range of bean, legume, and nut dishes, and possibly some egg, cheese, or dairy dishes. A strict vegetarian diet for a growing child is questionable, and parents should seek dietary advice from a nutritionist or medical professional. The other trace minerals and vitamins that children need for healthy bones, teeth, blood, and neural growth are usually found in a plentiful supply of fresh fruits and vegetables. On the other hand, fast foods and junk foods are high in fat and have led to an alarming trend of obesity in U.S. children by the age of 4 (Goran & Sun, 1998).

Question

What types of foods are a "must" for children and in what proportion of the daily diet?

When a Child Refuses Food   It is difficult for young children to understand when they hear, "You must eat everything on your plate" or "Don't waste food" or "You must eat your peas before you can leave the table." Like everyone else, young children will feel hungry, their bodies will tell them when to eat, and their hunger might not coincide with the family mealtime. Children can become difficult and disruptive at mealtimes if their parents impose strict eating behaviors, such as telling children how much to eat and requiring them to finish all of something and try everything. Do you remember as a child that there were certain foods you simply could not tolerate? Are there any you still cannot tolerate as an adult? Some foods, like broccoli, cabbage, brussels sprouts, spinach, cauliflower, olives, and onions, taste very bitter to the sensitive taste buds of young children. It might be better to reintroduce these foods to the child at several time intervals during development.

Leach (1998) cites a research study conducted at a nursery in London revealing that when children were offered trays of a wide range of suitably cooked and cut-up foods three times each day, they selected for themselves diets that were balanced over the long term. Some would prefer protein one day, fruits other days. Some days they ate more, while some days they ate less. A checkup with a pediatrician will determine whether the young child is growing adequately for gender, age, and body type. It is recommended, and soon seen by most parents, that mealtimes should be enjoyable, food should not be used as a reward, punishment, bribe, or threat. It has also been documented that dietary intervention programs can be successful when started early enough (Perry et al., 1998; Macaulay et al., 1997).

Variability in Eating Behaviors Among Children

Some children are larger or more energetic than others and might need to snack more between mealtimes. Other children are smaller and less energetic and might simply require fewer calories. Cultural background can affect how much a child eats. Gender, ethnicity, and even geographic location have all been examined as influential factors in determining how much energy is needed by children as young as 4 years (Goran et al., 1998). Food choices, cleanliness of food, and food rituals vary from family to family (some families eat with fingers or chopsticks, for instance). A child will need something to eat upon awakening to start the day, probably a midmorning snack, a lunch, a midafternoon snack, an evening meal, and maybe a before-bed snack. When hungry children must wait too long to eat, their blood sugar levels will dip, causing energy drain, lack of patience, and a "cranky" attitude. Nursery and day-care providers must be prepared to offer children adequate nutrition at various times of the day. Most preschool children and 5- and 6-year-olds in kindergarten classes need the energy boost of the midmorning or midafternoon snack to think clearly and participate in activities.

Food for Thought: Mom, I'm Still Hungry

Sherman (1997, p. 8) reveals the real-life devastation that poverty brings to families, when hard-pressed parents cannot afford adequate health care or food for their families. The following is a paraphrased version of Mary Tamper's testimony to the U.S. Senate Committee on the Budget in February 1990:

> In 1983 Mary Tamper's middle-income world came tumbling down when her husband, William, died of leukemia, leaving her to support her young son, Justin,

Young Children Need Nutritious Snacks Energetic young children may need several nutritious snacks a day between meals. These children are in a Montessori preschool where they are encouraged to prepare their own snacks.

on a meager salary as a teacher's aide. In the following months, Mary was sick frequently then found out she was pregnant with twins. One of the twins died *in utero* as a result of the strains on Mary, according to her physicians. The surviving twin contracted bacterial meningitis after birth, and the medical, food, housing expenses drained the family's savings. Mary, in desperation, turned to social service programs to supplement her small earnings. She received WIC (Women, Infants, and Children federal supplemental food and nutrition program), $770 per month in Social Security survivor's benefits, and $11 a month in food stamps. Her low income still put her over the eligibility threshold, so she was denied welfare and Medicaid benefits.

By 1986 she and two children were trying to live on $186 a month after paying her rent. "There were times when money ran short and we had no food," says Mary. WIC was not enough. "There were . . . times when my son would say, "Mom, I know you just fed me, but I'm still hungry." There was no food in the cupboard to feed him. By 1990 the family's Social Security benefits had been raised to $900 a month, she received a government housing subsidy, and there was still little money to go around. Has hunger hurt her boys? Mary says, "Yes, Justin, the older son, is slower to learn in school and not as easy-going as most other children—a result, Mary feels, of inadequate nutrition and the strains of growing up poor.

Tooth Development

In toddlers and young children the molars are still developing and can cause discomfort and irritability. Teething gels or cold foods seem to help. To prevent decay of both baby teeth and permanent teeth, children should not be given sweet foods or drinks before they go to bed to sleep. Factors that promote healthy tooth development include the appropriate diet mentioned above, training toddlers and young children to clean their teeth regularly, regular dental checkups, and fluoride in local water supplies. Many communities provide free dental checkups for young children at various locations (e.g., the revision authors' college has a dental hygiene program that provides free dental services for many children in the community while students earn their clinic hours toward state certification). Improper tooth development can impair a child's ability to eat and speak.

Self-Care Behaviors   An important aspect of a young child's development is training in self-care behaviors, such as daily bathing and shampooing, cleaning teeth, brushing the hair, wiping the nose, wiping the bottom and washing the hands after toileting, handwashing before and after eating, and dressing appropriately for the weather conditions (putting on boots, coat, and mittens when it is cold and snowy, for example). Again, cultures differ in how much independence they promote in children regarding self-care behaviors, and in the frequency of these behaviors. For instance, some cultures do not promote daily bathing and shampooing.

Toilet training is a big developmental milestone during early childhood, and children vary considerably in age of toileting mastery. On average, children demonstrate self-control of these bodily behaviors during the third year. Staying clean in this regard is easier than staying dry, because children eliminate from the bowel less often. Becoming able to announce the need to use the toilet is a milestone, and parents and caretakers need to understand that when a young child says, "I have to go," the child cannot wait long and facilities must be found quickly. Parents and caretakers should realize that extreme patience and a sense of humor help both them and the child get through this time of maturation and development. Some young children insist, "I want to do it myself!" and others plead, "You do it for me." As you shall see in upcoming chapters, though, children who master self-care skills and can self-regulate their own behaviors are likely to develop greater self-esteem and self-confidence, and this can lead to many other positive outcomes as the child matures.

Sleeping and Dreaming   There seem to be two camps of thought about young children's sleep habits. Some hold that children need a daily routine of one early afternoon nap with a reasonable evening bedtime established so the child will get 10 to 12 hours of sleep per night. Others believe that the child's sleep schedule can vary at times to meet the parents' needs, such as staying up later when at a friend's house. In actuality, sleep schedules are an individual family's decision, based on such things as how many parents or caretakers are in the family, whether there is outside child care, and the workday schedule of the parents. Today, with nearly 70 percent of mothers working outside the home, many young children are awakened early, transported to a child-care provider for the day, and picked up later to return home

for only a few hours of awake activity. Some families are comfortable living on a "schedule," while others appreciate variability in their lives.

One thing we can tell you for sure: An overtired child can be whiny and cranky and will probably be difficult or lethargic when awake. It is not fair to punish the child for being unruly or displaying uncooperative behavior, when the parents have caused this behavior in the first place by not providing enough rest, relaxation, or sleeptime for the growing child (who is expending a great deal more energy each day than during infancy or toddlerhood). Parents who establish a bedtime routine and stick to that routine are likely to have a child who will go to bed easier than those who reinforce a lot of rocking, getting up, walking around, playing, and getting into Mommy and Daddy's bed. Many children are relaxed by having a warm bath, a story read, or a quiet talk before bedtime, and they have fewer problems getting to sleep. Humorously, as illustrated in Figure 7-2, it isn't always the little one who goes to sleep first! Young children who are allowed to stay up late and run around the house at all hours are likely to be in control of the family over this sleep issue.

Though we do not know what causes nightmares or night terrors in preschool-age children, we can predict these are likely to occur at this stage. Daytime stress, anxiety, or fears in a young child's life, such as attending a new preschool, the birth of a new sibling, moving to a new home, separation of parents, or death of a grandparent, significant person, or pet can trigger bad dreams. Also, young children today are exposed to graphic displays of violence on TV and in other media that can be very disturbing. Some older siblings enjoy scaring younger ones about "the monster under the bed." A young child will not yet have the vocabulary to understand or express the feelings or vulnerability she or he is experiencing, and *nightmares* (a bad dream) and *night terrors* (screaming, profuse sweating, and no recall about this event because it occurs during deep sleep) can be a signal that the child is not coping well. Parents can reassure the child who is experiencing sleep disturbances by giving the child extra love and attention, talking about what is upsetting the child, and being patient and understanding.

Illness and Immunizations   With more children either in early childhood education programs or receiving out-of-home care in a nursery, day care, preschool, Head Start, and so on, children are continuously in contact with other children and at greater risk for contracting childhood diseases. Most states require proof of specific immunizations before a child can be enrolled in public school classes. Though in a few rare instances some children have more extreme reactions to certain immunizations, the vast majority will experience better health if their parents maintain the recommended inoculation schedule. Children in the U.S. public school system are

"I think he's finally asleep."

Figure 7-2 A nightly routine can help relax a child so bedtime is a calm time. Parents may find it relaxing, too!

checked annually by a health care professional for physical delays or disabilities, such as malnutrition, obesity, and curvature of the spine; sensory capabilities, such as vision and hearing; and overall health and self-care skills, with attention to bathing, body lice, and such.

The combination of proper nutrition for growth and development, required sleep for physical maturation and an increased alertness, and childhood immunizations protecting against specific diseases help to lay a healthy foundation for central nervous system development and mental functioning.

Question

What variables will a pediatrician probably examine to determine if a young child is healthy and developing?

Changing Demographics and Implications for Health of Minority Children

In 1928, Lewis Meriam (in Browne, Crum, & Cousins, 1997, p. 227) observed:

> The health of the Indians as compared with that of the general population is bad. Although accurate mortality and morbidity statistics are commonly lacking, the existing evidence warrants the statement that both the general death and infant mortality rate are high.

It is a sad and disturbing fact that his observation is just as accurate today as it was then (Browne, Crum, & Cousins, 1997). To understand the general health status of various racial and ethnic groups, it is important to look at changing demographics and their implications for minorities.

As we enter into the new millennium, the challenge facing our country is how to effectively reduce the disparities in health and health outcomes for the four major racial and ethnic populations Asian/Pacific Islanders, African Americans, Hispanic Americans, and Native Americans. According to national studies reported in *Trends in the Well-Being of America's Children and Youth* (U.S. Department of Health & Human Services, 1997d), over the past three decades the United States has become increasingly diverse racially and culturally. Table 7-2 shows statistics for 1996 and projections for the near future.

In 1980, 74 percent of U.S. children were non-Hispanic whites; as of 1996 that figure is reported to be 66 percent. And population experts project that over the next 20 years the non-Hispanic white child population will decrease to about 55 percent.

The relative size of the minority population is projected to change significantly as Hispanic and Asian groups grow rapidly; the black population is expected to remain close to its current relative size. This change in demographics is compounded by the fact that a significant proportion of the white population is growing older

Table 7-2   Minority Youth Population by Race and Ethnicity 1996–2020 (Projected)

Race and Ethnicity	1996	2010	2020
African American	16 percent	17 percent	17 percent
Hispanic American (Puerto Rican, Mexican, and Central, South, and Cuban Americans)	14 percent	19 percent	22 percent
Asian American (Chinese, Filipino, Japanese, Indian, Korean, Vietnamese, Hawaiian, Samoan, and Guamanian)	4 percent	6 percent	6 percent
Native American	1 percent	1 percent	1 percent
Total and Projected Totals	35 percent	43 percent	46 percent

Source: U.S. Department of Health and Human Services.

while the racial and ethnic minority populations have larger proportions of younger children and adolescents. Minority children accounted for approximately 30 percent of the population under 19 years of age in 1990, and in 1996 they made up approximately 35 percent of the child population. It is estimated that by the year 2020, minority children will constitute 46 percent of the child population in the United States (U.S. Department of Health & Human Services, 1997d).

The rapidly changing demographics in the United States have important implications, not only for the health of the individuals who compose these ethnic and racial groups, but for the overall population (see *Human Diversity:* "Health Beliefs and Practices Across Cultures"). With increasing numbers of individuals representing diverse groups, there will be large numbers of individuals speaking languages other than English. Failure to provide translation services will mean that members of minority and racial groups might have difficulty getting appropriate care and conversing with professionals about their health needs and health concerns. Such barriers can cause them to delay seeking services; this can mean that they don't receive health care until their medical condition has become more serious, thus leading to increased costs of these services and negative health outcomes. Tragically, in our (the revision authors') own community, a seriously ill Vietnamese baby died in the emergency room when the parents brought the baby in for treatment. The parents did not speak English, and no one in the ER could speak or understand Vietnamese. Our son-in-law is an ER physician in the Bronx in New York City; presently, if a translator cannot be found, the phone company is called, a telephone operator who speaks that particular language is put on the phone, and a three-way conversation is held.

Child Mortality Rates and Causes   A reduction in both early and late childhood mortality rates in the United States indicates an improvement in the health status of the child population. Even with decreasing mortality rates, since 1960 male children have died at a higher rate than female children: In 1993, for the age group 1 to 4 years, 49.5 males per thousand died versus 39.9 females per thousand; for the age group 5 to 9 years, 23.2 males per thousand died versus 19.0 females per thousand. In 1993, among children aged 1 to 4, injuries accounted for 44 percent of all deaths, followed by congenital anomalies, malignancies, diseases of the heart, and HIV or AIDS (Gardner & Hudson, 1996). In 1993, injuries accounted for 52 percent of deaths to children ages 5 to 14 (U.S. Department of Health & Human Services, 1997d).

Despite overall reductions in childhood mortality, there are also substantial differences across racial and ethnic groups (Kane, 1993). Among Hispanics, for example, Cuban children had lower death rates than their white counterparts, and Puerto Rican and Central and South American children had higher death rates in two recent studies. Asian and Pacific Islanders had significantly lower death rates than all other groups, true for both male and female children.

With their concern for the health and well-being of minority and poverty-stricken mothers and children, Maternal and Child Health (MCH) federal training programs are well-positioned to prepare professionals to play a unique role in research, service, and training. Four recommendations listed below, from Browne and colleagues (1997), can be used to improve health status and outcomes for minority children:

- Curricula in Maternal and Child Health (MCH) programs should be expanded to include courses in culturally sensitive communication and assessment that acknowledge cultural beliefs and values.
- MCH students, practitioners, and faculty must be given opportunities to understand the rapidly changing communities they work with, along with the cultural barriers to adherence to prevention and treatment programs.
- MCH professionals must be trained to create and validate culturally sensitive tools for data collection as well as approaches for interviewing individuals from ethnic and racially diverse populations.
- Minority researchers and students must be

Human Diversity

Health Beliefs and Practices Across Cultures

Significantly, all cultural subgroups have their own health beliefs and practices that shape their responses to wellness, illness, and restoration of health. Many of these practices and beliefs are inconsistent with Western practices in American hospitals and clinics. An elderly Vietnamese man in our (the revision authors') community died at home, and his family held a traditional burial service for him—unaware that by law a coroner must be called to determine the time and cause of death and that he could not be buried in the back yard. From their cultural background, they had no concept of a mortician being involved, nor could they afford the cost of a burial. Furthermore, many of these cultural subgroups rely on their traditional faith healers and do not seek standard medical care, such as immunizations for babies and children or regular dental checkups. Such beliefs and practices can impede interventions to improve health outcomes for the groups' children. Additionally, because the ethnic and racial populations are younger than the white population and the fertility rates for women of certain segments of the ethnic and racial communities are higher, there will also be higher rates of pediatric, reproductive, and other health problems that are commonly observed in mothers and children.

Childhood Checkups and Immunizations

Many childhood diseases are rare in countries where children are routinely protected by immunizations during regularly scheduled checkups with pediatricians and health care practitioners. However, a majority of the world's children live in countries where preventive immunizations are rare or nonexistent, and families seek their medical care from a nonmedical person such as a traditional faith healer. Because many young children in the United States today have parents who are immigrants from such countries, health care and child-care professionals must understand such countries' three basic myths regarding the cause of illness and appropriate cures (which more than likely do *not* include the concept of immunization) (Lecca et al., 1998):

- *Natural causes:* Illness can be caused by damp and cold. According to Chinese medical practice, illness is caused by the *yin* and *yang* being out of balance, which is cured with acupuncture. Illness can be caused by eating things that are poisonous or out of season; for instance, many practicing Muslims and Jews do not eat pork, Muslims fast during the month of Ramadan. Some members of Jehovah's Witness do not want their

Health Care and Minority Children While over the past several decades there has been a general improvement in child health status, increasing numbers of American children may come from ethnic groups that have alternative views regarding causes of illness and appropriate cures. Notice this Alaskan Eskimo toddler putting a feather in her mouth, which could lead to some type of illness.

children to have blood transfusions. Puerto Rican, South American, and Caribbean islanders utilize massage and natural folk and herbal remedies for common illnesses that they classify as hot or cold illnesses.

- *Supernatural causes:* Illness can be caused by someone, something, or a spiritual energy that is angry with the ill person and puts something bad on that person (hexes, curses, fixes).

Continued on page 204

Continued from page 203

- *Religious or spiritual causes:* Illness is caused by thinking or doing evil, by not praying enough, by not having faith, by lying, by cheating, by not respecting elders or religious leaders or God.

The Western medical model of wellness and illness assumes that widespread childhood immunizations prevent most serious illness and potential life-threatening complications. However, today more American families, for career or recreation, travel to countries that do not practice childhood im-

munization. And of the families that daily are moving into the United States from all over the world, many, both children and adults, have not been inoculated for childhood diseases and are unaware of, opposed to, or afraid of immunizations. Additionally, some U.S. families believe that childhood inoculations are not worth the risk, given the possibility that the child could develop fever or a mild form of the disease. Other families simply neglect taking their children for regular checkups, even though public clinics provide checkups and immunizations free of charge.

encouraged to continue and/or join in innovative and important research related to the health of women and children.

It is hoped that these recommendations and the discussion of minority health issues point to some of the crucial issues that need to be addressed in the health care field today.

Cognitive Development

I believe that every child has hidden away somewhere in his [or her] being noble capacities which may be quickened and developed if we go about it the right way, but we shall never properly develop the higher nature of our little ones while we continue to fill their mind with so-called basics. Mathematics will never make them loving, nor will accurate knowledge of the size and shape of the world help them to appreciate its beauties. Let us lead them during the first years to find their greatest pleasure in nature. Let them run in the fields, lear about animals, and observe real things. Children will educate themselves under the right conditions. They require guidance and sympathy far more than instruction.

Helen Keller

Preschool children who receive adequate nutrition and varied stimulation normally experience a rapid expansion of cognitive abilities. They become more adept at obtaining information, ordering it, and using it. Gradually, these abilities evolve into the attribute called *intelligence.* Whereas sensorimotor processes largely dominated development during infancy, a significant transition occurs after 18 months toward the more abstract processes of reasoning, inference, and problem solving. By the time children are 7 years old, they have developed a diversified set of cognitive skills that is functionally related to the elements of adult intelligence.

Intelligence and Its Assessment

For laypeople and psychologists alike, the concept of intelligence is a rather fuzzy notion (Sternberg, 1990). In some ways intelligence resembles electricity. Like electricity, intelligence "is measurable, and its effect, but not its properties, can be only imprecisely described" (Bischof, 1976, p. 137). Even so, David Wechsler (1975), a psychologist who devised a number of widely used in-

Early Intelligence Testing With the development of IQ tests in the early 1900s, seemingly a scientific instrument was at hand for evaluating the intellectual capabilities of the members of various ethnic groups. Immigrants in the United States from non-English-speaking nations tended to score rather poorly. With a score of about 100 considered "normal," the average score of Jews, Hungarians, Italians, and Poles was about 87. Some psychologists like Henry Goddard concluded that these groups were intellectually inferior—indeed, even "feeble-minded." Decisions were made to allow more Northern Europeans entry into the United States than Southern Europeans, based on early IQ testing.

telligence tests, has proposed a definition that has won considerable acceptance. He views **intelligence** as a global capacity to understand the world, think rationally, and cope resourcefully with the challenges of life. Seen in this light, intelligence is a capacity for acquiring knowledge and functioning rationally and effectively, rather than the possession of a fund of knowledge. Intelligence has captivated the interest of psychologists for a variety of reasons, including a desire to devise ways of teaching people to better understand and increase their intellectual abilities (Sternberg, 1986a).

Intelligence: Single or Multiple Factors? One recurrent divisive issue among psychologists is whether intelligence is a single, general intellectual capacity or a composite of many special, independent abilities. Alfred Binet (1857–1911), the French psychologist who in 1905 devised the first widely used intelligence test, viewed intelligence as a general capacity for comprehension and reasoning. Although his test used many different types of items, Binet assumed that he was measuring a general ability expressed in the performance of many kinds of tasks.

In England, Charles Spearman (1863–1945) quickly rose to eminence in psychological circles by advancing a somewhat different view. Spearman (1904, 1927) concluded that there is a general intellectual ability, the *g* (for "general") factor, that is employed for abstract reasoning and problem solving. He viewed the *g* factor as a basic intellectual power that pervades all of a person's mental activity. However, because an individual's performance across various tasks is not perfectly consistent, Spearman identified special factors (*s* factors) that are peculiar to given tasks, such as arithmetic or spatial relations. This approach is known as the **two-factor theory of intelligence** (see Brody, N. 1992). J. P. Guilford (1967) has carried the tradition further by identifying 120 factors of intelligence. Not all psychologists are happy, however, with such minute distinctions. Many prefer to speak of "general ability"—a mixture of abilities that can be more or less arbitrarily measured by a general-purpose intelligence test.

Multiple Intelligences In recent years psychologist Howard Gardner (1997, 1993a, 1993b, 1983) from Harvard has been conducting research with gifted children. On the basis of his research, he proposes that there are eight distinctive intelligences: *linguistic, logical-mathematical, spatial, musical, bodily-kinesthetic, interpersonal* (knowing how to deal with others), *intrapersonal* (knowledge of oneself), and *naturalistic* (nature smarts). Interpersonal or social intelligence is now referred to as *emotional intelligence* (see Goleman, 1995).

However, Gardner not only carves up intelligence into these separate types but also contends that the separate intelligences are located in different areas of the brain. When a person suffers damage to the brain through a stroke or tumor, all abilities do not break down equally. And youngsters who are *precocious* in one area are often unremarkable in others. In fact, people with retardation whose mental ability is lower in most areas will occasionally exhibit extraordinary ability in a specific area, most commonly mathematical calculation. Many autistic children display this ability. These observations led Gardner to say that the much-maligned intelligence quotient (IQ) ought to be replaced with an "intellectual profile." Gardner's proposed theory has been embraced by some teachers who work daily with children and readily see different capabilities of children (see "Following Up on the Internet" at the end of this chapter).

But critics such as Sandra Scarr (1985a) dispute Gardner, saying that he is really talking about talents or aptitudes, not intelligences. His critics have difficulty calling "intelligence" what people typically label human abilities or virtues.

Music, Spatial Skills, and Intelligence Gardner defines *spatial skills* as skills pertaining to the ability to form mental images, visualize graphic representations, and recognize interrelationships among objects (Gardner, 1997, 1983). Spatial skills are used to walk around objects in an environment and to solve mathematical and engineering problems later in life. Along similar lines, psychologist Fran Rauscher (1996) and neuroscientist Dee Joy Coulter (1995) have argued that singing, rhythmical movement, musical games, listening, and early musical instrument training are neurological exercises that introduce children to speech patterns, sensory motor skills, and vital rhythm and movement strategies. All of these nonverbal activities are independent of language, they suggest, and promote brain development in the same neural pathways as are used for spatial skills (Baney, 1998). Rauscher (1996) conducted a study in 1993 in which ten 3-year-old children took either singing or piano lessons. When tested later, the children's scores improved 46 percent on the Object Assembly Task of the Wechsler Preschool and Primary Scale of Intelligence–Revised. In a later study that took place over eight months with three groups of preschool children, the preschoolers taking piano lessons scored a significant 34 percent higher than the other groups not taking piano lessons. Heyge (1996) further states that music is essential to children's lives because it:

- Optimizes brain development
- Enhances multiple intelligences
- Facilitates genuine bonding between adult and child
- Builds social/emotional skills
- Promotes attention to task and inner speech
- Develops impulse control and motor development
- Communicates creativity and joy

In sum, these psychologists and neuroscientists recommend both incorporating music into children's lives as early as possible to build neural connections and revitalizing music programs in the public schools.

Intelligence as Process   Quite different from an "abilities" approach to intelligence are those perspectives that view intelligence as a *process*—they are not so much interested in *what* we know as in *how* we know. Proponents of this approach are less concerned with the "stuff" that allows people to think intelligently and more concerned with the operations involved in thinking. For instance, as we discussed in Chapter 2, Jean Piaget concerned himself with the stages of development during which given modes of thought appear. He focused on the continual and dynamic interplay between children and their environment through which children come to know the world and to modify their understanding of it. Piaget did not view intelligence in set or fixed terms, so he had little interest in the static assessment of individual differences in ability.

Intelligence as Information Processing   More recently, a number of cognitive psychologists have proposed an information-processing view of intelligence—a detailed, step-by-step analysis of how we manipulate information (Hunt, 1983; Sternberg, 1998, 1990, 1984). These psychologists are trying to unlock the doors to the mind by getting "inside" intelligence, seeking to understand the mental processes whereby people solve problems not only on tests but in everyday life. By way of illustration, try this question: If you have black socks and brown socks in your drawer, mixed in a four-to-five ratio, how many socks must you take out to ensure getting a pair of the same color? To find the answer, you must grasp what is important and ignore irrelevant details, a task that requires insight. The color ratio of socks is unimportant. (The answer is three: If the first is black and the second is brown, the third must be one or the other.)

In studying children in grades 4 through 6, Robert J. Sternberg (1986b) and his colleagues compared how children they had identified as gifted and those not so identified approached this problem. Three findings emerged: (1) The gifted children were better able to solve the socks problem because they ignored the irrelevant information regarding ratios. (2) Supplying nongifted children with the insights needed to solve this sort of problem increased their performance, whereas it had no effect on the gifted children because they already possessed the insights. And (3) insight skills can be developed by training. In a five-week training program, both gifted and nongifted youngsters achieved significant improvement in their scores, relative to the scores of an untrained control group. Moreover, these skills were evident in a follow-up study

a year later. In short, even though insight skills differentiate between the more and less intelligent, they are trainable in both groups. Sternberg says that this sort of understanding can be achieved only by studying the cognitive processes underlying intelligence—it is not supplied by global intelligence test scores.

> **Questions**
>
> *What is intelligence? How have the explanations of intelligence changed over time? Why is intelligence difficult to measure?*

Intelligence and the Nature-Nurture Controversy

Psychologists differ in the relative importance they attribute to heredity and environment, regarding intelligence. As we discussed in Chapter 2, some investigators of the nature-nurture issue have asked the "which" question, others the "how much" question, and still others the "how" question. And because they ask different questions, they come up with different answers.

The Hereditarian Position   Hereditarians tend to phrase the nature-nurture question primarily in terms of the "how much" question and seek answers in family resemblance studies (see Chapter 2) based on intelligence tests (Jensen, 1984). Psychologists have devised tests of intelligence, employing a single number called the *intelligence quotient*, or *IQ*, to measure intelligence. Today's intelligence tests provide an IQ score based on the test-taker's performance relative to the average performance of other individuals of the same age. Many psychologists believe that the assessment of intellectual abilities is one of their discipline's most significant contributions to society. But other psychologists say that it is a systematic attempt by elitists to "measure" people so that "desirable" ones can be put in the "proper" slots and the others rejected. Table 7-3 summarizes current data on IQ performance from family resemblance studies.

The data in Table 7-3 reveal that the median IQ correlation coefficient of separated identical (monozygotic) twins in three studies is + .72. Data are also shown from 29 studies of fraternal twins of the same sex who were reared together. The median IQ correlation coefficient of the fraternal twins is +.62. In sum, the identical twins reared in *different* homes are much more alike in IQ than the fraternal twins raised *together* (Jensen, 1972). Also note in Table 7-3 that as the biological kinship between two people increases (gets closer), the correlation between their IQ scores increases. On the basis of this and other evidence, hereditarians typically conclude that 60 to 80 percent of the variation in IQ scores in the general population is attributable to genetic differences and the

Table 7-3	Correlation Coefficients of IQ Scores Compared with the Degree of Family Relationship	
Correlations Between	Median Number of Studies	Correlation Coefficients
Unrelated Persons		
Foster parent and child	6	+.19
Children reared together	6	+.34
Collaterals		
Cousins	4	+.15
Half siblings	2	+.31
Siblings reared apart	2	+.24
Siblings reared together	69	+.47
Dizygotic twins, different sex	18	+.57
Dizygotic twins, same sex	29	+.62
Monozygotic twins, reared apart	3	+.72
Monozygotic twins, reared together	34	+.86
Direct Line		
Parent and child, different sex	12	+.39
Parent and child, same sex	14	+.40

Adapted from Thomas J. Bouchard, Jr., and Matthew McGue, "Familial Studies of Intelligence: A Review," *Science,* Vol. 212 (1981), pp. 1055–1059. Reprinted by permission of the author.

Cultural Bias in Intelligence Testing A "noncultural" or "culture-free" intelligence test is an impossibility. Membership in a particular culture influences what an individual is likely to learn or fail to learn. It would be invalid, for instance, to give these Spanish-speaking youngsters in Mexico the Stanford-Binet intelligence test, designed by and for English-speaking Americans.

remainder to environmental differences (Herrnstein & Murray, 1994).

The Environmentalist Rebuttal A number of scientists dispute the claim by Jensen and like-minded psychologists that differences in intelligence are primarily a function of heredity. Some disagree with the formulation of the nature-nurture question in terms of "how much" and insist that the question should be *how* heredity and environment interact to produce intelligence. And others such as Leon J. Kamin (1974, p. 1) go so far as to assert: "There exist no data which should lead a prudent man to accept the hypothesis that IQ test scores are in any degree heritable." Psychologists of Kamin's view, commonly called *environmentalists,* argue that mental abilities are learned. They believe that intellect is increased or decreased according to the degree of enrichment or impoverishment provided by a person's social and cultural environment (Blagg, 1991).

Kamin (1994, 1981, 1974) has vigorously challenged the adoption and identical-twin research that the advocates of heredity use to support their conclusions. He insists that it is improper to speak of individuals as being reared in differing environments simply because they were brought up in different homes. In some cases identical twins labeled as being "reared apart" were hardly separated at all: They were raised by relatives, or they

lived next door to one another, or they went to the same school. Similarly, environmentalists charge that studies of adopted children are biased by the fact that adoption agencies traditionally attempt to place children in a social environment that is religiously, ethnically, and racially similar to the one into which they were born.

Contemporary Scientific Consensus Most social and behavioral scientists believe that any extreme view in the nature-nurture controversy is unjustified at the present time. Estimates based on twin and adoption studies suggest that hereditary differences account for 40 to 80 percent of the variation found in the intelligence test performance of a population. Bouchard and colleagues (1990) found that 70 percent of the IQ differences among people is attributable to genetic factors, but other experts think that a 70 percent heritability estimate is too high. Jencks (1972), employing path analysis, which is a statistical technique used to partition the amount of variance within a group, estimated that 45 percent of IQ differences is due to heredity, 35 percent due to environment, and 20 percent due to gene-environment interaction. Jencks introduced the third element of gene-environment interaction because he felt that dividing IQ into only hereditary and environmental components oversimplifies the matter. Similarly, Loehlin and colleagues (1975) contend that we need to consider these three components:

- Genetic endowment when intellectual stimulation is held constant
- Environmental stimulation when genetic potential is held constant

- Covariance of heredity and environment—how the first two components vary relative to each other

If genes and environment reinforce each other, then the added component of variance cannot logically be assigned to either nature or nurture. Rather, it is a result of the association of their separate effects.

Question

What are the differences between the nature view of intelligence and the nurture view?

Piaget's Theory of Preoperational Thought

Jean Piaget, (1963, 1952) the Swiss developmental psychologist, pioneered the study of the development of intelligence in infants and children. He called the years between 2 and 7 the **preoperational period.** The principal achievement of that period is children's developing capacity to represent the external world *internally* through the use of symbols. *Symbols* are things that stand for something else. For example: Letters of the alphabet can be symbols, for in English *c-a-t* represents a 4-legged animal. And numbers, such as 3, are symbols for specific quantities of something. Here are some other familiar symbols—do you know what they represent? ☺ ♀ $ ♂ :-).

The ability to use symbols frees children from the rigid boundaries of the here and now. Using symbols, they can represent not only present events but past and future ones. The acquisition of language and numeration facilitates children's ability to employ and manipulate symbols.

Difficulties in Solving Conservation Problems   Piaget observed that although children make major strides in cognitive development during the preoperational period, their reasoning and thinking processes have a number of limitations. These limitations can be seen in the difficulties preschool children have when they try to solve conservation problems. **Conservation** refers to the concept that the quantity or amount of something stays the same regardless of changes in its shape or position.

For example, if a ball of clay is shown to a child and then rolled into a long, thin, snake-like shape, the child will say that the amount of clay has changed simply because the shape has changed. Similarly, if we show a child under age 6 two parallel rows of eight evenly spaced pennies and ask which row has more pennies, the child always correctly answers that both rows have the same amount. But if, in full view of the child, we move the pennies in one of the rows farther apart and again ask which row has more pennies, the child will reply that the longer row has more (see Figure 7-3). The child fails to recognize that the number of pennies does not change simply because we made a change in another dimension, the length of the row. Piaget said that the difficulties preschoolers have in solving conservation problems derive from the characteristics of *preoperational thought*. These characteristics inhibit logical thought by posing obstacles that are associated with centering, transformations, reversibility, and egocentrism.

Centration   Preoperational children concentrate on one feature of a situation and neglect other aspects, a process called **centration.** A preoperational child cannot understand that when the water that fills a tall, thin glass is poured into a short, wide glass, the *amount* of water

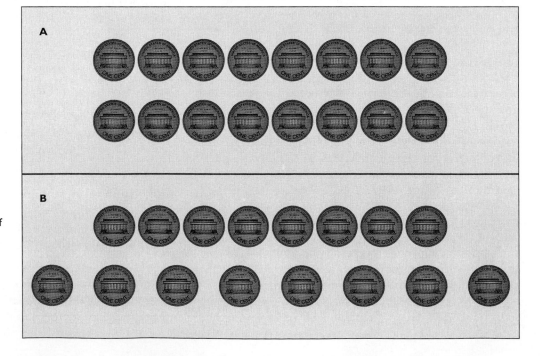

Figure 7-3 Conservation Experiment with Pennies
Children are first shown two rows of pennies arranged as in A. The experimenter asks if both rows contain the same amount of pennies. Then with the children watching, the experimenter spreads out the pennies in the bottom row, as in B. Children are once again asked if both rows contain the same number of pennies. Preoperational children will respond that they do not.

remains unchanged. Instead, the child sees that the new glass is half empty and concludes that there is less water than before; the child cannot attend simultaneously to both the amount of water and the shape of the container. To solve the conservation problem correctly, the child must *decenter*—that is, attend simultaneously to both height and width.

Likewise, in the case of the pennies, children need to recognize that a change in length is compensated for by a change in the other dimension, density. Thus, there is no change in quantity. Here, too, the ability to decenter—to explore more than one aspect of the stimulus—is said by Piaget to be beyond preoperational children.

States and Transformations   Another characteristic of preoperational thinking is that children pay attention to *states* rather than *transformations.* In observing water being poured from one glass into another, preschool children focus on the original state and the final state. The intervening process (the pouring) is lost to them. They do not pay attention to the gradual shift in the height or width of the water in the glasses as the experimenter pours the liquid. Preoperational thought fixes on static states. It fails to link successive states into a coherent sequence of events. Here's a more common example: If you offer a preoperational child a cookie, and the child complains that it is too small, simply break the cookie into 3 pieces and place it next to a whole cookie. The odds are the young child will think there is *more* in the cookie split into three pieces!

Their inability to follow transformations interferes with preschool children's logical thinking. Only by appreciating the continuous and sequential nature of various operations can we be certain that the quantities remain the same. Because preoperational children fail to see the relationship between events, their comparisons between original and final events are incomplete. Thus, according to Piaget, they cannot solve conservation problems.

Nonreversibility   According to Piaget, the most distinguishing characteristic of preoperational thought is the child's failure to recognize the **reversibility** of operations—that a series of operations can be gone through in reverse order to get back to the starting point. After we pour water from a narrow container into a wider container, we can demonstrate that the amount of water remains the same by pouring it back into the narrow container. But preoperational children do not understand that the operation can be reversed. Once they have carried out an entire operation, they cannot *mentally* regain the original state. Awareness of reversibility is a requirement of logical thought.

Egocentrism   Still another element that interferes with the preschool child's logical understanding of reality is **egocentrism**—lack of awareness that there are viewpoints other than one's own. According to Piaget, preoperational children are so absorbed in their own impressions that they fail to recognize that their thoughts and feelings might be different from those of other people. Children simply assume that everyone thinks the same thoughts they do and sees the world from the same perspective that they do.

Both Piaget (1963) and sociologist George Herbert Mead (1934; Aboulafia, 1991) pointed out that children must overcome an egocentric perspective if they are to participate in mature social interaction. For children to play their *role* properly, they must know something about other roles. Most 3-year-olds can make a doll carry out several role-related activities, revealing that the child has knowledge of a social role (for example, the child can pretend to be a doctor and examine a doll). Four-year-olds can typically act out a role, relating one social role to a reciprocal role (for instance, they can pretend that a patient doll is sick and that a doctor doll examines it, in the course of which both dolls make appropriate responses). During the late preschool years, children become capable of combining roles in more complicated ways (for example, being a doctor and a mother at the same time); most 6-year-olds can pretend to carry out several roles simultaneously.

> **Question**
>
> *How is preoperational thought different from the sensorimotor stage during infancy and the toddler years?*

Critiques of Piaget's Egocentric Child   More recent research by neo-Piagetians suggests, however, that although egocentricity is characteristic of preoperational thinking, preschool children are nonetheless capable of recognizing other people's viewpoints on their own terms. Even though, emotionally, toddlers can be quite *self-centered*, they are not necessarily *egocentric* in the sense of not understanding other perspectives (Newcombe & Huttenlocher, 1992). Increasingly researchers are uncovering many *sociocentric* (people-oriented) responses in young children. Indeed, some researchers have questioned the characterization of children as egocentric. Consider the following evidence.

Talking and Communicating   Catherine Garvey and Robert Hogan (1973) found that during the greater part of the time that children 3 to 5 years of age spend in nursery school, they interact with others, largely by talking. Furthermore, while some of their speech is *private speech* (directed toward themselves or nobody in particular), most of their speech is mutually responsive and adapted to the speech or nonverbal behavior of a partner

Sociocentric Behavior
Developmental psychologists are discovering that young children are considerably less egocentric and more sociocentric than earlier studies indicated. For example, one of the main priorities of a Montessori education is to promote prosocial behaviors.

(Spilton & Lee, 1977). Harriet Rheingold, Dale Hay, and Meredith West (1976) have shown that children in their second year of life share with others what they see and find interesting. Nursery school teachers know how to foster this type of social behavior by periodically scheduling a "show-and-tell time" when children bring special objects from home to talk about. Rheingold and her colleagues (1976, p. 1157) concluded:

> In showing an object to another person, the children demonstrate not only that they know that other people can see what they see but also that others will look at what they point to or hold up. We can surmise that they also know that what they see may be remarkable in some way and therefore worthy of another's attention. . . . That children so young share contradicts the egocentricity so often ascribed to them and reveals them instead as already able contributors to social life.

Altruism and Prosocial Behavior   Researchers have found evidence of altruistic and prosocial behavior even in very young children. Not uncommon are the following examples: A 2-year-old boy accidentally hits a small girl on the head. He looks aghast. "I hurt your hair. Please don't cry." Another child, a girl of 18 months, sees her grandmother lying down to rest. She goes over to her own crib, takes her own blanket, and covers her grandmother with it (Pines, 1979). If egocentricity is seen as the inability to psychologically "connect" with someone else, then these examples point to the child's ability to reach out and relate to another individual. If, on the other hand, egocentricity is seen as a "distortive interpretation of other people's experiences, and actions or persons or objects, in terms of one's own schemas" (Beard, 1969), then even these prosocial acts could be labeled "egocentric."

One of the better-known early childhood programs that has a set of altruistic priorities different from a traditional preschool is the Montessori "Children's Community" founded in 1907 by Dr. Maria Montessori. Her first school was made up of 60 inner-city children, most of whom came from dysfunctional families. She developed a child-centered philosophy of "education for life" directed toward the development of each child's interests, abilities, and human potential. But her vision expanded beyond an academic curriculum to promote prosocial behavior in young children. Montessori schools today give children the sense of belonging to a family and help them learn how to live with other human beings. By creating bonds among parents, teachers, and children, Montessori sought to create a community where children could learn to be a part of families, where they could learn to care for younger children, learn from each other and older people, trust one another, and find ways to be properly assertive rather than aggressive. Dr. Montessori envisioned her movement as essentially leading to a reconstruction of society. Montessori schools today are found all over the world, including Europe, Central and South America, Australia, New Zealand, India, Sri Lanka, Korea, and Japan (Seldin, 1996).

In conclusion, newly developing lines of research reveal that, in our efforts to understand children, we have been hampered by adult-centered concepts and by our preoccupation with the adult-child relationship—in other words, by *adult* egocentrism. Moreover, in recent years psychologists have moved increasingly away from Piaget's notion of broad, overarching stages to a more complex view of development (Case, 1991). Rather than searching for major overall transformations, they are scrutinizing separate domains such as causality, memory, creativity, problem-solving, and social interaction. Each domain has a somewhat unique and flexible schedule that is affected not only by age but also by the quality of the environment (Demetriou, Efklides, & Platsidou, 1993; Flavell, 1992). Chapter 9 will consider *prosocial behavior* at greater length.

The Child's Theory of Mind

Research in the **theory of mind** probes children's developing conceptions of major components of mental activity. When a child begins to comprehend that the mind exists, this paves the way to rudimentary distinctions such as that of being part of the environment while also being separate from it. Furthermore, the child can begin to understand that people can think about objects differently. For example, a researcher might show a 5-year-old a candy box and ask the youngster what is in it. "Candy," responds the child. But when the youngster opens the box, she is surprised to discover that it contains crayons rather than candy. The researcher then might inquire of the child, "What will the next child who has not opened the box think is in it?" "Candy!" the child answers, grinning at the trick. But when the researcher repeats the same procedure with a 3-year-old, the child typically responds to the first question as expected, "Candy!" but responds to the second, "Crayons!" Significantly, the 3-year-old also says "Crayons" when asked to recall what she herself had initially thought would be found in the box. The 3-year-old, unlike the 5-year-old, fails to understand that people can hold beliefs that are different from what they know and that are *false*. The 3-year-old believes that because she knows there is no candy in the box, everyone else automatically knows the same thing to be true (Wellman, 1990). In comparison with older youngsters, the thought of 3-year-olds is still confused. It seems that 3-year-olds struggle with, and typically fail to understand, what 4-year-olds and older children do understand: that people have internal mental states, such as beliefs, that represent or misrepresent the world, and people's actions stem from their mental representation of the world rather than from direct objective reality.

Piaget's sweeping account of children's reasoning was an early attempt to broach the topic of "mind"—the notion of an instrument that can calculate, dream, fantasize, deceive, and evaluate the thoughts of others. More recently, particularly over the past decade or so, a considerable amount of research has been done on children's developing understanding of their mental world (Feldman, 1992). This work has demonstrated that even 3-year-olds can distinguish between the physical and the mental and that they possess some understanding of what it means to imagine, think of, and dream of something (Flavell 1992). For example, if a 3-year-old is told that one child has ice cream and another child is only thinking of ice cream, then the 3-year-old will be able to say which child's ice cream can be seen by others, as well

as touched or eaten. This understanding is facilitated where youngsters enjoy a rich "database" derived from interactions with siblings, caregivers, and peers (Perner, Ruffman, & Leekam,1994).

Questions

How might a young child demonstrate prosocial behavior? What is meant by a "theory of mind"?

Implicit Understanding and Knowledge   Piaget's procedures have also tended to underestimate many of the cognitive capabilities of preschool children. Indeed, recent research has raised an important new issue concerning children's conceptual foundations for learning. Toddlers seem to possess a significant *implicit understanding* or *knowledge* of certain principles (Seger, 1994; Reber, 1993). Here we will consider two spheres of conceptual knowledge for illustrative purposes: causality and number concepts.

Causality   Piaget concluded children younger than 7 or 8 fail to grasp cause-and-effect relationships. When he would ask younger children why the sun and the moon move, they would respond that heavenly bodies "follow us about because they are curious" or "in order to look at us." These types of explanations for events led Piaget to emphasize the limitations associated with young children's intellectual operations.

But contemporary developmental psychologists who investigate the thinking of young children find that they already understand a good deal about causality. **Causality** involves our attribution of a cause-and-effect relationship to two paired events that recur in succession. Causality is based on the expectation that when one event occurs, another event, one that ordinarily follows the first, will again follow it. Apparently the rudiments for the causal processing of information are already evident among 3-month-old infants ("If I cry, Mom will come"). And by the time youngsters are 3 to 4 years old, they seem to possess rather sophisticated abilities for discerning cause-and-effect relationships (Gelman & Kremer, 1991).

The versatility of young children grasping causality has led some psychologists to conclude that human beings are biologically prewired to understand the existence of cause-and-effect relationships (Pines, 1983). Children appear to operate on an implicit theory of causality. Clearly, the ability to appreciate that a cause must always precede an effect would have enormous survival value in the course of evolution.

Number Concepts   Piaget also deemphasized children's counting capabilities, calling counting "merely verbal knowledge" and asserting that "there is no connection between the acquired ability to count and the actual operations of which the child is capable" (Piaget, 1965a, p. 61). Yet young children seem to have an implicit understanding of some number concepts. Preschool youngsters can successfully perform tasks requiring modified versions of counting procedures and can judge whether a puppet's performance in counting demonstrating the concepts of *more* and *less* is correct (Gelman & Meck, 1986). Counting is the first formal computational system children acquire. It allows youngsters to make accurate quantitative assessments of amount, rather than having to rely solely on their perceptual or qualitative judgments. The next time you are near a 2- to 3-year-old, ask, "How old are you?" Invariably the child will hold up fingers and say the number.

Disabilities in Cognitive Development

Some children are going to experience moderate to severe difficulties and or delays in cognitive development and language skills during early childhood. Children with central nervous system or genetic disorders (such as mental retardation, cerebral palsy, autism), sensory damage (such as blindness or deafness), motor skill delays (such as muscular dystrophy, paralysis, missing limbs), social neglect (such as abuse, neglect, institutionalization, homelessness, isolation), or serious illness and injuries will progress at a pace slower than is normally expected for their age. Those who receive early intervention services are likely to make more progress than at-risk children not enrolled in such programs. Additionally, as we mentioned in Chapter 6, children who have experienced social neglect because of moving from foster home to foster home or from community to community (as do many migrant workers' children) or due to deliberate isolation from society are also going to be cognitively delayed and will need extensive stimulation and interaction to develop their potential. Table 7-4 lists some early warning signs of developmental delays.

Language Acquisition

Young children oftentimes show a lag between comprehending (which involves *receptive language*) and producing language (which involves *expressive language*). Children younger than 1 year old demonstrate time and again that they understand what we say. Say "Where's Mommy?" and the child looks for Mommy. However, it takes several more months before most children can begin to express their own needs in more than one- or two-word expressions. A 3-year-old might utter the following sentence after knocking on the door, "Is everybody not home?" meaning to inquire, "Is anybody home?" It is important to remember that while children are acquiring language, they understand and use it in ways that represent how and what they know about the world at that particular point in their development.

Table 7-4	Early Signs of Developmental Delay in Preschool Children

Language

Pronunciation problems

Slow vocabulary growth

Lack of interest in storytelling

Memory

Trouble recognizing letters or numbers

Difficulty remembering things in sequences (e.g., days of the week)

Attention

Difficulty sitting still or sticking to a task

Motor Skills

Problems with self-care skills (fastening buttons, combing hair)

Clumsiness

Reluctance to draw

Other Functions

Trouble learning left from right

Difficulty categorizing things

Difficulty "reading" body language and facial expressions

From: Lisa Feder-Feitel, "Does She Have a Learning Problem?", *Child*, February 1997. Copyright © 1997 by Lisa Feder-Feitel. Originally published by Gruner & Jahr USA Publishing in the February 1997 issue of *Child* Magazine. Used with permission.

At this stage of language development, children move beyond two-word sentences such as "Doggy go" and begin to display a real understanding of the rules that govern language as well as master the different sounds within the language, which is known as **phonology.** Past tenses are used (first using "goed" and then "went" as past tense for *go*) as well as plurals ("girls" and "boys") and possessives ("Jim's" and "mine," although at first many children say "mines"), which are all examples of how a word can change form, or what is known as **morphology.** Around the age of 3, children will begin to properly ask the *wh-* questions (Why? What? Which? When? Who?), which shows an understanding of **syntax** (the ways words must be ordered in a sentence). And between the ages of 3 and 5, children learn what types of language they can use in different social contexts, or the **pragmatics** of language (see *Further Developments: "The Development of Linguistic Humor in Children"*).

Chomsky's Linguistic Theory Noam Chomsky (1980, 1965) proposes that a linguistic theory ought to be able to adequately explain language structure while taking into account the messy input children receive and from which they construct meaningful sentences. He

Further Developments

The Development of Linguistic Humor in Children

Around the age of 2 or 3, children begin to assimilate and experiment with the rules of language. It is not surprising that language becomes a vehicle for humor among children soon after they become somewhat comfortable with it. At each of its levels—phonology, morphology, semantics, syntax, and pragmatics—language is a rule-based system, and humor typically involves violating rules. Let's look at each of the rules and see what children find humorous in each case.

Phonology

Phonological humor consists of distorted and immature articulations, along with tongue twisters. If you think of children's cartoons, you know that two characters who use immature articulations are Tweety Bird and Elmer Fudd. Tweety Bird's famous line "I tought I taw a putty tat" probably derives much of its humor from the fact that it is as close to a baby's undifferentiated "da-da-da-da," or what is known as babbling, as one can get and still be a comprehensible sentence. Elmer Fudd, on the other hand, is best known for his "Come out of dat wabbit ho you wascally wabbit!" The humor in Fudd's case is the immature articulation based on systematic deviations from standard phonemes—for instance, replacing *r* with *w*. Another type of phonological humor is found in tongue twisters. These difficult-to-pronounce sentences derive their humor from the fact that the speaker unintentionally substitutes one phoneme for another when saying the sentences quickly. For fun, if you haven't done this in a while, try saying the following four sentences repeatedly, at a faster and faster rate:

- Rubber baby buggy bumpers.
- Bring a broad-backed black bath brush.
- Sherry sold sea shells at the seashore.
- Peggy Babcock

You will most likely have noticed that you made errors by *substituting* one phoneme for a neighboring phoneme, which is based on the principle of assimilation. The usual mistakes are these:

- *rubber* assimilated to *buggy* and pronounced "rugger"
- *baby* assimilated to *buggy* and pronounced "bagy"
- *backed* and *bath* assimilated to *black* and pronounced "blacked" and "blath"
- *Sold* assimilated to *Sherry* and pronounced "shold"
- *Peggy Babcock* assimilated reciprocally, resulting in "Pebby Bagcock"

Morphology

Morphological rules cover word changes involving plurality or tense. For example, the word *chair* must be changed (*s* added) when modified by the adjective *three*—three chairs. Humor based on morphology usually involves "play lan-

guages." Four types of morphological rule changes have been found to account for the creation of play languages: addition, subtraction, reversal, and substitution. *Pig Latin* invokes both reversal and addition, moving the first consonant to the end and adding *-ay*, so that "Many people enjoy speaking Pig Latin" becomes "Anymay eoplepay joyenay eakingspay igpay atinlay." One study found that children can segment sentences into words at age 4 and segment syllables into sounds at age 6 (Fox & Routh, 1975). So Pig Latin would not likely be spoken by children younger than 6.

Semantics

Semantics deals with the meaning of words as well as the rules for combining words together meaningfully. Semantic violations that contain humor for the child can involve absence of meaning in words. One of the best-known examples of words that lack meaning is the poem "Jabberwocky" by Lewis Carroll (in Philip, Neal, & Mistry, 1998):

> Twas brillig and the slithy toves
> Did gyre and gimble in the wabe:
> All mimsy were the borogroves,
> and the mome raths outgrabe

Children also find incongruous or impossible meanings quite funny, such as the following:

> One fine day in the middle of the night
> Two dead men got up to fight.
> Back to back they faced each other
> Drew their swords and shot each other.

Syntax

Syntax looks at the logical relation among words in a sentence. Syntactical humor seems to be the rarest among children, with very few examples. When given the sentence, "Kim and Amy is sisters" the observed child laughed and replied, "Kim and Amy are sisters" One explanation for the dearth of humorous syntactical rule violations is that awareness of correct word order is a rather late development among children, although it has been observed that children as young as 3 do find humor in sentences such as "The man jumped up to the ocean" and "I'm going to pillow on my sleep."

Pragmatics

Pragmatics concerns how people use language in different social contexts. Some of the implicit conversation rules are that people will

- be cooperative
- tell the truth

Continued on page 214

Continued from page 213

- offer new and relevant information, and
- request information they sincerely want

Pragmatical humor is easily seen when someone ignores the context of a request and gives a literal answer. For example,

Question: "Do you know what time it is?"
Answer: "Yes."

Another example is the use of double meanings of words to bring out a point that might not be heard without humor. A little girl was sitting at the dinner table eating spinach when she suddenly grabbed her stomach and said, "I'm sick . . ." then paused, looked at her worried parents, and continued, laughing, ". . . of eating this stuff."

By examining the humor that children (and adults) find in language, we can catch a glimpse of the connections between the linguistic, emotional, cognitive, and physical pieces that determine the interactions between members of the human race.

Source: Adapted from P. E. McGhee and A. J. Chapman, 1980, *Children's Humour,* pp. 59–86; W. Ruch, 1998, *The Sense of Humor,* New York: Aldine de Gruyter, pp. 330–331.

believes that humans are born possessing a **language acquisition device (LAD)** that takes all of the sounds, words, sentences an infant hears and produces a grammar that is consistent with this data (Lillo-Martin, 1997). Chomsky argues that this has to be the case because it would be impossible for an infant to learn language simply by induction—that is, by simply reusing the sentences it had already heard. If that were the case, then a youngster would not be able to utter a novel sentence, but we all know that youngsters come up with some extremely novel sentences.

Late Talkers   Provided the child's hearing is okay, and provided the child is in an environment to hear speech, there are a few reasons why a young child might not use words until 2, 2½, or even 3 years of age:

- The baby is a quiet baby.
- The baby was born prematurely and/or is experiencing health problems.
- The baby is a twin: Twins often develop their own private language.
- The baby is a male: Boy babies often talk a little later than girl babies.
- The baby is in a bilingual home.
- The child isn't motivated to use language because there are siblings who communicate the child's wishes.

Some children don't talk until later but surprise everyone with short sentences when they do begin to use language. Learning to use speech is empowering for children because it enables them to communicate their basic needs, relate to others, learn about and take command of their environment, and become normal social beings rather than isolates (Sachs, 1987).

Vygotsky's Perspective   Learning language and furthering cognitive development are not tasks that children can accomplish in the privacy of their cribs—they need to be in a social setting. Lev Vygotsky (1896–1934) first conceived of how language and thought are intertwined with culture and society thereby proposing that cognitive growth occurs in a sociocultural context dependent on a child's social interactions (Vgotsky, 1962). This led to the well-known concept of the **zone of proximal development (ZPD):** Tasks that are a little too hard for children to accomplish alone can be mastered by the children when they are helped by a more skilled partner.

Think of when you began to read. Your parents or teacher would give encouragement, suggestions, corrections, and praise as you worked through the process of learning to read. Little by little you began to read independently, but it was only through the interaction between you and an older, more accomplished person that you were able to foster this new skill. This same principle applies to learning a sport. There is usually someone there to show the child how to make a proper arm motion or a specific foot movement, just as the child is on the verge of developing the necessary hand-eye coordination to throw or hit a ball. Notice that there is a striking difference here between Vygotsky and Piaget: Piaget thought that children learn as independent explorers, whereas Vygotsky asserted that children learn through social interactions.

To summarize, Vygotsky has put forth a *sociocultural* view that stresses the social aspects of cognitive development that Piaget did not emphasize. Vygotsky asserts that children's minds develop as children engage in activities with more skilled others on tasks that are within the child's zone of proximal development. Furthermore, Vygotsky asserts that when a child and adult are engaged in activities, the child incorporates the language of the adult that pertains to the activity and then re-uses that language to transform her or his own thinking.

Questions

What are some specific cognitive capabilities of young children? What is meant by LAD? ZPD?

Vygotsky's Zone of Proximal Development Young children often learn from each other. Tasks that are a little too hard for a child to accomplish alone can often be mastered with help from a more skilled partner.

Language and Emotion

Several years ago Donna Shalala, Secretary of Health and Human Services, was visiting an exhibit of children's artwork displayed in the national Capitol building. Shalala stopped to ask a girl about five or six if she liked to do artwork. The little girl answered, "Yes." When Shalala then asked why the girl liked to do art, the young child responded, "Cause it lets the sad out."

Rodriguez, 1998

Young children's vocabularies generally are limited, consisting mainly of words that refer to things and action words, with a smaller percentage of words that express *affective* (emotional) states (James, 1990). Research has shown that when children are operating from an emotional state, such as fear, hurt, pain, or stress, they are not able to concentrate on learning tasks. Their energy is concentrated in the processing going on in the limbic (emotional) system of their brains. After reading the next few sentences, stop and try this yourself. Recall a recent hurtful experience. Try to remember the details, picture the faces of the people involved, and feel the emotions you felt when this event occurred. Close your eyes now, and give yourself several minutes.

Now answer this question: *What are the main principles of Chomsky's linguistic theory?* You read about this just a short while ago, but do you now feel a little confused or somewhat overwhelmed, after reliving that hurtful experience? Teachers who have done the exercise you just finished have often commented, "I just can't think" (Rodriguez, 1998). Obviously, then, this happens to children, too. Some of the healthy ways young children can communicate their feelings are through drawing and physical movement; such activities release the energy built up by the limbic system. This is one of the reasons why day-care centers and preschools usually have a variety of papers, paints, crayons, colored pencils; or small slides and climbing toys; or music time and dance time—to allow children to "express" themselves. Multisensory experiences also allow a child to learn to process information from a variety of sources, which helps them develop important skills for advanced learning.

Information Processing and Memory

Memory is a net; one finds it full of fish when he takes it from the brook; but a dozen miles of water have run through it without sticking.

Oliver Wendell Holmes, *The Autocrat of the Breakfast Table*

Memory is a critical cognitive ability. Indeed, all learning implies memory. In its broadest sense, **memory** refers to the retention of what has been experienced. Without memory we would react to every event as if we had never before experienced it. Furthermore, we would be incapable of thinking and reasoning—of any sort of intelligent behavior—if we could not use remembered facts. Hence, memory is critical to information processing.

Early Memory

The great majority of 5-year-old preschoolers can identify primary colors (89 percent), write or draw rather than scribble (84 percent), count to 20 (78 percent), and either read or tell connected stories while pretending to read (79 percent).

Building Knowledge for a Nation of Learners:
A Framework for Education Research,
Office of Educational Research & Improvement, 1997

During infancy and childhood, we learn a prodigious amount about the world (Rovee-Collier, 1987), yet by adulthood our memories of our early experiences have faded. This phenomenon is called *childhood amnesia.* As adults, generally we remember only fleeting scenes and isolated moments prior to the time we reached 7 or 8 years of age. Although some individuals have no recollections prior to 8 or 9 years of age, many of us can recall some things that took place between our third and fourth birthdays. Most commonly, first memories involve visual imagery, and most of the imagery is in color. In many cases we visualize ourselves in these memories from afar, as we would look at an actor on a stage (Nelson, 1982).

Just why early memories should wane remains an enigma (Newcombe & Fox, 1994). Sigmund Freud theorized that we repress or alter childhood memories because of their disturbing sexual and aggressive content. Others, particularly Piagetians and cognitive developmentalists, claim that adults have trouble recalling the events of childhood because they no longer think as children do (for instance, adults typically employ, as aids to memory, words and abstract concepts that do not mesh with the mental habits they employed as young children). Still others say that the brain and nervous system are not entirely formed in the young and do not allow for the development of adequate memory stores and effective retrieval strategies. Then there are those who contend that much learning during the first two years of life occurs in emotion-centered areas of the brain and that this learning is subject to limited recall in later years. Most types of earliest memories we are likely to recall are associated with the emotion of fear. What early experiences do you remember? Do you feel associated emotions?

Finally, Mark Howe and Mary Courage (1993) argue that the problem derives not from memory per se but from the absence of a personal frame of reference for making one's memories uniquely autobiographical (put in other terms, there is no developed "self" present as a cognitive entity). Nonetheless, our understanding of childhood amnesia continues to be relatively elusive, and theorizing about it remains largely speculative.

Questions

What are some of the explanations for childhood amnesia? Do you ever wonder why you cannot remember something from earlier in life (or from yesterday)? Does this bother you?

Information Processing

Memory includes recall, recognition, and the facilitation of relearning. In *recall* we remember what we learned earlier, such as the definition of a scientific concept or the lines of a play. (An essay question demands that you recall information.) In *recognition* we experience a feeling of familiarity when we again perceive something that we have previously encountered. (A matching type of quiz requires recognition.) In the *facilitation of relearning*, we find that we can learn material that is already familiar to us more readily than we can learn totally unfamiliar material.

On the whole, children's recognition memory is superior to their recall memory. In recognition the information is already available, and children can simply check their perceptions of an occurrence against their memory. Recall, in contrast, requires them to retrieve all the information from their own memory. When 4-year-olds are asked how many items from a list they think they will remember, they predict they will recall seven items—but they do recall fewer than four (Flavell, Freidrichs, & Hoyt, 1970).

Memory permits us to store information for different periods of time. Some psychologists distinguish sensory information storage, short-term memory, and long-term memory (see Figure 7-4). In **sensory information storage,** information from the senses is preserved in the sensory register just long enough to permit the stimuli to be scanned for processing (generally less than 2 seconds). This provides a relatively complete, literal copy of the physical stimulation. For instance, if you tap your finger against your cheek, you note an immediate sensation, which quickly fades away.

Short-term memory is the retention of information for a very brief period, usually not more than 30 seconds. For example, you might look up a number in the telephone directory and remember it just long enough to dial it, whereupon you promptly forget it. Another common short-term memory experience is your introduction to a stranger. How easily do you recall the person's

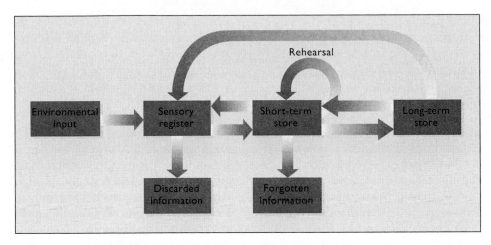

Figure 7-4 Simplified Flow Chart of the Three-Store Model of Memory Information flow is represented by three memory stores: the sensory register, the short-term store, and the long-term store. Inputs from the environment enter the sensory register, where they are selectively passed on to short-term storage. Information in the short-term store may be forgotten or copied by the long-term store. In some cases individuals mentally rehearse information in order to keep it in active awareness in short-term storage. Complicated feedback operations take place among the three storage components.

name 10 minutes later? Information typically is fleeting unless there is some reason or motivation to remember it longer.

Long-term memory is the retention of information over an extended period of time. A memory might be retained because it arose from a very intense single experience or because it is repeatedly rehearsed. Through yearly repetition and constant media reminders, you come to remember that Memorial Day is in late May, Labor Day is in early September.

Currently, Schacter and Tulving (1994) postulate that there are five major systems of human memory we use to operate in daily life:

- *Procedural memory* involves learning various kinds of motor, behavioral, and cognitive skills.
- *Working memory* is an elaboration on what is known as short-term memory, allowing one to retain information over short periods of time.
- *Perceptual representation (PRS)* is used in identifying words and objects.
- *Semantic memory* allows one to acquire and retain factual information about the world.
- *Episodic memory* allows one to remember events seen or experienced throughout life.

Metacognition and Metamemory

As children mature cognitively, they become increasingly active agents in their remembering process. The development of memory occurs in two ways: through alteration in the biological structuring of the brain (its "hardware") and through changes in types of informa-

tion processing (the "software" of acquisition and retrieving). Researchers have observed striking changes as a function of age, both in children's performance on memory tasks and in their use of memory strategies. As they grow older, children acquire a complex set of skills that enables them to control just what they will learn and retain. In short, they come to "know *how* to know," so that they can engage in deliberate remembering (Moore, Bryant, & Furrow, 1989). Individuals' awareness and understanding of their own mental processes is termed **metacognition,** and young children attempt to learn in ways that suit themselves. Listen carefully, and you will hear a young child say, "I can't do that" or the opposite, "Let me do that." These are signs of awareness of their own mental abilities. Overall, memory ability catapults upward from birth through age 5, and then it advances less rapidly through middle childhood and adolescence (Chance & Fischman, 1987).

Human beings require more than the factual and strategic information that constitutes a knowledge base. They must also have access to this knowledge base and apply strategies appropriate to task demands (*How am I going to remember to spell this word? to walk home from kindergarten? to throw this ball to home base?*) This flexibility in calibrating solutions to specific problems is the hallmark of intelligence. Flexibility reaches its zenith in the conscious control that adults bring to bear over a broad range of their mental functioning. This conscious control of strategies and awareness of reflection on these strategies continues to expand throughout adolescence and adulthood in both cognitive and sociomoral domains (Schrader, 1988).

Children's awareness and understanding of their memory processes is called **metamemory.** A common example of metamemory is the intentional approach children use to memorize their address and phone number. Because many young children spend a good deal of time away from home during the day, they are asked to learn this vital information as early as possible. Does the child repeat this over and over verbally, or does the child ask to look at the phone number and address on a sheet of paper? Research reveals that even 3-year-olds engage in intentional memory behavior. They appear to understand that when they are told to remember something, they are expected to store and later retrieve it. Indeed, even 2-year-olds can hide, misplace, search for, and find objects on their own (Wellman, 1990, 1977). By the time children enter kindergarten, they have developed considerable knowledge of the memory process. They are aware that forgetting occurs (that items get lost in memory), that spending more time in study helps them retain information, that it is more difficult to remember many items than a few, that distraction and interference make tasks harder, and that they can employ records, cues, and other people to help them recall things (Fabricius & Wellman, 1983). They also understand such words as *remember, forget,* and *learn* (Lyon & Flavell, 1994).

Question

If you ask a child, "Who taught you to walk?" the child will probably respond, "I taught myself." Why do children sometimes falsely remember things?

Memory Strategies

Children (and adults) might employ a variety of strategies to help themselves remember and recall information they are learning or desire to remember. Think about this for a few minutes, prior to reading the next section. If you had to remember the names of five people in your study group you were just introduced to this morning, how would you approach this task? Would you make a list and verbally repeat the list? Would you put names on flashcards and continue shuffling through the flashcards? Would you make a word out of the first letter of their names, such as taking Harriet, Angela, Paul, Peter, and Yvonne and making the word *HAPPY*? Then you could think, "We are a *happy* group" in order to remember their names. Or would you use some other approach?

Rehearsal as a Memory Strategy   One strategy that facilitates memory is *rehearsal,* a process in which we repeat information to ourselves. Many individuals who are adept at remembering people's names cultivate the talent by mentally rehearsing a new name several times to

themselves when they are introduced to a person. Researchers have demonstrated that children as young as 3 are capable of various rehearsal strategies. For instance, if 3-year-olds are instructed to remember where an object is hidden, they often prepare for future memory retrieval by extended looking at, touching, or pointing to the hiding place (Wellman, Ritter, & Flavell, 1975). As children grow, their rehearsal mechanisms become more active and effective (Halford et al., 1994). Some researchers believe that the process is facilitated through language, as children become increasingly skillful at verbally labeling stimuli. According to these investigators, the organizing and rehearsing process inherent in naming is a powerful aid to memory (Rosinski, Pellegrino, & Siegel, 1977). And as children come to process information in more sophisticated ways and learn how and when to remember, they become capable of making more decisions for themselves. Parents and teachers can cultivate children's decision-making skills.

Categorization as a Memory Strategy   One strategy that facilitates remembering is to sort information into meaningful *categories.* Sheila Rossi and M. C. Wittrock (1971) found that a developmental progression occurs in the categories children use to organize words for recall. In this experiment children ranging from 2 to 5 years old were read a list of twelve words (*sun, hand, men, fun, leg, work, hat, apple, dogs, fat, peach, bark*). Each child was asked to recall as many words as possible. The responses were scored in pairs in terms of the order in which a child recalled them: rhyming (*sun-fun, hat-fat*); syntactical (*men-work, dogs-bark*); clustering (*apple-peach, hand-leg*); or serial ordering (recalling two words serially from the list). Rossi and Wittrock found that rhyming responses peak at 2 years of age, followed by syntactical responses at 3, clustering responses at 4, and serial-ordering responses at 5. In many respects the progression is consistent with Piaget's theory, which depicts development as proceeding from concrete to abstract functioning and from perceptual to conceptual responding. Other research also confirms that changes occur in children's spontaneous use of categories during the developmental span from age 2 to adolescence (Farrar, Raney, & Boyer, 1992). Though children as young as 2 benefit from the presence of categories in recall tasks, older children benefit even more. For one thing, recall increases with age (Farrar & Goodman, 1992). For another, children of age 4 to 6, in comparison with older children, show less categorical grouping of items in recall tasks, fewer subordinate categories in tasks requesting that similar items be grouped together, and lower consistency in the assignment of items to selected categories (Best, 1993). Adolescents adopt even more sophisticated strategies, grouping items into logical categories, such as people, places, and things, or animals, plants, and minerals (Chance & Fischman, 1987).

Moral Development

Social feelings first appear during the preoperational stage. For the first time, feelings can be represented and recalled. The ability to recall feelings makes *moral* feelings possible. During the sensorimotor stage, children cannot reconstruct past events and experiences because they lack representation. Once the child has the capacity to reconstruct both the cognitive and affective past, the child can begin to exhibit consistent emotional behavior. If a child remembers what someone did in the past and also the emotional response linked to the action, the beginnings of moral decision making are in place. Piaget (1981) asserts that *reciprocity* of attitudes and values is the foundation for social interchange in children. **Reciprocity** leads to each child's valuing the other person in a way that allows him or her to remember the values that the interactions bring forth. The following scenario is an example of how this plays out in children:

> Two young children meet at the park and play. One of the children, Neiko, gives the other child, Kiri, some candy because she obviously wants some. They end up sharing their toys and having a nice time. By labeling Neiko's behavior as "good," Kiri can recall that scene the next time they encounter each other at the park. She might give Neiko something of hers or at least be nice to her owing to the recollection of warm feelings that seeing Neiko conjures up for Kiri.

When Piaget first studied moral development, he looked at the evolution of moral reasoning in children. He believed that moral feelings in young people pointed to what it was necessary to do and not just what was preferable. He also proposed that moral norms have three characteristics:

- They are generalizable to all situations.
- They last beyond the situations and conditions that engender them.
- They are linked to feelings of autonomy.

But, according to Piaget (1981):

> From two to seven years, none of these conditions is met. To begin with, norms are not generalized but are valid only under particular conditions. For example, the child considers it wrong to lie to his parents and other adults but not to his comrades. Second, instructions remain linked to certain represented situations analogous to perceptual configurations. An instruction, for example, will remain linked to the person who gave it, and finally, there is no autonomy. . . . Good and bad

are defined as that which conforms or fails to conform to the instructions one has received.

Piaget based his views on children's moral development by observing local boys playing marbles, which he considered to be a social game with a set of rules. By interviewing these boys, Piaget found two developmental stages in young children's knowledge of rules: the motor stage and the egocentric stage. In the motor stage, children are not aware of any rules. For example, a young child in this stage might build a nest out of the marbles and pretend to be a mother bird. The game of marbles is not understood. Between the ages of 2 and 5, the egocentric stage, children become aware of the existence of rules and begin to want to play the game with other children. But the child's egocentrism prevents the child from playing the game socially. The child continues to play alone without trying to compete. For example, a child will throw a marble at the pile and yell, "I won!" At this stage children believe everyone can win. Rules are seen as fixed and as coming from a higher authority.

Piaget (1981) also interviewed children to find out how they viewed intentionality versus accident. The children found it very difficult to separate the act from the reason for the act. For example, children were told the following two stories:

> Once there was a little girl named Heidi who was playing in her room when her mother called her to come to dinner. Since they were having guests, Heidi decided to help her mother and got a tray and then put 15 cups on the tray so she could carry them into the dining room. As she was walking with the tray, she tripped and fell and broke all of the cups.

> Once there was a little girl named Gretchen who was playing in the kitchen. She wanted some jam and her mother was not there. She climbed up on a chair and tried to reach the jam but it was too high. She tried for ten minutes and became angry. Then she saw a cup on the table so she picked it up and threw it. The cup broke.

Children younger than 7 usually see the first girl as having committed the worse act because Heidi broke more cups than Gretchen. The children judge the act based on the quantitative results of the action, with no appreciation of the intention behind the action. Morality, for preoperational children, is still based mainly on perception. Although young children have the rudiments of moral feelings, they are limited in their ability to comprehend justice as long as they do not understand the intentionality of an action.

Segue

The physical, sensory, and cognitive maturation and skills acquired during the early childhood years have profound implications for children's ability to function fully as members of society (Stipek, Recchia, & McClintic, 1992). We do not develop in isolation. To enter into sustained social interaction with others, we must impute meaning to the people around us. All of us, children and adults alike, confront the social world in terms of categories of people—we classify them as parents, family, cousins, adults, doctors, teachers, teenagers, business people, and so on. Society does not consist merely of so many isolated individuals. It is composed of individuals who are classed as similar because they play similar roles. And just as we impute meaning to others, we must also attribute meaning to ourselves. We have to develop a sense of ourselves as distinct, bounded, identifiable units. In Chapter 8 we will explore the "conceptions" of ourselves we acquire within emotional and social contexts.

Summary

Physical Development

1. Early childhood focuses on continued brain growth, physical development, the perfection of gross and fine motor skills, and maturation of the sensory systems. For good health, children need proper nutrition, exercise, and health care to support the developing body systems.

2. Visual, tactile, auditory, gustatory, linguistic, and kinesthetic sensory development rapidly occurs at this age, and children are anxious to use all of their senses in learning.

3. At the age of 5, the child's brain will weigh about 90 percent of its final, adult weight.

4. A "mixed diet" is important and means a variety of foods and beverages eaten in different combinations so that over a period of time the child will get the proper nourishment for brain and body growth.

5. Children should be routinely protected by immunizations during regularly scheduled checkups with pediatricians and other health care practitioners.

6. Minority children and children who recently immigrated to the United States are likely to come from cultural subgroups that have alternative perspectives on health, illness, and wellness. All children must meet immunization requirements in order to enter kindergarten and some preschools.

7. The leading cause of death for young children is accidents. More boys die than girls at all ages. Overall, the rate of deaths in early childhood has been decreasing over the past several decades.

Cognitive Development

8. Intelligence is a person's general problem-solving abilities. Psychologists have applied differing models to the study of intelligence. Some psychologists view intelligence as a composite of many special, independent abilities. Others depict it as a process deriving from the interplay between children and their environment. Still others portray intelligence as an information-processing activity.

9. Psychologists differ in the relative importance that they attribute to the roles of heredity and environment in fashioning an individual's intelligence. However, most social and behavioral scientists believe that any extreme view regarding the nature-nurture controversy is unjustified at the present time.

10. Jean Piaget calls the years between 2 and 7 the "preoperational period." The principal achievement of the preoperational period is the developing capacity of children to represent the external world internally through the use of symbols, such as words and numbers. During this period children have difficulty solving conservation problems. Logical thought is inhibited by obstacles associated with centering, transformations, reversibility, and egocentrism.

11. A considerable amount of research has been done on children's developing understanding of their mental world. "Theory of mind" research is socially significant because people seek to understand and predict the mental states and behaviors of others on the basis of it.

12. Recent research suggests that toddlers possess a significant implicit understanding of certain basic principles, including causality and number concepts. Such implicit understandings provide an important conceptual foundation for children's learning.

13. Children move beyond two-word sentences and begin to internalize the rules of language: phonology, morphology, syntax, semantics, and pragmatics.

14. The zone of proximal development (ZPD) involves tasks that are a little too hard for a child to accomplish alone but that the child can master with some help from a more skilled partner.

Information Processing and Memory

15. Memory is an integral component of cognition. Without memory we could not think or reason. Memory includes recall, recognition, and the facilitation of relearning.

16. As children develop, they become increasingly active agents in the remembering process. Strategies that facilitate remembering include grouping information into meaningful categories and rehearsing information.

17. As they grow older, children acquire a complex set of skills that enables them to control just what they will learn and retain. In short, they come to "know *how* to know," so that they can engage in deliberate remembering.

Moral Development

18. Piaget found in his study of the moral development of very young children that they do not understand the intentionality of actions, but as they mature cognitively and emotionally during early childhood, they come to understand more about moral reasoning.

Key Terms

causality *(211)*
centration *(208)*
conservation *(208)*
egocentrism *(209)*
intelligence *(205)*
language acquisition device (LAD) *(214)*
long-term memory *(217)*
memory *(215)*

metacognition *(217)*
metamemory *(218)*
morphology *(212)*
phonology *(212)*
pragmatics *(212)*
preoperational period *(208)*
reciprocity *(219)*
reversibility *(209)*
semantics *(213)*

sensory information storage *(216)*
short-term memory *(216)*
syntax *(212)*
theory of mind *(211)*
two-factor theory of intelligence *(205)*
zone of proximal development (ZPD) *(214)*

Following Up on the Internet

Web sites for this chapter focus on young children's physical, cognitive, language, and moral development. Please access the text web site at http://www.mhhe.com/crandell7 for up-to-date hot-linked Internet addresses for the following organizations and resources.

National Institute of Child Health and Human Development

Early Childhood.Com

Clearinghouse on Elementary and Early Childhood Education (ERIC EECE)

Children's National Day Care/School Directory

Early Childhood Quarterly Journal

Chapter 8

Early Childhood
Emotional and Social Development

Critical Thinking Questions

1. How does a young child develop a sense of identity, and is a child's gender inborn or does it come from sociocultural influences?

2. What other influences might affect a young child's growing sense of self in any society?

3. How would you explain the parenting style your parents used in raising you? Were they highly restrictive, permissive, or more democratic? In what ways has their approach affected your self-esteem? If you are a parent, do you parent in the same way?

4. When there is a divorce in a family or the death of a parent or sibling or a major move to a new home, people say of a young child, "Oh, she's young, she'll adjust" or "He's young, he's flexible." Do you think this is an accurate appraisal of young children's emotional capabilities?

Cognitive factors play an important part in setting the tone for the emotional life of youngsters, and social factors also have an impact on maximizing or minimizing intellectual ability. Through social interactions with family and others, children acquire guidelines that mentally or cognitively mediate their inner experience of emotion and their outer expression of it (Wintre & Vallance, 1994). Today, young children's development occurs within a broadening social world. Up to the 1980s most American children spent their early years in the confines of home with family members until they entered the social/academic world of school. Thus, much of the classic research on the child's sense of identity and the effectiveness of parenting styles was conducted with mainly intact, nuclear families in the 1950s to 1970s. Since the 1980s there has been a major exodus of mothers leaving the home to enter the ranks of the employed. Now young children are being cared for at nursery, day-care, and preschool programs in record numbers. It is within this broader context of social influences that we also examine the more recent research findings on the emotional and social development of young children from ages 2 to 6.

223

Emotional Development and Adjustment

Attitude determines altitude.

Reverend John Maxwell

Today's child psychologists and early childhood experts believe that children's emotions are central to their lives and should be central to the nursery school, preschool, and early elementary school curriculum. Hyson (1994, p. 2), an early childhood professional states, "A new generation of emotion researchers (developmental psychologists) has documented emotions' importance as organizers of children's behavior and learning. This knowledge base can provide a solid foundation for rebuilding an emotion-centered early childhood curriculum." She further states (1994, p. 4) that emotional development and social development are intertwined and "current theory and research support the belief that all behavior, thought, and interaction are in some way motivated by and colored by emotions. Thinking is an emotional activity, and emotions provide an essential scaffold for learning. In fact, children's feelings can support or hinder their involvement in and mastery of intellectual content." This next section examines the emotional development of young children from ages 2 to 6.

Timing and Sequence of Emotional Development in the Young Child

Many changes occur in emotional expression and emotional regulation during early childhood. The process of emotional development and emotional self-regulation might seem slow to parents and caretakers, yet children typically possess and express certain emotions as they mature.

Facial Expressions and Body Language   Very young infants express such emotions as happiness, sadness, distress, anger, and surprise (Izard & Malatesta, 1987). Other facial expressions that appear with greater frequency during toddlerhood and preschool years include pride, shame, shyness, embarrassment, contempt, fear, and guilt. These emotional expressions seem to require a certain level of cognitive ability and awareness of cultural values or social standards. Preschool children often display several emotions at once, such as anger and guilt, as when a child is stubbornly refusing to share something with another. With increased physical development, older children have better control of their facial muscles, and they can facially display more complex emotions (Hyson, 1994). Much developmental research has focused on the young child's face as the conduit for

expressing emotions, yet parents and other child caregivers would be wise to become good "readers" of children's whole body language. Young toddlers can use increased large- and fine-muscle control to express feelings in a more complex, deliberate way. Young children can jump, wave their arms, clap their hands together, or verbally express their delight (Hyson, 1994). Elementary-school-age children can demonstrate the socially approved action of "thumbs up" or a "high five."

Psychologists have also discovered that particular voice qualities, such as loudness, pitch, and tempo, can convey specific social-emotional messages, such as fear, anger, happiness, and sadness (Scherer, 1979). As children get older, sound and vocal quality continue to be important tools to convey feelings. Words accompany the emotion as well. "Don't touch my toys!" is conveyed both loudly and emotionally. Words also allow children to convey feelings about themselves and others: "I sad. Daddy make better." With progressive age, children can demonstrate and verbally express their more complex feelings (Ricks, 1979).

Symbolic Activity and Play   As children progress through the toddler and preschool years, they are able to demonstrate a complex array of emotions through play and manipulation of materials in their environment. Many early childhood professionals regard play as a primary way for children to communicate their deepest feelings (Erikson, 1977). **Play** may be defined as voluntary activities that are not performed for any sake beyond themselves. They are activities that people commonly view as being outside the serious business of life.

There are many forms of play, including pretend play, exploration play, games, social play, and rough-and-tumble play. In recent years pretend play has captivated the interest of many psychologists who have viewed make-believe or fantasy behavior as an avenue for exploring the "inner person" of the child and as an indicator of underlying cognitive changes. Thus, during the first two years children's play shifts from the simple manipulation of objects to an exploration of the objects' unique properties and to make-believe play involving ever more complex and cognitively demanding behaviors (Uzgiris & Raeff, 1995). For instance, many preschool youngsters engage in a form of make-believe play in which they create an *imaginary companion* who becomes a regular part of their daily lives—an invisible character who they name, refer to in conversations, and play with in an air of reality that lacks an objective basis (Taylor, Cartwright, & Carlson, 1993). As might be expected, the proportion of children who engage in pretend play with other youngsters increases with age.

One difference in play behavior is that boys tend to select games that require a larger number of participants than those selected by girls—but these differences in the

Differences in Play in Girls and Boys Girls tend to play in two-person groups, to be verbal during play, to hold hands or stay close to each other, and to display other signs of affection. Notice how closely these girls are sitting. In general, boys tend to play in larger groups, to engage in more boisterous, motor activities such as running, jumping, climbing, throwing objects, wrestling—often engaging in play with overtones of competition and dominance.

play activities of boys and girls are often fostered by parental and teacher socialization practices. The two genders have somewhat different play styles (Alexander & Hines, 1994). Boys' play has many rough-and-tumble physical qualities to it and strong overtones of competition and dominance. Girls engage in more intimate play than boys do, and two-person groups are conducive to intimate behavior. Girls are more likely than boys to disclose intimate information to a friend and to hold hands and display other signs of affection.

Self-Regulation of Emotions   Self-regulation is an area of emotional development where cultures differ and professionals disagree. Professional experts and parents often disagree regarding the age at which children can be expected to control their own emotions. Some believe children should be able to "let their feelings out," while others believe children need to learn to control their emotions and feelings. At best, self-regulation develops at each child's pace and likely proceeds gradually and unevenly. Emotional self-regulation is a difficult task that must be mastered in a healthy way for children to develop a positive relationship with others. Children who cry, whine, hit, bite, or scream to get their way certainly will not be well liked; nor will the sad or timid child who sits quietly in a corner away from others. Somewhere there is a healthy balance of emotional self-regulation for each child. Older children begin to regulate or control some aspects of their emotional life by inhibiting certain emotional expressions (such as outbursts they know will not be approved of). Deliberate holding back of feelings is difficult for children until after age 3 (Saarni, 1979). Individual cultural expectations also dictate the norms for the expression of emotions by boys and girls.

Cultural Expectations and Emotional Expression

Every society has its own cultural expectations or unwritten "rules" for expressing emotions. Young children also learn to convey culturally prescribed emotions through gestures and body language. In the past, U.S. social scientists used the white, middle-class child as a normative standard for comparison, but contemporary social scientists take a more cross-cultural approach to studying the influences on children's emotional development.

In general, American parents tend to promote freedom of expression, independence, and individuality, and allow a child to state or exhibit feelings more openly than in some other cultures. It is somewhat common to witness a young child in public having a "temper tantrum" or shouting for joy in a toy store, for example. In contrast, children from the Korean culture are encouraged to show a lack of emotion and to have a more compliant nature (Lynch & Hansen, 1992). Korean families stress obedience to authority and deference to elders (Rohner & Pettengill, 1985). Two- and 3-year-old Japanese children might show greater signs of distress at separating from their mothers, because the Japanese culture encourages extreme closeness between mother and child (Hyson, 1994). Lewis (1988) observed children in 15 Japanese nursery schools and learned that Japanese children are allowed to demonstrate anger as long as other children do not get hurt badly. However, as they grow older, Japanese children also encounter more social regulations that promote *ittaikan,* a feeling of oneness with the group (Weisz, Roghbaum, & Blackburn, 1984). Japanese children are prepared to accept a more rigorous school experience than American children. As early as kindergarten, young Japanese children face demanding school work 6 days a week for 240 days per year and must pass major exams to be promoted to the next grade (Fuligni & Stevenson, 1995).

According to Harwood (1992), Puerto Rican children are encouraged to have a close emotional bond with their mothers. African American children might be taught by caregivers to refrain from direct eye contact, which people from other cultural backgrounds might misinterpret as shyness or evasiveness (Hyson, 1994). The Hispanic American cultural stereotype for a male child is summed up in the word *macho,* derived from *machismo,* which means a boy is encouraged to restrain from expressing his feelings, to maintain an emotional distance while being strong and dominant. In contrast, *marianismo* is the Hispanic American cultural stereotype for a girl. She is expected to be feminine, to learn self-sacrifice, to provide joy in a family, and to subjugate her needs to prepare herself to live in a patriarchal subculture (Gil & Vazquez, 1996).

With greater *acculturation* into American society, though, traditional cultural gender-role expectations are being redefined. Caretakers of young children and parents who plan to adopt a child from a different background might want to become familiar with the styles of emotional expression prescribed by the child's group of origin.

Ekman (1972) found that preschoolers and school-age children learn about expectations of emotional self-regulation through observation and direct instruction. However, because this is a time of egocentric (centered on self) behavior for preschool children, repeated direct instruction may be necessary. It is common to hear parents of preschoolers and young elementary-age children express frustration: "How many times have I told you to. . . ." Caretakers of young children need to be patient and persevere by repeatedly explaining and demonstrating the desired behavior. Physical punishment, such as slapping or spanking children, at this stage simply teaches children to use physical measures themselves to control others' behaviors. Most children conform to their appropriate cultural norms over time.

Acquiring Emotional Understanding

Parents often talk to preschoolers about past events ("Remember when we did . . . and . . . happened?"). These reminiscences often are framed within an emotional context (how happy the child was, how much fun some activity was). As children get older and learn more about emotions, they become better able to match facial pictures with the correct emotion label. Older children are also able to identify why people feel the way they do—sad, happy, surprised, fearful, or angry (Stein & Jewett, 1986). Even 3-year-olds have their own ideas about what causes fear. Some studies show that

preschoolers have difficulty understanding that people can have several emotions at one time (as when a parent says "I love you, but I'm upset with what you did").

The Link Between Feeling and Thinking   When given familiar situations, preschoolers can accurately identify the commonly associated emotion; and when given an emotion, they can easily describe an appropriate eliciting situation (Lagattuta, Wellman, & Flavell, 1997). Although past research proposed that it is not until 8 or 9 years of age that children can begin to understand mental (rather than situational) causes of feelings, more recent research by Lagattuta and colleagues (1997) demonstrates that much younger children can understand the emotional consequences of past experiences when cued cognitively. The findings from their three recent research studies demonstrate that 3- and 4-year-olds have a decided tendency to explain emotions in terms of current external situations, and 4- and 5-year-olds have substantial knowledge about mind and emotion. More specifically, their research with young children demonstrated the following:

- Young children's prior experiences, desires, beliefs, and thoughts can affect their emotional reactions to current situations (e.g., they can relate to the loss of a pet in a story with photographs and elaborate on why the character in the story is unhappy).
- Young children have the ability to infer mental activity (thinking or remembering), particularly at 4, 5, and 6 years old.
- Young children's mental activity can influence their emotional arousal.
- Young children demonstrate knowledge about the sources of their thoughts.
- Young preschoolers can predict that a friend, who had never experienced a character's sad experience, would not feel sad.

Responding to Emotions of Others   Toddlers and preschoolers deliberately seek out information about the emotional reactions of others. They might appear to be fascinated by "getting a rise out of" their parents and caretakers (what at first might appear to be a game can become irritating to caretakers). By the preschool years, the emotional response to others' distress seems to motivate children to comfort and help their caregivers and peers (Hyson, 1994).

Forming Emotional Ties   Forming emotionally positive relationships with parents and significant others is a key developmental task in the development of a child's sense of self-awareness. Children in child-care settings nowadays have demonstrated that they can also develop close, affectionate bonds with caregivers in these settings (Howes & Hamilton, 1992). Emotional attachment is

Young Children Often Help Each Other George Mead would say that the child on the left can imagine himself in the role of the child on the right. Young children's prior experiences can affect their emotional reaction to a current situation. By the time children are 4- to 5-years-old, they are often motivated to help each other.

demonstrated by such behaviors as clinging, smiling, and crying. As children age, they are able to develop a close attachment to someone without necessarily having physical contact. Young children also demonstrate a close bond with siblings and other children in child-care settings (Hyson, 1994).

Emotions are central to children's lives and help them organize their experiences in order to learn from them and develop other behaviors. Though cultures differ in their norms for the expression of feelings by girls and boys, Table 8–1 shows the general progression of emotional development in children from age 2 through age 6.

Question

In early childhood, how do children increase their ability for emotional self-regulation?

The Development of Self-Awareness

Man can be defined as the animal that can say "I," that can be aware of himself as a separate entity.

Erich Fromm

As we have seen, emotional development is an essential component of a child's sense of self-awareness. Moreover, a child's own sense of self-worth or self-image is part of the overall dimension called **self-esteem.** Some children develop what is referred to as positive self-

esteem, while others develop a more negative view of themselves, referred to as low self-esteem. Enhancing children's self-esteem is an important goal for many parents and most preschool and kindergarten programs, and there is considerable evidence that children's self-esteem can have lifelong effects on their attitudes and behavior, performance in school, relationships with family, and functioning in society (Hong & Perkins, 1997). In this section, we examine the factors that influence a child's developing self-esteem.

The Sense of Self

Among the cognitive and social achievements of the child's early years is a growing self-awareness—the human sense of "I." At any one time, we are confronted with a greater quantity and variety of stimulation than we can attend to and process. Accordingly, we must select what we will notice, learn, infer, or recall. Selection does not occur in a random manner but depends on our use of internal cognitive structures—mental "scripts" or "frames"—for processing information. Of particular importance to us is the cognitive structure that we employ for selecting and processing information about ourselves. This structure is the **self**—the system of concepts we use in defining ourselves. It is the awareness we have of ourselves as separate entities who are able to think and initiate action. The self provides us with the capacity to observe, respond to, and direct our own behavior. The sense of self distinguishes each of us as a unique individual, different from others in society. It gives us a feeling of placement in the social and physical world and of continuity across time. And it provides the cognitive basis for our identities (Cross & Markus, 1991).

Neisser (1991) further differentiated between the ecological self and the interpersonal self. The *ecological self* is the self that picks up and acts on the perceptual information presented by objects in the environment (e.g., the family dog nudges the young child to play). The *interpersonal self,* in contrast, is that aspect of self that arises in interactions with other people (e.g., reacting happily when mother comes into the kindergarten room at the end of the day) (Pipp-Siegal & Foltz, 1997). Toddlers' abilities to develop categories of stimuli become more complex. Pipp-Siegel and Foltz (1997) conducted a study with 60 children and mothers at infancy and toddlerhood to determine toddlers' knowledge of self, mother, and inanimate objects. Their research findings show that the complexity of knowledge between knowing self and others changed as a function of age. Two-year-olds demonstrated an emergence of the interpersonal self, with the development of self-conscious emotions and an increased capacity for symbolic play.

The development of a *sense of self* as separate and distinct from others is a central issue of children's early years (Asendorpf & Baudonniere, 1993). This fundamental

Table 8-1   Progression of Emotional Development, Ages 2 to 6

Age Group	Emotional Expression
2-year-olds	Begin to show facial expressions of shyness, pride, shame, embarrassment, contempt, fear, and guilt.
	Parallel play is common in 2-year-olds—that is, playing independently near others without interaction.
	Play is more likely to be simple manipulation of objects.
	Deliberate holding back of feelings, called emotional self-regulation, is unlikely at this early age.
	Sometimes toddlers deliberately try to elicit emotional responses from caregivers.
	Definitely begin to use the word *no*.
	Toddlers can also develop close, affectionate bonds with caregivers, demonstrated by smiling, clinging, and crying.
	Two-year-olds demonstrate emergence of the interpersonal self (e.g., reacting happily when mother comes to pick up the child at the end of the day).
	"Watch me do this!"
	Young children conceive of the self strictly in physical terms, pointing to the head, for example.
	Plants and animals also have selves and minds.
	Gifted children master their environment quickly and thoroughly and appear to be self-directed.
3-year-olds	Increased control of large- and fine-muscles to express feelings in a more complex, deliberate way.
	Can now jump, wave arms, clap hands together, and verbally express delight.
	Voice qualities, such as loudness, pitch, and tempo vary, conveying specific feelings, (e.g., "Don't touch my toys!").
	Begin to demonstrate a complexity of emotions through play materials in their environment.
	Associative play is more common, with children beginning to interact with one another and sharing play materials.
	Playing cooperatively begins toward the end of this year.
	Some evidence of emotional self-regulation.
	Learn about expectations of emotional self-regulation from observing others, from modeling behaviors of others, and from direct instruction.
	Gifted children often display leadership qualities and often say "I can do it myself!"
3- to 4-year-olds	Have a tendency to explain feelings in terms of external events ("I sad. He hit me.")
	Are more likely to deliberately elicit emotional responses from caregivers.
	Are likely to want to help with tasks and to help others.
4-year-olds	Have better control of facial muscles and can display more complex emotions at one time.
	Begin to verbally express more complex feelings.
	Can accurately identify a commonly expressed emotion and can describe an eliciting situation.
	Boys seem to prefer to engage in more rough-and-tumble competitive play in larger groups, whereas girls seem to engage in two-person groups with more intimate behaviors, such as talking, holding hands.
	Some children this age have an imaginary friend.
	Four- and 5-year-olds show substantial knowledge about mind and emotion.
	A gender identity is usually established.
5- to 6-year-olds	Greater mastery of thinking and remembering.
	Evidence of an emerging distinction between the mental self and the physical self.
	Evidence of greater variety of temperaments and behaviors in a group.

cognitive change facilitates numerous other changes in social development. Youngsters come to view themselves as active agents who produce outcomes. They derive pleasure from being "self-originators" of behavior, and they insist on performing activities independently—resulting in behavior that is sometimes pejoratively labeled "the terrible twos." All the while toddlers increasingly focus on the outcomes of their activities rather than simply on the activities themselves. A common directive from young children is "Watch me do this!"

Toddlers also gain the ability to monitor their ongoing activities with respect to an anticipated outcome and to use external standards for measuring their task performance. This ability is first evident at about 26 months of age. By 32 months of age, they recognize when they err in performing certain tasks, and they correct their mistakes. For instance, in building a block tower, youngsters are not only better able to avoid errors in stacking blocks but manipulate and rearrange the blocks when their first efforts fail (Bullock & Lutkenhaus, 1988).

A concept of the self might also be necessary for self-conscious and self-evaluative emotions, because a notion of self seemingly precedes both self-conscious emotions (such as embarrassment) and self-evaluative emotions (such as pride or shame) (Harris et al., 1992). Having developed self-conscious and self-evaluative emotions—particularly negative emotions when engaging in some transgression—and the ability to evaluate objects and behavior in terms of some standard, youngsters gain the ability to inhibit their behaviors in the absence of caretakers: Social control becomes *self-control* (Stipek, Recchia, & McClintic, 1992).

During the preschool years, young children conceive of the self strictly in physical terms—body parts (the head), material attributes ("I have blue eyes"), and bodily activities ("I walk to the park") (Johnson, 1990). Children view the self and the mind as simply parts of the body (Damon & Hart, 1982). For the most part, children locate the self in the head, although they might also cite other body parts like the chest or the whole body. And they often say that animals, plants, and dead people also have selves and minds.

Between 6 and 8 years of age, children begin to distinguish between the mind and the body (Inagaki & Hatano, 1993). The emerging distinction between the mental and the physical allows children to appreciate the subjective nature of the self. They begin to recognize that people are unique not only because each of us looks different from other people but also because each of us has different feelings and thoughts. Hence, children come to define the self in *internal* rather than *external* terms and to grasp the difference between psychological and physical attributes.

The overall support and unconditional love of parents and caretakers provides the foundation for a child's ego, or developing self-concept. The **self-concept** or *self-image* is defined as the image one has of oneself. One theory is that a child's self-image develops as a reflection of what others think about the child. Parents' facial expressions, tone of voice, and patient or impatient interactions reflect how the parents value each child. Additionally, the child's own personality contributes to the child's growing self-awareness and to the parent's conceptions about the child. For example, a parent might recognize that her child is "shy" or "happy-go-lucky" or "strong-willed"—traits that appear to be inherent in this particular child's nature.

How Psychologists Measure a Child's Self-Esteem

Sometimes adults recognize a young child's lack of self-esteem by the child's withdrawn, sad, or antisocial behaviors and realize that the child might need intervention to boost self-esteem for better mental health and social functioning. Psychologists have devised a variety of psychometric instruments used to assess the accuracy of such observations. Harter and Pike (1984) developed the *Pictorial Scale of Perceived Competence and Social Acceptance in Young Children.* This has become a commonly used self-report measure, where children are asked to report on their own behavior, with separate scores obtained for cognitive and physical competence, maternal acceptance, and peer acceptance. An alternate approach to self-esteem measurement is the *Behavioral Rating Scale of Presented Self-Esteem in Young Children,* used to assess inferred self-esteem (Haltiwanger & Harter, 1988). Parents and teachers make inferences about children's self-esteem from observation of their behaviors. Ratings include such behaviors as initiative, preference for challenge, social approach/social avoidance, social-emotional expression, and coping skills. A lower outcome on these self-esteem measures typically associates behavioral difficulties with low self-esteem (Fuchs-Beauchamp, 1996).

Gifted Children and Their Sense of Self

More recently, Elizabeth Maxwell (1998), from the Gifted Development Center in Denver, reviewed information on the early childhoods of gifted children and their sense of self. Anecdotal information was provided by parents of at least 265 children with IQs of 160 or higher, and over 50 children who score at 180 or above. Maxwell (1998, p. 245) states, "It is difficult not to notice the assertive drive of gifted and highly gifted young children to master their environment as quickly and as thoroughly as possible, far beyond the age expectations of developmental timetables. They appear to be self-directed, their sense of self-awareness arriving early and thrust in the face of parents and the environment in general." Lovecky (1994) calls this *entelechy,* "a particular type of motivation, need for self-determination, and an inner strength and vital force directing life and growth to become all one is capable of being."

Gifted children are active learners with their own agendas, learn remarkably quickly, yet might be difficult to teach or to control. They do not automatically show respect to adults, are likely to see themselves as equal to adults, might demonstrate a strong will, and can present unique challenges to parents and teachers (Maxwell, 1998). On the other hand, many of these young children demonstrate *emotional self-efficacy*—that is, they are able to empathize with others' feelings and to detect moods and emotional subtleties at a very early age. They are likely to share easily with others, to have a strong sense of justice, and to display leadership qualities (Maxwell, 1998). A common highly charged directive from precocious children is "I can do it myself!" *Information You Can Use:* "Parental and Caretaker Assessment of Normal Three-Year-Old Development" provides a check-off list of behaviors that are assessed to

Information You Can Use

Parental and Caretaker Assessment of Normal 3-Year-Old Development

To assess whether a 3-year-old is developing normally, parents or other caretakers can observe the following behaviors in the child, preferably over a three-week period.

Behaviors to Assess	Yes	No	Follow-up Information
Physical Development and Health Related:			
Sleeping habits			
Does the child usually fall asleep easily and wake up rested (with occasional restless nights or nightmares)?			Children this age usually experience "morning eagerness" if well rested. Frequent insomnia or morning grouchiness for 3 or 4 weeks or more may indicate the child is coping with excessive stress or is not getting enough sleep.
Eating habits			
Does the child eat with a good appetite?			Occasional skipping meals, refusing certain foods is expected. Is the child distracted by the television or other activities during eating?
Toileting habits			
Does the child, on average over several weeks, exhibit bowel and bladder control?			Occasional accidents are normal, particularly if excessive intake of liquids, intestinal upset, or intense concentration on activities.
Cognitive Development:			
Curiosity			
Does the child exhibit curiosity, adventure, even mischief?			A child who does not exhibit curiosity may not be pushing against normal boundaries or may fear punishment. In contrast, frequent snooping where forbidden or prying for information may indicate a search for boundaries and limits.
Interest			
Does the child become involved in, absorbed, and interested in activities or projects outside of himself or herself (other than passive involvement with the TV)?			A preschooler who cannot become absorbed in an activity or who rarely stays with a project until completion may need help attending to a task.

determine if a child is experiencing delays, precocity, or typical development.

Question

What are some cognitive and social factors that promote a young child's growing self-awareness?

Gender Identification

What are little boys made of? Frogs and snails And puppy dogs' tails, That's what little boys are made of. What are little girls made of? Sugar and spice And all that's nice. That's what little girls are made of.

J. O. Halliwell, *Nursery Rhymes of England*, 1844

One of the attributes of self acquired early in life is *gender,* the state of being male or female. A major developmental task for the child during the first six years of life is to acquire gender identification. All societies appear to have seized on the anatomical differences between women and men to assign **gender roles**—sets of cultural expectations that define the ways in which the members of each sex should behave. Societies throughout the world display considerable differences in the types of activities assigned to women and men (Ickes, 1993; Murdock, 1935). In many societies, girls are socialized to identify with the nurturant, caregiving role of the mother, while boys are prepared to identify with the provisional-protector role of the father. In some societies, though, women do most of the manual labor; in others, as in the Marquesas Islands, cooking, housekeeping, and baby-tending are male occupations.

Behaviors to Assess	Yes	No	Follow-up Information
Variation in Playing			
Does the child's play vary? Does the child add elements to playing, even with the same materials (using same toys in a variety of ways)?			If the child plays with same objects in the same way every time, perhaps the child needs encouragement and modeling to learn to play with things in the environment in different ways.
Emotional-Social Development:			
Range of emotions			
Does the child display a range of emotions, such as affection, anger, enthusiasm, excitement, frustration, grief, joy, and love over a period of several weeks?			A child whose emotions don't vary, or who is always angry, may be having some difficulty. Expressions of sadness are not always a problem and may exhibit a sense of caring for others (unless the sadness is prolonged).
Friendship			
Has the child initiated and maintained satisfying relationships with one or more peers?			A child who is shy or fearful of peers or who frequently claims superiority over others may be seeking reassurance or may doubt his or her ability to meet parents' expectations.
Spontaneous affection			
Does the child express spontaneous affection and pleasure in being with one or more of those persons responsible for his or her care? Note: There is a cultural component to this question.			Demonstrations of affection vary among families and cultures. In culturally appropriate ways, a thriving child is likely to express affection toward caretakers and deep pleasure in being with them. Lack of affection might signal doubts about the feelings caretakers have toward the child.
Response to authority			
Does the child usually respond well and accept adult authority most of the time—but not all the time?			Occasional self-assertion, resistance, protest, and objections may indicate healthy socialization processes when followed by ultimate yielding to the adult. Unfailing acceptance of adult demands suggests excessive anxiety.
Enjoyment			
Does the child enjoy playing with others, going to new places, seeing new things?			If extreme shyness or strong dislikes prevent the child from having new experiences in life, help may be called for.

Lillian G. Katz, "Assessing the Development of Preschoolers," *ERIC Digest*, 1994. © 1994. Reprinted by permission.

Most people evolve gender identities that are reasonably consistent with the gender-role standards of their society. **Gender identity** is the conception that people have of themselves as being male or female. Over the past forty years, a good deal of research has explored the process by which children come to conceive of themselves in masculine or feminine terms and to adopt the behaviors that are considered culturally appropriate for them as males or females. It has also activated debate regarding the psychology of gender differences.

By the time most children are in preschool, they have acquired, through socialization experiences with parents, extended family, and media influence, a gender identity. How children are treated in any culture because of their gender significantly affects how children feel about themselves. You can see how most parents quickly plan to assimilate their child into the appropriate gender role beginning with the fetal ultrasound: What color paint and wallpaper is planned for the baby's room? What color and style of clothing is purchased prior to the birth? What types of toys and stuffed animals are purchased to put in the child's crib? What potential names are selected for the child-to-be? It would be fair to say that, in American society, gender roles have become more versatile over the past 30 years.

Hormonal Influences on Gender Behaviors

In recent years, there have been many studies regarding the influence of hormones (chemical messengers of the endocrine system) on gender behaviors (Eisenberg et al. 1997; Eagly, 1995). Both sexes have some male and female hormones, though the ratio of each varies in males and females. However, the prevalence of testosterone tends to make boys more physically active, more aggressive, and less likely to be able to sit still (Maccoby & Jacklin, 1974). Eleanor Maccoby (1980), after repeated reviews of research studies on sex behaviors, concluded: "The tendency of males to be more aggressive than

Acquiring Gender Identities The members of society typically take considerable pains to socialize their children in the roles they deem appropriate for males and females. Children pick up the gender expectations that permeate their environments, and adults also intervene to cue children when they engage in behavior thought to be inappropriate for their gender.

females is perhaps the most firmly established sex difference and is a characteristic that transcends culture." More recent research finds that females are less likely to inhibit aggression when aggression is socially expected and supported—as in school sports, professional teams, and Olympic competition, where females were not allowed until recently (Campbell, 1993). Today, many American communities begin teaching soccer and baseball to 5- and 6-year-old boys *and* girls.

The predominance of either female or male hormones influences the development of the fetal brain. Imaging techniques reveal that females tend to have a larger corpus callosum, which is the band of fibers and nerves that carries messages between the two hemispheres of the brain (Kolata, 1995a; 1995b). It is hypothesized that the slightly increased size of the corpus callosum allows the two hemispheres in female brains to communicate more easily. In general, boys tend to be more logical, analytical, spatial, and mathematical, whereas females tend to be more verbal at an earlier age, more "emotional," and more social (interested in "people-oriented" activities) (Halpern, 1997). Witness the number of preschool age boys who enjoy playing video games or sports that are based on spatial relations and the number of little girls who enjoy having a "tea party" or playing "school" and conversing with one another. Also, Halpern (1992) found that males are disproportionately represented among those manifesting dyslexia (reversal of letters and symbols) and stuttering problems.

Until many of these studies are replicated, we are not certain whether gender differences in the structuring of particular regions of the brain are the source of differences in gender identity, behavior, sexual orientation, or cognitive processes. Most psychologists agree that it is

premature to look only to biology for explanations of gender differences. Certainly each individual child's family experience and cultural socialization patterns also influence gender behaviors.

Social Influences on Gender Behaviors

The fact that hormonal and biological factors sometimes contribute to behavioral differences between men and women does not mean that environmental influences are unimportant. On the basis of his research with hermaphrodites (individuals having the reproductive organs of both sexes), Money (Money & Tucker, 1975, pp. 86–89) concludes that the most powerful factors in the shaping of gender identity are *environmental*:

> The chances are that society had nothing to do with the turnings you took in the prenatal sex development road, but the minute you were born, society took over. When the drama of your birth reached its climax, you were promptly greeted with the glad ritual cry, "It's a boy!" or "It's a girl!" depending on whether or not those in attendance observed a penis in your crotch. . . . The label "boy" or "girl," however, has tremendous force as a self-fulfilling prophecy, for it throws the full weight of society to one side or the other as the newborn heads for the gender identity fork [in the road], and the most decisive sex turning point of all. . . . [At birth you were limited to] something that was ready to become your gender identity. You were wired but not programmed for gender in the same sense that you were wired but not programmed for language.

Clearly, anatomy in itself does not provide us with our gender identity. Because of being labeled "boy" or "girl," a highly stylized treatment of the child is repeated countless times each day (Campbell, 1993). Boys receive more toy action vehicles, sport equipment, machines, toy animals, and military toys; girls receive more dolls, doll houses, and domestic toys. Boys' rooms are more often decorated with animal motifs; girls' rooms with floral motifs accompanied by lace, fringes, and ruffles. Although the sexual revolution has reshaped many nooks and crannies of U.S. life, it failed to reach very deeply into the toy box. Behind many parents' concerns about the type of toys their children play with are unexpressed fears about homosexuality, yet there is little evidence that children's toy preferences are related to their sexual orientation (Bailey & Zucker, 1995).

Although everywhere women give birth to children and men do not, there is little evidence to support popular notions that somehow biology makes women kinder and gentler beings or that nature equips them specifically for nurturing roles (Whiting & Edwards, 1988). Psychologist Jerome Kagan, who has spent more than 35 years studying children, speculates that any propensity women might have for caretaking can be traced to an early awareness of their role in procreation: "Every girl knows, somewhere between the ages of 5 and 10, that she is different from boys and that she will have a child—something that everyone, including children, understands as quintessentially natural. If, in our society, nature stands for the giving of life, nurturance, help, affection, then the girl will conclude unconsciously that those are the qualities she should strive to attain. And the boy won't. And that's exactly what happens" (Kagan, quoted by Shapiro, 1990, p. 59). Kagan's observations lead us to inquire as to the psychological processes that are at work in attuning youngsters to their *gender roles,* a matter to which we now turn our attention.

Theories Regarding the Acquisition of Gender Identities

Social and behavioral scientists have proposed a number of theories regarding the process by which children psychologically become masculine or feminine. Among these theories are the psychoanalytic, psychosocial, cognitive learning, and cognitive developmental approaches.

Psychoanalytic Theory   According to Sigmund Freud, children are psychologically bisexual at birth. They develop their gender roles as they resolve their conflicting feelings of love and jealousy in relation to their parents. A young boy develops a strong love attraction for his mother but fears that his father will punish him by cutting off his penis. The usual outcome of this Oedipal situation is for a boy to repress his erotic desire for his mother and identify defensively with the potential aggressor, his father. As a consequence of coming to feel identified with their fathers, boys later erotically seek out females. Meanwhile, Freud said, young girls fall in

love with their fathers. A girl blames her mother for her lack of a penis. But she soon comes to realize that she cannot replace her mother in her father's affections. So most girls resolve their Electra conflicts by identifying with their mothers and later by finding suitable men to love. These complexes are normally resolved by age 5 or 6. Though Freud's theory about gender identity remains controversial, it is common to hear a 4- or 5-year-old child proclaim, "When I'm older, I'm going to marry Daddy (or Mommy)."

Psychosocial Theory   Having accepted Freud's psychoanalytic theory of gender identity, Erikson shows how certain gender traits might be assimilated by social interaction. Erikson says that during the ages of 3 to 6, children strive exuberantly to do things and to test developing abilities. Erikson calls this psychosocial stage the conflict of **initiative versus guilt.** Children at this age will attempt to model their behavior after the adults and siblings in their environment. If Daddy is setting the table, then little Jimmy or Jill will want to help. If Daddy approves and lets them help, with supervision, then Erikson would say they are experiencing initiative. If, on the other hand, Jimmy and Jill decide they want to turn on the microwave, they might be scolded that they are too little yet, and they might experience a sense of guilt or inhibition. Erikson would say that there is a balance in this process of always wanting to learn and do and being told "No, you are too little" or "You can't, only grownups do this."

Children who learn how to regulate these opposing drives are developing what Erikson called the "virtue of purpose" (Erikson, 1982). Children might begin to say of themselves "I am good" or "I am bad" at this age, depending on the level of encouragement or discouragement they get from their parents and caretakers.

Preschool and kindergarten teachers were asked in a study to describe the behaviors of children with high self-esteem and low self-esteem (Haltiwanger & Harter, 1988). A finding from this study shows that children with high self-esteem are motivated to achieve. In contrast, some children exhibited a "helpless" behavior pattern: They did not attempt new tasks because they anticipated not succeeding, thus they simply did not try. Erikson described this as *inhibitory* behavior. When role-playing with dolls, the children with the "helpless" behavior pattern tended to scold the doll for failure and tell the doll it was "bad" (Burhaus & Dweck, 1995). (Children at this age tend to talk with their play things and typically repeat what they themselves have been told.)

Cognitive Learning Theory   Cognitive learning theorists take the view that children are essentially neutral at birth and that the biological differences between girls and boys are insufficient to account for later differences in gender identities. They stress the parts that *selective reinforcement* and *imitation* play in the process of acquiring a gender identity. Viewed from this perspective, children reared in a nuclear family setting are rewarded for modeling the behavior of the same-sex parent. And the larger society later reinforces this type of imitation through systematic rewards and punishments. Boys and girls are actively rewarded and praised, both by adults and by their peers, for what society perceives to be sex-appropriate behavior, and they are ridiculed and punished for behavior inappropriate to their sex (Smetana, 1986). Currently popular among some social and behavioral scientists is the "separate cultures" concept that children learn rules for social interaction from experience in largely sex-segregated peer groups in childhood and then carry this learning into their adult interactions (Maccoby, 1990).

Albert Bandura (1973; Bussey & Bandura, 1992, 1984) gives an additional dimension to cognitive learning theory. He points out that in addition to imitating the behavior of adults, children engage in *observational learning.* According to Bandura, children mentally encode a model's behavior as they watch it, but they will not imitate behavior they have observed unless they believe that it will have a positive outcome for them. He says that children discern which behaviors are appropriate for each sex by watching the behavior of many male and female models. They notice which kinds of behavior are performed by which sex in which kinds of situations. In turn, they employ these abstractions of sex-appropriate behavior as "models" for their own imitative actions. But not everything learned is performed. Thus, although boys might know how to wear a dress and apply makeup, few boys choose to perform these behaviors. Rather, they are most likely to perform those behaviors that they have coded as appropriate to their own sex. Consequently, the responses children select from their behavioral repertoires depend chiefly on the consequences they anticipate will follow from the behaviors (Bussey & Bandura, 1992).

Cognitive Developmental Theory   Still another approach, which is identified with Lawrence Kohlberg (1966; Kohlberg & Ullian, 1974), focuses on the part that cognitive development plays in children's acquisition of gender identities. This theory claims that children first learn to label themselves as "male" or "female" and then attempt to acquire and master the behaviors that fit their gender category. This process is called *self-socialization.* According to Kohlberg, children form stereotyped conceptions of maleness and femaleness—fixed, exaggerated, cartoonlike images—that they use to organize their environment. They select and cultivate behaviors that are consistent with their gender concepts.

Kohlberg distinguishes between his approach and cognitive learning theory in these terms. According to the cognitive learning model, the following sequence occurs: "I want rewards; I am rewarded for doing boy things; therefore I want to be a boy." In contrast, Kohlberg (1966, p. 89) believes that the sequence goes like this: "I am a boy; therefore I want to do boy things; therefore the opportunity to do boy things (and to gain approval for doing them) is rewarding."

Genital anatomy plays a relatively minor part in young children's thinking about sex differences. Instead, children notice and stereotype a relatively limited set of highly visible traits—hairstyle, clothes, stature, and occupation. Children use *gender schemes* or *models* to actively structure their experiences and to draw inferences and interpretations regarding gender behaviors (Bem, 1993). Viewed in this manner, children develop a rudimentary understanding of gender (*self-labeling*) and in turn invoke gender schemes to process information. These schemes begin developing rather early in life (Poulin-Dubois et al., 1994). When given a choice of toys, boys more often play with "boy" toys and girls more often play with "girl" toys (Lobel & Menashri, 1993). Apparently, by the time they are 3 years of age, 80 percent of U.S. children are aware of gender differences and can categorize tasks like driving a truck or delivering mail as "masculine" and cooking, cleaning, and sewing as "feminine" tasks (Fagot, Leinbach, & Hagan, 1986). Significantly, children tend to "forget" or distort information that runs counter to their developing gender schemes (Bauer, 1993).

Evaluation of Theories   The theories considered here stress the importance of children's knowledge about **gender stereotypes** (exaggerated generalizations about male or female behaviors) as powerful determinants of sex-typed behavior. Each emphasizes that behavioral differences between the sexes are at least in part perpetuated by the fact that children are more inclined to imitate the behavior of same-sex models than they are to imitate the behavior of opposite-sex models.

Each of the theories has some merit (Jacklin & Reynolds, 1993). Psychoanalytic theory has had historical importance in directing our attention to the significant part that early experience plays in fashioning an individual's gender identity and behavior. Erikson let us know that young children naturally exhibit a drive to act on their world with exuberance (with a purpose), and the word no doesn't have to be the only word young children hear if parents and caretakers are willing to supervise them patiently. Cognitive learning theory has contributed to our knowledge by highlighting the social and cultural components of gender role development and the importance of imitation in the acquisition of gender behaviors (Fagot, Leinbach, & Hagan, 1986). And cognitive developmental theory has shown how a gender scheme, or mental model, leads youngsters to sort incoming information on the basis of gender categories and then to adopt the gender-linked characteristics (Bigler & Liben, 1992). So rather than counterposing these theories in an either/or fashion, many psychologists prefer to see them as supplementing and complementing one another.

Mothers, Fathers, and Gender Typing

Social and behavioral scientists suggest that gender stereotypes arise in response to a society's division of labor by sex and serve to rationalize this division by attributing to males and females basic personality differences (Gilmore, 1990). It is hardly surprising, then, that parents have clear *stereotypes* regarding the behaviors they expect from female and male children (Jacobs & Eccles, 1992). Daughters more often than sons are described by both their mothers and fathers as "little," "beautiful," "pretty," and "cute." Furthermore, there are marked and relatively consistent differences in paternal and maternal reactions to female and male infants (Parke, 1995). Fathers are more likely than mothers to describe their sons as "firm," "large-featured," "well-coordinated," "alert," "strong," and "hard." And they are more likely than mothers to describe their daughters as "soft," "fine-featured," "awkward," "inattentive," "weak," and "delicate" (Rubin, Provenzano, & Luria, 1974).

Evidence suggests that in U.S. society, the father plays the critical role in encouraging "femininity" in females and "masculinity" in males in traditional terms (Weinraub et al., 1984). Fathers tend to be more concerned than mothers over their children's development of culturally appropriate sex roles, although this differ-

ence may be waning (Fagot, 1995). And fathers treat their sons differently from daughters (Lytton & Romney, 1991). Both fathers and mothers are more eager to push their sons toward masculinity than to push their daughters toward femininity. Parents generally express more negative reactions when boys make choices culturally defined as feminine than when girls make choices culturally defined as masculine. Additionally, fathers' fears of homosexuality, in themselves or in their sons, lead many men to inhibit displays of love and tenderness toward their sons (Parke, 1995).

Over the past 40 or so years, the following theory has dominated psychological inquiry and thinking regarding masculinity: Boys learn to be masculine by identifying with their own fathers. Mentally healthy men have a good, firm sense of themselves as masculine beings (Gilmore, 1990). However, because more fathers are now frequently absent from U.S. homes during their children's formative years, larger numbers of boys are growing into manhood lacking a secure sense of their masculinity. The opposing view that boys can learn to be masculine through socialization and interaction with other male models, without identifying with their fathers, is being questioned in face of current statistics about the increasing numbers of American children living without fathers. As mentioned earlier, though, it is difficult to avoid the many confounding effects of poverty when conducting this type of research on families and parenting. These factors cover a broad range of dimensions and reveal that the absence of fathers has far-reaching physical, cognitive, and emotional consequences for individual children as well as society (see *Further Developments:* "The Effects of Fatherlessness on Children" in Chapter 4).

Questions

Which biological and social factors play a role in a child's gender identity? What are the major theoretical views of children's acquisition of gender identity?

Family Influences

A young branch takes on all the bends that one gives it.

Chinese Proverb

Children are newcomers to the human group, strangers in an alien land. Genes do not convey *culture*, the socially standardized lifeways of a people. Clyde Kluckhohn (1960, pp. 21–22), a distinguished anthropologist, provides an illustration of this point:

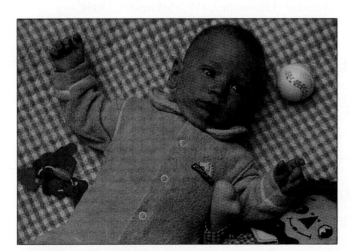

Gender Stereotypes Mothers and fathers typically react to male and female infants in distinct and consistent ways. In American society since blue is a color reserved for a newborn baby boy, we are likely to describe this child as "big," "strong," and "well-coordinated." However, a girl infant is likely to be described as "soft" and "delicate." How would you describe this infant? Infants are prepared for their society's gender role from birth. What other object in this picture symbolizes gender typing?

Some years ago I met in New York City a young man who did not speak a word of English and was obviously bewildered by American ways. By "blood" he was as American as you or I, for his parents had gone from Indiana to China as missionaries. Orphaned in infancy, he was reared by a Chinese family in a remote village. All who met him found him more Chinese than American. The facts of his blue eyes and light hair were less impressive than a Chinese style of gait, Chinese arm and hand movements, Chinese facial expression, and Chinese modes of thought. The biological heritage was American, but the cultural training had been Chinese. He returned to China.

The process of transmitting culture, of transforming children into bona fide, functioning members of society, is called socialization. Through **socialization,** children acquire the knowledge, skills, and dispositions that enable them to participate effectively in group life. Infants enter a society that is already an ongoing concern, and they need to be fitted to their people's unique social environment. They must come to guide their behavior by the established standards, the accepted dos and don'ts, of their society. Within the family setting, the child is first introduced to the requirements of group life. By the time a child reaches the age of 2, the socialization process has already begun. Developmental psychologists David P. Ausubel and Edmund V. Sullivan (1970, p. 260) observe:

> At this time parents become less deferential and attentive. They comfort the child less and demand more conformity to their own desires and to cultural norms. During this period [in most societies] the child is frequently weaned, is expected to acquire sphincter control, approved habits of eating and cleanliness, and do more things for himself. Parents are less disposed to

gratify his demands for immediate gratification, expect more frustration tolerance and responsible behavior, and may even require performance of some household chores. They also become less tolerant toward displays of childish aggression.

The magnitude of a child's accomplishment over a relatively short period is truly astonishing. By their fourth birthday most U.S. children have mastered the complicated and abstract structure of the English language and they can carry on complex social interactions in accordance with U.S. cultural patterns. At the same time, American children are being born into a greater diversity of family environments than ever before in U.S. history.

A Diversity of Family Environments for American Children

Many researchers are studying how children are being affected by the shifting trends in marriage, divorce, child-bearing, living arrangements, migration, education, work, income, and poverty (Wallman, 1998; Hernandez, 1997). The trend of *increased poverty* has critical far-reaching ill effects on the physical, cognitive, social, and emotional well-being of children, as we saw in Chapter 6. These current trends are related to six major demographic transformations over the past 100 years: (1) the shift from farm to industrial to service work for fathers; (2) a large reduction in family size; (3) greater educational attainment; (4) womens' participation in the labor force; (5) the increase in single-parent families; and (6) the subsequent rise in childhood poverty. These major trends have critical implications for child development in American society today (Hernandez, 1997; Schneider, 1993).

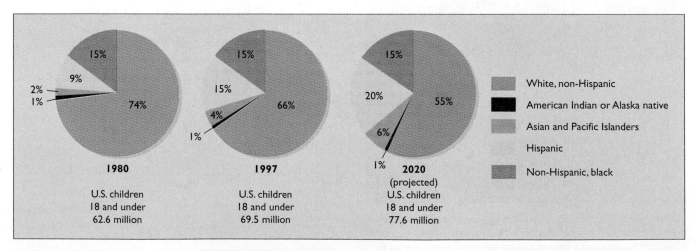

Figure 8-1 Actual and Projected Population and Ethnic Diversity of Children in the United States, 1980–2020. The number of children determines the demand for schools, health care, and other services and facilities that serve children and their families. In 1997, there were approximately 69.5 million children in the United States. By 2020, it is projected there will be 77.6 million children. In each age group there were nearly equal numbers of children: 0 to 5 years old, 23 million; 6 to 11 years old, 23 million; and 12 to 17 years old, 23 million. The proportion of Hispanic American children is increasing rapidly, as the proportion of white non-Hispanic children is decreasing.

Source: K. Wallman, *America's Children, 1998, Population and Family Characteristics,* U.S. Bureau of the Census, ChildStats.gov. at www.childstats.gov/ac1998/poptx.htm.

American children grow up in a diversity of developmental environments. Some are highly supportive and stimulating; too many are neglectful or abusive. Though children represent a smaller *percentage* (26 percent) of the overall population today than in 1960, they are a stable and substantial portion of the population and will remain so into the next century. In 1997, there were 69.5 million children in the United States. The age groups—0 to 5, 6 to 11, and 12 to 17 years—are nearly equal in size, about 23 million per group. The ethnic diversity of American children will continue to increase, with the proportion of Hispanic children increasing rapidly. It is projected by the year 2020, less than 50 percent of U.S. children will be non-Hispanic white (see Figure 8-1).

Some 38 percent of unwed mothers are teenagers. Nearly one in four children is born into poverty. More than 20 percent of children today live with one parent. Nearly two-thirds of children go home from school to an empty house, because their parent or parents are at work. What is happening to our children is of concern to us all, whether or not we are parents. Today's children will bear the responsibility for the next generation of children and a growing population of dependent adults. This state of affairs leads us to take a closer look at the significant role of family influences and parenting.

Determinants of Parenting

What we desire our children to become, we must endeavor to be before them.

Andrew Combe,
Physiological and Moral Management of Infancy

Until relatively recently, most socialization research focused on the processes whereby parental child-rearing strategies and behaviors shape and influence children's development. For the most part, psychologists and psychiatrists paid little attention to the parents themselves and the context in which they carried on their parenting. Also neglected was the part children play as active agents in their own socialization and in influencing the behavior of their caretakers. This focus has changed over the past two decades, resulting in a more balanced perspective (Lerner et al., 1995; Maccoby, 1992). Jay Belsky (1984) has provided a framework that differentiates among three major determinants of parental functioning: (1) the personality and psychological well-being of the parent; (2) the characteristics of the child; and (3) the contextual sources of stress and support operating within and upon the family.

The Parents' Characteristics Parenting, like other aspects of human functioning, is influenced by the relatively enduring characteristics, or personality, of a man or woman (Belsky, 1990). So, as you might expect, troubled parents are more likely to have troubled children (Dix, 1991). A 6-year study of 693 families found that 66 percent of children with emotionally troubled mothers had psychological problems; 47 percent had problems in families where only the father had symptoms; and 72 percent had problems when both parents were disturbed (this was double the rate for children with two healthy parents) (Parker, 1987). Other researchers find that constant parental irritability has an adverse effect on children's emotional well-being and on their development of cognitive skills (Crnic & Acevedo, 1995). Compared to nondepressed mothers, mothers suffering from psychological depression tend to be less affectionate, responsive, and spontaneous with their infants and less patient and more punitive with their older youngsters (Field, 1995). In contrast, children whose parents are happily married have more secure emotional ties to their parents and, as a result, seem to enjoy intellectual and other advantages over children of unhappily married people (Wilson & Gottman, 1995).

The Child's Characteristics Children's characteristics influence the parenting they receive (Sanson & Rothbart, 1995). These characteristics include such variables as age (Fagot & Kavanagh, 1993), gender (Kerig, Cowan, & Cowan, 1993), and temperament (for instance, aggressiveness, passivity, affection, moodiness, and negativity) (Sanson & Rothbart, 1995). And some children are simply more difficult to rear than others (Rubin, Stewart, & Chen, 1995). Anderson, Lytton, and Romney (1986) observed 32 mothers as they talked to and played with three different boys, ages 6 to 11. Half the mothers had wayward sons who had been referred to a mental health facility and diagnosed as having serious behavior problems; the other half were mothers of sons with no serious behavior problems. The researchers counted the mothers' positive and negative actions, their verbal requests, and the extent to which the boys complied with the women in the course of the play sessions. The mothers of the difficult and nondifficult youngsters did not differ in their behavior. The wayward boys were substantially less compliant than the other boys, regardless of how the women behaved or related to them. Overall, evidence suggests that parents and other adults typically react to disobedient, negative, and highly active youngsters with negative, controlling behavior of their own (Belsky, 1990).

Sources of Stress and Support Parents do not undertake their parenting in a social vacuum. They are immersed in networks of relationships with friends and relatives, and most are employed. These arenas of social interaction can be sources of stress or support, or both. For instance, difficulties at work commonly spill over to the home; arguments at work are likely to be followed by arguments between a husband and wife in the evening (Menaghan & Parcel, 1990). Yet one strength of the human condition is the propensity we have for giving and receiving support. Social support has a beneficial effect on us irrespective of whether we are under stress (Hashima & Amato, 1994). When we are integrated in social networks and groups, we have regular access to positive experiences and a set of stable, socially rewarded roles in the community. Support systems also buffer or protect us from the potentially harmful influence of stressful events (Cochran & Niego, 1995).

Not surprisingly, then, research undertaken among Japanese and U.S. mothers shows that the adequacy of a woman's mothering is influenced by the perception she has of her marital relationship. When a woman feels she has the support of her husband, she is more likely to involve herself with her infant (Wilson & Gottman, 1995). Likewise, researchers find that parents with little social support do a poorer job of parenting than do parents who are integrated in well-functioning support systems. They are likely to have more household rules and to use more authoritarian punishment techniques (Belsky, 1984). So let us turn our attention now to the suitability of various parenting practices.

The Search for Key Child-Rearing Practices

Parents wonder why the streams are bitter, when they themselves have poisoned the fountain.

John Locke

Most authorities agree that parenting is one of the most difficult tasks any adult faces (see Figure 8-2). Moreover, most parents are well intentioned and desire to succeed at parenting. Because this difficult task encompasses many years and consumes much energy, parents have often looked to pediatricians or experts in child psychology to provide them with guidelines for rearing mentally and physically healthy youngsters (Bornstein, 1995). But parents who turn to "authorities" can become immensely frustrated, for they will be confronted with an endless array of child-rearing books, conflicting information, gimmickry, and outright quackery (Young, 1990).

As we have already noted, until relatively recently psychologists assumed that socialization effects flow essentially in one direction—from parent to child. For some fifty years, roughly from 1925 to 1975, they dedicated themselves to the task of uncovering the part that different parenting practices have in shaping a child's personality and behavior. This research found three dimensions to be significant:

- The warmth or hostility of the parent-child relationship
- The control or autonomy of the disciplinary approach
- The consistency or inconsistency that parents show in using discipline

The Warmth-Hostility Dimension   Many psychologists have insisted that one of the most significant aspects of the home environment is the warmth of the relationship between parent and child (Kochanska & Aksan, 1995). Parents show warmth toward their children through affectionate, accepting, approving, understanding, and child-centered behaviors. When disciplining their children, parents who are warm tend to employ frequent explanations, use words of encouragement and praise, and only infrequently resort to physical punish-

BETTER OR WORSE/Lynn Johnston

Figure 8-2   Parenting Is One of the Most Difficult Tasks Any Adult Faces   Yet parents receive little formal training. It is common for new parents to be unsure of how to parent successfully. Societal and family expectations and the child's own characteristics also play a significant role in this process. Social support is likely to come from relatives, neighbors, coworkers, the media, and books. On the whole, most parents do quite well.

ment. Hostility, in contrast, is shown through cold, rejecting, disapproving, self-centered, and highly punitive behaviors (Becker, 1964). Wesley C. Becker (1964), in a review of the research on parenting, found that love-oriented techniques tend to promote children's acceptance of responsibility and to foster self-control through inner mechanisms of guilt. In contrast, parental hostility interferes with conscience development and breeds aggressiveness and resistance to authority.

The Control-Autonomy Dimension The second critical dimension is the range of restrictions that parents place on a child's behavior in such areas as sex play, modesty, table manners, toilet training, neatness, orderliness, care of household furniture, noise, obedience, and aggression toward others (Becker, 1964; Sears, Maccoby, & Levin, 1957). On the whole, psychologists have suggested that highly *restrictive* parenting fosters dependency and interferes with independence training (Bronstein, 1992; Maccoby & Masters, 1970). However, as Becker (1964, p. 197) observes in his review of the research literature, psychologists have had difficulty coming up with a "perfect" all-purpose set of parental guidelines:

> The consensus of the research suggests that both restrictiveness and permissiveness entail certain risks. Restrictiveness, while fostering well-controlled, socialized behavior, tends also to lead to fearful, dependent, and submissive behaviors, a dulling of intellectual striving and inhibited hostility. Permissiveness on the other hand, while fostering outgoing, sociable, assertive behaviors and intellectual striving, tends also to lead to less persistence and increased aggressiveness.

Combinations of Parenting Approaches Rather than examine the warmth-hostility and control-autonomy dimensions in isolation from one another, a number of psychologists (Becker, 1964) have explored their four combinations: warmth-control, warmth-autonomy, hostility-control, and hostility-autonomy.

Warm but restrictive parenting is believed to lead to politeness, neatness, obedience, and conformity. It also is thought to be associated with immaturity, dependency, low creativity, blind acceptance of authority, and social withdrawal and ineptness (Becker, 1964; Levy, 1943). Eleanor E. Maccoby (1961) found that 12-year-old boys who had been reared in warm and restrictive homes were strict rule enforcers with their peers. Compared with other children, these boys also displayed less overt aggression, less misbehavior, and greater motivation toward schoolwork.

Psychologists report that children whose homes combine *warmth with democratic procedures* (autonomy) tend to develop into socially competent, resource-ful, friendly, active, and appropriately aggressive individuals (Lavoie & Looft, 1973; Kagan & Moss, 1962). Where parents also encourage self-confidence, independence, and mastery in social and academic situations, the children are likely to show self-reliant, creative, goal-oriented, and responsible behavior. Where parents fail to foster independence, permissiveness often produces self-indulgent children with little impulse control and low academic standards.

Hostile (rejecting) and restrictive parenting is believed to interfere with the child's developing sense of identity and self-worth. Children come to see the world as dominated by powerful, malignant forces over which they have no control. The combination of hostility and restrictiveness is said to foster resentment and inner rage. These children turn some of the anger against themselves or experience it as internalized turmoil and conflict. This can result in "neurotic problems," self-punishing and suicidal tendencies, depressed affect, and inadequacy in adult role-playing (Whitbeck et al., 1992).

Parenting that combines *hostility with permissiveness* is thought to be associated with delinquent and aggressive behavior in children. Rejection breeds resentment and hostility, which, when combined with inadequate parental control, can be translated into aggressive and antisocial actions. When such parents do employ discipline, it is usually physical, capricious, and severe. It often reflects parental rage and rejection and hence fails as a constructive instrument for developing appropriate standards of conduct (Becker, 1964).

Discipline Consistency in discipline is the third dimension of parenting that many psychologists have stressed is central to a child's home environment. Effective discipline is consistent and unambiguous. It builds a high degree of predictability into the child's environment. Although it is often difficult to be consistent in how one punishes a child, research by Ross D. Parke and Jan L. Deur (1972) reveals that erratic punishment generally fails to inhibit the punished behavior. In the case of aggression, researchers have found that the most aggressive children have parents who are permissive toward aggression on some occasions but severely punish it on others (Sears, Maccoby, & Levin, 1957). Research suggests that parents who use punishment inconsistently actually create in their children a resistance to future attempts to extinguish the undesirable behavior (Parke, 1974).

Inconsistency can occur when the same parent responds differently at different times to the same behavior. It can also occur when one parent ignores or encourages a behavior that the other parent punishes (Belsky, Crnic, & Gable, 1995; Vaughn, Block, & Block, 1988). On the basis of his observation of family interaction patterns in the homes of 136 middle-class preschool boys, Hugh Lytton (1979) found that mothers typically initiate more

Further Developments

Sexual Abuse of Children

Most of us find the idea that parents use their children (of any age) for sexual gratification so offensive that we prefer not to think about it. Indeed, only in the past few decades have professionals and government officials come to view the sexual abuse of children as a major mental health problem. In 1974, the federal government adopted a more direct role in child-abuse policies with passage of the Child Abuse Prevention and Treatment Act (PL 93-247). This law established identification standards, reporting policies, and management policies for these maltreatment cases, while empowering individual states to investigate abuse and provide child protective services (CPS). Estimates of the prevalence of child sexual abuse range from 6 percent to 62 percent for females and from 3 percent to 31 percent for males (Fincham et al., 1994).

According to the National Committee to Prevent Child Abuse (NCPCA), in 1997 there were 1,054,000 confirmed cases of child abuse and neglect; of these cases, 8 percent (84,320) were confirmed sexual abuse cases. (These statistics are based on the *known* cases, not on estimates). This national agency also says that reports of *child abuse* have risen 41 percent between 1988 and 1997 (NCPCA, 1998). Of major concern is the alarmingly high 1997 statistic of deaths of children under 5 years old because of various forms of child abuse that often accompany sexual abuse: 79 percent of deaths of children under 5 years old reportedly occurred because of child neglect or abuse (statistics from 19 states only) (NCPCA, 1998).

Sexual abuse of children is commonly defined as sexual behavior between a child and an older person that the older person brings about through force, coercion, or deceit (Gelles & Conte, 1990). Sexual abuse often involves four types of experiences that alter a child's cognitive or emotional orientation to the world, distorting her or his self-concept, worldview, and affective capacities: (1) The child is exposed to sexual behaviors that are developmentally inappropriate; (2) the youngster experiences a loss of control and a sense of powerlessness; (3) the child is stigmatized for engaging in forbidden acts; and (4) the youngster feels betrayed because a trusted person has manipulated, used, or failed to protect him or her (Finkelhor, 1988; Wyatt & Powell, 1988).

Sexual abusers of children can be a parent, stepparent, sibling, other relative, trusted friend, neighbor, child-care worker, teacher, coach, or anyone else who has access to the child; males are reported to be the abusers in 80 to 95 percent of cases, with a dramatic increase in adolescent offenders (NCPCA, 1996a). Most research has dealt with the sexual abuse of females. A 1996 national survey found that girls are sexually abused three times more often than boys (NCPCA, 1996a). In cases of father-daughter incest, the fathers tend to be "family tyrants" who use physical force and intimidation to dominate their families (Herman & Hirschman, 1981). However, the daughter chosen as the object of the father's sexual attention might be exempt from other kinds of physical attack and might tolerate the father's sexual advances in order to continue to avoid other kinds of physical attack. Sexual contact usually begins when the child is between 5 and 12 years old (though instances of infant and toddler abuse have been reported), and it typically consists at first of fondling and masturbation. The behavior continues over time and might eventually proceed to intercourse or sodomy. One study of incest perpetrators found that almost all of them defined their incestuous impositions as love and care and their behavior as considerate and fair. However, their professed love, care, and sense of fairness were contradicted in many ways, including their refusal to stop when youngsters wanted them to stop (Gilgun, 1995).

Though children might be too young to know the sexual activity is "wrong," they will develop behavioral or physical problems resulting from the inability to cope with the over-

actions designed to control their son's behavior than fathers do. However, the boys were less inclined to obey their mothers than they were their fathers. But when their fathers were present, they were more likely to be responsive to their mothers' commands and prohibitions.

Many people have difficulty defining exactly where the line falls between discipline and abuse. Most Americans define child abuse and neglect as leaving young children home alone, living in a filthy home and lacking food, and hitting a child hard enough to cause bruises. However, they are ambivalent about spanking. According to a 1994 *USA Today*/CNN/Gallup poll, 67 percent of Americans agree that it is sometimes necessary to discipline a child with a "good, hard spanking." However, the percentage of Americans holding this opinion has dropped since 1986 (Poll: Discipline OK, abuse isn't, 1994).

Other evidence also points to the prevalence of spanking (Giles-Sims, Straus, & Sugarman, 1995). Being poor and being an unmarried mother seemingly increase the likelihood of spanking. Such findings support the notion that parents who are under stress are more likely to use corporal punishment. Child-rearing experts have long contended that children exposed to spanking are apt in later life to develop emotional and behavioral prob-

stimulation. Often there are no obvious physical signs of child abuse, only signs that a physician can detect, such as changes in the genital or anal area. Other behavioral signs are likely to include these:

- Unusual interest in, or avoidance of, all things of a sexual nature
- Sleep problems, nightmares
- Depression or withdrawal from friends or family
- Sexually inappropriate behaviors with others or knowledge beyond the child's years
- Statements that their bodies are dirty or damaged or fear there is something wrong in the genital area
- Refusal to go to school or delinquency
- Secretiveness
- Aspects of sexual molestation in drawings, games, fantasies
- Unusual aggressiveness
- Suicidal behaviors (even in a younger child)
- Other severe behavior changes

Sexually abused children are usually afraid to tell others about their experiences because the abuser will control the child by saying such things as, "Mommy won't love you anymore," or "Mommy will hurt you if she finds out," or "No one will believe you anyway since you are a child." Wives of men who molest their children usually are passive, have a poor self-image, and are overly dependent on their husbands. They frequently suffer from mental illness, physical disability, or repeated pregnancy. Under these circumstances, much of the housework and child care falls on the oldest daughter, who is the one most likely to be sexually victimized by the father. Although incest commonly involves the oldest daughter, the behavior is often repeated with younger daughters, one after another. Moreover, a father's incestuous behavior places his daughters at greater risk of sexual abuse by other male relatives and family intimates. Occasionally in the media there is a report of a drug-addicted parent who prostitutes his or her own child to pay for a drug habit. Female victims tend to show lifetime patterns of psychological shame and stigmatization.

Little is known or written about male victims, even though there has been an increase in the number of boy victims being identified (Hunter, 1990). The available evidence suggests that boys and girls respond differently to sexual victimization. Because boys are socialized to gain control of themselves and their environments, boys might feel their masculine competence has been compromised or destroyed when they are sexually abused. Or they attempt to "minimize" the importance of the abusive events. Complicating matters, male victims often face the additional problem of stigmatization as homosexual, because the large majority of abusers are male (Bolton, Morris, & MacEachron, 1989).

Sexual abuse should always be reported to the appropriate authorities (and many child-care workers are obligated to do so). Adults should never dismiss a child's complaint of sexual abuse; false complaints are quite rare. In turn, mental health professionals involved in interviewing children suspected of having been abused must be trained in objective, appropriate interview techniques.

Prevention Programs

By the early 1990s, a good many prevention programs had been designed that employed multiple modes of audiovisual technology (film, video, audiotape, and filmstrip) and format (storybooks, coloring books, songs, plays, and board games). The materials and programs are based on a number of assumptions: Many children do not know what constitutes sexual abuse, children must not tolerate sexual abuse, and children should inform responsible adults upon being sexually touched by an older person. Self-protection against sexual abuse is a complex process for any child (Berrick & Gilbert, 1991; Watts & Ellis, 1993).

lems, including an increased chance of being depressed, thinking about suicide, engaging in spouse beating, becoming violent and delinquent, and experiencing alienation and lower economic achievement (Giles-Sims, Straus, & Sugarman, 1995). T. Berry Brazelton, a renowned pediatrician, says he does not believe in spanking youngsters because it shows disrespect and teaches them that violence is an acceptable method for settling differences (Steinberg, 1995).

To read about another form of child abuse—sexual abuse—see *Further Developments:* "Sexual Abuse of Children."

Parenting Styles

It's not fun to be the boss of your parents.
Bruce Weber, *New York Times Magazine*, October 8, 1995

Diana Baumrind (1996, 1980, 1972, 1971), a developmental psychologist, examines the relationship between parental child-rearing styles and social competence in children of preschool and school age. From 1968 through 1980, she conducted research for The Family

Socialization Project, examining family socialization practices, parental attitudes, and factors of development at three crucial stages in a child's life: preschool, early childhood, and early adolescence. She included both parents and children from white, middle-class families in her sample, and collected data through questionnaires, personal interviews, and videotaped observations of family interactions at home. In her studies of white middle-class nursery-school children, Baumrind (1971) found that different types of parenting tend to be related to quite different behaviors in children. Among other findings in this longitudinal study, she distinguishes among **authoritarian, authoritative, permissive,** and **harmonious parenting** (see Figure 8-3).

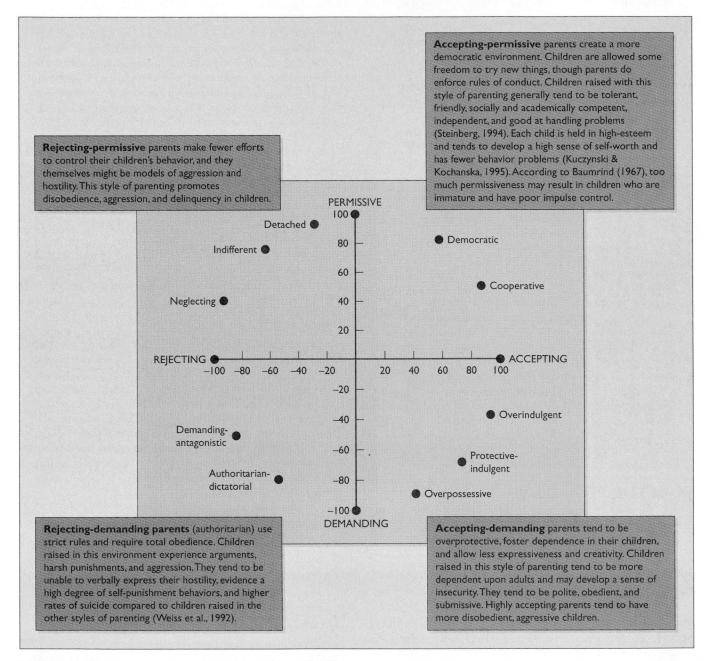

Figure 8-3 Baumrind's Research Expands on the Interaction of Parenting Dimensions Two parenting dimensions have been observed in all societies: the permissive-demanding dimension and the accepting-rejecting dimension (Rohner & Rohner, 1981). There are some research findings that parental choice of these dimensions—especially for the demanding style—may be influenced by the parent's genetic makeup; other findings suggest that the parent's own upbringing has a significant bearing on parental practices (Plomin, DeFries, & Fulker, 1988). Research studies also show that each parent is generally consistent in the chosen style of parenting.

Adapted from E. S. Schaefer, "A Circumplex Model for Maternal Behavior," *Journal of Abnormal and Social Psychology,* Vol. 59 (1959), p. 232; and M. L. Hoffman and L. W. Hoffman, *Review of Child Development Research* © 1964, Russell Sage Foundation. Used with permission of the Russell Sage Foundation.

Authoritarian Parenting The authoritarian parent attempts to shape, control, and evaluate a child's behavior in accordance with traditional and absolute values and standards of conduct. Obedience is stressed, verbal give-and-take is discouraged, and punitive, forceful discipline is preferred. More commonly, parents using this style of parenting are said to be operating from the *rejecting-demanding dimension*. The offspring of such *authoritarian* parents tended to be discontented, withdrawn, and distrustful.

Authoritative Parenting The authoritative parent provides firm direction for a child's overall activities but gives the child considerable freedom within reasonable limits. Parental control is not rigid, punitive, intrusive, or unnecessarily restrictive. The parent provides reasons for given policies and engages in verbal give-and-take with the child, meanwhile responding to the child's wishes and needs. (It may help you to distinguish *authoritative* from *authoritarian* by using the mnemonic device of gi*ve*-and-take with authorita*tive*.) *Authoritative* parenting was often associated with self-reliant, self-controlled, explorative, and contented children. In later research Baumrind (1996, 1994, 1991) found that an authoritative parenting style is especially helpful when parenting adolescents.

Baumrind believes that authoritative parenting gives children a comfortable, supported feeling while they explore the environment and gain interpersonal competence. Such children do not experience the anxiety and fear associated with strict, repressive parenting or the indecision and uncertainty associated with unstructured, permissive parenting. Laurence Steinberg and colleagues (1989) also found that authoritative parenting facilitates school success, encouraging a healthy sense of autonomy and positive attitudes toward work. Adolescents whose parents treat them warmly, acceptingly, democratically, and firmly are more likely than their peers to develop positive beliefs about their achievement and so are likely to do better in school. In addition, authoritative fathers and mothers seem more adept than other parents at "scaffolding." **Scaffolding** supports a child's learning through interventions and tutoring that provide helpful task information attuned to the child's current level of functioning (Pratt et al., 1988). On the basis of her research, Baumrind (1996, 1994, 1991) found a number of parental practices and attitudes that seem to facilitate the development of socially responsible and independent behavior in children:

- Parents who are socially responsible and assertive, and who serve as daily models of these behaviors, foster these characteristics in their children.
- Parents should employ firm enforcement policies geared to reward socially responsible and

independent behavior and to punish deviant behavior. This technique uses the reinforcement principles of conditioning. Parents can be even more effective if their demands are accompanied by explanations and if punishment is accompanied by reasons that are consistent with principles the parents themselves live by.
- Parents who are nonrejecting serve as more attractive models and reinforcing agents than rejecting parents do.
- Parents should emphasize and encourage individuality, self-expression, initiative, divergent thinking, and socially appropriate aggressiveness. These values are translated into daily realities as parents make demands upon their children and assign them responsibility.

Parents should provide their children with a complex and stimulating environment that offers challenge and excitement. At the same time, children should experience their environment as providing security and opportunities for rest and relaxation.

Permissive Parenting Permissive parents seek to provide a nonpunitive, accepting, and affirmative environment in which the children regulate their own behavior as much as possible. The children are consulted about family policies and decisions. The parents make few demands on the children for household responsibility or orderly behavior. The least self-reliant, explorative, and self-controlled children were those with permissive parents.

Harmonious Parenting Harmonious parents seldom exercise direct control over their children. They attempt to cultivate an egalitarian relationship, one in which the child is not placed at a power disadvantage. They typically emphasize humane values as opposed to the predominantly materialistic and achievement values they view as operating within mainstream society. The *harmonious* parents identified by Baumrind were only a small group. Of the 8 children studied from such families, 6 were girls and 2 were boys. The girls were extraordinarily competent, independent, friendly, achievement-oriented, and intelligent. The boys, in contrast, were cooperative but notably submissive, aimless, dependent, and not achievement-oriented. Although the sample was too small to be the basis for definitive conclusions, Baumrind tentatively suggests that these outcomes of harmonious parenting might be sex-related.

Discussion: Control and Autonomy Revisited Much research confirms Baumrind's findings and insights (Steinberg et al., 1994). The ability to "achieve one's goals without violating the integrity of the goals of the other" is undoubtedly a major component in the development of social competence (W. Bronson, 1974, p. 280). Clearly, disciplinary encounters between parents and their youngsters provide a crucial context in which children learn strategies for controlling themselves and for controlling others, so parents who model competent strategies are more likely to have children who also are socially competent (Kuczynski & Kochanska, 1995).

By way of illustration, consider Erik Erikson's (1963) notion that how toddlers resolve the stage of autonomy versus shame and doubt is linked to parental overcontrol. One indication that 2-year-olds are coming to terms with autonomy issues is their ability and willingness to say *no* to parents. The acquisition of *no* is a spectacular cognitive achievement because it accompanies youngsters' increasing awareness of the "other" and the "self" (Spitz, 1957). *Self-assertion, defiance,* and *compliance* are distinct dimensions of toddler behavior. For instance, if a mother tells her toddler girl to pick up her toys and place them in a box, and the child says, "No, want to play," the youngster would be *asserting* herself. If instead the toddler takes more toys from the box or heaves a toy across the room, she would be *defying* her mother. But if the child follows her mother's instructions, she would be *complying.*

Susan Crockenberg and Cindy Litman (1990) show that the way parents handle these autonomy issues has profound consequences for their youngsters' behavior. When parents assert their power in the form of negative control—threats, criticism, physical intervention, and anger—children are more likely to respond with defiance. Youngsters are less likely to become defiant when a parent combines a directive with an additional attempt to guide the child's behavior in a desired direction. This latter approach provides the child with information about what the parent wants, while inviting power sharing. For instance, if the parent asks the child to do something ("Would you pick up your toys, please?") or attempts to persuade the child through reasoning ("You made a mess, so now you'll have to clean it up"), the parent implicitly validates for the child that she is a distinct and separate person with her own individual needs. This approach is consistent with Baumrind's *authoritative* style of parenting and keeps the negotiation process going, allowing the toddler to "decide" to adopt the parent's goal. It seems that children are more willing to accept other people's attempts to influence their behavior if they perceive that they are participating in a reciprocal relationship where their attempts to influence others will also be honored (Kochanska & Aksan, 1995; Parpal & Maccoby, 1985).

Guidance alone seems less effective than guidance combined with control. An invitation to comply ("Could you pick up the toys now?") seems to afford the toddler a choice, and the child might feel free to turn it down in the absence of a clear and firm expression of parental wishes. This approach of guidance without control is consistent with Baumrind's *permissive* style of parenting and seems to be linked to less competent child behavior. When toddlers assert themselves and their parents follow with a power directive ("You better do what I say or I'll spank you!"), the children might interpret the behavior as an assertion of parental power and a diminution of their own autonomy, an approach in keeping with Baumrind's *authoritarian* style of parenting (Crockenberg & Litman, 1990).

Overall, it seems that parents who are most effective in eliciting compliance from their youngsters and deflecting defiance are quite clear about what they want their children to do, yet all the while they are prepared to listen to their children's objections and to make appropriate accommodations in ways that convey respect for their youngsters' individuality and autonomy (Gralinski & Kopp, 1993). At times, the process of achieving compliance can be somewhat extended and complex, with the parent explaining, reasoning, persuading, suggesting, accommodating, and compromising. These parental behaviors encourage and elicit competent behavior from the child (Crockenberg & Litman, 1990). Of course much also depends on the situation and the ability of youngsters to comprehend their parents' instructions (Grusec & Goodnow, 1994).

> **Questions**
>
> *According to Baumrind, what are the four main styles of parenting? How does each affect the outcome of the child's behavior?*

Gaining Perspective on Parenting

The parenting dimensions and styles that we have considered thus far in this chapter have focused on global patterns and practices. But they are much too abstract to capture the subtleties of parent-child interaction. In everyday life parents reveal a great variety of parenting behaviors, depending on many factors: the situation; the gender and age of the child; the parents' inferences regarding the child's mood, motives, and intentions; the child's understanding of the situation; the social supports available to the parent; the pressures parents feel from other adults; and so forth (Dix, Ruble, & Zambarano, 1989). For instance, parents can be warm or cold, restrictive or permissive, and consistent or inconsistent, depending on the setting and circumstances (Clarke-Stewart & Hevey, 1981). The child's response to being disciplined also modifies the parent's behavior and the parent's choice of future disciplinary measures. And the way the child *perceives* the actions of the parent can be more decisive than the parent's actions in themselves

(Grusec & Goodnow, 1994). Children are not interchangeable; they do not all respond in identical fashion to the same type of caretaker behavior (Kochanska, 1995).

The Harvard Child-Rearing Study   A follow-up on a classic study helps us clarify some of these matters. In the 1950s three Harvard psychologists carried out one of the most enterprising studies of child rearing ever undertaken in the United States. Robert Sears, Eleanor Maccoby, and Harry Levin (1957) attempted to identify those parenting techniques that make a difference in personality development. They interviewed 379 mothers of kindergartners in the Boston area and rated each mother on about 150 different child-rearing practices. Some 25 years later, a number of Harvard psychologists led by David C. McClelland (McClelland et al., 1978) contacted many of these children, who were then 31 years old, most of them married and with children of their own.

McClelland and his associates interviewed these individuals and administered psychological tests to them. They concluded that not much of what people think and do as adults is determined by the specific techniques of child rearing their parents used during their first five years. Practices associated with breast-feeding, toilet training, and spanking are not all that important. It is how parents *feel* about their children that *does* make a difference. What mattered was whether a mother liked her child and enjoyed playing with the child or whether she considered the child a nuisance, with many disagreeable characteristics. Furthermore, children of affectionate fathers were more likely as adults to show tolerance and understanding than were the offspring of other fathers. The Harvard researchers conclude:

> How can parents do right by their children? If they are interested in promoting moral and social maturity in later life, the answer is simple: they should love them, enjoy them, want them around. They should not use their power to maintain a home that is only designed for the self-expression and pleasure of adults. They should not regard their children as disturbances to be controlled at all costs. (McClelland et al., 1978, p. 53)

The Harvard Preschool Project   In the 1970s another group of Harvard researchers shed additional light on effective parenting. Burton L. White and his colleagues studied the development of overall competence in children (White & Watts, 1973). The goal of the research, christened the Harvard Preschool Project, was to find ways to optimize human development. The researchers intensively studied the mother-child interactions of 31 toddlers with different competence ratings. Burton White (1973, p. 242) concludes that a number of child-rearing practices appear to foster competence:

> [The effective mothers] talk a great deal to their children. . . . They make them feel as though whatever they are doing is usually interesting. They provide access to many objects and diverse situations. They lead the child to believe that he can expect help and encouragement most, but *not all* the time. They demonstrate and explain things to the child, but mostly on the child's instigation rather than their own. . . . They are secure enough to say "no" to the child from time to time without seeming to fear that the child will not love them. They are imaginative, so that they make interesting associations and suggestions to the child when opportunities present themselves. They very skillfully and naturally strengthen the child's intrinsic motivation to learn. They also give him a sense of task orientation, a notion that it is desirable to do things well and completely. They make the child feel secure.

One finding of considerable interest was that the most effective mothers did not devote the bulk of their day to child rearing. They were busy women, and some held part-time jobs. Instead of giving their children all of their time, these effective mothers tried to create an environment compatible with the burgeoning curiosity of their toddlers. They recognized that a youngster and a spotless, meticulously kept home are incompatible. Their children had access to small, manipulable, visually detailed objects (toys, plastic refrigerator containers, empty milk cartons, old shoes, magazines, and the like) and to materials that help to develop motor skills (scooters, tricycles, and so forth). Also, the effective mothers did not necessarily drop what they were doing to respond to their toddlers' requests. If they were busy, they said so,

How Parents Feel About Their Youngsters Makes a Difference   Increasingly psychologists are reaching the conclusion that how parents feel about their children makes a greater difference than the specific child-rearing techniques they employ. Parenting is not a matter of magical formulas but one of enjoying children and loving them.

giving the child a small but realistic taste of things to come in the larger world.

Summing Up   Clearly, parenting is not a matter of employing a surefire set of recipes or formulas. Cultures differ, parents differ, and children differ. Parents who employ identical "good" child-rearing techniques have children who grow up to be exceedingly different (Bornstein, 1995). Furthermore, situations differ, and what works in one setting can boomerang in another. As highlighted by the Harvard research, the essence of parenting is in the parent-child *relationship*. In their interactions, parents and children evolve ongoing accommodations that reflect each other's needs and desires. Parent-child relationships differ so much, both within the same family and among families, that in many respects each parent-child relationship is unique (Elkind, 1974). There is no mysterious, secret method you must master. It is the child that matters, not the technique—but child-rearing experts believe parents should not use corporal punishment. On the whole, most parents do very well.

> **Question**
>
> *What types of parenting styles seem to promote emotionally healthy children?*

The Changing Nature of American Families

*K*ids spend an average of 3 hours a day with their parents, 40 percent less time than they did in 1960.

New York Times Magazine, October 8, 1995

Single-Parent Families and Effects of Divorce   The American family structure has changed dramatically over the past few decades. Although sociologists tell us that the family tree is as deeply rooted as ever in the social landscape, it is sprouting varied branches. The normative *nuclear family* (mother, father, children) now takes a place alongside a variety of other family arrangements, such as cohabiting heterosexual couples, single-parent households, blended or bi-nuclear households, child-free married couples, lesbian and gay couples, and extended family households (such as grandparents raising grandchildren). Overall, family relationships are becoming more tangled as a result of people living longer and perhaps changing mates to suit the seasons of their lives. Consequently, increasing numbers of children are growing up with several sets of parents and an assortment of half brothers, half sisters, and stepsiblings (Newman, Roberts, & Syre, 1993; Ahlburg & DeVita, 1992).

According to the U.S. Bureau of the Census (1997a), about 69 percent of children currently live with both biological parents. About 27 percent of all U.S. children under 18 years of age live in single-parent households (19 percent with a separated parent, 4 percent with a widowed parent, and 4 percent with a parent whose spouse lives elsewhere). In contrast, in 1970, 12 percent of U.S. children under age 18 lived with one parent. The fastest increase in the number of single-parent families occurred during the 1970s; the number continued to rise during the 1980s, but at a slower rate than in the previous decade (Furstenberg & Cherlin, 1991). Figure 8-4 provides additional data on American households living in poverty.

In the past, many psychologists and sociologists viewed the single-parent family as a lamentable, defec-

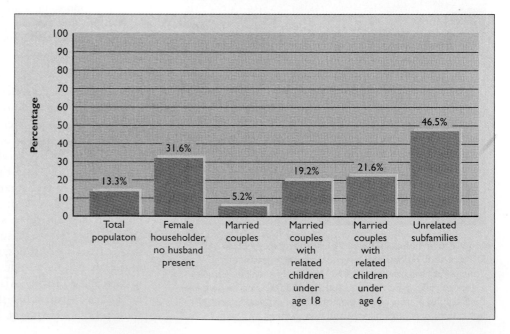

Figure 8-4 Proportion of the U.S. Population Living Below the Poverty Line, by Family Type, 1997

In 1997, the poverty line for a family of four was $16,400. For a family of three, it was $12,802. Nationwide the proportion of the population living below the poverty level declined from 13.7 percent in 1996 to 13.3 percent in 1997. This decline in the nation's overall poverty rate was mostly caused by a decline in poverty experienced by African Americans and Hispanic Americans.

Source: US Census Bureau. March 1998 Current Population Survey, Facts for Features.

tive, "broken" arrangement in need of fixing (Cavan, 1964). But over the past 30 years, professionals have increasingly come to recognize that the single-parent home is a different, but viable, family form (Hetherington & Stanley-Hagan, 1995). The nation's public and private schools, communities, and churches are beginning to assist the increasing number of single-parents families with before-school and after-school child-care programs and summer school enrichment or remedial classes for children of all ages. Some school districts have restructured the school day or extended the school year to 12 months with periodic breaks to meet the needs particularly of single-parent families.

Many single-parent families do have serious financial problems, although this is hardly evidence of pathology (Jayakody, Chatters, & Taylor, 1993). Of the 12 million U.S. youngsters—nearly 25 percent of the nation's children—who live below the federally defined poverty line, the vast majority are found in single-mother households (Polakow, 1993). One major difference between husband-wife families and one-parent families is that many of the former have two incomes, and the latter have at most one income. In addition, women's employment tends to pay less and is more sporadic than men's. Census Bureau figures reveal that 59 percent of women raising children without the children's father have been awarded child support, but only half of these mothers receive full payment, 24.9 percent receive partial payments, and 23.9 percent do not receive any payments (Pirog-Good, 1993). Significantly, 69 percent of never-married mothers live in poverty. About 45 percent of children who live with divorced single-parent mothers live at or below the poverty line (U.S. Bureau of the Census, 1997a).

The most common family arrangement in the period immediately following a divorce is for the children to live with their mothers and have only intermittent contact with their fathers. Divorce is a process that begins well before parents separate and continues long afterward (Guttman, 1993). E. Mavis Hetherington, Martha Cox, and Roger Cox (1977) did a two-year longitudinal study in which they matched a preschool child in a divorced family with a child in an intact family, on the basis of age, sex, birth-order position, and the age and education of the parents. In all, 48 children of divorced couples were paired with 48 children in intact families. They found that the first year after the divorce was the most stressful, for both parents. Many of the stressors that parents experience following divorce and the accompanying changes in their lifestyles are reflected in their relationships with their children (Webster-Stratton, 1989). Indeed, stressful events in the life of the mother are more predictive of adjustment in the children than are stressful events in the life of the child (Guidubaldi & Perry, 1985). Hetherington and colleagues (1976, p. 424) found that the interaction patterns between the divorced parents and their children differed significantly from those encountered in the intact families:

Single-Parent Families Although the single-parent family has traditionally suffered a bad reputation, It Is Increasingly recognized as a viable family form. Most children are resilient and can flourish in a wide variety of family arrangements so long as they are loved and their basic needs are met. Female single-parent families have been increasing since 1980. One of every three births in 1996 was to an unmarried mother.

> Divorced parents make fewer maturity demands of their children, communicate less well with their children, tend to be less affectionate with their children and show marked inconsistency in discipline and lack of control over their children in comparison to parents in intact families. Poor parenting seems most marked, particularly for divorced mothers, one year after divorce, which seems to be a peak of stress in parent-child relations. [Many of the mothers referred to their relationship with their child one year after divorce as involving "declared war" and a "struggle for survival."] Two years following the divorce, mothers are demanding more . . . [independent and] mature behavior of their children, communicate better and use more explanation and reasoning, are more nurturant and consistent and are better able to control their children than they were the year before. A similar pattern is occurring for divorced fathers in maturity demands, communication and consistency, but they are

becoming less nurturant and more detached from their children. . . . Divorced fathers were ignoring their children more and showing less affection [while their extremely permissive and "every day is Christmas" behavior declined].

Hence, many single-parent families had a difficult period of readjustment following the divorce, but the situation generally improved during the second year. Other researchers confirm this finding (Elias, 1987).

Hetherington has found that much depends on the ability of the custodial mother to control her children. Homes in which the mother loses control are associated with decline in children's IQ scores, poorer school grades, and a decrease in children's problem-solving skills. Children whose mothers maintain good control show no drop in school performance. Hetherington also found that when single-parent mothers lose control of their sons, a "coercive cycle" typically appears. The sons tend to become more abusive, demanding, and unaffectionate. The mother responds with depression, low self-esteem, and less control, and her parenting becomes worse (Hetherington & Stanley-Hagan, 1995). In contrast, mothers and daughters in mother-headed families often express considerable satisfaction with their relationships, except for early-maturing girls whose heterosexual and older-peer involvements frequently weaken the mother-child bond (Hetherington, 1989).

Hetherington (1989) undertook a follow-up study six years later, asking, "Which children are the winners, losers, and survivors six years after divorce?" (p. 11). Children from divorced or remarried families were overrepresented in these three categories:

- *Aggressive, insecure* children were prone both to impulsive, irritable outbursts and to sullen brooding periods of withdrawal at home and school. These children were unpopular with their peers, had difficulties in school, had few areas of satisfaction, and had low self-esteem, all of which contributed to their lonely, unhappy, angry, anxious, and insecure childhoods.
- *Opportunistic-competent* children seemed to adapt quite well. They were high in self-esteem, were popular with their peers and teachers, functioned at average or above-average levels in the classroom, and gave little evidence of behavior problems. They exhibited competence, flexibility, and persistence in dealing with demanding and stressful circumstances. However, in their social relationships, opportunistic-competent children were quite manipulative, attempting with considerable success to ingratiate themselves with people in power (parents and teachers and high-status peers).
- *Caring-competent* youngsters were high in self-esteem, were popular with peers and teachers,

functioned well in the classroom, and seemed to adapt well, with little evidence of behavior problems. They, too, exhibited competence, flexibility, and persistence in dealing with demanding and stressful circumstances. However, they were much less inclined toward manipulative behaviors and were disposed toward helping and sharing behaviors, often befriending neglected or rejected peers.

Joint-Custody Arrangements   Researchers find that the quality of the child's relationships with *both* parents is the best predictor of her or his postdivorce adjustment (Amato, 1993). Children who maintain stable, loving relationships with both parents appear to have fewer emotional scars—they exhibit less stress and less aggressive behavior, and their school performance and peer relations are better—than children lacking such relationships (Arditti & Keith, 1993). The importance for the child of maintaining a postdivorce relationship with both parents has led to the practice of awarding *joint legal custody* (Donnelly & Finkelhor, 1993). Under a **joint custody** arrangement, both parents share equally in the making of significant child-rearing decisions, and both parents share in regular child-care responsibilities. The child lives with each parent a substantial amount of time (for example, the child might spend part of the week or month in one parent's house and part in the other's). Joint custody also eliminates the "winner/loser" character of custodial disposition and much of the sadness, sense of loss, and loneliness that the noncustodial parent frequently feels.

But joint custody is not an answer for all children. Critics point out that parents who cannot agree during marriage cannot be expected to reach agreement on rules, discipline, and styles of parenting after marriage. And they say that alternating between homes interferes with a child's need for continuity in his or her life (Simon, 1991). Furthermore, the geographically mobile nature of contemporary society and the likelihood that parents will remarry render a good many joint-custody arrangements vulnerable to collapse (Simon, 1991). So probably it is not surprising that initial evidence suggests that joint-custody arrangements do not differ from sole-custody arrangements in children's adjustment to divorce (Donnelly & Finkelhor, 1992).

With respect to school achievement, social adjustment, and delinquent behavior, the differences are small or nonexistent between children from one- and two-parent homes of comparable social status (Amato, 1993). Some research suggests that children and adolescents from single-parent homes show less delinquent behavior, less psychosomatic illness, better adjustment to their parents, and better self-concepts than those from unhappy intact homes (Demo, 1992). Even so, neither an unhappy marriage nor a divorce is especially congenial for children. Each alternative brings its own sets of stressors (Cummings & Davies, 1994). Psychiatrists and clinical psychologists note that in many cases divorce reduces the amount of friction and unhappiness a child experiences, leading to better behavioral adjustments (White, 1994). Overall, research strongly suggests that the *quality* of children's relationships with their parents matters much more than the fact of divorce (Hetherington & Stanley-Hagan, 1995).

Young Children with Gay or Lesbian Parents   We do not know how many children have gay or lesbian parents, in part because we lack solid figures on the number of lesbian and gay adults. We do know, however, that the number of youngsters who are aware they have lesbian or gay parents is growing (Chira, 1993b). Usually the children are born to parents in heterosexual marriages who subsequently "come out." Most others are born to lesbians via artificial insemination. And there are a few hundred documented cases of adoption or foster parenting by lesbians and gays. Until recent years lesbian and gay parents have been largely invisible, mostly because of harassment and custody concerns, and because they wish to shield their children from turmoil. When lesbian and gay parents do face custody or visitation battles in court, the outcome varies from state to state and from court to court. In 1995, the Virginia Supreme Court denied a lesbian custody of her son, saying she was a poor mother and her live-in lesbian relationship could bring the child "social condemnation." Because of such difficulties, lesbian and gay parents have proven to be an especially difficult population to study (Hare & Richards, 1993). They are also a diverse population (Allen & Demo, 1995). However, there is no convincing evidence that the development of children with lesbian or gay parents is compromised in any significant way relative to the development of children with heterosexual parents who otherwise live under comparable circumstances—indeed, they seem to develop much the same (Patterson, 1992). The issue of gay and lesbian parenting is discussed further in Chapter 14.

Questions

In addition to the traditional nuclear family, what are some other types of family structures? What factors are associated with a young child's healthy emotional adjustment to parental divorce?

Sibling Relationships

*B*ig sisters are the crab grass in the lawn of life.

Charles M. Schulz, *Peanuts*

A child's relationships with sisters and brothers within the family are very important (Dunn, Slomkowski, & Beardsall, 1994). A child's position in the family and the number and sex of his or her siblings are thought to have major consequences for the child's development and socialization (Volling & Belsky, 1992). These factors structure the child's social environment, providing a network of key relationships and roles (E. Brody et al., 1994). An only child, an oldest child, a middle child, and a youngest child all seem to experience a somewhat different world because of the different social webs that encompass their lives, even though they receive the same style of parenting. Some psychologists contend that these and other environmental influences operate to make two children in the same family as different from one another as are children in different families (Daniels, 1986). These psychologists say that there is a unique *microenvironment* in the family for each child. In this view there is not a single family but, rather, as many "different" families as there are children to experience them. These psychologists conclude that the small degree of similarity in personality found among siblings results almost totally from shared genes rather than from shared experience. In

Sibling Interactions Are Important in All Societies
A child's birth order structures a somewhat different social web or environment for siblings. Indeed, scientists are increasingly coming to the conclusion that there is not a "single" family but as many "different" families as there are youngsters to experience them. Siblings typically maintain strong bonds throughout life.

short, the unique aspects of siblings' experiences in the family are more powerful in shaping their personalities than what the siblings experience in common. Many of the differences in the family environment are more obvious to children than to their parents (McHale et al., 1995), and much depends on how children perceive and interpret parental affection and discipline (Grusec & Goodnow, 1994).

The "pioneering function" of older brothers and sisters can persist throughout life, providing role models for how to cope with bereavement, retirement, or widowhood (Rosenthal, 1992). Indeed, by virtue of today's frequent divorces and remarriages, some children form stronger bonds with their siblings than they do with their parents or stepparents (Beer, 1989). Sibling relationships typically become more egalitarian but also less intense as youngsters move into later childhood and adolescence (Buhrmester & Furman, 1990).

Through the years research has been focused on firstborn children, for they appear to be fortune's favorites (Falbo & Poston, 1993; Cicirelli, 1978). Firstborns are overrepresented among students in graduate and professional schools (Goleman, 1985), at the higher-IQ levels (Zajonc, 1976), among National Merit and Rhodes Scholars, in *Who's Who in America* and *American Men and Women of Science,* among individuals on the cover of *Time* magazine, among U.S. presidents (52 percent), among the 102 appointments to the Supreme Court (55 percent were either only children or firstborns), among men and women in Congress, and in the astronaut corps (21 of the first 23 U.S. astronauts who flew on space missions were either only children or firstborn sons). However, these birth-order advantages do not hold in a good many families and for individuals from lower socioeconomic backgrounds (Ernst & Angst, 1983). Some research suggests that middle children tend to have lower self-esteem than firstborns and lastborns, perhaps a function of their less well-defined position within the family (Goleman, 1985), yet some researchers find that later-born children possess better social skills than firstborns (Jacklin & Reynolds, 1993).

Fewer effects related to birth order are found in the United States than in most other societies. A cross-cultural study of 39 societies reveals that firstborns are more likely than later-borns to receive elaborate birth ceremonies, to have authority over siblings, and to receive respect from siblings. In comparison with other sons, firstborn sons generally have more control of property, more power in the society, and higher social positions (Rosenblatt & Skoogberg, 1974). Moreover, older siblings act as caretakers for their younger siblings in a good many cultures (Dunn, 1983). Overall, sibling relationships in industrialized societies tend to be discretionary while those in nonindustrialized societies tend to be obligatory (Cicirelli, 1994).

A number of explanations have been advanced to account for differences between firstborn and later-born children. First, research reveals that parents attach greater importance to their first child (Clausen, 1966). There are more social, affectionate, and caretaking interactions between parents and their firstborn (Cohen & Beckwith, 1977). Thus, firstborns have more exposure to adult models and to adult expectations and pressures (Baskett, 1985).

A second explanation of the differences between first- and later-born children derives from **confluence theory,** a model devised by psychologist Robert B. Zajonc (1986; Zajonc et al., 1991) and his colleagues. Confluence theory gets its name from the view that the intellectual development of a family is like a river, with the inputs of each family member flowing into it. According to Zajonc, the oldest sibling experiences a richer intellectual environment than younger siblings do. He contends that each additional child "waters down" the intellectual climate by increasing the incidence of interactions with childish minds as opposed to adult minds. However, critics fault the confluence model, especially the mathematical operations that Zajonc has used in testing it (Retherford & Sewell, 1992).

A third explanation—the **resource dilution hypothesis**—extends the confluence model to encompass more resources than simply a rich intellectual environment. This theory says that in large families resources get spread thin, to the detriment of all the offspring. Family resources include such things as parental time and encouragement, economic and material goods, and various cultural and social opportunities (music and dance lessons and travel at home and abroad). Sociologists commonly employ the resource dilution hypothesis to explain the relationship they find between the number of siblings and educational attainment: Increases in the number of siblings are associated with the completion of fewer years of schooling and the attainment of fewer educational milestones (positions in student government, on the school newspaper, in drama groups, and so on). In this manner, family size is linked to greater or lesser degrees of achievement (Steelman & Powell, 1989).

A fourth explanation, which was first advanced by Alfred Adler, stresses the part that sibling power and status rivalry play in a child's personality formation (Ansbacher & Ansbacher, 1956). Adler viewed the "dethroning" of the firstborn as a crucial event in the development of the first child. With the birth of a sister or brother, the firstborn suddenly loses his or her monopoly on parental attention (Kendrick & Dunn, 1980). This loss, Adler said, arouses a strong lifelong need for recognition, attention, and approval that the child, and later the adult, seeks to acquire through high achievement. An equally critical factor in the development of the later-born child is the competitive race for achievement with older and more accomplished siblings. In many cases the rancor disappears when individuals get older and learn to manage their own careers and married lives (Dunn, 1986).

In sum, many unique genetic and environmental factors intervene in individual families, producing wide differences not just among families but among their members (Heer, 1985).

> **Questions**
>
> *In what ways do older siblings influence younger ones in terms of motivation, self-esteem, and social support? In general, how does a child's birth order in a family affect her or his personality?*

Peer Relationships

Man is a knot, a web, a mesh into which relationships are tied.

Antoine de Saint-Exupéry

We have seen that children enter a world of people, an encompassing social network. With time, specific relationships change in form, intensity, and function, but the social network itself stretches across the life span. Yet social and behavioral scientists mostly ignored the rich tapestry of children's social networks until the past two decades. They regarded social intimacy as centering on one relationship, that between the infant and mother, and treated young children's ties with other family members and with age-mates as if they did not exist or had no importance. However, a growing body of research points to the significance of other relationships in the development of interpersonal competencies. In this section we will explore children's peer relationships and friendships. **Peers** are individuals who are approximately the same age. Early friendship is a major source of a youngster's emotional strength, and its lack can pose lifelong risks (Newcomb & Bagwell, 1995).

Peer Relationships and Friendships

Friendship is a single soul dwelling in two bodies.

Aristotle

From birth to death we find ourselves immersed in countless relationships. Few are as important to us as those we have with our peers and friends (Dunn, 1993; Newcomb & Bagwell, 1995). Thus, not surprisingly, research reveals that peer relationships and friendships are

in their own right a meaningful experience in the lives of young children (Mueller & Cooper, 1986).

Children as young as 3 years of age form friendships with other children that are surprisingly similar to those of adults (Verba, 1994). And just as they do for adults, different relationships meet different needs for young children. Some child relationships are reminiscent of strong adult attachments; others, of relationships between adult mentors and protégés; and still others, of the camaraderie of adult coworkers. Although young children lack the reflective understanding that many adults bring to their relationships, they often invest in their friendships with an intense emotional quality (Selman, 1980). Moreover, some young children bring a considerable measure of social competence to their relationships and a high level of give-and-take (Farver & Branstetter, 1994). As we saw in Chapter 6, attachment theory predicts that the quality of the mother-child tie has implications for the child's close personal relationships. Researchers confirm that preschoolers with secure maternal attachments enjoy more harmonious, less controlling, more responsive, and happier relationships with their peers than do preschoolers with insecure maternal attachments (Turner, 1991).

A variety of studies reveal that with increasing age, peer relationships are more likely to be formed and more likely to be successful (Park, Lay, & Ramsey, 1993). Four-year-olds, for instance, spend about two-thirds of the time when they are in contact with other people associating with adults and one-third of the time with peers. Eleven-year-olds, in contrast, spend about an equal amount of time with adults and with peers (Wright, 1967).

A number of factors contribute to this shift in interactive patterns. First, as children grow older, their communication skills improve, facilitating effective interaction (Eckerman & Didow, 1988). Second, children's increasing cognitive competencies enable them to attune themselves more effectively to the roles of others (Verba, 1994). Third, nursery, preschool, and elementary school attendance offers increasing opportunities for peer interaction. And fourth, increasing motor competencies expand the child's ability to participate in many joint activities.

Children of preschool age assort themselves into same-sex play groups (Maccoby, 1990). In a longitudinal study, Eleanor E. Maccoby and Carol N. Jacklin (1987) found that preschoolers at 4½ years of age spent three times as much time playing with same-sex playmates as they did with opposite-sex playmates. By 6½ years of age, the youngsters were spending eleven times as much time with same-sex as with opposite-sex partners. Moreover, preschool girls tend to interact in small groups, especially two-person groups, whereas boys more often play in larger groups (Eder & Hallinan, 1978).

Human Diversity

Preparing for Children with Disabilities in Early Childhood Settings

One of the most fundamental needs of every child is to be accepted and to have a sense of belonging. However, research clearly shows us that children with disabilities or differences are not automatically accepted by their peers unless teachers, child-care workers, and parents take an active role in promoting their acceptance.

Early perceptions about people with disabilities or differences lay the groundwork for attitude formation. In fact, by the age of 5, children have already formed perceptions, positive or negative, about youngsters with disabilities (Favazza & Odom, 1996; Diamond, LeFurgy, & Blass, 1993). Without thoughtful planning and strategies to promote acceptance, these early attitudes are often negative (Favazza & Odom, 1996).

Teachers can address the three key influences in attitude formation by setting up a classroom to promote a positive, accepting attitude on the part of the class and positive self-esteem in children with a disability:

- *Indirect experiences.* Make available photographs, books, displays, and instructional programs that provide information about persons with disabilities. For example, it is not uncommon for children and their uninformed families to misperceive they can "catch" the other child's disability or disease (such as Down syndrome). Children with disabilities need to be visually represented in class settings and community activities, just as all children in the classroom need to

be represented in their surroundings. Favazza and Odom (1996) found in a study examining 95 preschool and kindergarten classes that most typically do not have persons with disabilities depicted in displays, books, or curricula that discourage stereotypic views.

- *Direct experiences.* Research has clearly demonstrated that positive experiences with individuals with disabilities can contribute to acceptance. Other children from the class might need to help the child with a disability during play time or snack time. All children in the group might need to learn sign language, for example. The teacher is crucial in modeling the helping behaviors first. Negative attitudes tend to arise in children who have little or no direct experience with the disability ("Climate and Diversity," 1997).

- *Primary social group.* The primary social group for a young child is her or his family. Children's attitudes are affected by parental attitudes, including parental silence, which the child experiences negatively. Teachers and classroom aides must educate themselves about particular children with disabilities. Get to know about disabilities—but also learn the specifics about that child. If possible, make arrangements for a successful adult with this disability to visit the classroom and talk with the children—for instance, invite an adult who is in a wheelchair, or who has a prosthetic limb, or who is working in a sheltered workshop.

From: Paddy C. Favazza. "Preparing for Children With Disabilities in Early Childhood," *Early Childhood Education Journal*, 25:4, © 1998. Reprinted by permission of Plenum Publishing Corporation.

Peer Reinforcement and Modeling

Children play an important part as reinforcing agents and behavioral models for one another, a fact that adults at times overlook. Much learning takes place as a result of children's interaction with other children (Azmitia, 1988). Here is a typical example (Lewis et al., 1975, p. 27):

Two four-year-olds are busily engaged in playing with "Playdoh." The room echoes with their glee as they roll out long "snakes." Each child is trying to roll a longer snake than the other. The one-year-old sibling of one of the children, hearing the joyful cries, waddles into the room. She reaches for the Playdoh being used by her sibling. The older child hands the one-year-old some Playdoh, and the child tries to roll her own snake. Unable to carry out the task and frustrated, the one-

year-old becomes fussy, at which point the older sibling gives the child a knife. The four-year-old then shows the sibling how to cut the snakes the other children have made.

Indeed, the old one-room school functioned, and functioned well, with the teacher teaching the older children and the older children teaching the younger boys and girls.

Seeing other children behave in certain ways can also affect a child's behavior. By means of imitate-in-turn and follow-the-leader games, youngsters gain a sense of connectedness, of other children being like themselves, and of successfully exerting social control over the behavior of others (Eckerman & Stein, 1990) (see *Human Diversity:* "Preparing for Children with Disabilities in Early Childhood Settings.") In addition, a study by Robert D.

O'Connor (1969) reveals that severely withdrawn nursery school children engage in considerably more peer interaction after they watch a 20-minute sound film that portrays other children playing together happily. And as we will see in Chapter 9, modeling has proved to be an important tool for helping children to overcome various fears and prepare for new experiences.

Aggression in Children

Much human aggression takes place in the context of group activity (Berkowitz, 1993). **Aggression** is behavior that is socially defined as injurious or destructive. Even young children display aggression. Across time children come to express aggression somewhat differently. With increasing age their aggressive behavior becomes less diffuse and more directed (Feshbach, 1970). The proportion of aggressive acts of an undirected, temper-tantrum type decreases gradually during the first three years of life, then shows a sharp decline after the age of 4. In contrast, the relative frequency of retaliatory responses increases with age, especially after children reach their third birthday. Verbal aggression also increases at 2 to 4 years of age (Egeland et al., 1990).

Girls and boys differ in how they express their aggression toward peers. Boys tend to harm others through physical and verbal aggression (hitting or pushing others and threatening to beat up others); their concerns typically center on getting their way and dominating other youngsters. Girls, in contrast, tend to focus on relational issues (establishing close, intimate connections with one another); they attempt to harm others by damaging their friendships or feelings of peer-group inclusion by such acts as spreading negative rumors, excluding the child from a play group, and purposefully withdrawing friendship or acceptance (Crick & Grotpeter, 1995).

Children have a considerable influence on one another in the expression of aggressive behavior (Herzberger & Hall, 1993). Gerry R. Patterson and colleagues (1967) have described how the process frequently operates in nursery school settings. They recorded aggressive interactions among 36 nursery school children over a 26-week period and found that when an aggressive response (e.g., a kick or a punch) was followed by crying, withdrawing, or acquiescing, the attacker was likely to aggress against the victim *again*. These reactions functioned as positive reinforcers for the aggressor. When aggressive behavior was followed by punishment such as retaliatory responses, efforts to recover the seized item, or teacher intervention, aggressors were more likely to pick a different victim for their future aggression or to alter their interactions with the original victim. Hence, the feedback provided for aggressors influences their subsequent behavior.

The researchers also found that whereas some children entered nursery school with a repertoire of aggressive behaviors, others were passive and unassertive at

Peer Aggression Many aggressive children attribute hostile intentions to their peers even under benign circumstances. Consequently, they are more likely than other children to respond with aggression in interpersonal settings.

first. But after the relatively unaggressive children learned to counteraggress and thus end other children's aggressive acts, they themselves began to aggress against new victims.

Even so, some children appear more prone than others to engage in aggressive behavior (Campbell et al.,1994). Some aggression derives from a developmental lag in children's acquisition of role-taking skills (Hazen & Black, 1989), but this is not the entire story. More aggressive children, particularly boys, report that aggression produces tangible rewards and reduces negative treatment by other children (Trachtenberg & Viken, 1994). Additionally, researchers find that youngsters who see and hear angry exchanges among adults become emotionally distressed and respond by aggressing against their peers (Vuchinich, Bank, & Patterson,1992).

Anthropologist Douglas P. Fry (1988) suggests that this observation might hold cross-culturally. He studied aggressive behaviors among children in two neighboring Zapotec-speaking communities of Mexico. Levels of violence, including homicide rates, were substantially higher in one of the Indian communities than in the other. Not surprisingly, youngsters from the more aggressive community engaged in considerably more actual fighting and rough play than did youngsters in the other community. Fry's research suggests that community differences in levels of aggression are perpetuated from one generation to the next as youngsters learn their community's patterns for handling and expressing aggression.

Questions

What role do early friendships play in a child's emotional development? What are some ways children can be encouraged to accept children with disabilities into the group? What are the research findings on aggressive behaviors in early childhood?

Preschools and Head Start

There is always one moment in childhood when the door opens and lets the future in.

Graham Greene, *The Power and the Glory*, 1940

With a majority of mothers working and with more children being born at risk during the 1990s, increasing numbers of children in the Western world are in early intervention programs, day care, and preschools long before they enter the public school arena (see Figure 8-5). That early childhood programs have a powerful impact on the development of young children is an understatement, and a flurry of research in emotional development of children is being undertaken with renewed vigor.

Edward Zigler (1994, p. x), says, "Today's children are spending less and less time with their parents, who are forced by economic necessity to work longer hours

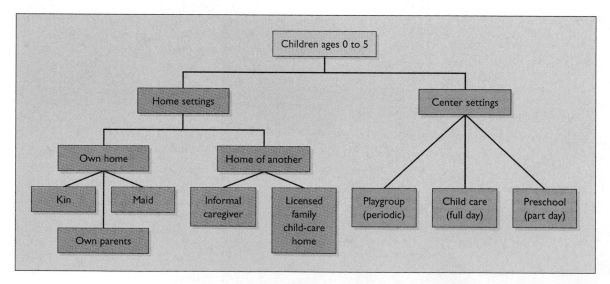

Figure 8-5 Early Childhood Care and Education Settings More young children in Westernized societies experience a diversity of child care settings, some informal and others regulated and licensed. The proportion of young children in out-of-home care while parents are working has risen. In 1993, 50 percent of toddlers aged 1 to 2 years were in nonparental care. By the time they enter kindergarten, 4 out of 5 have some preschool experience (West et al., 1995). Children in high-quality preschool classrooms display greater receptive language ability and pre-math skills. They also demonstrate more advanced social skills than those in lower quality classrooms (Helburn et al., 1995). The availability of quality child care has a direct impact on the employability of mothers (Bairraro & Tietze, 1993).

Source: Adapted from J. Bairraro, and W. Tietze, *Early Childhood Services in the European Community*, a report submitted to the Commission of the European Community Task Force on Human Resources, Education, Training, and Youth, October 1993.

and to place even their very young children in [nursery and] preschool settings. Thus the nature of the early childhood curriculum, its developmental appropriateness, and the emotional tone its practice creates are becoming increasingly important." The children in early childhood settings are just beginning to construct their personal social universes; they are experiencing for the first time many of their own emotional responses and those of others. A healthy goal at this stage of life should be to help young children learn desirable ways of expressing feelings and develop healthy patterns of understanding and regulating their emotions (Hyson, 1994).

Public kindergarten programs were originally established in the 1950s to support and nurture the young child's emotional and social development—to provide a healthy transition time from the emotional support at home for the cognitive tasks ahead. Zigler (1970) directed the earliest days of the *Head Start* program in the mid 1960s, with emotional development and the child's overall mental health being central to this program. The

Head Start program was an offshoot of the War on Poverty, which had been designed to provide children from low-income families with early intervention education in nursery school settings. Educators believed that appropriate services from outside the family could compensate for the disadvantages these youngsters experienced from the effects of poverty during their early years.

Significantly, in the 1960s and 1970s, Piaget's research promoted more extensive research in cognitive psychology. In the 1980s, lawsuits and charges of sexual abuse of young children in preschool and day-care programs forced such programs to focus on the young child's intellectual development, because caretaker displays of touching and physical affection and development of emotional connections with the children had become suspect. Many professionals and parents concerned about child development believe there has been an undue emphasis on formal academics (understanding symbols, time, number, volume, and space), while agreeing that a certain amount of intellectual preparation should be a component of these programs.

Much of what was said in Chapter 6 about child day-care centers also holds for preschools. Like child day-care programs, the preschool experience has been seen by a number of educators, child psychologists, and political leaders as a possible solution to many of the massive social problems of illiteracy, underachievement, poverty, and racism that confront Western nations.

In recent years, long-term data have become available that reveal that SES-disadvantaged children in such programs do indeed get a head start (Hofferth, 1997; Zigler & Styfco, 1994). Children who had participated in the preschool programs in the 1960s have performed as well as or better than their peers in regular school and have had fewer grade retentions and special-class placements. The project has also provided hundreds of thousands of young children with essential health care services and taught parents better parenting skills (Zigler & Styfco, 1994).

Although the improvement in children's IQ scores typically dissipates within several years, the children ultimately achieve a higher academic level than children lacking preschool instruction. Children who have been in compensatory education classes score significantly higher on mathematics achievement tests and have a better self-image than their peers in control groups. The programs have also produced dramatic "sleeper effects" on children later in adolescence and young adulthood. Youth who were in a Head Start program engage in less antisocial and delinquent behavior and are more likely to finish high school and get jobs or go to college than are their peers who were not in a program. Another significant impact of Head Start has been its effects on parents. The program has given parents access to community

Quality Head Start Programs Help   Children at-risk who have been enrolled in a quality early childhood program are 25 percent less likely to be retained a grade (Schweinhart, Barnes, & Weikert, 1993). Head Start has immediate positive effects on children's socioemotional development, including self-esteem, achievement motivation, and social behaviors (Schweinhart & Wikart, 1986). Seventy-two studies found Head Start to have sizable effects on children's cognitive development at the end of one year (Devaney, Ellwood, & Love, 1997). Researchers find that quality Head Start programs can make a long-term difference for poor youngsters that extends beyond the child's school years (Yoshikawa, 1995). Parental involvement in Head Start also contributes to their own positive growth.

resources, provided support for the entire family, and contributed to an improvement in parenting abilities and in later parent participation in school programs (Schweinhart, Barnes, & Weikert 1993).

In sum, the payoff of Head Start programs has been not only in education but in dollars, for in the long run its participants are less likely to need remedial programs as children and social support systems like welfare as adults. Apparently, children with preschooling learn how to extract a better education from the school system. Moreover, a program's effectiveness is increased by involving the parents, strengthening their ties to schools, and bringing them into partnership with the educational enterprise; many parents are themselves afraid of school or have had bad experiences with it, and these problems need be addressed lest the parents pass on their fears to their youngsters (Wasik et al., 1990). Nor will just any nursery school do—it must be a quality program. Many Head Start programs are marginal, staffed by people who are poorly paid and have little training, or run so poorly that they do virtually nothing to help youngsters (Zigler & Styfco, 1994). For information on studies that document the positive effects of Head Start for children at risk, follow the links for Research and Evaluation at the National Head Start Association Web site (see "Following Up on the Internet" at the end of the chapter).

Finally, as disadvantaged youngsters grow older, they continue to need the same comprehensive services and special attention as Head Start programs provided them in their early years. Our federal government is now providing some monies for Early Head Start programs for children from 0 to 3 and pre-kindergarten programs, after-school programs, and summer enrichment programs for this group of youngsters to continue developing their competencies to meet the academic curricula and behavioral standards of elementary schools.

Questions

Initially, what was the main goal of public kindergarten programs in the 1950s? Why has that goal changed over the years? What is the purpose of Head Start, and what population of children is served?

Media Influences on Children

For at least two decades, authorities have expressed concern over the effects of media violence on the behavior of the nation's youth (Kolbert, 1994; Slaby, 1994). In the United States, preschool children watch TV an average of 3 to 4 hours per day. Young children observe thousands of murder scenes, beatings, and sexual assaults in both televised and videotaped shows (Seppa, 1997).

Television

Average amount of time per day kids watch TV: 3 to 4 hours. Average amount of time kids spend doing homework: 26 minutes.

New York Times Magazine, October 8, 1995

Television has become a major socializing agent of American children, and studies have repeatedly found that television violence is as strongly correlated with aggressive behavior as any other behavioral variable that has been measured. The National Institute of Mental Health (Pearl, 1982), the American Academy of Pediatrics (1990), and many others have come to the same conclusion: Violence in the media contributes to childhood and adult aggression, especially for those already prone to aggressive behavior (Seppa, 1997; Huesmann & Miller, 1994). The nation's pediatricians ask parents to diminish their youngsters' television viewing by half or more because it also contributes to childhood obesity. Educators have also blamed TV for the nation's falling reading scores because they find a strong correlation between the lack of reading skills and excessive amounts of television viewing (Henry, 1993).

The ability of television to influence youngsters is compounded by several factors. For one thing, until they reach grade school, many children do not comprehend the motives behind the messages promoting products. For another, preschoolers and toddlers tend to believe what adults tell them. They lack the personal experience and cognitive development to question the accuracy of what they view (Wright et al., 1994). The harmful effects of violence are rarely portrayed in the media children watch.

According to the A. C. Nielsen Company, a firm that specializes in assessing the popularity of television programs, the television set stays on an average of 53 hours a week in homes with preschool children. This figure compares with 43 hours a week in the average U.S. household. Significantly, U.S. youngsters spend more time watching television than in any other activity except sleeping (Huston, Watkins, & Kunkel, 1989). Many of the programs directed toward child audiences are saturated with aggression, mayhem, and violence. For example, Bugs Bunny and Roadrunner cartoons are typical Saturday-morning fare and average 50 violent acts per hour. A survey by the Center for Media and Public Affairs, a nonprofit research organization, examined the programming available to a Washington, D.C., home wired for cable on April 7, 1994, and identified 2,605 acts of violence between 6 A.M., and midnight, the greatest number occurring in the early morning and in the afternoon, times when youngsters were most likely to be watching. Significantly, children watch high-quality informative programming such as "Sesame Street" less as they get older, and their cartoon and comedy viewing increases with age (Huston et al., 1990).

A variety of studies reveal that viewing media violence fosters aggressive behavior in a number of ways (Kolbert, 1994; Bandura, Ross, & Ross, 1963): (1) Media violence provides opportunities for children to learn new aggressive skills. (2) Watching violent behavior weakens children's inhibitions against behaving in the same way. And (3) television violence affords occasions for *vicarious conditioning,* in which children acquire aggressive behaviors by imaginatively participating in the violent experiences of another person. Research also demonstrates that media violence increases children's toleration of aggression in real life—a *habituation* effect—and reinforces the tendency to view the world as a dangerous place (Huesmann & Miller, 1994). Longitudinal research that has followed individuals over major portions of their lives, in some cases for up to 22 years, reveals that the viewing of large quantities of television violence by vulnerable youth is one of the best predictors of later violent criminal behavior (Slaby, 1994). In sum, an accumulating body of literature suggests that television provides a variety of entertainment for children but functions as a powerful negative socializer as well (Seppa, 1997; Huesmann & Miller, 1994).

Video Games and Computers

Many of the educational and psychological findings regarding television viewing extend to video games. In the early 1970s there was *Pong,* a simple electronic table tennis game. Then came *Pac-Man, Roadrunner, Space Invaders, Asteroids,* and others. By 1980 video games had developed such a considerable following among youngsters that physicians attributed a hand ailment—"space invaders cramp"—to excessive video game playing. A decade later, by 1990, Nintendo and Sega video games dominated children's play. Nearly one-third of U.S. homes owned a Nintendo or Sega set, and a survey of the 30 top-selling toys found that 20 were video-game related) (Provenzo, 1991). Also, in the late 1990s, many American families own a personal computer and have access to Internet and the World Wide Web. An Internet connection gives a child access to a vast array of educational and entertainment sites, but it also gives the child access to sites that are developmentally inappropriate

and sites that can be dangerous for children (for instance, some children have been victimized by pedophiles who logged on to children's chat rooms). Software manufacturers sell "filtering" software that allows parents to "block out" any site that contains the topics they select for blocking, but filters are not a fool-proof solution. To top it off, psychologists have now identified a psychological disorder labeled "Internet addiction," which affects children as well as adults.

Many parents are concerned that their child's preoccupation with video games and computers can lead to developmental drawbacks for the child. And there are ample grounds for concern (Hays, 1995; Sheff, 1993). Video games allow youngsters little opportunity to make decisions for themselves, to fashion their own fantasies, or to construct their own resolutions to problems. Most games do not reward individual initiative, creativity, or thought, nor do they allow sufficient freedom for youngsters to experiment with ideas, develop resourcefulness, and use their imaginations. Critics contend that video games provide very limited cultural and sensory stimulation and that many video games are appallingly violent and stereotypically sexist. In addition, children accustomed to the quick gratification that video games provide might be unwilling to put in the effort and long hours of practice that are necessary to play a musical instrument well or to excel at other endeavors.

Yet there is also room for optimism. Video game technology and multimedia educational software offer educators abundant opportunities for helping children, especially those with learning problems, expand their creative learning, thinking, and acting. Clearly, video games, computers, and Internet access, like television, are phenomena of tremendous import and significance in our society, and they demand our serious attention. And as we shall see when we turn to Part 5 and the next stage of development—middle childhood—computers are being used extensively in schools and homes to enhance instruction and train children with physical and cognitive disabilities.

Questions

How much television do you watch daily? Do you agree that television is a "silent" socializer?

Segue

In this chapter we discussed variables that can promote or detract from a young child's emotional and social development. Each child's whole development—physical, intellectual, emotional, and social—is unique. Even when adults attempt to rush the child

from one stage of development to another, these natural developmental processes will move along at the appropriate pace for each young child. A child's emotional self-regulation and self-image are influenced and reflected back by a variety of factors: the child's own developing sense of self; encouragement or discouragement from family members and caretakers; association, acceptance, or dissociation with peers; and, in Western cultures, the extent of media influence. Children need to feel accepted and loved by the significant people in their lives, if they are to maximize their own potential and to have a positive image of themselves. A healthy emotional-social foundation prepares young children for the demands of the middle childhood stage of development, from ages 7 to 12, which we will turn to next in Part 5.

Summary

Emotional Development and Adjustment

1. Recent research on emotions documents that all behavior, thought, and interactions are motivated by emotions. A child's emotions can hinder or support the child's engagement with and mastery of intellectual content.

2. Facial expressions, body language, and gestures appear with greater frequency during the preschool years, reflecting greater cognitive ability and awareness of cultural values and standards. Increased large- and fine-muscle control help the child express feelings in more complex ways. Voice qualities convey specific social-emotional messages.

3. Play allows a child to improve cognitive capacities and allows children to communicate their deepest feelings. There are many forms of play, including pretend play, exploration, games, social play, and rough-and-tumble play. Some young children create an imaginary companion. Boys tend to play in larger groups and seem to prefer rough-and-tumble, competitive play, whereas girls tend to prefer two-person groups conducive to verbal interaction and displays of affection.

4. Cultures and professionals vary in their views on the age at which children should be able to control their emotions. Self-regulation of emotions proceeds gradually at each child's own pace and is unlikely to appear until after age 3. Cultural norms vary for control of emotions by boys and girls. Many traditional gender roles are being redefined.

5. Young children demonstrate they can develop close, affectionate bonds with family members and other caregivers. One of the main goals of parents and preschool programs to enhance young children's self-esteem.

The Developing Sense of Self-Awareness

6. Emotional development is an essential component of self-awareness. A child's understanding of self as separate from others continues to develop during the preschool years.

7. Children come to view themselves as active agents in socializing with others. Self-control and social control are interrelated. While a 2-year-old is likely to view herself in terms of some aspect of her body, a 5- to 6-year-old can distinguish between her mind and her body. Some children might need adult intervention to raise their self-esteem and become better socialized.

8. Gifted children are likely to be self-directed and have a more mature self-awareness. They tend to be active learners and are able to empathize with others' feelings and emotional subtleties at earlier ages.

Gender Identification

9. Recognition of one's gender identity is a major developmental task during the first six years of life. All societies assign males and females specific gender roles. Socialization experiences help a young child develop a gender identity.

10. Hormones play a part in the display of gender behaviors. Testosterone tends to make boys more physically active, more aggressive, and less able to sit still. In general, boys also seem to be more logical, analytical, spatial, and mathematical, while girls tend to be more verbal, social, and somewhat more expressive emotionally. It might be premature to look only for biological factors to explain differences, because each child's family experience and cultural socialization patterns also influence gender behaviors. Some researchers propose that the most powerful factors influencing gender identity are environmental and not biological.

11. Psychoanalytic theory suggests that early experience plays a significant role in fashioning gender identity. Erikson argues that children exhibit a drive to act on their world with purpose. Cognitive learning theory proposes that modeling and imitation play a major role in gender identity. Cognitive developmental theory states that children acquire a mental model of a female or male and then adopt that model's gender-related characteristics. Mothers and fathers tend to treat their sons and daughters differently, promoting masculine or feminine traits.

Family Influences

12. Young children enter a society that is already an ongoing concern, and they need to be fitted to their society's cultural ways, in a process called "socialization."

13. Love-oriented parenting techniques tend to promote a child's conscience formation and responsibility. In contrast, parents who treat their child with hostility and rejection tend to interfere with the child's conscience development and to breed aggressiveness and resistance to authority.

14. Restrictive parenting tends to be associated with well-controlled, fearful, dependent, and submissive behaviors. Permissiveness, while fostering outgoing, sociable, assertive behaviors and intellectual striving, also tends to decrease persistence and increase aggressiveness. Effective discipline is consistent and unambiguous. The most aggressive children are those whose parents are permissive toward aggression on some occasions but severely punish it on other occasions.

15. Baumrind distinguishes among authoritarian, authoritative, permissive, and harmonious parenting. Authoritative parenting tends to be associated with children having self-reliance, self-control, an eagerness to explore, and contentment. In contrast, the offspring of authoritarian parents tend to be discontented, withdrawn, and distrustful.

16. One indication that 2-year-olds are coming to terms with autonomy issues is their ability and willingness to say *no* to parents. Self-assertion, defiance, and compliance are distinct dimensions of toddler behavior. The way parents handle these autonomy issues has profound consequences for their youngsters' behavior.

17. Divorce is a stressful experience for both children and their parents. Interactions between divorced parents and their children differ from those in intact families. Children often respond to their parents' divorce by pervasive feelings of sadness, strong wishes for reconciliation, and worries about having to take care of themselves.

18. Sibling relationships are normally significant throughout life. Firstborn children seem to be fortune's favorites. Three explanations have been advanced to account for differences between firstborns and later-borns: (a) Firstborns have greater exposure to adult models and to adult expectations and pressures, (b) they function as intermediaries between parents and later-borns, a role that appears to foster the development of verbal and cognitive skills, and (c) the "dethroning" of the firstborn by later siblings arouses a strong lifelong need for recognition, attention, and approval.

Peer Relationships

19. Peer relationships are in their own right a meaningful experience in the lives of young children. Such relationships become more likely to be formed and more successful with increasing age. Children are important to one another as reinforcing agents and behavioral models.

20. Play makes a number of major contributions to children's development. It functions as a vehicle of cognitive stimulation, allows children to handle the world on their own terms, provides for anticipatory socialization, fosters an individual sense of identity, and enables children to come to terms with their fears.

21. As children grow older, their aggression becomes less diffuse, more directed, more retaliatory, and more verbal.

Preschools and Head Start

22. Children who have participated in quality preschool and Head Start programs indeed do get a head start. They achieve a higher academic level than children who did not receive preschool instruction. The programs have also had positive "sleeper effects" on children later in adolescence and young adulthood.

Media Influences on Children

23. For the past two decades, research findings have suggested that television has become a major socializing agent of children and that televised violence is correlated with aggressive behavior in children. National child health organizations recommend limiting children's television viewing, and their use of video games and computers, in order to improve their mental health and academic performance.

Key Terms

aggression (253)	confluence theory (250)	gender stereotypes (234)
authoritarian parenting (243)	gender identity (231)	harmonious parenting (243)
authoritative parenting (243)	gender roles (230)	initiative versus guilt (233)

joint custody *(248)*

permissive parenting *(243)*

peers *(251)*

play *(224)*

resource dilution hypothesis *(250)*

scaffolding *(243)*

self *(227)*

self-concept *(229)*

self-esteem *(227)*

sexual abuse of children *(240)*

socialization *(236)*

Following Up on the Internet

Web sites for this chapter focus on early childhood emotional, self-esteem, gender, & identity issues. Please access the text web site at http://www.mhhe.com/crandell7 for up-to-date hot-linked Internet addresses for the following organizations, topics, and resources:

Directory of Children's Issues on the World Wide Web

Early Childhood Care and Development

American Library Association Links for Parents and Caregivers

Inclusion in the Preschool Setting

National Head Start Association

Part Five

Middle Childhood

7 to 12

Middle childhood, the time during the elementary school years, is a period of slower physical growth but faster intellectual development than what occurred during the preschool years. In Chapter 9 we will see that children experience greater cognitive sophistication, acquire more socialization skills, and understand and cope with their emotions better. Cognitive maturation allows the child to deal with fearful and stressful events or to model prosocial behaviors. In Chapter 10 we will see that children at this stage are industrious, inquisitive, and more socially aware than before. Peer groups begin to exert a stronger influence on preadolescent children than in earlier years. Youth become more aware of social standings such as popularity, acceptance, and rejection. Most children learn to control their emotions in school, in other group settings, or when around peers and friends. Some children, however, need help with academic performance or behavioral control.

Chapter 9

Middle Childhood
Physical and Cognitive Development

Critical Thinking Questions

1. What does it mean to say someone is gifted or a genius? For example, if Mozart had been taken at the age of 20 to an island in the Pacific where the inhabitants did not know or understand Western music, if he had no musical instruments and no means of expressing his musical talents except by humming or singing, would the islanders have considered him to be a genius? Or would they have thought he was disabled?

2. How do you perceive others? Think back to when you were a child and compare the perception you had of your parents then to the perception you have now. In what ways has your perception of them changed?

3. If you were going to try to teach a "hardened criminal" how to be moral, which of the following approaches would you use, and how would you implement your plan? (1) Place the criminal in a very moral environment and assume that morality would rub off. (2) Teach the criminal the basic principles of morality and then send the person to live in mainstream society. (3) Use examples of crime and deviancy to teach about morality and ethics.

4. Why don't we hold children to adult standards of morality? Why don't we treat adults like children and give them the same treatment for offenses?

In the middle childhood years, children get to rest up a bit after the dramatic physical development of early childhood and prepare for the onset of puberty and adolescence. From a developmental perspective, the changes during this time appear so smooth and uneventful that we might think nothing is happening. The greatest changes will be in cognitive growth, and we will discuss concrete operational thought, intelligence and its measurement, individual differences, and children with special learning needs.

These children capably use complex classifications and enjoy making collections of everything from sports cards to butterflies. They start paying attention to counts and amounts, what is bigger, who has more, and their knowledge of the physical world grows by leaps and bounds. We are also beginning to realize that a significant part of children's work during these years involves learning appropriate cultural and social skills. The patterns and habits of social interaction established now not only will affect the child's adolescence but also will carry into adulthood. We conclude this chapter with an examination of language skills and moral development, two significant issues in contemporary American society.

Physical Development

We went home and when somebody said, "Where were you?" we said, "Out," and when somebody said, "What were you doing until this hour of night?" we said, as always, "Nothing."

. . . but about this doing nothing: we swung on swings. We went for walks. We lay on our backs in backyards and chewed grass. I can't number the afternoons my best friend and I took a book apiece, walked to opposite ends of his front porch, sank down on a glider at his end, a wicker couch at mine, and read. We paid absolutely no attention to each other, we never spoke while we were reading, and when we were done, he walked me home to my house, and when we got there I walked him back to his house, and then he—

We strung beads on strings: we strung spools on strings; we tied each other up with string, and belts and clothesline. We sat in boxes; we sat under porches; we sat on roofs; we sat on limbs of trees. We stood on boards over excavations; we stood on tops of piles of leaves; we stood under rain dripping from the eaves; we stood up to our ears in snow. We looked at things like knives and immies and pig nuts and grasshoppers and clouds and dogs and people. We skipped and hopped and jumped. Not going anywhere—just skipping and hopping and jumping and galloping. We sang and whistled and hummed and screamed.

Writer Robert Paul Smith,
describing his elementary school years

Just as children have a range of play experiences during childhood, there are individual variations in physical development as well as variation due to gender or ethnic background. For example, African Americans tend to mature more quickly (as measured by bone growth, percent fat, number of baby teeth) than Americans of European descent. Asian Americans appear to have the slowest rate of physical change and are less likely to show signs of puberty during middle childhood. Although variations in size and maturity are normal, children who fall at one extreme or the other on the continuum can feel deficient because of their physical differences. These feelings are often reinforced by peers who rate attractiveness and popularity on the basis of physical competence and appearance (Hartup, 1983).

Growth and Body Changes

Children grow more slowly during the years of middle childhood than in early childhood or in adolescence. With adequate nutrition, the typical child gains about five or six pounds and grows approximately two inches per year. Girls and boys have similar growth patterns, except that girls tend to have more body fat, and they mature a bit faster than boys.

If we look at children during this time period, they appear thinner or slimmer because as they grow taller, their body proportions change. Muscles become bigger and stronger, and children can kick and throw a ball farther than in the earlier years. There is an increase in lung capacity that gives greater endurance and speed, of which children make full use. Some variations in height, strength, and speed are due partly to nutrition, particularly in developing countries, but most of the differences among children are the result of heredity. There are differences not only in size but also in rate of maturation. This is particularly noticeable at the end of middle childhood when some children begin to undergo the changes of puberty and find themselves quite different from their peers in shape, strength, and endurance.

Question

What are the expected growth changes for a child during the middle childhood years?

Motor Development

During their middle years, children become more skilled in controlling their bodies. Their rate of physical growth has slowed down temporarily, giving them time to feel comfortable with their bodies (unlike earlier years) and an opportunity to practice their motor skills and increase their coordination. Seven- or 8-year-olds might still have difficulty judging speed and distance, but they have improved their skills sufficiently to be successful in games like soccer and baseball. Jumping rope, skateboarding, rollerblading, and riding bicycles are also activities they can enjoy. Which specific skills are developed depends somewhat on the children's environment (e.g., whether they learn games for snow or the tropics, whether the culture favors soccer or football). Whatever activities a child chooses, they will lead to greater coordination, speed, and endurance. Gender differences are minimal during this time period, although girls tend to have greater flexibility and boys have greater forearm strength: Ask a girl to jump rope, and she can do it endlessly; ask a boy to jump rope, and he will be more awkward at it (which will frustrate him to no end!); but generally most boys will throw a ball better than most girls. Age and experience are much more important determinants than gender, and we can see this in team sports where girls and boys are equally likely to score goals or hit home runs and both enjoy doing cartwheels, somersaults, and other gymnastic maneuvers.

Brain Development

As the brain and nervous system develop during this time period, the capacity, speed, and efficiency of the child's mental processes typically increase as well. Memory span increases fairly steadily over the years of childhood, but more recent research has shown that children might not be developing an increase in their memory capacity, but instead might be making more efficient use of their storage space (Schneider & Pressley, 1989). One type of improved efficiency is simply faster response time (Kail, 1991). Faster response is due in part to the physical development that occurs in the brain, but also to the fact that children at this age are becoming adept at using more cognitive strategies to help them solve more complex tasks.

Studies utilizing imaging of the brains of young children have revealed evidence that children's brains appear to be organized differently than adult brains. A stroke in a young child of 6 or 7 might have no subsequent effect on the child's language development, whereas the same kind of stroke in an adult would normally cause permanent loss of language abilities. One theory is that in children the two hemispheres of the brain are "backups" for each other in the event of injury. Studies of children who have had one hemisphere removed indicate that these children continue to display excellent physical and mental development.

Question

What are the changes in body growth and brain maturation during these years?

Dyslexia Some children experience more difficulty with processing visual or auditory information. **Dyslexia** is a disorder in which an otherwise normally intelligent, healthy child or adult has extreme difficulty learning to read. This type of brain dysfunction affects at least 5 percent of school-age children. Neuroscientists are studying this brain disorder and so far have discovered from imaging, diagnostics, and limited autopsies of dyslexics that in the language area of the cortex in the left hemisphere, the layers are disorganized, whirled with primitive, larger cells. For you to understand what it must be like to have dyslexia, turn this book upside down and try to read the words fluently. We think you will find it frustrating and quite a chore. This is what reading tasks are like for children with dyslexia, who are at a higher risk of lower achievement in school, poor self-esteem, becoming school dropouts, and limiting their occupational opportunities. Interestingly enough, some of these children are also identified as *gifted* or *talented*.

Genius and Giftedness Neuroscientists and psychobiologists have found that more efficient brains have rich neuronal interactions and a multiplicity of synaptic connections. It has been hypothesized that child geniuses might have more complex synaptic connections in the association areas of the cerebral cortex or that their neurochemical transmissions might be more efficient. The areas in which some children have displayed genius range from piano playing to mathematical problem solving. A standard IQ score of 130 or above, or scoring in the 90th to the 99th percentile in reading or math on standardized achievement tests, usually qualifies a child for a pullout program, in-class enrichment, or advancing a grade.

Children who are identified as gifted during the middle childhood years of 7 to 12 typically need additional intellectual challenges to avoid boredom in their educational and home environments. Many, but not all, schools provide a special program for these students. Some colleges and universities offer special weekend and summer programs to stimulate the sharp minds of these students. On the East Coast, for example, Johns Hopkins University in Baltimore, Maryland, sponsors the Institute for the Academic Achievement of Youth (IAAY)–Center for Talented Youth (CTY) and conducts annual talent searches. The Johns Hopkins Talent Search Model looks for exceptional mathematical and/or verbal reasoning abilities among second-through eighth-graders in order to provide opportunities for full development of these children's mathematical and language abilities. Other national and international programs for highly intelligent, creative youth include, among others, the Odyssey of the Mind program, the Science Olympiad, the Invention Convention, National Geographic Kids Network, and MENSA programs. Those of you who are preparing to become educational psychologists, school psychologists, preschool to high school teachers, or future school administrators may want to find out more about the special needs of gifted and talented youth by contacting one of these special programs nearest you.

Health and Fitness Issues

Typically children are much healthier in the middle childhood years than at any time since birth. The rate of illness is lower, with most children in elementary schools reporting four to six acute illnesses in a year. The most common childhood illness is upper respiratory infection. Accidents, rather than illness, are the major cause of death or serious injury during this time period (and the most common cause is being struck by a moving car). Table 9-1 shows that mortality rates for children aged 10 to 14 have been decreasing over the years. The good news is that since 1960, overall child mortality rates have been declining for this age group for both male and female children and for black and white races. However, the mortality rate for black youth is still considerably higher than for white youth.

Children in middle childhood still need adult supervision and care. The greatest number of sport injuries for

Table 9-1   Decreasing Mortality Rates for Youth Aged 10 to 14 from 1960 to 1994 by Gender and Race

	1960	1965	1970	1975	1980	1985	1990	1991	1992	1993	1994
Gender											
Male	55.0	50.9	51.3	44.9	38.3	35.0	31.6	32.9	30.7	31.7	31.2
Female	32.6	29.7	29.5	25.3	22.9	20.6	20.2	18.2	18.2	19.2	18.8
Race											
White	41.4	38.6	38.4	33.7	29.8	27.0	24.3	24.2	24.2	22.8	23.0
Black	—	—	54.6	44.3	36.6	34.8	36.6	36.4	35.3	37.2	37.9

Source: National Center of Health Statistics (NCHS), unpublished data provided by the Statistical Resources Branch. P. Gardner and B. L. Hudson, "Advance Report of Final Mortality Statistics, 1993," National Center for Health Statistics, 1996. Published in *Trends in the Well-Being of America's Children and Youth,* 1997 edition (Washington, DC: U.S. Department of Health and Human Services).

children in this age group occur when they are playing basketball, football, or baseball, bicycling, and using playground equipment. Injuries from backyard trampolines have become "a national epidemic" (G. Smith, 1998). One possible explanation for the high incidence of injuries during these years is that middle childhood is the time when children test the limits of their bodies and skills but don't yet have cognitive awareness of the potential dangers of their activities.

There is a growing concern about violence perpetrated against or by this age group, particularly its older members. In 1995, juvenile violent crime arrests (children 10 to 17 years old) increased by as much as 28 percent, or by more than the state median in 17 states. It appears that youth are becoming involved in senseless violent crime at younger ages. Research on violence indicates that violent or aggressive behavior is often learned early in life, and steps need to be taken in families and communities to help children deal with their emotions without using violence. Consequently, many teacher education programs are training future teachers in *conflict resolution* methods. One of the strongest findings was that children "do what they see," and the recommendation was made that we take care not to be models for violence (American Academy of Pediatrics and American Psychological Association, 1996). The American Psychological Association (APA), in conjunction with other social science groups, produced a document designed to help parents work within the family, school, and community to reduce youth violence.

Obesity   Another risk to childhood health is obesity. Generally, **obesity** is defined as being 20 pounds over typical weight for height by gender. Ten to 20 percent of all U.S. children meet this criterion, and this is a substantial increase since 1960 (Wolf et al., 1993). Obesity is a physical and medical problem at any age, for the obese

person runs the risk of greater illnesses, including cardiac and respiratory disease. Being obese in childhood greatly increases the risk of being obese as an adult. This is particularly true if the child is overweight during the elementary school years (Kolata, 1992d).

Obesity appears to have three basic causes: heredity factors, activity, and diet. Studies of twins separated at birth by adoption and reunited in adulthood indicate that children have a genetic tendency toward fatness or thinness. Adult identical twins have much more similar body weights than fraternal twins (Stunkard, 1990). Likewise, adopted children reared by obese parents are not as likely to be obese as are the natural children of these parents. Activity and diet play some role in determining whether this genetic tendency will result in obesity. It is clear that obese adults and children take in more calories and exercise less than others of the same age, socioeconomic status, and gender who are not overweight. For example, Armstrong and colleagues (1998) report that the prevalence of obesity increases roughly 2 percent for each additional hour of television a child or teenager watches per day. Obesity is a hazard not only to physical health but also to mental health. In the elementary school years, overweight children are often picked on, teased, and rejected by their peers, partly because physically attractive children generally are better liked by their peers and are viewed as more competent. We, as the adult caretakers, have a responsibility to nourish our children with balanced meals, including fresh fruits and vegetables. Additionally, if we maintain a healthy lifestyle, exercise regularly, and eat low-fat, nutritious foods, our children will usually develop these habits also.

Anorexia and Bulimia   Though anorexia and bulimia are thought to mainly affect adolescents and young adults, there is evidence that a small percentage of middle school children, mainly females, are developing these eating disorders. Denying the body the nutrients it needs to grow and develop has far-reaching negative consequences for these young girls. It would be advantageous for parents, teachers, and school nurses to learn about the typical signs and causes of this disorder. We will discuss anorexia and bulimia in greater detail in Chapter 11.

Role of Play and Exercise   All children who attend public schools are required to participate in physical education classes during the regular school year, unless they have an illness or disability that prevents their participating. Additionally, most middle schools offer extracurricular sport activities for both boys and girls, such as indoor and outdoor soccer, basketball, track, football, baseball or softball, volleyball, and tennis. Extracurricular offerings depend upon the size and financial resources of the school district and the volunteerism of parents to assist with coaching teams. The number of elementary and middle-school girls who are now participating in sports has greatly increased recently. Armstrong and colleagues (1998) indicate that increased physical activity promotes overall fitness, improves cognitive functioning, and builds self-esteem. All children

should develop a regular routine of physical activity and maintain it throughout high school, and throughout life, to keep their immune and cardiovascular systems healthy. We now know that physical activity and play contribute to a positive attitude, reduce the effects of stress, and improve overall mental health (for us older kids, too!). Unfortunately, too many children are junior "couch potatoes," choosing to spend hours in front of the television instead of playing with others after school. Research has documented that children 6 to 11 years of age average more than 23 hours of television viewing per week (Armstrong et al., 1998).

Most children this age love to play, which usually means running and throwing balls as in touch football or dodge ball, wall ball, skateboarding, riding bikes, hopscotch, jumping rope, dancing, and gymnastics. Healthy girls and boys of this age have a difficult time sitting in a classroom all day without being able to expend their tremendous energy. For some—but not all—gym class is their favorite activity of the week because that is where they experience their greatest success!

> **Questions**
>
> *What are the common causes of childhood illness and death? What factors promote healthy development?*

Cognitive Development

An important feature of the elementary school years is an advance in children's ability to learn about themselves and their environment (Crick & Dodge, 1994). During this period they become more adept at processing information as their reasoning abilities become progressively more rational and logical (Schwanenflugel, Fabricius, & Alexander, 1994; Flavell, 1992).

Cognitive Sophistication

A crucial component of reasoning abilities is being able to distinguish fiction, appearance, and reality (Woolley & Wellman, 1993). Consider, for instance, what happens if you take hold of a joke sponge that looks like a solid piece of granite. Although it looks like a rock, you realize it is a sponge the moment you grasp it. But a 3-year-old is less certain. Very young children often do not grasp the idea that what you see is not necessarily what you get. But by the time they are 6 or 7 years old, most children appreciate the appearance-reality distinction that confronts us in many forms in everyday life (Flavell, Flavell, & Green, 1983).

Underlying the improvement in children's intellectual capabilities is a growing awareness and understanding of their own mental processes, what psychologists

Healthy Children Enjoy Expending Energy on Sports and Other Activities

term *metacognition* (Lyon & Flavell, 1994). In many respects Jean Piaget's developmental stages constitute sets of so-called **executive strategies** analogous to the tasks performed by a corporate executive. Executives select, sequence, evaluate, revise, and monitor the effectiveness of problem-solving plans and behavior within their corporations. In human mental functioning, executive strategies integrate and orchestrate lower-level cognitive skills. At higher stages the strategies are more complex and more powerful than at lower stages.

Piaget's Period of Concrete Operations   From Piaget's perspective, there is a qualitative change in children's thinking during middle childhood as children begin to develop a set of rules or strategies for examining the world. Piaget calls middle childhood the **period of concrete operations.** By an "operation" Piaget means an integration of the powerful, abstract, internal schemas such as *identity, reversibility, classification,* and *serial ordering.* The child begins to understand that adding something results in more, objects can belong to more than one category, and these categories have logical relationships. Such operations are "concrete" because children during these years are bound by immediate physical reality and cannot transcend the here and now. Consequently, during this period children still have difficulty dealing with remote, future, or hypothetical (or abstract) matters.

Despite the limitations of concrete operational thought, during these years children make major advances in their cognitive capabilities. For example, in the preoperational period before 6 or 7 years of age, children arrange sticks by size in their proper sequence by physically comparing each pair in succession. But in the period of concrete operations children "mentally" survey the sticks, then quickly place them in order, usually without any actual measurement. Because the activities of preoperational children are dominated by actual perceptions, the task takes them several minutes to complete. Children in the period of concrete operations finish the same project in a matter of seconds, because their actions are directed by internal cognitive processes: They can make the comparisons in their minds and do not have to physically place and see each stick side by side.

Conservation Tasks   The difficulty younger children had solving conservation problems is due to the rigidity of their preoperational thought processes. Let's look at how children use concrete operations to solve these kinds of problems when they are a bit older. **Conservation** requires recognition that the quantity of something stays the same despite changes in appearance. It implies that children are mentally capable of compensating in their minds for various external changes in objects. Elementary school children come to recognize that pouring

A Conservation Experiment   Place two balls containing the same amount of clay in front of a 4-year-old child and ask if they are the same size. Invariably the child will say yes. As the child watches, roll one of the balls into a long, sausage-like shape and again ask if the clay objects are the same size. The child will now claim that one of the clay objects is larger than the other. In this case, the child said that the sausage-shaped clay is larger than the clay ball. Not until children are a few years older will they come to recognize that the two different shapes contain the same amount of clay—that is, they will come to recognize the principle of the conservation of quantity.

liquid from a short, wide container into a long, narrow one does not change the quantity of the liquid; they understand that the amount of liquid is conserved. Whereas preoperational children fix ("center") their attention on either the width or the height of the container and ignore the other dimension, concrete operational children *decenter,* attending simultaneously to both width and height. Furthermore, concrete operational children assimilate *transformations,* such as to the gradual shift in the height or width of the fluid in the container as it is poured by the experimenter. And most important, according to Piaget, they attain *reversibility* of operations. They recognize that the initial state can be regained by pouring the water back into the original container. (Decentering, transformations, and reversibility were discussed in Chapter 7.)

Question

How does the acquisition of concrete operations allow a child to perform Piaget's conservation tasks?

Classification, or the concept of class inclusion, is usually understood by the age of 7 or 8. This development in children can be seen when using a game similar to "20 Questions." Children ages 6 to 11 were shown a group of pictures and were to ask questions to figure out which one the researcher "had in mind." ("Is it alive?" "Is it something you would play with?" etc.) The questions asked were divided into two basic types: strategy (categories) and no strategy (guessing). Six-year-olds typically resorted to guessing particular objects ("Is it a banana?"), older children asked categorical questions ("Is it something that you eat?") (Denney, 1985).

According to Piaget, children in the period of concrete operations develop the ability to use *inductive logic:* Given enough examples or provided with multiple experiences, children can come up with a general principle of how something operates. For example, if they are given enough problems of the form of *3 + 4 = ?* and *4 + 3 = ?,* they can induce that the order of the numbers in addition is not significant. They know that they will always get the same answer, 7. On the other hand, they are not very good at moving from the general to the specific, which is *deductive logic.* Starting with a concept or rule and then generating its application is difficult for them because it involves imagining something they might never have experienced. Despite the cognitive advances made in concrete operations, children in this period are still "concrete" and in some degree dependent on their own observations and experiences.

As is true of other cognitive abilities, children acquire some conservation skills earlier, some later (see Figure 9-1). Conservation of discrete quantities (number) occurs somewhat before conservation of substance. Conservation of weight (the heaviness of an object) follows conservation of quantity (length and area) and is in turn followed by conservation of volume (the space that an object occupies). Piaget calls this sequential development, with each skill dependent on the acquisition of earlier skills, *horizontal décalage.*

Horizontal décalage implies that repetition takes place within a single period of development, such as the period of concrete operations. For example, a child acquires the various conservation skills in steps. The principle of conservation is first applied to one task, such as the quantity of matter—the notion that the amount of an object remains unchanged despite changes in its position or shape. But the child does not apply the principle to another task, such as the conservation of weight—the notion that the heaviness of an object remains unchanged regardless of the shape that it assumes. It is not until a year or so later that the child extends the same type of conserving operation to weight.

The general principle is the same with respect to both quantity and weight. In each case children must perform internal mental operations and no longer rely on actual measurement or weighing to determine whether an object is larger or weighs more. However, children typically achieve the notion of the invariance of quantity a year or so before that of the invariance of weight (Flavell, Flavell, & Green, 1983).

Post-Piagetian Criticism The difference in acquisition times does not necessarily fit with Piaget's notion of stages. If, as Piaget says, each stage is a "cohesive whole," then we should find that children of any given age apply similar logic to a wide range of problems. A child who has concrete operations should be able to use operational logic on all tasks presented. Similarly, for children who are capable of operational logic, the amount of particular knowledge they have about the content of a task should not affect their use of these basic operations. But research has found that this is not really the case: It appears to make a difference how much experience or expertise a child has with the objects involved, for children can perform more complex operations on tasks they are familiar with as opposed to ones that are novel to them (Chi, Hutchinson, & Robin, 1989).

Other research has focused on whether the development of concrete operations can be accelerated, not just by general experience with objects, but through specific training on how to do conservation tasks. Jerome S. Bruner (1970) states: "The foundations of any subject may be taught to anybody at any age in some form." Learning theorists reject Piaget's stage formulations and disagree with Piaget's view that children below the age of 6 cannot benefit from experience in learning conservation because of their cognitive immaturity.

In the early 1960s any number of psychologists attempted to teach young children conservation skills. For the most part, they were unsuccessful. Subsequently, a number of researchers successfully used cognitive learning methods to train children in conservation. Furthermore, psychologists are finding that the content of a task decisively influences how a person thinks (Spinillo & Bryant, 1991). By altering the cognitive properties of a task, one often can elicit preoperational, concrete operational, or formal operational thinking from a child (Chapman & Lindenberger, 1988; Sternberg & Downing, 1982). Issues of this sort have stimulated research in *creativity.* For more on this subject, see *Further Developments:* "Creativity."

Questions

Is there a general consensus among psychologists and educators on how to define creativity? Do creative children tend to follow a single pattern, or do they work in different ways? What elements must be evident for a thought or invention to be considered creative? Do you think U.S. schools foster or stifle creativity?

Conservation Skill	Basic Principle	Test for Conservation Skills	
		Step 1	Step 2
Number (Ages 5 to 7)	The number of units in a collection remains unchanged even though they are rearranged in space.	 Two rows of pennies arranged in one-to-one correspondence	 One of the rows elongated or contracted
Substance (Ages 7 to 8)	The amount of a malleable, plastic-like material remains unchanged regardless of the shape it assumes.	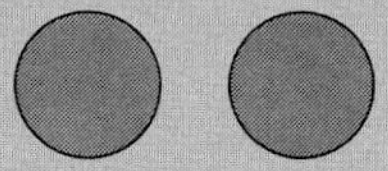 Modeling clay in two balls of the same size	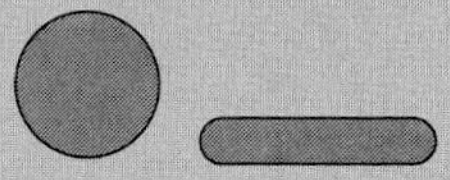 One of the balls rolled into a long, narrow shape
Length (Ages 7 to 8)	The length of a line or object from one end to the other end remains unchanged regardless of how it is rearranged in space or changed in space.	 Strips of cloth placed in a straight line	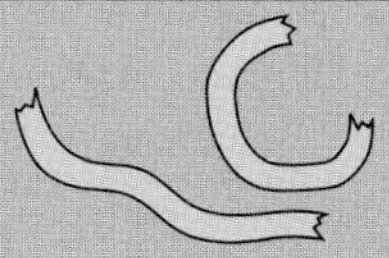 Strips of cloth placed in altered shapes
Area (Ages 8 to 9)	The total amount of surface covered by a set of plane figures remains unchanged regardless of the position of the figures.	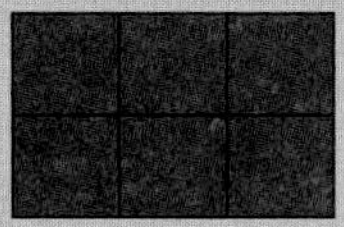 Square units arranged in a rectangle	 Square units rearranged
Weight (Ages 9 to 10)	The heaviness of an object remains unchanged regardless of the shape that it assumes.	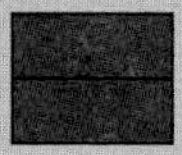 Units placed on top of each other	 Units placed side by side
Volume (Ages 12 to 14)	The space occupied by an object remains unchanged regardless of a change in its shape.	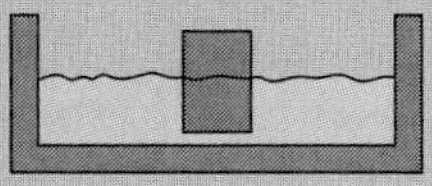 Displacement of water by object placed vertically in the water	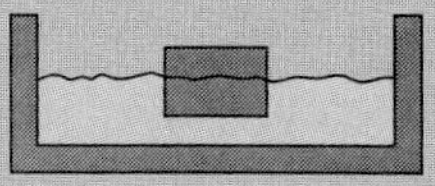 Displacement of water by object placed horizontally in the water

Figure 9-1 Sequential Acquisition of Conservation Skills During the period of concrete operations, Piaget says children develop conservation skills in a fixed sequence. For example, they acquire the concept of conservation of number first, then that of substance, and so on.

Further Developments

Creativity

We commonly value creativity as the highest form of mental endeavor and achievement. Whereas intelligence implies quick-wittedness in learning the predictable, creativity implies original and useful responses. Often we assume that high intelligence and creativity go hand in hand, yet psychologists find that high intelligence does not ensure creativity. Research reveals that differences between above-average and very high scores on intelligence tests have at best only a low association with creativity. Even so, although high intelligence does not guarantee creative activity, low intelligence seems to work against it. Above-average intelligence—although not necessarily exceptional intelligence—seems essential for creative achievement (Sternberg, 1988a).

In some cases too much brainpower can even get in the way of creativity. Psychologist Dean Keith Simonton (1991) looked into the lives of many renowned creators and leaders of the past century and concluded that the optimal IQ for creativity is about 19 points above the average of people in a given field. Nor is formal education essential. Many famous scientists, philosophers, writers, artists, and composers never complete college. Formal education often instills rote methods for doing things that blind people to offbeat but creative solutions. Albert Einstein describes the stifling effects that formal education had on his early scientific creativity: "The hitch in this was the fact that one had to cram all this stuff into one's mind for the examinations, whether one liked it or not. This coercion had such a deterring effect on me that, after I had passed the final examination, I found the consideration of any scientific problem distasteful for an entire year" (Einstein, 1949, p. 17).

Creative people seldom have bland personalities. Psycholinguist Vera John-Steiner (1986) interviewed 100 men and women active in the humanities, arts, and sciences, and she sifted through the notebooks, diaries, and biographies of creative individuals such as Albert Einstein and Leo Tolstoy. She finds that scientists and artists almost invariably mention that their talent and interest were revealed early in life and that they were often encouraged and nurtured by their parents or teachers. Even so, a good many creative individuals do not spend their childhood years basking in parental love and warmth. Instead, rather cool and even detached relationships are commonplace between parents and their creative sons and daughters. Perhaps such parents inadvertently encourage a rebellious attitude in their children that fosters independent thinking and action.

Most psychologists agree that a natural gift is not sufficient to produce creative effort. What seems to be required is the fortuitous convergence of innate talent and a receptive environment (Storfer, 1990; Greeno, 1989). According to one view, creativity requires that a person reorganize a tie or connection with some situation in the world, so creativity entails much more than a reorganization of the thinking apparatus. Creative accomplishment is exceedingly difficult precisely because it necessitates changing not merely the contents and organization of one's mind but an inclusive system of relationships involving the world, other people, or an established set of cultural concepts (Csikszentmihalyi, 1997). For this reason it may be much easier to stifle creativity than to stimulate it.

When people are inspired by their own interests and enjoyment, they are more likely to explore unlikely paths, take risks, and in the end produce something unique and useful. Dr. Salvador E. Luria of the Massachusetts Institute of Technology, mentor to several famous scientists and himself a Nobel laureate, says, "The most important thing is to leave a good person alone" (see Haney, 1985). Adds Dr. Mahlon B. Hoagland, scientific director of the Worcester Foundation for Experimental Biology, "I've often said that running a scientific institution is a lot like running an artist colony. The best an administrator can do is leave people alone to do what they want to do"(quoted in Haney, 1985, p. C-1). And although creative people may have innate talent, they must nurture their creativity with discipline and hard work.

There are a number of tips that parents and teachers can use to encourage creative thinking and originality in children:

- Respect children's questions and ideas.
- Respect children's right to initiate their own learning efforts.
- Respect children's right to reject, after serious consideration, the ideas of caretakers in favor of their own.
- Encourage children's awareness and sensitivity regarding environmental stimuli.
- Create "thorns in the flesh"—confront youngsters with problems, contradictions, ambiguities, and uncertainties.
- Give children opportunities to make something and then do something with it.
- Give youngsters opportunities to communicate what they have learned and accomplished.
- Use provocative and thought-producing questions.
- Encourage children's sense of self-esteem, self-worth, and self-respect.

In short, situational, cognitive, motivational, and personality characteristics all play a part in creativity.

Source: Creative Competitions, Inc.

Cross-Cultural Evidence   Over the past 40 years, children throughout the world have served as subjects for a variety of experiments designed to test Piaget's theory. Research has been conducted in over 100 cultures and subcultures, from Switzerland to Senegal and from Alaska to the Amazon (Feldhusen, Proctor, & Black, 1986). The results show that regardless of culture, individuals do appear to move through Piaget's hierarchical stages of cognitive development—the sensorimotor, preoperational, concrete operational, and formal operational periods—in the same sequence. Some cultural groups do not attain the state of formal operations, though. But cross-cultural research suggests that there is a developmental lag in the acquisition of conservation among children in non-Western, nonindustrialized cultures. Not entirely clear is whether this lag is due to genuine differences among cultures or to flaws in research procedures that use materials and tasks alien to some cultures.

The research also raises a question about whether the acquisition of conservation skills in the period of concrete operations occurs in the invariant sequence (horizontal décalage) postulated by Piaget. Children in Western nations, Iran, and Papua, New Guinea, exhibit the expected Piagetian pattern. Thai children, however, appear to develop conservation of quantity and weight simultaneously (Boonsong, 1968). And some Arab, Indian, Somali, and Australian aborigine subjects conserve weight before quantity (deLemos, 1969; Hyde, 1959).

Other cultural differences are also found. Children of pottery-making families in Mexico perform better on conservation of substance tasks than their peers from non-pottery-making families (Ashton, 1975). And Patricia M. Greenfield (1966) found in her studies involving Wolof children in Senegal, West Africa, that it made a difference whether the experimenter or the children themselves poured the water in the classic Piagetian experiment involving wide and narrow containers. Two-thirds of a group of children under 8 years old who themselves transferred the water achieved the concept of conservation. In contrast, only one-fourth of a group who had watched the experimenter pour the water then realized that the amount of water was the same. The children attributed to the experimenter's performance a "magical action" that they did not attribute to their own performance. It seems, then, that cultural and social factors play a role in children's cognitive development.

Information Processing: Another View on Cognitive Development

In chapter 7, we indicated that some theorists will ask children to report what they are doing intellectually while they perform a task. How does the child process the information she or he is given to solve a problem? How do these intellectual processes change with age? This is quite different from Piagetian theorists who ask what overall structure of logic the child uses and how these structures change over time. According to the information-processing view, we need to understand whether there is a change in the basic processing capacity of the system (hardware), and/or whether there is a change in the type of the programs (software) used to solve a problem. For instance, there are certain limitations to the number of operations a computer can perform at one time and to the speed at which it can perform them. Just like an executive, children at this age might be better at dividing tasks into more manageable segments. For example, with regard to memory, children at this age might become more efficient at using the mnemonic strategies of *rehearsal* and *categorization,* as discussed in Chapter 7.

Individual Differences   Much of our thinking so far has centered on how the typical child develops cognitively through this middle childhood period. We have described the changes in the use of concrete operations with advancing age and have shown that the developmental process is predictable to a certain extent, although there is still debate as to what mechanism explains this growth. We have also noticed variations in the rate at which children change and the overall amount of change that takes place in this period. Piaget believed that only 30 to 70 percent of adolescents and adults perform at the formal operation level (Piaget, 1963). Likewise, some children take quite a while to develop operational logic for any task, whereas other children learn some academic skills quite easily and yet struggle with other skills that appear quite similar operationally. School and home environments alone do not seem to account for the individual variability in conceptual ability.

Children's Perception of Others   The elementary school years are a time of rapid growth in children's cognitive understanding of the social world and of the requirements for social interaction (Crick & Dodge, 1994). Consider what is involved when we enter the wider world (Vander Zanden & Pace, 1984). We need to assess certain key statuses of the people we encounter, such as their age and sex. We must also consider their behaviors (walking, eating, reading), their emotional states (happy, sad, angry), their roles (teacher, sales clerk, parent), and social contexts (church, home, restaurant).

Accordingly, when we enter a social setting, we mentally attempt to "locate" people within the broad network of possible social relationships. By scrutinizing them for a variety of clues, we place them in *social categories.* For instance, if they wear wedding rings, we infer that they are married; if they wear business clothes during working hours, we infer that they are employed in white-collar jobs; if they are in a wheelchair, we infer that they are handicapped. Only in this manner can we decide what to expect of others and what they expect of us. In sum, we activate **stereotypes**—certain inaccurate, rigid, exaggerated cultural images—that guide us in identifying the mutual set of expectations that will govern the social exchange.

Research by W. J. Livesley and D. B. Bromley (1973), based on a sample of 320 English children between 7 and 16 years of age, traced developmental trends in children's perceptions of people. The study reveals that the number of dimensions along which children conceptualize other people grows throughout childhood. The greatest increase in children's ability to distinguish people's characteristics occurs between 7 and 8 years of age. Thereafter, the rate of change is generally much slower; the differences between 7-year-olds and 8-year-olds are often greater than the differences between 8-year-olds and 15-year-olds. This observation leads Livesley and Bromley (1973, p. 147) to conclude that "the eighth year is a critical period in the developmental psychology of person perception."

Children under 8 years of age describe people largely in terms of external, readily observable attributes. Their conception of people tends to be inclusive, embracing not only personality but also an individual's family, possessions, and physical characteristics. At this age children categorize people in a simple, absolute, moralistic manner and employ vague, global descriptive terms such as *good, bad, horrible,* and *nice.* Consider this account by a 7-year-old girl of a woman she likes (Livesley & Bromley, 1973, p. 214):

> She is very nice because she gives my friends and me toffee. She lives by the main road. She has fair hair and she wears glasses. She is forty-seven years old. She has an anniversary today. She has been married since she was twenty-one years old. She sometimes gives us flowers. She has a very nice garden and house. We only go in the weekend and have a talk with her.

When they are about 8 years old, children show rapid growth in their vocabularies for appraising people. Their phrases become more specific and precise. After this age children increasingly come to recognize certain regularities or unchanging qualities in the inner dispositions and overt behaviors of individuals. Here is a description of a boy by a 9-year-old girl (Livesley & Bromley, 1973, p. 130):

> David Calder is a boy I know. He goes to this school but he is not in our class. His behavior is very bad, and he is always saying cheeky [impudent] things to people. He fights people of any age and he likes getting into trouble for it. He is always being told off by his teachers and other people.

This suggests the rapid development that occurs during middle childhood in children's abilities to make psychological inferences about other people—about their thoughts, feelings, personality attributes, and general behavioral dispositions (Erdley & Dweck, 1993). As children's cognitive ability of "person perception" matures, their vocabulary increases, as does their ability to express their needs and to communicate with others.

Questions

How does Piaget describe children's cognitive development from age 7 to age 12? Are Piaget's theories about cognitive development applicable to children from other cultures?

Language Development in Middle Childhood

The use of the English language in our schools will be a hurdle to some of our new American schoolchildren, and our schools and communities must work together with these immigrants and their descendants to assimilate them into our culture. Part of the problem is their resistance to acculturation. Children of Hispanic or Asian origin are more likely to have difficulty speaking English because they are more likely to speak another language at home. In 1995, 5 percent of U.S. schoolchildren spoke a language other than English at home or had difficulty speaking English. This represents 2.4 million children, up from only 1.25 million in 1979 (Wallman, 1998). The percentage varies regionally. For example, 26 percent of Californian schoolchildren speak another language at home (Wallman, 1998). These statistics speak loudly about the dire need take action to prepare our future labor force and citizens of this country.

Learning any language is a lifelong process. Children from the ages 6 to 12 continue to acquire subtle phonological distinctions, vocabulary, semantics, syntax, formal discourse patterns, and complex aspects of pragmatics in their first language (Gleason, 1993a). During this time, children are usually in school and add to speech the cognitively complex systems of reading and writing. As they grow older, they grow wiser, and the complexity and cognitive level of their language increases as a reflection of their academic studies and life experiences. Let's take a closer look at the changes in language that take place during middle childhood—in particular, vocabulary, syntax, and pragmatics.

Vocabulary Anglin (1993) reports that a fifth-grader learns as many as 20 new words a day and achieves a vocabulary of nearly 40,000 words by age 11. Some of this increase is basic academic vocabulary, because children spend most of their day in school and are exposed to a wide range of subjects requiring them to understand and use specific new terms. The *concrete operational* child is also able to learn and understand words that might not be tied to their personal experience, such as *ecology* or *discrimination.* They are able to describe objects, people, and events in terms of categories and functions rather than just physical features. They are more likely to describe their family dog in terms of its breed, its personality traits, and the kinds of things it can do. They can compare their dog to other dogs and contrast their dog as a pet with other kinds of pets that people have. Children begin to understand that words have *multiple meanings,* and they can use this new skill to tell jokes and ask riddles. Third-, fourth-, and fifth-grade children typically love to tell and make up their own "knock-knock" jokes. Both in reading and speaking, we see them beginning to make hypotheses about what words mean, using the context of the speaker's situation or, when reading, using pictures or the surrounding known words.

Syntax and Pragmatics   During this time, children begin to formally study the *grammar* of their language, because they are now able to understand that one can apply rules to language (operational thought). We consider them "teachable" in that they might still speak incorrectly, using the wrong pronouns and nonagreement of subject and verb, but they can be taught the correct form and can recognize errors in syntax. The structure of their sentences becomes more complex—they use other conjunctions besides *and,* and use adverbs, adjectives, and prepositional phrases.

A second type of growth occurs in the child's pragmatic language. As discussed in Chapter 7, *pragmatic language* is use of language in a particular context or situation. In middle childhood, they recognize that they can (and should) modify their voice, volume, and even their vocabulary, depending on the context of the interaction. For example, they will adopt a more formal style of vocabulary and syntax when talking with a teacher or a neighbor, as compared to how they talk to their parents or close friends. This ability of *code switching,* or changing from one form of speech to another, indicates that the child has developed an awareness of the social requirements in a given context. It also allows the child a certain freedom in speech when talking with peers, enabling the child to express herself using casual, often emotional words. Think for a minute how you would describe your hectic evening last night to a friend, your boss, or your parents. Research has shown that children of all social strata engage in *code switching,* and their pronunciation, grammar, and slang all change in this process (Yoon, 1992).

Question

What changes do we see in the way children use and understand language from age 7 to age 12?

Bilingual Education in the United States   During the last 20 years, primarily as a result of increased immigration, children with *limited English proficiency (LEP)*

have had a major impact on U.S. schools. **LEP** is the legal term for students who were not born in the United States or whose native language is not English and who cannot participate effectively in the regular school curriculum because they have difficulty speaking, understanding, reading, and writing English. A 1997 federal report states there are 3.2 million LEP students nationwide (Crawford, 1997). Spanish, by far the most prevalent minority language, is spoken by nearly 3 out of 4 LEP students (Fleishman & Hopstock, 1993). Hispanic students in the United States are rapidly growing in numbers and historically are at an educational disadvantage. Over the past 20 years, the percentage of Hispanic students has risen from 6 percent to 14 percent. In states such as California, Texas, and Florida, Hispanic students constitute a majority of public school students in the larger urban areas. From 1990 to 1995, the number of LEP students mushroomed by 50 percent. Hispanics and LEP students are among the most educationally disadvantaged groups attending the nation's schools:

- Thirty-nine percent of Hispanic children live in families with an income below the poverty line, a rate twice as high as for white children.
- Hispanic children start elementary school with less preschool experience than white children (a federal report indicates that less than 15 percent of Hispanic children attend preschool programs).
- The Hispanic dropout rate is unacceptably high, at over 20 percent for those who enrolled in a U.S. school. Language limitation is one factor associated with the failure to complete high school.
- The demand for teachers qualified to serve LEP students far outstrips the supply. California estimates a shortage of nearly 21,000 bilingual or ESL teachers.
- Approximately one-third of U.S. students are classified as minority, but only 13.5 percent of their teachers are of minority status.

Yet our U.S. population is increasingly diverse. The most significant concentrations of diversity are in urban areas such as New York City, Chicago, Los Angeles, and Fairfax County, Virginia. The five most common language groups of LEP students are Spanish (72.9 percent), Vietnamese (3.9 percent), Hmong (1.8 percent), Cantonese (1.7 percent), and Cambodian (1.6 percent) (U.S. Department of Education, 1998). How do these children impact schools? The answer is complex, with educational, economic, social, and legal aspects.

Educating Children Whose Reading and Speaking Skills Are Below Standard   First, the *Education for All Handicapped Children Act,* PL 94-142, passed in 1975, states that every school-age child has the right to a free, equal, and appropriate education in the least restrictive environment. This law originally met with a great deal of public opposition but was amended in 1983 and strengthened again with passage of the *Individuals with Disabilities Education Act (IDEA),* which reinforced the original act's provisions. This federal law defines a learning disability as

> a disorder in one or more of the basic psychological processes involved in understanding or in using spoken or written language, which may manifest itself in an imperfect ability to listen, think, speak, read, write, spell, or to do mathematical calculations. (*General Information About Learning Disabilities, Fact Sheet #7,* 1997)

Initially, in 1968 Congress passed the first *Bilingual Education Act,* which requires schools to "rectify the language deficiency" in LEP children in order to guarantee these students an entrance into educational programs that are available to their English-proficient peers. An updated version of this act, passed in 1988, requires schools to gather data on the numbers of LEP students and services and to support evaluation and research on the effectiveness of bilingual education programs. This act requires that each school district, through its particular educational programs, present instruction such that non-English speakers have an opportunity to achieve (acquire language and math skills) equal to that of children who speak English. The language of instruction is a critical choice, and we must think about how language is learned and what variables affect speed of acquisition, the relationship of language and thinking, as well as the social context of language. Three approaches have been used to "rectify the deficiency" and open the doors to mainstream education.

English as a Second Language (ESL) The **ESL approach** is focused on teaching children English as quickly and efficiently as possible. For example, children might learn English in a separate class all day, every day, until they reach some standard of proficiency. Then the child will enter the regular classroom and receive full-time instruction in all academic areas in English. We can compare this approach to teaching English with the way foreign languages are taught in U.S. middle and high schools. That is, the child is taught specific vocabulary and the rules of grammar and has practice in speaking and interpreting English. The obvious question then is, When is a child ready to join the mainstream and exit the ESL class? One important variable is the level of competency and/or formal instruction the child has had in his or her primary language. If their first language is well developed, children might reach age- and grade-level norms in English in 3 to 5 years. If the child has received no formal education in the first language, it can take 7 to 10 years (Collier, 1997). In general, late-exit programs have resulted in higher academic achievement than early-exit programs. Crawford (1997) reports that children

from early-exit programs have rates of academic growth parallel to those for English-proficient children in regular classrooms, but their actual level of achievement is well below the national norms on standardized tests.

In addition to concerns about academic success, we often see these children devalue their first language, refusing to speak it at home and in the community. This limits the quality and quantity of their interactions with their families and excludes their development of cultural identity (Crawford, 1997).

Bilingual Education A second approach, **bilingualism,** provides instruction in both languages by teachers proficient in both. This approach seems ideal to some, who argue that because children receive instruction in their first language, they can continue to acquire both academic skills and language skills. Using their first language (which is more developed), they are able to express more complex ideas, read higher-level texts, and increase their basic vocabulary, which in turn allows them to learn a second language (English) more easily. It has been documented (Garcia, 1994; Collier, 1992;) that cognitive and academic development in the first language has an extremely important and positive effect on second-language schooling. A great transference of skills (literacy, concept formation, subject knowledge, learning strategies) can take place from the first to the second language, compared to having to learn these in a less developed language (Krashen, 1996). The National Research Council (1998) supports this approach:

> If language minority children arrive at school with no proficiency in English but speaking a language for which there are no instructional guides, learning materials, and locally available proficient teachers, then these children should be taught how to read in their native language while acquiring proficiency in spoken English, and then subsequently taught to extend their skills to reading English.

Bilingual programs have consistently shown the greatest gains for children in academic skills and in social and emotional skills. The child is able to interact with family and community with increasing complexity and greater meaning and can feel that her or his first language is on an equal basis with English. When children are tested in their second language, they typically reach and surpass native speakers' performance across all subject areas after 4 to 7 years in a quality bilingual program. Because they have not fallen behind in cognitive and academic growth during the 4 to 7 years it takes to build academic proficiency in a second language, bilingually schooled students typically sustain this level of academic achievement and outperform monolingually schooled students in the upper grades (Collier, 1997).

Most recently, bilingual programs have been under attack because they can be costly. School districts must recruit teachers who are proficient in at least two languages, and the pool of applicants is more limited. It can also be difficult to offer bilingual programs because of the diversity of languages spoken in a school (ESL programs can address all first languages, as the primary goal is to teach English). And bilingual programs are sometimes seen as prolonging the process of Americanization. One clear disadvantage of bilingual programs, as well as ESL programs, is that instruction takes place out of the mainstream, which can make the children feel separate from the rest of the school community. And informal English use and the social contexts of ordinary English are not as available to them as they would be in English-only classrooms.

According to Krashen (1996), a number of studies have shown bilingual education is effective if children are in well-designed programs and if teachers have the necessary preparation to work with these children. The National Board for Professional Teaching Standards recently completed a model of standards for bilingual and ESL teachers (U.S. Department of Education, 1998), and there is a national shortage of bilingual education teachers and teachers of all programs that serve LEP students. Krashen argues that bilingual education can be more effective if there are many more books available in both the first and second languages of children enrolled in these programs. He further states that "LES Spanish-speaking children have little access to books at home or at school"; there is an average of one book in Spanish per Spanish-speaking child in some school libraries in schools with bilingual programs, which is clearly not enough (Krashen, 1996).

In 1998, the California voters approved a referendum that all teachers must teach in English in the classroom. A reported 26 percent of California schoolchildren have low or no proficiency in the English language. Apparently the majority of voters in this state believed the bilingual approach was not successful and was going to continue to be a great tax burden. Significantly, many parents did not want their children segregated in schools by bilingual programs. The majority of California voters preferred, instead, an instructional approach called *total immersion.*

Total Immersion   In **total immersion** programs, children are placed in regular classrooms (with or without support in their first language), and English is used for all instruction. This eliminates "separate" education or classes and gives every child an opportunity to observe and learn social communication in English with peers. The theoretical basis of total immersion is that language is best learned not when it is an academic subject but rather when it is useful and that children are motivated to learn the language in order to understand what is going around them. However, although it is certainly true that

children first learn language in a natural setting or context, the process is quite lengthy, and learning is directed purposefully during this time to the child's linguistic and cognitive level. That is, we don't teach our 2 year-olds the history of the United States in the colonial period or the language of the scientific method. *Immersion* programs seem to work best when the children are young and the family has positive feelings about this approach.

During the middle school years, children begin to understand sentence structure and grammar, extend their vocabularies, write coherently, use school and public libraries to find information, give oral presentations in class, and use the Internet. They also utilize their language skills in the community through membership in organized clubs and activities or by helping neighbors, working as newspaper carriers, volunteering, and so forth. Children this age learn they have a right to voice their concern about societal issues such as pollution or recycling, and some write editorials to their local newspaper or to corporate America. A command of the English language is a symbol of empowerment for both these children and their parents.

Questions

Should children be "allowed" to speak a language besides English in school? Should teachers and other school personnel be required to speak a second language in order to be able to communicate with the children and parents who are not proficient in English? Should schools require all graduates to be proficient in a second language in order to be able to communicate with an increasingly diverse ethnic population?

Assessment of Intelligence

I am an individual . . . my own distinctive self with my own identity. I am like a snowflake that is different from all the others, with abilities and talents, emotions and opinions, ideals and beliefs . . . all shaped into a pattern never seen before.

Karen Raun, *Our Inward Journey,* 1979

When we think about *intelligence,* we think about the way individuals differ from one another in their ability to understand and express complex ideas, adapt effectively to their environment, learn from experience, and solve problems. (The concept of intelligence was defined in Chapter 7.) Although the children's differences in their ability to do those things may be tremendous, their abilities can also appear to vary due to the time at which we assess them, the methods that we use, and the specific tasks we ask them to perform (Daniel, 1997). Historically, what we know about differences in school-age children is based on academic performance in school and intelligence testing using *psychometric tests.* This makes sense because in school these differences play a major role in determining how children learn and what educational programs best meet their cognitive needs. Given that, two questions come to mind. First, what tests should we use to measure their cognitive abilities? Second, how should the information we get from the assessment be used to determine specific educational plans? For a cross-cultural perspective on academic performance, see *Human Diversity,* "The Academic Achievement of Japanese, Taiwanese, and U.S. Children."

Types of Intelligence Tests Intelligence tests come in many forms. Some use only a single type of item or question; examples include the *Peabody Picture Vocabulary Test* (a measure of children's verbal intelligence) and *Raven's Progressive Matrices* (a nonverbal, untimed test that requires inductive reasoning about perceptual patterns).

Other tests use several different types of items, including verbal and nonverbal, to measure a wide range of abilities instead of one specific construct such as spatial ability or verbal intelligence. The Wechsler scales and the Stanford-Binet test ask the test-taker to give meanings of words, complete a series of pictures, copy a block design, and complete analogies. Performance on these tests can be measured in several domains (*subscores*) as well as by an overall score relating to general intelligence. The scores are standardized, with a *mean* of 100 and a *standard deviation* of ±15. This permits us to make statements such as 95 percent of the population scored within 2 standard deviations of the mean (between 70 and 130) and compare the person's score to the norming population (the population the test-taker is being compared to).

In contrast to these psychometric tests, other tests give us insight about a person's ability in a specific type of task. If we want to measure spatial ability, we might measure the length of time it takes a person to find a certain place in a new town or unfamiliar setting, or we might measure linguistic ability by asking a person to give an extemporaneous talk.

Schools primarily use the more conventional measurements such as the *Stanford-Binet* or *WISC-R (Wechsler Intelligence Scale for Children–Revised)* because the reliability and validity of such tests are documented, the scores are stable at least for school-age children, and they correlate well with, and are predictive of, academic performance.

The EQ Factor: Emotional Intelligence Learning emotional self-regulation, a component of emotional intelligence, is not an easy task. As we discussed in Chapter 7, the development of emotional intelligence is receiving a great deal of attention from educational psychologists and other researchers concerned with the alarming trend of aggression and violence perpetrated by a growing number of children (and adults) in this country. More children in middle schools and high schools are bringing weapons to school and concealing them in lockers and book bags. Large urban and suburban schools have resorted to using metal detectors, 24-hour video security systems, and police officers to patrol hallways. When confrontations arise, as they always will, some youth have not been able to control their anger, rage, or fear. A few have taken tragic courses of action, killing innocent peers and school personnel. Drive-by shootings, in which youth in cars randomly shoot innocent children and adults for the thrill of it, have become all too common. Researchers who study emotional intelligence call this a type of "mind-blindness," where the powerful emotional limbic system short-circuits the frontal cortex, blocking all reasoning. Some psychologists call this "temporary insanity." The traditional intelligence tests that have been used for nearly a century do not begin to tap the social-emotional abilities that can predict success or healthy self-regulatory behaviors—important life skills. However, there are other psychometric instruments that assess the likelihood of deviancy or emotional disturbance in youth and adults.

Limitations of IQ Tests Many times children are included or excluded in specific educational programs based on their intellectual functioning, but the intellectual functioning of children with LEP often cannot be determined. For example, special education services for mental retardation or cognitive impairment are available only to children who are tested using a standard measure of intelligence (IQ test) and obtain a score (full scale) of less than 70 IQ. The intellectual functioning of a child who receives a score below 70 is assumed to be lower than 97 percent of the population tested. Yet we know without a doubt that a child's academic performance cannot be totally explained by this one score. In fact, other factors, such as the child's motivation, social and language skills, self-concept, and even family and ethnic values, play important roles in the child's success in school.

More recently, some schools have turned to a *multidisciplinary team assessment.* This more comprehensive look at the child might include teacher input on achievement and effort, parent and community reports of how a child functions, observations by a school psychologist of the child in the classroom, review of the child's portfolio or work samples, and recommendations from various professionals and others who have knowledge about the child (e.g., speech therapist, resource room teacher). These additional pieces of information provide a more global understanding of the child's abilities. Nevertheless,

Human Diversity

The Academic Achievement of Japanese, Taiwanese, and U.S. Children

Empirical data gathered from at least two extensive studies in 1984 and 1992 indicate that U.S. schoolchildren lag behind children in Japan and Taiwan from the day they enter school (see Table 9-2). These findings suggest that improving the quality of U.S. education requires changes in U.S. homes as well as in schools (Kurdek, Fine, & Sinclair, 1995; Grolnick & Slowiaczek, 1994). Cultural differences in family socialization seem to contribute to differences in school readiness. For instance, in soliciting their children's obedience, U.S. mothers assert their authority, an approach Japanese mothers avoid. Instead, Japanese mothers appeal to their children's feelings or to the consequences of disobedience. The Japanese strategy seems more effective in instilling adult values and norms and so better prepares children for school discipline. Having internalized their mothers' expectations, the Japanese youngsters are more likely than Americans to approach the classroom with "receptive diligence." Japanese culture instills discipline through notions of interpersonal harmony, knowing one's role, and role perfectionism. These polar views of children as autonomous and disciplined beings are reconciled in the cultural emphasis upon effort as being responsible for individual accomplishment (Hamilton et al., 1990).

American mothers also tend to expect less from their children academically and to be more satisfied with their children's education than are their Japanese or Taiwanese counterparts. And although an old adage depicts Americans as confirmed believers in hard work, U.S. mothers rate ability as a stronger factor than effort in their children's school success. In contrast, Japanese and Taiwanese mothers rate effort as more important than ability. American parents who think success depends more upon ability than effort are less likely to require their children to work hard at learning than are parents who value effort.

Stevenson and colleagues (1993), in a follow-up study with students from these same cities, determined that Asian children in first grade spend nearly equal amounts of time on language and math instruction, whereas American children spend much more time on language instruction in first grade.

Table 9-2   Mean Number of Questions Answered Correctly

	Math	Reading Comprehension	Percentage of Class Time Spent on Reading	Percentage of Class Time Spent on Math	Percentage of the Time Students Paid Attention in Class
First Grade					
United States	17.1	21.3	50.6	13.8	45.3
Japan	20.1	22.8	36.2	24.5	66.2
Taiwan	21.2	25.6	44.7	16.5	65.0
Fifth Grade					
United States	44.4	82.6	41.6	17.2	46.5
Japan	53.8	82.5	24.0	23.4	64.6
Taiwan	50.8	84.6	27.6	28.2	77.7

Table reprinted by permission of Harold W. Stevenson, University of Michigan, 1988.

it is useful to obtain *standard measures* as well to prevent bias or prejudice in the placement of the child in a special education program. It is well documented that certain groups of children are identified more frequently as being in need of special education services, even with the use of standard scores. For example, more African American children than white children are being placed in special education classes, even though there is no evidence of underlying differences in their abilities (Sternberg & Grigorenko, 1997). To ensure proper placement for all children, educational psychologists and educators must continue to pursue theoretical as well as practical knowl-

Are Japanese Schools Better Than American Schools?
Japanese schools have produced educational results that place their average 18-year-old on a par with the average American college graduate. Japanese schools provide a longer school year and a more orderly, focused, and faster-paced education than do their American counterparts.

American youngsters were more likely to engage in unproductive behaviors, wander around the room, ask irrelevant questions, and talk to peers (Stevenson, Chen, & Lee, 1993).

One explanation for the success of Asian Americans lies in the Confucian ethic that still infuses China, Korea, Japan, and Vietnam. The Confucian legacy—an ethical code rather than a religion—centers on tightly knit families, discipline, and a high respect for all forms of learning. Children have the strong support and encouragement of their families throughout their schooling. In turn, children work hard to bring honor to their families.

Clearly, then, cultural factors need to be taken into account in considering children's performance on cognitive tests. Asian children might score at the highest percentiles on standardized achievement tests, but they might not understand what it means to tackle a problem with personal *creativity*. Though Chinese culture promotes academic success through obedience to and respect for authority figures, personal discipline and sacrifice, and doing one's best, it does not foster personal creativity, which is an American trait. To complete a creative task, a child of Asian background might need to be given a framework, such as "Today we're all going to take this clay and make animals. Using your imagination, decide which animal you would like to make." In contrast, in American open-ended instruction, children might simply be told, "Today we're going to make things out of clay. Go to it!" Chinese cultural philosophy discourages individualism. Chinese children are taught total obedience to authority, which enables them to expend their energies in the classroom on more productive endeavors than challenging the authority (Zhang, 1995). Achievement consists of hard work and never giving up, according to Chinese philosopher Hsun Tzu (pronounced Sun Zi) (Zhang, 1995):

> Achievement consists of never giving up. If you start carving and then give up, you cannot even cut through a piece of rotten wood; but if you persist without stopping, you can carve or inlay metal or stone. Earthworms have no sharp claws or teeth, no strong muscles or bones, and yet above ground they feast on the mud, and below they drink at the yellow springs. This is because they keep their minds on one thing. If there is no dark and dogged will, there will be no shining accomplishment; if there is no dull and determined effort, there will be no brilliant achievement.

In light of Chinese cultural characteristics, the behaviors to be fostered are sharing, helpfulness, friendliness, cooperation, uniformity, and self-control. Punishment is to be conducted *privately* in order to save "face." Individual recognition and reward are to be avoided. The Chinese student also has a highly reflective response style that requires additional "wait time" in order to respond to a particular question. Yao (1985) found that Asian children learn best in a well-structured, quiet environment. Also, they are less likely to reveal their opinions, tend to hide their abilities, and seldom question a teacher. Children from Chinese, Taiwanese, and Japanese backgrounds have been taught that advancement comes from effort (Zhang, 1995).

edge about how children learn, the role of intellectual ability, and curriculum and instructional methods that best match children's learning needs.

Individual Cognitive Styles   Have you ever felt stupid or out of place in a classroom? Did you feel that you couldn't understand the information being presented, no matter how hard you concentrated? Have you ever wondered why some subjects come easier to you than others? Although intelligence has a role to play in all learning, it is becoming increasingly clear that not everyone learns in the same way. Some people learn best using visual information, others prefer to hear the spoken word. Yet others learn faster and retain the information longer when they involve all of their senses in hands-on learning. These differences in how individuals organize and process information have come to be called *cognitive styles*. **Cognitive styles** are powerful heuristics that cut across traditional boundaries between intelligence and personality and influence a person's preferred way of perceiving, remembering, and using information (Witkin, 1975, 1964; Sternberg & Grigorenko, 1997; Crandell, 1982, 1979). Psychologists have been researching cognitive styles for over 50 years and have identified a number of different cognitive style models, including Witkin's field-dependence versus field-independence, Kagan's impulsive versus reflective, and Hill's educational cognitive style (Messick, 1976). Research shows that when students are matched to instruction based on their preferred style of learning, they process the information faster, retain it longer, and are motivated to continue to learn (Dunn, Beaudry, & Klavas, 1989; Messick, 1984). Are you analytical or global? Do you work quickly or deliberately? Do you need to visualize and organize the task before starting? These are examples of cognitive style characteristics.

Putting theory into practice is difficult because teachers might no longer utilize one method of teaching (speaking, writing on the board). Teacher education programs are becoming more innovative, developing new delivery and testing strategies, requiring proficiency in a second language, and training in multimedia and multisensory approaches to accommodate the students' differing learning styles. The concept of *portfolio assessment* has been an outgrowth of learning styles: Any student is a whole being, not simply one who writes and speaks and calculates. Some are artistic, some are mechanical, some are musical, and some are quite adept socially. Proponents of portfolio assessment believe that a whole year of a child's academic life should not be reduced to a grade point average. These teachers keep a portfolio of the best of a child's performances for the year. The portfolio might include photographs of a class play or sports events, an original poem written by the child, notes on a car the child helped build for *Odyssey of the Mind* competition, in addition to the child's best essays and math exams of the year.

Questions

How do we measure individual differences in cognitive abilities? In what ways is the process of cognitive development universal? Are there individual and group differences in the way people process information?

Learning Disabilities (LDs)

Between 1977 and 1995 the number of students who participated in federal programs for children with all disabilities increased by 47 percent . . . and much of this rate increase is represented by the increased classification of children with learning disabilities.

National Center for Education Statistics,
The Condition of Education, 1997

Educators have come to employ the term **learning disabilities (LDs)** as an umbrella concept to refer to children, adolescents, and college students who encounter difficulty with school-related material despite the fact that they appear to have normal intelligence and lack a

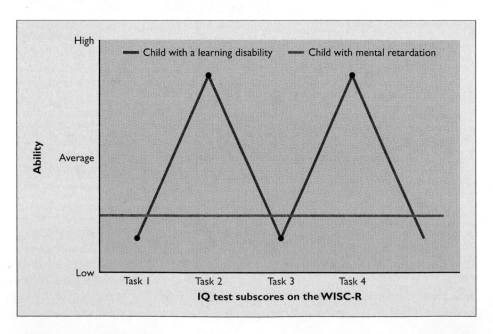

Figure 9-2 Typical Profile of Subtest Scores on the WISC-R IQ Test for a Child with a Learning Disability Versus a Child with Mental Retardation

demonstrable physical, emotional, or social impairment (some prefer the term *differently abled*):

> Learning disabilities (LD) is a disorder that affects people's ability to either interpret what they see and hear or to link information with different parts of the brain. These limitations can show up in many ways, as specific difficulties with spoken and written language, coordination, self-control, or attention. Such difficulties extend to schoolwork and can impede learning to read or write or to do math. Learning disabilities can be lifelong conditions that, in some cases, affect many parts of a person's life: school or work, daily routines, family life, and sometimes even friendship and play. In some people, many overlapping learning disabilities may be apparent. Other people may have a single, isolated learning problem that has little impact on other areas of their lives. (National Institute of Mental Health, 1993)

In practice, the notion implies that a discrepancy exists between a student's estimated ability and his or her academic performance. Youngsters are usually diagnosed as having a learning disability when their achievement level falls two or more grade levels below their ability, as predicted by a standardized IQ test score (see the hypothetical example in Figure 9-2).

Unlike those diagnosed as having mental retardation, the population of youngsters classified as having a learning disability is relatively unrestricted. In popula-

tion served, this converts to 796,000 youth in 1977 versus 2,508,000 youth in 1995. The U.S. Department of Education (1996) reports that total school enrollment declined 2 percent during this same time period. From 1977 to 1995, the percentage of children classified as having mental retardation declined from 26 percent to 10.5 percent (see Figure 9-3). In fact, as part of their efforts to secure adequate funding, school officials often reclassify pupils as having a learning disability in order to fit them into a program that will provide financial assistance.

Between 1977 and 1995, the percentage of children with specific learning disabilities as a percentage of total public K–12 enrollment rose from 2 percent to nearly 6 percent (see Table 9-3). (Note that this figure does not include statistics for preschool children with learning disabilities, who are serviced under separate legislation.) In contrast, the percentage of children classified as having learning disabilities, out of all children with disabilities (other children with speech or language impairments, mental retardation, serious emotional disturbance), rose tremendously from 21.6 percent in 1977 to 46.1 percent in 1995. Fortunately, the number of children with all other disabilities has been declining considerably since 1977 (see Figure 9-3). In 1992, males with specific learning disabilities represented 7.2 percent of total public school enrollment, and females classified with learning disabilities accounted for 3.3 percent (U.S. Department of Education, 1996). Public schools spend about $8,000 a

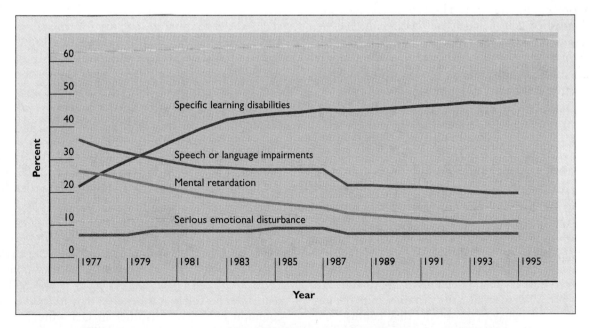

Figure 9-3   Children from Birth to 21 Who Were Served by Federally Supported Programs for Students with Disabilities, by Type of Disabilitiy: School Years Ending 1977–1995   This analysis includes students who were served under Chapter 1 of the Education and Improvement Act (ECIA) and Part B of the Individuals with Disabilities Education Act (IDEA). Data for 1995 are for children aged 3 to 21.

Source: U.S. Department of Education, Office of Special Education and Rehabilitative Services, Eighteenth Annual Report to Congress on the Implementation of the Individuals with Disabilities Education Act; and the National Center for Education Statistics, *Digest of Education Statistics,* 1996, in *The Condition of Education,* 1997.

Table 9-3   Children Who Were Served by Federally Supported Programs for Students with Learning Disabilities, by Percentage of K–12 Enrollment and by Percentage of All Children with Disabilities, 1977–1995

	1977	1979	1981	1983	1985	1987	1989	1991	1993	1994	1995
Specific learning disabilities as percentage of total public K–12 enrollment	1.8	2.7	3.6	4.4	4.7	4.8	4.9	5.2	5.5	5.6	5.7
Specific learning disabilities as percentage of all disabilities	21.6	29.1	35.3	40.9	42.5	43.8	43.7	44.7	45.9	45.6	46.1

Source: U.S. Department of Education, Office of Special Education and Rehabilitative Services, *Eighteenth Annual Report to Congress on the Implementation of the Individuals with Disabilities Education Act;* and the National Center for Education Statistics, *Digest of Education Statistics,* 1996.

year to educate a student with learning disabilities, compared to $5,500 for other youngsters (Roush, 1995).

A variety of youngsters are designated as "learning disabled." There are some whose eyes see correctly but whose brains improperly receive or process the informational input. Or instead of zeroing in on what is directly in front of their noses, they take a global view (Geiger & Lettvin, 1987). In either case, they might get letters mixed up, reading *was* as "saw" or *god* as "dog." Others have difficulty selecting the specific stimulus that is relevant to the task at hand from among a mass of sensory information. Still others might hear but fail to remember what they have heard by virtue of an auditory-memory problem. For instance, they might turn to the wrong page in the book or attempt the wrong assignment when they rely on oral instructions. The sources of such disabilities are varied and include combinations of genetic, social, cognitive, and neuropsychological factors (Roush, 1995). High-tech brain-imaging techniques, including PET, suggest that signal-processing regions like the thalamus might be implicated. The thalamus is a structure that functions like a telephone switching board; it takes incoming signals from the eyes, ears, and other sensory organs and routes them to different areas of the brain. However, whatever the source of the disability, youngsters with a learning disability have problems in reading *(dyslexia),* in writing *(dysgraphia),* or in mathematics *(dyscalcula).* Such difficulties might appear in 20 percent of the nation's schoolchildren (students identified as dyslexic make up about 80 percent of the LD population) (Roush, 1995).

Such disabilities are often called "the invisible handicap." They typically do not end with childhood. An estimated 15 to 20 percent of adults in the United States have thought and behavior problems that confuse and frustrate them in the course of their everyday activities (National Center for Learning Disabilities, 1997). However, learning disabilities do not automatically lead to low achievement. Many accomplished scientists (Thomas Edison and Albert Einstein), political leaders (Woodrow Wilson and Winston Churchill), authors (Hans Christian Andersen), artists (Leonardo da Vinci), sculptors (Auguste Rodin), actors (Tom Cruise and Whoopi Goldberg), athletes (Bruce Jenner), and military figures (George Patton) have had learning disabilities (Schulman, 1986).

The placement of children in a category like "learning disabled" is a serious matter. From one perspective, it could blight their lives by stigmatizing them, reducing their opportunities, lowering their self-esteem, and reducing their time for peer interactions. From another perspective, such classification can also open doors to services and experiences that can facilitate learning, improve a sense of self-worth, and foster social integration. But teachers and the public must be aware of the possible harmful effects of labeling so that the labels are not used for obscure, covert, or hurtful purposes (Harris et al., 1992). One thing we can say for sure is that up until 1975, children with disabilities were not offered a public education; most stayed at home with no prospects for the future. Over the past 25 years, the doors of opportunity and progress have opened. Each child and her or his family can determine to what extent they will take advantage of this opportunity.

Attention-Deficit Hyperactivity Disorder (ADHD). *Attention-deficit hyperactivity disorder (ADHD)* is the newest name applied to a complex disorder that has long puzzled professionals. The changing focus of research and treatment is reflected in the succession of diagnostic terms used to designate the disorder, including *brain-damaged child syndrome* (1940s), *minimal brain damage syndrome* (1950s), *hyperactive* or *hyperkinetic syndrome* (1960–1970), *attention-deficit disorder* (early 1980s), and the current *ADHD.* Today it is believed that ADHD is the most common behavioral disorder in U.S. youngsters. Some experts say it afflicts some 3.5 million youngsters, or up to 5 percent of those under age 18. It is two to three times more likely to be diagnosed in boys than in girls. Although a decade or so ago clinicians believed that the disorder faded with maturity, it is now one of the fastest-growing diagnostic categories for adults (Wallis, 1994).

Working with Children in Dyslexia Therapy Programs Dr. Paula Talla, shown above, finds that a subset of children with reading problems develop severe dyslexia because they have difficulty hearing and generating speech. The youngsters seemingly experience a "language fog." Many of the children can benefit from therapy programs that Talla and her coworkers developed. The therapy entails computer games in which the children practice thousands of sounds during each session.

Teachers had always considered me hyperactive, which meant I moved all the time. For awhile I thought I had something wrong with me that made me move. Now I understand. People who move a lot are taking in information through the part of the brain that controls sensory input and movement, allowing them to learn. They learn by doing. Educators now call these movers "haptic learners" because they process information primarily through the motor and sensory cortex of the brain.

Barbara Meister Vitale,
Free Flight: Celebrating Your Right Brain, 1986

Attention-deficit hyperactivity disorder—the designation employed by the American Psychiatric Association—involves a collection of vague and global symptoms. Typically youngsters who cannot stay in their seats, wait their turn, follow instructions, or stick with a chore are viewed as having ADHD. But the cutoff between exuberance and pathology is a matter of convention, and who is identified as having the disorder de-

pends on how the "scale" is set (Barkley, 1990). As noted above, clinicians now recognize ADHD in adults, and researchers are finding antisocial activity, conduct disorder, and substance abuse in many adults who were deemed hyperactive in childhood.

Significantly, clinicians report that some afflicted adults are successful, not in spite of ADHD, but in part because of it. These people, though easily distracted, often have the uncanny ability to "hyperfocus" and so become virtually immune to distractions. Under crisis circumstances or in difficult situations, such individuals often can make rapid decisions, act quickly, and bring boundless energy to bear. Ironically, while ADHD can put those with high IQs and good social skills at the top of their field (e.g., emergency room physicians, sales personnel, stock-market traders, and entrepreneurs), those with ADHD and without such attributes can land in prison given their zeal for risk taking and impulsive behaviors (Freeman, 1995).

There is little consensus on what causes ADHD or even whether it has a single cause. Experts have advanced a variety of theories to explain the disorder, including genetic defects, poor parenting, food additives, spicy foods, allergies, lead poisoning, fluorescent lights, insufficient oxygenation, and too much television. Most likely there are multiple routes to the development of ADHD in childhood (Carlson, Jacobvitz, & Sroufe, 1995). For some youngsters, intrusive and overstimulating care appears to play an important part. For others, organic considerations such as motor development or medical history may be operative. And for still other children, an interactive or transactive model combining genetic and environmental influences might best explain the development of the disorder. Increasingly, clinicians and researchers are concluding that ADHD is a heterogeneous disorder with disparate symptoms often being lumped together (Carlson, Jacobvitz, & Sroufe, 1995; Vatz, 1994). Though there are no established laboratory tests for diagnosing ADHD, a 1998 study conducted at Stanford University with 16 boys between the ages of 8 and 13 may pave the way for more accurate diagnosis. The boys underwent magnetic resonance imaging tests (MRI scans) while responding to a response task. Some had ADHD and had not taken Ritalin for 3 days, some did not have ADHD but were administered Ritalin prior to the MRI. Results indicate that subjects with ADHD experienced increases in blood flow to the attention-regulating basal ganglia (see Figure 9-4) (Rogers, 1998).

So whether a child is diagnosed as having ADHD, and whether a child is treated with medication, might be determined by behavioral/social considerations. Significantly, most nations of the world still treat unruly youngsters as behavior problems rather than as patients. France and England report ADHD rates of one-tenth the U.S. rate. The use of amphetamine drugs, primarily

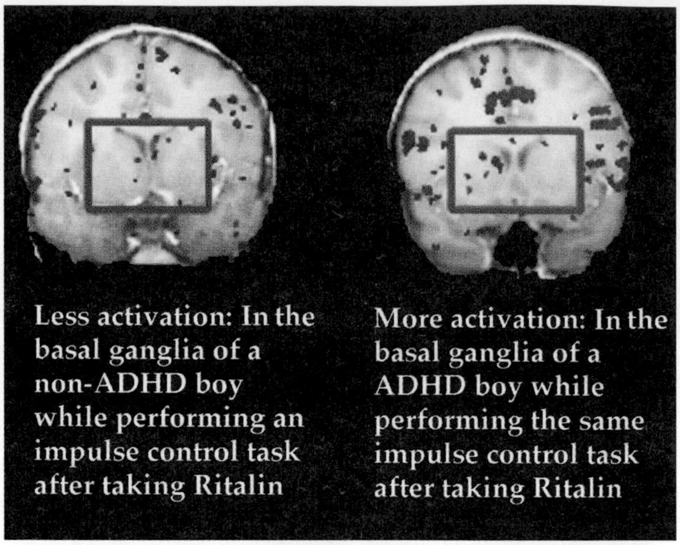

Less activation: In the basal ganglia of a non-ADHD boy while performing an impulse control task after taking Ritalin

More activation: In the basal ganglia of a ADHD boy while performing the same impulse control task after taking Ritalin

Figure 9-4 Brain Scans Give New Hope for Diagnosing Individuals with Attention-Deficit Hyperactivity Disorder (ADHD)

methylphenidate, better known as *Ritalin,* in the management of "problem" children is widespread and is increasing. In many cases parents, teachers, and peers report a significant improvement in symptoms. Physicians believe that the medication somehow allows hyperactive youngsters to attend to critical aspects of the learning situation and to filter out distractions. However, not all children respond to drug therapy. Moreover, as with any medication, there are sometimes adverse side effects—insomnia, gastrointestinal distress, dizziness, weight loss, and retardation of growth.

Many professionals and members of the public fear that the widespread enthusiasm for administering amphetamines to "hyperactive" children has led to abuses and to the acceptance of drug therapy for children who might not benefit from it. It becomes frighteningly easy for parents and teachers to control behavioral "problems" and "difficulties" under the guise of giving youngsters "medicine." The drugs then become "conformity pills" for rebellious youngsters. Some physicians have prescribed medication on the basis of reports from teachers or parents that a child is doing poorly in school, without giving the child a thorough medical evaluation. Some children benefit from amphetamine treatment, but the trend toward its indiscriminate use is cause for concern (Vatz, 1994).

Individual Education Plans (IEPs)   All students who are classified as having any disability, including a learning disability, are to be provided with an **individualized education plan (IEP)**. This plan, which is developed in a collaborative effort by the school psychologist, the child's teachers, an independent child advocate, and the parent(s) or guardian, is a legal document that ensures that the child with special learning needs will receive the needed educational support services in the least restrictive environment. The child's IEP is updated annually as a result of a formal assessment of the child's yearly academic and social-emotional performance. To see what is involved in developing an IEP, see *Information You Can Use:* "Individualized Education Plans (IEPs)."

Inclusion   Over the past two decades there have been appreciable changes in the policy and philosophy of educators toward youngsters with physical disabilities, learning disabilities, and mild mental retardation. We have shifted away from segregated or separate classrooms toward **inclusion**—the integration of students with special needs within the regular classroom programs of the school, in some cases with the assistance of an additional aide in the classroom. The inclusion movement received an impetus from federal IDEA legislation that requires youngsters with special needs be educated in the least restrictive environment possible. Though many have great hopes for inclusion, educators and psychologists warn that the mere physical presence of children with special needs in regular classrooms does not guarantee academic or social success. It works only where school officials, teachers, parents, and students are committed to making it work through designing, implementing, and supporting appropriate restructuring of the educational enterprise (Madden & Slavin, 1983).

Questions

What process is used to identify a child as "learning disabled"? Why do some children need an individual education plan? What do we mean by the concept of an inclusive education for children?

The Use of Methylphenidate (Ritalin) to Manage Children with ADHD Is Widespread Experts estimate there are more than 2 million children with this disorder. Many physicians, parents, and teachers believe that amphetamine medication helps hyperactive youngsters behave in more productive and less negative ways. The stimulant appears to increase the level of dopamine in the frontal lobe of the brain. Currently, U.S. Drug Enforcement Administration has classified it as a Schedule II controlled substance, in the same category as cocaine, methadone, and methamphetamines. In 1997, sales of the drug topped $350 million in the United States. Critics, including some parents, teachers, and physicians, are convinced Ritalin is overprescribed and simply a "quick fix" for a more complex attentional-behavioral problem.

What Do We Know About Effective Schools?

The principal goal of education is to create men [and women] who are capable of doing new things, not simply of repeating what other generations have done—men [and women] who are creative, inventive and discoverers.

Jean Piaget

Over the past 20 years, there has been an outpouring of reports decrying the state of the U.S. educational system. In 1983, *A Nation at Risk,* a report of the National Commission on Excellence in Education, set an alarmist tone that pointed to a "rising tide of mediocrity" in the schools. Other reports, criticisms, and recommendations followed. Yet there is little evidence that matters are better today than they were 20 years ago, despite society's

demands for better trained and more skilled citizens (Manegold, 1994). The reports point to many of the same indicators: poor achievement test scores; a long-term decrease in college entrance test scores; declines in both enrollments and achievement in science and mathematics; low levels of skill in communicating, writing, and thinking; the high costs incurred by business and the military for remedial and training programs; the substantial levels of functional illiteracy among the U.S. population; and the poor performance of U.S. students on international tests.

Educators, psychologists, and sociologists have examined what makes a school effective. Michael Rutter (1983) led a University of London team in a three-year study of students entering 12 London inner-city secondary schools. These schools—only a scant distance apart and with students of similar social backgrounds and intellectual abilities—had quite different educational results. The most important factor seemed to be the school "ethos" or "climate." The successful schools fostered expectations for order in the classrooms—they did not leave matters of student discipline to be worked out by individual teachers, and so teachers found that it was easier for them to do a good job of teaching in those schools. Moreover, the effective schools emphasized academic concerns:

- The teachers took considerable care in planning their lessons.
- High-achievement expectations permeated the classrooms.
- Students spent a high proportion of their time on instruction and learning activities.
- Homework was considered important.
- Students were encouraged to use the library.
- Schools that fostered respect for students as responsible people and held high expectations for proper behavior achieved better academic results.

Other researchers have also concluded that successful schools foster expectations that order will prevail and that learning is a serious matter (Johnson, 1994). Much of the success of private and Catholic schools has derived from their ability to provide students with an orderly environment and strong academic demands (Bryk, Lee, & Holland, 1993). Academic achievement is similarly high in the public sector when the policies and resulting behavior are like those in the private sector (McAdoo, 1995). Successful schools, then, possess "coherence"—things work together and have predictable relationships with one another. It seems that a growing number of U.S. parents are looking for many of these qualities in their youngsters' schooling. "Fed up" with public schools, some parents are turning to evangelical schools, private-tuition schools, charter or magnet schools, and

Information You Can Use

Individualized Education Plans (IEPs)

Every day more youngsters are determined to be eligible for an individual education plan. For 1995, the most recent year reported, 12.3 percent of the total K–12 student population in the United States was eligible for the services prescribed in an IEP (U.S. Department of Education, 1996). The IEP planning document is a written description of a program tailored to fit the eligible child's unique educational needs. The IEP is developed jointly by parents, educators, and other professionals. The children themselves may also sit in on the process and contribute to their own plans. Goals and objectives for a child, based on current levels of functioning, are outlined by everyone involved in planning and providing services. The IEP specifies the educational placement, or setting, and the related services necessary to reach those goals and objectives. The IEP also includes the date the services will begin, how long they will last, and the way in which the child's progress will be measured. The IEP is much more than just an outline of, or management tool for, a child's special education program. The development of an IEP gives caregivers the opportunity to work with educators as equal participants to identify a child's needs. The IEP is a commitment in writing to the educational program and resources the school agrees to provide. Periodic review of the IEP serves as an evaluation of the child's progress toward meeting the educational goals and objectives jointly decided upon by the caregivers and school professionals. In essence, the IEP is the cornerstone of special education.

IEP Description

Part 1 of the document describes the child applying for services, including basic information such as the child's name, age, and address, the child's educational level and behavioral performance, and subjective information about the effect of the disability on academic and nonacademic achievements. This section is also important in that it allows the caregivers to supply information about the child's typical behaviors in question and how the child approaches learning situations. Descriptive statements are essential so that everyone involved in teaching the child has accurate and complete information concerning the child's ability and potential.

Part 2 gives the caregivers and other members of the IEP team a chance to set goals and objectives that will help the child master the skills or behaviors that everyone wishes the child to attain. Goals are an expression of results to be achieved in the long run; objectives are the intermediate steps necessary to reach the long-range goals. *Annual goals* written on an IEP state what the child is expected to do by the end of the academic year. Each goal must be written as a positive statement that describes an observable event. A poorly written goal is vague (e.g., *Truett will learn to write*). A well-written goal will answer the questions Who? How? Where? When? (e.g., *Truett will write all of the letters of the alphabet unaided by May 7*). Caregivers often have questions concerning the goals and objectives put forth by the IEP committee. Common questions are these:

- How will I know if a goal is reasonable or appropriate?
- How do goals and objectives in the IEP relate to the instructional plans of educational personnel?
- Must goals and objectives be written on the IEP for

other government-sponsored alternative schools, and home schooling.

Since it became legal in 1993 for parents to home-school their children, the estimated number of children being home-schooled has increased 500 percent. In 1998 nearly 1.5 million school-age children were learning at home; this is approximately 1.5 percent of the total number of elementary and secondary school students (Kantrowitz & Wingert, 1998). Few people realize that a popular teen group, The Hansons, were home-schooled. The reasons more families are home-schooling include child safety issues, the ability to give greater attention to the child's learning and interests, reclaiming family closeness, teaching special-needs children (children who are highly gifted or have learning disabilities or emotional disturbances), and worries about unsavory peer influences in school environments. Critics of home schooling are concerned because to date only 37 states have statutes that set standards for home schooling. However, home-schoolers averaged 23 out of 36 on the 1998 ACT college admissions exam, compared to an average of 21 for traditional high school students (Kantrowitz & Wingert, 1998). Home-schooling newsletters, materials, seminars, and associations; Internet access; and sophisticated soft-

the parts of the child's program in the general education classroom?

- Is it necessary for parents/caregivers to learn to write goals and objectives?
- How much of my child's school day will be spent in pull-out learning activities, such as the Resource Room, rather than with peers?
- What do I do if I don't agree with my school district's plan for my child?

Part 3 describes related services to be provided at no cost to the parent/caregivers. These related services might include assistive technology, audiology, counseling services, medical services, occupational therapy, parent counseling and training, physical therapy, psychological services, recreation, rehabilitative counseling services, school health services, social work services in schools, speech therapy, an aide in the classroom, transportation methods, or out-of-school placement.

Part 4 describes the special education placement provided to the child. *Placement* refers to the educational setting in which the goals and objectives for the child's special education and related services can appropriately be met. For example, in the past, a child confined to a wheelchair would be placed in a room with all other children in wheelchairs, regardless of mental ability. Today the IEP guarantees that a child's ability is considered along with the disability.

Part 5 specifies when and for what duration the services will be in place. In general, the long-term duration of services will not exceed 1 year. This is because the services are reviewed annually to determine if they are still appropriate. In addition to specifying the long-term duration of the child's services, Part 5 of the IEP should include short-term, daily hours for each service (e.g., *Jessica will attend individual speech therapy twice a week for 30 minutes each session*).

Part 6 is the evaluation process, which allows the people who formulated the IEP to assess whether the IEP is working or needs modification. For example, suppose the short-term objective is this: *Kareem will choose two classmates with whom he will cooperatively plan and complete an art project to illustrate a story in the fifth-level reader, by January 10.* An evaluation of the child's services and program can occur by asking the following types of questions:

- Was the art project completed on time?
- Did Kareem read and comprehend the story? (There should be an objective, measurable assessment.)
- Did the teacher observe and note incidents to illustrate Kareem's cooperation with classmates?
- What indications show that Kareem made plans for the project?
- How did Kareem demonstrate pleasure in his accomplishments?

The IEP is an important document, and it is hoped that you now have a better understanding of how children with special education needs are able to get the services they require in order to better succeed in school. Today, students with learning disabilities or other disabilities are (1) eventually transitioning out to sheltered employment or regular employment, or (2) graduating from high schools, or (3) entering college programs where higher education provides special services to accommodate the student's individual learning needs. In the past these youth were not integrated into the mainstream of society and minimally sustained themselves on social services and welfare. Today, many have become productive, fully integrated members of society.

Anderson, Chitwood and Hayden, *Negotiating the Special Education Maze: A Guide for Parents and Teachers,* 1997. Bethesda, MD: Woodbine House. © 1997. Reprinted by permission.

ware equip parents better than ever before to meet the challenge of home-schooling their children.

Questions

What types of programs are mandated for children with learning or other disabilities? Why do you think more children are being classified as having a learning disability? What types of learning environments in schools have been found to promote effective learning? Why are more children being home-schooled?

Moral Development

Can you tell me, Socrates—can virtue be taught? Or if not, does it come by practice? Or does it come neither by practice nor by teaching, but do people get it by nature, or in some other way?

Plato, *Meno*

As human beings, we live our lives in groups. Because we are interdependent, one person's activities can affect the welfare of others. Consequently, if we are to live with one another—if society is to be possible—we must share certain conceptions of what is right and what is wrong. Each of us must pursue our interests, be it for food, shelter, clothing, sex, power, or fame, within the context of a moral order governed by rules. Morality involves how we go about distributing the benefits and burdens of a cooperative group existence (Wilson, 1993).

A functioning society also requires that its standards of morality be passed on to children—that moral development take place in its young. **Moral development** refers to the process by which children adopt principles that lead them to evaluate given behaviors as right and others as wrong and to govern their own actions in terms of these principles. If media interest is any indication, many Americans are quite concerned with the moral status of contemporary youngsters, and they look to the schools to teach values to fill what they deem to be a moral vacuum.

A century ago, Freud believed that children develop a conscience through feeling a sense of guilt for their actions. More recent theories come from the field of cognitive research.

Cognitive Learning Theory

To make your children capable of honesty is the beginning of education.

John Ruskin, *Time and Tide*, 1867

The discussion of cognitive learning theory in Chapter 2 emphasized the important part *imitation* plays in the socialization process. According to psychologists such as Albert Bandura (1986, 1977) and Walter Mischel (1977), children acquire moral standards in much the same way they learn any other behavior, and social behavior is variable and dependent on situational contexts. Most actions lead to positive consequences in some situations and not in others. Consequently, individuals develop highly discriminating and specific response patterns that do not generalize across all life circumstances (Bussey, 1992).

Studies carried out by cognitive learning theorists have generally been concerned with the effect that models have on other people's resistance to temptation (Bandura, Ross, & Ross, 1963). In such research, children typically observe a model who either yields or does not yield to temptation (see Figure 9-5). Walters, Leat, and Mezei (1963) conducted such an experiment. One group of boys individually watched a movie in which a child was punished by his mother for playing with some forbidden toys. A second group saw another version of the movie in which the child was rewarded for the same behavior. And a third group, a control group, did not see any movie. The experimenter took each boy to another

"How do you know when the yellow light means slow down and when it means to speed up?"

Figure 9-5 Children's Models Cognitive learning theorists have generally been concerned with the effect that models have on other people's resistance to temptations. Children are observant of "models" in their environment who either yield or do not yield to temptation.
DENNIS THE MENACE used by permission of Hank Ketchum and © by North America Syndicate.

room and told him not to play with the toys in the room. The experimenter then left the room.

The study revealed that boys who had observed the model being rewarded for disobeying his mother themselves disobeyed the experimenter more quickly and more often than the boys in the other two groups. The boys who had observed the model being punished showed the greatest reluctance of any of the groups to disobey the experimenter. In short, observing the behavior of another person does seem to have a modeling effect on children's obedience or disobedience to social regulations (Speicher, 1994). Of interest, other research reveals that dishonest or deviant models often have a considerably greater impact on children than do honest or nondeviating models (Grusec et al., 1979).

These findings provide a context for evaluating the prevalent notion that violent children's social cognitions, attitudes, and thinking patterns contribute to their violent behavior. Clinical psychologists have often viewed violence as symptomatic of "internal conflicts" or "antisocial personality traits" within individual children. In other words, they assume that the children have deficient or distorted cognitions that contribute to their violent behavior.

Cognitive Developmental Theory

Cognitive learning theorists view moral development as a cumulative process that builds on itself gradually and continuously, without any abrupt changes. In sharp contrast to this idea, cognitive developmental theorists like Jean Piaget and Lawrence Kohlberg conceive of moral development as taking place in stages, with clear-cut changes between them, such that a child's morality in a particular stage differs substantially from that child's morality in earlier and later stages. Although the learning and developmental perspectives are frequently counterposed to one another, they nonetheless provide complementary and interdependent analyses of human social interaction (Gibbs & Schnell, 1985).

Jean Piaget The scientific study of moral development was launched over 60 years ago by Jean Piaget. In his classic study *The Moral Judgment of the Child* (1932), Piaget said that there is an orderly and logical pattern in the development of children's moral judgments. This development is based on the sequential changes associated with children's intellectual growth, especially the stages that are characterized by the emergence of logical thought. In keeping with a constructivist perspective, Piaget believed that moral development occurs as children act upon, transform, and modify the world they live in. As they do, they in turn are transformed and modified by the consequences of their actions. Hence, Piaget portrayed children as active participants in their own moral development. In this respect Piaget differed from the cognitive learning theorists, according to whom the environment acts on and modifies children, and children are passive recipients of environmental forces. Cognitive learning theorists picture children as learning from their environment rather than, as Piaget would have insisted, in dynamic interaction with their environment.

Piaget provided a *two-stage theory of moral development.* The first stage, that of **heteronomous morality,** arises from the unequal interaction between children and adults. During the preschool and early elementary school years, children are immersed in an authoritarian environment in which they occupy a position decidedly inferior to that of adults. Piaget said that in this context children develop a conception of moral rules as absolute, unchanging, and rigid.

As children approach and enter adolescence, a new stage emerges in moral development—the stage of **autonomous morality.** Whereas heteronomous morality evolves from the unequal relationships between children and adults, autonomous morality arises from the interaction among status equals—relationships among peers. Such relationships, when coupled with general intellectual growth and a weakening in the con-

straints of adult authority, create a morality characterized by rationality, flexibility, and social consciousness. Through their peer associations, young people acquire a sense of justice—a concern for the rights of others, for equality, and for reciprocity in human relations. Piaget described autonomous morality as egalitarian and democratic, a morality based on mutual respect and cooperation.

Lawrence Kohlberg Lawrence Kohlberg refined, extended, and revised Piaget's basic theory of the development of moral values. Like Piaget, Kohlberg focused on the development of moral judgments in children rather than on their actions. He saw the child as a "moral philosopher." Like Piaget, Kohlberg gathered his data by asking subjects questions about hypothetical stories. One of these stories has become famous as a classic ethical dilemma:

> In Europe, a woman was near death from a special kind of cancer. There was one drug that the doctors thought might save her. It was a form of radium that a druggist in the same town had recently discovered. The drug was expensive to make, but the druggist was charging ten times what the drug cost him to make. He paid $200 for the radium and charged $2,000 for a small dose of the drug. The sick woman's husband, Heinz, went to everyone he knew to borrow the money, but he could only get together about $1,000, which is half of what it cost. He told the druggist that his wife was dying and asked him to sell it cheaper or let him pay later. But the druggist said, "No, I discovered the drug and I'm going to make money from it." Heinz got desperate and broke into the man's store to steal the drug for his wife. Should the husband have done that? (Kohlberg & Colby, 1990, p. 241)

On the basis of responses to this type of dilemma, Kohlberg identified six stages in the development of moral judgment. He grouped these stages into three major levels:

1. The preconventional level (Stages 1 and 2)
2. The conventional level (Stages 3 and 4)
3. The postconventional level (Stages 5 and 6)

These levels and stages are summarized in Table 9-4, together with typical responses to the story of Heinz. Study the table carefully for a complete overview of Kohlberg's theory. Note that the stages are based not on whether the moral decision about Heinz is pro or con but on what reasoning is used to reach the decision. According to Kohlberg, people in all cultures employ the same basic moral concepts, including justice, equality, love, respect, and authority; furthermore, all individuals, regardless of culture, go through the same stages of reasoning with respect to these concepts and in the same order (Walker, de Vries, & Bichard, 1984). Individuals differ only in how quickly they move through the stage

Table 9-4   Kohlberg's Stages of Moral Development

Level	Stage	Child's Sample Response to Theft of Drug
I Preconventional	1	• Heinz shouldn't steal the drug because he might be caught and go to jail. • Heinz should steal the drug because he wants it.
	2	• It is justified because his wife needs the drug and Heinz needs his wife's companionship and help in life. • Theft is condemned because his wife will probably die before Heinz gets out of jail, so it will not do him much good.
II Conventional	3	• Heinz is unselfish in looking after the needs of his wife. • Heinz will feel bad thinking of how he brought dishonor on his family; his family will be ashamed of his act.
	4	• Theft is justified because Heinz would otherwise have been responsible for his wife's death. • Theft is condemned because Heinz is a lawbreaker.
III Postconventional	5	• Theft is justified because the law was not fashioned for situations in which an individual would forfeit life by obeying the rules. • Theft is condemned because others may also have great need.
	6	• Theft is justified because Heinz would not have lived up to the standards of his conscience if he had allowed his wife to die. • Theft is condemned because Heinz did not live up to the standards of his conscience when he engaged in stealing.

Source: Lawrence Kohlberg, "The Development of Children's Orientations toward a Moral Order," *Vita Humana*, Vol. 6 (1963), pp. 11–33; "Stage and Sequence: The Cognitive-Development Approach to Socialization," in D. A. Goslin (ed.), Handbook of Socialization Theory and Research (Chicago: Rand McNally, 1969), pp. 347–480; "Moral Stages and Moralization," in T. Lickona (ed.), *Moral Development and Behavior: Theory, Research, and Social Issues* (New York: Holt, Rinehart & Winston, 1976), pp. 31–53.

sequence and how far they progress along it. Hence, it is Kohlberg's view that what is moral is not a matter of taste or opinion—there is a universal morality.

Research by Carol Gilligan (1982a) suggests that Kohlberg's moral dilemmas capture men's, but not women's, moral development. Gilligan's work reveals that women and men have differing conceptions of morality: Men have a morality of justice, the one described by Kohlberg, and women have a morality of care (Brown & Gilligan, 1992). Overall, it seems reasonable to conclude that there is considerable merit to Kohlberg's position that the course of moral development tends to follow a regular sequence, particularly in Kohlberg's first four stages. Even so, differences exist among individuals, both in the order and in the rate of attainment of given levels.

Question

What are the major differences between cognitive learning theory and cognitive developmental theory when trying to explain how children acquire morals?

Correlates of Moral Conduct

As discussed in the previous section, moral conduct tends to vary among individuals and within the same in-

dividuals in different situational contexts. Research on morality has focused on developmental transitions and universal processes more than on individual differences. Questions regarding individual differences are somewhat discomforting, for they imply that some people are more moral than others. And most of us are loathe to label youngsters as uncaring and unprincipled (Zahn-Waxler, 1990). Even so, a number of researchers have attempted to specify which personal and situational factors are most closely associated with moral behavior (Hart & Chmiel, 1992).

Intelligence   Maturity in several of the aspects of moral reasoning described by Piaget and Kohlberg tends to be positively correlated with IQ. The relationship between IQ and honesty disappears or declines when the context is nonacademic or when the risk of getting caught is low. Overall, being smart and being moral are not the same.

Age   Research provides little evidence that children become more honest as they grow older. There may be a small correlation between age and honesty, but it seems to be due to other variables that also correlate with increasing age, such as an awareness of risk and an ability to perform the task without the need to cheat (Burton, 1976).

Do Children of All Cultures Pass Through the Same Stages of Moral Development? Lawrence Kohlberg answers this question affirmatively. Based on the cross-cultural research he and his associates undertook, Kohlberg concluded that all youngsters pass through the same, unvarying sequence of stages. Piaget also believed that autonomous morality arises out of the reciprocal interaction that takes place among peers. Children's moral development is enhanced in different ways by one-on-one games, group competitions, and individual achievements.

Sex   In the United States, girls are commonly stereotyped as being more honest than boys, but research fails to confirm this popular notion. Hartshorne and May (1928) found, for instance, that girls tended to cheat more than boys on most of the tests that the researchers observed. Other studies undertaken during the intervening quarter part of a century show no reliable sex differences in honesty (Burton, 1976).

Group Norms   The Hartshorne and May (1928) research revealed that one of the major determinants of honest and dishonest behavior was the group code. When classroom groups were studied over time, the cheating scores of the individual members tended to become increasingly similar. This result suggests that group social norms were becoming more firmly established. Other research confirms the view that groups play an

important part in providing guideposts for the behavior of their members and in channeling their members' behavior.

Motivational Factors   Motivational factors are a key influence in determining honest or dishonest behavior. Some children have a high achievement need and a considerable fear of failure, and they are likely to cheat if they believe that they are not doing as well on a test as their peers.

Prosocial Behaviors

A man was going from Jerusalem to Jericho, and he fell among robbers, who stripped him and beat him, and departed, leaving him half dead. Now by chance a priest was going down that road; and when he saw him he passed by on the other side. So likewise a Levite, when he came to the place and saw him, passed by on the other side. But a Samaritan, as he journeyed, came to where he was; and when he saw him, he had compassion, and went to him and bound up his wounds, pouring on oil and wine; then he set him on his own beast and brought him next to an inn, and took care of him. And the next day he took out two denarii and gave them to the innkeeper, saying, "Take care of him; and whatever you spend, I will repay you when I come back."

Luke 10:30–35

Moral development is not simply a matter of learning prohibitions against misbehavior. It also involves acquiring **prosocial behaviors**—ways of responding to other people through sympathetic, cooperative, helpful, rescuing, comforting, and giving acts. Some psychologists distinguish between helping and altruism. *Helping* involves behavior that benefits or assists another person, regardless of the motivation that underlies the behavior. *Altruism*, in contrast, involves behavior that is carried out in order to benefit the other person, without the expectation of an external reward. Thus, we label a behavior altruistic only if we are fairly confident that it was not undertaken in anticipation of return benefits. Younger children report more self-oriented motives and older ones more genuine concern with others (Eisenberg, 1992). According to Piaget, a child younger than 6 or 7 is too egocentric to understand another person's point of view.

Although altruistic behaviors appear quite early, not all parents want their children to be Good Samaritans. Parents commonly teach their children not to be too generous and not to give away their toys, clothes, or other possessions. And in public places they might urge their children to ignore and not worry about some

Developing Prosocial Behaviors Initially, expectations regarding proper and improper behavior are external to children. As they grow older and immerse themselves in the life of their society, most children develop self-conceptions that regulate their conduct in accordance with the standards of the group. These expectations pertain not only to prohibited behavior—what people "ought not" do—but they also define what individuals "ought" to do. Moral development also involves acquiring prosocial behaviors—ways of responding to other people through sympathetic, cooperative, helpful, rescuing, comforting, and giving acts.

nearby person who is suffering or experiencing misfortune.

Research suggests that warm, affectionate parenting is essential for the development of helping and altruistic behaviors in children (Eisenberg, 1992). Yet nurturing parenting is not enough; parents must be able to convey with a certain intensity their own concern for other living things. If a cat is hit by a car, what matters in the development of a child's prosocial behavior is whether the parent appears to care about the cat—speaks about the cat's suffering and attempts to do something to alleviate it—or seems callous and unconcerned. Or should the child hurt someone else, what matters in the child's development is that the parent describe to the child how the other person feels. Yet there is a fine line between encouraging altruism and fostering guilt; parents should not inject too much intensity into such situations lest their youngsters become overanxious. Moreover, parental warmth and nurturing alone can encourage selfishness. So parents need to provide guidelines and set limits on what youngsters can get away with (see our discussion in Chapter 8 of what Diana Baumrind calls the authoritative style of parenting).

Altruism and helping are often associated with **empathy**—feelings of emotional arousal that lead an individual to take another perspective and to experience an event as the other person experiences it. Indeed, some psychologists deem empathy to be the foundation of human morality, particularly in inducing cooperation among strangers and provoking guilt in wrongdoers (Angier, 1995). However, the single most powerful predictor of empathy in adulthood was how much time the children's fathers had spent with them. It seems that

youngsters who saw their fathers as sensitive, caring beings were themselves more likely to grow up this way. In addition, mothers' tolerance of their children's dependency—reflecting their nurturance, responsiveness, and acceptance of feelings—was related to higher levels of empathic concern among 31-year-old adults (Koestner, Franz, & Weinberger, 1990).

Although in this section we have examined important personal and situational factors in moral and prosocial behavior, we need to emphasize that human behavior occurs in physical and social settings (see our discussion of the ecological approach in Chapter 1). Historically, when searching for answers to the question of what it takes to get people to live together harmoniously and without violence, sociologists have looked to larger social forces. Beginning with Émile Durkheim (1893/1964, 1897/1951), sociologists have stressed that people, including children, need to feel part of something. They must bond with some social entity, such as a family, church, neighborhood, or community. Moreover, children require a clear set of standards that tells them what is permissible and impermissible. In sum, children need to feel bonded to a larger social whole. Once they are part of this entity, then standards make a difference to them.

Questions

How does a child develop a conscience and establish prosocial values during middle childhood? Do all children develop these? Why or why not?

Segue

Children experience substantial growth in all areas during middle childhood, and in this chapter we have looked at the physical and cognitive aspects of this growth: Children become comfortable with using their bodies and minds in new and exciting ways, and their abilities and skills become more pronounced, as some children are advanced—or delayed—either physically or intellectually in comparison to their peers. Concepts like intelligence and morality start to become important because children are entering,

often without parental or caretaker supervision, into their first social institutions (e.g., sports teams, Girl Scouts, Boy Scouts, YMCA or YWCA, Boys and Girls Club, 4-H). Schools, religious institutions, and communities are important sources of interaction for children. Cognition and the ability to interact with others allow children a myriad of possibilities for emotional and social development, and we look at those significant issues in the next chapter.

Summary

Physical Development

1. Children grow more slowly during middle childhood and become more skilled in controlling their bodies.
2. The brains of young children appear to be organized differently from those of adults. Labels applied to some children during these years include *dyslexic* and *gifted.*
3. This is the period when children are healthiest. Risks include contagious illness, accidents, obesity, eating disorders, and sedentary lifestyles.

Cognitive Development

4. An important feature of the elementary school years is a marked growth in children's cognitive sophistication. During this time, which Piaget calls the *period of concrete operations,* children achieve mastery of conservation problems. They become capable of decentering, attending to transformations, and recognizing the reversibility of operations.
5. Considerable controversy exists about whether the development of conservation can be accelerated through training procedures. There is also some question about whether the acquisition of conservation skills in the period of concrete operations occurs in the invariant sequence—horizontal décalage—postulated by Piaget.
6. Using the computer as a model for the brain, information-processing theorists ask whether during childhood there are changes in the basic processing capacity of the system (hardware) or in the type of the programs (software) used to solve a problem.
7. Although the developmental process is somewhat predictable, children vary in their rate and overall amount of change.
8. The number of dimensions along which children

conceptualize other people grows throughout childhood. The greatest increase in children's ability to distinguish people's characteristics occurs between 7 and 8 years of age.
9. Children from the ages 6 to 12 continue to acquire subtle phonological distinctions, vocabulary, semantics, syntax, formal discourse patterns, and complex aspects of pragmatics in their first language.
10. For children whose first language is not English, three different approaches have been used in the United States to open the doors to mainstream education: ESL, bilingualism, and total immersion.
11. Schools primarily use conventional measurements, such as the *Stanford-Binet* test or WISC-R (*Wechsler Intelligence Scale for Children–Revised*), because the reliability and validity of such tests are documented, the scores are stable at least for school-age children, and they correlate well with and are predictive of academic performance.
12. A variety of youngsters are designated as having learning disabilities. LD youngsters have problems in reading (dyslexia), in writing (dysgraphia), or in mathematics (dyscalcula).

Moral Development

13. Cognitive learning theory views moral development as a gradual and continuous process. Children acquire moral standards primarily through imitating the observable values and behavior of others.
14. Cognitive developmental theorists such as Jean Piaget and Lawrence Kohlberg conceive of moral development as taking place in stages, with clear-cut changes distinguishing one stage from the next.
15. A number of researchers have attempted to specify which personal and situational factors are most

closely associated with moral behavior. Intelligence, age, and sex differences play only a small part in moral conduct. Group codes and motivational factors have a much larger role.

16. Moral development involves more than simply learning prohibitions against misbehavior. It also involves acquiring prosocial behaviors.

Key Terms

autonomous morality *(289)*
bilingualism *(275)*
cognitive styles *(280)*
conservation *(268)*
dyslexia *(265)*
empathy *(292)*
ESL approach *(275)*
executive strategies *(268)*

heteronomous morality *(289)*
horizontal décalage *(269)*
inclusion *(284)*
individualized education plan (IEP) *(284, 286)*
learning disabilities (LDs) *(280)*
LEP (limited English proficiency) *(274)*

moral development *(288)*
obesity *(266)*
period of concrete operations *(268)*
prosocial behavior *(291)*
stereotypes *(273)*
total immersion *(276)*

Following Up on the Internet

Web sites for this chapter focus on physical, cognitive, and moral development of youth in middle childhood. Please access the text web site at http://www.mhhe.com/crandell7 for up-to-date hot-linked Internet addresses for the following organizations, topics, and resources:

Office of Bilingual Education and Minority Languages Affairs and the U.S. Department of Education

Information About Learning Disabilities

Recording for the Blind and Dyslexic

National Association for Gifted Children (NAGC)

Coalition for Inclusive Education

Chapter 10

Middle Childhood
Emotional and Social Development

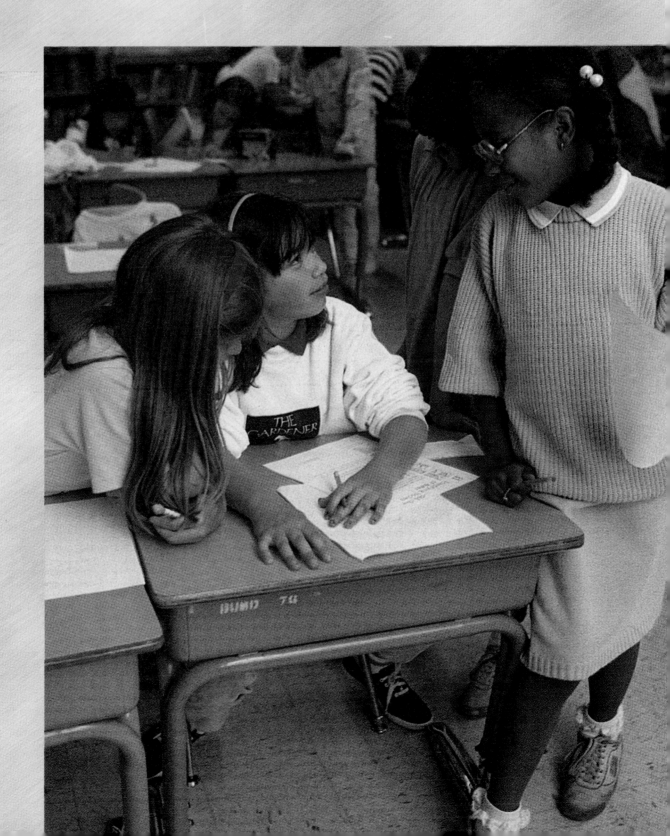

Critical Thinking Questions

1. Looking back to your childhood, do you remember disliking any particular subject or teacher when you were in elementary or middle school? On the positive side, was there a particular area of study or a teacher that you really liked? Do you remember how you felt when you were in those classes? Have those feelings carried over to how you react when you take courses in those subject areas today?

2. On a scale of 1 to 5 (with 1 being "no impact" and 5 being "most significant"), rate your mother's impact on your becoming who you are as an adult. Now do the same for your father. Is one parent becoming more influential as you get older?

3. Which would have had the greatest effect on altering your sense of self-esteem when you were 12: three friends telling you to "get lost," or your teacher telling you that she had just named her baby after you?

4. Have you experienced prejudice in any form? Do you remember committing prejudiced acts against others? Have these earlier behaviors and actions affected your attitude in any way today? Overall, did your school experience from first grade to sixth grade help or hurt your self-esteem?

As children progress into middle school, ability differences among peers become more apparent. A few excel, others experience difficulties. More immigrant children are in need of services to improve their English communication skills. Children with cognitive disabilities or with behavioral problems are assigned to special education services. Schools must meet the demands of inclusion, diversity, higher standards, and discipline.

Over time children engage in activities they enjoy, such as art, music, sports, carpentry, or cooking. Olympic Gold medalist Greg Louganis and actress Whoopi Goldberg were classified as dyslexic but excelled in other areas. They are successful as adults, but their self-esteem was affected during these years.

From the ages of 7 to 12, children form friendships through a variety of pursuits. Some youth, though, are classified as self-care children, are left on their own, and are vulnerable to victimization. School might be the only safe haven for these children, but they often find it too restrictive compared to the freedom they are used to. In this chapter we will examine the research on many of these emotional-social issues in middle childhood, many of which are alleviated by a supportive family.

The Quest for Self-Understanding

Life is with people.

Jewish Proverb

Erikson's Stage of Industry Versus Inferiority

According to Erikson's psychosocial model of development, children in middle childhood experience the fourth stage of the life cycle—**industry versus inferiority.** Think back to your own childhood, and you will probably remember this period as the time when you became interested in how things were made or how they worked. Erikson's notion of industry captures children's ability and desire to try their hands at building and working, with activities such as building models, cooking, putting things together and/or taking them apart, and solving problems of all sorts. A child who has difficulty can develop a sense of inferiority if compared to children who easily accomplish tasks. You can imagine two students in a math class: one who always has the right answer and another who tries but cannot come up with the right answer. At the beginning of the school year, the teacher will call on both students, but after a while the student who never gives the right answer will feel academically inferior to the other student and might decide to give up trying at math. In such situations, teachers are very important.

Erikson believed that good teachers were capable of instilling in students a sense of industry rather than a sense of inferiority. Likewise, if children are not given the opportunity to try their hand at constructing, acting, cooking, painting, fixing, and so forth—but are made to watch adults perform these tasks—they will develop a sense of remaining inferior to the adult who is able to accomplish these tasks, while they are relegated to observer status; imagine children who want to help bake a pizza and are told they may only watch because they might make a mess if they were allowed to actively participate. Extracurricular offerings that can *encourage* children to "jump right in" and try their hands at a variety of skills include clubs and activities such as Odyssey of the Mind, Science Olympiad, Invention Convention, talent shows, local science fairs, the school newspaper, cooking or computer instruction, scouting programs, and 4-H. Much of children's most exciting learning happens outside of the school classroom!

Self-Image

Self-image is the overall view that children have of themselves. When children are constantly praised or belittled, it is not uncommon for them to *internalize* this input and start to perceive themselves as "good" or "unworthy." Self-concept is a domain-specific assessment that children make about themselves. You might hear someone say, "I am a good athlete," or "I am horrible at math," which is very different from equating poor math performance with being a horrible person. For instance, Theresa is happy, confident, outgoing, and independent. She eagerly accepts new challenges and is not afraid to tackle problems. She thinks she is a nice, intelligent, friendly, caring child. In contrast, Lorri considers herself to be dumb, clumsy, and lacking confidence. She cannot handle being singled out for praise or criticism. She does not actively engage in new projects and tends to watch from the sidelines. Theresa is typical of the child with high self-image, whereas Lorri is a child with low self-image.

Self-Esteem

A man cannot be comfortable without his own approval.

Mark Twain, *What Is a Man?* 1906

In interacting with significant adults and peers, children get clues as to how others appraise their desirability, worth, and status. Through the accepting and rejecting behaviors of others, children continually receive answers to the questions "Who am I?" "What kind of person am I?" and "Does anyone care about me?" Central to much theory and research in social psychology is the notion that people discover themselves in the behavior of others toward them (Setterlund & Niedenthal, 1993).

The writings of social psychologists such as Charles Horton Cooley (1909, 1902) and George Herbert Mead (1934), and of neo-Freudian psychiatrists such as Harry Stack Sullivan (1953, 1947), are based on the view that our self-conceptions emerge from social interaction with others and that our self-conceptions in turn influence and guide our behavior. Considerable consensus exists among social and behavioral scientists that people have a strong and pervasive need for self-esteem (for instance, it buffers us against anxiety) (Greenberg et al., 1992). According to this social-psychological tradition, individuals' self-appraisals tend to be "reflected appraisals." If children are accepted, approved, and respected for who they are, they will most likely acquire positive, healthy attitudes of self-esteem and self-acceptance. But if the significant people in their lives belittle, blame, and reject them, they are likely to evolve negative or unhealthy attitudes toward themselves. On the whole, social psychological research has supported the overall postulate that we hold the keys to one another's self-conceptions and identities (Matsueda, 1992). However, the reflected-appraisal process often finds expression in the attitudes of *groups*—a general sense of how others view us—

Self-Appraisals as Reflected Appraisals
How we come to perceive ourselves is powerfully influenced by other people's definitions of us—how they respond to us plays a part in how we respond to ourselves. Children who are respected and approved of for what they are have more self-esteem and self-acceptance than children who aren't. This self-confidence is reflected in their achievements.

rather than necessarily in the attitudes of a specific, meaningful person.

But the fact is not simply that children can discover themselves only in the actions of others toward them. Children are not simply passive beings who mirror other people's attitudes toward them. They *actively* shape their self-conceptions as they go about their daily activities and test their power and competence in many situations. Through their interactions with others and through the effects they produce on their material environment, they derive a sense of their energy, skill, and industry.

Stanley Coopersmith (1967) studied the kinds of parental attitudes and practices that are associated with the development of high levels of self-esteem, using a sample consisting of 85 preadolescent boys. He found that three conditions were correlated with high self-esteem in these children.

- *First, the parents themselves had high levels of self-esteem and were very accepting toward their children:*

    The mothers of children with high self-esteem are more loving and have closer relationships with their children than do mothers of children with less self-esteem. . . . The child apparently perceives and appreciates the attention and approval expressed by his mother and tends to view her as favoring and supportive. He also appears to interpret her interest and concern as an indication of his significance; basking in these signs of his personal importance, he comes to regard himself favorably. This is success in its most personal expression—the concern, attention, and time of significant others.

- *Second, Coopersmith found that children with high self-esteem tended to have parents who enforced clearly defined limits:*

    Imposition of limits serves to define . . . the point at which deviation from . . . [group norms] is likely to evoke positive action; enforcement of limits gives the child a sense that norms are real and significant, contributes to self-definition, and increases the likelihood that the child will believe that a sense of reality is attainable. Consequently, such children are more likely to be independent and creative than those reared under open and permissive conditions.

- *Third, although parents of children with high self-esteem did set and enforce limits for their children's behavior, they showed respect for the children's rights and opinions:*

    Thus, within "benign limits" the children were given considerable latitude. The parents supported their children's right to have their own points of view and to participate in family decision making.

These findings from the Coopersmith study are reinforced by Baumrind's studies (1991, 1980, 1967), which found that competent, firm, accepting, and warm parenting is associated with the development of high self-esteem. By providing well-defined limits, the parents structure their children's world so that the children have effective standards by which to gauge the appropriateness of their behavior. And by accepting their children, the parents convey a warm, approving reflection that allows the children to fashion positive self-conceptions. On the whole, other researchers have confirmed Coopersmith's finding that warm and accepting parenting tends to be associated with children who have high self-esteem (Felson & Zielinski, 1989). So is "mattering," the feeling that we matter to others and that we make a difference in their lives.

Adding to the research on self-esteem is the work of Susan Harter (1983), whose *Self-Perception Profile for*

Children measures five domains of children's conceptions of self:

- Scholastic competency
- Athletic competency
- Physical appearance
- Social acceptance
- Behavioral conduct

Harter asked 8- to 12-year-olds to rate themselves in these domains and to specify how their competence in each domain affected their perceptions of themselves. What do you think the majority of children listed as the most important domain for their self-esteem? They listed physical appearance as most important, followed by social acceptance. We must keep in mind that the social-emotional support a child receives from parents, peers, and teachers also greatly influences their self-esteem. Additionally, more recent studies show that "girls who are involved in sports are more likely to attend college, have positive body images, perform better in school, and have higher test scores," says Megan Hougard, from the national Girl Scout Council (Bron-

ston, 1998). Since 1987 the number of American girls aged 6 to 11 participating in sports has risen 86 percent, according to the Sporting Goods Manufacturers Association. To bolster self-esteem in young girls, the Girl Scouts' new program, *GirlSports,* allows girls to earn badges for sports participation and has developed programs linking successful female athletes with girls' sports clinics. The U.S. Department of Health and Human Services also designed its Girl Power Web site and informational campaign to boost girls' self-esteem (Bronston, 1998).

In essence, all children's self-esteem will be affected by their own competencies, attributes, and behaviors and by the support they receive from immediate and extended family, friends, and society.

> **Questions**
>
> *What factors in an elementary-age child's environment are most likely to influence the child's sense of self-worth? Which of these factors, if any, are found to be more significant for boys or for girls?*

Self-Regulated Behaviors

Lying, stealing, destructive tendencies, fighting, and other delinquent behaviors are manifestations of emotional problems and low self-esteem. In educational and expanded societal settings, adults apply higher standards of behavior to fourth- through sixth-grade children than in earlier years. Children are expected to cooperate more on the school bus, in science labs, in group projects, on gym class teams, on the playground, in scouting activities, and in extracurricular groups. Children who repeatedly demonstrate that they cannot get along with their peers and cannot (or will not) control their overimpulsive or highly aggressive behaviors toward others are at risk for being classified by schools as *emotionally disturbed* or ED.

Public school district personnel are supposed to identify these youngsters (whom peers and teachers alike are afraid of), conduct an assessment of the child's needs, develop an individualized educational plan (IEP), and place the child into an appropriate educational setting where he (and a majority are boys, though some girls require these services) no longer will hurt or intimidate the teacher and classmates. At the heart of this situation is the child's ability—or inability, to use self-regulatory behaviors. In some children, especially those with *attention-deficit hyperactivity disorder* (see Chapter 9), features of their biochemistry make it very difficult for them to control a range of behaviors such as excessive impulsivity, short attention span, yelling out in class, jumping in and out of the chair, grabbing or hitting others,

More Girls Are Participating in Sports Research indicates that sports participation bolsters girls' self-esteem resulting in improved school attendance, higher grades, and more positive body images.

spilling things—childhood for them is something like being a human jumping bean! Some children have never been taught acceptable social or self-regulatory skills by the significant caretakers in their lives. And some parents are proud to encourage an aggressive child, thinking that the aggressiveness shows that the child can handle himself or herself in unpredictable situations—an ability that is often a necessity in some inner-city neighborhoods. In any case, these children are at high risk of being rejected by most peers, except those like themselves.

Understanding Emotion and Dealing with Fear, Stress, and Trauma

Cognitive factors also play an important part in setting the tone for the emotional life of youngsters. Recall from Chapter 6 that *emotion* involves the physiological changes, subjective experiences, and expressive behaviors that are involved in such feelings as love, joy, grief, and rage. As children interact with their mothers, fathers, siblings, peers, teachers, and others, they acquire guidelines that mentally or cognitively mediate their inner experience of emotion and their outer expression of it (Wintre & Vallance, 1994). For example, our society expects different emotional behavior from women than from men. And so parents, by responding more favorably to a daughter's sadness than to a son's sadness, encourage boys to suppress their sadness. Or by allowing a son to express his anger but encouraging a daughter to be "nice," parents suppress the daughter's expressiveness of "unladylike" feelings while allowing the boy to speak his mind (Fuchs & Thelen, 1988).

Questions

What do we mean by self-regulation? Why do parents treat girls and boys differently when they display emotions?

Children's knowledge of their own emotional experiences changes markedly from ages 7 through 12, as they increasingly attribute emotional arousal to internal causes:

- They come to know the social rules governing the display of emotion.
- They learn to "read" facial expressions with greater precision.
- They better understand that emotional states can be mentally redirected (for instance, thinking happy thoughts when in a sad state).
- They realize that people can simultaneously experience multiple emotions (Levine, 1995).
- They are able to identify their inner states and attach labels such as *anger, fear, sadness,* and *happiness* to them (Wintre & Vallance, 1994).

- They better understand how other people feel and why they feel as they do, and they become more adept at changing, containing, and hiding their feelings. For example, a 3-year-old as a guest for dinner at Grandma's house will readily say he doesn't like a particular food; the same child at 8 or 9, in the same situation, is likely to say he's not hungry so he doesn't hurt Grandma's feelings. Children also come to realize that the emotions they experience internally need not be automatically translated into overt action (rage need not become aggression), especially if they know there is someone who will listen, such as a friend, parent, or teacher.

Fear   Fear plays an important part in the lives of children of all ages. Psychologists define **fear** as an unpleasant emotion aroused by impending danger, pain, or misfortune. Some psychologists use the terms *fear* and *anxiety* interchangeably. Others define fear as a response to tangible stimuli (such as snakes or high places) and *anxiety* as a diffuse, unfocused emotional state (LaGreca & Wasserstein, 1995). However, distinguishing between fear and anxiety is often difficult. Psychologists also distinguish between fear and phobia. Whereas fear may be viewed as a normal reaction to threatening stimuli, a **phobia** is an excessive, persistent, and maladaptive fear response—usually to benign or ill-defined stimuli, such as a phobia of riding in an elevator or being afraid of spiders. Some children at this age develop what is called *school phobia.* This can occur under a variety of threatening circumstances, such as being victimized by bullies in school, while on a school bus, or when walking to and from school. Or if a parent is seriously ill at home over a period of time, the child might refuse to go to school for fear the parent will die while the child is at school. Occasionally, an insensitive teacher will ridicule or punish a child in front of peers for repeated failures in school, which can also give rise to phobia. Children who experience school phobia need professional counseling to help them cope with their excessive fear and anxiety.

As children move into the elementary school years, their fears change with their maturing cognitive and emotional understanding. Robinson and colleagues (1991) analyzed data from six normative-fear studies with child subjects from 1935 through 1988. They found that 4-year-olds are likely to have imaginary fears, including fears of dark rooms, large animals, other creatures, and monsters. Eight-year-olds tend to have unrealistic fears, such as fears of ghosts or tigers, and have other fears about being harmed bodily. Typically, preadolescents and adolescents have fears about school failures and violence in society. Gullone and King (1993) provided new normative data from a large Australian sample of children and found that children's fears are changing.

Fears about death and danger, including fear of nuclear war, were common in both this Australian sample and an American sample (Spence & McCathie, 1993). Two-thirds of U.S. elementary school children are afraid that "somebody bad" might get into their homes, and one-fourth are afraid that when they go outside, somebody might hurt them. More recent findings suggest that children seem more worried than in the past about physical harm or attack by others and about their health and that of their parents (Silverman, LaGreca, & Wasserstein, 1995). Children's fears are not completely irrational. For example, most of the children who say that they are afraid to go outside also say that in the past someone (child or adult) outside the home hurt them or made them afraid. Some children have legitimate concern that a parent might abuse them or a divorced parent might kidnap them. Note, however, that children show marked differences in their susceptibility to fear and in the sources of their fears. In more recent studies, girls seem to express more fear about wars, for example.

Few studies of children's fears have focused on any other ethnic group than white, middle-class children. More recently, though, some researchers have focused on Hispanic American or African American children as subjects, in particular those of lower socioeconomic status. Blakely (1994) interviewed 37 parents of Hispanic and non-Hispanic children from New York City. Findings from these interviews indicate that their children's greatest fears relate to social threats, such as kidnapping or sexual molestation. Hispanic children were more fearful than non-Hispanic children, and girls were more fearful than boys. The results of this study are somewhat questionable because it was a small sample and the children themselves were not interviewed. Silverman and colleagues (1995) surveyed second- through sixth-graders in an urban school and found that African American children report more fears than white or Hispanic children.

Patricia Owen (1998) examined the fears of 294 elementary-age children in San Antonio, Texas: 121 were middle-income SES, and 173 were low SES. All were of either Hispanic/Mexican or Anglo ethnicity: 163 were girls and 131 were boys. They were aged 7 to 9, in third or fourth grade, spoke English, and volunteered for this study. Owen administered the 48-item self-report instrument, the *Children's Fear Survey Schedule* (CFSS), with a few additions that were socially relevant, including items on divorce, street drugs, gunshots, gangs, being burned, and drive-by shootings. The findings indicate that children from low SES report a greater number of intense fears than did the middle SES children, though there were no significant ethnic differences. Table 10-1 lists the ten most common intense fears these children reported, by gender, SES, and ethnicity.

Girls and boys both reported significant fears of social violence. Hispanic and Anglo children differed minimally, and both reported significant fears of social violence. Children from low and mid SES differed little in their fears, which were significantly centered around social violence. Fears pertaining to death and social dangers were prominent in this study, which resemble the fears of children in past research. The pervasiveness of real-life fears (drive-by shootings, gangs, street drugs) is affecting children at younger ages, however, for children growing up on contemporary American society. Owen suggests further research on the coping and adaptive behaviors of young children exposed to these real-life dangers.

Table 10-1 Items with Highest Fear Intensity Rating for Children in San Antonio, Texas, by Gender, Socioeconomic Status (SES), and Ethnicity, 1998

Gender		SES		Ethnicity	
Girls	Boys	Low	Middle	Hispanic	Anglo
1. Drive-by shootings	1. To die	1. Drugs	1. Drive-by shootings	1. Drive-by shootings	1. Drive-by shootings
2. Kidnappers	2. Drive-by shootings	2. To die	2. Gangs	2. Burned	2. To die
3. Gangs	3. Nuclear weapons	3. Drive-by shootings	3. Gunshots	3. Gangs	3. Gunshots
4. Gunshots	4. Gangs	4. Earthquakes	4. Kidnappers	4. Nuclear weapons	4. Kidnapping
5. To die	5. Kidnappers	5. Nuclear weapons	5. To die	5. Gunshots	5. Gangs
6. Nuclear weapons	6. Earthquakes	6. Kidnappers	6. Nuclear weapons	6. Kidnapping	6. Death
7. Strangers	7. Burned	7. Gunshots	7. Drugs	7. Drugs	7. Drugs
8. Snakes	8. Death	8. Gangs	8. Burned	8. To die	8. Fire
9. Guns	9. Guns	9. Guns	9. Death	9. Snakes	9. Guns
10. Fire	10. Poison	10. War	10. Guns	10. Bad dreams	10. Nuclear weapons

From: Patricia R. Owen, "Fears of Hispanic and Anglo Children: Real-World Fears in the 1990s," *Hispanic Journal of Behavioral Sciences* (November 1998). Copyright © 1998. Reprinted by permission of Sage Publications, Inc.

Information You Can Use

Helping Children Cope with Their Fears

Fear is an inescapable and necessary human emotion. It fosters caution and prudence and increases our energy in times of danger. But fear can also outlast its usefulness or become misplaced, so that it interferes with healthy adaptations to life and, instead of helping us mobilize our resources, immobilizes us.

Many fears that children develop cannot be avoided, nor should they be. Rather, children can be encouraged to develop constructive mechanisms for coping with fear. Here are some techniques that psychologists have found useful for helping children deal with fear (Formanek, 1983):

- Create an accepting situation in which children feel at ease in sharing their fears with you. Help them to appreciate that adults, including yourself, also have fears.
- Encourage a child to acquire skills that will provide specific aid in dealing with the feared situation or object. Children are usually eager to shed their fears, and they are most successful when they themselves develop the competence to do so. Children who are afraid of the dark can be provided with the means to "control darkness," such as a night-light with a readily accessible switch next to their beds.
- Insofar as it is safe to do so, lead children by degrees into active contact with and participation in situations that they fear (an approach called *desensitization*). This technique has been successfully employed in the reduction of fears associated with snakes, spiders, heights, airplanes, and hospitals.
- Insofar as it is safe to do so, give children opportunities to encounter a feared stimulus in normal environmental

circumstances and to deal with it on their own terms. Permit them to inspect, ignore, approach, or avoid the stimulus as they see fit. And pair the feared stimulus with pleasant activities.
- Allow the child to observe the fearless actions of other children and adults when they encounter the same stimulus that the child fears. In one study a group of nursery school children who were afraid of dogs watched another child their own age happily playing with a dog. After eight sessions in which they saw the child romp with the dog, two-thirds of the children were willing to play with the experimental dog and with an unfamiliar dog as well (Bandura, Grusec, & Menlove, 1967).

In contrast, the following techniques are relatively ineffective and can complicate the child's difficulties:

- Coercing the child into contact with the feared situation by physical force, scolding, or ridicule ("Don't be such a baby!" or "Come on, pet the nice doggie. He won't hurt you!").
- Making fun of the child's fear.
- Shaming the child before others because of the fear.
- Ignoring the child's fear.
- Goading children into trying things they are not ready for, such as riding a roller coaster or diving off a diving board.

In sum, although caretakers cannot protect children from all fear (and it would be undesirable if they could), they can help children deal with their fears constructively.

Although fear can sometimes get out of hand and take on incapacitating and destructive qualities, it does serve an essential "self-preservation" function. If we did not have a healthy fear of fierce animals, fire, and speeding automobiles, few of us would be alive today.

> [The child] learns to anticipate "danger" and prepare for it. And he prepares for "danger" by means of anxiety! From this we immediately recognize that anxiety is not a pathological condition in itself but a necessary and normal physiological and mental preparation for danger. (Fraiberg, 1959, p. 11)

Of course, youngsters differ in their willingness to take risks. Some children seem much more drawn to the ex-

citement of risk than others. They seek out stimulation in ways that concern their parents and teachers. Indeed, we tend to categorize children by the frequency and appropriateness of the risks they take, describing a child as cautious or reckless and timid or bold. Risk taking finds different avenues of expression. Children who excel in sports, music, art, acting, or writing routinely take risks that their peers avoid—behaviors we deem "creative" or "courageous." But the pursuit of novelty, variety, and excitement can also lead children to seek out dangerous risks such as running away from home, stealing, experimenting with drugs, or playing with fire. To read about helping children who have fears, see *Information You Can Use: "Helping Children Cope with Their Fears."*

Question

Only 20 years ago, most American children worried about doing well in school, being liked by classmates, and meeting their parents' expectations, though a small number of children developed phobias. In comparison, what types of fears do contemporary American children, aged 7 to 12, report today?

Stress   The image of children in pain and anguish stirs our adult sense of vulnerability, concern, and indignation. Many of us are moved to intercede and attempt to heal. Yet all children must confront distressing situations. And, indeed, stress is an inevitable component of human life, so coping with stress is a central feature of human development. Psychologists view *stress* as a process involving the recognition of and response to a threat or danger. When we think and talk about stress, we usually mention its negative consequences—stomachache, tension headache, tight throat, aching back, short temper, crying jags, dizzy spells, sleepless nights, asthma attacks, and countless other unpleasant outcomes. Yet without some stress we would find life drab, boring, and purposeless. Stress can be beneficial if it contributes to personal growth and increases our confidence and skills for dealing with future events. (Many students know that fear can mobilize their study efforts the night before a major exam.) And so it is also with children. Hofferth's recently reported national study (1998) finds that for American children ages 3 through 12, parents report that "1 out of 5 children are fearful or anxious, unhappy, sad or depressed, or are withdrawn . . . and about 1 out of 25 have a behavior problem at school." Parents also reported "about 47 percent of children are rated excellent in health, friendships, and relationships."

Coping behaviors are a central aspect of stress. When we confront difficult circumstances, we typically seek ways for dealing with them. **Coping** involves the responses we make in order to master, tolerate, or reduce stress (Terry, 1994). There are two basic types of coping: problem-focused coping and emotion-focused coping (Folkman & Lazarus, 1985). *Problem-focused coping* changes the troubling situation, whereas *emotion-focused coping* changes one's appraisal of the situation.

Lacking specific information on how children respond to stress, professionals and social workers have often drawn inferences from the adult literature. But as psychologists have come to recognize that children are not miniature adults, they have increasingly turned to the study of stress and coping among children (Sorensen, 1993). In the course of this work, psychologists are finding that popular notions are not always accurate. For instance, such events as hospitalization, birth of a sibling, divorce, or war are not necessarily or universally stressful. Children's perception of such events greatly influences their stress reactions. Many of the stressors long cited by clinical psychologists as extraordinarily stressful are actually experienced by children as less stressful than such events as being ridiculed, getting lost, or receiving a poor report card (Terry, 1994).

Yet adults and children are alike in one important respect. People who feel *in control* of a situation experience a sense of empowerment (Whisman & Kwon, 1993). Individuals with a high sense of mastery believe that they can control most aspects of their lives. But those who are unable to gain mastery, to exert influence over their circumstances, feel helpless. Both children and adults with a low sense of mastery believe that their attempts at control are futile. Apparently a general sense of mastery moderates the negative effects of stress and encourages problem-focused as opposed to emotion-focused coping

Researchers find that an important moderator of our experience of stress is **locus of control**—our perception of who or what is responsible for the outcome of events and behaviors in our lives. When people perceive the outcome of an action as the result of luck, chance, fate, or powerful others, they believe in *external control.* When they interpret an outcome as the consequence of their own abilities or efforts, they believe in *internal control* (Weigel, Wertlieb, & Feldstein, 1989). Internal control typically increases with the age of a child. Scores on psychological measures of internal/external control tend to be relatively external at the third grade, with internality increasing by the eighth and tenth grades.

In evaluating matters of stress, coping, and locus of control, three factors stand out in the emerging body of research as being of particular importance:

1. The characteristics of the child
2. Developmental factors
3. Situation-specific factors

Let us examine each of these factors in turn:

- *Dispositional and temperamental differences among children play a central role in influencing their coping responses* (Kagan, 1983). Children differ in their sensitivity to the environment. Some show signs of arousal and distress to a much wider array of events than others, so they must cope with a greater number of stressful situations than do more stress-resistant youngsters. Moreover, children differ in the ways they react once they are aroused or threatened. For example, some become aggressive and enraged, others become withdrawn and sullen, and still others resort to daydreaming, fantasizing, and other escapist behaviors. Findings from the Healthy Environments, Healthy Children study (Hofferth, 1998) indicate that parents report girls experience more distress and withdrawn behavior than boys do.

- *Developmental factors also play a part.* In middle childhood, children's emerging sense of self makes them more vulnerable to events that threaten their self-esteem than they were earlier. For example, there is research evidence that children who change schools two or more times a year are likely to experience more stress, behavior problems, and problems in school (Hofferth, 1998). Also, as they move into middle school, students with learning problems or mild mental retardation experience more stressors as related to academics, peers, teachers, social support from family, and self-esteem (Wenz-Gross & Siperstein, 1998). These stressors are associated with the increased academic and social challenges for preadolescents and adolescents. Additionally, cognitive problem-solving skills differ as a function of age; as children get older, their ability to devise strategies for coping with stress expands, and they become more planful (Maccoby, 1983).

- *Situational factors influence how children experience and deal with stress.* Healthy parents often mediate many of the effects of stressful crises (Sorensen, 1993). A caretaker's irritability, anxiety, self-doubt, and feelings of incompetence are likely to intensify a child's fears of hospitalization or moving to a new school. On the other hand, support from the family can have a steeling effect, buffering the influence of stressors. Children's self-esteem is strengthened when they know they are accepted by their parents and others despite their difficulties or faults. And resourceful caretakers can help them to define and understand their problems and find ways to deal with them.

Questions

What types of physical symptoms might a child complain about who is experiencing stressful life events? What is the difference between external control and internal control? What are the three major factors in a child's coping responses to stress?

Trauma   Any extremely stressful event or trauma, even in an otherwise healthy environment, can affect a child's emotional and psychological well-being. Such events could include being personally attacked or raped; witnessing the violent abuse or death of a parent, sibling, friend, or pet; being evicted from a home; seeing mom or dad arrested and jailed; witnessing a home burning down; moving from foster home to foster home; and so

forth. Some children experience chronic child abuse or neglect; a smaller number of children experience "a single, sudden, episode of violence or victimization" (O'Maria & Santiago, 1998). Also, a significant number of children who witness domestic violence toward their mothers experience traumatic stress symptoms (Graham-Bermann & Levendovsky, 1998). In another study, Lehmann (1996) found that 56 percent of the children who were living in a battered women's shelter met the diagnostic criteria for post-traumatic stress disorder.

Children classified with **post-traumatic stress disorder (PTSD)** might exhibit a variety of symptoms including numbing and helplessness, increased irritability and aggressiveness, extreme anxiety, panic, and fears, exaggerated startle response, sleep disturbances, and bedwetting. To help these children reduce their stress, therapists and families need to restore a sense of security and safety for the child. Child-trauma experts suggest that art and play therapy illuminate the child's inner turmoil and distress, in contrast to conventional adult talk therapies. Some children need longer-term therapy before they experience a reduction in distress symptoms (O'Maria & Santiago, 1998).

"How-to" books and videos that supply programmatic strategies for assisting children living under stress often overlook the special nature of each child's individual experiences. Attentive caretakers need to understand they are dealing with the child's personality and developmental issues as well. Our consideration of emotional development, fear, and stress leads us to inquire how children encounter and manage morally relevant social situations, through family influences, a broadening social environment, and the world of school.

Questions

What changes in behavior will a child who has experienced a major trauma likely exhibit? What actions can adults take to help a traumatized child?

Continuing Family Influences

While it is important to see how children are faring on standardized tests for academic performance, as we saw in Chapter 9, children's experiences with family are also vital to their emotional health and social development. A recently updated study, the *Child Development Study* on the daily lives of American children, was conducted by the University of Michigan Institute of Social Research (Hofferth, 1998). Extensive data was collected in 1997 from a nationally representative sample of 3,586 children from 2,394 households. A majority of the children and parents in this study were white, but Asian, Hispanic, Black, and Native American children and families were also represented. Findings indicate that many children (ages 3 to 11) lead more structured lives as parents' lives have changed (Holmes, 1998).

The research project director, sociologist Sandra Hofferth, (1998) states that American children now spend less time at home and more time in structured social settings such as school and centers, more time doing household chores and less time eating, watching television, and playing than children did 16 years ago, when this study was last conducted. Children, generally, are spending more time on homework, more time with parents doing errands or household chores, and more time in organized activities. For example, in comparison to the findings in 1981, boys and girls now spend 50 percent more time in organized sports. Children's free time, or unstructured play, has declined sharply. The frequency of children and parents doing activities together rises slightly as children age. Forty percent of parents reported these shared activities with older children (Hofferth, 1998, p. 14):

- Washing or folding clothes
- Doing dishes
- Cleaning house
- Preparing food
- Looking at books and reading stories
- Talking about the family, working on homework
- Building or repairing something
- Playing on the computer or video games
- Playing board games, card games, or puzzles
- Playing sports or outdoor activities

Parents in this study rated 65 percent of the children under age 13 as extremely close or very close to their parents (including those living in stepfamilies, adopted families, and father figures in the home). As children grow older, closeness appears to decline, with parents reporting 59 percent of school-age children as close to them. Nearly 66 percent of parents reported very warm behaviors with their children, though the nature of the relationship changes as children mature and spend more time with peers. Whereas 80 percent of parents reported warmth with preschool children, 57 percent of parents of school-age children reported warmth factors (hugging, spending time together, joking, playing, talking, etc.).

Mothers and Fathers

According to this study, two-parent families with a male wage-earner and female homemaker spend an average of 22 hours per week in direct contact with their children. With both parents working in a family, parents spend 19 hours per week with their children. In general, single mothers spend 9 hours a week with their children. Much of this time together is on weekends. Children are increasingly being affected by the demands on their parents' time. Other major findings on family influences from Hofferth's (1998) study indicate that a warm relationship, parents' expectations for closeness, and parents' expectations for college completion are associated with more positive behaviors in children. Hispanic parents also rate their children more positively than non-Hispanic parents rate their children (Hofferth, 1998).

Working Mothers Over the past few decades, overall maternal employment rates increased from 53 percent in 1980 to 66 percent in 1995, with divorced mothers at 77 percent (U.S. Department of Health & Human Services, 1997d). In 1995, 70 percent of all employed mothers were working full time. Consequently, it is now common for children to either be in after-school programs or be self-care children. Research shows that mothers who work outside the home have better self-esteem, because they feel more economically secure, more confident in their ability to contribute to society, more competent, and generally more valued (Demo, 1992). And when a mother feels better about herself and her situation, she is more likely to be better able to nurture her children and will be a more effective parent. Also, Hofferth's (1998) findings indicate that the mother's verbal ability is associated with her children's higher verbal and math achievement scores, and in families with higher levels of income, children do better. Mothers who are warm and close to their children, who do activities with their children, who are involved in their children's schooling, and who expect their children to complete college are most apt to rate their children's behavior more positively (Hofferth, 1998).

However, a common feeling among many working mothers is a sense of guilt—feeling that the child is missing the mother, that the child is not receiving the best "maternal" care, and that the child is being harmed by not being home with the mother after school. On the other hand, children whose mothers work are encouraged to be more independent, and this independence benefits girls especially, as they become more competent, have more self-esteem, and perform better in academics (Bronfenbrenner & Crouter, 1983). But families with a working mother are not all cut from a single mold, and so it is impossible to say that what one family experiences will hold true for other families.

Questions

In general, what types of effects might children experience when their mother is employed? Are any of these effects benefits?

Caregiving Fathers Today there is substantial interest in the involvement of fathers in the lives of their children. It is only a stereotypical view that fathers are unconcerned about their children. According to recent studies, many men see their family role as being just as important as their working role (Aldous, Mulligan & Thoroddur, 1998; Cooksey & Fondell, 1996). Three hundred men were interviewed about their concerns regarding work, marriage, and parenting, and they were asked to evaluate the rewards of each of these domains. The men placed similar weight on the rewards stemming from work and parenting (Barnett 1992).

Cooksey and Fondell (1996) examined father-and-child living arrangement data from the *National Survey of Families and Households* to determine how fathers—both biological fathers and stepfathers—spend time with their children, what encourages them to do this, and the effects of fathers' time on children's well-being. Fathers were found to differ in the amount of time they want to spend with their children, in the activities they wish to share, and in the degree to which they take on the responsibilities of parenting. In family settings that include biological children, fathers consistently report higher frequencies of activities with children, particularly if the children are all males (Cooksey & Craig, 1998). Stepfathers who have no biological children of their own in the household report the least amount of activity with children. Fathers in father-stepmother families consistently report the highest amount of time spent with children. Fathers with only daughters, or only young children, or working long hours are less likely to take part in activities with their children. Hispanic fathers and more educated fathers are more likely to spend mealtimes with their children. Single-parent fathers spend significantly more time in activities with their children who reside with them. Black fathers were more likely than white fathers to spend time talking with their children, to spend time reading with their children, or to help with homework (Cooksey & Craig, 1998).

When no father or stepfather was present during the father's own childhood years, subjects were less likely to report participating in activities with their own children, which suggests the significance of a father as a role model for future behavior. Preteen children were found to have significantly better academic performance when their fathers share meals, spend leisure time, engage in activities, or assist with reading and homework with them. Overall findings indicate that when fathers are involved in children's lives, mothers and children experience many social-emotional benefits (Cooksey and Craig, 1998).

Most children still live with a biological father or a stepfather at least part of their childhood, and the presence or absence of a father is considered to be only one of many variables affecting a child's well-being and school achievement (Mott, 1994).

Question

How does a father's involvement, or lack of it, appear to affect children's lives? In general, in what ways do stepparents react differently to stepchildren?

Children Benefit from Father's Involvement Children benefit significantly when fathers spend time with their children, and increasing evidence suggests fathers serve as role models for future behaviors.

Sibling Relationships

Based on the recent, large national study on child well-being by Hofferth (1998), on average there are three children under 18 in a child's household. If you grew up with biological siblings, you probably remember experiencing a much more intense relationship with them than with your friends. Siblings do not have the luxury of choosing each other, as friends have, and so it is necessary to resolve conflicts and work on cooperation as a *modus vivendi*. Likewise, siblings know that an extremely angry confrontation will not bring about an end to the relationship. Even when they temporarily harbor the worst feelings for each other, siblings know that they must continue to live together. Sibling relationships are generally described as pleasant, caring, and often supportive (Jones & Costin, 1995). Older siblings normally play a role in helping younger siblings "learn the ropes"—be it with homework, coping with issues such as sex or drugs, or learning the values and morés of society (Cicirelli, 1994). However, there are times when some siblings experience conflicts in which aggression or abusive behavior can arise (Weihe, 1991). Parents play a substantial role in promoting prosocial behaviors in sibling relationships (Ross et al., 1994). Sibling relationships often improve when mothers are taught strategies for promoting child sharing and when children are directly taught prosocial-sibling behaviors (Kramer & Radey, 1997; Scarf & Grajek, 1982).

As more diverse family structures have evolved over the past two decades, more children are growing up with stepsiblings, half-brothers and half-sisters, adopted siblings, and nonrelated "siblings" in cohabiting households —which creates additional stress on all children (and adults) involved (see *Human Diversity:* "Adoption"). Sharing personal space and belongings with a sibling can cause conflicts in the best situations, but sharing personal space with children who are more or less strangers can create additional stress and anxiety for children whose parents have separated or divorced. The adults in a stepfamily might love each other, and the children might come to like each other over time, but this is not a sure thing. In the meantime, there can be disagreements and quarrels, many arising over discipline or sharing.

In many cultures older siblings take on the role of caregiver at a fairly early age and often become something of a surrogate parent. With the older sibling taking responsibility for younger siblings, the parents are able to work and pursue other activities. Research on birth order shows that the firstborn occupies a unique position; initially there are no siblings to vie for parental attention, but when siblings do come along, this can result in conflict between parents and the firstborn (Dunn, Kendrick, & MacNamee, 1982). Older siblings are generally more aggressive toward their younger siblings and but can also be more nurturing at times, whereas younger siblings are less so with their older siblings. Same-sex siblings seem to have more conflict in their interactions than siblings of different sex.

Question

What are the general findings about the role of siblings in a child's life?

Children of Divorce

Nearly half of all marriages in the United States end in divorce, though the divorce rate has been slowly declining during the 1990s. Some of these divorces follow second marriages. That divorce affects children is well documented—children experiencing divorce tend to have more problems in areas such as social interactions, behavior, and schoolwork. Each child reacts differently to the breakup of the family, depending on the child's age and temperament and the parents' competence in handling the situation (Hartup & van Lieshout, 1995). Schlesinger (1998) conducted a longitudinal study of 160 divorcing families that had children between 6 and 12 years of age. His major findings suggest the following:

- Children this age need to know what the separation is about.
- Children's concerns are very different from the concerns of parents during this time—and parents can be too occupied to notice.
- There are loyalty stresses for children, particularly when there is a high level of parental conflict. A recent study based on a national survey finds that high levels of parental conflict have a significant impact on children regardless of family structure (Vandewater & Lansford, 1998).
- To adjust to the divorce, children need a sense of safety and closeness to their parents and need their basic needs to be met.
- Children feel lower levels of conflict when their parents cooperate in matters concerning the children.

Wallerstein and Kelly (1987, 1980) found that children have six psychological tasks to complete after a divorce, and the ease of completing these tasks has much to do with how the parents handle the divorce. These are the six tasks:

- Accepting that the divorce is real
- Getting back into previous routines like school and other activities
- Resolving the loss of the family, which means having a "distant" or "absent" parent, restructured family traditions, and loss of security
- Resolving anger and self-blame, followed by forgiveness
- Accepting that the divorce will be permanent
- Believing in relationships

How divorce affects children's development is complex, with many confounding factors involved. Although most children eventually adjust, for some there are many troubled years after the divorce. In one study, children were interviewed 10 years after their parents had divorced (divorce having occurred when the children were between the ages of 6 to 8), and 75 percent of the girls and 50 percent of the boys were well adjusted. Nearly 75 percent of all the children lived with their mothers. Of the remaining children, about 4 percent lived with their fathers, the remaining lived with other relatives or in foster homes. Overall, children who maintain stable, loving relationships with both parents experience fewer emotional scars (Arditti & Keith, 1993; Kline et al., 1989). These are some factors to consider when examining the impact of divorce on children's subsequent development:

- *Age of the child at the time of the divorce.* Young children respond differently than older children, due to being in a different stage of development.

Human Diversity

Adoption

One alternative way to create a family with children is to adopt. Research shows that 11 to 24 percent of couples with infertility problems attempt to adopt (Mosher & Bachrach, 1996). Relatives and foster parents are also likely to adopt available children. Adoptions by single parents have risen from 2.5 percent to 5 percent in the 1980s to about 12 percent in the 1990s (Shireman, 1995). And co-parenting adoptions by lesbian and gay couples are on the rise (Collum, 1993). Though there are families adopting children with special needs, they are going to need a great deal of continual professional support from medical, educational, and mental health professionals, in contrast to families adopting a healthy child (Kramer & Houston, 1998).

By best estimates, 2.5 to 5 million children in the United States are adopted (Hollinger, 1998). Recent years have witnessed significant change in adoptive relationships. As news stories attest, there are many issues—birth parents versus adoptive parents, biological mothers versus biological fathers, adoptive children born at-risk and with disabilities, children's rights, surrogacy, and racial and religious compatibility (Waldman & Caplan, 1994). In 1992, the National Adoption Clearinghouse reported that 127,441 children of all races and nationalities were adopted in the United States (National Adoption Information Clearinghouse, 1996):

- 42 percent were stepparent or relative adoptions
- 37.5 percent of adoptions were managed by a private agency or lawyer
- 15.5 percent were adoptions of children from foster care
- 5 percent were adoptions of children from other countries

At the end of 1995, there were 483,000 children in foster care in the United States, and the number continues to mount. Many of these children are not babies or young children, and some come from dysfunctional homes or have established disabilities or behavioral concerns. Children under 18 were in foster care from 1982 to 1994 at a rate of 4.2 per thousand, which seems to be a low rate, but the *number* of children in foster care over this time increased *60 percent* during this period (see Figure 10-1). Both federal and state laws discourage the removal of children from their natural families unless it is necessary to ensure the child's safety. Consequently, the increase in the number of children placed in foster care indicates increased severe family dysfunction: neglect or physical or sexual abuse (U.S. Department of Health and Human Services, 1997d). Children in foster care tend to wait a long time for adoptive families, on average of 3.5 to 5.5 years (McKenzie, 1993). It is primarily younger children in foster care who are adopted, and adoptions tend to be more stable when a child in foster care is adopted at a younger age. The older the child, generally the more difficult the adoption is for the child and for the adoptive family.

Infant Adoption

Fewer women are relinquishing their infants for adoption. The rate for white women dropped from 19 percent in 1965–1972 to 3 percent in 1982–1988. The current rate for black women and Latina women is under 2 percent (Mosher & Bachrach, 1996). Domestic infant adoptions can take several years of wait time. Consequently, more families are turning to international adoptions because of their shorter wait time.

International Adoption

International adoptions are rising, though they are much more legalistic and complicated than domestic adoptions (Deacon, 1997). In 1997, 13,260 children from other countries were adopted by U.S. families; the number had risen yearly from 7,093 in 1990. The primary countries from which children were adopted internationally in 1997 were China (3,616 children), the Russian Federation (3,816), Korea (1,654), and Romania (554) (National Adoption Information Clearinghouse, 1998). The costs of international adoptions vary by agency and by country, ranging from a few thousand dollars (Haiti) to the mid teens (China, Guatemala, Russia, Romania) (Merrill, 1996). Transportation and other fees are additional. (We know someone who had to stay in China for 2 weeks while waiting for the release of her child.) Total wait time for international adoptions generally ranges from 6 months to 18 months.

Adoption Fees

The cost of adopting can be prohibitive for some families. The cost depends upon the type of service used. Public agencies charge either no fee or a minimal fee, though the family must pay attorney's fees. Religious agencies might charge from a few hundred dollars to several thousand dollars. Private agencies typically charge $10,000 to $20,000. Private adoptions can run from a few thousand dollars to the mid teens or higher if the family pays the medical costs for the pregnant mother and child. Some U.S. companies offer adoption benefits for employees, typically around $2,000. Some also cover the legal fees associated with adoptions, birth mother medical costs, and agency or placement fees (National Adoption Information Clearinghouse, 1998).

Becoming an Adoptive Parent

The root of the word *adopt* means "to choose" and "to take as one's own the creation of another." Adoption brings together the interconnected threads of somewhat paradoxical relationships. As Elinor B. Rosenberg (1992, p. 15) observes:

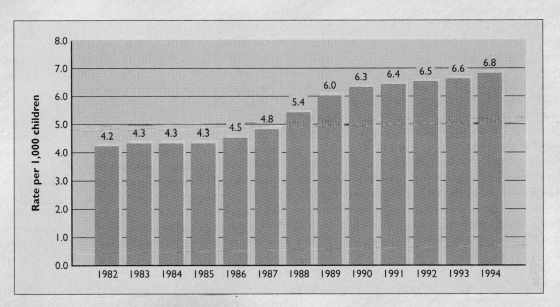

Figure 10-1 Children Living in Foster Care: 1982–1994 More children than ever are in foster care, though not all are available for adoption. A child is placed in foster care only when a court determines that the child's family cannot provide a minimally safe environment. In 1994, there were approximately 462,000 children in foster care in the United States.

Source: Tatara, Tashio. "U.S. Child Substitute Care Flow Data for FY 1993 and Trends in the State Child Substitute Care Populations," VCIS Research Notes, No. 11, August 1995. U.S. Bureau of the Census, Statistical Abstract of the United States, 1995 (Washington, DC: U.S. Government Printing Office, 1995).

"Birth parents are at once birth parents but not rearing parents; adoptive parents are rearing parents but not birth parents; adoptees are their adoptive parents' children but not their birth children, their birth parents' progeny but not their children by rearing." By virtue of the complexity of these relationships, members of the adoption circle often revisit issues of loss, separation, and insecurity in ways that differ from "regular" families. The traditional closed adoption system encourages birth parents to forget that a child was ever born, while denying any information to adoptees about their birth parents, thereby encouraging the adoptive family to live "as if" all are biologically related. But given conflicting societal messages, some adoptees decide to search for their birth parents, which typically is an emotional and painstaking business. The experiences of adult children who are reunited with their birth parents vary from happy reunions to distressing rejections (Lifton, 1994). The adoptive climate in contemporary U.S. society is much more open than in previous decades, and some states have passed legislation allowing for *mutual consent registries*—permitting parties to an adoption to meet later and allowing for a release of information if both the birth parent and the adult adoptee file formal consents of disclosure of their identities. As of February 1998, 24 states have mutual consent registries in place, and "search and consent" statutes were in place in at least 24 states (Hollinger, 1998).

Adoptive Child Well-Being

At one time it was believed that adoption invariably leads to emotional and psychological difficulties in adjustment. But a growing body of evidence demonstrates that there is no necessary relationship between adoption and psychopathology (Brodzinsky & Schechter, 1990). Indeed, a recent study of adoption found that adopted youngsters not only tend to be well adjusted, they may be better adjusted, on average, than children reared by their biological parents. A federally funded 4-year study by the Minneapolis-based Search Institute (1994), a nonprofit child research group, involving 715 families concluded that most adoptive families are thriving and that most adoptive adolescents show no signs that adoption has had a negative effect on their mental health, well-being, or identity formation. Overall, 52 percent of the adoptees said they thought about adoption less than once a month, or never, but 10 percent indicated they thought about adoption every day (10 percent of the adopted adolescents said they thought their parents would love them more if they were their biological children, 7 percent said it hurt them to know that they were adopted, and 6 percent indicated they felt unwanted).

We would do well to remind ourselves that first and foremost all children need and deserve the love and support of a family no matter what their origins in life.

- *Level of parental conflict during and after the divorce.* The conflict incurred during divorce can outweigh other positive factors on children's development.
- *Sex of child and custodial parent.* Children living with the same-sex parent were happier, more mature, and more independent, and had more self-esteem.
- *Nature of custody.* Sociologists find that children do better in mother-custody or joint-custody families than in father-custody families, when the mother is employed (Robinson, 1998).
- *Parental income.* Often the significant drop in income for mother-custody families creates considerable stress because the family must move to an apartment or lower-income housing. The children often must move out of their familiar neighborhoods, losing friends and acquaintances and frequently changing schools, along with experiencing changes in familiar family routines.

Questions

How do children learn to cope with parental separation and divorce? The stages a child goes through regarding a divorce apparently are similar to the stages of grief and bereavement over the loss of a loved one. How would you describe the stages a child aged 6 to 12 goes through to adapt to the changes in family structure and support?

Single-Parent Families

The family structure is associated with child well-being and with many future outcomes, such as rate of high school completion, substance abuse, criminal behavior, age at becoming a parent, lifetime earnings, and repeating the parents' marriage or nonmarriage pattern. An extensive amount of research comparing findings from the 1960s through the 1990s suggests that children born to single mothers, regardless of the age of the mother, are more likely to grow up in poverty, to spend their childhood without two parents, and to become single parents themselves (Ventura, 1995). However, child outcomes are likely to be more positive when single mothers are employed. Children growing up in homes with single-parent fathers are likely to fare better because males are more likely to be employed and generally earn higher wages, and single-parent fathers report they spend a great deal of time in activities with their children (McLannan & Sandefur, 1994). Here are some recent facts and figures from the recent national study *Trends in the Well-Being of America's Children and Youth* (U.S. Department of Health & Human Services, 1997d):

A Steep Increase in the Percentage of Single-Parent Families The single-parent family is under close examination by government researchers and social scientists. Some children live in single-parent families where the mother has never married; other children live with a divorced parent; some have a deceased parent; and some children are adopted by a single parent. Research suggests that children's well-being is most likely to suffer if the single-parent mother is unemployed. Recent findings from a large sample of men suggest that sons of working mothers do nearly as well professionally as those reared in two-parent homes.

- From 1960 to 1996, the proportion of children in two-parent families decreased from 88 percent to 68 percent. This change is evident in both white and black families, though the greatest decline (from 67 percent to 33 percent) was for black children living in two-parent families. Between 1980 and 1996, the proportion of Hispanic children living in two-parent families decreased from 75 percent to 62 percent. In 1996, 84 percent of Asian American children lived in two-parent households.
- Nonmarital childbearing increased considerably among mothers of all ages from 1960 (5.3 percent) to 1994 (32.6 percent). For teenage mothers aged 15 to 19, there was a dramatic change from 1960 (14.8 percent) to 1994 (75.5 percent). Nonmarital childbearing increased considerably for those aged 20 to 24, from 4.8 percent in 1960 to 44.9 percent in 1994.
- In 1996, 88 percent of single fathers were employed.

- In 1996, 77 percent of divorced mothers were employed, 83 percent on a full-time basis.
- In 1996, 66 percent of single mothers were employed.
- In 1996, 48 percent of never-married mothers were employed.

What do these numbers tell us about how living in a single-parent family affects children's development? Research studies on family structure indicate that children raised in a single-parent family headed by an unemployed mother are more likely to have problems in school, get into trouble, and have marital and parenting problems when they take on those responsibilities.

More recently, though, sociologist Timothy Biblarz from the University of Southern California reported research findings from a study of 22,761 adult men (Robinson, 1998). These male subjects were matched by occupational status, including income and education, to the family type in which they were raised. Sons of working single mothers did nearly as well professionally as those reared in two-parent homes. However, the sons of unemployed single mothers were more likely to be in the lowest-paying occupations. Biblarz states, "It seems that success—or lack of it—has more to do with finances than family structure" (Robinson, 1998). In another positive finding for single-parent families, children who report having positive interactions with nonresidential fathers or male role models have fewer problems in school and function better in both behavioral and cognitive realms (Coley, 1998). Furthermore, if children have a good relationship with the single parent and income stress is not a factor, they are inclined to be better adjusted than if they remain in a two-parent home that is a divided and hostile environment (Bray & Hetherington, 1993). Significantly, studies examining family-peer linkages have revealed that parenting styles, disciplining methods, parental support, and quality of child-parent attachment definitely influence the children's peer relationships—another important factor that promotes children's self-esteem (Stocker & Dunn, 1994). Also, because the likelihood of remarriage is relatively high over a period of time, children often enter a stepfamily structure, which can provide greater economic support but sometimes brings higher levels of emotional conflict.

Questions

Nearly every social institution, in Europe as well as in the United States, from the national to the local level, is studying the increasing number of children who live in either never-married single-parent families or divorced single-parent families, the majority of whom are headed by women. In what ways does this family structure affect children? What factors are associated with single parenting and children's well-being?

Stepfamilies

Between 75 and 80 percent of divorced parents remarry. These families are labeled *reconstituted* or *blended families,* contrasted with intact, natural families. Children can find it difficult to adjust to new parents who have their own children in tow, and the stress from trying to acclimate to the new family situation can lead to emotional and behavioral problems. Stepfathers are usually accepted by boys but can come between girls and their mothers. Therefore, girls are more likely to reject the stepfather (Bray & Hetherington, 1993). Stepparents usually adopt a *laissez-faire* attitude with their stepchildren in regard to discipline and find there may be fewer conflicts if the biological parent does the disciplining. A difference can be seen between men and women in the ways they bond with stepchildren, and stepmothers are more likely to slip into the day-to-day activities with their stepchildren whereas stepfathers are typically less involved with their stepchildren's activities (Hofferth, 1998).

Later Childhood: The Broadening Social Environment

Later childhood is a significant time when children refine their cognitive and social skills, become increasingly self-directed, and choose a large portion of their own social contacts. Many of these contacts are with peers with whom they form "chum" relationships. From 6 to 14 years of age children's conceptions of friendship show an increasing emphasis on reciprocity, intimacy, and mutual understanding (Youniss & Smollar, 1985). There is also a developmental change in what information they consider important to know about a friend.

Peer Relationships In their relationships with other youngsters of approximately the same age, young people acquire interpersonal skills essential for the management of adult life. Peer groups provide them with experiences in equalitarian relationships. In contrast, children are subordinates when dealing with adults.

The World of Peer Relationships

During later childhood, children begin focusing on a friend's preferences, such as her or his favorite games, activities, and people. As they enter adolescence, young people become increasingly concerned with a friend's internal feelings and personality traits. Hence, there is a progressive shift from concern with the observable and external qualities of a friend (he's my friend because he has a computer) to concern with a friend's internal psychological world (she's my friend because she and I like the same things and she's fun to be with). Clearly, peer relationships assume a vital role in children's development (Youniss & Smollar, 1985).

Developmental Functions of Peer Groups

Relationships with peers can help children develop interactional skills, such as communication, perspective taking, and conflict resolution (Dunn, 1993). Public approval or disapproval from peers is linked to self-worth in the middle childhood years (Harter, 1998). There are many different kinds of peer relationships and groups: a friendship, a school or neighborhood clique, a scout troop, a basketball or baseball team, a gang, and so on. Children may be simultaneously involved in a number of peer relationships, which provide them with a world of children, in contrast to a world of adults. Peer groups serve a variety of functions:

- Peer groups provide an arena in which children can exercise independence from adult controls. Because of the support they receive from their peer group, children can gain the courage and confidence they need to weaken their emotional bonds to their parents. By creating peer standards for behavior and then appealing to these standards, the peer culture also operates as a pressure group. The peer group becomes an important agency for extracting concessions for its members on such matters as bedtime hours, dress codes, choices of social activities, and amounts of spending money. It affirms children's right to a considerable measure of self-determination. Hence, the peer group furnishes an impetus for young people to seek greater freedom and provides them with support for behavior they would never dare attempt on their own. For this reason peer-group affiliations play a significant role in children's school motivation, performance, and adjustment.
- Peer groups give children experience with relationships in which they are on an equal footing with others. In the adult world, children occupy the position of subordinate, with adults directing, guiding, and controlling their activities. Group living calls for relationships characterized by sociability, self-assertion, competition, cooperation, and mutual understanding among equals (Edwards, 1994). By interacting with peers, children learn the functional and reciprocal basis for social rules and regulations. They practice "getting along with others" and subordinating their own interests to group goals. As discussed earlier, Jean Piaget views these relationships among status equals as the foundation for the stage in moral development that he terms *autonomous morality*.
- The peer group is the only social institution in which the position of children is not marginal. In it children can acquire status and realize an identity in which their own activities and concerns are supreme. Furthermore the "we" feeling—the solidarity associated with group membership—furnishes security, companionship, acceptance, and a general sense of well-being. And it helps children avoid boredom and loneliness during the unstructured hours when school is not in session.
- Peer groups are agencies for the transmission of informal knowledge, superstitions, folklore, fads, jokes, riddles, games, and secret modes of gratification. Peer groups are especially appropriate for the mastering of self-presentation and impression-management skills, because inadequate displays will often be ignored or corrected without severe loss of face. Upstairs, behind the garage, and in other out-of-the-way places, children acquire and develop many skills essential for the management of adult life.

Obviously, peer relationships are as necessary to children's development as family relationships are. The complexity of social life requires that children be involved in networks both of adults and of peers (Dunn, 1993). Sometimes, however, peer groups are in open conflict with adults, as in the case of delinquent gangs. A gang's behavior is oriented toward evading and flouting the rules and regulations of school and the larger adult-dominated society. Nondelinquent children also find themselves in conflict with parental expectations, as when they argue, "The other kids can stay out late; why can't I?" At the other extreme, the expectations of some peer groups may fully accord with those of adults. This agreement is usually true of scouting organizations, 4-H, and religious youth groups.

Questions

Why is it so important that children get along with peers? What are some of the major functions of children's peer groups?

Gender Cleavage

A striking feature of peer relationships during the elementary school years is **gender cleavage**—the tendency for boys to associate with boys and girls with girls. According to a recent large national study on the well-being of American children (Hofferth, 1998), the average child has four friends. For many children, same-gender friendships are closer and more intense in late childhood and early adolescence than in any other phase of the life span. Although the social distance between the genders is clearly present at preschool age, it increases greatly from the preschool to the school-age years and remains strong through middle childhood (Bukowski et al., 1993). Research reveals, for instance, that when first-grade children are given photographs of everyone in their class and asked to point out their best friend, 95 percent select a same-sex child. Moreover, when they are asked to point out their four best friends, 82 percent likewise choose children of the same gender. Whatever characteristics may differentiate the genders, the way boys and girls are socialized in Western nations magnifies the differences greatly. And in the early school years, peers intensify the pressure for *gender separation.*

Although first-grade children overwhelmingly name members of their own gender as best friends, girls and boys can be observed playing together on school playgrounds during recess. By the third grade, however, children have divided themselves into two gender camps. This separation tends to reach its peak around the fifth grade. Much of the interaction between groups of boys and girls at the fifth-grade level takes the form of bantering, teasing, chasing, name calling, and displays of open hostility. This "them-against-us" view serves to emphasize the differences between the genders and might func-

Gender Cleavage Often, same-sex friendships are closer during childhood and early adolescence than at any other time of life.

tion as a protective phase in life during which children can fashion a coherent gender-based identity. Some evidence suggests that gender segregation could be a universal process in human social development (Edwards & Whiting, 1988). Fifth- and sixth-grade boys are rather disgruntled about the issue of fairness. They are full of healthy energy, boisterous, and louder than most girls at this age. Consequently, teachers discipline them more, and boys loudly protest that teachers "favor the girls." Girls this age are more verbal and complain to teachers about the boys' behaviors. First-year middle school teachers be forewarned: You may feel that you're spending most of your time settling boys-against-girls disputes!

Developmental psychologist Eleanor E. Maccoby (1990, 1988) finds that gender segregation asserts itself in cultural settings in which children are in large enough numbers to permit choice. Indeed, children systematically frustrate adult efforts to diffuse their preferences for interacting with same-gender peers. For example, in modern coeducational schools, gender segregation is most marked in lunchrooms, playgrounds, on school buses, and other settings that are not structured by adults.

We should not assume, however, that *gender cleavage* is total, even in the preadolescent period. Whether working on class projects or playing handball, girls and boys intermingle (Thorne, 1993). Moreover, members of both genders show considerable interest in one another. Ten- and 11-year-old boys and girls talk a good deal about each other (for instance, who likes whom) and are often seen closely observing each other's actions.

Maccoby believes two factors give rise to gender segregation. First, boys and girls have differing styles for interacting with their peers, and so they segregate themselves into same-gender groups because they find play partners of the same gender more compatible. Second, girls have more difficulty influencing boys. Boys engage in a good deal of rough-and-tumble play—teasing, hitting, poking, pouncing, sneaking up on, mock fighting, piling on, chasing, holding, and pushing one another. Moreover, high levels of competitive and dominance-oriented behaviors prevail among boys (Dodge et al., 1990). Boys and men seem to evolve social structures—well-defined roles in games, dominance hierarchies, and team spirit—that allow them to function effectively in their preferred kind of milieu—group settings (Gurian, 1996).

Maccoby contends that rough-and-tumble play and competitive and dominance-oriented behaviors make many girls wary and uncomfortable. For instance, boys are more likely than girls to interrupt one another, use commands and threats, heckle a speaker, tell jokes or suspenseful stories, try to top another person's story, and call other youngsters names. In contrast, girls in their own groups are more likely than boys to engage in "collaborative speech acts"—they express agreement, pause to give another girl a chance to speak, acknowledge a point made by a previous speaker, smile, and provide nonverbal signals of attentiveness. In sum, boys' speech serves egoistic functions and is used to "stake out turf," whereas girls' conversation tends to be a socially binding undertaking. It is not that girls are unassertive among themselves; rather, girls pursue their ends by toning down coercive and dominance-type behaviors and using strategies that facilitate and sustain social relationships.

The prevalence of this childhood *gender cleavage* provided Sigmund Freud with his concept of the *latency period*. In Freud's view, once children no longer look on the parent of the opposite sex as a love object (thereby resolving their Oedipus or Electra conflict), they reject all members of the opposite sex until they reach adolescence. Hence according to Freud, the elementary school years are a kind of developmental plateau, one in which sexual impulses are repressed.

> **Questions**
>
> *Did you experience gender cleavage in elementary school? Who were your friends? Why are same-gender friends so prevalent during this stage of life?*

Popularity, Social Acceptance, and Rejection

Peer relationships often take on enduring and stable characteristics, in particular the properties of a **group,** defined as two or more people who share a feeling of unity and are bound together in relatively stable patterns of social interaction. Group members commonly have a psychological sense of oneness; they assume that their own inner experiences and emotional reactions are shared by the other members. This sense of oneness gives individuals the feeling that they are not merely in the group but of the group. A group's awareness of unity is expressed in many ways. One of the most important is through shared **values,** which are the criteria that people use in deciding the relative merit and desirability of things (themselves, other people, objects, events, ideas, acts, and feelings). Values play a critical part in influencing people's social interaction. They function as the standards, the social "yardsticks," that people use to appraise one another. In short, in fashioning accepting or rejecting relationships, people size up one another according to various group standards of excellence.

Peer groups are no exception. Elementary school children arrange themselves in ranked hierarchies with respect to a variety of qualities. Even first-graders have notions of one another's relative popularity or status. Consequently, children differ in the extent to which their peers desire to be associated with them. Recent research has also indicated that if children perceive their peer

group to be "different" in some way, the bonds will be much stronger between members (Bigler, Jones, & Lobliner, 1997).

One common measure used for assessing patterns of attraction, rejection, or indifference is *sociometry* (Ramsey, 1995; Newcomb, Bukowski, & Pattee, 1993). In a questionnaire or interview, the person is asked to name the three (or sometimes five) individuals in the group whom they would most like to sit next to (eat with, have as a close friend, go on a picnic with, live next to, have on their own team, or whatever) (for an example, see Figure 10-2). Researchers also use sociometry to classify children into categories based on their relative status within their peer group. These categories are used to identify children who are either extreme in status—popular, rejected, neglected, or controversial—or average in status or who differ along an acceptance or friendship continuum (Benenson, Apostolen, & Parnass, 1998). By asking people to name the individuals whom they would *least* like to interact with in a given context, teachers can identify children who are rejected by the group. Because these children are at risk for future adult developmental problems, proactive measures can be taken by school counselors or psychologists. The data derived from a sociometric study can be presented in a **sociogram,** which depicts the patterns of choice existing among members of a group at a given time. Sometimes adults overlook the impact of various ecological factors that influence early friendships and sociometric choices, such things as desk arrangements, ability groupings, and the scheduling of classes and recess.

Physical Attractiveness and Body Build   Researchers have found a good many qualities that make children appealing or unappealing in the eyes of their peers. Among the most important are physical attractiveness and body build. Studies using many different methods have reported a significant relationship between physical attractiveness and popularity (Stephan & Langlois, 1984). Furthermore, attractive strangers are typically rated as more socially desirable than are unattractive strangers. The marked agreement that is found among the members of a society on what constitutes "good looking" suggests that beauty is not in the eye of the beholder. Rather, physical attractiveness is *culturally defined* and is differently defined by different cultures.

Children begin to acquire these cultural definitions by about 6 years of age; by the age of 8 their criteria are the same as older people's. Before age 6, children's conceptions of physical attractiveness tend to be highly individualistic. Then, as children shift from the thought processes characteristic of the preoperational period to those of the period of concrete operations, they come to judge physical attractiveness in much the same manner as adults. Can you think why this might be true?

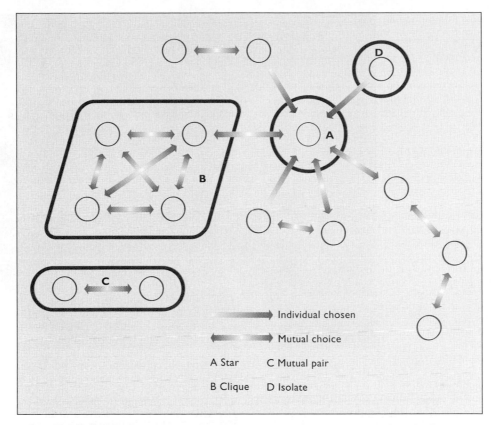

Figure 10-2 Sociometry Is a Technique Used for Identifying the Social Relations in a Group Each individual is asked to name three (sometimes five) group members with whom he or she would like to participate in a particular activity. The data gathered can be depicted graphically in a sociogram like this one. Different clusters of relationships typically emerge. The male peer group is likely to cluster as one big group of interconnected boys (as in B). The sociogram for girls is likely to be small clusters consisting of two or three close friends (as in C or A). This finding is consistent with observations about children in middle childhood and it appears to extend into adolescence. If a female has more than two best friends, these best friends were generally not best friends with one another. Note also that a few children are likely to be isolates (D) from the group and may need assistance with developing social skills.

Source: Bukowski, William & Cillessen, Antonius (eds.) *Sociometry Then and Now: Building on Six Decades of Measuring Children's Experiences with the Peer Group.* San Francisco: Jossey-Bass Publishers. (80). Summer 1998.

Stereotypes and appraisals of body configurations are also learned relatively early in life. "Lean and muscular," "tall and skinny," and "short and fat" are all bodily evaluations that influence the impressions people form of one another. Evidence suggests that negative attitudes toward "fatness" are already well developed among kindergarten children. Among boys, a favorable stereotype of the *mesomorph* (the person with an athletic, muscular, and broad-shouldered build) is evident at age 6. However, boys' desire to look like a mesomorph does not appear until age 7 and is not clearly established until age 8.

Behavioral Characteristics A variety of behavioral characteristics appear to be related to children's acceptance by their peers. Popular children tend to be described by their associates as active, outgoing, alert, self-

assured, helpful, good-natured, peppy, cheerful, and friendly. They are children who, although interested in others, do not too obviously or aggressively seek attention; they are active but not hyperactive; and they are confident but not boastful (Newcomb, Bukowski, & Pattee, 1993). Children who successfully gain entry into ongoing interaction apparently possess the ability to read the social situation and adapt their behavior to it on an ongoing basis (Putallaz & Sheppard, 1990). They take "a process view" in which they recognize that relationships take time and that they may have to work themselves into a group slowly. Several clusters of traits likewise tend to characterize children who are *unpopular* with their peers (Newcomb, Bukowski, & Pattee, 1993):

• Social isolates are physically listless, lethargic, and apathetic (or they might be experiencing an onagain, off-again chronic illness like asthma).

Standards of Popularity: Physical Attractiveness Among the most important traits influencing peer relationships and popularity are those having to do with physical attractiveness and body build. Children who meet their culture's standards enjoy a decided advantage in their peer relationships over youngsters who do not.

- Some children are so psychologically "introverted, timid, overdependent on adults, and withdrawn" that they do not have much contact with their peers.
- Children who are overbearing, aggressive, and egocentric are described by their peers and teachers as noisy, attention-seeking, demanding, rebellious, and arrogant. Such children have often been labeled "hyperactive" and placed on amphetamines in order to "control" their behavior.

Unpopular children require help from parents and teachers, and at times from professionals, because evidence suggests that they continue to have the same problems even when placed in totally new situations with unfamiliar peers (DeRosier, Kupersmidt, & Patterson, 1994). On the whole, the clusters of traits found among popular and unpopular youngsters in the United States hold cross-culturally in industrialized nations (Chen, Rubin, & Sun, 1992).

Significantly, early peer rejection in the first two months of kindergarten forecasts less favorable school perceptions among youngsters, higher levels of school avoidance, and lower performance levels. Children who experience early rejection are more likely to experience serious adjustment problems in later life. Early interventions by teachers and the school psychologist might help these children acquire more effective social skills.

Social Maturity   Children's social maturity increases rapidly during the early school years (French, 1984). In one school system, 50 percent of the first-graders said

they would rather play with younger children. This figure dropped to one-third among third-graders. Moreover, whereas 1 out of 3 first-graders said they would rather play alone, fewer than 1 in 5 third-graders expressed this preference. And although being with other children bothered 1 out of 3 first-graders, only 1 out of 5 third-graders reported this difficulty. In fact, some children go through school with few or no friends. For example, about 6 percent of third- through sixth-grade children in one school system were not selected by any classmate on a sociometric questionnaire. More recently, more than 10 percent of children from third through sixth grade reported feelings of loneliness and social dissatisfaction, and these feelings were significantly related to their sociometric status (Renshaw & Brown, 1993). Feelings of loneliness, rejection, and social isolation have profound consequences for child and adult self-esteem.

Peer influences operate in many ways. One of the most important is through the pressure that peer groups put on their members to conform to various standards of conduct. Although peer groups constrain the behavior of their members, they also facilitate interaction. They define shared goals and clarify acceptable, or unacceptable, means for pursuing these goals.

Questions

What personality characteristics does a "popular" child convey? In contrast, what traits or characteristics are likely to make a child unpopular with the group?

Racial Awareness and Prejudice

I have a dream that my four little children will one day live in a nation where they will be judged not by the color of their skin but by the content of their character.

Martin Luther King, Jr., Speech, June 15, 1963

A key aspect of many children's peer experiences involves relations with members of different racial and ethnic groups. A considerable body of research indicates that children as young as 3 can correctly identify racial differences between blacks and whites. By the age of 5 the vast majority of children can make such identifications accurately. Indeed, by age 5 youngsters already demonstrate a capacity for strong in-group bias and high levels of group cohesion. Children's perceptions and concepts about racial differences follow a developmental sequence similar to that of their perceptions and concepts about other stimuli (Clark, Lotto, & McCarthy, 1980). Their own social identities as members of particular ethnic groups evolve slowly with age, as they subjectively identify with a group and assimilate notions of ethnicity and belonging

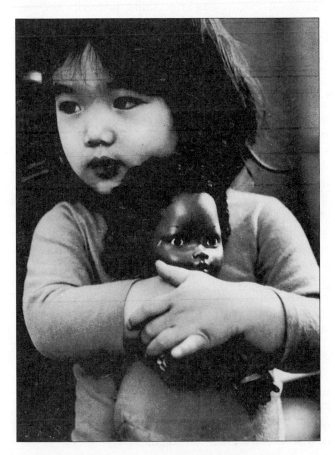

Racial Awareness The vast majority of children in our society can accurately identify a person's racial membership by the time they reach 5 years of age.

within their self-concepts (Hutnik, 1991). Labeling oneself as a member of an ethnic group is one of the earliest expressions of a child's social identity and typically is acquired by 8 years of age (Aboud, 1988, 1987; Spencer & Markstrom-Adams, 1990).

Some evidence suggests that white (dominant-group) children begin revealing own-group (in-group) biases around the age of 4. These biases apparently reach a peak between the ages of 5 and 7 (Aboud, 1988; Powlishta et al., 1994). Yet there is some doubt that children, especially younger grade-school children, show coherent, consistent **prejudice**—a system of negative conceptions, feelings, and action orientations regarding the members of a particular religious, racial, or nationality group. It is one thing to demonstrate that prejudice can develop in young children; it is quite another to say that prejudice is characteristic of young children (Powlishta et al., 1994; Westie, 1964). For instance, kindergartners and third-graders enrolled in an extraordinarily ethnically and racially diverse laboratory school located on the campus of the University of California at Los Angeles interacted and formed friendships independently of ethnic and racial memberships (Howes & Wu, 1990). At times, what adults perceive to be prejudice in children is instead a preference for other children who share similar subcultural practices and values and hence who provide a more comfortable "relational fit." Moreover, some question exists as to whether skin color is the principal determinant of racial prejudice (Sorce, 1979). Hair and eye characteristics might play an equal or even more important role (Swim et al., 1995).

Much research shows that how people act in an interracial group situation bears little or no relation to how they feel or what they think (Vander Zanden, 1987). The social setting in which individuals find themselves does much to determine their specific responses. Thus, in the United States today, a public show of blatantly racist and discriminatory behavior is commonly defined as counter to the nation's democratic ideals and as being "in poor taste." Interracial friendliness is promoted by policies that foster positive interracial contact at early ages in elementary schools and neighborhoods. For this reason, to be most effective, *desegregation* should begin at the earliest possible time, including daycare, preschool, and kindergarten. Many racial and ethnic attitudes are formed early, and adjusting to new environments and avoiding negative stereotypes is more difficult for older students than for younger ones. Indeed, the junior high school and middle school years can be the worst period in which to start desegregation (Hallinan & Williams, 1989). One of the most successful methods for reducing the hold of racism is grouping students into interracial learning teams, which, like sport teams, knit members together in common purpose that often leads to interracial friendships (Gaertner et al., 1990; Perdue et al., 1990).

Questions

How do children develop racial or ethnic prejudices? What types of behavioral interventions can be employed, and at what ages, to promote positive views among children with differences?

The World of School

I tell you we don't educate our children in school; we stultify them and then send them out into the world half-baked. And why? Because we keep them utterly ignorant of real life. The common experience is something they never see or hear. All they know is pirates trooping up the beaches in chains, tyrants scribbling edicts, oracles condemning three virgins to be slaughtered to stop some plague. Action or language, it's all the same; great sticky honeyballs of phrases, every sentence looking as though it has been plopped and rolled in poppyseed and sesame.

Gaius Petronius, Roman satirist, first century A.D.

The nature and mission of schools have been disputed through the ages. Contemporary Americans are no more in agreement about their schools than the citizens of Rome were 2,000 years ago. Controversy rages—about the content of school curricula, about teaching approaches (highly structured or informal), about busing to realize better racial mixes, about the amount and sources of school financing, about special programs for exceptional and disadvantaged youngsters, about the applications of academic freedom, and about school safety. A current approach is the development of *charter schools* or *magnet schools,* which are allowed more freedom in curriculum design and implementation.

School is the child's first big step into the larger society for most children, though home schooling is on the rise in the United States. Some U.S. children enter the world of school when they are enrolled in a day-care program, others when they attend preschool, still others when they enter elementary school. When they attend school on a regular basis, children are outside their homes and their immediate neighborhoods for several hours a day. In this new setting, teachers become a major source of potential influence. In school children also encounter other children of approximately the same age and grade level who will pass with them through a series of classrooms. Thus, school is a radical departure from a child's previous way of life. The school environment has a major influence on the development of a child's personality, intellectual capabilities, interpersonal skills, and social behavior ("The Standards: What Teachers Should Know," 1998).

Developmental Functions of Schools

What the best and wisest parent wants for his own child, that must the community want for all its children.

John Dewey

Schools came into existence several thousand years ago to prepare a select few to govern the many and to occupy certain professions. Over the past century or so, public schools have become the vehicle by which the entire population has been taught the basic skills of reading, writing, and arithmetic that an industrial, information-based, and service-oriented society requires. In contemporary society, we commonly think of schools as agencies that provide formal, conscious, and systematic training. In fact, education has become a crucial investment in the economy and a major economic resource. Higher education has also become a major military resource. Throughout the world, schools are increasingly seen as a branch of the state and as serving state purposes.

Elementary schools serve many functions. First, they teach specific cognitive skills, primarily the three Rs. School curricula—such "core" subjects as mathematics, natural science, and social science—are remarkably similar throughout the world. It seems that standardized models of mass education arose cross-culturally in conjunction with the rise of nation-states, which was closely tied to goals of national development, economic progress, and the formal integration of citizens within a larger social collectivity (Meyer, Ramirez, & Soysal, 1992). Additionally schools also inculcate more general skills, such as paying attention, sitting quietly, participating in classroom activities, and completing assignments.

Indeed, some argue that the authoritarian structure of the school mirrors the bureaucratic hierarchy of the workplace (Apple & Weis, 1983). So schooling functions to prepare students for future work conditions. Even the school grading system of A, B, C, D, and E (or F) has its parallel in the system of wage and salary scales as a device for motivating individuals. And like their counterparts in industry and service occupations, some students suffer boredom and alienation.

Second, schools have come to share with the family the responsibility for transmitting a society's dominant cultural goals and values. The schools perform a similar function in other societies. Like the United States, both China and Russia stress patriotism, national history, obedience, diligence, personal cleanliness, physical fitness, the correct use of language, and so on in their schools. With respect to basic social norms, values, and beliefs, all education indoctrinates students with what is called the "hidden curriculum."

Third, to one degree or another, schools function as a "sorting and sifting agency" that selects young people for upward social mobility. Families also influence the careers of their children by socializing them to higher educational and occupational aspirations and by providing them with the support necessary for achieving these aspirations (Sewell, 1981). Successful early schooling experiences seem to be particularly important in launching children into this process. Although early academic success does not ensure later success, early academic failure strongly predicts later academic failure (Temple & Polk, 1986).

Fourth, schools attempt to overcome gross deficits or difficulties in individual children that interfere with their adequate social functioning and participation. Often, schools work in consultation with parents, school psychologists, guidance personnel, physical therapists, speech therapists, and occupational therapists. In addition, schools meet a variety of needs not directly educational. They serve a custodial function, providing a daycare service that keeps children out from under the feet of adults and from under the wheels of automobiles (see *Further Developments:* "After-School Care and Supervision.") In higher grades they function as a dating and marriage market. And compulsory education, coupled with child labor laws, typically serves to keep younger children out of the labor market and hence out of competition with adults for jobs in the society.

Motivating Students

Most of us assume that behavior is functional and that people do certain things because the outcomes somehow meet their needs. This premise underlies the concept of motivation. **Motivation** involves the inner states and processes that prompt, direct, and sustain activity. Motivation influences the rate of student learning, the retention of information, and performance (Owen, 1995). Significantly, a gradual, overall decline occurs in various indicators of academic motivation—attention in class, school attendance, and self-perception of academic competence—across the early adolescent years (Eccles & Midgley, 1988). Here we will examine a small portion of the topic of motivation that is most relevant to our consideration of the schooling process.

Intrinsic and Extrinsic Motivation Mark Twain once observed that work consists of whatever we are obligated to do, whereas play consists of whatever we are not obligated to do. Work is a means to an end; play is an end in itself. Many psychologists make a similar distinction between extrinsic motivation and intrinsic motivation. **Extrinsic motivation** involves activity that is undertaken for some purpose other than its own sake. Rewards such as school grades, honor rolls, wages, and promotions are extrinsic, because they are independent of the activity itself and because they are controlled by

someone else. **Intrinsic motivation** involves activity that is undertaken for its own sake. Intrinsic rewards are those inherent to the activity itself and over which we have a high degree of control (Schrof, 1993).

As we have noted in earlier chapters, children want to feel effective and self-determining in dealing with their environment. Unhappily, formal education often undermines children's spontaneous curiosity and desire to learn. Children become more extrinsically motivated and less intrinsically motivated to do their schoolwork as they move into middle and junior high schools. At about this same age, many "turn off" to school and education (Eccles & Midgley, 1988). Most psychologists agree that punishment and pain impede classroom learning. But over the past two decades, they have also become aware that even rewards can be the enemies of curiosity and exploration (Ginsburg & Bronstein, 1993).

Research by Mark R. Lepper and David Greene (1975) suggests that parents and teachers can unwittingly undermine *intrinsic* motivations by providing youngsters with *extrinsic* rewards. They observed children in a school located on the campus of Stanford University and recorded which ones enjoyed drawing with felt-tipped pens of various bright colors. Then the felt pens were removed from the classrooms for two weeks. After this interval the children who had displayed interest in the pens were brought one by one to another room and asked to make drawings using the felt pens. Before beginning their drawing, one-third of the children were shown a "good player award" (a card with a gold star and a red ribbon) that they would receive upon completing their project. Another third were not told about the reward until they had finished drawing. The other third were neither told about nor given a reward.

A week later the teachers once again set out felt pens on a number of classroom tables. It was found that the children who had previously expected to receive an award for their pictures spent only half as much time playing with the felt pens as they had done before the experimental session. By contrast, children who had not received an award or had been given an award unexpectedly showed about the same interest in the felt pens after the experimental session as they had before it. Lepper and Greene take this and evidence from other studies they have conducted as showing that lavish praise, gold stars, and other extrinsic rewards serve to undermine children's intrinsic interest in many activities. They suggest that parents and educators should use extrinsic rewards only when necessary to draw children into activities that do not at first attract their interest. But even in these cases, extrinsic rewards should be phased out as quickly as possible.

Attributions of Causality Closely linked to the matter of intrinsic and extrinsic rewards is another matter—

Further Developments

After-School Care and Supervision

The number of American youngsters who routinely spend part of the day unsupervised has been mushrooming. According to a 1994 Census Bureau study, more than 1.6 million youngsters ages 5 through 14 are left home alone each day (Fields, 1994). Studies published by the Child Welfare League of America reveal that 42 percent of all American youngsters between the ages of 5 and 9 are home alone often or at least occasionally. For older children, the figure is about 77 percent. In some cases, older children become guardians for their younger brothers and sisters. For the most part, the United States as a nation has taken little notice of children left home alone except when disaster strikes—for instance, when a house fire tragically takes the lives of children, or in cases of gross negligence where parents take off for vacations and leave their children unattended (Creighton, 1993).

The consequences of lack of supervision torment many parents and concern the scientific community. Two types of questions are commonly asked: (1) What are the characteristics of families who leave their youngsters unattended? (2) Are "self-care" arrangements associated with increased social, emotional, or cognitive problems in children? Underlying these questions is an assumption that families typically use self-care only as a last resort and that unsupervised, self-care children are at greater risk for a wide range of behavior problems (Posner & Vandell, 1994).

Unhappily, the research we have in answer to these questions is less than definitive and even contradictory (Vandell & Corasaniti, 1988). Examining a sample of socially disadvantaged families, the *National Longitudinal Survey of Youth* found that type of after-school care was not related to children's gender, race, age, mothers' age, family marital composition, or geographic area. That is, there are many families using different types of after-school care, and no predictable pattern emerged.

With respect to social and cognitive development, the survey found indications that child self-care is associated with more behavioral problems than is adult-supervised care. However, when family income and family emotional support for a child are controlled (statistically allowed for),

the apparent differences in child functioning associated with type of after-school care dissipate. In sum, it seems that the critical contributor to development is not self-care per se, but a youngster's experiences within the family (Vandell & Ramanan, 1991).

The issue of after-school care and supervision is not limited to elementary school children. Many clinical psychologists believe that 12-, 13-, and 14-year-olds might be at even greater risk. Most child-care experts agree that children and adolescents today face perils that are more pernicious than a few decades ago (see Figure 10-1), particularly the increased risks associated with the accessibility of more addictive and hazardous drugs and the early age at which youngsters now begin experimenting with alcohol and sexual behaviors. One clinical psychologist observes that the lack of supervision can be more drastic for teenagers: "When a six-year-old runs away, he gets to the end of the block. When a 16-year-old runs away, she may wind up on Hollywood Boulevard prostituting herself" (Graham, 1995). In response to these concerns, over the past few years many schools across the United States—in partnership with local YMCAs, YWCAs, and Boys and Girls Clubs—have established constructive after-school programs and summer fun activities at both the elementary and the middle-school level. These programs provide safety and supervision, snacks, constructive activities, and adult care for children who would otherwise be on their own for perhaps several hours daily.

Overall, many American parents believe their children are much more competent than they really are. Although parents typically prefer not to leave a child alone at home, it is often difficult to avoid doing so, especially for families who cannot afford child care. When parents contemplate leaving a child alone, they need to consider carefully the child's maturity and the safety of the home and neighborhood. Many well-behaved children are unsupervised at times, but it seems that too many children are free to roam in parks and through neighborhoods at all hours, giving them open invitations to get into mischief, serious trouble, or harm.

people's perceptions of the factors that produce given outcomes (Hamilton, Blumenfeld, & Kushler, 1988). Consider the following experience. You have been watching a game involving your favorite football team. With five seconds left in the game and the score tied, a player on your team intercepts a pass and races for the goal line. As the player stumbles into the end zone, the

gun sounds, ending the game. Your team has won. Your friend, who preferred the other team, says, "Your guys were just lucky!" You indignantly respond, "Luck my eye! That was true ability." "Naw," exclaims another friend. "Your guys were more psyched up. They put out more effort." To which a fourth observer interjects, "Heck, it was an easy interception. No one was between him and the goal line!" Four differing explanations were set forth for the same event: luck, ability, effort, and the difficulty of the task (Weiner, 1993).

Youngsters also attribute their academic successes to these differing explanations. And it makes a considerable difference which of the explanations they employ. Educational psychologists find that when students attribute their successes to high ability, they are more likely to view future success as highly probable than if they attribute their success to other factors. By the same token, the attribution of an outcome to low ability makes future failure seem highly probable. The perception that one has failed because one has low ability is considerably more devastating than the perception that one has failed because of bad luck, lack of effort, or task difficulty (Gardner, 1991).

It seems that both success and failure feed on themselves. Students with histories of performing better than their peers commonly attribute their superior performances to high ability, and so they anticipate future success. Should they encounter periodic episodes of failure, they attribute them to bad luck or lack of effort. But youngsters with histories of low attainment typically attribute their successes to good luck or high effort and their failures to poor ability. Consequently, high attainment leads to attributions that maintain a high self-concept of ability, high academic motivation, and continued high attainment. It is otherwise for those youngsters with low attainment (Carr, Borkowski, & Maxwell, 1991).

Locus of Control   The research on attributions of causality has been influenced by the concept of locus of control. Earlier in this chapter we noted in our discussion of stress that **locus of control** refers to people's perception of who or what is responsible for the outcome of events and behaviors in their lives. Many studies have shown a relationship between locus of control and academic achievement (Smiley & Dweck, 1994). It seems that locus of control plays a mediating role in determining whether students become involved in the pursuit of achievement. Externally controlled youngsters tend to follow the theory that no matter how hard they work, the outcome will be determined by luck or chance; they have little incentive to invest personal effort in their studies, to persist in attempts to find a solution to a problem, or to change their behavior to ensure success. In contrast, internally controlled youngsters believe that their behavior accounts for their academic successes or failures and that they can direct their efforts to succeed in academic tasks. Not surprisingly, pupils who have an *internal sense of control* generally show superior academic performance (Carr, Borkowski, & Maxwell, 1991.) The impressive academic success of many Asian Americans seems related to their families' cultural belief

that "If I study hard, I can succeed, and education is the best way to succeed" (Sue & Okazaki, 1990). Low educational attainment is often associated with socioeconomic disadvantage and the effects of poverty, a topic that is being given extensive empirical scrutiny by behavioral scientists.

School Performance and Social Class

We must open the doors of opportunity. But we must also equip our people to walk through those doors.

Lyndon B. Johnson, Address, December 10, 1964

Repeatedly, studies have shown a close relationship between school performance and socioeconomic status (Hofferth, 1998; Hodgkinson, 1986; Matras, 1975). This relationship is evident regardless of the measure employed (occupation of main wage earner, family income, or parents' education). Studies have shown that the higher the social class of children's families, (1) the greater will be the number of the formal grades the children will complete, the academic honors and awards they will receive, and the effective offices they will hold; (2) the greater will be their participation in extracurricular activities; (3) the higher will be their scores on various academic achievement tests; and (4) the lower will be their rates of failure, truancy, suspensions, and premature dropping out of school. Among the hypotheses that have been advanced to explain these facts are the *middle-class bias* of schools, subcultural differences, and educational self-fulfilling prophecies.

Middle-Class Bias   Boyd McCandless (1970, p. 295) has observed that "schools succeed relatively well with upper- and middle-class youngsters. After all, schools are built for them, staffed by middle-class people, and modeled after middle-class people." Even when teachers are originally from a different social class, they still view their role as one of encouraging the development of a middle-class outlook on such matters as thrift, punctuality, respect for property and established authority, sexual morality, ambition, and neatness. Consequently, school districts nationwide are hiring teachers from a diversity of cultural backgrounds, especially those who are bilingual.

In some cases middle-class teachers, without necessarily being aware of their prejudice, find youngsters from lower socioeconomic status unacceptable—indeed, different or disobedient. Their students tend to respond by taking the attitude "If you don't like me, I won't cooperate with you." The net result is that some of these children fail to acquire basic reading, writing, and math skills and become disillusioned with the educational

"How do you know I'm an underachiever?
—Maybe you're just an overdemander!"

Figure 10-3 Achievement Is Relative
Drawing by Baloo, from *The Wall Street Journal*—Permission, Cartoon Features Syndicate.

system, which typically is the only avenue available to them to improve their well-being in the long run (Tapia, 1998) (see Figure 10-3).

Subcultural Differences   Children of different ethnic and social classes bring somewhat different experiences and attitudes into the school situation (Luster & McAdoo, 1994). Donald Hernandez (1997), in his extensive historic review of demographic trends of American childhood, states:

> Children in different racial and ethnic groups or with different immigration histories may live in family and neighborhood environments that differ sharply in
> (1) social organization, (2) economic opportunities and resources, or (3) behaviors, beliefs, and norms, including those pertaining to parent-child interaction, child-child interaction, nutrition, and child care and development with regard to play, reading, or learning new skills.

Middle-class parents generally make it clear to their children that they are expected to apply themselves to school tasks. But not all groups ascribe to this "middle-class" value. Therefore, children vary in their level of preparedness for school—for example, in their conceptions regarding a sense of time or punctuality, and use of books, pencils, drawing paper, numerals, and the alphabet.

Perhaps even more important, middle-class children are much more likely than disadvantaged youngsters to possess the conviction that they can affect their environments and their futures. With the more common violence in impoverished neighborhoods, some children say they are more concerned about making it through the day and are uncertain about their future. Members of social groups that face a job ceiling know that they do, and

this knowledge channels and shapes their children's academic behaviors (Dreeben & Gamoran, 1986). Moreover, youngsters from disadvantaged households are more likely to find themselves engulfed in recurrent school transfers; they fall behind time and again and find that they must start anew each time their parents move them. Finally, minority-group children who do not speak English are likely to find themselves educationally handicapped in schools where standard English is employed (Crawford, 1997).

Educational Self-Fulfilling Prophecies   Another explanation for social class difference is that children of lower socioeconomic status and minority children are frequently the victims of **educational self-fulfilling prophecies**—or teacher expectation effects (Jussim & Eccles, 1992). Some children fail to learn because those who are charged with teaching them do not believe that they will learn, do not expect that they can learn, and do not act toward them in ways that motivate them to learn. The hope is that as more college graduates from ethnically diverse backgrounds enter the teaching profession, they will have a better understanding of children's diverse backgrounds and serve as models to the younger generation.

Hence, James Vasquez (1998) writes that children brought up in Hispanic homes come to school with a sense of loyalty to the family, and the family is their basic support group throughout life. This is in direct contrast to the strong sense of individualism instilled in many other American youth today. Consider the differences in these two forms of hypothetical teacher praise: "This is good work, Maria. You should be proud of yourself." Or "Maria, this is good work. I'm going to send it home for your family to see." Vasquez also states that Hispanic youth thrive in an environment of cooperation with the group, which differs significantly from the mainstream value of individual competition and striving to be at the top at someone else's expense. Vasquez (1998) suggests a student-centered, three-step sequence for teachers to develop instructional strategies for adapting instruction to student cultural traits: (1) Identify individual student's cultural traits, (2) consider the *content* to be taught, the *context* in which it will be taught, and the *mode* or delivery method, and (3) write down and practice new instructional strategies (see Table 10-2).

Vasquez recommends that teachers avail themselves of educational literature, attend presentations, and read research to enhance their knowledge of the distinctive learning traits of many minority students.

Moreover, communities need to become involved, because schools and school districts do not provide equal learning opportunities, particularly facilities, equipment, and teacher quality. American inner-city school districts often claim they cannot afford to upgrade the infrastruc-

Table 10-2 Three-Step Procedure for Adapting Instruction to Cultural Traits		
Step 1: Observe/identify student trait	**Step 2: Consider content, context, and mode of delivery**	**Step 3: Verbalize or write out new instructional strategy**
Carlos does better when the material taught involves people interacting in a cooperative way.	*Content:* math concept of making change from $1, $5, and $10 *Context:* people are buying, or trading with each other. *Mode:* student dyads practice using age-appropriate objects	I will have students pair up and practice purchasing items and making change with imitation money and coins.

From: James A. Vasquez, "Distinctive Traits of Hispanic Students," *The Prevention Researcher,* Vol. 5, No. 1. Copyright 1998. Reprinted by permission.

ture of the schools because of lack of funds. Black and Hispanic students are often disadvantaged by schools in very much the same ways that their communities are disadvantaged in their interactions with major societal institutions (Cummins, 1986). To their disadvantage, many minority and lower-class youth respond by dropping out of school, some as early as middle school.

Questions

In what ways do contemporary American schools contribute to a child's overall well-being? In what ways do schools detract from a child's sense of self-worth or motivation?

Segue

Many interrelated factors are involved in children's healthy physical, cognitive, emotional, and social development throughout the elementary and middle school years. Many middle-class children have protective forces—an intact family structure, economic support, school curriculum and teacher support, friendships, community activities and recreational programs, and supervised after-school programs. Other children, particularly some at a socioeconomic disadvantage, do not develop in a healthy, resilient way because of inconsistent family supervision and the unsupportive nature of the social environments in their lives. The following factors are associated with healthy child development and competency: (1) good self-esteem, (2) optimism and a sense of hope, (3) a sense of resilience, (4) the ability to cope with fears and stress, (5) the ability to experience a range of emotions and self-regulation of emotions, (6) sociability, (7) cognitive abilities to problem-solve, (8) having regular chores at home, and (9) participating in school, church, or extracurricular activities (Determinants of Health in Children, 1996). These traits provide a solid foundation for thriving in junior high and high school environments during the adolescent stage of development, discussed in the next chapter.

Summary

The Quest for Self-Understanding

1. Erikson's psychosocial stage during middle childhood is industry versus inferiority. Children desire to try many new things and to develop their abilities. Those who are prevented from trying new activities, don't get the opportunity to try, or don't experience success in comparison to the group are likely to develop low self-esteem.

2. Children's self-concepts develop as they get feedback about their worth or status from the significant people in their lives. Children acquire positive, healthy self-esteem if they are accepted, approved, and respected. Harter identifies five domains that affect children's conceptions of self: scholastic competency, athletic competency, physical appearance, social acceptance, and behavioral conduct. More elementary-age girls

are now participating in sports; research findings indicate that girls who participate in sports develop healthier self-concepts.

3. In expanded social settings, children must also learn to regulate their own emotions, in order to get along with the group, (classmates, friends in the neighborhood, teammates, relatives). The peer group typically rejects children who cannot self-regulate their behaviors.

4. Children increasingly attribute emotional arousal to internal causes; they come to know the social rules governing the display of emotion; they learn to "read" facial expressions with greater precision; they better understand that emotional states can be mentally redirected; and they realize that people can simultaneously experience multiple emotions.

5. Fear plays an important part in the lives of young children. Children aged 5 to 6 often become afraid of imaginary creatures, the dark, and being alone or abandoned. Between 6 and 9 years of age children often fear unrealistic things, such as ghosts and monsters. Recent research indicates that across gender, ethnicity, and socioeconomic status, older children have more real-life fears of death or harm centered around social violence. Girls seem to experience more fears than boys.

6. All children experience stressful situations in response to perceived threats or dangers, yet they can be taught behavioral coping strategies to deal with stress. Two important aspects of coping with stress are a child's own sense of mastery and locus of control. Young children often believe the outcome of an action is the result of outside factors, whereas older children come to realize the outcome is moderated by their own efforts and abilities. Supportive caretakers also play a large role in buffering the effects of stress.

Continuing Family Influences

7. American children's lives have become more structured as parents' lives have changed. Children's unstructured play time has declined sharply.

8. Research findings indicate that more positive behaviors in children are associated with a warm parental relationship, parents' expectations for closeness, and parents' expectation for college completion.

9. More mothers than ever have become wage earners. Some research finds that women benefit from better economic security and higher self-esteem; other research findings suggest many working women feel guilty about their children spending time in after-school care or self-care situations. But working mothers are not all alike, and there is a great deal of variance in women's beliefs about the significance of working while raising children.

10. Most children still live with a biological father or stepfather during at least part of their childhood. But fathers differ in the amount of time they spend with children, in the activities they share with them, and in the degree to which they take on the responsibilities of parenting. Fathers in families with biological children spend the most time with children, while stepfathers with no biological children report the least amount of time spent with children. When fathers are involved in children's lives, both children and mothers benefit.

11. Results from a national study indicate that, on average, there are three children under 18 in a child's household. Sibling relationships are often more intense than relationships with friends. Siblings normally feel loyalty and support for each other but might also have conflict. Older siblings typically teach younger siblings about the practices and values of a society. More children today are living in stepfamilies. Blended family structures are likely to include stepsiblings, half-siblings, adopted siblings, and unrelated siblings. Sharing and discipline are two emotionally charged issues among siblings. Siblings in other cultures are more likely to take on the responsibility of caregiving at an early age.

12. Approximately half of U.S. marriages end in divorce, though this rate has been slowly declining. Children vary in their reactions to divorce, depending on children's age and temperament and parental competence in managing the divorce. Children's concerns about this loss are normally quite different from the parents' concerns. Studies find that children experience less stress when parents cooperate in matters concerning the children.

13. The number of children living in single-parent families has been increasing throughout the 1990s. The majority of single-parent families are headed by women, though the number of single-parent families headed by fathers is slowly rising. Single parents include those who have never married, those who are divorced or widowed, and those who have always been single and have adopted. The issue of single parenthood is highly charged in American society today because the parents in these families often need public assistance for the child's well-being. This is a tremendous cost to

society, and these children can lack necessary emotional and economic support.

14. Between 75 and 80 percent of divorced parents remarry, creating stepfamilies. Children can find it difficult to adjust to the authority of a new parent, who might or might not bring new children into the reconstituted family. The issue of who disciplines whose children, and how they are to be disciplined, can create frequent conflict in these new marital situations. In many stepfamilies, the divorced parents are also dealing with financial support issues with ex-spouses to adequately care for children. Child-stepparent attachment is likely to be a long, complex process.

Later Childhood: The Broadening Social Environment

15. Peer groups provide children with situations in which they are independent of adult controls, give them experience in egalitarian relationships, furnish them with status in a realm where their own interests reign supreme, and transmit informal knowledge.

16. Gender cleavage reaches its peak at about the fifth-grade level. Although same-gender friendships predominate during the elementary school years, children show a steady and progressive development of cross-gender interests as they advance toward puberty.

17. Elementary school children arrange themselves in hierarchies with regard to various standards, including physical attractiveness, body build, and behavioral characteristics. An observation of peer friendships and of children who are at risk of being social isolates can be done using a sociogram.

18. Children's self-conceptions tend to emerge from others' feedback regarding their desirability, worth, and status. Through their interactions with others and through the effects they produce on their material environment, they derive a sense of their energy, skill, and industry.

19. Over the elementary school years, conformity tends to increase with age in situations in which children are confronted with highly ambiguous tasks. But where the tasks are unambiguous, conformity tends to decline with age.

20. Children develop racial awareness between 3 and 5 years of age. However, whether children, especially younger school children, show coherent, consistent prejudice is doubtful.

The World of School

21. Schools teach specific cognitive skills (the three Rs), general skills associated with effective participation in classroom settings, and the society's dominant cultural goals and values. Throughout the 1980s and 1990s, American schools have been challenged by the increasing enrollment of immigrant and ethnically diverse children, as well as shifting to an inclusive education for children in special education. School buildings are used extensively today for after-school programs, while some children are unsupervised and care for themselves for long hours because more mothers and fathers are working. The issue of quality child care has become a priority in American society.

22. Although not all schools are effective, capable teachers can make a difference. Furthermore, effective schools differ from ineffective ones in important ways. Successful schools foster expectations that order will prevail and the view that learning is a serious matter.

23. Motivation influences rate of student learning, retention of information, and performance. Ideally, motivation comes from within (intrinsic motivation).

24. Overall, the higher the social class of children's families, the higher their academic achievement is likely to be. However, cross-cultural research shows that children of different racial and ethnic groups or immigration histories differ in behaviors, beliefs, and norms about children working, playing, reading, or learning new skills.

Key Terms

coping *(303)*

educational self-fulfilling prophecies *(322)*

extrinsic motivation *(319)*

fear *(300)*

gender cleavage *(313)*

group *(314)*

industry versus inferiority *(297)*

intrinsic motivation *(319)*

locus of control *(303, 321)*

motivation *(319)*

phobia *(300)*

post-traumatic stress disorder (PTSD) *(304)*

prejudice *(317)*

self-image *(297)*

sociogram *(314)*

values *(314)*

Following Up on the Internet

Web sites for this chapter focus on social-emotional issues, family and school influences, and continued identity development of children in middle childhood. Please access the text web site at http://www.mhhe.com/crandell7 for up-to-date hot-linked Internet addresses for the following organizations, topics, and resources:

Developmental Psychology Links

Girl Power

National Center on Fathers and Families

Panel Study of Income Dynamics:

National Adoption Institute

Youth Info, Department of Health & Human Services

Part Six

Adolescence

Adolescent youth generally experience rapid physical, intellectual, and emotional changes during the time of puberty. Chapter 11 presents maturity issues that accompany the adolescent growth spurt for males and females. Variations in timing of sexual maturation occur and influence both adolescent personality and behavior. Cognitive growth includes the ability to use logical and abstract thought and to plan for the future. Adolescents become more egocentric and influenced by peers than at any other time of development. Social pressures to belong or conform to a group cause some teens to experience anxiety, depression, or attempt suicide. In Chapter 12 we will discuss how teens continue to develop a sense of identity, to establish autonomy, and to explore vocational choices. Some teens begin to question authority, experiment with alcohol or other drugs, become sexually active or pregnant, and are likely to drop out of school. A small number exhibit antisocial behaviors and end up in the juvenile justice system. Many adolescents navigate through adolescence into early adulthood in a healthy fashion.

Critical Thinking Questions

1. During adolescence some children develop much faster than others. Should we separate students at this age into different schools or tracks based on how much they have developed physically or cognitively?

2. Two lifelong friends accidentally wear the exact same outfits to parties at ages 6, 16, and 60. At which age do you think this will have the most impact on them and why?

3. Think back to when you first became concerned with such things as the environment, your own nudity, politics, and whether you wanted to be seen publicly with your parents. Chances are, your awareness of these issues occurred during adolescence. Why would this be so?

4. If you were forced to get a tattoo tomorrow, but you could choose among (1) a word, (2) a face or a portrait of someone, or (3) an abstract design, which would you choose? Why? Which do you think a 13-year-old would choose: (1) picture of a cartoon character, (2) the "word" *phat,* or (3) a yellow "happy face" symbol? Why? Why would you choose or not choose the same thing?

For the first time since the 1970s the number of 12- to 19-year-olds is surging, and this increase is expected to continue until the year 2005. In the United States, adolescence is depicted as a carefree time of physical attentiveness and attraction, vitality, robust fun, love, enthusiasm, and activity. Although the majority of American teens are managing to get through their adolescent years with relatively few major problems, some find it to be a more difficult period. In contrast, in much of the world adolescence is not a socially distinct period in the human life span. Although young people everywhere undergo puberty, many assume adult status and responsibility by age 13 or younger.

This chapter focuses on the dramatic physical changes, cognitive growth, and moral challenges that adolescents experience as they make the transition from childhood to adulthood. During this transition they begin to experiment with what they see as "adult" behaviors, such as smoking, drinking, having sex, driving, and working. It is truly an exciting and sometimes frightening time, for it is probably the last time they will experience so many novel emotions and sensations in such a short period of time.

Physical Development

During adolescence young people undergo changes in growth and development that are truly revolutionary. After a lifetime of inferior size, they suddenly catch up to or surpass many adults in physical size and strength. Females typically mature earlier than males, and this becomes more evident in sixth and seventh grades, when many girls are taller than most boys. Accompanying these changes is the less evident development of the reproductive organs that signals sexual maturity. Remarkable chemical and biological changes are taking place that will, over time, fashion girls into women and boys into men.

Signs of Maturation and Puberty

Puberty is the period in the life cycle when sexual and reproductive maturation becomes evident. Puberty is not a single event or set of events but a crucial phase in a long and complex process of maturation that begins prenatally. However, unlike infants and young children, older children experience the dramatic changes of puberty through a developed sense of consciousness and self-awareness. So not only are they responding to biological changes, but their psychological states also have a significant bearing on those changes (Call, Mortimer, & Shanahan, 1995).

Hormonal Changes During Puberty

The dramatic changes that occur in children at puberty are regulated, integrated, and orchestrated by the central nervous system and glands of the endocrine system. The *pituitary gland,* a pea-sized structure located at the base of the brain, plays a particularly important role. It is called the "master gland" because it secretes into the bloodstream hormones that in turn stimulate other glands to produce their particular kind of hormone. At puberty some type of genetic timing triggers the pituitary gland to step up its production of the growth hormones, stimulates the manufacture of *estrogen* and *progesterone* in females, and stimulates cells of the testes in males to manufacture and secrete the masculinizing sex hormone, *testosterone.* A woman's eggs were created in an immature state while she was a fetus in her mother's womb. However, the hormonal changes of puberty will trigger the ovaries to release one mature egg (ovum) on a "monthly" cycle for about 30 years of her life, typically. Hence, puberty is a time when a system that was established prenatally becomes activated. Unlike females, males first begin to produce sex cells, called sperm, during puberty, and they continue to do so throughout life, unless their testes are affected by illness or are removed.

Biological Change and Cognitive Processes   In recent years researchers have taken a closer look at how biological changes influence cognitive processes during adolescence (Waber et al., 1985). The brains of children from 3 to 11 years of age seem to use twice as much energy as do the brains of adults. From 11 to 14 years of age this metabolic activity gradually falls to the adult level. Deep-sleep patterns also change. Children experience twice as much deep sleep as do adults. But again, from 11 to 14 years of age adolescents move into adult sleep patterns. Some researchers believe that the more metabolically active child's brain requires more deep sleep. The maturing brain seems to emerge as excessive or unused synapses are chipped away, possibly under the influence of hormonal changes. Such alterations in the brain could be associated with a growing capacity for *formal operational thought* (which we discuss later in this chapter).

Biological Change and Social Relations   Biological factors also have consequences for social relations during adolescence. Pubertal maturation, independent of changes in chronological age, is associated with increased emotional distance between youngsters and their parents (Steinberg, 1988). In addition, researchers at the National Institute of Mental Health have turned up evidence in support of a long-suspected link between hormones and adolescent behavior. They find that boys showing a profile of relatively low levels of testosterone and high levels of *androstenedione* are more likely than other boys to exhibit behavioral problems, including rebelliousness, talking back to adults, and fighting with classmates. The links between hormone levels and behavior tend to be considerably stronger and more consistent for boys than for girls. This research suggests that the activating influences of hormones may be reflected in the emotions of adolescents because neural tissues are the targets for some puberty-related hormones. This research, still in its infancy, indicates the need for continuing investigation of the role of hormonal processes in the behavioral development of adolescents (Graber, Brooks-Gunn, & Warren, 1995).

Even though puberty has a biological foundation, its social and psychological significance is a major determinant of how it is experienced by adolescents. For instance, boys might experience an increase in size and strength that encourages them to use aggression in achieving their goals. Or even though testosterone level is a strong predictor of sexual involvement in young women between the ages of 12 and 16, its effect is reduced or eliminated by having a father in the home or by sports participation; these environmental factors apparently reduce opportunities for sexual involvement and override hormonal effects on behavior (Udry, 1988). Hormone-level fluctuations in women have been associated with "mood swings," fluctuations that sometimes occur within a month, a week, or even a day. A pleasant, outgoing adolescent girl can become sullen and disagreeable and burst into tears and not even know why. Extreme or rapid changes in female hormone levels have also been linked to depression and unexpected behavior changes, though there is a great deal of individual variability in this. (For a discussion of PMS, or premenstrual syndrome, see Chapter 13.)

Questions

What physical changes are expected in girls and in boys during puberty? What seems to trigger the onset of this maturation?

Ethological Theory We have seen that biological factors have consequences for teenagers' social relationships. Some developmental psychologists take the argument a step further and contend that pubertal timing itself is an outcome of social experience or, at the very least, that social environmental events may influence the timing of pubertal change (Graber, Brooks-Gunn, & Warren, 1995; Wu & Martinson, 1993). Jay Belsky, Laurence Steinberg, and Patricia Draper (1991) have advanced an exceedingly controversial theory that some young mothers are responding to a pattern in human evolution that induces individuals who grow up under stressful circumstances to bear children early and often. In so doing, they address a matter that has troubled a good many U.S. policymakers, namely, the large numbers of inner-city young women in their early teens who are becoming mothers. The theory draws upon notions derived from ethology (see Chapter 2). According to this view, these teenage mothers are implementing a reproductive strategy that, from an evolutionary perspective, makes good sense. Belsky and his associates argue that youngsters growing up in dangerous conditions are "primed" to boost the chances of having their genes survive into the next generation by initiating sex early and entering motherhood early. One element of the theory is that girls reared in homes where there is a good deal of emotional stress, and especially where the father is absent, typically enter puberty at an earlier age than do girls reared in households where care and nurturance are relatively more abundant and predictable. Rather than being a biological given, puberty is said to be partially "set" by early experience. In this manner, human beings, like many other animals, adjust their life histories in response to environmental conditions in order to enhance their reproductive fitness. So experience shapes development. In sum, the model hypothesizes that young women who grow up under conditions of family stress experience behavioral and psychological problems that provoke earlier reproductive readiness.

In support of their theory, Belsky and his associates cite cross-cultural evidence that girls reared in father-absent households have an earlier onset of puberty than do girls whose fathers are present in the household.

But many developmental psychologists have expressed skepticism regarding these ethological formulations. For instance, Eleanor E. Maccoby (1991) favors a simpler explanation for the earlier pregnancies of girls from troubled homes: They receive less parental supervision. Other dissenters point out that girls tend to enter puberty at the same age as their mothers did by virtue of genetic factors (see Graber, Brooks-Gunn, & Warren, 1995). Girls who develop sexually at earlier ages are more likely to date and marry early, but they are also more likely to make the "worst" marital choices and terminate their marriages with divorce. So the girls whose parents are divorced might simply have had mothers who also tended to have undergone puberty at an earlier age. Significantly, Belsky has recently hinted that this "genetic transmission model may provide a more parsimonious account" than does his *sociobiological* model (Moffitt et al., 1992). Sociologists also distance themselves from these ethological formulations, contending that the more immediate cause of teenage sexuality and pregnancy is to be found in the lack of jobs and the presence of severe poverty in inner-city neighborhoods (Anderson, 1994).

The Adolescent Growth Spurt

During the early adolescent years, most children experience the **adolescent growth spurt,** evidenced by a rapid increase in height and weight. Usually, this spurt occurs in girls two years earlier than in boys. This means that many girls are taller than most boys in late middle school and junior high. The average age at which the peak is reached varies somewhat, depending on the people being studied. Among British and North American children it comes at about age 12 in girls and age 14 in boys. For a year or more, the child's rate of growth approximately doubles. Consequently, children often grow at a rate they last experienced when they were 2 years old. The spurt usually lasts about two years, and during this time girls gain about 6 to 7 inches and boys about 8 to 9 inches in height. By age 17 in girls and age 18 in boys, the majority of young people have reached 98 percent of their final height.

James M. Tanner (1972, p. 5), an authority on adolescent growth, writes that practically all skeletal and muscular dimensions of the body take part in the growth spurt, although not to an equal degree:

> Most of the spurt in height is due to acceleration of trunk length rather than length of legs. There is a fairly regular order in which the dimensions accelerate; leg length as a rule reaches its peak first, followed by the body breadths, with shoulder width last. Thus a boy stops growing out of his trousers (at least in length) a year before he stops growing out of his jackets. The

Adolescent Growth Spurt The rapid increase in height and weight that accompanies early adolescence tends to occur two years earlier in girls than in boys.

earliest structures to reach their adult status are the head, hands, and feet. At adolescence, children, particularly girls, sometimes complain of having large hands and feet. They can be reassured that by the time they are fully grown their hands and feet will be a little smaller in proportion to their arms and legs, and considerably smaller in proportion to their trunks.

Physical Growth   Asynchrony refers to the dissimilarity in the growth rates of different parts of the body. As a result of asynchrony, many teenagers have a long-legged or coltish appearance. Asynchrony often results in clumsiness and misjudgments of distances, which can lead to various minor accidents, such as tripping on or knocking over furniture, and to exaggerated self-consciousness and awkwardness in adolescents.

The marked growth of muscle tissue during adolescence contributes to differences between and within the sexes in strength and motor performance (Chumlea, 1982). A muscle's strength—its force when it is contracted—is proportional to its cross-sectional area. Males typically have larger muscles than females, which accounts for the greater strength of most males. Girls' performance on motor tasks involving speed, agility, and balance has generally been found to peak at about 14 years of age, though this statistic is based on past performance and does not reflect the increasing rate of female participation in junior high, high school, college, and professional sport competition. The performance of boys on similar tasks improves throughout adolescence.

At puberty the head shows a small acceleration in growth after remaining almost the same size for six to seven years. The heart grows more rapidly, almost doubling in weight. Most children steadily put on subcutaneous fat between 8 years of age and puberty, but the rate

drops off when the adolescent growth spurt begins. Indeed, boys actually tend to lose fat at this time; girls simply experience a slowdown in fat accumulation. Overall, the sequence of events in the pubertal process is similar across cultures and ethnicities (Brooks-Gunn & Reiter, 1990).

Maturation in Girls

In addition to incorporating the *adolescent growth spurt,* puberty is characterized by the development of the reproductive system (see Figure 11-1). The complete transition to reproductive maturity takes place over several years and is accompanied by extensive physical changes. As in the case of the adolescent growth spurt, girls typically begin their sexual development earlier than boys.

When puberty begins in girls, the breasts increase in size. The pigmented area around the nipple (the areola) becomes elevated, and the nipples begin to project forward. This change usually starts at about 9 or 10 years of age and is called the bud stage of breast development. In perhaps half of all girls the appearance of pubic down (soft hair in the pubic region) precedes this bud stage of breast development. Also early in puberty, hormonal action begins to produce an increase in fatty and supportive tissue in the buttocks and hip region. This perfectly normal developmental change prompts some girls in their early teen years to begin dieting or exercising to excess, which can lead to *anorexia* or *bulimia.* Many contemporary adolescent girls think their bodies should be as slender as in their childhood years. Another visible change in puberty is the growth of axillary (underarm) hair.

Menarche   The uterus and vagina mature simultaneously with the development of the breasts. However,

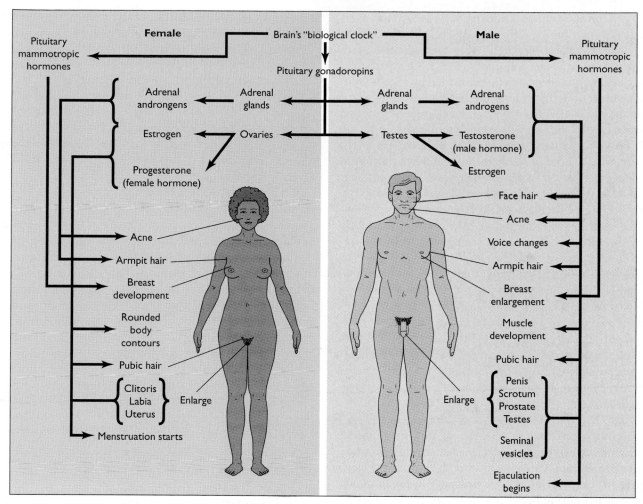

Figure 11-1 Effects of Sex Hormones on Development at Puberty At puberty the production of the pituitary gonadotropins (the follicle-stimulating hormone and the luteinizing hormone) stimulates the manufacture and secretion of the sex hormones. The release of these hormones affects a wide range of body tissues and functions.

From: HUMAN SEXUALITIES by John H. Gagnon. Copyright © 1977, Scott Foresman and Company. Reprinted by permission of Addison-Wesley Educational Publishers Inc.

menarche (me när´ key)—the first menstrual period—occurs relatively late in puberty, usually following the peak of the growth spurt. Early menstrual periods tend to be irregular, perhaps for a year or more. Furthermore, *ovulation* (the release of a mature ovum/egg) usually does not take place for 12 to 18 months after the first menstruation; hence, the girl typically remains sterile during this time.

The Earlier Onset of Menarche For over one hundred years, the average age of menarche in industrialized nations has shown a steady downward trend. The earlier onset of menarche appears to be caused largely by nutritional improvement. James M. Tanner (1972) points out that among well-nourished Western populations the onset of menarche occurs at about 12 to 13 years of age. In contrast, the latest recorded menarcheal ages are found among peoples with scarce food resources. In the highlands of New Guinea, the average age of onset is 17;

among the marginally nourished Bush people of the Kalahari, it is 16.5; in the Kikuyu poverty areas of Kenya, 15.9; and the Bantu poverty areas of South Africa, 15.5 (Worthman, 1986). Well-nourished girls of the African upper classes, however, have a median menarcheal age of 13.4.

Among European women in 1840, the average age for menarche was between 14 and 15 years, whereas today it is between 12 and 13 (Morabia & Costanza, 1998). It is important to note that a statistical average such as this does not mean that all girls will begin menstruating at this age. Some girls today begin menstruating as early as 8 or 9 years old, according to a study involving 17,000 girls aged 3 through 12 from across the United States. In this study nearly 10 percent of subjects were black and 90 percent were white. By 8 years of age, 48.3 percent of black girls and 14.7 percent of white girls had begun developing breasts, pubic hair, or both. Menstruation occurred at 12 years in blacks, on average, and at 13 in whites (Herman-Giddens et al., 1997). The average age of menstruation for white girls has stayed about the same for approximately 45 years, according to Herman-Giddens and colleagues (1997). This particular study began to investigate environmental estrogens (synthetic substances that mimic the female hormone estrogen) and their effects on sexual development. Another major study examining environmental factors and early puberty is to be concluded in 2001. Researchers at the Mount Sinai School of Medicine hope to have more conclusive information when they release their findings from "Environmental Exposures Related to Early Puberty." The potential association of cumulative environmental exposures to the onset of puberty, particularly early breast development, is being examined by researchers in many settings (Wolff et al., 1998–2001). However, though it appears that the early onset of menarche is becoming more prevalent, some girls do not menstruate until closer to 16 years old. Although girls show considerable variation in age at their first menstrual period, early menarche tends to be associated with a stout physique and late menarche with a thin physique. Menarche is also delayed by strenuous physical exercise; it occurs at about 15 years of age among dancers and athletes in affluent countries (Wyshak & Frisch, 1982).

Rose E. Frisch (1978) advances the hypothesis that menarche requires a critical level of fat stored in the body. She reasons that pregnancy and lactation impose a great caloric drain. Consequently, if fat reserves are inadequate to meet this demand, a woman's brain and body respond by limiting her reproductive ability. Frisch suggests that the improvement in children's nutrition contributes to an earlier onset of menarche because youngsters reach the critical fat/lean ratio, or "metabolic level," sooner. However, not all researchers are convinced that Frisch is correct; instead they speculate that changes in body fat are linked merely temporally rather than causally with menarche (Graber, Brooks-Gunn, & Warren, 1995).

The Significance of Menarche   Menarche is a pivotal event in an adolescent girl's experience. Most girls are both happy and anxious about their first menstruation (Paikoff & Brooks-Gunn, 1991). It is a symbol of a girl's developing sexual maturity and portends her full-fledged status as a woman. As such, menarche plays an important part in shaping a girl's image of her body and her sense of identity as a woman. Postmenarcheal girls report that they experience themselves as more womanly and that they give greater thought to their reproductive role. However, some researchers report an accentuation of conflict between the mother and the daughter shortly after menarche. This development is not necessarily negative, because it often facilitates the family's adaptation to pubertal change (Holmbeck & Hill, 1991).

In the above respects, menarche has positive connotations. But simultaneously, it is also portrayed within many Western nations as a hygienic "crisis" (Ruble & Brooks-Gunn, 1982). Menstruation is often associated with a variety of negative events, including physical discomfort, moodiness, and disruption of activities. The Westernized adolescent girl is often led to believe that menstruation is somehow unclean and embarrassing, even shameful, and such negative expectations of menstruation can prove to be self-fulfilling prophecies. Thus, preparedness for menarche is important. Indeed, the better prepared a woman feels she was as a girl, the more positive she rates the experience of menarche and the less likely she is to encounter menstrual distress as an adult (Petersen, 1983). It seems that most American girls discuss menarche with their mothers, but for the most part the content of the discussions focuses on practical concerns and symptoms rather than feelings. Fathers and daughters seldom, if ever, discuss the daughter's pubertal development. Overall, puberty creates discomfort and embarrassment for many American parents. Moreover, the issues confronting parents as they navigate their own life course influence their response to pubertal children. For instance, mothers might experience more problems with a daughter's menarcheal timing if they themselves are no longer menstruating (Paikoff & Brooks-Gunn, 1991). Cross-cultural studies on menstrual experiences are quite similar around the world, although interpretations, beliefs, and preferences are influenced by socializ-

Puberty Rites for Women   Although puberty rites are more common for boys than for girls, some societies seclude their young women at menarche. For example, upon their first menstrual period, Balese women in Zaire live apart from the village in a special hut and observe certain rituals and taboos.

ing factors such as country, religion, literacy, age, work environment, social status, and urban versus rural locale (Severy et al., 1993).

In Zimbabwe, for example, menstruation is traditionally associated with the desires of the body and is seen as being both morally and spiritually unhygienic. Traditionally, when a girl begins menstruation, she tells her grandmother, who informs the girl's mother. Her grandmother would show her how to take care of her pads and keep them ready for the next cycle. It is sometimes believed that if the pads fall into the hands of a witch, the witch can cause the girl harm through them (McMaster, Corme, & Pitts, 1997) (see *Human Diversity:* "Cultural Practices of Female Circumcision.")

Maturation in Boys

The first sign of puberty in males begins at about 11 to 11.5 years of age with an acceleration in the growth of the testes and scrotum, followed by the appearance of fine, straight hair at the base of the penis (see Figure 11-1). When boys are 13 to 16 years of age, their pubic hair multiplies, the testes and scrotum continue growing, and the penis lengthens and thickens. A boy's voice begins to change as the larynx enlarges and the vocal folds double in length (resulting at times in the embarrassing cracking of the adolescent boy's voice). When a boy is about 13 to 14 years of age, the *prostate gland* is producing fluid that can be ejaculated during orgasm. However, mature *sperm* are not present in the *ejaculatory fluid* until about a year later, although there is a wide variation among individuals. Also around 14 to 15 years of age, boys begin to have "wet dreams," involuntary emissions of seminal fluid during sleep. For most contemporary males, their first ejaculation elicits both very positive and slightly negative responses. Given greater openness about the matter today than in the past, most boys are somewhat prepared for the event (Gaddis & Brooks-Gunn, 1985).

Axillary and facial hair generally make their first appearance about two years after the beginning of pubic hair growth, though in some males axillary hair appears before pubic hair. The growth of facial hair begins with an increase in the length and pigmentation of the hair at the corners of the upper lip, which spreads to complete the mustache. Next, hair appears on the sides of the face in front of the ears and just below the lower lip, and finally it sprouts on the chin and lower cheeks. Facial hair is downy at first but becomes coarser by late adolescence.

Whereas girls develop fat deposits in the breasts and the hip region, boys acquire additional weight and size in the form of increased muscle mass. Furthermore, whereas the female pelvis undergoes enlargement at puberty, the most striking expansion in males takes place in the shoulders and rib cage (Chumlea, 1982). Some boys and girls experience dramatic body shape changes from

junior high through high school, which creates a time of adjustment. American society seems to equate male taller height with popularity, sex appeal, and success. Most boys want to be tall and talk about this a great deal during this growth time. Tall girls, however, might feel more self-conscious about their height during adolescence.

The Impact of Early or Late Maturation

Children show enormous variation in timing and rates of growth and sexual maturation. As Figure 11-2 demonstrates, some children do not begin their growth spurt and the development of secondary sexual characteristics until other children have virtually completed these stages (Tanner, 1973). A study of 781 girls in a middle-class Boston suburb revealed that age at menarche ranged from 9.1 to 17.7 years (Zacharias, Rand, & Wurtman, 1976). Thus, one cannot appreciate the facts of physical growth and development without taking account of individual differences.

The majority of young people in the United States move in chronological lockstep through elementary and secondary school. Consequently, fairly standardized criteria are applied to children of the same age with respect to their physical, social, and intellectual development. But because children mature at varying rates, they differ in their ability to meet these standards. Individual differences become most apparent at adolescence. Several studies over the past two decades confirm that whether adolescents mature early or late has important consequences for them in their relationships with both adults and peers.

Because of different rates of maturation, some adolescents have an advantage in the "ideals" associated with height, strength, physical attractiveness, and athletic prowess (Hayward et al., 1997; Stattin & Magnusson, 1990). Hence, some young people receive more favorable

Some Youth Are Favored by Early Maturation Because athletic excellence is usually prized among American adolescents, those boys who mature early are at a decided advantage in winning the admiration of their peers.

Human Diversity

Cultural Practices of Female Circumcision

In stark contrast to the experience of puberty by American teens, many young girls living in non-Westernized countries experience serious health risks or death before or during puberty, directly related to their developing sexuality. According to conservative estimates, nearly 110 million girls from countries in continental Africa, the Persian Gulf, and some parts of Indonesia and Malaysia are forced to undergo a radical "surgical" procedure called *female genital mutilation* (FGM), or variations of female circumcision called *excision* or *infibulation* (Hosken, 1998; Brady, 1998). The average age of girls who are forced to undergo these procedures is 4 to 7 years old (Brady, 1998). As more immigrants from these countries enter the United States, their practice of female mutilation continues. Health services professionals in the United States, Europe, and Australia are seeing evidence of this practice on female infants and young daughters of immigrants from Somalia, the Sudan, Ethiopia, Kenya, Nigeria, and some Muslims.

This "medical" procedure, often done on groups of girls at a time, is normally performed by older women, often reusing the same unsterilized, crude knife, piece of broken glass, razor blade, or scissors in unsanitary conditions (Brady, 1998).

> Infibulation or pharaonic circumcision [excision with infibulation] means that the entire clitoris and the labia

minora are cut away and the two sides of the labia majora are partially sliced off or scraped raw and then sewn together, often with catgut. In Sudan or Somalia, thorns are used to hold the two bleeding sides of the vulva together, or a paste of gum arabic, sugar and egg is used [or horsehair or catgut]. The introitus or entrance to the vagina is thus obliterated which is the purpose of the operation, except for a tiny opening in the back to allow urine, and later menstrual blood, to drain. The legs of the girl are tied together immediately after the operation, and she is immobilized for several weeks until the wound of the vulva has closed, except for a small opening that is created by inserting a splinter of wood or bamboo. (Hosken, 1998).

Bloodborne pathogens such as HIV and hepatitis B virus are easily transmitted during these procedures. Many girls die of hemorrhage or infection. Those who survive often have health complications of urinary tract infections, pelvic inflammatory disease, complications of pregnancy and delivery, or infertility. (An infertile woman can be divorced at will in these cultures and shamed for life.) Many need to undergo deinfibulation because fetal descent is obstructed, which can result in fetal death (Brady, 1998).

These centuries-old cultural practices from male-dominated societies are directly related to curbing the girls' developing sexuality and to guaranteeing that the girl is a

feedback regarding their overall worth and desirability, which in turn influences their self-image and behavior. For example, the value placed on manly appearance and athletic excellence means that early-maturing boys often enjoy the admiration of their peers. In contrast, late-maturing boys often receive negative feedback from their peers and hence can be more susceptible to feelings of inadequacy and insecurity (Meschke & Silbereisen, 1997).

Investigators at the University of California at Berkeley studied the physical and psychological characteristics of a large group of individuals over an extended period of time. On the basis of this work, Mary Cover Jones and Nancy Bayley (1950, p. 146) reached the following conclusion regarding adolescent boys:

> Those [boys] who were physically accelerated are usually accepted and treated by adults and other children as more mature. They appear to have relatively little need to strive for status. From their ranks came the outstanding student-body leaders in senior high

school. In contrast, the physically retarded [delayed] boys exhibit many forms of relatively immature behaviors: this may be in part because others tend to treat them as the little boys they appear to be. Furthermore, a fair proportion of these boys give evidence of needing to counteract their physical disadvantage in some way—usually by greater activity in striving for attention, although in some cases by withdrawing.

Boys who perceived themselves to be late maturing also tended to exhibit feelings of inadequacy, negative self-concept, and feelings of rejection. These feelings were coupled with a rebellious quest for autonomy and freedom from restraint (Mussen & Jones, 1957). Results of a study of college students by Donald Weatherley (1964) largely confirm the findings of Jones and Bayley. Late-maturing boys of college age were less likely than their earlier-maturing peers to have resolved the conflicts attending the transition from childhood to adulthood.

virgin at marriage. Several reasons are cited for why so many African and Middle Eastern groups still maintain this cruel practice (Hosken, 1998):

- This procedure has been handed down by ancestors to initiate the young girls (and males, in the case of male circumcision) into adult life. Girls and women must comply with the wishes of their ancestors. Some women view their genital mutilation as a source of pride in their households and community.
- The girl who has undergone infibulation will bring a higher price as a bride, because her virginity and purity are intact. Payment to the bride's father is still a prevalent practice in these cultures.
- In these cultures, young women are deemed incapable of controlling their sexual urges and might bring disgrace to their families. If a girl or young woman refuses to undergo this procedure, in many societies she will be considered a prostitute.
- Some men say the external genitalia of women are ugly if they are not removed.
- Others say that cleanliness, or purity, and better health are reasons for undergoing this procedure.
- Religious beliefs are also a significant factor. Generally, Muslims and some Christians in Africa also practice this procedure.
- It is believed in West Africa that the woman's clitoris represents "maleness" and the prepuce of the male's penis represents "femaleness," and that these tissues

must be removed before a person can be accepted as an adult into society.

Girls and women who experience the infibulation procedure, which nearly closes up the vaginal opening, will experience this procedure repeatedly during their reproductive lifetime: the scarred tissue will be cut open to allow for intercourse and birth of children, but sewn up again until children are weaned; then cut open again to allow for intercourse and conception, and then sewn up again; and so on.

There is no doubt that the forced practice of genital mutilation has terrible physical, health, and psychological consequences for these young women. Hoskens (1998), who has been an advisor to the World Health Organization and has witnessed this cruelty firsthand, states it is a violation of basic human rights because the procedure is performed on nonconsenting infants, girls, and women who are not anesthetized and are not allowed to heal in sanitary conditions. Furthermore, she protests that the procedure is now being carried out more in hospital settings using equipment provided by U.S. taxpayer monies. There seems to be strong agreement in Westernized countries that there must be an intense educational campaign to inform African and Middle Eastern families of the serious physical and emotional harm being done to their daughters and wives.

FGM has been outlawed in Sweden, the United Kingdom, Canada, France, Switzerland, Kenya, Senegal, and since 1996 in the United States. In each of these countries, it is now legally considered child abuse (Brady, 1998).

From: Margaret Brady, "Female genital mutilation." *Nursing 98,* 28 (9): 50. Used with permission. © Springhouse Corporation.

They were more inclined to seek attention and affection from others and readier to defy authority and assert unconventional behavior. More recent research also supports the finding that later maturation for boys is associated with a less positive self- and body-image (Sinkkonen, Anttila, & Siimes, 1998; Boxer, Tobin-Richards, & Petersen, 1983). Apparently for boys, being a late maturer is a psychological disadvantage.

A follow-up study of the early- and late-maturing males in the Berkeley sample was conducted when the men were 33 years old. Their behavior patterns were surprisingly similar to the descriptions recorded of them in adolescence (Jones, 1957). The *early maturers* were more poised, relaxed, cooperative, sociable, and conforming. *Late maturers* tended to be more eager, talkative, self-assertive, rebellious, and touchy.

In contrast, research with females has produced nearly the opposite results. Hayward and colleagues (1997) reported the findings of a longitudinal study of

growth and development with 1,463 ethnically diverse girls from San Jose, California. Subjects were in sixth, seventh, or eighth grade when the study began. Using self-report instruments, diagnostic interviews, and psychiatric assessment, the girls were evaluated over several years to determine the relationship between age of puberty and onset of any internalizing symptoms or disorders. Those girls who were defined as having gone through early puberty (the earliest 25 percent) were twice as likely to develop symptoms such as depression, panic attacks, and eating disorders as those girls who matured later. The findings further suggest that those girls who went through puberty early remained at increased risk for development of these specific disorders even after they entered high school (Greif & Ulman, 1982).

Still earlier research suggests that for girls, being "on time" is associated with a more positive set of self-perceptions than being "off time"—either early or late (Tobin-Richards, Boxer, & Petersen, 1983). And then

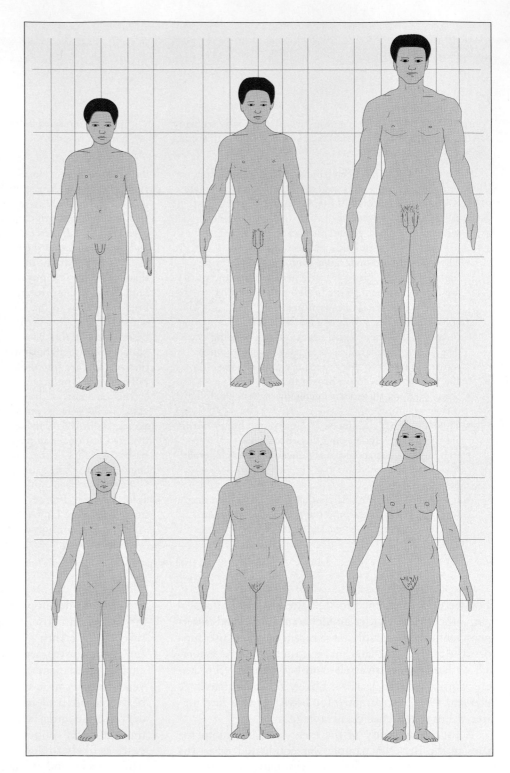

Figure 11-2 Variations in Adolescent Growth All the girls in the lower row are the same chronological age: 12.75 years. All the boys in the upper row are also the same chronological age: 14.75 years. Some persons of the same sex have completed their growth and sexual maturation when the others are just beginning the process.

there is the view that the early onset of menarche does not generate uniform reactions among young women; rather it appears to accentuate pretransition differences among them; for instance, stressful transitions merely intensify behavioral problems among girls who were predisposed to behavioral problems earlier in childhood (Hayward et al., 1997; Caspi & Moffitt, 1991).

So what are we to conclude? The answer may reside primarily in the ecological context: A dynamic interplay transpires between differing individual temperaments and the social environment to produce substantially differing outcomes among young women (Caspi et al., 1993). Additionally, the conflicting findings suggest that the situation for girls is more complicated than that for boys. The findings of more recent research point out the complex nature of variables that interact in producing a girl's reputation during adolescence. The discontinuity between rate of change in evaluations of prestige and rate of physical changes during adolescence means that, for girls, accelerated development is not a sustained asset throughout the adolescent period, as it is for boys.

Psychologists also point to still another factor. Early-maturing girls are more likely to be stout and to develop a stocky physique. In contrast, late-maturing girls are more likely to be thin and to acquire a slim, slight build that is more in keeping with the feminine ideal of many Western nations. Thus, over the long run (in contrast to the short run), late maturation in girls may be associated with factors other than maturation itself that function as assets in social adjustment (Brooks-Gunn & Petersen, 1983). In conclusion, an adolescent girl's subjective impression of timing is critical to understanding the outcomes of her puberty (Hayward et al., 1997; Brooks-Gunn, 1988).

Questions

What are the differences between girls and boys in physical growth and maturation during puberty? What are some of the factors that account for pubertal variability in males and females?

Self-Image and Appearance

The image that adolescents have of themselves is particularly susceptible to peer influences. Adolescents are quick to reject or ridicule age-mates who deviate in some way from the physical norm. Indeed, few words have the capacity to cause as much pleasure—and as much pain—to adolescents as does the word *popularity*. To say that many teenagers are preoccupied with their physical acceptability and adequacy is an understatement. These concerns arise at a time when the nature and significance of friendships are undergoing substantial developmental

change (Wong & Csikszentmihalyi, 1991). The ability to establish close, intimate friendships becomes more integral to social and emotional adjustment and well-being during adolescence than it was during preadolescence (Buhrmester, 1990).

Puberty brings with it an intensification of gender-related expectations, especially related to physical appearance. Adolescent girls often feel troubled about the development, size, and shape of their breasts. Although estimates vary, at least 150,000 women each year in the United States have breast implant surgery; 80 percent of these surgeries are for nonmedical reasons—and now we know that many of these women have become ill with a variety of symptoms because of leakage of silicone from the implants into their bodies. And adolescents, both boys and girls, express considerable concern about their facial features, including their skin and hair. Significantly, the media models (teen magazine cover girls, television actresses, and movie idols) that adolescent girls try to emulate are 9 percent taller and 16 percent thinner than are average American women, and so present an unrealistic ideal of beauty for women to imitate (Williams, 1992).

Weight A large proportion of adolescents want to change their weight—they perceive themselves as either "too thin" or "too heavy." Indeed, recent research findings from a nationally representative sample of 13,783 adolescents in the *National Longitudinal Study of Adolescent Health* indicate that 26.5 percent were obese (Popkin & Udry, 1998). In 1980, 1991, and 1994 the CDC reported findings from its series of National Health and Nutrition Examination Surveys to provide information about overweight children and adolescents. These findings are evidence of a steady increase in the percentage of overweight children and adolescents since the 1970s (see Table 11-1).

About 25 percent of all boys show a large increase in fatty tissue during early adolescence; another 20 percent show a moderate increase. During the adolescent growth spurt, however, the rate of fat accumulation typically

Table 11-1 Prevalence of Overweight Adolescents in the United States, 1976 to 1998

Study Results from Nationally Representative Samples of Adolescents	Adolescents Classified as Obese 12 to 17 Years Old
1976–1980 National Center for Health Statistics, CDC	5.7 percent
1988–1991 National Center for Health Statistics, CDC	10.8 percent
1988–1994 National Center for Health Statistics, CDC	12 percent
1998 National Longitudinal Study of Adolescent Health	26.5 percent

declines in both boys and girls. Whatever the source of obesity, the overweight adolescent is at a personal, health, and social disadvantage. In fact, a stigma is associated with obesity in the United States. Given this state of affairs, particularly for women, it is hardly surprising that adolescent girls are extremely sensitive regarding their body configurations and that even moderate amounts of fat generate intensely negative feelings and distorted self-perceptions. What makes such distortions so cruel is that our body image and our self-esteem are linked, so that if we dislike our bodies, we find it difficult to like ourselves. We examine the sensitive issue of weight concerns and obesity in the next section under nutrition and eating disorders.

> **Questions**
>
> *Which concerns about appearance or self-image did you have as an adolescent? Do you still worry about some of these things today? If so, which ones and why? Which ones no longer bother you?*

Health Issues in Adolescence

What does it mean to be healthy? You might think of yourself as being fairly healthy or as having been a healthy adolescent. Indeed, the majority of adolescents are free of disabilities, weight problems, and chronic illnesses, but a report from the early 1990s indicated that 20 percent of 10- to 18-year-olds suffer from at least one major health problem and that many youth are in need of counseling or other health services (Dougherty, 1993). What sorts of behaviors do adolescents engage in that lead to concerns about their health? It is during the adolescent years that many of us drink our first beer, experiment with other psychoactive substances, try smoking, and have our first sexual experiences with members of the opposite sex, same sex, or both sexes. Many adolescents are also stressed by attempting to earn money for college by working in part-time jobs, while attending high school full-time. Teens living in poverty are at higher risk of having health problems, and it is expected that the general health of adolescents will become worse over the next quarter century simply because more young people will be poor. Some of the negative health factors associated with poverty include poor nutrition, lack of or poor-quality medical and dental care, lack of necessary immunizations, higher risk of exposure to health hazards, and greater incidence of high-risk behaviors.

Many parents become slack about annual checkups for teenage children because they seem so healthy. Parents and caretakers can also be unaware of the recommended immunizations for older children and adolescents (see Table 11-2). Of course, adolescents themselves are also likely to refuse to go to the doctor because they associate this with "childhood" behavior. Public school health officials conduct annual hearing and vision exams, and students who participate on school sport teams are required to have an annual physical. But a large proportion of adolescents are likely to neglect regular health exams or immunizations. Fortunately, the majority are in excellent health. Many at this age believe that they can take care of themselves, even though they might be participating in risky social behaviors (alcohol or substance abuse, sexual activity, etc.).

As late teens approach the beginning of adulthood, they demand less supervision and more freedom to make their own decisions. They are learning to make more choices for themselves, basically by trial and error; and most are highly influenced by their risk-taking peers and youth-oriented media. Some of the most common health concerns for adolescents are discussed below. Though some people call this stage the beginning of the "age of reason," adolescents make many choices without fully realizing the real-life consequences of their actions.

Nutrition and Eating Disorders

What did you eat for lunch yesterday? A hamburger, a candy bar, pizza, French fries, or chips? Junk food is very tempting for most of us. Owing to our busy schedules, we do not want to sit down, relax, and eat if we can grab a snack or pick up a quick bite to eat. However, most "fast" food has very little nutritional value. Adolescents are usually deficient in calcium, iron, and zinc, which can lead to problems such as thinning of the bones (osteoporosis), anemia, and delayed sexual maturity (Rolfes & DeBruyne, 1997). Typically, three prevalent disorders stem from poor nutritional habits among today's teens: anorexia nervosa, bulimia, and obesity.

Anorexia nervosa **Anorexia nervosa** is an eating disorder that primarily affects females who have become obsessed with looking thin and terrified of becoming fat. Anorexics perceive food as being a threat to their bodies rather than a source of nutrition and as a result pursue a regimen of self-starvation. The cause of anorexia is unknown. Some suspect it involves a disturbance of the hypothalamus, others suggest that its cause might be traced to inadequate coping skills, whereby the person feels that the only thing under her (or in fewer cases, his) control is body weight. One scenario follows:

> Jeannette was always a bit chubby until she entered junior high, when she decided one day to lose a little weight. She lost 12 pounds in three weeks and received quite a few compliments from her friends, family, and teachers. She did not stop at 12 pounds and it was soon apparent to her family that something else was going on with their daughter besides her wanting to lose some

Table 11-2 Recommended Immunization Schedule for Adolescent Health

Disease	How often	Who should	Who shouldn't	Side effects
Flu	Yearly, in the fall	Teens with chronic illness or weakened immune systems, those who work or live with high-risk people	People allergic to eggs	Sore arm; rarely, flulike symptoms
Pneumococcal disease	Once, with a booster six years later for chronically ill people	Teens with chronic illness or weakened immune systems	Generally safe for most people	Mild soreness at injection site
Hepatitis A	Two doses, six months apart	Adolescents who are at increased risk of contracting the disease, such as those traveling to countries where it is common or live in communities where outbreaks are common; teens with chronic liver disease or who receive clotting factors	Generally safe	Mild soreness at injection site
Hepatitis B	Once; it takes three doses given over six months	Children aged 11 to 12 or older who were not immunized as infants	Generally safe	Mild soreness at injection site
Diptheria/tetanus	For immunized, three doses given within the space of a year; for partially immunized, booster every 10 years	Teens who did not get DPT shots as children or who did not receive all three DPT shots as children	Pregnant women in their first trimester; anyone who had a severe reaction to a previous shot	Possible redness and swelling at vaccination site; rarely, hypersensitivity, fever, and fatigue
Measles/mumps/rubella	Once, if immunized as a child; twice if not	Anyone born after 1957 who never had any of the diseases or were never vaccinated; anyone vaccinated before their first birthday	Pregnant teens or those who might become pregnant	Measles: high temperature and rash in 5% to 15% of people; mumps: rarely, swollen glands and fever; rubella: some join pain
Chickenpox	One dose for anyone under 13; two doses for 13 and above	Adolescents who have not been vaccinated and who have no reliable proof that they ever had chickenpox	Pregnant teens or those who may become pregnant	Mild soreness at injection site

Source: Centers for Disease Control and Prevention (CDC), Advisory Committee on Immunization Practices, June 1997, http://www.cdc.gov/nip

weight. Her mother describes a meal on a family trip that was undertaken in the hopes that it would divert their daughter's attention away from her constant dieting. "There was the way she would eat her food . . . how she would separate each bit of food, cut it into tiny, precise shapes, and repeatedly calculate calories on a small counter that she clicked. She had that calorie system down to a science, and she would continually pore over what she had on her plate, what she could eat, and how much she would gain." (Sacker & Zimmer, 1987)

Anorexia nervosa affects approximately 1 percent of the female population and is most common in countries such as the United States, France, Japan, and Australia—where attractiveness is equated with thinness and there is no scarcity of food. Anorexia and bulimia together affect about 3 percent of women over their lifetime, with cases of bulimia seemingly on the rise (Walsh & Devlin, 1998a, 1998b). (See Information You Can Use: "Understanding Anorexia and Bulimia.")

Questions

Why are more teenagers affected by anorexia and bulimia? How does their self-image become distorted and their body shape misperceived? What types of treatments are suggested to help victims become healthier?

Obesity **Obesity** is the most common eating disorder in the United States and the one most noticeable to visitors from other countries. A recent national sample finds that about 26.5 percent of U.S. adolescents are obese (Popkin & Udry, 1998). Obesity is not clearly defined. Various sources cite a range of being 20 to 30 percent heavier than one's ideal weight—although diverse factors such as body frame, status of adolescent growth spurt, and activity level make it impossible to generalize across the entire population. Research studies with nationally

Information You Can Use

Understanding Anorexia and Bulimia

Anorexia nervosa is a disorder in which the individual willfully suppresses appetite, resulting in self-starvation. Once considered to be quite rare, the incidence of anorexia nervosa has increased dramatically over the past 30 years. It occurs primarily in adolescent or young adult females of the middle and upper-middle classes. The victims have a fierce desire to succeed in their project of self-starvation; have a morbid terror of having any fat on their bodies; and deny that they are thin or ill, insisting that they have never felt better even when they are so weak they can barely walk. Simultaneously, these people might long for food and even have secret binges of eating (often followed by self-induced vomiting).

One explanation for the recent epidemic in cases of anorexia nervosa is the emphasis that Western societies place on slimness (Bordo, 1993; Macsween, 1993). From an historical perspective, the preoccupation with weight and thinness, especially among affluent women in the Western world, reflects a relatively recent but growing cultural trend (Attie & Brooks-Gunn, 1987). The development of eating problems apparently is one mode of accommodation that some girls make to pubertal change. As girls mature sexually, they experience a "fat spurt"; that is, they accumulate larger quantities of fat in subcutaneous tissues. Early maturers seem at greater risk for eating problems, partly due to the fact that they are likely to be heavier than their late-maturing peers (Graber et al., 1994). In some cases the refusal to eat is preceded by "normal" dieting, which could be prompted by casual comments by family or friends that the young woman is "putting on weight" or "getting plump." Furthermore, the victim's overestimation/misperception of her body size seems to increase with the severity of the illness. According to this interpretation, the disorder entails self-induced star-

Distorted Perceptions Among the Victims of Anorexia Victims of anorexia nervosa willfully starve themselves, denying that they are actually thin or ill, in the belief that they are too fat.

vation by women who desperately want to be beautiful but end up being grotesquely unattractive.

White and African American young women dramatically differ in how they view their bodies. Whereas 90 percent of white junior high and high school girls voice dissatisfaction with their weight, 70 percent of African American girls are satisfied with their bodies (64 percent of African American young women think it is better to be "a little" overweight than underweight). It seems that many African American teenagers equate a full figure with health and fertility and believe that women become more beautiful as they age. Significantly, anorexia and bulimia are relatively minor

---

representative samples of subjects confirm that there are many physical and psychological problems associated with obesity (Popkin & Udry, 1998; Gidding et al., 1996). Adolescents who are obese can have many psychological problems due to social prejudice and peer rejection.

How Is Obesity Determined? Obesity is the excess accumulation of body fat, or *adipose tissue.* A person can be overweight without being obese, as in the case of a bodybuilder with lots of muscle. Medical, government, and research professionals have not come to an exact consensus in their definition of obesity, perhaps because several factors can be examined to make the diagnosis—and overall weight is only one of those factors. One source states that doctors and scientists generally agree

that men with more than 25 percent body fat and women with more than 30 percent body fat are obese (Focus on Obesity, 1998). However, such variables as gender, height, body build, body mass index (BMI), percent of ideal body weight for height from published weight tables, skinfold measurements, and waist-to-hip ratios might also be considered (Gidding et al., 1996). A newer method is BIA, bioelectrical impedance analysis. BIA sends a harmless amount of an electrical current through the body, estimating total body water. A higher percentage of body water indicates more muscle and lean tissue. A mathematical equation is then used to estimate body fat and lean body mass (Focus on Obesity, 1998). Contemporary research scholars recommend using the *body mass index* assessment, which is weight in kilograms di-

problems among African American young women (Ingrassia, 1995).

Another explanation of the disorder is that it is an attempt to avoid adulthood and adult responsibilities. The young woman with anorexia invariably diets away her secondary sexual characteristics: her breasts diminish, her periods cease entirely (interestingly, menstruation ceases prior to pronounced weight reduction and hence cannot be attributed to starvation), and her body comes to resemble that of a prepubescent child. According to this view, such women are seeking a return to the remembered comfort and safety of childhood (Garner & Garfinkel, 1985).

Approximately two-thirds of victims of anorexia recover or improve, with one-third remaining chronically ill or dying of the disorder. Most authorities now recognize that anorexia nervosa usually has multiple causes and that it requires a combination of various long-term treatment strategies adjusted to the individual needs of the patient (Killian, 1994). A number of psychiatrists have suggested that a subgroup of male athletes—"obligatory runners"—resemble anorexic women (Brownell, Rodin, & Wilmore, 1992). These men devote their lives to running and are obsessed with the distance they run, their diets, their equipment, and their daily routines while ignoring illness and injury. Both anorexics and obligatory runners lead strict lives that assiduously avoid pleasure. Both groups are concerned about their health, feel uncomfortable with anger, are self-effacing and hard working, and tend to be high achievers from affluent families. And like anorexics, obligatory runners are exceedingly concerned about their weight and feel compelled to maintain a lean body mass.

Bulimia, a disorder often related to anorexia nervosa, is also called *binge-purge syndrome.* Between 1 and 3 percent of adolescent girls suffer from bulimia. Bulimia is characterized by repeated episodes of binging, particularly on high-calorie foods like candy bars, cakes, pies, and ice cream. The binge is followed by an attempt to get rid of the food through self-induced vomiting, taking laxatives, enemas, diuretics, or fasting (Crowther et al., 1992). Bulimic persons do not usually endeavor to become skeletally thin like anorexics, and they are ashamed and depressed about their eating habits and attempt to conceal their eating behaviors. Bulimics are typically within normal weight range and have healthy, outgoing appearances, whereas anorexics are skeletally thin. Although young women are the primary victims of binge-purge syndrome, young men in activities such as wrestling, male modeling, or acting might likewise engage in similar behavior to squeeze into a lower weight class, look good in photo shoots, or appear more slender on television.

This disorder can produce long-term side effects such as ulcers, hernias, hair loss, dental problems (stomach acid destroys the teeth), and electrolyte imbalance (resulting in heart attacks) (Keel et al., 1999). Like anorexia nervosa, bulimia calls for treatment. Some researchers believe that a hereditary form of depression might underlie some forms of both disorders, and indeed some patients respond to antidepressant medication (Graber et al., 1994). Here is an idea of how a bulimic person feels about this compulsion:

> The whole purge process was cleansing. It was a combination of every type of spiritual, sexual, and emotional relief I had ever felt in my life. Purging became the release for me. First, I felt a tremendous rush that you could really call orgasmic. Then, I relaxed completely and fell asleep. After a while, I was hooked. I actually believed I had to purge to fall asleep. (Sacker & Zimmer, 1987)

Bulimia is found in the same countries as anorexia, and the two disorders affect very similar populations, although more males suffer from bulimia than from anorexia. Up to 10 percent of bulimics are male, whereas 1 percent of anorexics are male, mostly owing to profession or lifestyle (Vollmer, 1999).

---

vided by height in meters squared (kg/m^2). Significant obesity in adults has been defined as ≥130 percent of ideal body weight for height (Gidding et al., 1996). Various measurement findings are compared with normative standards using either percentiles or percentages, which can further confound this vital issue.

Health Consequences   Obesity in adolescents often portends ill for their health in adulthood:

> At 5 feet 6 inches and 216 pounds, Tyshon represents an alarming new health trend: the sharp increase in the number of children with Type 2 diabetes, also known as adult-onset diabetes, an incurable and progressively damaging disease that can cause kidney failure, blindness and poor circulation. . . . Doctors long

believed that the disease occurred mostly during middle age or later. "Ten years ago we were teaching medical students that you didn't see this disease in people under 40, and now we're seeing it in people under 10," said Dr. Robin S. Goland, co-director of the Naomi Berrie Diabetes Center. . . . Type 2 diabetes was diagnosed in 10 to 20 percent of the center's new pediatric patients, compared with less than 4 percent in the hospital's clinic five years ago. (Thompson, 1998)

Obese adults are at greater risk for high blood pressure and heart disease, respiratory disease, diabetes, orthopedic disorders, gallbladder problems, breast and colon cancer, and the high costs of health care. Indeed, researchers find that adolescent obesity is even more strongly linked to health risks than being overweight in adult life (Guo et al., 1994). Boys in the top 25 percent of weight in relation to their height are more likely to experience any combination of these health risks before the age of 70. Overweight girls are found to be mainly at greater risk of developing arthritis, atherosclerosis, gall bladder disease, breast cancer, and diminished physical abilities in later life (Colditz, 1992; Brody, 1992a). Obesity also has social and economic consequences: 16-year-old young women who are in the heaviest 10 percent of their age group earn 7.4 percent less than do their non-obese peers (Hellmich, 1994). Compared with other women, those who are overweight during their teens and early twenties are 20 percent less likely to get married, are 10 percent more apt to live in poverty, and get an average of four months less schooling (Bishop, 1993). Thus, that obesity in childhood and adolescence is more common now than in the past is of considerable concern.

How Many Adolescents in the United States Are Obese? The obesity status of U.S. adolescents continues to increase. In the *National Longitudinal Study of Adolescent Health,* researchers examined a nationally representative sample of 13,783 adolescents (from 80 high schools across the United States) and found that 26.5 percent of American teenagers are currently obese (Popkin & Udry, 1998). For all groups, more obesity occurs among males than among females, except for blacks (27.4 percent for males and 34.0 percent for females). Asian American and Hispanic adolescents born in the United States are more than twice as likely to be obese as are the first-generation residents of the 50 states (Popkin & Udry, 1998). Results are indicated in Table 11-3.

Reasons cited for the increase of obesity in American youth include genetic and environmental factors. Studies of families and twins have clearly demonstrated a strong genetic component in resting metabolic rate, feeding be-

havior, and changes in energy level due to overfeeding. Environmental factors associated with obesity include socioeconomic status, race, region of residence, season, urban living, and being part of a smaller family (Gidding et al., 1996). Diet composition of children does not identify the cause of obesity in youth, because current dietary fat and saturated-fat intake of American children has decreased from previous years. Gidding and colleagues (1996) suggest obesity results, too, from an imbalance between energy intake and energy expenditure.

How Can We Prevent or Reduce Obesity?   Complicating matters, obesity has proven quite difficult to treat (O'Neill, 1995; Wadden & Van Itallie, 1992). In fact, a mounting body of evidence suggests that *dieting* can make matters worse, leading to counterproductive binge eating and a perpetual cycle of fruitless dieting (Brody, 1992b). Various psychiatrists used to argue that obesity is a response to psychological disorders (e.g., women who are fearful of men subconsciously gain weight to create a protective shell and keep men at a distance), but this view has little support among researchers. Another explanation that has attracted considerable interest is that fat babies and fat children develop a permanent excess of fat cells. This excess provides them with a lifelong storehouse of fat cells capable of being filled. When such individuals later become adults, the existing fat cells enlarge but are not thought to increase in number. Still another popular theory postulates the existence of a metabolic regulator or "set point" (Bennett & Gurin, 1982). According to this view, each of us has a built-in control system, a kind of fat thermostat that dictates how much fat we should carry. Some of us have a high setting and tend to be obese; others of us have a low setting and tend to be lean. Even if some lose the weight, long-term studies of weight reduction in children have shown that 80 to 90 percent return to their original weight percentile (Gidding et al., 1996).

Today many obese people challenge the prevailing social stereotypes and prejudices. Unlike the physically handicapped, obese people are held responsible for their condition. They are the object of much concern, criticism, and overt discrimination (Crandall, 1994; Crocker, Cornwell, & Major, 1993). Negative attitudes seem to intensify during adolescence, particularly among females. But increasingly, obese people are "fighting back" against discrimination by defining obesity as a disability, in an attempt to make it illegal to discriminate against persons who are obese.

Table 11-3   Adolescent Obesity by Gender and Ethnicity

Adolescents by Ethnicity	Total	Males	Females
Non-Hispanic White	24.2%	25.8%	22.6%
Non-Hispanic Black	30.9%	27.4%	34.0%
Hispanic	30.4%	31.6%	29.1%
Asian American	20.6%	25.7%	15.0%
Chinese American	15.3%	18.9%	10.9%
Filipino American	18.5%	22.6%	12.8%
Non-Hispanic American Indian	42.4%	44.4%	40.0%

From: Barry M. Popkin and J. Richard Udry. Adolescent obesity increases significantly in second and third generation immigrants: The National Longitudinal Study of Adolescent Health, *The Journal of Nutrition,* Bethesda, 1998, 128(4), 701–706. Reprinted by permission.

Note: Data is from the National Longitudinal Study of Adolescent Health, 1996 sample, body mass index (BMI) ≥ 85th percentile.

Question

Why is there an increase in the number of adolescents who have eating disorders, and what are the consequences of entering adulthood with these unhealthy behaviors?

Smoking and Chewing Tobacco

Since the early 1990s, smoking and "chewing" tobacco were on the rise among adolescents, though recent findings from a national PRIDE survey suggest that smoking among junior high students has somewhat declined—from 32 percent in 1996–1997 to 29 percent in 1997–1998 (Peterson, 1998). Furthermore, 90 percent of all smokers start their habit as teenagers. Most smokers begin smoking at an average age of just over 14. Smoking increases the risk of premature deaths from smoking-related diseases, with as many as 5.3 million teenagers succumbing to the ill effects of smoking. African American teenagers smoke far less than whites and Hispanics, with an 84 percent decline in African American teenage smokers over the last 20 years. Researchers do not know definitively why some adolescents are smoking more, but aggressive advertising by tobacco manufacturers is the probable reason. Contest prizes such as concert tickets, T-shirts, and hats are especially appealing to a young audience (Johnston & O'Malley, 1997). There is also statistical evidence of increased use of cigarette smoking in popular movies. Adolescents who "chew" are at risk of developing cancerous sores on the mouth and tongue and inside the mouth even after only a few years of use.

Alcohol and Other Substance Abuse

Recent findings from two extensive national studies yield contradictory findings about the extent of teenage alcohol and other substance abuse. PRIDE, a nonprofit drug prevention organization, surveyed 154,350 students in grades 6 to 12 and reports that tobacco, drug, and alcohol use by most teens declined this school year (Peterson, 1998). These findings suggest an across-the-board decline in alcohol, tobacco, and other drug use since the 1990–1991 school year. However, the *National Household Survey of Drug Abuse* found an increase in drug abuse by adolescents, led by a rise in marijuana smoking (Knox, 1998). This survey was conducted with a sample of 24,500 teens who were interviewed in their homes. The perception of many people is that marijuana is a "soft drug," but extensive studies on cannabis use suggest otherwise. In *National Drug Strategy Monograph Series No. 25* (1998), researchers report findings that adolescence is an important period of transition from childhood to adulthood in which regular cannabis intoxication can be expected to interfere with educational performance, the process of disengagement from dependence upon parents, the development of relationships with peers, and making important life choices (such as, dropping out of school, securing and maintaining employment).

Substance abuse is the harmful use of drugs or alcohol, lasting over a prolonged period, that puts self or others in hazardous situations. Although most adolescents do not abuse drugs, those who do run the risk of substance dependency and have a much higher chance of becoming adult drug addicts who often commit crimes to support their addictions. One recent study found that boys who exhibit a cluster of extreme personality characteristics (impulsivity, excitability, and low harm avoidance) by age 6 are far more likely to smoke, drink and use drugs upon reaching adolescence. And adult alcohol abuse is more than double in youngsters who begin drinking before age 15. Moreover, alcohol abuse significantly affects more adolescent boys than girls, although there are no gender differences found in drug use among high school students. Unfortunately, alcoholism is the third leading killer in the United States, and 25,000 people are killed each year by drunk drivers.

Teenage drinking does not just affect the health of adolescents. Inexperienced drinking and driving are related to many tragic automobile accidents, especially around "prom" time and graduation parties at the end of a school year. To the credit of some teens, more high school chapters of MADD (Mothers Against Drunk Driving) have been established around the country to promote safe teen driving and the use of a "designated driver." The serious issue of peer influence on substance use and abuse is discussed in Chapter 12.

Sexually Transmitted Diseases and HIV

Many adolescents frequently engage in sexual intercourse, and it is not uncommon for sexually active adolescents to have multiple partners. Government statistics show that of the 12 million cases of sexually transmitted diseases (STDs) that are estimated to occur each year, teenagers account for 3 million cases. By not using condoms, adolescents are vulnerable to **sexually transmitted diseases (STDs)**. Reasons given for not using condoms include these: *I was drunk; we decided on the spur of the moment; I was embarrassed to buy them; it spoils the romance.* It is not surprising, then, that U.S. teenagers have the highest rates of *gonorrhea, syphilis,* and *chlamydia* of the sexually active populations (DiClemente, 1990). *Chlamydia* is currently the most prevalent STD in the United States. It is caused by a parasite and can lead to infertility and blindness in women. An infected woman can pass on the chlamydia parasite to her infant as the baby passes through the birth canal. Five percent of female college students have been diagnosed with chlamydia. Fortunately, it is curable. *Syphilis* is a bacterial infection that can be passed on to a fetus through the placenta. Syphilis develops in four phases, beginning with incubation and ending in the final phase some five years later. Death is a possibility if left untreated. *Gonorrhea* is caused by bacteria and passes from one infected mucous membrane to another. The incidence of gonorrhea has steadily declined over the last 10 years in all populations *except* early-adolescent females, in whom the rate of infection has increased almost 50

Table 11-4 Percentage of High School Students Who Used a Condom During Their Last Sexual Intercourse, 1991 and 1997

	1991 (N = 12,272)	1997 (N = 16,272)
Grade		
9	53.3%	58.8%
10	46.3%	58.9%
11	48.7%	60.1%
12	41.4%	52.4%
Sex		
Male	54.5%	62.5%
Female	38%	50.8%
Race/Ethnicity		
Non-Hispanic White	46.5%	55.8%
Non-Hispanic Black	48.0%	64%
Hispanic	37.4%	48.3%

Source: The Centers for Disease Control and Prevention (CDC) conducted the *Youth Risk Behavior Survey* (YRBS) in 1991, 1993, 1995, and 1997. Reported in the *Morbidity and Mortality Weekly Report,* 47, no. 36 (Sept. 18, 1998), p. 751.

percent (Centers for Disease Control, 1997c). Other prevalent STDs are human *papilloma virus* (HPV) and *genital herpes simplex,* which is chronic, painful and extremely contagious. Genital herpes can be fatal to some persons with immune deficiencies and also to infants who contract the disease at birth. There is no cure for genital herpes simplex.

Have the national education/health campaigns been effective in informing adolescents to protect themselves against these STDs? Nationally, 62.5 percent of males and 50.8 percent of females in high school report having used a condom at last intercourse (Centers for Disease Control, 1998h). These figures indicate that among currently sexually active high school students, condom use has increased 23 percent from 1991 to 1997 (see Table 11-4).

This multi-year, national survey by the CDC (1998h) indicates that although fewer students are engaging in sexual behaviors that place them at high risk for STDs, AIDS, and pregnancy, many are still practicing high-risk behaviors. The prevalence of condom use *increased* for students in all grades, 9 to 12, both for males and females, and overall for all racial/ethnic groups. A dramatic 84 percent decline in U.S. syphilis rates occurred during this same time period, from 1990 to 1997 (after an alarming epidemic from 1986 to 1990). Factors associated with irregular or no contraceptive use include these: low academic skills and aspirations; being a younger teen; a tendency toward risk taking (for instance, alcohol and other

drug use); a strained relationship with parents; parental use of corporal punishment; and the absence of a committed relationship (Luster & Small, 1994a, 1994b).

To further compound the problem, many adolescents are not aware of how they can avoid STDs, cannot identify common symptoms, and do not know what action to take once they show signs of being infected. Most high schools and colleges have a staff nurse who can provide further information and be a safe resource person to those teens who suspect they have contracted a disease. Part of the nature of adolescence is to believe that bad things are not going to happen to you—that you have your whole future ahead. However, contracting or transmitting one of these diseases—particularly HIV—has life-threatening consequences. To further compound matters, teens are still getting pregnant, and the unborn child has a high risk of contracting HIV if the mother has this disease (see Chapter 3).

HIV and AIDS   It is hardly surprising that public health experts worry openly that sexually active adolescents are becoming the next AIDS "high-risk" group (Jadack et al., 1995; Sigelman et al., 1993). On the international level, an analysis of AIDS research by the United Nations Development Program finds that women under the age of 20 have higher rates of HIV infection than do older women or young men ("U.N. Finds Teenage Girls at High Risk of AIDS," 1993, p. A5). Significantly, a study of sexually active homeless and runaway girls revealed that even though 80 percent are fearful of contracting AIDS, "the need for intimacy and love—in effect, the need for a strong, secure bond with another person—outweighs the fear of the horrible suffering and sure death of AIDS" (Ravoira & Cherry, 1992).

AIDS became the most feared STD of the 1980s, with black and Hispanic adolescents being four times more likely to develop AIDS than other adolescents. HIV is now the *sixth leading cause of death* among persons aged 15 to 24 in the United States (Centers for Disease Control, 1998h). AIDS is caused by *human immunodeficiency virus* (HIV), which damages the human immune system and prevents the body from fighting infections. Some individuals can carry the virus for up to 10 years before developing AIDS. Most of the infected population in the United States in the 1980s were either gay males or intravenous drug users, but things have changed. Now HIV is spreading among the heterosexual community. An August 1998 report from the Centers for Disease Control and Prevention states that women aged 16 to 21 who had entered the federally funded Job Corps program (a training program for disadvantaged youth) between 1990 and 1996 were much more likely to be infected with HIV than their male counterparts. CDC epidemiologist Linda Valleroy ("Health Watch Rx for Risk," 1998) said the female infection rates were "alarm-

Public Health Campaigns to Combat AIDS Public health campaigns often attempt to alert individuals to the dangers associated with given social behaviors. Unfortunately, researchers find that many individuals are well aware of the dangers associated with their behavior but nonetheless persist in unsafe practices for a host of other reasons. For instance, by virtue of their intense yearning for love and intimacy, many homeless and runaway girls subject themselves to the known risk of contracting AIDS.

ing," with HIV infection rates 50 percent higher among young women than among young men. During the seven-year period of this study, though, rates for all women dropped from 4 per 1,000 in 1990 to 2 per 1,000 in 1996. Among men, HIV infection fell from 3 per 1,000 in 1990 to 1.5 per 1,000 in 1996 (Health Watch Rx for Risk, 1998). (In Chapter 15 we will discuss the increasing rate of HIV in heterosexual middle-aged women.)

Rates of adolescent sexual experience, sexual activity, and effective contraceptive practice are important determinants of changes in pregnancy rates. With sexual experience rates down, multiple-partner rates down, and condom use increasing, we would expect to see a decline in the teenage pregnancy rate.

Questions

What are the current trends in the incidence of sexually transmitted diseases in teenagers? What are some of the reasons teenagers give for using condoms? What are some of the reasons teenagers give for not using condoms?

Teenage Pregnancy

Recent reports from the U.S. Department of Health and Human Services (1997b) indicate that the national campaign to curb teenage pregnancy rates over the past several years has seen some success (see Figure 11-3). The sharpest drop in teen pregnancy is among adolescents 15 to 17 years of age. The birth rate in this group has dropped 20 percent in the last decade. Also, the birth rate among unmarried black women is at a 40-year low—74.4

per 1,000 births. The reasons given for lower rates were fear of getting AIDS, wanting to start on a career path, and avoiding the stigma of unwed motherhood. Additionally, more teenage girls are choosing to get the "shot" of Depo-Provera every three months at public health clinics as a more convenient, effective form of birth control; it has accounted for 19 percent of contraceptive use among black teens between 15 and 19 and slightly less than 10 percent among white and Hispanic teens (Freedman, 1998). Recent analysis of various approaches to curbing teenage pregnancy found that when young women are paired with older, nonparental mentors, the young women are more likely to practice safer sex methods (Blinn-Pike et al., 1998). The emotional, social, and socioeconomic consequences of teenage pregnancy are discussed in Chapter 12.

Stress, Anxiety, Depression, and Suicide

Rates of suicide and attempted suicide for contemporary adolescents are alarmingly high. Because stress, anxiety, and depression occur within the context of adolescent-parent relationships, peer-adolescent relationships, and boyfriend-girlfriend and other sexual relationships, we discuss the range of reactions to fear, loss, and attacks on personal sense of worth in Chapter 12. See *Further Developments:* "A Typical Teen's Day" to read about some of the stressors that can affect teens.

Body Art and Tattooing

Adolescents are currently "body-art" enthusiasts, with more young people getting tattoos and body piercings than previous generations did at their age. Tattoos and piercings have become widely acceptable and, in some circles, respectable. As far as being mainstream, a recent

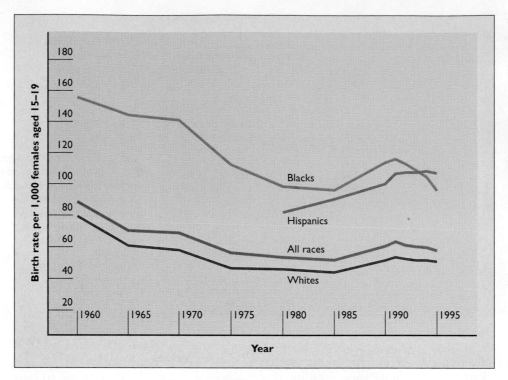

Figure 11-3 Teenage Birth Rates by Race/Ethnicity, 1960 to 1995

Source: Ventura, S. J., Martin, J. A. Mathews, T. J., Clarke, S. C., "Advance Report of Final Natality Statistics, 1994." *Monthly Vital Statistics Report,* Vol. 44, No. 11, Supplement, Hyattsville, Maryland: National Center for Health Statistics, 1996. Also previous issues of this annual report. Ventura, S. J., "Births of Hispanic Parentage, 1980." *Monthly Vital Statistics Report,* Vol. 32, No. 6. Supplement, Hyattsville, Maryland: National Center for Health Statistics, 1983. 1995 preliminary data from: Rosenberg, H. M., Ventura, S. J., Maurer, J. D., Heuser, R. L., and Freedom, M. A. "Births and Deaths: United States, 1995." *Monthly Vital Statistics Report,* Vol. 32, No. 6, Supplement 2. Hyattsville, Maryland: National Center for Health Statistics, 1996.

study found that many of the teenagers who were pierced or tattooed were the academic elite in their respective high schools. During 1997, tattoo parlors sprang up at the rate of over one a day, making tattooing one of the fastest growing businesses in the United States. Studies show that adolescents indulge in body art for the following reasons (Leonard, 1998):

- To differentiate themselves
- To commemorate an event
- To express intimacy (tattooing lover's name)
- Entertainment

Health risks from body piercing include hepatitis B and tetanus, as well as skin reactions from dye. Most teens do not realize that having a licensed dermatologist remove a single tattoo can cost $2,000 to $2,500. A poorly placed piercing can cause nerve damage. And a tongue stud can lead to speech problems and chipped teeth. Recently scarification, branding, and stretched earlobe holes have made their way into the body art scene; and as adolescents look for novel ways to express themselves, these become even more intriguing. It remains to be seen whether these adornments will be accepted in American work environments (see Figure 11-4). As we recall our own teen years, we recognize that

every adolescent cohort finds interesting visible ways to distinguish itself from the adult generation—including "slang" language, music style, hairstyles, length and fit of clothing, and use of body and facial adornments.

Question

Why do many young people choose adolescence as the time to begin experimenting with smoking, drinking, drugs, sex, and other risky behaviors?

Cognitive Development

During adolescence young people gradually acquire several substantial new intellectual capacities. They begin to reflect about themselves; their parents, teachers, and peers; and the world they live in. They develop an increasing ability to use *abstract thought*—to think about hypothetical and future situations and events. In our society they also must evolve a set of standards regarding family, religion, school, drugs, and sexuality; those who work at jobs during high school must also develop work standards.

Further Developments

A Typical Teen's Day

A detailed portrait of the day-to-day world of U.S. teenagers has emerged from a study undertaken by University of Chicago social scientists Mihaly Csikszentmihalyi and Reed Larson (1984), who sought to describe what it is like to be an adolescent from the inside. For a week they had 75 teenagers tell what they did, felt, and hoped for as they went about their daily rounds, from breakfast to school, from English class to lunch period, from the afternoons spent watching television to the wild parties on weekends. To do so, the researchers equipped the 39 boys and 36 girls with small electronic pagers of the sort used by doctors. A transmitter activated the pagers at random moments during the week, signaling the teenagers to fill out a questionnaire reporting their thoughts, feelings, and activities at the time. The subjects were randomly selected from the 4,000 students at a Chicago-area high school. Among the findings of the research were the following:

- Drastic mood swings seem to be a normal feature of adolescent life and do not necessarily signal deeply rooted psychological difficulties. From their reports, youth whose moods changed the most were as happy and as much in control of their lives as were their peers, and they seemed well adjusted in other spheres of life.
- Youth who spent more time with their families and less time with their peers achieved better school grades, were less likely to be absent from school, and were rated by their teachers as being more intellectually involved.
- Many of the adolescents reported that conflict with their siblings generated as much difficulty as generational tensions with their parents.
- Time spent at school and studying occupied about a quarter of the adolescents' waking time (about 38 hours per week). Another quarter of their time was spent alone. The Chicago researchers concluded that solitude is an important aspect of a healthy adolescence.

- The teenagers spent the largest part (42 hours) of their waking time in leisure activities: socializing (16 percent), watching television (7.2 percent), nonschool reading (3.5 percent), sports and games (3.4 percent), and listening to music (1.4 percent).
- On average, the adolescents spent 19 percent of their waking hours with family members, 23 percent with classmates, and 29 percent with friends.

However, these broad findings hide considerable diversity among the youth. Three days in the life of Greg, a disaffected, rebellious youth, and in the life of Kathy, a very directed student and accomplished violinist, reflect these differences. The substantial mood swings experienced by Greg and Kathy were not untypical. It usually took the teenagers about 45 minutes to come down from extreme happiness or up from deep sadness, whereas it usually takes adults several hours. Unhappiness frequently arose from the drudgery of schoolwork, jobs, and chores. The young people seemed to get the most satisfaction from meeting challenges that were appropriate to their skills and that provided them with meaningful rewards. Those youth who were active in sports and hobbies, for instance, felt stimulated to move on to new levels of challenge and accomplishment. But many also fell into patterns of aimless drifting, television viewing, and searching for short-term pleasures. And some simply removed themselves from work and play, relying on crutches (like smoking marijuana) rather than developing skills that would help them in later life. Coming to terms with ups and downs seems to be one of the greatest challenges that teenagers confront. When Csikszentmihalyi and Larson (1984) repeated the study with some of the adolescents, they found that many youth reported that their lives had become better. It was not so much that the reality of good and bad experiences had changed as that the young people had shifted their perspectives in ways that allowed them to better cope with the world.

Piaget: The Period of Formal Operations

Thinking is not a heaven-born thing. . . . It is a gift men and women make for themselves. It is earned, and it is earned by effort. There is no effort, to my mind, that is comparable in its qualities, that is so taxing to the individual, as to think, to analyze fundamentally.

Supreme Court Justice Louis D. Brandeis

Jean Piaget called adolescence the **period of formal operations,** the final and highest stage in the development of cognitive functioning from infancy to adulthood. This mode of thought has two major attributes. First, adolescents gain the ability to think about their own thinking—to deal efficiently with the complex problems involved in reasoning. Second, they acquire the ability to imagine many possibilities inherent in a situation—to generate mentally many possible outcomes of an event and thus to place less reliance upon real objects and events. *"If I don't*

VOCATIONAL COUNSELOR

DAVE CARPENTER

Figure 11-4 Body art, body piercing, and tattoos . . . now reaching mainstream culture. Is this an adolescent fad or a cultural change?

Used with permission of Dave Carpenter, PO Box 520, Emmetsburg, IA 50536, (712)852-3725.

"I imagine you'll be interested in one of the more highly visible occupations?"

come home at my curfew time with Dad's car, then . . . or . . . will happen." In sum, adolescents gain the capacity to think in logical and abstract terms.

Formal operational thought so closely parallels scientific thinking that some call it "scientific reasoning." It allows people to mentally restructure information and ideas so that they can make sense out of a new set of data. Through logical operations individuals can transfer the strategic skills they employ in a familiar problem area to an unfamiliar area and thus derive new answers and solutions. In so doing, they generate higher-level analytical abilities to discern relationships among various classes of events.

Formal operational thought is quite different from the concrete operational thought of the previous period. Piaget said that children in the period of concrete operations cannot transcend the immediate. They are limited to solving tangible problems of the present and have difficulty dealing with remote, future, or hypothetical matters. For instance, a 12-year-old will accept and think about the following problem: "All three-legged snakes are purple; I am hiding a three-legged snake; guess its color" (Kagan, 1972, p. 92). In contrast, 7-year-old children are confused by the initial premise because it violates their notion of what is real. Consequently, they can be confused and refuse to cooperate.

Likewise, if adolescents are presented with the problem: "There are three schools, Roosevelt, Kennedy, and Lincoln schools, and three girls, Mary, Sue, and Jane, who go to different schools. Mary goes to the Roosevelt school, Jane to the Kennedy school. Where does Sue go?" they quickly respond "Lincoln." The 7-year-old might excitedly answer, "Sue goes to Roosevelt school,

because my sister has a friend called Sue and that's the school she goes to" (Kagan, 1972, p. 93). Similarly, Bärbel Inhelder and Jean Piaget (Inhelder & Piaget, 1964, p. 252) found that below 12 years of age most children cannot solve this verbal problem:

> Edith is lighter than Suzanne.
> Edith is darker than Lily.
> Which is the darkest of the three?

Children under 12 often conclude that both Edith and Suzanne are light-complexioned and that Edith and Lily are dark-complexioned. Accordingly, they say that Lily is the darkest, Suzanne is the lightest, and Edith falls in between. In contrast, adolescents in the stage of formal operations can correctly reason that Suzanne is darker than Edith, that Edith is darker than Lily, and therefore Suzanne is the darkest girl.

Piaget suggested that the transition from concrete operational to formal operational thought takes place as children become increasingly proficient in organizing and structuring input from their environment with concrete operational methods. In so doing, they come to recognize the inadequacies of concrete operational methods for solving problems in the real world—the gaps, uncertainties, and contradictions inherent in concrete operational processes (Labouvie-Vief, 1986).

Not all adolescents, or for that matter all adults, attain full formal operational thought. Therefore, they fail to acquire its associated abilities for logical and abstract thinking. This lack of ability is shown, for instance, by people who score below average on standard intelligence tests. Indeed, as judged by Piaget's strict testing standards, less than 50 percent of U.S. adults reach the stage

The Development of Formal Operational Thought
Formal operational thought allows people to generate a higher level of analytical capability, allowing them to discern new relationships among various classes of events. It closely parallels the reasoning encountered in scientific thinking.

of formal operations. Some evidence suggests that secondary schools can provide students with experiences in mathematics and science that expedite the development of formal operational thought. And some psychologists speculate that various environmental experiences might be necessary to its development (Kitchener et al., 1993).

Furthermore, cross-cultural studies fail to demonstrate the full development of formal operations in all societies. For example, rural villagers in Turkey never seem to reach the formal operational stage, yet urbanized educated Turks do reach it (Kohlberg & Gilligan, 1971). Overall, a growing body of research suggests that full formal operational thinking might not be the rule in adolescence. Even so, considerable research confirms Piaget's view that the thought of adolescents differs from that of young children (Marini & Case, 1994; Pascual-Leone, 1988).

Adolescent Egocentricity

Every one believes in his youth that the world really began with him, and that all merely exists for his sake.

Goethe, 1829

Piaget (1967) said that adolescents produce their own characteristic form of **egocentrism,** a view expanded by the psychologist David Elkind (1970) in terms of two dimensions of egocentric thinking: (1) the *personal fable* and (2) the *imaginary audience.* As adolescents gain the ability to conceptualize their own thought, they also achieve the capacity to conceptualize the thought of others. But adolescents do not always make a clear distinction between the two. In turning their new powers of thought introspectively, adolescents simultaneously assume that their thoughts and actions are equally interesting to others. They conclude that other people are as admiring or critical of them as they are themselves. They tend to view the world as a stage on which they are the principal actors and all the world is the audience. According to Elkind, this characteristic accounts for the fact that teenagers tend to be extremely self-conscious and self-preoccupied: The preoperational child is egocentric in the sense that he is unable to take another person's point of view. The adolescent, on the other hand, takes the other person's point of view to an extreme degree.

As a result, adolescents tend to view themselves as somehow unique and even heroic—as destined for unusual fame and fortune. Elkind dubs this romantic imagery the **personal fable.** The adolescent feels that others cannot possibly understand what she or he is experiencing, and often this leads to the creation of a story or *personal fable*, which the adolescent tells everyone, although it is a story that is not true. If you have ever thought something like, "They will never understand the pain of unrequited love; only I have been through this torture," then you have created your own personal fable.

The **imaginary audience,** another adolescent creation, refers to the adolescent's belief that everyone in the local environment is primarily concerned with the appearance and behavior of the adolescent. The imaginary audience causes the adolescent to be very self-critical and/or extremely self-admiring. The adolescent really believes that everyone she or he encounters thinks solely about that individual night and day. Remember how devastating a pimple was in high school because you thought every eye would be glued to your affliction? You probably never thought that everyone else was too concerned and preoccupied with their own pimples to notice yours. Elkind believes that adolescents can eventually distinguish between real and imaginary audiences, and he also acknowledges that the adolescent's imaginary audience and personal fable are progressively modified and eventually diminished.

Other psychologists, such as Robert Selman (1980), also find that young adolescents become aware of their own self-awareness, recognizing that they can consciously monitor their own mental experience and control and manipulate their thought processes. However, only later in adolescence do they come to realize that some mental experiences that influence their actions are not accessible to conscious inspection. In brief, they become capable of distinguishing between conscious and nonconscious levels of experience. Hence, although they retain a conception of themselves as self-aware beings, they realize that their ability to control their own thoughts and emotions has limits. This gives them a more sophisticated notion of their mental self and what constitutes self-awareness.

The growing self-awareness of teenagers also finds expression in the increasing differentiation of the *self-concept* during adolescence. Adolescents provide different self-descriptions in different social contexts. The self-attributes teenagers assign themselves differ depending on whether they are describing their role in relation to their mother, father, close friends, romantic partners, or classmates, or their role as student, employee, or athlete. For instance, the self they depict with their parents might be open, depressed, or sarcastic; with friends—caring, cheerful, or rowdy; and with a romantic partner—fun-loving, self-conscious, or flirtatious. The cognitive-structural advances noted earlier in this chapter permit teenagers to make greater differentiations among role-related attributes. Simultaneously, the differing expectations of significant others in different social contexts compel adolescents progressively to differentiate the self with respect to varying social roles (Harter & Monsour, 1992). We will discuss adolescent egocentricity in its social context in Chapter 12.

Questions

How does a typical adolescent develop cognitively during the junior high to high school years? Why do adolescents tend to be more concerned with themselves than with others?

Educational Issues

For most teenagers, entering high school is a breath of fresh air, so to speak, from the more structured middle school or junior high. Students finally get an opportunity to select a few courses and a daily schedule along with enrolling in required state-mandated courses (mathematics, English, social studies, science). Students with an aptitude in vocational careers typically select from courses such as computer science, carpentry and woodworking, mechanical drawing, auto mechanics, machine shop, child care, veterinary care, food service, cosmetology, and pre-nursing. Many large high schools offer "schools within a school," such as a core of special courses for students who are highly gifted or talented in the fine arts along with a mentoring experience in typically music, drama, or computer art and design.

Of significant interest is the educational performance of those youth in junior high who were in the Head Start programs as preschoolers. Results from the *Extended Early Childhood Intervention and School Achievement: Age Thirteen Findings from the Chicago Longitudinal Study* indicate that at-risk children who participated in follow-up programs for three or more years after early intervention services had significantly higher reading achievement in the seventh grade, had a lower rate of being held back a grade, and were less likely to receive special education services (Reynolds & Temple, 1998). These researchers highly recommend the establishment of large-scale, community-based intervention programs that extend through the elementary years for youngsters earlier identified as being at risk.

Students who are high achievers can select advanced courses that challenge their problem-solving and critical thinking skills and allow them to earn college credits while in high school. Typically high schools have a variety of related extracurricular opportunities that allow adolescents real-world opportunities to try out potential career skills; these include working on the school newspaper or yearbook, participating on highly competitive sports teams, learning debate skills on mock trial teams, being an officer of a Spanish or French Club, participating on a community service club, and being a class officer.

Adolescents with low or limited intellectual capacity, while being included in some regular high school classes, are eligible to receive life-skills training in specific vocational programs with work experience. Their preparation during high school is called the "aging out" process, and these students are eligible to remain in academic and skills-training until they are 21 years old, should they choose to do so.

Students with high intellectual abilities who have had positive experiences during their school years typically enjoy the challenge of high school and anticipate entering college. However, those who have not experienced academic success or social acceptance in the earlier grades of school seem to begin to transition out, both physically and psychologically—poorer or more erratic school attendance, an unmotivated attitude toward schoolwork, disagreements with adults, increased substance abuse, and sometimes trouble with the law are strong signs of students at risk of dropping out. These youth are highly aware of when they can "quit" school, and some are at risk of dropping out well before the legal age by disappearing into the community or becoming a PINS youth (person in need of supervision). Many communities now have group homes for these teenagers, where they receive more structure and supervision than these youth may have experienced before.

Pregnant teenage girls are another group at risk of never earning a high school diploma. The *Even Start* program is federally sponsored and administered through many large urban high schools. The goals of this program are threefold: (1) to help the young mothers earn a high school diploma while learning effective par-

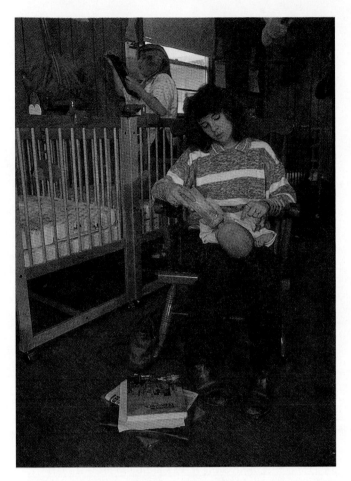

Student mothers in the Even Start program This federal program is offered in many large high schools to help teenage mothers get a high school diploma and learn to parent their babies and children.

enting skills, (2) to physically nurture and intellectually stimulate the infants and preschool children of these teenagers in a nearby setting where mothers can interact with their children during the school day, and (3) to nurture the self-esteem of mother and child and provide healthy role models who encourage the teenage mothers to develop their abilities and talents.

Effective Classroom Instruction   Today's classrooms are changing. Curriculum-specific expertise and technological preparedness are essential for using CD-ROMs, video laserdiscs, and multimedia presentations. Technical and specific computer expertise is needed to develop and access Internet web sites to gather and disseminate course information and homework assignments. Conflict-resolution skills and expertise are needed to diffuse potential violent behaviors. Today's students are also changing. They are a more culturally diverse population, some are more intellectually challenged than others, some have disabilities, and some have limited English proficiency. Teachers must come to the classroom well prepared to stimulate, motivate, educate, and evalu-

ate. Teachers who instruct with a "passion" for what they do shine like a light in the darkness, and high school students are astute enough to notice. Students must come to high school prepared to learn. High school classes are not an end in themselves, at least not for most students. High school instruction should be like a springboard, providing a firm foundation while launching the young adult onto new heights. Supporting the classroom teacher to provide the best education for every student are classroom aides, guidance counselors, school psychologists, school nurses, librarians, and school administrators.

Academic Standing and Global Comparisons   Even with all this professional and technological support, though, American high school seniors performed well below the international average in mathematics and science literacy in the largest international study of student achievement ever undertaken—the *Third International Mathematics and Science Study* (TIMSS), released in February 1998 by researchers at Boston College (Sullivan, 1998; Forgione, 1998) (see Table 11-5). This large-scale study included more than 500,000 students from 45 countries, and TIMSS assessed students in their last year in all types of schools and programs (Forgione, 1998). This research is evidence of a downward trend in the math and science skills of American youth in the years following the fourth grade, where American youngsters perform above average when compared to international peers. Researchers suggest that U.S. curricula in middle schools and high schools do not require many math or science courses. One significant finding was that students who use a calculator daily performed well above those who rarely used one. A gender gap was found as well: Boys outperformed girls in math and science literacy in nearly all countries tested.

Students from the Netherlands and Sweden fared best in overall mathematics and science literacy. French students performed highest in advanced mathematics. Those from Sweden and Norway performed highest in physics. American high school seniors showed a sharp dropoff in math and science skills after elementary school. Few American students take calculus or physics, compared to students from many other countries. Some factors appear to be related to the lower performance of U.S. high school seniors ("Building Knowledge," 1997).

- *Part-time jobs.* More U.S. students work part-time jobs and work more hours than students in countries that scored higher and lower than U.S. students.
- *Fewer hours of mathematics instruction per week.* The United States had significantly lower proportion than the international average of advanced students receiving five or more hours of mathematics instruction per week.

Table 11-5   U.S. Twelfth-Grade Mathematics and Science Achievement in International Context

Subject	Country	Score
Mathematics and Science Literacy (average score of 500)	Netherlands	559
	Sweden	555
	United States	471
Advanced Mathematics (average score of 501)	France	557
	United States	442
Physics (average score of 501)	Norway	581
	United States (the lowest ranking of all 45 countries)	423

Source: Pascal D. Forgione, Jr., U.S. Commissioner of Education Statistics, National Center for Education Statistics (NCES), *Pursuing Excellence: A Study of U.S. Twelfth-Grade Mathematics and Science Achievement in International Context,* and *The Release of U.S. Reports on Grade 12 Results from the Third International Mathematics and Science Study (TIMSS),* February 24, 1998. http://nces.ed/gov/timss/

U.S. researchers who participated in this study say there are no easy answers as to how American schools can improve educational achievement, and educators at the federal level suggest that TIMSS results are likely to raise the standard for teaching, curriculum, and evaluation of pupils in the United States. Standards are being raised around the nation in math and science, and students will be urged to take tougher courses (Forgione, 1998; Sullivan, 1998). Otherwise, American youth will be at a disadvantage for employment in the technologically driven global market.

Use of Computer Technology   The exciting diversity and richness of the world is opening up to our youth—and to all of us—through computer networks. Without a doubt, there is a high rate of job growth and opportunity for today's adolescent in the computer industry. The network wiring of schools has become a national goal, with funds coming from both the public and the private sector. However, some parents and faculty are concerned that computer games, Internet access, and chat rooms are a waste of time taking away from the 3 Rs or will lead to immoral, or dangerous, activities.

Modes of instruction using the computer, such as distance learning options, are expanding. Such instructional formats allow students with special interests to enroll in online classes that their high school or college might not be able to offer, such as Japanese, Russian, sign language, or Latin. Today all major research journals, magazines, and newspaper articles are available via the computer. High school and college libraries are now considered "Information Resource Centers." Many parents have purchased computers for home use and consider this expense an investment in their child's future. It is well known that nearly all occupations today utilize computers: Auto mechanics use computers to evaluate and calibrate auto engines; beauticians use computers to help a customer plan a "new" look; medical personnel use computers for diagnostics and imaging; farmers use computers to plan crops and costs; business personnel use computers to manage every aspect of business; astronauts use computers on the international space station; artists and animators use computer software to create cartoons and movies. We must prepare this adolescent cohort for its technologically oriented future.

Questions

Why do you think American high school seniors' academic performance in math and science ranked below average in the international TIMSS study? What would you recommend to improve the academic performance of U.S. teens?

Moral Development

The fundamental idea of good is thus, that it consists in preserving life, in favoring it, in wanting to bring it to its highest value, and evil consists in destroying life, doing it injury, hindering its development.

Albert Schweitzer, 1953

At no other period in life are people as likely to be as concerned with moral values and principles as they are during adolescence. A recurrent theme of American literature, from *Huckleberry Finn* to *Catcher in the Rye,* has been the innocent child who is brought at adolescence to a new awareness of adult reality and who concludes that the adult world is hypocritical, corrupt, and decadent. *Adolescent idealism,* coupled with *adolescent egocentricity,* frequently breeds "egocentric reformers"—adolescents who assume that it is their solemn duty to reform their parents and the world in keeping with their own highly personalized standards.

Some two and a half millennia ago, Aristotle came to somewhat similar conclusions about the young people of his time:

> [Youths] have exalted notions, because they have not yet been humbled by life or learned its necessary limitations; moreover, their hopeful disposition makes them think themselves equal to great things—and that means exalted notions. All their mistakes are in the direction of doing things excessively and vehemently. They love too much, hate too much, and the same with everything else.

The Adolescent as a Moral Philosopher

Significantly, young people have played a major role in many social movements that have reshaped the contours of history. In Czarist Russia the schools were "hotbeds of radicalism." In China, students contributed to the downfall of the Manchu dynasty and again to the political turmoil of 1919, the 1930s, and 1988 to 1989. And German students were largely supportive of different forms of right-wing nationalism from the mid nineteenth century on and showed support in student council elections for the Nazis in the 1930s (Lipset, 1989).

As we saw in Chapter 9, Kohlberg and his colleagues have found that in the course of moral development people tend to pass through an orderly sequence of six stages. These six stages of moral thought are divided into three major levels: the *preconventional,* the *conventional,* and the *postconventional.* Preconventional children are responsive to cultural labels of good and bad out of consideration for the kinds of consequences of their behavior-punishment, reward, or the exchange of favors. Persons at the conventional level view the rules and expectations of their family, group, or nation as valuable in

their own right. Individuals who pass to the postconventional level (and Kohlberg says most people do not) come to define morality in terms of self-chosen principles that they view as having universal ethical validity and application. (See Figure 11-5.)

The impetus for moral development results from increasing cognitive sophistication of the sort described by Piaget. Consequently, postconventional morality becomes possible only with the onset of adolescence and the development of formal operational thought—the ability to think in logical and abstract terms. Thus, postconventional morality depends primarily on changes in the structure of thought, rather than on an increase in the individual's knowledge of cultural values (de Vries & Walker, 1986). In other words, Kohlberg's stages tell us *how* an individual thinks, not *what* she or he thinks about given matters.

Questions

How does a typical adolescent's moral development change from sixth grade to twelfth grade? Is adolescent egocentrism the same as the egocentrism of a 4-year-old?

The Development of Political Thinking

If I do not acquire ideals when young, when will I? Not when I am old.

Maimonides, a Twelfth-Century Jewish Scholar

The development of political thinking, like the development of moral values and judgments, depends to a considerable extent on an individual's level of cognitive development. The psychologist Joseph Adelson and his colleagues have interviewed large numbers of adolescents between 11 and 18 years of age. Their aim has been to discover how adolescents of different ages and circumstances think about political matters and organize their political philosophies. Adelson (1972, p. 107) presents adolescents with the following premise:

> Imagine that a thousand people venture to an island in the Pacific to form a new society; once there they must compose a political order, devise a legal system, and in general confront the myriad problems of government.

Each subject is then asked a large number of hypothetical questions dealing with justice, crime, the citizen's rights and obligations, the functions of government, and so on. Adelson (1975, pp. 64–65) summarizes his findings as follows:

> The earliest lesson we learned in our work, and the one we have relearned since, is that neither sex, nor race, nor level of intelligence, nor social class, nor national origin is as potent a factor in determining the course of political thought in adolescence as is the youngster's

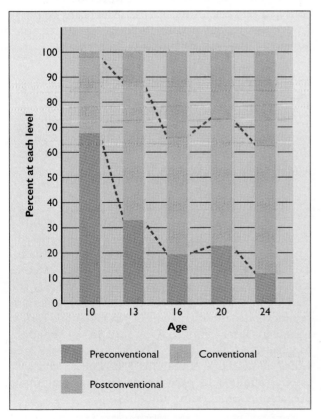

Figure 11-5 Age and Level of Moral Development The subjects in this study were urban middle-class male Americans. All percentages are approximate and are extrapolated from charts in the references cited below. The sizes of the samples studied are not stated in the originals.

Source: Daedalus, Journal of the American Academy of Arts and Sciences, from the issue entitled *Twelve to Sixteen: Early Adolescence,* Vol. 100, no. 4, Fall 1971, Cambridge, MA: Lawrence Kohlberg, "Continuities in Childhood and Adult Moral Development Revisited."

sheer maturation. From the end of grade school to the end of high school, we witness some truly extraordinary changes in how the child organizes his thinking about society and government.

Adelson finds that the most important change in political thought that occurs during adolescence is the achievement of increasing *abstractness*. This finding echoes Piaget, who described the hallmarks of formal operational thought in terms of the ability to engage in logical and abstract reasoning. Consider, for example, the answers given by 12- and 13-year-olds when they are asked, "What is the purpose of laws?" (Adelson, 1972, p. 108):

> They do it, like in schools, so that people don't get hurt.
> If we had no laws, people could go around killing people.
> So people don't steal or kill.

Now consider the responses of subjects two or three years older (Adelson, 1972, p. 108):

> To ensure safety and enforce the government.
> To limit what people can do.
> They are basically guidelines for people.
> I mean, like this is wrong and this is right and to help them understand.

An essential difference between the two sets of responses is that the younger adolescents limit their answers to concrete examples such as stealing and killing. Eleven-year-olds have trouble with abstract notions of justice, equality, or liberty. In contrast, older adolescents can usually move back and forth between the concrete and the abstract. In brief (Adelson, 1975, p. 68):

> The young adolescent can imagine a church but not the church; the teacher and the school but not education; the policeman and the judge and the jail but not the law; the public official but not the government.

Another difference between the political thinking of younger and older adolescents is that the former tend to view the political universe in rigid and unchangeable terms. Younger adolescents have difficulty dealing with historical causes. They fail to understand that actions taken at one time have implications for future decisions and events.

There is also a sharp decline in authoritarian responses as the child moves through adolescence. Preadolescents are arbitrary and even brutal in their views toward lawbreakers. They see issues in terms of good guys and bad guys, the strong against the weak, and rampant corruption versus repressive cures. They are attracted to one-person rule and favor coercive and even totalitarian modes of government. By late adolescence children generally have become more liberal, humane, and democratic in their political perspectives (Helwig, 1995; Gallatin, 1985).

Adelson finds some national variations among young people of different political cultures. Germans tend to dislike confusion and to admire a strong leader. British adolescents stress the rights of the individual citizen and the government's responsibility to provide an array of goods and services for its citizens. Americans emphasize social harmony, democratic practices, the protection of individual rights, and equality among citizens.

Question

Why and in what ways do adolescents generally become more politically aware as they progress through high school, graduate from high school, and enter college or the world of work?

Segue

We have seen how adolescents enter puberty and have discussed the physical and cognitive changes that are linked to this period in life. Maturation brings with it certain responsibilities and temptations that many adolescents find difficult to come to terms with; it is no surprise, then, that many adolescents begin to engage in behaviors that adults see as "destructive," although most adolescents come through these years unscathed. A review of the research on physical development and health issues in adolescence reveals that today's teens have many serious concerns they are dealing with on a daily basis. As we shall see in the next chapter, at a time when teens are presented with many challenges and choices, they become more reluctant to discuss these significant issues with parents or caretakers and turn to friends for information, advice, support, and comfort. In Chapter 12 we will turn our discussion to the individual adolescent's developing self-concept and self-esteem, family and peer influences, and continuing preparation for a healthy adult life.

Summary

Physical Development

1. During adolescence, young people experience the adolescent growth spurt, a very rapid increase in height and weight. The spurt typically occurs in girls two years earlier than in boys.

2. Adolescence is also characterized by the development of the reproductive system. The complete transition to reproductive maturity takes place over several years and is accompanied by extensive physical changes.

3. Children of the same chronological age show enormous variations in growth and sexual maturation. Whether they mature early or late has important consequences for them in their relationships with both adults and peers. Because of different rates of maturation, some adolescents have an advantage in height, strength, physical attractiveness, and athletic prowess.

4. Any difference from the peer group in growth and development tends to be a difficult experience for the adolescent, especially if the difference places the individual at a physical disadvantage or in a position of unfavorable contrast to peers.

5. Many teenagers are preoccupied with their physical acceptability and adequacy. These concerns take place during a time of substantial developmental change in the nature and significance of friendships.

Health Issues in Adolescence

6. Three disorders stem from poor nutritional habits: anorexia nervosa, bulimia, and obesity.

7. Substance abuse is the harmful use of drugs or alcohol, lasting over a prolonged period, that endangers self or others.

8. Unprotected sex often leads to the spread of STDs. AIDS is caused by a virus that damages the human immune system and prevents the body from fighting infections.

Cognitive Development

9. Jean Piaget called adolescence the period of formal operations. Its hallmarks are logical and abstract reasoning. Neither all adolescents nor all adults, however, attain the stage or acquire its associated abilities for logical and abstract thought

10. Adolescence produces its own form of egocentrism. In turning their new powers of thought upon themselves, adolescents assume that their thoughts and actions are as interesting to others as they are to themselves.

11. Students are divided into different ability groups in school during this period. Special programs help make the transition from school to work less problematic for many adolescents.

Moral Development

12. At no other period of life are individuals as likely to be concerned with moral values and principles as they are during adolescence. Some, but not all, adolescents attain Kohlberg's postconventional level of morality. In the process a number of young people go through a transitional phase of moral relativism.

13. During adolescence young people undergo major changes in the way they organize their thinking about society and government. Maturation appears to be the most potent source of these changes. As children move through adolescence, their political thinking becomes more abstract, less static, and less authoritarian.

Key Terms

adolescent growth spurt *(331)*
anorexia nervosa *(340, 342)*
asynchrony *(332)*
bulimia *(343)*
egocentrism *(351)*

imaginary audience *(351)*
menarche *(333)*
obesity *(341)*
period of formal operations *(349)*
personal fable *(351)*

puberty *(330)*
sexually transmitted diseases (STDs) *(345)*
substance abuse *(345)*

Following Up on the Internet

Web sites for this chapter focus on physical, cognitive, and moral maturation in adolescence. Please access the text web site at http://www.mhhe.com/crandell7 for up-to-date hot-linked Internet addresses for the following organizations, topics, and resources:

Adolescent Health from the American Medical Association

AIDS Prevention Education

Biological Changes in Adolescence for Males and Females

Survey of the Health of Adolescent Girls

TIMSS International Study Center at Boston College

Chapter 12

Adolescence
Emotional and Social Development

Critical Thinking Questions

1. When you were an adolescent, who had the most influence on how you saw the world and yourself? Was it your parents, your peers, or a best friend? Who has had the most lasting influence?

2. Are there very different developmental paths for different types of people? For example, does an African American lesbian go through the same developmental stages as a Caucasian heterosexual male? If so, what does this say about any theory?

3. How do you think the media influence adolescent identity formation?

4. What kinds of issues would you expect a teenager's diary to discuss—personal, political, moral, or cultural? What would you expect to find any different in a teenager's diary written in 1900?

The Western model of segregating youth from the "adult world" has given rise to a kind of youth culture. The obvious features of the youth culture revolve around various peer-group trademarks (such as style of music or the latest electronic gadgets), and the notion of a generation gap oversimplifies the relationship between youth and adults. Some difficult adjustments that teens must make revolve around their lifestyle choices and their sexual identity and expression. As teens begin to enter into the adult environment, they experience many exciting activities for the first time. Adults might encourage part-time employment and discourage sexual activity or substance use.

Most American adolescents make a successful transition into young adulthood. However, some teens are likely to engage in high-risk behaviors, such as substance abuse, indiscriminate sexual activity, pregnancies, abortion, suicide attempts, delinquency, self-mutilation practices, school failure, and little or no transition into employment. On the other hand, teens who continue to experience a high self-concept through tenth grade are likely to finish college, pursue advanced degrees, and continue building a foundation for their adult life.

In this chapter, we examine the influential factors that promote or demote an adolescent's self-worth and how—and with whom—an adolescent successfully navigates through this challenging stage of life.

Development of Identity

Experience: that most brutal of teachers. But you learn, my God do you learn.

C. S. Lewis

Over a period of several years or longer, a teenager spends a good deal of time focusing on the question "Who am I?" Through social interactions with family members, friends, classmates, teammates, teachers, coaches, advisors, and mentors, most adolescents come to a firmer understanding of their abilities and talents. Some want to make their own mark in the world using their unique talents; others decide to follow in a parent's footsteps or enter into a family-owned business. Many decide to enter college, which allows them time to postpone declaring a vocation or career. A minority decide to "drop out" of school and mainstream society to "find themselves," and some enter the world of work. As we shall see, there are several theories that attempt to explain why adolescence seems to be a pivotal point in an individual's life.

Hall's Portrayal of "Storm and Stress"

The notion that adolescence is a distinct and turbulent developmental period received impetus in 1904 with the publication of G. Stanley Hall's (1904) monumental work, *Adolescence.* Hall, one of the major figures of early U.S. psychology, depicted adolescence as a stage of **storm and stress,** characterized by inevitable turmoil, maladjustment, tension, rebellion, dependency conflicts, and exaggerated peer-group conformity. This view was subsequently taken up and popularized by Anna Freud (1936) and other psychoanalysts (Blos, 1962). Indeed, Anna Freud (1958) went so far as to assert: "The upholding of a steady equilibrium during the adolescent process is itself abnormal." Viewed from this Western perspective, the adolescent undergoes so many rapid changes (a convergence or "pileup" of life changes) that a restructuring of identity or self-concept is required if these changes are to be properly integrated into the individual's personality. Further complicating matters, biological and hormonal changes are thought to influence the adolescent's sense of emotional and psychological well-being and to generate in some youth, substantial mood swings, irritability, and restlessness (Buchanan, Eccles, & Becker, 1992). As you can see, adolescence was originally conceived as the "troubled waters" one had to pass over when voyaging from the more peaceful world of childhood to the demanding "real world" called adulthood. However, to this day non-Western cultures do not recognize an adolescent stage of development between youth and adulthood.

Sullivan's Interpersonal Theory of Development

One of the first theorists to propose that adolescents go through stages of development was Harry Stack Sullivan. He emphasized the importance of relationships and communication for teenagers in *The Interpersonal Theory of Psychiatry* (Sullivan, 1953). Sullivan's theory—in contrast to Freud's—explains the principal forces in human development as being social instead of biological. His social theory is enlightening when used to examine adolescent development and the impact on individuals of peer groups, friendships, peer pressure, and intimacy. In essence, Sullivan states that positive peer relationships during adolescence are essential for healthy development and that negative peer relationships will lead to unhealthy development, such as depression, eating disorders, and delinquency. We will focus on three periods of Sullivan's theory: preadolescence, early adolescence, and late adolescence.

Preadolescence   Preadolescence (which some now call the "tween" years) begins with a sudden powerful need for an intimate relationship with a same-sex playmate. It ends when the adolescent begins to experience a desire for genital sexuality. During this time personal intimacy involves interpersonal closeness but does not involve genital contact. Best friends, what Sullivan refers to as "chums," most likely have many of the same characteristics (same sex, social status, and age) and will share love, loyalty, intimacy, and the opportunity for self-disclosure—but they will not have a sexual relationship and will not experience what Sullivan calls the "lust dynamism." By having a "chum," the preadolescent gains insight into how others see the world, which helps diminish most forms of egocentric thought.

Early Adolescence   With the onset of puberty, most adolescents experience genital maturation. Sullivan (1953) says the intimate personal relationship that preadolescents had with their same-sex chums is challenged due to the emerging need for sexual intimacy with opposite-sex partners. Because the preadolescent has experienced intimacy with only someone of the same sex, the advent of early adolescence brings with it three separate needs: a need for sexual satisfaction, a continued need for personal intimacy, and a need for personal security (i.e., a need to be seen as socially acceptable by the potential sexual partners).

Security issues include self-esteem, value as an individual, and an absence of anxiety. For adolescents the new importance of their genitals as an indicator of their worth is enough to throw them into a state of disequilibrium. Remember what it was like when you first knew others were perceiving you as a "sexual" being? If you were to observe young adolescents today in malls or schools, the first thing you would probably notice are their various attempts to "catch the eye" of someone else, either by teasing, flirting, or some act of bravado. Sullivan says early adolescence leads to late adolescence when individuals have found a way to satisfy the genital drive they have acquired.

Late Adolescence   The period of late adolescence begins once the individual has established a method of satisfying sexual needs and ends with the establishment of a relationship that is both sexually and personally intimate. Love is the result of fusing intimacy and lust, and love with another person leads to a stable long-term relationship of adulthood. In late adolescence, the ability to sexually reproduce merges with the capacity for close interpersonal relationships.

Sullivan's theory attempts to get one step closer to the "nitty gritty" details of what adolescents do and experience on the journey to sexual adulthood. It also tries to explain why adolescents go through stages of development, in contrast to Hall, who wanted to paint a much more general picture of adolescence as a tumultuous period of life. Hall and Sullivan both wanted to explain certain aspects of adolescence in terms of how the youth will make a transition to becoming an adult. Hall looks at generalities, Sullivan looks at relationships. Neither emphasizes adolescent introspection or young people's psychological task of trying to make sense of the internal and external changes characteristic of the period we call adolescence. Erik Erikson, however, looks a little closer at the personal psychosocial tasks that teens struggle with during adolescence.

Erikson: The "Crisis" of Adolescence

*D*on't laugh at a youth for his affectations; he is only trying on one face after another to find a face of his own.

Logan Pearsall Smith, *Afterthoughts*, 1931

Erik Erikson's work has focused attention on the struggle of adolescents to develop and clarify their identity. His view of adolescence is consistent with a long psychological tradition that has portrayed adolescence as a difficult period (see Figure 12-1). As described in Chapter 2, Erikson divides the developmental life-span sequence into eight psychosocial stages. Each stage poses a somewhat different issue or significant challenge during development in which the individual must move in either a positive or a negative direction. A major task in self-development or ego adjustment becomes the focus of each psychosocial stage. Erikson's fifth stage covers the period of adolescence and consists of the search for **iden-**

Figure 12-1 Identity Issues Become a Focus of Adolescence

tity. He suggests that an optimal feeling of identity is experienced as a sense of well-being: "Its most obvious concomitants are a feeling of being at home in one's body, a sense of 'knowing where one is going,' and an inner assuredness of anticipated recognition from those who count" (Erikson, 1968a).

Erikson observes that adolescents, like trapeze artists, must release their safe hold on childhood and reach in midair for a firm grasp on adulthood. The search for identity becomes particularly acute because the adolescent is undergoing rapid physical change while confronting many imminent adult tasks and decisions. Recent empirical research has supported Erikson's view that adolescents do indeed go through identity exploration and a concomitant "crisis" (Kidwell et al., 1995). The older adolescent must often make an occupational choice or at least decide whether to continue formal schooling, seek employment, or simply "drop out." Other environmental aspects provide testing grounds for a concept of self: broadening peer relationships, sexual contacts and roles, moral and ideological commitments, and emancipation from adult authority.

Adolescents must synthesize a variety of new roles in order to come to terms with themselves and their environment. Erikson believes that, because adolescent identities are diffuse, uncrystallized, and fluctuating, adolescents are often at sea with themselves and others. This ambiguity and lack of stable anchorage can lead many adolescents to overcommit themselves to cliques or gangs, allegiances, loves, and social causes: To keep themselves together, they temporarily overidentify with the heroes of cliques and crowds, some to the point of apparently completely losing their sense of individuality. Yet in this stage not even "falling in love" is entirely a sexual matter. Adolescent love can be an attempt to arrive at a definition of one's identity by projecting one's diffused self-image on another and by seeing it reflected back and gradually clarified. This is why so much of young love is conversation. Clarification can also be sought by destructive means. Young people can become remarkably clannish, intolerant, and cruel in their exclusion of others who are "different," in skin color or cultural background, in tastes and talents, and often in entirely petty aspects of dress and gesture arbitrarily selected as the signs of being "in" or "out."

According to Erikson, this clannishness explains the appeal that various extremist and totalitarian movements have for some adolescents; in other words, you do not see many 75-year-old skinheads. In Erikson's view, every adolescent confronts a major danger: that he or she will fail to arrive at a consistent, coherent, and integrated identity. Consequently, adolescents might experience **identity diffusion**—a lack of ability to commit oneself, even in late adolescence, to an occupational or ideological position and to assume a recognizable station in life. Another danger is that adolescents might fashion a **negative identity**—a debased self-image and social role. Still another course taken by some adolescents is formation of a **deviant identity**—a lifestyle that is at odds with, or at least not supported by, the values and expectations of society. Other researchers have followed Erikson's lead.

James E. Marcia (1991, 1966) examined the development and validation of ego identity status in terms of achievement, moratorium, foreclosure, and diffusion. Marcia interviewed college students to find out how they felt about future occupations, religious ideology, and worldview. From these interviews Marcia found that students could be classified according to four types of identity formation:

1. **Identity diffusion.** A state in which the individual has few, if any, commitments to anyone or to a set of beliefs. Relativistic thought and emphasis on personal gratification are paramount. There is no core to the person that one can point to and state, "This person stands for X, Y, or Z." Those who are *identity diffused* do not seem to know what they want to do in life or who they want to be. *Example: Henri joins one cause this week and another next week. He is a strict vegetarian this month and an avid carnivore next month. He cannot tell you why he believes what he does except in very vague terms, such as "Because that's the way I am."*

2. **Identity foreclosure.** The avoidance of autonomous choice. Foreclosure is premature

identity formation. The adolescent accepts someone else's (such as parent's) values and goals without exploring alternative roles. *Example: Carmen wants to be a doctor and has wanted to be a doctor since her parents suggested it at age 7. Now at 18, she does not think twice about the idea because she has internalized her parents' expectations. You might hear her say, "Mom wants me to go to Harvard, so Harvard here I come!"*

3. **Identity moratorium.** A period of delay, during which adolescents can experiment with or "try on" various roles, ideologies, and commitments. It is a stage between childhood and adulthood when the individual can explore various dimensions of life without yet having to choose any. Adolescents might start or stop, abandon or postpone, implement or transform given courses of action. *Example: André joined the Peace Corps because he didn't quite know what he wanted to do after college, and he thought this would give him a chance to "find himself."*

4. **Identity achievement.** A period when the individual achieves inner stability that corresponds to what others perceive that person to be. *Example: Everyone agrees that when Jamella walks into the room she will handle the situation in a professional manner and refuse to divulge confidential information afterward. In fact, this is exactly what happens. Everyone knows that Jamella is trustworthy, and she sees herself in the same light.*

David Elkind (1994) has added a postmodern view of adolescence in which the psychologically harsh realities of the outside world cannot be kept at bay, and therefore identity formation can no longer be put off until late adolescence.

> **Questions**
> *In what respects do Hall, Sullivan, and Erikson view the developmental tasks of adolescence similarly? In what ways do these theorists differ in their views of adolescent identity formation?*

Cultural Aspects of Identity Formation

You have started out on the good earth. You have started out with good moccasins. With moccasin strings of the rainbow, you have started out. With moccasin strings of the sun's rays, you have started out. In the midst of plenty, you have started out.

Apache song sung at the "womanhood" rite marking puberty

Any number of social scientists have suggested that few people make the transition from childhood to adulthood more difficult than Western nations do (Chubb & Fertman, 1992; Elkind, 1979; Sebald, 1977). At adolescence boys and girls are expected to stop being children, yet they are not expected to be men and women. They are told to "grow up," but they are still treated like dependents, economically supported by their parents and frequently viewed by society as untrustworthy and irresponsible. According to this view, conflicting expectations generate an identity crisis among U.S. and European youth.

Many non-Western societies make the period of adolescence considerably easier, or at least more definitive. They ease the shift in status by providing **puberty rites**—initiation ceremonies that socially symbolize the transition from childhood to adulthood (Gilmore, 1990; Perlez, 1990). For example, adolescents in many African and Middle Eastern countries are subjected to various rituals and ceremonies, some of which are physically painful and psychologically harmful. Most youth are willing to endure these experiences in order to enter the higher status and privilege of adulthood in their culture. As we mentioned in Chapter 11, some circumcision rituals nowadays are performed on younger and younger youth, and some young people who have attended more formal schooling are refusing to go through these rituals that will scar them for life.

Western societies do provide a number of less obvious rites of passage. There are the Jewish Bar Mitzvah and Bat Mitzvah and the Christian confirmation ceremony. Securing a driver's license at age 16 or 17, voting at age 18, and entering the military also function as rites of passage. And graduation from high school and college, each affording a formal diploma and ritual, are special kinds of initiation ceremonies. But these are all rather mild versions of what youth must go through in many non-Western societies (Raphael, 1988).

Adolescence: Not Necessarily Stormy or Stressful?
Psychologist Albert Bandura (1964) believes that the stereotyped storm-and-stress portrait of adolescence most closely fits the behavior of "the deviant 10 percent of the adolescent population that appears repeatedly in psychiatric clinics, juvenile probation departments, and in the newspaper headlines." Bandura argues that the "stormy-decade myth" is due more to cultural expectations and the representations of teenagers in movies, literature, and other media than to actual fact. Daniel Offer (Offer, Ostrov, & Howard, 1981) likewise finds little evidence of "turmoil" or "chaos" in his longitudinal study of a sample of 61 middle-class adolescent boys. Most were happy, responsible, and well-adjusted boys who respected their parents. Adolescent "disturbance" tended to be limited mostly to "bickering" with their parents.

Like Bandura, Offer concludes that the portrayal of adolescence as a turbulent period comes from the work of such investigators as Erik Erikson who have spent their professional careers primarily studying disturbed adolescents. He concludes (Offer & Offer, 1975, p. 197): "Our data lead us to hypothesize that adolescence, as a stage in life, is not a uniquely stressful period." Offer's more recent study of some 6,000 adolescents in 10 nations (Australia, Bangladesh, Hungary, Israel, Italy, Japan, Taiwan, Turkey, the United States, and West Germany) lends cross-cultural support to this conclusion (Offer et al., 1988; Schlegel & Barry, 1991).

Adolescence can also be overrated as a time of major attitudinal change. Aspirations, self-concepts, and political attitudes (including those on racial issues) generally are well established by age 16. Though differences in these areas show up between individuals who achieve success in education (and later on the job) and those who do not, these differences are already largely established by the tenth grade (Bachman, O'Malley, & Johnston, 1978). In other words, young people who enter high school with high aspirations and positive self-concepts are likely to retain these advantages at least five years beyond high school. Hence, students in graduate and professional schools typically have high self-esteem that mirrors the positive self-images they possessed five years earlier. Similarly, the poor self-images of school dropouts are already established before these adolescents withdraw from school. Such individual differences are quite stable across time (Jessor, Turbin, & Costa, 1998).

Data from longitudinal studies first undertaken with youngsters between 1928 and 1931 by psychologists at the University of California at Berkeley and then followed by researchers for more than 50 years confirm these findings: Competent adolescents have more stable careers and marriages than less competent ones, and they experience less personality change over the adult years (Clausen, 1993; Fertman & Chubb, 1992). On the whole, many researchers find that the overall self-esteem of most individuals increases with age across the adolescent years (Chubb, Fertman, & Ross, 1997). Of course, there are exceptions. Changes in social environment, including changing schools, can interfere with those forces that otherwise bolster a child's self-esteem. Thus, the transition into a middle or junior high school can have a disturbing effect under certain circumstances, particularly for girls (Seidman et al., 1994). Indeed, the endangered self-esteem of adolescent girls is now the focus of considerable study.

Carol Gilligan: Adolescent Girls and Self-Esteem
Historically, females have not been studied by academics in most universities until very recently. However, according to *How Schools Shortchange Girls,* a study of 3,000 youngsters commissioned by the American

Human Diversity

A Contemporary Minority Female Perspective on Adolescence

The research of Carol Gilligan has been criticized for focusing on affluent white girls and therefore not being representative of girls of other ethnicities. In fact, young black women seem to be an exception to Gilligan's claim that adolescent girls lose their "voice": far more black than white women were still self-confident in high school. Apparently, many young black women identify with the strong, competent black women in their lives who hold full-time jobs and run their households. In addition, black parents often instill in their youngsters the belief that there is nothing wrong with them, only with the way the world treats them. And one African American researcher has another perspective: "young black girls have neither the time nor the opportunity to concern themselves with the contemplation of self-esteem" (Carroll, 1997). Carroll has interviewed young African American women in order to let them tell their own stories, stories that she feels cannot be captured in a research methodology that gives "voice" only to privileged groups of women. As exemplified below, it could be argued that Gilligan's theory applies only to those women and girls who have the "psychological leisure time" to entertain ideas about being authentic or false individuals.

Latisha, Fourteen, Oregon

I had stopped going to school for a month or so when I realized that I was starting to forget what I had learned during the time that I was in school. My cousin was in a program in the Urban League and she got me an application. I've only been in the program for two or three months. It's okay so far. The teachers are good, but some of 'em have attitude. When people are rude, just ignore you or say foul things or whatever, that's *attitude.* Guess it comes from their backgrounds. A lot of people say I got attitude, but I don't really see it. The only reason people be saying I have attitude is because I stand my own ground. Like when dudes try to push up on me—ask me if I got a boyfriend or whatever—I tell 'em to mind their own business 'cause I'm not givin' no play. Then they say I'm stuck up with attitude. But I feel like I'm a happy person, you know, try to be smilin' all the time and whatnot.

Most of the time, reason I don't wanna talk to none of these dudes is because they be gang-affiliated. Most of 'em just wanna get some anyway. It's not that I've had bad experiences with guys, you know, the boyfriends I've had are always friends first. That's real important, to be friends first. But all my homegirls have nothin' but bad things to say about their men. A friend of mine just got pregnant and her boyfriend asked if it was his. She told him yeah and he denied it and told her that all he wanted from her in the first place was just to have sex with her, he wasn't talkin' about no baby. See, all these niggas want these days is sex, and if we keep giving it to them, they gonna' think they can get it anytime they want it. It's disrespectful.

I think the difference between me and my homegirls who give it up so easy is that I'm aware of all them diseases and troubles having sex can bring on if you're not careful. Some of these girls out doin' what they doin' know about the diseases too, but they get attitude in full effect, and be sayin' how they can do whatever they want with their body. Strange thing is, I think lettin' dudes have sex with them

Association of University Women (AAUW), women emerge from adolescence with a poorer self-image, relatively lower expectations for life, and considerably less confidence in themselves and their abilities than men do (American Association of University Women Educational Foundation [AAUW], 1992). During their elementary school years, most girls are confident and assertive and feel positively about themselves. But by the time they reach high school, less than a third still feel this way. Boys also lose some measure of self-worth, but they end up far ahead of the girls. For instance, 67 percent of the elementary school boys said they always felt "happy the way I am." By high school, 46 percent of the boys still offered this response. With girls, the figure dropped from 60 percent to 29 percent (see *Human Diversity:* "A Contemporary Minority Female Perspective on Adolescence").

Developmental psychologist Carol Gilligan, who assisted in the design of the AAUW survey, finds that adolescence is a time when girls begin to doubt themselves (Brown & Gilligan, 1992). As we will see, Gilligan has argued that women hold a "responsibility or care perspective" on morality that leads them to be more concerned than men with human relationships and caring for others. Whereas men first define their identity as separate from others and then seek out intimate relationships, women focus primarily on close relationships and then wrestle with how to care for themselves (Brown & Gilligan, 1992; Gilligan 1982b).

Put in other terms, some social and behavioral scientists contend that males and females diverge in the degree to which they see themselves as separate from or connected with others. Women are more likely than men to develop what theorists call a *collectivist, ensembled,* or

makes them feel good about themselves in a way. If they have sex, then the dude is gonna keep being nice so he can get some more. It's only when something bad happens that dudes be showing their true colors. There are some gentlemens out there who will treat a girl nice and maybe stay with her, but you can't barely ever find 'em. Most dudes are all in gangs, or already dedicated to female, or just plain messed up. My father lives in California. I met him and I don't like him. When I was younger, he beat on my moms all the time. He beat her with a hammer and a gun and left her for dead. Us kids had to see all that, see her bleeding and rushed to the hospital. It hurt so much to see her like that. Now we consider our father dead, 'cause we don't got no love or respect for him. I think that's why my brothers are in gangs 'cause they need older male role models. Gang leaders be in gangs for a long time. My oldest brother told me the reason he be in and out of jail so much is 'cause he didn't have no guidance from a father figure. My mom is much better off now. She's going through a lot right now, and sometimes we don't see her for awhile 'cause she's been in and out of jail too. First time was for accessory to a burglary. She got three years for that. Then one time she was at the scene of a crime and there were no witnesses to testify that she didn't do anything. She's been in the wrong place at the wrong time a lot. But now she's tryin' to clean her life up, and we know she loves us.

My brother who's in this program with me is my role model. He's real smart. And also my aunt, who I stay with sometimes, is also a role model for me. She is a Christian and has a good job. She's always telling me I can be somebody. I want to get a good job and have a family when I'm ready. Her daughter, my cousin, is thirteen and she already got two babies. Yeah, she was pregnant with the first when she was eleven, had it when she was twelve. I don't want that. It's not a choice I'm gonna make anytime soon. Everyday I try and listen to what I know I got inside of me, you know, my personal strength, so I don't become a victim of society.

There are a whole lot of things that I want to do. I want to be an accountant, because accountants makes lots of money. I want to be a comic because making people laugh seems like a good profession and it feels good when people smile after I say something. I want to be an artist because I like to create other worlds. I know I have a long road ahead of me, but I don't feel disadvantaged because I know well enough to listen to myself and folks around me who are honestly trying to help me out. If I put my mind to it, I can do anything. When I wake up in the morning, I don't think about being black or about being a girl, I think about just making it through the day.

Sometimes, you know, I wonder what it would be like to be somebody else, somebody with lighter skin, longer, straighter hair, and pretty blue eyes. There don't seem to be no real famous dark, dark-skinned women with nappy hair out there. I know I'm pretty and all, but I also know I'd be prettier with lighter skin. That's what everybody says anyway. Besides, it would be so nice to be able to walk down the street and not have to worry about getting stopped by the police, just once not to worry 'bout some Caucasian policeman holdin' a gun to my head, askin' me 'bout which gang I'm in. Crazy.

I'm not scared of dying because I know the Lord will take care of me. I don't wanna die or nothin' but I know if I get killed it is because my time has come. If I could tell young black girls in America anything, I'd tell them to be hopeful and to know that there are always options. I would also tell them to take responsibility for their lives and to seek out people who will be supportive. The most important thing though, is to try and stay grounded, even when the world is spinnin' around you.

connected scheme or model for the self. In contrast, men are more likely to have an *individualist, independent,* or *autonomous scheme* for the self. So women are more likely to feel good about themselves, at least in part, from being sensitive to, attuned to, connected to, and generally interdependent with others. Men are more likely to feel positively about themselves by being independent, autonomous, separate, and better than others (Josephs, Markus, & Tafarodi, 1992). Adolescent girls who attend large, impersonal high schools can lose that connectedness that promotes meaningful relationships.

During the 1980s Gilligan and her colleagues undertook a series of projects designed to connect her earlier research on adult women to that of girls. She found that 11-year-old girls typically maintain the self-confident attitudes they had in the elementary school years: They retain honesty about what they like and what hurts in relationships, their belief in their own authority in the world, and their assured outspokenness. However, by age 15 or 16, they increasingly say, *"I don't know. I don't know. I don't know."* In brief, Gilligan finds that during adolescence girls begin to doubt the authority of their own inner voices and feelings and their commitment to meaningful relationships. Whereas as 11-year-olds they assert themselves and still speak their minds, in adolescence they come to fear rejection and anger, and so they mute their voices and repress their autonomy (also see Block & Robins, 1993). Western culture, Gilligan says, calls upon young women to buy into the image of the "perfect" or "nice" girl—one who avoids being mean and bossy and instead projects an air of calmness, quietude, and cooperation. Schools contribute to the problem by educating primarily for autonomy while negating the pursuit of rewarding relationships. Additionally,

Identity Formation in Adolescent Girls Some recent studies suggest that females are emerging from adolescence with a poorer self-image and less confidence in themselves and their abilities than they had in early adolescence.

school is seen as a place of unequal opportunity, where girls face discrimination from teachers, textbooks, tests, and their male classmates (sexual harassment) (Sadker & Sadker, 1994; American Association of University Women Educational Foundation, 1992).

More recently, Gilligan has turned her attention to developing programs that will help young women write authentic and meaningful scripts for their own lives and prevent them from "going underground" with their feelings. Meanwhile, Gilligan's thesis has not gone unchallenged. For instance, Christina Hoff Sommers (1994) attacks the credibility of virtually all gender-bias research as itself biased and lacking in substance.

Mary Pipher: Identity Formation in Adolescent Girls
In 1994, clinical psychologist Mary Pipher published a revealing exposé entitled *Reviving Ophelia: Saving the Selves of Adolescent Girls,* based on 20 years of counseling preadolescent and adolescent girls. From observing and documenting the significant changes she has witnessed in young women during the course of her clinical practice, she warns that our culture (schools, media, advertising industry) is destroying the identity and self-esteem of many adolescent girls, and she provides some recommendations for healthier identity formation. Her work not only supports Gilligan's theories, but she further states that girls are living in a whole new world in the 1990s. Theirs is a world of life-threatening experiences including living with anorexia, depression, self-

mutilation behaviors, STDs including genital herpes and warts, HIV, sexual violations and violence including rape, early and multiple pregnancies or abortions, earlier and more serious substance abuse, and higher incidences of suicide.

Something dramatic happens to early adolescent girls. Studies show a drop in girls' IQ scores and math and science scores. In early adolescence, girls become less curious and less optimistic; they lose their resiliency and their assertive "tomboyish" personalities and become more deferential, self-critical, and depressed. Their voices go "underground," their speech is more tentative and less articulate. Many vibrant, confident girls (particularly the brightest and most sensitive ones) become shy, doubting young women.

Michael Gurian: Identity Formation in Boys Michael Gurian is a counselor and therapist who has devoted a great deal of study to the identity development of boys. His 1996 book, *The Wonder of Boys: What Parents, Mentors and Educators Can do to Shape Boys into Exceptional Men,* describes what he thinks boys need to become strong, responsible, sensitive men. His theory about male identity development is centered around recognizing that brain and hormone differences basically control the way males and females operate. Some of the latest research in the 1990s on brain differences in female and male brains confirms the structural and behavioral differences that Gurian discusses.

Gurian states when a boy reaches puberty, the influence of testosterone on both brain and body increases. A male's body will experience five to seven surges of testosterone a day (Gurian, 1996, p. 10). A boy can be expected to bump into things a lot, be moody and aggressive, require a great deal of sleep, lose his temper, have a massive sexual fantasy life, and masturbate a lot. Most important, in Gurian's view, is that boys need a primary and extended family, relationships with mentors (wise and skilled persons, such as scouting and organized sports attempt to provide), and intense support from school and community. When positive role models and adult support are not available in our culture, adolescent boys are prey to gang activity, sexual misconduct, and crime. This is verified by the high rate at which young males are committing crimes (99 percent of violent crime is committed by males), being killed, or being incarcerated in our criminal justice system. Gurian states that "boys are acting out against society and parents because neither is providing them with enough modeling, opportunity, and wisdom to act comfortably within society" (Gurian, 1996, p. 54).

As we see, then, some theorists propose that the task of meeting the maturational challenges of adolescence varies by gender, though young females and males appear to prefer to work on meeting these challenges within the broader experience of a peer group.

Peers and Family

*T*rain up a child in the way he should go: and when he is old, he will not depart from it.

Proverbs 22:6

Historically we have had an image of adolescence as a time when the world of peers and the world of parents are at war with each other. However, we derive a quite different picture from psychological and sociological research. Western industrial societies have not only prolonged the period between childhood and adulthood, they also have tended to segregate young people. The notion of a **generation gap** has been widely popularized, implying misunderstanding, antagonism, and separation between youth and adults. The organization of schools into grades based on age means that students of the same age spend a considerable amount of time together. In both academic and extracurricular activities, the schools form little worlds of their own. Middle-aged and older people also tend to create a kind of psychological segregation through the stereotypes they hold of adolescents. They frequently define adolescence as a unique period in life, one that is somehow set apart from—indeed, even at odds with—the integrated web of human activity. Let us examine these matters more carefully.

The Adolescent Peer Group

*W*e are coming to live in a society that is segregated not only by race and class, but also by age.

Urie Bronfenbrenner, *Two Worlds of Childhood*, 1970

To the extent that young people are physically and psychologically segregated, they are encouraged to develop their own unique lifestyles (Brown & Huang, 1995). Some psychologists and sociologists say that Western societies, by prolonging the transition to adulthood and by segregating their youth, have given rise to a kind of institutionalized adolescence or **youth culture**—more or less standardized ways of thinking, feeling, and acting that are characteristic of a large body of young people. The first youth culture was identified after World War II. For the first time a large number of teens had free time, extra money, and unstructured energy. It was the time of football teams, cheerleaders, bobby socks, jukeboxes, and so forth, and television was not yet on the scene (Zoba, 1997). In the 1980s the second "watershed" of youth culture became "the most plugged-in generation ever"—and their electronic world has become their community (Sydney Lewis, in Zoba, 1997).

The *millenials* is a term coined by William Strauss, author of *Generations* and *The Fourth Turning*, to refer to the cohort born around 1980 and expected to graduate from college in the year 2000. Some sociologists and others have observed the following about this youth culture (Zoba, 1997; Machan, 1997):

- This generation's pulse runs fast. Bombarded by frequent images, they are in need of continual "bits." They process information in narrative images (like Nike commercials).
- The remote control symbolizes their reality: change is constant, focus is fragmented.
- They live for now.
- They are jaded by a "Been there, done that" attitude. Nothing shocks them.
- They take consumerism for granted.
- They don't trust adults—yet they say they want to be closer to their parents than previous generations.

Here is what "Millenials" say about themselves (Zoba, 1997; Machan, 1997):

- "My generation seems oblivious. We're just coasting."
- "Everybody is too feeble because everything is handed to us."
- "We don't do anything; we don't have any great achievements."
- "We feel like everything is changing and we have nothing to do with it, so we sit back and let it happen." (Perhaps this is why "Whatever" seems to be a stock response.)
- "No one's thinking for him/herself anymore."
- "No one has any sense of honor anymore."
- "We have nothing stable to grasp; no one to look up to; no one to believe in."
- "We're not standing for anything. We desperately need to be standing for something."

The most obvious features of the youth culture revolve around various peer-group trademarks: preferred music, dance styles, and idols; fashionable clothes and hairstyles; and distinctive jargon and slang. These features separate teenagers from adults and identify adolescents who share related feelings. Such trademarks facilitate a **consciousness of oneness**—a sympathetic identification in which group members come to feel that their inner experiences and emotional reactions are similar. Additionally, adolescents feel they lack control over many of the changes occurring in their lives. One way they "take back" control is by assuming the distinctive trademarks of the peer group: They cannot control getting acne, but they do have control over what music they listen to, what they wear, and how they wear their hair.

Among the central ingredients in the youth culture are various ideas about the qualities and achievements that reveal an individual's *masculinity* or *femininity*. Traditionally, for boys the critical signs of manhood are

A Distinctive Youth Culture Some psychologists and sociologists believe that the educational institution segregates young people within high schools and colleges and affords conditions conducive to a distinctive youth culture.

physical mastery, athletic skill, sexual prowess, risk taking, courage in the face of aggression, and willingness to defend one's honor at all costs. For girls the most admired qualities are physical attractiveness (including popular clothing), behaving properly and obeying rules, the ability to delicately manipulate various sorts of interpersonal relationships, and skill in exercising control over sexual encounters (Pipher, 1994; Sebald, 1986).

Overall, two qualities are essential for obtaining high status in today's adolescent society: (1) the ability to project an air of confidence in one's essential masculinity or femininity, and (2) the ability to deliver a smooth performance in a variety of situations and settings. Part of presenting a "cool" self-image is the display of the appropriate status symbols and behaviors (Sebald, 1986). On the other hand, other researchers' findings do not endorse the idea of a monolithic youth culture, instead suggesting that adolescents perceive their world as being made up of diverse peer groups with very distinct lifestyles (e.g., "the stoners," "the jocks," and "the computer whizzes") (Machan, 1997; Brown, Lohr, & Trujillo, 1990). Some psychologists and sociologists believe that the term *culture* in the expression *youth culture* clouds our understanding of adolescence by implying that there is a gap or break between generations.

The Developmental Role and Course of Peer Groups
Conformity to peer groups plays a prominent role in the lives of many teenagers, and peer pressure is an important mechanism for transmitting group norms and maintaining loyalties among group members. Although peers serve as major socialization agents in adolescence, peer pressure varies in strength and direction across grades. Clique membership seems to take on a growing significance for many sixth-, seventh-, and eighth-graders, but

then group membership drops off as the individual aspects of social relationships take on greater importance (Brown, Clasen, & Eicher, 1986; Brown, Eicher, & Petrie, 1986).

However, contemporary teenagers differ in a great many ways. Many of these differences arise from differences in socioeconomic, racial, and ethnic backgrounds (Perkins et al., 1998). Generally, every high school typically has several "crowds"—cliques that are often mutually exclusive. Additionally, a "cycle of popularity" seems to bring some teenagers together within relatively stable cliques (for instance, cheerleaders and athletes). In due course, however, many "outsiders" come to resent and dislike their "popular" counterparts, whom they define as "stuck up." Even so, leading-crowd members tend to exhibit higher self-esteem than "outsiders" do (Brown & Lohr, 1987). In sum, our search for similarities among young people should not lead us to overlook the differences that also exist among them (Wentzel & Erdley, 1993).

Questions
In what ways are peer groups important for identity development in adolescence?

Adolescents and Their Families

When I was sixteen, I thought my father was a damn fool. When I became twenty-one, I was amazed to find how much he had learned in five years.

Mark Twain

Overall, today's teenagers see their attitudes toward drugs, education, work, sex, and most other matters as closer to their parents' views than teenagers did in the 1970s (Machan, 1997). Indeed, a 1995 survey of youngsters ages 10 to 13 reveals that they want to be close to their parents: They are still insecure and afraid and therefore welcome the sense of safety and security that parents can supply. More particularly, 93 percent said their parents love them "all the time"; 61 percent indicated that talking to their parents had helped them deal with serious problems; 65 percent worried their parents might die; 47 percent were concerned that their parents might not be available to talk when they needed them; and 38 percent worried that their parents would get divorced (Painter, 1995). Nonetheless, parent-child relationships change at puberty. The amount of time spent with parents, the sense of emotional closeness, and the yielding to parents in decision making all decline during the pubertal period (Pipher, 1994; Paikoff & Brooks-Gunn, 1991).

Influence in Different Realms of Behavior Both the family and the peer group are anchors in the lives of most teenagers. Parents and peers provide adolescents with different kinds of experience, and the influence of the two groups varies with the issue at hand. When the issues pertain to finances, education, and career plans, adolescents overwhelmingly seek advice and counsel from adults, particularly their parents. Time with parents centers around household activities like eating, shopping, performing chores, and viewing television. And family interaction more closely parallels the goals of socialization dictated by the larger community. Contrary to some psychoanalytic formulations, adolescents do not seem to develop autonomy and identity by severing their ties with their parents. Rather, teenagers benefit in their development by remaining connected with their parents and by using them as important resources in their lives. This effect is most notable when the parenting style is authoritative (Lamborn, Dornbusch, Steinberg, 1996).

For issues involving the specifics of social life—including matters of dress, personal adornment, dating, drinking, musical tastes, and entertainment idols—teenagers are more attuned to the opinions and standards of their peer group (Lau, Quadrel, & Hartman, 1990). Time with peers is spent "hanging out," playing games, joking, and conversing. Teenagers report that they look to interaction with friends to produce "good times" (Larson, 1983). They characterize these positive times as containing an element of "rowdiness": They act "crazy," "out of control," "loud," and even "obnoxious"—deviant behavior that they describe as "fun." Such activities provide a spirited, contagious mood, a group state in which they feel free to do virtually anything. The extent and intimacy of peer relationships increases dramatically between middle childhood and adolescence (Larson & Richards, 1991).

Much of the similarity found in the attitudes and behaviors of friends is the result of people purposely selecting as friends individuals who are already compatible with them. Not surprisingly, therefore, adolescents who share similar political orientations, values, and levels of educational aspiration are more likely to associate with one another and then to influence one another as a result of continued association. Additionally, parents often seek to "nudge" their youngsters toward "crowds" that are consistent with their family's values (Brown et al., 1993). More directly, parents retain some control over their teenager's choice of peers through their selection of the neighborhood in which the family resides or the school (public or private) the youngster attends (Paikoff & Brooks-Gunn, 1991).

For many youth, the right to choose is often more important than the choice itself. It signals that their parents recognize their maturity and growing autonomy. Some evidence suggests that adolescents who believe that their parents are not providing them with sufficient "space"—that their parents are not relaxing their power and restrictiveness—are apt to acquire more extreme peer orientations and to seek out more opportunities for peer advice (Fuligni & Eccles, 1993). Additionally, psychological overcontrol, as well as behavioral undercontrol, places youngsters at greater risk for problem behaviors (Simons et al.,1994; Barber, 1992). Disagreements between parents and their teenage offspring occur primarily over differing interpretations of issues and the extent and legitimacy of the youngsters' personal jurisdiction (Smetana, 1995). Most conflict occurs over everyday matters such as chores and dress rather than over substantive issues such as sex and drugs (Barber, 1994).

The functional constraints provided by the family and the excitement by friends both have their part to play in development (Hunter & Youniss, 1982). Even so, parents and adolescents frequently differ in their perception of the extent to which continuity prevails in values, beliefs, and attitudes across generations. According to one view, teens exaggerate intergenerational differences out of a developmental need for emancipation, whereas parents minimize the differences out of a developmental need for validation. According to a second view, parents, after years of investing themselves in their offspring, have more at stake in maintaining the relationship than the children do.

Shift in the Family Power Equation We have seen that from early adolescence to late adolescence the cohesion or emotional closeness between parent and child ideally becomes transformed from one of considerable dependency to a more balanced connectedness that permits the youngster to develop as a distinct individual capable of assuming adult status and roles. Another shift also occurs in family structure. Power of hierarchical relations undergoes change (Cowen & Avants, 1988). Across adolescence, parents typically make increasingly less use of unilateral power strategies and greater use of strategies that share power with their youngsters (Smetana & Asquith, 1994).

However, much depends on the structural characteristics of the family. One-parent families tend to be less hierarchical (and more egalitarian) than two-parent families. Divorced, never-remarried mothers are more permissive, provide less chaperonage, and are less controlling than are mothers in intact, two-parent families, and their youngsters enjoy more independence, power, and responsibility in decision making. Overall, adolescents in divorced families have greater control over decisions regarding clothing, curfew, choice of friends, and the spending of money and have more responsibility for household tasks than do their counterparts in two-parent families (see Chapter 8).

Adolescents and Their Mothers One of the authors of this text worked in a middle school for five years and can attest that mothers get a "bad rap" among young adolescents. Boys who appear to be "close to" their mothers are teased unmercifully and goaded into fights by put-downs and vulgarities ("Your mother is so . . . that . . ."). Also, American adolescent girls receive more mixed messages about their relationship with their mothers than boys do about their fathers, normally. Growing up requires adolescent girls to reject the person with whom they have most closely identified—their mother. Daughters are socialized to fear becoming like their mothers. A great insult to most teenage girls is to say "Oh, you are just like your mother." Mixed messages bring inevitable conflict, especially today when mothers don't seem to understand the world their daughters are experiencing. Clinical psychologist Mary Pipher (1994, p. 107) relates an all-too-common first session with a teenage girl she calls "Jessica," whose mother (a single parent and social worker) had devoted her life to her daughter:

Dr. Pipher: How are you different from your Mom?

Jessica: (smirking) "I totally disagree with her about everything. I hate school; she likes school. I hate to work; she likes to work. I like MTV and she hates it. I wear black, and she never does. She wants me to live up to my potential, and I think she's full of shit. . . . I want to be a model. Mom hates the idea."

Fortunately, it appears that many young women return to a closeness with their mothers when they enter young adulthood, particularly after they have had their own children.

Question

In what areas do teens typically seek out family for advice and support, and when are they likely to seek out friends for advice and support?

Courtship, Love, and Sexuality

There are always two choices, two paths to take. One is easy. And its only reward is that it's easy.

Author unknown

One of the most difficult adjustments, and perhaps the most critical, that adolescents must make revolves around their developing sexuality. Biological maturity and social pressures require that adolescents come to terms with awakening sexual impulses. Consequently, sexual attraction and sexual considerations become dominant forces in their lives (Gullotta, Adams, & Montemayor, 1993). Indeed, first sexual intercourse is a developmental milestone of major personal and social significance and is often viewed as a declaration of independence from parents, an affirmation of sexual identity, and a statement of capacity for interpersonal intimacy.

Adolescent sexuality is also a matter that has commanded considerable popular and media concern. Not too long ago youngsters born outside of marriage were stigmatized as "bastards," and the word *illegitimate* was written on their birth certificates. In recent years the issue of extramarital pregnancies and births has become transformed in the public's mind into a symbol of social disorder and the source of such social ills as poverty and welfare dependence. Three images of teenage childbearing are currently vying for ownership of the nation's agenda (Nathanson, 1991):

- The *public health/preventive medicine perspective* sees the issue as a problem of unintended pregnancy that is best addressed by sex education, birth control, and abortion programs and services.
- The *conservative moral view* considers teenage pregnancy to be a problem of precocious sexual activity and advocates abstinence.
- The *conservative economic approach* defines the difficulty as residing with ethnic-minority adolescent mothers on welfare who require training to become economically self-sufficient.

Clearly, the matter of teenage pregnancy is a flash point for intense public passion and debate, exacerbating tensions of gender, race, and class. The matter is further complicated by its linkage to other vexing issues such as abortion, adoption, babies born at risk, special education services, health care for children with disabilities, welfare reform, absentee fathers, and political and taxation policies.

Differing Behavioral Patterns

Youth vary a good deal in the age at which they first experience intercourse (a large proportion of all sexual exposure of females before the age of 14 is involuntary). Contemporary earlier pubertal development applies a downward pressure on the age of sexual debut. Sociologist J. Richard Udry (1988) and his colleagues report strong evidence for a hormonal basis of sexual motivation and behavior, particularly in adolescent males. In addition, among both males and females, delinquent behavior, poor prospects for higher education, low levels of parental monitoring, along with smoking, alcohol and drug use, and early onset of sexual intercourse, tend to occur among the same teenagers (Luster & Small, 1994b; Metzler, Noell, & Biglan, 1992). Young people who remain virgins longer than their peers are more likely to value academic achievement, enjoy close ties with their parents, report stricter moral standards, begin dating later, and exhibit more conventional behavior with respect to alcohol and drug use. However, virgins are decidedly not "maladjusted," socially marginal, or otherwise unsuccessful. They report no less satisfaction and no more stress than nonvirgins, and they typically achieve greater educational success than nonvirgins. Furthermore, in many cases teenagers, especially girls, select as their friends individuals whose sexual behavior is similar to their own (Billy & Udry, 1985).

Aspects of family life also affect adolescent sexual behavior. Generally speaking, the earlier the mother's first sexual experience and first birth, the earlier the daughter's sexual experience. And teenagers with older, sexually active siblings are more likely to begin sexual intercourse at an earlier age (East, Felice, & Morgan, 1993). Living in poverty also tends to be associated with early sexual activity and early pregnancy. Unwed adolescent pregnancies are several times more likely among youth with poor academic skills and from economically disadvantaged families. Moreover, adolescents, especially daughters, from single-parent households typically begin sexual activity at younger ages than do their peers from two-parent families. A number of factors apparently contribute to the higher rates of sexual intercourse among adolescents in single-parent families (Whitbeck, Simons & Kao, 1994):

- There is often less parental supervision in single-parent households.
- Single parents are themselves often dating, and their sexual behavior provides a role model for their youngsters.

- Adolescents and parents who have experienced divorce tend to have more permissive attitudes about sexual activity outside of marriage.

Adolescent sexual behavior is shaped not only by individual characteristics but also by the surrounding social context. Communities characterized by limited economic resources, racial segregation, and social disorganization apparently provide young people with little motivation to avoid early childbearing. The opportunity structures available to many inner-city youth, particularly those relating to education and jobs, often lead the young people to conclude that legitimate pathways to social mobility are effectively closed to them. A concentration of poverty, crowded housing conditions, high levels of crime, unemployment, marital dissolution, and inadequate public services engenders a social climate of apathy and fatalism.

These communities also are likely to have fewer adult role models of economic and social success. Moreover, where few adult women are able to find stable, remunerative employment, the potential costs of sexual activity in terms of future occupational attainment appear minimal to some female adolescents. Indeed, a variety of social factors might encourage teenage girls to become pregnant. For instance, in some ethnic groups having a baby is frequently seen as symbolizing maturity and entrance into adulthood (getting one's own apartment, welfare check, and food stamps), and peers often ridicule adolescents who remain chaste. In sum, although the sexual behavior of adolescents is private and highly personal, it is shaped by the characteristics of the communities in which they live (Anderson, 1994; Billy, Brewster, & Grady, 1994).

Courtship

In the United States, *dating* traditionally has been the principal vehicle for fostering and developing sexual relations, or "courtship." Over the past quarter century, however, dating has undergone rapid change (Whyte, 1990). Traditionally, dating began with a young man inviting a young woman for an evening's public entertainment at his expense. The first invitation was often given during a nervous conversation on the telephone several days or even weeks in advance. Ideally, the man would call for the woman at the appointed hour in a car and return her by car.

Although the traditional pattern has not entirely been replaced, new patterns of courtship swept in on the wave of the youth movements of the late 1960s and early 1970s. The term *dating* itself became in many ways too stiff and formal to describe the "just hanging out" and "getting together" that took place among youth (Gross, 1990; Zeman, 1990). A more relaxed style came to govern the interaction between the sexes, including roving in groups through malls, informal get-togethers, group activities like "keggers," and spur-of-the-moment mutual decisions to go out for a pizza or to a movie. Today's Internet "chat" rooms also provide a casual meeting place for teens.

Many teenage males report that they do not date or, if they do, refuse to admit to it. It seems that growing numbers of young men "do not want to look soft" to their friends. This fear, and a desire to demonstrate their manhood, leads many teenage males to abuse or show disrespect to girls. A growing number of young men view sexual harassment as a game. Teenage males report that they gain popularity by yelling explicit propositions at or fondling girls who pass by, all the while competing with one another to demonstrate the most flair and audacity in "talking trash" and "making moves." Alarmingly, the abuse seems to be increasing as a group activity, a bonding ritual among young men (Gelman, 1993; Henneberger & Marriott, 1993). Additionally, fathering children is viewed by some young men as a sign of manhood, but one that does not necessarily mean marrying the mother or supporting the child.

Love

*A*mericans, who make more of marrying for love than any other people, also break up more of their marriages . . . but the figure reflects not so much failure of love as the determination of people not to live without it.

Morton Hunt, *The National History of Love,* 1967

In the United States nearly everyone is expected to fall in love eventually. Pulp literature, "soap operas," women's magazines, "brides only" and traditional "male" publications, movies, and popular music reverberate with themes of romantic ecstasy. In sharp contrast to the U.S. arrangement, consider the words of the elders of an African tribe. They were complaining to the 1883 Com-

The Concept of Romantic Love Social scientists have found it difficult to define the concept of romantic love. Some say it is recognized by physiological arousal, some say it is a unique chemical reaction that activates the brain's pleasure centers, and others say it is associated with a special transcendent feeling. From a cross-cultural view, the concept of romantic love is not universal.

mission on Native Law and Custom about the problems of "runaway" marriages and illegitimacy:

> It is all this thing called love. We do not understand it at all.

> This thing called love has been introduced. (Gluckman, 1955, p. 76)

These elders viewed romantic love as a disruptive force. In their culture marriage did not necessarily involve a feeling of attraction for the spouse-to-be; marriage was not the free choice of the couple marrying; and considerations other than love played the most important part in mate selection. In many non-Western countries of the world, marriage is often preplanned by the parents of the prospective bride and groom, and there is some type of monetary exchange for assurance of the virginity of the bride. Furthermore, in some Middle Eastern, African, and Asian countries, the wife then becomes the property of the husband, a practice that was abandoned in Western countries early in the twentieth century.

Clearly, then, different societies view romantic love quite differently (Cancian, 1987; Luhmann, 1986). At one extreme are societies that consider a strong love attraction to be a laughable or tragic aberration. At the other extreme are societies that define marriage without love as shameful. American society tends to insist on love; traditional Japan and China tend to regard it as irrelevant; ancient Greece in the period after Alexander, and ancient Rome during the Roman Empire, fell somewhere in the middle (Goode, 1959). It seems that the capacity for romantic love is universal, but its forms and extent to which the capacity gets translated into everyday life are highly dependent upon social and cultural factors (Goleman, 1992; Hendrick & Hendrick, 1992).

All of us are familiar with the concept of romantic love, yet social scientists have found it exceedingly difficult to define, so it is little wonder that many Americans—especially teenagers—are also uncertain about what love is supposed to feel like and how they can recognize the experience within themselves. Some social psychologists conclude that *romantic love* is simply an agitated state of physiological arousal that individuals come to define as love. The stimuli producing the agitated state might be sexual arousal, gratitude, anxiety, guilt, loneliness, anger, confusion, or fear. What makes these diffuse physiological reactions love, they say, is that individuals label them as love.

Some researchers reject the notion that love and other states of physiological arousal are interchangeable except for the label we give them. For instance, Michael R. Liebowitz (1983) says that love has a unique chemical basis and that love and romance are among the most powerful activators of the brain's pleasure centers. And they may also contribute to a special transcendent feeling—a sense of being beyond time, space, and one's own

body—that Liebowitz says parallels descriptions of psychedelic experiences. Intense romantic attractions can trigger neurochemical reactions that produce effects much like those produced by such psychedelic drugs as LSD, mescaline, and psilocybin. Just how we come to experience such changes in brain chemistry as feelings of love remains, for Liebowitz, an unanswered question.

Questions

Is the idea of romantic love a universal concept? How does the feeling called love manifest itself in physical, psychological, or emotional ways?

Sexual Attitudes and Behavior

Although we commonly equate adolescent sexuality with heterosexual intercourse, *sexual expression* takes a good many different forms. Furthermore, sexuality begins early in life and merely takes on more adult forms during adolescence.

Development of Sexual Behavior Both male and female infants show interest in exploring their own bodies, initially in a random and indiscriminate fashion. Even at 4 months of age babies respond to genital stimulation in a manner that suggests that they are experiencing erotic pleasure. When children reach 2 and 3 years of age, they will investigate their playmates' genitals and, if permitted, those of adults as well. But by this time strong social prohibitions come into effect, and children are socialized to restrain these behaviors.

Masturbation, or erotic self-stimulation, is common among children. In many cases children experience their first orgasm through self-stimulation. It might occur through the fondling of the penis or the manual stimulation of the clitoris or by rubbing against a bedcover, mattress, toy, or other object. Boys often learn about masturbation from other boys, whereas girls learn to masturbate primarily through accidental discovery (Kinsey et al., 1953; Kinsey, Pomeroy, & Martin, 1948).

A good many children also engage in some form of sex play with other children prior to adolescence. The activity is usually sporadic and typically does not culminate in orgasm. On the basis of his research in the 1940s and early 1950s, Alfred C. Kinsey and his associates (1948, 1953) found that the peak age for sex play among girls was 9, when about 7 percent engaged in heterosexual play and 9 percent in homosexual play. The peak age for boys was 12, when 23 percent participated in heterosexual play and 30 percent in homosexual play. But Kinsey believed that his reported figures were too low and that about a fifth of all girls and the vast majority of all boys had engaged in sex play with other children before reaching puberty.

Adolescent Sexual Expression Adolescent sexuality finds expression in a number of ways, including masturbation, nocturnal orgasm, heterosexual petting, heterosexual intercourse, and homosexual activity. Teenage masturbatory behavior is often accompanied by erotic fantasy. One study of 13- to 19-year olds found that 57 percent of the males and 46 percent of the females reported that they fantasized on most occasions while masturbating; about 20 percent of the males and 10 percent of the females rarely or never fantasized when masturbating (Sorensen, 1973). Many myths have attributed harmful effects to masturbation, but the physiological harmlessness of the practice has now been so thoroughly documented by medical authorities that there is no need to belabor the issue. Even so, some individuals might feel guilty about the practice for social, religious, or moral reasons.

Adolescent boys commonly begin experiencing *nocturnal orgasms,* or "wet dreams," between ages 13 and 15. Erotic dreams that are accompanied by orgasm and ejaculation occur most commonly among men in their teens and twenties and less frequently later in life. Women also have erotic dreams that culminate in orgasm, but apparently they are less frequent among women than among men.

Petting refers to erotic caressing that may or may not lead to orgasm. If it eventuates in sexual intercourse, petting is more accurately termed "foreplay." Although usually applied to heterosexual encounters, homosexual relations can involve similar techniques. There are few current statistics on the incidence of petting among teenagers because the practice is exceedingly prevalent,

and both public and scientific interests have moved beyond the issue (Katchadourian, 1984).

A little history . . . The past 30 years have seen substantial changes in U.S. attitudes toward teenage sexual activity (see Figure 12-2). An American societal change called the "sexual revolution" came about during the Vietnam War years, appearing mainly after 1965, along with availability of the birth control pill. Orgies, partner swapping, and love-ins were implicitly promoted in films, music, and advertising in the 1960s. The original Woodstock generation symbolized the idea of "free love" and communal living. This period altered the sexual landscape in that sexual attitudes and experimentation became much more relaxed until the advent of AIDS in the early 1980s.

Today greater openness and permissiveness prevail with regard to premarital sex, homosexuality, extramarital sex, and a variety of specific sexual acts. Television, movies, videos, Internet web sites, and magazines bombard young people and adults alike with sexual stimuli on an unprecedented scale. A growing proportion of teenagers are sexually active, and teens are beginning their sexual activity at earlier ages. According to a national sampling of high school students from all 50 states and the District of Columbia (CDC, 1997b), 60 percent of the country's adolescents have had sexual intercourse (and many have had multiple sex partners) by the time they are in their senior year of high school (see Table 12-1). More boys than girls have had sexual experience at every age level.

Figure 12-2 Dramatic Changes in Sexual Behavior Began about 1965

"But, Dad, in your day sex was still in the future."

Table 12-1   Percentage of High School Students Who Have Had Sexual Intercourse, by Grade Level, Sex, and Race/Ethnicity, 1991–1997

	1991 (N = 12,272)	1997 (N = 16,272)
Grade		
9th	39%	38%
10th	48.2%	42.5%
11th	62.4%	49.7%
12th	66.7%	60.9%
Sex		
Male	57.4%	48.8%
Female	50.8%	47.7%
Race/Ethnicity		
Non-Hispanic White	50%	43.6%
Non-Hispanic Black	81.4%	72.6%
Hispanic	53.1%	52.2%

Source: Centers for Disease Control and Prevention conducted the *Youth Risk Behavior Survey* (YRBS) in 1991, 1993, 1995, and 1997. Reported in the *Morbidity and Mortality Weekly Report, 47,* No. 36 (Sept. 18, 1998), p. 751.

Adolescent Sexual Activity Rates   The rates of teenage sexual experience increased throughout the 1980s, plateaued by the early 1990s (Besharov, 1993), and declined from 1991 to 1997 (see Table 12-1) (Centers for Disease Control, 1998h). Most recently, to determine trends in sexual risk behaviors among U.S. high school students, the Centers for Disease Control and Prevention (CDC) conducted the *Youth Risk Behavior Survey* (YRBS) in 1991, 1993, 1995, and 1997. Compared with findings from 1991 on the YRBS, the prevalence of students who had ever had sexual intercourse in 1997 decreased 11 percent. (However, note that adolescents who had dropped out of high school were not surveyed, nor were those in junior high school.) This is a definite reversal of the trend in sexual intercourse rates that occurred during the 1970s and 1980s (Centers for Disease Control, 1998h). Whereas in 1991, 57.4 percent of adolescent males surveyed reported ever having had sexual intercourse, by 1997, 48.8 percent surveyed reported having had sexual intercourse. During this same time period, 50.8 percent of adolescent females reported being sexually active in 1991, whereas in 1997, only 47.7 percent reported being sexually active.

Among male students, sexual experience decreased 15 percent, but sexual experience among female students did not show a significant decrease. From 1991 to 1997, adolescents in all racial and ethnic categories reported decreased rates of ever having had sexual intercourse. Sexual experience decreased 6.4 percent among white students and 11 percent among black students. Sexual experience among Hispanic students did not show a significant decrease. Notice, however, that by 12th grade, in 1997 in this national sample of students, almost 61 percent of all seniors reported having had sexual intercourse. This, coupled with having multiple sex partners, puts them at a high risk for STDs, AIDS, pregnancy and multiple pregnancies, and abortion.

Multiple Sex Partners   We are also seeing a downward trend in the number of teenagers who are having multiple sexual partners. The same *Youth Risk Behavior Survey* (YRBS) conducted in 1991, 1993, 1995, and 1997 by the Centers for Disease Control and Prevention (1998h) reveals that the prevalence of multiple partners overall decreased significantly from 1991 to 1997 (see Table 12-2). In 1997, 17.6 percent of male students reported having had multiple sex partners, a decrease from 23.4 percent in 1991. Prevalence of multiple sex partners among female students *increased* slightly.

Percentage of students who had had multiple sex partners *decreased* in all race and ethnic categories. However, the rates for multiple sex partners is still very

Table 12-2   Percentage of High School Students Who Have Had Multiple Sex Partners (4 or More Partners During Lifetime), by Grade Level, Sex, and Race/Ethnicity, 1991–1997

	1991 (N = 12,272)	1997 (N = 16,272)
Grade		
9th	12.5%	12.2%
10th	15.1%	13.8%
11th	22.1%	16.7%
12th	25%	20.6%
Sex		
Male	23.4%	17.6%
Female	13.8%	14.1%
Race/Ethnicity		
Non-Hispanic White	14.7%	11.6%
Non-Hispanic Black	43.1%	38.5%
Hispanic	16.8%	15.5%

Source: Centers for Disease Control and Prevention (CDC) conducted the *Youth Risk Behavior Survey* (YRBS) in 1991, 1993, 1995, and 1997. Reported in the *Morbidity and Mortality Weekly Report, 47,* No. 36 (Sept. 18, 1998), p. 751.

high for non-Hispanic black students, who are putting themselves at the highest risk for STDs, AIDS, multiple pregnancies, abortion, and high-risk births. About 70 percent of sexually active women ages 15 to 17 move on to a second sexual partner within 18 months of their first intercourse; sexually active women ages 18 and 19 do not move on to another partner quite as quickly—only half report having had more than one partner 18 months after their first intercourse (Lewin, 1992b).

Teenage Pregnancy

Ninety percent of the nation's prison inmates up to age 35 were born to mothers under 18.

Gurian, 1996, p. 183

Many Americans view unwed teenage parenthood as a major social problem associated with many negative outcomes for the teenage mother and child. According to the latest statistics from the National Center for Health Statistics, the number of unwanted births is declining. The birth rate for black teenagers aged 15 to 19 declined 21 percent from 1991 to 1996. Birth rates for other teens dropped between 5 to 12 percent during the same period (Centers for Disease Control, 1998f).

A decade or so ago, teenage pregnancy and childbearing were seen primarily as health problems. Although overall U.S. teen birthrates are declining, births to unwed teens are increasing: 76 percent of pregnant teenagers are single parents (Koch, 1998), up from 48 percent in 1980. This is a significant societal concern, for only 66 percent of adolescent mothers finish their high school education and can support their offspring (Luker, 1996). In the early 1990s, public concern was focused on teenage childbearing as one of the major sources of poverty in the United States; more recently researchers suggested that the real culprits are poverty and crime (Elliot et al., 1996). Yet our federal government spends $39 billion a year to support families begun by teen mothers (Koch, 1998). The birth of a child increases a young person's income requirements, and the lack of a diploma is a significant liability in today's job market.

Early childbearing also has implications for family life. Of those teenagers who do marry, pregnancy is often a major reason (one-third of brides under age 18 are pregnant). Teen marriages are highly unstable. They are two to three times more likely to end in divorce than marriages occurring after age 20, and most adolescent mothers spend at least some portion of their lives as single parents (60 percent of brides aged 17 years or less divorce within six years; 20 percent divorce within the first year) (Fiscella et al., 1998). The children of teen parents are also affected. Inept parenting, poor health care, child

neglect, and child abuse are comparatively more common among teenage parents. Moreover, children of younger parents tend to score lower than children of older parents on intelligence tests, and they typically do less well in school. In addition, children of younger parents are more prone to behavioral and adjustment disorders and are more likely to become teen parents themselves (Sommer et al., 1993). Also, among female adolescents, incidents of childhood sexual abuse increase the probability of having earlier sexual intercourse and earlier pregnancy (Fiscella et al., 1998).

Up to nine times as many teenagers give birth in the United States than in any other industrialized country (Koch, 1998) (see Figure 12-3). American youth are exposed to mixed messages about contraception, and birth control services are not effectively delivered to the nation's teenagers (Brooks-Gunn & Furstenberg, 1989). Each year, 1 out of every 10 teenage American females becomes pregnant, and 8 out of 10 of these pregnancies are unintended. Yet some teenagers unrealistically believe that having a baby will solve their problems, provide someone who will love them, and make them feel grown up. Only about 47 percent of teen pregnancies end in live birth; 40 percent are terminated by abortions, and 13 percent result in miscarriages.

Why Do Teenagers Become Pregnant?   Most teenagers do not consciously plan to become sexually active, and so they do not foresee their first sexual experience. Instead, they often experience their first sexual encounter not as a decision but rather as something that "just happened." Moreover, most teenagers wait almost a year after becoming sexually active before they seek medically supervised contraceptive care. Adolescents frequently have a sense of invulnerability and fail to associate consequences with actions. Other associative factors for female teenage sexual activity include coercion by male partners (Wulfert & Biglan, 1994); early childhood or physical trauma often leading to early onset of illegal drug use, affecting the judgment and increasing incidence of unprotected sex; procrastination about birth control; unrealistic expectations about having a baby; lack of impulse control; a conscious desire to seek welfare dependency; and a low sense of self-esteem or girls not wanting to lose their man (fear of abandonment) (DiMauro, 1995).

The Very Young Teen Mother   The younger the female, the greater the risk of unwanted pregnancy and STDs (Moore, Rosenthal, & Boldero, 1993). Of pregnancies that occur in girls under age 15, most occur in those aged 13 to 14. Very young women are likely to become pregnant from rape or incest. Early sexual activity is associated with parents' marital disruption, living with a single parent, early onset of puberty, school problems,

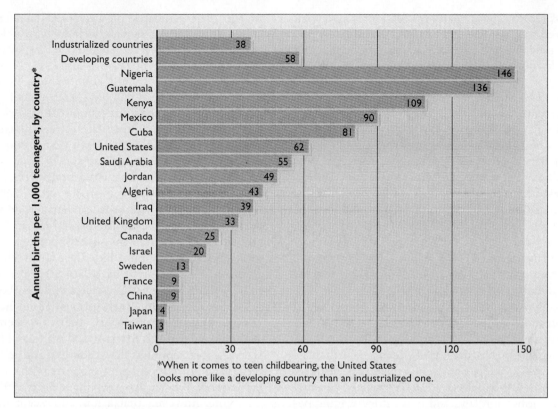

Figure 12-3 Teenage Pregnancies, 1994. Though there is a slight decline in the overall rate of teen pregnancy, the rates for non-marital births is increasing. Many Americans view unwed teenage parenthood as a major social problem, for it is associated with many negative outcomes for the mother, child, and society (Upchurch & McCarthy, 1990).

Copyright 1994, USA TODAY. Reprinted with permission.

delinquency, living in a disadvantaged neighborhood, lowered family income, inadequate supervision, sexually active siblings and friends, and chemical dependency problems (Waters, Roberts, & Morgan, 1997).

Sex Education, "Safe Sex," and Contraception Sex education advocates often operate on the mistaken notion of the "rational teenager" who, when given "the facts," will either abstain from sex or use contraception. These strategies tend to ignore the cognitive qualities of mind that are a prerequisite for developing complex interpersonal skills. When asked why they do not use birth control, 39 percent of teenagers give answers that can be summed up as "not wanting to" ("feels better without it," "don't think about it," "want to get pregnant"). Twenty-five percent give answers involving lack of knowledge or access; 24 percent are afraid or embarrassed to use contraception; 20 percent do not expect to need contraception or do not want to take the time; and 14 percent are not worried about pregnancy (some teenagers cite multiple reasons) (National Research Council, 1987).

Significantly, as adolescents become more sexually experienced, they tend to become more consistent contraceptive users. From 1988 to 1995, use of oral contraceptives declined but the use of condoms and long-acting contraceptive methods, such as the three-month contraception shot, increased (Centers for Disease Control, 1998f). In addition, teenagers who have high educational expectations and school success are more likely to use contraception effectively (Miller & Moore, 1990). Congress's 1996 abstinence-only law added more fuel to the fire in this highly charged issue. The new law not only prohibits teaching about contraception but also requires teenagers to be taught that adults in America are expected to refrain from extramarital sex because "it is likely to have harmful psychological and physical effects" (Koch, 1998).

Polls indicate that most American parents want their children to abstain from sex until they finish high school, yet the majority also want schools to teach sex education. Advocates of abstinence argue that teaching about "safe sex" is dishonest because condoms do not prevent transmission of several STDs and that teaching about birth control gives teenagers approval to have sex. They also say that decades of teaching our teenagers about "safe sex" has created a generation of teenagers who treat sex casually and irresponsibly and promotes easier access to confidential services that offer contraceptives and abortions. Critics in the medical community contend that not teaching about "safe sex" leaves our youth defenseless (Koch, 1998). Many professionals agree that parents need to play a more active role in combating the national epidemic of teen pregnancies, abortions, and STDs by discussing birth control and contraception.

Abortion   The number of reported legal abortions in the United States for 1995 (1,210,883) was the lowest annual total since 1975. From 1990 to 1995, the annual number of abortions has decreased overall by 15 percent (Abortion Surveillance: United States, 1996). The percentage of women under 19 who sought an abortion has decreased from 32.6 percent in 1972 to 20.1 percent in 1995. In 1995, the total number of legal abortions performed on teenage girls under age 15 was 5,949, whereas the total for girls 15 to 19 was 146,284 (Abortion Surveillance: United States, 1996). This lowered rate resulted from a lower proportion of pregnant women who had obtained an abortion. It is believed that four factors are involved: More pregnancies are being carried to term, attitudes about abortion have changed, the use of contraceptives has increased, and the number of unintended pregnancies has decreased. Teenage birth rates have decreased, access to abortion services has changed, and abortion laws that affect adolescents (e.g., parental consent or notification laws and mandatory waiting periods) have changed (Abortion Surveillance: United States, 1996; Koonin et al., 1998).

Sexual Orientation

A study of 34,706 Minnesota students in grades 7 through 12 found that more than one in four boys and girls enter adolescence "unsure" of their sexual orientation. However, by age 18, most youth deem themselves to be heterosexual or homosexual, with perhaps some 5 percent still "unsure" of their orientation (Painter, 1992b).

Research on homosexual behavior among adolescents is relatively limited (even less is known about the experiences of lesbian teenagers than about gay teenagers) (Herdt, 1989). What evidence there is suggests that growing up lesbian or gay is often a tortured journey toward self-acceptance. To be different is difficult at any age, and especially during adolescence, when conformity is celebrated and minor eccentricities can mean ostracism. In middle schools and high schools youngsters call one another a great many names, but few labels are more mortifying than "faggot." Overall, the victimization of lesbians and gay men through either verbal harassment or physical assault is the most common kind of bias-related violence. For youth who are lesbian or gay, the social pressures can be intense. By virtue of cumulative stress, lesbian, gay, and bisexual youth are at particularly high risk for suicide.

Estimates of attempted suicide among lesbian and gay adolescents range from 21 to 59 percent (consistently

higher than estimates of attempted suicide for high school students in general, which range from 8 to 13 percent) (Hershberger & D'Augelli, 1995). Especially troubling is the adolescent's fear of disclosing his or her sexual orientation to family and friends. Consequently, many young lesbians and gays keep their feelings hidden. Should they seek out school counselors, clergy, or physicians for help, they are often advised to "go straight." Such sentiments leave many young lesbians and gays feeling alienated and lonely and, not surprisingly, compromise their mental health (Hershberger & D'Augelli, 1995; Rotheram-Borus et al., 1995). Additionally, although heterosexual teenagers learn how to date and establish relationships, lesbian and gay youth are often precluded from such opportunities. Instead, they learn that they must hide their true feelings (Herdt & Boxer, 1993; Kantrowitz, 1986). Recent research also indicates that during adolescence children begin to develop *gender schemas* (perceptions) about heterosexual and homosexual personality traits. It seems that at around the age of 12, heterosexual males and females are seen as more feminine than their gay and lesbian counterparts, a perception that disappears with the advent of adolescence (Mallet, Apostolidis, & Paty, 1997).

Note, however, that a few adolescent homosexual experiences do not necessarily mean a lifetime of homosexuality. Genital exhibition, demonstration of masturbation, group masturbation, and related activities apparently are not uncommon among group-oriented preteen boys (Katchadourian, 1984). This prepubescent homosexual play generally stops at puberty. However, even though homosexual behavior during prepuberty does not necessarily lead to adult homosexuality, adult homosexuals typically report that their homosexual orientation had already been established before they reached puberty. Usually, the transition from homosexual experiences to predominantly heterosexual relationships occurs easily because many teenage boys do not regard their sexual contact with other boys as "homosexual" in the adult sense. But the question most teens in late adolescence are wrestling with is "What am I going to do with the rest of my life?"

Career Development and Vocational Choice

We cannot always build the future for our youth, but we can build our youth for the future.

Franklin D. Roosevelt, 1940

A critical developmental task confronting adolescents involves making a variety of vocational decisions. In the

United States, as in other Western societies, the jobs people hold have significant implications for their lives. The positions they assume in the labor force influence their general lifestyle, important aspects of their self-concept, their children's life chances, and most of their relationships with others in the community (Link, Lennon, & Dohrenwend, 1993). Additionally, jobs tie individuals into the wider social system and give them a sense of purpose and meaning in life.

Preparing for the World of Work

One focus of the transition from childhood to adulthood is preparation for finding and keeping a job in the adult years. Given the importance of the job-entry process, adolescents are surprisingly ill-prepared for making vocational decisions. Most teenagers have only vague ideas about what they are able to do successfully, what they would enjoy doing, what requirements are attached to given jobs, what the current job market is like, and what it will probably be like in the future. Complicating matters, many youth do not see a relationship between their current academic endeavors and their future employment opportunities. Youngsters given to antisocial behaviors are at risk for low academic achievement and school failure, and so they compromise their later job marketability. Of course, many teenagers work; but if they seek and gain work during their teen years, their working can have far-reaching consequences for their relationships with their parents, their schooling, their acceptance and status among their peers, and their standard of living and lifestyle (Lamborn, Dornbusch, & Steinberg, 1996).

Some teens encounter special difficulties in entering the job market—especially females, teens with disabilities, and teens from racial and ethnic minorities. Gender differences surface early, mirroring those of the adult work world (Greenberger & Steinberg, 1983). Young men are more likely than young women to be employed as manual laborers, newspaper deliverers, and recreation aides, whereas young women are more likely to work as clerical workers, retail sales clerks, child-care workers, health aides, and education aides. Gender segregation also occurs within industries. Among food service workers, young men more often work with things (they cook food, bus tables, and wash dishes), whereas young women more often work with people (they fill orders and serve as waitresses and hostesses).

Changing Employment Trends in the United States

For many U.S. teenagers a steady, decent-paying job is a distant hope. High rates of unemployment have traditionally been the lot of many young people, especially inner-city African Americans. Twenty-five years ago young people in the United States could find a job in manufacturing, construction, or sales and expect to make a career of it. But increasingly, young Americans without skills, and often those with them, cannot count on good wages and steady work. Even college graduates may be in for more difficult times. Labor Department economists calculate that although skill levels in the workplace rose over the past decade, the supply of college graduates rose even more rapidly. Should current job and educational trends continue, they estimate that some 30 percent of college graduates entering the workforce from 1990 to 2005 will work in jobs that do not require a college degree—which could result in one in four college graduates being "underemployed" (Koretz, 1992).

With urban public school systems failing to prepare many students for the more demanding jobs the economy is creating, African American and Hispanic young people are at growing risk of missing early work experience and of finding steady, well-paying jobs in adulthood (Winship, 1992). Some evidence suggests that racism is a factor contributing to the high level of unemployment for African Americans, because the unemployment rate of white high school dropouts is about equal to that of African American high school graduates. Poor job location—living in central cities when many jobs are in the suburbs—is another factor (O'Regan & Quigley, 1996; Stern & Eichorn, 1989).

Society pays a high price for the unemployment of its adolescents and young adults. Full-time employment is associated with low arrest rates for crime; unemployment is associated with high arrest rates. Low-quality employment with poor pay and poor hours is also associated with high youth arrest rates (Allan & Steffensmeier, 1989). Sociologists find that many other inner-city problems—educational underachievement, drug and alcohol abuse, and welfare dependency—are outgrowths of a more fundamental problem: no jobs (Anderson, 1994).

Balancing Work and School

Many of today's teens are already working while attending high school. A federally funded study of 12,000 students in grades 7 to 12 conducted by the University of Minnesota and the University of North Carolina shows teens who work 20 hours or more a week during the school year are more likely to be emotionally distressed, drink, smoke, use drugs, and have early sex. Among students in grades 9 to 12, nearly 18 percent work more than half time during the school year. This indicates that working takes a great toll on the time and energy of many high school students. The bottom line, say counselors, parents, and therapists, is that these youth struggle with balancing daily responsibilities. Good working

Part-Time Employment for Teenagers
The jobs that are open to teenagers typically involve repetitive operations that require few skills. Although it is commonly assumed that youth gain on-the-job training, they actually derive very little that is technically useful to them. But they do gain a "work orientation" and practical knowledge of how the business world operates.

conditions can boost a teen's morale and motivation to continue an education, but bad working conditions can hurt morale, lower self-esteem, and be a factor in dropping out of high school (Shellenbarger, 1997).

Dropping Out of High School

The U.S. high school dropout rate is estimated at 11 to 23 percent. However, the proportion is higher among the poor and minorities (perhaps 24 percent for blacks and 44 percent for Hispanics) (Freadhoff, 1992). And in some big-city school systems like Chicago and St. Louis, 52 percent of the young people do not graduate from high school (although many later admit regret at leaving school). Because of the technological orientation and requirements of contemporary society, young people who do not complete high school or who fail to acquire basic reading, writing, and mathematical skills find themselves at a serious disadvantage in the job market (Freadhoff, 1992; Krymkowski, 1991). Whereas in 1950, 34 percent of all jobs were open to workers without a high school diploma, by 1970 only 9 percent were open, and today the rate is even lower. Consequently, there are fewer jobs for high school dropouts. Many dropouts spend their time visiting and loafing, and some engage in a wide range of problem behaviors, such as using hard drugs, drinking, smoking, physical aggression, theft, and drug dealing (Anderson, 1990). Those who fail to move into stable employment by age 20 find it increasingly difficult to make the transition to an adult life of gainful employment (Harris, 1996).

School difficulties, both educational and social, are prominent in the history of most dropouts. In a comparative study of high school graduates and dropouts, Lu-

cius Cervantes (1965) found that dropouts tended to share a number of characteristics:

- Many had failed at least one grade.
- By seventh grade they were two years behind their classmates in reading and arithmetic; their attendance record was poor.
- Many were "underachievers."
- They had changed schools frequently.
- Many had behavior problems or were troubled emotionally.
- They tended to resent authority.

Usually, students who are most vulnerable to early school termination can be identified by the seventh grade. Boys and girls who are at high risk for dropping out reveal high levels of aggressive behavior and low levels of academic performance. And they are more likely to come from families in which they have to make decisions on their own and in which their parents show little interest in their education. Overall, research reveals that academic difficulties become cumulative, showing a gradual rise over the elementary school years and reaching a high point in the ninth and tenth grades (Luthar, 1995). Some 24 percent of high school dropouts leave school in their sophomore year; 47 percent in their junior year; and 29 percent in their senior year. According to the U.S. Education Department, dropouts cite school-related reasons over other reasons for leaving school (for instance, they were failing courses or simply did not like school). Additionally, more than 25 percent of young women who drop out of high school cite pregnancy as the reason; nearly 8 percent of male dropouts say it is because they have become fathers (U.S. Department of Education, 1994).

Not surprisingly, students who have difficulties with academic work and who view their assignments as incomprehensible and not germane typically find school frustrating and disheartening. Moreover, many adolescents fail to see that school attendance and continued effort are related to future employment. Many dropouts, especially those in the nation's big-city ghettos, regard the school experience as irrelevant to their personal, social, and vocational needs. For the most part, school authorities, police officials, and the public have argued that young people should be retained in school at least until they complete high school. The idea is that school will keep adolescents out of mischief and provide them with the skills necessary for gainful employment during adult life.

Many contemporary urban high schools have alternative education programs for this at-risk population of students who need life skills training as well as occupational skills. Our local urban high school has just instituted an agreement with the community college for mentors and transitional courses in certificate programs, such as office technologies, mechanical, electrical, civil technology, and computers. Additionally, more students are serving as interns in local businesses as part of their high school experience.

Questions

How are adolescents preparing for careers and vocations? Which students are most likely to drop out of high school? Why?

Risky Behaviors

*T*o me it seems that youth is, like spring, an overpraised season . . . more remarkable, as a general rule, for biting east winds, than genial breezes.

Samuel Butler

Even teenagers who do not perceive themselves as having "problems" might be seen by adult society as engaging in behaviors that are risky or deviant. Because adults control the channels of power, including legislative agencies, the courts, the police, and the mass media, they are in a better position than adolescents to make their definitions and values "stick" in the realm of everyday life. However, some risky behaviors are common for this age group, including drinking, illicit drug use, suicide, and delinquency. Such behaviors are called risky because typically they interfere with the person's long-term health, safety, and well-being.

Although adolescence is a period affording opportunities for positive development, many young people find themselves in social environments where they are exposed to considerable danger—"high-risk" settings. These youth confront social contexts—family, neighborhood, health-care system, schools, employment training, juvenile justice, and the child-welfare system—that are fragmented and not designed to meet the unique needs of adolescents. Poor and minority young people are especially ill-served. They are increasingly found in poor, deteriorating, inner-city neighborhoods riddled with crime, violence, and drugs and with underfunded and poorly designed schools and services (Commission on Behavioral and Social Sciences and Education, 1993; Takanishi, 1993).

Complicating matters, for at least an entire year between the ages of 16 and 23, more than 10 percent of young Americans are "disconnected": They are not in school, working, or in the military, nor are they married to someone who is in school, working, or in the military. Among young non-Hispanic white men, the rate is 7 percent; for Hispanics it is 13 percent, and for African Americans it is 23 percent. Female rates are even higher: 9 percent for whites, 21 percent for Hispanics, and 34 percent for African Americans it is 23 percent. Disconnected youth come disproportionately from families that are poor or on welfare, single-parent families, and families lacking a parent with a high-school degree. Many of the young people are high school dropouts, and high percentages of the women have given birth as teenagers (Otten, 1994).

Social Drinking and Drug Abuse

Nowadays, everyone talks about drugs, but the word itself is imprecise. If we consider a drug to be a chemical, then everything that we ingest is technically a drug. To avoid this difficulty, *drug* is usually arbitrarily defined as a chemical that produces some extraordinary effect beyond the life-sustaining functions associated with food and drink. For instance, a drug might heal, put to sleep, relax, elate, inebriate, produce a mystical experience or a frightening one, and so on. Our society assigns different statuses to different types of drugs. Through the federal Food and Drug Administration, the Bureau of Narcotics, and other agencies, the government takes formal positions on whether a given drug is "good" or "bad," and if "bad," how bad. Sociologists note that some drugs, like caffeine and alcohol, enjoy official approval. Caffeine is a mild stimulant that finds societal approval through the coffee break and coffee shop. Likewise, the consumption of alcohol, a central nervous system depressant, has become so prevalent in recreational and formal business settings that nonusers of the drug are often regarded as somewhat peculiar. And at least until a few years ago, the same held true for the use of nicotine (smoking), a drug usually categorized as a stimulant.

Whether or not they are culturally sanctioned, drugs can be abused. **Drug abuse** refers to the excessive or compulsive use of chemical agents to an extent that it interferes with people's health, their social or vocational functioning, or the functioning of the rest of society. Among adolescents, as among their elders, alcohol is the most frequently abused drug in the United States. According to a 1997 survey conducted by the Institute of Social Research at the University of Michigan, alcohol use was acknowledged by 46.8 percent of eighth-graders, 63.9 percent of tenth-graders, and 73 percent of high school seniors (Hofferth, 1998; "Teen Age Drug Use," 1998).

Nearly half of all college students engage in **binge drinking** (defined as downing five or more drinks in a row for men, or four or more in a row for women). College and university officials agree that binge drinking is a serious problem. It has far-ranging consequences, including potential death from alcohol poisoning. Binge drinkers are seven to ten times more likely than nonbinge drinkers to have unprotected sex, engage in unplanned sex or be raped while unconscious, get in trouble with campus police, damage property, or become hurt or injured (Carton, 1994). Alcohol was involved in more than two-thirds of all campus incidents last year that occurred in residence halls or involved violent behavior (Scrivo, 1998). Significantly, the percentage of college women who drink abusively has tripled in the last 20 years (from 10 percent in 1977 to 35 percent in 1993). Of great concern, 90 percent of all reported campus rapes occur when alcohol has been used by either the assailant, the victim, or both; 60 percent of college women in whom a sexually transmitted disease like herpes or AIDS was diagnosed

were intoxicated at the time of infection (Celis, 1994) (see *Information You Can Use:* "Determining Whether Someone You Know Has an Alcohol or Drug Problem").

After a decade of decline, illicit drug use began increasing again in 1993 among two age groups: teenagers and persons 35 and older. For instance, 13 percent of eighth graders said they used marijuana in 1994, compared with 9.2 percent in 1993 and 7.2 percent in 1992. In 1994, one in four high school sophomores and one in three seniors said they had smoked marijuana at least once within the last year. Nearly 7 percent of seniors acknowledged use of LSD, while 3.6 percent said they had used cocaine and 2 percent said they had used crack cocaine (Treaster, 1994; Janofsky, 1994).

According to federal health officials, as the nation entered the 1990s, more than a quarter of a million young people, mostly boys, used steroids to build their muscles and enhance their athletic performance. The nonmedical use of anabolic steroids is illegal. More than half of the users had started using by age 16, and 85 percent by age 17. Medical authorities say adolescents whose bodies are still developing are at special risk for adverse effects from steroids, including stunted growth, mood changes, long-term dependence on steroids, acne, fluid retention, breast development in males, masculinization in females, high blood pressure, and reversible sterility in males. Teenagers might find steroid use appealing because of their concerns about their appearance, peer approval, and "being large and strong enough to make the team" in competitive sports (Petraitis, Flay, & Miller, 1995).

Many psychologists say that coping with the presence of drugs in their social environment is now a developmental task that adolescents must reckon with, just as

Adolescent Binge Drinking Researchers find that an alarming number of high school and college men and women currently engage in binge drinking with far-reaching social consequences, including unprotected and unplanned sex, rape, traffic accidents and fatalities, encounters with law enforcement agencies, and death.

Information You Can Use

Determining Whether Someone You Know Has an Alcohol or Drug Problem

Q & A

How can I tell if my friend has a drinking or drug problem?

It might be difficult to tell because most people will try to hide their problems, but here is a quiz that you can ask yourself about their behavior (or your own):

1. Do they get drunk or high on a regular basis?
2. Do they lie about how much they drink or do drugs, or do they lie in general?
3. Do they avoid people in order to get drunk or high?
4. Do they give up activities they used to do, such as playing sports or hanging out with friends who do not get drunk or use drugs?
5. Do they plan drinking or taking drugs in advance, perhaps planning the day or activity around the act of drinking or getting high?
6. Do they have to increase the amount of alcohol or drug to maintain the same high?
7. Do they believe that drinking or drugs are necessary in order to have fun?
8. Are they frequently incapacitated because they are recovering from a "fun night"?
9. Do they pressure others to take drugs or condemn those who do not consume alcohol?
10. Do they take risks, such as careless driving, sexual risks, or otherwise act invincible?
11. Do they have blackouts where they have no recollection of what happened while they were drunk or high?
12. Do they talk of hopelessness or being depressed or committing suicide?
13. Do they sound selfish or uncaring about other people?

These are signs that a substance is taking control of someone's life. If you see these signs, chances are that your friend, or you, need help.

How serious can a drug or alcohol problem be?

People with serious drinking or drug problems usually start off saying that the experience is great, but things eventually change for the worse. They can develop serious psychological problems such as depression, thoughts of suicide, as well as physical problems such as liver damage, brain damage, if pregnant possible damage to the fetus, and of course overdose. Being under the influence of substances can cause people to engage in unsafe behaviors, such as driving while intoxicated, having unsafe sex, and trying dangerous "stunts."

What would cause my friend to have a problem like this?

Many things can lead to these sorts of problems. Sometimes these problems run in families, similar to heart disease or cancer. If there is a family history of substance abuse, it is more likely that your friend could develop a dependency. Some people take drugs or drink to mask other problems or to avoid things that bother them—stress, work pressures, feelings like they are different from others or they are not worthy, or unhappiness with their situation. They drink or get high to forget their problems. Even though they might forget their problems for a short while, they will more than likely become depressed when the high wears off and they have to face the same problem again and again (Schulenberg et al., 1997).

Why is it hard for individuals to help themselves?

It is very difficult for most people to admit that they have a serious problem. It is even more difficult when you are young and believe that nothing bad can happen to you. Denial and having to hide the problem from friends and family become problems themselves. It can seem easier to cut off everyone and isolate oneself than to constantly hide the evidence of the problem. Once someone withdraws, it is much more difficult to see that there really is a problem.

How can I help my friend?

You have several options:

- Approach him or her in a spirit of support.
- Encourage your friend to get professional help.
- Talk to someone you trust—counselor, teacher, doctor, clergy—this will give you another perspective on what you should do.
- Wait until your friend is sober to talk with him or her.
- Try not to accuse or blame your friend. Give examples of what you have noticed that worry you.
- Offer to accompany your friend to seek help.

What does my friend have to do to get help?

First, your friend must admit that there is a problem. This is difficult because it will mean admitting to having wasted part of her or his life. Your friend cannot solve this problem alone but will need professional help. Many people benefit from groups such as Alcoholics Anonymous (AA) or Narcotics Anonymous (NA). Both of these groups often have meetings on campus. Check with your own campus counseling center for meetings times. In the end, it is your friend who must make the decision to get help.

they must reckon with separation from parents, career development, and sexuality. Given the prevalence and availability of marijuana in the peer culture, it may not be surprising that psychologically healthy, sociable, and reasonably inquisitive young people would be tempted to try marijuana. But the findings also show that adolescents who use drugs frequently tend to be maladjusted, showing a distinct personality syndrome marked by interpersonal alienation, poor impulse control, and significant emotional distress (Johnson & Gurin, 1994; Petraitis, Flay, & Miller, 1995). For these youngsters, experimentation with drugs is highly destructive and easily leads to pathological functioning.

Why Do Teens Use Drugs?   Other researchers confirm that heavy and abusive use of drugs during adolescence is associated with increased loneliness, social isolation, disorganized and suicidal thought processes, and unusual beliefs—aspects that significantly interfere with problem solving and social and emotional adjustment (Johnson & Gurin, 1994; Petraitis, Flay, & Miller, 1995). And heavy drug use impairs competence in the crucial maturational and developmental tasks of adolescence and adulthood by generating premature involvement in work, sexuality, and family roles. In addition, teens see many of their peers using drugs without any apparent harmful effects, creating a climate of disbelief in antidrug campaigns. One place we see the effects of drug abuse is in the size of the nation's prison population, which passed the one million mark, having doubled from 1984 to 1994 and continues to climb.

A variety of factors have contributed to the illicit use of drugs by young people (Petraitis, Flay, & Miller, 1995). The recreational use of illegal drugs has become central to many adolescent peer groups in the past 25 years. Most adolescents who use illegal drugs move in peer groups in which drugs are a part of daily life. Another contributor in the use of illegal drugs by young people is that they see their parents use psychoactive drugs—such as alcohol, tranquilizers, barbiturates, and stimulants (Kandel, 1990). More than 80 percent of U.S. adolescents report that their family has a rule against illicit drugs, yet many mimic their parents' drug use and begin taking mood-changing drugs themselves. In this context adolescent drug use is a juvenile manifestation of adult behavior, so it is perhaps more accurate to view drug abuse as a society-wide problem.

Teenage Suicide

Although the world is full of suffering, it is also full of overcoming of it.

                                                    Helen Keller

Suicide ranks today as the third leading cause of death among adolescents in virtually every industrialized nation of the world. In the United States, the suicide rate among young people ages 10 to 19 increased by 120 percent from 1980 to 1995 (from 0.8 per 100,000 to 1.7 per 100,000). Suicide is still more common among white teens, though the suicide rate for black youth aged 10 to 19 has risen sharply. From 1980 to 1995, the black youth suicide rate rose an unprecedented 114 percent nationwide (representing 3,030 black youth over this 15-year period) (Centers for Disease Control, 1998g). Unreported suicides would push the figures much higher. Rates are highest in southern states, where they have risen 214 percent since 1980. The most vulnerable black youth are also younger. Among those aged 10 to 14, there was a 233 percent increase. Firearms were used in 66 percent of the cases of youth aged 10 to 19. The self-destructive behavior in all adolescents is alarming, but it is particularly alarming in the black community. "Traditionally, black people's rates of suicide were minimal," says Jewelle Gibbs, a specialist in African American issues at the University of California at Berkeley (McKenna, 1998).

In a recent study released by the U.S. Department of Health and Human Services, gay and lesbian youth were found to be two to six more times likely than nongay youth to attempt suicide. They account for 30 percent of the completed teen suicides (Centers for Disease Control, 1997c). Because stigma is often attached to suicide in Western countries, medical personnel frequently report a suicidal death as an accident or as a death from natural causes. The national *Youth Risk Behavior Surveillance Study* (CDC, 1997c) followed 10,904 high school students and found that during the previous 12 months:

- 24.1 percent had seriously thought about attempting suicide.
- 17.7 percent had made a specific plan to attempt suicide.
- 8.7 percent had already attempted suicide.
- 2.8 percent had made a suicide attempt that resulted in injuries requiring medical attention.

Comparison by gender reveals that female students (30.4 percent) were significantly more likely than male students (18.3 percent) to have thought seriously about suicide, to have made a specific plan to attempt suicide, or to have already attempted suicide. Comparison by race/ethnicity indicates that Hispanic students were significantly more likely than white students to have attempted suicide, and that white students were more likely than black students to have thought seriously about suicide (CDC, 1997c).

Risk Factors Associated with Suicide   A constellation of familial, biological, mental disorder, and environmental factors are associated with youth suicide. These factors include a sense of hopelessness, a family history of suicide, impulsiveness, aggressive behavior, social isolation, previous suicide attempt, easier access to alcohol, use of illicit drugs, low emotional support from a family, negative life events, and lethal suicide methods (McKeown et al., 1998). In many cases psychological *depression* underlies suicide and suicidal attempts (Garland & Zigler, 1993). **Depression** is an emotional state usually characterized by prolonged feelings of gloom, despair, and futility, profound pessimism, and a tendency toward excessive guilt and self-reproach. Other symptoms of depression include fatigue, insomnia, poor concentration, irritability, anxiety, reduced sexual interest, and overall loss of interest and boredom. At times, depression appears in the guise of other disorders, such as vague pains, headaches, or recurrent nausea. The rate of depression for adolescent girls is twice that of boys. A prime factor is the preoccupation many teenage girls have with their appearance (Ge et al., 1994; Nolen-Hoeksema & Girgus, 1994). The U.S. Education Department has named depression as a major cause of students' dropping out of college.

Suicide Prevention   The ability to screen for suicide risk is the most important part of suicide prevention. Second is linking at-risk individuals with community mental health services. The potential for suicide is on a continuum from low risk factors to high risk factors.

The strongest risk factor for completed suicide is the presence of a firearm in the home (McKeown et al., 1998). A broader societal focus on reducing substance abuse and poor family functioning includes both community and family interventions, including educating parents and educators (McKeown et al., 1998). Jessor and colleagues (1998) studied 1,493 Hispanic, white, and black high school students in a large, urban school district and found that the following protective psychosocial factors *enhance* adolescent health: regular physical exercise, healthy eating habits, dental care, safety behaviors, adequate sleep, religiosity, a commitment to school, having friends who take part in conventional youth activities and community volunteer work, an orientation toward parents, positive relationships with adults, church attendance, and involvement in prosocial activities.

Treatment of adolescents with suicidal tendencies usually involves a combination of psychotherapy and medication for depression (McKeown et al., 1998). The therapist seeks to help the teenager come to terms with his or her problems and acquire more effective techniques for coping with life and stressful circumstances. The therapist also attempts to foster self-understanding, a sense of inner strength, self-confidence, and a positive self-image. Dramatic progress has been made in recent years in the treatment of depression with such medi-

Crisis Intervention: The Hotline   Many communities have set up suicide hotlines that provide an instant, economical, and effective way to deal with emergency situations. Crisis hotlines are staffed by volunteers trained to listen, buy time, provide emotional support and sympathy, and give information about the community services that are available to deal with a caller's problems.

cations as Elavil, Tofranil, and Prozac (Shuchman & Wilkes, 1990).

Antisocial Behaviors and Juvenile Delinquency

Youthful "deviance" has been a common problem reported by societies throughout human history. The United States is no exception. Young men ages 15 to 29 are responsible for a significant proportion of the nation's crime. Men of this age group are also involved disproportionately with the criminal justice system. For instance, in California, young men between the ages of 20 and 29 represent 45 percent of those in prison, on parole, or on probation, although they constitute only 8 percent of the state's overall population (Fox, 1996).

A particularly troubling statistic is the rise in murder rates of and by adolescents. After dipping between 1980 and 1985, the nation's murder rates have again mounted, fueled in large measure by growing numbers of adolescent and young adult males dealing in cocaine, carrying assault weapons, and taking a more casual attitude about human life. Between 1988 and 1992 alone, the arrest rates for youths under age 18 for violent crimes increased significantly, with increases of 51 percent for homicide, 50 percent for robbery, 49 percent for aggravated assault, and 17 percent for forcible rape (Fox, 1996). Homicide is now the leading killer of African American males ages 15 to 24 and the second leading cause of death for whites

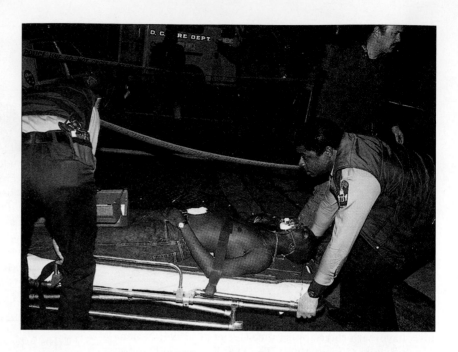

Violence Among American Youth Young African American males face especially difficult circumstances. The problems for young people are compounded by a deindustrializing economy where well-paying, secure jobs are scarce. Anger and defiance are readily translated into violence in an environment of labor market marginality. The leading cause of death among African American youth is homicide. Indeed, African American young men in New York City's Harlem are less likely to reach the age of 40 than are young men in Bangladesh.

after auto accidents (African American men are 14 times more likely to be murdered than the general population) (Hammond & Yung, 1993).

Even in places where young people commit crimes at about the same rate as they did in earlier years, the wider availability of firearms makes many offenses much more lethal. Each day more than 100,000 U.S. juveniles are held in custody in juvenile institutions and adult prisons. Mounting evidence suggests that the nation's juvenile justice system is antiquated, inadequate, and relatively incapable of dealing with the delinquency and violence wrought by children and teenagers. Psychologists, sociologists, and criminologists are increasingly coming to the conclusion that families and neighbors—not the police and wardens—hold the keys to cutting crime. In brief, society- and community-wide action is required.

Violence and Age   Age continues to be an important factor in violence. The youth of this nation have increasingly become both victims and perpetrators of violent acts. Violence by our youth is growing more rapidly than in any other subgroup. Between 1985 and 1994, the number of persons arrested for murder and non-negligent manslaughter increased by 150 percent for individuals under 18 years of age in comparison to only an 11.2 percent increase for persons 18 years of age and older (Children's Defense Fund, 1998). Chronic youthful offenders commit many more crimes than do chronic adult offenders—an average of 36 per year for juveniles and 12 a year for adults. From February 1997 through May 1998, there were seven instances where young male students under the age of 18 opened fire upon fellow stu-

dents and teachers, in all cases killing or wounding several people before being subdued.

Young people ages 12 through 17 are the most frequent victims of violent crime in the United States. Nationally, the second and third leading causes of death among youth under 21 are *homicide* and *suicide.* According to U.S. Justice Department statistics, children are beaten, raped, and robbed five times more often than adults age 35 and older. Although juveniles account for 10 percent of the population, they constitute 23 percent of all violent crime victims. Two-thirds of the victims between 12 and 19 years of age are attacked by someone aged 12 to 20. Of the violent personal crimes against juveniles in 1992, nearly 1.3 million, or 83 percent, were assaults and 229,000 were robberies (Office of Juvenile Justice and Delinquency Prevention, 1994). Significantly, nearly 1 in 13 high school students carries a weapon to school, and those who carry weapons are more likely than unarmed students to drink, smoke marijuana, and use cocaine (Manning, 1994). Almost half (48 percent) of African American male teenagers are killed by firearms (Children's Defense Fund, 1998). In addition, the rate of firearm deaths among youth under 15 is twelve times higher in the United States than in 25 other industrialized countries. For some thoughts on how to reduce firearm fatalities, see *Further Developments:* "What Can We Do to Protect Our Youth from Guns and Violence?"

Between 500,000 and 1 million teenagers run away from home each year, and many of these are forced into a life of crime to survive. More than 60 percent of homeless youth in shelters report having been physically or sexually abused by parents or other family members (Barden, 1990). Typically the runaways then must live on

Further Developments

What Can We Do to Protect Our Youth from Guns and Violence?

First we need to ask how kids are getting guns. Among inner-city youth, 53 percent admitted they got them from family and friends, while 37 percent said off the street. Also, 50 percent of U.S. households have a firearm, and at least one is stolen in each of the 340,000 burglaries committed each year in the United States. It is no wonder that young adolescents have access to guns (Children's Defense Fund, 1998).

The Brady Handgun Violence Prevention Act of 1993 and the Violent Crime Control and Law Enforcement Act of 1994 mandated, respectively, a five-day waiting period with background check for handgun purchasers and a moratorium on 19 assault weapons. Trigger locks, loading indicators, and lethality reductions are other possible policies that could be easily enacted.

Educationally, there are programs that emphasize conflict resolution, and street survival skills. We can also call for a tightening of security and protection in schools and manda-

tory sentences for crimes committed with a gun. Long-term solutions include reducing youth access to firearms. This will be difficult, given that so many U.S. homes have guns. Only 23 percent of firearm homicides are committed in the course of a felony or by a suspected felon, and 68 percent of U.S. felons admitted that their firearms came from friends, family, or acquaintances. This suggests that if private citizens possessed far fewer firearms, criminal access to guns would be greatly reduced. Many of these goals can be accomplished through these four policies (which some states have enacted):

1. Balance the number of gun dealers with the capacity of law enforcement officials to inspect dealers.
2. Establish a licensing and registration system that is similar to that used by motor vehicles.
3. Establish penalties for the illegal sale of firearms.
4. Regulate sale of guns over the Internet.

the street in a world filled with drug addicts, muggers, pimps, prostitutes, and pushers. If and when these youth become violent juvenile offenders, they pose a challenge to both the general public and the juvenile justice system, which no longer agree on how to treat violent adolescents (Tate, Reppucci, & Mulvey, 1995).

Childhood antisocial behaviors, such as juvenile delinquency, conduct disorders, and violent outbursts, are linked to a wide variety of troublesome adult behav-

iors, including criminal careers, general deviance, economic dependency, educational failure, employment instability, and marital discord. Even so, some juvenile delinquents later become law-abiding citizens. Apparently, they change not from fear of being arrested but because they realize that what was "fun" as a teenager is no longer appropriate behavior for an adult (Sampson & Laub, 1990).

Segue

In the United States the social boundaries of adolescence are rather ill defined. By tradition, the shift from elementary school to junior high school signaled entry into adolescence. But with the advent of the middle school, the transition became blurred. Nor is it entirely clear when a person leaves adolescence. Roughly speaking, adolescence is regarded as having ended when the individual assumes one or more adult roles, such as marriage, parenthood, full-time employment, or financial independence.

The extension of adolescence and youth has posed two related problems for society and for young people. Society has the problem of providing the young with a bridge to adult roles through appropriate socialization and role allocation.

As we shall discuss in Chapter 13, young adults have the problem of establishing a stable identity, achieving independence, and deciding on future alternatives. Many Western industrialized nations have put off entrance into adulthood for economic, educational, and other reasons. Today, college postpones full adult status for many socially and economically advantaged young people. Unemployment and underemployment produce a somewhat similar effect among less advantaged groups. At the same time, children are reaching puberty earlier than children did a century ago. Thus, physically mature people are told that they must wait 10 or more years before they can assume the full rights and obligations of adulthood.

Summary

Development of Identity

1. Hall's view is that adolescence is characterized by inevitable turmoil, maladjustment, tension, rebellion, dependency conflicts, and exaggerated peer-group conformity. Some social scientists believe this storm-and-stress view has been exaggerated by media influences and is not accurate today.

2. Sullivan's theory, in contrast to Freud's, explains the principal forces in human development as being social instead of biological. The three periods of Sullivan's theory are preadolescence, early adolescence, and late adolescence.

3. The crisis in Erikson's fifth stage of psychosocial development is the search for identity. Erikson suggests that an optimal feeling of identity is experienced as a sense of well-being.

4. Marcia proposes that identity formation can be further classified in terms of four different statuses: achievement, moratorium, foreclosure, and diffusion.

5. The United States and other Western countries seem to promote an extended adolescence. According to this view, conflicting expectations generate an identity crisis among U.S. and European youth. On the other hand, many non-Western societies mark the period of adolescence more clearly. They ease the shift in status by providing puberty rites—initiation ceremonies that socially symbolize the transition from childhood to adulthood. Some circumcision rituals marking adulthood are under attack by international human rights organizations as violating human rights.

6. Until very recently, only males were the subjects of research. An AAUW study released in 1992 entitled "How Schools Shortchange Girls" indicates that females emerge from adolescence with a poorer self-image, relatively lower expectations for life, and considerably less confidence in themselves and their abilities.

7. Gurian has studied identity development in boys and proposes accepting their biological differences and providing adolescent males with role models and opportunities to learn to act responsibly.

8. Western societies, by prolonging the transition to adulthood and by segregating their youth, have given rise to a kind of institutionalized adolescence or youth culture—more or less standardized ways of thinking, feeling, and acting that are characteristic of a large body of young people.

Peers and Family

9. Both the family and the peer group are anchors in the lives of most teenagers. However, the relative influence of the two groups varies with the issue. Adolescents, who are more segregated by age in high schools, are normally not at war with their parents. Teens often benefit from ties with their parents and seek advice from them on issues of finance, education, and careers.

10. Teens, who seem to develop a consciousness of oneness with their group members, seek advice from peers on topics such as clothes, hairstyle, dating, and music.

11. High status in adolescent society requires being able to project an air of confidence and a "cool" self-image in a variety of situations.

12. Conformity to a peer group and peer pressure play a prominent role in the lives of most teenagers. However, there are many different types of teen groups.

13. For adolescents, the demand to make their own choices is a signal of maturation and growth. Conflicts between teens and parents are more likely to arise over chores and appearance than over substantive issues.

14. Pipher proposes that girls get mixed messages about modeling their behavior after that of their mothers. Adolescent girls are socialized to psychologically distance themselves from their mothers at a time when they need their guidance and support the most.

Courtship, Love, and Sexuality

15. One of the most difficult adjustments adolescents must make as they make the transition into adulthood revolves around their sexuality, sexual orientation, and sexual expression. Sexual attraction and sexual considerations become dominant forces in their lives. Adolescent sexuality has received considerable public attention because it is associated with many negative individual and societal outcomes.

16. Youth vary a good deal in the age at which they first experience intercourse. Adolescent sexual behavior can be shaped by less parental supervision and by peer group pressures.

17. Social scientists find it difficult to define romantic love, and cross-cultural studies suggest it is not a universal concept.

18. Adolescent sexuality finds expression in a number of ways, including masturbation, nocturnal orgasm,

heterosexual petting, heterosexual intercourse, and homosexual activity.

19. The rates of teenage sexual experience increased throughout the 1980s, plateaued by the early 1990s, and declined by high school grades level, by gender, and by race and ethnicity by the late 1990s. Rates of having multiple sex partners have also declined by all grade levels and races and ethnic categories during this decade. As teens become more sexually experienced, they become more consistent contraceptive users.

20. Although teenage birth rates have been declining, childbearing by unwed teens is increasing and is associated with many negative outcomes for the mother, the child, and society.

21. More than one in four boys and girls enter adolescence "unsure" of their sexual orientation. However, by age 18, most youth deem themselves to be heterosexual or homosexual, with perhaps some 5 percent still "unsure" of their orientation.

Career Development and Vocational Choice

22. Adolescents are confronted with career development and vocational choice. College allows teens an opportunity to explore options and delay vocational decisions. Employment is associated with many positive outcomes as adults, but employment in high school can complicate the picture for teens. The high school dropout rate is higher among the poor and minorities, yet there are fewer employment opportunities for this group of teens.

Risky Behaviors

23. Some risk behaviors common for this age group are drinking, illicit drug use, suicide, and delinquency. Such behaviors are called "risky" because typically they interfere with a person's long-term health, safety, and well-being. The popularity of "binge drinking" puts many high school and college students at a high risk for death.

24. Familial, biological, mental disorder, and environmental factors are associated with substance abuse and suicide among youth. These factors include a sense of hopelessness, a family history of depression or suicide, impulsiveness, aggressive behavior, social isolation, previous suicide attempt, easier access to alcohol, use of illicit drugs, low emotional support from a family, negative life events, and lethal suicide methods.

25. Childhood antisocial behavior, such as juvenile delinquency, conduct disorder, and violent outbursts, is linked to a wide variety of troublesome adult behaviors, including criminal careers, general deviance, economic dependency, educational failure, employment instability, and marital discord.

Key Terms

binge drinking *(382)*
consciousness of oneness *(367)*
depression *(385)*
deviant identity *(362)*
drug abuse *(382)*

generation gap *(367)*
identity *(361)*
identity achievement *(362)*
identity diffusion *(362)*
identity foreclosure *(362)*

identity moratorium *(362)*
negative identity *(362)*
puberty rites *(363)*
storm and stress *(360)*
youth culture *(367)*

Following Up on the Internet

Web sites for this chapter focus on emotional and social issues, career development, and risky behaviors during adolescence. Please access the text web site at http://www.mhhe.com/crandell7 for up-to-date hot-linked Internet addresses for the following organizations, topics, and resources:

Adolescence: Biological, Social, and Emotional Changes

Teen Health Index

Teen Pregnancy

Risk Behaviors for Adolescents

The Working Adolescent

Understanding Depression and Suicide

Part Seven

Early Adulthood

Chapters 13 and 14 focus on the dynamic life stage of young adulthood, which extends from the late teens until the early forties, and discuss various theories of adult development. Contemporary young adults in the United States are ethnically diverse, better educated, and likely to delay marriage by cohabiting or living at home longer while establishing careers. Most young adults are physically active and more health conscious than ever. Recent findings suggest most are sexually conservative and are protecting themselves from AIDS and STDs. A smaller number engage in unhealthy behaviors, such as alcohol and drug abuse and unprotected sexual activity with multiple partners, leading to a higher risk of AIDS. Most are planning and preparing for their future by attending college or training for specific careers. Friendships and social relationships are of prime importance to young adults, who are searching for a compatible, intimate partner during this stage of life. Cohabitation prior to marriage is more common; the average age at marriage is now in the mid twenties. Though the media focus on teen pregnancies, most young adults today are delaying childbearing while establishing careers.

Chapter 13

Early Adulthood
Physical and Cognitive Development

Critical Thinking Questions

1. Do you think of yourself as an adult or as an adolescent? If you are already firmly in the adult category, when did you begin to think of yourself as an adult? What are some of the things you do that might fall into either category?

2. When you think of love, do you think of sex? When you think of sex, do you think of love?

3. What is sexuality for you? Is it possible for someone to be asexual?

4. If you witnessed a young woman with a baby stealing food from a supermarket, how would you react? Would it make a difference for you if the person were poor?

Today in American society, there is less demarcation between adolescence and adulthood than a few decades ago. Some social scientists have suggested that adolescence extends into the twenties, with more young adults living at home with parents during their expensive college years and when launching a career, or returning home after a short-lived marriage or as a single parent. Many middle-class young adults expect to maintain a lifestyle equivalent to or better than that of their parents while struggling through these early adult years.

Following the earlier lead of psychologists such as Erik Erikson, Charlotte Bühler, Carl Jung, Sidney Pressey, Robert Kegan, and others, the social science community has come to recognize that adulthood is not a single monolithic stage, not an undifferentiated phase of life between adolescence and old age. Developmentalists increasingly see individuals as undergoing change across the entire life span. The contemporary notion is of a process of becoming. Thus, adulthood is now seen as an adventure that involves negotiating ups and downs and changing direction to surmount obstacles. In this chapter, we examine several views of the stage called early adult development, typical physical and cognitive changes, the differing moral domains of men and women, and the impact of the diversity of values adults impose on society.

Developmental Perspectives

When we truly comprehend and enter into the rhythm of life, we shall be able to bring together the daring of youth with the discipline of age in a way that does justice to both.

J. S. Bixler, *Two Blessings of Joseph*

The category *adulthood* generally lacks the concreteness of *infancy, childhood,* and *adolescence.* Even in the scientific literature it has functioned as a kind of catchall category for everything that happens to individuals after they "grow up." Sigmund Freud, for instance, viewed adult life as merely a ripple on the surface of an already set personality structure; Jean Piaget assumed that no additional cognitive changes occur after adolescence; and Lawrence Kohlberg saw moral development as reaching a lifetime plateau after early adulthood. Many middle-aged American parents are asking themselves: *When is my son or daughter considered to be an adult, capable of working and living independently of Mom and Dad's resources?* There is no firm answer to their question today, and it seems to be quite an individual matter.

In the United States the beginning of adulthood is most often defined as the point at which a person leaves high school, attends college, takes a full-time job, enters the military, or gets married. However, becoming an adult is a rather different matter for different segments of the society. For instance, men in Western societies have traditionally emphasized such issues in development as autonomy, independence, and identity. In contrast, women, some ethnic minorities, and non-Western societies have typically assigned greater importance to issues of relatedness, such as closeness within a family (Guisinger & Blatt, 1994). And adulthood itself has different meanings for different age groups within the population (Krueger & Heckhausen, 1993). Because adulthood is not one experience but many experiences, people's conceptions of adulthood frequently differ.

Demographic Aspects of Adulthood

People's feelings, attitudes, and beliefs about adulthood are influenced by the relative proportion of individuals who are adults. In the United States, major population or demographic changes are under way that will have important social consequences. The *baby-boom generation* (those born between 1946 and 1964) represents a huge "age lump" passing through the population, a sort of demographic tidal wave. This age cohort—more than 79 million Americans—was responsible for the 70 percent jump in the number of school-age children from 1950 to 1970.

In the 1950s the baby boomers made the United States a child-oriented society full of new schools, suburbs, and station wagons. The baby boomers born before 1955 provided the nation with rock and roll, went to

A Young Couple Graduating from College More young adults than ever are attending college and staying in college longer waiting for a better employment picture or to attain better skills to be competitive in the workplace.

Vietnam and Woodstock, attended college in masses, and fueled the student, civil rights, women's movement, and peace movements of the 1960s and early 1970s. Later, as the baby boomers made their way into middle age, the contemplation—even celebration—of the middle years found expression in popular culture. It is hardly a coincidence that family sitcoms such as "The Mary Tyler Moore Show" of the 1970s focused on single life and work and played to what was then a younger audience. Or that "All in the Family" sharply contrasted the views of young boomers "Meathead" and Gloria with middle-aged Edith and Archie. Similarly, family situation comedy was popular in "The Cosby Show" and "Family Ties," which dominated the TV ratings in the 1980s. Television sitcoms about singles such as "Murphy Brown," "Seinfeld," and "Friends" were popular in the 1990s. What do you observe about the message of the popular contemporary sitcoms now that the baby boomers are in middle age and "Generation X" is coming on strong?

The baby boomers have brought about a rapid expansion of the nation's labor force. In the early 1970s there were about 52 million Americans between 20 and 39 years of age (25.8 percent of the population). By 1980 the number had risen to 72.4 million (32 percent of the population). The bumper crop of baby boomers were 33 to 51 years old in 1997 and constituted about 33 percent of the U.S. population (Stoneman, 1998). As the boomers have moved into their middle years, they have become more productive (many having attended college and acquired additional skills). Additionally, they earn more money and save more money—all of which will likely give a competitive edge to the United States in the world economic arenas in the 2000s. As the baby-boom genera-

tion moves into middle adulthood, their children—the next generation of young adults—have come of age.

Question

Think of your favorite television shows—who are the main characters and what issues are explored that demonstrate the values and characteristics of this generation?

Generation X or Twentysomethings

A new generation of young adults has higher college attendance rates, but their graduation rates are unchanged and incomes are down. They are more interested in the visual arts, less active in sports or sex, more scarred by divorce, and more likely to live with their parents. Their lives are defined by education, insecurity, and a slow transition to adulthood.

Zill & Robinson, 1995

Today's young adults in the cohort aged 18 to 27, who are more ethnically diverse than previous generations, are sandwiched between the baby-boom and "little boom" generations (see Figure 1-2 in Chapter 1). They are variously labeled by others, or by themselves, as the "Baby-Busters," "Generation X," "Twentysomethings" or "Boomerang Generation." These vague labels reflect the fact that in many ways the image of this generation has not as yet fully evolved, and there are various subgroups within this generation. Nonetheless, from opinion polls, research surveys, and anecdotal evidence, some characteristics of this cohort are emerging (see Table 13-1).

Table 13-1   Components of the American Dream, by Three Generations, 1996

	Percentage Saying This Is Part of Their American Dream		
	Elderly %	Baby-Boom Generation %	Generation X %
Having a happy marriage	77	74	70
Living in a decent, secure community	75	81	71
Owning own home	74	79	81
Having children	64	59	51
Unlimited opportunity to pursue dreams	63	71	70
College education for self	50	56	63
Being a winner	45	40	50
Living where everyone else shares my values	41	38	28
Becoming wealthy	33	37	46

This generation has been raised in a variety of family structures with many of their fathers and mothers working, and they share an appreciation for individuality and an acceptance of diversity in race, ethnicity, family structure, sexual orientation, and lifestyle (Stoneman, 1998; Ritchie, 1995; Zill & Robinson, 1995). At least 40 percent of today's young adults spent some time growing up in a single-parent family. "The National Survey of Children, a longitudinal study of people born in the late 1960s, found that 26 percent of this generation had received psychological treatment for either emotional, learning, or behavioral problems by the time they reached adulthood." Other surveys and research studies suggest the following about this generation (Zill & Robinson, 1995):

- They don't appreciate the exaggeration or hype of advertising and are more consumer savvy, because many of them have been doing the laundry, cleaning, and cooking since their early teens while their parents worked.
- They are going to two-year and four-year colleges in record numbers to attain the technological skills and education to adapt to today's work environments and earn decent wages (though they are likely to earn less than their parents).
- They are more likely to see a movie or use computers and multimedia than to read books (41 percent said they had not read any book not required for work or school in the past year, and 18- to 24-year-olds report watching an average of 3 hours of television per day).
- They are more likely to save money and to be concerned about wealth than their parents' generation.
- They are more likely to be less fitness oriented than the previous generation of young adults, particularly the women.
- They are less involved in sexual activity and are more likely to use birth control and practice safe sex.
- They are more likely to get along with their parents, siblings, and stepparents, to live with their parents longer, and to marry later. The percentage of women aged 25 to 29 who had never married tripled from 11 percent in 1970 to 33 percent in 1993; males in their twenties who have never married increased from 19 percent in 1970 to 48 percent in 1993.
- They are more likely to be cohabiting before marriage. The number of cohabiting couples rose from 500,000 in 1970 to 3.5 million in 1993.

Table 13-2 shows what this generation does for amusement or leisure when they aren't working, watching television, or using their computers.

Table 13-2   Leisure and Young Adults

Percentage of Young Adults, Ages 18 to 24, Who Participated in This Activity in the Past 12 Months

	1982	1992
Movies	87	82
Amusement or theme park	67	68
Amateur or professional sports event	65	51
Arts or crafts fair	35	37
Historic park or monuments	34	33
Stage plays	11	13
Live jazz performance	18	11
Classical music concert	11	10

As with prior generations, the process of leaving home remains an important part of the transition for today's young adults. Whereas in the past leaving home was frequently associated with the event of marriage, now moving out of the parental home comes from a desire to be independent.

> I want material success, but I also want balance in my life. My real dream is to live life as an entrepreneur. Not necessarily to start my own business, but to have this entrepreneurial focus in all aspects of my life. To really build my life and my lifestyle on my own terms, with my own skills, resources, and competencies (Stoneman, 1998).

Nearly half of young adults are remaining at home with parents much longer while going to college, or establishing a career, or returning after a divorce or breakup with a cohabiting partner (Ritchie, 1995). Generation Xers are marrying later than the previous generation did, delaying marriage until their education is complete and their careers have been established. Generally, those with a college education are also delaying the birth of children. Both partners in a marriage expect to work, to share in household chores, to have greater financial security, to own their own home, and to have more stable marriages than the previous generation (Ritchie, 1995).

Question

How would you concisely summarize the general traits and values of Generation X in contrast to the previous generation of young adults from the baby-boom generation, now entering midlife?

Conceptions of Age Periods

Historical evidence suggests that age distinctions were more blurred and chronological age played a less important role in the organization of U.S. society prior to 1850, compared to today. It seems that age consciousness has grown over the past 150 years in response to developments in education, psychology, and medicine (including the psychology of advertising). Public school systems used to impose strict age criteria for each grade. Psychological theories of development afforded a rationale for age legislation regarding child labor, school attendance, and pension benefits. And in medicine, pediatrics emerged as a specialty whereas old age became identified as gerontology. Popular culture, as expressed in the media, song lyrics, birthday celebrations, and advice columns of popular magazines, picked up and helped disseminate the notion that there are appropriate ages for experiencing various life events.

For the most part Americans perceive adults of all ages favorably. Nevertheless, older adults are viewed less favorably and as less desirable to be around than younger adults (unless you are an older adult, in which case you may prefer to be around people your own age). Such attitudes are influenced by a variety of factors. Adults who have had more formal education and more experience with a range of older adults have more positive attitudes toward older people than is true of the population generally. Adults who encounter burdens or conflicts associated with the elderly have more negative attitudes toward them.

College students tend to see young people as more adaptable, more capable of pursuing goals, and more active than older people. Overall, age, in and of itself, seems to be less important in determining people's attitudes toward the elderly than other types of information such as their personality traits. However, as shown in Figure 13-1, age groups differ on such issues as how old is old, whether they would like to be 100 years old, and how old they expect to live to be. People also evaluate the stages of the life cycle differently depending on their current age. It surprises many young people that their elders look back on the teenage years with little enthusiasm. And many older people see their retirement years as their best years.

Americans have some difficulty specifying the age at which an average man or woman becomes old. Much depends on the person's health, activity level, and related circumstances. Americans have little difficulty characterizing a person as a young adult or an elderly adult, but the boundaries between adjacent age categories are vague (Krueger, 1992). For instance, regarding the transition period between middle age and elderliness, the placing of an individual is only weakly connected to the chronological age of the person being judged. Even so, older adults hold more elaborate conceptions about development (its richness and differentiation) throughout the adult years than younger adults do.

A person's gender likewise makes a difference. Women are stereotyped as aging more quickly than men. But this tendency to assign women to older categories than comparably aged men becomes less pronounced when categorizing the elderly. Women are also less inclined than men to employ a "double standard of aging" (Kogan & Mills, 1992).

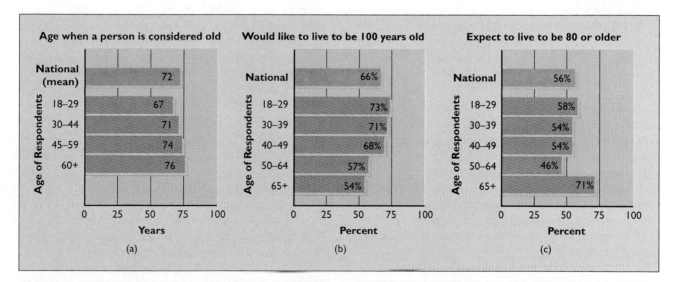

Figure 13-1 Differing Perspectives on Growing Old With the number of people age 100 years or older doubling every decade since 1950 in the industrialized world, it is of interest to note differences in attitudes toward old age. Younger Americans are more likely than older Americans to indicate that they wish to live to be 100. However, older Americans are more likely to believe they will live to an older age.

Source: This article is from The American Enterprise, a national magazine of politics, business and culture.

According to surveys of adult Americans, two-thirds perceive themselves as being younger than they actually are, though those under 30 years of age often perceive themselves as being older than they actually are (Riley & Staimer, 1993). It seems that younger people desire to be grown up and to dissociate themselves from potential social stigmas and disadvantages attached to being "too young." Once individuals reach middle age, however, they think of themselves as being 5 to 15 years younger than they are. Indeed, people frequently say that they feel about 30 or 35 years old, regardless of their actual age. The thirties seem to have eternal appeal. In addition, consistent with the notion that "you're only as old as you feel," aging people's own conceptions of their age seem to be better predictors of their mental and physical functioning than their chronological age is.

> **Question**
> *Historically, how did social scientists come to distinguish various life stages during adulthood, and what are those stages?*

Age Norms and the Social Clock

We commonly associate adulthood with **aging**—biological and social change across the life span. **Biological aging** refers to changes in the structure and functioning of the human organism over time. **Social aging** refers to changes in an individual's assumption and relinquishment of roles over time. So the life course of individuals is punctuated by **transition points**—the relinquishment of familiar roles and the assumption of new ones.

Age, as reckoned by society, is a set of behavioral expectations associated with given points in the life span. Many behaviors are prescribed for us in terms of society's dos and don'ts. Conformity with these expectations generally has favorable results; violation, unpleasant ones. Such dos and don'ts are termed **social norms**—standards of behavior that members of a group share and to which they are expected to conform, such as the unacceptability of pushing and shoving when entering or exiting a subway car. Social norms are enforced by positive and negative sanctions. For example, even though the expected social norm of marrying before having children has relaxed, many single-parent mothers and children still experience many negative outcomes related to living in poverty.

Social norms that define what is appropriate for people to be and to do at various ages are termed **age norms.** A societal "Big Ben" (the age norms) tends to define the "best age" for a man or woman to marry, to finish school, to settle on a career, to hold a top job, to become a grandparent, and to be ready to retire. Individuals tend to set their personal "watches" (their internalized age norms) by society's Big Ben (Kimmel, 1980). Examples are compulsory school attendance laws, the minimum voting age in election laws, the age at which youth may purchase alcoholic beverages, and the age at which individuals become eligible for Social Security benefits.

Age norms can also represent informal expectations about the kinds of roles appropriate for people of various ages. At times, such expectations are only vague notions about who is "too old," "too young," or "the right age" for certain activities. The appeal "Act your age!" pervades a great many aspects of life. Variations on this theme are often heard in such remarks as, "She's too young to wear that style of clothing" and "That's a strange thing for a man of his age to say." Do you have any difficulty accepting the reality of 11-year-olds having babies or 63-year-old women having babies?

Here is an illustration of how age norms and social norms can diverge: A woman of 24 does not fit the age norm for being a full professor at a university, but she can meet the social norms set by the other faculty by fulfilling her professorial obligations, such as researching, publishing, teaching, and advising.

Age grading at the social level—the arranging of people in social layers that are based on periods in the life cycle—creates a **social clock,** a set of internalized concepts that regulate our progression through the age-related milestones of the adult years. The social clock sets the standards that individuals use in assessing their conformity to age-appropriate expectations. Table 13-3 shows adults' responses in 1965 and 1996 to questions regarding the best age at which to marry, to have children, to become grandparents, to be settled in one's career, to have reached the top, and to retire. Note that those responding in 1996 tend to think women and men should marry later, become grandparents later, and retire at younger ages, compared to the survey respondents of 1965.

Likewise, people describe what personality characteristics are salient in particular age periods. For example, they think it appropriate to be impulsive in adolescence but not in middle age. And they readily report whether they themselves are "early," "late," or "on time" with regard to family and occupational events. Such an internal sense of social timing can act as a "prod" to speed up accomplishment of a goal or as a "brake" to slow down passage through age-related roles. Not being "on time" can have differing outcomes. For instance, World War II brought substantial social disruptions. There were particular social disruptions in the lives of men who were mobilized beyond the age of 30, and these disruptions were associated with an increased risk of adverse change in the trajectories of the men's health across the adult years (Elder, Shanahan, & Clipp, 1994). In contrast, compared to "on time" fathers, "late" fathers are

Table 13-3   Notions of the "Right Time" for Life Events, American and Australian Adult Respondents, 1965 and 1996

Question	Percentage Responding There Is No Appropriate Age		1965 Age Range	1996 Average Age
Best age for a man to marry	Men	9	20–25	25.21
	Women	15	20–25	25.96
Best age for a woman to marry	Men	12	19–24	23.04
	Women	12	19–24	24.48
When most people should become grandparents	Men	6	45–50	52.89
	Women	21	45–50	54.05
Best age for most people to finish school and go to work or university	Men	12	20–22	18.68
	Women	27	20–22	18.92
When most people should be ready to retire	Men	6	60-65	60.03
	Women	17	60-65	62.66

Source: Neugarten, Moore, & Lowe (1965) and Peterson, C. (1996).

more highly involved with their youngsters and have more positive feelings regarding their involvement.

Any period of transition or crisis in life can initiate a *life review* process and an assessment of where one stands with respect to age-related milestones (Suedfeld & Bluck, 1993). Exposure to death has the power to do this at all ages, even when one is not of an age when one's own death is imminent. Serious illness—one's own or that of another—can also instigate such a life review.

Although the members of a society tend to share similar expectations about their life cycle, some variations do occur. Social class is one important factor. The lower the social class, the more rapid the pacing of the social clock tends to be. The higher the social class, the later individuals generally leave school, acquire their first job, get married, begin parenthood, secure their top job, and begin grandparenthood. And new generations can reset the social clock. For instance, young women currently prefer earlier ages for educational and occupational events and later ages for family events than did women of earlier generations. It is not uncommon today for professional women to wait until their late thirties or even their forties to have a first child (as did Candace Bergen's character "Murphy Brown").

Neugarten and Neugarten (1987) believe that the distinctions between life periods are blurring in the United States. They note the appearance of the category "young-old," retirees and their spouses who are healthy and vigorous, relatively well off financially, and well integrated within community life. A young-old person could be 55 or 85. The line between middle age and old age is no longer clear. What was once considered old now characterizes a minority of elderly people—the "old-old," a particularly vulnerable group who often are in need of special support and care. Neugarten observes that in-

creasingly we have conflicting images rather than firm stereotypes of age: the 18-year-old who is married and supporting a family, but also the 18-year-old college student who still lives at home; and the 70-year-old in a wheelchair, but also the 70-year-old running a marathon.

A more fluid life cycle affords new freedoms for many people. But even though some timetables may be losing their significance, others are becoming more compelling. Young adults might feel they are failures if they have not "made it" in a corporation or law office by the time they are 35. And a young woman might delay marriage or childbearing in the interests of a career but then hurry to catch up with parenthood in her late forties or even early fifties today, even though she might expect to live to her late seventies or early eighties.

Question

What evidence do we see that traditional age-grade norms seem to be "blurring," as Neugarten states?

Age-Grade Systems

In a number of African societies, age norms are embodied in an age-grade system (Foner & Kertzer, 1978). Members of each grade are alike in chronological age or life stage and have certain roles that are age-specific. For instance, the Latuka of Sudan distinguish among five age grades: children, youths, rulers of the village, retired elders, and the very old. In such societies the individuals of each age grade are viewed as a corporate body and move as a unit from one age grade to another. For example, among the African Tiriki, uninitiated boys may not engage in sexual intercourse, they must eat with other

Distinctions Between Life Periods Are Blurring It is quite common for contemporary middle-aged Americans to be attending college to upgrade their skills or to earn a first or an advanced degree. Do you have notions concerning the "best age" at which to graduate from college, settle on a career, marry, have children, advance in a job, become a grandparent, or retire?

children and with women, and they are permitted to play in the women's section of the hut. After initiation they may engage in sexual intercourse, are expected to eat with other men, and are forbidden to enter the women's section of the hut. In India, young women who marry become the caretakers of everyone in the household, and the older mother or mother-in-law in the home finally enjoys higher social status and a more leisurely life.

Furthermore, age grades differ in the access they afford their members to highly rewarded economic and political roles. In Western societies, in contrast, people's chronological age is but a partial clue to their social locations. Class and ethnic factors cut across lines of age stratification and provide additional sources of identity.

On the surface, societies with age-grade systems seem to provide an orderly method for role allocation and reallocation. But in practice, the transition process is often less than orderly. Conflicts frequently arise between age grades, essentially a version of the time-honored struggle between the "ins" and the "outs." The desire of people to gain access to or hold on to various privileges and rewards fuels social discord and individual grievances. The rules governing transition might not be clear. Even when they seem to be unambiguous, the rules are always open to different interpretations and to "bending" in one or another group's favor. The continuing debate in the United States over Social Security funding and medical benefits reflects this type of tugging between the young and the elderly.

All societies are faced with the fact that aging is inevitable and continuous. Hence, they all must make provision for the perpetual flow of one cohort after another by fitting each age group into an appropriate array of so-

cial roles. Societies with age-grade systems attempt to achieve the transition by establishing points in the life course for entering roles and leaving them. Another solution, more closely approximated in Western societies, is to allow "natural" forces to operate: Younger people assume adult roles when they are ready to do so, while older people give up roles when they are ready to do so or when they become ill or die.

In the United States no collective rituals mark the passage from one age grade to another. High school and college graduation ceremonies are an exception, but even here not all individuals in a cohort graduate from either high school or college. And there is some flexibility in the operation of the age system, in that intellectually gifted children skip grades in elementary school and bright youths are allowed to enter college after only two or three years of high school. At the older age levels, too, flexibility often operates within some companies and some occupations with regard to early retirement; for instance, one can retire from the police work or the military after 20 years of service. With corporate downsizing and mergers, however, it is more common to see more middle-aged college students retraining for a second career in the workforce. Nevertheless, age norms serve as a counterpart of the age-grade system in broadly defining what is appropriate for people to be and to do at various ages.

Life Events

People locate themselves across the life span in terms of social timetables. They also do so in terms of **life events.** But some events are largely independent of age or stage,

Further Developments
The Effects of a Violent Crime

Rape is commonly defined as sexual relations obtained through physical force, threats, or intimidation. About one in four women report that a man they were dating persisted in attempting to force sex on them despite their crying, pleading, screaming, or other resistance (Celis, 1991). In many respects, each rape and coping experience is unique, yet there are common ingredients in such stressful life events. For instance, researchers find that victims show a significant increase in health-related problems in the year following the rape. Moreover, the vast majority of rape victims blame themselves (Frazier, 1990; King & Webb, 1981). Controversy rages, however, as to whether self-blame is a healthy or unhealthy response.

According to one view, self-blame can be a highly adaptive strategy in the aftermath of a rape. Victims who blame themselves might cope successfully as long as they direct the blame at specific, controllable behaviors that they define to be "foolish" or "careless." The women can feel that if they change these behaviors they can avoid being raped in the future. According to another view, blaming oneself for negative events such as rape is associated with increased depression and malfunctioning. Most research supports the second view (Frazier, 1990; Katz & Burt, 1988).

Societal attention to the problems that surround rape is a recent phenomenon. It seems that behavior in which some people are victimized is accorded public concern only when the victims have enough power to demand attention. Indeed, for centuries rape was a crime in which the victims were stigmatized for their victimization (a "fallen woman") and blamed for their participation in the act (Allison & Wrightsman, 1993; Fairstein, 1993). Marital rape was viewed as legal, acquaintance rape was typically unrecognized, and women's accusations of rape were generally presumed to be unreliable. And when military histories have been written, the plight of raped women as casualties of war has been deemed not serious enough for inclusion in scholarly works (Brown-

miller, 1993). Yet rape during war—as under contemporary circumstances in the former Yugoslavia—is commonplace and considered an act of the victors. Significantly, the current concern with rape has paralleled the emergence of the women's movement.

Rape myths abound in U.S. life. Psychologist Martha R. Burt (1980) found that over half of her representative sample of Minnesota residents agreed with such statements as "In the majority of rapes, the victim was promiscuous or had a bad reputation," and "A woman who goes to the home or apartment of a man on a first date implies she is willing to have sex." The threat of rape affects women whether or not they are its actual victims, limiting their freedom, keeping them off the streets at night, and at times imprisoning them in their homes. Women who carry the highest burden of fear are those with the fewest resources—the elderly, ethnic minorities, and those with low incomes.

Date rape is also a serious social problem. A three-year study of 6,200 female and male students on 32 campuses found that 15 percent of all women reported experiences that met legal definitions of forcible rape. More than half of the incidents were *date rapes*. Some 73 percent of the women forced into sex avoided using the term *rape* to describe their experiences, and only 5 percent reported the incident to authorities (Lewin, 1991). Acquaintance rape is a paralyzing event, so outside the realm of normal events that many women are left without a way of understanding it. Instead, they feel shame and guilt and attempt to bury the experience. Rape is a stressful event with lifelong consequences that can have unique meanings because of the extent to which other people blame victims of rape. Many campuses and rape crisis centers now sponsor speakers and programs directed at preventing date rape. Those few women who falsely accuse a man of rape do a great injustice to the man involved but also leave real victims of rape more vulnerable to being disbelieved.

including losing a limb in an automobile accident, winning a lottery, undergoing a "born again" conversion, or being alive when Pearl Harbor was attacked or when we first landed on the Moon. We often employ major events as reference points or time markers in our lives, speaking of "the time I left home for school," "the day I had my heart attack," and "when I met my first love." Such life events define transitions. Events that produce traumatic or reflective consequences include the death of a loved one, divorce, being fired from a job, or being the victim of a traumatic crime such as rape. These life events cause

one to ask questions about oneself or society. Some of the effects of rape on the lives of women are discussed in *Further Developments:* "The Effects of a Violent Crime."

Life events may be examined in a great many ways (Suedfeld & Bluck, 1993). For instance, some are associated with internal growth or aging factors like puberty or old age. Others, including wars, national economic crises, and revolutions, are the consequences of living in society. And still others derive from events in the physical world such as fires, storms, tidal waves, earthquakes, or avalanches. And there are those events that have a strong inner or psychological component, including a profound religious experience, the realization that one has reached the zenith of one's career, or the decision to leave one's spouse. Any of these events we might view as good or bad, a gain or a loss, controllable or uncontrollable, and stressful or unstressful.

Often in thinking about a life event, we ask ourselves three questions: *Will it happen to me? If so, when will it occur?* and *If it happens, will others also experience it or will I be the only one?* The first question concerns the probability of an event, such as getting married or being injured playing football. If we believe that the probability of an event's occurring is low, we are unlikely to attend to or anticipate it in advance. For instance, most of us are more likely to devote thought and preparation to marriage than to a serious football injury. The second question involves the correlation between chronological age and an event, such as the death of one's spouse or the suffering of a heart attack. Age-relatedness matters because it influences whether or not we are caught unexpectedly by the event. The third question concerns the social distribution of an event, whether everyone will experience it or just one or a few persons. This question is important because it will largely determine whether people will organize social support systems to assist us in buffering the change. In the United States, for example, we have extensive, organized social support systems for ushering children through formal schooling, for getting and staying employed, for getting married, and for making the transition to retirement.

The Search for Periods in Adult Development

Many psychologists have undertaken the search for regular, sequential periods and transitions in the life cycle. They depict adulthood as being, like childhood, a sort of stairway, a series of discrete, steplike levels. The metaphor of the life course as being divided into stages or "seasons" has captured the imagination of philosophers and poets as well as other writers. One of the most popular versions of the stage approach is contained in Gail Sheehy's best-selling books *Passages* (1976) and *New Passages* (1995). She views each stage as posing problems that must be resolved before the individual can successfully advance to the next stage. In these passages from one stage to the next, each person acquires new strengths and evolves an authentic identity. Such an identity has many of the qualities that Abraham Maslow associates with the self-actualized person.

Other psychologists have taken exception to the stage approach. Some believe that an individual's identity is fairly well established during the formative years and does not fundamentally change much in adulthood. Ac-

cording to this view, people might change jobs, addresses, even faces, but their personalities persist, much like adult height and weight do, with only minor changes. As we noted in our treatment of developmental continuity and discontinuity in Chapter 2, this approach views aging as a continuous yet dynamic process. The discontinuity approach stresses the differences between stages and the uniqueness of the issues for each stage.

> **Question**
>
> *How does the stage perspective of adult development differ from the view that adult development is a continual process of change over the life course?*

Physical Changes and Health

Our physical organism changes across the life span. But in and of themselves such changes may be less important than what people make of them. As we pointed out earlier, cultural stereotypes and social attitudes have a profound effect on our perceptions of biological change and our experience of it. Whereas the physical changes associated with puberty are comparatively easy to identify, later changes (with the possible exceptions of prostate problems or menopause) are less easy to pinpoint as marking stages in adulthood.

Physical Performance

To most individuals, getting older means losing a measure of physical attractiveness, vigor, and strength. Yet although physical changes take place throughout adulthood, for the most part they have only minimal implications for an individual's daily life in early adulthood. The years from 18 to about 30 are peak years for speed and agility. Most Olympic athletes fall between these ages. However, in some events, such as women's gymnastics, women of 18 are deemed "old." For long-distance running and most positions in baseball, the peak age of performance is about 28, whereas top tennis players reach their highest levels of performance at age 24. Golfers, in contrast, peak at about 31 years of age. In swimming events, men typically peak at 18 or 19 years of age and women at 16 or 17. Studies confirm that maximum physical strength is highest in the twenties, after which it declines (Rikli & Busch, 1986; Welford, 1977). Some research suggests that hand-grip strength in the thirties is 95 percent of what it was in the twenties, falling to 91 percent in the forties, 87 percent in the fifties, and 79 percent in the sixties. Back strength also diminishes to about 97 percent in the forties, 93 percent in the fifties, and 85 percent in the sixties. But such averages mask the considerable variation found among individuals. Complex sensorimotor coordination also begins slowing in the thirties. (Welford, 1977).

One area in which individuals in early and middle adulthood are most likely to note changes is in their vision. Between 30 and 45 years of age, individuals tend to experience some loss in the power and elasticity of their eye lenses. Through early and middle adulthood, individuals who are nearsighted tend to become more nearsighted, and those who are farsighted tend to become more farsighted. Aging also produces deterioration in the human auditory and vocal systems, but these changes are generally of little consequence for those in early and middle adulthood.

Physical Health

A recent national survey by the Centers of Disease Control (1998d) found that nearly 87 percent of adult Americans said their health was good or excellent. Alaskans gave themselves the highest overall health rating (91.6 percent rated their health as good to excellent); West Virginians reported the poorest health status (76.6 percent). Adult health is a function of a wide variety of factors, including heredity, nutrition, exercise, previous illness, access to health insurance, and the demands and constraints of the social environment. For the most part we assess people's health by how well they are able to function in their daily lives and adapt to a changing environment. Health, then, has a somewhat different meaning for a young pregnant woman, a nursing home resident, a college professor, a presidential candidate, a high school basketball player, an airline pilot, a construction worker, and a surgeon (Van Mechelen et al., 1996).

Extrapolating from the results of interviews with the more than 10,000 American women who participated in the 1995 National Survey of Family Growth, 47 percent of U.S. married women are dependent upon their husbands' health insurance for health care coverage, and 38 percent have health care coverage with their own employers. About 9 percent of married women are covered by Medicaid, 3 percent are covered by military health insurance, 5.5 percent pay for their own coverage, and 9 percent—or 2.7 million—have no health coverage. Also, 21 percent of married Hispanic women are not covered by any medical insurance, compared with 8 percent of non-Hispanic white women (U.S. Department of Health & Human Services, 1997c).

For single women, 34 percent have health insurance through their employer, 25 percent were covered under parents' medical policies, and 23 percent are covered by Medicaid. About 14 percent—or 4.3 million single women—have no health insurance coverage of any kind. One-third of single women rely solely upon Medicaid for their health coverage. Sixty-eight percent of unmarried mothers rely upon Medicaid for medical coverage (Hofferth, 1998). Obviously, any government policies that incorporate cutbacks to the Medicaid system could have negative health outcomes for many U.S. women and children. On the other hand, full-time employment could provide much-needed health care coverage. Like other Americans, the vast majority of young adults enjoy good to excellent health. Infectious diseases—particularly colds, upper respiratory infections, and sexually transmitted infections—are among their most common illnesses. Employers can expect to lose more work days due to sickness among young adults than among older adults. (A business with 100 employees can expect to lose about 10 months of work days per year; if the employees are 100 young adults, the expectation is nearly 16 months). Motor vehicle accidents tend to take a higher toll in death and disfigurement among young people than among other age groups. For the adult of any age who is concerned with both physical and mental health, exercise affords both physical and psychological benefits (Wiley et al., 1996).

Dieting, Exercise, and Obesity Being healthy is rarely something that just happens to you, especially as you get older. Having and maintaining good health consists of choosing certain patterns of behavior and eschewing others. Diet and exercise are two important components of staying healthy, and these have become daily obsessions for many Americans. After completing high school, many young adults enter into new routines that affect how active they are—some enter into environments that include time for exercise, others take on responsibilities that prevent them from being as active as they were during adolescence. Today, more people walk, jog, cycle, rollerblade, swim—engage in *aerobic exercises*—than in the last 40 years. Young adult women report that walking and swimming are their two most common physical activities. And it is therefore no surprise that most people today understand the word *aerobic*, related to cardiovascular fitness, whereas very few people would have recognized the word in 1960.

Physical Activity and Health Across Cultures Public health programs throughout the Western world have goals to increase leisure-time physical activity of the general public in order to promote their health and fitness. Results of the European Health and Behaviour Survey conducted with more than 16,000 adults aged 18 to 30 enrolled in universities in 21 European countries indicate that physical activity levels vary across different countries, that participants are generally well aware of the benefits of exercise, and that men are more likely than women to engage in frequent leisure-time physical activity (Steptoe et al., 1997) (see Tables 13-4 and 13-5.) Seventeen percent of the young men and 23 percent of the

Table 13-4   Percentage of Young Adults Who Were Most Physically Active (Leisure Physical Activity) by Gender and Country

Men		Women	
% Most Physically Active	Country	% Most Physically Active	Country
87%	Hungary	95%	Hungary
84	Finland	92	Finland
83	Switzerland	85	Sweden
83	Iceland	83	The Netherlands
80	W. Germany	79	W. Germany

From: A. Steptoe, et al. Leisure-Time Physical Exercise: Prevalence, Attitudinal Correlates, and Behavioral Correlates Among Young Europeans from 21 Countries. *Preventive Medicine*, 26, 845–854. Copyright 1997. Reprinted by permission of Academic Press.

Table 13-5   Percentage of Young Adults Who Were Least Physically Active (Leisure Physical Activity) by Gender and Country

Men		Women	
% Least Physically Active	Country	% Least Physically Active	Country
52%	Portugal	29%	Greece
55	Greece	35	Portugal
58	Spain	36	Spain
64	Scotland	58	France
66	England	58	Italy

From: A. Steptoe, et al. Leisure-Time Physical Exercise: Prevalence, Attitudinal Correlates, and Behavioral Correlates Among Young Europeans from 21 Countries. *Preventive Medicine*, 26, 845–854. Copyright 1997. Reprinted by permission of Academic Press.

young women reported persistent health problems. More than 70 percent of the respondents in Denmark, Finland, the Netherlands, and Norway were aware of the benefits of exercise and cardiovascular fitness; but fewer than 40 percent in Belgium, Greece, Italy, and Poland were aware of these benefits. Lack of physical exercise was associated with cigarette smoking among both men and women: 41 percent of inactive men were smokers, and 34 percent of inactive women were smokers. Alcohol consumption levels were low: 27 percent of men and 13 percent of women drank more than one drink per day, and 35 percent of men and nearly 50 percent of women were nondrinkers. Many more women (44 percent) than men (17 percent) in this study participate in leisure physical activity out of a desire to lose weight.

Data from this extensive self-report study comes from university students, who tend to be healthier and more informed about health than the general population. This study also found an association between lack of physical exercise and incidence of depression. An interesting finding was that despite public fitness education campaigns in England, Scotland, and Ireland, low rates of leisure physical activity were reported in those countries. The planning of health services and preventive programs is especially important to the European Union at the turn of the century as the European Community Act allows for more population mobility and inclusion of Eastern European countries into the European Union is being considered. Apparently, these findings also suggest that U.S. students studying abroad in northern European countries are likely to encounter higher rates of leisure physical activity among their university peers there. The opposite would be true in southern Europe.

Some individuals combine their exercise program with some form of dieting; others use diet as the primary means of maintaining health. What we eat can affect us in many ways, but most commonly our diet affects how we look and feel and how prone we are to sickness. People who consume lots of fruits and vegetables decrease their chances of contracting cancer or suffering from stroke (Nutrition, 1996). Excessive cholesterol in the blood vessels can eventually clog them, leading to heart attacks. Several studies have linked cholesterol levels to risks of heart disease and have shown that controlling one's intake of cholesterol is a key factor to lowering one's risk for heart problems (Pinkowish, 1996).

Simple ways to lower cholesterol are to eat fiber found in beans, fruits, and vegetables; eat fewer eggs; cut down on saturated fats found in milk, cheese, and meat; and cook with polyunsaturated fats such as sunflower, safflower, or olive oil.

Adolescents are, for the most part, active and resilient. Only after they begin the transition to adulthood and the concomitant changes in lifestyle do the effects of poor diet and little exercise begin to show. It is therefore no coincidence that dieting is common in the United States (as one European was overheard remarking to another, "I think all Americans are on a diet"). In truth, typically around 40 percent of U.S. women and 24 percent of U.S. men are dieting (Brody, 1996). Some argue that diets work only in tandem with exercise, education, and behavior modification. And one major concern regarding diets is the constant on-again, off-again pattern of weight loss/gain that dieting produces. And as discussed in Chapter 11, dieting might lead to eating disorders such as bulimia or anorexia.

Information You Can Use

Benefits of Aerobic Exercise

Most health experts agree that your cardiovascular fitness is best promoted by engaging in exercise that raises your heart rate to 60 percent of your maximum heart rate a minimum of three times a week. To find out what your maximum heart rate is, simply subtract your current age from 220 and then multiply by 0.6. A few sample ages are shown in Table 13-6.

Some experts contend that high-quality exercise on a regular basis significantly reduces the chances of heart attack. The amount of exercise recommended for these results include swimming or running for 25 minutes each day, cycling for 50 minutes at greater than 10 mph, walking for 45 min-

Health and Aerobic Exercise The physical and mental health benefits of regular aerobic exercise are well known, but more young females today are turning away from strenuous exercise in favor of walking, swimming, and cycling.

Table 13-6   Sample Calculations of Target Heart Rate During Exercise

Current Age	220 – Current Age	Multiply by 0.6	Heart Rate During Exercise
20	220 – 20 = 200	200 * 0 .6 = 120	120
30	220 – 30 = 190	190 * 0 .6 = 114	114
40	220 – 40 = 180	180 * 0 .6 = 108	108
50	220 – 50 = 170	170 * 0 .6 = 102	102

Table 13-7   Examples of Moderate and Strenuous Exercise

Moderate (Activities That Raise Aerobic Levels to 3 to 6 Times Nonactive Levels)	Strenuous (Activities That Raise Aerobic Levels to More Than 6 Times Nonactive Levels)
Walking at 3–4 mph	Walking at 3–4 mph or uphill 5 times a week
Easy cycling	Fast cycling for 1 hour
Leisure swimming	Swimming laps 3 times a week
Golf without cart	Stair-climber 2–3 hours a week
Table tennis	Tennis or racquetball 3 days a week
Canoeing 2–4 mph	Canoeing faster than 4 mph
Mowing yard with power mower	Mowing yard with push mower

utes at 4 mph pace, or doing aerobics for 30 minutes. According to the American Heart Association (1998), exercise also benefits us in the following ways:

- Maintains desired body weight
- Strengthens heart and lungs
- Protects against stroke, diabetes, cancer, and osteoporosis
- Lowers blood pressure
- Relieves anxiety

But too much of even a good thing can be bad; overexertion does not produce extra benefits and can even be harmful. If you are just beginning to exercise or want to start an exercise program, consult your physician and ease into exercise slowly. In a recent survey, awareness of the benefits of exercise was consistently associated with engaging in physical exercise and having a desire to lose weight. Also, 52 percent of men and 54 percent of women were aware that exercise decreases the risk for heart disease (Steptoe et al., 1997). About 37 percent of American adults exercise strenuously at least three times a week. For some examples of moderate and strenuous exercises, see Table 13-7.

In contrast to many other societies of the world, Western societies put a premium on being thin, and being overweight, or even perceiving oneself as overweight, can place a Westerner in a precarious emotional state. Added to the emotional stress of being overweight are the potential physical problems of high blood pressure, gallstones, diabetes, stroke, and heart disease ("Executive Summary," 1998). Approximately 70 percent of adults age 25 to 50 are over their ideal weight. Thirty-three percent of these are considered obese (generally considered to mean being more than 20 percent over the ideal body weight for one's sex, body frame, and age). Because our *metabolism* slows down and the likelihood that we will gain weight increases as we age, preventative measures in the early adult years can play a significant role in keeping us healthy as we grow older.

HIV and AIDS   Findings from personal interviews with women aged 15 to 44 in the National Survey of Family Growth (1999) indicate that 48 percent of women in early adulthood have at some time in their lives had an HIV test related to a blood donation, prenatal care, hospitalization, surgery, or an insurance application, or by request. One serious finding was that 28 percent of unmarried women reported that their male partner(s) in the last 12 months were having sexual relations with other women. About 33 percent of unmarried women who were sexually active over the past year reported using condoms every time they had sexual relations; 33 percent said they used condoms sometimes; and 33 percent said they had unprotected sex. Do you think that the serious health threat of HIV/AIDS over the past two decades changed adult sexual behavior in the United States?

Reported Changes in Adult Sexual Behavior over the Past Decade   Studies of individual responses to the threat of HIV/AIDS and other sexually transmitted diseases (STDs) have found substantial evidence of behavioral changes in high-risk groups, (e.g., homosexuals, intravenous drug users, pregnant mothers, and STD patients). One of the few studies conducted to date with the general U.S. adult population, The National Health and Social Life Survey, was done in 1992 with a general population of adult Americans (Feinleib & Michael, 1998). Findings indicate that nearly 30 percent of the survey participants aged 18 to 59 reported an increase in self-protective strategies to avoid AIDS (see Table 13-8). If the 30 percent figure seems low, it is because a majority of the respondents in this survey were married and did not consider themselves at risk for contracting HIV or STDs. Younger male nonmarried adults, nonwhites, and those living in urban areas were more likely to have changed their behaviors because of the threat of HIV/AIDS. Those who had had only one sexual partner during their lifetime were not changing their behaviors,

Table 13-8   Changes in Sexual Behavior to Avoid HIV/AIDS

30 percent of adult survey participants report changing their sexual behaviors in the following ways:

Type of Sexual Behavior Change	Percent
Reducing number of partners	12%
Using condoms more frequently	9
One partner or monogamy	8
Abstaining from sex altogether	3

From: Joel A. Feinleib and Robert T. Michael. Reported Changes in Sexual Behavior in Response to AIDS in the United States. *Preventive Medicine, 27,* 400–411. Copyright 1998. Reprinted by permission of Academic Press.

whereas those who had had multiple partners were much more likely to have changed their behaviors. However, 25 to 30 percent of those with the greatest exposure to risk reported making no changes whatsoever in their sexual behaviors. These young adults constitute a major concern for U.S. health practitioners.

Practicing Safe Sex   Many individuals falsely believe that various forms of birth control protect them against STDs and HIV/AIDS transmission. Some young adults who are already using another form of birth control believe that requesting condom use is awkward and indicates a belief that the partner may carry an STD. Recent research shows, however, that women who identify themselves as using condoms to prevent pregnancy were more likely to use a condom at last intercourse than were other women (Critelli & Suire, 1998). One of the reported reasons why sexually active young adults might not use condoms is that they have taken other measures to avoid pregnancy. Research findings also suggest that once a relationship becomes established and partners trust one another, condom use is replaced by oral contraceptives, which seem more convenient and effective (Feinleib & Michael, 1998). Monogamy is frequently cited as a reason for not using condoms, and many young adults consider monogamy to be an effective way to prevent transmission of STDs. However, many relationships between young adults do not last. Short-term, serial monogamy is not an effective way to prevent transmission of STDs, especially given that in a survey of college men, about 25 percent reported that they lie about their sexual history in order to obtain sex (Fischer, 1996). Knowledge of risks does not necessarily lead to changed behaviors (Gupta & First, 1998).

What health issues are currently of paramount importance varies across cultures. Russia has seen almost a 4,500 percent increase of diagnosed HIV infections in the last decade. These data point to the toll the disease will have on young adults, because they are the ones most likely to be using drugs (Williams, 1995). Rates of HIV/AIDS transmission and death are very high in African countries, where many myths about transmission of the disease abound and condoms and medicines to curb the course of the disease and prolong life are often unaccepted or unavailable. Also, cultural female and male circumcisions are often conducted on groups of adolescents at one time using one unsterilized "surgical" device (Duke, 1999). Additionally a recent report by the Chinese Ministry of Health indicates that HIV-infected people have been found in all 31 Chinese provinces (Lili, 1999).

Socioeconomic Status, Ethnicity, and Gender

In the United States today, people who are poor and lack higher education have a higher death rate than people who are wealthy and better educated (Stevens, 1996). The reasons are not surprising—poverty increases the likelihood that one will experience inadequate or poor nutrition, poor housing, insufficient prenatal care, limited access to health care facilities, and less education. And many studies confirm that racial and ethnic minorities and single parents are more likely to live in poverty. Poor people who do not have health insurance cannot afford good medical care, and it has been shown that with less education one is more susceptible to heart disease, hypertension, and other health ailments (Pincus & Callahan, 1994). Does this mean that if someone living in poverty won the lottery tomorrow, good health would follow? Better medical care would be within reach with the boost in income, but poor habits with respect to eating, smoking, and drinking have social links—and these might remain entrenched in the person's lifestyle.

We all know that women have a longer life expectancy than men, but it is not clear how this is related to specific health problems. It has been suggested that women's greater longevity is due to women's having two X chromosomes and larger amounts of the hormone estrogen. But as women's patterns of work and recreation have become more like men's, their health patterns and concerns have become similar. For example, women have not had the same decrease in smoking behaviors as men, which shows up as a 182 percent increase in deaths among women from lung cancer over the past 25 years (American Cancer Society, 1997). This figure is expected to rise, as recent studies with adolescent girls indicate that more are smoking (because they have discovered that another way to keep their weight down is to smoke instead of eating). There is also an association between gender and depression; women resort much more often than men to dysfunctional ways of coping such as self-blame, venting anger on others, and seeking comfort in sweets or alcohol (Hänninen & Aro, 1996).

In young adulthood the gender gap in depression becomes obvious; women are more likely than men to experience depression during young adult life, except that the likelihood of depression for young black males is similar to that for women. Research suggests that the general gap between women and men with respect to depression relates to gender inequities in numerous areas of adult life, such as employment opportunities, pay and authority in the workplace, the burden of child care and housework, and situations that women and men find themselves in as young adults, sometimes for the first time. One reason the gap is not disappearing is that men still get more pay outside the home while the balance of power within the home still rests with men (Mirowsky, 1996). The median wage and salary income for all male workers in 1997 was $26,939, compared to $16,849 for women—about 63 percent of men's average income (U.S. Bureau of the Census, 1998a).

Changes in Drug Use over Time

The end of high school is a major transition period for most adolescents as they enter into the world of young adulthood. The usual roles that young adults take on include college student, civilian employee, and member of the armed forces—though approximately 10 percent of young adults simply "drop out" of mainstream society (some are incarcerated, institutionalized, chronically ill, or disabled). Common experiences for young adults have included completing college, securing full-time employment, being promoted in the workplace, marrying, working toward advanced degrees, and parenting. Each of these experiences affects drug use during the post high school years, so it is useful to consider the timing of these experiences during young adulthood and how they are interrelated. Data collected from high school classes between 1976 to 1997 were compiled in annual Monitoring the Future Surveys (Johnston, Bachman, & O'Malley, 1997). Results indicate that one of the most fundamental choices facing young adults as they leave high school is whether they should go on to college. More than half of the respondents went on to college. Overall, their experience indicates that young adults going on to college are

- less likely to marry or more likely to marry later (the average age of marriage for these young adult women is 25 years old and for men it is 26.8 years old),
- more likely to leave home,
- more likely to live in dormitories or other student housing, and
- more likely to not be employed or to be employed only part-time.

In contrast, results on respondents who were high school graduates but *not* going to college indicate that these young adults are

- continuing to live at home for a while,
- more likely to marry at younger ages,
- employed, and
- for the women, more likely to be homemakers.

How do these differences compare in terms of new freedoms and responsibilities? College students must learn to manage time and deal with the pressures of exams and school requirements, while they also have more flexibility in their schedules, housing arrangements, and day-to-day lives. It is important to note that these differences do translate into how frequently individuals use drugs during young adulthood. Mapped over time, data on annual marijuana use and marital status show the following:

- Overall, women use marijuana less than men do.
- Married individuals use marijuana less than single individuals do.
- Young individuals use marijuana more than older individuals do.

The data taken at five points in time—at ages 18, 19–20, 21–22, 23–24, 25–26—show that among single men, those aged 19–20, 21–22, and 23–24 had the highest rates of marijuana use. Among single women, those aged 19–20 and 21–22 had the highest rates of marijuana use (Johnston & O'Malley, 1997). Among married men and women, those under age 20 were most likely to use marijuana. These findings show that getting married strongly correlates with a lower rate of use of marijuana. Although this survey did give reasons for this trend, perhaps you can suggest some.

Questions

Is there any particular segment of the young adult cohort that is most at risk for negative health outcomes? What are some recommendations to maintain or improve physical health during this period of the life span?

Mental Health

Mental health is also a matter of concern at any age. We know that poor health, alcohol and drug use, and depression are generally interrelated. According to a survey sponsored by the National Institute of Mental Health (NIMH), approximately 52 million adults in the United States—more than one in four—suffer from a mental disorder at some point during a year (Goleman, 1993). Of these, only 28 percent seek professional help, and general practitioners rather than psychiatrists or other psychotherapists treat the bulk of those individuals. In a given year, close to 9 million Americans develop a mental health problem for the first time; another 8 million suffer a relapse; and the remaining 35 million experience continuing symptoms.

Overall, two elements stand out in any consideration of mental health. First, from a social perspective, mental health involves people's ability to function effectively in their social roles and to carry out the requirements of group living. Second, from a psychological perspective, mental health involves a subjective sense of well-being—happiness, contentment, and satisfaction. Yet adequate social functioning and a sense of psychological well-being are not so much states of being as processes. Mental health requires that people continually change and adapt to life experiences, and people who have good mental health are commonly taken to have found a comfortable fit between themselves and the world—they "have it all together." People who do not "have it all together" can experience anxiety, stress, or depression. Let's take a closer look at these issues.

In intensive interviews with 20,000 men and women from households selected to represent the entire population, the mental disorders that were diagnosed ranged from mildly impairing problems like a phobia of heights or enclosed places to incapacitating bouts of depression or chronic problems like schizophrenia. Some 20 million Americans report having at least one phobia severe enough to interfere with their daily functioning. Alcoholism is also prevalent, with close to 14 million Americans reporting an alcohol problem. Many people report having more than one disorder; for instance, 6 million people report having a substance-abuse disorder along with one or more other mental disorders (Hankin et al., 1998).

Some 18 million Americans suffer from some form of depression each year. And cross-cultural studies consistently find that twice as many adult women as adult men are depressed, particularly women aged 18 to 24 (Hankin et al., 1998). Findings suggest that women experience more depression because they tend to resort to emotion-focused coping instead of problem-solving coping. In a recent study conducted by Hänninen and Aro (1996) in Finland with nearly 1,700 young adult females and males, the focus was on functional versus dysfunctional coping strategies. An effective strategy among women was "thinking the problem over with a friend," whereas men tended to use effective coping strategies such as "trying to find something relaxing to do," "tackling the problem more persistently than before," and reassuring themselves that "there's no reason to get upset." Women were more likely to engage in dysfunctional coping strategies, such as "venting anger on other people," "blaming oneself for what happened," and "seeking comfort in sweets." Men were more likely to resort to a dysfunctional approach of "going out for a few beers." Overall, dysfunctional ways of coping with stress exacerbate stress and depression. A variety of studies reveal that depression and other mental illnesses become more prevalent as income and education levels drop (Hankin et al., 1998). Additionally, young adults between the ages of 25 and 34 report a higher incidence of psychiatric disorder than do elderly people. One partial explanation seems to be that older people sometimes overlook the mental difficulties they had earlier in life. But it also seems to be true that modern young people face substantial stressors, such as drugs, alcohol, eating disorders, competition in the academic and work world, single parenthood, and fear about the future (Miller, 1994).

Lifetime alcohol use increases the risk for major depression for both men and women across studies. Both cultural and acculturation factors appear to play a role in alcohol use, drug use, and depression, according to findings from a recent study with young adult males from Puerto Rico, the Dominican Republic, and Colombia (Zayas, Rojas, & Malgady, 1998). Johnson and Gurin (1994) reported that among Puerto Rican men, depression is strongly associated with alcohol consumption. Research findings are mixed from examining the association between the strains of Hispanic acculturation to American society and alcohol use and mental health problems: Some findings indicate a positive relationship, others find a negative relationship. Some studies, however, suggest that Hispanic men begin heavy alcohol consumption in their early and mid twenties and drink more heavily than men in the general population (Johnson & Gallo-Treacy, 1993). A study of young adult Hispanic males found that Mexican American and Puerto Rican men between the ages of 25 and 34 years old report more alcohol consumption than do men in older age groups (Black & Markides, 1994). More community-based research is needed to understand the associations among immigration experience, sociocultural norms of drinking or drug use, and acculturation experiences (Zayas, Rojas, and Malgady, 1998).

Psychological disorders and disturbances result from both individual vulnerability and environmental stresses (see Figure 13-2). Some people are genetically so susceptible that it is exceedingly difficult for them to find an environment that is sufficiently low in stress to prevent them from having a breakdown. Hereditary predisposition most commonly takes the form of a defect in the metabolism of one or more neurotransmitters. At the other end of the continuum are those people who are so resilient and resistant to stress that few, if any, environments would trigger severe disturbance in them. For example, some political prisoners, despite years of torture and solitary confinement, manage to retain their sanity. In sum, people differ greatly in their vulnerability to mental disorder and disturbance (Wiebe, 1991).

Stress   In the course of our daily lives, most of us experience one or more demands that place physical and emotional pressure on us (Zirkel, 1992). We commonly term these experiences "stress." According to a 1993 national survey of Americans undertaken by Princeton

"I can't get used to being a grownup."

Figure 13-2 Making the Transition to Full Adulthood Can Be More Problematic Today

Survey Research Associates (1994), 6 in 10 Americans feel under great stress at least once a week. Nineteen percent say they feel great stress virtually every day, 15 percent several days a week, and 27 percent once or twice a week. Only 10 percent of Americans report never feeling stressed. Adults who sleep six or fewer hours each night are more likely than those who get more sleep to feel great stress every day (43 percent in contrast to 14 percent for those who sleep seven to eight hours each night). The survey also finds that younger adults experience more feelings of distress than do older adults. The portion of Americans who report stress several times a week declines gradually from 39 percent of those aged 18 to 29, to 23 percent of those aged 65 and older. What might be some reasons why elderly Americans report less stress per week?

Gender Differences in Stress   A variety of research studies conducted over the past two decades consistently find that women, especially married women, are more likely to feel—or at least to admit to feeling—stress than men are (though men might be less likely to report being distressed because they want to present themselves in a better light) (Almeida & Kessler, 1998). There appear to be two major views about this gender difference. *Rumination theory* states that women are more likely to ruminate (dwell upon) their negative emotions, which prolongs distress, whereas men tend to respond to distress in a behavioral fashion, distracting themselves from the stress (Nolen-Hoeksema, Morrow, & Frederickson, 1993). The *gender-role perspective,* however, suggests that women are more distressed than men because their roles as nurturant, empathic caretakers expose them to more daily stressors in contrast to male roles, which are generally more instrumental (Mirowsky & Ross, 1989). Several studies confirm that women were more distressed over family events and men were more distressed by financial or work-related problems (Almeida & Kessler, 1998; Conger et al., 1993).

The Americans most likely to report being stressed are women aged 30 to 44; women in this age group were 6 percent more likely than men of the same age to say they experience some stress. This is not surprising, because women in contemporary American society are likely to be wage earners, caretakers of family members, housekeepers, and caretakers of elderly relatives. Women also are more likely than men to indicate that stress has affected their health (Daley et al., 1998). In fact, stressful life events impact heavily on all aspects of anyone's routines, and there is documented evidence of increased accidents or injuries when an individual is under psychological stress (Almeida & Kessler, 1998).

Stress Reported by Nontraditional versus Traditional College Students

Nontraditional students have entered college in record numbers over the past decade. A nontraditional student is defined as a student who has major multiple roles (e.g., spouse, parent, employee, student) and at least a one-year hiatus between high school and college. The average reported absence time between high school and college entry is 10 years. Traditional college students are typically 18 to 23 years old and have come to college directly from high school. In 1997 the U.S. Department of Education estimated that 46 percent of all college students were older than 24, which means that many were of nontraditional status. Nontraditional students typically report role strains because of lack of time and more role demands. Dill and Henley (1998) recently conducted a study in which nontraditional and traditional students completed a survey on major and daily life events that they perceived to be stressful (positive, neutral, and negative life events). Their findings suggest that in comparison to nontraditional students, traditional students are likely to

- attend class more often
- belong to campus organizations and find social and peer events more important
- report more problems with roommates
- worry more about their academic performance

- spend more time relaxing with friends, though their friends often drink and use drugs
- be happy about becoming independent, though a social network is very important, and
- feel more pressure about expectations from their parents

In contrast to traditional students, nontraditional students are likely to

- find doing homework more desirable
- experience a greater impact from bad classes or poor teaching
- report more responsibility and obligations at home and have little time for friends
- report concerns about family or friends recovering from illness
- indicate they enjoy going to classes but find that other responsibilities interfere sometimes
- have overall satisfaction with their role as student, and
- have specific financial aid problems

Overall, both groups of students experience stress, but high levels of stress are more common among nontraditional students. Most colleges have established clubs for nontraditional students, but many cannot find the time to participate.

Stages of Stress Reaction

According to a classic study undertaken by Hans Selye (1956), our bodies respond to stress in several stages. The first stage is the *alarm reaction.* The nervous system is activated; digestion slows; heartbeat, blood pressure, and breathing rate increase; and the level of blood sugar rises. In brief, the body pulsates with energy. Then, the *stage of resistance* sets in. The body mobilizes its resources to overcome the stress. During this phase the heart and breathing rates often return to normal. But the appearance of normality is superficial, because the pituitary hormone ACTH remains at high levels. Finally, if some measure of equilibrium is not restored, a *stage of exhaustion* is reached. The body's capacity to handle stress becomes progressively undermined, physiological functioning is impaired, and eventually the organism dies.

By virtue of having been linked to various disorders, including heart disease, high blood pressure, ulcers, asthma, and migraine headaches, stress has acquired a bad name. Yet stress is a factor in everyone's life. Indeed, without some stress we would find life quite drab, boring, and stagnant. Therefore, psychologists are increasingly concluding that stress in and of itself is not necessarily bad. Much depends on how we react to and cope with the various stresses in our lives. Even so, most of us agree some events are more stressful than others, with the death of a loved one and divorce being the most severe.

More often than not, stress resides neither in the individual nor in the situation alone but in how the person perceives a particular event (Terry, 1994). Not surprisingly, some individuals seem to be more stress-resistant than others by virtue of the attitudes they bring to their lives, and consequently they enjoy better health (Wiebe, 1991). Psychologists Kobasa, Maddi, and Kahn (1982) find that *hardiness* is associated with an openness to change, a feeling of involvement in what one is doing, and a sense of control over events. Take the matter of a person's attitude toward change. Should a man lose his job, for example, he can view it as a catastrophe or as an opportunity to begin a new career more to his liking. Likewise, stress-resistant individuals get involved in life rather than hanging back on its fringes: They immerse themselves in meaningful activity. Furthermore, psychologically hardy people believe that they can actively influence many of the events in their lives and that they have an impact on their surroundings. Other researchers also find that *good self-esteem* and a *sense of control* are important buffers against the harmful effects of stress (Brandtstadter & Rothermund, 1994).

Because we are social beings, the quality of our lives depends in large measure on our interpersonal relationships. One strength of the human condition is our propensity for giving and receiving support from one another under stressful circumstances. Social support consists of the exchange of resources among people, based on their interpersonal ties. Group and community supports affect how we respond to stress through their

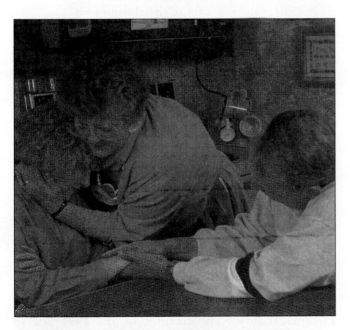

Significance of Social Support One of the protective factors for good mental health is an established support network of friends and family when experiencing negative life events. Black women are especially strong in this regard and have much lower rates of suicide than any other adult group.

health-sustaining and stress-buffering functions. Those of us with strong support systems appear better able to cope with major life changes and daily hassles. As we will see in Chapters 18 and 19, people with strong social ties live longer and have better health than those without such ties. Studies covering a range of illnesses, from depression to arthritis to heart disease, reveal that the *presence of social support* helps people fend off illness, and the absence of such support makes poor health more likely (Turner, Wheaton, & Lloyd, 1995). Social support cushions stress in a number of ways:

- Friends, relatives, and coworkers may let us know that they value us. Our self-esteem is strengthened when we feel accepted by others despite our faults and difficulties.
- Other people often provide us with informational support. They help us to define and understand our problems and find solutions to them.
- We typically find social companionship supportive. Engaging in leisure-time and recreational activities with others helps us to meet our social needs while simultaneously distracting us from our worries and troubles.
- Other people may give us instrumental support—financial aid, material resources, and needed services—that reduces stress by helping us resolve and cope with our problems.

Adults who say they feel little or no stress tend to be regular exercisers and nonsmokers and to be in good health on the whole. Additional factors affect a person's reaction to stress, because much stress is associated with relationships (family, coworkers, friends, neighbors, and others).

Suicide in Young Adulthood   In the past, young black males and females had low rates of suicide, yet the rates for young black men have been rising steadily since the late 1980s (Gibbs, 1997). Of black men, those between the ages of 25 and 34 have the highest rate for suicide, and substance abuse is often involved. White males, however, have higher rates of suicide than black males during adolescence and again in late adulthood following retirement. Black females of all ages were found to have the lowest rate of suicide when compared to others by sex and race (Gibbs, 1997). Protective factors that reduce suicide risk for black females include religious beliefs and learning to cope with high levels of stress by forming strong social support networks among extended families in neighborhoods. Gibbs (1997) suggests young adults should make every effort to graduate from high school, go to college or enroll in job training programs, and increase their opportunities for employment to mitigate the effects of the stress and depression in their lives.

Human Diversity

Gender Aptitude Quiz

You are not the same person you were just half an hour ago—emotions, psychology, physicality—these are constantly in flux. We all change careers, attitudes, and relationships. How about gender? Some believe that gender is as malleable as some of the other parts of our lives, but here is a quick quiz to see how your aptitude on gender rates.

1. **Why are you taking this quiz?**
   a) I'm not taking this quiz. I'm just looking through the questions.
   b) This kind of thought-provoking discussion is interesting, even though it doesn't apply to me.
   c) I like to question things I've taken for granted. It changes how I think.
   d) This gender question seems like a real Pandora's box to me. Let's throw open the cover and see what flies out.

2. **How many genders do you really think there are?**
   a) Two. What a pointless question.
   b) I'm going to guess that there are several genders and two sexes.
   c) 3? 4? 5? Wait . . . 6? Um . . . I don't know.
   d) An infinite number. What a pointless question!

3. **What exactly do you feel the basis of gender to be?**
   a) Genitalia, hormones, chromosomes, and brain structure. I mean, duh!
   b) Gender may be a social construct, but it's based on, or influenced by, biology.
   c) Gender is how we act, not what we are: we are all human after all.
   d) It's a lot of heretofore unnamed social, biological, and psychological factors masquerading as a bipolar system.

4. **Which of the following most nearly expresses your ideas about gender and sexual preference?**
   a) Birds and bees do it, so face it, heterosexuality is natural.
   b) I'm straight, but sometimes I lust in my heart for . . . umm . . . other things.
   c) Well, in theory, I imagine I could be attracted to anyone, no matter what gender.
   d) As long as no one's getting hurt, any pleasurable activity between two (or more) consenting adults is just fine.

5. **How have you lived your gendered life so far?**
   a) I've been living my biological destiny.
   b) I may have been living a biologically destined gender, but I don't want to anymore.
   c) I've been experimenting; using bits of some genders here and bits of other genders there.
   d) Let's just say that in the battle of the sexes, I'm a conscientious objector.

6. **Has anything you've read about gender recently made you want to stretch your own gender?**
   a) No, I'm happy the way I am. Always have been, always will be.
   b) Not really, but I might be more lenient when it comes to other people.

Questions

It is common for all of us to experience greater stress or depression about some issue at a few points in our lives, but which young adults are most likely to experience more serious bouts of depression and mental illness? What types of activities are known to mitigate the effects of significant stress, depression, and other mental illness?

Sexuality

As we all make, or made, the transition to young adulthood, sexuality takes on added importance because we need to position ourselves as competent, independent, caring individuals. "How do I fulfill my sexual needs?" and "How does sexuality fit in with my idea of who I am?" are some of the questions we ask when we are deciding whether sexual relations for us will be casual, monogamous, or simply another form of "fun." At this time, gender roles are likely to become more complex and challenging. Also, the impact of AIDS has given rise to more serious caution and changes in sexual behaviors for a majority of young adults.

Heterosexuality The findings of the 1994 Sex in America Survey of sexual patterns (Michael, 1994) included these:

- One-third of Americans have sex more than twice a week.

c) Yes, I might consider that but it's a little intimidating.

d) Yup. Now I'm more determined than ever.

7. **When you see someone on the street whose gender is unclear, how do you react?**

a) Honestly? With some combination of revulsion, pity, and bewilderment.

b) I try to figure out if it's a man or a woman.

c) I mentally do a makeover so the person can pass better as one or the other.

d) I probably notice the person is staring at me, trying to figure out what I am.

8. **Which of the following most nearly matches your definition of the word transgender?**

a) It's a disorder that results in men cutting off their penises.

b) Being born in the wrong body, or having the wrong sex for your gender.

c) Changing from one gender to another, or just looking like you have done that.

d) Transgressing gender, breaking any rule of gender in any way at all.

9. **Who gets to say exactly what gender you are?**

a) It's not up to anyone to say, it's a biological fact, man or woman.

b) Gender may be a biological fact, but biologists are constantly refining their definitions, and that makes me nervous.

c) We are force fed gender by a conspiracy of science, law, and the media. Oliver Stone should make a movie about this.

d) I do.

10. **OK, what is your gender?**

a) Oh, please, obviously I'm a real man/woman.

b) Well, I am a biological man/woman.

c) I was afraid I'd be asked this question. I'm just not sure anymore.

d) Fiddle dee, tomorrow is another gender.

TO CALCULATE YOUR GENDER APTITUDE

Give yourself
5 points for every A answer
3 points for every B
1 point for every C
0 for D

0–15 GENDER FREAK

This was child's play for you wasn't it? But aren't you glad to see it has finally made it into a mainstream text?

16–25 GENDER OUTLAW

You have gone too far to make the climb back up to "real" man or woman.

26–35 GENDER NOVICE

You are not always taken for normal, are you? Is gender stuff kind of scary to you? Maybe it's time to look it in the face.

36–45 WELL GENDERED

This gender as construction stuff sounds weird but the world is changing. Maybe someone you know is exploring gender, keep your eyes open, and you might want to study this stuff a bit.

46–50 YOU ARE CAPTAIN KIRK!

Reprinted with permission from the *Utne Reader* (October 1998). © 1998 Kate Bornstein. First appeared in *Utne Reader*.

- One-third have sex a few times a month.
- One-third have sex a few times a year or do not have sex.
- Married couples have more sex and achieve orgasm more often than others.
- Favorite sexual acts were vaginal sex, watching partner disrobe, and oral sex.

Furthermore, by age 22, 90 percent of young adults have engaged in sex, and a majority of these adults have had multiple partners. Gender differences have all but disappeared with regard to premarital sex, with females having almost as many premarital sexual experiences as males. Also, there are no gender differences in attitudes about homosexuality, masturbation, or sexual satisfac-tion. Promiscuity does not appear to be rampant; the majority of young adults claim that they have had only one partner during the last year. Many of these young adults have changed their sexual behaviors owing to the threat of AIDS. Some of the recent modifications include having casual sex less frequently, having fewer partners, and practicing abstinence.

Gay, Lesbian, and Bisexual Attitudes and Behaviors
Sexual orientation is no longer thought of in terms of either/or—individuals are not locked into heterosexual or homosexual categories. Bisexuality is a case in point, and in recent surveys around 5 percent of males and 2 percent of females describe themselves as homosexual (Hershberger & D'Augelli, 1995; Kurdek, 1998).

Historically homosexuality as been seen as an aberration, and it was only in the mid 1970s that the American Psychiatric Association declassified homosexuality as no longer being considered a disorder. In the United States, only one state—Hawaii—recognizes the legality of marriage between same-sex adults. And more recently, the realization that gender is a social "construction" has led many people to question the rigid gender roles that males and females have traditionally held. The new option "transgendered" is now available for those uncomfortable with the limitations of being a "woman" or a "man." Some reject the notion of a singular gender identity as being able to account for how a person experiences the world and accept a much more fluid conception of gender containing aspects of what have traditionally been called male and female characteristics. The transgender survey in *Human Diversity:* "Gender Aptitude Quiz" addresses this issue in a light way.

How do gay men and lesbians manage in the United States when they are given minority status and the majority insists on defining them? During the 1960s and 1970s, collectives of men and women created new social structures that were characterized as "open communities" and became havens for people who chose not to define themselves by narrow definitions of sexuality. The gay man made a transition from queen, fairy, closet, to clone, hot man, and dancer (Chauncey, 1994). The 1970s portrayed gay men as bar hoppers looking for hot, quick sex. The caring professional who was looking for a stable committed relationship was not seen as a legitimate role for gay men, until recently. If one looks at recent research on lesbian couples, the picture is quite different from the stereotypes of the 1960s and 1970s (Klinger, 1996). Recent survey findings regarding lesbians include the following (Klinger, 1996):

Sexual Orientation Studies indicate that lesbian couples tend to be committed to their relationship and endorse a high degree of equality between partners.

- Sixty to 80 percent of lesbians are in committed relationships.
- Almost 100 percent of lesbian couples are sexually exclusive.
- Most lesbian couples endorse a high degree of equality between partners.

In short, there are many more similarities in human relationships, regardless of sexual preference, than differences.

Question

What is social science research finding out about the changes in sexual attitude and behavior of the current generation of young adults?

Cognitive Development

The varied experiences of adult life pose new challenges and require that we continually refine our reasoning capabilities and problem-solving techniques. Be it in the realm of interpersonal relationships, working, parenting, managing homes, or participating in church or community volunteerism—or even when vacationing—we confront new circumstances, uncertainties, and difficulties that call for decision making and resourceful thought. Consequently, we must learn to identify problems, analyze them by breaking them down into their relevant components, and devise effective coping strategies. The following is a true story of a vacation that challenged the problem-solving and critical thinking skills of two of your authors and demonstrates how well stressful events become etched in our memories. One of the things we learn when things go wrong is what we would do differently the next time the situation arises. We're sure you have your own version of our misadventure:

In 1985 we decide to vacation at a rental condo in sunny Myrtle Beach during Easter break. While packing the rental car, which was supposed to be a three-seat station wagon, we discover there was no third seat and that the car could not accommodate everyone. The rental car dealer says he "thought" there was a third seat when he rented the car to us—not to worry, though, because a passenger van will be there in the morning. When picking up the van, we discover it has about 110,000 miles on it, and there's no time to service it. The embarrassed dealer gives us the van for free. Off we go a day later than planned to start our much-needed vacation.

Driving well into the night in extremely high winds, we arrive at a motel in the Carolinas. The next morning the motel has no power because the high winds knocked out some lines. With four hungry, irritable teenagers in tow, we attempt to get something to eat and discover that the rental van now will drive only in reverse! After driving around the parking lot in reverse in a van that resembles a large brown elephant, we call the dealer to see what he wants us to do with his van. He says he will reimburse us for repairs. It is Saturday, and most garages are closed. Luckily, about three hours later a mechanic arrives, fixes the problem, and we're on our way. Our first night in a beautiful new condo, at 3 A.M., the fire alarm goes off, and all of the pajama-clad residents have to evacuate the building into the parking lot while firemen check the building. Some teenager thought that setting off the alarm was a fun prank! As adults, we are never on "vacation" from utilizing our reasoning and critical thinking skills!

Post-Formal Operations

For Jean Piaget, the *stage of formal operations* constituted the last stage in cognitive development. Piaget depicted adolescence as opening a new horizon in thought. During this period adolescents gain the ability to think about their own mental processes, to imagine multiple possibilities in a situation, and to mentally generate numerous hypothetical outcomes. In brief, adolescence opens to teenagers the prospect of thinking in logical, abstract, and creative ways.

A number of psychologists have speculated about whether a fifth and qualitatively higher level of thought follows formal operations (Demtrious, 1988; Soldz, 1988). Common to the various formulations is the notion that **post-formal operational thought** is characterized by these three features:

- First, adults come to realize that knowledge is not absolute but relativistic. They recognize that there are no such things as facts, pure and simple, but deem facts to be constructed realities—attributes we impute to experience and construe by the activities of the mind. (*There are many times in the world of work when people have different visions about a project and need to learn to collaborate.*)

- Second, adults come to accept the contradictions contained in life and the existence of mutually incompatible systems of knowledge. This understanding is fostered by the adult's expanding social world. In the larger community the adult is confronted by differing viewpoints, contrary people, and incompatible roles. And she or he is constantly required to select a course of action from among a multitude of possibilities. (*There are times when we must respect and do what our older in-laws tell us to do—even if we believe it might not be the best course of action.*)

- Third, because they recognize that contradiction is inherent in life, adults must find some encompassing whole by which to organize their

experience. In other words, adults must integrate or synthesize information, interpreting it as part of a larger totality. (*Sometimes we have to look at the larger picture and realize we need to work to support ourselves and our families—even if we are not satisfied with the atmosphere at work.*)

Here, then, is a working model of post-formal operational thought. Future research will be needed to determine the appropriateness of models that set forth a fifth stage in cognitive development.

Or perhaps future research will show that a Piagetian model has limited usefulness. Critics of Piaget are increasingly challenging the assumptions underlying his theory. For instance, evidence suggests that younger and older adults might merely differ in their cognitive competencies: Younger adults seem to place greater reliance on rational and formal modes of thinking; older adults seem to develop a greater measure of subjectivity in their reasoning, and they place greater reliance on intuition and the social context in which they find themselves (Labouvie-Vief, DeVoe, & Bulka, 1989). Further, more complex cognitive tasks place greater demands on working-memory resources that decline with increased age (Salthouse, 1992a; 1992b). In any event, most psychologists now acknowledge that cognitive development is a lifelong process, a viewpoint that has gained widespread acceptance only in the past 20 years.

Thought and Information Processing

Adult thinking is a complex process. We would be little more than glorified cameras and projectors if information handling were limited to storage and retrieval. Psychologist Robert J. Sternberg (1997) has studied how we think by examining what is involved in information processing. He views **information processing** as the step-by-step mental operations that we use in tackling intellectual tasks. He examines what happens to information from the time we perceive it until the time we act on it. The various stages or components of this process are highlighted by an analogy problem: *Washington* is to *one* as *Lincoln* is to *(a) five, (b) ten, (c) fifteen, (d) fifty.*

In approaching this problem, we first encode the items, identifying each one and retrieving from our long-term memory store any information that might be relevant to its solution. For instance, we might encode for "Washington" such attributes as "president," "depicted on paper currency," and "Revolutionary War leader." Encodings for "Lincoln" might include "president," "depicted on paper currency," and "Civil War leader." Encoding is a critical operation. In this example, our failure to encode either individual as having his portrait on paper currency will preclude our solving the problem.

Next, we must infer the relationship between the first two terms of the analogy: "Washington" and "one."

We might infer that "one" makes reference to Washington's having been the first president or to his being portrayed on the one-dollar bill. Should we make the first linkage and fail to make the second, we will again be stymied in solving the problem.

Then, we must examine the second half of the analogy, which concerns "Lincoln." We must map the higher-order relationship that links "Washington" to "Lincoln." On all three dimensions the men share similarities: both were presidents, both are depicted on currency, and both were war leaders. Should we fail to make the connection that both Washington and Lincoln are portrayed on currency, we will not find the correct answer.

In the next step we must apply the relation that we infer between the first two items ("Washington" and "one") and the third item ("Lincoln") to each of the four alternative answers. Of course, Washington appears on the one-dollar bill and Lincoln on the five-dollar bill. Here we could fail to recognize the relationship because we make a faulty application (we might mistakenly recall Lincoln as appearing on the fifty-dollar bill).

We then attempt to justify our answer. We check our answer for errors of omission or commission. We might recall that Lincoln was the sixteenth president, but if we are uncertain, we might select "fifteen," figuring we are somewhat amiss in our recollection. Finally, we respond with the answer that we conclude is most appropriate.

Sternberg (1998) finds that the best problem solvers are not necessarily those who are quickest at executing each of the above steps. In fact, the best problem solvers

spend more time on "encoding" than poor problem solvers do. Good problem solvers take care to put in place the relevant information that they might need later for solving the problem. Consequently, they have the information that they require in later stages. Thus, expert physicists spend more time encoding a physics problem than beginners do, and they are repaid by their increased likelihood of finding the correct solution.

Cognitive Development in College Students   It is important that college professors—and college students—understand that students reason differently depending on their level of experience and intellectual development. Over the past few decades several social scientists have spent time examining how college students mature intellectually over time. They observed that social interactions with faculty, peers, parents, and other adults especially influenced student's cognitive development from the freshman to the senior year. William Perry (1981, 1968), studied male students at Harvard at two time intervals in 1971 and 1979. Though his theory is criticized because his subjects were all males, others who have conducted similar studies with both genders find there is some validity in his theory, in which he attempted to extend Piaget's theories of cognitive development. Perry theorized that freshmen college students generally use more dualistic thinking, as in labeling things "right or wrong" or "good or bad." When asked for an opinion, they were likely to say to the professor: "You tell us. You're the teacher." Over time and with a variety of experiences in college, they begin to use "multiplicity" in their thought. That is, they begin to recognize that there are diverse perspectives on the same subject, and they become willing to listen to them. As they continue to develop cognitively, they transition to "relativistic" thought. Using this type of thinking, they realize that they—and others—must support and defend their position in some rational way. This happens as students move into their junior and senior years, making more decisions such as declaring a major, planning a career, choosing a religion, firming up relationships, taking a stand in politics, securing jobs, and so forth.

The Developmental Instruction Model proposed by Knefelkamp (1984) aided in operationalizing Perry's model. Their model of college student cognitive development involves four components of challenge and support: structure, diversity, experiential learning, and personalism. Each exists within a continuum, such as more structure to less structure, less diversity to more diversity, low involvement in learning to experiential learning, and moderate levels of personalism to high levels of personalism in instruction. College freshman typically appreciate more structure and social support during their first year, and they are more likely to call home for advice, rely upon others, and learn through traditional

Adult Problem Solving The best problem solvers are more likely to gather relevant information and spend more time encoding to find the best solution.

classroom instruction. This model has implications for instructional methods, because it suggests that freshmen might not be ready to "discuss their views" in a class—they might feel they are supposed to be told what to think, unlike a senior, who is definitely ready for this type of cognitive challenge. As students progress through college, they cognitively transition into making and defending their own decisions, appreciating diverse views, experiencing internships in their field, and interacting with professors and mentors for more personalized instruction. This model suggests that colleges and universities must provide students with resources and materials to be able to support and defend their cognitive growth over time.

Questions

What features characterize maturing adult cognition? In light of the above information about adult thought and problem solving (encoding), should time components for tests such as the SAT and the GRE be modified?

Moral Reasoning

As we saw in Chapter 9, Lawrence Kohlberg identified six stages in the development of moral reasoning and grouped them into three levels:

1. Preconventional (Stages 1 and 2)
2. Conventional (Stages 3 and 4)
3. Postconventional (Stages 5 and 6)

Of particular significance, Kohlberg's cognitive-developmental theory stresses the universal and invariant character of the sequence, which he contends derives from the inherent structuring of thought in the stages described by Piaget. We pointed out that the existence of Kohlberg's sixth stage—a stage characterized by noble ideals of brotherhood and the community good—is debatable (also see Kincheloe & Steinberg, 1993).

Given the "political" quality of many of the features that characterize Kohlberg's stages, it is not surprising that researchers should also have looked for evidence of a linkage between moral reasoning and political attitudes. Overall, individuals with liberal leanings are more likely to reason at the postconventional level, while conservatives are more likely to reason at the conventional level. This finding has led a number of psychologists to investigate the possibility that individual differences in adult moral reasoning reflect differences in their political ideologies. They find that right-wing and politically moderate individuals increase their principled-reasoning scores if they are given the task of responding in a manner that they believe characterizes left-wingers. Such results suggest that variations in adult moral reasoning are more a product of political position than of developmental status. In other words, people seem to have some ability to differentiate among moral reasons and to use them in their own interests. In practice, people often draw upon ideological "scripts" of the political left or right and use these "moral reasons" to advance their political cause (Sparks & Durkin, 1987).

For more than two decades, Carol Gilligan (1982a; Taylor, Gilligan, & Sullivan, 1995) has conducted thoughtful and systematic research involving Kohlberg's framework. She finds that as they move through their young adult years, men and women take somewhat different approaches to the moral dilemmas employed by Kohlberg in his research. Indeed, women tend to score lower than men on Kohlberg's scale of moral development. Gilligan contends that the lower scores result from bias in Kohlberg's approach.

According to Gilligan, men and women have different moral domains. Men define moral problems in terms of right and rules—the "ethic of justice." In contrast, women perceive morality as an obligation to exercise care and to avoid hurt—the "ethic of care." Men deem autonomy and competition to be central to life, so they depict morality as a system of rules for taming aggression and adjudicating rights. Women consider relationships to be central to life, so they portray morality as protecting the integrity of relationships and maintaining human bonding. In sum, whereas men view development as a means of separating from others and achieving independence and autonomy, women view it as a means of integrating oneself within the larger human enterprise.

These two ethical views provide somewhat different bases for finding one's identity and integrating the self. Gilligan calls upon developmental psychologists to recognize that the feminine moral construction is as credible and mature as its masculine counterpart. The full response to Gilligan's proposal is not in yet. Not all researchers support her contention that women and men differ in their orientation for moral reasoning. Some researchers have found only limited support for Gilligan's assertion that women are more attuned to issues of care in moral conflicts and men more attuned to issues of justice (Pratt et al., 1991). Still others stress that the realm of care is not an exclusively female realm, nor are justice and autonomy exclusively in the male realm. Indeed, orientations toward justice and care are frequently complementary (Gilgun, 1995).

Question

What is a major difference between Kohlberg's and Gilligan's views of moral reasoning?

Segue

In this chapter we have examined the physical and cognitive changes that young adults go through upon graduating from high school and embarking on the road to adulthood. It is a time when young adults are beginning to understand that there are age norms and to understand the dos and don'ts of each age. Young adulthood is considered to be a time of increased stressors, including having to make "adult" choices such as whether to drink, smoke, or have sex. The young adult needs to worry about other aspects of healthy development, including regular exercise, proper nutrition, and safe sex. How these changes and choices affect the individual we will see in the next chapter, when we look at the emotional and social dimensions of becoming an adult—most importantly Erikson's idea that becoming intimate with another is the "crisis" of young adulthood.

Summary

Developmental Perspectives

1. People's feelings, attitudes, and beliefs about adulthood are influenced by the relative proportion of individuals who are adults. At the present time the post–World War II baby-boom generation has brought about a rapid expansion in the nation's labor force. Their children have become a surplus of well-educated individuals in keen competition for managerial and professional positions.

2. The contemporary generation of young adults have been given such labels as *Generation X*. Studies indicate this is a cohort interested in video entertainment and less exercise, a college education, wealth and home businesses, stability in relationships (though more are cohabiting), less sex because of the threat of STDs, more accepting of the diversity around them, and more willing to live at home longer to finish college, establish a career, or raise a child as a single parent.

3. For the most part, people in the United States perceive adults of all ages favorably. Age norms may also represent informal expectations about the kinds of roles appropriate for people of various ages. Also, age grading is blurring.

4. People pass through a socially regulated cycle from birth to death just as surely as they pass through the biological cycle.

5. Turning points are times at which individuals change direction in the course of their lives. Some life events are related to social clocks, including entering school, graduating from school, starting to work, marrying, and having children. Other life events are unexpected and individualistic, such as winning the lottery or being the victim of a crime.

Physical Changes and Health

6. The majority of young adults in the U.S. indicate they are in good health.

7. Factors such as socioeconomic status, ethnicity, and gender might not seem obviously connected, but they interrelate to impact on health. People who live in poverty generally experience more negative health outcomes. Men are more likely than women to choose exercise as a leisure activity.

8. The most common mental disorders range from mildly impairing problems like a phobia of heights or enclosed places to incapacitating bouts of depression or chronic problems like schizophrenia. The rate of depression for women generally is twice that for men. Suicide rates for young black men are rising. Black women have the lowest rates of suicide.

9. In the course of our daily lives, most of us experience one or more demands that place physical and emotional pressure on us. We commonly call these experiences "stress." Nontraditional students typically experience more stressors than traditional college students. One buffer against stress is having a social network of friends and family.

10. In the transition to young adulthood, sexuality takes on added significance, because we need to position ourselves as competent, independent, caring individuals. "What is sex for me?" and "How does sexuality fit with my idea of who I am?" are some of the questions we ask when we are deciding whether sexual relations for us will be casual, monogamous, or simply another form of fun. At this time, gender roles are likely to become more complex and challenging. Most people define themselves as heterosexual, homosexual, or bisexual.

Cognitive Development

11. Some psychologists have speculated about whether a fifth, and qualitatively higher, level of thought

follows formal operations. Common to the various formulations is the notion that post-formal operational thought is characterized by three features: accepting that knowledge is not absolute but relativistic; accepting the contradictions contained in life and the existence of mutually incompatible systems of knowledge; and finding an encompassing whole by which to organize experience.

12. College students typically progress in their cognitive development as they proceed through the college years and into graduate school. As they progress, they learn to make important decisions, support and defend their positions, experience internships and fieldwork, and accept that there is a diversity of views on any one issue. Students at all levels generally benefit from personalized instruction.

Moral Reasoning

13. The two main theories of moral reasoning are Kohlberg's and Gilligan's. Kohlberg's is based on a justice orientation, defining moral problems in terms of right and rules—the "ethic of justice." Gilligan argues that Kohlberg's theory reflects male moral reasoning and that women, in contrast, perceive morality as an obligation to exercise care and to avoid hurting others—the "ethic of care."

Key Terms

age grading *(398)*
age norms *(398)*
aging *(398)*
biological aging *(398)*

information processing *(415)*
life events *(400)*
post-formal operational thought *(415)*

social aging *(398)*
social clock *(398)*
social norms *(398)*
transition points *(398)*

Following Up on the Internet

Web sites for this chapter focus on physical changes, cognitive maturation, and moral reasoning in early adulthood. Please access the text web site at http://www.mhhe.com/crandell7 for up-to-date hot-linked Internet addresses for the following organizations, topics, and resources:

Women of the World: Reproduction and Contraception

National Institutes of Health: Adult Health (English or Spanish)

World Wide Web Virtual Library—Social Sciences

Psychology Virtual Library

Chapter 14

Early Adulthood
Emotional and Social Development

Critical Thinking Questions

1. Which would be the more difficult way for you to spend the rest of your life—living with one person you love and not being able to work, or working with stimulating people at a job you love without ever finding someone to love?

2. Do you believe in or have you experienced "love at first sight"?

3. Would you be willing to marry someone your parents chose for you as a life partner, as is the custom in many parts of the world? Why or why not?

4. Which of your personal characteristics do you think would be hardest for your potential partner to ignore?

The process of leaving the parental home during early adulthood has become increasingly complex and variable, with many young people experiencing numerous living arrangements in the course of assuming adult status. Contemporary young men and women are also more open than in the past concerning their sexual orientation and behaviors. American society has granted greater latitude to those individuals who practice a lifestyle that is less constrained by the "traditional" standards of a nuclear family with one man, one woman, and 2.3 children. For most young adults, the period from the late teens until the early forties is a time of establishing intimate relationships, preparing for and building up a position in the work world, and looking forward with hopes and dreams for the future.

Theories of Emotional-Social Development

To love and to work.

Sigmund Freud *(When asked for capacities that are characteristic of a mature person)*

Central to any lifestyle are the bonds we forge with other people. Much of our identity is linked with our relationships with other people—in relatively stable sets of expectations that sociologists term **social relationships.** For example, if someone asks you who you are, you might answer, "I am the daughter (or son) of _____, or "I am the husband (or wife) of _____," or "I am an employee of _____." Two common types of bonds are *expressive ties* and *instrumental ties.* An **expressive tie** is a social link formed when we invest ourselves in and commit ourselves to another person. Many of our needs can be satisfied only in this fashion. Through association with people who are meaningful to us, we gain a sense of security, love, acceptance, companionship, and personal worth. Social interactions that rest on expressive ties are termed **primary relationships.** We view these relationships—with friends, family, and lovers—as ends in themselves, valuable in their own right. Such relationships tend to be personal, intimate, and cohesive. For example, one of the longest lasting primary relationships people normally have is with siblings.

In contrast, an **instrumental tie** is a social link that is formed when we cooperate with another person to achieve a limited goal. At times, this relationship can mean working with people we disagree with, as in the old political saying, "Politics makes strange bedfellows." More commonly, it merely means that we find ourselves integrated in complex networks of diverse people, such as the division of labor extending from the farmers who grow the grain, to the grocers who sell bread, to those who serve us sandwiches. Social interactions that rest on instrumental ties are called **secondary relationships.** We view such relationships as means to ends rather than as ends in their own right. Examples are our casual contacts with the cashier at the supermarket, the clerk in the registrar's office, or a gas station attendant. Secondary relationships are everyday touch-and-go contacts in which individuals need have little or no knowledge of one another. As we progress through our adulthood, our days are filled with social contacts among our primary and instrumental ties.

Question

Are the ties you have with family expressive or instrumental?

Erikson: Psychosocial Stages

Each ten years of a man's life has its own fortunes, its own hopes, its own desires.

Goethe, *Elective Affinities,* 1809

As we saw in Chapters 2 and 13, some social scientists have undertaken the search for regular, sequential stages and transitions in the life cycle. Erik Erikson, who pioneered a theory of the psychosocial stages of development, identified eight life-span stages, three of which apply to adulthood: *early adulthood,* which involves intimacy versus isolation; *middle adulthood,* which involves generativity versus stagnation; and *late adulthood,* which involves integrity versus despair. In this chapter, we examine the early adulthood stage.

The principal developmental task confronting young adults in the stage of **intimacy versus isolation** is to reach out and make connections with other people. Erikson refers to this stage as the first stage "beyond identity," which was the stage associated with adolescence. Individuals must cultivate the ability to enter into and establish close and intimate relationships with others. Should they fail to accomplish this task, they confront the hazards of leading more isolated lives devoid of society-sanctioned, meaningful bonds (e.g., they might join the ranks of young adult gang members, prisoners, the institutionalized, the unemployed, prostitutes, cults, the homeless, the drug-addicted). In young adulthood, should the individual fail to come to terms with the critical developmental task of the previous *identity* stage, the person might temporarily drop out of college or settle for a highly stereotyped interpersonal relationship. Erikson expanded on Freud's succinct dictum "To love and to work," explaining that by love he meant the generosity of intimacy as well as genital pleasure. A general work productiveness should not preoccupy the individual to the extent that there would be a loss in one's right or capacity to be a sexual and a loving being. Isolation is the inability to take chances with one's identity by sharing true intimacy, and such inhibition is often reinforced by fears of the outcomes of intimacy, such as having children.

The experience of loneliness is not reserved for singles. Research studies have found that there is a cultural component as well as a gender component to the experience of loneliness. Personal inadequacies, lack of social contacts, unfulfilling intimate relationships, relocation or significant separation, and social marginality are major factors promoting loneliness (Rokach, 1998). Additionally, Sadler (1978) presents a model of loneliness that many first-generation American immigrants must be experiencing—that of **cultural dislocation,** defined as a feeling of homelessness and alienation from a traditional way of life. Immigrants from South American and

Asian cultures typically value family and extended family for personal support; in contrast North Americans emphasize self-reliance, competitiveness, independence, and autonomy (Jylha & Jokela, 1990). Additionally, immigrants who settle in large metropolitan areas often experience unemployment, fear of crime, social prejudice, and large apartment complex living—all of which promote reluctance to interact or get involved with others (Rokach, 1998). Rokach (1998) found that women across cultures tend to derive most of their self-worth from family and offspring, and if they do not have these kinds of relationships, they can become lonely. However, men often invest heavily in their work.

The psychiatrist George E. Vaillant and his associates (Vaillant & Milofsky, 1980) found support for Erikson's formulations when they followed up a group of 392 white lower-class youth and 94 highly educated men who were first studied in the 1940s as subjects in the Grant Study. They concluded, as Erikson contends, that the post-childhood stages of an individual's life cycle must be passed through sequentially. Failure to master one of Erikson's stages typically precluded mastery of later stages. However, the men varied enormously in the age at which they mastered a given stage. In fact, one man in six was still struggling in his forties with issues characteristic of adolescence, such as "What do I want to be when I grow up?" When the researchers examined subsequent employment patterns, the men who had been least emotionally mature as boys were much more likely to have experienced significant unemployment by the time they had reached their mid forties. Clearly, then, people do not march in lockstep across the life span in developing their identities—optimal psychological functioning is a lifelong challenge (Pulkkinen & Ronka, 1994).

Question

What constitutes a meaningful adult life, according to Erikson?

Levinson: Phases in Adult Male Development

A number of Yale researchers, led by psychologist Daniel Levinson, have also approached adulthood from a stage perspective. Levinson and his colleagues (1978) constructed a framework for defining phases in the lifespan development of adult males. They studied 40 men in their mid thirties to mid forties who were blue- and white-collar workers in industry, business executives, academic biologists, and novelists, and concluded that men go through six periods from their late teens or early twenties to their late forties. Levinson and his associates say that the overriding task throughout a man's adulthood is the creation of a *life structure*. A man must peri-

odically restructure his life by creating a new structure or reappraising an old one. He must formulate goals, work out means to achieve them, modify long-held assumptions, memories, and perceptions regarding himself and the world, and then initiate the appropriate goal-seeking behaviors. *Transition periods* tend to loom within two or three years of, and on either side of, the symbolically significant birthdays—20, 30, 40 and 50. The man and his environment interact to move him developmentally through a series of new levels of life organization. This approach focuses on the underlying set of developmental tasks confronting men rather than on the timing of major life events. Following are summaries of the levels, which are depicted in Figure 14-1.

Leaving the Family   The process begins in the individual's late teens or early twenties when he leaves the family. This phase is a period of transition between his adolescent life, which was centered in the family, and his entry into the adult world. Young men might choose a transitional institution, such as the military or college, to start them on their way, or they might work while continuing to live at home. During this period a roughly equal balance exists between "being in" the family and "moving out." Crossing the family boundary is the major developmental task. He must become less financially dependent, enter new roles and living arrangements, and achieve greater autonomy and responsibility. The period lasts about three to five years.

Getting into the Adult World   This period begins with a man's shifting away from his family of origin. Through adult friendships, sexual relationships, and work experiences, he arrives at an initial definition of himself as an adult. This definition allows him to fashion a temporary life structure that links him to the wider society. During this period men explore and tentatively begin committing themselves to adult roles, responsibilities, and relationships that reflect their evolving set of priorities. A man might lay the groundwork for a career; he might develop one career and then discard it; or he might drift aimlessly, precipitating a crisis at about age 30, when the pressures become strong to achieve more order and stability within his life.

Settling Down   This period usually begins in the early thirties. The man establishes his niche in society, digs in, builds a nest, and makes and pursues longer-range plans and goals. By this time he has often evolved a dream, a vision of his own future. In succeeding years there can be a major shift in life direction, when he revives the dream and experiences a sense of betrayal, disillusionment, or compromise with respect to it. Careers such as those of professional athletes can interfere with satisfying the developmental tasks associated with this phase. Being "on the road" to compete does not allow most professional

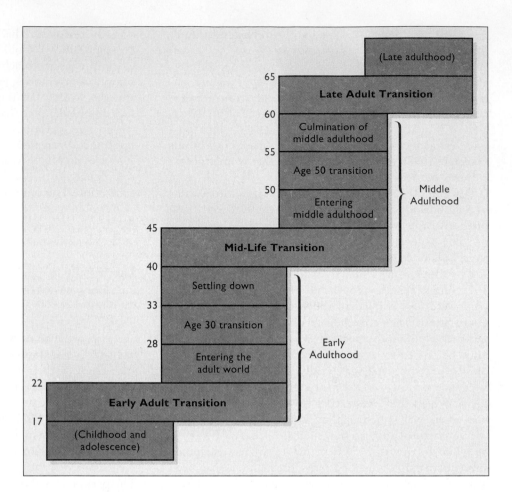

Figure 14-1 Periods in Adult Male and Female Development
Daniel Levinson and colleagues conceive of adult development as a succession of periods requiring the restructuring of critical aspects of a person's assumptions regarding self and the world. Levinson's initial study, *Seasons of a Man's Life*, was done with male subjects, but in 1996 similar findings on adult female development were published in *Seasons of a Woman's Life*.

From: *THE SEASONS OF A MAN'S LIFE* by Daniel J. Levinson. Copyright © 1978 by Daniel J. Levinson. Reprinted by permission of Alfred A. Knopf, Inc.

athletes opportunities to develop meaningful intimate relationships; and their high-profile status makes them targets of close scrutiny by the media. Though they might enjoy the limelight, the status, and the high salaries, they are likely to have fewer friendships and to lack a strong intimate relationship with someone. As we all know, several high-profile athletes have admitted to having hundreds of "one-night" stands and are now experiencing the consequences of such a life.

Becoming One's Own Man   This period tends to occur in the mid to late thirties. It is the high point of early adulthood and the beginning of what lies beyond. A man frequently feels that no matter what he has accomplished so far, he is not sufficiently independent. He might long to get out from under the authority of those over him. He commonly believes that his superiors control too much and delegate too little; he impatiently awaits the time when he will be able to make his own decisions and get the enterprise "really going." If a man has a **mentor**—a teacher, experienced coworker, boss, or the like—he will often give him up now. At this time, men want to be affirmed by society in the roles that they most value. They will try for a crucial promotion or some other form of recognition. Work and family have traditionally been separate spheres of living for most men, and traditionally

many men invested much of their lives in their work role. Because of the American corporate downsizing, takeovers, and massive layoffs of the 1990s, some men have come to a more balanced understanding of their life priorities—which is not surprising given that many have arrived at work to find a termination slip and a cleaned-out desk, even after 25 to 30 years of dedicated service.

Levinson: Stages in a Woman's Life

Although there is a growing interest in adult development, studies dealing with phases in adult female development have lagged behind studies of men (Guisinger & Blatt, 1994; Gilligan, Rogers, & Tolman, 1991). Such research is clearly called for. For example, although Erik Erikson (1968a) says the formation of identity in adolescence is followed by the capacity for intimacy in early adulthood, many women describe the opposite progression, with the sense of identity developing more strongly in midlife (Kahn et al.,1985; Baruch & Barnett, 1983). Women have a longer life span than men, they are increasing their educational attainments, and perhaps 70 to 77 percent of women participation in the labor force made obsolete much of previous research and theory.

Levinson received much criticism for not including women in his original study, so he and his colleagues embarked on another study that resulted in the book *Seasons of a Woman's Life* (1996). These studies confirm that entry into adulthood is similar for men and women, in that both faced the four developmental tasks and the *age-30 transition*. Levinson was surprised to find that while men see themselves tied to a future in terms of their job, women are much more interested in finding ways to combine work and family. None of the professional women they studied found that they could balance the demands of work, family, and their own well-being satisfactorily, feeling that they had sacrificed either career or family in the struggle to maintain both. The women also stated that they had more trouble than men finding someone "special" who would stay with them during their personal and career growth.

The *age-30 transition* marked another developmental difference between women and men, in that women tend to reprioritize their goals around this age (see Figure 14-2). For example, women who began a career early on gravitate toward marriage and family, and those who started off as wives and mothers entered into occupations around the age of 30. Growing numbers of women enter or reenter the labor force, change jobs, undertake new careers, or return to school. Of equal significance are the growing numbers of women who rear their children first

in two-parent, then in one-parent, and then again in two-parent households. Numerous combinations of career, marriage, and children occur with respect to both timing and commitment, and each pattern has different ramifications. Some variations in life arrangements also include returning to the parental home for a period.

New Social Definitions for Women

Until the past few decades in the United States, a woman's life was seen primarily in terms of her reproductive role—bearing and rearing children, menopause, and "the empty nest" being the major events of the woman's adult years. In many countries of the world, this view of women has not changed. Indeed, people commonly equate the female life cycle with the family life cycle. Not surprisingly, the major psychosocial transitions in the lives of contemporary U.S. women now aged 60 and over were more likely to be associated with phases of the family cycle than with chronological age (Moen, Dempster-McClain, & Williams, 1992). But today, with 90 percent of all women working for pay at some point in their lives, employment outside the home is playing an increasingly important role in women's self-esteem and identity. Although the participation of white and nonwhite women in the labor force has increased from 1890 to the present, proportionately more nonwhite women have been employed outside their own homes than white women. Figure 14-3 provides data on the employment of American women.

Family and Work An accumulating body of evidence corroborates Levinson's conclusions that women differ from men in the ways they approach tasks and the outcomes they achieve. In large measure these differences derive from the greater complexity of women's visions for their future and the difficulties they encounter in carrying them out. Unlike men, most women do not report dreams in which careers stand out as the primary component; women are more likely to view a career as insurance against not marrying or a bad marriage and difficult economic circumstances. Instead, most women's dreams contain an image in which they are immersed in a world centered in relationships with others, particularly husbands, children, friends, and colleagues.

Most American women say they have attempted to balance work and family. Even today, many single mothers and married women tend to work only in the economic interest of the family and only after all the

"Nothing personal, but I tend to avoid men who are still under construction."

Figure 14-2 But Are We All Under Construction?

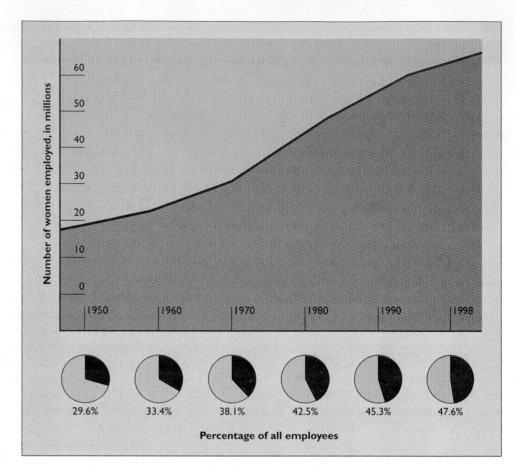

Figure 14-3 Women in the U.S. Workforce. The figures provide data on the number of women who were employed and the percentage of all employees who were women between 1950 and 1998.

Source: New York Times, October 11, 1991 and the U.S. Census Bureau, Current Population Reports, July 27, 1998a.

other needs of its members have been addressed. Furthermore, unlike men, whose likelihood of marrying and having a family correlates with career success, more successful women are less likely to marry and have a family (Fraker, 1984). Clearly, that a choice must be made between work and family is more apparent for women. Consequently, many women struggle to maintain a balance between the demands of their career and the needs of their family.

The struggle for greater equality between the sexes has increased women's roles and workloads (Haslett, Geis, & Carter, 1992). Consequently, some women have difficulty fulfilling all their work and family obligations, giving rise to role conflict and role overload. **Role conflict** ensues when they experience pressures within one role that are incompatible with the pressures that arise within another role, such as the conflicting demands made upon them as a parent, spouse, and paid worker. **Role overload** occurs when there are too many role demands and too little time to fulfill them. Women who encounter these types of role strains, and not all women do, are more likely to experience a diminished sense of well-being and a decrease in work and marital satisfaction (Tiedje et al., 1990). We should emphasize, however, that most women manage their multiple responsibilities quite

well and derive many benefits from participation in the labor force, especially if they receive social support on the job and at home. It is also hypothesized that those women who view themselves as being "feminine and committed to marriage" do not anticipate experiencing the role conflict that many women face (Livingston, Burley, & Springer, 1996). Single-parent families headed by a female, however, face dramatic socioeconomic and psychosocial challenges that dual-earner families do not.

Reentering the Paid Labor Force   Levinson finds that around age 40 men reconsider various of their commitments and often attempt to free themselves from a previously central male mentor. Levinson labeled this the BOOM (Becoming One's Own Man) phenomenon. In contrast, some women enter the world of work, then leave full-time employment entirely while raising children or work part-time, and then reenter full-time employment in their late thirties or early forties. Recent findings suggest that many contemporary women have more mentors, either male or female, to help them navigate to upper administrative positions in their careers (Van Collie, 1998). Moreover, women confront an additional problem: Although men can act as mentors, for women, attempts at cross-gender mentoring are vulnerable to disruption by sexual attractions (Roberts & Newton, 1987). Furthermore, despite current legal and social trends, there is still evidence of job discrimination against women, particularly middle-aged women, who might have stayed at home to raise children or support their husband's career.

Stocktaking Psychiatrist Kathleen M. Mogul (1979) also finds that the "stocktaking," or the reevaluation of the situation that might lead to personal questions and changes that men do in their forties, can occur earlier among women. Childless women in their late thirties and forties often have a "last chance" feeling with regard to motherhood, and options for contemporary women include adoption, in vitro fertilization, or remaining childless. On the other hand, those who became mothers earlier experience a decrease in their absorption of the burdens of child care. And the larger number of middle-aged women who are having children and simultaneously work outside the home also begin to reflect upon their coming life pattern.

Differing Adult Experiences Some psychologists and sociologists believe that the adult experience is different for women than for men (Pugliesi, 1995; Gilligan, 1982a, 1982b). As we noted in Chapter 11, Gilligan questions the traditional psychological assumption that boys and girls both struggle to define a distinct identity for themselves during adolescence. Instead, she contends that girls must struggle to resist the loss of psychological strengths and positive conceptions of themselves that they had possessed in childhood. Therefore, a woman's development is not necessarily a steady progression; women tend to recover in adulthood the confidence, assertiveness, and positive sense of self that Western society pressured them to compromise during their adolescence (Gilligan, 1982a, 1982b).

Part of the difficulty, Gilligan says, is that women often find it difficult to commit themselves to competitive success because they are socialized toward the achievement of cooperation, mutuality, and consensus. Many women focus on preserving rather than using relationships. It was otherwise for the men interviewed by Levinson, for whom "friendship was largely noticeable by its absence," and work typically fostered distance between self and others. Indeed, life in contemporary bureaucracies and corporations frequently rewards those who relate to others not as persons but as objects to be manipulated to get ahead. Significantly, the ten most popular sports for men are mostly competitive activities, such as softball, basketball, billiards, and pool, while women's top ten tend to be noncompetitive sports, such as walking, swimming, aerobics, running, hiking, and calisthenics.

According to Deborah Tannen (1994), Western nations socialize the two genders differently (also see Chapter 7). Males often gather in groups that are hierarchical and teach boys how to dominate and jockey for the spotlight, often by versing boys in displays of ridicule and putdowns. In contrast, female groups are structured chiefly around pairs of good friends who share secrets and forge intimacy—basically the same social pattern we see both girls and boys engage in from the preschool years. In large institutional settings such as companies, universities, hospitals, and government agencies, women tend to be consensus builders. So when found in positions of authority, they are inclined to ask others for their opinions. Tannen says that men often misinterpret this behavior as a show of indecisiveness. Moreover, women tend to hesitate to call attention to their accomplishments or seek recognition. Yet Tannen also challenges the notions that women are more indirect than men and that tentativeness reflects low confidence. She notes a parallel in Japanese culture, where it is deemed boorish for a higher-status person to be direct or to be singled out for praise and recognition over others.

Yet it is easy to overstate the differences between women and men (Yoder & Kahn, 1993). There is neither the "normative (or interchangeable) woman" nor the "normative (or interchangeable) man." Gender is intertwined with race, class, sexual orientation, and countless other variables of human identity (Riger, 1992; Spelman, 1988). Indeed, women and men are more similar than different, and most of their apparent differences are culturally and socially produced.

> **Question**
>
> *What are some of the differing career experiences for contemporary women?*

A Critique of the Stage Approach

Any stage theory of adult development suggests basic principles for identifying orderly changes that occur in people's lives as well as individual variations from these broad tendencies. However, one drawback of earlier investigations of stage theory was that the subjects were predominantly male, white, and upper-middle class and were born before and during the Great Depression of the 1930s. What held for these Depression Era males might not hold for today's 40-year-old men and women who were born in the more optimistic years following World War II or those twenty-somethings entering young adulthood today.

Furthermore, even chance events play a large part in shaping adult lives. American careers and marriages often result from the happenstance of meeting the right, or wrong, person at the right, or wrong, time. Coming of age at a certain point in time and experiencing certain decisive economic, social, political, or military events has a profound impact on people's lives. But even though the events could be random, their consequences are not (Cooney & Hogan, 1991). For example, in non-Western cultures, such as India, China, Japan, and countries of the Middle East and Africa, many marriages are arranged by families to remove the element of chance. And many psychologists and sociologists reject the notion that one must resolve certain developmental tasks in one stage before going on to the next (Rosenfeld & Stark, 1987), pointing out that critical transitions—"passages" or "turning points"—need not be characterized by "crisis," stress, and turmoil. We actively welcome and embrace some new roles. Consider, for instance, the excitement associated with one's first "real" job, serious love, or first child.

For his part, Levinson acknowledges that a life-cycle theory does not mean that adults, any more than children, march in lockstep through a series of stages. He recognizes that the pace and degree of change in a person's life are influenced by personality and environmental factors (war, a death in the family, poor health, a sudden windfall). Hence, Levinson does not deny that very wide variations occur among people in any one life period. He uses the analogy of fingerprints. If we have a theory of fingerprints, we have a basis for order, in that we can identify individuals because we know the basic principles around which fingerprints vary. Let's now turn to the variety of ways young adults resolve the "crisis" of intimacy versus isolation.

Question

What are some drawbacks of Levinson's stage theories regarding women and men?

Establishing Intimacy in Relationships

One can live magnificently in this world, if one knows how to work and how to love.

Count Leo Tolstoy, 1856

As Tolstoy suggests in the above quotation, love and work provide the central themes of adult life. Both love and work place us in a complex web of relationships with others. Indeed, we can experience our humanness only within our relationships with other people. Of equal significance, our humanness must also be sustained through such relationships, and fairly steadily so. Like other periods of human life, young adulthood, from the late teens until the late thirties or early forties, can be understood only within the social context in which it occurs. In addition to relationships in our immediate family, our earliest social relationships are those we have with friends.

Friendships

What is a friend? One old saying goes, "A friend: One soul, two bodies." Think of your friends and see if the following descriptors are true for you—

- You like to spend time with your friends.
- You accept your friends for who they are, and you are not overly concerned to change them.
- You trust your friends.
- You respect your friends.
- You would help your friends and would expect your friends to help you.
- You can confide in your friends.
- You can let down your guard with friends because you do not feel vulnerable with them.

Our friends become our major source of socializing and support during our adult years. We tend to want to spend our free time with those who are experiencing many of the same life events as we are. Women who have babies and young children tend to develop friendships with other women who have babies and young children, often at work or in the neighborhood. Single friends might begin to feel "left out" and will gravitate toward finding new friends who are single and share common interests. Men tend to develop friendships within the spheres of work and recreation, and often the topics of their conversations revolve around such events. And with the growing use of the Internet at work and at home, more studies are focusing on the isolation of our young adult "technocrat" generation, who develop "friendships" in chat rooms in cyberspace rather than across the hall in an apartment building. Recent findings indicate that greater use of the Internet leads to declines in communication with family, a smaller social circle, and increased risk of depression and loneliness (Kraut et al., 1998). Friends normally provide us with much-needed social support when life gets us down.

Love

It is not easy to clearly define what being in love is. We have, however, made the observation that most people find it easy to state whether or not they are or were in love.

Jürg Willi, M.D., Chair of the Psychiatric Outpatient Department, University Hospital, Zurich, Switzerland

As mentioned earlier, the concept of romantic love is not universal, as is evident in the practice of arranging marriages between people who have never even met before or only know each other superficially. Yet many Westerners describe themselves as being "in love" when they marry, so what do they mean? Why are certain marriages and relationships happier and longer lasting than others? Is being in love significant for happiness and satisfaction in marriage?

Traditionally, love has been divided into what are known as *romantic* and *companionate* types of love. **Romantic love** is what we think typically of when we say we are "in love" with someone. **Companionate love** is usually understood as the kind of love you have for a very close friend. The latter is usually manifested in the words, "I love you . . . as a friend, but nothing more." Even though most people can differentiate between the two types of love, psychologists have tried to measure companionate love using a variety of measurements.

Sternberg's Triangular Theory of Love   Robert J. Sternberg (Aron & Westbay, 1996; Sternberg, 1988b) has proposed that companionate love consists of two other types of love: intimacy and commitment. According to his **triangular theory of love,** love is made up of these three elements:

- Passion (sexual as well as physical attraction to someone)
- Intimacy (having a close, warm and caring relationship)
- Commitment (intent and ability to maintain the relationship over an extended period of time and under adverse conditions)

A relationship that does not have all three components of a complete, consummate love—*passion, intimacy,* and *commitment*—is, says Sternberg, an emotional attachment of one of these seven kinds:

- Infatuation—only passion is evoked
- Fatuous love—a relationship that has passion and commitment, but not intimacy
- Companionate love—a relationship having intimacy and commitment, without passion
- Romantic love—a relationship with intimacy and passion, but lacking commitment
- Nonlove—none of the three components are present
- Liking someone—intimacy is present, but passion and commitment are absent
- Empty love—a relationship consisting only of commitment

However, when all three aspects of his triangular model exist in a relationship, Sternberg calls the emotional bond **consummate love** (Sternberg & Hojjat, 1997). This theory is best thought of as an attempt to explain the real complexities involved in initiating and maintain-

ing a meaningful relationship with another person. Research indicates that lovers' definitions and communications of commitment, intimacy, and passion remain stable among different age cohorts (Reeder, 1996).

Significance of Romantic Love   In a study on the significance of romantic love for relationship quality and duration, Willi (1997) surveyed more than 600 adults in Switzerland and Austria, ranging in age from 18 to 82. Most were married, some were single, some were divorced, and a few were widowed. Based on the following definition of romantic love, most indicated they had fallen in love 2 to 5 times, some indicated 6 to 15 times; 2 percent indicated they had never been in love; and 1 percent indicated they had been in love more than 16 times.

> Being in love does not simply mean a fleeting or simplistic feeling, but something that for a longer period of time leads to an intensive, erotic attraction and inner fulfillment through the idealization of the relationship with the partner. (Willi, 1997, p. 172)

In this study, men did not fall in love more frequently than women, and being in love did not lead more frequently to a relationship for men (contrary to popular thought). An interesting finding was that for one-third of the sample, a relationship developed with one's great love but did not lead to living together. The married in this study indicated most frequently that they had lived or continue living with their great love (62 percent). For 13.5 percent, the greatest love of their life did not lead to a relationship, mainly because their love was not reciprocated. Those who were married to their great love described themselves as significantly happier than other married respondents did, and their divorce rate was the lowest (6 percent). On a scale of 1 to 5 (1 = not happy, 5 = very happy), those married without children rated themselves significantly happier (4.4) than did those married with children (3.9) (Willi, 1997).

The greatest satisfaction was reported for all groups in the areas of sexual fidelity as well as in the security of the partnership. The least amount of satisfaction overall for marrieds was reported in the areas of tenderness, sex life, and conversation. Singles with stable partnerships and those who were divorced and with new partners were more satisfied with their communication than were marrieds. Overall, 76 percent of the married respondents characterized their partnership as happy to very happy. There were no gender differences in frequency and intensity of being in love, though women were more likely to admit that they had been more intensely in love with someone else than they were with their current partner.

In sum, being in love seems to be a special relationship that clearly distinguishes itself from other kinds of relationships, but it does not necessarily lead to the

Table 14-1	Subjects Reporting Love at "First Sight" or Love Building Up Gradually Over Time

N = 605

Love at first sight or first day (13 percent described a "lightening like" love)	25%
Within first 8 weeks	45
Not until later	30

Sex and Human Life by Penegelley © 1978, Reprinted by permission of Prentice-Hall, Inc., Upper Saddle River, NJ.

relationship that one might expect. The findings indicate that being in love on the first day or at "first sight" leads to a stable partnership just as often as does love that builds up gradually over time. Also, one does not necessarily need to have married one's great love in order to be happy or satisfied in a relationship, though marrying one's great love is associated with marital duration.

Question
What are the characteristics of romantic love?

Diversity in Lifestyle Options

People in modern complex societies typically enjoy some options in selecting and changing their lifestyles. A **lifestyle**—the overall pattern of living whereby we attempt to meet our biological, social, and emotional needs—provides the context in which we come to terms with many of the issues discussed in the previous chapter. More particularly, lifestyle affords the framework by which we work out the issues of *intimacy versus isolation* that Erik Erikson described. **Intimacy** involves our ability to experience a trusting, supportive, and tender relationship with another person. It implies a capacity for mutual empathy and for both giving and receiving pleasure within an intimate context. Comfort and companionship are among the ultimate rewards to be found in a close relationship.

A striking aspect of American society over the past 30 years has been the rapid expansion in alternative lifestyles. Much of the turmoil of the late 1960s revolved around living arrangements, including communal living and cohabitation. From the various liberation movements (African American, Hispanic, Native American, women's, gay, lesbian, feminist, and youth), there has come a broader acceptance of pluralistic standards for judging behavior. Individuals are now permitted greater latitude in tailoring for themselves a lifestyle that is less constrained by traditional standards of what a "respectable" person should be like.

Leaving Home

Leaving home is a major step in the transition to adulthood. Prior to this transition, the two generations typically form a single family. For many decades, marriage was the major reason for leaving home—young people left their parent's home to form a new family, signaling their attainment of adulthood. But over the past two or three decades, the process of leaving the parental home has become increasingly complex, with many young people experiencing numerous living arrangements in the course of assuming adult status (Goldscheider & Goldscheider, 1993a).

More than four out of five young adults leave home between the ages 15 and 23, at least for a short time period. Overall, in the United States, Germany, Denmark, Australia, and Britain, contemporary young people are leaving home later than earlier cohorts did. And when they do leave home, they are more likely to live independently and less likely to marry, with women leaving home at a younger age than men to both destinations (Buck & Scott, 1993). Of U.S. men and women, the percentage who spent at least part of their twenties living at home was roughly equal—about 40 percent.

Most young Americans do not leave the parental home until around age 18. About this age, however, nonfamily living occurs with greater frequency. Living in college dormitories and military barracks spikes sharply around age 18, followed by a sharp upturn in housemate living around ages 19 and 20. Even so, the pathways out of the parental home are quite varied. Overall, young adults fan out in a good many directions, with no one pattern of nonfamily living being dominant. Other common arrangements include dormitory living, housemate arrangements, cohabitation, marriage, and living alone. Significantly, despite these changes, the parental home remains the primary residence for most young people aged 15 to 23, accounting for about two-thirds of the total time (Thornton, Young-DeMarco, & Goldscheider, 1993).

In the contemporary United States, young adults can leave home whenever they or their parents desire. Because separate residences are usually more expensive than coresidence, parents often have a strong voice in the matter because they can subsidize new housing arrangements or they can withhold financial assistance. In most families the children are "ahead" of their parents in *expecting* to leave home. Consequently, parents can and do use their resources in influencing their children's nest leaving, either to forestall departures that are "too early" or to expedite those that are "too late." The timing of marriage typically rests on the decision of the young adult, but leaving the parental home for a nonmarriage situation usually involves a joint decision of child and parents.

Two factors have contributed to making living arrangements negotiable between maturing children and their parents. First, the growth of premarital residential independence has been spurred by an increased emphasis on autonomy and individualism that has led some young people to deem the various costs of leaving home well "worth it." Second, young people are marrying at later ages (between 1960 and 1998, the median age at first marriage rose from 22.8 to 26.8 years for men and from 20.3 to 25 years for women). The negotiations between generations can be problematic because, given the rapid social and economic change of recent years, the experiences of parents and their children are likely to be quite different (Goldscheider & Goldscheider, 1993a, 1993b). Parents might not understand the current trend of being an undergraduate for five or more years interrupted by, or followed immediately by, a trip to Europe. Significantly, there are ethnic and religious variations in leaving home. Hispanic Americans are much more likely than non-Hispanic whites to live at home until marriage. The same pattern prevails among Asian American students. Protestant fundamentalists and students who attended Catholic schools are also more likely to live at home until they marry (Goldscheider & Goldscheider, 1993a).

Some sociologists believe that the transition to adulthood is particularly problematic for today's young people. They argue that the amount and duration of parental support have been increased by such factors as postponed careers, recurrent recessions, low beginning salaries, rising housing costs, high divorce rates, high levels of nonmarital childbearing, and damaged lives resulting from drug abuse (Amato, Rezac, & Booth, 1995; Goldscheider & Goldscheider, 1993a). These factors complicate the many transitions that often are associated with young adulthood. Between the ages of 18 and 30, individuals have traditionally married, begun families, and launched careers. Indeed, rates of childbearing, first marriage, divorce, remarriage, and relocating for family reasons are higher during this period than at any other time in life (Rindfuss, 1991). Moreover, the early years of a career are frequently unstable, requiring job changes and moving for job-related reasons. Further complicating matters, today's generations have confronted a gender revolution that has confused many roles and behaviors. Given these circumstances, significant numbers of young adults remain dependent upon their parents in many ways well into their late twenties (Cooney & Uhlenberg, 1992).

Living at Home

In U.S. and English textile communities in the late 1800s and early 1900s, different generations often resided in the same household, providing a good deal of assistance to one another (Hareven, 1987). This extended family arrangement is still common in many Asian American, Hispanic American, African American, and Russian American homes (Rokach, 1998). These societal norms began to change in mid-twentieth century, such that U.S. youth were encouraged to leave home and make their own way in the world. Indeed, two decades ago it was not deemed acceptable to live at home after one reached 20 years of age. The goal of raising independent children was grounded historically in the expanding employment opportunities that prevailed in the United States for much of the twentieth century (Schnaiberg & Goldenberg, 1975).

Nowadays, many young people remain in the parental home or return home when circumstances become difficult. But economic pressures need not be severe: Some youth merely wish to save the rent money and instead spend it on a car or save it to purchase a house. Among the middle classes, comfort is another reason to live in the parental nest. The young people can support themselves, but not in their own home with a comparable standard of living. So instead they borrow a slice of their parent's prosperity. In addition, some young people, particularly men, get another advantage— "maid service" from their mothers. Women are less likely to be doted on by their mothers, and both parents are likely to keep a "shorter leash" on their daughters than on their sons, if they live at the parents' home. Even so, contemporary young adults experience greater equality while living with parents than previous generations did.

Living with parents can be anywhere from highly successful to disastrous (Aquilino & Supple, 1991). Those who are pleased with the arrangement say that it affords them the benefits historically available in extended family living. There is a sense of warmth, closeness, and emotional support at a time of widespread personal alienation. But the most common complaint voiced by members of both generations concerns the loss of privacy. Couples report that they feel uncomfortable fighting in front of family members. Young adults, especially the unmarried, complain that parents cramp their sex lives or their music playing, treat them like children, and reduce their independence. Parents often grumble that their peace and quiet is disturbed, the phone rings at odd hours, they lie awake at night worrying and listening for the adult child to return home, meals are rarely eaten together because of conflicting schedules, and too much of the burden of baby-sitting falls on them. Higher expenses might compel parents to relinquish long-awaited vacations, and a need for space means they must postpone a move to a smaller, less expensive retirement home.

Usually, the happiest refilled nests are those with ample space and open, trusting communication. Those that are most difficult involve grown children who lack resilience and maturity and who have drug or mental health problems. Parents might treat such a 28-year-old like a 15-year-old, and the 28-year-old behaves like one. Family therapists express concern that those who stay at home do not have opportunities to fully develop their sense of individuality. Staying at home tends to aggravate tendencies toward excessive protectiveness in parents and toward a lack of self-confidence in youth. The resulting tensions lead some families to seek professional counseling.

Staying Single

Census data reveal that single status among both men and women under 35 years of age has sharply increased in recent years, to 9 percent of the adult population (U.S. Bureau of the Census, 1998a). This increase has resulted in part from the tendency of young people to postpone marriage; just 60 percent of Americans over age 18 are married, down from a peak of 74 percent in 1960. Figure 14-4 shows rates of women and men living alone in the United States. Many men and women have simply postponed marriage; the changes in data suggest that a growing proportion of Americans will never marry at all. The Census Bureau (1998a) recently raised its estimate of the number of Americans who will never marry from 5 percent to 10 percent. Even so, the population remaining single today is smaller than it was at the turn of the century, when fully 42 percent of all U.S. adult men and 33 percent of adult women were never married. This same societal phenomenon is occurring in Japan as well, as a broad spectrum of college-educated adult women ("thirty-somethings") are spending time pursuing a career before settling down to marry (Butler, 1998). In the 25 to 29 range, 48 percent of Japanese women were not married in 1995, compared to 18 percent in 1970. Roughly 70 percent of unmarried Japanese women in their thirties live at home with their parents. Japanese so-

ciety expects married women to stay at home full-time to care for husband and family (Butler, 1998).

According to the 1990 census, nearly a quarter of the 94 million U.S. households consisted of one person, up from 17 percent in 1970. However, the singles population is not a monolithic group. Half the single men, but only 23 percent of the single women, have never married. And more than two-thirds of adult singles live with a friend, relative, or "spouse equivalent." Furthermore, 40 percent are age 65 and older. There are nearly twice as many single men as single women under age 35, and almost three times as many single women as men who are aged 65 or older. For instance, half of the women living alone, but only 14 percent of the men, are widowed. Although there are fewer unmarried African American men than unmarried African American women in all age groups, young unmarried white and Hispanic men outnumber unmarried women of the same groups. A variety of factors have fueled the increase in nonfamily households: the deferral of marriage among young adults, a high rate of divorce, and the ability of the elderly to maintain their own homes alone. Singlehood is also a reclaimable state. A person may be single, then choose to marry or cohabit, and perhaps later divorce and become single again (U.S. Department of Commerce, 1997).

Until a few decades ago, social stigma was attached to the terms *spinster* and *bachelor,* and remaining single was actively discouraged. Over the past generation, the notion that individuals must marry if they are to achieve maximum happiness and well-being has been increasingly questioned (Blakeslee, 1991a). Many Americans no longer view "singlehood" as a residual category for the unchosen and lonely. Even so, in Western culture the nuclear family composed of a husband, a wife, and their offspring continues to be the measure or standard against which other family forms and lifestyles are judged (Ganong, Coleman, & Mapes, 1990).

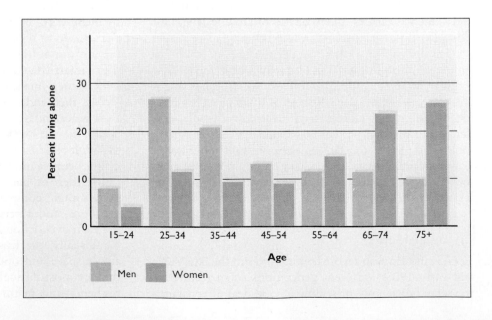

Figure 14-4 Adults of All Ages Living Alone Adult men are more likely than women to live alone, but women generally live longer than do men.

Source: U.S. Bureau of the Census, *Current Population Reports,* Series P-20, No. 450. 1990.

Variable Paths for Leaving Home The pathways from the parental home are varied. Young adults can take up residence by themselves, as has this young man, or they can live in a college dormitory or military barracks, have a housemate, cohabit, or marry. Many of today's young adults are remaining in the parental home until their late twenties or early thirties or returning home after a divorce or separation.

Single individuals (both nevermarried and divorced) find that as their numbers have grown, singles communities have arisen in most metropolitan areas. Single adults can move into a singles apartment complex, go to a singles bar, take a singles trip, join a singles consciousness-raising group, and so on. If they choose, they can lead an active sex life without acquiring an unwanted mate, child, or reputation. And homeownership is substantially increasing among single people (48 percent of single women own homes, compared with 36 percent of single men) (U.S. Bureau of the Census, 1998a). Although staying single can offer greater freedom and independence than married life does, it can also mean greater loneliness. The impersonal nature of singles bars has led them to be labeled "meat racks," "body works," and other nicknames that signify a sexual marketplace. Still, many singles remain wary about marriage and look to their work and other interests for their primary life satisfactions.

Cohabiting

The number of adults who are sharing living quarters with an unrelated adult has increased substantially over the past 35 years, to 4.1 million couples in 1998 (marriage is distinguished from other types of intimate relationships by its state-sanctioned, and often church-sanctioned, status) (U.S. Department of Commerce, 1998). "Cohabitation is more common among Blacks than among whites and is most prevalent among those with low levels of education" (Brown & Booth, 1996). It is easier for unmarried couples to live together today because of more permissive codes of morality, whereas in the recent past cohabitation was looked upon as morally wrong. Now university officials, landlords, and other

agents of the establishment tend to ignore the matter. Its increase has been fueled by lenient contemporary attitudes toward premarital sex and by fears, sparked by high divorce rates, that marriages fail.

The rise in cohabitation is associated with the decline in marriage and seems most attractive to young adults and to those who have been divorced. Nearly 50 percent of Americans in their twenties and thirties have cohabited (Brown & Booth, 1996). The high proportion of married couples who live together prior to marriage suggests that premarital cohabitation might become institutionalized as a new step between dating and marriage (Brown & Booth, 1996; Manning, 1993). Cohabiting between marriages is even more prevalent: 58 percent of recently remarried couples have cohabited, though plans to remarry are less common among this group of cohabitants (Wu & Balakrishnan, 1994).

Although the media have at times labeled cohabiters "unmarried marrieds" and their relationships "trial marriages," the couples usually do not see themselves in this category. College students typically view cohabitation as part of the courtship process rather than as a long-term alternative to marriage. Nor does research support the notion that cohabitation before marriage is associated with later marital success. It seems that the people who flout convention by cohabiting also tend to flout traditional conventions regarding marital behavior, have a lower commitment to marriage as an institution, and are more likely to disregard the stigma of divorce (Brown & Booth, 1996; Thomson & Colella, 1992).

In many cases living together as an alternative to marriage is not radically different from marriage (Manning, 1995; Wu, 1995). In fact, as with many marriages, biological children have been found to exert a stabilizing influence on a cohabiting relationship (Brown & Booth,

1996). The partners typically fall into traditional gender roles and engage in many of the same activities that married couples do. Like married men, cohabiting men are more likely to initiate sexual activity, make most of the spending decisions, and do far less of the housework than their working women partners (Sullivan, 1997). Moreover, cohabiting couples typically encounter many of the same sorts of problems found among married couples. Surprisingly, the incidence of interpersonal disagreements, fights, and violence is higher among cohabiting than among married couples (Brown & Booth, 1996). Commonly, the cohabiting couples view themselves as less securely anchored, so they feel more tentative about their capacity to endure difficult periods, and they are less likely than married couples to practice fidelity (Steinhauer, 1995). The average cohabitation lasts about two years and ends either through dissolution or through marriage.

Many cohabitors intend to marry, but of those who have no plans to marry, typically at least one of the partners is poor marriage material. That is, typically they exist on low incomes, receive welfare, or have resident children. Previously married cohabitors are also less likely to report plans to marry than are their never-married counterparts (Brown & Booth, 1996).

Separating is not as easy for an unmarried couple as popular belief has it. The emotional trauma can be every bit as severe as among married couples undergoing divorce. And in some cases there are legal complications associated with apartment leases and jointly owned property. Overall, the dissolution of cohabitation resembles divorce in that the partners experience similar processes of disengagement, emotional distress, and adjustment (Cole, 1977).

> **Question**
>
> *What differentiates cohabiting couples from married couples?*

Living as a Lesbian or Gay Couple

Homosexuality is assuredly no advantage but it is nothing to be ashamed of, no vice, [no] degradation, it cannot be classified as an illness. . . . Many highly respectable individuals of ancient and modern times have been homosexuals, several of the greatest men among them (Plato, Michelangelo, Leonardo da Vinci, etc.). It is a great injustice to persecute homosexuality as a crime, and cruelty too.

Sigmund Freud,
Letter to the mother of a homosexual, April 9, 1935

As mentioned in Chapter 13, *sexual orientation* refers to whether an individual is more strongly aroused sexually by members of his or her own sex (homosexual), the opposite sex (heterosexual), or both sexes (bisexual). Most people assume that there are two kinds of people, whom they label "heterosexual" and "homosexual." In reality, however, a more accurate view is that a *heterosexual orientation* and a *homosexual orientation* are on a continuum (in other words, one could be "more heterosexual" or "more homosexual"). Some individuals show varying degrees of orientation, including a *bisexual orientation* (Weinberg, Williams, & Pryor, 1994). In brief, human sexuality is quite versatile.

Because there are so many gradations in sexual orientation and practice, some experts on sexuality take the position that there are various sexual behaviors—female-female (lesbian), male-male (gay), and male-female (heterosexual)—not heterosexual or homosexual *individuals* per se (Kitzinger & Wilkinson, 1995). Complicating matters, some evidence suggests that some women seem to broaden their sexual experience as they get older, so that some heterosexual women become lesbian and some lesbians become heterosexual (Kitzinger & Wilkinson, 1995).

Additionally, it is important to distinguish between "orientation" and "behavior"—between erotic attraction and what a person actually does. A gay, lesbian, bisexual, or heterosexual person might or might not elect to engage in sexual behavior (about 1 in 10 American men and women report that they have not had sex with anyone in the past five years). And genital sexuality need not simply and mechanically equate with affection and loyalty. For instance, many militantly straight men reserve their deepest feelings for one or more male-bonded groups, such as their football team or military unit.

How common are homosexual behaviors? In his early data from the late 1940s, Alfred Kinsey found that among single males 20 to 35 years old, those who had exclusively homosexual experience ranged in various samples from 3 to 16 percent, and among females that age, from 1 to 3 percent (Kinsey et al., 1953). However, surveys by the National Opinion Research Center (NORC) in 1970 and 1988 suggest that the percentage of U.S. men who have had at least one homosexual experience is lower than Kinsey's estimates. Researchers with the NORC point out, though, that many men have been reluctant to admit same-gender sexual experiences, given the history of discrimination and oppression gays have encountered (Fay et al., 1989). In all studies, however, the proportion of persons who are exclusively homosexual throughout their entire lives is quite small (perhaps from 1 to 2 percent of adults) (Barringer, 1993b; Crispell, 1993b).

In the Western world, individuals practicing homosexual behavior have experienced a history of oppression by a culture that has long regarded the behavior as deviant. Indeed, until 1973, the American Psychiatric Association included homosexuality in its manual of pathological behaviors. As a result, many people who were erotically attracted to the same sex remained "in the closet," hiding their sexual orientation. In 1975, the American Psychiatric Association reversed its position and no longer lists homosexuality as a disorder. Today the issue of whether or not sexual orientation is a matter of individual *choice* occupies center stage and is hotly debated by psychiatrists, clinical psychologists, biologists, and others. The public is also divided on the issue, particularly with the issue of same-sex marriages and adoption by gay men or lesbians. In 1994, pollsters asked Americans, "Are homosexual relationships between consenting adults morally wrong?" Some 53 percent responded yes, and 41 percent responded no (Henry, 1994).

Lesbians and gay men are a varied group (Kurdek, 1995). They come from all racial and ethnic backgrounds, work in all occupational fields, have varied political outlooks, and have varied religious affiliations. Just as there is no such thing as a "heterosexual lifestyle," there is no such thing as a "lesbian lifestyle" or "gay lifestyle." Some lesbians and gay men "pass" for heterosexuals, are married, have children, and in most respects seem indistinguishable from the larger population. Just as some heterosexual adults do, some seek transitory relationships, whereas others establish relatively durable relationships. And still others enter homosexual unions repeatedly, each time hoping they have found an ideal lover. Lesbians tend to form more lasting ties than gay men do, and they are less often detected and harassed—though over the past few years the news media have sensationalized child custody cases involving the "fitness" of lesbian mothers.

More recently, researchers examined the social and psychological adjustment of a sample of nearly a thousand gay men and lesbians living in the San Francisco Bay area. On the whole, they found that adults practicing homosexual behavior resemble adults practicing heterosexual behavior in their reports about their physical health and their feelings of happiness or unhappiness. Overall, the quality of relationships has been the same for heterosexual and homosexual couples across time when one looks at intimacy, autonomy, problem solving, commitment, and equality (Kurdek, 1998). Fidelity typically is defined not by sexual behavior but rather by the partners' emotional commitment to each other. For many couples the passion of the sexual encounter dwindles rapidly after two or three years, and outside sexual activity increases. Gay male couples are more likely to break up over other incompatibility issues, such as how money is spent, than over the issue of sexual faithfulness. Household duties tend to be sorted out according to partners' skills and preferences and seldom on the basis of stereotyped roles of "husband" and "wife" (Kurdek, 1993).

In recent years there has occurred a marked decrease in casual sex among individuals practicing either homosexual or heterosexual behavior. The main reason can be traced to fear of AIDS and herpes. The AIDS epidemic has created anxiety and caution within gay male communities throughout the United States because many victims of the lethal disease have been gay men. Lesbians have a low incidence of any kind of sexually transmitted disease and are not among the high-risk groups for AIDS. This disease renders the body's immune system ineffective, leaving the victim susceptible to cancers, various infections, and eventually to an early death—though regimens of new medicines are, in general, prolonging the lives of AIDS patients. Gay communities worldwide have witnessed the untimely death of thousands of men over the past decade but have responded by practicing safer sex methods to prevent the transmission of HIV. However, heterosexuals of all ages are mistaken if they think this disease affects only those in the homosexual community.

By virtue of some continuing hostility from the larger community, individuals practicing homosexual behavior have often had to live double lives, "gay" at home and "straight" on the job. But shifts in public attitudes and gay rights laws have prompted many to live openly, particularly in large cities such as San Francisco, Los Angeles, and New York. The same does not hold true in many smaller communities. The brutal murder of a young gay college student in Wyoming in 1998 prompted a public outcry by a great many people—a national level of condemnation of crimes against gays and lesbians that we have not witnessed in the past. Furthermore, gay and lesbian organizations and political caucuses have been vigorously championing gay and lesbian rights measures, particularly the right to be covered under a partner's health insurance, the right to adopt, and the right to marry. An extension of this effort is the creation of gay and lesbian clubs on many college campuses, in response to which a small number of students have protested that student activity fees should not pay for such activities. Additionally, more colleges are offering—and requiring students to take—courses on gay and lesbian issues.

Question

What changes have occurred recently in homosexual behaviors?

Getting Married

A lifestyle practice that apparently exists in all societies is **marriage**—a socially and/or religiously sanctioned union between a woman and a man with the expectation that they will perform the mutually supportive roles of wife and husband. After studying extensive cross-cultural data, anthropologist George P. Murdock (1949) concluded that reproduction, sexual relations, economic cooperation, and the socialization of offspring are functions of families throughout the world. We now recognize that Murdock overstated the matter, because in some societies, such as Israeli kibbutz communities, the family does not perform all four of these functions (Gough, 1960). What Murdock describes are commonly encountered tendencies in family functioning in most cultures.

Societies differ in how they structure marriage relationships. Four patterns are found: *monogamy* (one husband and one wife), *polygyny* (one husband and two or more wives), *polyandry* (two or more husbands and one wife), and *group marriage* (two or more husbands and two or more wives). Although monogamy exists in all societies, Murdock discovered that in some societies other forms are not only allowed but preferred. Of 238 societies in his sample, only about one-fifth were strictly monogamous.

Polygyny has been widely practiced throughout the world. The Old Testament reports that both King David and King Solomon had several wives. In his cross-cultural sample of 238 societies, Murdock found that 193 (an overwhelming majority) permitted husbands to take several wives. In one-third of these polygynous societies, however, less than one-fifth of the married men had more than one wife. Usually only the rich men in a society can afford to support more than one family.

In contrast with polygyny, polyandry is rare among the world's societies. And in practice, polyandry has not usually allowed freedom of mate selection for women—it has often meant simply that younger brothers have sexual access to the wife of an older brother. For example, in some cultures, the father must pay a fee to secure a virginal bride for his son; if the father is unable to afford wives for each of his sons, he might secure a wife for only his oldest son. Consider this account of polyandrous practices among the Todas, a non-Hindu tribe of India:

> The Todas have a completely organized and definite system of polyandry. When a woman marries a man, it is understood that she becomes the wife of his brothers at the same time. When a boy is married to a girl, not only are his brothers usually regarded as also the husbands of the girl, but any brother born later will similarly be regarded as sharing his older brother's right. . . . The brothers live together, and my informants seemed to regard it as a ridiculous idea that there should even be disputes or jealousies of the kind that might be expected in such a household. . . . Instead of adultery being regarded as immoral . . . according to the Toda idea, immorality attaches rather to the man who grudges his wife to another. (Rivers, 1906, p. 515)

Fraternal Polyandry Ethnic Tibetans along the Tibet-Nepal border in central Asia practice polyandry. A number of brothers jointly take a wife. Although the eldest brother is typically the dominant figure in the household, all the brothers share the work and participate as sexual partners with the wife. Here a 12-year-old bride stands with three of her five husbands-to-be. The grooms, ages 19, 17, and 7, are brothers. Two other brothers, ages 14 and 21, were away. The cultural ideal calls on the wife to show all the brothers equal affection and sexual accommodation.

Anthropologists disagree on whether group marriage genuinely exists in any society as a normatively encouraged lifestyle. There is some evidence that it might take place among the Marquesans of the South Pacific, the Chukchee of Siberia, the Kaingang of Brazil, and the Todas of India. On occasion, as among the Todas, polyandry slides into group marriage when a number of brothers share more than one wife.

As in the past, *monogamy* remains the dominant lifestyle in the United States. Of Americans aged 35 or older, over 90 percent have been married at least once. But nearly half of all marriages end in divorce. Four out of five divorced people remarry; an estimated 44 percent of all current marriages are remarriages. About 45 percent of those who divorce and remarry divorce a second time (Nock, 1995). Hence, rather than having recourse to polygyny, polyandry, or group marriage, many Americans have maintained a legal monogamous arrangement through *serial monogamy*—a pattern of marriage, divorce, and remarriage.

In sum, Western countries have not given up on marriage (Nock, 1995). Surveys show that American and European adults depend very heavily on their marriages for their psychological well-being, and, in fact, married couples who described themselves as "in love with their great love" were very satisfied with their relationship (Willi, 1997; Contreras, Hendrick & Hendrick, 1996). In cross-cultural studies of arranged and love marriages, marital satisfaction was higher in love marriages (see

Human Diversity: "Arranged Marriages or Love Matches: Which Are Better?"). Also, marriage seems to enhance mental health; married couples report better health than do the individuals in their cohort who remain unmarried (Horwitz, 1996). Indeed, history reveals marriage to be a very resilient institutional arrangement. Not surprisingly, marriage is the most prevalent American lifestyle (see Figure 14-5), although increasing numbers of Americans have come to define marriage as something that can be ended and reentered, no longer viewing it as a permanent institution. Moreover, marriages differ. Marriage encompasses a wide range of interaction patterns, each of which involves a somewhat different lifestyle. We will return to this topic in Chapters 16 and 18.

Questions

How does the structure of "marriage" vary across cultures? Is marriage as an American institution less stable than it was previously?

Family Transitions

The relationships in families are the juice of life, the longings and frustrations and intense loyalties. We get our strength from those relationships, we enjoy them, even the painful ones. Of course, we also get some of our problems from them, but the power to survive those problems comes from the family, too.

Urie Bronfenbrenner (1977, p. 47)

We hear a good deal nowadays about the "demise of the American family." Yet for many Americans the family remains a central and vital institution (see *Further Developments:* "Is the U.S. Family Disintegrating or Merely Changing?"). Over the course of their lives, most Americans find themselves members of two family groups. First, a person belongs to a *nuclear family* that often consists of oneself and one's father, mother, and siblings. This group is called the individual's *family of orientation.* Second, because over 90 percent of Americans marry at least once, the vast majority of American adults are members of a nuclear family in which they are one of the parents. This group is called the individual's *family of procreation.*

Various psychologists and sociologists have sought to find a framework for describing the transitions that occur across a person's life span that are related to these shifts in family patterns (Cowan & Hetherington, 1991). One tool that they have devised is the concept of the **family life cycle**—the sequential changes and realignments that occur in the structure and relationships of

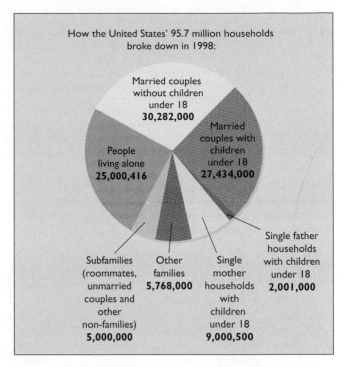

Figure 14-5 Household Compositions The U.S. Census Bureau expects nonfamily households, singles, and cohabiting couples to grow to almost one-third of the total households by 2000. Fairly high divorce rates and marriages at later ages are cited as factors. An increasingly elderly population of widows, who often outlive their husbands, continue to contribute to the growth of singles households.

Source: U.S. Department of Commerce, *Current Population Reports,* Oct. 1998.

family life between the time of marriage and the death of one or both spouses. The family life-cycle model views families, like individuals, as undergoing development that is characterized by identifiable phases or stages.

The Family Life Cycle

In the United States, families have traditionally had a fairly predictable natural history. Major changes in expectations and requirements are imposed upon a husband and wife as their children are born and grow up. The sociologist Reuben Hill (1964) describes the major milestones in a nine-stage **family life cycle:**

1. Establishment—newly married, childless
2. New parents—until first infant is 3 years old
3. Preschool family—oldest child is 3 to 6 years old, possibly younger siblings
4. School-age family—oldest child is 6 to 12 years old, possibly younger siblings
5. Family with adolescent—oldest child is 13 to 19 years old, possibly younger siblings
6. Family with young adult—oldest child is 20 years old or more, until first child leaves home

Human Diversity

Arranged Marriages or Love Matches: Which Are Better?

Everybody knows that love can have no place between husband and wife . . . for what is love but an inordinate desire to receive passionately a furtive and hidden embrace? But what embrace between husband and wife can be furtive, I ask you, since they may be said to belong to each other and may satisfy all of each other's desires without fear that anybody will object . . . love cannot exert its powers between two people who are married to each other.

When do you think this was written? Last year? The beginning of the twentieth century? You will probably be a bit surprised to find that Andreas Capellanus wrote these lines in 1174 in *The Art of Courtly Love.* It was to be another 400 years before society in general began to accept the idea that passionate love and marriage were not at opposite ends of the spectrum. Historically, owing to the belief that love and marriage did not belong under the same roof, parents wielded great power in that they chose mates for their children through arranged marriages. Even today in some countries of the world, girls as young as 11 or 12 are married off by their parents to eligible males. Following the Islamic custom of prearranged marriages, Houston Rockets' star Hakeem Olajuwon wed 18-year-old Dalia Asafi in a traditional ceremony in Houston in 1996. "There is no dating process, no boyfriends and girlfriends in Islam," Olajuwon said. "Families meet, talk, get to know one another. Then the marriage is arranged" (Thomas, 1996). The justification given for arranging marriages is that children are too young to understand the real purpose of marriage—an alliance between families. Think about Romeo and Juliet—the problem for those two lovers was that the *families* objected to their uniting as a couple. Or recall the heartfelt question Tevye, a poor Jewish milkman in Czarist Russia, sings to his wife Golde in *Fiddler on the Roof:* "Do you love me?" But even in arranged marriages, love can sometimes be taken into consideration, and the extent to which this occurs can be seen if one looks at contemporary marriage arrangement practices around the world. For example in some cultures:

- Parents choose the partner, no discussion or objection allowed.
- Parents choose the partner, discussion or objection allowed.
- Parents, child, relatives, and friends come to a group decision.
- Child chooses prospective partner and parents give approval.
- Arranged or love marriages are both alternatives.

Those who favor marriage or relationship based on love argue that it is cruel to force young people to enter into relationships with people they do not love or even like. Furthermore, only the individual can know who that perfect someone will be. Those in favor of arranged marriages argue that love is a kind of madness (a type of irrational thought) that interferes with one's ability to make a good choice for the future. They also point out that "love" is not as discerning as some would like to believe—most people fall in love with someone they meet in the tiny circle of individuals they are acquainted with. Look at what some young people in India said about the pros and cons of different kinds of marriages (Sprecher & Chandak, 1992):

---

7. Family as launching center—from departure of first child to departure of last child
8. Postparental family—after children have left home, until father retires
9. Aging family—after retirement of father

As viewed by Hill and other sociologists, the family begins with the husband-wife pair and becomes increasingly complex as members are added, creating new roles and multiplying the number of interpersonal relations. The family then stabilizes for a brief period, after which it begins shrinking as each of the adult children is launched. Finally, the family returns once again to the husband-wife pair and then terminates with the death of a spouse. However, in the contemporary United States some individuals do not form families, and many families do not pass through these stages. Because this view of the family life cycle revolves around the reproductive process, the approach is not particularly helpful in understanding childless couples. Nor does the family life-cycle approach apply to single-parent families, divorced couples, and stepfamilies (Hill, 1986). Significantly, in its original formulation, the family life-cycle approach made no reference at all to the mother's participation in the paid labor force. Newer versions of the scheme have recognized that a career woman and mother are increasingly one and the same person.

Critics such as Glenn H. Elder, Jr. (1985, 1974) contend that in contemporary society, many behaviors do not occur at the usual ages or in the typical sequence as-

Arranged Marriages

Advantages	Disadvantages
1. Support from families	1. Not knowing each other well
2. Quality and stability of marriage	2. Problems with dowry
3. Compatible or desirable backgrounds	3. Incompatibility/ unhappiness
4. Time to learn to adjust to marriage	4. Limited choice
5. Happiness of parents and family	5. Family and in-law problems
6. Approval of society	
7. Easy to meet a partner	
8. Excitement of the unknown	
9. Parents know best	

Dating and Love Marriages

Advantages	Disadvantages
1. Getting to know the other person	1. Sex, pregnancy, and immoral behavior
2. Stimulation and fun	2. Disapproval from parents
3. Socialization concerning the opposite sex	3. Cause anguish
4. Broadening outlook	4. Short-lived
5. Sex	5. Disapproval from society
6. Love and romanticism	6. Negative effect on studies
7. Freedom of choice	7. Waste of time and money
8. Leads to a good choice	8. Bad reputation
	9. Risky
	10. Immature relations

One study reported that in India arranged marriages might be better than love marriages, but overall arranged marriages do not seem to work out for the best (Gupta & Singh, 1982). Several studies in which men and women from different countries were asked to assess their marriages found:

- Japanese women were less happy in arranged marriages; the longer the marriage, the unhappier.
- Chinese women were far happier if allowed to choose a partner.
- Indian women did not end up liking or loving their husbands as the marriage progressed.

As times change, so do institutions and customs. For some this is terrifying and for others it is liberating. What is your perception of marriage now? What do you hope it will be 10 years from now, or what was your view 10 years ago? Before reading this you probably thought that the conception of love that we hold today was universal and had remained unchanged since time immemorial.

Today, in the United States, love is seen as an integral part of marriage. Whether love is going to be stronger at the beginning of the marriage or the end seems to be the only difference for those preferring arranged or love marriages. But if we compare the idea that love has a place in marriage with the quote at the beginning of this box, we see that human relational patterns are subject to vast change and unpredictable invention.

sumed by the family life-cycle model. At times, decisive economic, social, political, or military events intervene to alter the normal course of events. In addition, it is important whether parents are in their early twenties or their late thirties during the childbearing stage of the cycle. And a retarded or handicapped child often stays with the family long after the normatively established "launching" period.

But despite its shortcomings, the family life-cycle model provides a clear picture of family change, particularly for families that remain intact and in which children are present. Each change in the roles of one family member affects all other family members, because they are bound together in a network of complementary roles—a set of mutually contingent relationships. Consequently, each stage in the family life cycle requires new adaptations and adjustments. Of particular importance are the events surrounding parenthood. Accordingly, let us examine more closely the significance of pregnancy and parenthood for young adults.

Pregnancy

Within the life cycle of a couple, particularly a woman, the first pregnancy is an event of unparalleled importance (Ruble et al., 1990). The first pregnancy signals that a couple is entering into the family cycle, bringing about new role requirements. As such, the first pregnancy functions as a major marker or transition and confronts a couple with new developmental tasks.

Further Developments

Is the U.S. Family Disintegrating or Merely Changing?

The family, in its old sense, is disappearing from our land, and not only our free institutions are threatened but the very existence of our society is endangered.

Boston Quarterly Review, 1859

Some 90 percent of U.S. men and women consider marriage the best way to live. So concern about the future of the family is hardly surprising, given the directions in which family life has been moving in recent decades. However, opinions differ as to the significance and meaning of the changes. There are those who say that the family is a durable feature of the human experience, a resilient institution rooted in our social and animal nature. But because the institutional structure of society is always changing, the family must change to reflect this fact. Indeed, there are those who argue that the traditional family is no longer appropriate for modern times because they see the traditional family structure as flawed by unhealthy conformity and male domination. Others contend that the family is in crisis and they fear its impending death—noting that divorce rates have soared, birth rates have fallen, the proportion of unwed mothers has increased, single-parent households have proliferated, mothers of young children have entered the labor force in large numbers, and the elderly and some segments of the younger adult population are relying more on the government than on the family for financial support.

Not surprisingly, family issues have entered the political arena with a vengeance. Pessimistic conservatives usually decry what they see as the lack of traditional family values and issue urgent calls for their revival. All the while, optimistic liberals endorse the proliferation and flexibility in family structures and call for additional government assistance programs. As sociologist and U.S. Senator Daniel Patrick Moynihan (1985) has noted, conservatives—fearing government intervention and interference—like to talk about family values but not new government initiatives,

whereas liberals—fearing a "blame the victim" mentality—prefer to talk about public policy initiatives but not family values.

Much of the debate over the "state of the American family" might be misinformed and misguided because it uses a stereotyped image of the white middle-class family of the 1950s and 1960s as its point of departure for either praise or criticism of subsequent changes. Such an image is informed less by reality than by old television series such as "Ozzie and Harriet," "Father Knows Best," and "Happy Days." More particularly, the notion that family life is disintegrating implies that at an earlier time in history the family was a more stable and harmonious institution than currently. Yet historians have never located a golden age of the family (Coontz, 1992; Cherlin, 1983). Their research reveals that the marriages of seventeenth-century England and New England were based on family and property needs, not on choice and affection. Loveless marriages, the tyranny of husbands, and the beating and abuse of children were commonplace (Shorter, 1975). Additionally, families were riddled by desertion and death to an unimaginable degree. In fact, because of fewer deaths, disruptions of marriages up through the completion of child rearing have been *declining* in the United States since 1900 (Uhlenberg, 1980).

The idea that families should consist of a breadwinner husband, a homemaker wife, and their dependent children is more recent. In the late 1890s, the rural, preindustrial family was a largely self-sufficient unit, meeting most of its consumer needs. Husbands, wives, children, and lodgers were all expected to participate in gainful work. Later, with the onset of industrialization, more and more family members sought work for wages in factories and workshops. This trend led Karl Marx and his coworker Friedrich Engels to deplore the employment of female and child workers at minuscule wages to run factory machines, compelling able-bodied men to accept "children's work at children's wages." Throughout the Western world the nascent labor movement pressed for the

---

Pregnancy requires a woman to marshal her resources and adjust to a good many changes. Unfortunately, in many cases a woman's earliest experiences of pregnancy might be somewhat negative; she might be an unmarried, young adolescent; she might have morning sickness, vomiting, and fatigue. **Pregnancy** can also compel a woman to reflect on her long-term life plans, particularly as they relate to marriage and a career. And pregnancy can cause her to reconsider her sense of iden-

tity. Many women seek out information on birth and motherhood to help them prepare for the new role. The woman's partner faces many of these same concerns. He might have to reappraise his conception of age, responsibility, and autonomy. Similarly, pregnancy frequently contributes to changes in the couple's sexual behavior. Few events equal pregnancy in suddenness or significance, and many couples experience the initial phase of pregnancy as being somewhat disruptive.

establishment of a "living wage," an income sufficient for a male breadwinner to support a wife and children in modest comfort.

Americans began sorting jobs into male and female categories during the nineteenth century. The domestic sphere was defined as the "special place" of women. If women were in the labor force outside the home, it was expected that they would stop working after marriage or make a lifetime commitment of secular celibacy as a nurse or schoolteacher. The restriction of large numbers of married women to domestic activities took place only after industrialization was well established (Carlson, 1986; Cherlin, 1983).

Contrary to popular myth, family life prior to the 1950s was hardly orderly (Coontz, 1992; Mintz & Kellog, 1988). Family members expected children to leave school, postpone leaving home, or put off marriage to help the family meet an unexpected economic crisis or a parental death. At the turn of the century, young people married relatively late because they were often obligated to help support their parents and siblings. The economic prosperity of the post–World War II years contributed to a sharp drop in the average age at which marriage occurred. Today's young adults, however, are reversing the trend and marrying at later ages (U.S. Bureau of the Census, 1998a). The view that family life should meet emotional needs—accompanied by the transformation of the family into a highly private institution—did not spread beyond the middle class until the twentieth century. The development of affectionate and private bonds within a small nuclear family were accelerated in the early 1900s with the decline in the boarding and lodging of nonfamily members, the growing tendency for unmarried adults to leave home, and the fall in fertility (Laslett, 1973).

All in all, reports of the death of the U.S. family are greatly exaggerated. Indeed, if you literally believe what Americans say about family life, it might actually be flourishing. Nearly two-thirds of married Americans rate their own marriages as "very happy." For more than 50 years pollsters have been asking Americans about their families, and Americans consistently respond that the most important as-

Reports of the Death of the American Family Are Greatly Exaggerated Throughout human history the family has been subject to a wide variety of societal pressures. It has proven itself to be an adaptive, resilient institution that satisfies the needs of many people. Although new challenges confront today's families, it seems that people continue to prefer and vitally need the kinds of relationships that a healthy family life can provide.

pect of their lives is the relationship among parents, children, and siblings. They contend that families are more important than work, recreation, friendships, or status. But what people say and what they do might not necessarily coincide. Surveys also reveal that in practice many Americans put their work, possessions, and personal freedom ahead of their family responsibilities. And there has been a substantial decline in the ideal of marital permanence. The goal of "having a happy marriage" currently ranks well above "being married to the same person for life" and even farther above merely "being married" (Glenn, 1992).

Overall, even though family life often presents challenges and frustrations, it is something that most people vitally need, particularly when many other spheres of life are becoming "depersonalized." In sum, for most Americans the family remains a vital, adaptive, resilient human institution.

On the broader social level, relatives, friends, and acquaintances commonly offer judgments on numerous matters, including whether the woman stands in a proper social relationship with the father-to-be (or will remain a single parent). An employed woman may have to confront changing relationships at work, as her employer and colleagues reappraise their ties with her. If the mother-to-be should withdraw from the paid workforce in preparation for childbirth, she could find that her do-

mestic situation also alters: the more egalitarian values and role patterns of dual-career couples tend to give way to the stereotyped gender-role responsibilities found in traditional nuclear families.

Researchers have identified four major developmental tasks confronting a pregnant woman. First, she must come to accept her pregnancy. She must define herself as a parent-to-be and incorporate into her life frame a sense of impending parenthood. This process requires developing an emotional attachment to her unborn child, something that is now easier to do earlier in the pregnancy, due to the marvels of fetal ultrasound. Women typically become preoccupied with the developing fetus and, especially around the time that they begin to detect clear movements of the child in the uterus, ascribe personal characteristics to it.

Second, as a woman's pregnancy progresses, she must come to differentiate herself from the fetus and establish a distinct sense of self. She might accomplish this task by reflecting on a name for the infant and imagining what the baby will look like and how it will behave. This process is expedited when her increasing size brings about alterations in her clothing and she assumes a "pregnancy identity."

Third, a pregnant woman typically reflects on and reevaluates her relationship with her own mother. This process often entails the woman's reconciliation with her mother and the working through of numerous feelings, memories, and identifications.

Fourth, a woman must come to terms with the issue of dependency. Her pregnancy and impending motherhood often arouse anxiety concerning her loss of certain freedoms and her reliance on others for some measure of support, maintenance, and help. Such concerns are frequently centered on her relationship with her husband or partner.

The accomplishment of these developmental tasks is often expedited by childbirth-training classes, which teach women what to expect during pregnancy and labor. The knowledge and techniques the pregnant woman gains from the classes gives her a measure of "active control" and self-help. Finally, when husbands or partners also participate in the training classes, mothers-to-be find additional social support and assistance. Both preparation in pregnancy and a partner's presence are positively associated with the quality of a woman's birth experience. Indeed, much that happens before birth influences what transpires between parent and child after birth.

Transition to Parenthood

Psychologists and sociologists who view the family as an integrated system of roles and statuses have often depicted the onset of parenthood as a "crisis" because it involves a shift from a two-person to a three-person system (Rubenstein, 1989). The three-person system is thought to be inherently more stressful than the two-person system. Sociologist Alice S. Rossi (1968, p. 35) also finds other reasons the transition to parenthood could pose a crisis:

> The birth of a child is not followed by any gradual taking on of responsibility, as in the case of a professional work role. It is as if the woman shifted from a graduate student to a full professor with little intervening apprenticeship experience of slowly increasing responsibility. The new mother starts out immediately on 24-hour duty, with responsibility for a fragile and mysterious infant totally dependent on her care.

It is also important to note that not all marriages change in exactly the same way, and important individual differences surface in the ways spouses respond to parenthood (Alexander & Higgins, 1993; Levy-Shiff, 1994). A continuing study of more than 250 families by Jay Belsky and

Pregnancy: A Time of Developmental Change Pregnancy is a major marker or transition in the life of a woman or a couple. She is confronted with new developmental tasks, coming to terms with her pregnancy, taking care of herself, and redefining her relationships with her partner and other people who are important to her.

his colleagues is providing a rich array of insights about the transition to parenthood (Belsky & Rovine, 1990). This research shows that having children does not turn good marriages into bad ones or bad marriages into good ones. Yet overall, after the birth of a baby, couples typically experience a modest decline in the overall quality and intimacy of their marital life. Husbands and wives show each other less affection, and they share fewer leisure activities. The decline in marital satisfaction tends to be greater for wives than for husbands. But on the positive side, there is an increase in a couple's sense of partnership and mutual caretaking. And with the addition of a second child, fathers often become more involved in the work of the home, taking on more household tasks and significantly increasing their interaction with the first-born youngster (Stewart, 1990). One father observes, "It took only one child to make my wife a mother, but two to make me a father" (Stewart, 1990, p. 213).

Couples who are most likely to report marital problems in early parenthood are those who held the most unrealistic expectations of parenthood. Belsky believes the most successful couples seemingly shift to rose-colored glasses, focusing on what is going well rather than on what is going poorly; for instance, they downplay the fact that she has not lost all the weight or that he does not help out enough. One of the biggest post-birth stumbling blocks is the division of labor. Overall, parents typically move to more stereotyped gender roles after the arrival of a baby. Consequently, the wife frequently assumes a heavily disproportionate share of the division of household chores, and this can lead her to have negative feelings toward her husband. It seems that role consensus and a shared division of responsibilities between husband and wife are especially important for the maintenance of ongoing intimate relationships and marital satisfaction (Levy-Shiff, 1994). Some couples are also helped along by the evolution of family rituals, such as highly stylized religious observances, a certain type of greeting when the spouse returns home, a routine dinnertime, or a special weekend activity (Fiese et al., 1993).

Although having a baby might not save a marriage, a study by researchers associated with the Rand Corporation found that a first baby can stabilize marriages (Waite, Haggstrom, & Kanouse, 1985). The researchers followed 5,540 new parents and 5,284 nonparents for three years; the couples were matched on such factors as age and years married. The study showed that by the time their children were 2 years of age, the parents had a divorce rate under 8 percent. The nonparents had a rate of more than 20 percent. But why should babies make a marriage more stable? The Rand researchers speculate that people who decide to have children might be happier together to begin with, that children are a deterrent to divorce because they add complexity and expense to a breakup, and that parents acquire a bond with their part-

ners through their children. After the birth of a baby, some parents initially experience a short period called the *postpartum blues* (see *Information You Can Use:* "Postpartum Blues").

Question

How does parenthood change the nature of a marriage in both positive and negative ways?

Lesbian Parenthood

It has been estimated that in the past 1 to 5 million lesbians have had children in the context of traditional heterosexual relationships, although these percentages are still highly debatable (Patterson, 1992; Falk, 1989). Lesbians also choose other pathways to parenthood, including donor inseminations, adoption, and stepparenting. But many lesbian mothers are hesitant to be open about their sexual orientation for fear of losing their children in custody disputes, and it is still common that lesbian mothers in custody disputes often are unsuccessful in keeping their children (Tasker & Golombok, 1997; Brophy, 1992). In many courts of law, lesbian mothers have been deemed "unfit" as parents on a number of grounds (Arnup, 1995). But in fact there is no evidence that lesbian mothers are emotionally unstable or that they might sexually abuse their children. As Patterson (1992) reminds us, the majority of child molesters are heterosexual males.

Research shows that there are few differences between heterosexual and lesbian mothers, as it is motherhood—not sexuality—that emerges as the dominant identity marker for these women (Lewin, 1993). Lesbian women choose to become parents for many of the same reasons as heterosexual parents. But they face certain issues that heterosexual mothers do not face, such as internalized homophobia and societal disapproval (Flaks et al., 1995). Both of these factors place additional stress on the lesbian mother that can lead her to develop feelings of self-doubt, ambivalence, or a sense of having to over achieve. Lesbian couples with children reported greater sexual and interpersonal satisfaction than those lesbian couples who remained childless. Furthermore, lesbian couples tend to divide housework and child care more equitably than other couples (Parks, 1998).

Even though lesbian parents face social stigma, legal battles, and economic disadvantages, research indicates that both parents and children are healthy, secure, and effective in coping with the challenges they face in their years together (Parks, 1998). Findings from Tasker and Golombok's (1997) longitudinal study of the psychological well-being of children raised to adulthood in lesbian homes indicate the following:

Information You Can Use

Postpartum Blues

About two or three days after delivery, some new mothers experience what is commonly called the *postpartum blues.* Symptoms include irritability, waves of sadness, frequent crying spells, difficulty sleeping, diminished appetite, and feelings of helplessness and hopelessness. Generally, the episode is mild and lasts only a short time, up to a few weeks (O'Hara, 1995). Similar symptoms often appear in women who adopt a child, and some new fathers also report that they feel "down in the dumps." One study reported that 89 percent of new mothers experienced some symptoms that have been traditionally associated with the postpartum blues and 62 percent of the fathers had similar symptoms (Collins et al., 1993). Further, if a woman has had a *postpartum depression,* she is 50 percent more likely to have another case with the next baby (Wisner & Wheeler, 1994).

Various explanations have been advanced for the postpartum blues. Some experts believe that a woman's hormonal changes associated with childbirth and metabolic readjustment to a nonpregnant state can influence her emotional and psychological state. Following childbirth, dramatic changes occur in the levels of various hormones, and changes can occur in thyroid and adrenal hormones. All these changes can contribute to depressive reactions. Among the evidence supporting a biological explanation is research showing that postpartum depression is as common in rural Africa as it is in industrialized Western nations (Brody, 1994b).

Other explanations of a more psychological nature emphasize the adjustments required of a woman in her new role as a mother (Hopkins, Marcus, & Campbell, 1984). Some women experience a sense of loss of independence, of being tied down and trapped by the new infant. Other women feel guilty about the anger and helplessness they feel when their infants cry and cannot be comforted. And some women also feel overwhelmed by the responsibility of caring for, rearing, and shaping the behavior of another human being. More particularly, women with temperamentally difficult babies can find that child care severely taxes their emotional and psychological resources, contributing to depression. Although mothers can do little to alter their infant's basic temperament, they can more successfully cope with stress by developing a network of supportive people (Cutrona & Troutman, 1986).

In a small number of cases, the birth of a child can catalyze severe mental illness in women who are predisposed to schizophrenia or bipolar (manic-depressive) disorder. Women who have the more devastating forms of mental illness might even commit infanticide or suicide. Women should not be afraid or feel ashamed to seek help. Treatment options include psychotherapy, antidepressant medication, participation in a support group, family counseling, and, if circumstances warrant, hospitalization (Brody, 1994b). Overall, contemporary parents seem to prepare better than parents of previous generations for coping with the changes parenthood brings. They have to juggle a whole new set of questions about their parenting role, work roles, demands on their time, communication patterns, privacy, and the companionship aspects of their relationship. But on the whole, they report enormous satisfaction in parenthood.

- These children are no more likely to have a gay or lesbian sexual orientation as adults than are their peers raised by heterosexual parents.
- Men and women raised by lesbian mothers were no more likely to experience anxiety or depression than were their peers raised in heterosexual homes.
- Fear of group stigmatization and the experience of being teased or bullied are central elements of how children feel about growing up in a lesbian-mother family.

Question

What are the findings of the first longitudinal study on the psychological effects of children growing up in a lesbian-parent home?

Employed Mothers

I say that the strongest principle of growth lies in the human choice.

Elliot (1992)

One aspect of motherhood that some women experience as a challenge is balancing motherhood and career. Despite the "sexual revolution," employed women still bear the brunt of being the primary parent and the primary housekeeper (South & Spitze, 1994). Economist Sylvia Ann Newlett (quoted in Castro, 1991, p. 10) notes:

In the U.S. we have confused equal rights with identical treatment, ignoring the realities of family life. After all, only women can bear children. And in this country, women must still carry most of the burden of raising them. We think that we are being fair to everyone by stressing identical opportunities, but in fact we are punishing women and children.

Over the past several decades more and more mothers with young children have found employment outside the home. In 1998, 77 percent of mothers with children under 6 years of age were in the labor force (either on a full- or part-time basis) in comparison to 20 percent in 1960 (U.S. Bureau of Labor Statistics, 1998). Even among mothers of infants, over half are back at work before their baby celebrates a first birthday (68 percent who are college graduates do so). All the while, rising educational levels for women have increased both their ability and motivation to obtain employment outside the home. Moreover, a great many families find it an economic necessity that the woman be employed; most divorced, single-parent, and widowed mothers must work to avoid poverty. Even two-parent families find a second income is often required to maintain an acceptable standard of living. And the growing prevalence of working mothers has increased the full-time homemaker's isolation from other adults (Menaghan & Parcel, 1990). As with men, a job in the workplace allows women to participate in the larger society and to be compensated for one's skills.

Serious concern is frequently voiced about the future of U.S. children as more and more mothers enter the workforce. Many people fear that the children of a working mother suffer a loss in terms of supervision, love, and cognitive enrichment. Much of the earlier research on maternal employment and juvenile delinquency was based on this assumption: Mothers were working, children were unsupervised, and thus, they became delinquents. But the matter is not that simple. In a classic study of lower-class boys, Sheldon and Eleanor Glueck (1957) found that sons of regularly employed mothers were no more likely to be delinquent than sons of nonemployed mothers. However, inadequate supervision does appear to be associated with delinquency, whatever the mother's employment status (Bank et al., 1993).

Research findings are contradictory regarding the effects of maternal employment during a child's first year, with some studies reporting negative cognitive and social outcomes (Baydar & Brooks-Gunn, 1991; Belsky & Eggebeen, 1991) and others finding only minimal negative outcomes (Parcel & Menaghan, 1994a). A positive outcome is that mothers' working provides positive role models for children. For older youngsters, however, an accumulating body of research suggests that there is

"I'm tired of this full-time job. I want a part-time job."

Figure 14-6 The Division of Labor A job in the workplace permits participation in the larger society and compensation for one's skills. Additionally, a shared division of responsibilities between husband and wife improves marital satisfaction.

Source: The New Yorker Collection 1997 Joseph Farris from cartoonbank.com. All Rights Reserved.

little difference in the development of children whose mothers work and that of children whose mothers remain at home (Hofferth & Phillips, 1991).

Much depends on whether the mother, regardless of employment, is satisfied in her situation (see Fig 14-6). Parcel and Menaghan (1994b) conclude that the working mother who obtains personal satisfaction from employment, who does not feel excessive guilt, who has high-quality child care, and who has adequate household arrangements is likely to perform as well as or better than the nonemployed mother. Mothers who are *not* working and would like to, and working mothers whose lives are beset by harassment and strain, are the ones whose children are most likely to show maladjustment and behavioral problems. How much time parents spend with their children is not so predictive of young children's development as are the attitudes and behaviors their parents take toward them.

More significantly, many psychologists and sociologists are asking, "Under what conditions do *both* mothers' and fathers' working experiences affect children's cognitive and social development?" (Parcel & Menaghan, 1994a; 1994b). Clearly, contemporary youngsters are affected by the father's as well as the mother's working. Further, having two working parents impacts youngsters in many direct and interactive ways. For instance, parents who have heavy work schedules are less able to spend time with their children fostering the development of their cognitive and social skills. And fathers who work less than full-time might not be able to set an appropriate example of self-discipline and to help their children establish suitable standards of self-control (Parcel & Menaghan, 1994a; 1994b).

Social and behavioral scientists are examining the complex interactions of the home environments (marital status and number of children), parental characteristics (age, schooling, cognitive skills, and self-concepts), parental occupations (levels of substantive complexity, supervision, and autonomy), child characteristics (gender and temperament), family resources (income), child-care arrangements, and larger social and economic conditions. For example, a mother's job complexity is associated with enhanced reading scores for a single child, but the positive effects of job complexity decrease with additional children (Parcel & Menaghan, 1994a; 1994b).

Separation and Divorce

Divorce is widespread, and a divorce affects everyone involved with the family. The experience does not affect all couples in the same ways; for instance, some spouses continue to have sex even after they are separated or divorced and get along fairly well. But for the most, the effects are negative. People who have gone through separation or divorce have increased chances of psychiatric

disorders, depression, alcoholism, weight loss or weight gain, and sleep disorders. The effects on young children can also continue into young adulthood, with children of divorce having a decreased capacity for intimacy (Westervelt & Vandenberg, 1997). And parental divorce is associated with an increased risk of offspring divorce, especially when wives or both spouses have gone through the experience of their parents' divorce (Amato, 1996). Economically, most women suffer a major decline in standard of living, while men generally gain a moderate increase in their standard of living (Peterson, 1996). Traditionally women are granted either sole custody or joint custody of children and might be awarded a regular child-support stipend by the courts, which might or might not be paid.

Well-Being of Children and Young Adults Whose Parents Have Divorced   Gohm and colleagues (1998) recently conducted a multinational study investigating the effects of parental divorce on the subjective well-being of adult children. They collected international data during 1995 to 1996 as part of a larger study of issues related to cultural differences in subjective well-being. Participants were 6,820 college students (2,625 males and 4,118 females) from 39 countries (14 Asian countries, 13 European countries, 5 African countries, 4 from South American, and Australia, the United States, and Puerto Rico). In some cases, parental divorce increased the well-being of adult offspring, and for some being raised in a second marriage was better than being raised by a single parent (Chase-Lansdale, Cherlin, & Kiernan, 1995). Family stress theory suggests, however, that remarriage is probably beneficial if the remarriage does not result in high levels of marital conflict. Analysis of 37 studies of young adult offspring found no support for gender differences in the impact of divorce (Amato & Keith, 1991). However, "within original family marriages, marital conflict has clear, long-term, negative consequences for the well-being of offspring . . . and is consistent across gender" (Gohm, Oishi, & Darlington, 1998).

Offspring raised in a remarried household where parental conflict is low report life satisfaction that is similar to that of offspring from average marriages and of offspring from single-parent, divorced homes. Offspring raised in divorced, single-parent households report greater life satisfaction and more positive emotional life (higher subjective well-being) than offspring raised in high-conflict marriages—though this finding varied across cultures (Gohm, Oishi, & Darlington, 1998). "The subjective well-being of offspring in remarried households is not much greater than that of offspring in a high-conflict marriage, and sometimes it is lower" (Gohm, Oishi, & Darlington, 1998). In fact, Wilson and Daly (1997) report a considerably higher incidence of child abuse and homicide in remarriage families, particularly for children under age 3, in comparison to statistics for intact families. The extended social network found in collectivist cultures (those with extended kinship patterns) appears to provide greater psychological and emotional support for children experiencing the trauma of marital conflict and divorce.

Single-Parent Mothers

From 1980 to 1997 the number of single-parent families in the United States increased by nearly 60 percent: from 6.2 million to 10 million. The U.S. Census Bureau reported in October 1998 that 42.2 percent of today's mothers have never married and that 27.3 percent of all parents—or 10 million—are female single parents, up from 19 percent in 1980. Additionally, as shown in Figure 14-7, increasing numbers of women in Western nations are bearing children outside of marriage. Single

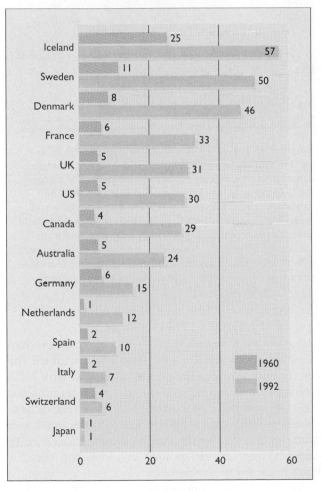

Figure 14-7 Percentage of Births That Occur Outside of Marriage Some sociologists and demographers predict that by the year 2000 at least 40 percent of U.S. babies will be born outside of marriage. *Note:* Germany is West Germany for 1960. Data for Canada and Spain are from 1991. Data for Japan are from 1997.

Source: This article is from *The American Enterprise,* a national magazine of politics, business, and culture.

parenthood is evident at all socioeconomic levels. Though the numbers are highest for women who live in poverty and among African American women, the most rapid rate increase is seen in nonmarried motherhood among women who have attended college for at least a year and among women with professional or managerial jobs. Some women who have earned professional training and good jobs are less threatened with disgrace if they have a child when not married, and many also are adopting as singles (Pollitt, 1993). Significantly, nonmarried childbirth, divorce, and various cohabiting arrangements have also changed the meaning of "family." As noted earlier in this chapter, the traditional nuclear family is currently not a reality for a growing number of Americans.

In single-parent families the responsibilities fall upon one adult rather than two, so single parents must allocate their time to cover both their own and their children's physical, social, and psychological needs. The matter is complicated by the fact that many schools and workplaces have inflexible hours, and these hours do not coincide—though more businesses now have on-site day care and more U.S. schools are providing both before-school care and after-school care, as well as all-day summer programs for youth. Single mothers frequently suffer from a lack of free time, spiraling child-care costs, loneliness, and the unrelenting pressures of attempting to meet the demands posed by both home and work, and, for some, college classes as well. Being a single parent calls for a somewhat different kind of parenting. Frequently, single parents find themselves making "the speech," as one mother termed it, explaining:

> I sat down with my three children and said, "Look. Things are going to have to be different. We're all in this together and we're going to have to be partners. I'm earning a living for us now. I'm doing it all. I need your help, if this household is going to work." (McCoy, 1982, p. 21)

Women heading single-parent families typically have lower incomes and lower levels of social support than women in two-parent families, leading to greater stress (Simons et al., 1993). Disruptions due to substantial income changes, residential relocation, unpredictable financial support from an ex-spouse, and household composition changes are more likely. Not surprisingly, female heads generally report much lower self-esteem, a lower sense of effectiveness, and less optimism about the future than their counterparts in two-parent settings. However, recently divorced, separated, and widowed women experience more major life-event disruptions than women who have been single for three or more years. And though many women do not choose single parenting, most are proud of their ability to survive under adverse circumstances (Lewin, 1992a).

For many single mothers, kin networks are important sources of financial, emotional, and child-care support (Jayakody, Chatters, & Taylor, 1993). Many single mothers are dependent on government programs, local charitable organizations, and churches for survival. More recently federal Welfare to Work legislation is providing job training for healthy single mothers to become employed within a recommended time period (Polakow, 1993). There is a critical difference between the married poor with children and the single-parent poor: On average, the married poor move out of poverty, the single-parent poor do not (Weiss, 1984). Social stigma often is attached to the single mother as a result of her unwed or divorced status, and she might be further discriminated against as a female head of household. In many respects a single father is in a better position than a single mother, because he is frequently viewed as a person who is doing something extraordinary.

According to the U.S. Bureau of the Census (1997a), about 56 percent of women raising children without the children's father have been awarded child support, but only half of these receive full payment. In the United States, divorce courts often demand the father's monetary obligation but not his presence, though more courts are awarding joint custody. And as men start seeing their children less, they often start paying less. Among never-married mothers, only a small number get any financial help from their children's father. It is hardly surprising, then, that nearly three-fourths of never-married mothers aged 15 to 24 live in poverty. And children living with divorced, single-parent mothers typically experience a dramatic decline in their standard of living compared with their predivorce, two-parent household (Weitzman, 1990).

Some families headed by women survive these hardships with few ill effects. But a disturbing number of children and parents are saddled with problems (Grych & Fincham, 1992). Children living in single-parent families are much more likely to be enrolled below the grade that is modal for their age and to be experiencing school difficulties than are children living with both parents. Some studies also show that juvenile delinquency rates are twice as high for children from single-parent households as they are for children from two-parent households. Lack of parental supervision and chronic social, health, and psychological strains are often associated with poverty (Bank et al., 1993).

Social isolation can create a sense of vulnerability for single mothers. Yet most women report having a partner, boyfriend, friends, or relatives who provide them with assistance on a fairly regular basis (Parish, Hao, & Hogan, 1991; Richards, 1989). In nearly half of all single-parent families, the parent remarries within five years. This new marriage results in a "blended" or "reconstituted" family, which can produce complicated kinship

Single Parenthood Is on the Rise Even though 27.5 percent of all American youngsters currently live in single-parent households—and the proportion is steadily rising—the United States lags behind most other industrialized nations in making safe, high-quality child care available and accessible to single parents employed outside the household. Some large corporations have on-site day care, and many public schools offer before-school and after-school programs, as well as summer supervision. Many single-parent moms live at a lower socioeconomic status than single-parent dads.

networks (see Chapter 16). Where both partners have been previously married, each has to deal with the former spouse of the current partner as well as with his or her own former spouse and several sets of grandparents (some of them being ex-in-laws). Adding to the difficulties are stepparent-stepchildren relationships, one's own children's reactions to the current spouse, one's own reactions to the current spouse's children, and the children's reactions to one another.

Single-Parent Fathers

The movie *Kramer vs. Kramer* broke new ground in the 1970s by portraying what is now becoming a way of life for increasing numbers of U.S. men: men rearing children alone. In 1998, there were 2.1 million father-child families in the United States. The number of men who become single parents as a result of their wife's death has declined. But overall the number of single fathers has grown slowly, as more men are awarded custody of children in divorce proceedings. Indeed, 20 years ago a father was awarded custody only if he could demonstrate in court that the mother was totally "unfit" for parenthood, and single men adopting children was out of the question.

Although the expectations attached to the father role in a two-parent family are fairly explicit, they are not so explicit for a father in the single-parent family. A number of studies have shown that even though single fathers are confronted with adjustment requirements, most of them raise their children successfully (Risman, 1986). But juggling work and child care commonly poses difficulties for single fathers, as it does for single mothers, especially for those with preschool youngsters. Compared with single mothers, however, fathers often make more money and have greater economic security and job flexibility (Meredith, 1985). Overall, the single father is neither the extraordinary human being nor the bumbling "Mr. Mom" depicted in popular stereotypes. Concerning child supervision, fathers tend to gravitate toward nurseries and child-care centers, where they feel that the staff has a professional commitment to children.

Generally, single fathers seem better prepared for the physical aspects of parenting—shopping, cooking, cleaning, taking the child to the doctor, and the like—than for dealing with their children's emotional needs. Men who adeptly juggle work schedules to stay home and nurse a sick child report that they fall apart in the face of a healthy temper tantrum. They view their children's strong displays of emotion as "irrational," especially when they cannot trace those emotions to some specific event in the children's lives. Single fathers also tend to express more anxiety over the sexual behavior of their daughters than that of their sons. And many are concerned about the absence of adult female role models within the home. Many single fathers admit that they have had to learn to deal with their children's emotional needs and to develop their own nurturing skills (Seligmann, 1992).

Some suggest that the main difficulty for many fathers in making the transition to becoming a single parent is losing companionship rather than becoming a single parent (Smith & Smith, 1981). Though dating is often an important part of the single father's lifestyle, he is in no hurry to marry again. Indeed, half were uncertain if they want to remarry and were committed to remaining single for the present. In sum, single-parent fathers and single-parent mothers both find that their greatest difficulties lie in balancing the demands of work and parenthood. Let's now examine the role that work plays in the lives of adults.

Question

What are the proposed stages of a traditional family life cycle, and how are these stages out of sync with the reality of many contemporary relationships?

Work

Originality and the feeling of one's own dignity are achieved only through work and struggle.

Fyodor Dostoevsky, *Diary of a Writer,* 1873

The central portion of the adult life span for both men and women today is spent at work, and nowadays Americans are working longer than ever. Economist Juliet B. Schor (1993) finds that over the past 20 years the time Americans spend on the job has been rising—a reversal of a century-long trend toward a shorter workday. Today, only Japanese workers put in more time. Schor (1992) contends that Americans are working longer to achieve the equivalent of a 1970 standard of living. It seems that Americans have chosen to put in more time at work (now averaging about 47 hours a week). The net result for Americans, says Schor, is a decline in their happiness and an erosion of their collective ability to care for children, cook, sleep, visit, and enjoy life. A 1995 survey confirms that a majority of Americans believe they do not have enough time for leisure and family. However, given a choice, 26 percent would work more hours for more income, while more than half would work the same hours for the same income (*USA Today*/CNN/Gallup, 1995).

The work experience of Americans has also undergone a significant change over the past 150 years. More than 70 percent of the labor force worked on the farm in 1820; today less than 5 percent are engaged in agriculture. Today, employment in the service industries is approaching 70 percent, the same percentage as for farming a century and a half ago.

For a majority of young Americans, the transition to adult occupational roles is postponed by college. Most youth view a college education primarily as a means to a better job rather than as a vehicle for broadening their intellectual horizons. One of the most significant developments in higher education in recent years is that U.S. colleges are enrolling more adult "nontraditional" students, those who have been out of high school for a year or more, are over 23, and have put off getting a degree. Over the past 15 years the number of nontraditional students enrolled in colleges and universities grew by over 70 percent. *Part-timers* (those taking less than a full-time course load) now account for nearly half of all U.S. college students. The adult nontraditional students are more likely to come from working-class backgrounds, have family and work responsibilities, do not want to waste any time in getting a degree, and strive to do their best. These adults aspire to prepare themselves for a better job or a career.

The Significance of Work for Women and Men

People work for a great many reasons. "Self-interest" in its broadest sense, including the interests of one's family and friends, is an underlying motivation of work in all societies. However, self-interest is not simply the accumulation of wealth. For instance, among the Maori, a Polynesian people of the Pacific, a desire for approval, a sense of duty, a wish to conform to custom, a feeling of emulation, and a pleasure in craftsmanship also contribute to economic activity (Hsu, 1943).

Even in the United States, few activities seriously compete with work in providing basic life satisfaction (Menaghan & Parcel, 1990; Weiss, 1990). In a study conducted more than 40 years ago (Morse & Weiss, 1955) and since replicated several times, a representative sample of U.S. men were asked whether they would continue working if they inherited enough money to live comfortably (Gallup Poll Monthly, 1991; Opinion Roundup, 1980). About 80 percent said they would. The reasons are not difficult to discover. Work, in addition to its economic functions, structures time, provides a context in which to relate to other people, offers an escape from boredom, and sustains a sense of identity and self-worth. Perhaps not surprisingly, then, only one in four million-dollar lottery winners quits working after hitting the jackpot. Sociologist Harry Levinson (1964, p. 20) observed:

> Work has quite a few social meanings: When a man works he has a contributing place in society. He earns the right to be the partner of other men. . . . A man's work . . . is a major social device for his identification as an adult. Much of who he is, to himself and others, is interwoven with how he earns his livelihood.

Much the same assessment has been made regarding the meaning of work for women. Although paid work is becoming an economic necessity for an increasing number of women, one of the central themes of the women's movement has been the symbolic meaning of a paid job (Stern, 1991). For many, though not all, contemporary women, exclusive commitment to the unpaid work of homemaker and mother implies being cut off from the full possibilities of self-fulfillment. Some women view parenting children as the most important job of all and are willing to make sacrifices to continue as full-time homemakers, but a paid job has increasingly come to be seen as the "price of admission" to independence in the greater society and as a symbol of self-worth (Lewin, 1995a; Yankelovich, 1981).

The Americans with Disabilities Act (ADA) legislation (P.L.101-336) passed by Congress in 1990 recognizes the significance of work for all adults and opens the doors for more meaningful participation in higher education, work, and society for adults with disabilities. Corporations, municipalities, industries, and colleges are making buildings more accessible; transit companies are providing buses with wheelchair lifts and rescheduling routes to get people to school or a job; and communities across the United States are redesigning buildings, doorways, restrooms, and sidewalks for adults with disabilities. American corporations that receive federal contracts must demonstrate that they have hired persons with disabilities, and everyone benefits from providing meaningful participation in society for this segment of our adult population.

For *all* of us, our work is an important socializing experience that influences who and what we are. Sociologists Kohn and colleagues (1990) found that college-educated people are more likely to acquire jobs that require independent judgment and lead to higher rankings in the socioeconomic system. By virtue of the intellectual demands of their work, the college-educated evolve an intellectual prowess that carries over to their private lives. They might even seek out intellectually demanding activities in their leisure pursuits (such as community theater, coaching, mentoring, serving on boards of agencies). Typically, people who engage in self-directed work come to value self-direction more highly, to be more open to new ideas, and to be less authoritarian in their relationships with others. As parents, they pass these characteristics on to their children.

In the United States it is a blunt and ruthlessly public fact that to do nothing leads to a sense of worthlessness and purposelessness. We witness this daily in our inner cities where there are high numbers of unemployed, unskilled, or homeless workers with a sense of hopelessness (Toro et al., 1991). And we witness this with the thousands of skilled, college-educated, but unemployed workers in their thirties, forties, and fifties who lost their jobs in the early to mid 1990s because of corporate takeovers, layoffs, and downsizing. The longer a person is unemployed, cannot find a job, and cannot support a family, the more worthless (or depressed) that person feels, whether man or woman. To be unemployed—especially for a man, whom society views as the main wage earner—is to be a social outcast. Anthropologist Elliot Liebow (1967), in a study of "street-corner men" living in a Washington, D.C., ghetto, found that the inability to gain steady, remunerative, and meaningful employment undermines an individual's self-respect and self-worth. Liebow concludes that the street-corner man is attempting to achieve the goals and values of the larger society, and when he fails to do so, he tries to conceal this failure as best he can from himself and others, often by escaping through alcoholism or substance abuse. In essence, for most people work is a truly defining activity necessary for their healthy development.

Question

What kind of meaning do people derive from their work?

Segue

We have looked at love and the institutional extension of love—marriage—as they affect different adults. As individuals go through changes, or stages, they reevaluate themselves, the interactions they have with others, and the meanings that love and work have for them at the moment. For some people this evaluation can lead to transitions like marriage, divorce, quitting a job, sexual reorientation, cohabitation, and, more likely than not, remarriage. Early adulthood is seen as the time when individuals start out on their first truly

independent journey—it is a time when they are able to leave parents in order to do what they "really want to do." But we have also seen that sometimes, as society changes, individuals make changes—perhaps postponing their dreams, or at least modifying them. Also, these changes lead to physical and cognitive changes, as adults in their middle years become more sedentary, experience midlife changes, and reassess their lives. We will explore these significant issues in the next chapter.

Summary

Theories of Emotional-Social Development

1. Love and work are the central themes of adult life. Both place us in a complex web of relationships with others. Relationships derive from two types of bonds: expressive ties and instrumental ties. Relationships that rest on expressive ties are called primary relationships; those that rest on instrumental ties are called secondary relationships.

2. The metaphor of the life course as being divided into stages or "seasons" has captured the imagination of philosophers and poets as well as other writers. Theorists tend to be divided over the idea that development is divided into discrete intervals.

3. Erikson's first stage of adult development is called the crisis of intimacy versus isolation.

4. Levinson and his associates say that the overriding task throughout adulthood is the creation of a *life structure*. Men and women must periodically restructure their lives by creating a new structure or reappraising an old one.

5. Stage theory typically has four characteristics. First, qualitative differences in structure are said to take place at given points in development. Second, the theory posits an invariant sequence or order. Third, the various ingredients making up a distinct structure appear as an integrated cluster of typical responses to life events. And fourth, higher stages displace or reintegrate the structures found at lower stages.

Establishing Intimacy in Relationships

6. Friends become our major source of socializing and support during our adult years. Research on love attempts to explain the real complexities involved in initiating and maintaining a meaningful, intimate relationship with another person.

7. Individuals in modern complex societies generally enjoy some options in selecting and changing their lifestyles. A striking aspect of U.S. society over the past 30 years has been the rapid expansion in lifestyles. Greater latitude is permitted individuals in tailoring for themselves lifestyles that are less constrained by traditional standards of what a "respectable" person should be like.

Diversity in Lifestyle Options

8. Leaving home is a major step in the transition to adulthood. For a good many decades, marriage was the major reason for leaving home. But over the past few decades, the process of leaving the parental home has become increasingly complex, and many young people experience numerous living arrangements in the course of assuming adult status.

9. Norms have dictated that U.S. youth leave home and make their own way in the world. But lately adult children have been making their way back to the parental home in increasing numbers. Family therapists express concern that those who stay at home or return home do not have opportunities to

fully develop their sense of individuality. The practices aggravate tendencies toward excessive protectiveness in parents and aggravate tendencies toward a lack of self-confidence in youth.

10. Census data reveal a sharp increase in the percentage of men and women under 35 years of age who are single. This increase has resulted in part from the tendency of young people to postpone marriage. More than two-thirds of adult singles live with someone else, such as a friend, a relative, or a "spouse equivalent."

11. The number of couples who are not married but live together has increased substantially over the past two to three decades, and those who follow this lifestyle do so more openly than they used to. Cohabitation is not restricted to the younger generation. It is becoming increasingly prevalent among the middle-aged and elderly who are divorced or widowed.

12. Sexual orientation refers to whether an individual is more strongly aroused sexually (erotic attraction) by members of his or her own sex (homosexual), the opposite sex (heterosexual), or both sexes (bisexual). Individuals show varying degrees of orientation, so homosexuals (lesbians and gay males) are a varied group.

13. Marriage is a lifestyle found in all societies. It remains the dominant lifestyle in the United States. About 4 in 10 marriages end in divorce, and many Americans practice serial monogamy.

Family Transitions

14. Families, like individuals, undergo development. In the United States most families have traditionally had a fairly predictable natural history, but today there is more variability. Major changes in expectations and requirements are imposed on a husband and wife as their children are born and grow up.

15. Within the life cycle of a couple, particularly for the woman, the first pregnancy is an event of unparalleled importance. It signals that a couple is entering the family cycle, bringing about new role requirements. As such, the first pregnancy functions as a major marker or transition and confronts a couple with new developmental tasks.

16. Many psychologists and sociologists who view the family as an integrated system of roles and statuses depict the onset of parenthood as a "crisis" because it involves a shift from a two-person to a three-person system. But many researchers have questioned this perspective. Their research suggests that relatively few couples view the onset of parenthood as especially stressful.

17. There are few differences between heterosexual and lesbian mothers, as motherhood, and not sexuality, is the dominant identity marker for these women. Although lesbian women choose to become parents for many of the same reasons as heterosexual women, they face certain issues that heterosexual women do not face: internalized homophobia, societal disapproval, and fear of losing their children.

18. One aspect of motherhood that appears particularly stressful is balancing motherhood and career. Many people fear that the children of a working mother will suffer a loss in terms of supervision, love, and cognitive enrichment. But researchers are finding that the working mother who obtains personal satisfaction from employment, who does not feel excessive guilt, and who has adequate household arrangements is likely to perform as well as or better than the nonworking mother.

19. Unfortunately, divorce is widespread and affects everyone involved. It does not affect all couples in the same ways, but most effects are negative.

20. Single-parent mothers frequently suffer from a lack of free time, spiraling child-care costs, loneliness, and the unrelenting pressures of attempting to fill the demands posed by both home and work. Not uncommonly, most find themselves in difficult economic circumstances. Nearly one-half live below the poverty level.

21. Increasing numbers of men are becoming single-parent fathers. Studies show that even though single fathers are confronted with some unique adjustment requirements, most of them are successful in raising their children. Like single-parent mothers, single-parent fathers find that one of their greatest difficulties is balancing the demands of work and parenthood.

Work

22. Work plays an important part in the lives of all adults, not only because of the money it brings in but also because people's self-definitions and sense of self-worth are tied to their work.

Key Terms

companionate love *(429)*

consummate love *(429)*

cultural dislocation *(422)*

expressive tie *(422)*

family life cycle *(437)*

instrumental tie *(422)*

intimacy *(430)*

intimacy versus isolation *(422)*

lifestyle *(430)*

marriage *(435)*

mentor *(424)*

pregnancy *(440)*

primary relationship *(422)*

role conflict *(426)*

role overload *(426)*

romantic love *(429)*

secondary relationship *(422)*

social relationship *(422)*

triangular theory of love *(429)*

Following Up on the Internet

Web sites for this chapter focus on the diverse social and work roles of contemporary young adults. Please access the text web site at http://www.mhhe.com/crandell7 for up-to-date hot-linked Internet addresses for the following organizations, topics, and resources:

Erikson and Young Adulthood

Single Parent Resource Center

Single Fathers

Unmarried Couples and the Law (Cohabitation)

American's with Disabilities Act—PL101-336, 1990

PartEight

Middle Adulthood

If we think of middle age as roughly the years between 45 and 65, middle-aged Americans compose about a fifth of the population. Some were born at the end of the Great Depression and the beginning of World War II. But the majority were born shortly before or during the 1950s and lived out their youth during the Korean War, the Elvis–Bob Dylan–Beatles era of rock and roll, the space race, the cold war, school integration, the Vietnam conflict, the women's liberation movement, "free love" and the birth control pill, the original Woodstock, and an era of opportunities for a college education. They entered the job market and made their way in the labor force during the economic golden years of the late 1960s and early 1970s. Now one American is turning 50 every 7.6 seconds (Kuczynski, 1998). Each one is experiencing midlife changes, and many are feeling the psychological pressures associated with the shift from the smokestack industrial age to the age of the information superhighway. As you read about their development through this life stage, it will become evident they are redefining middle adulthood.

Chapter 15

Middle Adulthood
Physical and Cognitive Development

Critical Thinking Questions

1. If you temporarily lost your sight for a week and had no way of knowing how you looked to others, would you feel comfortable in public?

2. Would you feel comfortable knowing that your partner was going to tell an interviewer (anonymously) all of the details of your sex life together? Would you want to listen to a recording of the interview?

3. Do you think you would like to have a baby at 55 years of age? Why or why not?

4. Assume you have just retired and money is no problem—what would you truly want to do with your time? Would you pursue some creative or new endeavor, or would you accept your life just as it is?

Until recently the middle years, from about age 45 through age 65, were a time of peak earnings and well-being for many adults. Middle-aged Americans were often portrayed as being the settled "establishment" of our society, the power brokers and decision makers. However, this scenario does not depict those middle-aged Americans living in poverty or from ethnic minority communities who have experienced many more difficulties. Additionally, with the changing U.S. socioeconomic conditions (due to corporate mergers, takeovers, downsizing, and international relocations), newly unemployed men and women are returning to college in ever-larger numbers to retrain in new areas of expertise or to become entrepreneurs. Others are being forced into early retirement. And some are becoming parents for the first time. Having learned to cope with the many contingencies of childhood, adolescence, and young adulthood, middle-aged women and men have a substantial repertoire of strategies for dealing with the physical and intellectual challenges of middle adulthood.

Defining (or Defying) Middle Age

With the life expectancy of Americans now reaching at least 76 or so years, the middle of life falls statistically at age 38—but what we define as middle age typically comes considerably later. Indeed, most of us have difficulty identifying which years of life are the middle years. Does middle age begin at 40, 45, or 50? And does middle age end at 60, 65, or 70? It seems that chronological age had a more definitive meaning several decades ago. For instance, in 1900 Americans could expect to live, on average, to age 47; only 3 percent of the population lived past 65. With Americans living longer today, the distinctions between life periods seem to be blurring.

By analogy to machines, a human body that has been functioning for a number of decades tends to work less efficiently than it did when it was "new." At age 50 or 60, the kidneys, lungs, heart, and other sensory organs are less efficient than they were at 20. Yet across middle adulthood the physical and cognitive changes that typically occur are, for the most part, not precipitous.

Sensory and Physical Changes

In most cases sensory and physical changes in midlife are so gradual that people are often not aware of any changes until they take stock on a birthday, at the wedding of a child, the first time they are called "Grandma" or "Grandpa," at a retirement celebration, at the death of a parent, or at some other significant life event. After adolescence, most integrated bodily functions decline at the rate of about 1 percent a year. Overall, middle-aged individuals report that they are not appreciably different from what they were in their early thirties. They mention that their hair has grayed (and often thinned), they have more wrinkles, they are paunchier, they may have "lost a step," they tire more easily, and they rebound less quickly. Even so, except for those in poor health, they find that, on the whole, they carry on in much the manner that they did in their younger years. The best preventatives against loss of strength and vitality and some minor sensory changes are an engaged lifestyle, proper nutrition, and regular exercise, along with a healthy dose of humor about how life marches along.

Vision

At least 75 million Americans over age 40 are experiencing **presbyopia** (prez-bee-OH-pee-uh), a normal condition in which the lens of the eye starts to harden, losing its ability to accommodate as quickly as it did in youth (*presbys* is Greek for "old man" or "elder"). Symptoms include getting headaches or "tired" eyes while doing close work. With many baby boomers reaching middle age, more people are likely to find they need contacts, bifocals, or half glasses to read the newspaper, computer screens, restaurant menus, or small printed numbers on price tags, wristwatches, and prescription dosages. If this problem is not corrected, people find they can read printed material only by holding it farther and farther away from their eyes, until eventually they cannot see to read even at arm's length. Adaptation to darkness and recovery from glare also take longer, making night driving somewhat more taxing. Distance acuity, contrast sensitivity, visual search, and pattern recognition are also diminished (Madden, 1990; "Learning to see," 1998).

A number of disorders that affect sight become more common with the normal aging process. **Glaucoma,** increased pressure caused by fluid buildup in the eye, can damage the optic nerve and lead to blindness, if left untreated. The disorder has no symptoms in the early stages and can be detected only by a professional eye examination. **Cataracts,** or clouding of the lens, typically occur in 30 to 50 percent of people over age 65, but in some instances they appear among individuals in their late fifties and early sixties. The condition can usually be remedied by surgical removal of the affected lens. Then this vision impairment can be corrected by eyeglasses, a contact lens, or an artificial lens placed in the eye at the time of the operation. **Floaters,** annoying floating spots, are particles suspended in the gel-like fluid that fills the eyeball, and generally they do not impair vision. However, a severe "floating" problem accompanied by flashes of light could indicate the more serious problem of retinal detachment, which if detected early enough can be treated with laser surgery. **Dry eye,** stemming from diminished tear production, can be uncomfortable and can usually be eased with drops. Those who look at computer screens or read documents for long hours are more likely to experience this discomfort. The thinning of the layers of the retina and/or rupturing of tiny blood vessels is **macular degeneration,** the first signs of which are faded, distorted, or

Macular Degeneration One of the eye disorders midlife adults might experience is macular degeneration, the first sign of which is faded, distorted, or blurred central vision. Eating plenty of fruits and vegetables and reducing fats in the diet can promote clarity of vision and overall health of the eyes.

blurred central vision (Browder, 1997). In a University of Wisconsin study, men and women aged 45 to 84 who ate the most saturated fat were 80 percent more apt to show early signs of age-related eye degeneration. Just as saturated fat contributes to clogged arteries and reduces blood flow to the heart, it might reduce the amount of blood reaching the eye. A diet rich in fresh fruits and vegetables helps preserve vision (Browder, 1997).

Hearing

Changes in hearing usually begin at about age 30. In **presbycusis** (prez-bee-KOO-sis) typically the ability to hear high-pitched sounds, such as speech, declines, but the magnitude of the change varies appreciably among individuals. There appears to be a genetic predisposition to hearing loss, because age-related hearing loss tends to run in families. The baby boomers (including President Clinton) were the first generation to listen to highly amplified music over a period of years and are now experiencing premature hearing loss requiring hearing aids (Cleveland, 1998). By age 50 about one in every three men and one in every four women have difficulty understanding a whisper. However, only a small number of the population at 50 have substantial hearing problems.

Some of the more common causes of conductive hearing loss include cochlear damage due to prolonged exposure to loud noise, lack of good muscle tone in the middle ear (caused by stress or poor diet), and overgrowth of the cochlear bone, which results in the stapes (stirrup) bone becoming fixed. People who have jobs associated with high noise levels—such as miners, truck drivers, heavy equipment operators, air-hammer operators, some industrial workers, and rock concert performers—are particularly at risk. Also, some individuals in their sixties report that they "take in" information more slowly than they did earlier in life. *Audiology testing* determines the extent of hearing loss. To conduct a simple test, rub your fingers together next to each ear. If you cannot hear this slight sound, you might have the beginning of hearing loss. A treatment regimen could include cleaning your ears, using an amplification device (hearing aid), or surgical intervention. A standard hearing aid for one ear can cost $1,000 or more; for individuals who cannot afford this, hearing loss reduces their quality of life (Applied Medical Informatics, 1997).

Taste and Smell

Taste buds, which detect salty, sweet, sour, and bitter tastes, normally are replaced nearly every 10 days. However, in people who are in their forties, taste buds are replaced at a slower rate, and smell receptors begin to deteriorate, affecting the sense of taste. Women have a better sense of taste than men do because they generally have more taste buds, and scientists believe estrogen increases a woman's taste sensitivity. Salty and sweet are the first

tastes to change, and one's appetite may become especially partial to sweet and salty foods (Chaikivsky, 1997).

From age 50 on, the sense of smell typically starts to decline. Probably half of all people who are 65 have had a noticeable loss of their sense of smell. Tastes and flavors are almost entirely detected in the nose, so food won't taste as delicious. Valery Duffy, a taste-and-smell researcher at the University of Connecticut at Storrs, observes that women with a weakened sense of smell are more likely to gain weight as they attempt to satisfy their yearning for flavors or compensate for the loss of flavors with the gratifying texture of fat. One can compensate for the loss of flavors by holding food in the mouth longer or cooking with more flavorful seasonings. Dr. Duffy also suggests as a preventative measure that people get the flu shot each year, because each time one catches the flu, the virus can diminish one's sense of smell (Browder, 1997).

Question

What are some of the typical changes in the senses in middle adulthood?

Appearance

Sometimes I think of the alternatives to looking older, and I wonder what it would be like to have my face frozen the way I was in my thirties, and I think—that would be ridiculous! That's not me, that doesn't reflect the years I've lived and all the things I've experienced. I don't want to deny my experiences, and I feel that if I dislike my aging looks, I'm denying all the wonderful parts of my life. I don't want to do that.

A 48-year-old woman, in Doress-Worters & Siegal,
The New Ourselves Growing Older (1994), p. 39

During midlife, the gums of the teeth begin to recede, which for some leads to a condition called *periodontal disease.* This in turn leads to loss of teeth, and caps, bridges, and false teeth might become necessary. Those who cannot afford dental work are likely to have difficulty eating nutritious foods, and their health might decline (Chaikivsky, 1997; Browder, 1997). Some dental experts suggest that regular flossing and brushing and scheduled dental visits can help people keep their own teeth into their nineties. Losing teeth or loosening teeth might sound trivial, yet it takes time to become comfortable with a new facial appearance created by extensive dental work—and these procedures are quite costly.

As one ages, the skin becomes dryer, thinner, and less elastic after years of exposure to ultraviolet rays. As the skin loses collagen, fat, and oil glands, it becomes more wrinkled. Skin cells grow more slowly with age, and the outer layer of skin is not shed and replaced at the same rate as in younger years. With aging, cells lose some of their ability to retain water, causing dryness. Soaps, antiperspirants, perfumes, and hot baths can aggravate this condition or cause itching. As the skin loses tone and elasticity, it sags and wrinkles, especially in areas where there is frequent movement, such as the face, neck, and joints. Some people also develop "droopy" eyelids as a result. Additionally, darker patches of skin ("age spots") caused by many years of exposure to the sun begin to appear before the age of 60 (Dickinson, 1997).

Scaly patches of skin and any changes in skin color should be checked by a dermatologist, because these could be signs of *basal-cell carcinoma,* which can be treated. However, a study of 37,000 patients with basal-cell carcinoma researchers at the Danish Epidemiology Science Center suggests that this malady could be a marker for other, more serious cancers. People who smoke tend to have more wrinkles than nonsmokers of the same age, complexion, and history of sun exposure (National Institute on Aging, 1996b). Regular head-to-toe examinations of the skin, which is the largest organ of the body, should be conducted by a medical professional as a necessary component of a checkup.

The facial appearance of aging men and women in American society is part of the so-called "double standard of aging" (see Figure 15-1). As men age, they are often considered "mature" or "sophisticated" or more attractive than when they were younger. However, there are very few kind expressions for the way many older women look. American men and women spend billions of dollars each year on "wrinkle creams," bleaching products, skin lotions, facials, and electrolysis for unwanted hair removal and other dermatology procedures (National Institute on Aging, 1996b).

Women vary greatly in how they respond to looking older. There are indications that those who were hugged, kissed, and cuddled when younger tend to accept their bodies as they are. Also, women whose families provided opportunities for physical activity or who were athletes in high school and college feel more positive about their looks than women without such experiences. When self-esteem is high, body image is often positive. Considering the relationship between athletics and self-esteem, it is no surprise that studies show that men tend to overrate their body image, while women tend to underrate theirs. A woman who was considered unusually attractive when younger may find it more difficult as she begins to look older, as compared to a woman who never set great store on her looks. Women who are concerned about looking older seem to experience this concern up to their middle and late sixties. In fact, some women *like* getting older; they feel freer and more confident. What they don't like is *looking* older (Doress-Worters & Siegal, 1994). In general, compared to women, men have better-looking skin as they age. As men take whiskers from their face each day, they also slough off dead skin cells, leaving a more youthful facial appearance.

There are many socioeconomic and behavioral implications of the aging of skin. A new cosmetic industry is cropping up that is catering to the needs of Black Americans. Sam Fine, makeup artist to black actors, actresses, and musicians, recently published *Fine Beauty: Beauty Basics and Beyond for African-American Women* (1998). He states that there are 40 to 50 skin shades in African American women alone. Fine has developed a line of cosmetic products to help this segment of the American population become more comfortable with their aging skin (Belluck, 1998).

More American women and men are taking advantage of a European health habit: having facials and full body massages to maintain a healthier, more youthful appearance. Use of cosmetic surgery (face lifts, tummy tucks, breast lifts, liposuction) by both men and women

Figure 15-1 The Double Standard of Aging Women in Western cultures are likely to be more concerned than men about their changing physical appearance during midlife. Studies show that men are likely to overrate their image, whereas women tend to underrate theirs.

has also increased as baby boomers have entered their middle years. Wearing hats, long sleeves, and long pants, using more protective sunscreens, using a humidifier in the home, and staying out of the hottest sun of the day from 11 A.M. through 2 P.M. are all highly preventative measures to avoid skin damage. Millions of retirees migrate to southern and western states with sunny climates, where skin cancer rates are high. Proper nutrition and plenty of drinking water, regular exercise, cosmetics and creams such as those that contain alpha hydroxy, and limited sun exposure can help keep the skin more supple and healthy (Atkins, 1996).

Hair color change to gray or white is usually the obvious physical change that marks the change in chronological age. During midlife or sooner, the color, thickness, and texture of hair undergo changes, which can include thinning, balding, and graying. Thinning hair is often associated with aging males, many of whom develop hair loss and balding. Many women are also likely to find their hair thinning as they experience *perimenopausal* symptoms and hormonal changes.

Some hair products, such as Rogaine, are marketed as stimulating the scalp to prevent hair loss. Hair coloring and hair restoration techniques can help maintain a more youthful appearance. Many women and men in middle adulthood who are at the prime of their professions want to look their best and consider it essential to retain a more youthful appearance. On the other hand, a growing number of adults in midlife are very accepting of their changing looks and find emotional comfort and satisfaction in this transitional stage of life. Many are so satisfied with their busy new social roles of grandmother and grandfather, mentor at work, and community or church volunteer that they barely focus on "looks."

Body Composition

One of the major concerns of some individuals in midlife is body composition, or proportion of muscle to fat. Around age 30, muscles begin to atrophy, which can diminish strength, agility, and endurance. Men are less likely to notice this because their muscles tend to be larger. By the time a person reaches the fifties, muscle loss is likely to become more evident—and weight gain more pronounced. According to the *Baltimore Longitudinal Study of Aging*, muscle mass declines, on average, 5 to 10 percent each decade. This can become more evident in daily activities like carrying groceries or getting out of a chair. Muscle loss also leads to an increase in body fat. By age 50, one should eat *240 fewer calories a day* to maintain the same weight carried at age 30 (Doress-Worters & Siegal, 1994).

Americans are preoccupied with weight because currently in this culture "thin is in." Emphasis on "ideal" weight on weight charts creates damaging pressure on women (and men) to be thin. What seems to be overlooked is that each person has a unique size, shape, and body chemistry strongly determined by heredity. One's weight is a function of several additional factors: energy input (food consumption), energy used (activity level), and the body's rate of using that energy (metabolism). Dieting to lose weight can become an unhealthy way of life. Each time a person goes on a low-calorie diet, the body reacts as if it is being starved and tries to preserve energy by decreasing metabolism. Rate of metabolism slows each time a person diets. After a diet, unless the person dramatically increases exercise, the body adds more fat to protect itself from food deprivation. This explains why 90 to 99 percent of dieters regain their weight, and perhaps even gain back more, within five years.

Thin people, who are more likely to be malnourished, are also more susceptible to certain diseases, including lung diseases, osteoporosis, fatal infections, ulcers, and anemia. As more baby boomers join the ranks of the middle-aged population, it is likely that two more severe eating disorders will be prevalent. *Anorexia*, common among young women, might be increasing among older women. *Bulimia* seems to be found among all age groups. According to Doress-Worters and Siegal in *The New Ourselves, Growing Older* (1994), the extent of eating disorders among middle-aged and older women is not known, and funding for research on this topic has been cut. Studies show that older people who are "significantly underweight" die sooner than those who are not (Doress-Worters & Siegal, 1994).

Americans' obsession with dieting programs and the escalation in the number of cosmetic surgeries confirm that some of today's adults would rather acquiesce to than ignore current norms. Goodman (1996) examined the social, psychological, and developmental factors that precipitated cosmetic surgery with 24 female subjects ranging in age from 29 to 75. She found that age, or cohort membership, differentiated women's attitudes and behaviors regarding their bodies. Younger women in the group (ages 29 to 49) were uniformly preoccupied with their bodies, the classic area of discontent being breast size—either too big or too small. Older women in the group (over age 50) assigned more importance to their faces than to their figures; their classic complaints involved symptoms of aging—wrinkles, "saggy" jowls, and "droopy" eyelids. Davis (1990) has suggested that television and other media images ushered in the age of "commercialized feminity" for women growing up in the 1950s. Women were encouraged to aspire to an "ivory" complexion, blonde Breck-girl hair, and a Marilyn Monroe figure. (Marilyn Monroe, by the way, was a perfect size 12. Today's models are size 6). Breasts and cleavage became focal points in media imagery of women.

On a more realistic note, scientists now know that most muscle aging is preventable and reversible with a regular exercise program of resistance training, or weight lifting. In a Tufts University study, women in their sixties increased muscle strength in their legs, backs, abdomens, and buttocks by 35 to 76 percent. After one year of strength training, these women emerged both physiologically stronger by 15 to 20 years and psychologically more youthful (Browder, 1997). A middle-aged person with an active lifestyle can still take advantage of fitness programs that support these health-oriented goals.

Question

What are some of the aspects of natural aging that the baby-boom generation has challenged?

With the normal aging process after age 35, bones become less supple and more brittle and begin to lose their density. It is generally held that women are at greater risk of **osteoporosis** (a disorder of thinning bone mass and microarchitectural deterioration of bone tissue) than men because men have 30 percent more bone mass at age 35 than women, and they lose bone more slowly as they age. Age-related decreases in bone density accelerate with menopause in women, so the earlier a woman experiences menopause, the higher the risk. Osteoporosis is a complex condition. Usually it takes years for it to advance to the stage where it can be detected, because the skeletal structure is not visible and one does not feel the slowing in the replacement of older bone cells with newer bone cells. The NIH has estimated that 25 million Americans are affected by this type of bone degeneration; although osteoporosis is more prevalent in women, about one-fifth of osteoporotic fractures occur in men, and occasionally men suffer from severe osteoporosis in middle age (National Institute on Aging, 1996a).

As with muscle mass, men start out with greater bone density than women. Consequently, women are likely to experience bone fractures up to a decade earlier than men. A major contributing factor to bone loss in women is the major decline in the production of the hormone estrogen as a woman experiences *perimenopause, menopause,* and *postmenopause.* Most physicians today recommend that, as a preventative measure, women begin taking calcium supplements by the time they are 35 years old. And a painless medical test, using a scan of the spine or hip, is now suggested for adults in middle age or older. One can also add calcium and Vitamin D to one's diet by eating more milk products, egg yolks, leafy green vegetables, certain shellfish, tofu, and other soybean products. For people who are lactose intolerant and cannot digest milk products, other foods can supply the additional calcium. We shall discuss the serious health consequences of bone loss for elderly adults in Chapter 17.

Adequate body weight and fat tissue offer some protection from osteoporosis. Bones that carry body weight must work to produce new bone tissue, and fat tissue helps maintain some estrogen in the body after menopause. Diuretics such as caffeine and alcohol can cause loss of calcium and zinc in the urine. It is also suspected that smoking interferes with the body's production of estrogen which in turn affects the onset of osteoporosis. Weight-bearing exercises such as walking, jogging, jumping rope, and dancing make bones work harder, strengthen the muscles and ligaments supporting the skeleton, and slow bone loss. For a person at a higher risk of osteoporosis, a nutritionist can help plan a preventive diet tailored to individual needs, and a physical therapist can suggest an exercise program to strengthen muscles and ligaments at any age.

Many studies are being conducted to examine risk of osteoporosis across cultures. Black women have 10 percent more bone mass than white women, and they might have more *calcitonin,* the hormone that strengthens bones. However, the consequences of a hip fracture for blacks are far greater, possibly because of the effects of more poverty, inadequate resources for health care, and the greater likelihood of other underlying disease. One study showed that Japanese women continue to gain bone mass until the age of 39 and are also likely to have fewer fractures because they sleep on futons close to the floor, wear broader shoes (rather than spiked heels), and use less alcohol and tranquilizers. Another study showed a lower rate of hip fractures in Hong Kong compared to the white U.S. population. The fairer your complexion, the greater your risk for osteoporosis (Doress-Worters & Siegal, 1994).

Another disorder that surfaces in middle age is **rheumatoid arthritis,** an inflammatory disease that causes pain, swelling, stiffness, and loss of function of the joints (shoulder, knee, hip, hands, etc.) (NIH, 1998b). People with this condition may experience fatigue and occasional fever in addition to the typical symptoms. Arthritis symptoms vary from person to person and can last a few months and then disappear for years. Some people have mild forms of this disease, others live with serious disability. Scientists classify rheumatoid arthritis as an autoimmune disease—the person's own immune system attacks her or his own body tissue. Although there is no single test for this disease, especially in its early stages, lab tests and X rays can determine the extent of bone damage and monitor the progression of this disease. Current treatment strategies include pain relief, a balance of rest and exercise, and patient education, and most people with this disease can lead active and productive lives (National Institute of Arthritis and Musculoskeletal and Skin Diseases, 1998). Dr. Neil Gordon, director of physiology at the Cooper institute in Dallas, has found that higher levels of vascular strength, cardiopulmonary fitness, and increased flexibility are associated with a lowered incidence of arthritis (Physical Fitness in Midlife, 1993).

Hormones

Human growth hormone (HGH), a powerful hormone that has been used to treat children afflicted by dwarfism, has become a trendy anti-aging potion among some of the social elite, as well as with Hollywood stars and executives. Doctors say thousands of Americans are currently using the hormone, which costs $200 to $400 a week, in hopes of reversing the droops, dwindles, and sags of middle age (Kuczynski, 1998). Currently, the National Institute of Aging is funding medical studies to investigate earlier findings that HGH and other hormones can slow, stop, or possibly reverse the changes associated with aging. HGH, also known as *somatotrophin,* is the most abundant hormone secreted by the pituitary gland and consequently impacts the production of all other endocrine hormones in the body. Daily HGH secretion diminishes with age; a 60-year-old might secrete 25 percent of the HGH secreted by a 20-year-old. Daniel Rudman, an endocrinologist, pioneered the original research on HGH in 1985, hypothesizing that the changes in body composition that become apparent around age 35 had to do with declining hormone levels (Rudman et al., 1991).

Some middle-aged and older Americans believe they have found the "fountain of youth" by having regular hormone injections. Hormones are protein messengers of the endocrine system that circulate throughout the body to all organs, causing natural reactions that affect not only ovaries and testes but many other vital life functions such as memory, protein synthesis, cell repair, metabolism, ability to sleep, body temperature, water balance, and sexual functioning. Hormone production decreases as one ages. Scientists are trying to decide whether the body suffers without hormones or simply has no use for them in later years. Consequently, a great deal of research has been and is being conducted (Kuczynski, 1998).

The physicians administering these hormones call themselves "anti-aging specialists." The American Academy of Anti-Aging Medicine, founded in 1993, currently boasts a membership of more than 4,300 U.S. doctors who specialize in youth preservation. Their president states, "We're not about growing old gracefully. We're about *never* growing old" (Kuczynski, 1998). Daniel Rudman, M.D., in a 1990 study, found that some men over the age of 60 given HGH for six months had improved energy, less body fat, and more muscle mass.

Owen Wolkowitz, M.D., of the University of California at San Francisco, administered another hormone—DHEA (dehydroepiandrosterone)—to a small group of depressed men and women between the ages of 50 and 75. The findings suggest that the hormone not only lifted depression but improved memory as well. Some researchers also believe that DHEA blocks the decline of the body's immune system. Like HGH, DHEA is abundant in the body when one is age 20, but it continues to decrease with time. At 80 years of age, humans usually produce only 10 to 20 percent of DHEA produced in a 20-year-old (DHEA Center, 1997; DHEA Prohormone Complex, 1997).

Question
Under what circumstances might a doctor recommend HGH or DHEA to a patient?

Menopause and Female Midlife Change

Archie: Edith, if you're gonna have a change of life, you gotta do it right now. I'm gonna give you just 30 seconds. Now come on, *change!*

Edith: Can I finish my soup first?

"All in the Family," 1972

Females born during the baby boom are now beginning to experience **menopause**—a process culminating in the cessation of menstrual activity (Love & Lindsey, 1997; Reichman, 1996; Sheehy, 1992). The cessation of menstruation typically takes two to four years, with intermittent missing of periods and the extension of the intervals between periods. This time period is referred to as premenopause or **perimenopause.** A woman is said to have gone through menopause when she no longer menstruates for one year. Many American women, known as "midlife Mommies," are racing the biological clock to have children before their bodies cease to produce mature, viable eggs. Menopause is the culminating sign of the **climacteric,** characterized by changes in the ovaries and in the various hormonal processes over 2 to 5 years prior to complete cessation of menstruation. Probably the most significant change is the profound drop in the production of the female hormones (particularly *estrogen*) by the ovaries.

In Western countries, the average age range for complete cessation of menses is around 45 to 55, and the average age is about 51, but it can occur as early as the thirties or as late as the sixties (Sheehy, 1998). A cross-cultural study with nearly 20,000 women from Europe, the Americas, Asia, Australia, and Africa found the median age of natural menopause to range from 49 to 52 years (Morabia & Costanza, 1998). Bromberger (1997) reported study findings that black women enter menopause sooner than white women: blacks at an average age of 49.3, whites at an average age of 51.5. However, she states that stress might be a factor. Younger women also undergo an early menopause if their ovaries are removed by a surgical procedure called *hysterectomy.* *Premature menopause,* occurring naturally or induced before age 40, affects about 4 percent of the female population (Midlife Passages, 1998b).

Women in Japan and other Asian countries rarely complain about menopausal symptoms. A possible explanation has emerged, pointing to a new dietary approach to treating menopausal symptoms: Soybeans, and products made from soybeans, are rich in large amounts of chemicals in plants that produce natural estrogen. Dietary habits in Asian countries might also account for the lower rates of heart disease and breast cancer there. The dietary approach appeals to many women who seek a more "natural" alternative to hormone replacement therapy (Brody, 1997).

Aging changes that can accompany the hormonal changes of menopause include incontinence (involuntary leaking of urine), heart disease, and *osteoporosis* (bone thinning). These changes are considered normal reactions to the body's reduced production of sex hormones. Every woman's experience with menopause is different, and most have minimal symptoms and continue to function quite well. Most effects can be reduced with exercise and diet modifications, disappear with treatment, or diminish over time. Menopause reminds a woman to continue with a good health program as she progresses through midlife. Family history might provide clues as to when a woman is likely to go through menopause, because the timing of natural menopause has a genetic component.

Much controversy surrounds **hormone replacement therapy (HRT),** a regimen often recommended to menopausal women by physicians to maintain cardiovascular fitness, slow bone loss, and slow memory loss. Many women are skeptical about taking the hormone treatment advice of their physicians because their mothers or grandmothers were subjects of massive estrogen replacement therapy (ERT) treatment in the 1960s to 1970s, and many developed cancer of the ovaries or uterus, leading to early deaths. However, with an HRT that combines estrogen and progesterone and is prescribed and monitored by a physician, many women assert they feel and look much better.

Today, considerable controversy persists as to whether hormone replacement therapy (HRT) increases a woman's risk of developing breast cancer. Some scientific evidence suggests that women who continue to use HRT for five or more years after menopause are 30 to 40 percent *more likely* to develop breast cancer than are women who have never used menopausal hormones (especially if there already exists a hereditary risk). Although the inclusion of *progestin* (a form of the hormone progesterone) in HRT apparently protects the uterus against estrogen-induced endometrial cancer, it does not protect against breast cancer. Some authorities believe that estrogen's benefits to the heart more than outweigh any risk to the breast (Love & Lindsey, 1997; Brody, 1995c).

Clearly, women face a conflict with respect to HRT. Each case must be judged on its own merits, and each woman should be medically evaluated for her individual risk of developing heart disease, osteoporosis, and breast cancer (Brody, 1995a). Nearly 4 in 10 women going through menopause currently use HRT (Painter, 1994a). Critics of hormone therapies also contend that the medical and drug establishment has turned a normal aging event into a "disease" requiring medication, a phenomenon they call "the medicalization of menopause"; the powerful pharmaceutical industry, they argue, has much to gain by making every woman feel that she is "diseased" instead of aging normally (Brody, 1995c).

Today women openly discuss this normal aging event and share information on alternative approaches, the media is educating the public, and up-to-date access to discussion of issues is available through Internet newsgroups. Although psychoanalytic literature stresses the adverse psychological impact of no longer being able to bear children, less than 5 percent of all menopausal women report that they feel distressed by this fact

Women and Menopause Contemporary middle-aged women openly share information about menopause and traditional or alternative therapies. Contrary to popular opinion, scientific studies have not linked menopause to depression or dramatic change for most women, and cross-cultural studies find that women in other cultures experience heightened social status after menopause. Many American women, freed of reproductive concerns, find it a marker to start another productive phase in their lives.

(Healy, 1985). Many women are relieved that they no longer have to worry about pregnancy, and some report an improvement in their sex lives. And when menopause occurs on time during midlife, it is not likely to be a source of psychological distress (Reichman, 1996; Lennon, 1982). Although conventional wisdom has linked menopause with depression, scientific studies have failed to establish a causal relationship between them. Indeed, research surveys consistently reveal that women are considerably more likely to suffer from depression in their twenties and thirties than at midlife (Elias, 1993).

Women are more likely to feel upset if they view menopause as signaling the end of their attractiveness, usefulness, and sexuality. Such feelings can be heightened by our youth-oriented culture, which tends to devalue older people. From cross-cultural studies, we know that physical and emotional symptoms of menopause are rarer in societies such as Japan and India, where postmenopausal women gain greater power and heightened social status than they enjoyed during their reproductive years (Elias, 1993).

Evidence suggests that the climacteric typically does not cause problems that were not already present in a woman's life. Women with preexisting or long-standing difficulties might be more susceptible to problems and react more adversely during the climacteric than less vulnerable women (Greene, 1984). Many women express satisfaction that their children are grown and have left home, opening all sorts of new possibilities for them (Reichman, 1996; Goleman, 1990b). As Judith Reichman, M.D., (1996) notes, "Menopause is the completion of a life cycle that started before we were born. It does not constitute a stop but actually is the start of the next phase of our lives."

One of the largest and potentially most beneficial studies being conducted is the Women's Health Initiative. Currently researchers from the National Institutes of Health (NIH) are studying 164,500 women of various ages and racial and ethnic backgrounds across the United States. This scientific investigation, which will not be completed until the year 2005, is expected to find out whether a low-fat diet, HRT, calcium, and vitamin D might prevent heart disease, breast and colorectal cancers, bone fractures, and memory loss (Weinstein, 1997). To find out more about the Women's Health Initiative, see "Following Up on the Internet" at the end of this chapter.

Reproduction After Menopause

For the first time in history (excluding the Biblical account of Sarah's giving birth when she was over 80 years old), women can potentially "conceive" and carry a baby to term after menopause. This reproductive process involves special medical procedures to inject the mother-to-be with hormones for a period of time to build up her uterus to provide a viable home for an implanted embryo. A postmenopausal woman no longer has any eggs (ova), so she must seek out a donor to provide healthy eggs. Then her husband's (or donated) sperm is used to fertilize an egg to create a zygote, which will be grown into an embryo before it is transplanted into the woman's uterus in an *in vitro* fertilization procedure. That an Asian American woman gave birth at age 63 was sensationalized in all the media in 1997 (the birth occurred in 1996). The woman had lied about her age, because the California infertility clinic she attended had an upper limit of 55 years old. It was also reported that this couple spent over $50,000 and a great deal of time and effort to have this child (Kolata, 1997a).

Worldwide, some postmenopausal women have borne children using this method of procreation. The age of the mother has become a controversial issue, while the age of the father is rarely discussed, which perhaps is due to the double standard of aging we discussed earlier. When men in their seventies and eighties father children, they are lauded for their virility. Ethicists are debating such questions as these: Are older mothers yet another example of the way society encourages women to clutch at eternal youth? Are they a laudable example of the way technology can overcome the barriers of age? If women's bodies are not capable of making babies after menopause except through medical intervention, is there something inherently wrong in creating a pregnancy that would not have occurred naturally? Will the children created so late in their parents' lives be able to have a childhood, or will they as teenagers become responsible for aging, possibly senile parents? Will the parents live long enough to raise this child? (Kolata, 1997a; 1997b).

> ### Questions
> *What is menopause, and how does it affect the woman's life? How is it possible for a woman to give birth to a child after menopause?*

Male Midlife Change

Men obviously do not undergo menopause, but in midlife they are likely to experience enlargement of the **prostate gland,** a walnut-sized gland at the base of the urethra (the tube that emerges from the bladder and carries urine to the outside world via the penis). About 10 percent of men aged 40 already have recognizable enlargement of the prostate; the condition is virtually universal in men at age 60. The exact reasons for the enlargement of this gland are unclear, but research is being conducted to shed some light on this in the next few years (National Cancer Institute, 1998a). Hormonal changes associated with aging, along with a high-fat diet low in fruit and vegetables, are thought to be implicated in the process (Giovannucci et al., 1993). As the prostate enlarges, it puts pressure on the urethra, which contributes to decreased force in the urinary stream, difficulty in beginning urination, and an increased urge to urinate. Although the condition is not dangerous itself, it can contribute to bladder and kidney disorders and infections and urges to urinate that interrupt sleep. Should serious obstruction to the outflow of urine occur, the tissue causing the enlargement of the prostate can be removed. Only rarely does a patient become impotent after the operation.

A more serious problem is cancer of the prostate. The National Cancer Institute reports that "prostate cancer is the most common malignant cancer in North American men," and it is now the second leading cause of cancer death in men, claiming nearly 40,000 lives each year (National Cancer Institute, 1998c). This type of cancer is rarely seen in men under 50; incidence increases with each decade of life. Japanese men have a low incidence of prostate cancer, however, perhaps due to dietary, genetic, or screening factors. Black males have been found to have a higher incidence of prostate cancer than white males, although any male with a family history of prostate cancer is at an increased risk of this disease. Other lifestyle factors associated with this disease include alcohol consumption, vitamin or mineral interactions, dietary habits, promiscuity, and genital warts (National Cancer Institute, 1998c). A recent double-blind study conducted with U.S. males found that those with a higher intake of selenium (a trace nutrient in grains, fish, and meat) apparently had a lower risk of prostate cancer—but NIH researchers caution that further research needs to be conducted (Giovannucci, 1998).

A major risk with prostate cancer that is not discovered and treated early is that it will metastasize and spread to the bones. Current screening methods include the DRE (digital rectal exam) and PSA (prostate-specific antigen), which are conducted during physical checkups for middle-aged and elderly males. Fortunately, most cancers of the prostate, particularly in older men, grow at a slow pace. By age 80 the majority of men have prostate cancer, but most outlive it and die from something else.

Although women's levels of the female hormone estrogen plunge during the climacteric, men experience a much slower drop over a period of many years in levels of the male sex hormones (primarily testosterone) (Tsitouras, Martin, & Harman, 1982) (see Figure 15-2). Cultural stereotypes have frequently depicted men in their forties and fifties as suddenly undergoing severe psychological disturbances, leaving their wives for women young enough to be their daughters, quitting their jobs to become beachcombers, or beginning to drink to excess. Such difficulties are commonly attributed to a "male menopause." Medical authorities are quick to point out, however, that what has never been— menses—cannot cease to be.

What part hormonal changes could play in a male midlife crisis that only some men experience is a matter of lively and sometimes heated medical controversy (Angier, 1992). Some authorities attribute a decline in testosterone levels with a concomitant decrease in muscle mass and strength, the buildup of body fat, the loss of bone density, flagging energy, lowered sperm output, decreasing strength and mobility, fading virility, and heightened mood swings. But human growth hormone also influences these functions and declines with age. Researchers find it difficult to sort out which hormone plays which role, or, as is most likely, how the two hormones combine to affect men's vigor. Older men also typically lose the circadian rhythms that affect testosterone fluxes in their younger counterparts, in whom hormone levels usually peak prior to rising in the morn-

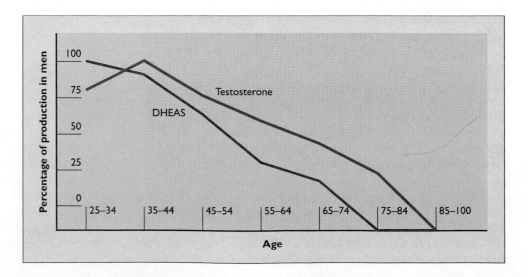

Figure 15-2 Testosterone Production in Men Men experience a much slower drop in production of sex hormones over a period of many years in comparison to the female climacteric around age 50 in middle age.

From: *Newsweek,* Geoffrey Cowley. ©1996 Newsweek, Inc. All rights reserved. Reprinted by permission.

ing, which is one reason why young men often awake with erections.

Although most medical authorities currently do not believe that men undergo a hormonal midlife change comparable to menopause, some nonetheless speculate that as the baby-boom generation enters midlife, interest in treating aging men for profit will encourage the definition of a "clinical syndrome" and its "medical treatment" (Angier, 1992). Researchers at Yale also find that men's concerns about sexuality are a major problem area; such concerns include worries about waning virility and declining physical attractiveness. A rather high incidence of **impotence,** the inability to have or sustain an erection, can reinforce their concerns.

One recent federally funded study carried out in Massachusetts found that about half of American men over age 40 experience potency problems. Although aging can contribute to declining male virility, impotence often results from drugs, medical conditions such as diabetes, cardiovascular disorders, and cancer, and social habits including smoking, alcohol consumption, and lack of exercise. Stress, depression, grief, illness, and accidents can also give rise to temporary impotence (Altman, 1993). The jury is still out on the new pharmaceutical product Viagra, which when it came on the market for male impotence sold 200,000 new prescriptions in the first four weeks and which has a potential market of 30 million American men (Comarow, 1998). As of this writing, nearly 100 men have died using Viagra; these men ranged in age from 48 to 80, with most in their sixties and seventies. Some doctors have been highly criticized for prescribing Viagra for women without any proof of its safety from clinical trials with women.

Psychologists and sociologists typically dismiss notions of a male menopause as a social myth deriving from an overmedicated culture that is obsessed with health and pathologically fearful of aging and death. They look mainly to social explanations for understanding changes in a man's life that produce a crisis in his self-concept. The middle years call for readjustments and reassessments, some of which can be unpleasant. Most commonly, troubling events are spread over one or two decades.

Researchers have followed a sample of unusually accomplished, self-reliant, and healthy young men from their first year as Harvard University students, mostly from the classes of 1942 to 1944, until their late forties. The men judged to have had the *best* outcomes in their late forties regarded the period from 35 to 49 as the happiest in their lives and the seemingly calmer period from 21 to 35 as the least happy. But the men *least* well adapted at midlife longed for the relative calm of young adulthood and regarded the storms of midlife as unusually painful (Rosenfeld & Stark, 1987).

Men's responses to middle age are varied. Many seem to move calmly through it; others have a stormy

Varied Responses by Males to Midlife Many men navigate through middle adulthood with a healthy lifestyle, seeming to be ten years younger than they are. Some, however, find it more difficult to adapt, particularly if they lead a less engaged lifestyle.

passage. Middle age can be a time of new personality growth, a period when a man moves toward a new kind of intimacy in his marriage, greater fulfillment in his work, and more realistic and satisfying relationships with his children. But for fewer men, it can be a time of developmental defeat, leading to such problems as depression, alcoholism, obesity, and a chronic sense of futility and failure (Sheehy, 1998).

Question

Is there any validity to the notion of a male "menopause" or midlife change?

Health Changes

The incidence of various health problems increases with age. However, people can maximize their chances for leading healthy and long lives by altering their lifestyles

Changing Unhealthy Habits A nutritious diet and regular exercise are keys to healthy living at all stages of life. At midlife, one's metabolism slows, often leading to weight gain unless one reduces caloric intake or increases activity level. Although many Americans recognize the importance of these factors, some have considerable difficulty changing unhealthy habits. Many programs assist people in achieving behavioral changes to reach their health goals.

to include a variety of health-conscious practices such as regular exercise and eating a healthy diet. Nazarid's 1989 study of the experience of 10,000 devout Mormons residing in California is illuminating. The individuals observe lifestyle habits recommended by the Church of Jesus Christ of Latter-Day Saints, which include abstaining from tobacco, alcohol, caffeine, and drugs, eating meat sparingly, consuming plenty of herbs, fruits, and grains, securing ample sleep, and engaging in regular physical activity. Middle-aged Mormon men have a cancer death rate only 34 percent of that of middle-aged non-Mormon white men and only 14 percent of their death rate for cardiovascular disease. A 25-year-old Mormon man has a life expectancy of 85 years, whereas the average American white 25-year-old male can expect to reach age 74. Middle-aged Mormon women have a cancer mortality rate 55 percent of that of middle-aged non-Mormon white women and a death rate only 34 percent of theirs for cardiovascular disease. A 25-year-old Mormon woman has a life expectancy of 86 years; a non-Mormon white female has a life expectancy of 80 years. Studies of other nonsmoking, health-conscious religious groups such as the Seventh-Day Adventists also show low rates of mortality from cancer and heart disease (Nazario, 1989). Of course, health-promoting habits afford no guarantees but merely weight the odds in a person's favor.

Social habits and lifestyles affect our health in many ways. C. Everett Koop, former Surgeon General, estimates that half of U.S. deaths in 1990 were caused by smoking, drinking, sexually transmitted diseases, drug abuse, poor nutrition, guns, and motor vehicles. He observes that many of the old epidemics of the eighteenth century have been replaced by "self-induced diseases." Moreover, a mounting body of evidence suggests that strong social ties are conducive to health (Wolf & Bruhn, 1993). Groups provide the structure by which we involve ourselves in the daily affairs of life. Not surprisingly, therefore, living alone can be hazardous to one's health. For example, men and women who lack social and emotional support are more than twice as likely to die following a heart attack as are people with a caring family and friends (Friend, 1995).

Social support can be of tremendous help when we are ill. For instance, patients are far more likely to live at least six months after heart surgery if they draw strength from religion and participation in social groups (Elias, 1995). Religious beliefs typically takes on much more significance for middle-aged adults and the elderly. Research studies have shown that patients who have faith that they will heal often do heal more quickly than those who despair over their illness (Elias, 1995). Fortunately, more adults are smoking less, eating more sensibly, and exercising more. Yet many continue their unhealthy ways. *Information You Can Use:* "A Schedule of Checkups for Midlife Women and Men" highlights potential bodily changes and gives a time schedule for diagnostic exams to help maintain a healthy, active life.

Sleep

Most Americans are famous for filling every hour of the day with activity, to the point of being sleep deprived. However, as one proceeds through middle adulthood, several factors can also affect the length and quality of sleep. Use of mild stimulant drugs, such as caffeine and

Information You Can Use

A Schedule of Checkups for Midlife Women and Men

Health Checkups	Diagnosis or Potential Health Concerns	Recommendations
Vision	Presbyopia, glaucoma, cataracts, macular degeneration, dry eyes, watery eyes, "droopy" eyes, other eye conditions.	*If 40 or over,* do you have to hold this book farther and farther away to be able to read this print? *Over 60,* have an eye exam at *least once every 2 years.* This should include dilating the pupils to get a good view of the retina and optic nerve, essential in detecting diseases of the eye. www.nei.nih.gov/publications/cataract/htm
Hearing	Presbycusis, nerve deafness, or other hearing disorders. About 35% of adults age 65 and older have a hearing loss. It is estimated that 50% of those 75 and older have a hearing loss.	*Check yourself:* Can you hear a bird chirping nearby? Can you hear the telephone ringing? Are you asking others to repeat themselves? Do you have difficulty hearing in a crowded room? Can you hear the sound of rubbing your fingers together near each ear? What is the typical sound level of your work environment? Do you have a familial background of hearing loss? An audiology exam may be in order.
Skin Appearance	Look for patches of dry, scaly skin, darkening of a mole, or other skin protrusion; any bleeding or discharge from skin, sunburn.	The American Academy of Dermatology suggests that people, especially those who live in warm climates, *have a yearly skin exam* as part of a regular checkup. In between checkups, be alert for any changes in your skin. www.nih.gov/nia/health/pubpub/skin.htm
Dental Health	Cleaning and examination; X rays for periodontal disease.	*Minimum once per year* for teeth cleaning and *every few years for dental X rays.*
Weight	Sources vary on recommendations. Some say 20% over, others say 30% over recommended weight for body size puts a person at a higher risk of heart disease, cancer, and other illnesses.	Look at muscle-to-fat ratio; consider body frame and lifestyle. Start physician-monitored exercise regimen if overweight. Make an appointment with a nutritionist.
Bone Density	Osteoporosis and other bone disorders can be diagnosed with a painless bone densitometry scan. *Osteoporosis is preventable.* Follow a diet rich in calcium and vitamin D and a lifestyle of regular weight-bearing exercise. *Those at higher risk:* small body frame, low body weight, sedentary lifestyle, anorexics and dieters.	Men and women aged 25 to 65 should have 1,000 mg calcium daily. Women near menopause or postmenopause should have 1,500 mg calcium daily. Calcium-rich foods include low-fat dairy products (cheese, yogurt, milk); canned fish with edible bones, such as salmon and sardines; dark leafy-green vegetables such as kale, broccoli, collard greens; breads made with calcium-fortified flour; juices fortified with calcium; calcium supplements with vitamin D as needed. www.nih.gov/nia/health/pubpub/osteo.htm
Cervical Cancer (women)	A pap smear, or cervical biopsy if at higher risk.	A *pap smear* every 1 to 3 years to detect cervical cancer. Request lab where a person examines smear; some are examined electronically. Should be done more often if at higher risk: multiple sex partners, sequential sex partners, IV drug user, have unprotected sex, previous history of STD(s), familial history of cancer, heavy smoker, postmenopausal and on HRT.
Menopause (women)	Blood test for level of FSH (follicle-stimulating hormone); begins in only 4% of women under 40. Typical age range is 45 to 55; average age is 51.	Typically if female age 40 or over and if menses are irregular or atypical, or if any other signs of menopause are indicated (hot flashes or hot flushes, etc.).
Breast Cancer (can affect both females and males)	Self-exam. Mammogram: an X-ray picture of the breast, which takes only a few seconds per breast.	If you are in your forties or older, *having a mammogram every 1 to 2 years could save your life.* Look for any unusual lump or swelling lasting for a period of time; some lumps (cysts) normally occur at a certain time each month; check breasts monthly while in the shower by running hands over breasts lightly to detect any changes or lumps; also check nearby lymph glands for changes. A doctor should do a breast exam during a yearly

Continued on page 470

Continued from page 469

Health Checkups	Diagnosis or Potential Health Concerns	Recommendations
Breast Cancer (can affect both females and males) (continued)	With aging after 40, the chances of getting breast cancer get higher. After skin cancer, breast cancer is the most frequently diagnosed cancer in U.S. women.	exam. Mammograms are recommended once every 2 years after 40 years old or more frequently if there is a familial or personal history of breast cancer. rex.nci.nih.gov/mammog_web/ or 1 (800)-4-CANCER (National Cancer Institute Information Service)
Cardiovascular Fitness (Age 18 & older)	Hypertension or high blood pressure. | Systolic	Diastolic
Normal	<130	<85
High normal	130–139	85–89
High		
Stage 1	140–159	90–99
Stage 2	160–179	100–109
Stage 3	180+	110		Have blood pressure taken regularly if overweight or have familial or personal history of hypertension. *To lower blood pressure:* • Maintain healthy weight (lose weight if overweight) • Be more physically active. • Choose foods low in salt (sodium). • Drink alcoholic beverages only in moderation. www.nhlbi.nih.gov/nhlbi/cardio/hbp/gp/prevhbp/blood.htm
Blood Cholesterol	High 239+ Borderline 200–239 Desirable below 200.	A blood test to check for LDL and HDL for everyone aged 20 and older, every 5 years; men with no risk factors can wait until age 35; women with no risk factors can wait until age 45 (see Table 15-1).
HIV/AIDS	Blood tests to confirm absence or presence of HIV virus; gynecological cervical tests in women.	*Higher-risk factors include* unprotected sexual behaviors, homosexual male sexual behaviors, history of sexual abuse (rape), history of multiple partners, history of sequential partners, IV drug user sharing needles, history of STD(s), extreme fatigue, recurrent female pelvic diseases, female over 50 having unprotected sexual relations.
Prostate Gland (men)	Prostatitis can be caused by several types of bacteria. Symptoms include burning sensation, discharge, difficulty beginning a stream of urine. Treatment regimen is antibiotics. Prostate enlargement is more typical beginning in late 40s and symptoms are similar to prostatitis. Prostate cancer is detected by biopsy.	See a physician (a urologist perhaps, depending upon the severity of the symptoms). The physician inserts a gloved finger into the rectum to feel the prostate through the wall of the bowel. This exam is *part of a routine physical for men over 40* to detect signs of prostatic cancer—a slow-growing cancer that almost never affects young men but is seen in most elderly men (Nickel, 1997).
Testicular Cancer (men)	Testicular cancer usually affects men between the ages of 15 to 35. If testicular cancer is detected early, it is very curable.	A monthly testicular self-exam (TSE) is performed to identify a number of conditions, but primarily cancer. Regularly examine the testes during a warm shower when the heat of the shower can relax the scrotum. Rotate each testicle between the thumb and forefinger, feeling for a round, firm surface. If you discover a small, painless lump on the surface of the testicle that does not appear to be epididymis, consult a physician immediately. Testicular cancer is almost always painless, so do not wait for the lump to grow or pain to develop.

nicotine, can interfere with sleep needs, as can prescription drugs and over-the-counter drugs such as pain relievers, cold remedies, antihistamines, appetite suppressants, decongestants, and drugs for asthma, high blood pressure, and heart and thyroid problems. Even common foods containing sugar (especially eaten late at night) and common beverages, such as alcohol, soft drinks, cocoa, and tea, and not getting enough protein in one's diet can keep one awake or prevent sleep for an extended period of time. Insomnia can also result from worry or depression. Changes in one's circadian rhythms can affect sleep patterns as well. It is quite common to be able to go to sleep when desired but to wake up only a few hours later and not be able to get back to sleep. Women experiencing *perimenopause* often experience this pattern, accompanied by "night sweats," because of ongoing changes in their biochemistry (Doress-Worters & Siegal, 1994). Another factor interfering with sleep for many midlife adults is the "waiting-up-for-teenagers-to-get-home-safely" syndrome!

Medications formulated to help induce sleep, including sleeping pills, tranquilizers, and antianxiety drugs, can actually cause more sleeplessness after discontinuation of their use. Additionally, these types of medications can become addictive. Because they also are central nervous system depressants, their use is likely to affect a person's alertness, aggravate memory loss, and make a person unsteady upon arising, which could increase the likelihood of falls or more severe injury. Use of sleeping pills for a shorter duration or situationally is not likely to become addictive. Anyone with serious sleep disturbances should consult a sleep clinic at a local hospital for assistance.

Cardiovascular Fitness

There are several risk factors associated with cardiovascular (heart and circulatory system) health, including high blood pressure, smoking, a family history of heart disease, being male, being diabetic, and being obese. Two cardiovascular diseases are heart disease, the number one cause of death in the United States, and stroke, the third most common cause of death. Blood pressure readings and cholesterol screenings are predictors of cardiovascular fitness.

Blood Pressure   About one in every four American adults has high blood pressure, called **hypertension,** which is a risk factor for other diseases affecting the heart, kidneys, and brain. Hypertension has no warning signs or symptoms. Blood pressure is the force of the blood pushing against the walls of arteries. Each time the heart beats (about 60 to 70 times a minute at rest), it pumps out blood to the arteries. Blood pressure is at its greatest when the heart contracts and is pumping the blood, which is called *systolic pressure.* When the heart is at rest, in between beats, blood pressure falls, which is *diastolic pressure.* Blood pressure is always noted as these two numbers, and both are important. They are usually written one above or before the other, such as 120/80, with the top number systolic and the bottom diastolic. A reading of 140/90 is considered high normal; 120/80 is even better for the heart and blood vessels (National Institute of Health (NIH), 1996). Blood pressure can fluctuate during a regular day, depending upon activity level. During sleep, for instance, blood pressure goes down. Some people have blood pressure that stays up all or most of the time; if left untreated, this can lead to serious medical problems such as arteriosclerosis (hardening of the arteries), heart attack (reduced blood flow that reduces oxygen supply to the heart), enlarged heart, kidney damage, or stroke (NIH, 1996).

Who Is at Risk?   Anyone can develop high blood pressure, but it is more common in African Americans than in whites. In the early and middle adult years, men have high blood pressure more often than women; but as men and women age, more women have high blood pressure than men. After menopause, women have high blood pressure as often as men of the same age. The percentage of men and women with hypertension increases rapidly in older age groups. Some midlife adults have hypertension, more than half of all Americans over age 65 have high blood pressure, and older African American women who live in the Southeast are more likely to have high blood pressure than those in any other region of the United States. Heredity can make some individuals more prone to developing high blood pressure, and some children have hypertension (NIH, 1996).

The NIH, (1996), promotes several lifestyle changes to help reduce the risk of hypertension, including these:

- Maintain a healthy weight, and lose weight if overweight.
- Eat foods high in starch and fiber, limiting serving sizes.

Coming to Terms with Physical Changes in Middle Age Though anyone can develop high blood pressure and high cholesterol readings, some Americans are more at risk for these effects and need to maintain a healthy weight, follow a high-fiber, low-fat diet, increase their activity level, eat less salt, drink alcohol only in moderation, and learn stress management skills. The southeastern United States has high rates of adult hypertension.

- Increase activity level.
- Choose foods lower in salt.
- Drink alcoholic beverages only in moderation.
- Learn stress management skills.

Medications can help reduce high blood pressure. Regular blood pressure screenings several times a year are recommended during middle adulthood and are available for free in many community settings, such as neighborhood clinics, pharmacies, churches, and schools. Adults with a high risk for hypertension can purchase diagnostic instruments to take their blood pressure readings at home.

Cholesterol is a white, waxy fat that occurs naturally in the body and is used to build the cell walls and make certain hormones. *Cholesterol* has been the "buzz word" in health throughout the 1990s. Too much of it in one's diet can clog arteries and eventually choke off the supply of blood to the heart. High cholesterol is a leading risk factor for heart disease. The National Cholesterol Education Program suggests being tested from age 20 on, with men 35 and older and women 45 and older being tested every five years. Cholesterol level can be determined with a simple, inexpensive blood test. HDL is considered the "good" cholesterol that cleans the blood vessels; LDL is the "bad" cholesterol that builds up and clogs arteries. See Table 15-1 for blood cholesterol levels. It is never too late to eat more fresh fruits and vegetables and to eat fewer foods with saturated fat (NIH, 1999).

Changing one's diet to reduce consumption of saturated fats and increasing activity level by exercising regularly (brisk walking, running, swimming, cycling, dancing, jumping rope, skating, aerobics) can raise the beneficial HDL cholesterol and lower the damaging LDL cholesterol.

Smoking, especially over a long period of time, can lead to many health deficits. It has been documented in many empirical studies that smoking can lead to heart disease, lung disease, kidney and bladder disease, and many types of cancers. A variety of smoke-reduction programs (and medications) are available to those who wish to make this lifestyle change to promote better health and well-being, not just for oneself but for one's family as well, since a great deal of research has documented that secondhand smoke is possibly even more harmful than ever believed (American Cancer Society, 1997). Cigarette packages are now required to be labeled *This product is likely to be harmful to your health.* People who have quit at any age usually state that food tastes better, breathing is easier, endurance for activity is better, and they feel more energetic—and, as a boon to people around them, they no longer reek of stale smoke. It is never too late to quit. Smoking-related illnesses and diseases cost Americans billions of dollars in medical costs every year, along with the emotional devastation associated with the loss of loved ones who would not give up the habit.

Question

How do smoking, cholesterol, and hypertension contribute to cardiovascular disease?

Cancer

For women between the ages of 40 to 60, breast cancer is the leading cause of death. Colon cancer and lung cancer are also leading causes of death for both males and females. Researchers estimate that over 180,000 women are diagnosed with breast cancer each year in the United States. Ovarian and cervical cancer are also more prevalent at this stage of life (Midlife Passages, 1998a). Some inborn factors might predispose a person to developing cancer, but virtually all experts agree that about 80 percent of all cancers are caused by environmental factors such as toxic substances in cigarettes, air, water, and food, medical treatment, and workplace hazards (Doress-Worters & Siegal, 1994).

Smoking is the number one controllable cause of cancer, causing at least 30 percent of cancer deaths (Doress-Worters & Siegal, 1994). Anyone who has a history of cancer is at a higher risk for developing further cancers. Other risk factors are these:

- *Poverty.* Of persons diagnosed with cancer, those who have financial resources live longer compared to poorer people.
- *Longevity.* Those who live longer have a greater likelihood of getting some type of cancer. Cancer incidence begins to rise at about age 35 for women and a little later for men.
- *Gender and race.* More men than women, and more blacks than whites, die from cancer.
- *Family.* Heredity plays a role in the incidence and types of cancers diagnosed. Inform any health provider of a familial disposition to cancer (Doress-Worters & Siegal, 1994).

As compared to many years ago when cancer was whispered to be "the *C* word," many communities now have

Table 15-1 Blood Cholesterol Levels

Risk	Total Cholesterol	HDL ("good")	LDL ("bad")
High	above 239	less than 35	above 159
Borderline	200–239	n/a	130–159
Desirable	below 200	above 60	below 130

cancer support groups available to help cancer patients and their families cope with this serious ordeal. Researchers around the world are making scientific breakthroughs every day in the search to understand and cure cancer.

The Brain

The *Baltimore Longitudinal Study of Aging* found that more than 25 percent of persons in their seventies showed no decline in memory or reasoning skills, and many showed little decline into their eighties (Browder, 1997). Recent research suggests that people can slow the process of brain cell loss by staying intellectually active, continuing to problem-solve and using challenging thought processes. People in midlife should continue the path of active learning: learn and use new words, play games of Scrabble, compete along with "Jeopardy" contestants, demonstrate creativity, try a new hobby, take a college course or continue education in some other fashion, mentor someone, join a speakers' group, join an Elderhostel program—in other words anything to keep yourself intellectually engaged with the world.

A **stroke,** or "brain attack," occurs when blood circulation to the brain fails. Brain cells can die from decreased blood flow and the resulting lack of oxygen. Both blockage and bleeding can be involved. Stroke is the third leading killer of Americans and is the most common cause of disability. Each year more than 500,000 Americans have a stroke, with about 145,000 dying. Strokes occur in all age groups, in both sexes, and in all races of every country. For African Americans the death rate from stroke is almost twice that of the white population. High blood pressure, cigarette smoking, heart disease, history of stroke, and diabetes are considered risk factors for stroke (National Institute of Neurological Disorders and Stroke [NINDS], 1998a). This brain disorder will be discussed in more detail in Chapter 17.

Parkinson's disease belongs to a group of motor system disorders that are likely to show up in late middle age, usually affecting people over 50. The average age of onset is 60 years old. The four primary symptoms are tremor or trembling in hands, arms, legs, jaw, and face; rigidity or stiffness of limbs and trunk; slowness of movement; and postural instability or impaired balance and coordination. About 50,000 Americans are diagnosed with Parkinson's disease each year, and they include prominent people such as Pope John Paul and actor Michael J. Fox. Parkinson's affects men and women in nearly equal numbers, and it knows no social, economic, or geographic boundaries. Some studies show that African Americans and Asians are less likely than whites to develop this disease. There is no cure at this time for Parkinson's, but a variety of medications can provide dramatic relief (NINDS, 1998b).

Alzheimer's disease (AD) sometimes develops in middle adulthood, but this brain disorder is more likely to show up after age 65. AD begins slowly, and at first the only symptom could be mild forgetfulness. Minimal memory loss is also a symptom of the complex hormonal changes many women experience during menopause, and many middle-aged women worry that they are developing Alzheimer's when they find themselves forgetting things. This form of dementia and memory loss will be addressed in more detail in Chapter 17.

Consumption of alcohol (a depressant), at any age, slows down brain activity, which in turn affects alertness, judgment, coordination, and reaction time. It is well known that drinking increases the risk of accidents and injury. Some research has shown that it takes less alcohol to affect older people than younger ones. Over time, heavy drinking damages the brain and central nervous system, as well as the liver, heart, kidneys, and stomach. Alcohol is often harmful (even fatal) when mixed with prescription or over-the-counter medications. An older body cannot absorb or dispose of alcohol or other drugs as easily as a younger body. Some people are likely to develop a drinking problem later in life because of situational factors, such as a job layoff, forced retirement, failing health, break up of a marriage, or loss of friends and loved ones. However, chronic drinkers have been drinking for many years. Once a person decides he or she needs help, there are many treatment avenues available to help change this brain-destroying behavior (National Institute on Aging, 1998).

Question

What are some potential brain disorders associated with aging, and why are they so life-threatening?

Midlife Men and Women at Risk for HIV/AIDS

In the past few years, AIDS cases rose faster in middle age and older people than in people under 40.

National Institutes of Health (1997)

Fact: According to a May 1998 National Institutes of Health press release, since the AIDS epidemic began in 1981 *more than 630,000 Americans have been diagnosed with AIDS*—and nearly 400,000 American men, women, and children have lost their lives to this disease. An estimated 650,000 to 900,000 Americans are believed to be living with HIV. For many years, homosexual males and intravenous drug users seemed to be the most affected populations. Then the virus spread to the heterosexual population, especially to sexually active adolescents. The number of cases among midlife and older women has been steadily increasing, mainly among postmenopausal women who no longer use birth control. Because of their age or monogamous marital status, these women do not see themselves as being at risk of STDs. Generally health-care providers fail to ask middle-age adults about their sex histories or current practices and often do not test for HIV/AIDS. However, anyone who is sexually active is at risk for the HIV virus, leading to AIDS. Yet women in midlife are among those least likely to practice safe sex and to get tested. The immune system gets weaker with age, but the decline is faster in older AIDS patients (National Institute on Aging, 1994).

Risk Factors   A link between child sexual abuse and risk for HIV infection has been proposed by several researchers, and recent findings strongly confirm that association (Cassese, 1993; Paone & Chavkin, 1993). In the Women's Interagency HIV Study (WIHS), data from 1,560 women in New York City, Chicago, Washington, DC, and Los Angeles revealed that 40 percent reported a history of child sexual abuse (Cook, 1997). For these women, a history of sexual abuse, physical abuse, or domestic abuse was highly correlated with engaging in behavior that put them at risk for HIV. Significantly, childhood sexual abuse was associated with use of IV drugs, exchange of sex for drugs, paying money for shelter, multiple sexual partners, and having sex with a person at high risk for HIV. Another risk factor is adolescent or adult sexual assault. It is estimated that more than 30 percent of all females and nearly 15 percent of all males in the United States have been victims of childhood sexual abuse. One in four women have been raped, and one in five women have experienced domestic abuse (Denenberg, 1997a). (See *Further Developments:* "Why Are More Midlife Women Contracting HIV/AIDS?") A survey finding released in August 1996 by the XI International Conference on AIDS in Vancouver, Canada, identified fatigue as the most difficult side effect of treatment for people living with HIV/AIDS. The patients surveyed said that fatigue negatively affects their outlook on life, willingness to continue aggressive treatment, ability to work, and their role as a family member or partner (Spiller & Reeves, 1996).

Female-to-Male Transmission of HIV/AIDS   Of women and men engaging in heterosexual sex, the women are much more likely than the men to become infected by an HIV-positive partner, according to a 10-year-study conducted by researchers at the University of California, San Francisco. The probability of HIV-positive women infecting their male partners with the virus was found to be significantly low. The risk factors for HIV infection among heterosexuals are (1) unprotected anal receptive sex, (2) lack of condom use,

(3) injection drug use, (4) sharing of tainted injection equipment, and (5) the presence of a sexually transmitted disease (STD) (Padian et al., 1997).

Female-to-Female Transmission of HIV/AIDS   Through December 1996, 85,500 women were reported with HIV/AIDS. Of these 1,648 were reported to have had sex with women. However, the vast majority had other risks—such as injection drug use, sex with high-risk men, receipt of blood or blood products, alternative insemination, and needle use for piercing and tattooing (Denenberg, 1997b). Of the 333 (out of 1,648) who were reported to have had sex only with women, 97 percent of these women also had another risk—injection drug use. Information on whether the woman had sex with women is missing in half of the 85,500 case reports, possibly because the physician did not elicit the information or the woman did not volunteer it (Centers for Disease Control, 1997a). Although female-to-female transmission of HIV is apparently rare, female sexual contact should be considered a possible means of transmission among women who are having sexual relations with women. Women who are lesbian need to know that exposure of a mucous membrane to vaginal secretions and menstrual blood is potentially infectious. A major preventative is accurate knowledge of a partner's HIV status and use of effective barriers (CDC, 1997b).

The focus of research at present is to develop women-initiated methods to avoid sexually transmitted diseases, including HIV. Technical products for safer sex are referred to as "barriers," and women and men need to learn the nature of each type of barrier, its relative effectiveness and availability, and how to use it. Mechanical barriers (such as male and female condoms, the diaphragm, the cervical cap) and chemical barriers (such as bacteria- and virus-killing microbicides) can be used alone or in combination (Segal, 1997).

Ron Stall of the Center for AIDS Prevention Studies at the University of California, San Francisco, points out that a major problem is a lack of information on patterns of sexual behavior among midlife and older people (Marks, 1998). Support for HIV/AIDS research at the National Institutes of Health in 1998 totaled $1.6 billion, an increase of 50 percent since fiscal year 1993. Since 1997, three separate research initiatives have been funded through NIH that will expand our understanding of the risk and coping strategies for this age of Americans: "Sexual Risk Behavior and Behavior Change in Heterosexual Women and Men" (Ehrhardt, 1997); "Health Outcomes—Transition to Midlife for African American and Caucasian Women" (Glazer, 1998); and "Late Middle Age and Older Adults Living with HIV/AIDS" (Segal, 1997). Several national health agencies are also funding research to investigate prevention issues relevant to AIDS in middle-aged and older populations (NIH, 1998a). HIV-related behavior and behavior changes must

Further Developments
Why Are More Midlife Women Contracting HIV/AIDS?

Janet is an intelligent, aware woman who educated herself on HIV and AIDS and the modes of HIV transmission. After she had had months of recurrent yeast infections and debilitating fatigue, she developed symptoms of pneumonia, and her doctor ordered an HIV test. Her HIV-positive diagnosis came as a surprise to her doctor and was a shock to Janet. She had never considered herself in any way at risk for HIV. Janet is 68 years old (Solomon, 1995).

Older unmarried women have a variety of sexual options available to them. Postmenopausal women often see themselves as free from the responsibility of birth control—and therefore, mistakenly, as free from needing to protect themselves from disease protection. Certain biological changes that accompany aging make midlife and older women particularly vulnerable to HIV transmission. During menopause, the vaginal wall becomes thinner and more likely to tear. As women age, vaginal acidity decreases, which can facilitate urinary tract infections, and changes in the immune system can contribute to increased vulnerability to HIV. Drug use, thought to be primarily a behavior of younger people, also occurs among older men and women. The cohort who are reaching middle age now grew up in the late 1950s and 1960s during the era of "free love" and an American culture more accepting of drug use. Researchers who visited "shooting galleries," where needles are commonly shared, found women over 50 there regularly. Some women who were drug users in their thirties and forties are now finding they are HIV-positive many years after they stopped using (Solomon, 1995). Once diagnosed, the progression of HIV in older people is not clearly understood. Research up to 1997 basically focused on women through childbearing years, usually up to age 45 to 49 (Buttenwieser, 1994).

Once HIV/AIDS is diagnosed, older women experience further problems in getting the care they need. Weight and hormonal differences between men and women make it difficult to prescribe proper dosages of the *antiretroviral drugs* for women. Women who are also taking hormonal contraceptives should be aware that protease inhibitors will significantly lower blood levels of those drugs, reducing the effectiveness of the contraceptive, and pregnancy can occur. Many HIV-positive women report changes in their menstrual cycles, including longer, shorter, heavier, irregular, or painful periods. **Amenorrhea,** the absence of a menstrual cycle, is three times more likely to occur in HIV-positive women. Originally this was thought to be premature menopause in which ovarian function wanes and eventually ceases. However, the levels of FSH (follicle-stimulating hormone) detected in HIV-positive women experiencing amenorrhea do not indicate true menopause (Marks, 1998).

HIV Manifestations in Women
Based on recent statistics from the Centers for Disease Control (1998a), nearly 20 percent of the newest cases of HIV/AIDS are in women, as compared to 7 percent of cases in 1985.

Studies have found that the most common reasons infected women first sought medical attention were recurrent vaginal yeast infections, enlarged lymph nodes, and extreme fatigue, followed by bacterial pneumonia. Cervical cancer was added to the list of AIDS-defining conditions in 1993. A *Pap smear* is typically used to detect cellular changes in the cervix that can indicate a risk of developing cancer. HIV-positive women have a 30 to 40 percent chance of having an abnormal Pap smear after testing positive. Invasive cervical cancer is usually more serious in women who are HIV-positive than it is in HIV-negative women.

On a psychological level, midlife or older women who are HIV-positive are likely to keep this diagnosis to themselves until they become seriously ill. Some cannot afford the medical care or adequate services or believe that they do not have the time to take care of themselves if they are already involved in caring for others. Dr. Alexandra Levine, with the Women's Interagency HIV Study, says that American women are very likely to experience a sense of profound isolation as a result of knowing they are HIV-positive (Marks, 1998). In her study with 2,000 participants, a significant number said they had never met or spoken to another woman living with AIDS. To compound matters, people with HIV or AIDS who need elder care might face discrimination in facilities such as nursing homes. Most must rely on in-home care, which for some is not an option either. Also, based on cross-cultural studies of attitudes and health, it is likely that many ethnic minority adults infected with HIV are likely to seek support from each other and not request assistance from health agencies or community resources.

be viewed within a human development framework. In March 1998, President Clinton outlined a proposal to Congress requesting $400 million to be used in 30 model communities over the next five years to eliminate dispar-ity in health care between whites and people of color, with HIV/AIDS a major targeted area. Clinton cited the fact that minority women and children account for 75 percent of all AIDS cases (Bornhoeft, 1998).

Question

Why is HIV/AIDS becoming more prevalent among the middle aged?

Stress and Depression

Midlife is often associated with change and adaptation—children leaving home, elderly parents moving in, potential divorce or remarriage, change of jobs, retirement, relocation, children returning home with grandchildren, and so on. Midlife is also associated with losses—forced early retirement, traditional retirement, physical changes or health decline, death of a spouse, parents' loss of physical well-being, and so forth. Samuels (1997) reports in *Midlife Crisis: Helping Patients Cope with Stress, Anxiety, and Depression* that depression and substance abuse are common but often underrecognized and undertreated in middle age adults. According to Samuels, major depression in midlife is common; approximately 2.2 percent of midlife adults experience major depression. Depression in older life is associated with increased mortality or suicide, even after controlling for physical illness and disability. Dr. Samuels recommends that physicians pay attention to change and loss in a patient's life as serious predictors of depression, by using various rating scales during a checkup, because suicide risk increases with age particularly in adult males. People with mood disorders are likely to self-medicate with alcohol or other substances of abuse (about 1 in 8 older adults has a problem related to alcohol abuse), and one-third of these develop alcohol problems late in life that are related to the psychological stress of aging (Samuels, 1997). An analysis of the adult's history, a full physical, and neurological and mental status examinations can help assess whether a patient is suffering from multiple stressors and depression. An individualized treatment regimen should be established, which might include changes in lifestyle or antidepressant medications. The goal is to give the patient a renewed sense of control and to eliminate stressors or develop coping strategies for those that cannot be eliminated. Depression is a treatable illness, and many organizations provide support and educational materials for midlife patients faced with complex problems.

Sexual Functioning

Americans have well-established stereotypes regarding the sexual lives of various age groups (Laumann et al., 1994). They think of young adults as the most sexually active—as desiring, attempting, and achieving the most sex. They view the middle-aged as the most sexually knowledgeable and skilled, but they consider the old as asexual or sexless. A display of erotic interest by older people is considered unnatural and undignified. For example, what our youth-oriented society considers virility in a 20-year-old male it views as lechery in a 65-year-old, who is likely to be labeled a "dirty old man." Despite such stereotypes, classic research on human sexuality by William H. Masters and Virginia E. Johnson (1966) revealed that sexual effectiveness need not disappear as human beings age. Like their other activities, people's sexual performance might not have the same physical energy in the later years of life as in the earlier years. But Masters and Johnson found that many healthy men and women do function sexually into their eighties or beyond. Although time takes its toll, it need not eliminate sexual desire nor bar its fulfillment.

Sexual arousal in human beings is the product of a complex interaction of affective, cognitive, and physiological processes (Marx, 1988; Morokoff, 1985). But what happens in far too many cases is that older people come to accept social definitions of their sexlessness; they become victims of the myth. Believing they will lose their sexual effectiveness becomes a self-fulfilling prophecy, and some older people do lose it even though their bodies have not lost the capacity for sexual responsiveness. The belief that they are sexless may be reinforced in some men when they are unable to attain or maintain an erection during a number of sexual attempts. This is a common occurrence among men of any age group and can be associated with stress, illness, or overindulgence in alcohol. Indeed, fear of failure is not uncommon among older men. Often, a woman is unaware of her partner's fears and mistakes his caution for disinterest.

Changes nonetheless do occur with age. Men over 50 find that it takes longer for them to achieve an erection. The erection of older men, particularly those over 60, is generally not as firm or full as when they were younger, and maximum erection is achieved only just before orgasm. With advancing age comes a reduction in the production of sperm and seminal fluid, in the number of orgasmic contractions, and in the force of the ejaculation. And the frequency of sexual activity typically declines with advancing age. Yet Painter (1992a) reports findings that 61 percent of married couples in their fifties have sex at least once a week.

Overall, researchers find that the general level of sexual activity of the individuals when they were between 20 and 39 years of age correlated highly with the frequency of their sexual activity in later life (Elias, 1992; Tsitouras, Martin, & Harman, 1982). Hence, if men have maintained elevated levels of sexual activity from their earlier years, and if acute or chronic ill health does not intervene, they are able to continue some form of active sexual expression into advanced age. However, if aging males are not stimulated over long periods of time, their responsiveness can be permanently lost (see Figure 15-3) (Masters & Johnson, 1966). Physicians find that many

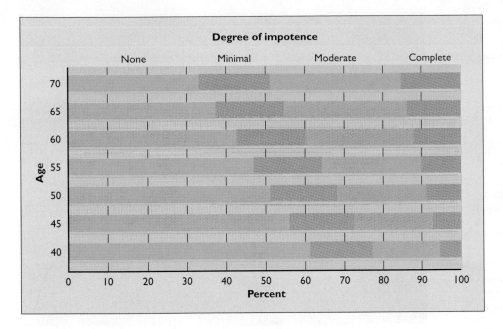

Figure 15-3 Aging and Impotence The figure shows the incidence of impotence, by age and degree, in a study carried out among men in Massachusetts who answered a series of questions about their sexual potency during the previous six months.

Lawrence K. Altman, "A Study on Impotence Suggests Half of Men over 40 May Have a Problem." *The New York Times*, December 22, 1993. Copyright © 1993 by The New York Times. Reprinted by permission.

medical problems manifest themselves sexually. Medical problems that affect male sexual performance include diabetes, which occurs in about 10 percent of men over age 50, and Peyronie's disease, a scarring of the tissue inside the shaft of the penis, more common with men who have diabetes or high blood pressure. Psychological depression is also associated with a decrease in sexual desire. In recent years physicians have become aware of the sexual side effects of medications for heart disease, high blood pressure, and coronary artery disease. Blocking agents used to control high blood pressure, for instance, reduce the flow of blood into the pelvic area. In a younger person it might not matter much, but in older men the result is often impotence.

Masters and Johnson (1966) also found no reason that menopause or advancing age should interfere with the sexual capacity, performance, or drive of women. Basically, older women respond as they did when they were younger, and they continue to be capable of sexual activity and orgasm. Older women tend to lubricate more slowly than they did earlier in life, and the vaginal walls become thinner, which means that the tissues can be easily irritated and torn with forceful sexual activity. Male gentleness and the use of artificial vaginal lubricants can do much to minimize this difficulty. Like aging men, older women also typically have fewer orgasmic contractions; younger women average 5 to 10 contractions, whereas older women average 3 to 5 contractions).

It appears that American men are becoming more aware of their own sensitivity and humanness. Privately, they are increasingly coming to recognize, and even approve of, their feelings of tenderness, dependence, weakness, pain, and so on. But a good many of them, especially older men, are not yet able to talk freely about

these traditionally "unmasculine" emotions. While many men are becoming more aware of their own sensitivity and humanness, over the past two decades women have become more aware of their own sexuality. A growing number of women are tired of the traditional pattern of sexual relations, which focused on male erection, male penetration, and male orgasm. Women are increasingly admitting to themselves what they like sexually and are asking their partners for it.

The need to learn about the prevalence of various sexual behaviors and about how the prevalence varies with age and other factors is not strictly an academic issue. Only by studying sexual behavior can we hope to understand and combat the epidemic of sexually transmitted diseases, including the AIDS virus. Admittedly, it is difficult to measure many sorts of private behavior, particularly behavior that is potentially awkward or embarrassing to report to others (Barringer, 1993a; Lord, 1994). Given this caveat, one of our best sources for data on the sexual behavior of adult Americans is the *National Health and Social Life Survey* conducted in 1992. It contains responses obtained from interviews with a representative sample of over 3,000 men and women, ranging in age from 18 to 59 (Laumann et al., 1994). Among key findings of the survey:

- Married and cohabiting couples had the most sex. Forty percent of marrieds and 56 percent of cohabiting couples had intercourse twice a week or more, and they enjoyed their sex lives more than did singles who lived alone.
- Only 2.7 percent of men and 1.3 percent of women reported they had homosexual sex over the past year.

The Sexual Behavior of Middle-Aged Americans There is a stereotypical view that aging men and women are sexless, but research indicates that many men and women are sexually active well into middle age and older years, though frequency of sexual activity often declines with advancing age. Married and cohabiting couples have the most sex. However, drug treatments for certain medical problems can have side effects that affect sexuality.

- Some 20 percent of men and 31 percent of women have had only one sex partner since age 18; 21 percent of men and 36 percent of women have had 2 to 4 sex partners; 23 percent of men and 20 percent of women have had 5 to 10 sex partners; 16 percent of men and 6 percent of women have had 10 to 20 partners; and 17 percent of men and 3 percent of women have had 21 or more partners.
- Extramarital sex is the exception, not the rule, among Americans. Nearly 75 percent of married men and 85 percent of married women say they have never been unfaithful.

Since the 1970s the media has moved sex from the private to the public arena. The "more liberated" career woman depicted in the media also introduced the necessity of competing and succeeding at a job on the basis of sex appeal and changing oneself to match the media image of beauty. The youngest group in the 1992 National Health and Social Life Survey seems to bear the brunt of society's confusion over changing sex roles and orientations, as women are more publicly protesting abuse in the home, sexual harassment or discrimination in the workplace, date rape, and other mistreatment of women. Yet powerful ads continue to erroneously portray men and women as preoccupied with a youthful appearance and sexual gratification (Laumann et al., 1994).

Question

What does research say in general about sexual frequency, sexual performance, and sexual health among middle-aged adults?

Cognitive Functioning

As noted earlier in this chapter, as people age, they change physically in several different ways. Some physical changes, such as graying hair and facial wrinkles, are obvious and can be quickly verified by a nearby mirror. But what about cognition? Are there parallel changes in aging adults' intellectual abilities as well? And, if so, how and in what ways? Research studies designed to answer this question have yielded mixed results: Yes, no, and it depends.

Research Findings: A Methodological Problem

Most of the data on age-related cognitive differences is based on IQ scores and have been tabulated using a *cross-sectional* research method. Results of studies that have employed the cross-sectional method indicate that overall composite IQ reaches a peak for most people when they are in their twenties, remains stable for a couple of decades, and then takes a dramatic downward drop (Schaie, 1994). Remember from Chapter 1 that cross-sectional studies employ the "snapshot" approach. Researchers administer tests of intelligence to a large group of individuals of different ages at about the same time and compare their performance. Consequently, a major weakness of cross-sectional studies is that "uncertainty regarding comparability" is always a problem. That is, we can never be sure that the reported age-related differences between subjects are not the product of other variables or events. For example, people in their fifties might have a lower average score than those in their twenties—not because of their age difference but because they might have less experience taking standardized tests.

As psychologists Schaie & Willis (1993) have pointed out, cross-sectional studies of adult aging do not allow for generational differences in performance on intelligence tests. Because of increasing educational opportunities and other social changes, successive generations of Americans perform at progressively higher levels. Hence, the measured intelligence (IQ) of the population is increasing. When comparing adults from different generations—80-year-olds with 40-year-olds, for instance—you are comparing people from vastly different environments. Thus, cross-sectional studies tend to confuse generational differences with differences associated with chronological age.

Using the *longitudinal research method,* researchers study the same individuals over a period of years, more like a case history. When this technique is used, the results are quite different: overall or global IQ tends to rise until the mid fifties and then gradually decline. In fact, most adults do not experience a decline in general intelligence functioning in middle age and show little decline throughout their sixties, seventies, and even beyond. However, the longitudinal method also presents a problem for the researcher. Whereas the cross-sectional method tends to magnify or overestimate the decline in intelligence with age, the longitudinal method tends to minimize or underestimate it. One reason is that some people drop out of the study over time; the probability is higher that the eldest will die over the course of a 10-year study. Generally, it is the more able, healthy, and intelligent subjects who remain in the study, and those who perform poorly on intelligence tests tend to be less available for longitudinal retesting. Consequently, the researchers are left with an increasingly smaller or biased sample as the subjects are retested at each later period (Brooks-Gunn, Phelps, & Elder, 1991; Botwinick, 1977; Siegler & Botwinick, 1979).

Question

What are the problems with cross-sectional and longitudinal research on intelligence conducted with subjects over the lifespan?

The Varied Courses of Cognitive Abilities

As we saw in Chapter 7, intelligence is not a unitary concept in the same sense that a chemical compound is a single entity. People do not have intelligence as such but, rather, *intelligences.* Thus, different abilities can follow quite different courses as a person grows older (Neisser et al., 1996). Many traditional measures of intelligence focus on abilities that are useful in academic environments. For instance, tests that measure *verbal* abilities tend to show little or no decline after the age of 60,

whereas those that measure *performance* do seem to show a decline (Schaie, 1989):

- *Verbal scores* are usually derived from tests in which people are asked to do something verbally, such as define a series of words, solve arithmetic story problems, or determine similarities between two objects.
- *Performance scores* are commonly based on people's ability to do something physically, such as assemble a puzzle or fill in symbols to correspond to numbers.

Fluid Versus Crystallized Intelligence   Some psychologists distinguish between **fluid intelligence** (the ability to make original adaptations in novel situations) and **crystallized intelligence** (the ability to reuse earlier adaptations on later occasions) (Cattell, 1971, 1943). Fluid intelligence (Horn, 1976) is generally tested by measuring an individual's facility in reasoning, often by means of figures and nonword materials (letter series, matrices, mazes, block designs, and picture arrangements). Presumably, fluid intelligence is "culture-free" and based on the individual's genetic and neurological structures. Crystallized intelligence is commonly measured by testing an individual's awareness of concepts and terms in vocabulary and general-information tests in areas such as science, mathematics, social studies, and English literature. Crystallized intelligence is acquired in the course of social experience. It is the scores on tests of crystallized intelligence that are most influenced by formal education. Often crystallized intelligence *increases* with age, or at least it does not decline, whereas fluid intelligence declines with age in later life (Gilinsky & Judd, 1994).

In the earliest published account of the theory, Cattell (1943) argued that fluid ability is "a purely general ability to discriminate and perceive relations between any fundamentals, new or old." Fluid ability was hypothesized by Cattell to increase until adolescence and then slowly decline. Further, fluid intelligence was thought to be the cause of the general (*g*) factor found among ability tests administered to children and among the "speeded or adaptation-requiring" tests administered to adults. Crystallized intelligence, on the other hand, was thought to increase with age. The important psychological distinction in the theory was between process (fluid intelligence) and product (crystallized intelligence) (Cattell, 1963).

So, does intelligence really decline with age? When the data are gathered and examined within a framework of fluid versus crystallized intelligence, the answer is yes and no. It depends on how you define intelligence. In the *Seattle Longitudinal Study,* which tested the mental abilities of more than 5,000 adults over a period of 35 years,

Human Diversity

Strategies for Midlife College Students

Preparation and planning are half the battle of being a successful college student. Midlife students might save some time, money, and frustration when pursuing a college education if they will read the following suggestions, which are based upon years of professionally counseling and advising nontraditional college students:

Investigate your career options

Do you need to complete a degree, or is there a one-year certificate program that will prepare you for the job market or launch you into a new career? Have you taken a career interest inventory to see if your interests match those of people who are currently working in a specific field?

Visit the College's Career Placement Office

Start here first. This is a valuable college resource, but most students don't make use of its resources until graduation. This office has your chosen career's employment statistics and employment outlook. Will there be jobs in your locale in a few years? Is your area flooded with certain professionals? When there is a surplus of workers for a field, starting salaries are low and it is difficult to secure a job. Are you willing to move to a new locale? This office has data from professional organizations in your field. Look at the "bigger picture" of national or international trends. Is this job being taken over by computers? You might need to reevaluate your plans or plan a dual major. This office has annual statistics on job placement of graduates in academic departments. How many transferred for advanced degrees or became employed after graduation? Check with a Chamber of Commerce for employment data, too.

Make an appointment with someone working in your field of choice

Find out how a successful person in the field managed the educational path. Will you need an associate's degree (2

years), a bachelor's degree (4 years), or graduate school training? Plan ahead to reach your goal. Is there an internship or field experience required for entrance into this field? Is it a paid or volunteer experience? Actual work experience makes a person more competitive when job hunting the next time around. Does this career require passing a state certification exam before you can be hired for a job?

Minimize the financial strain of career preparation

There are web sites, books, and consultants that will help you find sources of grants and scholarships (money you do not have to pay back) and loans (money to be paid back in the future). Start with the financial aid office at the college. Funding is available if you take the time to search. Private colleges offer grants, scholarships, or work-study monies if you meet the criteria. Just ask! Funding is available for minorities (women included here), those with disabilities, retraining, and so forth. Do the organizations you are currently affiliated with offer any scholarships? If you are retraining in the same career area, ask if there is a subsidy. Adopt the philosophy "Where there's a will, there's a way!"

Have your eyes checked

You are going to use your eyes for extensive reading, watching videos, reading information on the board, peering through microscopes, and reading from computer screens.

Begin or maintain a regular exercise regimen

Top researchers on aging and the brain have discovered that "there's a simple way to ward off slowness, stay in shape" (Jaret, 1996). With no exercise and a sedentary lifestyle, neurons get less nourishment and can't move electrical impulses as fast—the mind slows, leading to memory problems. Exercise energizes! As an American college student, you can take physical education classes or at least utilize the fitness center on campus that you have paid to use when you paid your tu-

---

results clearly showed no uniform patterns of age-related changes across all intellectual abilities (Schaie, 1994, p. 306). The majority of participants showed no statistically significant reduction in most abilities until after age 60, and then only in certain abilities. The only tests on which ability declined with age were related to speed of performance. It was found that fluid intelligence tended to decrease in young adulthood, but these deficits were offset by crystallized abilities, which remained stable or increased into middle age, followed by slight declines (Schaie, 1996). Indeed, this study clearly demonstrates

that as people age, they learn to make good use of their abilities that remain intact and learn to compensate for abilities that diminish with age (Sternberg, 1997).

Maximizing Cognitive Abilities

Until recently, exceptional performance among healthy adults had not been extensively studied by researchers. However, in the last several years, interest in exceptional achievement and performance has proliferated (Schultz & Heckhausen, 1996; Gardner, 1993a). Different ap-

ition. Many fitness centers and P.E. departments offer swimming classes, line dancing, Tai Chi, low-impact aerobics, ballroom dancing, and similar activities. By committing to this exercise regimen as part of your college schedule, you are more likely to stick with it! Hot line for exercisers: American Council on Exercise: 1-800-529-8227. Here you can get help with choosing a program that suits your lifestyle and level of fitness.

Be aware of the role of previous experience and expertise

Start with courses on subjects you have some familiarity with. Jumping into courses or full programs in subjects in which you have no background can be overwhelming—not that it can't be done. Scenario: Suppose you are an interior decorator and you have decided to become an engineer. Start with some liberal arts courses in order to experience some success then start taking on engineering courses. If you have been technically oriented for the past 25 years, the opposite is true. Start with a course similar to your background training, such as a formal computer course or a drafting or design course. Then work your way into the liberal arts arena.

Get organized

Make your transition in over a period of time, if possible. Do not take on too much too soon. Become familiar with the physical campus territory as well as the academic program and the requirements outlined in the college catalog. Make an appointment with the chairperson of your department to find out if you are on the right track and in the right degree program. Be sure to sign up for prerequisite courses first (e.g., take general psychology before taking human development). Organize and save all official paperwork in one place.

Save your college catalog

Be sure to get and keep your own copy of the college catalog. It is your contract with the college for your degree program. You might need to show course descriptions and degree requirements to someone at the next college you attend or to state certification personnel. *Hint:* Most colleges have a disclaimer statement in the catalog stating that you, the student, are responsible for following the degree plan printed in the catalog.

Get connected with other middle-aged adult learners

Is there a group or club on campus for adult students? If not, start one with the midlife students in your classes. Social support has been proven to promote mental, physical, and emotional health. You will also give each other important advice. Many adults who have come from the work world are used to feeling connected to a network of support.

Find out about distance learning classes

If you have a minimal amount of time to take courses, are a disciplined and dedicated student, or if you are commuting a great distance to take classes, you might be able to sign up for classes you can take using a home computer, saving both time and money! Check with your campus registrar.

Learn to use a computer—This is essential

Most colleges have minicourses on computer operating systems or the Internet. Credit courses can teach you how to use the computer to do spreadsheets, word processing, database, art, Internet, and other information retrieval.

Do not focus on short-term losses, but look at the bigger picture of where you are going

Don't let a low grade on a quiz or test get you down. As with any new venture, there are peaks and valleys. If you are having difficulty, sign up in your counseling center for a free tutor (a student who can help you through the "bumps in the road.") None of us excel at everything. Adult learners tend to be very hard on themselves and tend to expect all A's. Relax a little. If you tend to be an auditory learner and like working in small groups, set up a discussion group with other students in a class before major exams. Use your text, class notes, and the study guide that accompanies your text if you are a visual learner. Ask the instructor for a copy of an old exam so you can benefit from the practice effect.

proaches are being employed to explore the behavior of expert performers. One approach is to study the individual characteristics of exceptional performers. This approach is being spearheaded by Howard Gardner (1993b), whose work on multiple intelligences was discussed earlier in Chapter 7. Gardner proposes that exceptional performance in later life is dependent upon early identification of talent and nurturing of that talent by providing the individual with task-related practice over long periods of time (Gardner, 1993a). Much of Gardner's research findings are based on advances in brain physiology and achievements of savants, prodigies, and geniuses in specific domains. The important aspect of talent, according to Gardner (1983), is not the innate ability of the individual, rather the capacity to learn material relevant to one of the eight intelligences. Gardner believes that innate intellectual abilities, in order to persist throughout life, must be exercised or practiced regularly and over long periods of time.

Maintaining Expert Performance Most elite or expert performers in most domains are engaged in their domain of expertise essentially full-time from childhood to late adulthood. For example, millions of people are active in sports, music, visual arts, and chess, but only a small number reach the highest levels of performance—these are distinguished by the *length of time* and *duration* they commit to practicing their skill.

It is extremely rare for performers to reach their optimal performance before adulthood, but it has been found that performance does not necessarily continue to improve in those who keep exercising their skills across the life span. Rather, as Leman (1953) first noted, peak ages for performance seem to fall in the twenties, thirties, and forties. In vigorous sports, such as professional basketball, it is rare for elite athletes above age 30 to reach their personal best. Michael Jordan and Cal Ripkin are noted exceptions to this rule. Similar age distributions, centered around age 30, are also found for fine motor skills and some cognitive activities (Simonton, 1988). Typically the probability of producing an outstanding work declines with age. However, in novel writing, history, and philosophy, the optimal ages are in the forties and fifties. One common hypothesis on aging and expertise is that experts generally age more slowly than other performers. Recent research on expert performance has shown this not to be the case in chess (Charness & Gerchak, 1996), typing (Bosman, 1993), and music (Krampe, 1994). The superior performance of older experts is found to be restricted to relevant tasks in their preferred domains of expertise.

The Role of Deliberate Practice   The most marked age-related decline is generally observed in perceptual-motor performance as displayed in different sports. High levels of practice are necessary to attain the physical readiness that is found in mature performers, and the effects of practice appear to be particularly large when intense practice overlaps with physical development during childhood and adolescence. Most of these adaptations require that practice be maintained. When older master athletes are compared with young athletes at a similar level, many physiological measurements do not differ between them. However, at least some physiological functions, such as maximum heart rate, show an age-related decline independent of past or current practice. The ability to retain superior performance in sports appears to depend critically on maintaining practice during adulthood (Ericsson, 1990).

Evidence on the role of early and maintained practice in retaining cognitive aspects of expertise is much less extensive. Some abilities, such as the acquisition of a second language, especially accents and pronunciation, appear easier to acquire at young rather than adult ages. Expert performers remain highly active in their domains of expertise, though with increasing age, they typically reduce their intensive work schedule (Ericsson & Charness, 1994).

The traditional view of talent, which concludes that successful individuals have special innate abilities and basic capacities, is not consistent with the reviewed evidence. Efforts to specify and measure characteristics of talent that allow early identification and successful prediction of adult performance have failed. Differences between expert and less-accomplished performers reflect acquired knowledge and skills or physiological adaptations developed through training (Ericsson & Charness, 1994). Consequently, there is no reason to believe that developed expertise in human performance is limited to traditional domains, such as sports. Through systematic practice and training, similar changes can be expected in several everyday activities, such as thinking, problem solving, and communication (Jaret, 1996). What it takes is desire, practice, and commitment, whether in a sport, such as tennis, or in an academic pursuit, such as taking a human development course in a college continuing education program. A successful life course is achieved when adults continue to stay both physically and mentally active.

Cognition and Dialectical Thinking

If we look at cognition instead of intelligence, we see that a slightly different debate has emerged over the years concerning the existence of post-formal operations. Recall that Piaget maintained that cognitive development stops around the age of 15 with what he called formal operations, the ability to perform abstract reasoning. Critics assert that the thought processes of adults are qualitatively different from the logical problem-solving characteristics upon which Piaget focused. These post-formal operations let adults enter the realm of dialectical thought, which does not insist on a single correct answer to any given problem or dilemma, instead searching for complex and changing understandings of the processes or elements involved in the problem.

One proponent of dialectical thinking sums it up this way: **Dialectical thinking** is an organized approach to analyzing and making sense of the world one experiences that differs fundamentally from formal analysis. The latter involves the effort to find fundamental fixed realities-basic elements and immutable laws; the former attempts to describe the fundamental process of change and the dynamic relationship through which the change occurs (Basseches, 1980). An example of post-formal thought would be the understanding that a family quarrel might be nobody's "fault" and that the solution does not involve one of the parties "giving in" or changing their view of the situation. Instead, the dialectical approach consists of understanding the merits of different or opposing points of view and looks at the possibility of integrating them into a workable solution. This dialectical approach to problem solving is apparent when you think of situations where you have been on both sides of the same issue—for example being a child arguing with your parents, and then being a parent arguing with your child. One reason why post-formal operations are attributed to older individuals is that life experience is probably necessary in order to see the "bigger" picture. Schaie has proposed a four-stage model of cognitive development that is quite different from Piaget's in that it covers the entire life span and is not confined to preadulthood. Schaie's (1994) four stages are as follows:

1. *Acquisitive stage.* What should I know? (childhood-adolescence)
2. *Achieving stage.* How should I use what I know in career and love? (young adult)
3. *Responsible/executive stage.* How should I use my knowledge in social and family responsibilities? (middle adulthood)
4. *Reintegrative stage.* What should I know? (old age)

Even though the questions in Stages 1 and 4 are worded the same, in Stage 4 the acquisition of information is guided by one's interests, attitudes, and values. An older adult is willing to expend effort on a problem she or he faces in everyday life.

This interest in problem solving might shed light on another aspect of cognition—creativity. J. P. Guilford (1967) distinguished two sorts of thinking—convergent and divergent. **Convergent thinking** is very much like formal operations—the application of logic and reasoning to arrive at a single correct answer to a problem. **Divergent thinking** is more open-ended, and multiple solutions are sought, examined, and probed, thereby leading to what are deemed creative responses on measures of creativity. Creativity goes beyond problem solving and penetrates into problem generation. A creative person not only solves problems but sees problems of which others are not yet even aware. Creativity seems to demand imagination, motivation, and a supportive environment—characteristics that probably do not gel for the individual until middle adulthood. This might explain why it is during middle adulthood that the peak period of creative productivity occurs (Kastenbaum, 1993). The major problem with assessing creativity is the difficulty in defining criteria that capture creativity as well as originality, utility, and productivity (Aiken, 1998). The history of science and art indicates that Michelangelo, Verdi, Goethe, Picasso, and Monet created highly original work throughout the life span. From a psychological perspective, creativity results in flow: "the state in which people are so involved in an activity that nothing else seems to matter: the experience itself is so enjoyable that people will do it at great cost, for the sheer sake of doing it" (Csikszentmihalyi, 1993). During such "flow" experiences, people become so absorbed in an activity they disregard distracting concerns such as matters of the self, material gain, security, or personal advancement. Think of artists slaving away day and night, going without food or sleep in order to finish a masterpiece. It is almost as if they are possessed by the activity, unable to see beyond or outside of it.

Question

How can middle-age adults maximize cognitive abilities?

Moral Commitments

Moral development in certain individuals leads to something akin to flow, in that they become committed to doing "good" activities to the extent that, like saints,

Middle Age and Divergent Thinking Divergent thinking leads to creative responses, which demand imagination, motivation, and a supportive environment. Middle adulthood is considered the peak period of creative productivity. Creative experience is described as the "flow," where people become absorbed in an activity.

they have "dedicated the totality of their psychic energy into an all-encompassing goal [which they] follow unto death" (Csikszentmihalyi, 1997). Recently, two of Lawrence Kohlberg's colleagues set out to find individuals dedicated to acts of morality who they designated as *moral exemplars* (Colby & Damon, 1992). After an extensive search throughout the United States, they found 23 people who demonstrated all of the following characteristics:

- A sustained commitment to moral ideals that include respect for humanity
- Consistency between ideals and actions
- A willingness to risk one's self-interest for the sake of one's moral values
- Being an inspirational force for others, who then became active in moral work
- Humility about one's work and importance, unconcerned with ego

Most of the moral exemplars were in their forties, fifties, and sixties, and they were a diverse collection of individuals. Five of the exemplars were chosen to be interviewed in depth to discover what makes a moral exemplar "tick." Of the five:

- Two did not finish high school, one had a bit of college, one finished college, and one had a Ph.D.
- All were religious.
- Politically they were conservative (1), moderate (2), and liberal (2).
- Vocationally there were a minister, a businessperson, an innkeeper, a civil rights advocate, and a charity worker.

While interviewing these exemplars, Colby and Damon found that they showed enormous ability to critically examine old habits and assumptions and adopt new strategies for dealing with problems, and they constantly took up new and interesting challenges—while maintaining a lasting dedication to their particular values, goals, and moral projects. This amounted to the ability to remain morally stable without becoming cognitively or behaviorally stagnant. Charleszetta Waddles is one of the exemplars, and through her activities you can get a hint of how moral acts affect people on a day-to-day basis. Mother Waddles, as she likes to be called, runs the Perpetual Mission for Saving Souls of All Nations in Detroit. The mission serves 100,000 people a year by offering food, clothing, legal services, tutoring, and emergency assistance among other services. Her ultimate goals are to lift the poor out of their condition, help them believe in themselves, and encourage them to take full responsibility for their lives. She will do whatever is in her power to help anyone who comes to her mission—be it a woman who needs money for an eye operation, a young girl who needs advice about her pregnancy, or a

man who has lost everything in a fire and has no place to go. Oftentimes she has used most of her $900 monthly income to pay for these services herself when she was unable to find resources elsewhere. Mother Waddles, with an eighth-grade education, has run her mission for 30 years and during that time she raised ten children (Colby & Damon, 1992).

Based on Mother Waddles and other exemplars' experiences, Colby and Damon (1992) have suggested that it is *moral commitment* that develops in adulthood—as opposed to moral cognition, as might be expected from looking at Kohlberg's thesis. Interestingly enough, Colby and Damon found that most of the exemplars were between the third and fourth stages in Kohlberg's model of moral development (see Chapter 9). In essence, the development of moral commitment results when adults are socially influenced to transform their personal goals, especially when they have a "sense of continued openness to change and growth, an openness that is not the usual expectation in most adult lives" (Colby & Damon, 1992). This capacity to change while keeping a sense of personal stability harks back to the dialectical approach to cognition. It seems that adults have an easier time, perhaps owing to life experience, being able to engage in the dialectical thinking that allows personal aspirations and social needs to inform each other. We will see in Chapter 19 that often the impetus for this dialogue between the self and "other" comes from religious affiliations.

Segue

Significantly, today's baby boomers are looking, acting, and feeling younger at middle age than did prior generations. Indeed, the defining characteristics of middle age are starting later and lasting longer, so that age 50 has taken on the connotations that age 40 had only a few decades ago. T-shirts proclaim "*50 Never Looked So Good!*", some 50-year-olds are following the Grateful Dead on the Internet, and Cher, The Rolling Stones, and Tina Turner are still performing professionally and producing hits. (And Elvis, who still has fans, would have turned 65 in January 2000.) Midlife baby boomers are actively revolutionizing cultural attitudes toward how the "middle-aged" are viewed—not unlike how this generation of Americans initiated sweeping social, political, and cultural changes in the 1960s.

In the next chapter we will examine the changing self-concept of adults in midlife, who are often reevaluating intimate relationships, reestablishing life priorities, becoming grandparents, shifting occupational ventures, or preparing for retirement.

Summary

Sensory and Physical Changes

1. The human body, after functioning for a number of decades, tends to work less efficiently than it did when it was "new." Seventy-five million Americans over age 40 are experiencing presbyopia, a normal condition in which the lens of the eye starts to harden, losing its ability to accommodate as quickly as it did in youth. Other symptoms of ocular "wear and tear" include glaucoma, cataracts, floaters, dry eye, and macular degeneration.

2. By age 50 about one in every three men and one in every four women have difficulty understanding a whisper. However, only about 5 percent of the population at 50 can be deemed to have substantial hearing problems.

3. By the mid forties, one may notice a gradual drop in the ability to taste; and by age 60, one will be tasting only one-half of what was possible in one's twenties. Half of those who reach the age of 65 are predicted to have a noticeable loss of the sense of smell.

4. The appearance of aging men and women in American society is part of the so-called "double standard of aging." As many men age, they are considered "mature" or "sophisticated" or even more attractive than when they were younger. However, there are very few kind expressions for the way an older woman looks.

5. One of the major concerns of some midlife adults is body composition, or proportion of muscle to fat. Muscle mass declines an average of 5 to 10 percent each decade.

6. The powerful hormone HGH (human growth hormone), which was developed to treat children afflicted by dwarfism, has become a trendy anti-aging potion. Hormones are protein messengers of the endocrine system that circulate throughout the body to all organs causing natural reactions, affecting not only ovaries and testes but many other vital life functions like memory, protein synthesis, cell repair, metabolism, ability to sleep, body temperature, water balance, and sexual functioning.

7. Menopause is one of the most readily identifiable signs of the climacteric, characterized by changes in the ovaries and in the various biological processes associated with these changes. Probably the most significant change is the profound drop in the production of the female hormones (particularly estrogen) by the ovaries. The average age at complete cessation of menses is around 45 to 55 years in Western countries.

8. Cultural stereotypes have frequently depicted men in their forties and fifties as suddenly undergoing severe psychological disturbances, leaving their wives for women young enough to be their daughters, quitting their jobs to become beachcombers, or beginning drinking to excess. Their difficulties are commonly attributed to a "male menopause." Psychologists and sociologists typically dismiss notions of a male menopause as a social myth deriving from an overmedicated culture that is obsessed with health and pathologically fearful of aging and death.

Health Changes

9. About 35 million Americans—1 in 7—have disabilities ranging from arthritis, diabetes, and emphysema to mental disorders. However, people can maximize their chances for leading healthy and long lives by altering their lifestyles to include a variety of health-conscious practices such as regular exercise and eating a healthy diet.

10. There are several risk factors associated with cardiovascular (heart and circulatory system) fitness or the lack of fitness. These risk factors include high blood pressure, smoking, a family history of heart disease, being male, being diabetic, or being obese.

11. A stroke, or "brain attack," occurs when blood circulation to the brain fails. Brain cells can die from decreased blood flow and the resulting lack of oxygen. Both blockage and bleeding can be involved. Stroke is the third leading killer of Americans and is the most common cause of disability. Each year more than 500,000 Americans have a stroke, with about 145,000 dying. Strokes occur in all age groups, in both sexes, and in all races of every country.

12. Women are much more likely than men to become infected by an HIV-positive heterosexual partner.

13. Depression and substance abuse are common but are often underrecognized and undertreated in middle age adults. Depression in older life is associated with increased mortality (or suicide), even after controlling for physical illness and disability.

14. Far too many older people accept social definitions of themselves as sexless; they become victims of the myth. Believing that they will lose their sexual effectiveness becomes a self-fulfilling prophecy, and some older people do lose it even though their bodies have not lost the capacity for sex.

Cognitive Functioning

15. Psychologists distinguish between fluid intelligence (the ability to make original adaptations in novel situations) and crystallized intelligence (the ability to reuse earlier adaptations on later occasions). A debate has emerged over the years concerning the existence of post-formal operations, contrary to Piaget's model.

16. People in midlife can maximize their cognitive abilities by practicing them regularly over time. Even experts must be committed to practice in their domains of expertise. Schaie proposed a model of cognitive development over the life span that includes the acquisitive stage, the achieving stage, the responsible/executive stage, and the reintegrative stage. Guilford distinguishes between convergent thinking (using standard logic and reasoning) and divergent thinking (creative). Further, middle adulthood is considered the peak time of creative thinking.

17. Some people in midlife have been identified as moral exemplars; that is, they consistently demonstrate sustained commitment to moral ideals that include respect for humanity and doing good for others.

Key Terms

amenorrhea *(475)*

cataract *(458)*

cholesterol *(472)*

climacteric *(463)*

convergent thinking *(483)*

crystallized intelligence *(479)*

dialectical thinking *(482)*

divergent thinking *(483)*

dry eye *(458)*

floaters *(458)*

fluid intelligence *(479)*

glaucoma *(458)*

human growth hormone (HGH) *(463)*

hormone replacement therapy (HRT) *(464)*

hypertension *(471)*

impotence *(467)*

macular degeneration *(458)*

menopause *(463)*

osteoporosis *(462)*

perimenopause *(463)*

presbycusis *(459)*

presbyopia *(458)*

prostate gland *(465)*

rheumatoid arthritis *(462)*

stroke *(473)*

Following Up on the Internet

Web sites for this chapter focus on physical changes, health risks, and intellectual functioning of adults in midlife. Please access the text web site at http://www.mhhe.com/crandell7 for up-to-date hot-linked Internet addresses for the following organizations, topics, and resources:

Women's Health Initiative Study sponsored by the NIH

Health Touch Online

National Cancer Institute

National Institutes of Health—Alphabetical Index

Elderhostel

Chapter 16

Middle Adulthood
Emotional and Social Development

Critical Thinking Questions

1. Suppose that each stage of a person's life could be defined in terms of a consuming purpose. For instance, suppose that the child's purpose is to play, the adolescent's purpose is to explore, and the young adult's purpose is to settle. What would the middle-age adult's consuming purpose be?

2. What would you think if your 55-year-old doctor had her or his tongue pierced? Would you judge this person to be acting immaturely?

3. Who do you think is happier in midlife—people who are married with a job and now raising grandchildren, or people who remained single and have the resources and time to do whatever they desire? Why?

4. Suppose that pay were inversely related to job popularity. Would you collect garbage for $150,000 a year or be a surgeon for $47,000 a year? When thinking of a career in this way, what would motivate you more—economic security or job satisfaction?

Transition and adaptation are central features in middle adulthood, as they are in other phases of life. Midlife is a time of looking back and at the same time looking forward. Some of the changing aspects of middle age are associated with the family life cycle (see Chapter 14). Many, but not all, parents at midlife today enter the "empty nest" period of life. Some couples at midlife are just beginning their families and are raising young children. Others are caring for both growing children or grandchildren and elderly parents, and we refer to these as the "sandwich generation." There have also been changes in the workplace. Because of advances in technology and international competition, job obsolescence confronts many blue-collar workers, and studies find this has had a particularly negative economic impact on African American and Hispanic American adults in midlife. Most midlife adults in white-collar and professional occupations are reaching the upper limits of their careers and realizing that they now must settle for lateral occupational shifts; others at midlife must find new inner resources to deal with unemployment or early retirement. In this chapter we examine the changing emotional-social context of middle adulthood from individual, social, occupational, and cultural perspectives.

Theories of the Self in Transition

> Now in their fifties, many American men and women are confronted with the fact that there are time limits to their lives. A powerful reminder is the symbolic meaning attached to the number 50. In terms of the life span, age 50 is roughly two-thirds of the way through life, but because 50 marks a half century, the 50th birthday carries a strong symbolic connotation that many men and women see as marking their entry into the "last half of life." . . . They begin counting the number of birthdays left to them rather than how many they have reached. So the fifties become a time for more introspective reflection and stock taking (Sheehey, 1998).

Traditionally, developmental psychologists have focused their attention on the many developmental changes that occur during infancy, childhood, and adolescence—and presently on old age. Underlying most of their approaches, however, is the notion that each individual has a relatively unique and enduring set of psychological tendencies and reveals them in the course of interacting with the environment. Only more recently has *middle adulthood* been examined to determine how and in what ways adults continue to develop and mature (Brim, 1992). Much of the available research on the midlife experience typically focused on physical and intellectual changes in the white middle-class population, was limited by its use of cross-sectional data, and focused on clinical populations (those who were having physical or mental health problems). Until more recently, only a few studies centered on emotional and social development of a diversity of adults in midlife, often identified as 35 to 65 years old (Lachman & James, 1997, p. 3). With the increase in longevity in the United States and other Western nations, many of the events that used to occur in old age (e.g., grandparenthood, retirement) now occur during midlife.

Significantly, in recent years a *life-span perspective* of development has emerged that views middle adulthood as a period of both continuity and transition in which individuals must adapt to new life situations and make a variety of role transitions—in the family, at work, in the community, and so on. Recent cross-cultural and gender development research shows that there are multiple paths that individuals take during this transitional time before late adulthood (Lachman & James, 1997). The adaptations along these paths can lead to changing aspirations, changing timetables, giving up goals, or taking on new ones (Brim, 1992). That is why it is more common today to see two 48-year-olds who are at different phases of their life cycle—one a first-time parent and the other a grandparent or one launching a new career and the other retired. Consequently, Helson (1997, p. 23) says that "middle age has different meanings in different times and places for different individuals." With this recent perspective in mind, we examine some research that is multidisciplinary, some that address midlife from a gender perspective, and some that look at the issues of midlife for adults in other cultures. Interestingly enough, we also learn from recent findings that adults in some societies experience neither a "middle adulthood" stage nor a "midlife crisis" (Lachman & James, 1997).

Maturity and Self-Concept

Most personality theorists emphasize the importance of maturity to individuals as they move through life. **Maturity** is our capacity to undergo continual change in order to adapt successfully and cope flexibly with the demands and responsibilities of life. Maturity is not some sort of plateau or final state but a lifetime process of becoming (Waterman, 1993). It is a never-ending search for a meaningful and comfortable fit between ourselves and the world—a struggle to "get it all together." Gordon W. Allport (1961, p. 307) identifies six criteria that psychologists commonly employ for assessing individual personalities:

> The mature personality will have a widely extended sense of self; be able to relate warmly to others in both intimate and nonintimate contacts; possess a fundamental emotional security and accept himself; perceive, think, and act with zest in accordance with outer reality; be capable of self-objectification, of insight and humor; and live in harmony with a unifying philosophy of life.

Underlying these elements of the mature personality is a positive self-concept (Hattie, 1992; Ross, 1992). **Self-concept** is the view we have of ourselves through time as "the real me" or "I myself as I really am." Self-concept has considerable impact on behavior (Dunning & Cohen, 1992; Greenberg et al., 1992). Indeed, much of our significant behavior can be understood as an attempt to approach or avoid various of our "possible selves" (Cross & Markus, 1991). For instance, a middle-aged man whose feared self includes being a heart attack victim, and who worries about how to avoid becoming that self, might undertake an exercise and diet regimen. The human ability to preserve, enhance, promote, defend, and revise notions about the self (our self-schemas) helps us to explain how older people, experiencing the frailties of advancing age, nonetheless function so well, especially on the subjective level. It seems that aging people do not simply react to aging processes but instead make cognitive shifts and behavioral adjustments that preserve their mental health and behavioral functioning despite their losses (Heidrich & Ryff, 1993).

Mounting evidence suggests that sad people and happy people are each biased in their basic perceptions of themselves and the world (Baumgardner, 1990). People bring to the world somewhat different cognitive templates or filters through which they view their experiences (Feist et al., 1995) (see Figure 16-1). And the way they structure their experiences determines their mood

Figure 16-1 Happy and Sad People View Their Experiences Through Somewhat Different Cognitive Filters

© The New Yorker Collection 1994 Edward Koren from cartoonbank.com. All Rights Reserved.

"I'm somewhere between O. and K."

and behavior. If we see things as negative, we are likely to feel and act depressed. If we see things as positive, we are likely to feel and act happy. That is why some view middle age as a time of feeling secure and settled, whereas others view it as a time of being bored or in a rut (Helson, 1997). Such perceptions tend to reinforce and even intensify people's feelings about their self-worth and their adequacy in the larger world. It is how individuals take the idea of self and interact in social situations that interests the theorists we will next encounter.

Stage Models

Erikson posits that the midlife years are devoted to resolving the "crisis" of **generativity versus stagnation.** Erikson (1968b, p. 267) views **generativity** as "primarily the concern in establishing and guiding the next generation." Adults express generativity through nurturing, teaching, mentoring, and leading—by promoting the overall interests of the next generation while contributing to the world of politics, art, culture, and community. As a group, they seek to benefit the larger society and facilitate its continuity across generations. To do otherwise is to become self-centered and to turn inward, resulting in "psychological invalidism," which leads to rejectivity and the unwillingness to care for others. McAdams and de St. Aubin (1992) also see generativity as springing from two deeply rooted desires: the communal need to be nurturant and the personal desire to do something or be something that transcends death. You will recognize the two poles of this particular crisis in stereotypical portrayals of adults. The generative teacher thinks of each pupil as "one of my own children" and tries to *generate* a love of life in each of the students. The individual

whose resolution is weighted toward stagnation is usually depicted as a "humbug" sort, Scrooge being probably the best-known example. Stories that show an "old grump" turning into an avuncular, caring person depict the transition from stagnation to generativity.

Erikson's "crisis" view has been criticized by others, such as Costa and McCrae (1980). In their studies they find no evidence that psychological disturbance is any more common during midlife than during other periods of life. Other critics of this normative stage approach say midlife is more a time of productivity and altruism, not a time of turmoil.

Various psychologists have elaborated on Erikson's formulations. Robert C. Peck (1968) has taken a closer look at midlife and suggests that it is useful to identify more precisely the tasks confronting midlife individuals. Peck defines these four tasks:

- *Valuing wisdom versus valuing physical powers.* As we saw in Chapter 15, when individuals progress through middle age, they experience a decline in their physical strength. Even more importantly, in a culture that emphasizes looking youthful, people lose much of their edge in physical attractiveness. But they also enjoy new advantages. The sheer experience of longer living brings with it an increase in accumulated knowledge and greater powers of judgment. Rather than relying primarily on their "hands" and physical capabilities, they must now more often employ their maturity and wisdom in coping with life. In Japanese and Chinese cultures this is definitely the case, and it is not uncommon for public officials to be in their seventies and eighties.

- *Socializing versus sexualizing in human relationships.* Allied to midlife physical decline, although in some ways separate from it, is the sexual climacteric. In their interpersonal lives individuals must now cultivate greater understanding and compassion. They must come to value others as personalities in their own right rather than chiefly as sex objects.
- *Cathectic flexibility versus cathectic impoverishment.* This task concerns the ability to become emotionally flexible. In doing so, people find the capacity to shift emotional investments from one person to another and from one activity to another. Many middle-aged individuals confront the death of their parents and the departure of their children from the home, and they must widen their circle of acquaintances to embrace new people in the community. And they must try on and cultivate new roles to replace those that they are relinquishing. Those who do not are likely to experience a sense of isolation and loneliness.
- *Mental flexibility versus mental rigidity.* As they grow older, some people too often become "set in their ways." They might become "closed-minded" or unreceptive to new ideas. Those who have reached their peak in status and power are tempted to forgo the search for novel solutions to problems. But what worked in the past might not work in the future. Hence, they must strive for mental flexibility and, on an ongoing basis, cultivate new perspectives as provisional guidelines to tackling problems.

Peck's formulations provide a more positive, dynamic image of middle age than those portraying midlife as a time of turmoil and crisis as people reevaluate their lives. Most adult Americans also see middle age as a time when people deepen their relationships and intensify their acts of caring. A variety of studies confirm that people in their fifties become more altruistic and community-oriented than they were at age 25; for instance, many volunteer to serve on the boards of schools, universities, and community agencies. This might be seen as striving for symbolic immortality by furthering a worthy cause or group, catalyzed by an awakened confrontation with one's own mortality. In general, those individuals who have consolidated their marriages and careers have a secure base from which to reach out to assist others (Peterson & Klohnen, 1995; McAdams, de St. Aubin, & Logan, 1993). Those without a secure base are likely to flounder and experience dissatisfaction with middle age.

Question

How is Erikson's stage theory different from Peck's task theory of midlife personality development?

Trait Models

Until relatively recently, most psychologists believed that personality patterns are established during childhood and adolescence and then remain relatively stable over the rest of the life span. This view largely derived from Sigmund Freud's psychoanalytic theory. As described in Chapter 2, Freud traced the roots of behavior to personality components formed in infancy and childhood—needs, defenses, identifications, and so on. He deemed any changes that occur in adulthood to simply be variations on established themes, for he believed that an individual's character structure is relatively fixed by late childhood.

Likewise, clinical psychologists and personality theorists have typically assumed that an individual gradually forms certain characteristics that become progressively resistant to change with the passage of time. These patterns are usually regarded as reflections of inner traits, cognitive structures, dispositions, habits, or needs. Indeed, almost all forms of personality assessment assume that the individual has stable *traits* (stylistic consistencies in behavior) that the investigator is attempting to describe (Goldberg, 1993). For example, developmentalists Robert McCrae and Paul Costa, Jr. (1990) found that a few dimensions recur across their many investigations, thereby proposing that the "big five" traits—*extraversion, neuroticism, openness, agreeableness,* and *conscientiousness*—constitute the core of personality. Most models concur on the existence of extraversion-introversion and neuroticism (emotional stability/instability), although there is less agreement on what narrower traits constitute these dimensions. What lies beyond extraversion and neuroticism, however, is a more perplexing and contentious issue, with some theorists identifying five, eleven, and even sixteen factors (Block, 1995).

Moreover, we all typically view ourselves as having a measure of consistency, and we anticipate and adapt to many events without appreciably changing our picture of the entirety of our lives. So we see ourselves as essentially stable even though we assume somewhat different roles in response to changing life circumstances (Tomkins, 1986).

Situational Models

The trait approach to personality views a person's behavior in terms of recurring patterns. In contrast, proponents of situational models view a person's behavior as the outcome of the characteristics of the situation in which the person is momentarily located. Cognitive learning theorist Walter Mischel (1985, 1977, 1969) provides a forthright statement of the situational position. Mischel says that behavioral consistency is in the eye of the beholder and hence is more illusory than real. Indeed, he wonders if it makes sense to speak of "personality" at all, suggesting that human beings might not have personalities. Mischel concludes that we are motivated to believe that the world around us is orderly and patterned because only in this manner can we take aspects of our daily lives for granted and view them as predictable. Consequently, we perceive our own behavior and that of other people as having continuity. Even so, Mischel admits that a fair degree of consistency exists in people's performance on certain intellectual and cognitive tasks. However, he notes that the correlation between personality test scores and behavior seem to reach a maximum of about .30, not a particularly high figure. People's behavior across situations is highly consistent only when the situations in which the behavior is tested are quite similar. When circumstances vary, little similarity is apparent (see Figure 16-2).

Question

Which theory, trait or situational, explains why you might not mind waiting in a slow movie line one day but might become agitated doing so another time?

"Another thing I remember about the 1960 release is that back then you identified with the slaves."

Figure 16-2 The Least Continuity Is Found in Attitudes Across the Life Span

Interactionist Models

In recent years psychologists have come to recognize the inadequacies of both trait and situational models. Instead, they have come to favor an *interactionist approach* to personality (Field, 1991). They claim that behavior is always a joint product of the person *and* the situation. Moreover, people seek out congenial environments—selecting settings, activities, and associates that provide a comfortable context and fit—and thereby reinforce their preexisting bents (Setterlund & Niedenthal, 1993). And through their actions, individuals create as well as select environments (Rausch, 1977). By fashioning their own circumstances, they produce some measure of stability in their behavior (Heidrich & Ryff, 1993).

Psychologists have employed a variety of approaches for specifying the form of interaction that transpires between a person and a situation. One approach is to distinguish between those people for whom a given trait predicts behavior across situations and those people for whom that trait is not predictive of behavior. For example, individuals who report that their behavior is consistent across situations with respect to friendliness and conscientiousness do indeed exhibit these traits in their behavior. In contrast, individuals who report that their behavior is inconsistent across situations with respect to these qualities reveal little consistency in their behavior with regard to them (Bem & Allen, 1974). Moreover, people vary in their consistency on different traits (Funder & Colvin, 1991). Hence, if we are asked to characterize a *friend,* we typically do not run through a rigid set of traits that we use to inventory all people. Instead, we select a small number of traits that strike us as particularly pertinent and discard as irrelevant the hundreds or so other traits. Usually, the friendship traits—such as trustworthiness, kindness, and acceptance—that we select are those that we find hold for the person *across a variety of situations.*

Gender and Personality at Midlife

In the long years liker must they grow; The man be more of woman, she of man.

Alfred, Lord Tennyson, *The Princess*

In recent years many psychologists have moved away from the traditional assumption that masculinity and femininity are inversely related characteristics of personality and behavior. A variety of studies have examined the dynamics of gender and personality traits during middle adulthood.

Levinson's Theory of Male Midlife Development

Psychologist David J. Levinson (1986) and his associates

at Yale University studied stages of male adult development and formed their *life structure theory* (Levinson et al., 1978). Central to this theory are the roles of family and one's occupation in life. Analysis of their subjects' intensive biographies, described in *The Seasons of a Man's Life* (Levinson et al., 1978) led them to conclude that men often experience a turning point in their lives between the ages of 35 and 45. More recently, Levinson concluded that men's inner struggles occur with renewed intensity in the mid fifties (Goleman, 1989a). These Yale researchers believe that a man cannot go through his middle adult years unchanged, because he encounters the first indisputable signs of aging and reaches a point at which he is compelled to reassess the fantasies and illusions he has held about himself. Levinson identified several substages from 40 to 65 in his initial study with men in the early 1970s (and with women in the early 1980s) with both beginnings and endings:

- *Midlife transition (age 40 to 45).* This is a developmental bridge between early adulthood and middle adulthood. People come to terms with the end of their youth in the late thirties and try to create a new young-old way of being. Working at *individuation* at midlife is an especially important task. Initiation into midlife involves making choices.
- *Entry life structure for middle adulthood (age 45 to 50).* People establish an initial place in a new generation and a new season of life. Even if they are in the same job, marriage, or community, they create important differences in these relationships. They try to establish a new place in midlife.
- *Age 50 transition (age 50 to 55).* This is a time to reappraise and explore one's self within this new midlife structure. Developmental crises are common in this period, especially for persons who have made few life changes in the previous 10 to 15 years.
- *Culminating life structure for middle adulthood (age 55 to 60).* A time for realizing midlife aspirations and goals.
- *Late adult transition (age 60 to 65).* This requires a profound reappraisal of the past and a readiness for the next era of adulthood. This involves termination of midlife and readiness for late adulthood.

The realization that the career elevator is not going up any longer, and might indeed be about to descend, can add to a man's frustrations and stresses. There is historical evidence that in the early 1900s in the United States, management arbitrarily, and overnight, replaced men in their forties and fifties with younger men; and it was then impossible for these older men to find another job. Thus, the fears that many midlife men have about losing

their jobs or being replaced in their jobs is legitimate today, as corporations continue to "merge," "restructure," and "downsize." A man who has not accomplished what he had hoped to with his life now becomes keenly aware that he does not have much more time or many more chances to do it. And even many who have reached their career goals—say, owning their own company, having a high executive position, or earning a full professorship—often find that what they have achieved is less rewarding than they had anticipated. A man might ask himself, "If I am making so much money and doing so well, how come I don't feel better and my life seems so empty?" On the other hand, a blue-collar worker of more modest means might ask, "How much longer am I going to have to put up with this dull job?" Even so, research shows that, for most people, aspirations gradually come to match attainments without severe upset and turmoil (Bridgman, 1984).

Levinson concluded that this age-graded sequence of periods in adult development violates conventional wisdom and the findings that there are no comparable sequences of periods for personality development, cognitive development, and family development. However, based on his research findings, he concludes that this sequence exists. He further states it is a much more complex matter to determine the "satisfactoriness" of individual lives within these stages (Levinson, 1996, p. 29). A developmental crisis occurs when someone is having great difficulty in meeting the tasks in each stage (e.g., makes an abrupt career change, gets divorced, has an affair, or "acts like an adolescent"). Levinson concluded prior to his death in 1994 that though the genders experience different life circumstances, and women work on the developmental tasks of each period with different resources than men have, women and men both go through these life stages.

Question

According to Levinson's research findings, what are the stages of a man's life?

To gild refined gold, to paint the lily,
To throw perfume on the violet,
To smooth the ice, or add another hue
Unto the rainbow, or with taper-light
To seek the beauteous eye of heaven to garnish,
Is wasteful and ridiculous excess.

William Shakespeare, "King John"

Levinson's Theory of Women's Midlife Development
A white-male-centered view of adult life has been domi-

nant since the earliest of psychological and sociological studies, and before that in the history and literature of various cultures. Though Levinson chose to study adult male development in his initial study in the 1970s, he said that one couldn't study one gender without studying the other, for both influence each other in complex ways (e.g., relationships of husbands and wives, fathers and daughters, mothers and sons, ex-husbands and ex-wives). From 1980 to 1982, Levinson used the method of intensive biographical interviewing with 45 females, ranging in age from 35 to 45. The subjects, selected randomly, constituted three equal groups: (1) primarily homemakers, (2) women with careers in the corporate world, and (3) women with careers in the academic world. The research findings indicate there were apparent gender differences in adult life and development, particularly in the realms of domestic and public life, within a marriage, in the division of "women's work" and "men's work," and in the male and female psyche. Let's look at some of Levinson's findings.

Primarily Homemakers   Women who live primarily as homemakers in a traditional marriage are likely to center their lives in the domestic sphere, to engage solely in women's work, and to accept and value themselves and other women as "feminine." These women are also more likely to be marginal to the public world, to be limited as "provisioners" and authorities in the family, and to have difficulty engaging in "men's work." Men's lives are usually centered in the public sphere (their territory, under their control), whereas women typically participate in the public sphere in more segregated, marginal, and subordinate ways (e.g., as secretary to a lawyer, nurse attending a physician—not their territory, not under their control), (though many more women have entered the male-dominated professions since this study was conducted in the early 1980s). The domestic sphere consists of a household and its surrounding social world. In many societies of the world, (e.g, Japan, India, China), women's lives continue to be centered in the domestic sphere. The home is the key source of their identity, meaning, and satisfaction.

Traditional Marriage   In many societies, marriage is not about being in love. It is, "first of all, about building an enterprise in which the partners can have a good life" and "take care of each other" in their own fashion (Levinson, 1996). Arranged marriages are still common in many countries. In a **traditional marriage,** the women are homemakers and the men are provisioners (involving themselves much more in their work world than in their family). *Patriarchy,* the rule of the father or husband, is a universal theme to some degree in most societies of the world. Consequently, the man is still the dominant authority over the family members. A wife has the responsibility of the housekeeping, family life, and care of the children and husband (still a prevalent way of life in Japan, China, and India). A wife might enter the public sphere marginally in her attempts to further her husband's career (e.g., entertain colleagues), work on the family farm or business, or contribute time to the local school or church. In the United States a prime example of this is the thousands of hardworking farm wives who were never employed outside the farm, have not personally contributed to Social Security, and are ineligible for Social Security benefits. Women who have lived their lives in a traditional marriage and then become divorced or widowed during midlife are often faced with a major transition:

> "I knew from the beginning that when he was gone [her husband of 39 years, dying of emphysema], I'd have to go to work," she says. That wasn't a comforting thought. The emotions she felt at the prospect of losing her husband were devastating enough; the idea of having to construct a life on her own was often overwhelming. . . . Work was not a part of the blueprint Vinita had drawn for herself. [Widowed at 56, Vinita Justus, long a homemaker, faced the daunting necessity of joining the workforce]. (Coburn, 1996, p. 57)

As homemaker, women get to create their own home base as center for their life. It becomes the place where they rest, play, love, enjoy privacy and leisure, and have strong affectional relationships with family members. Domestic work can be hard, though, particularly as women age and are still facing endless household, childcare, eldercare, and husband-care responsibilities (see *Human Diversity:* "Life Without a Middle Age"). However, taking care of each other has different meanings for men and women in a traditional marriage. A man readily accepts nourishment and support from his wife the homemaker, but he has much more difficulty discussing with her his possible inadequacies at work, and she might find his work world quite alien to her own. And a woman is dependent upon her husband to provide for the family. The traditional homemaker and spouse are in a *mutual-care arrangement.*

Questions

According to Levinson's findings, what are a woman's needs at midlife? What does Levinson mean by the traditional mutual-care arrangement between a husband and a wife?

However, many cultural changes since the 1940s and 1950s in the United States have reduced women's involvement in the family and increased their involvement in the outside world. Today, homemaking is not a permanent, full-time occupation for most American

Human Diversity

Life Without a Middle Age

Perceptions of age are always mediated by cultural models of the life course, and the Western view of middle age differs significantly from the views of those in non-Western cultures.

India

For centuries in India, well-defined extended family structures in each caste system identified the roles of family members throughout life. The three-generation family life was the norm. When a woman married, she became the person of lowest status in her in-laws' home, with her mother-in-law becoming the "keeper of the traditions" and controlling the behavior of the younger family members. The young bride was expected to take over all household responsibilities, while the elder mother-in-law could now spend more time with her friends and attend to more religious obligations. The elders in the home were to be accorded respect, care, and support by the young. The young were to defer to the elders and care for them during their declining years. With this age-old view of growing older, those in middle adulthood looked forward to receiving respect and care and relinquishing home responsibilities. Tikoo (1996) says it is difficult to give a definite age at which one is middle-aged in India, because the average life expectancy in India is only 58 years; "middle age" in that context would be younger than American middle age. However, Adler and colleagues (1989) grouped their Indian subjects into "young" (16–35), "middle" (34–54), and "old" (55–80).

In India, aging is not unwelcome, and much folklore highlights the significance of grey hair and wrinkles as signs of one's wisdom and experience (Tikoo, 1996). Tikoo and colleagues surveyed 56 adult men and women (Hindu) with the 97-item self-report Men's Adult Life Experience Survey. Findings in 11 developmentally related domains suggested very few gender or age differences, "raising questions as to the meaning or relevance of midlife crisis in India" (Tikoo, 1996).

However, other recent findings (Kumar & Suryanarayna, 1989) indicate that along with modernization in certain locales in India, there has been an evident decline in extended family life, including a decline in the respect, prestige, and care given the elderly by the young. Young adults in India are more mobile today, and some elderly no longer receive the care and support they would have commanded in earlier years.

Japan

In Japan the three-generation household still exists in nearly 20 percent of homes. Women are trained to be good wives and mothers and are formally educated for their role in soci-

ety (Lock, 1998, p. 59). The dominant image of females is nurturer. Formerly about 98 percent of Japanese women married; by 1998, that rate had dropped to 88 percent. The divorce rate is increasing but is still very low (1.45 per thousand). Traditionally women were expected to marry in their mid twenties and produce two children within five or six years. By their mid thirties, their families were expected to be complete. Birth control pills are not available in Japan, but woman have access to other birth control devices. Women who remain unmarried by age 25 are described as "unsold merchandise" (*urenoki*) or "overripe fruit" (*tô ga tatsu*), not unlike the U.S. concept of the "old maid."

However, modern, young Japanese women are delaying marriage; the average age of marriage is now 27 years old. Virtually all women who give birth are married or in long-term common-law arrangements. Young women are choosing to delay marriage mainly because of the cultural expectation that they will leave the workforce entirely once they become pregnant the first time. "It is believed that women cannot work effectively with young children in the household" (Butler, 1998). Also, the traditional view is that a woman's life is meaningful only in terms of what she accomplishes for others.

The ideal life course of the people of a village, region, or the nation is age-graded; that is, "the life cycle is predominantly a social process involving community rituals in which people born the same year participate together" (Lock, 1998). Biological aging has always been subordinate to social maturation of a family and a community. The female's life cycle has been shaped more by biology than a male's. Middle age, referred to as the *prime of life* (*sônen*), is simply a part of the life cycle that begins with marriage and ends at the ritual of turning 60—old age. There is no word in Japanese for menopause, nor is menopause noted by the family or community, though the end of a woman's reproductive cycle (*tenki*) has traditionally been recognized to occur in the "seventh stage" in a woman's life. Japanese physicians have not adopted the "disease" approach to managing menopause. Hormone replacement therapy is not available, but women do use herbal concoctions. Japanese women in midlife also have low rates of osteoporosis and heart disease (Lock, 1998, p. 63).

Interviews with Japanese women in their fifties reveal that many spend time and energy cultivating various art forms (dance, flower arranging, classical poetry, calligraphy, archery). These traditional art forms are thought to be paths to spiritual awareness, self-development, and self-discipline. Lifelong socialization "is based on an ideology that accepts

Continued on page 496

Continued from page 495

the possibility of human perfectibility over time, a condition that transcends and continues beyond the unavoidable decline in the body" (Lock, 1998). A woman is recognized as fully mature—a complete adult—after she has raised her own children to adulthood. Grandmothers devote a good deal of time monitoring the household, caring for and educating grandchildren, and caring for aging in-laws.

Confucianism encourages respect for the elderly in life and in the afterlife. To the Japanese, age denotes wisdom and authority. Overall, people strive for a cooperative life in which the "self" is subordinate or nonexistent. In classical Buddhist thinking, individual discipline fosters a path to maturity and otherworldly transcendence, which can take many incarnations to accomplish. Individual aging therefore is not viewed as a bad thing or a disease, though caring for growing numbers of bed-ridden elderly has become a more pressing social problem.

From: Margaret Lock. Deconstructing the change: Female Maturation in Japan and North America. *Welcome to Middle Age!* Richard A. Shweder, editor. Copyright 1998. Reprinted by permission of the University of Chicago Press.

women. Many young women live with the awareness that they will become the primary provisioner for themselves and their children. Those who want to remain homemakers experience social pressures from dominant cultural values to become educated and employed. Many women have chosen, or have been forced by divorce or widowhood, to be employed as caregivers: in child care, teaching, nursing, social work, psychology, clerical, service, clergy, and sales. Only since the 1970s in this country have women been able to enter the higher-status, higher-salaried occupations. Women's entry into the occupational sphere has changed the division of labor and has violated the traditional division of authority in the home. For many midlife women, balancing their investments in a career and a family has become very difficult.

The Single Successful Career Woman Reflecting back on their youth, many women in midlife today would admit that they once were afraid to become this "pitied" creature. Typically parents harped at their daughters that they would become an "old maid" or "spinster" if they chose a career path instead of the normative security of marriage and family. The view of the single woman as the pitied woman, the odd person out, has deep historical roots, and there were few successful career women as models when today's middle-aged women were young.

Midlife Transition Career Women Today's midlife women also bought the myth in the late 1960s and 1970s that a woman could have any career she wanted, as well as marry, have children, and have time for recreation and leisure. These women earned advanced degrees in record numbers, all while caring for home and family or supporting themselves. When something has had to "give" in the family, it has usually been the woman's career. Many midlife women today have made transitions in and out of employment, or have accepted adjunct or temporary work positions, depending upon demands of

birthing, caring for children, and caring for elderly parents. These same women thought their so-called equality in the workplace also meant equality at home, as in sharing domestic and child-care chores equally. But there has been a great imbalance at home in the division of labor, and women still perform most of the domestic chores for their needy family while they work at a demanding job. Levinson's study revealed that each woman in midlife transition recognized that her efforts to combine love and marriage, motherhood, and full-time career had not given her as much satisfaction as she had hoped. These women were exploring new ways to live in middle adulthood.

The career women and female faculty members in Levinson's study in their early forties went through a major reappraisal of their careers and made significant changes in career and other aspects of their lives. A major component of this struggle was coming to terms with the myth of the successful career woman. Two of the seven female faculty interviewed were not full professors, though they had attained a doctorate by the age of 40. The full professors found that there were even more demands placed upon them to continue teaching, conduct scholarly work and administrative work, to contribute to professional organizations and activities in their field, and so on. All of the female midlife subjects reported a sexist attitude at work (both in education and in business settings).

> Debra (44): My primary dissatisfaction in my life now is my job, and I ought to look for a change in my job situation. I am very ambitious, and so much of my present frustration in my life really has to do with the feeling now of getting nowhere with my career. I'm very unhappy with my work, but I stay because of the money. It enables my family to live a certain lifestyle that I'm not totally prepared to give up, especially with my children about to go to college. I feel trapped in my job economically (Levinson, 1996, p. 381).

Midlife Transition Career Woman Some midlife women today have changed employment status while raising young children and later when caring for aging parents and in-laws. Other women in the baby-boom generation with a career, a marriage, a family, and other personal commitments find life challenging.

Levinson says that the traditional roles of men and women in marriage have been split, and no clear alternative has yet been established. Thus, several alternative lifestyles have emerged. With one out of every two marriages ending in divorce, many women are finding themselves to be single parents who have to be both provisioner and homemaker. And men are also confused and uncomfortable with their changing roles; changing diapers as first-time dads in midlife was not in their "blueprint" either. Critics of Levinson's work cite the small sample of subjects (approximately 45 in each study). They point out that men or women in their fifties and sixties were not interviewed, though his life structure theory extends to these ages, and they say it is unlikely that such a midlife "crisis" occurs within the narrow age range of 40 to 45.

Ravenna Helson and the Mills Longitudinal Study Subjects at Midlife Helson and Wink (1992) studied personality change of middle-aged women who were subjects in the *Mills Longitudinal Study* (women born in the late 1930s who graduated from a private college between 1958 and 1960). Findings indicate that these women experienced turmoil in their early forties, followed by an increase in stability by age 52. Over this same decade, the women had decreased negative emotions, increased decisiveness, and increased comfort and stability through adherence to personal and social standards. By age 52 they had fewer caregiving responsibilities toward their children and more toward their elderly parents. Three-fourths of this sample were menopausal or postmenopausal (Helson, 1997, p. 26).

Helson's findings support those of Neugarten and Datan (1974). Their study in the 1970s found that middle-aged men reported that their period of greatest productivity and reward was after they turned 50. Women homemakers experienced more adjustments as children departed home and were more likely to describe middle age as a period of "mellowness and serenity." But blue-collar workers, both men and women, were likely to report this as a time of decline. (Time, place, and socioeconomic status are factors here, because women who were traditional homemakers did not have a role in the occupational world while some blue-collar women were wage earners at this stage.) Mitchell and Helson (1990) describe women's *prime of life* as being in their early fifties, with a sense of accomplishment from occupational achievement and/or launching of children, convergence of personality resources, and a new, freer lifestyle.

Continuity and Discontinuity in Gender Characteristics Carl Jung (1933/1960), the influential Swiss psychoanalyst, was one of the first to suggest that gender differences tend to diminish or even "cross over" in later life. Some research confirms that men and women move in *opposite* directions across the life span with respect to assertiveness and aggressiveness, so that patterns of a later-life "unisex" tend to emerge (Hyde, Krajnik, & Skuldt-Niederberger, 1991).

David Gutmann (1987) pursued this possibility by comparing the male subjects in Neugarten's Kansas City study with men in a number of other cultures: the subsistence, village-dwelling, lowland and highland Maya of Mexico; the migratory Navajo herdsmen of the high-desert plateau of northeastern Arizona; and the village-dwelling Galilean Druze herdsmen and farmers of Israel. Gutmann found that the younger men (aged 35 to 54) in all four cultures relied on and relished their own internal energy and creative capabilities. They tended to be competitive, aggressive, and independent. On the other hand, the older men (aged 55 and over) tended to be more passive and self-centered. They relied on supplication and accommodation to influence others. Gutmann concluded that this change from *active* to *passive mastery* seems to be more age- than culture-related. Other researchers similarly report that, on the whole, older men are more reflective, sensual, and mellow than younger men (Zube, 1982).

Gutmann (1987) has continued to study personality and aging in a wide range of cultures. In subsequent research he reports that his earlier finding has been confirmed—that around age 55, men begin to use passive instead of active techniques in dealing with the demands of their environment. Women, however, appear to move in the opposite direction, from *passive* to *active mastery* (see Figure 16-3). They tend to become more forceful, domineering, managerial, and independent. Those who study the effects of lowered levels of estrogen and progesterone—and thus an increasing influence of testosterone—in postmenopausal women are beginning to document similar findings. Sheehy writes (1992) about women's *postmenopausal zest*:

> Today's pioneering women in postmenopause in advanced societies eventually give up the futile gallantry of trying to remain the same younger self. Coming through the passage of menopause, they reach a new plateau of contentment and self-acceptance, along with a broader view of the world that not only enriches one's personality but gives one a new perspective on life and humankind. Such women—and there are more and more of them today—find a potent new burst of energy by their mid-fifties (p. 237).

Gutmann's approach to gender behaviors—a functional, role-based theory—is controversial. It has been criticized on theoretical and methodological grounds (McGee & Wells, 1982). Even so, it has provided a starting point for research on gender-role orientation across the life span. The findings of Gutmann and other researchers seem to suggest that in later life people tend toward *androgynous* responses—thought to be associated with increased flexibility and adaptability and hence with successful aging (Wink & Helson, 1993). **Androgyny**, the incorporation of both male-typed and female-typed characteristics within a single personality, provides an alternative perspective. Androgynous individuals do not restrict their behavior to that embodied in cultural stereotypes of masculinity and femininity.

Questions

What do we know about men's and women's lives and transition through middle adulthood? Do studies of midlife development include ethnic minorities and those who are not of the middle class?

Personality Continuity and Discontinuity

*T*urning 30 scared and depressed me. . . . Fifty fascinated me. I'm more comfortable with who I am and what I do than I have ever been.

Sylvia, a married college professor in the
Our Boomers-Turning-50 Happiness Index

As noted earlier, Robert R. McCrae and Paul T. Costa, Jr. (1990) find considerable continuity in a person's personality across the adult years. They have tracked individuals' scores over time on standardized self-report personality

Figure 16-3 Moving from Passive to Active Mastery. Gutmann reports that at around age 55 men begin to use *passive mastery* in dealing with demands in their environment, whereas women tend to move toward *active mastery*. Sheehy calls this women's *postmenopausal zest*.

"Hi. You've been randomly selected to participate in a sex survey upstairs in fifteen minutes."

scales. On such personality dimensions as *warmth, impulsiveness, gregariousness, assertiveness, anxiety,* and *disposition to depression,* a high correlation exists in the ordering of persons from one decade to another. An assertive 19-year-old is typically an assertive 40-year-old and later an assertive 80-year-old. Likewise, "neurotics" are likely to be "complainers" throughout life (they might complain about their love life in early adulthood and decry their poor health in late adulthood). Although people can "mellow" with age or become less impulsive by the time they are in their sixties, the relation of individuals to one another regarding a given trait remains much the same; when tested, most persons drop the same few standard points.

In sum, for many facets of our personality there is strong evidence of continuity across the adult years (Caspi, Elder, & Bem, 1987; Costa, McCrae, & Arenberg, 1980). This element of stability makes us adaptive; we know what we are like and hence can make more intelligent choices regarding our living arrangements, careers, spouses, and friends. If our personality changed continually and erratically, we would have a hard time mapping our future and making wise decisions.

This conclusion is supported by much of the work exploring Erikson's thesis regarding generativity (Peterson & Klohnen, 1995). As the Neugartens note (1987), the psychological realities confronting the individual shift with time. Middle age, for instance, often brings with it responsibilities for aging parents and for one's own minor children. With these obligations comes the awareness of oneself as the bridge between the genera-

tions—the so-called **sandwich generation** (see Table 16-1) (Lang & Brody, 1983). As a perceptive woman observed in Neugarten's study (1968, p. 98):

It is as if there are two mirrors before me, each held at a partial angle. I see part of myself in my mother who is growing old, and part of her in me. In the other mirror, I see part of myself in my daughter. I have had some dramatic insights, just from looking in those mirrors. . . . It is a set of revelations that I suppose can only come when you are in the middle of three generations.

In a sense, individuals in their forties and fifties are catching up with their own parents, so they might experience increased identification with them and a greater awareness of their own approaching senescence (Stein et al., 1978). Some of the issues of middle age are related to increased stocktaking, in which individuals come to restructure their time perspective in terms of time-left-to-live rather than time-since-birth.

The Social Milieu

To study an individual life, we must include all aspects of living. A life involves significant interpersonal relationships—with friends and lovers, parents and siblings, spouses and children, bosses, colleagues, and mentors. It also involves significant relationships with groups and institutions of all kinds: family, occupational world, religion, and community.

Levinson (1996)

Table 16-1 The Sandwich Generation: Caught in the Middle—The Needs of Middle-Aged Adults, Their Children, and Their Aging Parents

What Kids Want and Need	What Middle-Aged Adults Want and Need	What Aging Parents Want and Need
Independence	Help	Acceptance
Respect	Appreciation	Independence
Sounding board	Pressure off	Respect
Separate entity	"My turn"	Control
Patience	Independence	Sharing
Guidelines	Listening ear	Involvement
Flexibility	Acceptance	Emotional support
Acceptance	Time with my own generation	Interpersonal relationships
Security	Solitude	Interaction
Money	Space	Inclusion
Support	Unconditional love	Control of one's own life
Make choices	Control of one's own life	
Unconditional love		
Control of one's own life		

Source: Herbert G. Lingren, 1996, "The Sandwich Generation: A Cluttered Nest" [on-line], NebGuide, Cooperative Extension, University of Nebraska–Lincoln, www.ianr.unl.pubs/family/g1117.htm.

Close and meaningful social relationships play a vital part in human health and happiness. Through our association with others—family, friends, acquaintances and co-workers—we achieve a sense of worth, acceptance, and psychological well-being. This social network is often referred to as a **social convoy,** the company of other people who travel with us from birth to death.

Familial Relations

As we noted in Chapter 14, the vast majority of adult Americans have a profound wish to be part of a couple and to make the relationship work.

Married Couples   A 1995 survey of 1,000 men and women born in 1946 (the first of the baby-boom generation to turn 50 years old) revealed that the majority (80 percent) were married and had a healthy attitude about turning age 50 and about their future; (see Table 16-2) (Callum, 1996). The figure of 80 percent married does not reveal how many were remarriages, but marriage seemed to be the relationship of choice for these middle-aged adults. More than half of the survey participants had children still living at home, for many had started their families later than their parents did (and this was the first generation to use the birth control pill). Most were better educated, were employed full-time and in decent financial shape, and seemed to be happy with their sex lives. Forty-seven percent reported that they now had sex less often but that sex was now better. One finding among both men and women was that religion had become a source of community, spirituality, and comfort. Many said they had a personal responsibility to make the world a better place.

As they have grown older, these first boomers have found themselves to be more conservative in their views, and they enjoy their material success and comforts. They look forward to spending leisure time with their spouses. Still, the majority are concerned about developing health problems, about maintaining their lifestyle after retirement, and some worry about not being able to retire.

"... And as part of your community service, you're to help your wife with the cooking and cleaning around the house."

Figure 16-4 Most Men Do Little or No Housework, Which Can Cause Friction Between Marital Partners and Cohabiting Couples
From *The Wall Street Journal*—Permission, Cartoon Features Syndicate.

Most indicate they don't feel their age—that they feel much younger (Callum, 1996).

An earlier study of 12,000 U.S. couples was undertaken by sociologists Blumstein and Schwartz (1983). They surveyed a nationally representative sample of adults in four types of couple relationships: married, cohabiting, gay male, and lesbian. Three hundred couples were then selected for in-depth interviews and were interviewed again 18 months later. As with most studies of the early 1980s, this sample was composed of white, well-educated couples. These couples turned out to be more conventional than the researchers had initially expected. Although 60 percent of the wives worked outside the home, only 30 percent of the men and 39 percent of the women believed that both spouses should work. Wives with full-time jobs still did most of the housework. In general, husbands so objected to doing housework that the more they did of it, the more unhappy they were. If the man did not contribute what the woman felt was his fair share of the housework, the relationship was sometimes imperiled (see Figure 16-4). Much of the same pattern held for cohabiting couples (Huston & Geis, 1993; Suitor, 1991).

Men, both straight and gay, placed a considerable premium on power and dominance. Heterosexual men apparently took pleasure in their partner's success only if it was not superior to their own. Gay males likewise tended to be competitive about their career success. In

Table 16-2	Marital Status of First Baby-Boomers Born in 1946 Turning 50

(Survey of 1,000 middle-age adults)

Married	80%
Divorced (currently)	10
Single	6
Widowed	3
Separated	1

From: Myles Callum. Our Boomers-Turning-50 Happiness Index. *New Choices.* February 1996. Copyright © 1996. Reprinted with permission.

contrast, lesbians did not feel themselves particularly threatened by their partner's achievements, perhaps because women are not socialized to link their self-esteem so highly with success in the workplace (Kurdek, 1994a).

Most married couples in this large sample pooled their money, though some wives did not pool their money. Regardless of how much the wife earned, married couples measured their financial success by the husband's income. In contrast, cohabitors and gay couples appraised their economic status individually rather than as a unit. Recent findings by Forste and Tanfer (1996) concur with previous studies that found that cohabitors are more likely to value independence and equality.

Typically, one partner in a marriage expressed a desire for "private time." Early in marriage, husbands were more likely than wives to assert that they needed more time on their own. But in long-standing marriages, it was the wives who more often asserted that they needed more time to themselves. Women with retired husbands who "hung around" the house were especially troubled by the constant presence of their partner.

Sociologists Jeanette Lauer and Robert Lauer (1985) looked into the question of what makes successful marriages. They surveyed 300 happily married couples, asking why their marriages survived. The most frequently cited reason was having a positive attitude toward one's spouse. The partners often said "My spouse is my best friend" and "I like my spouse as a person." A second key to a lasting marriage was a belief that marriage is a long-term commitment and a sacred institution. Marriage itself increases commitment between partners, regardless of prior cohabitation status (Forste & Tanfer, 1996).

Question

What are some of the factors associated with long-lasting and satisfying marriages of middle-aged men and women?

Extramarital Sexual Relations   Traditionally, Western society has strongly disapproved of extramarital sexual relations (referred to now as "EMS"). Opposition has mainly been the result of two beliefs: that marriage provides a sexual outlet and therefore the married person is not sexually deprived and that extramarital involvement threatens the marriage relationship and therefore imperils the family institution. These beliefs find expression in religious values that brand extramarital sexual activity as sinful and label it adultery.

In comparison to Kinsey and colleagues' (1953) studies, rates of EMS reported in the 1970s, 1980s, and early 1990s have varied (see Table 16-3) (some studies investigated incidence of EMS over the past year, while others investigated incidence of EMS over a lifetime). It appears that the incidence of self-reported EMS increased for respondents in the early 1980s, with a decline in rates of self-reported EMS into the 1990s by both men and women—although men are more likely to have engaged in extramarital sexual relations over a lifetime. In the 1990s, some researchers question whether having multiple sex partners might be a "disappearing practice" among married Americans (Allegeier & Allegeier, 1995, p. 430). Perhaps public information campaigns about the risks of HIV and STDs have had an impact upon a large portion of married American adults. Another factor to consider, however, is that more couples are "cohabiting" in long-term arrangements and would not be considered "married" for these national "representative" surveys. We do not know their degree of "faithfulness" to each other in these less formal arrangements.

Clements (1994) reported greatest prevalence of EMS within the middle-aged cohort—those aged 45 to 60. Lifetime prevalence of EMS appeared to increase with age among men up to the oldest age group (age 70 and older), at which point it decreased. Laumann and colleagues (1994, p. 216) found that lifetime prevalence of EMS increased steadily with age for men but showed a curvilinear relationship for women, for whom the highest

Table 16-3   Extramarital Sex (EMS): Prevalence in U.S. National Surveys, 1953 to 1994

Research Studies on EMS Using Large-Scale Samples of Respondents (Self-Report and Face-to-Face Interview Methods Used)	Percentage of Married Male Respondents Who Reported Engaging in EMS	Percentage of Married Female Respondents Who Reported Engaging in EMS
Kinsey et al. (1953), p. 417	33%	20%
Laumann et al. (1994) national survey; lifetime experience of engaging in EMS	24.5%	15.0%
Clements (1994) *Parade Magazine* telephone survey of 1,049 adults, ages 18 to 65	19%	15%

incidence was among women in their forties. Women 60 and older were least likely to report ever having engaged in EMS. The large majority of currently married men (78 percent) and women (88 percent) consistently denied EMS both during the past year and during their lifetime (Wiederman, 1997a). Lifetime rates of EMS were twice as high among those who had been divorced or legally separated, compared to respondents who had never divorced or separated (at a statistically significant level).

Sex does not seem to be the major lure for extramarital affairs. Loneliness, emotional excitement, and wanting to prove that one is not getting old tend to be more frequently cited reasons (Elias, 1986). Additionally, many people report that they seek a new partner because they crave companionship and someone to make them feel special (Hall, 1987). For their part, evolutionary psychologists contend that human beings are designed to fall in love but not to stay in love. According to this view, males are "programmed" to maximize the spread of their genes into the future by copulating with many women; women, in contrast, are more given to fidelity because they can have only one offspring a year (Wright, 1994).

Separation and Divorce   Many Americans follow a path leading from marriage to divorce, remarriage, and widowhood. Marital separation, divorce, and remarriage have also become more common for those in midlife and older as well. Apparently, women currently in their late forties are a somewhat unique group. Their generation was the trendsetters, attending college and entering the workforce in extraordinary numbers and shaping new social standards. These changes had vast consequences for traditional husband-wife relationships. Although U.S. divorce rates had been on an upswing during the late 1960s through the 1970s, it appears that they stabilized in the 1980s and declined in the early 1990s. The divorce rate was 23 per 1,000 married women in 1980, and fell to 19.8 per 1,000 married women in 1995 (Rubenstein, 1997).

The first two-decade empirical study of women at midlife—the *National Longitudinal Survey of Mature Women*—was conducted with several thousand women from 1967 through 1989. Hiedemann, Suhomlinova, and O'Rand (1998) examined the data on the 2,000 women who remained in a first marriage and were biological mothers and looked at risk factors for separation and divorce. On average, the women had been married for 17 years. Hiedemann and colleagues' analysis finds an association between the following factors and the risk of separation or divorce:

- Educational attainment reflects the wife's economic independence and the couple's economic status. College-educated wives are more likely to remain married. Women who do not have a high school diploma face higher risks of marital disruption.

- Marriage at young ages increases the risk of marital disruption.
- The longer a marriage survives, the less likely it is to be disrupted.
- A larger family and purchase of a home tend to indicate greater emotional investment in a marriage.
- Controlling for education and work experience, an increase in a wife's economic independence in the form of employment or wages appears to increase the probability of marital disruption.
- The last child's departure from the home has a strong effect on marital stability. The empty-nest experience increases marital disruption for those who reach this phase early in their marriage (around 20 years), but it tends to decrease disruption for those who reach this phase later in their marriage (30 years or more).

Hiedemann and colleagues (1998) suggest that because the baby-boom cohort is expected to live longer after they experience the empty nest, the study of their marriage status should continue.

Friedberg (1998) conducted a 50-state analysis of divorce rates from 1960 to 1990. She examined the rates of divorce from the early 1960s—a time when one had to prove "grounds for divorce" (adultery or cruelty); liberal "no fault" and "unilateral" divorce legislation passed in nearly all states in the early 1970s. She has determined from her study that divorce rates would have been about 6 percent lower if states had not adopted such liberal policies, accounting for 17 percent of the overall increase between 1968 and 1988. Since early 1996, at least eight states have considered returning to more conservative "mutual consent" policies. Proponents of tightening the current divorce requirements argue that making divorce more difficult will strengthen families (Friedberg, 1998).

There is little doubt, from reviewing many research studies conducted over the past 20 to 30 years, that the place of marriage in family life has declined. Cohabitation outside of marriage has become common, and rates of separation and divorce are still high (with one out of every two marriages ending). These changes have occurred among Americans of all races and ethnic groups, but they have been more pronounced among African Americans (Cherlin, 1998). African Americans, who used to marry at younger ages, now marry later. More importantly, they are less likely to marry at all than whites. The marriage rate for white women declined from 95 percent in the 1950s to nearly 90 percent in the late 1980s. During the same time period, the marriage rate for African American women declined from about 88 percent to 70 to 75 percent (Bennett, Bloom, & Craig, 1989). In 1970, married-couple households far outnumbered those headed by unmarried women among African Americans, but by 1993, 53 percent were headed by single women (U.S. Bureau of the Census, 1994). Socioeco-

The Role of Grandparents With separation, divorce, and remarriage rates high, grandparents provide stability and support in grandchildren's lives. Extended kinship ties are more common in ethnic minority families.

nomic factors such as lower incomes, higher rates of unemployment, and lower education are suspected to be the major sources of this racial difference.

Though some African Americans have long, stable marriages, they are the minority and are among the more prosperous (Cherlin, 1998). These intact families are often unreported in the discussions of African American and poverty effects, and the strengths of African American families are often overlooked. Extended kinship ties in three-generation African American families are common, and grandparents are very involved in the upbringing of grandchildren (Cherlin, 1998). In times of adversity, they typically have an extensive social-support system. Economic fluctuations have had a significant impact on the employment/unemployment of poor and middle-class African American families (Cherlin, 1998). Without jobs, African American men fear they cannot provide for a family, and women are reluctant to marry them.

Question

What are some of the reasons why some adults at midlife separate or divorce?

Life as a Single Although returning to single life after divorce has become more common, it is a very difficult experience. In many cases divorce exacts a greater emotional and physical toll than almost any other life stressor, including the death of a spouse (Kurdek, 1991). Studies over the past 20 years document that, compared with married, never-married, and widowed adults, both parties to a divorce have more financial strain, social isolation, physical distress, and increased parenting responsibilities (Wu & Penning, 1997). Study after study documents that individuals who are divorced or separated are overrepresented among psychiatric patients, compared to those who are married and living with their spouses (Stack, 1990).

The trauma of divorce tends to be the greatest for women who are older, who have been married longer, who have two or more children, whose husband initiated the divorce, and who still have positive feelings for their husband or want to punish him. Some middle-aged and older women lack the educational background, skills, and employment experience to reenter or advance within the paid labor market (Wu & Penning, 1997). Additionally, the older the woman is at the time of divorce, the more likely it is that she will not remarry and that she will live many years without support and assistance in her elderly years (Wu & Penning, 1997).

The **displaced homemaker** is a woman whose primary activity has been homemaking and who has lost her main source of income because of divorce from or death of her husband. Nearly 90 percent of displaced homemakers aged 65 and older are widowed, whereas nearly 90 percent of displaced homemakers under age 35 are divorced or separated. About a third of displaced homemakers aged 35 to 64 have children living with them who are under age 18 (for older women, the children might be grandchildren). A majority (53 percent) of Hispanic displaced homemakers have young children at home, compared with 36 percent of African Americans and 17 percent of whites (Demos, 1994). Forty percent work part-time or seasonally, but 60 percent do not work at all. Many of these women find themselves ill equipped to deal with the financial consequences of divorce. They frequently find themselves cut off from their former husband's private pension plan and from medical insurance. Consequently, some 57 percent of displaced homemakers live in households at or below the official poverty threshold (Demos, 1994).

Even though, at midlife or beyond, divorce is an especially disruptive experience—often frightening, frustrating, and depressing—for some, it can simultaneously be exciting and liberating (Bursik, 1991). Many women report that they have a more positive self-image and higher self-esteem than they did during their dissatisfying marriage. And nearly two-thirds say that the process helped them gain control for the first time in their lives (in marriage, they report, they had to "knuckle under"). Nor are the women necessarily lonely, because most have supportive female friends. Moreover, many find new activities and careers (Peterson, 1993).

In contrast, men seem to be unwilling converts to single life (Gross, 1992). Overall, divorced men have much narrower support networks than do divorced women. Even when they have ample money, many men find it difficult to put together new lives that can sustain them. American men have typically depended on women to create social lives for them. Not only do men suffer from want of regular companionship, they miss the amenities of established domestic life. And some men begin questioning their worth and competence. Although men might find it easier than women to remarry (statistics favor men), they still do not find it easy to find love, comfort, and a feeling of at-homeness. If men do not remarry, their rates of car accidents, drug abuse, alcoholism, and emotional problems tend to rise, especially five to six years after their divorce. Whereas a decade or so ago many Americans were willing to take a chance on divorce, in recent years they have become more conservative and more realistic in their marital expectations. More and more couples are finding it better to make up than to break up.

As women become better educated and establish occupational and economic independence, the gap of well-being between marrieds and singles is diminishing. For example, the factor of economic strain due to remaining single or returning to single status can become less of an issue. Some even suggest that the stigma of being single might be decreasing, with more unmarried persons engaging in sexual relationships without negative social stigma attached. Mature single women are beginning to report more advantages to remaining single, such as personal autonomy and growth (Marks, 1996).

Remarriage   Many divorced people eventually remarry. About five of every six divorced men and three of every four divorced women marry again. In fact, nearly half of all recent marriages are remarriages for one or both partners (see Figure 16-5). And some of these remarriages are third and fourth marriages. This social pattern is known as "conjugal succession" or "serial marriage." Although lifelong marriage still remains an ideal, in practice marriage has become for most Americans a conditional contract (Bumpass, Sweet & Martin, 1990). The net result is that only about 4 of 10 adult Americans are currently married to their first spouses; the others are single, cohabiting, or remarried. The findings of a recent study suggest that remarried couples in middle and later life might indeed have lower risks of marital disruption, particularly if both couples have been previously married (Wu & Penning, 1997).

Men are more likely than women to remarry, for a number of reasons. For one thing, men typically marry younger women, and thus they have a larger pool of potential partners from which to choose. Moreover, men are more likely to marry someone who was not previously married and often marry women with less education than themselves. The likelihood that a woman will remarry declines with age and with increasing levels of education.

In recent years there has been an upward trend in second divorces, indicating that those who divorce once might have fewer of the personality traits or social skills that make a marriage work (Martin & Bumpass, 1989). And having undergone a divorce, they might feel less threatened by the experience. However, individuals who report the quality of their new marriage to be good also report a renewed feeling of well-being (Coleman & Ganong, 1990).

Stepfamilies   Many remarriages create stepfamilies. Sixty percent of remarried persons are parents, and their new partners become stepparents. Almost one-fifth of all married couples with children have at least one stepchild under age 18 living with them. A little more than a third of children born today will live in a stepfamily before they are 18. So remarriage can mean having to accommodate "strangers in the home" (Beer, 1989). Steppar-

Figure 16-5 Nearly Half of All Recent Marriages Are Remarriages

"I've been to *all* his weddings."

ents are probably the most overlooked group of parents in the United States (Pasley & Ihinger-Tallman, 1994). Professionals have by and large studied intact, original families or single-parent families. And for their part, a good number of stepparents feel stigmatized. Images of wicked stepmothers, cruel stepbrothers and stepsisters, and victimized stepchildren found in such tales as *Cinderella* still abound (Dainton, 1993).

Children typically approach a parent's remarriage with apprehension rather than joy. Remarriage shatters their fantasy that their mother and father will get together again someday. And the new spouse may seem to threaten the special bond that often forms between a child and a single parent. Moreover, after having dealt with divorce or separation and single parenthood, children are again confronted with new upheaval and adjustment. Matters are complicated because people often expect instant love in the new arrangement. Many women assume, "I love my new husband, so I will love his children and they will love me, and we all will find happiness overnight." Such notions invite disappointment because relationships take time to develop (Hetherington, Stanley-Hagan, & Anderson, 1989). Complicated scenarios arise, like the following, which would not occur in a traditional family (Fishman, 1983, p. 365):

> My ex-husband has money and he bought our son Ricky a car. That's great! But when my stepson David needs transportation, he is not permitted to borrow Ricky's car because my ex is adamant about not wanting to support someone else's child. Ricky would love to share the car with his stepbrother, but he can't risk angering his father. Besides, David is sensitive about being the "poor" brother and would not drive it anyway. It just burned us up [her and her current husband]. So we scraped together some money we

could ill afford and bought an old junker for David. Of course, we fixed it up so it runs safely, and now we've put that problem behind us.

Most stepparents attempt to recreate an intact-family setting because it is the only model they have. The more complex the social system of the remarriage (for instance, stepsiblings, stepgrandparents, and in-laws from a previous marriage), the greater the likelihood of difficulties. One woman in a stepfamily with seven children tells of this experience (Collins, 1983, p. 21):

> Our first-grader had a hard time explaining to his teacher whether he had two sisters or four, since two of them weren't living in our home. The teacher said, "Justin seems unusually confused about his family situation," and we told her, "He's absolutely right to be confused. That's how it is."

With stepfamilies there is yet another dimension—the absent natural parent, whose existence can pose loyalty problems for the children. They wonder, "If I love my stepparent will I betray my *real* parent?" In most cases children are happier when they can maintain an easy relationship with the absent parent. Not uncommonly, however, ex-husbands resent having lost control over raising their children and fail to maintain strong relationships or make child-support payments.

Given these tensions, it is hardly surprising that stepparents report significantly less satisfaction with their family life than do married couples with their biological children. Research shows that 17 percent of remarriages that involve stepchildren on both sides wind up in divorce within three years, compared with 6 percent of first-time marriages and 10 percent of remarriages without stepchildren (White & Booth, 1985). Moreover, one divorce and remarriage does not end the

Stepfamilies A growing number of American households are composed of stepfamilies. Because half of all remarried persons are parents, their new partners become stepparents. On the basis of the research currently available, it seems fair to conclude that children may turn out well or poorly in either a stepfamily or a natural family.

Information You Can Use

Adaptation in Stepfamilies

Each stepfamily presents its own set of problems and requires its own unique solutions because of the complex mix of the members of the blended family. Several matters of everyday living are common sources of friction for stepfamilies, but a little prudence can go a long way.

Food

Food preferences are quite salient, especially to the stepchildren. Mealtimes are normally times when a family is together, so there are more opportunities during mealtimes for disruptions to occur. Finding out what family members enjoy eating—and their definite dislikes—can make for a more pleasant experience when the family comes to the table. Stepparents who cook should be prepared for negative or rejecting comments for awhile, because their stepchildren are accustomed to their own parents' cooking. Preparing each child's favorite meal on his or her birthday or a special occasion is a subtle way of saying to each child: "You *are* special. I care about you." Table manners and hygiene should be taught carefully over a period of time. A simple "Don't eat your peas with your fingers" can cause a flare-up! A recently popular expression (and the title of a book stepparents might consider purchasing) is "Don't sweat the small stuff—and it's all small stuff."

Division of Labor

To resolve the problem of who is going to do which chores, many families find they must put up charts allocating duties. This practice seems to work. Find out what each child is good at doing and make tasks appropriate for age level. Even 2-year-olds enjoy setting the table. Children generally resist biological parents when it comes to chores, so expect friction when a stepparent steps in and is too controlling or demanding. The best rule of thumb for teaching chores to any child is to do the chore with them in the beginning. Large tasks might need to be broken down into smaller tasks that are manageable over a few days (picking up dirty clothes one day, vacuuming the next). Even children who are in the home only on the weekend or during vacations still need to have responsibilities in both homes—it actually gives them a sense of belonging ("I had to clean my room at Dad's"). Typically,

children develop more confidence the more they know how to do. Biological parents who have not seen their children all week might have a tendency to want to do everything for their children—but a child of 6 who gets away with no responsibility becomes a 16-year-old adolescent who can be impossible to live with. A home is not a hotel. A stepparent is not a "servant," and neither is a stepchild.

Personal Territory

Changes in living arrangements pose turf problems. The stepparent who has moved into the spouse's home finds that areas of the house are already designated for use and intrusions are deeply resented. This matter is more easily handled by moving to a new house or making personal spaces within the old one. New paint, fresh curtains, a new rug selected by children can personalize their space. If possible, enlist the children's opinions in some of these decisions. Bring something from the other home to put in his or her new room. Even new husbands and wives do not like sitting in the ex-spouse's chair or sleeping in the ex-spouse's bed. Even biological siblings fight over what belongs to whom, so expect these types of spats among stepsiblings. In the beginning, it is best if the biological parent manages discipline in this area. Stepchildren, even young ones, seem to feel in the beginning that stepparents have no right to tell them what to do (anymore than we would want a stranger telling us what to do). Expect to hear "You're not my Mom (or Dad)!" Be prepared to discuss the underlying *feelings* behind any sharp remarks. Acknowledge those feelings and get them into the open for age-appropriate discussions ("I know you're *angry* that I'm asking you to put your clothes away. It isn't easy having to pack and unpack every weekend, is it? I understand it's a lot of work for you. Perhaps we could do this together.")

Financial Matters

The biological parent might be making child-support payments to another household, so there can be financial strains in supporting the "new" family. Where monies come from, and for whom and how funds should be used, are matters that must be mutually decided. Stepfamily money practices

---

family transitions for a good many youngsters. About half of those whose parents divorce will experience a second divorce (Brody, Neubaum, & Forehand, 1988). To succeed, the stepfamily must develop workable solutions that leave some of the "old" ways of doing things (traditions, rituals, and customs) intact while

fashioning new ones that set the stepfamily apart from the previous family. Although some strains are associated with stepfamilies, so is a good deal of positive adaptation (Kurdek, 1994b). To read more about life in a stepfamily, see *Information You Can Use:* "Adaptation in Stepfamilies."

often come under closer scrutiny by outsiders, such as lawyers, courts, the social services system, and the IRS. The family must keep track of medical expenses, insurance monies, orthodonture payments, clothing and school-related expenses, and so on. Over time, the ex-spouse is likely to initiate court-ordered audits of a family's income for support readjustment (particularly if someone gets a bonus or raise in pay), or a tax audit might be required—dependents are often split equally between ex-spouses by the courts for tax purposes. Limited financial resources require careful management. Use of a computer software program for managing family finances can make the time-consuming chore of recording and tracking income and expenses easier. You will have fewer problems in court if your records are organized and documented. This can save a family thousands of dollars. The demands of an ex-spouse could seem unreasonable, but this is generally up to the courts to decide. Make sure separation and divorce documents are precise about who is responsible for what expenses. There are legal consequences if a biological parent does not follow through with court-ordered support payments.

Discipline

Perhaps the touchiest point of all is discipline, and this is likely to be the area of most dispute between parent and stepparent. Children who are accustomed to one type of discipline have to adjust to another—often on a weekly basis. Most professionals agree that the parenting styles of both adults must yield to compromise so that the children are presented with a united front. Children must not have the opportunity to pit one parent against the other. In natural families, parenting techniques evolve gradually as the parents and children move through the family life cycle. But in blended families there is no time for such evolution. A solid understanding of child and adolescent development helps parents communicate effectively. Whatever forms of discipline are decided on, their administration must be fast and firm, and such decisions are best worked out prior to taking the marriage vows. As stated earlier in this text, discipline should be administered as promptly after the infraction as possible, especially for younger children who cannot remember what they did earlier, and the type and degree of discipline should match the misbehavior. As in any family,

fairness should prevail. The most effective discipline teaches the child what she or he should do rather than focuses on what was done wrong. ("You didn't pick up your room, so you will have to do it now rather than go to your friend's house" instead of "You are grounded for the weekend!") Fathers in an intact family do not always notice whether bedrooms are clean, chores are done, homework is done, children have brushed their teeth and showered, and so forth. These tasks have traditionally been considered within the mother's domain. Stepmoms, like many women, tend to pick up on such details, which fathers might not see as significant at the time. The stepfamily must reevaluate what is *really important* in the family over the long run.

Changes over Time

As children become teenagers, they naturally want to spend more time with friends. Both sets of parents need to support the adolescent's interests in school and outside activities. Adolescence is a time when teens need to express themselves and experience some independence, and parents who force children to adhere to a rigid visitation schedule spend much of their time with an unhappy teen. Allow teens to make more choices, just as they do in intact families. Do not, however, let the teen get away with *less supervision*. At this stage, manipulation of both sets of parents is more likely. A child who has two homes might think, "I can always go to the other home," even though they might have 2 parents at each home telling them what to do. Try not to dwell on the negative; that type of attention is a reinforcer for some children. Look for and praise the positive. Make them be responsible and face the consequences of their actions.

Support Groups

More communities now have support groups for stepparents. Those who are more experienced with this type of family arrangement can understand the special stresses and emotions that young stepparents experience. Many school districts also have support groups for children. Children in intact families often say "I love you, Mom [or Dad]." It can be years before a stepparent hears anything close to this. Recognize that children do not realize the sacrifices parents make to raise them (until they themselves become parents).

From: White and Booth, The Quality and Stability of Remarriages: The Role of Stepchildren. *American Sociological Review* 50, 689–698. Copyright 1985. Reprinted by permission of the American Sociological Association.

Question

What are some of the particular stressors for adults who remarry into stepfamilies?

Adult Children and Grandchildren   The term **empty nest** is applied to that period of life when children have grown up and left home. Yet today, the nest might not be so empty. Many adult children are living at home longer while attending college or when establishing their careers or returning home with grandchildren during a separation or divorce or through longer periods of unemployment. Additionally, more middle-aged adults (primarily women) are working and caring for either their elderly parents and in-laws with the assistance of health care aides and adult day care or raising grandchildren whose parents are incapable of parenting. Middle-aged parents of young adults with disabilities (such as mental retardation, head injuries and paralysis, and other multiple handicaps) today have difficult, time-consuming decisions concerning placement of their children into group homes or supported apartments. They also need to become the legal guardians of their adult child and assist with sheltered employment or supervised employment. These are often difficult, heart-wrenching decisions for aging parents. We normally expect that our adult children will be capable of making their own life decisions.

But for some older married couples, the nest stays empty after the children depart. This can be especially stressful for a woman who has been a traditional homemaker, for whom being a mother has been a central ingredient in her life and identity. Clinical psychologists and psychiatrists have emphasized the emotional difficulties women face when their children leave home, dubbing the problems the **empty-nest syndrome** (Bart, 1972). A parent who has found her or his meaning in life primarily in the children often experiences a profound sense of loss when the children are no longer around. One mother who had been completely wrapped up in a selfless nurturing of her four children told one of your authors a few months after the last child had left home for college, "I feel such a hole in my life, such a void. It is like I'm a rock inside. Just a vast, solid emptiness. Really, I have nothing to look forward to. It is all downhill from here on!"

Some women object to the overemphasis on physical changes and the "empty nest" in women's midlife maturation. Recent cross-sectional and longitudinal studies find that many women do not regret the end of fertility, experience minimal menopausal symptoms, feel happy when their children are successfully launched, and feel satisfaction when their children continue to stay in touch with them (Helson in Lachman & James, 1997, p. 29). Couples often report that they view the empty-nest period "as a time of new freedom." One 42-year-old woman whose two children were off at college and who had returned to Ohio State University to secure her degree in accounting told one of your authors:

> Sure, I experienced a throb or two when Ida [her
> youngest child] left home. But, you know, I had
> expected it to be a lot worse. I really like the freedom I
> have now. I can do the things I want to do when I want
> to do them, and I don't have to worry about getting
> home to make dinner or to do household chores. I'm
> now back in school, and I love it.

Often, both parents and children adjust to the empty nest on a gradual basis. Children might live at home for a period after securing their first job. Or they might go off to college and return home for vacations. Most parents adapt to the empty-nest period quite well. Ryff and colleagues (1994) interviewed a sample of midlife parents and confirmed that parents' self-evaluations are influenced by the perceptions they have of the lives of their grown children. Overall, midlife adults' sense of self-acceptance, purpose in life, and environmental mastery was strongly linked with their assessments of their children's adjustment.

Caring for Elderly Parents   Psychologists call middle-aged adults the *sandwich generation* because many find themselves with responsibilities for their own children on the one side and for their elderly parents on the other. At the very time when they are launching their own children and looking forward to having more time for themselves, many encounter new demands from their parents. Some midlife couples find that they no sooner reach the empty-nest stage of the family life cycle than the nest is refilled with either an elderly parent or a grown son or daughter who returns home after a divorce or the loss of a job (see Chapter 14). Studies show that 30 to 40 percent of Americans in their fifties help their chil-

The Sandwich Generation   Middle age often brings with it responsibilities for one's own children, grandchildren, and aging parents. The tasks fall disproportionately upon women who, in the American gender division of labor, are the persons assigned primary responsibility for family caretaking.

dren, and a third help their parents financially or in other ways (Kolata, 1993a). According to a 1988 U.S. House of Representatives report, the average U.S. woman will spend 17 years raising children and 18 years helping aged parents. And because many U.S. couples delayed childbirth during the 1980s, more couples will find themselves "sandwiched" between child care and elder care (Beck, 1990a, 1990b; Begley & Check, 1998).

Aging parents sometimes require increased time, emotional energy, and financial aid from their adult children. Despite the profound changes in the roles of family members, the grown children (particularly the daughters and daughters-in-law) still bear the primary responsibility for their aged parents (Brody et al., 1994). The sense of obligation is strong even when the emotional ties between the parent and child were previously weak (Cicirelli, 1992). In 80 percent of cases, any care an elderly person will require will be provided by her or his family. This assistance might be supplemented by help with income and health care costs through Social Security, Medicare, and Medicaid programs, though in 1992 the federal health insurance program for elderly people (Medicare) covered only one-third of all paid home health care. Despite the fact that the vast majority of adult children provide help to their parents, Americans continue to echo the myth that "nowadays, adult children do not take as much care of their parents as they did in past generations."

Not surprisingly, many "women in the middle" are subjected to role-overload stresses similar to those experienced by younger women in relation to work, child care, and other household responsibilities. Their difficulties are often compounded by their own age-related circumstances, such as lower energy levels, the onset of chronic ailments, and family losses (E. M. Brody, 1990). And large numbers of these middle-aged women will end up caring for their husbands in the years ahead because women typically marry older men (Day, 1986). Some believe these facts in part explain the appeal that religion has for many middle-aged and elderly women. According to the American Association of Retired Persons, about 14 percent of contemporary caregivers to the elderly have switched from full- to part-time jobs, and 12 percent have left the workforce (Beck, 1990a). One 60-year-old woman, whose 90-year-old father shares her home along with her husband and her 30-year-old daughter who moved back home to save money, says (Langway, 1982, p. 61):

> At a time of my life I should have less to do, I have greater demands put on me. I have my own getting older to cope with. Sometimes I get angry, not at him [her father], but at what age brings with it. Whenever he gets ill, I panic: now what? Still I couldn't live with myself if I resorted to a nursing home.

Despite the changing roles of women, when it comes to the elderly, the old maxim still seems to hold: "A son's a

son till he takes a wife, but a daughter's a daughter for the rest of her life." Thus, adult daughters and daughters-in-law often face complex time-allocation pressures. They must juggle competing role demands of employed worker, homemaker, wife, mother, grandmother, and caregiving daughter.

The motivations, expectations, and aspirations of the middle-aged and the elderly differ to some extent because of their different life periods and cohort memberships. At times these differences can be a source of intergenerational strain (Scharlach, 1987). However, resentment and hostility are usually less where the financial independence of the generations enables them to maintain separate residences. Both the elderly and their adult offspring seem to prefer intimacy "at a distance" and opt for residing independently as long as possible. Consequently, the elderly parents who need to call on children for assistance are apt to be frail, greatly disabled, gravely ill, or failing mentally. When middle-aged adults express reluctance to take on primary care for an ailing parent, they are not necessarily being "hard-hearted." Rather, they recognize that their marriage or emotional health could be endangered by taking on those caretaking responsibilities (Miller & McFall, 1991). Simultaneously, this realization can produce strong feelings of guilt (Pruchno et al., 1994). As noted earlier, most adult children would rather make sacrifices to care for their parents than place them in a nursing home (E. M. Brody, 1990), and recent statistics indicate that fewer elderly adults are now being placed in nursing homes (U.S. Department of Health & Human Services, 1997a).

On the other hand, female caregivers who find their caregiving role to be satisfying or rewarding can derive positive health benefits from the endeavor. Of course, the outcomes are frequently otherwise for those who find caregiving stressful or unfulfilling (Stephens & Franks, 1995). In contrast to women, most men apparently find caregiving stressful and are likely to invest themselves more in their occupational role (Allen, 1994).

Questions

What is meant by the expression the sandwich generation? *How does being in this generation impact the relationships of middle-age adults with their parents, grown children, and grandchildren?*

Friendships

I get by with a little help from my friends.
I get high with a little help from my friends.
I'm gonna try with a little help from my friends.

The Beatles, *A Little Help from My Friends,* 1967

Most people distinguish friend from relative and coworker. This does not mean that a relative could not be a friend. Friends are people who seek each others' company and with whom one can talk and share activities and for whom one has warm feelings (Fehr, 1996). As we have noted throughout this text, close and meaningful social relationships play a vital part in human happiness and health. In study after study, researchers report that people who have friends to whom they can turn for affirmation, empathy, advice, assistance, and affection are less likely to develop diseases like cancer and respiratory illness and more likely to survive health challenges like heart attacks and major surgery. Indeed, it seems it is more important to have at least one person with whom we can share open and honest thoughts and feelings than it is to have a substantial network of superficial friendships (Brody, 1992c).

So what do we know about the baby-boom generation entering midlife and their friendships? Adams and Blieszner (1998) have been studying the generation born between 1946 and 1964 and its patterns of friendship. They find this is the first American cohort to relocate from their home communities in large numbers as they finished college and sought employment in the 1960s to 1970s. Consequently, this cohort has had more opportunities than any other generation to develop a diversity of, and greater tolerance within, friendships. With the development of television, this generation was also the first to have almost instant knowledge of major current events (e.g., the assassination of President Kennedy, our landing on the moon, the civil rights movement, the Vietnam War, the women's movement) and developed a high level of cohort identity.

Women in midlife are more likely than men to have intimate friendships (Adams & Blieszner, 1998). Men often have many acquaintances with whom they share experiences, but they frequently have few or no friends. The social contacts of many elderly men are restricted to their wives, their children, and their children's families. Male relationships are often limited to group settings involving their sports team, occupational colleagues, or "brothers" in a fraternal group, such as Rotary or Knights of Columbus. Numerous social constraints limit the development of closeness among men (Weiss, 1990). Men have also tended to disenfranchise themselves from the house, the primary center of security, warmth, and nurturance. They are more likely to construe the house as a "physical structure." In contrast, women tend to define the house more in terms of a "personalized place" affording "relationships with others." Hence, men commonly socialize outside the home.

Moreover, women tend to maintain family contacts and emotionally invest themselves in the family to a greater degree than men do throughout the lifetime of a family. Women's friendships often take up where their marriages leave off. A woman's best friends typically compensate her for the deficits of intimacy she encounters within her marriage. And shared experiences of child rearing promote a moral and social dialogue among women that gives their friendships depth and intimacy. Significantly, women in the baby-boom generation were the first to give friendships equal status with romantic relationships with men and have adopted a "code of loyalty" to their women friends in contrast to the behavior of older women (Adams & Blieszner, 1998). Thus, women's networks of interpersonal ties work to satisfy their emotional needs and add coherence to lives pulled in multiple directions by the fragmenting demands of modern life (O'Connor, 1992). The nature and quality of friendships for this generation have implications for their networks of support in old age and the types of communities they will choose to live in after they retire. As we shall see in the following section, the relationships we establish with co-workers have less stability than in past generations, as significant changes are occurring in the American workplace.

Question

How do women's and men's friendships in midlife differ in value and purpose?

The Workplace

Today, new currents are at play in the workplace. These currents involve new technologies, new industries, new markets, job migration, population shifts, and continuing education. Employees, especially those in their forties and fifties, experience continual demands to retrain and upgrade their job skills. Today's midlife employee can no longer rely on the foundation of education and training she or he acquired 20 or 30 years ago.

Many nontraditional students in their mid thirties and forties have experienced corporate "downsizing," "takeovers," and "restructuring." Some have been forced into unemployment, disability, or early retirement; others are making career changes in order to reenter the occupational sphere. A few make dramatic career changes by personal choice. This experience has a significant impact on both economic security and self-esteem. Old jobs in manufacturing, in mining, and on the farm are disappearing. Some have applied the label *Second Industrial Revolution* to the vast changes being brought about by computers and other electronic innovations. In some respects old terms like *blue-collar, white-collar,* and *service economy* no longer define with precision the realities of the "information age" workplace.

Job Satisfaction

Without work all life goes rotten. But when work is soulless, life stifles and dies.

Albert Camus

Levinson says that in midlife, employed men and women experience a different relationship with their occupational status. They are asking themselves, "Is my work satisfying?" There is much greater emphasis on self-fulfillment and satisfaction than on the collection of external rewards. Others ask themselves, "In what ways have I made a contribution or formed a 'legacy' to something outside of myself or my family?" People often want to be appreciated for their contributions at work and are likely to offer to mentor others who are coming up in the ranks. However, not everyone experiences satisfaction in their work, and some take "psychological retirement" (Levinson, 1996, p. 375). Some whose work performance is minimally adequate accept "early retirement." Some with midlife dissatisfaction with work may become depressed, consume more alcohol or other drugs, become accident prone or regularly absent, experience marital and family discord, or search for youthful forms of excitement that typically are considered inappropriate in middle age (Levinson, 1996, p. 375). Those in their early forties might see that the anticipated promotion or advancement isn't coming and make career changes more to their liking. Some today have launched their own consulting businesses. Typically, "the higher one's position [at work], the fewer the attractive alternatives and the greater the potential fall" (Levinson, 1996).

In the quotation that opens this section, Albert Camus captures the importance of work in giving meaning and satisfaction to our lives. Significantly, Americans, especially women, have been entering the labor force in ever-increasing numbers since the 1960s, so that today about 67 percent of the working-age population holds jobs or seeks them. Yet work that is not fulfilling can erode and undermine much of our humanness. Some psychologists and sociologists have applied the concept of *alienation* to such troubles and have sought solutions to these problems in programs designed to decrease alienation. The word **alienation** commonly implies a pervasive sense of powerlessness, meaninglessness, normlessness, isolation, and self-estrangement (Erikson, 1986). Alienation can be expressed in **job burnout.** In terms of psychological effects, the problem that confronts most people in occupational life today is that they cannot gain a sense of *self-actualization* in their work. Hence, the most potent factors in job satisfaction are those that relate to workers' self-respect, their chance to perform well, their opportunities for achievements and growth, and the chance to contribute something personal and quite unique.

Work affects an individual's personal and family life in many ways (Crouter & Manke, 1994). For instance, jobs that permit occupational self-direction—initiative, thought, and independent judgment in work—foster people's intellectual flexibility. Individuals with such jobs become more open approaching and weighing evidence on current social and economic issues. Individuals who enjoy opportunities for self-direction in their work are more likely to become more self-confident, less authoritarian, less conformist in their ideas, and less fatalistic in their nonwork lives than other individuals are. In

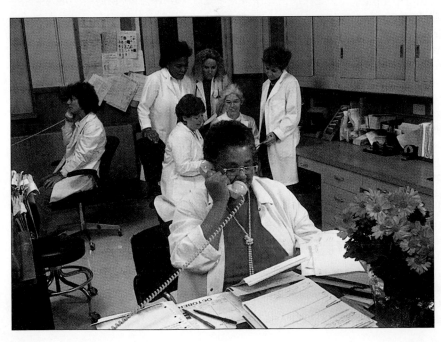

Job Satisfaction and Job Burnout
Research suggests that a crucial key to job satisfaction is the ability to exercise substantial control over one's work environment. People yearn for personal achievement and opportunities for self-direction in their work.

turn, these traits lead, in time, to more responsible jobs that allow even greater latitude for occupational self-direction. In sum, the job affects the person and the person affects the job in a reciprocal relationship across adult life (Kohn et al., 1990).

Research on work satisfaction has consistently shown that older people are more satisfied with their jobs than younger people are (MOW International Research Team, 1987; Kalleberg & Loscocco, 1983). Two major hypotheses have been advanced as explanations for this tendency. According to one interpretation, the "now" generation of workers subscribes to a set of leisure-oriented values that are different from those of the past. For example, the work environment at Microsoft is more casual and very different from the conservative environment of the "old" IBM, yet both companies are achievement and profit oriented (to date, both have been phenomenally successful). Proponents of this view cite as key features of the "new values" a willingness to question authority, a weakening of materialistic standards, and a demand that work be fulfilling and enriching. These values contradict those of an industrial order founded on deference to authority and responsiveness to such traditional rewards as income and promotion. Although polls show that Americans *believe* that people currently take less pride in their work than they did a decade ago, it seems premature to conclude that the traditional U.S. work ethic has fallen by the wayside.

A second interpretation of age differences in job satisfaction looks to life-cycle effects (Wright & Hamilton, 1978). Proponents of this hypothesis say that older workers are more satisfied with their jobs because on the whole they have better jobs than younger workers do. In the usual career pattern, a person begins at or near the bottom and, where possible, moves up. Young people typically begin their careers when they have relatively few pressing responsibilities. Usually they are unmarried, or without children, or both. They require little more than a start—a job that is "good enough" for the immediate present, supplies sufficient money to meet short-term needs, and affords some opportunity for advancement. But as workers develop new needs (marry, have children, and grow old), they also accumulate the experience, skills, and seniority that allow them to find positions that are progressively more satisfying.

Using data gathered by the University of Michigan's Survey Research Center (based on national surveys of the economically active U.S. labor force), sociologists find age differences in the rewards that workers look for from their jobs and in certain of their work values (Kalleberg & Loscocco, 1983). For instance, older workers seek job security, fringe benefits, and convenient hours, while young workers seek promotions and advancement. These differing concerns can be explained by life-cycle effects.

Questions

Are middle-aged workers likely to be satisfied with their jobs? Why or why not?

Midlife Career Change

Most of us start off our careers with the assumption that we will spend our lives in one line of work (chances are, our parents and grandparents did). This perspective is most characteristic of white-collar professionals, such as physicians, lawyers, accountants, engineers, and college professors. Such individuals spend their late adolescence and early adulthood acquiring special skills and credentials. They—and their family, friends, and associates—assume that they will spend their remaining years successfully pursuing a career that constitutes a lifetime commitment. Their work is expected to produce a considerable sense of fulfillment and, with few exceptions, unfold in an orderly progression of steps from an entry position to eventual retirement. Each advancement is thought to bring new levels of satisfaction and well-being. But according to studies by psychologist Seymour B. Sarason (1977), this view is excessively optimistic and for many people quite unrealistic. A survey shows that over half of the Americans polled had switched careers at least once, and 43 percent said a future switch is somewhat or very likely (Giese, 1987).

Recent surveys on emotional well-being and aging, and an examination of suicide statistics, have indicated that those who enter "the professions" are also likely to become dissatisfied with their careers, but many feel they cannot change occupations because the financial rewards, security, and prestige are so great. This serious

Midlife Career Change At midlife many people take stock of themselves and reassess where they are going. Some middle-aged workers today have been forced to take early retirement or have returned to college to retrain for a new occupation.

conflict between changing to more satisfying work and living at a different status or remaining in a career that gives no satisfaction might contribute to rise in the suicide rate for people in their forties and older.

The Bureau of Labor Statistics gathers statistics on workers who tell that they are doing different work than they did one year earlier. About one man in eight aged 25 to 34 typically changes his occupation in a year. The same holds true for women, except that women are more likely to change jobs now than in the 1960s. Both men and women switch careers for a variety of reasons. Some find that their career has not provided the fulfillment they had expected or that it no longer challenges them. One professor who left academic life at age 43 for a career in marketing told one of the authors:

> I just got fed up with teaching. It no longer turned me on. I would wake up in the morning and dread the day ahead. When driving over to the university, I would develop waves of nausea. I began thinking to myself, "This is no way to spend the rest of my life." I had always been interested in marketing, and I had done consulting for a number of years before I left the university. It took me about five years before I got the business really rolling but now I'm doing pretty well. I like being my own boss and I like making money—big money. I never could do that as a prof.

At midlife many people take stock of themselves and reassess where they are going and what they are doing with their lives (see *Further Developments:* "Working"). Some look to formal education to provide them with new skills. Others build on contacts, interests, skills, or hobbies.

Unemployment and Forced Early Retirement

Most people find unemployment a painful experience. As mentioned earlier in this chapter, job insecurity might be mounting in the United States, due to automation, corporate downsizing, and globalization of the economy (Zachary & Ortega, 1993). There has been a rise in the number of workers employment agencies call "contingent workers," "flexible workers," and "assignment workers"—what some labor economists by contrast call "disposable" and "throwaway" workers (Kilborn, 1993). In fact, many U.S. college faculty are part-time workers (adjuncts).

Sociologists and psychologists find that unemployment has adverse effects on physical and mental health (Feather, 1990). Based on data from the 1970s, M. H. Brenner (1976) of Johns Hopkins University calculated that a rise of 1 percent in the national rate of unemployment, when sustained over a six-year period, is associated with a 4.1 percent increase in suicide, a 5.7 percent increase in homicide, and a 4.3 percent increase in first-time male admissions to state mental institutions. More

recent research confirms these findings (Hamilton et al., 1990). The worst psychological effects of job loss, however, can be minimized if opportunities exist for reemployment (Hamilton et al., 1993). Unemployment also increases the financial and role strains of parents, intensifies conflict between parents and their children, and undermines children's school achievement and health (Conger et al., 1993; Flanagan & Eccles, 1993). George Clem, a 31-year-old unemployed manufacturing worker in Jackson, Michigan, observed: "I've lost everything I ever had—it's all gone. I've lost my job. I've lost my home. I thought I had my future assured, but now I know I have no future." A woman in the same community said: "Emotionally, you begin to feel worthless. Rationally, you know you're not worthless, but the rational and the emotional don't always meet" (Nelson, 1983, p. 8).

National unemployment figures include people who are currently eligible for unemployment benefits. We really do not know the occupational status of people who are no longer eligible for unemployment and are unemployed on a long-term basis. However, studies of workers reveal that their behavioral and emotional reactions to unemployment typically pass through several stages (Kaufman, 1982):

- Initially, they undergo a sequence of shock, relief, and relaxation. Many had expected that they were about to lose their jobs. Hence, when the dismissal comes, they may feel a sense of relief that at last the suspense has ended. On the whole, they remain confident and hopeful that they will find a new job when they are ready. For the first month or two, they maintain normal relationships with their family and friends.

- The second stage centers on a concerted effort to find a new job. If these unemployed persons have been upset or angry about losing their jobs, the feeling tends to evaporate as they marshal their resources and concentrate their energy on finding a new job. This stage can last for up to four months. But individuals who have not found another job during this time move into the next stage.

- The third stage lasts about six weeks. Their self-esteem begins to crumble, and they experience high levels of self-doubt and anxiety. Those nearing retirement age find their outlook particularly bleak (Love & Torrence, 1989).

- The fourth stage finds unemployed workers drifting into a state of resignation and withdrawal. They become exceedingly discouraged and convinced that they are not going to find work, so they either stop looking for work or search for it only halfheartedly and intermittently. Some come through the stage and look back on it as a

Further Developments

Working

Most people feel that they would lose a large part of themselves if they lost their job or if they could not work. Work is not only an activity, it is a *defining* activity. When two people begin a conversation, occupation is usually one of the first topics to come up, as if one can decipher who another person is by knowing what they do for 40 hours each week. For many people, work is a search for meaning—making sense of the world, themselves, and others by giving a sense of order to what might otherwise be a very unstructured existence. Let's look at two people in midlife (Terkel, 1974) and see what they think about their vocations. As you read about them, pay attention to some of the themes that have been discussed in our chapters on midlife, and watch how these individuals articulate their sense of generativity, physical changes, family relations, health, and cognition.

Hub Dillard—Heavy Equipment Operator

Hub is a 48-year-old crane operator who is waiting for his turn to land a "soft job" within the profession (usually at the age of 55). "There's no job in construction which you could call an easy job. I mean, if you're out there eating dust and dirt for eight, ten hours a day, even if you're not doing anything, it's work. The difficulty is not in running a crane. Anyone can run it. But making it do what it is supposed to do, that is the big thing. It only comes with experience. Some people learn it quicker and there are some people can never learn it. What we do you can never learn out of a book. You could never learn to run a hoist or a tower crane by reading. It's experience and common sense. This is a boom crane. It

goes anywhere from 80 to 240 feet. You're setting iron. Maybe you're picking up fifty, sixty ton and maybe you have some ironworkers up there. You have to be real careful that you don't bump one of these persons, where they would be apt to fall off. At the same time, they're putting bolts in holes. If they want a half-inch, you have to be able to give them a half-inch. I mean, not an inch, not two inches. These holes must line up exactly or they won't make their iron. And when you swing, you have to swing real smooth. You can't have your iron swinging back and forth, oscillating. If you do this, they'll refuse to work with you, because their life is at stake.

It's not so much the physical, it's the mental. The average . . . workingman [lives to] . . . 72. The average crane operator lives to 55. They don't live the best sort of life. There's a lot of tension. We've had an awful lot of people have had heart attacks. Yeah, my buddy. Before I had this heart attack, I sure wanted a drink. Sure, it relaxes. There's a lot of times you have to take another man's word for something and a lot of people get hurt. I was hurt because I took another man's word. When the crane went over backwards and threw me out, a 500 pound weight went across my leg and crushed my ankle and hip. I was in the hospital, had three operations on my leg and was out of work 18 months. My wife worked a little bit and we managed. My father was a crane operator since 1923. We lived on a farm and he was away from home a lot. So I said I'd never do this. When I got out of the service, I went to school and was a watchmaker. I couldn't stay in a pack. It was the same thing, every day and every night.

"cleansing" experience. They might make a conscious decision to change careers or to settle for some other line of work. And they might look for other sources of self-esteem, including their family, friends, and hobbies.

However, individuals who undergo long-term unemployment often find that their family life deteriorates (Larson, Wilson, & Beley, 1994). Unemployment and health benefits end for most Americans when they lose their jobs, and many lose their pensions as well if they did not have enough vested time in the company or agency. Financial pressures mount. They are unable to keep up their mortgage payments, or they fall behind in the rent. They see their cars and furniture repossessed. It is little wonder that they feel that they are losing control of their lives. Child abuse, violence, family quarreling,

alcoholism, and other evidence of maladjustment mounts. The divorce rate soars among the long-term unemployed. Many men feel emasculated when confronted by an involuntary change of roles in the family, and they lash out with destructive reactions (remember, traditionally across all cultures males are prepared to be the *provisioners* of a family). The unemployment scenario has been more prevalent in African American and other ethnic minority families, especially for unskilled laborers without an education.

A recent study of forced early exits (Table 16-4) and average age at retirement (Table 16-5) for men and women in Germany, Japan, Sweden, and the United States found that from the late 1960s to the middle 1990s there were significant declines in the age of retirement. Both men and women are retiring earlier, yet they are expected to live longer, in many cases increasing the de-

It was inside. And being a farm boy . . . So I went to work with my father, construction work, and stayed with it ever since. I have one son doing this work. But this youngest one, he's pretty intelligent, I'd like to see him be a professional man if he will. Of course, I wanted the other one too. But . . . there's so many changes now. When I started, to build a road a mile it took you two or three months. Now they build a mile a day. There's a certain amount of pride. I don't care how little you did. You say, 'I worked on that bridge, or I worked on that building.' Maybe it don't mean anything to anybody else, but there's a certain pride knowing you did your bit."

Babe Secoli—Checker in a Supermarket for 30 Years

"We sell everything here, millions of items. From potato chips to pop—there are items I've never heard of that we sell here. I know the price of every one. Sometimes the boss asks me and I get a kick out of it. You sort of memorize the prices. It just comes to you. I know a half gallon of milk is 64 cents; a gallon, $1.10. I don't have to look at the keys on my register. I'm like the secretary who knows her typewriter. The touch. My hand fits. The number nine is my big middle finger. I use my three fingers—my thumb, my index finger, and my middle finger. The right hand. And my left hand is on the groceries. They put down their groceries. I got my hips pushing on the button and it rolls around on the counter. When I feel I have enough groceries in front of me, I let go of my hip. I'm just movin' the hips, the hand, and the register. You just keep going, one, two, one, two. If you've got that rhythm, you are a fast checker. Your feet are flat on the floor and you're turning your head back and forth. Somebody talks to you. If you take your hand off the item, you're gonna forget what you were ringin'. It's the feel. When I'm pushin' the items through, I'm always having my hand on the items. If somebody interrupts me to ask the price, I'll answer while I'm movin'. Like playin' a piano.

I'm eight hours a day on my feet. When I get home I get my second wind. As far as standing there, I'm not tired. It's when I'm roamin' around trying to catch a shoplifter. When I see one, I'm ready to run for them. I'm a checker and I'm very proud of it. There's some that say, 'A checker—ugh!' To me, it's like somebody being a teacher or a lawyer. I'm not ashamed that I wear a uniform and nurse's shoes and that I got varicose veins. I'm making an honest living. Whoever looks down on me, they're lower than I am. What irritates me is when customers get cocky with me, 'Hurry up,' or 'Cash my check quick.' I make my mistakes, I'm not infallible. I apologize. I catch it right there and then. I tell my customers, 'I overcharged you two pennies on this. I will take it off your next item.' It hurts my feelings when they distrust me. I wouldn't cheat nobody, cause it isn't going in my pocket.

Years ago it was more friendlier, more sweeter. Now there's like tension in the air. A tension in the store. The minute you walk in you feel it. I've never been late except for that big snowstorm. I've never thought of any other work. I'm a couple of days away, I'm very lonesome for this place. When I'm on vacation, I can't wait to go, but two or three days away, I start to get fidgety. I can't stand around and do nothing. I have to be busy at all times. I look forward to coming to work. It's a great feeling. I enjoy it somethin' terrible."

Excerpts from *Working* by Studs Terkel is reprinted by permission of Donadio & Olson, Inc. Copyright 1974 by Studs Terkel.

pendency on government-subsidized pension or social security systems.

Many communities now have career centers where job hunters can search the Internet to get professional assistance in writing and preparing resumes and letters of application, to update and practice interviewing skills, and so on. These services also can help persons with long-term unemployment find retraining opportunities.

Question

Why are many middle age American workers unemployed or forced into early retirement, and what are these men and women doing about their occupational status?

Dual-Earner Couples

As of the last official U.S. census, of the 45.1 million married couples drawing paychecks in the United States in 1990, 65 percent were dual-earner pairs. The figure continues to grow as more women take jobs outside the home (Rubin & Riney, 1993). Today various sources report that 70 to 77 percent of adult women work, which means that more than 60 million women are part of the U.S. labor force. In 1950, women made up nearly 25 percent of the labor force (it was socially acceptable for single women to work to support themselves then). Both the women's movement of the 1960s and the passage of Title IX of the Civil Rights Act of 1964 allowed more women to make gains in employment and enter traditionally male-dominated universities and occupations (Barnett, 1997). Today women make up nearly 50

Table 16-4

Comparison (in Percent) of Labor-Force Early Exits (Retirement), Men and Women Aged 50 to 54, in Germany, Japan, Sweden, and the United States, 1965–1970 and 1990–1995

Years	Germany	Japan	Sweden	United States
Men				
1965–1970	3.1	4.4	4.2	7.1
1990–1995	10.9	.5	10.5	11.9
Women				
1965–1970	5.2	12.3	–3.9*	–4.3*
1990–1995	11.6	14.1	9.3	11.1

Gendell, Murray. (1998). Trends in Retirement Age in Four Countries, 1965–1995. *Monthly Labor Review.* Bureau of Labor Statistics, U.S. Department of Labor. Vol. 121, No. 8, 20–30.

*Net accessions rather than net exits. This means more women in the age group were entering the labor force than exiting from it at this time.

Table 16-5

Average Age at Labor Force Exit of Men and Women in Germany, Japan, Sweden, and the United States, 1965–1970 and 1990–1995

Years	Germany	Japan	Sweden	United States
Men				
1965–1970	64.7	66.6	65.7	64.1
1990–1995	60.3	65.2	62.0	62.2
Women				
1965–1970	63.0	63.8	65.5	65.3
1990–1995	59.9	62.9	62.0	62.7

Gendell, Murray. (1998). Trends in Retirement Age in Four Countries, 1965–1995. *Monthly Labor Review.* Bureau of Labor Statistics, US Department of Labor. Vol. 121, No. 8, 20–30.

percent of the labor force, and a majority of working women are married (U.S. Bureau of Labor Statistics, 1998).

Men and women typically see and experience inequalities in family work quite differently. Much depends on how fair the man and the woman judge the division of household labor to be (Mederer, 1993).

Research by Thoits (1986) and Verbrugge (1989) concludes, however, that there is a positive association between occupying a greater number of roles and increased psychological well-being. These findings are contrary to the traditional belief that multiple roles could not be good for married women with children. More recent studies support the findings that women who are wives and mothers and employees show no more signs of distress than women who occupy fewer of these roles (Barnett, 1997). Moreover, women who have children report increased distress only if they *decrease* their commitment to the labor force (Barnett, 1997). Examining cross-sectional and longitudinal studies on job distress for men and women, Barnett and colleagues (1995) found that people who report positive job experiences report low psychological distress. If the job experience deteriorates, psychological distress increases. Even today, women who enter male-dominated professions might have to continually fight subtle (or not so subtle) sexism obstacles (see Figure 16-6) (Kolbert, 1991). Also, we do know that work has different meanings for men and women (Barnett, 1997).

The dynamics of family decision making are currently undergoing change in the United States. In the process dual-earner couples are evolving new patterns and traditions for family life (Orbuch & Custer, 1995). Though work outside the home creates new sources of conflict for many couples, it also gives them new sources of personal fulfillment (Paden & Buehler, 1995). Recent studies also indicate that women's adult career commitments have a positive impact on their children's career

"I love being a partner Mr. Jenkins! There's just one problem."

Figure 16-6 Gender Roles Change but Old Ways and Attitudes Persist

From *The Wall Street Journal*—Permissions, Cartoon Features Syndicate.

goals and aspirations (Vandewater & Stewart, 1997). In addition, workplace policies on parental leave (for child care and parental care) and other institutional arrangements (e.g., flextime policies) that can provide relief to dual-earner couples are becoming top social and political issues.

Segue

Perhaps more than at any other life stage, there are evident, dynamic changes for adults in midlife. This is a time of looking back on one's life and looking forward and making plans for retirement years without the daily structure of an occupation. Our self-esteem and personality traits affect our family and occupational choices and overall happiness at this stage of life. Those who feel overwhelmed with responsibilities of taking care of children, grandchildren, and parents will be psychologically distressed. Those who learn to manage their time and adapt to changing family and occupational responsibilities usually have a positive outlook and are often much happier as they enter and go through these years of life. Adaptability, flexibility, and a sense of humor all contribute to successful passage through this stage of life. Fewer women nowadays are displaced homemakers because many are engaged in meaningful, satisfying occupational pursuits along with their spouses. Many of those who are dissatisfied with their employment status are taking charge of their careers, going back to college (even in their fifties and sixties), and entering new occupations. Those who are unemployed or experience forced early retirement require more understanding of their emotional distress. They may justifiably feel "old before their time."

Summary

Theories of the Self in Transition

1. Maturity is the capacity to undergo continual change in order to adapt successfully and cope flexibly with the demands and responsibilities of life. Maturity is not some sort of plateau or final state but a lifetime process of becoming.

2. Self-concept is the view we have of ourselves through time as "the real me" or "I myself as I really am." Self-concept in part derives from our social interactions because it is based on feedback from other people.

3. Erik Erikson posits that the midlife years are devoted to resolving the "crisis" of generativity versus stagnation, where generativity is "primarily the concern in establishing and guiding the next generation."

4. Robert C. Peck suggested that midlife individuals confront four tasks: making the transition from valuing physical powers to valuing wisdom; socializing instead of sexualizing relationships; become emotionally flexible; becoming "open" rather than "closed minded."

5. Trait models are based on the assumption that an individual gradually develops certain characteristics that become progressively resistant to change with the passage of time.

6. Situational models view a person's behavior as the outcome of the characteristics of the situation in which the person is momentarily located.

7. According to interactionist models of personality, behavior is always a joint product of the person and the situation; people seek out settings, activities, and associates that are congenial for them and thereby reinforce their preexisting bents.

8. According to Levinson's male stage theory, a man cannot go through his middle adult years unchanged, because he encounters the first indisputable signs of aging and is compelled to reassess the illusions he has held about himself.

9. Levinson's study of women in midlife found that combining love/marriage, motherhood, and full-time career had not given women as much satisfaction as they had hoped for, and women were exploring new ways to live in middle adulthood.

10. The women in the Mills Longitudinal Study experienced turmoil in their early forties, followed by an increase in stability by age 52, when they had fewer negative emotions, increased decisiveness, and increased comfort and stability through adherence to personal and social standards.

11. In later life people tend toward androgynous responses, which are thought to be associated with increased flexibility and adaptability and hence with successful aging. Androgynous individuals do not restrict their behavior to that embodied in cultural stereotypes of masculinity and femininity.

12. On the whole, the greatest consistency in personality appears in various intellectual and

cognitive dimensions, such as IQ, cognitive style, and self-concept. The least consistency is found in the realm of interpersonal behavior and attitudes.

The Social Milieu

13. Most American adults in midlife have a desire to be or to remain married. Eighty percent of baby boomers born in 1946 are currently married. Most enjoy spending time with their spouse. Having a positive attitude toward one's spouse is associated with longevity and satisfaction with marriage.

14. Rates of extramarital sexual activity have varied since Kinsey's original self-report sexual behavior surveys in the early 1950s. Peaks of extramarital sexual activity occurred in the 1980s; married couples in the 1990s report less EMS. STDs and AIDS might be factors in a lower rate of EMS in the 1990s.

15. Although divorce is becoming more common, it is hardly routine. Divorce can exact a greater emotional and physical toll than almost any other life stress, including death of a spouse. Compared with married, never-married, and widowed adults, the divorced have higher rates of psychological difficulties, accidental death, and illness.

16. Many Americans follow a path leading from marriage to divorce, remarriage, and widowhood. Marital separation, divorce, and remarriage have also become more common for those in midlife and older. The effects of divorce are diverse—displaced homemakers, liberated souls, and depressed individuals are all outcomes of divorce.

17. Most divorced people eventually remarry. About five of every six divorced men and about three of every four divorced women marry again.

18. Half of remarried persons are parents, and, for better or worse, their new partners become stepparents. A good number of stepparents feel stigmatized. Boys particularly seem to benefit when their mothers remarry.

19. Middle-aged adults constitute a sandwich generation, with responsibilities for their own teenage children on the one side and for their elderly parents on the other side. Eighty percent of care for elderly persons is provided by their families. Responsibility for the elderly falls most commonly on daughters and daughters-in-law.

20. Friendships are very important in middle age. However, women are more likely than men to have intimate friendships in midlife. The nature and quality of friendships for today's generation of middle aged adults have implications for their networks of support in old age.

The Workplace

21. Work fills many needs. At midlife, some individuals ask themselves, "Is my work satisfying?" while others ask themselves, "In what ways have I made a contribution or formed a 'legacy' to something outside of myself or my family?" Job satisfaction is associated with opportunities to exercise discretion, accept challenges, and make decisions. People appear to thrive on occupational challenges. However, not everyone experiences satisfaction in their work, and some take "psychological retirement."

22. The past three decades have brought considerable economic and social change in the status of women. Many women are employed either full time or part time today.

23. Career changes are now occurring more often in midlife. Women and men switch careers for a variety of reasons.

24. Unemployment is difficult for many reasons. Financial pressures mount. Child abuse, violence, family quarreling, alcoholism, and other evidences of maladjustment mount. The divorce rate soars among the long-term unemployed.

25. There has been an increase in dual-earner families. Today more than 70 percent of adult women work either full-time or part-time, which means that more than 60 million women are part of the U.S. labor force. In 1950, women made up only 25 percent of the labor force.

Key Terms

alienation *(511)*	generativity *(490)*	sandwich generation *(499)*
androgyny *(498)*	generativity versus stagnation *(490)*	self-concept *(489)*
displaced homemaker *(503)*	job burnout *(511)*	social convoy *(500)*
empty nest *(507)*	maturity *(489)*	traditional marriage *(494)*
empty-nest syndrome *(508)*		

Following Up on the Internet

Web sites for this chapter focus on successful midlife transitions of middle-age adults. Please access the text web site at http://www.mhhe.com/crandell7 for up-to-date hot-linked Internet addresses for the following organizations, topics, and resources:

MacArthur Foundation Research Network on Successful Midlife Development (MIDMAC)

The Wisconsin Longitudinal Study

Institute on Aging and Love

Midlife Mommies

The Sandwich Generation

The Riley Guide: Employment Opportunities and Job Resources on the Internet

Part Nine

Late Adulthood

Late adulthood brings a broader range of physical, cognitive, and social changes than any other stage in life. At left, Sarah Knauss, the world's oldest person at 118, heads six generations of descendants, including a daughter who is 95 and a great-great-great-grandson who is 3. In Chapter 17, we will see there is no consensus as to when old age begins and that across the world humans are living longer than ever. Many assume that "old age" starts around retirement and when eligibility for public or private pensions begin. The "beginning of old age" is changing, though, for millions of elderly are now living well into their eighties and beyond. Recent research disputes the myths that portray old age as an undesirable time of life mainly characterized by debilitation. As we will discuss in Chapter 18, though the elderly experience changes in their cognitive functioning, some of the "oldest old" continue to care for themselves and remain socially engaged into their eighties and older. Many, like Sarah, continue to share their maturity and wisdom with members of their extended family. As the elderly experience physical, cognitive, and social changes, they call on their religion and inner reserves of faith and spirituality for comfort and satisfaction with having lived a meaningful life.

Chapter 17

Late Adulthood
Physical and Cognitive Development

Critical Thinking Questions

1. If a new "wonder" drug were invented that could stop the aging process at age 45 and guarantee another 20 years of life as a 45-year-old, would you take the drug, knowing that you would be unable to experience what we call late adulthood? What do you think you might miss out on?

2. You can easily articulate why children, young adults, and middle-aged adults are important to society as potential and productive members, but why are elderly adults, those 65 and older, important to society?

3. Do you think men's and women's differences will be stronger or weaker in late adulthood? If they no longer work or have caretaker responsibilities that help define them, what might they find as other commonalities?

4. What is the longest you would want people to live? Would you really enjoy being married to the same person for 100 years or keeping track of all your great-great-great-great-grandchildren's names and birthdates? How do you predict humans would benefit from greater longevity?

This chapter looks at the physical and cognitive functioning of adults in the final stages of life. Longevity is a relatively new area of study, and we will see where the research stands. Most physical and health changes are due to biological aging, though some 80-year-olds stay active and proclaim they feel like 60-year-olds. Because more people around the world are living longer, all industrialized countries are experiencing a "graying" of their population and reexamining policies for their citizenry. Issues such as when the elderly should retire, where the funding for old-age pensions is going to come from, and who is going to pay the bill are serious concerns all over the world. Significantly, specialists in aging predict that in future years, human life will be extended by many more years.

Though most elderly experience a slowing in cognitive functioning, they also experience a newfound sense of inner peace, finding hope and meaning in their faith and religious rituals. We will examine this last stage, not as the final chapters of a book, but more as an explorer's tentative steps toward the unknown.

Aging: Myth and Reality

At 15 I set my heart upon learning.
At 30 I had planted my feet firm upon the ground.
At 40 I no longer suffered from perplexities.
At 50 I knew what were the biddings of heaven.
At 60 I heard them with a docile ear.
At 70 I could follow the dictates of my own heart; for what I
desired no longer overstepped the boundaries of right.

Confucius (quoted in Levinson et al., 1978, p. 326)

Many of us have a half-conscious, irrational fear that some day we will find ourselves old. It is as if we will suddenly fall off a cliff—as if what we will become in old age has little to do with what we are now. But at no point in life do people stop being themselves and suddenly turn into "old people." Aging does not destroy the continuity of what we have been, what we are, and what we will be. In some cultures, such as Japan, aging is viewed positively, but our culture views aging ambivalently, even negatively. The older the old become, especially as they reach quite advanced age, the more likely they are to be unfavorably stereotyped. Take a minute to think about words commonly used to describe people who are in this later stage of development: *slow, tired, miserly, grouchy, wise, doting, serene,* and *incompetent.* Using these words contributes to a myth of old age, which tends to become a monolithic myth about *all* older people.

The truth about aging is far more optimistic than myths would have us believe. Many losses of function once thought to be age-related, particularly declines in cognitive ability and mobility, are overgeneralized and exaggerated. Decline is now known to take place later in life than formerly believed, and it can be modified—and in some areas even prevented—with adequate medical care and a healthy lifestyle (Schaie, 1995). **Ageism** is stereotyping and judging a group of people solely on the basis of their age. Just as sexism or racism denies the variabilities found in any race or gender, ageism focuses on a narrow range of descriptors of older adults, which are negative for the most part (Perls, 1995).

Most of the signs of aging are viewed as unattractive in both males and females but even more unappealing in women (Harris, 1994). The elderly are often depicted as frail, critical, and dependent; yet at the same time they are also portrayed as emphasizing deep and meaningful friendships, enjoying hobbies, being family oriented, valuing companionship, and exuding serenity (Kite, Deaux, & Miele, 1991). Moreover, although our society negatively stereotypes elderly people as a whole, we have quite different attitudes toward different kinds of older adults. Americans have a favorable view of the elderly in roles such as sage, John Wayne conservative, grandpar-

ent, liberal matriarch/patriarch, and golden ager. On the other hand, we have an unfavorable view of the elderly in roles such as recluse, nosy neighbor, shrew/curmudgeon, and vagrant (Hummert et al., 1994).

We will see in this chapter and the next that aging does affect physical, cognitive, emotional, and social development, but we have all heard many inconsistent views on when these changes occur, how much change occurs, and why these changes occur. Ongoing research in **gerontology,** the study of aging and the special problems associated with it, and **geropsychology,** the study of the behavior and needs of the elderly, can help us separate myths from reality.

Aging is not a disease, and the ravages of aging are somewhat of a myth (Costa, Yang, & McCrae, 1998). Ignorance, superstition, and prejudice have surrounded aging for generations. To dispel some of the mystery, we will look at the demographics of aging and some specific myths in relation to current research.

Older Adults: Who Are They?

A grandmother becomes palsied. Her grown child gives her a wooden bowl that trembling hands cannot break. The old woman dies, and the bowl is discarded. But the granddaughter retrieves it; the bowl, she knows, will be needed again.

A Yiddish folk anecdote

The time at which old age is said to begin varies according to period, place, and social rank. One researcher reported, for instance, that the Arawak of Guyana (in South America) seldom lived more than 50 years and that between the thirtieth and fortieth years in the case of men, and even earlier in the case of women, "the body, except the stomach, shrinks, and fat disappears, [and] the skin hangs in hideous folds" (Im Thurn, 1883). Life expectancy for the Andaman Islanders of the Bay of Bengal rarely exceeded 60 years (Portman, 1895), and the Arunta women of Australia were regarded as fortunate to reach 50 (Spencer & Gillen, 1927). In addition, the Creek Indians of North America were considered lucky if they lived to see gray hair on the heads of their children (Adair, 1775).

Future Growth The U.S. Bureau of the Census (1998c) reported that in the United States in 1990, 35,808 people were 100 years or older—double the number in 1980—and that in 1998, 66,000 people were 100 and over. Likewise, as of 1997 in the United States, there were 24.3 million over 70, 8.5 million over 80, and 1.4 million over 90 (Neuharth, 1997). As you can see, the number of elderly continues to climb as over 5,000 Americans reach their 65th birthday every day (see Figure 17-1). But the

85-and-older age group, several million strong, is the fastest-growing age segment of our population (Powell & Whitla, 1994). Future projections call for over 60 million people living beyond 70 years by the year 2050 in the United States (Neuharth, 1997).

The growing population of older Americans also will differ significantly from current generations of older adults in some important ways. The current population is predominantly female and white. In 1997, there were 20 million older women and 14 million older men (U.S. Department of Health and Human Services, 1998). In the coming years the largest increases in population will be among nonwhites, particularly Hispanic Americans. The older white population is projected to grow by approximately 8 percent; the projected increase in the older African American population is over 20 percent. Even more dramatic, the projected increase in Hispanic Americans over the age of 65 is approximately 60 percent. By the year 2050, Hispanics will constitute more than 15 percent of older U.S. adults.

These increases in the percentages of elderly Americans who are nonwhite will be accompanied by socioeconomic shifts as well. For example, national statistics indicate that older African Americans and Hispanics report lower median incomes than older whites. Rates of poverty are significantly higher for older African Americans (33 percent) and Hispanics (22 percent) compared to whites (11 percent). Lower socioeconomic status (SES) is associated with increased risks for illness and disability, which in turn are predictive of longer hospitalizations, greater home health care and hospice care, and greater per capita health care expenditures. Not only are there likely to be increases in those population subgroups at highest risk for negative health outcomes, but these populations also are least likely to have private health care insurance and will require public health care (Seeman & Adler, 1998).

The risks for disease and disability are higher at older ages for all persons, though many older adults enjoy good health and experience little or no functional impairment. Research evidence suggests that we increase our chances for more successful aging through lifestyle choices that include personal control and efficacy, regular exercise, and participation in social activities (Schulz & Heckhausen, 1996; Schaie, 1994). According to Seeman and Adler (1998), the two biggest challenges are (1) to convince older adults of the value of such behaviors, and (2) to develop programs and policies that not only encourage such lifestyle changes but also promote and facilitate their adoption by all socioeconomic and ethnic segments of the growing population of older Americans. Beyond health care, Powell & Whitla (1994) predict the likely effects of these population changes will include:

- An increased dependency ratio in the nation. Today, 75 percent of the adult population is in the labor force; that figure is projected to be less than 60 percent in 2020.
- An increased demand by the aged for various resources such as Social Security, welfare, adult day care, medical and rehabilitative facilities and services, and recreational centers.
- The emergence of older people as a political force and social movement. The American Association of Retired Persons (AARP) is a powerful lobbying group in Washington.

Some 30 percent of the annual federal budget is dedicated to this population, and this figure is expected to rise to about 60 percent over the next several decades. So there is a growing concern over the so-called graying of the budget. Some are concerned that age divisiveness might appear in U.S. politics. See *Further Developments:* "Generational Tensions: How Should We Allocate Resources?" to take a closer look at such highly charged issues.

Myths

The facts of aging are often hidden by a great many myths that have little to do with the actual process of growing old. Let's examine each one in light of recent findings from research.

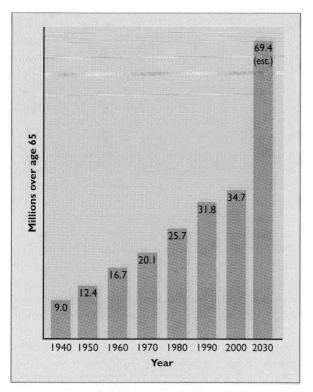

Figure 17-1 Americans Age 65 and Over From 9.0 percent of the population in 1940, older Americans may grow to 13.1 percent of the population by the year 2000 and to 21.1 percent by 2030.

Source: U.S. Department of Commerce, 1997.

Myth: Most persons age 65 and over live in hospitals, nursing homes, and other such institutions.

Fact: Only relatively recently have people lived long enough to require long-term care. This increased longevity is primarily due to improved sanitation, nutrition, and medical care (Shute, 1997). Although Census Bureau data reveal that the chances of needing nursing-home care do increase with age, only about 5 percent of Americans over 65 will require placement in a nursing home (see Figure 17-2) (U.S. Department of Health and Human Services, 1997a).

Myth: Many of the elderly are incapacitated and spend much of their time in bed because of illness.

Fact: In the United States about 3 percent of the elderly who live at home are bedridden and about 9 percent are housebound. An additional 5 percent are seriously incapacitated, and another 11 to 16 percent are restricted in mobility. By contrast, one-half to three-fifths function without any limitation; 37 percent of those 85 and older report no incapacitating limitation on their activity (U.S. Department of Health and Human Services, 1998). A person who becomes severely ill or disabled in advanced old age most likely will not linger another four or five years but will enter a relatively short terminal decline of 90 to 120 days. Indeed, given contemporary trends toward greater longevity, many demographers believe that the baby-boom generation will, on average, spend less time in nursing homes and fewer years being severely disabled than did its parents and grandparents (Manton, Stallard, & Liu, 1993). In sum, notions of gradual aging may well be replaced with perceptions of vigorous adulthood across the life span followed

by a brief, precipitous **senescence** (mental decline in old age) or period of physical decline.

Myth: Most elderly people are "prisoners of fear" who are "under house arrest" by virtue of their fear of crime.

Fact: Overall, those 65 and older have the lowest victimization rates of any age group. For crimes of violence, including robbery, assault, and rape, 4 out of every 1,000 individuals 65 or older are victimized, compared with rates of 8.5 for people 50 to 64, 27.2 for those 25 to 49, and 64.6 for those aged 12 to 25 (Otten, 1993b).

Myth: Most people over 65 find themselves in serious financial straits.

Fact: As a group, Americans over 65 today are in better financial shape than were those who were over 65 three decades ago. Overall, the poverty level for those 65 and older dropped from 35.2 percent in 1959 to 10.5 percent in 1997. Roughly 9 percent of the white elderly, 23.8 percent of the Hispanic elderly, and 26 percent of the African American elderly had 1997 incomes below the poverty level (U.S. Bureau of the Census, 1998b). Social Security and retirement benefits are the primary sources of income for most elderly persons. Indeed, since 1970, overall income levels of the elderly have climbed relative to the rest of the population because Social Security benefits rose 46 percent, after adjustment for inflation, while the buying power of people earning wages and salaries fell 7 percent. In contrast, the rate of poverty among the nation's children has worsened. Moreover, the elderly pay a smaller share of their income to taxes. So although the elderly might have lower incomes than most U.S. householders, they have a higher net worth—often consisting of home ownership—and they tend to be more satisfied with their financial circumstances than other Americans are.

The economic gap between men and women widens in retirement. Although many women entered the workforce in recent decades, women's average 1998 monthly Social Security benefits were $632, compared with an average of $843 for men (U.S. Bureau of the Census, 1998b). Generally women earn less than men do, and they are more likely to have worked part-time or left the workforce for several years for family reasons. For most women who receive Social Security benefits through their husbands, the benefits—equal to half their husbands' monthly check—are nonetheless more than the amount they would receive on the basis of their own work record. Moreover, whereas 33 percent of men over age 65 have private

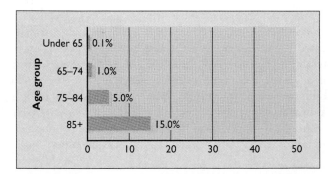

Figure 17-2 Percentage of Age Groups Living in a Nursing Home, 1995 Although a small percentage (4 percent) of the 65+ population lived in nursing homes in 1995, the percentage increased dramatically with age, ranging from 1 percent for persons 65 to 75 years, to 5 percent for persons 75 to 84 years, and 15 percent for persons 85 and older.
Source: U.S. Department of Health and Human Services, 1997a.

pensions, only 13 percent of women over 65 do. Some 31 percent of African American women aged 65 and over live in poverty, as do 22 percent of older Hispanic women and 14 percent of older white women (Lewin, 1995b).

Myth: Most grown children live away from their elderly parents and basically abandon them.

Fact: Most middle-aged children take care of their elderly parents rather than abandon them. According to the National Alliance for Caregiving, nearly one in four American households cares for an elderly relative or friend, helps support their elderly parents financially, or provides other types of support (Cooper, 1998). They found that at least 60 percent of people age 18 and over who have living parents either lived with them or lived within one hour away. Family members are still the most important source of help for older people (Kolata, 1993a; 1993b).

Additional myths will be considered in the course of this chapter. Although the problems cited above are not true for most older adults, they are realities for some segments of the older population. But by and large there is a great gap between the actual experiences of most older people and the difficulties attributed to them by others. Many of the elderly are resilient and very much alive and not hopeless, inert masses teetering on the edge of senility and death. Generalizations that depict the elderly as an economically and socially deprived group can do them a disservice, for such stereotypes give younger people a clear conscience about distancing themselves from the elderly and treating the elderly as if they have inferior status.

Growth in the relative size of the older population is not unique to the United States. In nearly all industrial societies, the number of elderly people is growing at an unprecedented rate (Zarit, Johansson, & Jarrott, 1998). Reductions in fertility and mortality in East Asian countries have created similar demographic changes. By the year 2025, the number of those 65 or older in Japan will double. In the People's Republic of China and Latin America, this number will triple. A near quadrupling is projected for Korea and Malaysia (Powell & Whitla, 1994).

According to the U.S. Bureau of the Census (1998a), the average life expectancy of American women was 79 years in 1998; for men it was 73 years. Projections for 2010 show life expectancy will be 81 years for women, and 74 years for men. The life span for a girl born in 1900 was about 48 years; for a girl born in 2000 it will be about 81. White U.S. males born in 1992 can expect to live 73.2 years; African American men have a life expectancy of 65.5 years. Japanese adults live, on average, nearly 3 years longer but surpass the average Egyptian

by 18 years (see Table 17-1). While the life span for infants is continuing to increase dramatically, the increase in life expectancy for adults, especially males, is less spectacular. A child born in 1997 could expect to live 76.5 years, about 29 years longer than a child born in 1900. The major part of the increase in this longevity statistic is due to the fact that the U.S. has experienced lower death rates for children and young adults. Longevity statistics are averaged to include babies born stillborn or die at birth, those who die in infancy and childhood, and ages of all adult deaths.

Generally, adults who are 65 or older are considered to be in the final stage of human development. Why should this be so? Considering how much these adults vary in their activities, health, and welfare, it is difficult to say when old age begins. Probably the simplest and safest rule is to consider individuals old whenever they become so regarded and treated by their contemporaries (Golant, 1984). Indeed, although our society is getting older, the old are getting younger. The activities and attitudes of a 70-year-old today closely approximate those of a 50-year-old two decades ago. Their quality of life, especially for the "youngest old," ages 60 to 75, is much greater than that of their parents. A large number remain physically and mentally active, even continuing to work full- or part-time.

Questions

How will U.S. population demographics change due to age? What effect will these changes have on our society's social and health policies in the next few years?

Table 17-1 Average Life Expectancy at Birth Across Cultures, 1998

Country	Age	Country	Age
Japan	80.0	Turkey	72.8
Australia	79.9	Mexico	71.6
Canada	79.2	China	69.6
Sweden	79.2	Philippines	66.4
Switzerland	78.9	Russia	65.0
Iceland	78.8	Brazil	64.4
Italy	78.4	India	62.9
Israel	78.4	Egypt	62.1
Britain (UK)	77.2	Zaire	49.3
Germany	77.0	Cambodia	48.0
United States	76.1	Angola	47.9
Argentina	74.5	Kenya	47.6

Source: U.S. Bureau of the Census, "International Data Base (IDB)." Retrieved January 18, 1999, from the World Wide Web: www.census.gov/ipc/www/idbnew.html

Further Developments

Generational Tensions: How Should We Allocate Resources?

The proportion of elderly in our society is growing, and this fact will affect every American and every U.S. institution. Over the past five decades, a variety of social policies have been put in place that have allowed the elderly both to disengage from economically productive activities and to experience an improved standard of living. Congress set up the Social Security program in 1935 and since then has significantly expanded its benefits (62 percent of the beneficiaries are retired workers; 8 percent are disabled workers; and the remaining recipients are family members). Medicare provides the elderly with national health insurance, while the Supplemental Security Income program provides them with a guaranteed minimum income. And the Older Americans Act supports an array of services specifically intended for older persons.

Government spending on the elderly has risen steadily. In fiscal 1965 (the year Medicare was enacted), spending on the elderly accounted for 16 percent of federal outlays. By 1995, benefits for the elderly surpassed 33 percent of federal government expenditures. Currently, most programs for the elderly seem on sound financial footing. For instance, contributions to Social Security exceed outlays. As the baby-boom generation moves into its peak earning years and the number of retired people remains relatively small, the funds should keep swelling. Even so, the Social Security system is headed for difficulty. In 1965 the number of births dropped below 4 million a year, inaugurating the baby-bust generation. Today, 3.3 workers toil to support a single Social Security beneficiary. By 2010, when the first wave of boomers nears 65, the figure will fall to 2.9 workers. And after 2030 the support ratio withers to 1.9 workers per beneficiary (see Figure 17-3).

At the same time, lifestyle and medical advances since 1960 have added five years to the average American's life expectancy and retirement. By the year 2008, on the basis of current trends, the Social Security system is projected to pay out more than it takes in. Moreover, no society in the history of the world has experienced a situation in which over 20 percent of its population were 65 years of age and over. Not surprisingly, many Americans—especially those aged 18 to 30—have lost confidence in the Social Security system. Only

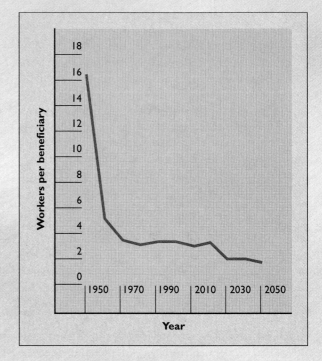

Figure 17-3 Workers per Social Security Beneficiary As the baby-boom generation continues to retire, there will be fewer workers to support Social Security beneficiaries.
Source: U.S. Bureau of the Census, 1998b.

20 percent of working Americans expect to get back all of what they paid into Social Security; 70 percent expect to get back less than half their contribution; and about 18 percent expect to get nothing back from Social Security in retirement (Merline, 1994).

There are those who argue that Social Security, as currently constructed, is a mechanism for redistributing income from the younger to the older generation. For example, the elderly who retired in the early years of Social Security had to pay the program's taxes for only a few years before retiring, and the initial taxes were low. The maximum annual So-

Women Live Longer Than Men

The gap between the life expectancy rates for men and women has been increasing since 1920. Because of women's greater longevity, in the U.S. population of people over 65 the ratio of women to men is currently 4 to 1. Centenarians are also more likely to be women than

men, with from 31 to 43 men per 100 women. Although at birth there are about 105 males for every 100 females, the male death rate is consistently higher, so that by their early twenties, females start outnumbering men. Health statistics typically reveal that women have higher rates than men of acute illnesses and of most nonfatal chronic conditions. More women than men report hav-

"Son, they say your generation doesn't have as much to look forward to as my generation did. Do you want to talk about it?"

Figure 17-4 How Should the United States and Other Countries Allocate Their Resources Among Generations?

Many people do not realize that Social Security was never intended to provide for all their needs when they retired or became disabled. Pension planners typically view Social Security as one leg of a three-legged stool, with private pensions and individual savings providing the other two kinds of support. Critics of the current system contend that most people could do much better if they were able to put their Social Security contributions into a private pension plan. But this argument fails to consider the disability and survivor's benefits that workers or their families might begin drawing at an early age. Moreover, economists say that we cannot make Social Security voluntary. Low-cost people would get out because they could get better investment returns and benefits elsewhere. The high-cost people would stay in, and the system would collapse.

Given these issues, the question of how the nation should allocate its resources among generations is making its way into political discourse. Some contend that the generation currently in power is passing the bill for its needs and upkeep to its successors. Growing numbers of Americans are expressing concern that our nation is mortgaging our children's and their children's futures (see Figure 17-4). For instance, the national debt more than doubled during the 1980s. Since 1973, yearly earnings for workers younger than 35 have declined 15 percent, adjusted for inflation. Younger families are finding it increasingly difficult to purchase their own homes. And the percentage of children who live below the poverty line has increased 50 percent since 1980. Overall, then, the nation's policy makers and public officials are confronting some difficult choices.

cial Security tax, including both employer and employee shares, was $189 in 1958 and still was only $348 in 1965. However, today's younger workers must pay Social Security taxes of several thousand dollars a year for their entire working careers. Some economists estimate that, adjusted for inflation and investment potential, payroll taxes paid by a middle-income 25-year-old will exceed benefits by nearly $40,000. In contrast, today's 70-year-old retiree who had the same inflation-adjusted salary at age 25 would receive $20,000 more in benefits than he or she had paid in taxes (U.S. Bureau of the Census, 1998b).

ing disabilities and functional limitations (Merrill et al., 1997; Harris, 1993).

According to the U.S. Bureau of the Census, in 1990 some 30.2 percent of women aged 75 and over needed help with such basics as eating, dressing, bathing, preparing meals, managing money, and getting around outside; in contrast, 17.2 percent of the men needed such assis-

tance (Barringer, 1992). But men have higher rates of the leading fatal conditions, which parallel their higher mortality. There is some evidence that men also have an incidence of acute illness that is at least equal to that of women when such factors as health-reporting behaviors, environmentally acquired risks, and various psychosocial aspects of illness are taken into account (Verbrugge, 1989).

Human Diversity
Women Survivors

Jean Calmut, a citizen of France and the world's oldest living person at that time, died in August 1997 at the age of 122. She loved wine and candy and smoked until she was 117. She rode her bicycle daily until she was 100 and lived alone until the age of 110, when she entered a nursing home.

(Neuharth, 1997)

The oldest living person on record at the time of this writing is Sarah Knauss, 118, from Allentown, Pennsylvania (Vitez, 1999).

[Knauss's family photo of six generations begins Part 9]

Do 122 and 118 sound extraordinary to you? Jean Calmut and Sarah Knauss reached an old age that anti-aging researchers predict more people will be able to achieve in the next several decades. But what about the quality of life of the "oldest old" humans?

In the mid 1980s the National Institute of Aging began a program to research the oldest population—those over 85—of whom the majority are women (72 percent in 1996) (US-DHHS, 1998). These older women are extending life to its chronological limit, and they hold the secret of what normal aging is at the extreme of old age. Why do the oldest women live so long? What accounts for their ability to survive?

Medical and scientific literature draws attention to genetic solutions, rarely, if at all, acknowledging social and psychological factors. The broad view currently is to describe the oldest women as victims—victims of a system that leaves them with a high risk of disabilities and with neither a husband nor an adequate income. Furthermore, most of these women live alone or in nursing homes. All of these factors—lower income, living alone, disability—lead to a lower quality of life for those who have triumphed in the long journey facing us all. To live through the normal struggles of life and survive into advanced old age is seldom considered heroic or laudable. Women who live long lives usually are pitied as having unfortunately outlived their friends and family.

What Realities Face These Oldest of the Old?

According to the 1990 U.S. Census, many more women than men over the age of 75 reported that they suffered from diseases that were not life-threatening but were disabling. For example, twice as many women as men report suffering from arthritis. This leads to more women entering nursing homes because of their higher rates of disability and not because they live longer. More times than not, aging will turn women into widows—and many widows suffer financially when their husbands die. Retirement income is set up to reward the primary wage earner and the living spouse. The benefits from Social Security drop by one-third when the husband dies. This drop can be extremely harsh for women, especially if they survive their husbands for many years.

Who Are These Women?

Women who are 85 in 1999 were born in 1914, during World War I. This cohort had fewer children than previous cohorts,

---

Genetic differences might play a part in women's greater longevity (Epstein, 1983). Women seem to have an inherent sex-linked resistance to some types of life-threatening diseases. Apparently, a woman's hormones give her a more efficient immune system. Estrogen appears to be protective against cardiovascular disease, because premenopausal women have a substantially lower risk of heart disease than men of comparable ages.

Lifestyle differences also contribute to gender differences in life expectancies. A major factor is the higher incidence of smoking among men. One study examined 17,000 Seventh-Day Adventists (nonsmoking vegetarians) and found that women outlived men by three years. Most medical experts believe that smoking accounts for about half of the gender difference in longevity (Enstrom, 1984). Hence, the rising incidence of smoking among female teenagers may mean that women will lose some of their statistical advantage in the future, and researchers need to examine this group more closely. For some of the data on the oldest group of women, see *Human Diversity:* "Women Survivors."

Question

Why do more women than men survive into old age, and how are these women in their eighties, nineties, and beyond faring?

Women Survivors Women outlive men four to one in old age, but the quality of life for many elderly women is generally poor. They are likely to live alone or in nursing homes, without husbands and without an adequate income.

the Great Depression. African American women have always been part of the labor force, but usually in jobs like private housekeeping and child care that are not covered by Social Security. These women are now having a much harder time in general than those who received some benefits or whose husbands jobs were covered by Social Security. In ural areas in 1925, education was not a priority for girls, and even if a girl did go to school, eighth grade was considered enough school for almost anyone. The lack of education among today's oldest women diminished their employment opportunities later in life.

How Are These Women Situated?

Research shows that this population is quite diverse in terms of physical ability:

- 58 percent needed help getting to the store or doctor's office.
- 50 percent were living in their homes, with various levels of disability.
- 44 percent lived alone (75 percent of these women live in poverty).
- 31 percent lived with relatives (usually daughters).
- 25 percent are in nursing homes.
- 25 percent feel "great" and lead independent lives.

Fewer social norms exist for the oldest segment of our population, and their daily lives generally are free of the worries that plague younger individuals. For women who enter the oldest years without severe disability and with adequate financial resources, these years can be a time of independence, personal mastery, and self-assurance. Future research might do well to focus on helping this group of women develop their potential, which should ultimately benefit us all.

which might mean they now have fewer available adult caregivers in old age. Having lived through the Great Depression, followed by World War II, these women faced many hardships. For a large number, their childbearing years were "taken up" by larger social issues, such as World Wars and

Source: Adapted from S. Bould & C. Longino, *Handbook on Women and Aging* (Westport, CT: Greenwood Press, 1997).

Health

Since I came to the White House, I got two hearing aids, a colon operation, skin cancer, a prostate operation, and I was shot. The damn thing is, I've never felt better in my life.

Ronald Reagan, on turning 76

Although 47 percent of the general public believe that the elderly have poor health, only 18 percent of those 65 and over consider poor health to be a serious problem for them (AARP, 1996). Indeed, the incidence of self-reported acute illnesses (upper respiratory infections, injuries, digestive disorders, and the like) is lower among

the elderly than among other segments of the population. However, the incidence of chronic diseases (heart conditions, cancer, arthritis, diabetes, varicose veins, and so on) rises steadily with advancing years.

Despite the higher incidence of chronic health problems among the elderly, most do not consider themselves to be seriously handicapped in pursuing their ordinary activities. A recent survey (CDC, 1998d) reported health-related quality-of-life measures by age, gender, general health status, and activity limitation and found that the mean number of days in a 30-day period that the participants did not have good physical health tended to increase with age (there was an 1.8-day increase for the 60- to 74-year-olds, compared to the 40- to 59-year-olds).

Information You Can Use

Exercise and Longevity

Many of us think of Juan Ponce de Leon (1460–1521) as the Spaniard who undertook a fruitless search for the fountain of youth. Yet contemporary medical researchers are finding that the explorer might not have failed in his endeavor, although he very likely made the discovery without even realizing it. During the sixteenth century, when few men survived to age 53, Ponce de Leon was hiking Florida's uncharted coastal terrain. Eight years later he suffered a mortal wound from an arrow. Yet historical accounts reveal that Ponce de Leon was a man younger than his years right up to his end (Scheck, 1994).

In the past several decades, study after study has pointed to how modern-day Ponce de Leons are benefiting by staying active during a time in life when they are supposedly past their prime. Medical science has long recognized the rewards to be reaped from exercise by patients with high blood pressure and coronary artery disease. But new information reveals countless other benefits afforded by exercise. With proper counseling, exercise can reduce the risk of falls by restoring coordination and balance and make fractures less likely by strengthening the musculoskeletal system. And physical activity lowers the risk of developing diabetes, osteoporosis, colon cancer, breast cancer, and depression. Exercise not only can add vigor to the later years; it is also an excellent mood enhancer (Hill, Storandt, & Malley, 1993; McAuley, Lox, & Duncan, 1993).

Exercise gurus have long urged seniors to walk, run, bike, hike, dance, or paddle their way to better endurance and better aerobic capacity (respiratory and circulatory function). But evidence suggests that simply maintaining an independent lifestyle also requires strength. Exercise specialists say that "pumping iron"—lifting weights—can be enormously

beneficial. Aerobic activities such as walking, running, and cycling require moving large groups of muscles hundreds of thousands of times against relatively little resistance other than gravity. Strength training entails the working of small groups of muscles only a few times against high, and gradually increasing, resistance. As people enter midlife, their muscles become less bulky as the size and number of their muscle fibers decrease (for instance, from age 30 to age 80, men typically lose 40 percent of their leg muscle strength, and this correlates with a similar loss of muscle mass) (Woollacott, 1993). Exercise cannot replace lost muscle fiber, but it can restore the robustness of the muscles that remain. Young people who are bedridden lose muscle in much the same manner as do the elderly, leading some experts to believe that *atrophy* is due to disuse as well as age.

Strength training can benefit even the very old and frail. Maria A. Fiatarone and her colleagues at the Harvard Medical School analyzed the results of two months of high-intensity workouts by ten frail men and women 86 to 96 years of age. On average, regular conditioning boosted the power of their knee extensors by 174 percent, their heel-to-toe walking speed by 48 percent, and the size of their midthigh muscles by 9 percent (Fiatarone & Evans, 1993).

Does exercise adds years to life? This is still being debated. Medical authorities generally concur that moderate exercise is sufficient to contribute to a healthier life, but recent evidence suggests that exercise must be strenuous to add to the life span. Harvard researchers followed 17,321 healthy male alumni over a 26-year period and found that only vigorous exercise increased longevity (Brody, 1995d). The risk of dying for men who expended more than 1,500 calories a week in such activities as running, cycling, and swimming

In mental health, the older age group reported fewer days than the 40- to 59-year-old group. There was very little difference in the mean number of days that usual activity was limited.

Most of the conditions that create chronic disease increase with advanced age. Over time, it is more likely that a person will accumulate more risk factors for a variety of illnesses. Not only will the number of risk factors increase, but the efficiency of the body's systems is reduced by primary aging or the irreversible changes that occur over time. An older adult is more susceptible to disease, will take longer to recuperate from an illness, and is more likely to have other complications associated with a disease.

The *Baltimore Longitudinal Study,* which began in 1958, has contributed much to our understanding of the health of older Americans (Fozard et al., 1994). The study follows 950 volunteers ranging in age from the early twenties through the late eighties. Every two years the individuals undergo two and a half days of exhaustive physical tests. The study's findings that many people over 65 tend to be overweight caused researchers to re-examine traditional weight tables that equate health with thinness. However, recent research suggests that the increases in so-called desirable weights for men made over the last three decades are unjustified and that higher weights are likely to result in higher death rates, particularly from heart disease.

Cultivating Healthy Living Habits
An accumulating body of scientific evidence points to the important part good habits play in fostering a healthy old age. Researchers with the MacArthur Foundation Consortium on Successful Aging report that we are largely responsible for our own old age. Staying active both physically and socially contributes to successful aging. Indeed, such practices apparently play a larger role than genes in laying the foundations for a healthy and independent old age.

was as much as 25 percent lower than the risk for men who expended fewer than 150 calories a week in such activities (even so, most exercise physiologists believe there is a point at which too much exercise is detrimental to health). The Harvard study defined as *vigorous* any activity that raised the metabolic rate in a 10-second interval to six or more times the rate at rest. Activities such as the following would achieve the level of caloric expenditure associated with the lowest death rates:

- Walking for 45 minutes at 4 to 5 mph five times a week
- Playing tennis for one hour three times a week
- Swimming laps for three hours a week
- Cycling at 10 mph for one hour four times a week
- Running for 30 minutes at 6 to 7 mph five to six times a week
- In-line skating for two and a half hours a week

In sum, older age is not uniformly associated with declines in physical performance or health. Effective interventions, such as appropriate exercise, can promote successful aging (Seeman et al., 1994; Svanborg, 1993).

Also, since it has been found that 60 percent of those over age 60 have high blood sugar levels in glucose tolerance tests, medical authorities are questioning many diagnoses of diabetes in older people. Perhaps a reduction in insulin production in the elderly is a normal occurrence and not a sign of disease. As a result of this research, the American Diabetes Association and the World Health Organization have lowered the statistical range used to diagnose diabetes. Observes Dr. Reubin Andres, one of the researchers with the project, "Think of it, several million people 'cured' by the stroke of a pen" (Fozard et al., 1994). Proper exercise appropriately and carefully pursued throughout life—even into the eighties and beyond—can significantly deter the deterioration of bodily functions that traditionally accompany aging (see *Information You Can Use:* "Exercise and Longevity"). But, researchers find that only 22 percent of adults are active enough; 54 percent are somewhat active; and 24 percent are sedentary (Hellmich, 1995). Disuse is thought to account for about half the functional decline that typically occurs between ages 30 and 70. The demonstrated benefits of exercise include increased work capacity, improved heart and respiratory function, lower blood pressure, increased muscle strength, denser bones, greater flexibility, quicker reaction times, clearer thinking, and reduced susceptibility to depression. Although exercise by the elderly is not without risk, recent studies show that age-associated declines can be delayed by fitness-promoting exercise.

Nutrition and Health Risks

Good nutrition is another controllable factor contributing to health in old age. Although energy requirements decrease with advancing age, elderly people do not require fewer nutrients than younger adults. Indeed, they may require more. Recent research suggests that vitamins and minerals are metabolized differently as people age and that the recommended dietary allowances could be inaccurate for elderly persons (Brody, 1994a; Altman, 1992).

Middle-aged and older adults might require calcium supplements to prevent broken bones in later life (Brody, 1993b). Americans annually suffer more than 275,000 hip fractures, 500,000 vertebral fractures, and 200,000 wrist fractures because of **osteoporosis**—a condition associated with a slow, insidious loss of calcium that results in porous bones (see Figure 17-5).

Not every elderly person experiences osteoporosis, but according to the National Institute on Aging, in the United States one out of every two women and one in eight men over the age of 50 will have an osteoporosis-related fracture, with white women, Asian American women, and thin persons with small bone structure being at a higher risk (National Institute on Aging, 1996a). Along with internal bones becoming more "porous," the discs between the vertebrae of the spine become less dense, causing vertebrae to become closer to each other. A person's appearance, over a period of many years, begins to change as he or she loses height.

Hip fractures cost close to $13 billion annually in the United States. Such fractures are associated with a one-year mortality of 25 percent in 80-year-old women and have important social implications: More than 25 percent of affected women must give up their independent status and enter nursing homes (NIA, 1996a).

Calcium supplements seem to slow or stop bone loss, but by themselves they do not increase bone mass (see Figure 17-5). However, bone mass sometimes increases when fluoride is taken with the calcium supplements. The therapy should be supervised by a physician, because overdoses of fluoride can be highly toxic. Hormone replacement therapy has also been found effective if started before or within 10 years of menopause. Although there is no absolute cure for osteoporosis, if treatment is begun early enough, its progress can be slowed and later fractures prevented. Women who remain physically active, have good leg and arm muscles, and exercise into the seventh and eighth decade seem to have less of a problem; women who lead a sedentary life are more likely to develop the problem.

Brief blackouts are also a major hazard among the elderly that can result in broken hips, bleeding inside the skull, and other injuries. Medical researchers find that many elderly people have a 20-point drop in their blood pressure when they stand up. Eating also lowers the blood pressure in the elderly for an hour after meals. When the two factors coincide, an elderly person may experience a fainting spell.

Figure 17-5 Osteoporosis: Bone Loss Osteoporosis is associated with a loss of calcium. Some evidence suggests that poor nutrition, even at a young age, can contribute to bone loss later in life.

© 1991. Reprinted by permission of The Columbus Dispatch.

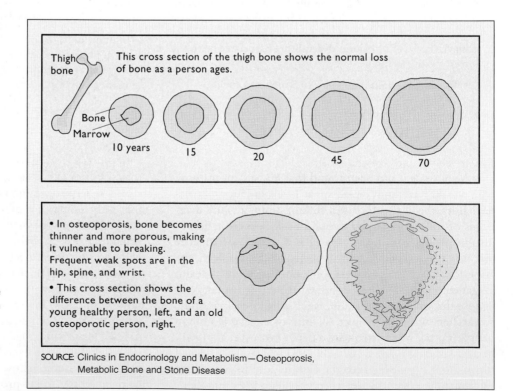

Thigh bone

This cross section of the thigh bone shows the normal loss of bone as a person ages.

Bone Marrow

10 years   15   20   45   70

• In osteoporosis, bone becomes thinner and more porous, making it vulnerable to breaking. Frequent weak spots are in the hip, spine, and wrist.

• This cross section shows the difference between the bone of a young healthy person, left, and an old osteoporotic person, right.

SOURCE: Clinics in Endocrinology and Metabolism—Osteoporosis, Metabolic Bone and Stone Disease

Drug Dosages and Absorption Effects Some health problems of older Americans result from overmedication and mixing medications. Elderly persons do not absorb drugs as readily from the intestinal tract, their livers are less efficient in metabolizing medications, and their kidneys are 50 percent less efficient than those of a younger person in excreting chemicals. Hence, a person over age 60 is two to seven times more likely to suffer adverse side effects than a younger patient (Kolata, 1994b).

Although older people need higher doses of some medications, they need lower doses of others. For instance, the aging brain and nervous system are unusually sensitive to antianxiety drugs such as Valium and Librium, and these can produce confusion and lethargy in the elderly. Sedatives such as phenobarbital often have a paradoxical effect on the elderly, inducing excitement and agitation rather than sleep. Close to a quarter of all Americans 65 years of age or older are given prescriptions for drugs that medical authorities have established should not be prescribed for older people, either because safer alternatives are available or because the drugs are simply not needed (Kolata, 1994b).

These facts are quite dismaying when we realize that people over 65 take more than 25 percent of all prescription drugs. Indeed, the average healthy elderly person takes at least eleven different prescription medicines in the course of a year. When taken in combination, the medications can produce severe secondary reactions. Some problems arise because the elderly have different doctors treating them for different conditions, and each doctor prescribes several potent medications, unaware of the other medications that have been prescribed. Many pharmacy chains now have computerized networks that share information among pharmacies in order to help prevent harmful mixing of medications, and potential drug effects are now printed out for the patient.

Mental Health and Depression Most older adults adapt well to the changes and losses they are confronted with in late adulthood and have good mental health. However, as with all other stages of life, a small percentage develop depressive symptoms that can lead to more serious depressive disorders or mental illness if not treated or discussed (Kasl-Godley, Gatz, & Fiske, 1998). As with depression in earlier life stages, symptoms can include loss of energy, fatigue or sleep disorders, loss of appetite, or loss of interest in sexual activity. Other factors associated with depression include diagnosed health problems, cognitive dysfunction, strained interpersonal relations, stressful life events, and genetic makeup. Depression can result from a physical illness or its treatment. People living with chronic pain or a fatal diagnosis have a much higher risk of depression.

Generally, women have more major depressive disorders than men, though this difference diminishes and might even reverse in very old age (Kasl-Godley, Gatz, & Fiske, 1998). At this writing, little research has been done to determine an association between depressive disorders and race, ethnicity, or culture. However, certain factors are associated with depressive disorders and higher rates of suicide: institutionalized and inpatient clients, being male, being 75 years of age or older, substantial health and mobility problems, and cognitive impairment. The suicide rate for older adults is the highest of any age group (see Chapter 19) (Kasl-Godley, Gatz, & Fiske, 1998).

Depressed elderly adults often do not seek treatment. The older adult might not see a physician, or the physician might focus on a physical condition and miss the depressive symptoms, especially if the patient does

The Elderly with Animal Companions Go to the Doctor Less Psychologist Judith M. Siegel finds that the elderly who have animal companions visit their doctor less often than do those who lack animal companions, and those with dogs have the fewest visits of all. Pets provide their owners with companionship and an object of attachment and seem to help their owners in times of stress. Animal companionship is a way of life for nearly one-third of adults age 70 or older.

not mention them. Allowing depression to go untreated until it becomes more severe can lead to use of potent medications or electroconvulsive therapy. Recent findings from a longitudinal study suggest that a combination of medications and psychotherapy yielded the best results for elderly patients suffering from major depression (Winslow, 1999). Social support, coping styles, perceived control of life factors, and cognitive appraisals play a protective role against depression (Kasl-Godley, Gatz, & Fiske, 1998). Another protective factor is having a pet.

Questions

How do we know there is a relationship between nutrition and aging? To what extent are the elderly likely to experience depression, and what approaches are suggested to help them cope with this disorder?

Biological Aging

To me, old age is always fifteen years older than I am.

Bernard Baruch

Biological aging refers to changes that occur in the structure and functioning of the human organism over time (see Figure 17-6). *Primary aging,* or time-related changes, is a continuous process that begins at conception and ceases at death. As human beings advance from infancy through young adulthood, biological change typically enables them to make a more efficient and effective adaptation to the environment. Beyond this period, however, biological change generally leads to impairment in the ability to adapt to the environment; and ultimately, it jeopardizes survival. Improvements in the conditions of life and advances in medicine, however, have allowed more people to live longer and have facilitated successful aging (Baltes & Carstensen, 1996).

Physical Changes Some of the most obvious changes associated with aging are in physical characteristics. The hair grows thinner, turns gray, and becomes somewhat coarser. The skin changes texture, loses its elasticity and moistness, and gathers spot pigmentation. Some of the subcutaneous fat and muscle bulk built up during earlier adulthood begins to decrease; this decrease, coupled with the loss of the elasticity of the skin, produces skin folds and wrinkling.

Other changes are noticeable in body height, shape, and weight. As the vertebrae begin to settle closer because the "cushioning" material in between them becomes thinner, reducing the height of the spine, there are also muscular changes that result in a loss of flexibility, making it harder to stand straight. Body-shape changes are primarily the result of the redistribution of fat away from the arms, legs, and face and onto the torso. Aging is associated with decreases in both size and strength of muscles, particularly such muscles as the knee extensors and knee flexors of the lower limb (Overend et al., 1992).

There are also other physiological changes, such as a decline in the capacity for physical work and exercise (sport performance declines after the twenties or early thirties) (Sinaki, 1996). From age 30 to age 70, maximum oxygen intake declines 60 percent, and maximum ventilatory volume declines 57 percent during exercise. Because oxygen is needed to combine with nutrients for the release of chemical building blocks and energy, the older person generally has less staying power and lower reserves. Furthermore, at age 75 the heart pumps about 65 percent as much blood as at age 30; the brain receives 80 percent as much blood, but the kidneys only 42 percent as much. However, the nerve fibers that connect directly with the muscles show little decline with age—nerve impulses travel along single fibers in elderly people only 10 to 15 percent slower than in young people. Even so, psychomotor performance is slower and less consistent in the elderly (Kallman, Plato, & Tobin, 1990).

Collagen, a substance that constitutes a very high percentage of the total protein in the body, appears to be implicated in the aging process. Collagen is a basic structural component of connective tissue. Loose connective tissue resembles styrofoam packing material. It supports and holds in place blood vessels, nerves, and internal organs while simultaneously permitting them some freedom of movement. It also holds muscle cells together and binds skin to underlying tissue. Over time, collagen fibers become thicker and less elastic. In early life these changes are fundamental to development. But once set in motion, the process apparently is not halted, contributing to a loss of elasticity in the skin, hardening of the arteries, and stiffening of the joints. Thus, over time, collagen speeds the destruction of the organism it helped to build.

Sensory Changes Our sensory abilities, such as hearing, sight, taste and smell, also change with age. It is important to understand both the magnitude of the changes and the course of action that can reduce the amount of impairment. Recent studies have shown a strong link between sensory and cognitive functioning in aging. A person whose memory ability has declined with age is more likely to have impaired hearing as well. This relationship does not imply direct causation, but there is connection between these abilities (Stevens, Cruz, Marks & Lakatos, 1998). Baltes and Lindenberger (1997) found comparable links among visual acuity, auditory thresholds, and intelligence.

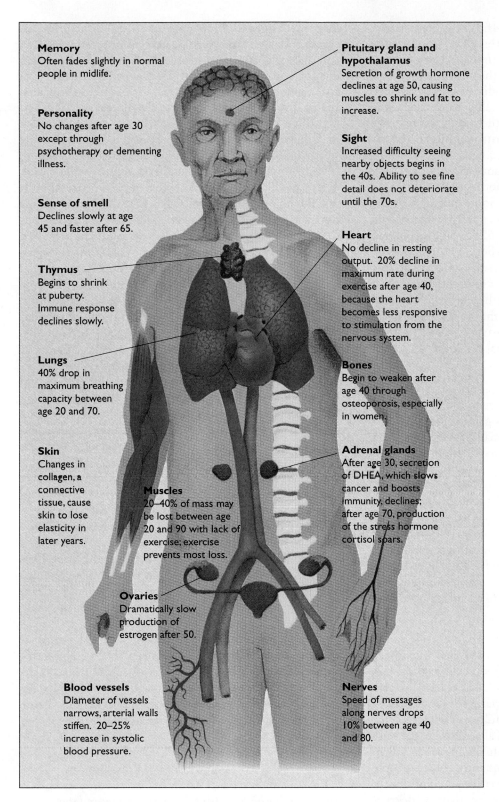

Memory
Often fades slightly in normal people in midlife.

Personality
No changes after age 30 except through psychotherapy or dementing illness.

Sense of smell
Declines slowly at age 45 and faster after 65.

Thymus
Begins to shrink at puberty. Immune response declines slowly.

Lungs
40% drop in maximum breathing capacity between age 20 and 70.

Skin
Changes in collagen, a connective tissue, cause skin to lose elasticity in later years.

Muscles
20–40% of mass may be lost between age 20 and 90 with lack of exercise; exercise prevents most loss.

Ovaries
Dramatically slow production of estrogen after 50.

Blood vessels
Diameter of vessels narrows, arterial walls stiffen. 20–25% increase in systolic blood pressure.

Pituitary gland and hypothalamus
Secretion of growth hormone declines at age 50, causing muscles to shrink and fat to increase.

Sight
Increased difficulty seeing nearby objects begins in the 40s. Ability to see fine detail does not deteriorate until the 70s.

Heart
No decline in resting output. 20% decline in maximum rate during exercise after age 40, because the heart becomes less responsive to stimulation from the nervous system.

Bones
Begin to weaken after age 40 through osteoporosis, especially in women.

Adrenal glands
After age 30, secretion of DHEA, which slows cancer and boosts immunity, declines; after age 70, production of the stress hormone cortisol soars.

Nerves
Speed of messages along nerves drops 10% between age 40 and 80.

Figure 17-6 Tendencies in Rates of Aging The rate of aging varies considerably among individuals and even among different organs in the same person. Nonetheless, a number of tendencies in aging occur in a predictable fashion.
© 1992. Reprinted by permission of The Columbus Dispatch.

Vision and Hearing Some of the visual changes associated with aging begin to unfold during middle age. We saw in Chapter 15 that many adults in their forties need to wear bifocals and develop "dry" eyes, requiring drops of "liquid tears" occasionally. One should have annual eye exams to check for pressure buildup in the fluid of the eye, or *glaucoma,* which must be treated to prevent blindness. Blurred detail vision for most is not a problem until some time in the seventies or eighties, and this can be a sign of a *cataract,* or hardening of the lens of the eye so that it cannot accommodate as efficiently. Fortunately, cataract surgery has become routine. The "oldest old" are also more prone to *retinal detachment*, a serious condition in which the retinal layer at the back of the eyeball begins to "peel away." If the condition is caught early enough, laser surgery might be able to reattach the retina. A person with other health complications (e.g., high blood pressure, diabetes, or stroke) might find vision difficulties appearing as well.

Hearing loss is a hazard in some occupations, such as working near large engines or motors or working with sound equipment for rock bands. Some hearing loss appears to have a genetic component. Hearing loss occurs in about 25 percent of people aged 65 to 75 years old and in 50 percent of those over age 75. Severe hearing loss or loss of vision diminishes the person's quality of life, making the person more dependent upon others to meet even basic needs. Loss of vision or hearing can even lead to one of the most severe things that can happen to an older person—losing his or her driver's license and the freedom of choice and mobility that it allows.

Taste and Smell   Older people frequently report that they are losing their ability to enjoy food (de Graaf, Polet, & van Staveren, 1994). This problem is related to a decline in the taste buds (the small protuberances on the surface of the tongue). Persons 70 to 85 years old have, on average, only one-third as many taste buds as young adults (Bartoshuk et al., 1986). Olfactory sensitivity (smell—the ability to distinguish oranges from lemons or chocolate from cheese) also declines among older adults and helps explain why many complain about their food (Ship et al., 1996). Cross-sectional studies reveal that a large number of older individuals have higher detection thresholds for smell, diminished intensities, and an impaired ability to identify and discriminate odors (Ship & Weiffenbach, 1993).

Touch and Temperature Sensitivity   Touch sensitivity also decreases with age, but older people differ considerably in this respect (Kenshalo, 1986). The elderly are also less sensitive to changes in temperature (Richardson, Tyra, & McCray, 1992). Young adults can detect a temperature drop of only 1 degree Fahrenheit in the surrounding air. Elderly individuals can fail to notice a drop of 9 degrees Fahrenheit. Consequently, older people tend to be susceptible to **hypothermia**—a condition in which body temperature falls more than 4 degrees Fahrenheit and persists for a number of hours. This can be life-threatening, especially because the aging body becomes less able to maintain an even temperature in winter weather. Early symptoms of hypothermia include drowsiness, mental confusion, and eventually a loss of consciousness.

Sleep Changes   Complaints of sleep difficulties increase with age. More than 50 percent of persons aged 65 and older report regular problems with sleep. Older men display greater age-related sleep changes than older women do (Reynolds et al., 1991). Most frequently, men report difficulties with daytime sleepiness, napping, and nighttime awakenings, whereas women report difficulty in falling asleep, staying asleep, and getting adequate sleep (Middelkoop et al., 1996). It seems important to keep in mind that older people do report being sleepy during the day, and they do take naps, so apparently it is not the amount of sleep needed that declines with age but rather the ability to sleep. There are many reasons that the ability to sleep decreases with age, but the two main ones are changes in circadian rhythms and the presence of sleep disorders (Ancoli-Israel, 1997).

Sleep patterns change across the life span (Morin et al., 1993). For instance, older people have less Stage 3 and Stage 4 sleep (deep sleep) and less rapid-eye-movement (REM) sleep (dream sleep). The sleep/wake cycle is controlled by our biological clock, or circadian rhythm. The average younger adult gets sleepy around 10 or 11 P.M. and sleeps for about 8 to 9 hours, waking between 6 and 8 A.M. As we age, our circadian clock advances, causing *advanced sleep phase syndrome*. People with advanced sleep phase syndrome get sleepy early in the evening; and if they were to go bed at that time, they would sleep for about 8 hours and wake up at 4 to 5 A.M. But the tendency, of course, is not to go to bed so early, rather delay sleep until the "usual" time of 10 or 11 P.M. The problem is that the person still wakes up at 4 or 5 A.M. Now they have had only 5 or 6 hours sleep, so they will feel tired and maybe nap during the day to "catch up." According to Sonia Ancoli-Isreal (1997), director of the Sleep Disorder Clinic, University of California at San Diego, sunlight treatment is the best stabilizer of circadian rhythms, and she recommends bright light exposure in the early evening or late afternoon.

Sleep apnea is a disorder in which the person occasionally stops breathing during sleep. The person gasps for air and might jump up to breathe again. The likelihood of this disorder increases with age. Treatment is continuous positive airway pressure provided by a machine or a tongue-retaining device in milder cases.

Questions

What are some common physical and sensory changes in late adulthood? How do sleep changes in later years manifest themselves?

Sexuality   "The need for companionship, tenderness, love and yes, sex, remains as important as ever, but the rules of the game have changed" (Rimer, 1998, A1). For most of us, our sex life changes over time. In surveys of older people's attitudes toward sexuality, most of those interviewed felt that sex was important for both physical and emotional health (see Figure 17-7). Wiley and Bortz (1996) found that 92 percent of the men and women as a whole reported that ideally they would wish to have sex at least once per week, and this desire was not diminished with older respondents, but fewer than half the men and women reported having sexual activity at least once per week. This decrease in activity is well reported, so the question is why sexual activity becomes infrequent (Levine, 1998).

Three major factors are partner availability, difficulty with sexual arousal, and the overall health of older adults. As mentioned earlier, after age 75 women outlive men four to one. And older women find that older men simply want younger women (Rimer, 1998). Sexual arousal difficulties occur in both men and women, but male impotence is the frequently occurring problem. Women also experience physical changes after the onset of menopause (lack of vaginal lubrication and thinning of the walls of the vagina), which can make sexual intercourse uncomfortable or painful. Other health factors that can contribute to less frequent sexual activity are medications that diminish the desire or inhibit erection in males; pain from chronic illness; and mobility problems.

Questions

What physical changes are considered a normal part of sexual aging for women and men? What are some factors that might account for a decrease in sexual activity?

Biological Theories of Aging

We have discussed some of the physical changes associated with aging. Now we need to ask what is driving the changes—why do they occur? There are many competing theories to explain the process. We will review several alternative explanations that recently have received the most attention. According to the *wear-and-tear theory,* there is a natural limit to the human life expectancy—about 85 years—and little can be done to push the figure upward. The other view holds that there are no absolute biological limits to how long human beings can live. Rheumatologist James Fries (1997, 1989) is the leading proponent of the notion of inborn limits. He contends that the human body is biologically destined to fall apart after 85 years—frailty rather than disease being the primary killer of people at very old ages. Around age 85, give or take seven years, the tiniest insults—a fall that would be trivial to a 20-year-old, a spell of hot weather, or a mild case of the flu—are sufficient to cause death. Fries likens the process to a sun-rotted curtain: You sew up a tear in one place and it promptly tears somewhere else. In the opposing camp are those who argue that old people do die from such causes as osteoporosis or athereosclerosis, but these diseases—once deemed to be the inevitable hallmarks of old age—now can be prevented or delayed.

As the population becomes generally healthier, people will enter old age in better shape and so life spans can be expected to increase. Demographers project that by the year 2080 the average life expectancy for men will be 94 and for women 100. Others like demographer S. Jay Olshansky (1992) believe that, while small gains may continue, it will become harder and harder to maintain the pace, so that any major increase is highly unlikely. As

"And this one is my grandma and her current lover."

Figure 17-7 The Contemporary Elderly Are Enjoying Healthier and More Zestful Lives Than Prior Generations

in economics, a curve of diminishing returns operates. It is not so much that we are programmed to die, says Olshansky. Rather, we are not programmed to survive very long past the end of our reproductive period. Olshansky calculates that in order to get an average life expectancy of 85 from today's average of 77.5, we would need to reduce death rates by 55 percent at each age interval (equivalent to the complete elimination of heart disease and cancer). And once life expectancy reaches age 85, a practical limit is reached because there are too few people over 85 to influence the overall statistics.

In evaluating these perspectives, it is useful to examine the more prominent theories that seek to explain the biological process of aging. So far researchers have not reached a consensus. Indeed, the process of aging may be too complex for any one-factor explanation.

Genetic Preprogramming   This is also referred to as "mean time to failure." Engineers contend that every machine has a built-in obsolescence and that its lifetime is limited by the wear and tear on the parts. In the same way, aging is viewed as a product of the gradual deterioration of the various organs needed for life (Hayflick, 1980). Most significantly, DNA repair capacity declines with age and DNA damage accumulates (Warner & Price, 1989).

Aging Effects of Hormones   Hormones can promote or inhibit aging, depending on the conditions. Reducing the secretion of some hormones (e.g., pituitary hormones) in rodents depresses their body metabolism and delays the aging of their tissues. Caloric restrictions likewise seem to slow aging processes. However, a reduction in the secretion of many hormones also occurs with age in rodents and humans. Increasing these hormones (e.g., growth hormone and DHEA) enhances metabolism and stimulates organ functioning (Joseph & Roth, 1993; Angier, 1990).

Accumulation of Copying Errors   According to this theory, human life eventually ends because body cells develop errors in copying. The prints taken from prints are thought to deteriorate in accuracy with the number of recopying events (Lumpkin et al., 1986; Busse, 1969).

Error in DNA   Another line of evidence suggests that alterations (mutations) occur in the DNA molecules of the cells—that is, errors creep into the chemical blueprint—that impair cell function and division (Wareham et al., 1987; Comfort, 1976; Busse, 1969).

Autoimmune Mechanisms   Some scientists believe that aging has a marked impact on the capabilities of the immune system. They are convinced that the body's natural defenses against infection begin to attack normal cells because the information is blurring or because the normal cells are changing in ways that make them appear "foreign" (Miller, 1989; Schmeck, 1982).

Accumulation of Metabolic Wastes   Biologists have suggested that organisms age because their cells are slowly poisoned or hampered in functioning by waste products of metabolism. Such waste products accumulate, leading to progressive organic malfunctioning (Chown, 1972; Carpenter, 1965). For instance, researchers have found significant changes with age in the amounts and kinds of metals in certain organs, including the lens of the eye. Additionally, molecules that are the normal by-products of cells' use of oxygen—called "free radicals"—react with virtually every other molecule they encounter, wreaking havoc on vital cellular machinery. In due course, the injuries are so substantial that cells no longer function properly, organ systems fail, arthritis cripples joints, emphysema undermines lungs, cataracts cloud the eyes, diseases like cancer and heart disease occur, and finally the organism dies (Kolata, 1994a; Merz, 1992). Vitamins C and E and beta carotene are believed by some to be helpful in mopping up free radicals, and so they are taken as dietary supplements. Biologists have identified specific genetic traits that are associated with the improvement of an organism's defense system against free radicals. This knowledge has allowed them to breed fruit flies that live the human equivalent of 150 years (McDonald, 1992).

Stochastic Processes   *Stochastic* implies that the probability of a random happening increases with the number of events. Radiation, for instance, could alter a chromosome through a random "hit" that either kills a cell or produces a mutation in it. The chances for such an event obviously increase, the longer one lives.

Longevity Assurance Theory   The theories outlined above focus on cell-destroying mechanisms. In sharp contrast to these approaches, George Sacher (Brues & Sacher, 1965) offers what he terms a "positive" theory of aging because he portrays evolution as having prolonged life among some species. Thus, instead of asking why organisms age and die, he asks why they live as long as they do. Sacher observes that the life spans of mammals vary enormously, from about 2 years for some shrews to more than 60 years for great whales, elephants, and human beings. He says that in long-living species natural selection has favored genes that repair cells while weeding out genes that impair cell functioning. Individuals who are the bearers of cell repair genes are more likely to survive and thus pass on their favorable genes to their offspring.

In support of this explanation, Sacher notes that when researchers exposed cells from seven species to ultraviolet light, the amount of DNA repair that occurred was in direct proportion to the life span of the species.

Because animals with large brains produce small litters, evolution has favored them with longevity genes that lengthen the life span and make up for the losses in reproductive potential (Lewin, 1981).

Given what we know about physical changes in late adulthood, there appear to be three primary ways of slowing the aging process: (1) Individuals can make behavior changes in diet and lifestyle that are known to further life expectancy; (2) Medical scientists can develop ways to replace the body's growth factors, hormones, and chemical defense systems that diminish over time and affect youthfulness; and (3) Medical scientists can change the genetics of aging with drugs and gene therapies. Clearly, the first option is the principal and perhaps only one currently and readily available to us.

Questions

What are the major theories about why biological aging occurs? In what ways might we slow the aging process?

Cognitive Functioning

There is a wicked inclination in most people to suppose an old man decayed in his intellect. If a young or middle-aged man, when leaving a company, does not recollect where he laid his hat, it is nothing; but if the same inattention is discovered in an old man, people will shrug their shoulders and say, "His memory is going."

Samuel Johnson

Sooner or later almost all adults worry about whether they are thinking, remembering, and making decisions with the same acuity as when they were younger. Even as early as 40, we can begin to notice occasional mental lapses. It could be as simple as having difficulty remembering where we parked the car or retrieving a certain word or name in conversation. We might wonder if these are normal consequences of aging or if they foretell a more serious process such as Alzheimer's disease. Attitudes or perceptions about aging appear to be important in determining whether people are honored for their wisdom or cast aside for their incompetence. Not all societies value old age in the same way. The Bible celebrates the wisdom of the elderly King Solomon, and Eastern cultures have long revered their elders. We have seen in the United States a less positive attitude, although much depends upon the person and the historic period in which she or he lives.

Overall, psychological literature supports the view that aging often brings a decline in intellectual ability (Giambra et al., 1995; Schaie, 1994). However, there is less decline—or even little or no decline—for people with favorable lifestyles and good health (Schaie, Willis, & O'Hanlon, 1994). Indeed, a growing body of research suggests that disease, including depression, metabolic disorders, hardening of the arteries, chronic liver and kidney failure, amnesia, or Alzheimer's disease—not age in itself—underlies much of the decline and loss of cognitive and intellectual functioning among the elderly.

Researchers looking at the physiology of aging brains are surprised at their flexibility and resilience (Cerella et al., 1993). For instance, contrary to what was established scientific opinion only a few years ago, investigators now find that older brains rejuvenate and rewire themselves to compensate for losses. According to Stanley Rapoport, chief of the neurosciences lab at the National Institute on Aging, as brains age, neighboring brain cells (neurons) help pick up the slack; indeed, responsibilities for a task can actually shift from one region to another (Schrof, 1994). Overall, although there is some decline in cognitive functioning for some people in their seventies and more in people in their eighties, many people seem not to be affected. Let's take a closer look at some of these issues.

The Varied Courses of Different Cognitive Abilities

Not only do the declines in cognitive functioning appear later in life than we might have expected, but cognitive abilities vary in how they are affected by aging. The *Seattle Longitudinal Study* (see Chapter 15) found that many abilities start a dramatic downward trend beginning in the twenties, with an even greater one as advanced age is reached. There are important exceptions.

The cross-sectional data for verbal and numeric abilities indicate a peak in midlife with relatively little change into early old age but a significant decline in the eighties. Take a minute to review some weaknesses of a cross-sectional design. The major one, of course, is that you don't know if your groups are comparable. For instance, would your 80-year-olds, if measured at the age of 40, be similar to your current 40-year-olds? In the Seattle study, both longitudinal and cross-sectional data were collected and analyzed.

The only ability that shows profound linear decrement is *perceptual speed*. This decline is the result of progressive slowing of neural impulses throughout the central nervous system. Another important finding is the extent of individual variation; that is, some very old people are capable of quite quick responses (Schaie, 1995; Powell & Whitla, 1994). Even more interesting is the fact that most other abilities show a gain from young adulthood into midlife. Intellectual competence generally peaks in the forties and fifties, because people continue to gain experience and without significant physiological loss to offset the gain. Speed of numeric computation declines significantly with age when followed longitudinally, but in contrast, verbal ability does not peak until the sixties and declines only modestly after that age.

Some researchers suggest that cognitive functioning should be assessed in less traditional ways than standard intelligence testing. Critics of traditional intelligence testing suggest that aspects of adult functioning, such as social or professional competence and the ability to deal with one's environment, should also be considered (Berg & Sternberg, 1992). Many psychologists are developing new measures of adult intelligence and revising our notions of adult intelligence.

For example, psychologist Gisela Labouvie-Vief (Adams et al., 1990) is investigating how people approach everyday problems in logic. She notes that researchers usually find that the elderly do poorly on measures of formal reasoning ability. But Labouvie-Vief contends that this poor performance results from differences in the way younger adults and older adults approach tasks. Older adults tend to personalize the tasks, to consider alternative ways to answer a question, and to examine affective and psychological components associated with a problem solution. She says that reasoning by intuition rather than by principles of formal logic is not an inferior mode of problem solving—merely a different one. In other research (cited in Meer, 1986), Labouvie-Vief has found that when older people are asked to give summaries of fables they have read, they excel at recalling the metaphoric meaning of a passage. In contrast, college students try to remember the text as precisely as they can. Other researchers also find that the lower performance of older people stems from the fact that they might view some things as unimportant and hence selectively ignore what younger people may attempt to capture (Hess & Flannagan, 1992; McDowd & Filion, 1992).

Cognitive functioning depends to some extent on whether the elderly use their abilities. You most likely have heard the expression *use it or lose it*. For instance, people can perform such complex cognitive tasks as playing chess or the cello well into old age at the same time as they are losing many simpler abilities. Many elderly persons find that what they have been doing, they can keep on doing. Recently John Glenn, in his late seventies, returned to space as an astronaut to conduct research on aging. Erik Erikson, Jean Piaget, Dr. Benjamin Spock, Arthur Rubinstein, Eubie Blake, Martha Graham, George Burns, Andrés Segovia, Pablo Picasso, George Bernard Shaw, Arthur Miller, James Michener, Bertrand Russell, and Senator Margaret Chase Smith are examples of people who continued to excel at the same high standards of performance well into advanced age. Moreover, Schaie and Willis (1993, 1986) found, in a long-term study of a sample of older adults, that individualized training resulted in an improvement in spatial

orientation and deductive reasoning for two-thirds of those they studied. Nearly 40 percent of those whose abilities had declined returned to the level they had been at 14 years earlier. In fact, cognitive-training techniques can in many cases reverse declines (Schaie, 1994; Willis & Nesselroade, 1990). In brief, much of our fate is in our own hands, and "use it or lose it" is an underlying principle.

Older adults do slow down in their performance of many tasks (Verhaeghen, Marcoen, & Goossens, 1993), and slower reflexes are a disadvantage in tasks such as driving a car, but for many speed of activity is relatively unimportant (Meer, 1986). Developmentalist K. Warner Schaie (1994), whose pioneering work has done much to shape our understanding of cognitive functioning across the life span, finds that a variety of factors reduce the risk of cognitive decline in old age:

- Good health and the absence of chronic diseases
- Environmental circumstances characterized by above-average education, a history of stimulating occupational pursuits, above-average income, and the maintenance of an intact family
- A complex and stimulating lifestyle, including extensive reading habits, travel, and a continuing pursuit of educational opportunities

Intellectual Functioning Among the Elderly
Although some aspects of cognitive functioning diminish after age 60, the decline is often of little consequence and greatly exaggerated. Dr. Marvin Glock, professor emeritus from Cornell University, continues to teach communication skills courses in his eighties.

- A personality that is flexible and adaptable at midlife
- Marriage to a spouse with high cognitive capabilities

In sum, from a review of the research, these conclusions seem to be the most reasonable: A decline in intellectual ability tends to occur with aging, particularly very late in life. Some aspects of intelligence, mainly those that are measured by tests of performance and fluid ability, appear to be more affected by aging than others. But older people can learn to compensate, and they can still learn what they need to, although it can take them a little longer. Other aspects of intelligence, notably crystallized intelligence, might increase, at least until rather advanced age. There are also considerable differences among people, some faring poorly and others faring quite well. One of the major factors in maintaining or improving mental capabilities is using them. Too often the expectation of decline becomes a self-fulfilling prophecy. Those who expect to do well in old age seem to remain involved in the world about them and thus do not become ineffective before their time.

Questions

What factors have been identified with optimal cognitive aging? What can we do to mitigate the decline in certain intellectual abilities?

Overestimating the Effects of Aging

What happens in psychological aging is complex, and we are only beginning to understand it. What is clear, however, is that psychologists have taken too negative a view of the impact of aging on intellectual functioning. One reason for this is that researchers have relied too heavily on cross-sectional studies. As we explained in Chapter 1, cross-sectional studies employ the snapshot approach; they test individuals of different ages and compare their performance. Longitudinal studies, in contrast, are more like case histories; they retest the same individuals over a period of years (Holahan, Sears, & Cronbach, 1995).

Psychologists such as Baltes and Schaie (1976; Schaie, 1994) have pointed out that cross-sectional studies of adult aging do not allow for generational differences in performance on intelligence tests. Because of increasing educational opportunities and other social changes, successive generations of Americans perform at progressively higher levels. Hence, the measured intelligence (IQ) of the population is increasing. When individuals who were 50 years old in 1993 are compared with those who were 50 in 1973, the former score higher almost any kind of cognitive task. But because the people who were 50 years old in 1993 were 30 in 1973, a cross-sectional study undertaken in 1973 would falsely suggest that they were "brighter" than those who were 50 in 1993. This result would lead to the false conclusion that intelligence declines with age. When you compare people from different generations—80-year-olds with 40-year-olds, for instance—you are comparing people from different environments. Thus, cross-sectional studies tend to confuse generational differences with differences in chronological age.

Other factors have also contributed to an overestimation of the decline in intellectual functioning that occurs with aging. Research suggests that a marked intellectual decline, called the **death drop** or the *terminal decline phenomenon*, occurs just a short time before a person dies (Johansson & Berg, 1989). Because relatively more people in an older age group can be expected to die within any given span of time, compared to a younger group, the average scores of older age groups are depressed relatively more as a result of the death-drop effect than are the average scores for younger age groups.

Whereas the cross-sectional method tends to magnify or overestimate the decline in intelligence with age, the longitudinal method tends to minimize or underestimate it. One reason is that some people drop out of a longitudinal study over time. Generally it is the more able, healthy, and intelligent subjects who remain available. Those who perform poorly on intelligence tests tend to be less available for longitudinal retesting. Consequently, the researchers are left with an increasingly smaller, biased sample as the subjects are retested at each later period.

Question

What are some reasons cited for the apparent overestimation of declines in intellectual functioning in old age?

Memory and Aging

Growing older is difficult on a personal level for many people. They might have trouble adjusting to changes in how they look or what they can do, and the limitations in daily life, even if minor, can cause them concern. One concern that appears to be paramount for nearly all older adults is memory. No other criterion is used more often to evaluate how we are doing. And for good reason! No other cognitive skill is as pervasive in everything we do, from remembering what we need at the grocery store to remembering how to get there (Cavanaugh, 1998b). Memory loss is also a preoccupation of American culture, reflected in the prevalence of cartoons and jokes about memory and aging. One of the common complaints of middle-aged and older adults is that their memory is "not as good as it used to be."

Samples of adults ranging in age from 40 to 80 suggest that anywhere from 43 percent to 79 percent of the respondents say they had experienced some deterioration of memory in the previous year (Aiken, 1998). Clearly, large percentages of middle-aged and older adults believe they have some difficulty with their memory (Lachman & James 1997). But with sufficient motivation, time, instruction, and a suitable environment, older adults can continue to expand their interests, abilities, and outcomes. According to Aiken (1998), these are some of the characteristics of older learners that need to be taken into account in any learning situation for them:

- Preference for a slower instructional pace
- Inclination to make more errors of omission due to cautiousness
- More disrupted by emotional arousal
- Less attentive
- Less willing to engage with material that is irrelevant to their own lives
- Less likely to use imagery

Older adults often say that the first sign of cognitive aging they noticed was difficulty remembering people's names. Recalling names seems to become harder as people age. But by the same token, the pool of names they know also becomes larger, so they might have difficulty remembering, not so much because their memories are not as good as they once were, but because they have more things stored in their memory and searching therefore takes longer.

Although the memory for names seems to decline regularly over a lifetime, vocabulary memory remains stable and can even increase slightly (Schaie,1994; Powell & Whitla, 1994). Moreover, a progressive loss of memory does not necessarily accompany advancing age (Jennings & Jacoby, 1993). Instead, some memory loss is found in an increasing *proportion* of older people with each advance in chronological age, but some elderly retain a sound memory regardless of age. Nor are all aspects of memory equally affected by aging (Hultsch, Hertzog & Dixon, 1990; Smith et al., 1990). For instance, age-related decreases are more severe for recall tasks than for recognition tasks, yet simple short-term or primary memory shows little decline until late adulthood (Cavanaugh, 1998b) (see Chapter 7). Short-term memory includes remembering such things as whether you took a prescribed pill this morning, whether you just went to the grocery store, and whether a friend just paid a visit. Older adults have more difficulty with this type of memory, particularly if they are taking medications. Younger adults are sometimes amazed by an older adult's long-term memory, such as remembering something from their youth 85 years ago, reciting poetry they learned in grade school, or remembering specific events and names from childhood.

Question

What types of memory is an older adult likely to retain, and what types of memory is an older adult most likely to have difficulty with?

Phases in Information Processing   When information is remembered, three things occur: (1) **encoding,** the process by which information is put into the memory system; (2) **storage,** the process by which information is retained in memory until it is needed; and (3) **retrieval,** the process by which information is regathered from memory when it is required. These components are assumed to operate sequentially. Incoming signals are transformed into a "state" (or "trace"). A trace is a set of information; it is the residue of an event that remains in memory after the event has vanished. When encoded, the trace is said to be placed in storage. To remember that stored information, the individual actively searches for the stored material.

Information processing has been likened to a filing system (Vander Zanden & Pace, 1984). Suppose you are a secretary and have the task of filing a company's correspondence. You have a letter from a customer criticizing a major product of your firm. Under what category are you going to file the letter? If the contents of the letter involve a defect in a product, will you decide to create a new category—"product defects"—or will you file the letter under the customer's name? The procedure you used for categorizing the letter must be used consistently for categorizing all other correspondence you receive. You cannot file this letter under "product defects" and the next letter like it under the customer's name and hope to have an efficient system.

Encoding involves perceiving information, abstracting from it one or more characteristics needed for classification, and creating corresponding memory traces for it. As in the case of the filing system, the way in which you encode information has an enormous impact on your ability to retrieve it. If you "file" an item of experience haphazardly, you will have difficulty recalling it. But encoding is not simply a passive process whereby you mechanically register environmental events on some sort of trace. Rather, in information processing you tend to abstract general ideas from material. Hence, you are likely to have a good retention of the meaning or gist of prose material but poor memory for the specific words.

Memory Failure   Memory failure can occur at any phase in information processing. For instance, difficulty can occur in the encoding phase. Returning to the example of the office filing system, you might receive a letter from a customer and accidentally place the letter with trash and discard it. In this case the letter is never encoded because it is not placed in the filing cabinet. It is unavailable because it was never stored. This difficulty is more likely to be experienced by older than by younger people. Older individuals are not as effective as younger ones are in carrying out the elaborate encoding of information that is essential to long-term retention. For instance, the elderly tend to organize new knowledge less well and less completely than they did when they were younger (Hess, Flannagan, & Tate, 1993; Hess & Slaughter, 1990). Thus, overall, older adults process information less effectively than younger adults (Verhaeghen, Marcoen, & Goossens, 1993).

Memory failure can also stem from storage problems. For instance, when filing, you might place the letter in the filing cabinet but by mistake put it in the wrong folder. The letter is available but it is not accessible because it was improperly stored. Apparently, this problem occurs more frequently with older adults than with younger adults (Earles & Coon, 1994; Mantyla, 1994). But other factors are also involved. **Decay theory** posits that forgetting is due to deterioration of the memory traces in the brain (Salthouse, 1991). The process is believed to resemble the gradual fading of a photograph over time or the progressive obliteration of the inscription of a tombstone. **Interference theory** says that retrieval of a cue becomes less effective as more and newer items come to be classed or categorized in terms of it (Kausler, Wiley, & Lieberwitz, 1992). For example, as you file more and more letters in the cabinet, more items compete for your attention, and your ability to find a letter is impaired by all the other folders and letters.

Faulty retrieval of knowledge is a third major cause of memory loss. Older persons can suffer a breakdown in the mechanisms and strategies by which stored information is recalled (Cavanaugh, 1998b). John C. Cavanaugh (1998b) indicates that attention is an important aspect of information processing. The ability to focus on what we need to do (selective attention), perform more than one task at the same time (divided attention), and sustain attention in order to accomplish long-term tasks all can result in cue overload—a state of being overwhelmed or engulfed by excessive stimuli—and failure to process retrieval information effectively. (For instance, you may file a letter under a customer's name but later lack the proper cue to activate the category under which you filed it).

Some researchers suggest that the elderly might be subject to greater inertia or failure of a "selector mechanism" to differentiate between appropriate and inappropriate sets of responses (Allen et al., 1992). Also, retrieval time becomes longer with advancing age (Cavanaugh, 1998b; Salthouse & Babcock, 1991). Overall, older adults have more difficulty with memory than younger adults do. This fact has practical implications. Older people are more likely to be plagued by doubts as to whether or not they carried out particular activities—"Did I mail that letter this morning?" "Did I close the window earlier this evening?" (Kausler & Hakami, 1983). And they are more likely to have difficulty remembering where they placed an item or where buildings are geographically located (Pezdek, 1983).

Question

What are some of the theories associated with memory difficulties in old age?

Learning and Aging

Psychologists are finding that the distinctions they once made between learning and memory are becoming blurred. Learning parallels the encoding process whereby individuals put into memory material that is presented to them. Psychologist Endel Tulving (1968) says that learning constitutes an improvement in retention. Hence, he contends, the study of learning is the study of memory. Clearly, all processes of memory have

Aging and Learning Older people continue to learn throughout life, but they might need a little more time than a younger adult might need. They are less motivated to learn materials that are irrelevant to their lives. Engaging in mental tasks helps prevent memory decline.

consequences for learning. If people do not learn (encode) well, they have little to recall; if their memory is poor, they show few signs of having learned much. Not surprisingly, therefore, psychologists find that younger adults do better than older adults on various learning tasks (Crook, Larrabee, & Youngjohn, 1993). This fact has given rise to the old adage "You can't teach an old dog new tricks." But this adage is clearly false. Both older dogs and older human beings can and do learn. They would be incapable of adapting to their environment and coping with new circumstances if they did not.

Research suggests that both younger and older individuals benefit when they are given more time to inspect a task. Allowing people ample time gives them more opportunity to rehearse a response and establish a linkage between events, and it increases the probability that they will encode the information in a fashion that facilitates later search and recall. Older adults benefit even more than younger ones when more time is made available for them to learn something.

Older people often give the impression that they have learned less than younger people have because they tend to be more reluctant to venture a response. At times, the elderly do not provide learned responses, especially at a rapid pace, although they can be induced to do so under appropriate incentive conditions. And when tested in a laboratory setting, older adults seem to be less motivated to learn arbitrary materials that appear to be irrelevant and useless to them. Complicating matters, today's young adults are better educated than their older counterparts. Furthermore, another hidden bias is that many elderly individuals take medications that can diminish mental functions. All these factors suggest that we should exercise caution when appraising the learning potential of the elderly, lest we prematurely conclude that they are incapable of learning new things.

Alzheimer's Disease

My wife refused to believe I was her husband. Every day we went through the same routine: I would tell her we had been married for thirty years, that we had four children. She listened, but she still thought she lived in her hometown with her parents. Every night when I got into bed she'd say, "Who are you?"

Husband of an Alzheimer's patient

In disclosing that he had developed Alzheimer's disease, former President Ronald Reagan said he wanted to make more Americans aware of this illness, which has eluded a medical cure. Until recently, most everyone, including physicians, accepted the view that senility is the penalty people pay for living longer than the Biblical three score and ten years. **Senility** is typically characterized by progressive mental deterioration, memory loss, and disorientation regarding time and place. Irritability and other personality changes usually accompany the intellectual decline (Cohen et al., 1993).

In persons over 65, about 20 to 25 percent of all senility results from **multiinfarcts** (better known as "little strokes"), each of which destroys a small area of brain tissue. Another 50 percent is due to **Alzheimer's disease**—a progressive, degenerative disorder that involves deterioration of brain cells. Autopsies of victims show microscopic changes in brain structure, especially in the cerebral cortex. Areas involved in cognition, memory, and emotion are riddled with masses of proteins called "plaques" and tangles of nerve cells. Some of the nerve cells look like infinitesimal bits of braided yarn. Apparently, the clumps of degenerating nerve cells disrupt the passage of electrochemical signals across the brain and nervous system. One hallmark of the disease seems to be a breakdown of the system that produces the neurotransmitter *acetylcholine*.

Alzheimer's disease affects some 4 million Americans. Although an estimated 3 percent of the elderly between 65 and 74 years of age suffer from the disease, the figure rises to 18.7 percent for those aged 75 to 84 and mushrooms to 47.2 percent for those 85 and older. The National Institutes of Health estimates that 60 percent of nursing-home patients over age 65 suffer from the disease (U.S. Department of Health and Human Services, 1997a). It is also the fourth leading cause of death in the United States.

Unfortunately, as yet there is no cure for Alzheimer's disease. But new scientific findings offer hope in the midst of despair. Pharmaceutical companies are testing more than 100 compounds that might relieve or delay the symptoms of the disease. The drug tacrine (also known as THA or Cognex) can alleviate some the cognitive symptoms during the early and middle stages of the disease. Other medications can help control behavioral symptoms such as sleeplessness, agitation, wandering, anxiety, and depression.

The disorder has a devastating impact not only on its victims but on their relatives as well. Many family members complain bitterly about the lack of skilled nursing facilities for Alzheimer's patients, the failure of government insurance programs to pay for needed care during the prolonged period of deterioration, the absence of counselors and programs to assist families in coping with the demands of patients, and the ignorant, indifferent, and even callous attitudes of many physicians (Lyman, 1993; Kolata, 1991a). Not surprisingly, family members caring for Alzheimer's patients are at especially high risk for depressive disorders (Chesla, Martinson, & Muwaswes, 1994). Families find their loved one progressively regressing, eventually unable to perform the simplest tasks. One woman, Marion Roach (1983, p. 22), told of her experiences with her 54-year-old mother, who suffered from the disease:

Caring for Victims of Alzheimer's Disease
Although experts suggest a variety of guidelines to assist those caring for loved ones suffering from Alzheimer's disease, the task nonetheless remains exceedingly taxing and stressful. More nursing homes today have added special facilities to accommodate the needs of older adults with this disease.

In the autumn of 1979, my mother killed the cats. We had seven; one morning, she grabbed four, took them to the vet and had them put to sleep. She said she didn't want to feed them anymore. . . . Day by day, she became more disoriented. She would seem surprised at her surroundings, as if she had just appeared there. She stopped cooking and had difficulty remembering the simplest things. . . . Until she recently began to take sedation, she would hallucinate that the television or the toaster was in flames. She repeats the same few questions and stories over and over again, unable to remember that she has just done so a few moments before.

The disease typically proceeds through a number of phases (American Psychiatric Association, 1994). At first, in the "forgetfulness phase," individuals forget where things are placed and have difficulty recalling events of the recent past. Later, in the "confusional phase," difficulties in cognitive functioning worsen and can no longer be overlooked. Finally, in the "dementia phase," individuals become severely disoriented. They are likely to confuse a spouse or a close friend with another person. Behavior problems surface: Victims may wander off, roam the house at night, engage in bizarre actions, hallucinate, and exhibit "rage reactions" of verbal and even physical abuse. In time, they become incontinent and unable to feed or otherwise care for themselves. Victims show a marked decrease in life expectancy in comparison with age-matched men and women. Patients usually die of infections, often from pneumonia. The course of the illness varies enormously with different individuals—from under 3 years to over 20 years before death ensues.

Alzheimer's researchers resemble the blind men studying the elephant: Each grabs onto a different part of the disease and comes to a different conclusion as to its causes. One popular hypothesis relates Alzheimer's disease to increased levels of a toxic brain chemical—*beta amyloid protein*—that is believed to result from some biochemical blunder. The beta amyloid protein is a major component of the plaques found on the nerve endings of the brain cells in patients suffering from Alzheimer's (Marx, 1993; Hardy & Higgins, 1992). Other researchers believe beta amyloid is a side effect of some other damage.

Another hypothesis looks to a link between zinc and the disease (zinc ions can cause certain brain proteins to convert into an insoluble form that in turn accumulates into plaque clumps) (Kaiser, 1994). Still another hypothesis postulates the existence of a defect in the immune system of Alzheimer's victims. And another hypothesis relates Alzheimer's disease to a puzzling infectious agent known as a "slow virus." Such brain disorders as kuru and Creutzfeldt-Jakob disease are caused by slow viruses and are accompanied by distinctive brain lesions or plaques that bear a close resemblance to those that characterize Alzheimer's disease. Kuru, a disease occurring in New Guinea and once believed to be of hereditary origin, is a slow-acting viral infection transmitted from person to person by ritual cannibalism. Creutzfeldt-Jakob disease might simply be a form of pre-senile dementia that strikes at an earlier age than Alzheimer's disease. Medical researchers find that the brains of patients suffering from "slow virus" diseases have a huge deficit in a key enzyme, choline acetyltransferase, a substance used in the manufacture of material employed by the brain to transmit nerve signals from cell to cell (Price et al., 1991).

Because Alzheimer's disease tends to run in families, some investigators have looked to a genetic source of the disease. In fact the strongest risk factor identified is having at least one first-degree relative with symptoms of dementia. A family history of Parkinson's disease, which also causes dementia, was found 3½ times more frequently in the families of patients with Alzheimer's disease (Polymeropoulos & Golbe, 1997). Duke University scientists have found a connection between a gene called apolipoprotein E and the risk of contracting Alzheimer's; the gene produces proteins that are necessary to shuttle cholesterol through the bloodstream and that are thought to be essential in protecting brain cells from collapse (Travis, 1993). Additional evidence of the importance of genetic or chromosomal factors is found in studies of Down syndrome showing that many adults with the syndrome eventually succumb to Alzheimer's lesions (National Down Syndrome Society, 1998; Blakeslee, 1994). Still other researchers have located on chromosome 14 a genetic defect linked to an inherited form of Alzheimer's that develops unusually early at age 45 (Marx, 1992). It might also be true that Alzheimer's disease can spring from a variety of causes and that it is a group of closely associated disorders rather than a single illness (Stipp, 1990).

Senility, a lack of consistency and a deterioration in cognitive functioning, is one of the most serious conditions that a physician can diagnose in a patient. The prognosis is grim, and the effectiveness of current treatments is uncertain. Consequently, it is incumbent upon professionals who treat the elderly to do a full battery of tests to make certain a treatable cause for a patient's symptoms has not been overlooked. Often, underlying physical diseases that can make an elderly person seem senile go unnoticed and untreated. Such individuals are simply dumped into the wastebasket category of "senile" by families and physicians who have accepted the conventional wisdom that senility is an inevitable part of the aging process. Common problems often mistakenly diagnosed as senility include tumors; vitamin deficiencies (especially B_{12} or folic acid); anemia; depression; such metabolic disorders as hyperthyroidism and chronic liver or kidney failure; and toxic reactions to prescription or over-the-counter drugs (including tranquilizers, anticoagulants, and medications for heart problems and high blood pressure). Many of these conditions can be reversed if they are identified and treated early in the course of the illness.

Questions

How does Alzheimer's disease affect the person's brain functioning? What are its effects on a person's quality of life? What does it mean when a person is classified as "senile"?

Moral Development

Morals, values, and beliefs are often transmitted through formal religious organizations and have varying influences on people's lives. The purpose of organized religions—such as Christianity, Judaism, Islam, Buddhism, Hinduism, Confucianism, and Taoism—is to give spiritual meaning to life. These values, beliefs, and rituals bring people together, provide a source of strength and hope in difficult times, and provide a source of continuity throughout life.

One of the continuous projects facing us is the need to make sense of our lives and, above all, to give our lives meaning. If you think back over the major theorists we have reviewed throughout this book, each one is concerned with explaining how our actions make sense as we understand the meaning-making system within which they occur. Let us now examine a developmental theory whose locus is *faith*. We introduce it at this time because it is well documented that the elderly are more likely to be involved in religion and religious activities than are the young (Payne & McFadden, 1994).

Religion and Faith

As people age and their support systems are reduced through loss (deaths of their spouse, children, siblings, friends, neighbors, and acquaintances) or movement to a new location (nursing home, living with adult children, or foster care), the elderly might be particularly comforted by the meaningfulness afforded through faith in something outside of the self. Many elderly find strength, solace, and meaning in attending religious services.

James Fowler's Theory of Faith Development   James Fowler, an ordained minister and professor, has attempted to combine theology and developmental psychology to explain the different trajectories that faith takes in individuals. A system of faith continually changes in response to cognitive development, maturation, and both religious and life experiences. Although Fowler posits seven stages of faith development, we will examine only Stage 5 in detail, as this stage is most likely to be reached after midlife. The stages are described in Table 17-2.

As we progress through the different stages of faith, we begin and continue to ask the fundamental questions regarding existence: What is life all about? Who is the ultimate authority over my life? How do I give meaning to my life? What is the purpose of my life? Is my life really over when I die? The answers to these questions revolve around what Fowler calls our *Master Story*, which contains the crucial ideas we draw upon to give meaning to our lives. The Master Story will become increasingly ex-

Table 17-2   Fowler's Seven Stages of Faith Development

Stage 0	Primal Faith (birth to age 2)	Really more like a basic trust in caretakers.
Stage 1	Intuitive-Projective Faith (2–8 years)	Beginning to understand cause-and-effect relationships, which are given meaning through the stories and images one gets from caretakers.
Stage 2	Mythic-Literal Faith (childhood and beyond)	Separation of fantasy and real world, but meaning is carried and restricted by the narrative it occurs in.
Stage 3	Synthetic-Conventional Faith (adolescence and beyond)	Belief that everyone has basically the same received beliefs; e.g., that all Catholics understand their religion in the same way. Not yet a personal belief.
Stage 4	Individuative-Reflective Faith (young adulthood and beyond)	A personal faith that can be reflected upon.
Stage 5	Conjunctive Faith and the Interindividual Self (midlife and beyond)	Seeks truth of faith in a dialectical, multidimensional manner. (Few people get to this stage.)
Stage 6	Universalizing Faith	Committed to selfless, universal goals. Gandhi and Mother Teresa are examples.

"Seven Stages of Faith Development" from STAGES OF FAITH: THE PSYCHOLOGY OF HUMAN DEVELOPMENT AND THE QUEST FOR MEANING by James W. Fowler. Copyright © 1981 by James W. Fowler. Reprinted by permission of HarperCollins Publishers, Inc.

plicit as we mature, and in many instances can be summed in a few words, such as "It's all about me," or might be in the form of a communal ritual such as the Passover Haggadah for the devout Jewish community.

Fowler's seven stages are usually called "soft stages" because they do not involve the strict adherence to logical structures associated with a "hard stage" theory like Piaget's. Furthermore, transitions between stages are usually experienced during developmental milestones such as adolescence, young adulthood, midlife, and old age.

For individuals in mid and late adulthood (who typically are at Fowler's Stage 5), life's paradoxes are numerous and not easily brushed aside. The possibility for appreciating opposites, polarities, and diverse perspectives due to having experienced these aspects of life firsthand contributes to a higher level of meaning-making for some of those in their "winter" years. Older adults understand that all narratives point to the existence of a "grand narrative," which in turn suggests that every religious perspective is a "vehicle for grasping truth" (Fowler, 1983). Put another way, the individual in Stage 5 realizes that religions are much more similar than they are different—they all have the same seed of truth at the core. The moral aspect of faith becomes apparent if we

assume, as Fowler does, that we invest faith in an aggregate of ideas, values, attitudes, and behaviors that are significant to us.

Those at the higher stages of faith development will be more likely to invest in goals focused more on others and less on self, much like Gandhi. Fowler worked with Kohlberg for a time, and at one point Kohlberg entertained a seventh stage for his model that reflected an element of faith. Additionally, in 1997 Joan Erikson, at 93 years old, published the last of Erik Erikson's works. Before Erik died (in his early nineties), they collaborated on an update of *The Life Cycle Completed* in which they incorporated a ninth psychosocial stage called "Very Old Age." Their ninth stage outlines the critical role that hope and faith play in the lives of those in their eighties, nineties, and beyond for valuing wisdom.

Questions

Why do elderly adults become more interested in religious and spiritual observances? How would you explain Fowler's fifth stage of faith development?

Seque

In many societies around the world, the "oldest old" are a growing population of adults who are living longer with a higher quality of life than past generations had. Philosophical concerns are being raised about extending life and the allocation of resources

for the elderly. Our understanding of biological aging continues to grow, and we now know that longevity is generally associated with healthy genes, a lifestyle of regular exercise, sensible health habits, regular medical checkups, and a network of social support.

Recent research on longevity and aging has led to a medical field of anti-aging.

Concerning intellectual functioning, those elderly who remain engaged in mental activities are less likely to experience declines in cognitive functioning, though certain types of memory decline before other types. Though it can take a longer for older people to remember recent events, they often are a wealth of information about the past. Additionally, older adults are typically concerned about staying healthy, remaining independent and not dependent on others, and not being left alone in their last years. Eventually, most cope with continual loss—loss of friends and loved ones, loss of health, loss of one's lifelong home, and loss of independence—which can lead to depression and mental illness. For coping with these losses, many find a sense of peace by having or developing a faith in some type of spiritual life after this physical one. And we shall see in Chapter 18, a network of social relationships through family support, social contacts, and adult day care can do a great deal to mediate these losses—improving quality of life.

Summary

Aging: Myth and Reality

1. At no point in life do people stop being themselves and suddenly turn into "old people." Aging does not destroy the continuity between what we have been, what we are, and what we will be.

2. There are a great many myths about aging that have little to do with the actual process of growing old. Included among these myths are those that portray a large proportion of the elderly as abandoned, institutionalized, incapacitated, in serious financial straits, and living in fear of crime.

3. The gap between the life expectancy rates of men and women has been increasing since 1920. On the average, women live several years longer than men. Women seem to be more durable organisms because of an inherent sex-linked resistance to some types of life-threatening disease. Lifestyle differences also contribute to gender differences in life expectancies. A major factor is the higher incidence of smoking among men.

Health

4. Despite the higher incidence of chronic health problems among the elderly, most elderly do not consider themselves to be seriously handicapped in pursuing their ordinary activities. Most of the conditions that create chronic disease increase with advanced age. Some of the health problems experienced by older Americans are the product of side effects associated with medication.

5. Some of the most obvious changes associated with aging are related to an individual's physical characteristics. The hair grows thinner, the skin changes texture, some of the bulk built up during earlier adulthood begins to decrease, and some individuals experience a slight loss in stature. Sensory abilities also decline with age. Aging is likewise accompanied by various physiological changes. One of the most obvious is a decline in the individual's capacity for physical work and exercise. Sleep patterns also change.

6. The incidence of self-reported acute illnesses (upper respiratory infections, injuries, digestive disorders, and the like) is lower among the elderly than among other segments of the population. However, the incidence of chronic diseases (heart conditions, cancer, arthritis, diabetes, varicose veins, and so on) rises steadily with advancing years.

7. The elderly experience muscle loss, decreases in both size and strength of muscles, a decline in oxygen intake, and reduced heart efficiency.

8. Sensory abilities, such as hearing, sight, taste, touch, and smell, decline with age.

9. Touch and temperature sensitivity also decrease. Some elderly individuals might fail to notice a temperature drop of up to 9 degrees Fahrenheit. Consequently, older people tend to be susceptible to hypothermia—a fall in body temperature of more than 4 degrees Fahrenheit that persists for a number of hours and potentially fatal.

10. More than 50 percent of persons aged 65 and older report regular problems with sleep. Older people have less Stage 3 and Stage 4 sleep (deep sleep) and less rapid-eye-movement (REM) sleep (dream sleep). Sleep apnea, a disorder in which the person occasionally stops breathing during sleep, increases with age.

11. Some factors leading to decreased sexual activity are difficulty with sexual arousal, the overall health of older adults, and the availability of a partner.

12. Many theories seek to explain the biological process of aging by focusing on cell-destroying mechanisms. Many of the mechanisms overlap. Although the effects of aging are often confounded

with the effects of disease, aging is not the same thing as disease.

Cognitive Functioning

13. A decline in adult intelligence becomes more evident after age 60, mainly with slowing of response time and declines in short-term memory. However, different abilities follow quite different courses in aging individuals. Aspects of intelligence that are measured by tests of performance and fluid ability appear to be the most affected by aging.

14. Psychologists have traditionally taken too negative a view of the impact of aging on intellectual functioning, partly because researchers have relied too heavily upon cross-sectional studies.

15. Memory is often affected by aging. But to assume that a progressive loss of memory necessarily accompanies advancing age is incorrect. Memory loss among the elderly has many causes, related to the acquisition, retention, and retrieval of knowledge.

16. Senility is typically characterized by progressive mental deterioration, memory loss, and disorientation regarding time and place. In persons over 65, about 20 to 25 percent of all senility results from multiinfarcts. Another 50 percent is due to Alzheimer's disease—a progressive, degenerative disorder that involves deterioration of brain cells. The disorder has a devastating impact not only on its victims but also on their relatives.

Moral Development

17. Morals, values, and beliefs are often transmitted to members of society through formal religious organizations. As people age and their support systems (spouse, children, friends, acquaintances) are reduced through loss or movement to a new location (nursing home, living with adult children, foster care), the elderly may be particularly comforted by the meaningfulness afforded through faith in something outside of the self.

18. James Fowler has attempted to combine theology and developmental psychology to explain the different trajectories that faith takes in individuals throughout the life course. His theory involves a seven-stage model of faith development that covers religious as well as nonreligious perspectives. Stages 5 and 6 are predominant in the late adulthood stage of life.

Key Terms

ageism *(524)*	gerontology *(524)*	retrieval *(544)*
Alzheimer's disease *(546)*	geropsychology *(524)*	senescence *(526)*
collagen *(536)*	hypothermia *(538)*	senility *(546)*
death drop *(543)*	interference theory *(545)*	sleep apnea *(538)*
decay theory *(545)*	multiinfarct *(546)*	storage *(544)*
encoding *(544)*	osteoporosis *(534)*	

Following Up on the Internet

Web sites for this chapter focus on the changing health needs, declining cognitive functioning, and growing spiritual needs of those in late adulthood. Please access the text web site at http://www.mhhe.com/crandell7 for up-to-date hot-linked Internet addresses for the following organizations, topics, and resources:

Administration on Aging

The Women's Health and Aging Study

NIH Resource Directory for Older People

International Year of Older Persons

Web Resources on Gerontology

Spreading Eternal Youth—Extensive links

Chapter 18

Late Adulthood

Emotional and Social Development

Critical Thinking Questions

1. Do you think that a 75-year-old and a 25-year-old could fall in love with each other and have a high-quality life together? Why or why not?

2. If society painted an exciting or respectful picture of old age, do you think most elderly individuals would view their own lives differently? How does the media's portrayal of old age affect people's self-perceptions?

3. Nursing homes exist largely because most adults work and cannot take care of those in need on a full-time basis. If the time comes for you to help your elderly parent(s), do you think you would quit your job to support those who sacrificed for you? Why or why not?

4. If, starting tomorrow, you could no longer work, attend school, or have any family or close friends, and you did not know how many more years you had to live—what would you do or think about? How do you think you might feel about living a life "in limbo," away from the mainstream of life?

As Americans make the transition into the new century, we are experiencing a paradigm shift from the Industrial Age to the Information Age, which is significantly impacting individuals, societies, and nations. Advances in medicine, technology, education, the social sciences, and human services allow more and more people to reach old age with fewer ailments and to live a higher-quality life. Predictions about the increasing number of elderly show that national programs will be stretched to their limit in the future and that families will need to learn how to cope with caring for aging relatives (Kaetz, 1998).

Significantly, the number of American minority elderly is increasing more rapidly, but many of these elderly hold positions of respect within their ethnic communities and are taken care of by family members. Only about 6 percent of elderly Americans are placed in institutional care facilities to live out their later years if they develop a serious physical or mental health condition, such as Alzheimer's disease. Today home health care and alternative living arrangements are available for elderly persons who need special assistance to maintain independent living and for persons who are caring for an elderly relative in their home. According to the social model of living, the elderly will have a higher quality of life if they can maintain independence, control, and social engagement.

Social Responses to Aging

Our national denial and denigration of age has prevented us from viewing it as a new period of human life.

Betty Friedan, Winter 1994

It has become increasingly clear that social and behavioral factors influence the quality of aging, as well as the physical and health factors discussed in Chapter 17. In the United States in 1992, nearly 70 behavioral science organizations and federal agencies formulated a national research agenda called The Human Capital Initiative to address the serious challenges and opportunities that increasing numbers of diverse elderly pose to our society. In 1993, these organizations compiled the *Vitality for Life* proposal, which called for research in four major aspects of aging:

1. Behavior change to prevent damage to body systems and maintain health
2. Psychological health of the oldest old
3. Maximizing and maintaining productivity
4. Assessing mental health and treating mental disorders

Research in these areas can improve our understanding of the needs of this large group of adults and enhance their quality of life. Another significant factor that is improving life for the elderly is the current celebration of cultural differences that has replaced traditional pressures for immigrants to assimilate to an "American" way of life. A greater respect and caring for the elderly of all ethnic and cultural backgrounds is surfacing in American society.

Some American elderly are experiencing greater satisfaction in their later years. "Industrialized societies have demonstrated remarkable efficacy and flexibility in extending longevity for many people and in providing the economic and social resources for them to lead a more satisfying life in old age" (Baltes & Baltes, 1998). Many who are living longer are also staying physically active, mentally engaged, and socially involved longer. Even now, the sight of vigorous people in their eighth and ninth decades is not unusual. Given present trends in mortality, a larger segment of the population is expected to live into their eighties, with a high rate of mortality in their mid eighties. On the whole, people find that adapting to aging is simply a continuation of the lifelong process of coping with life (Rowe & Kahn, 1987). Recent research results from *The Berlin Study on Aging* (Baltes & Baltes, 1998) show that the oldest old do typically experience declines in health, intellectual functioning, mobility, and social contacts. This does not mean, though,

that they are not coping effectively with their late-life circumstances.

Much of the existing research up through the mid 1990s on adulthood focused on the physical and cognitive development of persons aged 20 to 60. A few research efforts have focused on the cognitive functioning of the "youngest old," those aged 60 to 75. There is scant research on those 80 and older, particularly in the areas of emotional and social functioning. However, worldwide there is a flurry of research getting under way.

Because we know that physical and intellectual declines are common in the elderly, can we predict a constricted range of emotions in elderly adults? If so, what emotional states can we expect the elderly to experience? If not, what life circumstances help some elderly maintain a rich range of emotions, especially those that promote constructive coping? Some studies indicate that elderly adults typically experience a "flatness" of emotional states along with more limited physical, cognitive, and social functioning (Baltes & Baltes, 1998; Myers & Diener, 1995). However, researchers in Stockholm, Sweden, as part of the Kungsholmen Project, recently reported findings of both positive and negative affect (emotional components of subjective well-being) in 105 people, 90 to 99 years of age, who were not cognitively impaired (Hilleras et al., 1998). *Positive affect* was defined as the extent to which a person feels active, alert, and enthusiastic. *Negative affect* was defined as the extent to which a person feels guilt, anger, and fear.

False Stereotypes

With our culture's preoccupation with youthfulness, many Americans prefer to ignore late adulthood or have distorted notions about it. Most likely, though, you have heard of "the graying of America," or "the golden years," or "the silent revolution." But specifically to whom do these phrases refer? Gerontologists tend to segment late adulthood into the periods of "young old" (65 to 75 or 80) and "old old" (80 and above) (Baltes & Smith, 1997). Historical gerontologist and chair of the National Council on Aging, W. Andrew Achenbaum (1998) reminds us why we historically have held more negative views of old adulthood:

1. Old age has always been considered to be the last stage of existence before death, and no one wants to be reminded about mortality.
2. Old age is undefined: There are few rites of passage comparable to those celebrated in youth. Not all elderly are married to celebrate golden anniversaries, nor are all grandparents, nor do all elderly retire.
3. There is a growing, diverse composition of elders with varied physical, cognitive, behavioral, and

socioeconomic characteristics. Discussion about a "typical" older adult is difficult, yet many are dependent upon government entitlements and family support.

These stereotypic views weigh heavily on older adults' self-esteem: A large portion of our elderly suffer from illnesses associated with physiological aging and are likely to be overmedicated. Many suffer from depression, and suicide rates are highest among elderly males. With emerging empirical research on our elderly population, though, we are finding that growing into late adulthood is not necessarily the unhealthy burden that it was thought to be. Recall that worldwide the fastest-growing segment of the population is older adults—more specifically, those over 80. Overall, by the year 2025, it is estimated that nearly 20 percent of the U.S population will be over age 65, compared to the current 12 percent. The projected growth of the population aged 80 and older, of whom the majority are women, is even more dramatic, with the size of this group increasing at a much higher rate than any other segment of U.S. society (*Vitality for Life,* 1993). Achenbaum (1998) predicts that female and Hispanic attitudes about age and aging will become a larger determining factor in overall American perceptions.

Questions

Why has little empirical research been done with the elderly? Why are behavioral scientists studying more issues in late adulthood now?

Positive and Negative Attitudes

The Indian Summer of Life should be a little sunny and a little sad, like the season, and infinite in wealth and depth of tone.

Henry Brooks Adams, *The Education of Henry Adams*

A common stereotype is that adults in old age are unhappy, but some research disputes this myth. Latten (1989) found that older adults tend to report higher levels of life satisfaction than younger adults do. Participants recognized that there are different life demands than in younger adulthood but that old age has its own demands and rewards. *Positive affect* has also been found to be related to favorable life events, positive health status and functional ability, availability of social contacts, and higher levels of educational attainment (Hilleras et al., 1998). However, correlates of positive and negative affect in the very elderly have rarely been researched.

Using the Positive and Negative Affect Schedule (PANAS), Hilleras and colleagues (1998) surveyed their elderly subjects to examine whether similar patterns of positive and negative affect were correlated with personality traits. Factors associated with affect in this study were grouped into categories of personality, social relationship, subjective health, activities, life events, religiousness, and sociodemographic variables.

Factors associated with positive affect included:

- Social relationships, such as contact with friends, living with others, attending church, and participating in clubs/society/organizations
- Reading and following news
- An extroverted personality
- Death of a close friend
- Definite beliefs and definite disbeliefs
- Living with other persons

Factors associated with negative affect included:

- Neuroticism
- Own major illness
- Money problems
- Living alone

The Stockholm study also demonstrated that positive affect and negative affect were not correlated with each other. That is, the very elderly who scored high on positive affect (active, alert, and enthusiastic) did not necessarily score low on negative affect (guilt, anger, and fear). This means these elderly subjects responded with a full range of emotions on this scale, not unlike younger adults. One observation, though, was that subjects took a long time to answer each question during the interview process. Findings from this study also underscore that the emotional health of the "oldest old" is quite diverse and that further research is needed on the affective health of this population of adults.

Over the past few decades, researchers in geropsychology and gerontology have posed several theories about personality development and adjustment for those living into the highest ranges of adulthood.

Self-Concept and Personality Development

What might seem unusual today will be usual tomorrow, as our population continues to move to the South and the West, grows older, and becomes more diverse.

Paul Campbell, U.S. Bureau of the Census, December 1996

Recall from Chapter 17 that genetics, biochemistry, and lifestyle factors are variables that play a large role in physical health and longevity. In this chapter, we continue to examine research findings from social and emotional perspectives: Is there such a thing as an ideal personality type that promotes a high quality of life and longevity? Is there continuity in personality across the life span, or are there dramatic changes along the way?

Psychosocial Theories

Unfortunately, many of the theories proposed on social-emotional development in the earlier years of life never extended to adulthood, especially not to late adulthood. However, as societies around the world are experiencing the demographic shift to an older population, many gerontologists, geropsychologists, and sociologists are studying the concepts of "successful aging," "satisfaction with aging," "productive aging," and "optimal aging."

Personality Change and Stability Our personalities are complex and multifaceted. Some of the components change markedly over time, while others show considerable continuity. This photo is of a retired teacher, 80 years old. The portrait in the background shows her as a young adult.

There are several proposed psychosocial theories of late adulthood at this time.

Erikson: Integrity Versus Despair   A recurrent theme in Erikson's work is that psychosocial development occurs in stages across the entire life span. According to him, the elderly confront the issue of **integrity versus despair.** In this eighth stage, Erikson says individuals recognize that they are reaching the end of life. Provided they have successfully navigated the previous stages of development, they are capable of facing their later years with optimism and enthusiasm. With retrospection, they can take satisfaction in having led an active, full, and complete life. This recognition produces contentment and compensates for decreased physical potency and performance. They find a new unification in their personality, producing a sense of *integrity.*

In late adulthood, Erikson himself demonstrated a life of *ego integrity.* At age 87 Erikson published *Vital Involvements in Old Age,* reporting the results of his personal interviews with American men and women in their eighties and nineties. His last works examined why some elderly live hopeful, productive lives, despite failing health and alertness, and why others, though relatively robust, give in to loneliness, narcissism, and despair (Woodward, 1994). At 92 he was formulating and personally experiencing a last stage of personality development in which each individual confronts his life in relation to existence itself. In 1997, Joan Erikson (in her nineties) published an update of Erikson's *The Life Cycle Completed,* incorporating a ninth stage of psychosocial development. She writes of the challenges faced by elderly adults facing lost autonomy and more limited life choices. At the end of his life, Erikson viewed hope and faith as playing critical roles in developing *wisdom* in this life stage.

Those who appraise their lives as having been wasted, believing that they missed opportunities they should have taken years ago (a different career path, marrying or marrying a different person, retiring later or sooner, having children) experience a sense of despair. They realize that time is running out and that it is too late to make up for past mistakes. They view their lives with a feeling of disappointment, loss, and purposelessness. Consequently, Erikson says, they approach death with regrets and fear.

Some evidence also suggests that aspects of personality change among the elderly are developmental. The Berkeley Studies, a longitudinal study of adults begun in 1928, suggest a decline in extraversion in old age (Field & Millsap, 1991). The decline might reflect a developmental increase in what has variously been termed "inner-directedness" and "interiority" (Field, 1991; Neugarten, 1973). Erikson's notion of wisdom as "a detached concern with life itself, in the face of death itself" suggests a

Information You Can Use

Reminiscence: Conducting a Life Review

For a person in late adulthood, conducting a life review is a significant experience. A **life review** is a reminiscence and sharing of family history from one generation to another. This is especially difficult today as more families are geographically separated from one another or have such busy schedules they see each other infrequently. The older person gains feelings of self-worth, continuity in personality, and happiness about preserving family history. During this connection among the generations, individuals discover interesting life experiences and memories about each other, too. "Life reviews affirm the importance of life experiences and achievements and, for some individuals, give new meaning to life" (Moyer & Oliveri, 1996). As Erikson has indicated, adults in this stage of life strive to integrate past psychological themes into a new level of meaning, and they have an individual and wiser perspective on the world and life that is different from the perspective they had when they were younger. A life review allows the older adult to take a look at her or his lifetime as a whole and promotes a sense of integrity.

Butler (1971) described a life review as a normative process engaged in by anyone, regardless of age, when they are aware that they are approaching death. The person looks back at his or her entire life, tries to resolve issues that were unresolved, tries to accept negative experiences, celebrates the good experiences, and gains a sense of accomplishment and closure with life.

Sometimes natural reminiscence occurs at family reunions or funerals. However, as people age and significant people in their life move, become disabled, or die, they have fewer contacts that promote natural reminiscence. Reminiscence can be incorporated into private or group settings and sometimes in therapeutic settings. John Kunz (1991) suggests that practitioners who conduct intake interviews at nursing homes, group homes, or adult day-care facilities should be respectful of their elderly interviewees and should expect the interview to take some time, because an elderly person can have more history to tell, and relating that history can be therapeutic. When reminiscence is done in a group, such as

an adult day-care setting, those of the same age often benefit from each other, as long as no one or two people dominate the reminiscence.

Taking the time to listen to others helps them know they are important. Here are some suggestions that might spark a reminiscence with someone you love or know.

Items that prompt sharing

- Photographs (from childhood, or family reunions, birthdays, anniversaries)
- Family scrapbooks, journals, books
- Newspaper clippings and old magazines
- Mementos from life accomplishments (trophies, badges, ribbons, plaques, etc.)
- Special personal belongings (keepsakes, jewelry, etc.)

Some Good Communication Skills

- Be an active listener
- Maintain eye contact
- Ask questions that encourage the person to explain or expand on things that seem important
- Watch nonverbal cues for posture, eye contact, and expressions (comfort or discomfort)
- Accept what is said, and allow the person to continue talking without jumping in with your own life experiences
- Realize there may be times of silence or tears (this is normal; allow the person time to regather thoughts and continue)

The sharer of the life review can help record it. There are many ways to preserve a life review, including preparing a scrapbook, an audiocassette or videotape, a newspaper article, a letter, or a book for family members or as a contribution to a local museum, historical society, or library. These are forms of remembrance of this special person. The American Association of Retired Persons has an interactive bulletin board on the Internet called "Through the Years," where people can tell the world about their memories of the twentieth century and be a part of history.

Source: Adapted from John Kunz, *Communicating with Older Adults,* 1996, and John S. Kunz, Reminiscence Approaches Utilized in Counseling Older Adults, in *Illness, Crisis, and Loss,* 1991, 48–54.

similar inner-directedness (Erikson, Erikson, & Kivnick, 1986). Indeed, some psychologists have wondered whether it is possible to attain Erikson's stage of *ego integrity* without the inward focus associated with interiority (Ryff, 1982).

Question

According to Erikson, what psychosocial tasks confront those in late adulthood?

Peck's Psychosocial Tasks of Later Adulthood   Psychologist Robert C. Peck (1968) provides a somewhat related view of personality development during the later years that is more focused than Erikson's. Peck says that old age confronts men and women with three challenges, or tasks, which we will examine here.

Ego Differentiation Versus Work-Role Preoccupation
The central issue here is presented by retirement from the workforce. Men and women must redefine their worth in terms of something other than their work roles. They confront this question: "Am I a worthwhile person only insofar as I can do a full-time job; or can I be worthwhile in other, different ways—as a performer of several other roles and also because of the kind of person I am?" (Peck, 1968, p. 90). The ability to see themselves as having multiple dimensions allows individuals to pursue new avenues for finding a sense of satisfaction and being worthwhile.

Body Transcendence Versus Body Preoccupation   As people age, they might develop chronic illness and a substantial decline in their physical capabilities. Those who equate pleasure and comfort with physical well-being can feel that this decrease in health and strength is the gravest of insults. They can either become preoccupied with their bodily health or find new sources of happiness and comfort in life. Many elderly persons suffer considerable pain and physical unease and yet manage to enjoy life greatly. They do not succumb to their physical aches, pains, and disabilities but find human relationships and creative mental activities to be sources of fulfillment.

Ego Transcendence Versus Ego Preoccupation   Younger individuals typically define death as a distant possibility, but this privilege is not accorded the elderly. They must come to terms with their own mortality, and their adaptation need not be one of passive resignation. Rather, the elderly can come to see themselves as living on after death through their children, their work, their contributions to culture or community, and their friendships. Thus, they perceive themselves as transcending a mere earthly presence.

Common to the approaches taken to psychosocial development by Erikson, Peck, and many other psychologists is the notion that life is never static and seldom allows a prolonged respite (Shneidman, 1989). Follow-up research on Terman's gifted men and women documents the importance to good mental health and psychological well-being of establishing and maintaining appropriate goals and commitment throughout adulthood, including late adulthood (Holahan, 1988). Both the individual and the environment constantly change, necessitating new adaptations and new life structures.

Question

What are Robert Peck's views about personality development during late adulthood?

Vaillant's Theory of Emotional Health   As noted earlier regarding the Stockholm study findings, our personality also seems to make a difference in promoting or detracting from our emotional health and well-being. These recent results support earlier longitudinal research. At five-year intervals, researchers have followed some men who graduated from Harvard University in the early 1940s (Goleman, 1990c). Psychiatrist George Vaillant and colleagues conducted a follow-up study of 173 of these men at age 65 who had participated in the Grant Study (Vaillant & Milofsky, 1980).

The project provides insights into which personality factors matter, for better or worse, in later life. The investigators viewed emotional health among the elderly as the "clear ability to play and work and to love" and to achieve satisfaction with life. An ability to handle life's blows without passivity, blame, or bitterness proved especially important. At age 65 the subjects reported that their emotional health was not grounded in a happy childhood, or a satisfying marriage, or professional recognition and accomplishment. Those men who developed a *sense of resilience* to absorb the shocks and changes of life were best able to enjoy life. Their self-awareness allowed them to control their first impulses and to respond calmly using what researchers called *mature adaptive mechanisms*—instead of lashing out in anger or blame—promoting emotional satisfaction and well-being.

Men who in college were rated by a psychiatrist as being good practical organizers of coursework—rather than as having a theoretical, speculative, or scholarly bent—were among those making the best emotional adjustment in their later years. So were those who as college sophomores had been described as "steady, stable, dependable, thorough, sincere, and trustworthy." These two traits—*pragmatism* and *dependability*—seemed to matter more than traits such as spontaneity and the ability to make friends easily, which had seemed important for psychological adjustment during the college years. Many factors of early life, even a relatively bleak childhood (such as being poor, orphaned, or a child of divorce), had little effect on well-being at age 65 for the Harvard men. This finding was supported more recently in a four-year study reported by Suh and Diener (1998). They found that emotional well-being is strongly determined by enduring personality characteristics rather than by external life circumstances. In the Grant Study, though, severe psychological depression earlier in life was associated with persistent problems. Being close to one's siblings while in college was strongly linked to later emotional health. As other researchers have also found, the Harvard project revealed that people are extraordinarily adaptable and that over a half-century most people retain the capacity to recover from adversity and get on with their lives.

A Trait Theory of Aging

Most of us go about our daily lives "typing" or "pigeonholing" people on the basis of a number of traits that seem particularly prominent in their behavior. They reveal these traits in the course of interacting with others and with the environment. On the basis of this observation, a number of psychologists, including Bernice L. Neugarten, R. J. Havighurst, and S. Tobin (1968), attempted to identify major personality patterns, or *traits*, that have relevance for the aging process. They studied several hundred persons aged 50 to 80 in the Kansas City area over a six-year period. From their research results, they identified four major personality types:

- Integrated
- Armor-defended
- Passive-dependent
- Disintegrated

The *integrated* elderly are well-functioning individuals who reveal a complex inner life, intact cognitive abilities, and competent egos. They are flexible, mellow, and mature. However, they differ from one another in their activity levels, and the researchers identified three subgroups of integrated elderly: The *reorganizers* are capable people who place a premium on staying young, re-

maining active, and refusing to "grow old." As they lose one role in life, they find another, continually reorganizing their patterns of activity. The *focused* elderly display medium levels of activity. They are selective in what they choose to do and center their energy on one or two role areas. The *disengaged* elderly also show integrated personalities and high life satisfaction. But they are self-directed people who pursue their own interests in a calm, withdrawn, and contented fashion, with little need for complex patterns and networks of social interaction.

The *armored-defended* elderly are striving, ambitious, achievement-oriented individuals, with high defenses against anxiety and with the need to retain tight control over events. Here, too, there are differences: The *holders-on* view aging as a threat and relentlessly cling as long as possible to the patterns of middle age. They take the approach "I'll work until I drop dead." They are successful in their adaptation as long as they can continue their old patterns. The *constricted* elderly structure their world to ward off what they regard as an imminent collapse of their rigid defenses. They tend to be preoccupied with "taking care of themselves," but in their preoccupation they close themselves off from other people and experiences.

Passive-dependent elderly form a third group: The *succorance-seeking* have strong dependency needs and elicit responsiveness from others. They appear to do well as long as they have one or two people on whom they can lean and who meet their emotional needs. The *apathetic* elderly are "rocking chair" people who have disengaged from life. They seem to "survive" but with medium to low levels of life satisfaction.

Finally, there are those elderly who show a *disintegrated* pattern of aging. They reveal gross defects in psy-

Differing Responses to Aging
Bernice Neugarten and her associates have identified a number of personality patterns that influence how people respond to the aging process. "Reorganizers" search for settings that permit them to remain integrated by carrying on an active life; the "disengaged" seek integration through more solitary activities.

chological functions and an overall deterioration in their thought processes. Their activity levels and life satisfaction levels are low.

Neugarten, Havighurst, and Tobin (1968) concluded that personality is an important influence in how people adapt to aging. It has major consequences for predicting their relationships with other people, their level of activity, and their satisfaction with life.

Other Models of Aging

How successfully people age depends on a complex interaction of variables, including physical and mental health, financial security, and individual and cultural perceptions about aging. Human beings also need social settings to develop and express their humanness. Accordingly, psychologists and sociologists have advanced a number of theories that describe changes in the elderly in the United States in terms of the changes in their self-perception and social environments:

- **Disengagement theory of aging.** A view of aging as a progressive process of physical, psychological, and social withdrawal from the wider world. Consequently, the elderly can face death peacefully, knowing that their social ties are minimal, that they have said all their goodbyes, and that nothing more remains for them to do.
- **Activity theory of aging.** The view that the majority of healthy older persons maintain fairly stable levels of activity as long as possible and then find substitutes for the activities they are forced to relinquish. The amount of engagement or disengagement that occurs among the elderly appears to be more a function of past life patterns, socioeconomic status, ability to use the English language, and health than of anything inherent in the aging process.
- **Role exit theory of aging.** The view that retirement and widowhood terminate the participation of the elderly in the principal institutional structures of society—the workforce and the family— diminishing the opportunities open to the elderly for remaining socially useful. The loss of occupational and marital statuses are regarded as particularly devastating, because these positions are *master* statuses or *core* roles, anchoring points for adult identity. The social norms that define the behavioral expectations for old age are weak, limited, and ambiguous. Furthermore, the elderly have little motivation to conform to an essentially "roleless role," a socially devalued status.
- **Social exchange theory of aging.** The theory that people enter into social relationships because these provide rewards—economic sustenance, recognition, a sense of security, love, social

approval, gratitude, and the like. In the process of seeking such rewards, however, they also incur costs—they have negative, unpleasant experiences (effort, fatigue, embarrassment, etc.), or they are forced to abandon other positive, pleasant experiences in order to pursue the rewarding activity. A relationship tends to persist only as long as both parties receive profit (total reward minus total cost) from it. As applied to old age, social exchange theory suggests that the elderly find themselves in a situation of increasing vulnerability because of the deterioration in their bargaining position (Schulz, Heckhausen, & Locher, 1991). In industrial societies, skills become increasingly outmoded through technological change, and as a consequence of the decline in power available to the elderly, older workers exchange their position in the labor force for the promise of Social Security and Medicare; that is, they "retire."
- **Modernization theory.** The view that the status of the aged tends to be high in traditional societies and lower in urbanized, industrialized societies (Cowgill, 1974, 1986). This theory assumes that the position of the aged in preindustrial, traditional societies is high because the aged tend to accumulate knowledge and control through their years of experience. Modernization theorists believe that industrialization undermines the importance of traditional knowledge and control, and some evidence supports this contention (Kertzer & Schaie, 1989). However, Japan is one exception to modernization theory, for the Japanese values of filial piety and ancestor worship have mediated the impact of economic factors on the treatment of the elderly (Martin, 1989).

Although social exchange theory and modernization theory are helpful in drawing attention to elements of exchange that influence the position of the elderly in a society, they fall short of providing a complete explanation (Ishii-Kuntz & Lee, 1987). Indeed, at least for those under 75, aging seems to be accompanied by stability and by continuity in levels of participation in voluntary associations such as religious, civic, service, patriotic, and fraternal or sororal organizations (Lam & Power, 1991). Older people in other cultures often remain engaged in life into old adulthood, where there is no official "retirement" date or age. See *Human Diversity:* "Elderly Hispanic Americans" for further discussion.

Question

Which models of aging deal with internal aspects and which deal with external aspects?

Selective Optimization with Compensation

We have succeeded in adding years to our lives, but we are uncertain of our ability, as a society and as individuals, to add life to the years we have gained.

Paul Baltes and Margret Baltes, *Successful Aging,* 1990

Paul and Margret Baltes endorse the life-span model they call **selective optimization with compensation.** At all stages, and especially in old age, we are adjusting our standards of expectation. A great deal depends on how each elderly person perceives old age. The aging adult who views late adulthood as a time of gaining knowledge and wisdom will have a healthier, more positive self-concept; the aging adult who views old age as a time of physical debilitation and loss of control will form a more negative self-concept. We will enjoy more successful aging if we recognize our capabilities and compensate for losses and limitations. Late adulthood brings many changes in life: That is, we demonstrate "the ability to adjust and to transform reality so that the self continues to operate well if not better than in earlier years. Because of this remarkable power of the self, older adults on average are not at all more depressed or anxious than younger ones" (Baltes & Baltes, 1998). Findings from the recent *Berlin Aging Study* suggest that people aged 70 to 80 continue to have a purpose in life, and for the most part they live in the present with much engagement, mastering the tasks of everyday life (Smith & Baltes, 1997).

The Third Age   Some research indicates that some aspects of personality actually improve during what the Balteses call the *Third Age.* These include emotional intelligence and wisdom. Emotional intelligence includes the ability to understand the causes of emotions (their expression, underlying feelings, and the associated nonverbal cues). **Wisdom** is "expert knowledge about life in general and good judgment and advice about how to conduct oneself in the face of complex, uncertain circumstances" (Baltes & Baltes, 1998). The Balteses suggest that "our store of wisdom benefits from the ability to engage ourselves with others in discussions of life dilemmas and from having a personality that is open to new experiences and strives toward excellence in matters of human lives" (Baltes & Baltes, 1998, p. 14).

The Balteses propose that on a sociocultural level the elderly have potential that is inactivated. Can we take the adult focus on productivity and transform it into other forms of productivity for society? Are there better ways to prepare for old age? What are the emerging biomedical and genetic interventions to repair and prevent conditions of illness?

Aging and Wisdom   Paul and Margret Baltes suggest that a few aspects of personality improve during what they call the "Third Age," including emotional intelligence and wisdom, having expert knowledge of life in general, and good judgment and advice.

The Fourth Age   Paul and Margret Baltes advance the theory that as people move into the *Fourth Age* (the late eighties and beyond), they will face increasingly difficult obstacles and become more vulnerable. People of these advanced ages require more cultural, technical, and behavioral resources to attain and maintain high levels of functioning because they have biological deficits. Baltes and Baltes (1998) point to the recent findings of the Berlin Aging Study, a study focusing on the functioning of those aged 70 to 100. The results indicate that the oldest old do experience more demands and stressors and have fewer mental, emotional, and social reserves to cope with and compensate for these conditions. The prevalence of Alzheimer's disease, in particular, was found to be 2 to 3 percent in those aged 70 to 80; 10 to 15 percent in those 80 to 90; and about 50 percent in those over 90. But Baltes and Baltes (1998) suggest that in old age people can still find effective strategies for life management. By carefully selecting, optimizing, and compensating, we are able to minimize the negative consequences of old age. They remind us

Human Diversity

Elderly Hispanic Americans

"This is the age of aging" (Fried & Mehrotra, 1998). Until recently, research on aging and late adulthood studied mainly white subjects, who in 1990 made up nearly 90 percent of American elderly. However, the number of ethnic minority elderly is projected to grow much more rapidly than the number of elderly whites over the next 50 years. To date there are limited studies on issues that enhance the lives of our elderly, especially for those of minority backgrounds. It is especially important that we address the needs of our growing, more diverse elderly American population in the related fields of gerontology, gerontological nursing, geropsychology, adult development and aging, human services, health care and health education, public health, health policy, and family studies. Because demographic projections suggest that within the next 20 years about one-third of the American population will be of Hispanic origin (Mexican American, Cuban American, Puerto Rican American, etc.), we shall focus here on recent research findings regarding the Hispanic American elderly (Angel & Hogan, 1994). Hispanics are those persons whose families originated in Central or South America or the Caribbean, and they can be Black, White, or Asian—but they share the Spanish language (Fried & Mehrotra, 1998).

Acculturation

The relationship between diversity and psychological distress was studied with immigrant elders from Cuba, Mexico, and Puerto Rico. Using data from the National Survey of Hispanic People, Krause and Goldenhar (1992) conducted a phone survey with 1,339 elderly and found an association between financial strain and level of acculturation. These findings indicate that Hispanic Americans who are more acculturated experience less social isolation, fewer financial problems, and fewer symptoms of depression. Puerto Rican American elderly subjects reported more depressive symptoms than Cuban American subjects.

Mental Health and Well-Being

Minority elders usually seek help for emotional issues from other members of their ethnic group and not from social service agencies. Zamanian and colleagues (1992) surveyed a sample of 159 older Mexican Americans to determine the relationship between depressive symptoms and acculturation. They conclude from their findings that "retention of aspects of Mexican culture without concomitant attempts to incorporate aspects of the dominant culture results in the most vulnerable position to depression" (Zamanian et al., 1992). Because there is a relationship between depression and suicide, this is of great concern. Additionally, minority caregivers are less likely to use agencies offering services for elderly with dementia and Alzheimer's disease (Braun et al., 1995). In comparison with other older adults, Hispanic American elders have poorer physical and mental health, are more likely to lose their family supports through family movement and relocation, and are likely to underuse formal support systems (Garcia, 1985).

Health Care

Gelfand (1994) reminds us that there are large differences among Hispanic American groups in terms of the health of older adults. He cites that Puerto Rican American men have a higher rate of cardiovascular disease than Cuban or Mexican American men; Mexican Americans have the highest rate of cerebrovascular disease, while Cuban Americans have higher rates of cancer. The health beliefs of the Mexican American elderly are more likely to derive from folk medicine than from professional medical practice. One such folk belief system of some Mexican Americans in the Southwest is *curanderismo* (Gafner & Duckett, 1992). This involves a blend of Catholicism, medieval medicine, and the medicine of indigenous Indians. Some of the features of this belief system are (1) the belief that God heals the sick through per-

that the indomitable human spirit, when confronted with the challenges of living, comes up with many solutions. This will be a worldwide challenge of the new century.

A Life-Span Model of Developmental Regulation

Schulz and Heckhausen (1996, p. 708) have recently proposed a life-span theory with the construct of *control* as the central theme for characterizing human development

from infancy to old age. Implicit in their view is the thesis that successful aging includes the development of primary control throughout the life course. "Primary control targets the external world and attempts to achieve effects in the immediate environment external to the individual, whereas secondary control targets the self and attempts to achieve changes directly within the individual" (Schulz & Heckhausen, 1996, p. 708).

Both primary and secondary control can involve cognition and action. Schulz and Heckhausen (1996) also

sons, called *curanderos,* who are blessed with the *don,* a special gift; (2) the reality that a number of conditions can be cured; (3) belief in mystical diseases such as "*susto,* meaning loss of spirit of fright"; (4) the belief that illness and health exist on material, mental, and spiritual levels; and (5) the use of proscribed rituals and herbs for healing (Gafner & Duckett, 1992). *Curanderos* are readily available, do not require medical and insurance forms, and speak the same language and share the same beliefs as the patient.

Providers of Care

Adult children provide the informal help network. A recent study (Garcia et al., 1992) of four generations of Hispanic American women revealed beliefs that adult children should care for their elderly, children should share their homes with the elderly, and contact should be maintained between children and their elderly parents. Because older Hispanics are viewed as wise, knowledgeable, and deserving of respect, they continue to expect and receive help provided by their adult children and not professional help (Cox & Gelfand, 1987). This view is especially burdensome for Hispanic Americans who are caring for elderly with dementia and Alzheimer's disease (Cox & Monk, 1993). The use of professional services contradicts the norms of filial responsibility, so reaching out for professional care is an extremely difficult choice.

Community Services

Hispanic American elders are likely to differ in their use of community services. Mexican Americans tend to be concentrated in urban areas, where many live in ghettos. However, they do not typically use senior services such as transportation, senior centers, meals-on-wheels, church-based assistance, homemaker assistance, or routine telephone checks. Use of such services increases, though, when Hispanic Americans are involved in the delivery of such services. It is important to involve leaders of ethnic groups in the design and implementation of community services so that they do not violate or disrespect cultural values and practices (Fried & Mehrotra, 1998).

Retirement/Standard of Living

Older Mexican Americans tend to have significantly lower incomes than older White Americans. Most Mexican American elderly were unskilled or semiskilled laborers employed for many years on farms or ranches, in factories, or in part-time seasonal jobs (Fried & Mehrotra, 1998). The traditional retirement criteria (age 65, with income from retirement sources, and with a view of oneself as being retired) do not apply to Mexican Americans, who experience "retirement" differently and have access to fewer sources of retirement benefits (Zsembik & Singer, 1990). Some who migrated to the United States late in life are not eligible for Social Security benefits, do not know how to apply for benefits, and might work "under the table" for employers who do not report employee contributions (Garcia, 1993).

Religion and Spirituality

Older Americans of Mexican heritage practice religion that is a blend of different rituals and doctrines from both Aztec and European influences. The Aztecs believed that all things in nature are to be respected and valued. Catholicism is practiced by many elderly Mexican Americans, but there is a trend toward conversion to other religions. Religion remains an important aspect of the lives of these elderly, and many attend weekly services (Villa & Jaime, 1993). Faith, however, represents a way of being and living and is required for spiritual growth. The concept of *fe* embodies a way of life and incorporates varying degrees of faith, hope, spirituality, beliefs, and cultural values. *Fe* is personally defined by each individual, and *fe* helps individuals cope with poor health, poverty, and death. *Fe* helps elderly Mexican Americans maintain a positive attitude toward life. Anyone who works with this population of elderly needs to understand the significance of *fe* as a valid coping mechanism (Villa & Jaime, 1993).

Question

What are some pertinent concepts to understand about elderly Hispanic Americans and their attitudes toward health care and mental health?

state that because primary control is directed outward, engaging the external world, it is preferred and has greater adaptive value to the individual. It enables individuals to explore and shape their environment to fit their particular needs and optimize their developmental potential. Primary control also provides the foundation for diversity and selectivity. Adelman says diversity (sampling different performance domains) is optimal throughout the life course: "The principle of diversity has important implications for socializing agents respon-

sible for childhood development. Early in their development, children should be exposed to a variety of domains of functioning so they are challenged, develop diverse skills, and have the opportunity to test their genetic potential" (Adelman, quoted in Schulz & Heckhausen, 1996, p. 706).

The principle of *selectivity* is that individuals must selectively invest time, effort, abilities, and skills—and selectivity must work hand in hand with diversity so that the potential for high levels of functioning in some domains is attained, while other broad, generalizable skills are developed. Some of the major challenges individuals face throughout life involve assessing the trade-offs for a given investment of time and effort and making decisions about whether to continue within a given domain or switch to another (Schulz & Heckhausen, 1996).

For example, as a college student right now, you are investing time and effort by preparing yourself for some career (developing skills in English, perhaps a second language, mathematics, social sciences, life sciences, history, as well as specific courses in your major area of study). At the same time as you are taking a diversity of courses, you are also becoming more selective as you prepare yourself for your chosen career. During adolescence and early adulthood, we develop a broad range of secondary control strategies (strategies focused inward), including changing aspiration levels, denial, egotistic attributions, and reinterpretation of goals.

Selectivity continues throughout adulthood, and diversity gradually decreases. The trade-off of selectivity over diversity is a hallmark of middle adulthood and old age. For example, most people who can no longer play tennis or basketball because of age are still able to play golf, a sport that does not require the same level of physical exertion. Increasing age-related biological and social challenges to primary control put a premium on secondary (self-directed) control strategies. The older person's own competencies and motivation will lead into experiences of either failure outcomes or positive outcomes. Failure experiences have the potential to undermine their self-concept, and therefore the elderly develop various strategies of compensation, such as using hearing aids, canes, and other adaptive devices. Those elderly who are able to engage and impact their environments for the longest time would be judged most successful. Schulz and Heckhausen (1996) encourage future investigation of their position on successful life-course development. Associations among personal control, choice, and healthy psychological adjustment have been studied since the early 1900s with younger adults.

The Impact of Personal Control and Choice

As psychologists since Alfred Adler (1870–1937) have noted, a sense of control over one's fate makes a substantial contribution to most people's mental health. It seems that most individuals prefer and benefit from control most of the time. However, there are exceptions to this rule (Reich & Zautra, 1991). Psychoanalyst Erich Fromm (1941) argued that a good many people do not wish to be masters of their own fate and hence are attracted to totalitarian leaders and movements. Witness the large number of people who voluntarily sign up for a life in the military, for instance. Perceived control and the desire for control can also decrease with age

(Mirowsky, 1995). But even though one's sense of control might erode with age, there often comes the wisdom to "shuck off" a sense of responsibility for matters that seem beyond one's control (Schulz, Heckhausen, & Locher, 1991). Many older people simply put their ailments on the back burner (Jacobson, 1992–1993). Hence one need not "feel in control over things" so long as "things seem under control." As we shall see, generally those elderly who are forced into institutionalized living arrangements have serious health conditions and lose their sense of control and purpose as others take control over such daily decisions as when to bathe, when to eat, when to sleep, what to eat, and what to wear.

A Sense of Purpose   The important contribution that a sense of responsibility, usefulness, and purpose makes to successful aging is highlighted by the research of social psychologists Langer and Rodin (1976). They investigated the impact that feelings of control and personal choice had among residents of a high-quality nursing home in Connecticut. Forty-seven residents on one floor of the home heard a talk by the nursing-home administrator in which he stressed the residents' responsibility for caring for themselves and shaping the home's policies and programs. At the conclusion of his talk, he presented each resident with a plant "to keep and take care of as you'd like." He also informed them that a movie would be shown on two nights, saying, "You should decide which night you'd like to go, if you choose to see it at all." Forty-five residents on another floor of the four-story building were also given a talk by the administrator. This time, however, he emphasized the responsibility that the staff felt for the residents. He gave them plants with the comment, "The nurses will water and care for them for you." Finally, he told them that they would be seeing a movie the following week and would be notified later about which day they were scheduled to see it.

The residents were individually interviewed by a trained researcher one week before and again three weeks after the administrator's talk. To prevent bias, the researcher was not told the purpose or the nature of the experiment. The questions she asked the residents dealt with how much control they felt they had over their lives and how happy and active they believed themselves to be. She also rated each resident on an 8-point scale for alertness. On the same two occasions, a questionnaire was filled out by each member of the nursing staff. The nurses were asked to evaluate each resident in terms of her or his overall activity, happiness, alertness, sociability, and dependence. Like the interviewer, the nurses were not aware of the nature and purpose of the study. According to the various ratings made before the administrator's talks, the two groups of residents were quite similar in their feelings of control, alertness, and satisfaction. Three weeks after the talks, however, the differences were marked.

Despite the high-quality care given them, nearly three-fourths of those in the second group (the group in which the staff retained control and took primary responsibility) were rated as having become more debilitated—over a mere three-week period. In contrast, 93 percent of those in the first group (the group in which the residents were encouraged to make their own decisions and were given decisions to make) showed overall improvement. Langer and Rodin (1976, p. 197) conclude that their findings support the view that "some of the negative consequences of aging may be retarded, reversed, or possibly prevented by returning to the aged the right to make decisions and a feeling of competence."

Rodin and Langer (1976) found that the group of residents who had been given responsibilities during the experimental period were healthier than comparable residents who had been treated in the conventional manner. The two treatment groups also showed different death rates. By the time of the follow-up, twice as many members of the conventional-treatment group, compared with the responsibility-induced group, had died. In sum, medical practitioners would be well-advised to incorporate patient choice in their caregiving.

Question

How do elderly individuals benefit from personal control and choice?

Faith and Adjustment to Aging

Carter Catlett Williams (1998), as a "Third Ager" and social work consultant in aging, suggests there is a possibility of far greater richness in the elder years than the professional literature on aging suggests. According to Williams, old age is an adventure to be explored, a God-given mystery, not something to fear or despise. From working with the elderly, she says there is an "inner life that beckons," remembering the joys and sweet sorrows of the past about people, places, and events that have shaped each one of us and continue to shape who we will become.

More researchers from around the world are also documenting through empirical studies that there is a healing association between religion and health. Dale Matthews (1998) relates findings from hundreds of empirical studies on what he calls the catalyst of the "faith factor" and its powerful benefits to overall well-being and attitude for people of all ages. He says that many centuries ago Western cultures separated medicine from religion, and we are just now coming around to recog-

nize the powerful relationship between the two. "Scientific studies show that religious involvement helps people *prevent* illness, *recover* from illness, and—most remarkably—*live longer*. The more religiously committed you are, the more likely you are to benefit" (Matthews, 1998, p. 19). A classic 1972 survey by Comstock and Partridge of 91,909 subjects living in Washington County, Maryland, showed that those who attended church at least once a week had significantly lower death rates from coronary-artery disease (50 percent reduction); emphysema (56 percent reduction); cirrhosis of the liver (74 percent reduction); and suicide (53 percent reduction) (Matthews, 1998). In a study of 1,077 college students at Northern Illinois University, one-sixth of the subjects who were more religious had better health, less sickness, fewer medical visits, and fewer injuries than other subjects who were less religious or nonreligious. These healthier subjects also had significantly lower rates of alcohol, tobacco, and substance abuse, and they exercised and used seatbelts regularly (Oleckno & Blacconiere, 1991).

Gardner and Lyon (1982) have studied the effects of the healthy lifestyle of members of the Church of the Latter-Day Saints (Mormons). Church doctrine includes abstaining from alcohol, tobacco, coffee, and tea, and includes morals, honesty, payment of tithes, and regular church attendance. Those who were more religiously involved had significantly fewer cases of cancer. Studies of Seventh-Day Adventists in the Netherlands found that male Adventists lived an average of nine years longer than men in the general population. These healthier, longer-living believers adopt better health practices *because of* their religious commitment (Berkel & deWaard, 1983). A study of 2,754 female and male subjects in Tecumseh, Michigan, found that those women who attended church more often lived longer than those who did not (House, Robbins, & Metzner, 1982). Thomas Oxman, of Dartmouth Medical School, followed 232 elderly patients who had all undergone open-heart surgery. Those who said they attended church regularly had a death rate of 5 percent. The death rate was three times as high for nonchurchgoers (Oxman, Freeman, & Manheimer, 1995). Many subjects in this study acknowledged the healing power of prayer.

In a study of 760 midwestern women, those who attended church more than once a month experienced significantly less depression and anxiety than those who attended less frequently or not all (Hertsgaard & Light, 1984). This finding was confirmed in a study of 451 African American subjects that examined the correlation between regular church attendance and incidence of depression. "People with high levels of religious involvement reported significantly less depression" (Matthews, 1998, p. 25). Several other similar studies confirmed that religious involvement helps prevent depression at particularly stressful times such as severe illness or loss of a loved one. "Widows and widowers with an authentic sense of personal relationship with God cope better with the loss of their spouses than do their nonreligious peers or religious individuals who do not experience an active awareness of the presence of God in their lives" (Rosick, 1989, p. 25–26). Religious involvement appears to greatly enhance the quality of life for young and old and in many walks of life.

Question

What do researchers find about the role of religious faith in late adulthood?

Familial Roles: Continuity and Discontinuity

Though the majority of older persons lived in a family setting in 1997, the social world of old age differs significantly from that of early and middle adulthood (*Profile of Older Americans,* 1998). Changes in physical vigor and health and in cognitive functioning have social consequences. And shifts in marital roles and work profoundly affect the lives of elderly people through the behavioral expectations and activities that they allow. Hence, the "social life space" of aging adults provides the context in which elderly men and women, like their younger counterparts, define reality, formulate their self-images, and generate their interaction with other individuals.

Throughout history, there has been obvious continuity in intergenerational relations in families: Parents cared for and nurtured children, launched them into independent lives, and typically transferred resources to children and grandchildren upon death. But Pillemer and Suitor (1998) forecast that the contemporary baby-boom generation will experience special issues and challenges, unlike previous generations, due to the *plurality of family forms:* Divorce, remarriage, single parenting, stepparenting, single lifestyle, and alternative family forms create a wide field of kinship bonds. Also, in general younger adults today are more dependent on the labor market for their livelihood than dependent upon family resources, so intergenerational support is more a choice than an obligation. However, for elderly who require long-term care, past cultural experience demonstrates that sustained, stable support requires very close kin. The current elderly appear to have large resources for stable family support, including greater likelihood of an intact marriage and relatively large numbers of living children and grandchildren (Pillemer & Suitor, 1998).

Love and Marriage

You meet somebody that attracts your eye. First comes the physical attraction. . . . Then you begin to feel that you need that person. My marriage has lasted because the need has persisted. . . . And it's also, you know, from time to time—the other thing.

Actor Carroll O'Connor on the longevity of his 47-year marriage to his wife Nancy,
Life Magazine, February 1999, p. 42

The parents of the baby-boom generation are more likely to still be married because they had the highest rates of marriage for any cohort of the twentieth century (Pillemer & Suitor, 1998). Additionally, caregiving research indicates that the spouse is the first person married elders turn to when they need care. Until age 75 most American householders are part of a married couple (see Table 18-1). Accordingly, social and behavioral scientists have asked, "What is the nature of marriage in older age?" It seems that most Americans believe that marriages that do not terminate in divorce begin with passionate love and evolve into cooler but closer companionship. However, researchers paint a somewhat different picture. Both marital satisfaction and adjustment begin declining quite early in marriage. The speed and intensity of this decline vary from one study to another. In the middle and later stages of the family life cycle, the evidence is less clear (Vaillant & Vaillant, 1993). Some investigators find a continual decline (Swensen & Trahaug, 1985). Usually, however, they report a U-shaped curve, with a decline in satisfaction during the early years, a leveling off during the middle years, and an increase in satisfaction during the later years (Glenn, 1990).

Though the quality of marriage varies from couple to couple, most elderly husbands and wives report greater happiness and satisfaction with marriage during their later years than at any other time except for the

Table 18-1   Marital Status of Persons 65 and Older, 1995

	Women %	Men %
Married	42	74
Widowed	46	16
Single (never married)	4	4
Divorced	7	6

Source: "Marital Status and Living Arrangements: March 1997 Update," *Profile of Older Americans, Current Population Reports,* PPL-90, U.S. Bureau of the Census, retrieved January 14, 1999, from the World Wide Web: www.aoa.dhhs.gov/aoa/stats/profile

Satisfaction with Marriage Among the Elderly Most elderly couples report greater happiness and satisfaction with marriage than at any time since the newlywed phase. Many say their marriage improved during late adulthood. The couple shown are celebrating their fiftieth wedding anniversary.

newlywed phase. Many say that companionship, respect, and the sharing of common interests improve during later adulthood (Powers & Bultena, 1976). Sexual activity continues to play an important role in relationships of older men and women, according to results of the 1998 Healthy Sexuality and Vital Aging study. According to this national random survey of 1,300 older Americans conducted by the National Council on Aging (1998) and released in September 1998, nearly half of all Americans aged 60 and older indicated they engage in sexual activity at least once a month. Four out of 10 indicated they would like to have sex more frequently than they do (56 percent of men vs. 25 percent of women). Only 4 percent said they wanted sex less frequently. Men and women in their sixties are more sexually active than those in their seventies, partly because many women no longer have a partner because they are widowed. Lower sexual satisfaction was also attributed to medical conditions and treatments that prevent the elderly from having sex and medications that seem to reduce sexual desire. This study indicates that for many older Americans, sex remains an important and vital part of their lives (National Council on Aging, 1998). In 1998, the "little blue pill" Viagra came on the market to improve the sex lives of men clinically diagnosed with impotence and their partners (Weber et al., 1998).

A number of other factors appear to contribute to the improvement of marriage in the later years. For one thing, children are launched. Parenthood requires a heavy commitment and is often the source of role overload and strain (Schumm & Bugaighis, 1986). Children also interfere with the amount of communication between spouses and limit the time they have available for companionable activities (Fitzpatrick, 1988). And children often create new sources of conflict while intensifying existing sources (Udry, 1971). Furthermore, in later life problems with such issues as in-laws, money, and sex have often been resolved or the stresses associated with them have dissipated. And as we discussed in Chapter 15, older adults tend to be more androgynous in their roles than younger adults are. However, retirement can create new strains for a couple. One retiree points out the following:

> A husband and wife may each have a dream of what retirement would be, but those dreams don't necessarily mesh. They've got to sit down and talk—outline their activities, restructure their time, and define their territories. I discovered that my wife was very afraid that after I retired she'd have to wait on me hand and foot and would lose all her freedom (Brody, 1981, p. 13).

Some women report that they feel "smothered" having their retired husbands about the house so much of the time. Men who attempt to increase their involvement in household tasks might be seen by their wives as intruders. Even so, some wives welcome the participation of their husbands because it relieves them of some responsibilities. The loss of privacy and independence is often offset by opportunities for nurture and companionship. And wives mention having the "time available to do what you want" and the greater flexibility in schedules as advantages of retirement. Yet perhaps the most important factors influencing a wife's satisfaction with her husband's retirement are her own and her husband's good health and adequate finances (Brubaker, 1990).

Over the past few decades as women have entered the labor force in growing numbers, couples increasingly have had to confront the issue of whether or not the husband and wife will retire at or about the same time (Pienta, Burr, & Mutchler, 1994). One study of retirement decisions—based on the National Longitudinal Survey of Mature Women—found that among couples in which both the husband and wife had retired, about a quarter of the men and women did so in the same year. Another 44 percent of the women retired before their husbands and 30 percent retired later (Jacobs & Furstenberg, 1986). Therefore, the stereotyped image of retirement in which the husband leaves his job and joins his wife at home no longer fits many family situations.

Marriage also seems to protect people from premature death (Rogers, 1995). Married individuals are healthier than unmarried individuals, and death rates are consistently higher among single and socially isolated people (even after adjustments are made for age, initial health status, smoking, physical activity, and obesity). Significantly, although popular folklore depicts marriage as a blessed state for women and a burdensome trap for men, it is men rather than women who researchers find receive marriage's greatest mental and physical benefits (Anderson & McCulloch, 1993). For instance, women are more likely to be named as confidants by their husbands than they are to name their husbands as confidants. Women are more likely to select as confidants adult daughters and female friends (Antonucci, 1994).

Questions

Why does the happiness of couples who have been married for many years typically have a U-shaped curve? What are the findings about the well-being of elderly singles?

Widows and Widowers   A majority of elderly men—72 percent—live with their spouses. But because women tend to outlive their husbands, there are far more widows than there are widowers. Only 42 percent of elderly women still reside with their husbands. This statistic demonstrates particular gender issues that are of great concern worldwide: (1) Elderly women are at much higher risk of living in poverty, especially in nonindustrialized countries where they may have never earned a pension, (2) most are not able to afford adequate health care services, and (3) they are much more likely to experience neglect and elder abuse. Approximately five times as many women age 65 and older live alone compared to men 65 and older who live alone (U.S. Bureau of the Census, 1998a).

In the Healthy Sexuality and Vital Aging study, more older men (61 percent) than women (37 percent) said they were sexually active (National Council on Aging, 1998). When older people are not sexually active, it is usually due to lack of a partner or to a medical condition. In this same study, 90 percent of the respondents identified a high moral character, a pleasant personality, a good sense of humor, and intelligence as the important qualities in a partner. More women than men were likely to seek financial security from a partner, and more women than men sought a partner who observed a religious faith. Men were more likely to seek a partner who was attractive and interested in sex (National Council on Aging, 1998). The National Council on Aging (NCOA) announced its new *Love & Life: A Healthy Approach to Sex for Older Adults* program with four major goals: (1) to educate older adults that sex can be a natural part of life, (2) to educate about sexual dysfunction, disease, and the aging process, (3) to generate discussion between physicians and their elderly patients about sexual needs, and (4) to coordinate education and training on sexuality workshops for older adults.

Remarriage and the Elderly   Remarriage in later life is another recent trend. Nearly half a million people over the age of 65 in the United States remarry each year (U.S. Department of Health & Human Services, 1998). With more people entering late adulthood, it is likely that more remarriages are occurring during this stage of life. With factors of increased longevity and more later-life remarriages, those who are currently in middle age could conceivably have several groups to whom they feel obligated: grandparents, parents, children, grandchildren, in-laws, stepchildren, and stepparents. Whereas intergenerational studies have demonstrated that children sense obligation toward elderly parents, there is little basis for obligations to extend to stepparents acquired through remarriage. In fact, due to the conditional nature of step relationships that are formed when stepchildren are adults, these adult stepchildren might not be perceived as having responsibilities to help their stepparents (Ganong et al., 1998).

Singles

The elderly who remain single tend to have more emotional and physical pathology than the married elderly. This appears to be because the married have continuous companionship with a spouse, and spouses normally provide interpersonal closeness, nurturing, emotional gratification, and support in dealing with life's hassles and stress (Venkatraman, 1995). Indeed, married men and women are typically happier and less stressed than are the unmarried (Mastekaasa, 1992).

Studies of intergenerational social supports demonstrate that the people most likely to be listed as key network members at all times are parents and children. If singles who have never married have no children, they are lacking a main component of a significant social network in their later years. If singles have been married and then divorced, their social network can be unstable. As mentioned above, death rates are consistently higher among single and socially isolated people—even after adjustments are made for age, initial health status, smoking, physical activity, and obesity (Pillemer & Suitor, 1998). Males are particularly vulnerable for being socially isolated, whereas older women typically establish a social support network with other females and with family members. With greater numbers of divorced or widowed singles in late life, there are more remarriages, which create elderly stepparent/stepchild relationships. This new family structure can create strain on adult children and less stable intergenerational relationships.

Question

Are men or women more likely to be widowed in late adulthood, and do widowed adults tend to remarry or remain single?

Lesbian and Gay Elderly

Older lesbians, gay men, and bisexuals have been stereotyped as "lonely and pathetically miserable" (Cruikshank, 1991). History-graded influences on development have affected different cohorts of lesbians, gay men, and bisexuals (Slater, 1995). For instance, in recent history homosexuals were either ignored or scorned by social scientists, but today's "gay community" has been granted civil rights and protection under the law in many states. It was erroneously believed in the past that older gay and lesbian adults lead a lonely existence. Today we know that they are not inevitably alone and despondent. Some have children and grandchildren as a result of previous heterosexual unions. Kehoe's (1989) study found that 27 percent of older lesbians reported previous marriages. Friend (1980) found that older gays and lesbians have reconstituted families in the form of friendship and support networks.

Friend's (1991) theory of successful aging for gay men and lesbians has as its defining feature the achievement of a positive identity as an openly gay man or lesbian. Optimally, the person will reconstruct a new meaning for the feelings and experiences associated with being lesbian or gay and reject the larger group value held by a homophobic culture (Reid, 1995). Or less optimally, the older adult will (1) conform to the larger culture and keep his or her identity secret or (2) remain closeted and continue with heterosexual relationships. In contrast, in a four-year longitudinal study of 47 Canadian gay men aged from 50 to 80, Lee (1987) found wealth, health, and lack of loneliness to be associated with higher life satisfaction. Lee concludes that, in his sample, successful aging includes being skillful in avoiding life stressors—

including dealing successfully with the stress of coming out. Quam and Whitford (1992) studied older gay men and lesbians and found that those who maintained a high level of involvement in the gay community had higher levels of life satisfaction and acceptance of the process of aging. Research on aging in lesbians is sparse, but a few studies indicate that they are more concerned about privacy and being identified as lesbian (Martin & Lyon, 1992). Adelman (1991) discusses the damaging effects on older lesbians who were forced to keep their sexual orientation secret from coworkers and family members for extended periods of time. Kehoe (1989, 1986) surveyed two groups of women ranging in age from 65 to 85 years and found that, overall, older lesbians accept aging, are in good or excellent health, feel positive about being lesbian, and have a high level of life satisfaction.

Questions

Overall, how do gay and lesbian elderly rate their life satisfaction? What factors contribute to their well-being?

Children or Childlessness

The notion that most elderly people are lonely and isolated from their families and other meaningful social ties is false (Hannson & Carpenter, 1994). Moreover, the elderly are often involved in exchanges of mutual aid with their grown children as both providers and receivers. Many times, the elderly parent helps the adult child by performing child care and other home-related roles, whereas the adult child helps the parent with heavy housework, shopping, bureaucratic mediation, and transportation. Among the American middle class, gifts from living parents play a substantial role in bolstering the living standards of their adult children. For instance, economists have estimated that one in four home buyers receive assistance from their parents or other relatives (Zachary, 1995).

Also, in 80 percent of cases, care for the elderly is provided by their families. This is especially true for elderly ethnic minorities. Despite substantial social change (increased geographic mobility, divorce, and women's participation in the labor force), which has been thought by many to weaken intergenerational family cohesion, adult children, particularly daughters, remain major sources of instrumental support to their parents (Silverstein, Parrott, & Bengston, 1995). Significantly, older Americans are less likely than their adult children to believe that when elderly parents can no longer take care of themselves, the best solution is for them to move in with their children. The elderly value their privacy and independence (Cicirelli, 1992; Cherlin & Furstenberg, 1986).

Those with adult children prefer to live near but not with them, what psychologists term "intimate distance." In sum, the elderly are not so isolated from kin and friendship networks as is commonly believed (Aldous, 1987).

Grandparenting and Great-Grandparenting

Grandparents in America are like volunteer firefighters: They are required to be on the scene when needed but otherwise keep their assistance in reserve.

(Cherlin & Furstenberg, 1986, p. 183)

Child psychologists emphasize that both children and their grandparents are better off when they spend a good deal of time in each other's company (Smith, 1991) (see Figure 18-1). Yet at the turn of the twentieth century, surviving grandparents were in short supply. Now, with the increase in adult life expectancies, more children and adults have living grandparents, stepgrandparents, and great-grandparents. Presently in the United States, grandparents are likely to have five or six grandchildren, on average, and grandparents are healthier, more active, and better educated than they used to be, and many have more money and leisure time (Szinovacz, 1998).

One in ten Americans aged 30 to 44 is a grandparent, as are half of those aged 45 to 59 and 81 percent of those

Being a Grandparent Regardless of culture, grandparents can be an important developmental resource for their grandchildren. And they can reap important psychological satisfactions from spending time with their grandchildren.

aged 65 and older. Grandparenthood, for most, is now seen as a sequential phase of life rather than an overlapping phase. In the past, people often became grandparents while still raising their own youngest children, whereas today the youngest has usually left home before people become grandparents (Szinovacz, 1998). In 1998, grandmothers outnumbered grandfathers by more than 9 million, because older women typically outlive older men (Crispell, 1993a). However, many middle-aged and older women are working and are unlikely to be available to baby-sit full-time for their grandchildren. Of older people with adult children, 94 percent are grandparents and 46 percent are great-grandparents. So almost half of all persons 65 and over in the United States who have living children are members of four-generation families. The likelihood of being a great-grandparent increases with age, so that among persons 80 and over, almost three-fourths are great-grandparents (Szinovacz, 1998). Great-grandparents are now more common than grandparents were at the turn of the century.

Another trend over the past few decades is that nearly 25 percent of grandparents will be stepgrandparents, through either their own or their children's divorces (Szinovacz, 1998). The more complex family structures created by multiple divorces involving children have become more common. Overall, 3.3 percent of grandparents lived in households with grandchildren,

3-19

"Megan says she has a great grandmother, but I told her we have the GREATEST grandmother."

Figure 18-1 Both Children and Grandparents Can Benefit from Their Relationship.
© 1982 Bill Keane, Inc., Dist. By Cowles Synd., Inc. Reprinted with special permission of King Features Syndicate.

Table 18-2 Prevalence of Grandparenthood in the United States, 1992 to 1994

	All	Male				Female			
		Black	White	Hispanic	All	Black	White	Hispanic	All
Youngest child age 40+	94.6	83.5	94.3	86.1	93.6	99.3	94.7	100	95.1
Those 65 and over	81.4	84.5	83.7	89.7	83.7	78.7	79.3	90.3	79.6
Mean number of grandchildren	5.2	5.7	4.6	7.2	4.9	5.9	5.3	6.6	5.4
3-generation families	80	79.5	77.8	88.1	78.4	84.2	80.9	82.9	81.3
4-generation families	16.1	18.6	13.5	16.8	14.2	25.7	17.1	16.1	17.8

Republished with permission of the Gerontological Society of America, 1030 15th Street, NW, Suite 250, Washington, DC 20005. *Grandparents Today: A Demographic Profile* (Table), M.E. Szinovacz, *The Gerontologist*, Vol. 38, No. 1. Reproduced by permission via Copyright Clearance Center, Inc.

and nearly 5 percent live in households with three or more generations. Consequently, today's grandparents live through more life transitions than grandparents did in the past, and a majority of grandparents have more roles than grandparents did in the past (Szinovacz, 1998). Table 18-2 is based on data collected in a national study from more than 10,000 subjects in the United States between 1992 and 1994. Of those with children aged 40 and over, nearly 95 percent are grandparents. Over 80 percent of today's families contain three generations, and 16 percent are four-generation families (see Table 18-2) (Szinovacz, 1998).

Despite stereotypes about grandparents, research suggests that grandparents vary considerably in the ways they approach their roles (Thomas, 1986). Much depends on their age and health, their race and ethnic background, and geographic distance. In some cases, due to death, divorce, remarriage, or other family disruption, grandparents are obligated to take on a larger role in the lives of their grandchildren and stepgrandchildren. More than 4 percent of grandparents have either become surrogate parents or assumed primary caretaking responsibility for a grandchild while the mother works (Szinovacz, 1998). Grandparents tend to be more involved with grandchildren when the children's mothers are divorced or unmarried (Denham & Smith, 1989). Countless grandparents have stepped into the breach to rescue grandchildren whose parents have faltered due to drugs, abuse, or crime, and these grandparents can find their grandchildren to be a source of much stress (Hayslip et. al., 1998). The American Association of Retired Persons (AARP) has special resources on their Web site for grandparents who have custodial responsibility for grandchildren (see "Following Up on the Internet" at the end of this chapter). So although some grandparents are emotionally or geographically remote, a majority are very closely involved with their grandchildren (Jendrek, 1993).

In contemporary society, the most common kind of relationship between grandparents and grandchildren is companionate, where the grandparent and grandchild are essentially good pals. Grandparents might play, joke, and watch television with their grandchildren, but they are less inclined to discipline them. By the same token, grandparents also want freedom and fulfillment—to spend their leisure time as they please and to have close but not constant association with their children and grandchildren (Hayslip et. al., 1998; Smolowe, 1990).

Although the grandparenting role has different meanings for different people, some themes recur: For many individuals grandparenting is a source of biological renewal or continuity. Being a grandparent instills a sense of extension of self and family into the future and is often a source of emotional self-fulfillment. It generates feelings of companionship and satisfaction between

Growing Numbers of U.S. Youngsters Are Raised by Their Grandparents In many contemporary American homes, grandparents play a substantial role in the rearing of their grandchildren. Indeed, in a good many cases, grandparents assume the role of custodial parents.

the grandparent and a child that were often absent in the earlier parent-child relationship. Kornhaber and Woodward (1981) conducted lengthy interviews with 300 grandchildren and as many grandparents. They concluded that for those grandchildren and grandparents who are close, the relationship provides mutual rewards. The children derive unadulterated doses of tender, loving care. They may also acquire a storyteller, family historian, mentor, wizard, confidant, and role model for the grandchild's own old age.

Questions

In what ways has the role of grandparent been redefined over the course of the twentieth century? How do grandparents and grandchildren benefit from a relationship with each other?

Siblings

The longest lasting relationships people normally have are with siblings.

John Cavanaugh,
Friendships and Social Networks Among Older People

Siblings normally play a significant role in the lives of the elderly (Connidis, 1994; Cicirelli, Coward, & Dwyer, 1992). They provide a continuity in family history that is uncommon to most other family relationships—siblings might be the only members of the family of origin that remain. A shared family history frequently affords a foundation for interaction that supplies companionship and a support network as well as validation for an older person's reminiscences of family events. As they advance into old age, many siblings report that they think more often about one another and find that their acceptance, companionship, and closeness deepens. The most frequent contact occurs among siblings in relatively close physical proximity. Siblings are especially important kin for those with few or no children (Connidis, 1994). Five types of sibling interactions have been identified, ranging from extremely close to distant: congenial, loyal, intimate, apathetic, and hostile (Gold, Woodbury, & George, 1990). The majority of elderly report congenial and loyal relationships with siblings. The sister-sister relationship tends to be the most potent sibling relationship, followed by the sister-brother and brother-brother relationship (Campbell, Connidis, & Davies, 1999).

Question

How does the presence or absence of children, grandchildren, and siblings affect the elderly adult's emotional well-being?

Social and Cultural Support

Numerous findings from scientific research support the notion that frequent social contact appears to be associated with good health, facilitation of coping with stress, and greater life satisfaction (Cavanaugh, 1998a; Matthews, 1998). A functioning network of social ties is

Siblings and Social Support The longest lasting relationships we normally have are with our siblings, who provide a continuity in family history that is uncommon to most other family relationships. A shared family history frequently affords a foundation for interaction that supplies companionship, closeness, and a support network as well as a validation for an older person's memories of family events.

necessary in times of stress, illness, and aging—and includes both emotional support and practical assistance as needed. Indeed, researchers such as Herbert Benson state, "Our need for contact with other human beings is vital" (Matthews, 1998, p. 250).

Friendships

Across ethnic-group lines, older adults' life satisfaction depends more on the quality and quantity of contacts with friends than it does on contacts with younger members of their own family.

T. Antonucci,
Handbook of Aging and the Social Sciences, 1985

Overall, in terms of companionship, friends are more important and satisfying to older people than their offspring are (Cavanaugh, 1998a; Dykstra, 1990). In fact, some research suggests that greater loneliness exists among the single elderly who live with relatives than among those who live alone. An elderly widow who lives with her daughter's family can be quite lonely if she has little contact with associates her own age.

As in other aspects of human affairs, individuals differ enormously in what they view as adequate or inadequate contact with other people. Furthermore, it is not a certain absolute degree of isolation that produces feelings of loneliness in old age but, rather, the fact of becoming socially *more* isolated than one had previously been.

> Solitude need not be experienced as loneliness, while loneliness can be felt in the presence of other people. For instance, persons residing in nursing homes often complain of loneliness, even though they are surrounded by people and, at a superficial level, are interacting with them. Loneliness is the awareness of an absence of meaningful integration with other individuals or groups of individuals, a consciousness of being excluded from the system of opportunities and rewards in which other people participate (Busse & Pfeiffer, 1969, p. 188).

The quality of a relationship is more important than mere frequency of contact (Field et al., 1993). Overall, maintaining even one meaningful, stable relationship is more closely associated with good mental health and high morale among the elderly than is a high level of social interaction. A confidant serves as a buffer against gradual diminishing of social interaction and against the losses associated with widowhood and retirement (Sugisawa, Liang, & Liu, 1994). Moreover, being able to give to and for others seems to be very rewarding for older people and seems to reinforce their sense of well-being and independence (Roberto & Scott, 1986).

Retirement/Employment

Retirees have discovered that the secrets to happiness are as varied as their own interests.

Sandra Dallas, *Retirement Can Be Ideal*

Retirement is a relatively recent notion. Indeed, retirement as we understand it today did not exist in preindustrial American society—nor does it exist in some societies today. Older people formerly were not sidelined from mainstream employment, and life expectancies were much shorter. In 1900 the average American male had a life expectancy of 46.3 years and spent about 3 percent of his lifetime in retirement; today the average life expectancy of Americans has surpassed 75 years, with retirement averaging more than 10 years, or at least 13 percent of one's lifetime. Simultaneously, older Americans are leaving the labor force at younger ages by choice or by forced early retirement (Elder & Pavlko, 1993) (see Figure 18-2). Consequently, Americans now spend a substantially greater proportion of their lives in retirement than ever before. Even so, *retirement* has become a somewhat ambiguous term because retirement is frequently a transitional process in which a person leaves a "career" job and engages in other paid activities before withdrawing completely from the labor market (Hayward, Crimmins, & Wray, 1994; Parnes & Sommers, 1994).

The proportion of males aged 65 and over who were gainfully employed dropped from 68 percent in 1900, to 42 percent in 1940, to only 16 percent today (Bossé, 1998). Of equal social significance, the proportion of men aged 55 to 64 who were employed declined from 89 percent in 1947 to 69 percent in 1990, a drop of 20 percentage points. The U.S. Bureau of Labor Statistics estimates that by the year 2000 only one in four men 60 years of age and over will be working (Lewin, 1990).

Three-fourths of men and more than four-fifths of women retiring on Social Security currently leave their jobs before they are 65, and in companies with high early-retirement pension benefits, the retirement age drops below 60 (Bossé, 1998). In government, nearly two out of three civil servants retire before age 62. Workers who retire at age 60 or 62 can expect to have 15 to 20 years of life remaining. So retirement appears to be a factor of mounting significance in the lives of U.S. men. And it is taking on added significance in the lives of women, as more and more women enter the labor force (in 1998 about 8 percent of women over 65 hold jobs or are seeking work, down from 9.5 percent in 1971) (U.S. Bureau of the Census, 1998b).

People who return to work after tasting retirement usually do so rather quickly or not at all. Some

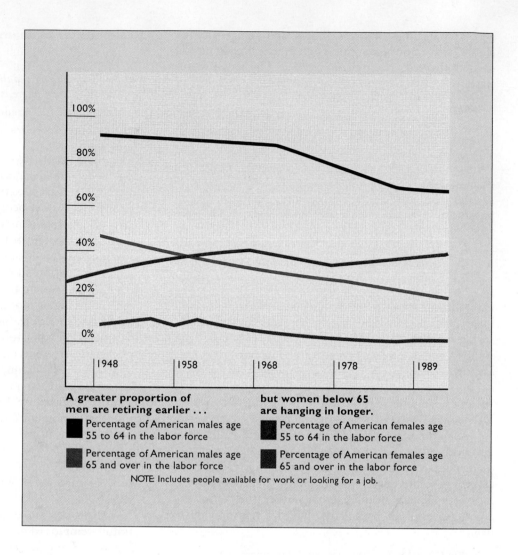

Figure 18-2 Elderly Americans, Particularly Men, Have Shown an Increased Willingness to Retire Early
Over the past several decades, the nation's older workers have been retiring at an earlier age.
Copyright, May 14, 1990, *U.S. News & World Report.*

individuals with good health, a strong commitment to work, and a corresponding distaste for retirement can and do work well into their seventies and early eighties (Parnes & Sommers, 1994). Of men aged 55 or older who retire, about one-third return to the workforce (see Table 18-3). The self-employed, professionals, sales workers, and farm laborers have higher than average rates of returning. But nearly three-fourths of all those who resume work do so in the first year of retirement; an additional 18 percent join them in the second and third years. Significantly, about two-thirds of postretirement work is full-time (Bossé, 1998; Otten, 1993a).

Involuntary Retirement In 1978 Congress passed legislation banning compulsory retirement for most workers before age 70 and in 1986 passed legislation largely abolishing mandatory retirement at any age. Many Americans view the practice of compelling workers to retire as a curtailment of basic rights. Prior to 1978 about half of the nation's employers had policies requiring employees to step down at 65. Business organiza-

tions like the U.S. Chamber of Commerce supported forced retirement; they argued that abolishing mandatory retirement would severely disrupt companies' personnel and pension planning and would keep younger people from advancing. For the most part, however, these fears have not been realized. Despite the changes in the law, there has not been a reported increase in the labor force participation of older Americans. People have increasingly wanted and chosen to retire at earlier ages, so the new laws have affected only a small minority of workers (Dentzer, 1990). Nevertheless, high unemployment in many industries (especially in mining and manufacturing), coupled with the prospect of an extended layoff or a difficult search for a new job, has led many older workers to opt for early retirement. And in other cases some older workers have been crowded out of their jobs or have been given special incentives for leaving to make room for younger workers. However, women have generally not worked enough years to be eligible for the generous pension inducements offered by many firms (Uchitelle, 1991).

Table 18-3	Americans 65 and Over in the Workforce	
Year	Men %	Women %
1900	68	9.1
1940	42	6.0
1994	17	9.2

Raymond Bossé, "Retirement and Retirement Planning," *Clinical Geropsychology.* Copyright © 1998 by the American Psychological Association. Reprinted with permission.

Contemporary employers, although often pleased with the work habits of older employees, still harbor many negative attitudes toward older workers that are rooted in concerns about higher health insurance costs and older employees' inexperience with new technology such as computers and robotics.

Seemingly, businesses looking to elderly workers to make up for labor shortages might have to offer rather substantial financial inducements to lure some of them out of retirement; many value their free time and independence more than they do the income or other benefits of a job (Parnes & Sommers, 1994). Surveys undertaken by the American Association of Retired Persons (AARP) show that from a quarter to a half of older workers and retirees would delay retirement if they could work fewer hours. Such findings seem to suggest that it does not make sense to operate on a system where people go from working all the time to one where they do not work at all (Lewin, 1990).

Retirement Satisfaction By Western norms, people are integrated into the larger society by their work roles. Work is seen as an important aspect of identity and self-esteem and as providing people with many personal satisfactions, meaningful peer relationships, and opportunities for creativity—in sum, the foundation for enduring life satisfactions. The loss of these satisfactions through retirement has been viewed traditionally as stressful, inherently demoralizing, and leading to major problems in older age (Mowsesian, 1987).

In recent years the negative view of retirement has been challenged. Probably no more than a third of Americans find retirement stressful, either as a transition or as a life stage (Bossé, 1998; Midanik et al., 1995). Moreover, some research suggests that it is money that is most missed in retirement and that when people are assured an adequate income, they will retire early (Beck, 1982). One longitudinal survey of 5,000 men found that most men who retire for reasons other than poor health are "very happy" in retirement and would, if they had to

do it over again, retire at the same age. Only about 13 percent of whites and 17 percent of blacks said they would choose to retire later if they could choose again. Other studies have similarly found that when individuals are healthy and their incomes are adequate, they typically express satisfaction with retirement (Bossé, 1998). Significantly, more and more retirees are also going back to college.

Social scientists are increasingly coming to recognize that preretirement lifestyle and planning play an important part in retirement satisfaction (Bossé, 1998). Positive anticipation of retirement and concrete and realistic planning for this stage in life are related to adjustment in retirement. On the whole, voluntary retirees are more likely to have positive attitudes and higher satisfaction in retirement than those who are forced into early retirement. However, other factors, not always controllable, also influence retirement satisfaction. In sum, those with better health and higher socioeconomic status seem to make a better adjustment to retirement.

Questions

Do the majority of retirees find life after retirement to be stressful? Which elderly are most likely to be dissatisfied with retirement?

Living Arrangements

A change in living (housing) arrangements represents a significant life event for an older person (VandenBos, 1998). Several factors are likely to initiate a change (or changes) in living arrangements for the elderly in the United States, including retirement, change in economic conditions, death of a spouse, and declining physical and/or mental capacities. Potential living arrangements include living at home alone, living at home with assisted-living services, living with adult children with adult day-care options, institutional care, retirement communities, group homes and community shelters, or homelessness. The majority of elderly adults prefer to remain independent in their own homes as long as they can (VandenBos, 1998).

Living Alone at Home and Assisted-Living Services Nine million Americans over the age of 65 live alone, but 2 million indicate they have no one to turn to if they need help (see Table 18-4). Of those living at home, 80 percent are women. Nearly half of those aged 85 and older live alone (Profile of Older Americans, 1998). Large numbers of the elderly are now cared for more appropriately, as well as more economically, in their homes—with the help of visiting nurses and meals-on-wheels services, but lack of a caregiver is a serious problem for many of these elderly. Depending on the state,

Table 18-4   Living Arrangements of Persons 65 and Older, 1995

	Men %	Women %
Living with spouse	72	40
Living with other relatives	8	17
Living alone or with nonrelatives	20	43

Source: *Profile of Older Americans, Household and Family Characteristics: March 1997, Current Population Reports,* P20-509 (Washington, DC: U.S. Bureau of the Census), retrieved on January 14, 1999, from the World Wide Web: www.aoa.dhhs.gov/aoa/stats/profile

home-care assistance may be available under the following programs: Medicare, Medicaid, the Older Americans Act, and Title 20 of Social Security. Most elderly fear being forced to leave their home because of frailty, illness, widowhood, a decaying neighborhood, or rural isolation and fear of becoming dependent upon their relatives or institutional care. Although the philosophy today within the field of aging is to promote independence, many elderly eventually develop health or cognitive problems that make them unable to safely remain in their homes.

Living with Children and Adult Day Care   For the elderly who are living with relatives, senior day-care centers give busy family members a break from caregiving and a chance to work or catch up on tasks. **Adult day care** involves long-term care support to adults who live in the community, providing health, social, and support services in a protective setting during any part of the day. Socialization, recreation, nutrition, and professional supervision are provided for a few hours or on a daily basis, as needed. Some elderly from ethnic minority families with debilitating illness or dementia expect to live with their adult children, who are expected to care for their parent or grandparent without seeking assistance from the greater community—to do so would be shameful or indicate failure on the part of the adult child to meet filial obligations (see the *Human Diversity* box earlier in this chapter). This puts a great psychological and emotional burden on the adult children or grandchildren.

Institutional Care   In the United States there are 1.6 million elderly people—5 percent of all those 65 and over—in some 22,000 nursing homes (Goldstein, Damon, & Taeuber, 1993). The figure is expected to reach 2 million by the year 2000, 2.8 million by 2020, and 4.4 million by 2040. Like individuals, each caregiving environment has a distinct personality, and some social climates are friendlier, more oriented toward independence, and better organized than others. Certainly many

facilities are doing their best to provide the elderly with decent care. Ethnic minorities are significantly underrepresented in nursing home populations in the United States (Fried & Mehrotra, 1998). Language and bureaucratic barriers are often cited as reasons, as well as other cultural norms that enforce intergenerational responsibility.

Even so, nursing homes are hardly "homes." Becoming a resident of a nursing home shifts control of one's life from the individual to the "total institution" and often brings physical and chemical restraint (Mor et al., 1995). Once in a nursing home, the elderly typically become physically, emotionally, and economically dependent on the facility for the rest of their lives (Wolinsky et al., 1992). Staff practices all too often foster patient dependency (Baltes, Neumann, & Zank, 1994). Indeed, some of the characteristics frequently encountered among the institutionalized aged, including depression, feelings of helplessness, and accelerated decline, are partly attributable to the loss of control over their lives (Zarit, Dolan, & Leitsch, 1998). Under such circumstances of forced dependency, the elderly come to see themselves as powerless—as passive objects manipulated and buffeted by the environment. In many nursing homes, the majority of residents share a small bedroom with others; eat mass-prepared, high-carbohydrate meals in a large, linoleum-floored dining hall; and watch television in a common recreation room.

Many of the nation's skilled-care nursing homes fail to meet federal standards for clean food and do not administer drugs properly or on a timely basis. Inspection reports on code violations in nursing homes can be difficult to locate, and the decision to place a relative in a nursing home is usually done in a few days under emergency circumstances through recommendations of hospital personnel. Not having enough time to get all the facts to make good decisions about placement, most people select a home nearby so they can visit their loved one. The elderly poor, especially blacks, are often placed in public facilities—frequently state mental hospitals—that generally provide inferior care. Frail and disabled elderly are often pushed into passive roles; some patients are tied to their beds or wheelchairs or given powerful tranquilizing drugs on the premise that "a quiet patient is a good patient." Recent research suggests that it is rarely necessary to restrain nursing-home patients suffering from mental confusion or extreme physical weakness (Brody, 1994c). Most nursing homes in the United States are owned by private proprietors and are operated for profit. In European countries, in contrast, long-term institutional care is usually under government sponsorship, with local public control operating in accordance with national guidelines (Landsberger, 1985).

Inadequate, unskilled staffing is a major problem in many nursing homes. People with little education and minimal skills are often hired as aides, orderlies, janitors, and kitchen help because the jobs are viewed as unattractive, low paying, and rather stressful. With assisting many who may have dementia or multiple care needs, such personnel may not develop a strong commitment to their jobs or to the nursing home and may abuse the patients (Zarit, Dolan, & Leitsch, 1998).

As the nursing-home industry has evolved in the United States, nursing homes have become a place of last resort for a variety of problems. They are used by terminally ill patients who require intensive nursing care, by recuperating individuals who need briefer convalescence, and by less ill but infirm aged who lack the social and financial resources necessary to manage in the community. Consequently, nursing-home residents differ greatly in degree of physical impairment and mental disorientation. Indeed, about 15 to 40 percent of the elderly in nursing facilities could be more appropriately cared for in private homes, and some are now being moved into adult foster care (Roberts et al., 1991).

Government regulations have cast nursing homes in the role of miniature hospitals (Winslow, 1990). The quality of life they afford is largely determined by bureaucratic requirements regarding Medicare and Medicaid and by state regulations mandating standards for licensing. Even chronically ill persons moving into a nursing home continue in a "sick role" long after it is no longer functional, forfeiting independence, autonomy, and quality of life (Cohn & Sugar, 1991). For suggestions on factors to consider when choosing a nursing home, see *Further Developments:* "Selecting a Nursing Home."

Retirement Communities More retirement communities are being developed for middle-class and well-to-do elderly, particularly in the warm southern coastal states and golf resort communities in the southwest. Of concern to politicians in many northern states is the mass migration of their newly retired elderly, who have been productive citizens and provided a stable tax base in many communities. Some retirement communities are known as "trailer" communities, where the elderly can live comfortably with fewer home-care responsibilities and continue to lead active lives with senior citizens like themselves. Some are townhouses and condos that border lush golf courses and other prime real estate. The nature of such communities, however, is unlike a typical community where people of all ages live side by side. Some elderly especially miss being with children and younger adults daily.

Adult Group Homes Some communities are experimenting with newer alternatives such as sheltered housing, assisted-living facilities, and continued-care retirement communities—protected communities arranged in apartment complexes or detached cottages—that provide such supportive services as centrally prepared meals, housekeeping, laundry, transportation, recreational options, and health care (Jeffrey, 1995). This is projected to be a more common living arrangement over the next decade, allowing the elderly to maintain much of their self-reliance while providing a sense of security and support (Zarit, Dolan, & Leitsch, 1998).

Questions

What are some alternatives for assisted-living arrangements for elderly adults who need supervision or care? What are some factors that should be considered when selecting a nursing home?

Elder Abuse

Our study suggests that elder mistreatment is as dangerous to the health and well-being of older adults as the many chronic diseases we associate with death and disability.

Dr. Mark S. Lachs, New York Hospital–Cornell University Medical College, August 1998

Elder abuse and neglect are both acts of commission and omission that cause unnecessary suffering to older persons (Wolf, 1998). It is estimated that one million Americans age 60 and older are abused in domestic settings each year. Another study reports that as many as 5 to 6 percent of persons 65 and older have experienced elder abuse. Because most cases never come to the attention of the authorities, the full scope of the problem isn't known (Wolf, 1998). The old stereotype of a younger male abusing a frail elderly female victim is inaccurate. Pillemer and Finkelhor (1988) found that spousal abuse was more prevalent than abuse by adult children. Furthermore, a recent survey conducted by Harris (1996) revealed that for more than half, the physically abused spouses had been mistreated for years.

Elder abuse is the form of family violence about which we know the least—for it often goes unreported—and the risk factors for elder mistreatment are inconsistently reported in scarce research studies. Findings of a 13-year study were recently reported by Lachs and colleagues (1998). Risk factors were analyzed for both reported and verified elder abuse and neglect in a Connecticut cohort of 2,812 adults 65 and older from diverse ethnic, racial, and social backgrounds. Research methodology included inception interviews, yearly phone interviews, and in-person interviews once a year for three years as well as psychometric evaluation of functional and cognitive impairment. The legal definitions of abuse include the following:

Further Developments

Selecting a Nursing Home

Americans have an enormous fear of growing old, becoming infirm, losing their minds, and being placed in a nursing home, and the vast majority of spouses and adult children typically postpone the arrangement as long as they can (Montgomery & Kosloski, 1994). When they do place a spouse or an elderly parent in a nursing home, it is usually at the point of desperation (Newman et al., 1990). Adult children find the decision to be exceedingly excruciating, and they commonly feel ambivalence, shame, and guilt. One 49-year-old East Coast public relations woman tells of her anguish after her 80-year-old mother had suffered a number of small strokes (Moore, 1983, p. 30):

> When it was clear that she couldn't go on living alone, we hired round-the-clock nurses at $400 a week. But first one didn't show up; then my mother didn't like another. Each time it was something else. So we brought her to our house for a while, but it was extremely hard on me and the rest of the family. We talked about a nursing home. And though she didn't want to go, it became apparent that it was the only way. She started in a minimum-care facility. But she became more and more confused, so they decided to switch her to their skilled-care facility—a decision in which I had no choice. . . . It's very sad. She keeps asking: "When can I get out of here and get on with my life?"

Clearly, the prospect of nursing-home care can be traumatic for both the aging parents and their adult children. Should a nursing home be required, the person who is going to be placed there should be involved in the planning, when possible. A number of nursing homes should be visited before a decision is made. A list of licensed nursing homes can generally be secured from the local Social Security office or county nursing service. Nursing homes can choose their patients, and they often give preference to those who demand the least attention and care. Many good nursing homes have a waiting list. There are a number of things to look for in a nursing home:

The Facility

- Is the nursing home licensed by a governmental agency? Are recent inspection reports available?
- What arrangements exist for the transfer of a patient to a hospital, should it be required?
- Are the rooms clean, relatively odor-free, and comfortable?
- What are the visiting hours? Who is welcome?
- Does each room have a window, and does the room open to a corridor?

Safety

- Is the building fire-resistant? Does it have a sprinkler system? Are emergency exit routes clearly posted?
- Are there ramps for wheelchairs and handicapped persons?
- Are the hallways well lighted?
- Are the floor coverings nonskid and safe?
- Are there grab bars in appropriate settings, including bathrooms?

- *Abuse.* Physical abuse typically means intentionally inflicting, or allowing someone else to intentionally inflict, bodily injury or pain and includes such harmful behaviors as slapping, kicking, biting, pinching, and burning, as well as sexual abuse. It might also include inappropriate use of drugs and physical restraints.
- *Psychological abuse.* This includes verbal harassment, intimidation, denigration, and isolation. It might include repeated threats of abandonment or of physical harm.
- *Neglect.* This includes failure of a caretaker to provide the goods, services, or care necessary to maintain the health or safety of a vulnerable adult. Neglect can be repeated conduct or a single incident that endangers the person's physical or psychological well-being (Morris, 1998).

- *Exploitation.* Taking advantage of an older adult for monetary gain or profit (Lachs et al., 1997).

The findings from the 13-year study were as follows: 47 members of the 2,812 cohort had experienced abuse that was substantiated: 9 (19 percent) were cases of abuse; 8 (17 percent) were cases of exploitation; and 30 (64 percent) were cases of neglect by another party. Adult children were the most common perpetrators, followed by spouses, then by grandchildren and paid caregivers (Lachs et al., 1997). This current study discovered an association between the following risk factors and elder mistreatment: poverty, minority status, functional and cognitive impairment, worsening cognitive impairment, and living with someone (spouse or family member). Age was an additional risk factor, but gender was not. With respect to social network factors, living alone was

Staff

- Are a physician and a registered nurse on call at all times?
- What provision is made for patient dental care?
- Are the staff patient with questions, and are they happy to have visitors inspect the facility?
- Is there a physical therapy program staffed by a certified therapist?
- Are most of the patients out of bed, dressed, and groomed? (Be suspicious if you see many patients physically restrained in their beds or chairs or if they appear to be overtranquilized.)
- Are staff members on the floor actually assisting residents?

Activities

- Does the facility offer a recreation program?
- Are the grounds well maintained, and are patients encouraged to get outdoors when weather permits?
- How are patients kept busy? Are patients left idle and shown no care or interest?
- Are activities scheduled outside the facility and in the community?

Food

- Are the meals adequate and appetizing? Are they served at the proper temperature?
- Does a dietician prepare the menu?
- Is the menu posted, and does it accurately describe the meal?
- Does someone notice if a resident is not eating?

- What provision is made for patients who have difficulty feeding themselves?
- What arrangements are made for patients requiring special diets?

Atmosphere

- Are patients accorded privacy in receiving phone calls and visits and in dressing?
- Can residents send and receive mail unopened?
- Are patients allowed to have personal belongings, including items of furniture?
- May residents have plants?
- Is there an outdoor area where residents can sit?
- Do residents of the home recommend it? What do they have to say about it?
- Are patients permitted to wear their own clothing?
- Are the patients well groomed? If they cannot bathe themselves, are they bathed daily and as needed?
- What arrangements are made so that patients can follow their own religious practices?
- Are patients treated with warmth and dignity?

Cost

- Is the facility approved for Medicare/Medicaid reimbursement?
- Will the patient's own insurance policy cover any of the expenses?
- Is the cost quoted inclusive, or are there extra charges for laundry, medicines, and special nursing procedures?
- Are advance payments required? Are the payments refunded should the patient leave the home?

Sources: Adapted from W. Andrew Achenbaum, "Perceptions of Aging in America," *National Forum: Phi Kappa Phi Journal, 78,* No. 2 (Spring 1998), 30–33; and "Nursing Homes: When a Loved One Needs Care," *Consumer Reports, 60* (August 1995), 518–528.

significantly associated with protecting older adults from abuse (80 percent of mistreated subjects lived with someone) (Lachs et al., 1997).

Caregiver Burnout   Caregivers of the elderly with cognitive and functional impairments—even those who are loving and dedicated—are at high risk to experience "caregiver burnout," (Marks & Lambert, 1998). It is most likely that women will become the unappreciated caretakers of disabled elders (Marks & Lambert, 1998). In one study, 20 percent of caregivers said their tasks were so frustrating and difficult that they were afraid they might hurt the patient (Lachs et al., 1998). Pillemer says caregivers must first recognize there's a problem, must not feel guilty, but must seek help. Caretakers should address these questions: "Am I depressed?" "Do I fly off the handle?" "Do I resent my relative?" "Am I

denying my relative social activity?" "Do I threaten nursing home placement?" Pillemer (Lachs et al., 1998) recommends the following strategies to cope with caregiving:

- Join a support group that shares the same problems.
- Continue activities you enjoy.
- Seek professional help for parents' physical needs.
- Get more information about caregiver burnout from area agencies on aging.
- Investigate adult day care or adult respite options in your community.

Elder Abuse in the Long-Term-Care Community
Providers now recognize that elder abuse and neglect must be prevented—for they have a devastating impact

Caregiver Burnout Women are generally the ones who provide caretaking. Caregivers who take care of a loved one with cognitive impairment or Alzheimer's disease are likely to experience frustration, depression, and difficulties coping and should seek support or relief for themselves.

on the victim, the confidence of other residents, their families, and the morale of staff, and they arouse outrage of the public and law enforcement officials. The state of Massachusetts has instituted a pilot training program of video training, conferences, and workshops to educate staff on recognizing and preventing elder abuse. The American Health Care Association (AHCA), which represents 11,000 long-term care facilities nationwide, is distributing this video and lobbying Congress for a federal database for CORE (criminal offenders record information) investigations to screen out employment candidates with prior convictions, pending charges of elder abuse, theft, or other serious crimes. In September 1997 in a significant case for patient rights, the Arizona Supreme Court upheld that families of elderly persons who are abused by their caretakers may recover damages for the victims' pain and suffering—even after their deaths (*Denton v. American Family Care Corp.*) (Cassens, 1998).

Questions

In what ways are the elderly victimized by abuse, and who is most likely to perpetrate that abuse? What can be done to prevent elder abuse?

Policy Issues and Advocacy in an Aging Society

The original Older Americans Act was passed in 1965, in response to the needs of a growing number of older people. This legislation established the Administration on Aging (AOA), an agency of the U.S. Department of Health and Human Services and headed by the Assistant Secretary for Aging. This national agency advocates and administers programs for senior citizens. Also, it sponsors research and training programs on aging and seeks to educate older people and the public through its regional, state, and area offices about benefits and services available to help this vulnerable population. Nationwide there are 660 Area Agencies on Aging. AOA programs help older persons remain in their homes and offer opportunities for older Americans to enhance their health and remain active in their families and communities through volunteer and employment programs. Supportive services include:

- Information and referral, outreach, case management, escort and transportation
- In-home services, such as personal care, chores, home-delivered meals, home repair
- Community services, such as senior centers, adult day care, elder abuse prevention, congregate meals, health promotion, employment counseling and referral, and fitness
- Caregiver services, such as counseling, education, and respite

As of fall 1998, the Older Americans Act was up for Congressional renewal (S.2295 and H.R. 4344, The McCain-Mikulski and Defazio-Lobiondo Reauthorization of the Older Americans Act). Many of the same issues concerning the elderly are still pressing, and delay in passing this act would affect many vital services for the elderly. The 1999 public policy agenda issues for seniors continue to be economic security (pension plans, retirement planning, and Social Security), health care issues (Medicare and HMOs), access to community-based services, maintaining or securing employment (second careers), education and volunteer issues, and adequate, safe housing (National Council on the Aging, 1998). The

American Association of Retired Persons (AARP), the Gray Panthers, and the National Council on Aging (which sponsors Meals on Wheels and foster grandparents) are private, nonprofit advocacy organizations that promote a higher quality of life for our elderly citizens.

Segue

As you have now read and discussed the research findings about the status of America's elderly citizens, note the remarkable progress that has occurred in improving the longevity and the quality of life for our senior citizens. Many of our elderly are reaping the benefits of recent years of research in the physical and social sciences. It seems it was only a few years ago we heard the word *gerontology* for the first time, yet now countless researchers are studying longevity, aging, and the quality of life for all of us when we reach late adulthood. In October 1998 John Glenn reentered the world of space exploration at the age of 77 to provide American scientists with never-before-gathered data on the effects of aging. He passed rigorous physical and cognitive tests to be selected. His research in space aptly symbolizes that the study of aging is truly in its infancy—and "the sky's the limit" for senior citizens living around the world. In our concluding chapter, we shall examine several end-of-life scenarios for people of all ages and the range of coping strategies employed by those loved ones who remain.

Summary

Social Responses to Aging

1. Industrialized societies have extended longevity for many people and provided economic and social resources for them to lead more satisfying lives in old age. Consequently, many of our American elderly are experiencing greater satisfaction in their later years.

2. With our culture's preoccupation with youthfulness, most Americans want to ignore late adulthood or have distorted perceptions of it. These stereotypic views have negative effects on older adults' self-esteem.

3. Among the elderly, positive emotions have been found to be related to favorable life events, positive health status and functional ability, availability of social contacts, and higher levels of educational attainment.

Self-Concept and Personality Development

4. The elderly confront the issue of integrity versus despair. Provided they have successfully navigated the previous stages of development, they are capable of facing their later years with optimism and enthusiasm; those who appraise their lives as having been wasted experience a sense of despair.

5. Peck stresses the changes people go through, from their reflections on retirement, recurring illness, and mortality.

6. Emotional health among the elderly is described as the "clear ability to play and work and to love," and to achieve satisfaction with life. An ability to handle life's blows without passivity, blame, or bitterness is especially important.

7. Neugarten and colleagues identified four major personality types, or *traits,* related to the aging process: integrated, armored-defended, passive-dependent, and disintegrated.

8. Other theories of aging concentrate on disengagement, activity levels, social usefulness, social exchange, and the modernization of society.

9. Some research indicates that a few aspects of personality, including emotional intelligence and wisdom, actually improve during what the Balteses call the "Third Age," though decline is evident in the "Fourth Age."

10. Research from around the world suggests that there is a healing association between religion and health. One study of 2,754 male and female participants found that women who attended church lived longer than those who did not.

Familial Roles: Continuity and Discontinuity

11. The quality of marriage varies from couple to couple. Most elderly husbands and wives report greater happiness and satisfaction with marriage during their later years than at any other time except for the newlywed phase.

12. Only 42 percent of elderly women still reside with their husbands. This statistic demonstrates a particular gender issue that is of great concern worldwide: Elderly women are at much higher risk of living in poverty.

13. About four of every five of the elderly have living children, and the elderly are not so isolated from kin and friendship networks as is commonly believed, due to exchanges of mutual aid and proximity of kin.
14. Child psychologists emphasize that both children and their grandparents are better off when they spend a good deal of time in each other's company.
15. Siblings often play a significant role in the lives of the elderly. They provide a continuity in family history that is uncommon to most other family relationships and might be the only members of the family of origin who are still alive.

Social and Cultural Support

16. Some research suggests that the single elderly who live with relatives are lonelier than those who live alone. An elderly widow who lives with her daughter's family might be quite lonely if she has little contact with associates her own age.
17. Retirement as we understand it today did not exist in preindustrial American society—nor does it exist

in some societies today. Older people formerly were not sidelined from mainstream employment. Many of today's workers can expect to have 15 to 20 years of life remaining after retiring.
18. Nine million Americans over the age of 65 live alone, but 2 million indicate they have no one to turn to if they need help; other arrangements are living with families, nursing homes, retirement communities, and adult group homes.
19. It is estimated that a million Americans age 60 and older are abused in domestic settings each year. Another study reports that elder abuse is as prevalent as 32 per 1,000 adults. Because most of the cases never come to the attention of the authorities, the full scope of the problem isn't known.
20. The quality of life for American elderly has been improved by public policy issues, the Older Americans Act, AARP, the Gray Panthers, and community supports.

Key Terms

activity theory of aging *(560)*
adult day care *(576)*
disengagement theory of aging *(560)*
elder abuse *(577)*

integrity versus despair *(556)*
life review *(557)*
modernization theory *(560)*
role exit theory of aging *(560)*

social exchange theory of aging *(560)*
selective optimization with compensation *(561)*
wisdom *(561)*

Following Up on the Internet

Web sites for this chapter focus on the changing demographics and quality of life for those in late adulthood. Please access the text web site at http://www.mhhe.com/crandell7 for up-to-date hot-linked Internet addresses for the following organizations, topics, and resources:

Vitality for Life Report

APA Division 20 on Aging

National Institute on Aging (NIA)

American Association of Retired Persons (AARP)

Organizations Representing Professionals in the Field of Aging

Professional Opportunities in the Field of Aging

Part Ten

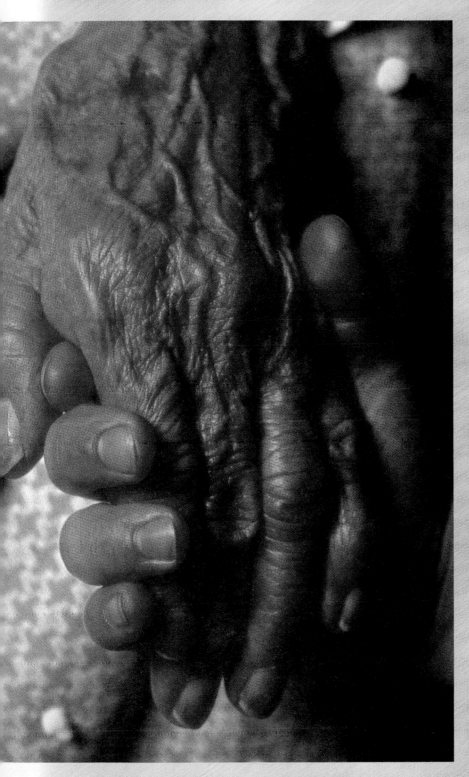

The End of Life

Over the past 30 years, the old American practice of "hiding death" has largely been abandoned. With considerable truth it can be said that Western society has "rediscovered" death. An outpouring of scholarly (and not so scholarly) works in sociology, psychology, and thanatology, television documentaries, paperback books, and feature articles in newspapers and magazines have drawn attention to every aspect of the topic. Simultaneously, controversy has swirled in the mass media, in the political arena, in legislative chambers, in courtrooms across the country, and in our own homes about such matters as the right to die, clinical death, adolescent homicides, the death penalty, mass suicide, and life after death. People are more willing to discuss dying and death, and such discussions have paved the way for families to experience more meaningful closure to life for loved ones. Our gradual cultural embracing of death as a natural stage of development is leading toward a deeper understanding of the meaning of life.

Chapter 19 Dying and Death

Critical Thinking Questions

1. If death is a natural part of life—if the natural progression of life is birth, growth, maturity, aging, and death—then why do so many people regard death as bad?

2. If you had a life-threatening illness, would you want the physician to tell you? Why or why not?

3. Why is it so difficult for us to console someone who is grieving after the loss of a loved one?

4. Would you fill out a do-not-resuscitate order or living will for yourself? Why or why not?

In the cycle of life, each person comes into this world alone and then—typically many years later—exits alone. During the time in between, called a lifetime, each of us ponders about this mystery of death many times: when experiencing the actual loss of loved ones, when viewing media portrayals of dying and death, or perhaps when conducting one's daily life as a soldier or police officer, fireman, medical professional, clergy, or hospice volunteer.

Throughout the ages, people from every culture have wondered about this great mystery by taking risks and "teasing" death, by wrestling with or writing about their fears, by devising elaborate rituals in preparation, and by building monuments in recognition. Some even welcome or promote their own "final exit." We conclude our final chapter of human development by discussing cultural awareness of issues in thanatology, cross-cultural views of death and grief, the role of religious beliefs, the stages of dying, and adjusting to the death of a loved one. At best, it can be said that facing death is as individual as our own personal philosophies about how we have lived our lives.

The Quest for "Healthy Dying"

A cultural revolution is underway in our society, bringing with it sweeping changes in our outlook and behaviors concerning death.

Hannelore Wass

Until the last few decades, death was a taboo subject in Western society. Dying was something to be kept out of sight and out of mind. Typically the terminally ill were whisked away to a hospital and our senile relatives and oldest old to nursing homes to live out their remaining days. Even medical schools avoided the topic of dying and death, and physicians did their best to ignore it in dealing with dying patients and their families—for their professional oath is to maintain and prolong life. Indeed, some social scientists claim that the United States is a "death-denying culture." Americans stress youth, beauty, and physical fitness. Less than 25 percent have wills, and many are made uncomfortable by discussions of death.

Death confronts us both subtly and blatantly in diseases such as Alzheimer's and cancer and in epidemics, such as AIDS. On a nationwide scale, it tore at our hearts and brought us to our knees when we heard of the Challenger disaster, the Oklahoma City bombing, the explosion of Flight 800 over Long Island Sound, and the tragic Columbine High School shootings. We conceal it under a variety of names—*abortion, stillbirth, mortality, morbidity, homicide, suicide, natural disasters, drive-by shootings, school-yard shootings, occupational fatalities, auto fatalities,* and so forth.

We are reminded of its inevitability in rituals invoking collective memories, such as memorials surrounding the deaths of famous persons—such as Princess Diana, Mother Teresa, Martin Luther King, and President John Kennedy—that invoke larger moral principles. As the traveling AIDS quilt or the traveling Vietnam Memorial make their way around the country, whole communities come together in deep sorrow for the thousands of lives lost. And most of us experience some forms of preliminary finality in our lives in the "endings" of marriages by divorce or widowhood, the end of careers or group memberships by retirement or resignation, and a passing over of family traditions such as when the children instead of the parents prepare the annual Thanksgiving meal. If we don't have children, we might decide it is time to go to the community or church Thanksgiving dinner with friends and acquaintances.

Even our everyday conversations are sprinkled with expressions about death: "My back is killing me!" "I was so embarrassed I could have died!" "You scared me to death!" Some people risk death on a daily basis: police, firemen, soldiers, the flag person for a construction crew on a superhighway, stunt people who allow themselves to be set afire or make death-defying leaps out of buildings, those who buy and sell drugs on the street, those who have sex with multiple partners, and others. Our television movies, popular action films, the evening news, and the local newspapers sensationalize death, making it seem less than real. Death and the true meaning of life have been themes in great literature from the earliest recorded texts, such as the Bible, the *Iliad* and the *Odyssey* to Shakespeare's *Romeo and Juliet* and *Hamlet.* We personify death by giving it names like "the Grim Reaper," "the Gentle Comforter," or the "Gay Deceiver" (Kastenbaum, 1997).

More recent death themes in literature and the media include the ideas of near-death experiences, past lives, life after death, communicating with the dead, and the presence of angels (e.g., the "Touched by an Angel" television series). Some of these works have made it to best-sellers lists, including *Life After Life* by Dr. Ray Moody, Jr., *Embraced by the Light* by Betty J. Eadie, *Conversations with God (Book 1, Book 2,* and *Book 3)* by Neale Donald Walsch, and *Talking to Heaven: A Medium's Message of Life After Death* and *Reaching to Heaven* by James Van Praagh. Though they are not grounded empirically, at the very least these works are pricking our consciousness and making us confront what we think about life and death. More sophisticated medical technology has permitted many people to come back from the brink of death, and Koerner (1997) reports that nearly one-third of these people—as many as 15 million Americans—report having had mystical experiences, with vivid images of an afterlife, when they were near death.

Brian L. Weiss—psychiatrist, experimental psychologist, distinguished university chairperson, and author of *Many Lives, Many Masters*—says that after many years of being immersed in the traditional, conservative aspects of his profession, scientists still have much to learn about the human mind that now seems to be beyond our comprehension. "Rigorous scientific study into the mystery of the mind, the soul, death, continuation of life after death is in its infancy in our society" (Weiss, 1988, p. 11).

Question

Perspectives on the causes and meaning of death have varied throughout history. In contemporary American society, what events are prompting a more open discussion of death and dying?

Thanatology: The Study of Death and Dying

If you can begin to see death as an invisible, but friendly, companion, on your life's journey—gently reminding you not to wait till tomorrow to do what you mean to do—then you can live your life rather than simply passing through it.

Elisabeth Kübler-Ross (1993, p. 47)

Over the past 20 years, public and professional awareness of the dying person's experience has increased dramatically. "Death with dignity!" has become a major rallying cry. Interest in the field of **thanatology**—the study of death (*thanatos* is the Greek word for "death")—has grown, and more colleges and health care settings are offering courses and programs in death, dying, and bereavement. Death awareness advocates assert that the power to control one's own dying process is a basic human right. They point out that in the United States the majority of those who die spend part of their final year in a nursing home or hospital, often in pain and alone. So allowing someone to die naturally often involves a team of professionals who must make a conscious decision whether to continue medical treatment. Some professionals draw a distinction between ordinary and extraordinary treatment, contending that ordinary measures such as nutrition and hydration should be continued, whereas extraordinary treatment such as dialysis with a kidney machine or artificial maintenance of blood circulation may be halted if the case is hopeless. *Death-with-dignity* advocates insist that aggressive medical care, the norm that life must be maintained at all costs, prevents people from dying quickly and naturally.

The "healthy dying" quest has led some *thanatologists* to expound on the good, acceptable, or self-actualized death (Kastenbaum, 1997, 1979). According to this view, it is not sufficient that death be reasonably free of pain and trauma. Instead, it is argued, individuals suffering a terminal illness should be able to select the particular style of exit that they believe to be consistent with their total lifestyle, such as a romantic death, a brave death, or a death that integrates and confirms the person's unique identity.

Overall, it seems that more and more Americans are trying to take back control of the time, place, and circumstances of their death. We are granting powers of attorney and making advance directives, purchasing cemetery plots and customized gravestones, executing wills and living wills, and so on. Some choose to leave the hospital to die at home. Many are giving family members or close friends the power to terminate medical treatment when they themselves can no longer do so. In a number of states, death-with-dignity advocates have secured the passage of laws that provide for the drawing up of a **living will**—a legal document that states an individual's wishes regarding medical care (such as refusal of "heroic measures" to prolong his or her life in the event of terminal illness) in case the person becomes incapacitated and unable to participate in decisions about his or her medical care (see *Information You Can Use:* "An Example of a Living Will"). The 1994 deaths of Jacqueline Kennedy Onassis and Richard Nixon, both of whom rejected medical treatment that could have prolonged their lives, have accelerated changes in Americans' approach to death (Scott, 1994). Other developments—including the considerable sales of Derek Humphry's 1991 do-it-yourself suicide manual *Final Exit* and widespread controversy revolving around Dr. Jack Kevorkian's practice of "assisted suicide"—suggest that some Americans want still more control at the end of life. More recently, some people who were healthy and functional formed a communal group, Heaven's Gate, and for reasons unrelated to health chose to end their lives together on their own terms. These people and events then bring more prominence to the right-to-die movement.

Question

How would you describe the study of thanatology, in your own words?

The Right-to-Die Movement

More than two decades have passed since the New York Times *startled many of its readers with the story of Jo Roman, a 62-year-old artist with terminal cancer, who took her own life with an overdose of Seconal after gathering intimates around her for a "last day." With her family and friends she completed a "life-sculpture" that consisted of a coffinlike pine box containing photograph albums, scrapbooks, paintings, and other cherished items, and drank champagne toasts in a rite of farewell.*

(Johnston, 1979)

"Healthy dying" has become something of a consumer demand in the United States. Public opinion surveys show that 8 out of 10 Americans believe patients should be allowed to die under certain circumstances, and about half say incurably ill people have the moral right to commit suicide. Only 15 percent say that doctors and nurses should always do everything possible to save a patient's life. Fifty-nine percent would want their doctors to stop administering life-sustaining treatment if they had a terminal illness and were in a great deal of physical pain. Significantly, a third of adult Americans can imagine

Information You Can Use

An Example of a Living Will

To my family, my physician, my lawyer and all others whom it may concern: Death is as much a reality as birth, growth, maturity, and old age. It is the one certainty of life. If the time comes when I can no longer take part in decisions for my own future, let this statement stand as an expression of my wishes and directions, while I am still of sound mind.

If at such a time the situation should arise in which there is no reasonable expectation of my recovery from extreme physical or mental disability, I direct that I be allowed to die and not be kept alive by medications, artificial means, or "heroic measures." I do, however, ask that medication be mercifully administered to me to alleviate suffering even though this may shorten my remaining life.

This statement is made after careful consideration and is in accordance with my strong convictions and beliefs. I want the wishes and directions here expressed carried out to the extent permitted by law. Insofar as they are not legally enforceable, I hope that those to whom this Will is addressed will regard themselves morally bound by these provisions.

_____   _____
Signed   Date

_____   _____
Witness   Witness

_____

Copies of this request have been given to _____

_____

Source: From Judy Oaks & Gene Ezell, _Dying and Death: Coping, Caring, Understanding_ (Scottsdale, AZ: Gorsuch Scarisbrick, 1993), p. 197.

themselves taking the life of a loved one who was suffering terribly from a terminal illness (_Gallup Poll Monthly_, 1991; Times Mirror Center for the People and the Press, 1990).

Much criticism is currently leveled at how modern technology is applied to the terminally ill. Critics contend that too much is done for too long a period at too high a cost, all at the expense of basic human considerations and sensitivities, and that the terminally ill become the property of health care institutions, which override individual autonomy, endurance, dignity, and personhood. Modern medicine has come to be seen by some as the enemy rather than the friend of the terminally ill (Nuland, 1994). Additionally, the argument is often made that ending "futile" medical care for dying patients could be an important cost-saving step in overhauling the nation's health care system. _Futile care_ is sometimes defined as "any clinical circumstance in which the doctor and consultants conclude that further treatment cannot, within a reasonable possibility, cure, palliate, ameliorate, or restore a quality of life that would be satisfactory to the patient" (Snider & Hasson, 1993, p. 1A).

Physician-Assisted Suicide (PAS) Requests for physician-assisted suicide are not new, and most physicians are likely to face this challenge. Physicians point out they have a problem in deciding what medical measures they should undertake. At times, patients and their families are so distraught and frightened that they will not or cannot say what they want done. Cicirelli (1997) reports from a study he conducted with subjects aged 60 to 100 that one-third of the respondents wanted a family member, physician, or close friend to make their end-of-life decision. The decision the family must make often presents an _avoidance-avoidance conflict_ in which people's emotionally-charged thoughts tend to freeze: _Do we attempt to save our loved one at all costs, even though this person may never regain consciousness or be the person he or she used to be? How can we let this person die whom we love so dearly?_

Physicians often withhold antibiotics from terminally ill and senile patients who develop respiratory infections. It is often the nurses who shoulder much of the responsibility for making such decisions. The nurses might decide not to call the physician after the onset of fever, or they might influence the physician to opt for nontreatment. Significantly, the American Hospital Association estimates that 70 percent of the 6,000 deaths occurring in the United States every day are somehow timed or negotiated by patients, families, and doctors who arrive at a private consensus not to do all that they can do and instead allow a dying patient to die (Malcolm, 1990).

A national survey of 1,902 physicians (Meier, 1998) found that of those who have been asked by a patient to assist with a suicide, some admit having already assisted a suicide. It was also found that if assisted suicide were legal, more physicians would be willing to assist (see Table 19-1).

Taboo has long silenced medical personnel who have sped up the death of an incurably ill patient or helped such individuals commit suicide. When pressed on the issue, the medical profession typically concedes that throughout history some physicians have helped their patients end their own lives. In recent years, a number of physicians have publicly confided that they have sped up the death of an incurably ill patient or helped such individuals commit suicide. They have injected overdoses of narcotics or written prescriptions for drugs potent enough to end their patients' suffering (Quill, 1993). However, such "hidden practices" are risky for patients and can potentially damage a medical practitioner's reputation. By the same token, however, terminally ill patients who take their own lives often find that they must die alone so as not to place their caregivers or families in legal jeopardy.

Some of the most painful decisions physicians confront concern newborns. With today's neonatal technology, about half the infants born weighing 750 grams (1 pound 10 ounces) can be saved. However, there is a high risk that they will have serious physical or mental handicaps. Another dilemma concerns infants born with life-threatening complications, such as a blockage of the intestinal tract or a heart defect, that require surgical correction. A somewhat similar issue is posed by newborns with meningomyelocele, a condition in which the spinal cord is deformed and protrudes outside the body. If untreated by surgery, the infants develop a spinal infection and die; if treated, they often suffer paralysis and incontinence. All these cases entail complex bioethical decisions as to whether severely handicapped infants should be treated so that they can survive.

The American Medical Association (AMA) said in 1986 that doctors could ethically withhold all means of life-prolonging medical treatment, including food and water, from patients in irreversible comas even if death were not imminent. The withholding of such therapy should occur only when a patient's coma is beyond doubt irreversible and there are adequate safeguards to confirm the accuracy of the diagnosis. Although the opinion of the 271,000-member association did not make such an action mandatory for doctors, it did open the way for them to withdraw life-prolonging treatment with less fear of being taken to court and to use the opinion as a defense if they are challenged. Though technological developments continue to outpace the nation's legal framework, the overall trend seems to be toward basing decisions less on strictly legalistic interpretations regarding specific treatments and more on balancing benefits on a case-by-case basis. The debate on these issues has spread to the treatment of patients with acquired immune deficiency syndrome (AIDS) (see *Further Developments:* "The Worldwide Epidemic of AIDS").

Probably most controversial of all is the issue of **euthanasia,** or "mercy killing." The practice of euthanasia goes back to ancient history. *Passive euthanasia* allows death to occur by withholding or removing treatments that would prolong life. In *involuntary euthanasia,* someone in the family or a legally empowered person decides to withhold or remove medical treatments when the patient is medically considered brain dead. In *voluntary euthanasia,* the patient grants permission to remove treatments that would prolong life. Some people have prepared and signed a legal document known as a "living will" that states the person's wishes about medical treatment in case she or he becomes unable to participate in decisions regarding medical care (see *Information You Can Use:* "An Example of a Living Will").

Table 19-1	Physician Responses to Assisting a Patient with Suicide	
Survey Question		**Percentage Answering Yes**
Received request from patient for some type of suicide.		18
Received request from patient to provide lethal injection.		11
Would prescribe medication to patient to use in suicide under present conditions.		11 (would rise to 36% if legal)
Would administer lethal injection if asked under present conditions.		7 (would rise to 24% if legal)
Have assisted a suicide at least once.		6

D. Meier, "A National Survey of Physician Assisted Suicide and Euthanasia in the United States," *New England Journal of Medicine, 338* (April 23, 1998), 1193–1201. Cited in Arthur Zucker, "Treatment of the Terminally Ill," *Death Studies, 22* (December 1998), 784–786.

Further Developments

The Worldwide Epidemic of AIDS

AIDS (acquired immune deficiency syndrome) has become an obsession among many people, leading to a plague mentality in some quarters. Recent statistics are staggering: Worldwide more than 30 million people are living with HIV infection. In 1997 nearly 6 million people became infected with HIV, with more than 90 percent of the new infections occurring in developing countries (Cohen & Fauci, 1998). The stigma of AIDS has taken on an irrational life of its own. In the United States some fearful parents have withdrawn their youngsters from school upon learning that a classmate has HIV. Adults with AIDS have been denied jobs, housing, schooling, dental care, health services, and insurance; those with jobs are often harassed and threatened by coworkers. Movies such as *Philadelphia* brought this issue to light. Even people who know that the disease is not spread casually engage in discriminatory acts. Nor does a clean bill of health necessarily place a person beyond suspicion: An insurance company refused a policy to a Colorado man who voluntarily provided the results of an HIV test that came out negative; the company said the fact that he got tested at all made him too great a risk.

AIDS has been likened to the Black Death—the plague that killed a quarter to half of Europe's population in a three-year span from 1347 to 1350. But AIDS would have to kill 60 to 120 million Americans in a three-year period to have a similar impact. And unlike the many plagues of old, AIDS does not spread through casual contact: It cannot be contracted from doorknobs, drinking glasses, toilet seats, or social kissing.

During the 1980s, most persons with AIDS in the United States fell into three categories: homosexual men, intravenous drug users, and hemophiliacs. About 40 percent of persons diagnosed with AIDS are black or Hispanic, although these minorities make up only about a fifth of the population. In the 1990s nearly half of the minority victims are heterosexuals, chiefly intravenous drug users or their sex partners. The inequality is even more striking among babies, though pregnant women with HIV in the United States can be given medications to stay healthy and reduce the risk of transmission at birth ("Perinatal Transmission: A Mother's Story," 1998). In 1996, rates of heterosexual female AIDS cases accounted for 20 percent of newly reported AIDS cases—ranging from teens to women in midlife (Centers for Disease Control, 1996).

Scientists have learned a good deal about how the HIV virus which causes AIDS infects cells. But efforts to devise a vaccine or a treatment have been complicated by the fact that AIDS is caused by dozens of strains of the virus, and each strain mutates frequently. The worldwide scientific community is in agreement on a number of key matters: (1) Although today's medicines are prolonging patients' lives, an effective vaccine or a definitive cure for HIV infection is years off; (2) at present, behavioral remedies are the world's lone "vaccine"; and (3) even were a medical vaccine or a cure found, behavioral interventions would still be required (Aggleton et al., 1994).

Within the United States, the prestigious Institute of Medicine of the National Academy of Sciences a few years ago recommended a massive national educational campaign centering on two precepts. First, random sex of any sort is dangerous: When a person engages in a sexual act, the person is not just sexually connected with that partner; he or she is sexually connected to everybody that partner has had sexual relations with in the past 10 years. Second, for all sex outside monogamous relationships with well-known partners, a condom should be used. Significantly, the distinguished scientists rejected ideas for a massive national quarantine program or mandatory blood testing to detect HIV. The panel

---

The Netherlands is the only country where euthanasia is officially condoned, if not exactly legal (Steinfels, 1993). Polls in the United States show large majorities of Americans expressing support for the proposition that physicians should help terminally ill patients commit suicide or give them lethal injections if they request them. But when voters or legislators have faced concrete proposals legalizing such practices, the measures have usually been defeated; the exception is that Oregon residents passed a Death with Dignity Act in October 1997 (Caplan, 1999). As of October 1998, eight people have taken their lives under the law's provisions. The average age of the patients was 71, and 9 of the 10 who asked for lethal prescriptions suffered from cancer (2 more people requested the injection but died before its administration). Of those who requested the lethal medication, five were men and five were women. Nine physicians were involved in making the requests for the lethal injection on behalf of the 10 patients. The average time of death was 40 minutes after taking the drug with no reported complications (Prager, 1998). Moreover, most religious groups maintain their traditional opposition to the practice. For example, the Roman Catholic Church regularly reaffirms its condemnation of euthanasia, while stating that individuals in certain circumstances have the right to renounce extraordinary and burdensome life-support systems (Steinfels, 1992).

A number of concerns and attitudes have converged in the right-to-die movement. For one thing, Americans

Mysterious and Incurable Diseases Often Foster a Plague Mentality During the terrible Black Death that swept Europe in the fourteenth century, thousands of Jews were massacred in countless communities because gentiles blamed them for the disease. The painting depicts Jews being burned to death.

argued that those in high-risk groups were not likely to comply with quarantine or mandatory screening programs and that it would be neither ethical nor in keeping with American civil liberties to compel them to do so. Indeed, the number of exposed people in some cities is already so high that a quarantine would be impractical (Hilts, 1991).

Social and cultural factors contribute to risk behavior, and modifying these factors must be a central element in HIV risk reduction. Such factors include social pressures, cultural expectations, cultural scripts influencing sexual negotiation, subcultures of drug and alcohol use, economic resources affecting the availability of condoms and clean syringes, laws and regulations that marginalize some groups and thus limit their access to information and services, and political and religious ideologies that seek to restrict information about the

full range of safer sex behaviors. Consequently, diverse social and behavioral programs are required to bring about changes in risk behaviors (for instance, contexts in which "safe sex" becomes normative). Some of the most successful interventions have been community-based programs initiated in well-educated gay communities. The Centers for Disease Control and Prevention (CDC) reported for the first time in 1996 that there has been a marked decrease in deaths among people with AIDS. Advances in HIV therapeutics, including combinations of antiretroviral agents, have prolonged the life span of those diagnosed with AIDS. Unfortunately, few people worldwide have access to these medicines for long-term treatment. "Development of a safe and effective vaccine for HIV infection remains the 'holy grail' of AIDS research" (Cohen & Fauci, 1998).

From: C. M. Parkes, P. Laungani, and B. Young, *Death, and Bereavement Across Cultures.* Copyright 1996. Reprinted by permission of Routledge.

tend to favor a quick transition between life and death. But this belief that "the less dying, the better" has come up against an altered biomedical technology, in which people are increasingly approaching death through a "lingering trajectory." Many people also have a profound fear of being held captive in a state between life and death, as "vegetables" sustained entirely by life-support equipment. Coupled with these concerns are a pervasive nonacceptance and intolerance of pain and a growing expectation that one should be without pain and discomfort in the ordinary course of life. Increasing numbers of Americans have become dependent on aspirin and other analgesics to control even relatively minor discomfort.

Some physicians have cautioned that the current preoccupation with the issues of patient autonomy and death with dignity can lead doctors and patients to make clinically inappropriate decisions. They warn of the perils of taking patients at their word. Depressed patients are particularly likely to tell doctors that they do not wish to live any longer and to ask that they be allowed to die. Psychiatric illness, including symptoms of deep despair and hopelessness, can severely distort rational decision making. Concern has also been expressed that legalizing physician-assisted suicide for the terminally ill would benefit only a few and would open the door to widespread abuse. For instance, in 1994 an influential New York State advisory panel, the New York State

Task Force on Life and the Law, released a report saying that the practice might lead some doctors to prescribe deadly drugs rather than improve the substandard care they are giving patients who are dying, in severe pain, or seriously depressed (Rosenthal, 1994). Further, research commissioned by the Dutch government suggests that in more than a thousand cases per year, physicians actively cause or hasten death without the patient's request (Hendin, 1994). Some ethicists fear that most at risk will be patients unable to speak for themselves—the tens of thousands of elderly people suffering from dementia and other diseases in nursing homes and the severely disabled of all ages.

In 1983 a U.S. presidential commission made a public statement on euthanasia confirming that decisions on whether to continue life-sustaining medical treatment should generally be left to mentally competent patients. Family members would be permitted to make similar decisions for mentally incompetent patients. But the commission said that ending a patient's life intentionally could not be sanctioned on moral grounds. Even so, doctors would be allowed to administer a pain-relieving drug that hastens death provided that the sole reason for giving the drug is to relieve the pain (Schmeck, 1983; Senate Special Committee, 1997). Related developments are "living will" laws that afford protection against dehumanized dying and confer immunity upon physicians and hospital personnel who comply with a patient's wishes.

More recently, Linda Emanuel (1998) proposed a model of response to the patient who requests physician-assisted suicide (PAS) (see Table 19-2). She describes a series of steps to assess a patient's competency for such a decision, which include evaluation for depression and other psychiatric conditions and decision-making competence, listing goals of care, providing full information to the patient, physician consultation with professional colleagues, following care plans, removing all unwanted life-support interventions, and securing maximum relief of suffering. The patient would ultimately have the right to decline nutrition and hydration and any regular oral intake (Emanuel, 1998).

Questions

What is our nation's stance on the issues of euthanasia or physician-assisted suicide? Are there other nations with a different perspective on this issue?

Table 19-2   Emanuel's Eight-Step Approach to the Patient Who Requests Physician-Assisted Suicide

Step	Suggested Procedure
1	**Assess for depression:**
	If Yes, then treat for depression.
	If No, then see Step 2.
2	**Assess for decision-making capacity:**
	If No, then assess for treatable causes; if none, seek proxy.
	If Yes, continue with Step 3.
3	**Engage in structural deliberation including advance care planning. Affirm other form of comfort and control:**
	Request for PAS may be dropped. Provide care as discussed in patient/proxy.
	Request for PAS may continue. See Step 4.
4	**Establish and treat root cause(s) of request: physical, personal, or social. Hospice philosophy, palliative care skills, as appropriate:**
	Request dropped. Provide care as discussed in patient/proxy.
	Request continues. See Step 5.
5	**Ensure full information on consequences, risk, and responsibilities; attempt to dissuade from PAS:**
	Request dropped. Provide care as discussed with patient/proxy.
	Request continues. See Step 6.
6	**Involve consultants or institutional committee as appropriate:**
	Request dropped. Provide care as discussed with patient/proxy.
	Request continues. See Step 7.
7	**Review adherence to goals and care plan, supporting removal of unwanted intervention and providing full-comfort care:**
	Request dropped. Provide care as discussed with patient/proxy.
	Request continues. See Step 8.
8	**Decline PAS (all states except Oregon) explaining why and affirming alternatives.**
	Provide care as discussed with patient/proxy.

From: Linda Emanual, "Facing Requests for Physician-Assisted Suicide: Toward a Practical and Principled Clinical Skill Set," *Journal of the American Medical Association* (JAMA), 1998, Vol. 280, No. 7. Copyright 1998. Reprinted by permission.

Suicide   In many countries, suicide is one of the leading causes of death (Oaks & Ezell, 1993). Suicidologists report that in many societies, because of the stigma associated with suicide, many cases are reported as accidents or death from undetermined cause. In addition to the emotional burden for those left behind, there is an economic burden because most insurance companies do not honor the life insurance policies of those who take their own lives. The social isolation of survivors is also greater than for any other cause of death, as was expressed by a bereaved mother of a child who died from a self-inflicted gunshot wound:

> People I had known for years stopped speaking to me. They couldn't even look me in the eyes in the grocery store. All of a sudden we had no friends. People didn't call, visit, or invite us over. They treated us like we had some dreaded disease (Oaks & Ezell, 1993, p. 209).

To commit suicide is to kill oneself. The National Institute of Mental Health (NIMH) has defined three concepts pertaining to suicide (Oaks & Ezell, 1993, p. 212):

- *Suicide ideas.* This pertains to the observation or inference that someone might be moving in the direction of taking his or her life. The subject might seem to be suggesting it, writing about it, talking about it—but does not carry it out.
- *Suicide attempts.* This pertains to the situation in which a person performs an overt life-threatening behavior with the intent of taking his or her own life (e.g., cutting wrists, jumping off a bridge, overdosing on pills). This may become a repetitive behavior with the intent of communicating a message to a loved one and may be a call for help rather than true intent to end one's life.

- *Completed suicide.* This pertains to all persons who were successful in taking their own lives, based on the circumstances surrounding the death.

Various cultural attitudes toward suicide still waver between the extremes of the early Christians, who labeled suicide a sin and denied burial rites to persons who had killed themselves, to considering the victim as insane or weak, to considering suicide the honorable choice when confronted with capture, defeat, or disgrace (e.g., Japanese kamikaze pilots in World War II). The idea that suicide is a sin comes from the Christian view that God made each of us and it is up to God, and only God, to determine our death. Criminologists, on the other hand, are likely to take the position that the person who commits suicide is mentally ill and has a flawed, deviant mind, as in criminal cases when an ex-spouse murders his own family and then takes his own life.

Who Commits Suicide and Why?   How do we classify the thousands of suicides committed every year mainly by children, adolescents, and elderly men (see Figure 19-1)? Although more females attempt suicide, more males succeed at suicide (Oaks & Ezell, 1993). The most recent statistics reported were for 1995, and young black

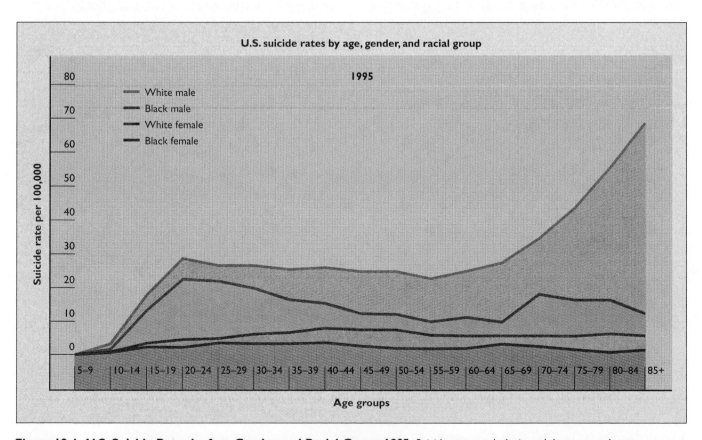

Figure 19-1 U.S. Suicide Rates by Age, Gender, and Racial Group, 1995 Suicide rates peak during adolescence and young adulthood and then again in late adulthood, particularly for white males during retirement.

Source: Centers for Disease Control and Prevention, National Center for Health Statistics, National Institute of Mental Health. Posted January 14, 1998.

males from their mid teens through their early thirties had higher rates of suicide. For males, common causes are pressures associated with peer acceptance, job goals, loneliness, and poor health. For elderly males, loneliness, lack of a sense of purpose or usefulness, loss of strength, illness and overmedication, and a realization that more loved ones have passed on than are left in life all contribute to suicidal behaviors.

> His parents were moving to Houston because the father had been transferred by his employer. He said he was not going with them. He was an honor student, involved in sports and extracurricular activities. He showed his friends his gun, but no one took him seriously enough to tell school personnel. He walked to the front of the classroom, put the gun in his mouth, and pulled the trigger. Postvention assistance included individual counseling for some students and teachers, participation in support groups led by professional counselors, and a memorial service held at the school. The event was so traumatic, it was weeks before the school atmosphere returned to normal (Martin & Dixon, 1986, p. 265)

Suicide rates for white males climb rapidly as men age from their forties into their eighties. White-collar workers have higher rates of suicide than laborers do. "Highly educated professionals, such as dentists, psychiatrists, doctors, lawyers, and mental health workers have rates

three times the national average" (Oaks & Ezell, 1993, p. 216). Christmastime and springtime have the highest rates of suicide. Those who are divorced, widowed, and separated commit suicide more than those who remain married and have children. A much higher number of suicides are committed by white males during their post-retirement years in comparison to any other group. Yet since the early 1980s, the rates of teenage suicide have increased significantly.

More black youth are committing suicide than ever before, particularly in southern states (CDC, 1998g). The U.S. Department of Health and Human Services (USDHHS) has identified high-risk areas of the country for the main causes of death. Centers for Disease Control (1998a) revealed that among whites, suicide rates are highest in the western states and in nonmetropolitan areas throughout the country. Using a firearm is the most common method of suicide in the United States; women are more likely than men to use poison.

Kastenbaum (1991) believes that suicide victims view suicide as a reunion with God or a loved one, as a rest or refuge, as getting back at or hurting someone, as a penalty for failure, as getting attention, or as loss of a life-sustaining drive. Each year over 400,000 people in the United States attempt suicide. Several thousand succeed and leave unbelievable shock and grief behind. Suicide peaks during adolescence and during late adulthood.

Table 19-3   Warning Signs of Suicide

Signs Among Elderly	Behavioral Signs Typical for Teens	Environmental Signs	Verbal Cues
Severe physical illness	Lack of energy or increased fatigue	Previous suicide attempts by a family member or friend	Direct statements that may need immediate attention:
Chronic pain	Acting bored or disinterested		"I want to die."
Marked change in body image	Tearful sadness	Problems at school	"I don't want to live anymore."
Loss of significant emotional ties	Difficulty concentrating or making decisions, confusion	Family violence	"Life sucks and I want to get out."
Decrease in level of socialization	Silent or withdrawn	Sexual abuse	
	Angry and destructive behaviors	Major family change	"I won't be a problem much longer."
Lack of a religious faith	Less interest in usual activities		"Nothing matters anymore."
Physical disability	Giving away prized possessions		
Cognitive deficiencies	Poor school performance		
Change in or loss of familiar surroundings	Dwelling on death in creative activities, such as music, poetry, and artwork		
Isolation	Difficulty sleeping or change in sleeping patterns		
Loss of functional capacities	Increased thrill-seeking and risky behaviors		
Reduction of responsibilities	Increased use of drugs and alcohol		
	Change in appearance or cleanliness		
	Change in appetite or eating habits		
	Suddenly cheerful after a depression		

From: Gary J. Kennedy (Ed.), *Suicide and Depression in Late Life.* © 1996. Reprinted by permission of John Wiley & Sons, Inc.

The most vulnerable times are during "rites of passage," such as graduation, anniversaries, birthdays, retirement, or death of a spouse or child. Teens also commit suicide after a "crisis" of rejection or humiliation, and 30 percent of teen suicides are related to issues of sexual identity. If someone exhibits several of the following signs together, it is time to talk to the person, demonstrate love and concern, convince the person to get counseling, and lock up any weapons in the home. Table 19-3 lists warning signs for suicide.

Questions

What are some common behavioral changes noted prior to suicide? What life events are potential triggers to suicide attempts?

The Hospice Movement

Hospice neither hastens nor postpones death. It affirms life, recognizes dying as a part of the normal process of living, and focuses on maintaining the quality of remaining life.

P. North, 1998

Some have called death the undiscovered country. At its border, *hospice programs* provide an affordable alternative for dealing with pain and the end of life in a dignified, graceful fashion. In medieval times a hospice was a place where sick and weary travelers could seek comfort and care before continuing on their journey. The **hospice** of today likewise provides comfort and care but with the knowledge that the recipients are nearing the end of their life's journey—that they are dying. (The word *hospice* is Latin for "host" or "guest.") The approach is modeled after St. Christopher's Hospice in England, in operation since 1967. It entails a variety of programs designed to afford an alternative to conventional hospital care for the terminally ill, especially cancer patients.

Polls show that most people are less terrified of dying than they are of dying an agonizing, impersonal, and undignified death among machines and strangers. Although few people can be accorded a "perfect death," most can be free of pain; and the fear of terror, isolation, and chaos can be replaced by calm and control. Many hospice proponents believe that rather than debating euthanasia and assisted suicide, the nation should have a national debate about care for the dying (Chase, 1995). As of 1995, there were over 2,000 hospices in the United States. A typical hospice staff includes nurses, clergy, social workers, physicians, and a host of volunteers.

The hospice program takes a positive attitude toward dying. It does not discontinue such medical treat-

ments as chemotherapy and radiation when they are conducive to the comfort of the patient. But the emphasis falls on "comfort-care" rather than on attempts to prolong life. Comfort-care involves an aggressive treatment of symptoms, both physical and emotional, through the use of psychological, religious, and nutritional counseling, antidepressive medications, and high-dose morphine preparations (designed to free patients from the severe and recurrent pain that frequently accompanies terminal cancer). Significantly, a study of patients at major cancer centers reveals that nearly half are reluctant to report their pain. Others are reluctant to take pain medication for fear that when their pain becomes truly intolerable, it will then be ineffective. And others fear that narcotic painkillers will result in addiction. Many physicians share these misconceptions. Yet much can be done to ease the pain and suffering of patients (and their families) with cancer, AIDS, and degenerative neuromuscular disorders (Lang & Patt, 1994). Children, too, can be well served by hospice care (see Papadatos & Papadatos, 1991).

Most hospice programs are centered about caring for the dying person at home. According to a Gallup poll, 86 percent of those surveyed said that if they were terminally ill and had only six months to live, they would prefer to receive care and die in their own home or that of a family member (Baer, 1993). More hospice services are offered now through hospitals as well as in the patient's home. In a hospital hospice setting, families are included as part of the treatment, visiting hours are unlimited, and day beds are available in the patient's room so family members may stay overnight to be with a loved one as needed. Physicians, nurses, social workers, and volunteers provide emotional and spiritual assistance as well as medical care—either in the patient's home or in this special unit of the hospital.

But the concept of hospice is not so much a place as it is a program or a mode of care. It seeks to give dying patients greater independence and control over their lives so that they do not have to surrender themselves to the care of impersonal bureaucratic organizations. Consequently, hospice services can be quite personal. One hospice had a client whose one wish was to walk again, and so the hospice arranged for a physical therapist who helped the man regain his ability to walk. Another client wanted to make a last trip to Hawaii, and the hospice set it up (Walters, 1991). In sum, the hospice movement undertakes to restore dignity to death.

Advocates of the hospice approach say that it is difficult for physicians and nurses taking care of patients in hospital settings to accept the inevitability of death. Hospitals are geared to curing illness and prolonging life. An incurable case is an embarrassment, evidence of medical failure. Consequently, hospice proponents say, an alternative-care arrangement is required that accepts the

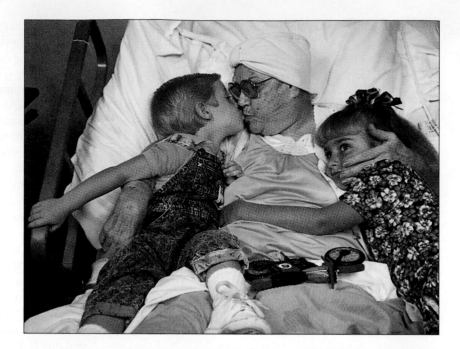

Hospice A hospice provides comfort and care for terminally ill patients. Here grandchildren spend time with their grandmother in a hospice setting.

inevitability of death and provides for the needs of the dying and their families.

Hospice programs hope to make dying less emotionally traumatic for both patient and loved ones. Indeed, much effort is directed toward helping family members face the problems that surround terminal illness. The visiting home-care staff also assist the family with changing bedding, securing all necessary medical equipment, informing the hospice nurse or physician when the patient is in great discomfort and may need to change pain medication, and might sit for long hours with the patient or family members during reminiscence, or discuss many spiritual questions about death and dying. They also give the immediate family members time for a brief respite, if needed. Each patient and family is treated with dignity and warmth. A bereavement follow-up service maintains contact with the family in the period following the loved one's death. Most major health insurance plans, including Medicare, now cover virtually the entire cost of treatment in hospices that meet standards set by the U.S. Department of Health and Human Services. However, there are some patients (and families) who feel anxious about death and dying and opt to be in a typical hospital setting getting aggressive medical treatment for their last days of life.

Question

What is the philosophy of the hospice approach, and who is most likely to be eligible for these services?

The Dying Process

The significance a person gives to dying is based on the personal philosophy he or she has acquired during life. The attitude may be spiritual or materialistic, life affirming or nihilistic, or unformulated or uncertain.

Dewi Rees, *Death and Bereavement*, p. 94

Throughout history, human societies seemingly have given death their most elaborate and reverent attention (Ariès, 1978, 1981; Williams, 1990). Some of the world's most gigantic constructions, its most splendid works of art, and its most elaborate rituals have been associated with death. More than 500,000 years ago, ceremonial rituals for burying the dead were being employed by Peking Man. And we still are awed by the Egyptian pyramids (witness our fascination with King Tut), the huge European burial mound of Silbury Hill, the towering Pyramid Tomb of the High Priest in the Central American Yucatan forest, the beehive tombs at Mycenae, the Taj Mahal, and megalithic burial remains of inhabitants of northwestern Europe.

Monotheistic world religions—Judaism, Christianity, and Islam—put an end to the practice of equipping the elite dead with wives, concubines, slaves, jewels, armor, and other luxuries in hopes of guaranteeing their enjoyment and comfort in the next world. However, Christianity did nothing to impede the elaboration of rituals surrounding the act of dying, mourning rituals, funeral rites, and rituals to appease and assist the souls of the departed. Today, though, because we so rarely have contact with them and because they tend to take a low profile, the funeral industry has become a corporate conglomerate of immense proportions.

Defining Death

The earliest biblical sources, as well as English common law, considered a person's ability to breathe independently to be the prime index of life. This view coincided with the physiological state of the organism. In the past, absence of spontaneous breath or heartbeat resulted in the prompt death of the brain. Conversely, destruction of the brain produced prompt cessation of respiration and circulation.

Over the past 40 years, technological advances have rendered these traditional definitions of death obsolete. In 1962, Johns Hopkins University physicians developed a coordinated method of resuscitation by artificial ventilation, heart compression (CPR), and electric shock (defibrillation) by which many victims of cardiac arrest can be saved. Another advance has been the invention of the mechanical respirator, which sustains breathing when the brain is no longer sending out the proper signals to the lungs. Still another technological innovation has been the use of an artificial pacemaker to induce regular heartbeat after failure of the heart's own electrical conduction system. And dialysis machines prolong the lives of patients suffering kidney failure.

These life-extending technologies have compelled courts and legislatures to grapple with new definitions of death. The need to accept *brain death* as one standard has now been acknowledged in most states and endorsed by the American Medical Association, the American Bar Association, and in 1981 by the President's Commission for the Study of Ethical Problems in Medicine. **Brain death** occurs when the brain receives insufficient oxygen to function (Oaks & Ezell, 1993, p. 3). The first stage of total brain death is *cerebral death* (the cerebral cortex is no longer functioning). A problem arises when a patient shows no cerebral activity (thought) but has function in the lower brain stem (which controls breathing, pulse, blood pressure). As medical procedures are invented that prolong life, the definition of death changes. A series of landmark court decisions have upheld the validity of the brain-death criterion. Many of these cases involved transplantation of organs (such as cornea, kidney, lung, liver, or heart) and have included cases of alleged murder in which organs were removed from the victims for transplantation purposes with the permission of the families. In their trials defendants have alleged that death resulted not from the assailants' violent attacks on the victims but from the removal of the organs. But the courts have invariably affirmed the pretransplantation death of the victims on the basis of brain-death criteria.

As one might expect, definitions as to what constitutes death afford an ethical minefield and have fueled a continuing debate among medical professionals, bioethicists, and legal scholars. For instance, although individuals in persistent vegetative states (such as Nancy Cruzan and Karen Ann Quinlan) were deemed to be legally alive, courts have ruled that families can allow persistent vegetative patients to die by withholding food and water. But because the patients still retain a working brain stem, which controls breathing and primitive reflexes, the courts have typically held that it is not permissible to take their organs for donation to help others. The President's Commission recommended that persons be declared dead only when their entire brain, including the brain stem, ceased to function, to allow a safety margin. If one allowed people to be declared dead while they could still breathe, it was feared, some patients who conceivably could recover from their injuries might be inappropriately declared dead (Kolata, 1992b).

Confronting One's Own Death

In one manner or another, everyone must adapt to the fact of their own dying and death (see Figure 19-2). Indeed, a realistic acceptance of death may well be the hallmark of emotional maturity. People differ considerably, however, in the degree to which they are consciously aware of death (Rees, 1997). Some individuals erect formidable defenses to shield themselves from facing the reality that they too must die. Death is a highly personal matter, and its meaning tends to vary from individual to individual. Some elderly visualize death as the dissolution of bodily life and the doorway to a new life, a passing into another world. Those with Western religious convictions often express the belief that in death they will be reunited with loved ones who have died. For such individuals death is seen as a transition to a better state of being; few say that death will entail punishment. There are also people who look with resignation on death as "the end"—the cessation of being (Rees, 1997).

Researchers tend to agree that only a relatively small proportion of the elderly express fear of death (Rees, 1997). This is not to say that the problem of such fear does not exist for older people. As one elderly person put it:

> No, I'm not afraid to die—it seems to me to be a
> perfectly normal process. But you never know how
> you will feel when it comes to a showdown. I might get
> panicky. (Jeffers & Verwoerdt, 1969, p. 170)

And a 90-year-old man put it this way: "I'm afraid of the unknown, and if I had my druthers, I'd rather not do it" (quoted by Chase, 1995, p. B1). One of the most frequent questions asked of health practitioners about dying (incidentally, also asked about giving birth by

Figure 19-2 Coming to Terms with One's Own Mortality

© The New Yorker Collection 1993 Roz Chast from cartoonbank.com. All Rights Reserved.

mothers about to have a baby) is "How long is it going to take, and is it going to hurt?" Generally, younger people show greater fear of death than persons 65 and over (Riley, 1983).

Physicians who care for the elderly report that patients frequently tell them, "I am not so much afraid of death as I am of dying." A 1991 national survey found that two-thirds of Americans want to live to age 100, but 75 percent are worried about losing control of their lives. Nearly 80 percent said that they feared ending up in a nursing home more than dying quickly from a disease ("Most want long lives," 1991). It seems that a good many Americans fear painful, lonely, debilitating illness, such as cancer and senility. More than a century ago Canadian physician Sir William Osler, in a study of some 500 deaths, found that only 18 percent of the dying suffered physical pain and only 2 percent felt any great anxiety. Osler concluded: "We speak of death as the king of terrors, and yet how rarely the act of dying appears to be painful" (Ferris, 1991, p. 44).

A Life Review   Psychologist Robert N. Butler (1971) suggests that the elderly tend to take stock of their lives, to reflect and reminisce about it—a process he calls the **life review.** Often the review proceeds silently without obvious manifestations and provides a positive force in personality reorganization. In some cases, however, it finds pathological expression in intense guilt, self-deprecation, despair, and depression. Reviewing one's

life can be a response to crises of various types—such as retirement, the death of one's spouse, or one's own imminent death. In Butler's opinion, the life review is an important element in an individual's overall adjustment to death, a continuation of personality development right to the very end of life. Some people are helped along in the process if they write autobiographical accounts of their lives or if a close family member (often a grandchild) does a videotaped interview. Life reviews can give new meaning to people's present lives by helping them understand the past more fully (Birren, 1987). To the extent that the elderly are able to achieve a sense of lifetime integrity, competence, and continuity, the life-review process contributes to their successful aging by increasing their self-understanding, personal meaning, self-esteem, and life satisfaction (Staudinger, Smith, & Baltes, 1992). (Reminiscence and conducting a life review are discussed in more detail in Chapter 18.)

Changes Before Death   Various investigators report that systematic psychological changes occur before death, even several months ahead of the event, in what is sometimes called the *death drop.* These changes do not appear to be a simple result of physical illness. Individuals who become seriously ill and recover apparently do not exhibit similar changes. Morton Lieberman and Annie S. Coplan (1970) report, for instance, that individuals whom they later found to be a year or less away from death showed poorer cognitive performance, lower introspective orientation, and a less aggressive and more docile self-image on personality tests than did those who were three or more years away from death. A number of researchers also report a decline in measured intelligence and the complexity of information processing for those who die within a year as compared with those who die a number of years later (White & Cunningham, 1988). Psychomotor performance tests, depression scales, and self-report health ratings likewise have predictive value and can alert the physician to the patient's decline (Botwinick, West, & Storandt, 1978). Duke University researchers, however, have generally found less evidence of a death drop than that reported by other investigators (Palmore & Cleveland, 1976). Furthermore, a critical loss in intellectual functioning is less predictive of further survival for people beyond 80 years of age than for the young-old (Steuer et al., 1981).

Not all dying people panic when realizing their death is impending. Some become nostalgic, want to see people they have not seen for a long time, and become more sensitive to those who will be left behind. Some take the opportunity to enhance relationships and forgive long-held grudges and misunderstandings.

Questions

Is there a universal experience of dying or is it highly individualistic? Do you think there is a cultural component to how one views death?

Near-Death Experiences

I think we have reached an era of transition in our society. We have to have the courage to open new doors and admit that our present-day scientific tools are inadequate for many of these new investigations.

R. Moody, *Life After Life,* 1976

The theme of "a beautiful death" has been linked by some with belief in a life hereafter. The "life after life" movement—supported and researched by Dr. Elisabeth Kübler-Ross, Dr. Ray Moody, Dr. Brian L. Weiss, Dr. Bruce Greyson, Betty Eadie, and others—claims evidence of a spiritual existence beyond death. Some persons pronounced clinically dead who have been resuscitated by medical measures or by some miraculous intervention have told of having left their bodies and undergone otherworldly experiences before they were resuscitated. Plato wrote about such an event centuries ago in the *Republic,* says Dr. Bruce Greyson, psychiatrist and NDE researcher, and such reports seem to be prevalent across all cultures (Koerner, 1997). A **near-death experi-**

ence (NDE) commonly is precipitated by medical illness, traumatic accident, surgical operation, childbirth, or drug ingestion. An estimated 20 to 40 percent of people who survive a brush with death report a near-death experience (Irwin, 1985). "I had a floating sensation . . . and I looked back and I could see myself on the bed below (Moody, 1976). The "typical" report runs along the following lines (Rees, 1997; Morse, 1992): The dying individuals feel themselves leave their bodies and watch, as spectators, from a few yards above their bodies, the resuscitation efforts being made to save them. Then, about a third report passing through a tunnel and entering some unearthly realm. "I went through this dark black vacuum at super speed" (Moody, 1976). Half say they see guides or the spirits of departed relatives, a religious figure, or a "being of light." "From the moment the light spoke to me, I felt really good—secure, and loved" (Moody, 1976). Many individuals report having approached a sort of border, seemingly representing the divide between earthly and next-worldly life. But they are told they cannot cross the divide now. They are told they must go back to earth, for the time of their death has not yet arrived. They resist turning back, for they are overwhelmed by intense feelings of love, joy, and peace. In some metaphysical manner they are then reunited with their physical body and they live.

A near-death experience can be a catalyst for spiritual awakening and development in the months and years following the episode. Many individuals develop firmer beliefs in God and an afterlife, become less materialistic and more spiritual, feel a greater love for other people, and spend more time searching for the meaning of life (Rees, 1997). The more NDEs are openly discussed, the more thousands of people claim to have experienced such an otherworldly experience. However, a neuroscientist in a research setting at Laurentian University in Sudbury, Ontario, has induced many of the characteristics of an NDE by stimulating the brain's right temporal lobe, the area of the brain above the right ear that is responsible for perception (Koerner, 1997).

Skeptics such as psychologist Ronald K. Siegel (1981) suggest that the visions reported by dying people are virtually identical to the descriptions given by individuals experiencing drug-induced hallucinations, who often report hearing voices and seeing bright lights and tunnel imagery. The episodes, Siegel says, derive from intense arousal of the central nervous system and disorganization of the brain's normal information-processing procedures. Another major skeptic is Daniel Alkon (Koerner, 1997), chief of the Neural Systems Laboratory at the National Institutes of Health, who says anoxia (oxygen deprivation to the brain) can induce such mental states. In England, Karl Jansen has zeroed in on the shifting levels of a brain neurotransmitter called ketamine (Koerner, 1997). A related interpretation holds that at death various forces combine to sever the connections between consciousness and somatic processes while the brain remains active. Dying people, aware that they are dying, may turn their thoughts to the possibility of re-uniting with loved ones and the broader meaning of life. And experiences of intense joy, profound insight, and love may be produced by endorphins (molecules that act as both neurotransmitters and hormones), which were designed in the course of evolution to blot out over-whelming pain (Irwin, 1985).

Another skeptic, psychologist Robert Kastenbaum (1977), suggests that the current fascination with "life after life" is simply another "mind trip." He cites cases of some heart-attack victims who, when pronounced clini-cally dead and then resuscitated, have no recollection of an out-of-body experience. And he tells of people in res-piratory failure (choking on a bone or undergoing an acute episode of emphysema) who later report feeling as though they had been in direct hand-to-hand combat with death. Kastenbaum (1977, p. 33) expresses concern lest death be "romanticized" and cautions:

> Death seems to be less demanding, perhaps more friendly, than life. . . . I do not believe the frustrated adolescent, the unemployed worker, the grieving widow or the ailing old person needs to be offered the invitation to suicide on quite so glittering a silver platter.

Hence, there is considerable controversy about near-death experiences, and researchers on both sides of this issue continue to investigate NDEs in as scientific a manner as possible.

Questions

What common elements seem to be present in people's descriptions of near-death experiences? Are there any scientific explanations for these sensations?

Religious Beliefs

Notions of life after death have old roots. Ancient Greek philosophy frequently mentioned Hades, the Bible speaks of the "kingdom of Heaven," and Eastern reli-gions speak of realms of life we'll inhabit after this one. Many Eastern religions believe in an afterlife and rein-carnation of the souls of animals and people and teach their followers not to harm animals or people. In the New Testament, Jesus spoke about the kingdom of Heaven as a place of eternal reward:

> I go to prepare a place for you, and if I go to prepare a place for you I will come again, and receive you to myself; that where I am you may be also (John 14:2–3).

Christian theologians traditionally have viewed the book of Revelation as providing the most vivid and familiar biblical images of a mystical heaven: images of "pearly gates and streets of gold, of a vast white throne, and of throngs of saints and angels gathered around God at the culmination of history" (Sheler, 1997). With the new mil-lennium being ushered in, a proliferation of writings about being "born again," the signs of the Apocalypse, the second coming of Christ, and the "Rapture" have a powerful influence on some Christians and others in their views of an afterlife. Strikingly, "nearly 80 percent of Americans—of various religious faiths and of none— say they believe in life after death, and two thirds are cer-tain there is a heaven" (Sheler, 1997). In his book *Teach-ing Your Children About God,* Rabbi David Wolpe reminds us of the experience of birth into the unknown world of joyous waiting arms, as is the classic view of the afterlife—a birth into a place that we humans can only try to imagine (Sheler, 1997, p. 66). Agnostics are unsure about the existence of an afterlife: atheists believe that death is the end of their existence.

The Jewish position has been that no one knows the nature of life after death, and therefore speculation is pointless and to be discouraged. Judaism rejects the con-cept of hell, but Orthodox Jews do believe that in a Mes-sianic era there will be a resurrection—a time when the souls and bodies of the dead will be reunited.

Most Buddhists believe that the body retains "life" for eight to twelve hours after death, and so it is possible

Religious Healing Many societies assign responsibility for the treatment of illness to faith healers and shamans, individuals who work with supernatural power. Modern medicine is taking a closer and more sympathetic look at folk healers such as this !Kung shaman in South Africa who places his arms around a patient in a healing trance. Although folk medicine has generated much skepticism, people who believe they are victims of an evil spirit not only think they are ill but are ill. So they often respond to the supernatural entreaties of traditional healers. Faith healing is not uncommon in the United States.

to speak and act in a meaningful way with the deceased. That the person still exists after death has implications for Buddhists because their idea of rebirth hinges on the fact that people experience changes in consciousness every moment. For the Buddhist, every instant involves the death and rebirth of consciousness. In the moment just before death, each person recollects parts of their past lives and glimpses of the future. Then the mind drops into a nonaware state, only to reawaken into one of the six realms of the universe: Hell, Hungry Ghost, Animal, Human, Jealous Gods, or Gods. None of these realms is very desirable, and Buddhism teaches a doctrine of liberation from karma into a state of nirvana. This detailed picture of the after-death state is very different from the uncertain picture in the West about the nature of afterlife and after-death experience.

Dying

Social scientists observe that modern societies attempt to control death by turning over its management to large, bureaucratic organizations (Rees, 1997). Only a few generations ago, most people in the United States died at home surrounded by family and closest friends. Before the twentieth century, doors on homes were built wide enough to allow a coffin to pass through them, and parlors were designed to hold mourners. The family assumed the responsibility for laying out the corpse and otherwise preparing for the funeral. Today, in general, the nursing home or hospital typically cares for the terminally ill and manages the crisis of dying. A mortuary establishment—euphemistically called a "home"—prepares the body and makes the funeral arrangements or undertakes the cremation. The exceptions to this are evident among pockets of immigrants in this country who are unaware that they are required by law to notify various authorities of any death—and some of these families still attempt to manage death and burial according to their cultural customs.

This bureaucratization of death minimizes the average person's exposure to death. The dying and the dead are segregated from others and placed with specialists for whom contact with death has become a routine and impersonal matter (Kamerman, 1988). But as we become increasingly less exposed to death, we have less opportunity to learn and pass along the lessons of how to cope adequately with it. Often neither the person who is dying nor the family and friends understand how to deal with death. These trends concern many thanatologists, who point out that people are often ill equipped to do the "grief work" so essential to coming to terms with the death of a loved one. Sometimes grief seems to be even more prolonged on a massive scale when a prominent person such as John Kennedy or a celebrity dies an early or tragic death, such as Elvis, John Lennon, the Mexican-American pop singer Selena, or Princess Diana. Witness the proliferation of web sites set up in their memory. And they point out that funerals, memorial services, and other forms of grieving rites give continuity to both family and community life.

Stages of Dying Over the past 30 years, Elisabeth Kübler-Ross (1969, 1981) has contributed a good deal to the movement to restore dignity and humanity to death and to reinstate the process of dying to the full course of human life. She notes that when medical personnel and the family attempt to hide the fact that a patient is dying, they create a barrier that prevents everyone from preparing for death. Furthermore, the dying patient generally sees through the make-believe. Kübler-Ross finds that it is better for all parties if their genuine emotions are respected and allowed expression. In this fashion, dying can afford a new opportunity for personal growth. Indeed, surveys reveal that four out of five individuals would want to be told if they had an incurable disease. And today, physicians are sharing with patients and family members information as to the likely medical outcome of an illness.

Thanatologists are finding that dying, like living, is a process. Though there are different styles of dying, just as there are different styles of living, some elements are common to the death experience. Kübler-Ross (1969) has observed that dying persons typically pass through five stages. Not everyone goes through all the stages, some slip back and forth between stages, and some experience several stages at the same time. These are the five stages according to Kübler-Ross:

1. *Denial.* Individuals resist acknowledging the reality of impending death. In effect, they say "No!" to it.
2. *Anger.* Dying people ask the question "Why me?" They might look at the persons around them and feel envy, jealousy, and rage over their health and vigor. During this phase a dying person often makes life difficult for others, criticizing friends, family, and medical personnel with little justification.
3. *Bargaining.* Dying individuals often begin to bargain with God, fate, or the illness itself, hoping to arrange a temporary truce. For instance, a dying person might say, "Just let me live long enough to attend my son's marriage," or "Allow me to get my business in order." In turn the patient promises to be "good" or to do something constructive during his or her remaining time alive. The "bargain" generally is successful for only a short period, because the advance of the illness itself invalidates the "agreement."
4. *Depression.* Dying people begin to mourn their own approaching death, the loss of all the people and things they have found meaningful, and the plans and dreams never to be fulfilled—they experience what Kübler-Ross terms "preparatory grief."

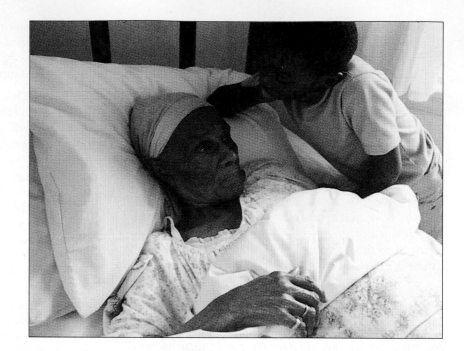

Dying at Home For the most part, people want to die at home among loved ones and surroundings that are familiar and comforting.

5. *Acceptance.* The dying have by this time mourned their impending loss, and they begin to contemplate the coming of the end with a degree of quiet expectation. In most cases they are tired and quite weak. They no longer struggle against death but make their peace with it.

Says Kübler-Ross about the fifth stage:

> Acceptance should not be mistaken for a happy stage. It is almost void of feelings. It is as if the pain had gone, the struggle is over, and there comes a time for "the final rest before the long journey," as one patient phrased it (Kübler-Ross, 1969, p. 113).

Kastenbaum's Trajectories of Death Psychologist Robert Kastenbaum (1975) points out that although Kübler-Ross's theory has merit, it neglects certain aspects of the dying process. One of the most important is the nature of the disease itself, which generally determines pain, mobility, the length of the terminal period, and the like:

> Within the realm of cancer alone, for example, the person with head or neck cancer looks and feels different from the person with leukemia. The person with emphysema, subject to terrifying attacks in which each breath of air requires a struggle, experiences his situation differently from the person with advanced renal failure, or with a cardiovascular trajectory [condition]. Although Kübler-Ross's theory directs welcome attention to the universal psychosocial aspects of terminal illness, we also lose much sensitivity if the disease process itself is not fully respected (Kastenbaum, 1975, p. 43).

Other factors that must be considered are differences in sex, ethnic-group membership, personality, developmental level, and the death environment (a private home or a hospital). Kastenbaum believes that Kübler-Ross's stages are very narrow and subjective interpretations of the dying experience. He claims that her stages are cast with an exaggerated salience and are isolated from the total context of the individual's previous life and current circumstances. And he is concerned lest this stage approach encourage an attitude in which, for instance, medical personnel or the family are able to say, "He is just going through the anger stage," when there may be concrete, realistic factors that are arousing the patient's ire (Kastenbaum & Costa, 1977).

> **Questions**
>
> *According to Elisabeth Kübler-Ross, what are the five stages of dying? What does Kastenbaum mean by the "trajectories" of death?*

Causes of Death

The most recent comprehensive statistics on causes of death available at the time of publication of this text come from the *1993 National Mortality Followback Survey* (NMFS)—the first study since the 1980s to examine detailed patterns of mortality by supplementing the information provided by birth certificates with interviews of next of kin. This survey, published by the Centers for Disease Control (1998a), allowed for the examination of trends in mortality, differences by income and education, risk factors and causes of death, and health care utilization in the last year of life. This survey was based on examination of nearly 23,000 records of individuals aged 15 years and older who died in 1993. This sample of subjects was drawn from all states except South Dakota (where state law restricts the use of death certificate information). Here is a summary of the major findings:

Where does death occur?

- 56 percent of deaths occur in a hospital, clinic, or medical center.
- 21 percent of people died at home.
- 19 percent died in a nursing home.

What diseases did people have?

- 25 percent had had a heart attack and about an equal number had angina; more than 40 percent had hypertension.
- Other frequent conditions were cancer and arthritis, each reported for about one-third of decedents.
- 15 percent suffered from a memory impairment.

How many smoked or used drugs or alcohol?

- 50+ percent of all decedents smoked cigarettes at some point in their lives.
- About 25 percent of all decedents used alcohol during the last year of their life.
- 29 percent of drinkers used alcohol every day.
- About 2 percent used marijuana during their last year of life.
- Less than 1 percent were reported to have used other types of illicit drugs.

How was their last year of life?

- 58 percent of those with functional limitations received help at home from a spouse.
- 50 percent reported functional limitations due to physical or mental conditions during the last year.
- 46 percent had daughters who provided care.
- 39 percent of the decedents took pain medication in the last year of life.
- 31 percent received help from visiting nurses.
- 10 percent were in bed most of their last year of life due to illness or injury.

For those who died of homicide, suicide, or fatal motor vehicle accident:

- 33 percent of decedents involved in fatal vehicle crashes were not wearing seatbelts.
- 19 percent had an alcoholic beverage within 4 hours of death.
- 17 percent had taken drugs or medication within 24 hours of death.
- Of the 36,000 firearm-related deaths, 72 percent involved handguns.

Who gets and pays for health care?

- 75 percent of decedents were covered by Medicare.
- 50+ percent were covered by private insurance or HMOs.
- For 46 percent, Medicare was the principal source of medical payments.
- 20 percent had private medical insurance.
- 10 percent paid for their own medical care.
- 10 percent relied upon Medicaid as the source of payment.
- 10 percent of decedents were reported to have never visited a doctor during the last year of life.
- Nearly 10 percent had made 50 visits or more for health care during the previous year.

Questions

Do you think the decedent's families would give honest answers to the survey questions about past drug and alcohol use and who cared for a dying relative? Would you give honest answers to this type of survey?

Other studies reported by the CDC for 1988 to 1992 indicate that overall death rates for heart disease, the nation's leading cause of death, are now higher in the southeastern United States than in the Northeast, which previously was the region with the highest rates. Deaths by heart disease have declined in all areas during the past 30 years. Cancer—lung, prostate, and colorectal—is the second leading cause of death. HIV death rates are highest in states on the East and West Coasts and in nearly every urban area of the United States. Homicide, suicide, and motor vehicle injury deaths are also major public health problems in well-defined geographic patterns. Homicide rates are high for young black adults in urban areas, but for young white men the high rates are in southern and southwestern states. For young adults, motor vehicle death rates are higher in the southeastern states and generally in less densely populated areas (Centers for Disease Control, 1998a). Other sources indicate that among children, the primary causes of death are in the understudied areas of accidents and injuries, which are now classified as "behavior-related" injuries and under closer research scrutiny (Midlife Passages, 1998a). Table 19-4 lists the major causes of death or disability by various age groups.

Questions

Is there a single major cause of death of adults in the United States? How does the major cause of death change across the life span?

Table 19-4   Causes of Death or Disability, by Developmental Stage*

Adolescence	Early Adulthood (ages 20–40)	Mid Adulthood (ages 40–60)	Late Adulthood (ages 60–80)	Advanced Age (ages 80–100)
Auto accidents	Auto accidents	Breast cancer	Breast cancer	Breast cancer
Sexually transmitted diseases	Sexually transmitted diseases	Obesity	Obesity	Osteoporosis
AIDS	AIDS	Diabetes	Diabetes	Diabetes
Poor diet or obesity	Obesity	Smoking	Smoking	Smoking
Homicide	Homicide	Hypertension	Hypertension	Hypertension
Stress	Stress	Lack of exercise	Lack of exercise	Lack of exercise
Suicide	Suicide	Colon cancer	Colon cancer	Colon cancer
Depression	Depression	Stress	Depression	Stress
Smoking	Drug and alcohol abuse	Lung cancer	Heart disease	Strokes
Drug and alcohol abuse	Smoking	Depression	Alzheimer's disease	Depression
Lack of exercise	Lack of exercise	Heart disease	Strokes	Heart disease
Unplanned pregnancy		Medicine and alcohol abuse/misuse	Osteoporosis	Alzheimer's disease
			Medicine and alcohol abuse/misuse	Other cancers
				Medicine abuse/misuse

From: U.S. Preventative Health Services.
*Causes are rank-ordered, with most frequent cause listed first.

Grief, Bereavement, and Mourning

When a person is born, all his dear ones rejoice. When he dies they all weep. It should not be so. When a person is born, there is as yet no reason for rejoicing over him, because one knows not what kind of a person he will be by reason of his conduct, whether righteous or wicked, good or evil. When he dies, there is cause for rejoicing if he departs with a good name and departs this life in peace.

The Talmud

We know for certain that there is life after death—for the relatives and friends who survive. As they go on with their lives, they must come to terms with the death of the loved one and make many types of adjustment. First, there is psychological coping, often termed "grief work" (mourning, talking about, and acknowledging the loss). Second, there are numerous procedural details that must be attended to, including funeral arrangements and legal routines (dealing with attorneys, settling the estate, filing for insurance, pension, and Social Security benefits, and the like). Third, there is the social void produced by the death of a family member, which requires revising life patterns and family roles (for instance, housekeeping, marketing, securing a livelihood). Where the death of a loved one is anticipated, individuals frequently experience **anticipatory grief**—a "state of emotional limbo, unable to resolve the loss because it has not yet occurred, and unable to avoid the authoritative diagnosis that

death will occur" (Stephenson, 1985, p. 163). Life after the death of a loved is still difficult, even when one feels "prepared" for the inevitable.

Adjusting to the Death of a Loved One

It is sweet to mingle tears with tears; griefs, where they wound in solitude, wound more deeply.

Seneca, *Agamemnon*, first century A.D.

Bereavement is a state in which a person has been deprived of a relative or friend by death. **Grief** involves keen mental anguish and sorrow over the death of a loved one. **Mourning** refers to the socially established manner of displaying signs of sorrow over a person's death (for instance, wailing, wearing black, hanging flags at half-mast, and the like).

Expressing Anguished Feelings In *Macbeth* Shakespeare proclaims: "Give sorrow words. The grief that does not speak, whispers the o'erfraught heart and bids it break." And a Turkish proverb declares, "He that conceals his grief finds no remedy for it." Many contemporary clinicians and psychologists agree with these statements. Sympathetic assistance can be all-important in the process of expressing anguished feelings. Rather than uttering platitudes ("She lived a full life"), psychologists suggest, well-meaning individuals can offer emotional

Celebrating a Death: Dr. Benjamin Spock   The noted author of the classic book on child rearing died in 1998 at the age of 94. He lived an energetic life and spoke out on important parenting issues as well as other social causes. He requested a "New Orleans style of sendoff" for his own funeral, led by his widow, Mary Morgan.

support and a ready ear. Today in most communities there are support groups associated with nearly every cause of mortality to help other people go through their grief work. Bereavement is less an intellectual process than one of coming to terms with one's feelings. A social worker who found herself "on the fringe of madness" when her husband died of a heart attack eight years earlier, leaving her with three young children, said: "Being strong and bucking up is a lot of baloney. You don't feel like bucking up. It is a real process that you must be allowed to go through" (Gelman, 1983, p. 120). People who receive the support and comfort of family and friends typically have a lower incidence of mental and physical disorders following bereavement. And for most people the expression of grief following the death of a loved one is an important component in recovery. But there are exceptions. Those whose grief includes an intense yearning for and a high degree of dependency on the deceased person tend to have a harder time recovering from their loss (Cleiren, 1993).

Culture and Grief Work   Unfortunately, cultural expectations, social values, and community practices at times interfere with necessary grief work. Dying is often left to medical technology and commonly takes place in a facility outside the home. Funerals are often brief and simplified, and mourning is thought of as a form of mental pathology (the cultural ideal is the self-contained widow who displays a "stiff upper lip"). Yet thanatologists say that expressions of grief and mourning rituals are therapeutic for survivors. Such traditions as the Irish *wake* and the Jewish *shivah* help the bereaved come to terms with the loss and reconstruct new life patterns with

family and friends (see *Human Diversity,* "Cross-Cultural Perspectives on Death"). For those with a religious commitment, church organizations can provide significant support systems (Romanoff & Terenzio, 1998).

In Japan, however, the bereaved are not expected to sever their relationship with the dead by the grief work

Bereavement Ceremonies   Throughout the world societies have evolved funerals and other rituals to assist their members in coming to grips with the death of a loved one. The ceremonies highlight the finality of death. Here mourners in Ghana express their grief in such a ritual.

that is promoted in the West. Japanese culture encourages the survivors to maintain a relationship with the deceased in the form of a family shrine. The family altar can be used to make direct contact with the spirit of the deceased as though one were calling them on a spiritual "telephone." In this way ancestors can be accessed by all family members and so they continue to have a relationship with those who remain on earth (Rees, 1997).

Consequences of Grief   Bereavement and grief often have a much greater impact than is evident in the period immediately following the death. The survivor is more vulnerable to physical illness and mental illness, even to death, especially if the death was sudden and unexpected. Bereaved people have a higher than average incidence of illness, accidents, mortality, unemployment, and other indices of a damaged life (Kastenbaum & Costa, 1977). Psychological effects, especially depression, are even more serious than the effects on physical health (Bodnar & Kiecolt-Glaser, 1994). Even one year after death, some 20 to 26 percent of bereaved persons are still depressed (Norris & Murrell, 1990). In addition, significant clinical depression at the time of a spouse's death substantially increases the risk for psychological complications during the bereavement process, and survivors of spouses who committed suicide are at even greater risk (Gilewski et al., 1991). An authority on bereavement has this to say about grief:

> It isn't a problem or an illness that can be solved or cured. That's why it is really inappropriate for someone to say: "You can get over this. You can recover." In fact, people can't recover in the sense of going back to the way things were. You must make changes in your life in order to go on. The death and the bereavement wizen you, weather you and make you look at life differently (Silverman, 1983, p. 65).

Adjusting to Violent and Premature Death   Violent and premature deaths often result in the most severe grief reactions. Suicide is one of the most difficult types of death for survivors to handle. They may have difficulty acknowledging that the death was a suicide, and even when they do, they can feel guilt and shame that impair normal mourning. Suicide survivors frequently feel that they are somehow to blame—for not seeing the signs and for not fulfilling the stated or unstated needs of the person who died. They also must bear the thought that the dead person did not believe they were worth living for. And their guilt can be intensified if survivors experience a sense of relief after an ordeal of mental illness or of suicide threats and attempts (Murphy et al., 1999).

Adjusting to the Death of a Parent   The bereavement experienced by adult children following the death of an elderly parent has until recently been largely ignored by the media and the academic community—though it has always been of concern of those in the legal profession (see Figure 19-3). As noted in prior chapters, ties between a parent and an adult child often remain strong across the life span. Giving care to frail older parents, particularly by daughters, tends to extend earlier patterns of intergenerational exchange (Bodnar & Kiecolt-Glaser, 1994). From an early age youngsters realize that they can lose their parents, and by midlife they have usually seen friends lose parents. Only 1 in 10 children has lost a parent by age 25; but by 54 years of age, 50 percent of adult children have lost both parents, and by 62 years of age, 75 percent have lost both parents) (Winsborough, Bumpass, & Aquilino, 1991). Although "anticipatory orphanhood" can acquaint adults with the idea that their parents will die, the sense of preparation is seldom equal to the reality of death (Moss & Moss, 1983–1984). Consequently, the loss of a parent through death can be followed by feelings of considerable distress (Umberson & Chen, 1994).

The bereavement accompanying a parent's death is a complex emotional, cognitive, and behavioral process. Most men and women experience their mother's death as a profound loss. Women seem to be more emotionally affected than men are by the death of their fathers (Douglas, 1990–1991). Bereavement can be especially difficult for adult children who provided care prior to the parent's death—particularly those who experienced an intensification of their bonds during caregiving (Pratt, Walker, & Wood, 1992). For middle-aged adults, a parental death is often an important personal and symbolic event that heightens a person's awareness of her or his own mortality, and bereavement can stimulate personal growth and development by fostering a greater sense of personal identity and stronger ties to family

"Judging from her will, your aunt was very interested in efforts to save the whale."

Figure 19-3 Living on After One's Death
From *The Wall Street Journal*—Permission, Cartoon Features Syndicate.

members (Bass & Bowman, 1990). Yet bereavement feelings are often complicated and contradictory. For instance, caregiving daughters frequently report feeling psychological strength in coping with their mothers' deaths while simultaneously reporting feelings of shock, anger, and guilt (Pratt, Walker, & Wood, 1992).

Bereavement seems to have a three-step progression for healthy adults who have lost a parent, according to a recent study by Petersen and Rafuls (1998). Based on their study of six adults who had recently lost a parent, they suggest that a transition in family values occurs, which they call a model for receiving the scepter of family values. Each new generation takes over the responsibility, role, and authority of the previous generation in this three-step transformation (Petersen & Rafuls, 1998):

Step 1. Going back to the origins. Duty and obligation become the most important response the first few days after death, along with an unconscious emotional "moving away" from one's own family.

Step 2. Reevaluation phase. A period of preoccupation permeated with a formal sadness, with release in one or two intense periods of crying and very different from depression found with other grief. Participants dealt with this stage quietly, internally, without sharing, yet thought about the deepest meanings of life. Also, within relationships, the most evident consequence during this stage was reduced lovemaking by a couple.

Step 3. Assuming leadership. Partners brought the second phase to a conclusion by reminding spouses that they were needed by their own families. These phases seemed to bring about a sense of strength never experienced by these subjects before; they reordered their priorities, and began to appreciate the richness of life around them.

Phases in the Bereavement Process   In standard bereavement by adults, the individual typically passes through a number of phases (Malinak, Hoyt, & Patterson, 1979). The first phase is characterized by shock, numbness, denial, and disbelief. The most intense feelings of shock and numbness usually last several days, although the process of struggling with denial and disbelief can persist many days, even months.

The second phase involves pining, yearning, and depression. It usually reaches its peak within 5 to 14 days but can continue longer. Weeping, feelings of hopelessness, a sense of unreality, feelings of emptiness, distance from people, lack of vigor and interest, and preoccupation with the image of the deceased are quite common during this stage. Other symptoms might include anger, irritability, fear, sleeplessness, episodes of impaired recall or concentration, lack of appetite, and weight loss. Not uncommonly, the bereaved might idealize the dead person, maintaining an element of "reverence" despite a recognition of the deceased's human faults and failings. In fact, survivors of a bad marriage can become hopelessly stuck in grief. They mourn not only for the marriage that was but also for the marriage that might have been and was not. In cases of pathological grief following the death of a spouse, three factors are typically present: an unexpected death, ambivalence regarding the marriage, and overdependence on the spouse (Rees, 1997). Recovery from grief seems to be quicker and more complete when a marriage was happy.

The third phase of bereavement involves emancipation from the loved one and an adjustment to the new circumstances. In this period the individual mobilizes his or her resources, attempts to become reconnected with people and activities, and seeks to establish a new equilibrium that will permit some element of satisfaction and comfort. Some people may complete the psychological and emotional work of this stage in about six to eight weeks, others in a matter of months; but for still others the process may continue for years.

The fourth phase is characterized by identity reconstruction. The person crystallizes new relationships and assumes new roles without the loved one. At this stage approximately half of the survivors report realizing some benefit or experiencing some growth from bereavement. These gains include an increased sense of self-reliance and strength, a greater caring for friends and loved ones, and a more general quickening to life and deepening appreciation of existence.

Individual Variations in the Bereavement Process
People differ greatly in how they handle the death of a loved one, their specific symptoms of grief, and the intensity and duration of the symptoms (Johnson, Lund, & Dimond, 1986). The reactions cannot be neatly plotted in a series of well-defined stages, nor is the progression from the time of death to the resolution of bereavement necessarily a straight line. And although shock, anger, and depression are common reactions to loss, not everyone experiences them, no matter how deeply they cared about the person they lost. For instance, a man who lost his wife six months earlier might say he is ready to remarry, and a young mother might laugh with friends only days after losing her child. Between a quarter and two-thirds of those who lose a loved one do not show great distress. Indeed, the absence of extreme distress can be a sign of psychological strength and resilience (Goleman, 1989b). New interests and strong social networks seem to promote adjustment (Norris & Murrell, 1990).

However, some aspects of grief work do not end for a significant proportion of bereaved individuals. They still feel themselves strongly affected by the deceased

Human Diversity

Cross-Cultural Perspectives on Death

In some societies, one is required to deal with death during one's entire lifetime—a survivor might be expected to dress, behave, and participate in a certain manner within society that reflects the loss she or he has suffered perhaps 20 years previously. For most Americans, one is either dead or living, but for people in other cultures, there are gradations of dead. For example, in some cultures, a person might well participate in her own death ritual, or someone whom we would describe as dead—with no vital signs—might be considered an active member of the community. These differences are easier to understand if we take it as our premise that for all cultures, death is a transition from one stage to another. In some cultures, such as our own, the demarcation between life and death is based on medical criteria and is therefore very narrow and precise. In other cultures, this transition is founded on social and emotional ties that might cause the line between life and death to be less clear cut—if the deceased is considered as one who is getting ready for an extended journey, it might take several months to make sure all of the preparations are ready.

Variations among cultures in defining death can be traced to how each culture makes sense of what occurs after death. If death is thought of as a gift, that the deceased will be given something greater than what is here on earth, then the rituals, emotions, and bereavement will be shorter and less doleful. If, on the other hand, death is thought to be caused by a malevolent act, such as a curse, then the ritual might be full of anger and thoughts of retribution. We might have trouble understanding or appreciating the grief of someone from another culture if the manifestation of that sorrow also includes joking and laughter, murderous rage, wailing and physical mutilation, or catatonic unresponsiveness for months. If we look at death in other cultures, we may better understand why our culture's view of death has taken its present form. All forms of bereavement and ritual are highly prescribed by the society in which we live. Parkes, Laungani, and Young (1996) have gathered many examples of cultural differences surrounding death and dying, a few of which are described here. The following two examples show us that different cultures handle death in different ways and have different attitudes regarding death. In India, death is not considered a taboo subject, while Muslims tend to not discuss it openly. Some cultures look at death as the end of life, and others see it as the beginning of a new life. Whether or not one believes in an afterlife, one thing is certain, the dead live on in the memories of those who survive them.

Hinduism

A Brahmin gentleman has been sick for some years and he does not have long to live. Maintaining Hindu custom, his wife undertakes severe fasts in order to help him recover. She knows that her actions will propitiate the gods, and they will help her husband. Even though she begins to suffer physically from not eating, she sits by a family shrine singing devotional songs, reciting mantras, and reading from holy texts. The eldest son will become head of the family when his father dies. The man's three children take turns spending the night with him in the hospital so that he is never alone. On the day that the old man dies, the hospital room is full of people who are paying their respects. They begin to wail and lament loudly until a doctor comes in and puts his hands on the eldest son's shoulders. Everyone calms down and looks at the eldest son, acknowledging that he is now head of the family and must take control of the situation. He persuades everyone to leave and begins to make preparations for the funeral. He knows that the hospital will only sign the death certificate and release the body after he has paid the bill in full. As soon as that is done, the deceased is taken home to await the funeral. At home the women are in charge, and after they take a ritually cleansing bath, they begin to prepare for the funeral. All of the women will wear white saris, and the wife of the deceased will wear a white sari for one year. The red mark, called a *sindoor*, which was adopted upon her marriage, is wiped off the wife's head, indicating that she is a widow. Everyone sleeps on the floor for the next 12 days, including the deceased. For the next 12 days, cooks and cleaners are hired to supply food and to clean and to feed the guests. Everyone will participate in two prayer meetings every day, at sunrise and sunset.

The day after the death, the sons' heads are shaven so they can offer their hair to their departed father. The dead man

person, say they are upset on the yearly anniversary of the person's death, and experience an emotional void in their lives. Their sense of loss may overshadow other experiences, changing the way they interpret even positive events because the events remind them of their loss and their inability to share the experiences with the loved one (Zautra, Reich, & Guarnaccia, 1990). At unguarded moments people can stumble into "little ambushes" of grief. One widower tells of being out driving and spotting a woman with a hairstyle similar to his wife's:

> I said to myself, "There's Nola." Then I laughed out loud and told myself, "How silly of me. Nola is dead." Then my next thought was, "I must go home and tell

is lying on the floor, covered by a white sheet, but with his face left uncovered. After a few hours, flies have begun to settle on his uncovered face. Two days later the three sons wash the body and then oil and anoint it. The body is then placed on a bier and taken to a crematorium where it is then ceremoniously cremated. Although everyone else will soon leave, the family will wait by the fire until they hear the skull crack—it is believed that when the skull cracks open, the soul is released. The ashes will be put in the Ganges to ensure spiritual salvation. For the rest of the 12-day period, the women gather at four o'clock to wail and cry. The oldest son has one more task to perform: He must feed 350 needy mendicants after the final funeral rite of throwing the ashes into the river.

Islam

One essential belief among Muslims is that although life on Earth is finished, life after death will continue and so preservation of the body is necessary. For that reason, cremation is forbidden by Islamic law. When death approaches, the dying are helped to sit up or at least turn their faces toward Mecca. If the dying person cannot pronounce a confession of faith, those present will say the prayer, and the dying person, too weak to repeat the words, will lift a finger to sign their presence. This ritual is important because it signifies that the dying person understands that death is not the end and that now she or he is entering the world of the divine. For the survivors it means defying the definiteness of death by taking the first step toward continuation.

As soon as death has occurred, the body is laid out on a hard surface. The feet are bound together, the arms straightened along the sides, the eyes closed, and the chin wrapped up with a piece of cloth. The body is washed—men washing a man's body and women washing a woman's body. All washers follow a fixed pattern. They clean themselves three times, then they clean the corpse's nose, ears, and neck. Then they wash the right forearm and hand, followed by the left. The entire body will be cleaned, the feet coming last. Everything is done with the utmost care, because a little life is supposed to linger in or around the corpse through which the dead still can feel and hear. After being ritually cleansed, the dead are shrouded from head to toe in clean white cloth. Women mourners usually cry, scratch their faces, beat their chests, and begin a long dirge that details circumstances of the departed's life and death. Men are usually constrained in an attempt to separate one's life from that of the deceased. Concentration on reciting the Koran serve this purpose of restraining oneself. For example, when King Hussein of Jordan died, the Koran was recited on radio and television for three days while all news and entertainment were strictly forbidden.

The greatest service one can render is to bury the dead immediately—If possible on the same day. Shortly before burial a ceremony takes place in which people can make their last private farewell. This is followed by a prayer for the deceased. It is forbidden to touch the body, as this could undo its purity. Everyone present will forgive the dead person's sins and begin posing questions, "Was [the person] a good person?" To which all present will answer "good!" Then six men will carry the bier to the burial site. The deceased will be lowered into the bottom of the grave and positioned as if sleeping. The corpse is turned a bit on its right, with eyes facing Mecca and feet pointing south. The grave is quickly filled in while the Koran is read. As soon as the last foot has finished tamping the grave, it is believed the deceased will wake up and receive a visit. This is the visit for which the body has been prepared. Two angels enter the grave and ask the following five final questions:

- Who is your God? (My God is Allah)
- Who is your prophet? (My prophet is Mohammed)
- Which is your book? (My book is the Koran)
- Who is your Imam? (My Imam is _____)
- Which is your prayer direction? (My prayer direction is Mecca)

Because the dead person is not able to answer, the Imam "helps" the deceased by calling his or her name. After they leave the gravesite, the mourners linger within earshot to hear the Imam calling the dead person's name. The dead are supposed to react by bumping their heads against a piece of wood that was put into the grave for this purpose. If the Imam calls once, everyone knows that the person was indeed a "good" person, but if the Imam calls 15 or 20 times, everyone knows that this person was "no good." Children do not partake of the ceremony and are not encouraged to ask questions (Parkes, Laungani, & Young, 1996).

Nola how silly I just was." It all happened in a fraction of a second (Gelman, 1983, p. 120).

In sum, we are increasingly coming to realize that there is not a universal prescription for how best to grieve and that people handle grief in a great many different ways (Goleman, 1989b).

Questions

Why do so many people think that the one-year anniversary of the death of a loved one is the time to be "over it"? Is there a standard time frame in many societies when grief work should be over?

Widows and Widowers

Two are better than one, because they have a good reward for their toil. For if they fall, one will lift up his fellow; but woe to him who is alone when he falls and has not another to lift him up. Again, if two lie together, they are warm; but how can one be warm alone?

Ecclesiastes 4:9–11

Three out of four American men 65 years old and older are married and living with their wives. One out of three women in the same age group are married and living with their husbands. Of those over the age of 75, two-thirds of the men are living with a spouse, whereas less than one-fifth of the women are. Hence, women 65 and older are much more likely to be widowed than married. This situation results from the fact that the life expectancy of women tends to be seven or more years longer than that of men and from the tradition that has ordained that women marry men older than themselves (of interest, women married to men younger than themselves tend to live longer than would otherwise be expected, whereas women married to older men tend to die sooner than would be expected) (Klinger-Vartabedian & Wispe, 1989). Consequently, remarriage tends to be a male prerogative: After age 65 men remarry at a rate eight times that of women (Horn & Meer, 1987).

Of the more than 800,000 people widowed each year, some 20 to 26 percent still suffer serious depression a year or more later. Their eating habits frequently are altered, resulting in poorer health and nutrition (Rosenbloom & Whittington, 1993). Use of alcohol, drugs, and cigarettes rises. Health problems in survivors tend to be worse among those who were already in poor physical or mental health, those who are alcohol or drug abusers, and those who lack a social support network. Survivors with strong social support or who remarry seem to suffer fewer health problems.

The death of a spouse seems particularly stressful for men (Osterweis, Solomon, & Green, 1984). The death rate for widowers is seven times as high as the rate for married men of comparable age. Widowers die four times as often from suicide, three times as often from motor vehicle accidents, and six times as often from heart disease. But over half of all widowers between 46 and 65 remarry, and substantial numbers remarry beyond the age of 65 (among people over 65, there are nine bridegrooms to every bride). Because healthy widowers presumably remarry relatively rapidly, the statistics showing higher mortality among widowed men might apply primarily to the less healthy widowers (Greenberg, 1981).

One problem confronted by widowers is the cultural dictate that men are not supposed to feel emotion and pain or say "I need help," due to which men traditionally have had difficulty expressing emotion. Furthermore, many widowers have trouble cooking and caring for themselves. They develop poor eating habits, which, together with other poor health practices and feelings of loneliness and emptiness, often lead to heavy drinking, sleeplessness, and chronic ailments. Overall, U.S. men seem to have a more difficult time living alone than do U.S. women, and they are far less likely to receive help from others (Coombs, 1991).

We know quite a bit more about widows than about widowers, largely owing to research by sociologist Helena Znaniecki Lopata (1981). The women studied by Lopata lived in metropolitan Chicago and were interviewed by the National Opinion Research Center. Lopata was able to distinguish three categories of widows on the basis of the extent of their involvement in different types of social relationships. At one extreme were women, primarily better educated and belonging to the middle class, who were strongly involved in the role of wife when their husbands were still alive. They built many other roles on the husband's presence as a person, a father, and a partner in leisure-time activities. There was a strong tendency for these women to idealize the late husband, often to the point of sanctification (Futterman et al., 1990). At the other extreme were women, primarily lower- or working-class and living in black or ethnic neighborhoods, who belonged to sex-segregated worlds and were immersed in kin, neighboring, or friendship relationships with other women. Between these two extremes were women who led multidimensional lives in which the husband was involved in only part of the total set of relations. The adjustments confronting the women tended to vary with the degree to which their social relationships revolved about or were integrated with those of their husbands.

The main conclusion Lopata drew from her data was that the higher the woman's education and socioeconomic class, the more disorganized her self-identity and life became with her husband's death—but by the same token, the more resources she had to form a new lifestyle once her grief work was accomplished. Other research has suggested that the negative long-term consequences of widowhood seem to derive from socioeconomic deprivation rather than from widowhood itself (Bound et al., 1991).

Of considerable interest, Lopata found that about half of the widows lived entirely alone, and most of these women said they much preferred to do so. Only 10 percent had moved in with their married children. One reason this figure was so low is that the widows cherished their independence, which they did not wish to jeopardize by giving up their own homes and moving into an

Coping with Death The adjustments confronting widows tend to vary with the degree to which their social relationships revolved about or were integrated with those of their husbands. Typically, lower socioeconomic status is a significant factor in coping. Children sometimes have greater difficulty coping with the death of a loved one than do adults.

unfamiliar network of relations. Furthermore, the widows anticipated problems in their role as mothers of grown-up children. They said that if they could not criticize or speak up, they would feel inhibited, but that if they did speak up, their children would become upset. They regarded their relationships with grandchildren as presenting similar problems.

Question

What are the differences in the ways widows and widowers cope with the loss of a spouse?

The Death of a Child

The loss of a child is also frequently associated with depression, anger, guilt, and despair; recovery can take a very long time. Parents must come to recognize the limits of their protective powers. Women who find their primary satisfactions in the mother role might feel useless without someone to care for. Guilt can be especially in-

tense after a death from sudden infant death syndrome (SIDS). Parents of children who die from cancer might find that their grief intensifies during the second year. Parents who feel they did all they could do to care for a child during the final illness seem to make a more speedy recovery. Some evidence suggests that parents who chose hospital care rather than home care are more depressed, socially withdrawn, and uncomfortable afterward (Murphy et al., 1999).

Loss by Miscarriage   Parents who have lost a child by miscarriage often receive no recognition of their loss from others, yet their grief work might continue for a lifetime. Support groups can help because other people who have lost a child through miscarriage will understand their feelings of loss. Statements such as "At least you can have another one" or "This child wasn't meant to be" do not comfort people experiencing this loss. Every anniversary of the loss is a reminder of the child, who usually has been named and placed in a gravesite but didn't have the opportunity for life. The mother is likely to feel a great deal of guilt, thinking that if only she had done something differently, the child would be here. Most fathers suffer in silence and make efforts to go on with their lives. Repetitive miscarriages can bring about deep depression and feelings of failure in the mother.

Loss by Murder or Violence   Nicholas Green was a 7-year-old American boy on vacation in Italy in 1994 with his parents, Reg and Maggie Green, when he was fatally shot by highway robbers during a night time drive toward their destination. Within hours, his parents made the difficult decision to donate his organs to seven Italian citizens, who survived because of this gift. To this day, Reg Green says he thinks of his son a hundred times a day, and though it does not take away the pain of the loss of his son, he does have a sense of peace knowing that his son helped people in life-threatening situations. John Walsh and his wife lost their son Adam to a vicious child killer, and their lives have never been the same. The loss of a child is one of life's greatest agonies, even when parents know it is inevitable due to illness or disease; but when someone senselessly takes the life of a child, the bereavement process goes on indefinitely. Some parents start a crusade in honor of their child, such as the campaign to enact Megan's Law, which now requires that community residents be notified if a released child molester moves into the community. No parents want their child to have lived in vain, and humanitarian efforts—such as organ donation, a national database of convicted child molesters, hot lines and web sites for teenage runaways, stricter laws and penalties for those who commit the crimes—help these parents keep the memories of their child alive.

Segue

Perhaps the most difficult transition anyone has to make after the death of a loved one is the journey from grief and sorrow to remembrance and honoring. Our loss seems to overwhelm us and, fearing that we will dishonor the one we love by forgetting, we nurture grief, clinging to it with a desperation that comforts us. But in time, this grieving must give away to honoring and remembering. So often are the words "get on with your life" spoken to the one who is left behind, alone. To move on is not to forget. It is to remember.

It is to remember all that your loved one gave you and all that you shared with your loved one. To move on is to celebrate those gifts and to know your loved one lives on in your memories. It is to realize how deeply he or she touched your life.

Ted Menten, *Gentle Closings: How to Say Goodbye to Someone You Love*, p. 136

Summary

The Quest for "Healthy Dying"

1. Until very recently, death was a taboo topic in Western society. During the past 35 years, however, this pattern has been reversed. Considerable controversy swirls about such matters as the right to die, clinical death, the death penalty, and life after death.

2. Thanatology is the study of death and dying, and interest has grown in this field.

3. Over the past 20 years, public and professional awareness of the dying person's experience has increased dramatically. "Death with dignity!" has become a major rallying cry. The death awareness movement asserts that a basic human right is the power to control one's own dying process. Much criticism is currently leveled at the way modern technology is applied to the terminally ill. According to this view, too much is done for too long a period at too high a cost, all at the expense of basic human considerations and sensitivities.

4. Cultural attitudes toward suicide still waver among extremes. Early Christians called suicide a sin and denied burial rites to those who killed themselves. Some consider suicide victims to be insane or weak. Others consider suicide the honorable choice when confronted with capture, defeat, or disgrace.

5. The hospice approach involves a variety of programs designed to provide an alternative to conventional hospital care for the terminally ill, especially cancer patients. The emphasis of the movement falls on "comfort care" rather than on attempts to prolong life. Comfort care involves an aggressive treatment of symptoms, both physical and emotional, through the use of counseling, antidepressive medications, and high-dose morphine preparations. Most hospice programs are centered about caring for the dying person at home.

The Dying Process

6. The earliest Biblical sources, as well as English common law, considered a person's ability to breathe independently to be the prime index of life. Over the past 35 years, technological advances have rendered this traditional definition of death obsolete. Such advances have included methods to resuscitate victims of cardiac arrest, mechanical respirators, and artificial heart pacemakers. A growing acceptance in medical and legal circles of the need for an additional criterion of death has resulted in a legal definition that includes the absence of spontaneous brain function.

7. A realistic acceptance of death could well be the hallmark of emotional maturity. People differ considerably, however, in the degree to which they are consciously aware of and think about death. Furthermore, death is a highly personal matter, and its meaning tends to vary from individual to individual. Researchers agree that only a relatively small proportion of the elderly express a fear of death.

8. Elisabeth Kübler-Ross identifies five stages through which dying persons typically pass: denial, anger, bargaining, depression, and acceptance.

9. In some cases persons pronounced clinically dead but then resuscitated by medical measures have told of having left their bodies and undergone otherworldly experiences. Some individuals have interpreted such experiences as scientific evidence of a spiritual existence beyond death. Skeptics rejoin that the visions reported by dying people are hallucinations associated with the intense arousal of the central nervous system and disorganized brain functioning.

Grief, Bereavement, and Mourning

10. Bereavement and grief have a considerably greater impact than is evident in the period immediately

following the death of a loved one. The survivor is more vulnerable to physical and mental illness, even to death. The bereaved adult typically passes through a number of phases. The first phase is characterized by shock, numbness, denial, and disbelief. The second phase entails pining, yearning, and depression. The third phase involves emancipation from the loved one and an adjustment to the new circumstances. The fourth phase is identity reconstruction.

11. Cultures differ in how they perceive death and dying. Every culture has a unique approach for dealing with death, but each will involve an understanding of death, spiritual beliefs, rituals, expectations, and taboos.

12. The major causes of death vary from country to country. Heart disease and cancer are major causes of adult death in the United States.

13. Loss of parents, spouse, or children all have a significant impact on a person. Typically parents and adult children retain strong ties throughout life. The loss of a child is one of life's most intensely agonizing experiences, one that many people never quite get over.

14. Women 65 and over are much more likely to be widowed than married. The difficulty women have in adjusting to widowhood tends to vary with the degree to which their social relationships revolved about or were integrated with those of their husbands.

Key Terms

anticipatory grief *(604)*

bereavement *(604)*

brain death *(597)*

euthanasia *(589)*

grief *(604)*

hospice *(595)*

life review *(598)*

living will *(587)*

mourning *(604)*

near-death experience (NDE) *(599)*

thanatology *(587)*

Following Up on the Internet

Web sites for this chapter focus on controversies about "healthy" dying, the dying process, and the coping of survivors. Please access the text web site at http://www.mhhe.com/crandell7 for up-to-date hot-linked Internet addresses for the following organizations, topics, and resources:

Thanatology Forum

Euthanasia

DeathNet

Sociology of Death

Bereavement Resources Directory

Glossary

A

abortion   Spontaneous or induced expulsion of the fetus prior to the time of viability, occurring most often during the first 20 weeks of the human gestation period.

accommodation   In Piaget's cognitive theory, the process of changing a schema to make it a better match to reality.

activity theory of aging   The theory that as an elderly person's level of activity declines, so also do that person's feelings of satisfaction, contentment, and happiness.

adolescent growth spurt   Rapid increase in height and weight during the early adolescent years.

adult day care   A program of long-term care and support to adults who live in the community, providing health, social, and support services in a protective setting during any part of the day.

afterbirth   Placenta and the remaining umbilical cord expelled from the uterus through the vagina after childbirth.

age cohort   (also called *cohort* or *birth cohort*) A group of persons born in the same time interval.

age grading   Arranging people in social layers based on place in the life cycle.

age norms   Social standards that define what is appropriate for people to be and to do at various ages over the life span.

age strata   Social layers within societies that are based on chronological age and serve to differentiate people as superior or inferior, higher or lower.

ageism   Stereotyping and judging a group of people solely on the basis of their age.

aggression   Behavior that is socially defined as injurious or destructive toward a person or a group of persons.

aging   The process of biological and social change across the life span.

alienation   A pervasive sense of powerlessness, meaninglessness, isolation, and estrangement from others.

allele   A pair of genes found on corresponding chromosomes that affect the same trait.

Alzheimer's disease   A progressive, degenerative neurological disorder involving deterioration of brain cells that can occur in late adulthood.

amenorrhea   The absence of a menstrual cycle in a female who was normally menstruating, other than during pregnancy. This is more common among females who are anorexic or bulimic and those in perimenopause, and in some females who participate in endurance sportslike marathons.

amniocentesis   A commonly used invasive procedure conducted between the 14th to 20th week of gestation to determine the genetic status of the fetus. It involves withdrawing and analyzing amniotic fluid.

androgyny   The presence of both male-typed and female-typed characteristics.

anorexia nervosa   A potentially life-threatening eating disorder that affects primarily females, and a smaller percentage of males, in which the person becomes obsessed with looking thin and seriously alters her or his eating behaviors in order to lose weight. Excessive exercise is also associated with anorexia.

anoxia   Oxygen deprivation caused when the umbilical cord becomes squeezed or wrapped around the baby's neck during delivery.

anticipatory grief   A state of emotional limbo where the death of a loved one is anticipated and individuals feel unable to resolve the loss because it has not yet occurred and are unable to avoid the diagnosis that death will occur.

Apgar scoring system   A standard scoring system developed by anesthesiologist Virginia Apgar to objectively appraise the normalcy of a baby's condition at birth, based on five criteria.

assimilation   In Piaget's cognitive theory, the process of taking in new information and interpreting it in such a way that it conforms to a currently held model of the world.

assisted reproductive technologies (ARTs)   Scientific technological options used to increase a woman's chance of becoming pregnant when conception does not occur through heterosexual intercourse.

asynchrony   Dissimilarity in the growth rates of different parts of the body. For example, during adolescence the hands, feet, and legs grow before the trunk of the body.

attachment   An affectional bond that one individual forms for another and that endures across time and space.

authoritarian parenting   A parenting style distinguished by attempts to shape, control, and evaluate a child's behavior in accordance with traditional and absolute values and standards of conduct.

authoritative parenting   A parenting style distinguished by firm direction of a child's overall activities but allowing the child freedom to make some decisions within reasonable limits and supervision.

autonomous morality   The second stage of Piaget's two-stage theory of moral development which arises from the interaction among status equals relationships among peers.

autosomes   The 22 pairs of chromosomes similar in size and shape that each human being normally possesses in addition to the sex chromosomes.

B

behavior modification   The application of learning theory and experimental psychology to alter behavior.

behavioral theory   A psychological theory that focuses on observable behavior—what people do and say—and how their environment shapes their development across the life span.

bereavement   Being deprived of a relative or friend by death.

bilingualism   Instruction in both first and second languages by teachers proficient in both; proficiency in two languages.

binge drinking   Downing five or more alcoholic drinks in a row for men, or four or more in a row for women.

biological aging   Changes in the structure and functioning of the human organism over time.

birth   The transition from dependent existence inside the woman's uterus to life as a separate organism.

birthing centers   Primary care facilities, other than hospitals, typically located in urban centers, that are used for low-risk childbirth.

birthing rooms   A homelike atmosphere in a hospital or other setting where labor and delivery can occur.

blastocyst   The gel-like fluid-filled ball of cells produced after the zygote goes through early meiotic cell division. The blastocyst gradually moves into the uterus and implants itself into the uterine wall to nourish itself and becomes the embryo.

brain death   Cessation of neural activity when the brain receives insufficient oxygen to function.

bulimia   Also called binge-purge syndrome. A serious eating disorder characterized by repeated episodes of binging—particularly on high-calorie foods like candy bars, cakes, pies and ice cream—and after eating purging by forced vomiting, taking laxatives, enemas, diuretics, or fasting.

C

caretaker speech   The simplified form of language used by adults when they are talking to infants and young children.

case-study method   A special type of longitudinal study that focuses on a single individual rather than a group of subjects.

cataracts   A clouding of the lens of the eye that impairs vision, typically seen in older adults; in rarer cases found in infants or young people.

causality   A cause-and-effect relationship between two paired events that recur in succession. Piaget concluded that children younger than 7 or 8 fail to grasp cause-and-effect relationships.

centration/centering   The process whereby preoperational children, ages 2 to 7, concentrate on only one feature of a situation and neglect other aspects. It is characteristic of preoperational thought.

cephalocaudal development   The process of development that commences with the brain and head areas and then proceeds down the body to the feet.

cesarean section (C-section)   A surgical delivery technique by which the physician enters the uterus through an abdominal incision and removes the infant.

child abuse   Intentional physical attack on or sexual abuse or injury to a child.

cholesterol   A white, waxy substance found naturally in the body that the body uses to build the cell walls and make certain hormones. Certain foods can cause cholesterol buildup in the bloodstream that can cause cardiovascular problems.

chorionic villus biopsy (CVS)   An invasive procedure performed between 9½ and 12½ weeks of gestation to determine genetic characteristics of the fetus. The physician inserts a thin catheter through the vagina and cervix and into the uterus, removing a small plug of villous tissue.

chromosomes   The long threadlike structures made up of protein and nucleic acid that contain the genes, which transmit hereditary materials found in the nuclei of all cells.

classical conditioning   A type of learning in which a new, previously neutral stimulus, such as a bell, comes to elicit a response, such as salivation, by repeated pairings with an unconditioned stimulus, such as food.

climacteric   A time in a woman's life characterized by changes in the ovaries and in the various biological processes over 2 to 5 years prior to complete cessation of menstruation.

cognition   The act or process of knowing.

cognitive development   A major domain of development that involves changes in mental activity, including sensation, perception, memory, thought, reasoning, and language.

cognitive learning   The process of observing other people and learning new responses or behaviors without first having had the opportunity to make the responses ourselves.

cognitive stages   Sequential periods in the growth or maturing of an individual's ability to think, to gain knowledge, and to be aware of one's self and the environment.

cognitive styles   Consistent individual differences in how a person organizes, processes, recalls, and uses information.

cognitive theory   A theory that attempts to explain how we go about representing, organizing, treating, and transforming information as we modify our behavior.

colic   An uncomfortable condition of unknown origin that can cause a baby to cry for at least an hour or more (typically every day about the same time). Colic typically disappears after the first several months of development.

collagen   A basic structural component of connective tissue in body cells that appears to be implicated in the aging process. During late adulthood the body has less collagen, thus a person's skin appears more wrinkly.

communication   The processes by which people transmit information, ideas, attitudes, and emotions to one another.

companionate love   The kind of love a person has for a very close friend.

conceptualization   A grouping of perceptions into classes or categories on the basis of certain similarities. Related ideas can come together to create a concept.

confluence theory   The view that the intellectual development of a family is like a river, with the inputs of each family member flowing into it.

consciousness of oneness   A sympathetic identification in which group members come to feel that their inner experiences and emotional reactions are similar.

conservation   Understanding that the quantity or amount of something stays the same regardless of changes in its shape or position. According to Piaget, preoperational children (2 to 7 years old) do not have the concept of conservation. Children are typically in elementary school before they learn this concept.

consummate love According to Sternberg, the kind of love present when all three aspects of his triangular theory of love exist in a relationship.

continuum of indirectness The notion that the role played by hereditary factors is more central in some aspects of development than in others.

control group In an experiment, a group of subjects who are similar to the subjects in the experimental group but do not receive the independent variable (treatment). The results obtained with the control group are compared with the results of the experimental group.

convergent thinking The application of logic and reasoning to arrive at a single, correct answer to a problem.

coping The responses, behaviors, and actions one takes in order to master, tolerate, or reduce stress.

correlation coefficient The numerical expression of the degree or extent of relationship between two or more variables or conditions.

critical period For the developing embryo, the time of development when each organ and structure is most vulnerable to damaging influences. Also a relatively short period of time in which specific development or imprinting normally takes place.

cross-cultural method A study in which researchers compare data from two or more societies and cultures. Culture, rather than individuals, is the subject of analysis.

cross-sectional method Investigation of development by simultaneously comparing people from different age groups.

crowning The stage during childbirth when the widest diameter of the baby's head is at the mother's vulva.

crystallized intelligence The ability to use on later occasions knowledge that was acquired earlier in life. Crystallized intelligence often shows an increase with age.

cultural dislocation A sense of homelessness and alienation from a traditional way of life.

culture The social heritage of a people—those learned patterns of thinking, feeling, and acting that are transmitted from one generation to the next.

D

death drop A marked intellectual decline that occurs shortly before a person dies.

decay theory A theory of cognitive decline according to which forgetting occurs due to deterioration in the memory traces in the brain.

delivery Childbirth. The process that begins when the infant's head passes through the mother's cervix and ends when the baby has completed its passage through the birth canal.

deoxyribonucleic acid (DNA) The active biochemical substance of genes that programs the cells to manufacture vital protein substances.

dependent variable In an experiment, an objective measure of the subject's behavior—the variable that is affected by the independent variable.

depression A state of mind characterized by prolonged feelings of gloom, despair, and futility, profound pessimism, and a tendency toward excessive guilt and self-reproach.

development The orderly and sequential changes that occur with the passage of time as an organism moves from conception to death.

developmental psychology That branch of psychology that investigates how individuals change over time while remaining in some respects the same.

deviant identity A lifestyle that is at odds with, or at least not supported by, the values and expectations of society.

dialectical thinking An organized approach to analyzing and making sense of the world one experiences that differs fundamentally from formal analysis.

disengagement theory of aging A view of aging as a progressive process of physical, psychological, and social withdrawal from the wider world.

displaced homemaker A woman whose primary life activity has been homemaking and who has lost her main source of income because of divorce or widowhood.

divergent thinking Open-ended thought in which multiple solutions are sought, examined, and probed, thereby leading to what are deemed creative responses on measures of creativity.

dominant character In genetics, the property of an allele (gene) that completely masks or hides the other paired allele, as in *AA* or *Aa.*

drug abuse Excessive or compulsive use of chemical agents to an extent that interferes with an individual's health and social or vocational functioning, or the functioning of the rest of society.

dry eye Diminished tear production that can be uncomfortable and can usually be eased with eye drops; typically a condition in late adulthood.

dyslexia A type of brain dysfunction exhibited by extreme difficulty in learning to read in an otherwise normally intelligent, healthy child or adult.

E

eclectic approach An approach to studying behavior in which psychologists select from the various theories and models those aspects that provide the best fit for the descriptive and analytical task at hand.

ecological approach Bronfenbrenner's system of understanding development, according to which the study of developmental influences must include the person's interaction with the environment, the person's changing physical and social settings, the relationship among those settings, and how the entire process is affected by the society in which the settings are embedded.

ecological theory A theory of development proposed by Bronfenbrenner that proposes that, in order to understand development, researchers must look at the relationship between the developing individual and the changing environment(s).

educational self-fulfilling prophecies Teacher expectation effects whereby some children fail to learn because those who are charged with teaching them do not believe that they will learn, do not expect that they can learn, and do not act toward them in ways that motivate them to learn.

egocentrism A lack of awareness that there are viewpoints other than one's own. There are two characteristic forms of egocentric thinking in adolescents: the personal fable and the imaginary audience.

elder abuse Acts of commission and omission that cause unnecessary pain or suffering to older persons.

embryo The developing organism from the time the blastocyst implants itself in the uterine wall until the organism becomes a recognizable human fetus.

embryonic period The second stage of prenatal development, the period from the end of the second week to the end of the eighth week of gestation.

emotions The psychological changes, subjective experiences, and expressive behaviors that are involved in such feelings as love, joy, grief, anger, and many others.

emotional intelligence A relatively new concept, proposed by Goleman, that includes such abilities as being able to motivate oneself, persisting in the face of frustrations, controlling impulses and delaying gratification, empathizing, hoping, and regulating of one's moods to keep distress from overwhelming one's ability to think.

emotional-social development (also called **psychosocial development**) A major domain of development that includes changes in an individual's personality, emotions, and relationships with others.

empathy Feelings of emotional arousal that lead an individual to take another person's perspective and to experience an event as the other person experiences it.

empty nest That period of the family life cycle when children have grown up and left home.

empty-nest syndrome A variety of emotions parents can experience associated with their children growing up and leaving home.

encoding A cognitive process that involves perceiving information, abstracting from it one or more characteristics needed for classification, and creating corresponding memory traces for it.

entrainment A kind of biological feedback system across two organisms in which the movement of one influences the other.

epigenetic principle According to Erikson, the principle that each part of the personality has a particular range of time in the life span when it must develop if it is going to develop at all.

equilibrium In Piaget's theory, this is the result of balance between the processes of assimilation and accommodation.

ESL approach Instructional methods focused on teaching English to children with limited English proficiency.

ethology The study of the behavior patterns of organisms from a biological perspective.

euthanasia The act of terminating an ill or injured person's life for reasons or mercy. Also called assisted suicide and mercy killing.

event sampling A research technique of recording a class of behaviors observed at particular time intervals.

executive strategies Strategies for integrating and orchestrating lower-level cognitive skills.

exosystem Bronfenbrenner's third level of environmental influence, consisting of the social structures that directly or indirectly affect a person's life.

experiment A rigorous study in which the investigator manipulates one or more variables and measures the resulting changes in the other variables to attempt to determine the cause of a specific behavior.

experimental group In an experiment, the group that receives the independent variable (compared with the control group).

experimental method A rigorously objective scientific technique that allows a researcher to attempt to determine the cause of a behavior or event.

expressive tie A social link formed when we invest ourselves in and commit ourselves to another person.

extraneous variables Factors that can confound the outcome of an experiment, such as the age and gender of the subjects, the time of day the study is conducted, the educational levels of the subjects, the setting for the experiment, and so on.

extrinsic motivation The kind of motivation at work when an activity is undertaken for some reason other than its own sake. Rewards such as school grades, wages, and promotions are examples of extrinsic motivation.

F

failure to thrive An infant or child being severely underweight for its age and sex.

fallopian tubes Two passages from the ovaries to the uterus that carry the ova from the ovary to the uterus. Fertilization, if it occurs, typically occurs in the fallopian tubes.

family life cycle The sequential changes and realignments that occur in the structure and relationships of family life between the time of marriage and the death of one or both spouses.

fear An unpleasant emotion aroused by impending danger, loss, pain, or misfortune.

fertilization The union (or fusion) of an ovum and a sperm that usually occurs in the upper end of the fallopian tube and that results in a new structure called the zygote.

fetal period The third stage of prenatal development, extending from the end of the eighth week until birth.

fetoscopy A procedure that allows a physician to examine the fetus directly through a lens after inserting a very narrow tube into the uterus.

fetus The developing organism during the fetal period while in the mother's womb.

fixation According to psychoanalytic theory, the tendency to stay at a particular psychosexual stage of development.

floaters Floating spots that actually are particles suspended in the gel-like fluid that fills the eyeball but generally do not impair vision.

fluid intelligence A cognitive ability to make original adaptations in novel situations. Fluid intelligence is generally tested by measuring an individual's facility in reasoning. Fluid intelligence usually declines with age in later life.

fraternal (dizygotic) twins Siblings that develop from two eggs are fertilized by two different spermatozoa, develop separately in the womb at about the same time, and are born at the same time.

fusion (fertilization)   The process where a sperm enters and unites with an ovum to form a fertilized egg.

G

gametes   Reproductive cells (sperm and ova).

gender cleavage   The tendency for boys to associate with boys and girls with girls.

gender identity   The conception that people have of themselves as being female or male.

gender roles   A set of cultural expectations that define the ways males and females should behave.

gender stereotypes   Exaggerated generalizations about female or male behaviors.

generation gap   Mutual antagonism, misunderstanding, and separation between youth and adults.

generativity   The concern of an older generation in establishing and guiding the next generation.

generativity vs. stagnation   The "crisis" that, according to Erikson, the midlife years are devoted to resolving. Generativity is the concern to guide the next generation through nurturing and mentoring. Failure to do otherwise is to become self-centered, which results in psychological invalidism, which leads to rejection.

genes   Small units of heredity located on chromosomes that transmit inherited characteristics from biological parents to children.

genetic counseling   A process whereby physicians and specialists counsel couples about concerns they may have about inherited diseases in their family history.

genetics   The scientific study of biological inheritance.

genotype   The genetic makeup of an organism.

germinal period   The first stage of prenatal development, which extends from conception to the end of the second week.

gerontology   The study of aging and the special problems associated with it.

geropsychology   The study of the changing behaviors and psychological needs of the elderly.

glaucoma   A condition of vision impairment characterized by increased pressure caused by fluid buildup within the eye that can damage the optic nerve and lead to blindness if left untreated.

grief   An experience involving keen mental anguish and sorrow over the death of a loved one.

group   Two or more people who share a feeling of unity and are bound together in relatively stable patterns of social interaction.

growth   The increase in size that occurs with age.

H

harmonious parenting   A parenting style distinguished by an unwillingness to exert direct control over children, in an attempt to cultivate an egalitarian relationship.

heredity   The genes we inherit from our biological parents, which help shape our physical, intellectual, social, and emotional development.

heteronomous morality   The first stage of Piaget's two-stage theory on moral development which arises from the unequal interaction between children and adults.

heterozygous   In biological inheritance, the arrangement in which two paired alleles (genes) are different.

hierarchy of needs   A key concept of Abraham Maslow's humanistic theory, which indicates that basic needs must be met before self-development and self-esteem needs can be fulfilled.

holistic approach   The humanistic approach according to which the human condition must be viewed in its totality, and each person is a whole rather than a mere collection of physical, social, and psychological components.

holophrase   Using single words to convey complete thoughts or sentences; characteristic of the early stages of language acquisition in young children.

homozygous   In biological inheritance, the arrangement in which the two paired alleles (genes) are the same.

horizontal décalage   According to Piaget, a type of sequential development in which each skill is dependent upon the acquisition of earlier skills.

hormone replacement therapy (HRT)   A medical regimen often recommended by physicians to women after menopause to maintain cardiovascular fitness, and a slowing of bone loss and memory loss.

hospice   A program or mode of providing comfort, care, and pain relief to persons dying of cancer or other long-term illness and comfort to the relatives of the patient. Care may be conducted in a hospice center or in the patient's home.

human genome   A map of the genetic makeup of all the genes on their appropriate chromosomes being studied extensively by researchers in the Human Genome Project.

human growth hormone (HGH)   A powerful hormone originally administered to treat children afflicted by dwarfism but is now taken as a trendy anti-aging potion by some of the social elite.

humanistic psychology   A psychological theory, deriving from Abraham Maslow, Carl Rogers, and others, proposing that humans are different from all other organisms in that they actively intervene in the course of events to control their destinies and shape the world around them.

hypertension   High blood pressure.

hypothermia   A condition in which body temperature falls more than 4 degrees Fahrenheit below normal and persists at this low level for a number of hours.

hypothesis   A tentative proposition that can be tested in a research study; forming an hypothesis is one of the initial steps in the scientific method.

I

identical (monozygotic) twins   Twins resulting from the division of a single fertilized ovum.

identity   According to Erikson, identity is defined as a sense of well-being and is achieved by being comfortable in one's body, by knowing where one is going, and by being recognized by significant others.

identity achievement   One of James Marcia's four identity types. The individual is able to achieve inner stability that corresponds to how others perceive that person.

identity diffusion   According to Erikson, an inability to commit to an occupational or ideological position and to assume a recognizable station in life, experienced by some adolescents.

identity foreclosure   One of James Marcia's four identity types. Identity foreclosure is characterized by the individual's avoidance of autonomous choice.

identity moratorium   One of James Marcia's four identity types that occurs when the adolescent experiments with various roles, ideologies, and commitments.

imaginary audience   The adolescent's belief that everyone in the local environment is primarily concerned with the appearance and behavior of the adolescent.

implantation   The process in which the blastocyst completely buries itself in the wall of the uterus.

impotence   A male's inability to have or sustain an erection.

imprinting   A process of attachment that occurs only during a relatively short period and is so resistant to change that the behavior appears to be innate.

in vitro fertilization (IVF)   Fertilization that occurs outside the body, typically in a petri dish in a medical lab environment, followed by implantation of a fertilized egg into a woman's uterus in an attempt to accomplish pregnancy.

inclusion   The integration of students with special needs within the regular classroom programs of the school.

independent variable   The variable that is manipulated by the researcher during an experiment in order to observe its effects on the dependent variable. This is often referred to as the treatment variable.

individualized education plan (IEP)   A plan developed in a collaborative effort by the school psychologist, the child's teachers, an independent child advocate, and the parent(s) or guardian and is a legal document that ensures that the child with special learning needs will be provided with the needed educational support services in the least restrictive learning environment.

industry vs. inferiority   The fourth stage of Erikson's psychosocial model of development in which children either work industriously and are rewarded or fail and develop a sense of inferiority.

infancy   The period of child development during the first two years of life.

infant mortality   The death of an infant within the first year of life.

information processing   The application of mental operations that we use in tackling intellectual tasks in a step-by-step fashion.

informed consent   An ethical standard, established by the American Psychological Association, that requires the researcher to inform each subject about the research study and obtain from each subject his/her voluntary, written consent to participate in a research study.

initiative vs. guilt   According to Erikson, the psychosocial stage in early childhood (about ages 3 to 6) when children strive exuberantly to do things and to test their developing abilities, sometimes reaching beyond their competence.

instrumental tie   A social link that is formed when we cooperate with another person to achieve a limited goal.

integrity vs. despair   Erikson's psychosocial stage of late adulthood in which individuals recognize they are reaching the end of life. People take satisfaction in having led a full and complete life, or view their lives with feelings of disappointment, loss, and purposelessness—depending upon how they navigated the previous stages of development.

intelligence   According to Wechsler, a global capacity to learn exhibited by an understanding of the world, rational thought, and resourceful coping with the challenges of life.

interference theory   The theory that retrieval of a cue becomes less effective as more and newer items come to be classed or categorized in terms of it.

intimacy   The ability to experience a trusting, supportive, and tender relationship with another person.

intimacy vs. isolation   Erikson's stage of psychosocial development when young adults reach out and attempt to make close connections with other people. Failure to do so can lead to more isolated lives devoid of meaningful bonds.

intrinsic motivation   The motivation at work when activity is undertaken for its own sake.

J

job burnout   The development of feelings of dissatisfaction and lack of fulfillment regarding work that once was fulfilling and satisfying.

joint custody   A legal custody arrangement where both parents share equally in significant child-rearing decisions and share in regular child-care responsibilities.

L

labor   In childbirth, the stage when the strong muscle fibers of the mother's uterus rhythmically contract and push the infant downward toward the birth canal.

language   A structured system of sound patterns (words and sentences) that have socially standardized meanings that enable people to categorize objects, events, and processes in their environment and communicate with each other about them.

language acquisition device (LAD)   An inborn language-generating mechanism in humans, hypothesized by Chomsky. Central to Chomsky's theory of language development is the idea that young children, just from having heard words and sentences spoken, can produce speech.

learning   The more or less permanent change in behavior that results from the individual's experience in the environment across the entire life span.

learning disabilities (LD)   The classification used for children, adolescents, and college students who encounter difficulty with school-related material, despite the fact that they appear to have normal intelligence and lack a demonstrable physical, emotional, or social impairment.

life events   Turning points at which individuals change direction in the course of their lives.

life review   A reminiscence and sharing of family history from one generation to another.

lifestyle   The overall pattern of living whereby one attempts to meet biological, social, and emotional needs.

lightening   The repositioning of the infant that occurs a few weeks prior to birth, shifting the infant downward and forward in the uterus to lighten the mother's discomfort and ensure that the baby will be born head first.

limited English proficiency (LEP)   The legal educational term used to describe students who were not born in the United States and/or whose native language is not English and who cannot participate effectively in the regular school curriculum because they have difficulty speaking, understanding, reading, and writing English.

living will   A legal document that states the individual's wishes regarding medical care in case the individual becomes incapacitated and unable to participate in medical care decisions (such as a refusal of heroic measures to prolong her or his life in the event of terminal illness).

locomotion   The infant's ability to walk, which typically evolves between 11 and 15 months of age, and is the climax of a long series of early motor development preceded by crawling and creeping.

locus of control   An individual's perception regarding who or what is responsible for the outcome of events and behaviors in her or his life. An important moderator of an individual's experience of stress.

longitudinal method   A research approach in which scientists study the same individuals at different points in their lives to assess developmental changes that occur with age.

long-term memory   The retention of information in memory over an extended period of time.

M

macrosystem   Bronfenbrenner's fourth level of environmental influence, consisting of the overarching cultural patterns of a society that find expression in family, educational, economic, political, and religious institutions.

macular degeneration   A thinning of the layers of the retina and/or rupturing of tiny blood vessels in the eye, producing faded, distorted, or blurred central vision; more common in late adulthood.

marriage   A legally, socially, and/or religiously sanctioned union between a woman and a man with the expectation that they will perform the mutually supportive roles of wife and husband.

maternal blood sampling   A blood test that can detect birth defects by analyzing fetal cells shed into the pregnant women's bloodstream.

maturation   A component of development that involves the more or less automatic unfolding of biological potential in a sequence of physical changes and behavior patterns.

maturity   In human beings, the capacity to undergo continual change in order to adapt successfully and cope flexibly with the demands and responsibilities of life.

mechanistic model   A model of development that represents the universe as a machine composed of elementary particles in motion. Human development is portrayed as a gradual, chainlike sequence of events.

meiosis   The process of cell division in reproductive cells that produces gametes with one-half of the organism's normal number of chromosomes.

memory   The cognitive capacity to retain information that has been experienced.

menarche   The first menstrual period.

menopause   The normal aging process that culminates in the cessation of female menstrual activity. In Western countries, the typical age range of menopause is from 45 to 55 years old.

menstrual cycle   A series of hormonal changes within a woman, beginning with menstruation and ovulation, typically with 28 days per cycle.

menstruation   The maturing of an ovum and the process of ovulation and eventual expulsion of an unfertilized ovum from the body through the vagina.

mentor   A teacher, experienced co-worker, boss, or the like who shares expertise and guides someone in new learning.

mesosystem   Bronfenbrenner's second level of environmental influences, consisting of the interrelationships among the various settings in which the developing person is immersed.

metacognition   An individual's awareness and understanding of his or her own mental processes.

metamemory   An individual's awareness and understanding of her or his memory processes.

microsystem   Bronfenbrenner's first level of environmental influence, which consists of the network of social relationships and the physical settings in which a person is involved each day.

midwifery   The legalized provision of prenatal care and delivery by midwives.

miscarriage   Expulsion of the zygote, embryo, or fetus from the uterus before it can survive outside the mother's womb.

mitosis   The process of ordinary cell division, which results in two new cells identical to the parent cell.

modernization theory   The theory that the status of the elderly tends to be high in traditional societies and lower in urbanized, industrialized societies.

moral development   The process by which children adopt principles and values that lead them to evaluate behaviors as "right" or "wrong" and to govern their own actions in terms of these principles.

morphology   The study of word formation and changes, as when a child learns to say "Jim's and mine" instead of "Jim's and mines."

motherese   The speech adults tend to use with infants and young children; language that characteristically is simplified, redundant, and highly grammatical.

motivation   The inner emotional or cognitive states and processes that prompt, direct, and a person's sustain activity.

mourning   The culturally or socially established manner of expressing sorrow over a person's death.

multifactorial transmission   The interaction of environmental factors with genetic factors to produce traits. For example, a child's inborn predisposition for musical talent can be nurtured or defeated by environmental forces, such as parents or teachers.

multiinfarcts   "Little strokes" that destroy small areas of brain tissue.

multiple mothering   An arrangement in which responsibility for a child's care is dispersed among several caregivers.

N

natural childbirth   A form of childbirth in which the woman is awake, aware, and unmedicated during labor and delivery.

natural selection   Darwin's evolutionary theory that organisms best adapted to their environment survive and pass on their genetic characteristics to offspring.

naturalistic observation   A research method that involves carefully watching and recording behavior as it occurs in natural settings. The researchers must be careful not to disturb or affect the events under investigation.

near-death experience (NDE)   An experience such as having spiritually left one's body, having undergone otherworldly experiences, and having been "told to come back"—commonly reported by person's experiencing a medical illness, a traumatic accident, a surgical operation, childbirth, or drug ingestion.

negative identity   A diminished self-image, often associated with a diminished social role.

neglect   A type of abuse committed by a caretaker in failing to provide adequate social, emotional, and physical care to maintain the health or safety of a vulnerable person.

neonate   A newborn baby in the first month of life.

nonnormative life events   In the timing-of-events model, a set of unique turning points at which people change some direction in their life.

normative age-graded influences   In the timing-of-events model, a set of influences that include physical, cognitive, and psychosocial changes at predictable ages.

normative history-graded influences   In the timing-of-events model, historical events, such as wars, epidemics, and economic depressions, that affect large numbers of individuals about the same time.

norms   In child development, standards used for evaluating a child's developmental progress relative to the average for the child's age group.

O

obesity   Being at least 20 percent over the recommended weight for one's sex, height, and body structure.

object permanence   The understanding that objects continue to exist when they are out of sight; Piaget said this cognitive capacity is mastered by the end of infancy.

obstetrician   Physician who specializes in conception, prenatal development, birth, and the woman's postbirth care.

operant conditioning   A type of learning in which the consequences of a behavior alter the strength of that behavior.

organismic model   A model of development that views human beings as an organized configuration. Human development is characterized by discrete, steplike states.

osteoporosis   A serious bone-thinning disorder that is a "silent" disease. An adult's bones typically begin thinning during the midthirties, but weight-bearing exercise and calcium in the diet can prevent or slow its progression. Osteoporosis most often develops in women in late adulthood, particularly if they fall or have not taken hormone replacement therapy.

ovaries   The primary female reproductive organs. The ovaries produce mature ova and the female sex hormones, estrogen and progesterone. The ovaries are a pair of almond-shaped structures that lie in the pelvis.

ovulation   The discharge of an ovum from a follicle in the ovary into the fallopian tube.

ovum   The female gamete (sex cell), or egg.

P

parent-infant bonding   A process of interaction and mutual attention that occurs over time that builds an emotional bond between parent and infant.

peers   Individuals of approximately the same age.

penis   A man's external reproductive organ.

perimenopause   The period of two to four years before the cessation of menstruation, with sometimes heavy flow and the extension of intervals between periods.

period of concrete operations   Piaget's cognitive stage of middle childhood during which children demonstrate a qualitative change in cognitive functioning and develop a set of rules or strategies for examining the world.

period of formal operations   Piaget's highest stage in the development of cognitive functioning, generally reached during adolescence by a majority of people. Adults with cognitive disabilities might never reach the stage of formal operations, characterized by the ability to think abstractly and to plan for the future.

permissive parenting   A style of parenting distinguished by a nonpunitive, accepting, and affirmative environment in which the child regulates her or his own behavior as much as possible.

personal fable   Romantic imagery in which adolescents tend to view themselves as somehow unique and even heroic—as destined for unusual fame and fortune.

phenotype   The observable (expressed) characteristics of an organism.

phobia   An excessive, persistent, and maladaptive fear response—usually to benign or ill-defined stimuli.

phonology   The study of the sounds involved in a given language.

physical development   Changes that occur in a person's body, including changes in weight and height; in the brain, heart, and other organ structures and processes; and in skeletal, muscular, and neurological features that affect motor, sensory, and coordination skills.

placenta   A structure formed from uterine tissue and the trophoblast of the blastocyst. It functions as an exchange terminal, permitting entry of food materials, oxygen, and hormones into the fetus and allowing exit of carbon dioxide and metabolic wastes.

play   Voluntary activities that are performed for their own sake.

polygenic inheritance   The determination of traits by a large number of genes in combination, and not by a single gene. Personality, intelligences, aptitudes, and abilities are examples of polygenic inheritance.

post-formal operational thought   A fifth stage of cognitive development proposed by neo-Piagetians. This stage is characterized by three features: Adults come to realize that knowledge is not absolute but relativistic, recognize that contradiction is inherent in life, and must find some encompassing whole by which to organize their experience.

postmature infant   A baby delivered more than 2 weeks after the usual 40 weeks of gestation in the womb.

postpartum depression (PPD)   Symptoms of depression experienced by some new mothers. These include feeling unable to cope, thoughts of not wanting to take care of the baby, or thoughts of wanting to harm the baby.

post-traumatic stress disorder (PTSD)   A person's delayed response to severe stress or prolonged stress. Symptoms might include numbing and helplessness, increased irritability and aggressiveness, extreme anxiety, panic and fears, exaggerated startle response, sleep disturbances, and bed-wetting.

pragmatics   Rules governing the use of language in different social contexts.

prejudice   A system of negative conceptions, feelings, and actions regarding the members of a particular religious, racial, or nationality group.

premature infant   By common standards, a baby weighing less than 5 pounds 8 ounces at birth or having a gestational age of less than 37 weeks.

prenatal diagnosis   A determination of the health and condition of an unborn fetus.

prenatal period   The period of time from conception to birth.

preoperational period   Piaget's stage of cognitive development of children aged 2 to 7. The principle achievement of this stage is the developing capacity to represent the external world internally through the use of symbols.

presbycusis   The inability to hear high-pitched sounds, more common in elderly adults.

presbyopia   A normal condition in which the lens of the eye starts to harden with age, losing its ability to accommodate as quickly as it did in younger years.

primary relationships   Social interactions based on significant expressive ties, such as with parents, spouse, siblings, and children.

prosocial behavior   Being sympathetic, cooperative, helpful, rescuing, comforting, and generous; learning such behavior is considered to be an aspect of moral development.

prostate gland   A small gland in males at the base of the urethra that produces prostate fluid for semen. It generally enlarges as males reach their fifties or sixties, creating urinary difficulties.

proximodistal development   Development that proceeds outward from the central axis of the body toward the extremities.

psychoanalytic theory   Theory based on Freud's view that personality is fashioned progressively as the individual passes through various psychosexual stages of development.

psychoprophylactic method   A preparatory technique used to encourage women in childbirth to relax and concentrate on their breathing when a contraction occurs during labor.

psychosexual stages   The stages of personality development that Freud believed all human beings pass through: oral, anal, phallic, latency, and genital.

psychosocial development   An individual's development within a social context over the life course.

puberty   The period during early adolescence when sexual and reproductive maturation become evident.

puberty rites   Cultural initiation ceremonies that socially symbolize an adolescent's transition from childhood to adulthood.

R

random sampling   A sample for which each member of the population had an equally likely probability of being chosen; a technique to insure that the sample under study represents the larger population.

receptive vocabulary   An understanding of spoken words, exhibited by infants before they have developed an expressive vocabulary (speaking their first words).

recessive character   A gene that can determine a trait in an individual only if the other member of that gene pair is also recessive, as in *aa*.

reciprocity   A process that leads to each child's valuing the other person in a way that allows the child to remember the values that their interactions bring about. Piaget asserts that reciprocity of attitudes and values is the foundation of social interchange in children.

reflex   A relatively simple, involuntary, and unlearned response to a stimulus.

reinforcement   One event's strengthening the probability of another event's occurrence. A concept popularized by B. F. Skinner.

releasing stimuli   Biologically preadapted behaviors and features in infants that activate parenting.

reproduction   The process by which organisms create more organisms of their own kind.

resource dilution hypothesis   The theory that in large families resources get spread thin, to the detriment of all the offspring.

response   A term used by behavioral theorists to break down behavior into units.

retrieval   The cognitive process by which information is gathered from memory when it is required for recall or recognition.

reversibility   The child's failure to recognize that operations can be done in reverse to regain to an earlier state. According to Piaget, the most distinguishing characteristic of preoperational thought.

rheumatoid arthritis   An inflammatory disease that causes pain, swelling, stiffness, and loss of function of the joints; most common in the elderly.

right to privacy   An ethical standard established by the American Psychological Association requiring that researchers keep confidential all their records of behaviors or information from research participants.

role conflict   A type of stress that occurs when people experience pressures within one role that are incompatible with the pressures that arise within another role (for example, having to be at work and also having to take care of a sick child).

role exit theory of aging   A theory that views retirement and widowhood as life events terminating the participation of the elderly in the principal institutional structures of society—the job and the family.

role overload   A type of stress that occurs when people have too many role demands and too little time to fulfill them.

romantic love   Typically what we think of when we say we are "in love" with someone.

rooming in   An arrangement in a hospital whereby the newborn stays in a bassinet beside the mother's bed, allowing the mother and other family members to become acquainted with the newborn.

S

sandwich generation   People in middle adulthood who are caring for growing children at the same time as they are helping elderly parents and relatives.

scaffolding   Helping the child to learn through intervention and tutoring that is geared to the child's current level of functioning.

schemas   Piaget's term for mental structures that people evolve to deal with events in their environment.

scientific method   A systematic and formal process for conducting research, including selecting a researchable problem, formulating a hypothesis, testing the hypothesis, arriving at conclusions, and making the findings public.

secondary relationships   Social interactions based on instrumental ties, such as a relationship with a mechanic, or a teacher in a classroom, or a clerk at a store.

selective optimization with compensation   The life-span model endorsed by Paul and Margret Baltes. Older people cope with aging through a strategy that involves focusing on the skills most needed, practicing those skills, and developing ways to compensate for other skills.

self   The system of concepts we use in defining ourselves, the awareness of ourselves as separate entities who think and initiate action. The self provides us with the capacity to observe, respond to, and direct our behavior.

self-actualization   Maslow's concept from humanistic psychology that each person needs to fulfill his or her unique potential to the fullest.

self-concept (self-image)   The image a person has of herself or himself.

self-esteem   An overall dimension of one's sense of self-worth or self-image.

self-image   The overall view people have of themselves, which can be positive or negative.

semantics   The rules of meaning in a language, by which words have meaning and are combined to express complete thoughts.

senescence   The process of growing old, which affects us both physically and cognitively.

senility   A deterioration in cognitive functioning in late adulthood characterized by a lack of consistency in personality and/or behavior.

sensitive period   A brief period when specific events or experiences affect an organism more than they do at other times; an ethological concept.

sensorimotor   Piaget's first stage of cognitive development, lasting from birth to about two years. Infants use actions—looking, grasping, and so on—to learn about their world. The major tasks of the period revolve around coordinating motor activities with sensory inputs.

sensory information storage   The preservation of sensory information in the sensory register just long enough to permit the stimuli to be scanned for processing, generally less than 2 seconds.

separation anxiety   An infant's fear of being separated from the caregiver, demonstrated by distress behaviors.

sequential methods   A combination of the longitudinal and cross-sectional methods of research.

sex chromosomes   The 23rd pair of chromosomes which determine the baby's sex.

sex-linked traits   Traits other than gender that are affected by genes found on the sex chromosomes. For example, hemophilia is a sex-linked characteristic carried on the X chromosome.

sexual abuse of children   Sexual behavior between a child and an older person that the older person brings about through force, coercion, or deceit.

sexually transmitted diseases (STDs)   Diseases transmitted through sexual intercourse (e.g., gonorrhea, syphilis, chlamydia, HIV). Using condoms greatly reduces the risk of contracting an STD.

shaken-baby syndrome (SBS)   Serious brain damage or death that occurs when a baby's head is violently shaken back and forth or strikes something, resulting in bruising or bleeding of the brain, spinal cord injury, and eye damage.

short-term memory   The retention of information in memory for a very brief period, usually no more than 30 seconds.

sleep apnea   A sleep disorder in which the person occasionally stops breathing during sleep.

small-for-term infant   A low-birthweight infant that has developed over the usual 40 weeks of gestation in the womb but is born weighing less than is expected.

social aging   Changes in an individual's assumptions and roles as she or he ages.

social clock   A set of social concepts regarding the appropriate ages for reaching milestones of the adult years.

social convoy   The company of other people who travel with us from birth to death.

social exchange theory of aging   The theory that people enter into social relationships in order to derive rewards (economic, social, emotional). People also incur costs in social relationships. Relationships persist as long as both parties profit.

social norms   Standards and expectations that specify what constitutes appropriate and inappropriate behavior for individuals at various periods in the life span.

social referencing   The practice whereby an inexperienced person relies on a more experienced person's interpretation of an event to regulate his or her own behavior.

social relationships   Bonds forged with other people in relatively stable social circumstances.

social survey method   A research method used to study the incidence of specific behaviors or attitudes in a large population of people.

socialization   The process of transmitting culture to children in order to transform them into well-functioning members of society.

sociocultural theory   The theory that psychological functions as thinking, reasoning, and remembering are facilitated through language and anchored in the child's interpersonal relationships.

sociogram   A type of graph depicting the patterns of peer friendships and relationships existing among members of a group at a given time.

sperm   The male gamete (sex cell).

states   In child development, an infant's first line of defense. By regulating their internal states, infants shut out certain stimuli or set the stage to actively respond to their environment. Their continuum of alertness ranges from sleep to vigorous activity, and includes such behaviors as crying, sleeping, eating, and eliminating.

stereotypes   Exaggerated cultural understandings that guide us in identifying the mutual set of expectations that will govern the social exchange.

stillbirth   Fetal death in the womb or during labor and delivery.

stimuli   A term used by behavioral theorists to break down the environment into units.

storage   The retention of information in memory until it is needed.

storm and stress   Hall's notion that adolescence is a stage of inevitable turmoil, maladjustment, tension, rebellion, dependency conflicts, and exaggerated peer-group conformity.

strange situation   A research technique consisting of a series of eight episodes in which researchers observe infants in an unfamiliar playroom in order to study attachment.

stranger anxiety   A wariness or fear of strangers. This anxiety is first exhibited by infants at about 8 months, peaks around 13 to 15 months, and decreases thereafter.

stroke   A life-threatening blockage of blood flow to the brain.

substance abuse   The harmful use of drugs or alcohol, lasting over a prolonged period, that can harm the user or others.

sudden infant death syndrome (SIDS)   Sudden death of an infant during sleep, due to unknown causes, also called "crib death." One of the leading causes of infant death during the first several months of life. Making sure that babies sleep on their backs, and not on their stomachs, helps reduce the risk of SIDS.

syntax   Rules governing the proper ordering of words to form sentences.

T

telegraphic speech   The use of two- or three-word utterances to express complete thoughts, characteristic of young children's speech.

temperament   The relatively consistent, basic dispositions inherent in people that underlie and modulate much of their behavior.

teratogen   Any agent that contributes to birth defects or anomalies.

teratology   The study of teratogens and birth defects.

testes   A pair of primary male reproductive organs, normally lying outside the body in a pouchlike structure called a scrotum.

thanatology   The study of death and dying.

theory   A set of interrelated statements intended to explain a class of events.

theory of mind   In child development research that probes children's developing conceptions of major components of mental activity.

time sampling   An observational technique that involves counting the occurrences of a specific behavior over systematically spaced intervals of time.

total immersion   The instructional approach of placing children of all language backgrounds together in regular classrooms and using English for all instruction (with or without support in their first language).

traditional marriage   The style of marriage in which women are homemakers and men are providers.

transition points   Periods in development when the individual relinquishes familiar roles and assumes of new ones.

triangular theory of love   Sternberg's theory that companionate love consists of two other types of love: intimacy and commitment.

two-factor theory of intelligence   Spearman's view that intelligence is a general intellectual ability employed for abstract reasoning and problem solving.

U

ultrasonography   A noninvasive diagnostic procedure that allows physicians to see inside the body—for instance, to determine the size and shape of the fetus and placenta, the amount of amniotic fluid, and the appearance of fetal anatomy.

umbilical cord   A connecting lifeline carrying two arteries and one vein linking the embryo to the placenta.

uterus   A hollow, thick-walled, muscular organ in a female that can house and nourish a developing embryo.

V

vagina   A muscular passageway in the female reproductive system that is capable of considerable dilation, allowing for intercourse or birth of a baby.

values   The criteria individuals use in deciding the relative merit and desirability of things (such as themselves, other people, objects, events, ideas, acts, feelings).

W

wisdom   Expert knowledge about life in general and good judgment and advice about how to conduct oneself in complex, uncertain circumstances.

Y

youth culture   Characteristics of a large body of young people that become standardized ways of thinking, feeling, and acting.

Z

zone of proximal development (ZPD)   Vygotsky's concept that children develop through participation in activities slightly beyond their competence when helped by a more skilled partner.

zygote   A single fertilized ovum (egg).

References

A

Abortion Surveillance Preliminary Analysis: United States 1996. (1996). *Morbidity and Mortality Weekly Reports, 47,* 1025–1034.

Aboud, F. (1987). The development of ethnic self-identification and attitudes. In J. Phinney & M. Rotheram (Eds.), *Children's ethnic socialization: Pluralism and development.* Newbury Park, CA: Sage.

Aboud, F. (1988). *Children and prejudice.* Oxford: Blackwell.

Aboulafia, M. (Ed.). (1991). *Philosophy, social theory, and the thought of George Herbert Mead.* Albany: State University of New York Press.

Abramov, I., Gordon, J., Hendrickson, A., Hainline, L., Dobson, V., & LaBossiere, E. (1982). The retina of the newborn human infant. *Science, 217,* 265–267.

Abrams, S., Prodromidis, M., Scafidi, F., & Field, T. (1995). Newborns of depressed mothers. *Infant Mental Health Journal, 16,* 233–239.

Achenbaum, W. (1998). Perceptions of aging in America. *National Forum: Phi Kappa Phi Journal, 78,* 30–33.

Acredolo, L. P., & Goodwyn, S. (1988). Symbolic gesturing in normal infants. *Child Development, 59,* 450–466.

Acredolo, L. P., & Hake, J. K. (1982). Infant perception. In B. B. Wolman (Ed.), *Handbook of developmental psychology.* Englewood Cliffs, NJ: Prentice Hall.

Adair, J. (1775). *The history of the American Indians.* London: E. D. Dilly.

Adams, C., Labouvie-Vief, G., Hobart, C. J., & Dorosz, M. (1990). Adult age group differences in story recall style. *Journal of Gerontology, 45,* P17–P27.

Adams, R. G., & Blieszner, R. (1998, Spring). Baby boomer friendships. *Generations, 22,* 70–75.

Adamson, L. (1996). *Communication development during infancy.* Boulder, CO: Westview Press.

Adelman, K. (1991). The toughest thing. *Washingtonian, 26,* 23.

Adelson, J. (1972). The political imagination of the young adolescent. In J. Kagan & R. Coles (Eds.), *Twelve to sixteen.* New York: Norton.

Adelson, J. (1975). The development of ideology in adolescence. In S. E. Dragastin & G. H. Elder, Jr. (Eds.), *Adolescence in the life cycle: Psychological change and social context.* New York: Wiley.

Adler, L. L. (Ed.). (1989). *Cross-cultural research in human development: Life span perspectives.* New York: Praeger.

Aggleton, P., O'Reilly, K., Slutkin, G., & Davies, P. (1994). Risking everything? Risk behavior, behavior change, and AIDS. *Science, 265,* 341–345.

Ahlburg, D., & De Vita, C. J. (1992). New realities of the American family. *Population Bulletin, 47,* 2–44.

Aiken, L. (1998). *Human development in adulthood.* New York: Plenum.

Ainsworth, M. D. S. (1967). *Infancy in Uganda: Infant care and the growth of attachment.* Baltimore: Johns Hopkins University Press.

Ainsworth, M. D. S. (1983). Patterns of infant-mother attachment as related to maternal care. In D. Magnusson & V. Allen (Eds.), *Human development: An interactional perspective.* New York: Academic Press.

Ainsworth, M. D. S. (1992). A consideration of social referencing in the context of attachment theory and research. In S. Feinman (Ed.), *Social referencing and the social construction of reality in infancy.* New York: Plenum.

Ainsworth, M. D. S. (1993). Attachment as related to mother-infant interaction. *Advances in Infancy Research, 8,* 1–50.

Ainsworth, M. D. S. (1995). On the shaping of attachment theory and research: An interview with Mary Ainsworth (Fall 1994). *Monographs of the Society for Research in Child Development, 60,* 3–21.

Ainsworth, M. D. S., & Bell, S. M. (1974). Mother-infant interaction and the development of competence. In K. Connolly & J. Bruner (Eds.), *The growth of competence.* New York: Academic Press.

Ainsworth, M. D. S., Bell, S. M., & Stayton, D. J. (1974). Infant-mother attachment and social development. In M. P. M. Richards (Ed.), *The integration of a child into a social world.* Cambridge: Cambridge University Press.

Ainsworth, M. D. S., Blehar, M. C., Waters, E., & Wall, S. (1979). *Patterns of attachment: A psychological study of the strange situation.* New York: Halsted.

Ainsworth, M. D. S., & Wittig, B. A. (1969). Attachment and the exploratory behavior of one-year-olds in a strange situation. In B. M. Foss (Ed.), *Determinants of infant behavior* (Vol. 4). London: Methuen.

Alberts, N. , & Dowling, C. (1999, February). The science of love. *Life,* 38–51.

Albright, J., & Kunstel, M. (1999, February 2). Grim odds for Chinese girl babies: Modern sex screening, old traditions are accelerating the purging of female children. *Atlanta Constitution,* A08.

Aldous, J. (1987). New views on the family life of the elderly and the near-elderly. *Journal of Marriage and the Family, 49,* 227–234.

Aldous, J., Mulligan, G. M., & Thoroddur, B. (1998). Fathering over time: What makes the difference? *Journal of Marriage and the Family, 60,* 809.

Alexander, G. M., & Hines, M. (1994). Gender labels and play styles: Their relative contribution to children's selection of playmates. *Child Development, 65,* 869–879.

Alexander, M. J., & Higgins, E. T. (1993). Emotional trade-offs of becoming a parent: How social roles influence self-discrepancy effects. *Journal of Personality and Social Psychology, 65,* 1259–1269.

Allan, E. A., & Steffensmeier, D. J. (1989). Youth, underemployment, and property crime: Differential effects of job availability and job quality on juvenile and young adult arrest rates. *American Sociological Review, 54,* 107–123.

Allegeier, R., & Allegeier, E. (1995). *Sexual interactions.* Boston: Houghton-Mifflin.

Allen, K. R., & Demo, D. H. (1995). The families of lesbians and gay men: A new frontier in family research. *Journal of Marriage and the Family, 57,* 111–127.

Allen, P. A., Madden, D. J., Weber, T., & Crozier, L. C. (1992). Age differences in short-term memory: Organization or internal noise? *Journal of Gerontology: Psychological Sciences, 47,* P281–P288.

Allen, S. M. (1994). Gender differences in spousal caregiving and unmet need for care. *Journal of Gerontology: Social Sciences, 49,* S187–S195.

Alley, T. R. (1983). Growth-produced changes in body shape and size as determinants of perceived age and adult caretaking. *Child Development, 54,* 241–248.

Allison, J. A., & Wrightsman, L. S. (1993). *Rape: The misunderstood crime.* Newbury Park, CA: Sage.

Allman, W. F. (1991, August 19). The clues in the idle chatter. *U.S. News & World Report,* 61–62.

Allport, G. W. (1955). *Becoming: Basic considerations for a psychology of personality.* New Haven, CT: Yale University Press.

Allport, G. W. (1961). *Pattern and growth in personality.* New York: Holt, Rinehart & Winston.

Almeida, D. M., & Kessler, R. C. (1998). Everyday stressors and gender differences in daily stress. *Journal of Personality and Social Psychology, 75,* 670–680.

Altman, L. K. (1992, November 6). Vitamin array is found to be benefit to elderly. *New York Times,* A13.

Altman, L. K. (1993, December 22). A study on impotence suggests half of men over 40 may have problem. *New York Times,* B7.

Altman, L. K. (1994a, August 17). High HIV levels said to raise newborns' risk. *New York Times,* B6.

Altman, L. K. (1994b, February 21). In major finding, drug limits HIV infection in newborns. *New York Times,* A1, A8.

Amato, P. R. (1993). Children's adjustment to divorce: Theories, hypotheses, and empirical support. *Journal of Marriage and the Family, 55,* 23–38.

Amato, P. R. (1996). Explaining the intergenerational transmission of divorce. *Journal of Marriage and the Family, 58,* 628–640.

Amato, P. R., & Keith, B. (1991). Parental divorce and the well-being of children: A meta-analysis. *Psychological Bulletin, 110,* 26–46.

Amato, P. R., Rezac, S. J., & Booth, A. (1995). Helping between parents and young adult offspring: The role of parental marital quality, divorce, and remarriage. *Journal of Marriage and the Family, 57,* 363–374.

America's children: Key national indicators of child well-being 1998. (1998). ChildStats.gov. Retrieved August 31, 1998, from the World Wide Web: http://www. childstats.gov/ac1998/ac98.htm

American Academy of Family Physicians. (1998). HIV and pregnancy: What is the risk to my baby? *American Family Physician.* Retrieved February 1, 1998, from the World Wide Web: http://www. aafp.org/afp/980201ap/quantum.html

American Academy of Pediatrics. (1990, April 16). [News release]. Provided by Dr. Victor Strasburger of the University of New Mexico School of Medicine.

American Academy of Pediatrics and the American Psychological Association. (1996). Raising children to resist violence: What you can do. Retrieved February 13, 1999, from the World Wide Web: http:// www.apa.org/pubinfo/apa-aap.html

American Association of Retired Persons. (1996). Taking control of your health: Keeping in shape. AARP Webplace. Retrieved February 12, 1999, from the World Wide Web: http://www.aarp.org/ health/letsgetmoving.html

American Association of University Women Educational Foundation. (1992). *How schools shortchange girls: The AAUW report.* Washington, DC: AAUW.

American Cancer Society. (1997). Cancer facts and figures. Retrieved March 5, 1999, from the World Wide Web: http://www. cancer.org/statistics/97cff/97tobacc.html

American Heart Association. (1998). Benefits of daily physical activity. Retrieved March 6, 1999, from the World Wide Web: http://www.amhrt. org/health/lifestyle/physical_activity/ beneact.html

American Psychiatric Association. (1994). *Diagnostic and statistical manual of mental disorders* (4th ed.). Washington, DC: Author.

American Psychological Association. (1982). *Ethical principles in the conduct of research with human participants.* Washington, DC: Author.

Anastasi, A. (1958). Heredity, environment, and the question "how?" *Psychological Review, 65,* 197–208.

Ancoli-Israel, S. (1997, January). Sleep problems in older adults: Putting myths to bed. *Geriatrics, 52,* 20.

Anderson, E. (1994, May). The code of the streets. *Atlantic Monthly,* 81–94.

Anderson, J. R. (1990). *The adaptive character of thought.* Hillsdale, NJ: Erlbaum.

Anderson, K. E., Lytton, H., & Romney, D. M. (1986). Mothers' interactions with normal and conduct-disordered boys: Who affects whom? *Developmental Psychology, 22,* 604–609.

Anderson, T. B., & McCulloch, B. J. (1993). Conjugal support: Factor structure for older husbands and wives. *Journal of Gerontology: Social Sciences, 48,* S133–S142.

Anderson, W. (1997). *Negotiating the special education maze.* Bethesda, MD: Woodbine House.

Angel, J. L., & Hogan, D. P. (1994). The demography of minority aging population. In *Minority elders: Five goals toward building a public policy base* (pp. 9–21). Washington, DC: Gerontological Society of America.

Angier, N. (1990, July 5). Human growth hormone reverses effects of aging. *New York Times,* A1, A12.

Angier, N. (1991, March 29). Older women at no greater risk in bearing children with defects. *New York Times,* 9.

Angier, N. (1992, May 20). Is there a male menopause? Jury is still out. *New York Times,* B1, B7.

Angier, N. (1995, May 9). Scientists mull role of empathy in man and beast. *New York Times,* B7, B9.

Anglin, J. M. (1993). Vocabulary development: A morphological analysis. *Monograph of the Society for Research in Child Development, 58,* Serial No. 238.

Ansbacher, H. L., & Ansbacher, R. R. (1956). *The individual psychology of Alfred Adler.* New York: Basic Books.

Antonucci, T. C. (1985). Personal characteristics, social support, and social behavior. In R. Binstock & E. Shanas (Eds.), *Handbook of aging and the social sciences* (pp. 94–128). New York: Van Nostrand Reinhold.

Antonucci, T. C. (1994). A life-span view of women's social relations. In B. F. Turner & L. E. Trolls (Eds.), *Women growing older: Psychological perspectives.* Thousand Oaks, CA: Sage.

Apgar scoring for newborns. (1998). Childbirth.org Retrieved July 31, 1998, from the World Wide Web: http://www. Childbirth.Org./articles/apgar.html

Apgar, V. (1953). Proposal for a new method of evaluation of the newborn infant. *Anesthesia and Analgesia, 32,* 260–267.

Apple, D. (1956). The social structure of grandparenthood. *American Anthropologist, 58,* 656–663.

Apple, M. W., & Weis, L. (1983). *Ideology and practice in schooling.* Philadelphia: Temple University Press.

Applied Medical Informatics. (1997). Hearing loss and sound therapy. Retrieved May 14, 1998, from the World Wide Web: http://www.intouchmag.com/ soundtherapy1.html

Aquilino, W. S., & Supple, K. R. (1991). Parent-child relations and parent's satisfaction with living arrangements when adult children live at home. *Journal of Marriage and the Family, 53,* 13–27.

Archer, S. L. (1992). A feminist's approach to identity research. In G. R. Adams, T. P. Gullota, & R. Montemayor (Eds.), *Adolescent identity formation.* Newbury Park, CA: Sage.

Arditti, J. A., & Keith, T. Z. (1993). Visitation frequency, child support payment, and the father-child relationship postdivorce. *Journal of Marriage and the Family, 55,* 699–712.

Ariès, P. (1962). *Centuries of childhood* (R. Baldick, Trans.). New York: Random House.

Ariès, P. (1978). *Western attitudes toward death: From the Middle Ages to the present.* Baltimore: Johns Hopkins University Press.

Ariès, P. (1981). *The hour of our death.* New York: Knopf.

Armstrong, C., Salles, J., Alcaraz, J., Kolody, B., & McKenzie, T. (1998). Children's television viewing, body fat, and physical fitness. *American Journal of Health Promotion, 12,* 363–368.

Arnup, K. (1995). Living in the margins: Lesbian families and the law. In K. Arnup (Ed.), *Lesbian parenting: Living with pride and prejudice* (pp. 378–398). Charlottetown, Canada: Gynergy.

Aron, A., & Westbay, L. (1996). Dimensions of the prototype of love. *Journal of Personality and Social Psychology, 70,* 535–551.

Aronson, E., Brewer, M., & Carlsmith, J. M. (1985). Experimentation in social psychology. In G. Lindzey & E. Aronson (Ed.), *Handbook of social psychology* (3rd ed., Vol. 2). New York: Random House.

Asendorpf, J. B., & Baudonniere, P. (1993). Self-awareness and other-awareness: Mirror self-recognition and synchronic imitation among unfamiliar peers. *Developmental Psychology, 29,* 88–95.

Ashmead, D. H., McCarty, M. E., Lucas, L. S., & Belvedere, M. C. (1993). Visual guidance in infants' reaching toward suddenly displaced targets. *Child Development, 64,* 1111–1127.

Ashton, P. T. (1975). Cross-cultural Piagetian research: An experimental perspective. *Harvard Educational Review, 45,* 475–506.

Atkins, A. (1996, September). Who ages better—Men or women? *New Choices, 36,* 22–25.

Atkinson, M. P., & Blackwelder, S. P. (1993). Fathering in the 20th century. *Journal of Marriage and the Family, 55,* 975–986.

Attie, I., & Brooks-Gunn, J. (1987). Weight-related concerns in women: A response to or a cause of stress? In R. C. Barnett, L. Biener, & G. K. Baruch (Eds.), *Gender and stress.* New York: Free Press.

Ausubel, D. P., & Sullivan, E. V. (1970). *Theory and problems of child development* (2nd ed.). New York: Grune & Stratton.

Aviezer, O., Van IJzendoorn, M. H., Sagi, A., & Schuengel, C. (1994). "Children of the Dream" revisited: 70 years of collective early child care in Israeli kibbutzim. *Psychological Bulletin, 116,* 99–116.

Azmitia, M. (1988). Peer interaction and problem solving: When are two heads better than one? *Child Development, 59,* 87–96.

B

Bachman, J. G., O'Malley, P., & Johnston, J. (1978). *Youth in transition. Vol. 6: Adolescence to adulthood—changes and stability in the lives of young men.* Ann Arbor, MI: Institute for Social Research, University of Michigan.

Baer, K. (1993, April). A guide to hospice care [Special suppl.]. *Harvard Health Letter.*

Bailey, J. M., & Zucker, K. J. (1995). Childhood sex-typed behavior and sexual orientation: A conceptual analysis and quantitative review. *Developmental Psychology, 31,* 43–55.

Baillargeon, R., & DeVos, J. (1991). Object permanence in young infants: Further evidence. Child Development, 62, 1227–1246.

Bairraro, J., & Tietze, W. (1993). Early childhood services in the European community. Commission of the European Community Task Force on Human Resources, Education, Training, and Youth. Retrieved October 15, 1998, from the World Wide Web: http://futureofchildren.org

Bakeman, R., Adamson, L. B., Konner, M., & Barr, R. G. (1990). !Kung infancy: The social context of object exploration. *Child Development, 61,* 794–809.

Baldwin, D. A., & Markman, E. M. (1989). Establishing word-object relations: A first step. *Child Development, 60,* 381–398.

Baldwin, J. M. (1895). *Mental development in the child and the rare methods of processes.* New York: Macmillan.

Baldwin, J. M. (1897). *Social and ethical interpretations of mental development: A study in social psychology.* New York: Macmillan.

Baltes, M. M., & Carstensen, L. L. (1996). The process of successful ageing. *Ageing and Society, 16,* 397.

Baltes, M. M., Neumann, E. M., & Zank, S. (1994). Maintenance and rehabilitation of independence in old age: An intervention program for staff. *Psychology and Aging, 9,* 179–188.

Baltes, P. B., & Baltes, M. M. (1990). Psychological perspectives on successful aging: The model of selective optimization with compensation. In P. B. Baltes & M. M. Baltes (Eds.), *Successful aging: Perspectives from the behavioral sciences.* New York: Cambridge University Press.

Baltes, P. B., & Baltes, M. M. (1998). Savoir Vivre in old age: How to master the shifting balance between gains and losses. *National Forum: Phi Kappa Phi Journal, 78,* 13–18.

Baltes, P. B., & Lindenberger, U. (1997). Emergence of a powerful connection between sensory and cognitive functions across the adult life span: A new window to the study of cognitive aging? *Psychology and Aging, 12,* 12–21.

Baltes, P. B., & Schaie, K. W. (1976). On the plasticity of intelligence in adulthood and old age. *American Psychologist, 31,* 720–725.

Baltes, P. B., & Smith, J. (1997). A systemic-wholistic view of psychological functioning in very old age: Introduction to a collection of articles from the Berlin Aging Study. *Psychology and Aging, 12,* 395–409.

Bandura, A. (1964). The stormy decade: Fact or fiction? *Psychology in the Schools, 1,* 224–231.

Bandura, A. (1973). *Aggression: A social learning analysis.* Englewood Cliffs, NJ: Prentice Hall.

Bandura, A. (1977). *Social learning theory.* Englewood Cliffs, NJ: Prentice Hall.

Bandura, A. (1986). *Social foundations of thought and action: A social cognitive theory.* Englewood Cliffs, NJ: Prentice Hall.

Bandura, A. (1989a). Human agency in social cognitive theory. *American Psychologist, 44,* 1175–1184.

Bandura, A. (1989b). Regulation of cognitive processes through perceived self-efficacy. *Developmental Psychology, 25,* 729–735.

Bandura, A., Grusec, J. E., & Menlove, F. L. (1967). Vicarious extinction of avoidance behavior. *Journal of Personality and Social Psychology, 5,* 16–23.

Bandura, A., Ross, D., & Ross, S. (1963). Imitation of film-mediated aggressive models. *Journal of Abnormal and Social Psychology, 66,* 3–11.

Baney, C. (1998). Wired for sound: The essential connection between music and development. *Early Childhood News.* Retrieved February 12, 1999, from the World Wide Web: http://www.earlychildhoodnews.com/wiredfor.htm

Bank, L., Forgatch, M. S., Patterson, G. R., & Fetrow, R. A. (1993). Parenting practices of single mothers: Mediators of negative contextual factors. *Journal of Marriage and the Family, 55,* 371–384.

Barber, B. K. (1992). Family, personality, and adolescent problem behaviors. *Journal of Marriage and the Family, 54,* 69–79.

Barber, B. K. (1994). Cultural, family, and personal contexts of parent-adolescent conflict. *Journal of Marriage and the Family, 56,* 375–386.

Barden, J. C. (1990, February 5). Toll of troubled families: Flood of homeless youth. *New York Times,* A1, B7.

Barinaga, M. (1994). Looking to development's future. *Science, 266,* 561–564.

Barkley, R. A. (1990). *Attention-deficit hyperactivity disorder: A handbook for diagnosis and treatment.* New York: Guilford Press.

Barnes, M. L., & Sternberg, R. J. (1997). A hierarchical model of love and its prediction of satisfaction in close relationships. In R. J. Sternberg and M. Hojjat (Eds.), *Satisfaction in close relationships* (pp. 79–101). New York: Guilford Press.

Barnett, R. C. (1992). Multiple roles, gender, and psychological distress. In L. Goldberger & S. Breznitz (Eds.), *Handbook of stress* (pp. 427–445). New York: Free Press.

Barnett, R. C. (1997). Gender, employment, and psychological well-being: Historical and life course perpsectives. In M. E. Lachman & J. B. James (Eds.), *Multiple paths of midlife development* (pp. 325–343). Chicago: University of Chicago Press.

Barnett, R. C., Raudenbush, S. W., Brennan, R. T., Pleck, J. H., & Marshall, N. L. (1995). Change in job and marital experience and change in psychological distress: A longitudinal study of dual-earner couples. *Journal of Personality and Social Psychology, 69,* 839–850.

Barr, H. M., Streissguth, A. P., Darby, B. L., & Sampson, P. D. (1990). Prenatal exposure to alcohol, caffeine, tobacco, and aspirin: Effects on fine and gross motor performance in 4-year-old children. *Developmental Psychology, 26,* 339–348.

Barringer, F. (1992, November 10). Among elderly, men's prospects are the brighter. *New York Times,* A1, A9.

Barringer, F. (1993a, April 25). Measuring sexuality through polls can be shaky. *New York Times,* 12.

Barringer, F. (1993b, April 25). Sex survey of American men finds 1% are gay. *New York Times,* A1, A9.

Bart, P. B. (1972). Depression in middle-age women. In V. Gornick & B. K. Moran (Eds.), *Women in sexist society.* New York: New American Library.

Bartoshuk, L. M., Rifkin, B., Marks, L. E., & Bars, P. (1986). Taste and aging. *Journal of Gerontology, 41,* 51–57.

Baruch, G., & Barnett, R. C. (1983). Adult daughters' relationships with their mothers. *Journal of Marriage and the Family, 45,* 601–606.

Baskett, L. M. (1985). Sibling status effects: Adult expectations. *Developmental Psychology, 21,* 441–445.

Bass, D., & Bowman, K. (1990). The impact of an aged relative's death on the family. In K. F. Ferraro (Ed.), *Gerontology: Perspectives and issues.* New York: Springer.

Bass, L., & Jackson, M. S. (1997). A study of drug abusing African-American pregnant women. *Journal of Drug Issues, 27,* 659–671.

Basseches, M. (1980). Dialectical schemata: A framework for the empirical study of the development of dialectical thinking. *Human Development, 23,* 400–421.

Bates, E., Bretherton, I., & Snyder, L. (1988). *From first words to grammar: Individual differences and dissociable mechanisms.* New York: Cambridge University Press.

Bauer, P. J. (1993). Memory for gender-consistent and gender-inconsistent event sequences by twenty-five-month-old children. *Child Development, 64,* 285–297.

Bauer, P. J., & Mandler, J. M. (1989). One thing follows another: Effects of temporal structure on 1- to 2-year-olds' recall of events. *Developmental Psychology, 25,* 197–206.

Baumgardner, A. H. (1990). To know oneself is to like oneself: Self-certainty and self-affect. *Journal of Personality and Social Psychology, 58,* 1062–1072.

Baumrind, D. (1967). Child care practices anteceding three patterns of preschool behavior. *Genetic Psychology Monographs, 75,* 43–88.

Baumrind, D. (1971). Current patterns of parental authority. *Developmental Psychology Monographs, 4,* 1.

Baumrind, D. (1972). Socialization and instrumental competence in young children. In W. W. Hartup (Ed.), *The young child* (Vol. 2). Washington, DC: National Association for the Education of Young Children.

Baumrind, D. (1980). New directions in socialization research. *American Psychologist, 35,* 639–652.

Baumrind, D. (1991). The influence of parenting style on adolescent competence and substance use. *Journal of Early Adolescence, 11,* 56–95.

Baumrind, D. (1994). The social context of child maltreatment. *Family Relations, 43,* 360–368.

Baumrind, D. (1996). The discipline controversy revisited. *Family Relations, 45,* 405–414.

Baumwell, L., Tamis-LeMonda, C. S., & Bornstein, M. H. (1997). Maternal verbal sensitivity and child language comprehension. *Infant Behavior and Development, 20,* 247.

Baydar, N., & Brooks-Gunn, J. (1991). Effects of maternal employment and child-care arrangements on preschoolers' cognitive and behavioral outcomes: Evidence from the Children of the National Longitudinal Survey of Youth. *Developmental Psychology, 27,* 932–945.

Bayley, N. (1935). *The development of motor abilities during the first three years.* Washington, DC: Society for Research Development.

Bayley, N. (1936). *The California infant scale of motor development: Birth to three years.* Berkely: University of California Press.

Bayley, N. (1956). Individual patterns of development. *Child Development, 27,* 45–74.

Bayley, N. (1965). Research in child development: A longitudinal perspective. *Merrill-Palmer Quarterly, 11,* 184–190.

Beard, R. M. (1969). *An outline of Piaget's developmental psychology for students and teachers.* New York: New American Library.

Beck, M. (1990a, July 16). Aging: Trading places. *Newsweek,* 48–54.

Beck, M. (1990b, July 2). A home away from home. *Newsweek,* 56–58.

Beck, S. H. (1982). Adjustment to and satisfaction with retirement. *Journal of Gerontology, 37,* 616–624.

Becker, W. C. (1964). Consequences of different kinds of parental discipline. In M. L. Hoffman & L. W. Hoffman (Eds.), *Review of child development research* (pp. 169–208). New York: Russell Sage Foundation.

Beer, W. R. (1989). *Strangers in the house: The world of stepsiblings and half-siblings.* New Brunswick, NJ: Transaction.

Begley, S., & Check, E. (1998, September 7). The parent trap. *Newsweek,* 52–59.

Behrens, M. L. (1954). Child rearing and the character structure of the mother. *Child Development, 20,* 225–238.

Beilin, H. (1990). Piaget's theory: Alive and more vigorous than ever. *Human Development, 33,* 362–365.

Beilin, H. (1992). Piaget's enduring contribution to developmental psychology. *Developmental Psychology, 28,* 191–204.

Belkin, L. (1992, March 25). Births beyond hospitals fill an urban need. *New York Times,* A1, A15.

Bell, S. M. (1970). The development of the concept of object as related to infant-mother attachment. *Child Development, 41,* 291–311.

Belluck, P. (1998, April 12). Say black? Not so fast. *New York Times,* 9, 1.

Belsky, J. (1984). The determinants of parenting: A process model. *Child Development, 55,* 83–96.

Belsky, J. (1990). Parental and nonparental child care and children's socioemotional development: A decade in review. *Journal of Marriage and the Family, 52,* 885–903.

Belsky, J. (1993). Etiology of child maltreatment: A developmental-ecological analysis. *Psychological Bulletin, 114,* 413–434.

Belsky, J. (1996a). Infant attachment, security, and affective-cognitive information processing at age 3. *Psychological Science, 7,* 111–114.

Belsky, J. (1996b). Parent, infant, and social-contextual antecedents of father-son attachment security. *Developmental Psychology, 32,* 905–913.

Belsky, J., Crnic, K., & Gable, S. (1995). The determinants of coparenting in families with toddler boys: Spousal differences and daily hassles. *Child Development, 66,* 629–642.

Belsky, J., & Eggebeen, D. (1991). Early and extensive maternal employment and young children's socioemotional development: Children of the National Longitudinal Survey of Youth. *Journal of Marriage and the Family, 53,* 1083–1110.

Belsky, J., & Rovine, M. (1987). Temperament and attachment security in the strange situation: An empirical rapprochement. *Child Development, 58,* 787–795.

Belsky, J., & Rovine, M. (1990). Patterns of marital change across the transition to parenthood: Pregnancy to three years postpartum. *Journal of Marriage and the Family, 52,* 5–19.

Belsky, J., & Rovine, M. (1993). Non-maternal care in the first year of life and the security of infant-parent attachment. In R. Pierce (Ed.), *Life-span development: A diversity reader.* Dubuque, IA: Kendall, Hunt.

Belsky, J., Steinberg, L., & Draper, P. (1991). Childhood experience, interpersonal development, and reproductive strategy: An evolutionary theory of socialization. *Child Development, 62,* 647–670.

Belsky, J., Youngblade, L., Rovine, M., & Volling, B. (1991). Patterns of marital change and parent-child interaction. *Journal of Marriage and the Family, 53,* 487–498.

Bem, D., & Allen, A. (1974). On predicting some of the people some of the time: The search for cross-situational consistencies in behavior. *Psychological Review, 81,* 506–520.

Bem, S. L. (1993). *The lenses of gender: Transforming the debate on sexual inequality.* New Haven, CT: Yale University Press.

Bem, S. L. (1998). *An unconventional family.* New Haven, CT: Yale University Press.

Bendersky, M., & Lewis, M. (1994). Environmental risk, biological risk, and developmental outcome. *Developmental Psychology, 30,* 484–494.

Benedict, H. (1976). *Language comprehension in 10 sixteen-month-old infants.* Unpublished doctoral dissertation, Yale University.

Benenson, J., Apostolen, N., & Parnass, S. (1998). The organization of children's same-sex peer relationships in sociometry, then and now: Building on six decades of measuring children's experiences with the peer group. In W. Bukowski & A. Cillesser (Eds.), *New Directions for Child Development, 80.*

Bengtson, N. G. (1985). *Grandparenthood.* Beverly Hills: Sage.

Bennett, N. G., Bloom, D. E., & Craig, P. H. (1989). *The divergence of black and white marriage patterns.* New Haven, CT: Economic Growth Center, Yale University.

Bennett, W., & Gurin, J. (1982). *The dieter's dilemma.* New York: Basic Books.

Benoit, D., & Parker, K. C. H. (1994). Stability and transmission of attachment across three generations. *Child Development, 65,* 1444–1456.

Berg, C. A., & Sternberg, R. J. (1992). Adults' conceptions of intelligence across the adult life span. *Psychology and Aging, 7,* 221–231.

Berkel, J., & deWaard, F. (1983). Mortality pattern and life expectancy of Seventh-day Adventists in the Netherlands. *International Journal of Epidemiology, 12,* 455–459.

Berkowitz, L. (1993). *Aggression: Its causes, consequences and control.* New York: McGraw-Hill.

Bernstein, I. L. (1990). Salt preference and development. *Developmental Psychology, 26,* 552–554.

Berrick, J. D., & Gilbert, N. (1991). *With the best of intentions: The child sexual abuse prevention movement.* New York: Guilford Press.

Besharov, D. J. (1990). *Recognizing child abuse: A guide for the concerned.* New York: Free Press.

Besharov, D. J. (1993). Teen sex. *American enterprise, 4,* 52–59.

Best, D. L. (1993). Inducing children to generate mnemonic organizational strategies: An examination of long-term retention and materials. *Developmental Psychology, 29,* 324–336.

Bettelheim, B. (1989). *Uses of enchantment: The meaning and importance of fairy tales.* New York: Vintage Books.

Bettes, B. A. (1988). Maternal depression and motherese: Temporal and intonational features. *Child Development, 59,* 1089–1096.

Bigler, R., Jones, L. C., & Lobliner, D. B. (1997). Social categorization and the formation of intergroup attitudes in children. *Child Development, 68,* 530–543.

Bigler, R. S., & Liben, L. S. (1992). Cognitive mechanisms in children's gender stereotyping: Theoretical and educational implications of a cognitive-based intervention. *Child Development, 63,* 1351–1363.

Bijou, S. W., & Baer, D. M. (1965). A social learning model of attachment: Socialization—The development of behavior to social stimuli. In *Child development II.* New York: Appleton-Century-Crofts.

Biller, H. B. (1974). *The father-infant relationship: Some naturalistic observations.* Unpublished manuscript, University of Rhode Island.

Biller, H. B. (1976). The father and personality development: Paternal deprivation and sex-role development. In M. E. Lamb (Ed.), *The role of the father in child development.* New York: Wiley.

Biller, H. B. (1993). *Fathers and families: Paternal factors in child development.* Westport, CT: Auburn House.

Billy, J. O. G., Brewster, K. L., & Grady, W. R. (1994). Contextual effects on the sexual behavior of adolescent women. *Journal of Marriage and the Family, 56,* 387–404.

Billy, J. O. G., & Udry, J. R. (1985). Patterns of adolescent friendship and effects of sexual behavior. *Social Psychology Quarterly, 48,* 27–41.

Birnholz, J. (1981). The development of human fetal eye movement patterns, *Science, 213,* 679–681.

Birnholz, J. C., & Benacerraf, B. R. (1983). The development of human fetal hearing. *Science, 222,* 516–518.

Birren, J. E. (1987, May). The best of all stories. *Psychology Today, 21,* 91–92.

Bischof, L. J. (1976). *Adult psychology* (2nd ed.). New York: Harper & Row.

Bishop, J. E. (1993, September 30). Obese adolescents found less likely to marry, more likely to earn less. *Wall Street Journal,* B16.

Bisping, R., Steingrueber, H. J., Oltmann, M., & Wenk, C. (1990). Adults' tolerance of cries: An experimental investigation of acoustic features. *Child Development, 61,* 1218–1229.

Bjorklund, D. F., & Green, B. L. (1992). The adaptive nature of cognitive immaturity. *American Psychologist, 47,* 46–54.

Black, S. A., & Markides, K. S. (1994). Aging and generational patterns of alcohol consumption among Mexican Americans, Cuban Americans, and mainland Puerto Ricans. *International Journal of Aging and Human Development, 39,* 97–103.

Blagg, N. (1991). *Can we teach intelligence? A comprehensive evaluation of Feuerstein's instrumental enrichment program.* Hillsdale, NJ: Erlbaum.

Blakely, K. S. (1994). Parents' conceptions of social dangers to children in the urban environment. *Children's Environments, II,* 16–25.

Blakeslee, S. (1986, June 24). Rapid changes seen in young brain. *New York Times,* 17, 20.

Blakeslee, S. (1989, February 14). Crib death: Suspicion turns to the brain. *New York Times,* 17, 19.

Blakeslee, S. (1991a, August 28). Bachelorhood after 40: It may be a state of mind. *New York Times,* B1, B6.

Blakeslee, S. (1991b, January 1). Research on birth defects turns to flaws in sperm. *New York Times,* 1, 16.

Blakeslee, S. (1994, September 28). Study of birth defects holds clues to aging. *New York Times,* B7.

Blanchard, R. W., & Biller, H. B. (1971). Father availability and academic performance among third-grade boys. *Developmental Psychology, 4,* 301–305.

Blass, E. M., & Ciaramitaro, V. (1994). A new look at some old mechanisms in human newborns: Taste and tactile determinants of state, affect, and action. *Monographs of the Society for Research in Child Development, 59* (Serial No. 239).

Blinn-Pike, L., Kuschel, D., McDaniel, A., & Mingus, S. (1998). The process of mentoring pregnant adolescents: An exploratory study. *Family Relations, 47,* 119–127.

Block, J. (1995). A contrarian view of the five-factor approach to personality description. *Psychological Bulletin, 117,* 187–215.

Block, J., & Robins, R. W. (1993). A longitudinal study of consistency and change in self-esteem from early adolescence to early adulthood. *Child Development, 64,* 909–923.

Bloom, L. (1970). *Language development: Form and function in emerging grammar.* Cambridge, MA: MIT Press.

Bloom, L. (1993). *The transition from infancy to language: Acquiring the power of expression.* New York: Cambridge University Press.

Blos, P. (1962). *On adolescence: A psychoanalytic interpretation.* New York: Free Press of Glencoe.

Blumstein, P., & Schwartz, P. (1983). *American couples.* New York: Morrow.

Bock, R., & Dubois, R. (1998, July 21). SIDS rate drops as more babies are placed to sleep on their backs or sides. National Institute of Child Health and Human Development, NIH. Retrieved August 5, 1998, from the World Wide Web: http://www.nih.gov/news/pr/july98/nichd-21.htm

Bodnar, J. C., & Kiecolt-Glaser, J. K. (1994). Caregiver depression after bereavement: Chronic stress isn't over when it's over. *Psychology and Aging, 9,* 372–380.

Bolton, F., Jr., Morris, L. A., & MacEachron, A. (1989). *Males at risk: The other side of child sexual abuse.* Newbury Park, CA: Sage.

Boonsong, S. (1968). *The development of concentration of mass, weight, and volume in Thai children.* Unpublished master's thesis, College of Education, Bankok, Thailand.

Bordo, S. (1993). *Unbearable weight: Feminism, western culture and the body.* Berkeley: University of California Press.

Bornhoeft, M. A. (1998). Strategies for preventing HIV in women. *The Body, XI.* Retrieved September 10, 1998 from the World Wide Web: http://www.thebody.com/bp/mar98/news.html#international

Bornstein, K. (1998, September/October). Just who do you think you are? *Utne Reader,* 61.

Bornstein, M. H. (1989). Sensitive periods in development: Structural characteristics and causal interpretations. *Psychological Bulletin, 105,* 179–197.

Bornstein, M. H. (1995). Parenting infants. In M. H. Bornstein (Ed.), *Handbook of parenting* (Vol. 1). Hillsdale, NJ: Erlbaum.

Bornstein, M. H., & Marks, L. E. (1982, January). Color revisionism. *Psychology Today,* 64–73.

Bornstein, M. H., & O'Reilly, A. W. (Eds.). (1993). *The role of play in the development of thought.* San Francisco: Jossey-Bass.

Bosman, E. A. (1993). Age-related differences in motoric aspects of transcription typing skills. *Psychology and Aging, 8,* 88–102.

Bossé, R. (1998). Retirement and retirement planning. In I. H. Nordhus, et al., (Eds.), *Clinical geropsychology* (pp. 155–159). Washington, DC: American Psychological Association.

Botwinick, J. (1977). Intellectual abilities. In J. E. Birren & K. W. Schaie (Eds.), *Handbook of the psychology of aging.* New York: Van Nostrand.

Botwinick, J. (1978). *Aging and behavior: A comprehensive integration of research findings* (2nd ed.). New York: Springer.

Botwinick, J., West, R., & Storandt, M. (1978). Predicting death from behavioral test performance. *Journal of Gerontology, 33,* 755–762.

Bouchard, T. J., Jr., Lykken, D. T., McGue, M., Segal, N. L., & Tellegen, A. (1990). Sources of human psychological differences: The Minnesota study of twins reared apart. *Science, 250,* 223–228.

Bouchard, T. J., Jr., & McGue, M. (1981). Famial studies of intelligence: A review. *Science, 212,* 1055–1059.

Bould, S., & Longino, C. (1997). *Handbook on women and aging.* Westport: Greenwood Press.

Bound, J., Duncan, G. J., Laren, D. S., & Oleinick, L. (1991). Poverty dynamics in widowhood. *Journal of Gerontology, 46,* S115–124.

Bower, T. G. R. (1974). *Development in infancy.* San Francisco: Freeman.

Bower, T. G. R. (1976, November). Repetitive processes in child development. *Scientific American, 235,* 38–47.

Bowlby, J. (1969). *Attachment.* New York: Basic Books.

Bowlby, J. (1988). *A secure base: Clinical applications of attachment theory.* London: Routledge.

Boxer, A. M., Tobin-Richards, M., & Petersen, A. C. (1983). Puberty: Physical change and its significance in early adolescence. *Theory into Practice, 22,* 85–90.

Bradley, R. H., Whiteside, L., Mundfrom, D. J. H., Casey, P. H., Kelleher, K. J., & Pope, S. K. (1994). Early indications of resilience and their relation to experiences in the home environments of low birthweight, premature children living in poverty. *Child Development, 65,* 346–360.

Brady, M. (1998). Female genital mutilation. *Nursing, 28,* 50–51.

Brain facts: A parent's guide to early brain development. (1999). I Am Your Child. Retrieved February 4, 1999, from the World Wide Web: http://iamyourchild. org/docs/bf-0.html

Braine, M. D. S. (1963). The ontogeny of English phrase structure: The first phase. *Language, 39,* 1–14.

Brandtstadter, J., & Rothermund, K. (1994). Self-percepts of control in middle and later adulthood: Buffering losses by rescaling goals. *Psychology and Aging, 9,* 265–273.

Braun, K. L., Takamura, J. C., Forman, S. M., Sasaki, P. A., & Meininger, L. (1995). Developing and testing outreach materials on Alzheimer's disease for Asian and Pacific Island Americans. *Gerontologist, 35,* 122–126.

Bray, J. H., & Hetherington, E. M. (1993). Families in transition: Introduction and overview. *Journal of Family Psychology, 7,* 3.

Brazelton, T. B. (1978). Introduction. In A. J. Sameroff (Ed.), Organization and stability of newborn behavior: A commentary on the Brazelton Neonatal Behavior Assessment Scale [Monograph]. *Monographs of the Society for Research in Child Development, 43* (177), 1–13.

Brazelton, T. B. (1998, December 6). Early bonding is important to parents, infants, *Houston Chronicle,* 9.

Brazelton, T. B., Nugent, J. K., & Lester, B. M. (1987). Neonatal Behavioral Assessment Scale. In J. D. Osofsky (Ed.), *Handbook of infant development* (2nd ed.) New York: Wiley.

Brenner, M. H. (1976). Estimating the social costs of national economic policy: Implications for mental and physical health and criminal aggression (Paper No. 5). *Report to the Congressional Research Service of the Library of Congress and Joint Committee of Congress.* Washington, DC: Government Printing Office.

Brenton, M. (1977). What can be done about child abuse? *Today's Education, 66,* 51–53.

Bridgman, M. (1984, December 28). Midlifers accept situations and look ahead. *Columbus, Ohio, Dispatch,* C1.

Brim, G. (1992). *Ambition: How we manage success and failure through our lives.* New York: Basic Books.

Broder, T. (1996, December 10). He's a new dad at 103. *Weekly World News.* Retrieved August 13, 1997, from the World Wide Web: http://cbs1.cornell.edu/hss315/ articles/103.html

Brody, D. J., Pirkle, J. L., Kramer, R. A., Flegal, K. M., Matte, T. D., Gunter, E. W., & Paschal, D. C. (1994). Blood lead levels in the U. S. population: Phase I of the third national health and nutrition examination survey, NHANES III, 1988–1991. *Journal of the American Medical Association, 272,* 277–283.

Brody, E. M. (1990). *Women in the middle: Their parent-care years.* New York: Springer.

Brody, E. M., Litvin, S. J., Albert, S. M., & Hoffman, C. J. (1994). Marital status of daughters and patterns of parent care. *Journal of Gerontology: Social Sciences, 49,* S95–S103.

Brody, G. H., Neubaum, E., & Forehand, R. (1988). Serial marriage: A heuristic analysis of an emerging family form. *Psychological Bulletin, 103,* 211–222.

Brody, G. H., Stoneman, Z., Flor, D., McCrary, C., Hastings, L., & Conyers, O. (1994). Financial resources, parent psychological functioning, parent co-caregiving, and early adolescent competence in rural two-parent African-American families. *Child Development, 65,* 590–605.

Brody, J. E. (1981, May 27). Planning to prevent retirement "shock," *New York Times,* 13.

Brody, J. E. (1991, October 1). A quality of life determined by a baby's size. *New York Times,* A1, A13.

Brody, J. E. (1992a, November 5). Adolescent obesity linked to ailments in adults. *New York Times,* A8.

Brody, J. E. (1992b, November 23). For most trying to lose weight, dieting only makes things worse. *New York Times,* A1, A8.

Brody, J. E. (1992c, February 5). Maintaining friendships for the sake of good health. *New York Times,* B8.

Brody, J. E. (1993a, April 28). Modern certified midwives are leading a revolution in high-quality obstetric care. *New York Times,* B7.

Brody, J. E. (1993b, August 18). The time to head off osteoporosis, the nemesis of many older women, is in the teen-age years. *New York Times,* B7.

Brody, J. E. (1994a, May 11). For many older Americans, an increase in age can often mean a decrease in nutrition. *New York Times,* B6.

Brody, J. E. (1994b, October 26). New research on postpartum depression holds hope for the mothers caught in its grip. *New York Times,* B7.

Brody, J. E. (1994c, November 8). Restraints for elderly. *New York Times,* B17.

Brody, J. E. (1995a, January 18). Estrogen therapy: Questions are answered and raised. *New York Times,* B8.

Brody, J. E. (1995b, March 1). Infant mortality and premature birth. *New York Times,* B7.

Brody, J. E. (1995c, June 15). New clues in balancing the risks of hormones after menopause. *New York Times,* A1, A12.

Brody, J. E. (1995d, April 19). Study says exercise must be strenuous to add to lifespan. *New York Times,* A1, B7.

Brody, J. E. (1997, August 27). Diet may become one reason complaints about menopause are rare in Asia. *New York Times Health,* C8.

Brody, L. (1996, February 1). The diet years: Sure, it's a $35-billion industry, but you can't call it "new." Take a trip with us through the long and storied history of calorie counting. *Los Angeles Times,* E1.

Brody, N. (1992). *Intelligence* (2nd ed.). San Diego: Academic Press.

Brodzinsky, D. M., & Schechter, M. D. (Eds.). (1990). *The psychology of adoption.* New York: Oxford University Press.

Bromberger, J. T. (1997). A hot flash. *Heart and Soul, 23,* 21.

Bronfenbrenner, U. (1977, May). Nobody home: The erosion of the American family. *Psychology Today, 10,* 40–47.

Bronfenbrenner, U. (1979). *The ecology of human development: Experiments by nature and design.* Cambridge, MA: Harvard University Press.

Bronfenbrenner, U. (1986, February). Alienation and the four worlds of childhood. *Phi Delta Kappan, 67,* 430–436.

Bronfenbrenner, U. (1997). Systems vs. associations: It's not either/or. *Families in Society, 78,* 124.

Bronfenbrenner, U., & Crouter, A. C. (1983). Evolution of environmental models of developmental research. In P. Mussen and W. Kessen (Eds.), *Handbook of child psychology.* New York: Wiley.

Bronson, G. (1977, March 1). Long exposure to waste anesthetic gas is peril to workers, U.S. safety unit says. *Wall Street Journal,* 10.

Bronson, G. (1997). The growth of visual capacity: Evidence from infant scanning patterns. *Advances in Infancy Research, 11,* 109–141.

Bronson, G. W. (1972). Infants' reactions to unfamiliar persons and novel objects. *Monographs of the Society for Research in Child Development, 37* (3).

Bronson, G. W. (1994). Infants' transitions toward adult-like scanning. *Child Development, 65,* 1243–1261.

Bronson, W. (1974). Mother-toddler interaction: A perspective on studying the development of competence. *Merrill-Palmer Quarterly, 20,* 275–301.

Bronstein, R. F. (1992). The dependent personality: Developmental, social, and clinical perspectives. *Psychological Bulletin, 112,* 3–23.

Bronston, B. (1998, October 19). Girl power. *Times-Picayune,* D1.

Brooks-Gunn, J. (1988). Antecedents and consequences of variations in girls' maturational timing. *Journal of Adolescent Health Care, 9,* 365–373.

Brooks-Gunn, J., & Furstenberg, F. F., Jr. (1989). Adolescent sexual behavior. *American Psychologist, 44,* 249–257.

Brooks-Gunn, J., Klebanov, P. K., Liaw, F., & Spiker, D. (1993). Enhancing the development of low-birthweight, premature infants: Changes in cognition and behavior over the first three years. *Child Development, 64,* 736–753.

Brooks-Gunn, J., & Petersen, A. C. (Eds.). (1983). *Girls at puberty: Biological and psychosocial perspectives.* New York: Plenum.

Brooks-Gunn, J., Phelps, E., & Elder, G. H., Jr. (1991). Studying lives through time: Secondary data analyses in developmental psychology. *Developmental Psychology, 27,* 899–910.

Brooks-Gunn, J., & Reiter, E. O. (1990). The role of pubertal processes in the early adolescent transition. In S. S. Feldman & G. R. Elliott (Eds.), *At the threshold: The developing adolescent.* Cambridge, MA: Harvard University Press.

Brophy, J. (1992). New families, judicial decision-making, and children's welfare. *Canadian Journal of Women and the Law, 5,* 484–497.

Browder, S. (1997). Which body parts wear out the fastest . . . and what you can do to prolong their vitality. *New Choices, 37,* 52–55.

Brown, B. B., Clasen, D. R., & Eicher, S. A. (1986). Perceptions of peer pressure, peer conformity dispositions, and self-reported behavior among adolescents. *Developmental Psychology, 22,* 521–530.

Brown, B. B., Eicher, S. A., & Petrie, S. (1986). The importance of peer group ("crowd") affiliation in adolescence. *Journal of Adolescence, 9,* 73–96.

Brown, B. B., & Huang, B. (1995). *Pathways through adolescence.* Hillsdale, NJ: Erlbaum.

Brown, B. B., & Lohr, M. J. (1987). Peer-group affiliation and adolescent self-esteem. *Journal of Personality and Social Psychology, 52,* 47–55.

Brown, B. B., Lohr, M. J., & Trujillo, C. (1990). Multiple crowds and multiple lifestyles: Adolescents' perceptions of peer group stereotypes. In R. E. Muuss (Ed.), *Adolescent behavior and society.* New York: McGraw-Hill.

Brown, B. B., Mounts, N., Lamborn, S. D., & Steinberg, L. (1993). Parenting practices and peer group affiliation in adolescence. *Child Development, 64,* 467–482.

Brown, L. M., & Gilligan, C. (1992). *Meeting at the crossroads: Women's psychology and girls' development.* Cambridge, MA: Harvard University Press.

Brown, R. (1973). *A first language.* Cambridge, MA: Harvard University Press.

Brown, R., & Herrnstein, R. J. (1975). *Psychology.* Boston: Little, Brown.

Brown, S. L., & Booth, A. (1996). Cohabitation versus marriage: A comparison of relationship quality. *Journal of Marriage and the Family, 58,* 668–678.

Brown, T. (1996). Values, knowledge, and Piaget. In E. Reed, E. Turiel, & T. Brown (Eds.), *Values and knowledge.* Mahwah, NJ: Erlbaum.

Browne, D. C., Crum, L., & Cousins, D. S. (1997). Minority health. In J. B. Kotch (Ed.), *Maternal and child health: Programs, problems, and policy in public health* (pp. 227–252). Gaithersburg, MD: Aspen.

Brownell, K. D., Rodin, J., & Wilmore, J. H. (1992). *Eating, body weight, and performance in athletes: Disorders of modern society.* Philadelphia: Lea & Febiger.

Brownmiller, S. (1993, January 4). Making female bodies the battlefield. *Newsweek,* 37.

Brubaker, T. H. (1990). Families in later life: A burgeoning research area. *Journal of Marriage and the Family, 52,* 959–981.

Brues, A. & Sacher, G. (Eds.). (1965). *Aging and levels of biological organization.* Chicago: University of Chicago Press, 1965.

Bruner, J. S. (1970, December). A conversation with Jerome Bruner. *Psychology Today, 4,* 51–74.

Bruner, J. S. (1983). *Child's talk: Learning to use language.* New York: Norton.

Bruner, J. S. (1991). *Acts of meaning.* Cambridge, MA: Harvard University Press.

Bruner, J. S., Goodnow, J. J., & Austin, G. A. (1956). *A study of thinking.* New York: Wiley.

Bruner, J. S., Oliver, R. R., & Greenfield, P. M. (1966). *Studies in cognitive growth.* New York: Wiley.

Bryan, T., & Pflaum, S. (1978). Social interactions of learning disabled children: A linguistic, social, and cognitive analysis. *Learning Disabilities Quarterly, 1,* 70–79.

Bryk, A. S., Lee, V. E., & Holland, P. B. (1993). *Catholic schools and the common good.* Cambridge, MA: Harvard University Press.

Buchanan, C. M., Eccles, J. S., & Becker, J. B. (1992). Are adolescents the victims of raging hormones: Evidence for activational effects of hormones on moods and behavior at adolescence. *Psychological Bulletin, 111,* 62–107.

Buck, N., & Scott, J. (1993). She's leaving home: But why? An analysis of young people leaving the parental home. *Journal of Marriage and the Family, 55,* 863–874.

Bühler, C., & Massarik, F. (1968). *The course of human life: A study of goals in the humanistic perspective.* New York: Springer.

Buhrmester, D. (1990). Intimacy of friendship, interpersonal competence, and adjustment during preadolescence and adolescence. *Child Development, 61,* 1101–1111.

Buhrmester, D., & Furman, W. (1990). Perceptions of sibling relationships during middle childhood and adolescence. *Child Development, 61,* 1387–1398.

Building knowledge for a nation of learners: A framework for education research. (1997). Office of Educational Research & Improvement and the National Educational Research Policy & Procedures Board. Washington, DC: U.S. Department of Education.

Bukowski, W., & Cillessen, A. (1998). *Sociometry then and now: Building on six decades of measuring children's experiences with the peer group.* San Francisco: Jossey-Bass.

Bukowski, W. M., Gauze, C., Hoza, B., & Newcomb, A. F. (1993). Differences and consistency between same-sex and other-sex peer relationships during early adolescence. *Developmental Psychology, 29,* 255–263.

Bullock, M., & Lutkenhaus, P. (1988). The development of volitional behavior in the toddler years. *Child Development, 59,* 664–674.

Bumpass, L., Sweet, J., & Martin, T. C. (1990). Changing patterns of remarriage. *Journal of Marriage and the Family, 52,* 747–756.

Burhaus, K. K., & Dweck, C. S. (1995). Helplessness in early childhood. *Child Development, 66,* 1717–1738.

Bursik, K. (1991). Adaptation to divorce and ego development in adult women. *Journal of Personality and Social Psychology, 60,* 300–306.

Burt, M. R. (1980). Cultural myths and supports for rape. *Journal of Personality and Social Psychology, 38,* 217–230.

Burton, R. V. (1976). Honesty and dishonesty. In T. Lickona (Ed.), *Moral development and behavior: Theory, research, and social issues.* New York: Holt, Rinehart & Winston.

Bushnell, E. W. (1985). The decline of visually guided reaching during infancy. *Infant Behavior and Development, 8,* 139–155.

Buss, D., Haselton, M., Schackelford, T., Bleske, A., & Wakefield, J. (1998). Adaptations, exaptations, and spandre. *American Psychologist, 53,* 533–548.

Busse, E. W. (1969). Theories of aging. In E. W. Busse & E. Pfeiffer (Eds.), *Behavior and adaptation in late life.* Boston: Little, Brown.

Busse, E. W., & Pfeiffer, E. (1969). Functional psychiatric disorders in old age. In E. W. Busse & E. Pfeiffer (Eds.), *Behavior and adaptation in late life.* Boston: Little, Brown.

Bussey, K. (1992). Lying and truthfulness: Children's definitions, standards, and evaluative reactions. *Child Development, 63,* 129–137.

Bussey, K., & Bandura, A. (1984). Influence of gender constancy and social power on sex-linked modeling. *Journal of Personality and Social Psychology, 47,* 1292–1302.

Bussey, K., & Bandura, A. (1992). Self-regulatory mechanisms governing gender development. *Child Development, 63,* 1236–1250.

Butler, R., & Ruzany, N. (1993). Age and socialization effects on the development of social comparison motives and normative ability assessment in kibbutz and urban children. *Child Development, 64,* 532–543.

Butler, R. N. (1971, December). The life review. *Psychology Today, 5,* 49–51f.

Butler, S. (1998). Japan's baby bust. *U.S. News & World Report, 125,* 42–44.

Buttenwieser, S. (1994, May/June). AIDS update: Women over 50. *Ms,* 62–63.

C

Cadman, J. (1997, July/August). Early treatment for HIV-infected infants. *GMHC Treatment Issues.* Retrieved July 29, 1998, from the World Wide Web: http://www.thebody.com/gmhc/issues/julaug97/infants.html

Cahill, S. E. (1990). Childhood and public life: Reaffirming biographical divisions. *Social Problems, 37,* 390–402.

Cairns, R. B. (1983). The emergence of developmental psychology. In P.H. Mussen (Series Ed.) & W. Kessen (Vol. Ed.), *Handbook of child psychology:* Vol.

1. *History, theory, and methods* (4th ed., pp. 41–102). New York: Wiley.

Calhoun, G., Jr., & Alforque, M. (1996). Prenatal substance afflicted children: An overview and review of the literature. *Education, 117,* 30–38.

Call, K. T., Mortimer, J. T., & Shanahan, M. J. (1995). Helpfulness and the development of competence in adolescence. *Child Development, 66,* 129–138.

Callum, M. (1996, February). Our boomers-turning-50 happiness index. *New Choices,* 24–27.

Calvert, K. (1992). *Children in the house: The material culture of early childhood, 1600–1900.* Boston: Northeastern University Press.

Campbell, A. (1993). *Men, women, and aggression.* New York: Basic Books.

Campbell, L., Connidis, I., & Davies, L. (1999). Sibling ties in later life: A social network analysis. *Journal of Family Issues, 20,* 114–148.

Campbell, P. (1996). Warmer, older, more diverse. *Census Brief.* Washington, DC: U. S. Bureau of the Census.

Campbell, S. B., Pierce, E. W., March, C. L., Ewing, L. J., & Szumowski, E. K. (1994). Hard-to-manage preschool boys: Symptomatic behavior across contexts and time. *Child Development, 65,* 836–851.

Campos, J. J., Barrett, K. C., Lamb, M. E., Goldsmith, H., & Stenberg, C. (1983). Socioemotional development. In M. M. Haith & J. J. Campos (Eds.), *Infancy and developmental psychobiology.* Vol. 2 of P. H. Mussen, Handbook of child psychology. New York: Wiley.

Campos, J. J., Mumme, D. L., Kermoian, R., & Campos, R. G. (1993). A functionalist perspective on the nature of emotion. *Monographs of the Society for Research in Child Development, 59* (Nos. 2–3, Serial No. 240).

Cancian, F. M. (1987). *Love in America: Gender and self-development.* New York: Cambridge University Press.

Candland, D. K. (1993). *Feral children and clever animals: Reflections on human nature.* New York: Oxford University Press.

Cantor, D. S., Fischel, J. E., & Kaye, H. (1983). Neonatal conditionability: A new paradigm for exploring the use of interoceptive cues. *Infant Behavior and Development, 6,* 403–413.

Caplan, A. (1999). The nation waits and watches—Oregon physician-assisted suicide may open Pandora's box. Retrieved January 27, 1999, from the World Wide Web: http://www.med.upenn.edu/~bioethic/PAS/oregon.html

Caplan, F. (1973). *The first twelve months of life.* New York: Putnam.

Carlsmith, L. (1964). Effect of early father-absence on scholastic aptitude. *Harvard Educational Review, 34,* 3–21.

Carlson, A. C. (1986). What happened to the "family wage"? *Public Interest, 83,* 3–17.

Carlson, E. A., Jacobvitz, D., & Sroufe, L. A. (1995). A developmental investigation of inattentiveness and hyperactivity. *Child Development, 66,* 37–54.

Caron, A. J., Caron, R. F., Caldwell, R. C., & Weiss, S. J. (1973). Infant perception of the structural properties of the face. *Developmental Psychology, 9,* 385–399.

Carpenter, D. G. (1965). Diffusion theory of aging. *Journal of Gerontology, 20,* 191–195.

Carr, M., Borkowski, J. G., & Maxwell, S. E. (1991). Motivational components of underachievement. *Developmental Psychology, 27,* 108–118.

Carroll, L. (1998). Jabberwocky. In N. Philip (Ed.), *The new Oxford book of children's verse* (pp. 887–88). New York: Oxford University Press.

Carroll, R. (1997). *Sugar in the raw: Voices of young black girls in America.* New York: Crown.

Carton, B. (1994, December 7). Binge drinking at nation's colleges is widespread, a Harvard study finds. *Wall Street Journal,* B7.

Case, R. (Ed.). (1991). The mind's staircase: *Exploring the conceptual underpinnings of children's thought and knowledge.* Hillsdale, NJ: Erlbaum.

Casper, L. M., & Bryson, K. R. (1998). *Co-resident grandparents and their grandchildren: Grandparent maintained families.* Population Division Working Paper No. 26, Population Division. Washington, DC: U.S. Bureau of the Census.

Caspi, A., Elder, G. H., Jr., & Bem, D. J. (1987). Moving against the world: Life-course patterns of explosive children. *Developmental Psychology, 23,* 308–313.

Caspi, A., Elder, G. H., Jr., & Bem, D. J. (1988). Moving away from the world: Life-course patterns of shy children. *Developmental Psychology, 24,* 824–831.

Caspi, A., Lynam, D., Moffitt, T. E., & Silva, P. A. (1993). Unraveling girls' delinquency: Biological, dispositional, and contextual contributions to adolescent misbehavior. *Developmental Psychology, 29,* 19–30.

Caspi, A., & Moffitt, T. E. (1991). Individual differences are accentuated during periods of social change: The sample case of girls at puberty. *Journal of Personality and Social Psychology, 61,* 157–168.

Cassens, D. (1998). Expanded damages in elder abuse cases, *ABA Journal, 84,* 39.

Cassese, J. (1993). The invisible bridge: Child sexual abuse and the risk of HIV infection in adulthood. *SIECUS Report, 21,* 1–7.

Cassidy, J., & Berlin, L. J. (1994). The insecure/ambivalent pattern of attachment: Theory and research. *Child Development, 65,* 971–991.

Castro, J. (1991, August 26). Watching a generation waste away. *Time,* 10–12.

Catherwood, D., Crassini, B., & Freiberg, K. (1989). Infant response to stimuli of similar hue and dissimilar shape: Tracing the origins of the categorization of objects by hue. *Child Development, 60,* 752–762.

Cattell, R. B. (1943). The measurement of adult intelligence. *Psychological Bulletin, 40,* 153–193.

Cattell, R. B. (1963). Theory of fluid or crystallized intelligence. *Journal of Educational Psychology, 54,* 1–22.

Cattell, R. B. (1971). *Abilities: Their structure, growth, and action.* Boston: Houghton Mifflin.

Causes of death and disability. (1998). *Midlife Passages.* Retrieved September 15, 1998, from the World Wide Web: http://www.midlife-passages.com/newpage2.htm

Cavan, R. S. (1964). Structural variations and mobility. In H. T. Christense (Ed.), *Handbook of marriage and the family.* Chicago: Rand McNally.

Cavanaugh, J. C. (1998a). Friendships and social networks among older people. In I. H. Nordhus et al., (Eds.), *Clinical geropsychology* (pp. 137–140). Washington, DC: American Psychological Association.

Cavanaugh, J. C. (1998b). Memory and aging. *National Forum: Phi Kappa Phi Journal, 78*(2), 34–37.

Celis, W., III. (1991, January 2). Growing talk of date rape separates sex from assault. *New York Times,* A1, B7.

Celis, W., III. (1994, June 8). More college women drinking to get drunk. *New York Times,* B8.

Centers for Disease Control and Prevention. (1986). Public health guidelines for enhancing diabetes control through maternal- and child-health programs. *Morbidity and Mortality Weekly Reports, 13,* 201–208.

Centers for Disease Control and Prevention. (1995). Suicide among children, adolescents, and young adults—United States, 1980–1992. *Morbidity and Mortality Weekly Report, 44,* 289–291.

Centers for Disease Control and Prevention. (1996, February). *HIV/AIDS trends provide evidence of success in HIV prevention and treatment: AIDS deaths decline for the first time.* Retrieved September 15, 1998, from the World Wide Web: http://www.cdc.gov/od/oc/media/pressrel/aids-d1.htm

Centers for Disease Control and Prevention. (1997a, April). *Atlas of United States mortality.* Atlanta: USDHHS.

Centers for Disease Control and Prevention. (1997b). *HIV/AIDS and women who have sex with women.* Retrieved June 12, 1998, from the World Wide Web: http://www.thebody.com/cdc/wsw.html

Centers for Disease Control and Prevention. (1997c). *Teen sex down new study shows.* Retrieved September 10, 1998, from the World Wide Web: http://www.cdc.gov

Centers for Disease Control and Prevention. (1997d). *Youth risk behavior surveillance: United States 1997.* Atlanta: CDC, Division of Media Relations.

Centers for Disease Control and Prevention. (1998a). *1993 National Mortality Followback Survey* (NMFS). National Center for Health Statistics. Retrieved October 1, 1998, from the World Wide Web: http://www.cdc.gov/nchswww/about/major/nmfs/nmfs.htm

Centers for Disease Control and Prevention. (1998b). *Children on losing end of access to new drugs.* CDC National Center for HIV, STD, and TB Prevention. AIDS Alert. Retrieved July 29, 1998, from the World Wide Web: http://www.thebody.com/cdc/children1.html

Centers for Disease Control and Prevention. (1998c). Epidemic of congenital syphilis—Baltimore, 1996–1997. *Morbidity and Mortality Weekly Reports, 47,* 904–907.

Centers for Disease Control and Prevention. (1998d). Health-related quality of life and activity limitation: Eight states, 1995. *Morbidity and Mortality Weekly Report, 47* (February 27), 134–139.

Centers for Disease Control and Prevention. (1998e). Primary and secondary syphilis: United States 1997. *Morbidity and Mortality Weekly Report, 47,* 493–504.

Centers for Disease Control and Prevention. (1998f). State-specific pregnancy rates among adolescents—United States, 1992–1995. *Morbidity and Mortality Weekly Report, 47* (June 26), 497–504.

Centers for Disease Control and Prevention. (1998g). Suicide among black youths—United States, 1980–1995. *Morbidity and Mortality Weekly Reports, 47,* 193–196.

Centers for Disease Control and Prevention. (1998h). Trends in sexual risk behaviors among high school students—United States, 1991–1997. *Morbidity and Mortality Weekly Report, 47,* 749–752.

Centers for Disease Control and Prevention. (1999). *1996 Assisted reproductive technology success rates: National summary and fertility clinic reports.* Retrieved February 4, 1999, from the World Wide Web: http://www.cdc.gov/nccdphp/drh/art96.htm

Cerella, J., Rybash, J., Hoyer, W., & Commons, M. L. (Eds.). (1993). *Adult information processing: Limits on loss.* San Diego: Academic Press.

Cervantes, L. (1965). *The dropout: Causes and cures.* Ann Arbor: University of Michigan Press.

Chaikivsky, A. (1997, June). Getting older will leave a bad taste in your mouth. *Esquire,* 100.

Chamberlain, D. (1998). Pregnancy, birth, and very early parenting. Retrieved July 15, 1998 from the World Wide Web: http://www.birthpsychology.com/ressources/index.html

Chance, P., & Fischman, J. (1987, May). The magic of childhood. *Psychology Today, 21,* 48–58.

Chapman, M. (1988). Contextuality and directionality of cognitive development. *Human Development, 31,* 92–106.

Chapman, M., & Lindenberger, U. (1988). Functions, operations, and decalage in the development of transitivity. *Developmental Psychology, 24,* 542–551.

Charlesworth, R., & Hartup, W. W. (1967). Positive social reinforcement in the nursery school peer group. *Child Development, 38,* 993–1002.

Charlesworth, W. R. (1992). Darwin and developmental psychology: Past and present. *Developmental Psychology, 28,* 5–16.

Charness, N., & Gerchak, Y. (1996). Participation rates and maximal performance: A log-linear explanation for group differences, such as Russian and male dominance in chess, *Psychological Science, 7,* 46–51.

Chase, M. (1995, February 27). Gently guiding the gravely ill to the end of life. *Wall Street Journal,* B1.

Chase-Lansdale, P. L., Cherlin, A. J., & Kiernan, K. E. (1995). The long-term effects of parental divorce on the mental health of young adults: A developmental perspective. *Child Development, 66,* 1614–1634.

Chauncey, G. (1994). *Gay New York: Gender, urban culture, and the making of the gay male world, 1890–1940*. New York: Basic Books.

Chavkin, W., & Breitbart, V. (1997). Substance abuse and maternity: The United States as a case study. *Addiction, 92*, 1201–1205.

Chen, X., Rubin, K. H., & Sun, Y. (1992). Social reputation and peer relationships in Chinese and Canadian children: A cross-cultural study. *Child Development, 63*, 1336–1343.

Cherlin, A. (1983). Changing family and household: Contemporary lessons from historical research. *Annual Review of Sociology, 9*, 51–66.

Cherlin, A. J. (1998). Marriage and marital dissolution among black Americans. *Journal of Comparative Family Studies, 29*, 147–158.

Cherlin, A. J., & Furstenberg, F. F., Jr. (1986). *The new American grandparent: A place in the family, a life apart*. New York: Basic Books.

Chesla, C., Martinson, I., & Muwaswes, M. (1994). Continuities and discontinuities in family members' relationships with Alzheimer's patients. *Family Relations, 43*, 3–9.

Chess, S., & Thomas, A. (1996). *Temperament: Theory and practice*. New York: Brunner/Mazel.

Chi, M., Hutchinson, J., & Robin, A. (1989). How inferences about novel domain-related concepts can be constrained by structural knowledge. *Merrill Palmer Quarterly, 35*, 27–62.

Chiesa, M. (1992). Radical behaviorism and scientific frameworks: From mechanistic to relational accounts. *American Psychologist, 47*, 1287–1299.

Children's Defense Fund. (1998). Safe start: Facts on youth, violence, and crime. Retrieved August 17, 1998, from the World Wide Web: http://www.childrensdefense.org/safestart_facts.html

Chira, S. (1993a, September 22). Census data show rise in child care by fathers. *New York Times*, A10.

Chira, S. (1993b, September 30). Gay and lesbian parents grow more visible. *New York Times*, A1, B5.

Chira, S. (1994, April 8). Broad study says home-based day care, even if by relatives, often fails children. *New York Times*, A9.

Cholesterol. (1998, July/August). *Health Magazine* (supplement).

Chomsky, N. (1957). *Syntactic structures*. The Hague: Mouton.

Chomsky, N. (1965). *Aspects of a theory of syntax*. Cambridge, MA: MIT Press.

Chomsky, N. (1968). *Language and mind*. New York: Harcourt Brace Jovanovich.

Chomsky, N. (1975). *Reflections on language*. New York: Pantheon Books.

Chomsky, N. (1980). *Rules and representations*. New York: Columbia University Press.

Chown, S. M. (Ed.). (1972). *Human aging*. Baltimore: Penguin Books.

Chubb, N. H., & Fertman, C. I. (1992). Adolescents' perceptions of belonging in their families. *Families in Society, 73*, 387–394.

Chubb, N. H., Fertman, C. I., & Ross, J. L. (1997). Adolescent self-esteem and locus of control: A longitudinal study of gender and age differences. *Adolescence, 32*, 113–129.

Chugani, H. T., & Phelps, M. E. (1986). Maturational changes in cerebral functions in infants determined by FDG positron emission tomography. *Science, 231*, 840–843.

Chumlea, W. C. (1982). Physical growth in adolescence. In B. B. Wolman (Ed.), *Handbook of developmental psychology*. Englewood Cliffs, NJ: Prentice Hall.

Cicirelli, V. G. (1978). The relationship of sibling structure to intellectual abilities and achievement. *Review of Educational Research, 48*, 365–379.

Cicirelli, V. G. (1981). *Helping elderly parents: The role of adult children*. Boston: Auburn House.

Cicirelli, V. G. (1992). *Family caregiving: Autonomous and paternalistic decision making*. Newbury Park, CA: Sage.

Cicirelli, V. G. (1994). Sibling relationships in cross-cultural perspective. *Journal of Marriage and the Family, 56*, 7–20.

Cicirelli, V. G. (1997). Relationship of psychosocial and background variables to older adults' end-of-life decisions. *Psychology and Aging, 12*, 72–83.

Cicirelli, V. G., Coward, R. T., & Dwyer, J. W. (1992). Siblings as caregivers for impaired elders. *Research on Aging, 14*, 331–350.

Clark, D. L., Lotto, L. S., & McCarthy, M. M. (1980). Factors associated with success in urban elementary schools. *Phi Delta Kappan, 61*, 467–470.

Clark, E. V., Gelman, S. A., & Lane, N. M. (1985). Compound nouns and category structure in young children. *Child Development, 56*, 84–94.

Clarke-Stewart, K. A., Gruber, C. P., & Fitzgerald, L. M. (Eds.). (1994). *Children at home and in day care*. Hillsdale, NJ: Erlbaum.

Clarke-Stewart, K. A., & Hevey, C. M. (1981). Longitudinal relations in repeated observations of mother-child interaction from 1 to 2½ years. *Developmental Psychology, 17*, 127–145.

Clausen, J. A. (1966). Family structure, socialization and personality. In L. W. Hoffman & M. L. Hoffman (Eds.), *Review of child development research* (Vol. 2). New York: Russell Sage Foundation.

Clausen, J. A. (1993). *American lives: Looking back at the children of the Great Depression*. New York: Free Press.

Cleiren, M. P. H. D. (1993). *Bereavement and adaptation: A comparative study of the aftermath of death*. Washington, DC: Hemisphere.

Clements, M. (1994, August 7). Sex in America today: A new national survey reveals how our attitudes are changing. *Parade Magazine*, 4–6.

Cleveland, J. (1998). Ear-Splitting music really did split ears. *Senior Connection*. Retrieved June 12, 1998, from the World Wide Web: http://www.seniornews.com/senior-connection/article1032.html

Climate and diversity of educational institutions. (1997). Education of students with disabilities. *The Condition of Education 1997*. Washington, DC: National Center for Education Statistics.

Coburn, M. F. (1996). It was scary, but exciting. *New Choices, 36*, 56–58.

Cochran, M., & Niego, S. (1995). Parenting and social networks. In M. H. Bornstein (Ed.), *Handbook of parenting* (Vol. 3). Hillsdale, NJ: Erlbaum.

Cohen, D., Eisdorfer, C., Gorelick, P., Paveza, G., Luchins, D. J., Freels, S., Ashford, J. W., Semla, T., Levy, P., & Hirschman, R. (1993). Psychopathology associated with Alzheimer's disease and related disorders. *Journal of Gerontology: Medical Sciences, 48*, M2555–M260.

Cohen, J., & Tronick, E. Z. (1987). Mother-infant face-to-face interaction: The sequence of dyadic states at 3, 6, and 9 months. *Developmental Psychology, 23*, 68–77.

Cohen, M. E., Hoffman, H. S., Kelley, N. E., & Anday, E. K. (1988). A failure to observe habituation in the human neonate. *Infant Behavior and Development, 11*, 297–304.

Cohen, O. J., & Fauci, A. S. (1998). HIV/AIDS in 1998—Gaining the upper hand? *Journal of the American Medical Association, 280*, 87–88.

Cohen, S. E., & Beckwith, L. (1977). Caregiving behaviors and early cognitive development as related to ordinal position in preterm infants. *Child Development, 48*, 152–157.

Cohn, J., & Sugar, J. A. (1991). Determinants of quality of life in institutions: Perceptions of frail older residents, staff and families. In J. E. Birren, J. E. Lubben, J. C. Rowe, & D. E. Deutchman (Eds.), *The concept and measurement of quality of life in the frail elderly*. New York: Academic Press.

Colby, A., & Damon, W. (1992). *Some do care*. New York: Free Press.

Colby, K. M., & Stoller, R. J. (1988). *Cognitive science and psychoanalysis*. Hillsdale, NJ: Erlbaum.

Colditz, G. A. (1992). Economic costs of obesity. *American Journal of Nutrition, 55*, 503–507.

Coldren, J. T., & Colombo, J. (1994). The nature and processes of preverbal learning. *Monographs of the Society for Research in Child Development, 59* (4, Serial No. 241).

Cole, C. L. (1977). Cohabitation in social context. In R. W. Libby & R. N. Whitehurst (Eds.), *Marriage and alternatives*. Glenview, IL: Scott, Foresman.

Cole, R. A. (1979, April). Navigating the slippery stream of speech. *Psychology Today, 12*, 77–87.

Coleman, M., & Ganong, L. H. (1990). Remarriage and stepfamily research in the 1980s: Increased interest in an old family form. *Journal of Marriage and the Family, 52*, 925–940.

Coley, R. (1998). Children's socialization experiences and functioning in single mother households: The importance of father and other men. *Child Development, 69*, 219–230.

Collier, V. P. (1992). A synthesis of studies examining long-term language minority student data on academic achievement. *Bilingual Research Journal, 16*, 187–212.

Collier, V. P. (1997). Acquiring a second language for school. *Direction in Language and Education, 1*, 4.

Collins, G. (1983, October 24). Stepfamilies share their joys and woes. *New York Times*, 21.

Collins, N. L., Dunkel-Schetter, C., Lobel, M., & Scrimshaw, S. C. M. (1993). Social support in pregnancy: Psychosocial correlates of birth outcomes and postpartum depression. *Journal of Personality and Social Psychology, 65*, 1243–1258.

Collum, C. S. (1993). Co-parenting adoptions: Lesbian and gay parenting. *Gay and Lesbian Issues*. Retrieved December 15, 1998, from the World Wide Web: http://nac.adopt.org/adopt/gay/gay2.html

Colombo, J. (1993). *Infant cognition: Predicting later intellectual functioning*. Newbury Park, CA: Sage.

Comarow, A. (1998, May 4). Viagra tale: How one man sought an impotence cure—And found one. *U.S. News & World Report*, 64–66.

Comfort, A. (1976). *A good age*. New York: Crown.

Commission on Behavioral and Social Sciences and Education. (1993). *Losing generations: Adolescents in high-risk settings*. Washington, DC: National Academy Press.

The condition of education. (1997). National Center for Education Statistics. Retrieved February 7, 1999, from the World Wide Web: http://nces.ed.gov/pubs/ce/index.html

Condom use increasing among U.S. women. (1997, Oct.). *Population Today, 25*, 8.

Condon, W. S., & Sander, L. W. (1974a). Neonate movement is synchronized with adult speech: Interactional participation and language acquisition. *Science, 183*, 99–101.

Condon, W. S., & Sander, L. W. (1974b). Synchrony demonstrated between movements of the neonate and adult speech. *Child Development, 45*, 456–462.

Conger, R. D., Conger, K. J., Elder, G. H., Jr., Lorenz, F. O., Simons, R. L., & Whitbeck, L. B. (1993). Family economic stress and adjustment of early adolescent girls. *Developmental Psychology, 29*, 206–219.

Conger, R. W., Lorenz, F. O., Elder, G. H., Simons, R. L., & Ge, X. (1993). Husband and wife differences in response to undesirable life events. *Journal of Health and Social Behavior, 34*, 71–88.

Connidis, I. A. (1994). Sibling support in older age. *Journal of Gerontology: Social Sciences, 49*, S309–S317.

Contreras, R., Hendrick, S., & Hendrick, C. (1996). Perspectives on marital love and satisfaction in Mexican American and Anglo-American couples. *Journal of Counseling and Development, 74*, 408–415.

Conway, K. (1995). Miscarriage experience and the role of support systems: A pilot study. *British Journal of Medical Psychology, Part 3*, 259–267.

Cook, J. A. (1997, May 4–7). Minority women in U.S. need better access to potent HIV drug regimens. National Conference on Women and HIV. Pasadena, CA: HIV/AIDS Information Center, Journal of the American Medical Association.

Cooksey, E. C, & Craig, P. H. (1998). Parenting from a distance: The effects of paternal characteristics on contact between nonresidential fathers and their children. *Demography, 35*, 187–200.

Cooksey, E. C., & Fondell, M. F. (1996). Spending time with his kids: Effects of family structure on fathers and children's lives. *Journal of Marriage and the Family, 58*, 693.

Cooley, C. H. (1902). *Human nature and the social order*. New York: Scribner's.

Cooley, C. H. (1909). *Social organization*. New York: Scribner's.

Coombs, R. H. (1991). Marital status and personal well-being: A literature review. *Family Relations, 40*, 97–102.

Cooney, T. M., & Hogan, D. P. (1991). Marriage in an institutionalized life course: First marriage among American men in the twentieth century. *Journal of Marriage and the Family, 53*, 178–190.

Cooney, T. M., & Uhlenberg, P. (1992). Support from parents over the life course: The adult child's perspective. *Social Forces, 71*, 63–84.

Coontz, S. (1992). The way we never were: *American families and the nostalgia trap*. New York: Basic Books.

Cooper, C. R., & Denner, J. (1998). Theories linking culture and psychology: Universal and community-specific processes. *Annual Review of Psychology, 49*, 559–584.

Cooper, M. H. (1998, February 20). Caring for the elderly: Is adequate long-term care available? *CQ Researcher, 8*, 145–168.

Cooper, R. P., & Aslin, R. N. (1990). Preference for infant-directed speech in the first month after birth. *Child Development, 61*, 1584–1595.

Coopersmith, S. (1967). *Antecedents of self-esteem*. San Francisco: Freeman.

Copper, R. L., Goldenberg, R. L., Das, A., Elder, N. Swain, M., Norman, G., Ramsey, R., Cotroneo, P., Colllins, B. A., Johnson, F., Jones, P., & Meier, A. M. (1996). The preterm prediction study: Maternal stress is associated with spontaneous preterm birth at less than 35 weeks' gestation. National Institute of Child Health and Human Development Maternal-Fetal Medicine Units Network. *American Journal of Obstetrics and Gynecology, 175*, 1286–1292.

Corby, B. (1993). *Child abuse: Toward a knowledge base*. Buckingham, PA: Open University Press.

Corn, A., Ferrell, K.A., Spungin, S.J., & Zimmerman, G. (1996). *What we know about teacher preparation programs in blindness and visual impairment*.

Washington, DC: National NASDE Policy Forum.

Costa, P. T., Jr., & McCrae, R. R. (1980). Still stable after all these years: Personality as a key to some issues in adulthood and old age. In P. B. Baltes & O. G. Brim, Jr. (Eds.), *Life-span development and behavior* (Vol. 3, pp. 65–102). New York: Academic Press.

Costa, P. T., Yang, J., & McCrae, R. (1998). Aging and personality traits: Generalizations and clinical implications. In I. H. Nordhus et al., *Clinical Geropsychology.* Washington, DC: American Psychological Association.

Costa, P. T., Jr., McCrae, R. R., & Arenberg, D. (1980). Enduring dispositions in adult males. *Journal of Personality and Social Psychology, 38,* 793–800.

Cottle, M. (1998, July/August). Who's watching the kids? *Washington Monthly, 30,* 16–25.

Coulter, D. J. (1995). Music and the making of the mind. *Early childhood connections: The Journal of Music and Movement-Based Learning.* Cited in Early Childhood News. Retrieved February 12, 1999, from the World Wide Web: http://earlychildhoodnews.com/wiredfor.htm

Cowan, P. A., & Hetherington, M. (Eds.). (1991). *Family transitions.* Hillsdale, NJ: Erlbaum.

Cowen, G., & Avants, S. K. (1988). Children's influence strategies: Structure, sex differences, and bilateral mother-child influence. *Child Development, 59,* 1303–1313.

Cowgill, D. O. (1974). Aging and modernization: A revision of the theory. In J. F. Gubrium (Ed.), *Late life.* Springfield, IL: Thomas.

Cowgill, D. O. (1986). *Aging around the world.* Belmont, CA: Wadsworth.

Cowley, G. (1996, September 16). Attention aging men. *Newsweek,* 68–75.

Cox, C., & Gelfand, D. E. (1987). Familial assistance, exchange and satisfaction among Hispanic, Portuguese, and Vietnamese ethnic elderly. *Journal of Cross-Cultural Gerontology, 2,* 241–255.

Cox, C., & Monk, A. (1993). Hispanic culture and family care of Alzheimer's patients. *Health and Social Work, 18,* 92–100.

Crandall, C. S. (1994). Prejudice against fat people: Ideology and self-interest. *Journal of Personality and Social Psychology, 66,* 882–894.

Crandell, T. L. (1979). *The effects of educational cognitive style and media format on reading procedural instruction in picture-text amalgams.* Dissertation

Abstracts. Ann Arbor, MI: University Microfilms International.

Crandell, T. L. (1982, June). Integration of illustrations and text in reading. In B. A. Hutson (Ed.), *Advances in Reading Language Research, Vol. 1.* Greenwich, CT: Jai Press.

Crano, W. D. (1998). The leniency contract and persistence of majority and minority influence. *Journal of Personality and Social Psychology, 74,* 1437–1450.

Crano, W. D., & Aronoff, J. (1978). A cross-cultural study of expressive and instrumental role complementarity in the family. *American Sociological Review, 43,* 463–471.

Cratty, B. J. (1970). *Perceptual and motor development in infants and children.* New York: Macmillan.

Cravens H. (1992). A scientific project locked in time: The Terman genetic studies of genius, 1920s–1950s. *American Psychologist, 47,* 183–189.

Crawford, J. (1997). *Best evidence: Research foundations of the Bilingual Education Act.* Washington, DC: Clearinghouse for Bilingual Educators.

Creighton, L. L. (1993, December 20). Kids taking care of kids. *U.S. News & World Report,* 26–33.

Crick, N. R., & Dodge, K. A. (1994). A review and reformulation of social information-processing mechanisms in children's social adjustment. *Psychological Bulletin, 115,* 74–101.

Crick, N. R., & Grotpeter, J. K. (1995). Relational aggression, gender, and social-psychological adjustment. *Child Development, 66,* 710–722.

Crispell, D. (1993a). Grandparents galore. *American Demographics, 15,* 63.

Crispell, D. (1993b). Sexual surveys: Does anyone tell the truth? *American Demographics, 15,* 9–10.

Crispell, D. (1994). Child-care choices don't match moms' wishes. *American Demographics, 16,* 11–13.

Critelli, J. W., & Suire, D. M. (1998). Obstacles to condom use: The combination of other forms of birth control and short-term monogamy. *Journal of American College Health, 46,* 215–219.

Crnic, K., & Acevedo, M. (1995). Everyday stresses and parenting. In M. H. Bornstein (Ed.), *Handbook of parenting* (Vol. 4). Hillsdale, NJ: Erlbaum.

Crockenberg, S., & Litman, C. (1990). Autonomy as competence in 2-year-olds: Maternal correlates of child defiance, compliance, and self-assertion. *Developmental Psychology, 26,* 961–971.

Crocker, J., Cornwell, B., & Major, B. (1993). The stigma of overweight: Affective consequences of attributional ambiguity. *Journal of Personality and Social Psychology, 64,* 60–70.

Crook, C. K., & Lipsitt, L. P. (1976). Neonatal nutritive sucking: Effects of taste stimulation upon sucking rhythm and heart rate. *Child Development, 47,* 518–522.

Crook, T. H., III, Larrabee, G. J., & Youngjohn, J. R. (1993). Age and incidental recall for a simulated everyday memory task. *Journal of Gerontology: Psychological Sciences, 48,* P45–P47.

Cross, S., & Markus, H. (1991). Possible selves across the life span. *Human Development, 34,* 230–255.

Crouter, A. C., & Manke, B. (1994). The changing American workplace: Implications for individuals and families. *Family Relations, 43,* 117–124.

Crowther, J. H., Tennenbaum, D. L., Hobfoll, S. E., & Stephens, M. A. P. (Eds.). (1992). *The etiology of bulimia nervosa: The individual and family context.* Washington, DC: Hemisphere.

Cruikshank, M. (1991). Lavender and gray: A brief survey of lesbian and gay aging studies. In J. A. Lee (Ed.), *Gay midlife and maturity* (pp. 77–87). New York: Haworth Press.

Csikszentmihalyi, M. (1993). *The evolving self.* New York: HarperCollins.

Csikszentmihalyi, M. (1997). *Creativity: Flow and the psychology of discovery and invention.* New York: HarperPerennial.

Csikszentmihalyi, M., & Larson, R. (1984). *Being adolescent: Conflict and growth in the teenage years.* New York: Basic Books.

Cummings, E. M., & Davies, P. T. (1994). *Children and marital conflict: The impact of family dispute and resolution.* New York: Guilford Press.

Cummins, J. (1986). Empowering minority students: A framework for intervention. *Harvard Educational Review, 56,* 18–36.

Cutrona, C. E., & Troutman, B. R. (1986). Social support, infant temperament, and parenting self-efficacy. *Child Development, 57,* 1507–1518.

D

Dainton, M. (1993). The myths and misconceptions of the stepmother identity. *Family Relations, 42,* 93–98.

Daley, S. E., Hammen, C., Davila, J., & Burge, D. (1998). Axis II symptomatology, depression, and life stress during the transition from adolescence to adulthood. *Journal of Consulting and Clinical Psychology, 66,* 595–603.

Damon, W., & Hart, D. (1982). The development of self-understanding from infancy through adolescence. *Child Development, 53,* 841–864.

Dangerous bumps ahead. (1999, Jan./Feb.). *Men's Health, 14,* 40.

Daniel, M. (1997). Intelligence testing: Status and trends. *Journal of the American Psychological Association, 52,* 1038–1045.

Daniels, D. (1986). Differential experiences of siblings in the same family as predictors of adolescent sibling personality differences. *Journal of Personality and Social Psychology, 51,* 339–346.

Daniels, D., & Plomin, R. (1985). Origins of individual differences in infant shyness. *Developmental Psychology, 21,* 118–121.

Dannemiller, J. L. (1989). A test of color constancy in 9- and 20-week-old human infants following simulated illuminant changes. *Developmental Psychology, 25,* 171–184.

Davis, D. M. C. (1990). Portrayals of women in prime-time network television: Some demographic characteristics. *Sex Roles, 23,* 325–332.

Davis, K. (1949). *Human society.* New York: Macmillan.

Dawe, H. C. (1934). An analysis of two hundred quarrels of preschool children. *Child Development, 5,* 139–157.

Day, A. T. (1986). *Who cares? Demographic trends challenge family care for the elderly.* Washington, DC: Population Reference Bureau.

Deacon, S. A. (1997). Inter-country adoption and the family life cycle. *American Journal of Family Therapy, 25,* 245–259.

DeAngelis, T. (1997, September). There's new hope for women with postpartum blues. *APA Monitor,* American Psychological Association. Retrieved July 31, 1998 from the World Wide Web: http://www.apa.org/monitor/sep97/hope.html

DeCasper, A. J., & Carstens, A. A. (1981). Contingencies of stimulation: Effects on learning and emotion in neonates. *Infant Behavior and Development, 4,* 19–35.

DeFrain, J., Millspaugh, E., & Xiaolin, X. (1996). The psychosocial effects of miscarriage: Implications for health professionals. *Families, Systems, and Health, 14,* 331–347.

de Graaf, C., Polet, P., & van Staveren, W. A. (1994). Sensory perception and pleasantness of food flavors in elderly subjects. *Journal of Gerontology: Psychological Sciences, 49,* P93–P99.

Dein, S. (1997). ABC of mental health: Mental health in a multiethnic society, *British Medical Journal, 315,* 473–476.

DeLaguna, G. (1929). Perception and language. *Human Biology, 1,* 555–558.

deLemos, M. M. (1969). The development of conservation in Aboriginal children. *International Journal of Psychology, 4,* 255–269.

Demetriou, A., Efklides, A., & Platsidou, M. (1993). The architecture and dynamics of developing mind. *Monographs of the Society for Research in Child Development, 58* (5–6, Serial No. 234).

Demo, D. H. (1992). Parent-child relations: Assessing recent changes. *Journal of Marriage and the Family, 54,* 104–117.

Demos, R. (1994). Stuck at home. *American Demographics, 16,* 6.

Demtrious, A. (Ed.). (1988). *The Neo-Piagetian theories of cognitive development: Toward an integration.* Amsterdam: North-Holland.

Denenberg, R. (1997a). Childhood sexual abuse as an HIV risk factor in women. *The body: An AIDS and HIV information resource, 11.* Retrieved July 11, 1997, from the World Wide Web: http://www.thebody.com/gmhc/issues/julaug97/abuse.html

Denenberg, R. (1997b). HIV risks in women who have sex with women. *The body: An AIDS and HIV information resource, 11.* Retrieved September 10, 1998, from the World Wide Web: http://the body.com/gmhc/issues/julaug97/wsw.html

Denham, S. A., Renwick, S. M., & Holt, R. W. (1991). Working and playing together: Prediction of preschool social-emotional competence from mother-child interaction. *Child Development, 62,* 242–249.

Denham, S. A., Zoller, D., & Couchoud, E. A. (1994). Socialization of preschoolers' emotional understanding. *Developmental Psychology, 30,* 928–939.

Denham, T. E., & Smith, C. L. (1989). The influence of grandparents on grandchildren: A review of the literature and resources. *Family Relations, 38,* 345–350.

Denney, N. (1985). A review of lifespan research with the twenty questions task (TQT). *International Journal of Aging & Human Development, 21,* 161–173.

Dennis, W., & Dennis, M. G. (1940). The effect of cradling practices upon the onset of walking in Hopi children. *Journal of Genetic Psychology, 56,* 77–86.

Dentzer, S. (1990, May 14). Do the elderly want to work? *U.S. News & World Report,* 48–50.

DeRosier, M. E., Kupersmidt, J. B., & Patterson, C. J. (1994). Children's academic and behavioral adjustment as a function of the chronicity and proximity

of peer rejection. *Child Development, 65,* 1799–1813.

Determinants of health in children. (1996): National Crime Prevention Council of Canada. Retrieved February 7, 1999, from the World Wide Web: http://www.crime-prevention.org/ncpc

Devaney, B. L., Ellwood, M. R., & Love, J. M. (1997). Programs that mitigate the effects of poverty on children. *Future of Children, 7,* 88–112.

Devereux, E. C., Shouval, R., Bronfenbrenner, U., Rodgers, R. R., KavVenaki, S., Kiely, E., & Karson, E. (1974). Socialization practices of parents, teachers, and peers in Israel: The kibbutz versus the city. *Child Development, 45,* 269–281.

DeVito, J. A. (1970). *The psychology of speech and language.* New York: Random House.

de Vries, B., & Walker, L. J. (1981). Moral reasoning and attitudes toward capital mental sequelae of maltreatment in infancy. In R. Rizley & D. Cicchetti (Eds.), *Developmental perspectives on child maltreatment.* San Francisco: Jossey-Bass.

DeWitt, P. M. (1993). The birth business. *American Demographics, 15,* 44–49.

DeWolff, M. S., & IJzendoorn, M. H. (1997). Sensitivity and attachment: A meta-analysis on parental antecedents of infant attachment. *Child Development, 68,* 571–591.

DHEA Center. (1997). DHEA Center. Retrieved November 16, 1997, from the World Wide Web: http://www.dheacenter.com/dheaquo.htm

DHEA prohormone complex. (1997). Be Healthy Retrieved September 10, 1998, from the World Wide Web: http://Be-Healthy.simplenet.com/pendo.htm

Diamond, K., LeFurgy, W., & Blass, S. (1993). Attitudes of preschool children toward their peers with disabilities: A year-long investigation in integrated classrooms. *Journal of Genetic Psychology, 154,* 215–221.

Dickinson, B. (1997, June). Are you a marked man? *Esquire,* 100.

Dick-Read, G. (1944/1994). *Childbirth without fear.* New York: Harper.

Dickstein, S., & Parke, R. D. (1988). Social referencing in infancy: A glance at fathers and marriage. *Child Development, 59,* 506–511.

DiClemente, R. J. (1990). Adolescents and AIDS: Current research, prevention strategies, and policy implications. In L. Temoshok and A. Baum (Eds.), *Psychosocial perspectives on AIDS: Etiology, prevention, and treatment* (pp. 51–64). Hillsdale, NJ: Erlbaum.

Dietary Guidelines Advisory Committee. (1995). Guidelines for Americans 1995. *Nutrition Reviews, 53,* 376+.

Dill, P. L., & Henley, T. B. (1998). Stressors of college: A comparison of traditional and nontraditional students. *Journal of Psychology, 132,* 25–32.

DiMauro, D. (1995). *Sexuality research in the United States: An assessment of social and behavioral sciences.* New York: Social Science Research Council.

Dix, T. (1991). The affective organization of parenting: Adaptive and maladaptive processes. *Psychological Bulletin, 110,* 3–25.

Dix, T., Ruble, D. N., & Zambarano, R. J. (1989). Mothers' implicit theories of discipline: Child effects, parent effects, and the attribution process. *Child Development, 60,* 1373–1391.

Dobzhansky, T. (1962). *Mankind evolving.* New Haven: Yale University Press.

Dodge, K. A., Bates, J. E., & Pettit, G. S. (1990). Mechanisms in the cycle of violence. *Science, 250,* 1678–1683.

Dodge, K. A., Price, J. M., Coie, J. D., & Christopoulos, C. (1990). On the development of aggressive dyadic relationships in boys' peer groups. *Human Development, 33,* 260–270.

Donaldson, E. B. (1998). *Adoption in the United States: National adoption statistics.* The Evan B. Donaldson Adoption Institute: Research Resources. Retrieved December 15, 1998, from the World Wide Web: http://www.adoptioninstitute.org/research/ressta.html

Donate-Bartfield, E., & Passman, R. H. (1985). Attentiveness of mothers and fathers to their baby's cries. *Infant Behavior and Development, 8,* 385–393.

Donnelly, D., & Finkelhor, D. (1992). Does equality in custody arrangement improve the parent-child relationship? *Journal of Marriage and the Family, 54,* 837–845.

Donnelly, D., & Finkelhor, D. (1993). Who has joint custody? Class differences in the determination of custody arrangements. *Family Relations, 42,* 57–60.

Doress-Worters, P. B., & Siegal, D. L. (1994). *The new ourselves, growing older: Women aging with knowledge and power.* Boston Women's Health Book Collective. New York: Simon & Schuster.

Dougherty, D. (1993). Major policy options from a report to Congress on adolescent health. *Journal of Adolescent Health, 14,* 499–504.

Douglas, J. (1990–1991). Patterns of change following parent death in midlife adults. *Omega: Journal of Death and Dying, 22,* 123–138.

Dowd, J. M., & Tronick, E. Z. (1986). Temporal coordination of arm movements in early infancy: Do infants move in synchrony with adult speech? *Child Development, 57,* 762–776.

Dreeben, R., & Gamoran, A. (1986). Race, instruction, and learning. *American Sociological Review, 51,* 660–669.

Duffy, F. H., Als, H., & McAnulty, G. B. (1990). Behavioral and electrophysiological evidence for gestational age effects in healthy preterm and full term infants studied two weeks after expected due date. *Child Development, 61,* 1271–1286.

Duhaime, A. C., Christian, C. W., Rorke, L. B., & Zimmerman, R. A. (1998). Nonaccidental head injury in infants—The "shaken-baby syndrome." *New England Journal of Medicine, 338,* 1822–1829.

Duke, L. (1999, February 16). Ignorance feeds deadly South African AIDS epidemic: Spurred by myth and social morés, infection reaches crisis proportions. *The Washington Post,* A01.

Dunn, D. S., Goldbach, K. R. C., Lasker, J. N., & Toedter, L. J. (1991). Explaining pregnancy loss: Parents' and physicians' attributions. *Omega, 23,* 13–23.

Dunn, J. (1983). Sibling relationships in early childhood. *Child Development, 54,* 787–811.

Dunn, J. (1986). Growing up in a family world: Issues in the study of social development of young children. In M. Richards and P. Light (Eds.), *Children of social worlds: Development in a social context.* Cambridge, MA: Harvard University Press.

Dunn, J. (1988). *The beginnings of social understanding.* Cambridge, MA: Harvard University Press.

Dunn, J. (1993). *Young children's close relationships: Beyond attachment.* Newbury Park, CA: Sage.

Dunn, J., Kendrick, C., & MacNamee, R. (1982). The reaction of first-born children to the birth of a sibling: Mothers' reports. In S. Chess & A. Thomas (Eds.), *Annual Progress in Child Psychiatry and Child Development* (pp. 143–165). New York: Brunner/Mazel.

Dunn, J., & Munn, P. (1985). Becoming a family member: Family conflict and the development of social understanding. *Child Development, 56,* 480–492.

Dunn, J., & Plomin, R. (1990). *Separate lives: Why siblings are so different.* New York: Basic Books.

Dunn, J., Slomkowski, C., & Beardsall, L. (1994). Sibling relationships from the preschool period through middle childhood and early adolescence. *Developmental Psychology, 30,* 315–324.

Dunn, J., Slomkowski, C., Beardsall, L., & Rende, R. (1994). Adjustment in middle childhood and early adolescence: Links with earlier and contemporary sibling relationships. *Journal of Child Psychology and Psychiatry, 35,* 491–504.

Dunn, R., Beaudry, J. S., & Klavas, A. (1989). Survey of research on learning styles. *Educational Leadership, 46,* 50–58.

Dunning, D., & Cohen, G. L. (1992). Egocentric definitions of traits and abilities in social judgment. *Journal of Personality and Social Psychology, 63,* 341–355.

Durkheim, E. (1951). *Suicide.* New York: Free Press. (Original work published 1897).

Durkheim, E. (1964). *The division of labor in society.* New York: Free Press. (Original work published 1893).

Dykstra, P. A. (1990). *Next of (non) kin: The importance of primary relationships for older adults' well-being.* Lisse, The Netherlands: Swets & Zeitlinger.

E

Eagly, A. H. (1995). The science and politics of comparing women and men. *American Psychologist, 50,* 145–158.

Earles, J. L., & Coon, V. E. (1994). Adult age differences in long-term memory for performed activities. *Journal of Gerontology: Psychological Sciences, 49,* P32–P34.

Early child care and self-control compliance and problem behavior at twenty-four and thirty-six months. (1998). The NICHD Early Child Care Research Network, Bethesda, MD: National Institutes of Health.

East, P. L., Felice, M. E., & Morgan, M. C. (1993). Sisters' and girlfriends' sexual and childbearing behavior: Effects on early adolescent girls' sexual outcomes. *Journal of Marriage and the Family, 55,* 953–963.

Eccles, J. S., & Midgley, C. (1988). Stage/environment fit: Developmentally appropriate classrooms for early adolescents. In R. E. Ames & C. Ames (Eds.), *Research on motivation in education* (Vol. 3). San Diego: Academic Press.

Eckerman, C. O., & Didow, S. M. (1988). Lessons drawn from observing young peers together. *Acta Paediatrica Scandinavica, 77,* 55–70.

Eckerman, C. O., & Stein, M. R. (1990). How imitation begets imitation and toddlers' generation of games. *Developmental Psychology, 26,* 370–378.

Eder, D., & Hallinan, M. T. (1978). Sex differences in children's friendships.

American Sociological Review, 43, 237–250.

Edmonds, M. H. (1976). New directions in theories of language acquisition. *Harvard Educational Review, 46,* 175–198.

Edwards, C. A. (1994). Leadership in groups of school-age girls. *Developmental Psychology, 30,* 920–927.

Edwards, C. P., & Whiting, B. B. (1988). *Children of different worlds.* Cambridge, MA: Harvard University Press.

Egeland, B. (1993). A history of abuse is a major risk factor for abusing the next generation. In R. J. Gelles & D. R. Loeske (Eds.), *Current controversies on family violence.* Newbury Park: Sage.

Egeland, B., Jacobvitz, D., & Sroufe, L. A. (1988). Breaking the cycle of abuse. *Child Development, 59,* 1080–1088.

Egeland, B., Kalkoske, M., Gottesman, N., & Erickson, M. F. (1990). Preschool behavior problems: Stability and factors accounting for change. *Journal of Child Psychology and Allied Disciplines, 31,* 891–909.

Eggs for Sale. (1998). *Commonweal, 125* (March 27), 5–6.

Ehrhardt, A. (1997). *Sexual risk behavior and behavior change in heterosexual women and men.* National Institutes of Health. Retrieved June 10, 1998, from the World Wide Web: http://www.nih.gov/ silk/arisabst/fy1997/P0004818.html

Eibl-Eibesfeldt, I. (1989). *Human ethology.* New York: Aldine de Gruyter.

Eimas, P. D. (1985, January). The perception of speech in early infancy. *Scientific American, 252,* 46–52.

Einstein, A. (1949). Autobiography. In P. Schilpp (Ed.), *Albert Einstein: Philosopher-scientist.* Evanston, IL: Library of Living Philosophers.

Eisenberg, N. (1992). *The caring child.* Cambridge, MA: Harvard University Press.

Eisenberg, N., Fabes, R. A., Shepard, S. A., Murphy, B. C., Guthrie, I. K., Jones, S., Friedman, J., Poulin, R., & Maszk, P. (1997). Contemporaneous and longitudinal prediction of children's social functioning from regulation and emotionality. *Child Development, 68,* 642–664.

Eisenberg, N., Shepard, S. A., Fabes, R. A., Murphy, B. C., & Guthrie, I. K. (1998). Shyness and children's emotionality, regulation, and coping: Contemporaneous, longitudinal, and across-context relations. *Child Development, 69,* 767–790.

Ekman, P. (1972). Universal in cultural differences in facial expressions of emotion. In J. K. Cole (Ed.), *Nebraska symposium on motivation* (Vol. 19). Lincoln: University of Nebraska Press.

Ekman, P. (1980). *The face of man: Expressions of universal emotions in a New Guinea village.* New York: Garland STPM Press.

Ekman, P. (1994). Strong evidence for universals in facial expressions: A reply to Russell's mistaken critique. *Psychological Bulletin, 115,* 268–287.

Elder, G. H., Jr. (1974). *Children of the great depression.* Chicago: University of Chicago Press.

Elder, G. H., Jr. (1985). Perspectives on the life course. In G. H. Elder, Jr. (Ed.), *Life course dynamics: Trajectories and transitions, 1968–1980.* Ithaca, NY: Cornell University Press.

Elder, G. H., Jr., & Pavlko, E. K. (1993). Work careers in men's later years: Transitions, trajectories, and historical change. *Journal of Gerontology: Social Sciences, 48,* S180–S191.

Elder, G. H., Jr., Shanahan, M. J., & Clipp, E. C. (1994). When war comes to men's lives: Life-course patterns in family, work, and health. *Psychology and Aging, 9,* 5–16.

Elias, M. (1986, April 17). Sex isn't the main lure for extra-marital affairs. *USA Today,* D1.

Elias, M. (1987, March 16). Divorce is easier on well-off kids. *USA Today,* D1.

Elias, M. (1989, August 9). Inborn traits outweigh environment. *USA Today,* 1D, 2D.

Elias, M. (1992). Late-life love. *Harvard Health Letter, 18,* 1–3.

Elias, M. (1993). Mind and menopause. *Harvard Health Letter, 19,* 1–3.

Elias, M. (1995, January 30). Social life and religion boost heart survival. *USA Today,* D1.

Elkind, D. (1970, April). Eric Erikson's eight stages of man. *New York Times Magazine,* 24.

Elkind, D. (1974). *A sympathetic understanding of the child from birth to sixteen.* Boston: Allyn & Bacon.

Elkind, D. (1979, February). Growing up faster. *Psychology Today, 12,* 38–45.

Elkind, D. (1987, May). Superkids and super problems. *Psychology Today, 21,* 60–61.

Elkind, D. (1994). *Parenting your teenager.* New York: Ballantine Books.

Elkind, D. (1995). *Ties that stress: The new family in balance.* Cambridge, MA: Harvard University Press.

Elliot, D. S., Wilson, W. J., Huizinga, D., Sampson, R. J., et al. (1996). The effects of neighborhood disadvantage on adolescent development. *Journal of Research in Crime and Delinquency, 33,* 389–426.

Ellis, D. G. (1992). *From language to communication.* Hillsdale, NJ: Erlbaum.

Emanuel, L. (1998). Facing requests for physician-assisted suicide: Toward a practical and principled clinical skill set. *Journal of the American Medical Association. 280,* 643–647.

Engen, T. (1991). *Odor sensation and memory.* New York: Praeger.

Engen, T., Lipsitt, L. P., & Kaye, H. (1963). Olfactory responses and adaptation in the human neonate. *Journal of Physiology and Psychology, 56,* 73–77.

Engen, T., Lipsitt, L. P., & Peck, M. B. (1974). Ability of newborn infants to discriminate sapid substances. *Developmental Psychology, 10,* 741–744.

Engle, L. (1998). When HIV is kid stuff: Body positive. The Body: An AIDS and HIV Information Resource (February). Retrieved July 29, 1998, from the World Wide Web: http://www.thebody.com/bp/ feb98/kid.html

Enstrom, J. E. (1984). Smoking and longevity studies. *Science, 225,* 878.

Epstein, S. H. (1983, October). Why do women live longer than men? *Science, 83,* 4, 30–31.

Erdley, C. A., & Dweck, C. S. (1993). Children's implicit personality theories as predictors of their social judgments. *Child Development, 64,* 863–878.

Ericcson, K. A. (1990). Peak performance and age: An examination of peak performance in sports. In P. B. and M. M. Baltes (Eds.), *Successful Aging: Perspectives froom the behavioral sciences.* New York: Cambridge University Press.

Ericcson, K. A., & Charness, N. (1994). Expert performance: Its structure and acquisition. *Journal of the American Psychological Association, 49,* 725–747.

Erikson, E. H. (1959). Identity and the life cycle. *Monograph, Psychological Issues* (Vol. 1). New York: International Universities Press.

Erikson, E. H. (1963). *Childhood and society.* New York: W. W. Norton.

Erikson, E. H. (1964). Inner and outer space: Reflections on womanhood. *Daedalus, 93,* 582–606.

Erikson, E. H. (1968a). *Identity: Youth and crisis.* New York: W. W. Norton.

Erikson, E. H. (1968b). Life cycle. In D. L. Sills (Ed.), *International encyclopedia of the social sciences* (Vol. 9). New York: Free Press and Macmillan.

Erikson, E. H. (1974). *Dimensions of a new identity.* New York: Norton.

Erikson, E. H. (1977). *Toys and reasons: Stages in the ritualization of experience.* New York: Norton.

Erikson, E. H. (1982). *The life cycle completed: A review.* New York: W. W. Norton.

Erikson, E. H., & Erikson, J. M. (1997). *The life cycle completed* (ext. version). New York: W. W. Norton.

Erikson, E. H., Erikson, J. M., & Kivnick, H. Q. (1986). *Vital involvement in old age.* New York: W. W. Norton.

Erikson, K. *(1986).* On work and alienation. *American Sociological Review, 51,* 1–8.

Ernst, C., & Angst, A. (1983). *Birth order.* New York: Springer-Verlag.

Executive summary of the clinical guidelines on the identification, evaluation, and treatment of overweight and obesity in adults. (1998). *Journal of the American Dietetic Association, 98,* 1178–1191.

Eyer, D. E. (1992, November 24). Infant bonding: A bogus notion. *Wall Street Journal,* A14.

F

Fabricius, W. V., & Wellman, H. M. (1983). Children's understanding of retrieval cue utilization. *Developmental Psychology, 19,* 15–21.

Faden, R. R., & Kass, N. E. (1993). Genetic screening technology: Ethical issues in access to tests by employers and health insurance companies. *Journal of Social Issues, 49,* 75–88.

Fagot, B. I. (1995). Parenting boys and girls. In M. H. Bornstein (Ed.), *Handbook of parenting* (Vol. 1). Hillsdale, NJ: Erlbaum.

Fagot, B. I., & Kavanagh, K. (1993). Parenting during the second year: Effects of children's age, sex, and attachment classification. *Child Development, 64,* 258–271.

Fagot, B. I., Leinbach, M. D., & Hagan, R. (1986). Gender labeling and the adoption of sex-typed behavior. *Developmental Psychology, 22,* 440–443.

Fairstein, L. A. (1993). *Sexual violence: Our war against rape.* New York: William Morrow.

Falbo, T., & Poston, D. L., Jr. (1993). The academic, personality, and physical outcomes of only children in China. *Child Development, 64,* 18–25.

Falk, P. (1989). Lesbian mothers: Psychosocial assumptions in family law. *American Psychologist, 44,* 941–947.

Fantz, R. L. (1963). Pattern vision in newborn infants. *Science, 140,* 296–297.

Fantz, R. L. (1966). Pattern discrimination and selective attention as determinants of perceptual development from birth. In A. H. Kidd & J. F. Rivoire (Eds.), *Perceptual development in children.* New York: International Universities Press.

Fantz, R. L., Fagan, J. F., & Miranda, S. B. (1975). Early visual selectivity. In L. B.

Cohen & P. Salapatek (Eds.), *Infant perception: From sensation to cognition* (Vol. 1). New York: Academic Press.

Fantz, R. L., & Miranda, S. B. (1975). Newborn infant attention to form of contour. *Child Development, 46,* 224–228.

Farrar, M. J., & Goodman, G. S. (1992). Developmental changes in event memory. *Child Development, 63,* 173–187.

Farrar, M. J., Raney, G. E., & Boyer, M. E. (1992). Knowledge, concepts, and inferences in childhood. *Child Development, 63,* 673–691.

Farver, J. A. M., & Branstetter, W. H. (1994). Preschoolers' prosocial responses to their peers' distress. *Developmental Psychology, 30,* 334–341.

Favazza, P. C. (1998). Preparing for children with disabilities in early childhood classrooms. *Early Childhood Special Education, 25,* 255–258.

Favazza, P. C., & Odom, S. L. (1996). Use of acceptance scale to measure attitudes of kindergarten-age children. *Journal of Early Intervention, 20,* 232.

Fay, R. E., Turner, C. F., Klassen, A. D., & Gagnon, J. H. (1989). Prevalence and patterns of same-gender sexual contact among men. *Science, 243,* 338–348.

Feather, N. T. (1990). *The psychological impact of unemployment.* New York: Springer-Verlag.

Feder-Feitel, L. (1997, February). Does she have a learning problem? *Child,* 50–53.

Fehr, B. (1996). *Friendship processes.* Thousand Oaks, CA: Sage.

Feinleib, J. A., & Michael, R. T. (1998). Reported changes in sexual behavior in response to AIDS in the United States. *Preventive Medicine, 27,* 400–411.

Feist, G. J., Bodner, T. E., Jacobs, J. F., Miles, M., & Tan, V. (1995). Integrating top-down and bottom-up structural models of subjective well-being: A longitudinal investigation. *Journal of Personality and Social Psychology, 68,* 138–150.

Feldhusen, J. F., Proctor, T. B., & Black, K. N. (1986). Guidelines for grade advancement of precocious children. *Roeper Review, 9,* 25–27.

Feldman, C. F. (1992). The new theory of theory of mind. *Human Development, 35,* 107–117.

Felson, R. B., & Zielinski, M. A. (1989). Children's self-esteem and parental support. *Journal of Marriage and the Family, 51,* 727–735.

Fernald, A. (1985). Four-month-old infants prefer to listen to motherese. *Infant Behavior and Development, 8,* 181–195.

Fernald, A. (1990). Intonation and communicative intent in mothers' speech to infants: Is the melody the message? *Child Development, 60,* 1497–1510.

Fernald, A., & Morikawa, H. (1993). Common themes and cultural variations in Japanese and American mothers' speech to infants. *Child Development, 64,* 637–656.

Ferris, T. (1991, December 15). A cosmological event. *New York Times Magazine,* 44.

Fertman, C. I., & Chubb, N. H. (1992). The effects of a psychoeducational program on adolescents' activity involvement, self-esteem, and locus of control. *Adolescence, 27,* 517–526.

Feshbach, S. (1970). Aggression. In P. H. Mussen (Ed.), *Carmichael's manual of child psychology* (3rd ed., Vol. 2). New York: Wiley.

Feyereisen, P., & de Lannoy, J. (1991). *Gestures and speech: Psychological investigations.* New York: Cambridge University Press.

Fiatarone, M. A., & Evans, W. J. (1993). The etiology and reversibility of muscle dysfunction in the aged. *Journals of Gerontology, 48* (Special Issue), 77–83.

Field, D. (1991). Continuity and change in personality in old age: Evidence from five longitudinal studies. *Journal of Gerontology: Psychological Sciences, 46,* P271–P274.

Field, D., & Millsap, R. E. (1991). Personality in advanced old age: Continuity or change? *Journal of Gerontology: Psychological Sciences, 46,* P299–P308.

Field, D., Minkler, M., Falk, R. F., & Leino, E. V. (1993). The influence of health on family contacts and family feelings in advanced old age: A longitudinal study. *Journal of Gerontology: Psychological Sciences, 48,* P18–P28.

Field, T. (1995). Psychologically depressed parents. In M. H. Bornstein (Ed.), *Handbook of parenting* (Vol. 4). Hillsdale, NJ: Erlbaum.

Field, T. (1998). *Depressed mothers and their newborns.* Miami: University of Miami School of Medicine.

Field, T. M., Scafidi, F., Pickens, J., Prodromidis, M., Pelaez-Nogueras, M., Torquati, J., Wilcox, H., Malphurs, J., Schanberg, S., & Kuhn, C. (1998). Polydrug-using adolescent mothers and their infants receiving early intervention. *Adolescence, 33,* 117–143.

Fields, C. M. (1981, September 9). Minors found able to decide on taking part in research. *Chronicle of Higher Education,* 7.

Fields, G. (1994, May 20). 1.6 million kids home alone. *USA Today,* A1.

Fiese, B. H., Hooker, K. A., Kotary, L., & Schwagler, J. (1993). Family rituals in the early stages of parenthood. *Journal of Marriage and the Family, 55,* 633–642.

Fincham, F. D., Beach, S. R. H., Moore, T., & Diener, C. (1994). The professional response to child sexual abuse: Whose interests are served? *Family Relations, 43,* 244–254.

Fine, S. , & Chance, J. (1998). *Fine beauty: Beauty basics and beyond for African-American women.* New York: Riverhead Books.

Finkelhor, D. (1988). The trauma of sexual abuse: Two models. In G. E. Wyatt & G. J. Powell (Eds.), *Lasting effects of child sexual abuse.* Newbury Park, CA: Sage.

Fiscella, K., Kitzman, H. J., Cole, R. E., Sidora, K. J., & Olds, D. (1998). Does child abuse predict adolescent pregnancy? *Pediatrics, 101,* 620–624.

Fischer, G. J. (1996). Deceptive, verbally coercive college males: Attitudinal predictors and lies told. *Archives of Sexual Behavior, 25,* 5.

Fischer, K. W., & Silvern, L. (1985). Stages and individual differences in cognitive development. *Annual Review of Psychology, 36,* 613–648.

Fishman, B. (1983). The economic behavior of step-families. *Family Relations, 32,* 359–366.

Fitzpatrick, M. A. (1988). *Between husbands and wives: Communication in marriage.* Newbury Park, CA: Sage.

Flaks, D. K., Ficher, I. Masterpasqua, F., & Joseph. G. (1995). Lesbians choosing motherhood: A comparative study of lesbian and heterosexual parents and their children. *Developmental Psychology, 31,* 105–114.

Flanagan, C. A., & Eccles, J. S. (1993). Changes in parents' work status and adolescents' adjustment at school. *Child Development, 64,* 246–257.

Flavell, J. H. (1992). Cognitive development: Past, present, and future. *Developmental Psychology, 28,* 998–1005.

Flavell, J. H., Flavell, E. R., & Green, F. L. (1983). Development of the appearance of reality distinction. *Cognitive Psychology, 15,* 95–120.

Flavell, J., Freidrichs, A., Hoyt, J. (1970). Developmental changes in memorization processes. *Cognitive Psychology, 1,* 324–340.

Fleischman, H. I., & Hopstock, P. J. (1993). *Descriptive study of services to limited English proficient students, Vol. 1. Executive summary.* Arlington, VA: Development Associates.

Focus on obesity. (1998). *Medical Sciences Bulletin.* Retrieved February 7, 1999, from the World Wide Web: http://pharminfo.com/pubs/msb/obesity.html

Fogel, A., & Thelen, E. (1987). Development of early expressive and communicative action: Reinterpreting the evidence from a dynamic systems perspective. *Developmental Psychology, 23,* 747–761.

Folkman, S., & Lazarus, R. S. (1985). If it changes it must be a process: Study of emotion and coping during three stages of a college examination. *Journal of Personality and Social Psychology, 48,* 150–170.

Folven, R. J., & Bonvillian, J. D. (1991). The transition from nonreferential to referential language in children acquiring American Sign Language. *Developmental Psychology, 27,* 806–816.

Foner, A., & Kertzer, D. (1978). Transitions over the life course: Lessons from age-set societies. *American Journal of Sociology, 83,* 1081–1104.

Forgione, P. D. (1998). *Pursuing excellence: A study of U.S. twelfth-grade mathematics and science achievement in international context.* National Center for Education Statistics (NCES). Retrieved December 10, 1998, from the World Wide Web: http://nces.edu/gov/timss/

Formanek, R. (1983, August 22). How children's fears are changing. *U.S. News & World Report,* 43–44.

Forste, R., & Tanfer, K. (1996). Sexual exclusivity among dating, cohabiting, and married women. *Journal of Marriage and the Family, 58,* 33–48.

Fowler, J. (1983). *Stages of faith: The psychology of human development and the quest for meaning.* San Francisco: Harper & Row.

Fox, B., & Routh, D. (1975). Analysing spoken language into words, syllables, and phonemes: A developmental study. *Journal of Psycholinguistic Research, 4,* 331–342.

Fox, J. A. (1996, March). *Trends in juvenile violence: A report to the United States Attorney General on current and future rates of juvenile offending.* Washington, DC: U.S. Department of Justice, Bureau of Justice Statistics.

Fozard, J. L., Vercruyssen, M., Reynolds, S. L., Hancock, P. A., & Quilter, R. E. (1994). Age differences and changes in reaction time: The Baltimore Longitudinal Study of Aging. *Journal of Gerontology: Psychological Sciences, 49,* P179–P189.

Fraiberg, S. H. (1959). *The magic years.* New York: Scribner's.

Fraker, S. (1984, April 16). Why women aren't getting to the top. *Fortune,* 40–45.

Frazier, P. A. (1990). Victim attributions and post-rape trauma. *Journal of Personality and Social Psychology, 59,* 298–304.

Freadhoff, C. (1992, August 28). America goes back to school. *Investor's Business Daily,* 1–2.

Freedman, A. M. (1998, October 14). Why teenage girls love "the shot"; Why others aren't so sure. *Wall Street Journal,* A1, A14.

Freeman, L. L. (1995, May 22). When success and distraction go together. *Investor's Business Daily,* 1–2.

Freeseman, L. J., Colombo, J., & Coldren, J. T. (1993). Individual differences in infant visual attention: Four-month-olds' discrimination and generalization of global and local stimulus properties. *Child Development, 64,* 1191–1203.

French, D. C. (1984). Children's knowledge of the social functions of younger, older, and same-age peers. *Child Development, 55,* 1429–1433.

Freud, A. (1936). *The ego and the mechanisms of defense.* New York: International Universities Press.

Freud, A. (1958). Adolescence. *Psychoanalytic Study of the Child, 13,* 255–278.

Freud, S. (1930/1961). *Civilization and its discontents.* London: Hogarth Press.

Freud, S. (1940). An outline of psychoanalysis. In J. Strachey (Ed. and Trans.), *The standard edition of the complete psychological works of Sigmund Freud.* London: Hogarth Press.

Fried, S. B., & Mehrotra, C. M. (1998). *Aging and diversity: An active learning experience.* Washington, DC: Taylor & Francis.

Friedberg, L. (1998). Did unilateral divorce raise divorce rates? Evidence from panel data. *American Economic Review, 88,* 608–627.

Friedman, H. S., Tucker, J. S., Schwartz, J. E., Tomlinson-Keasey, C., Martin, L. R., Wingard, D. L., & Criqui, M. H. (1995). Psychosocial and behavioral predictors of longevity: The aging and death of the "Termites." *American Psychologist, 50,* 69–78.

Friedman, M. A., & Brownell, K. D. (1995). Psychological correlates of obesity: Moving to the next research generation. *Psychological Bulletin, 117,* 3–20.

Friend, R. A. (1980). Gayaging: Adjustment and the older gay male. *Alternative Lifestyles, 3,* 231–248.

Friend, R. A. (1991). Older lesbian and gay people: A theory of successful aging. In J. A. Lee (Ed.), *Gay midlife and maturity* (pp. 99–118). New York: Haworth.

Friend, T. (1995, January 19). Emotional ties can help heal heart patients. *USA Today,* D1.

Fries, J. F. (1989). The compression of morbidity: Near or far? *Milbank Quarterly, 67*, 208–231.

Fries, J. F. (1997). Can preventive gerontology be on the way? *American Journal of Public Health, 87*, 1591–1593.

Frisch, R. E. (1978, June 30). Menarche and fatness. *Science, 200*, 1509–1513.

Fromm, E. (1941). *Escape from freedom.* New York: Avon.

Fry, D. P. (1988). Intercommunity differences in aggression among Zapotec children. *Child Development, 59*, 1008–1019.

Fry, D. P. (1993). The intergenerational transmission of disciplinary practices to conflict. *Human Organization, 52*, 176–185.

Fuchs, D., & Thelen, M. H. (1988). Children's expected interpersonal consequences of communicating their affective state and reported likelihood of expression. *Child Development, 59*, 1314–1322.

Fuchs-Beauchamp, K. (1996). Preschoolers inferred self-esteem. *Journal of Genetic Psychology, 157*, 204–210.

Fuligni, A. J., & Eccles, J. S. (1993). Perceived parent-child relationships and early adolescents' orientation toward peers. *Developmental Psychology, 29*, 622–632.

Fuligni, A. J., & Stevenson, H. W. (1995). Time use and mathematics achievement among American, Chinese, and Japanese high school students. *Child Development, 66*, 830–842.

Funder, D. C., & Colvin, C. R. (1991). Explorations in behavioral consistency: Properties of persons, situations, and behaviors. *Journal of Personality and Social Psychology, 60*, 773–794.

Furstenberg, F. F., Jr., & Cherlin, A. J. (1991). *Divided families.* Cambridge, MA: Harvard University Press.

Futterman, A., Gallagher, D., Thompson, L. W., Lovett, S., & Gilewski, M. (1990). Retrospective assessment of marital adjustment and depression during the first 2 years of spousal bereavement. *Psychology and Aging, 5*, 277–283.

G

Gaddis, A., & Brooks-Gunn, J. (1985). The male experience of pubertal change. *Journal of Youth and Adolescence, 14*, 61–69.

Gaertner, S. L., Mann, J. A., Dovidio, J. F., Murrell, A. J., & Pomare, M. (1990). How does cooperation reduce intergroup bias? *Journal of Personality and Social Psychology, 59*, 692–704.

Gafner, G., & Duckett, S. (1992). Treating the sequelae of a curse in elderly Mexican Americans. In T. L. Brink (Ed.), *Hispanic aged mental health* (pp. 145–153). New York: Haworth Press.

Gallatin, J. (1985). *Democracy's children: The development of political thinking in adolescents.* Ann Arbor, MI.: Quod.

Gallup Poll Monthly. (1991, January). Princeton, NJ: Gallup Organization.

Ganong, L. H., Coleman, M., & Mapes, D. (1990). A meta-analytic review of family structure stereotypes. *Journal of Marriage and the Family, 52*, 287–297.

Ganong, L., Coleman, M., McDaniel, A. K., & Killian, T. (1998). Attitudes regarding obligations to assist an older parent or stepparent following later-life remarriage. *Journal of Marriage and the Family, 60*, 595–610.

Garcia, A. (1993). Income security and elderly Latinos. In M. Sotomayer & A. Garcia (Eds.), *Elderly Latinos: Issues and solutions for the 21st century* (pp. 17–28). Washington, DC: National Hispanic Council on Aging.

Garcia, E. (1994). *Understanding and meeting the challenge of student cultural diversity.* Boston: Houghton Mifflin.

Garcia, J. (1985). A needs assessment of elderly Hispanics in an inner-city senior citizen complex: Implications for practice. *Journal of Applied Gerontology, 4*, 72–85.

Garcia, J., Kosberg, J., Mangum, W., Henderson, J., & Henderson, C. (1992, November). *Caregiving for and by Hispanic elders: Perceptions of four generations of women.* Paper presented at the annual meeting of the Gerontological Society of America. Washington, DC.

Gardner, H. (1983). *Frames of mind: The theory of multiple intelligences.* New York: Basic Books.

Gardner, H. (1991). *The unschooled mind: How children think and how schools should teach.* New York: Basic Books.

Gardner, H. (1993a). Lessons in life from creative geniuses: Freud, Einstein, Picasso, Stravinsky, Eliot, Graham and Gandhi. *Boardroom Reports, 22*, 13.

Gardner, H. (1993b). *Multiple intelligences: The theory in practice.* New York: Basic Books.

Gardner, H. (1997). Multiple intelligences as a partner in school improvement. *Educational Leadership, 55*, 20–21.

Gardner, J. W., & Lyon, J. L. (1982). Cancer in Utah Mormon women by church activity level. *American Journal of Epidemiology, 116*, 258–165.

Gardner, P., & Hudson, B. L. (1996). *Advance report on final mortality statistics, 1993.* Hyattsville, MD: National Center for Health Statistics.

Garland, A. F., & Zigler, E. (1993). Adolescent suicide prevention: Current research and social policy implications. *American Psychologist, 48*, 169–182.

Garner, D. M., & Garfinkel, P. E. (Eds.). (1985). *Handbook of psychotherapy for anorexia nervosa and bulimia.* New York: Guilford Press.

Garvey, C., & Hogan, R. (1973). Social speech and social interaction: Egocentrism revisited. *Child Development, 44*, 562–568.

Ge, X., Lorenz, F. O., Conger, R. D., Elder, G. H., Jr., & Simons, R. L. (1994). Trajectories of stressful life events and depressive symptoms during adolescence. *Developmental Psychology, 30*, 467–483.

Geber, M., & Dean, R. (1957a). Gesell tests on African children. *Pediatrics, 20*, 1055–1065.

Geber, M., & Dean, R. (1957b). The state of development of newborn African children. *Lancet, 1*, 1216–1219.

Geiger, G., & Lettvin, J. Y. (1987). Peripheral vision in persons with dyslexia. *New England Journal of Medicine, 316*, 1238–1243.

Gelfand, D. E. (1994). *Aging and ethnicity: Knowledge and services.* New York: Springer.

Gelfand, D. M., & Teti, D. M. (1990). The effects of maternal depression on children. *Clinical Psychology Review, 10*, 329–353.

Gelles, R. J., & Conte, J. R. (1990). Domestic violence and sexual abuse of children: A review of research in the eighties. *Journal of Marriage and the Family, 52*, 1045–1058.

Gelman, D. (1983, November 7). A great emptiness. *Newsweek*, 120–126.

Gelman, D. (1993, April 12). Mixed message. *Newsweek*, 28–29.

Gelman, R., & Meck, E. (1986). The notion of principle: The case of counting. In J. Hiebert (Ed.), *Conceptual and procedural knowledge: The case of mathematics.* Hillsdale, NJ: Erlbaum.

Gelman, S. A., & Kremer, K. E. (1991). Understanding natural cause: Children's explanations of how objects and their properties originate. *Child Development, 62*, 396–414.

Gendell, M. (1998). Trends in retirement age in four countries, 1965–1995. *Monthly Labor Review, 121*, 20–30.

General information about learning disabilities, Fact sheet no. 7. (1997). National Information Center for Children and Youth with Disabilities. Retrieved December 16, 1998, from the World Wide Web: http://www.nichcy.org

Gerwitz, J. L. (1972). *Attachment and dependency.* Washington, DC: Winston.

Gesell, A. (1928). *Infancy and human growth.* New York: Macmillan.

Giambra, L. M., Arenberg, D., Zonderman, A. B., Kawas, C., & Costa, P. T., Jr. (1995). Adult life span changes in immediate visual memory and verbal intelligence. *Psychology and Aging, 10,* 123–139.

Gibbs, J. C., & Schnell, S. V. (1985). Moral development "versus" socialization. *American Psychologist, 40,* 1071–1080.

Gibbs, J. T. (1997). African-American suicide: A cultural paradox. *Suicide and Life-Threatening Behavior, 27,* 68–79.

Gibson, E. J. (1969). *Principles of perceptual learning and development.* New York: Appleton-Century-Crofts.

Gibson, E. J., & Walk, R. D. (1960, April). The "visual cliff." *Scientific American, 202,* 64–71.

Gidding, S., Leibel, R. L., Daniels, S., Rosenbam, M., Horn, L. V., & Marx, G. (1996). Understanding obesity in youth. *Circulation, 94,* 33–83.

Giese, W. (1987, June 15). Four of 5 of us say we enjoy going to work. *USA Today,* 1, 2.

Gil, R. M., & Vazquez, C. I. (1996). *The Maria paradox: How Latinas can merge old-world traditions with new-world self-esteem.* New York: Putnam.

Giles-Sims, J., Straus, M. A., & Sugarman, D. B. (1995). Child, maternal, and family characteristics associated with spanking. *Family Relations, 44,* 170–176.

Gilewski, M. J., Farberow, N. L., Gallagher, D. E., & Thompson, L. W. (1991). Interaction of depression and bereavement on mental health in the elderly. *Psychology and Aging, 6,* 67–75.

Gilgun, J. F. (1995). We shared something special: The moral discourse of incest perpetrators. *Journal of Marriage and the Family, 57,* 265–281.

Gilinsky, A. S., & Judd, B. B. (1994). Working memory and bias in reasoning across the life span. *Psychology and Aging, 9,* 356–371.

Gilligan, C. (1982a). *In a different voice: Psychological theory and women's development.* Cambridge, MA: Harvard University Press.

Gilligan, C. (1982b, June). Why should a woman be more like a man? *Psychology Today, 16,* 68–77.

Gilligan, C., Rogers, A. G., & Tolman, D. L. (Eds.). (1991). *Women, girls & psychotherapy: Reframing resistance.* New York: Haworth Press.

Gilmore, D. D. (1990). *Manhood in the making: Cultural concepts of masculinity.* New Haven, CT: Yale University Press.

Ginsburg, G. S., & Bronstein, P. (1993). Family factors related to children's intrinsic/extrinsic motivational orientation and academic performance. *Child Development, 64,* 1461–1474.

Giovannucci, E. (1998). Selenium and risk of prostate cancer. *Lancet, 352,* 755–756.

Giovannucci, E., Rimm, E. B., Colditz, G. A., Stampfer, M. J., Ascherio, A., Chute, C. C., & Willett, W., C. (1993). A prospective study of dietary fat and risk of prostate cancer. *Journal of the National Cancer Institute, 85,* 1571–1579.

Glazer, G. (1998). *Health outcomes— Transition to midlife for African American and Caucasian women.* National Institutes of Health. Retrieved September 10, 1998, from the World Wide Web: http://www.nih.gov/silk/arisabst/fy1997/P0007298.html

Gleason, J. B. (1993a). *The development of language.* New York: Macmillan.

Gleason, J. B. (1993b). The neglected role of fathers in children's communicative development. *Seminar in Speech and Language, 14,* 314–324.

Glenn, N. D. (1990). Quantitative research on marital quality in the (1980)s: A critical review. *Journal of Marriage and the Family, 52,* 818–831.

Glenn, N. D. (1992). What does family mean? *American Demographics, 14,* 30–37.

Glover, V. (1997). Maternal stress or anxiety in pregnancy and emotional development of the child. *British Journal of Psychiatry, 171,* 105–106.

Gluckman, M. (1955). *Custom and conflict in Africa.* Oxford: Blackwell.

Glueck, S., & Glueck, E. (1957). Working mothers and delinquency. *Mental Hygiene, 41,* 327–352.

Gohm, C. L., Oishi, O., & Darlington, J. (1998). Culture, parental conflict, parental marital status, and the subjective well-being of young adults. *Journal of Marriage and the Family, 60,* 319–334.

Golant, S. M. (1984). *A place to grow old: The meaning of environment in old age.* New York: Columbia University Press.

Gold, D., Woodbury, M., & George, L. (1990). Relationship classification using grade of membership analysis: A typology of sibling relationships in later life. *Journal of Gerontology: Social Sciences, 45,* S43–S51.

Goldberg, L. R. (1993). The structure of phenotypic personality traits. *American Psychologist, 48,* 26–34.

Goldfarb, W. 1945. Psychological privation in infancy and subsequent adjustment. *American Journal of Orthopsychiatry, 15,* 247–255.

Goldin-Meadow, S., & Feldman, H. (1977). The development of language-like communication without a language model. *Science, 197,* 401–403.

Goldin-Meadow, S., & Mylander, C. (1984). Gestural communication in deaf children. *Monographs of the Society for Research in Child Development 49* (Serial No. 207).

Goldscheider, F. K., & Goldscheider, C. (1993a). *Leaving home before marriage: Ethnicity, familism, and generational relationships.* Madison: University of Wisconsin Press.

Goldscheider, F. K., & Goldscheider, C. (1993b). Whose nest? A two-generational view of leaving home during the 1980s. *Journal of Marriage and the Family, 55,* 851–862.

Goldsmith, H. H. (1997). Toddler and childhood temperament. *Developmental Psychology, 33,* 891–905.

Goldsmith, H. H., & Alansky, S. A. (1987). Maternal and infant temperamental predictors of attachment. *Journal of Consulting & Clinical Psychology, 55,* 805–816.

Goldstein, A., Damon, B., & Taeuber, C. M. (1993, September). *We the American elderly.* U.S. Bureau of the Census. Retrieved February 12, 1999, from the World Wide Web: http://www.census.gov/apsd/wepeople/we-9.pdf

Goleman, D. (1985, May 28). Spacing of siblings strongly linked to success in life. *New York Times,* 17, 18.

Goleman, D. (1986, December 2). Major personality study finds that traits are mostly inherited. *New York Times,* 17, 18.

Goleman, D. (1989a, February 7). For many, turmoil of aging erupts in the 50's, studies find. *New York Times,* 17, 21.

Goleman, D. (1989b, August 8). New studies find many myths about mourning. *New York Times,* 17.

Goleman, D. (1990a, March 6). As a therapist, Freud fell short, scholars find. *New York Times,* B5, B9.

Goleman, D. (1990b, February 6). In midlife, not just crisis but care and comfort, too. *New York Times,* B1, B8.

Goleman, D. (1990c, January 16). Men at 65: New findings on well-being. *New York Times,* 19, 23.

Goleman, D. (1992, November 24). Anthropology goes looking for love in all the old places. *New York Times,* B1.

Goleman, D. (1993, March 17). Each year, more than 1 in 4 U.S. adults suffers a mental disorder. *New York Times,* B7.

Goleman, D. (1995). *Emotional intelligence: Why it can matter more than IQ.* New York: Bantam.

Goode, E. E. (1991, June 24). Where emotions come from. *U.S. News & World Report,* 54–62.

Goode, W. J. (1959). The theoretical importance of love. *American Sociological Review, 24,* 38–47.

Goodman, M. (1996). Culture, cohort, and cosmetic surgery. *Journal of Women and Aging, 8,* 55–58.

Goran, M. I., Nagy, T. R., Gower, B. A., Mazariegos, M., et al. (1998). Influence of sex, seasonality, ethnicity, and geographic location on the components of total energy expenditure in young children: Implications for energy requirements. *American Journal of Clinical Nutrition, 68,* 675–682.

Goran, M. I., Shewchuk, R., Gower, B. A., Nagy, T. R., et al. (1998). Longitudinal changes in fatness in white children: No effect of childhood energy expenditure. *American Journal of Clinical Nutrition, 67,* 309–316.

Goran, M. I., & Sun, M. (1998). Total energy expenditure and physical activity in prepubertal children: Recent advances based on the application of the doubly labeled water method. *American Journal of Clinical Nutrition, 68,* 944S–949S.

Gorsky, R. D., Farnham, P. G., & Straus, W. L. (1996). Preventing perinatal transmission of HIV—Costs and effectiveness of a recommended intervention. *Public Health Reports, 111,* 335–341.

Goslin, D. A. (1969). *Handbook of socialization theory and research.* Chicago: Rand McNally.

Gottlieb, G. (1991). Experiential canalization of behavioral development: *Theory. Developmental Psychology, 27,* 4–13.

Gough, E. K. (1960). Is the family universal? The Nayar case. In N. W. Bell & E. F. Vogel (Eds.), *A modern introduction to the family.* New York: Free Press.

Gould, C. G. (1983, April). Out of the mouths of beasts. *Science 83, 4,* 69–72.

Graber, J. A., Brooks-Gunn, J., Paikoff, R. L., & Warren, M. P. (1994). Prediction of eating problems: An 8-year study of adolescent girls. *Developmental Psychology, 30,* 823–834.

Graber, J. A., Brooks-Gunn, J., & Warren, M. P. (1995). The antecedents of menarcheal age: Heredity, family environment, and stressful life events. *Child Development, 66,* 346–359.

Graham, E. (1995, May 9). Working parents' torment: Teens after school. *Wall Street Journal,* B1, B5.

Graham-Bermann, S., & Levendovsky, A. (1998) Traumatic stress syndrome in children of battered women. *Journal of Interpersonal Violence, 13,* 111–128.

Gralinski, J. H., & Kopp, C. B. (1993). Everyday rules for behavior: Mothers' requests to young children. *Developmental Psychology, 29,* 573–584.

Granrud, C. (Ed.). (1993). *Visual perception and cognition in infancy.* Hillsdale, NJ: Erlbaum.

Gratch, G., and Schatz, J. A. (1988). Evaluating Piaget's infancy books as works-in-progress. *Human Development, 31,* 82–91.

Greenberg, J. (1981, July 3). Study finds widowers die more quickly than widows. *New York Times,* 1, 7.

Greenberg, J., Solomon, S., Pyszczynski, T., Rosenblatt, A., Burling, J., Lyon, D., Simon, L., & Pinel, E. (1992). Why do people need self-esteem? Converging evidence that self-esteem serves an anxiety-buffering function. *Journal of Personality and Social Psychology, 63,* 913–922.

Greenberger, E., & Steinberg, L. (1983). Sex differences in early labor force experience: Harbinger of things to come. *Social Forces, 62,* 467–486.

Greene, J. G. (1984). *The social and psychological origins of the climacteric syndrome.* Brookfield, VT: Gower.

Greene, M. F. (1998, May). How to raise septuplets. *Life, 21,* 39–48.

Greenfield, P. M. (1966). On culture and conservation. In J. Bruner, R. R. Olver, & P. M. Greenfield (Eds.), *Studies in cognitive growth.* New York: Wiley.

Greeno, J. G. (1989). A perspective on thinking. *American Psychologist, 44,* 134–141.

Greenspan, S. I. (1991). *Infancy and early childhood: The practice of clinical assessment and intervention with emotional and developmental challenges.* Madison, CT: International Universities Press.

Greenspan, S., & Greenspan, N. T. (1985). *First feelings.* New York: Viking Press.

Greif, E. B., & Ulman, K. J. (1982). The psychological impact of menarche on early adolescent females: A review of the literature. *Child Development, 53,* 1413–1430.

Griswold, R. L. (1993). *Fatherhood in America: A history.* New York: Basic Books.

Grolnick, W. S., & Slowiaczek, M. L. (1994). Parents' involvement in children's schooling: A multidimensional conceptualization and motivational model. *Child Development, 65,* 237–252.

Gross, D. (1990, February 14). Roses are red, violets blue. . . . *New York Times,* A19.

Gross, J. (1992, December 7). Divorced, middle-aged and happy: Women, especially, adjust to the 90's. *New York Times,* A8.

Grusec, J. E. (1992). Social learning theory and developmental psychology: The legacies of Robert Sears and Albert Bandura. *Developmental Psychology, 28,* 776–786.

Grusec, J. E., & Goodnow, J. J. (1994). Impact of parental discipline methods on the child's internalization of values: A reconceptualization of current points of view. *Developmental Psychology, 30,* 4–19.

Grusec, J. E., Kuczysnki, L., Rushton, J. P., & Simutis, Z. M. (1979). Learning resistance to temptation through observation. *Developmental Psychology, 15,* 233–240.

Grych, J. H., & Fincham, F. D. (1992). Interventions for children of divorce: Toward greater integration of research and action. *Psychological Bulletin, 111,* 434–454.

Guidubaldi, J., & Perry, J. D. (1985). Divorce and mental health sequelae for children: A two-year follow-up of a nationwide sample. *Journal of the American Academy of Child Psychiatry, 24,* 531–537.

Guilford, J. P. (1967). *The nature of human intelligence.* New York: McGraw-Hill.

Guillemin, R. (1982). Growth hormone-releasing factor from a human pancreatic tumor that caused acromegaly. *Science, 218,* 583–587.

Guillen, M. A. (1984, April). The I and the beholder. *Psychology Today, 18,* 68–69.

Guisinger, S., & Blatt, S. J. (1994). Individuality and relatedness: Evolution of a fundamental dialectic. *American Psychologist, 49,* 104–111.

Gullone, E., & King, N. J. (1993). The fears of youth in the 1990s: Contemporary normative data. *Journal of Genetic Psychology, 154,* 137–153.

Gullotta, T. P., Adams, G. R., and Montemayor, R. (Eds.). (1993). *Adolescent sexuality.* Newbury Park, CA: Sage.

Guo, S., Wu, G., Wang, N., et al. (1994). Follow-up study on children with obesity and hypertension. *Chinese Medical Journal, 77,* 18.

Gupta, A. K., & First, E. R. (1998). Community Outreach Health Information System (COHIS). Boston University Medical Center. Retrieved August 10, 1998, from the World Wide Web: http://gopher1.bu.edu/COHIS/aids/risks.htm

Gupta, U., & Singh, P. (1982). An exploratory study of love and liking and type of marriage. *Indian Journal of Applied Psychology, 19,* 92–97.

Gurian, M. (1996). *The wonder of boys: What parents, mentors, and educators can do to shape boys into exceptional men.* New York: Putnam.

Gutmann, D. L. (1987). *Reclaimed powers: Toward a new psychology of men and women in later life.* New York: Basic Books.

Guttman, J. (1993). *Divorce: Theory and research.* Hillsdale, NJ: Erlbaum.

H

Haith, M. M., & Benson, J. B. (1998). Infant cognition. In W. Damon (Ed.), *Handbook of child psychology* (5th ed). New York: Wiley.

Haith, M. M., & Goodman, G. S. (1982). Eye-movement control in newborns in darkness and in unstructured light. *Child Development, 53,* 974–977.

Haith, M. M., Bergman, T., & Moore, M. J. (1977). Eye contact and face scanning in early infancy. *Science, 198,* 853–854.

Hajela, D. (1998, September 3). Morning-after contraceptive pill to hit market this month. *Press & Sun-Bulletin,* C1.

Halford, G. S., Maybery, M. T., O'Hare, A. W., & Grant, P. (1994). The development of memory and processing capacity. *Child Development, 65,* 1338–1356.

Hall, D. G. (1991). Acquiring proper nouns for familiar and unfamiliar animal objects: Two-year-olds' word-learning biases. *Child Development, 62,* 1142–1154.

Hall, G. S. (1904). *Adolescence: Its psychology, and its relations to physiology, anthropology, sociology, sex, crime, religion, and education.* New York: Appleton.

Hall, T. (1987, June 1). Infidelity and women: Shifting patterns. *New York Times,* 20.

Hallinan, M. T., & Williams, R. A. (1989). Interracial friendship choices in secondary schools. *American Sociological Review, 54,* 67–78.

Halpern, D. F. (1992). *Sex differences in cognitive abilities* (2nd ed.). Hillsdale, NJ: Erlbaum.

Halpern, D. F. (1997). Sex differences in intelligence: Implications for education. *American Psychologist, 52,* 1091–1102.

Haltiwanger, J., & Harter, S. (1988). *A behavioral measure of young children's presented self-esteem.* Unpublished manuscript, University of Denver.

Halverson, H. M. (1931). An experimental study of prehension in infants by means of systematic cinema records. *Genetic Psychology Monographs, 10,* 107–286.

Hamilton, V. L., Blumenfeld, P. C., & Kushler, R. H. (1988). A question of standards: Attributions of blame and credit for classroom acts. *Journal of Personality and Social Psychology, 54,* 34–48.

Hamilton, V. L., Blumenfeld, P. C., Akoh, H., & Miura, K. (1990). Credit and blame among American and Japanese children: Normative, cultural, and individual differences. *Journal of Personality and Social Psychology, 59,* 442–451.

Hamilton, V. L., Broman, C. L., Hoffman, W. S., & Renner, D. S. (1990). Hard times and vulnerable people: Initial effects of plant closing on autoworkers' mental health. *Journal of Health and Social Behavior, 31,* 123–140.

Hamilton, V. L., Hoffman, W. S., Broman, C. L., & Rauma, D. (1993). Unemployment, distress, and coping: A panel study of autoworkers. *Journal of Personality and Social Psychology, 65,* 234–247.

Hammond, W. R., & Yung, B. (1993). Psychology's role in the public health response to assaultive violence among young African-American men. *American Psychologist, 48,* 142–154.

Haney, D. Q. (1985, February 3). Creativity is fragile and easily stifled. *Columbus, Ohio, Dispatch,* C1.

Hankin, B. L., Abramson, L.Y., Moffitt, T. E., Silva, P. A. (1998). Development of depression from preadolescence to young adulthood: Emerging gender differences in a 10-year longitudinal study. *Journal of Abnormal Psychology, 107,* 128–140.

Hänninen, V., & Aro, H. (1996). Sex differences in coping and depression among young adults. *Social Science & Medicine, 43,* 1453–1460.

Hansson, R. O., & Carpenter, B. N. (1994). *Relationships in old age: Coping with the challenge of transitions.* New York: Guilford Press.

Hardy, J. A., & Higgins, G. A. (1992). Alzheimer's disease: The amyloid cascade hypothesis. *Science, 256,* 184–185.

Hare, J., & Richards, L. (1993). Children raised by lesbian couples. *Family Relations, 42,* 249–255.

Hareven, T. K. (1982). *Family time and industrial time.* New York: Cambridge University Press.

Hareven, T. K. (1987). Historical analysis of the family. In M. E. Sussman & S. K. Steinmetz (Eds.), *Handbook of marriage and the family.* New York: Plenum Press.

Harkness, S. (1992). Cross-cultural research in child development: A sample of the state of the art. *Developmental Psychology, 28,* 622–625.

Harlow, H. F. (1971). *Learning to love.* San Francisco: Albion.

Harnish, J. E., Dodge, K. A., & Valenta, E. (1995). Mother-child interaction quality as a partial mediator of the rules of maternal depressive symptomatology and socioeconomic status in the development of child behavior problems. *Child Development, 66,* 739–753.

Harris, C. (1993). Osteoarthritis: How to diagnose and treat the painful joint. *Geriatrics, 48* (8), 42–46.

Harris, K. M. (1996). Poverty, paternal involvement, and adolescent well-being. *Journal of Family Issues, 17,* 614–640.

Harris, K. M., & Morgan, S. P. (1991). Fathers, sons, and daughters: Differential paternal involvement in parenting. *Journal of Marriage and the Family, 53,* 531–544.

Harris, M. B. (1994). Growing old gracefully: Age concealment and gender. *Journal of gerontology: Psychological sciences, 49,* P149–P158.

Harris, M. J., Milich, R., Corbitt, E. M., Hoover, D. W., & Brady, M. (1992). Self-fulfilling effects of stigmatizing information on children's social interactions. *Journal of Personality and Social Psychology, 63,* 41–50.

Harris, S. B. (1996). For better or worse: Spouse abuse grown old. *Journal of Elder Abuse & Neglect, 8(1),* 1–33.

Hart, B., & Risley, T. R. (1992). American parenting of language-learning children: Persisting differences in family-child interactions observed in natural home environments. *Developmental Psychology, 28,* 1096–1105.

Hart, D., & Chmiel, S. (1992). Influence of defense mechanisms on moral judgment development: A longitudinal study. *Developmental Psychology, 28,* 722–730.

Harter, S. (1983). Causes and consequences of low self-esteem in children and adolescents. In R. Baumeister (Ed.), Self-Esteem: *The puzzle of low self-regard* (pp. 87–1170). New York: Plenum Press.

Harter, S. (1998). A model of the effects of perceived parent and peer support on adolescent false self-behavior. *Child Development, 67,* 360.

Harter, S., & Monsour, A. (1992). Developmental analysis of conflict caused by opposing attributes in the adolescent self-portrait. *Developmental Psychology, 28,* 251–260.

Harter, S., & Pike, R. (1984). The pictorial scale of perceived competence and social acceptance for young children, *Child Development, 55,* 1969–1982.

Hartshorne, H., & May, M. A. (1928). *Studies in the nature of character.* Vol. 1: *Studies in deceit.* New York: Macmillan.

Hartup, W. W. (1983). Peer relations. In P. H. Mussen & E. M. Hetherington, *Handbook of child psychology* (Vol. 3). New York: Wiley.

Hartup, W. W., & Van Lieshout, C. F. (1995). Personality development in social context. *Annual Review of Psychology, 46,* 655–687.

Harwood, R. L. (1992). The influence of culturally derived values on Anglo and Puerto Rican mothers' perceptions of attachment behavior. *Child Development, 63,* 822–839.

Hashima, P. Y., & Amato, P. R. (1994). Poverty, social support, and parental behavior. *Child Development, 65,* 394–403.

Haslett, B. J., Geis, G. L., & Carter, M. R. (1992). *The organizational woman: Power and paradox.* Norwood, NJ: Ablex.

Hattie, J. (1992). *Self-concept.* Hillsdale, NJ: Erlbaum.

Hawley, T. L., & Disney, E. R. (1992, Winter). Crack's children: The consequences of maternal cocaine abuse. *Social Policy Report: Society for Research in Child Development, 6.*

Hayes, S. C., & Hayes, L. J. (1992). Verbal relations and the evolution of behavior analysis. *American Psychologist, 47,* 1383–1395.

Hayflick, L. (1980, January). The cell biology of aging. *Scientific American, 242,* 58–65.

Haynes, H., Rovee-Collier, C., & Perris, E. E. (1987). Categorization and memory retrieval by three-month-olds. *Child Development, 58,* 750–767.

Hays, L. (1995, April 24). PCs may be teaching kids the wrong lessons. *Wall Street Journal,* B1, B7.

Hayslip, B., Shore, R. J., Henderson, C., & Lambert, P. (1998). Custodial grandparenting and the impact of grandchildren with problems on role satisfaction and role meaning. *Journal of Gerontology, 53B,* S164–S173.

Hayward, C., Killen, J. D., Wilson, D. M., Hammer, L. D., Litt, I. F., Kraemer, H. C., Haydel, F., Varady, A., & Taylor, C. B. (1997). Psychiatric risk associated with early puberty in adolescent girls. *Journal of the American Academy of Child and Adolescent Psychiatry, 36,* 255–262.

Hayward, M. D., Crimmins, E. M., & Wray, L. A. (1994). The relationship between retirement life cycle changes and older men's labor force participation rates. *Journal of Gerontology: Social Sciences, 49,* S219–S230.

Hazen, N. L., & Black, B. (1989). Preschool peer communication skills: The role of social status and interaction context. *Child Development, 60,* 867–876.

Health watch Rx for risk: Poor young black females the group hit fastest by AIDS. (1998, August 28). *Atlanta Constitution,* 1.

Healy, M. (1985, August 27). Menopause doesn't harm health. *USA Today,* D1.

Heer, D. M. (1985). Effects of sibling number on child outcome. *Annual Review of Sociology, 11,* 27–47.

Heidrich, S. M., & Ryff, C. D. (1993). Physical and mental health in later life: The self-system as mediator. *Psychology and Aging, 8,* 327–338.

Heimann, M. (1989). Neonatal imitation, gaze aversion, and mother-infant interaction. *Infant Behavior and Infant Development, 12,* 495–505.

Heinrichs, C., Munson, P. J., Counts, D. R., Cutler, G. B., Jr., & Baron, J. (1995). Patterns of human growth. *Science, 268,* 442–445.

Heinstein, M. I. (1963). Behavioral correlations of breast-bottle regimens under varying parent-child relationships. *Monographs of the Society for Research in Child Development, 28* (4).

Helburn, S., Culkin, M. L., Morris, J., Moran, N., Howes, C., Phillipsen, L., Bryant, D., Clifford, R., Cryer, D., Peisner-Feinberg, E., Burchinal, M., Kagan, S. L., & Rustici, J. (1995). *Cost, quality, and child outcomes in child care centers.* Denver: University of Colorado, Economics Department.

Helfer, R. E., & Kempe, C. H. (1977). *The battered child.* Chicago: University of Chicago Press.

Hellmich, N. (1994, July 13). Size, gender, income linked. *USA Today,* B1.

Hellmich, N. (1995, February 1). Exercise many options for life. *USA Today,* D1.

Helson, R. (1997). The self in middle age. In M. E. Lachman & J. B. James (Eds.), *Multiple paths of midlife development* (pp. 21–43). Chicago: University of Chicago Press.

Helson, R., & Wink, P. (1992). Personality change in women from the early 40s to early 50s. *Psychology and Aging, 7,* 46–55.

Helwig, C. C. (1995). Adolescents' and young adults' conceptions of civil liberties: Freedom of speech and religion. *Child Development, 66,* 152–166.

Hendin, H. (1994, December 16). Scared to death of dying. *New York Times,* A19.

Hendin, J., Rutenfrans, C., & Zylicz, Z. (1997). Physician-assisted suicide and euthanasia in the Netherlands. *Journal of*

the *American Medical Association, 277,* 1720–1722.

Hendrick, S., & Hendrick, C. (1992). *Romantic love.* Newbury Park, CA: Sage.

Henneberger, M., & Marriott, M. (1993, July 11). For some, youthful courting has become a game of abuse. *New York Times,* 1, 14.

Henry, T. (1993, September 16). TV gets blame for poor reading. *USA Today,* D1.

Henry, W. A., III. (1994, June 27). Pride and prejudice. *Time,* 54–59.

Herdt, G. (Ed.). (1989). *Gay and lesbian youth.* New York: Haworth Press.

Herdt, G., & Boxer, A. (1993). *Children of horizons: How gay and lesbian teens are leading a new way out of the closet.* Boston: Beacon Press.

Herman, J., & Hirschman, L. (1981). Families at risk for father-daughter incest. *American Journal of Psychiatry, 138,* 967–970.

Herman-Giddens, M. E., Slora, E. J., Wasserman, R. C., Bourdony, C. J., Bhapkar, M. V., Koch, G. G., & Hasemeier, C. M. (1997). Secondary sexual characteristics and menses in young girls seen in office practice. *Pediatrics, 99,* 505–512.

Hernandez, D. J. (1993a). *America's children: Resources from family, government, and the economy.* New York: Russell Sage Foundation.

Hernandez, D. J. (1993b). *We, the American children.* U.S. Bureau of the Census, WE-10. Washington, DC: Government Printing Office.

Hernandez, D. J. (1997). Child development and the social demography of childhood. *Child Development, 68,* 149–169.

Herodotus. (1964). *The histories* (A. de Selincourt, Trans.). London: Penguin Books.

Herrnstein, R. J., & Murray, C. (1994). *The bell curve: Intelligence and class structure in American life.* New York: Free Press.

Hershberger, S. L., & D'Augelli, A. R. (1995). The impact of victimization on the mental health and suicidality of lesbian, gay, and bisexual youths. *Developmental Psychology, 31,* 65–74.

Hertsgaard, D., & Light, H. (1984). Anxiety, depression, and hostility in rural women. *Psychological Reports, 55,* 673–674.

Herzberger, S. D., & Hall, J. A. (1993). Consequences of retaliatory aggression against siblings and peers: Urban minority children's expectations. *Child Development, 64,* 1773–1785.

Hess, T. M., & Flannagan, D. A. (1992). Schema-based retrieval processes in young and older adults. *Journal of Gerontology: Psychological Sciences, 47,* P52–P58.

Hess, T. M., & Slaughter, S. J. (1990). Schematic knowledge influences on memory for scene information in young and older adults. *Developmental Psychology, 26,* 855–865.

Hess, T. M., Flannagan, D. A., & Tate, C. S. (1993). Aging and memory for schematically vs taxonomically organized verbal materials. *Journal of Gerontology: Psychological Sciences, 48,* P37–P44.

Hetherington, E. M. (1989). Coping with family transitions: Winners, losers, and survivors. *Child Development, 60,* 1–14.

Hetherington, E. M., & Stanley-Hagan, M. M. (1995). Parenting in divorced and remarried families. In M. H. Bornstein (Ed.), *Handbook of parenting* (Vol. 3): Hillsdale, NJ: Erlbaum.

Hetherington, E. M., Cox, M., & Cox, R. (1976). Divorced fathers. *Family Coordinator, 25,* 417–427.

Hetherington, E. M., Cox, M., & Cox, R. (1977, April). Divorced fathers. *Psychology Today, 10,* 42–46.

Hetherington, E. M., Stanley-Hagan, M., & Anderson, E. R. (1989). Marital transitions: A child's perspective. *American Psychologist, 44,* 303–312.

Heyge, L. L. (1996). Music makes a difference. *Early Childhood Connections: The Journal of Music and Movement-Based Learning.* Cited in *Early Childhood News.* Retrieved February 12, 1999, from the World Wide Web: http://www.earlychildhoodnews.com/wiredfor.htm

Hiedemann, B., Suhomlinova, O., & O'Rand, A. (1998). Economic independence, economic status, and empty nest in midlife marital disruption, *Journal of Marriage & the Family, 60,* 219–231.

Hill, R. (1964). Methodological issues in family development research. *Family Process, 3,* 186–206.

Hill, R. (1986). Life cycle stages for types of single parent families: Of family development theory. *Family Relations, 35,* 19–29.

Hill, R. D., Storandt, M., & Malley, M. (1993). The impact of long-term exercise training on psychological function in older adults. *Journal of Gerontology: Psychological Sciences, 48,* P12–P17.

Hilleras, P., Herlitz, A., Jorm, A., & Winblad, B. (1998). Negative and positive affect among the very old: A survey on a sample age 90 years or older. *Research on Aging, 20,* 593–610.

Hilts, P. J. (1991, January 11). U.S. abandons idea of carrying out household survey on cases of AIDS. *New York Times,* A10.

Hirshberg, L. M., & Svejda, M. (1990). When infants look to their parents: Infants' social referencing of mothers compared to fathers. *Child Development, 61,* 1175–1186.

Hodgkinson, H. L. (1986, December). Reform? Higher education? Don't be absurd! *Phi Delta Kappan, 68,* 271–274.

Hofferth, S. L. (1997). *Findings about early childhood development, poverty, and child care environments.* Paper presented June 30 at the American Sociological Association and NIH Behavioral and Social Sciences Research Division. Retrieved August 5, 1998, from the World Wide Web: http://www1.od.nih.gov/obssr/asaschof.htm

Hofferth, S. L. (1998). *Healthy environments, healthy children: Children in families. A report on the 1997 panel study of income dynamics. Child Development Supplement.* Ann Arbor: University of Michigan Survey Research Center. Retrieved December 5, 1998, from the World Wide Web: html://www.isr.umich.edu/src/child-development/fullrep.html

Hofferth, S. L., & Phillips, D. A. (1991). Child care policy research. *Journal of Social Issues, 47,* 1–13.

Hoff-Ginsberg, E. (1986). Function and structure in maternal speech: The relation to the child's development of syntax. *Developmental Psychology, 22,* 155–163.

Hoff-Ginsberg, E., & Shatz, M. (1982). Linguistic input and the child's acquisition of language. *Psychological Bulletin, 92,* 3–26.

Hoffman, M. L. (1971). Father absence and conscience development. *Developmental Psychology, 4,* 400–406.

Holahan, C. K. (1988). Relation of life goals at age 70 to activity participation and health and psychological well-being among Terman's gifted men and women. *Psychology and Aging, 3,* 286–291.

Holahan, C. K., Sears, R. R., & Cronbach, L. J. (1995). *The gifted group in later maturity.* Stanford, CA: Stanford University Press.

Hollinger, J. H. (1998). *Adoption law and practice. Volume 1: 1998 supplement.* New York: Matthew Bender.

Holmbeck, G. N., & Hill, J. P. (1991). Conflictive engagement, positive affect, and menarche in families with seventh-grade girls. *Child Development, 62,* 1030–1048.

Holmes, S. A. (1998, November 11). Children study longer and play less, a report shows. *New York Times,* Late edition, 16.

Hong, E., & Perkins, G. (1997). Children's responses to self-concept questionnaire administered in different contexts. *Child Study Journal, 27,* 111.

Hopkins, J., Marcus, M., & Campbell, S. B. (1984). Postpartum depression: A critical review. *Psychological Bulletin, 95,* 498–515.

Horan, P. M., & Hargis, P. G. (1991). Children's work and schooling in the late nineteenth-century family economy. *American Sociological Review, 56,* 583–596.

Horn, J. C., & Meer, J. (1987, May). The vintage years. *Psychology Today, 21,* 76–84.

Horn, J. L. (1976). Human abilities. A review of research and theory in the early (1970)s. *Annual Review of Psychology, 27,* 437–485.

Horn, M. (1993). Childhood and children. In M. K. Cayton, E. J. Gorn, & P. W. Williams (Eds.), *Encyclopedia of American social history* (Vol. 3). New York: Scribner's.

Horn, W. F. (1998). *Why fathers count: The National Fatherhood Initiative.* Retrieved July 20, 1998, from the World Wide Web: http://www.fatherhood.org/fathers-count.html

Horwitz, A. V. (1996, November). Becoming married and mental health: A longitudinal study of a cohort of young adults. *Journal of Marriage and the Family, 58,* 895–907.

Hosken, F. P. (1998). *Female genital mutilation: Strategies for eradication.* Retrieved December 22, 1998, from the World Wide Web: http://nocirc.org/symposia/first/hosken.html

House, J. S., Robbins, C., & Metzner, H. L. (1982). The association of social relationships and activities with mortality: Prospective evidence from the Tecumseh Community Health Study. *American Journal of Epidemiology, 116,* 123–140.

Howe, M. L., & Courage, M. L. (1993). On resolving the enigma of infantile amnesia. *Psychological Bulletin, 113,* 305–326.

Howes, C., & Hamilton, C. (1992). Children's relationships with caregivers, *Child Development, 63,* 859–866.

Howes, C., & Wu, F. (1990). Peer interactions and friendships in an ethnically diverse school setting. *Child Development, 61,* 537–541.

Hsu, F. L. K. (1943). Incentives to work in primitive communities. *American Sociological Review, 8,* 638–642.

Huesmann, L. R. & Miller, L. S. (1994). Long-term effects of repeated exposure to media violence in childhood. In L. R. Huesmann (Ed.) *Aggressive behavior.* New York: Plenum Press.

Hultsch, D. F., Hertzog, C., & Dixon, R. A. (1990). Ability correlates of memory performance in adulthood and aging. *Psychology and Aging, 5,* 356–368.

Hummert, M. L., Garstka, T. A., Shaner, J. L., & Strahm, S. (1994). Stereotypes of the elderly held by young, middle-aged, and elderly adults. *Journal of Gerontology: Psychological Sciences, 49,* P240–P249.

Humphrey, T. (1978). Function of the nervous system during prenatal life. In U. Stave (Ed.), *Perinatal physiology.* Hillsdale, NJ: Erlbaum.

Humphry, D. (1991). *Final exit.* Eugene, OR: Hemlock Society.

Hunt, E. (1983). On the nature of intelligence. *Science, 219,* 141–146.

Hunter, F. T., & Youniss, J. (1982). Changes in functions of three relations during adolescence. *Developmental Psychology, 18,* 806–811.

Hunter, M. (Ed.). (1990). *The sexually abused male. Vol. 1: Prevalence, impact, and treatment.* Lexington, MA: Lexington Books/DC Heath.

Huston, A. C., Watkins, B. A., & Kunkel, D. (1989). Public policy and children's television. *American Psychologist, 44,* 424–433.

Huston, A. C., Wright, J. C., Rice, M. L., Kerkman, D., & St. Peters, M. (1990). Development of television viewing patterns in early childhood: A longitudinal investigation. *Developmental Psychology, 26,* 409–420.

Huston, T. L., & Geis, G. (1993). In what ways do gender-related attributes and beliefs affect marriage? *Journal of Social Issues, 49,* 87–106.

Hutchings, D. E. (Ed.). (1989). *Prenatal abuse of licit and illicit drugs.* New York: New York Academy of Sciences.

Hutnik, N. (1991). *Ethnic minority identity: A social psychological perspective.* Oxford, England: Clarendon Press.

Huttenlocher, J., Haight, W., Bryk, A., Seltzer, M., & Lyons, T. (1991). Early vocabulary growth: Relation to language input and gender. *Developmental Psychology, 27,* 236–248.

Hyde, D. M. (1959). *An investigation of Piaget's theories of the development of number.* Unpublished doctoral dissertation, University of London.

Hyde, J. S., Krajnik, M., & Skuldt-Niederberger, K. (1991). Androgyny across the life span: A replication and longitudinal follow-up. *Developmental Psychology, 27,* 516–519.

Hyson, M. C. (1994). *The emotional development of young children: Building an emotion-centered curriculum.* New York: Teachers College Press.

I

Ickes, W. (1993). Traditional gender roles: Do they make, and then break, our relationships? *Journal of Social Issues, 49,* 71–85.

Im Thurn, E. F. (1883). *Among the Indians of Guiana.* London: Kegan Paul, Trench & Trubner.

Inagaki, K., & Hatano, G. (1993). Young children's understanding of the mind-body distinction. *Child Development, 64,* 1534–1549.

Ingrassia, M. (1995, April 24). The body of the beholder. *Newsweek,* 66–67.

Inhelder, B., & Piaget, J. (1964). *The early growth of logic in the child.* New York: W. W. Norton.

Internet Health Resources. (1996). Embryo and semen freezing. *University of Southern California IVF Program.* Retrieved December 5, 1996, from the World Wide Web: http://www.ihr.com/ucsfivf/embfreez.html

Irwin, H. J. (1985). *Flight of mind: A psychological study of the out-of-body experience.* Metuchen, NJ: Scarecrow Press.

Ishii-Kuntz, M., & Lee, G. R. (1987). Status of the elderly: An extension of the theory. *Journal of Marriage and the Family, 49,* 413–420.

Istvan, J. (1986). Stress, anxiety, and birth outcomes: A critical review of the evidence. *Psychological Bulletin, 100,* 331–348.

Izard, C. E. (1991). *The psychology of emotions.* New York: Plenum.

Izard, C. E., & Ackerman, B. (1995, Winter). What's new in the new functionalism? What's missing? *SRCD Newsletter,* 1–2, 8, 10.

Izard, C. E., & Malatesta, C. Z. (1987). Perspectives on emotional development: Differential emotions theory of early emotional development. In J. D. Osofsky (Ed.), *Handbook of infant development* (2nd ed.). New York: Wiley.

Izard, C. E., Hembree, E. A., & Huebner, R. R. (1987). Infants' emotion expressions to acute pain: Developmental change and stability of individual differences. *Developmental Psychology, 23,* 105–113.

Izard, C. E., Libero, D. Z., Putnam, P., & Haynes, O. M. (1993). Stability of emotion experiences and their relations to traits of personality. *Journal of Personality and Social Psychology, 64,* 847–860.

J

Jacklin, C. N., & Reynolds, C. (1993). Gender and childhood socialization. In A. E. Beall & R. J. Sternberg (Eds.), *The psychology of gender.* New York: Guilford Press.

Jackson, J. F. (1993). Multiple caregiving among African Americans and infant attachment. The need for an emic approach. *Human Development, 36,* 87–102.

Jacobs, J. A., & Furstenberg, F. A. (1986). Changing places: Conjugal careers and women's marital mobility. *Social Forces, 64,* 714–732.

Jacobs, J. E., & Eccles, J. S. (1992). The impact of mothers' gender-role stereotypic beliefs on mothers' and children's ability perceptions. *Journal of Personality and Social Psychology, 63,* 932–944.

Jacobsen, L., & Edmondson, B. (1993). Father figures. *American Demographics, 15,* 22–27.

Jacobsen, T., Edelstein, W., & Hofmann, V. (1994). A longitudinal study of the relation between representations of attachment in childhood and cognitive functioning in childhood and adolescence. *Developmental Psychology, 30,* 112–124.

Jacobson, S. (1992–1993, December/January). Attitude: Mind over matters. *Modern Maturity,* 37–38.

Jacobson, S. W., Jacobson, J. L., O'Neill, J. M., Padgett, R. J., Frankowski, J. J., & Bihun, J. T. (1992). Visual expectation and dimensions of infant information processing. *Child Development, 63,* 711–724.

Jadack, R. A., Hyde, J. S., Moore, C. G., & Keller, M. L. (1995). Moral reasoning about sexually transmitted diseases. *Child Development, 66,* 167–177.

James, J. S. (1998). Preventing mother-infant transmission worldwide: What is needed? An interview with Joseph Saba, M.D. *The Body: An AIDS and HIV Information Resource, 294.* Retrieved July 29, 1998, from the World Wide Web: http://www.thebody.com/atn/294.html

James, S. L. (1990). *Normal language acquisition.* Boston: Allyn & Bacon.

Janofsky, M. (1994, December 13). Drug use rising among teen-agers, study says. *New York Times,* A1, A12.

Jaret, P. (1996, March/April). Think fast: What gives you speedier reactions, quicker recall? Step this way. *Health,* 44–46.

Jayakody, R., Chatters, L. M., & Taylor, R. M. (1993). Family support to single and married African American mothers: The provision of financial, emotional, and child care assistance. *Journal of Marriage and the Family, 55,* 261–276.

Jeffers, F. C., & Verwoerdt, A. (1969). How the old face death. In E. W. Busse &

E. Pfeiffer (Eds.), *Behavior and adaptation in late life.* Boston: Little, Brown.

Jeffrey, N. A. (1995, April 19). More seniors see security in retirement villages. *Wall Street Journal*, C1, C13.

Jencks, C. (1972). *Inequality: A reassessment of the effect of family and schooling in America.* New York: Basic Books.

Jendrek, M. P. (1993). Grandparents who parent their grandchildren: Effects on lifestyle. *Journal of Marriage and the Family, 55,* 609–621.

Jennings, J. M., & Jacoby, L. L. (1993). Automatic versus intentional uses of memory: Aging, attention, and control. *Psychology and Aging, 8,* 283–293.

Jensen, A. R. (1972, Summer). The heritability of intelligence. *Saturday Evening Post,* 149.

Jensen, A. R. (1984, March). Political ideologies and educational research. *Phi Delta Kappan, 65,* 460–462.

Jensen, K. (1932). Differential reactions to taste and temperature stimuli in newborn infants. *Genetic Psychological Monographs, 12,* 363–479.

Jessor, R., Turbin, M. S., & Costa, F. M. (1998). Protective factors in adolescent health behavior. *Journal of Personality and Social Psychology, 75,* 788–800.

Johansson, B., & Berg, S. (1989). The robustness of the terminal decline phenomenon: Longitudinal data from the Digit-Span Memory Test. *Journal of Gerontology, 44,* P184–186.

Johnson, C. N. (1990). If you had my brain, where would I be? Children's understanding of the brain and identity. *Child Development, 61,* 962–972.

Johnson, D. (1994, September 21). Study says small schools are key to learning. *New York Times,* B12.

Johnson, J. S., & Newport, E. L. (1989). Critical period effects in second language learning: The influence of instructional state on the acquisition of English as a second language. *Cognitive Psychology, 21,* 60–69.

Johnson, M., & Puddifoot, J. (1996). The grief response in the partners of women who miscarry, *British Journal of Medical Psychology, 69,* 313–327.

Johnson, P. B., & Gallo-Treacy, C. (1993). Alcohol expectancies and ethnic drinking differences. *Journal of Alcohol and Drug Education, 38,* 80–88.

Johnson, P. B., & Gurin, G. (1994). Negative affect, alcohol expectancies and alcohol-related problems. *Addiction, 89,* 581–586.

Johnson, R. J., Lund, D. A., & Dimond, M. F. (1986). Stress, self-esteem and coping during bereavement among the elderly. *Social Psychology Quarterly, 49,* 273–279.

John-Steiner, V. (1986). *Notebooks of the mind: Explorations of thinking.* Albuquerque: University of New Mexico Press.

Johnston, L. (1979, June 17). Artist ends her life after ritual citing "self-termination" right. *New York Times,* 1, 10.

Johnston, L. D., Bachman, J. G., & O'Malley, P. M. (1997). Drug use among American teens shows some signs of leveling after a long rise. *Monitoring the future survey: 1997.* University of Michigan Survey Research Center. Retrieved March 6, 1999, from the World Wide Web: http://www.health.org/pressrel/dec97/10.htm

Johnston, L. D., & O'Malley, P. M. (1997). The recanting of earlier-reported drug use by young adults. In L. Harrison (Ed.), *Validity of self-reported drug use: Improvising the accuracy of survey estimates.* (National Institute on Drug Abuse Research Monograph). Rockville, MD: National Institute on Drug Abuse.

Jones, D. C., & Costin, S. E. (1995). Friendship quality during preadolescence and adolescence: The contributions of relationship orientations, instrumentality, and expressivity. *Merrill-Palmer Quarterly, 41,* 517–535.

Jones, M. C. (1957). The later careers of boys who were early- or late-maturing. *Child Development, 28,* 113–128.

Jones, M. C., & Bayley, N. (1950). Physical maturing among boys as related to behavior. *Journal of Educational Psychology, 41,* 129–148.

Jones, S. S. (1985). On the motivational bases for attachment behavior. *Developmental Psychology, 21,* 848–857.

Joseph, J. A., & Roth, G. S. (1993). Hormonal regulation of motor behavior in senescence. *Journals of Gerontology, 48* (Special Issue), 51–55.

Josephs, R. A., Markus, H. R., & Tafarodi, R. W. (1992). Gender and self-esteem. *Journal of Personality and Social Psychology, 63,* 391–402.

Josselson, R. (1988). *Finding herself: Pathways to identity development in women.* New York: Jossey-Bass.

Jung, C. G. (1933). *Modern man in search of a soul.* New York: Harcourt, Brace & World.

Jung, C. G. (1960). The stages of life. In H. Reed, M. Fordham, & G. Adler (Eds.), *Collected works* (Vol. 8). Princeton, NJ: Princeton University Press. (Original work published 1931)

Jussim, L., & Eccles, J. S. (1992). Teacher expectations II: Construction and reflection of student achievement. *Journal of Personality and Social Psychology, 63,* 947–961.

Jylha, M., & Jokela, J. (1990). Individual experiences as cultural—A cross-cultural study on loneliness among the elderly. *Ageing and Society, 10,* 295–315.

K

Kaetz, J. (1998, Spring). In this issue. *National Forum: Phi Kappa Phi Journal, 78,* 2.

Kagan, J. (1972). A conception of early adolescence. In J. Kagan & R. Coles (Eds.), *Twelve to sixteen: Early adolescence.* New York: Norton.

Kagan, J. (1983). Stress and coping in early development. In N. Garmezy & M. Rutter (Eds.), *Stress, coping, and development.* New York: McGraw-Hill.

Kagan, J. (1984). *The nature of the child.* New York: Basic Books.

Kagan, J. (1989). *Unstable ideas: Temperament, cognition, and self.* Cambridge: Cambridge University Press.

Kagan, J. (1993). On the nature of emotion. *Monographs of the Society for Research in Child Development, 59* (2–3, Serial No. 240).

Kagan, J. (1994, October 5). The realistic view of biology and behavior. *Chronicle of Higher Education,* A64.

Kagan, J. (1997). Attention, emotion, and reactivity in infancy and early childhood. In P. J. Lang (Ed.), *Attention and orienting: Sensory and motivational processes.* Mahwah, NJ: Erlbaum.

Kagan, J., & Klein, R. E. (1973). Cross-cultural perspectives on early development. *American Psychologist, 28,* 947–961.

Kagan, J., & Moss, H. A. (1962). *Birth to maturity.* New York: Wiley.

Kagan, J., & Snidman, N. (1991). Temperamental factors in human development. *American Psychologist, 46,* 856–862.

Kagan, J., Kearsley, R. B. & Zelazo, P. R. (1978). *Infancy: Its place in human development.* Cambridge, MA: Harvard University Press.

Kahn, S., Zimmerman, G., Csikszentmihalyi, M., & Getzels, J. W. (1985). Relations between identity in young adulthood and intimacy at midlife. *Journal of Personality and Social Psychology, 49,* 1316–1322.

Kail, R. (1991). Developmental change in speed of processing during childhood and adolescence. *Psychological Bulletin, 109,* 490–501.

Kaiser, J. (1994). Alzheimer's: Could there be a zinc link? *Science, 265,* 1365.

Kakar, S. (1998). The search for middle age in India. In R. A. Schweder (Ed.), *Welcome to middle age! (and other cultural fictions)* (pp. 75–98). Chicago: University of Chicago Press.

Kaler, S., & Freeman, B. J. (1997). Analysis of environmental deprivation: Cognitive and social development in Romanian orphans. *Journal of Psychology and Psychiatry and Allied Disciplines, 35,* 769–781.

Kalleberg, A. L., & Loscocco, K. A. (1983). Aging, values, and rewards: Explaining age differences in job satisfaction. *American Sociological Review, 48,* 78–90.

Kallman, D. A., Plato, C. C., & Tobin, J. D. (1990). The role of muscle loss in the age-related decline of grip strength: Cross-sectional and longitudinal perspectives. *Journal of Gerontology, 45,* M82–M88.

Kalter, H., & Warkany, J. (1983a). Congenital malformations. I. *New England Journal of Medicine, 308,* 424–431.

Kalter, H., & Warkany, J. (1983b). Congenital malformations. II. *New England Journal of Medicine, 308,* 491–497.

Kamerman, J. B. (1988). *Death in the midst of life: Social and cultural influences on death, grief, and mourning.* Englewood Cliffs, NJ: Prentice Hall.

Kamin, L. J. (1974). *The science and politics of IQ.* Hillsdale, NJ: Erlbaum.

Kamin, L. J. (1981). Commentary. In S. Scarr (Ed.), *IQ: Race, social class, and individual differences.* Hillsdale, NJ: Erlbaum.

Kamin, L. J. (1994, November 23). Intelligence, IQ tests, and race. *Chronicle of Higher Education,* B5.

Kandel, D. B. (1990). Parenting styles, drug use, and children's adjustment in families of young adults. *Journal of Marriage and the Family, 52,* 183–196.

Kane, H. (1993). *Child mortality continues to fall.* Retrieved February 11, 1999, from the World Wide Web: http://www.ai.rcast. u-tokyo.ac.jp/~dobashi/workhtml/ vi939697.html

Kantrowitz, B. (1986, January 13). Growing up gay. *Newsweek,* 50–52.

Kantrowitz, B., & Wingert, P. (1998, October 5). Learning at home: Does it pass the test? *Newsweek,* 64–70.

Kasl-Godley, J. E., Gatz, M. & Fiske, A. (1998). Depression and depressive symptoms in old age. In I. H. Nordhus et al. (Eds.), *Clinical geropsychology* (pp. 211–218). Washington, DC: American Psychological Association.

Kastenbaum, R. (1975). Is death a life crisis? On the confrontation with death in theory and practice. In N. Datan & L. H. Ginsburg (Eds.), *Lifespan developmental psychology: Normative life crisis.* New York: Academic Press.

Kastenbaum, R. (1977, September). Temptations from the ever after. *Human Behavior, 6,* 28–33.

Kastenbaum, R. (1979). "Healthy dying": A paradoxical quest continues. *Journal of Social Issues, 35,* 185–206.

Kastenbaum, R. (1991). Where do we come from? What are we? Where are we going? An annotated bibliography of aging and humanities by Donna Polisar, Larry Wygant, Thomas Cole &Cielo Perdomo. *International Journal of Aging & Human Development, 33,* 247.

Kastenbaum, R. (1993). *Encyclopedia of adult development.* Phoenix, AZ: Onyx Press.

Kastenbaum, R. (1997). Final acts of love: Families, friends; and assisted dying / Life beyond 85 years: The aura of survivorship / Letting go: Morrie's reflections on living while dying / The psychology of growing old. *Gerontologist, 37,* 698–701.

Kastenbaum, R., & Costa, P. T., Jr. (1977). Psychological perspectives on death. In M. R. Rosenzweig & L. W. Porter (Eds.), *Annual review of psychology* (Vol. 28). Palo Alto, CA: Annual Reviews, Inc.

Katchadourian, H. A. (1984). *Fundamentals of sexuality* (4th ed.). New York: Holt, Rinehart & Winston.

Kato, D. (1996, March 31). Growing pains: Female youngsters learn to cope with early menstruation. *Houston Chronicle,* 6.

Katz, L. (1994). Assessing the development of preschoolers. *ERIC Digest.* Retrieved August 30, 1998, from the World Wide Web: http://136.242.172.58/db/riecije/ ed372875.htm

Katz, S., & Burt, M. (1988). Self-blame: Help or hindrance in recovery from rape? In A. Burgess (Ed.), *Rape and sexual assault.* New York: Garland.

Kaufman, H. G. (1982). *Professionals in search of work: Coping with the stress of job loss and underemployment.* New York: Wiley.

Kaufman, J., & Zigler, E. (1987). Do abused children become abusive parents? *American Journal of Orthopsychiatry, 57,* 186–192.

Kaufman, J., & Zigler, E. (1993). The intergenerational transmission of abuse is overstated. In R. J. Gelles and D. R. Loseke (Eds.), *Current controversies on family violence.* Newbury Park, CA: Sage.

Kausler, D. H., & Hakami, M. K. (1983). Memory for topics of conversation: Adult age differences and intentionality. *Experimental Aging Research, 9,* 153–157.

Kausler, D. H., Wiley, J. G., & Lieberwitz, K. J. (1992). Adult age differences in short-term memory and subsequent long-term memory for actions. *Psychology and Aging, 7,* 309–316.

Keefer, C. H., Tronick, E., Dixon, S., & Brazelton, T. B. (1982). Special differences in motor performance between Gusii and American newborns and a modification of the neonatal behavioral assessment scale. *Child Development, 53,* 754–759.

Keel, P. K., Mitchell, S. E., Miller, K. B., Davis, T. L., & Crow, S. J. (1999). Long-term outcome of bulimic nervosa. *Archives of General Psychiatry, 56,* 63.

Kegan, R. (1988). *In over our head.* Boston: Harvard University Press.

Kegan, R. (1994). *In over our heads: The mental demands of modern life.* Cambridge, MA: Harvard University Press.

Kehoe, M. (1986). Lesbians over 65: A triple invisible minority. *Journal of Homosexuality, 12,* 139–152.

Kehoe, M. (1989). *Lesbians over sixty speak for themselves.* New York: Haworth Press.

Keller, W. D., Hildebrandt, K. A., & Richards, M. E. (1985). Effects of extended father-infant contact during the newborn period. *Infant Behavior and Development. 8,* 337–350.

Kelly, G. A. (1955). *The psychology of personal constructs.* New York: Norton.

Kendrick, C., & Dunn, J. (1980). Caring for a second baby: Effects on interaction between mother and firstborn. *Developmental Psychology, 16,* 303–311.

Keniston, K. (1970, Autumn). Youth: A "new" stage in life. *American Scholar,* 586–595.

Kennell, J. H., Voos, D. K., & Klaus, M. H. (1979). Parent-infant bonding. In J. D. Osofsky (Ed.), *Handbook of infant development.* New York: Wiley.

Kenshalo, D. R., Sr. (1986). Somesthetic sensitivity in young and elderly humans. *Journal of Gerontology, 41,* 732–742.

Kerig, P. K., Cowan, P. A., & Cowan, C. P. (1993). Marital quality and gender differences in parent-child interaction. *Developmental Psychology, 29,* 931–939.

Kertzer, D. I., & Schaie, K. W. (Eds.). (1989). *Age structure in comparative perspective.* Hillsdale, NJ: Erlbaum.

Kidwell, J., Dunham, R., Backo, R., & Pastorino, E. (1995). Adolescent identity exploration. *Adolescence, 30,* 785–793.

Kilborn, P. T. (1993, March 15). New jobs lack the old security in time of "disposable workers." *New York Times,* A1, A6.

Killian, K. D. (1994). Fearing fat: A literature review of family systems understandings and treatments of anorexia and bulimia. *Family Relations, 43,* 311–318.

Kimble, G. A. (1984). Psychology's two cultures. *American Psychologist, 39,* 833–839.

Kimmel, D. C. (1980). *Adulthood and aging: An interdisciplinary, developmental view* (2nd ed.). New York: Wiley.

Kincheloe, J. L., & Steinberg, S. R. (1993). A tentative description of post-formal thinking: The critical confrontation with cognitive theory. *Harvard Educational Review, 63,* 296–320.

King, H. E., & Webb, C. (1981). Rape crisis centers. *Journal of Social Issues, 37,* 93–104.

Kinsey, A. C., Pomeroy, W. B., & Martin, C. E. (1948). *Sexual behavior in the human male.* Philadelphia: Saunders.

Kinsey, A. C., Pomeroy, W. B., Martin, C. E., & Gebhard, P. H. (1953). *Sexual behavior in the human female.* Philadelphia: Saunders.

Kipnis, D. (1994). Accounting for the use of behavior technologies in social psychology. *American Psychologist, 49,* 165–172.

Kisilevsky, B. S. (1995). The influence stimulus and subject variables on human fetal responses to sound and vibration. In J. P. Lecanuet et al. (Eds.), *Fetal development.* Hillsdale, NJ: Erlbaum.

Kitchener, K. S., Lynch, C. L., Fischer, K. W., & Wood, P. K. (1993). Developmental range of reflective judgment: The effect of contextual support and practice on developmental stage. *Developmental Psychology, 29,* 893–906.

Kite, M. E., Deaux, K., & Miele, M. (1991). Stereotypes of young and old: Does age outweigh gender? *Psychology and Aging, 6,* 19–27.

Kitzinger, C., & Wilkinson, S. (1995). Transitions from heterosexuality to lesbianism: The discursive production of lesbian identities. *Developmental Psychology, 31,* 95–104.

Klaus, M. H., & Kennell, J. H. (1976). *Maternal-infant bonding: The impact of early separation or loss on family development.* St. Louis: Mosby.

Kline, M., Tschann, J. M., Johnston, J. R., & Wallerstein, J. S. (1989). Children's adjustment in joint and sole physical custody families. *Developmental Psychology, 25,* 430–438.

Klinger, R. L. (1996). Lesbian couples. In R. P. Cabaj & T. S. Stein (Eds.), *Textbook of homosexuality and mental health* (pp. 339–352). Washington, DC: American Psychiatric Press.

Klinger-Vartabedian, L., & Wispe, L. (1989). Age differences in marriage and female longevity. *Journal of Marriage and the Family, 51,* 195–202.

Kluckhohn, C. (1960). *Mirror for man.* Greenwich, CT: Fawcett.

Knefelkamp, L. L. (1984). *A workbook for the practice-to-theory-to-practice model.* Unpublished manuscript. University of Maryland, College Park

Knox, N. (1998, February 22). Drug abuse problems are on the rise as the labor pool shrinks. *The New York Times,* Section 3, 11.

Kobasa, S. C., Maddi, S. R., & Kahn, S. (1982). Hardiness and health: A prospective study. *Journal of Personality and Social Psychology, 42,* 168–177.

Kobre, K. (1998a). Crack babies in infancy (Part 1). The Gannett Foundation and San Francisco State University. Retrieved July 29, 1998, from the World Wide Web: http://www.gigaplex.com/photo/kobre/crack/crack1/htm

Kobre, K. (1998b). Crack babies in infancy (Part 2). The Gannett Foundation and San Francisco State University. Retrieved July 29, 1998, from the World Wide Web: http://www.gigaplex.com/photo/kobre/crack/crack2/htm

Kobre, K. (1998c). Crack's next generation: How the children of crack addicts grow up. The Gannett Foundation and San Francisco State University. Retrieved July 29, 1998, from the World Wide Web: http://www.gigaplex.com/photo/kobre/crack/crack1/htm

Koch, K. (1998, July 10). Encouraging teen abstinence. *CQ Researcher, 8,* 577–600.

Kochanska, G. (1995). Children's temperament, mothers' discipline, and security of attachment: Multiple pathways to emerging internalization. *Child Development, 66,* 597–615.

Kochanska, G., & Aksan, N. (1995). Mother-child mutually positive affect, the quality of child compliance to requests and prohibitions, and maternal control as correlates of early internalization. *Child Development, 66,* 236–254.

Koerner. B. I. (1997). Is there life after death? *U.S. News & World Report, 122,* 58–64.

Koestner, R., Franz, C., & Weinberger, J. (1990). The family origins of empathic concern: A 26-year longitudinal study. *Journal of Personality and Social Psychology, 58,* 709–717.

Kogan, N., & Mills, M. (1992). Gender influences on age cognitions and preferences: Sociocultural or sociobiological? *Psychology and Aging, 7,* 98–106.

Kohlberg, L. (1963). The development of children's orientations toward a moral order. I: Sequence in the development of human thought. *Vita Humana, 6,* 11–33.

Kohlberg, L. (1966). A cognitive-developmental analysis of children's sex-role concepts and attitudes. In E. E. Maccoby (Ed.), *The development of sex differences.* Stanford, CA: Stanford University Press.

Kohlberg, L., & Colby, A. (1990). *Measurement of moral judgment.* New York: Cambridge University Press.

Kohlberg, L., & Gilligan, C. F. (1971). The adolescent as philosopher: The discovery of the self in a postconventional world. *Daedalus, 100,* 1051–1086.

Kohlberg, L., & Ullian, D. Z. (1974). Stages in the development of psychosexual concepts and attitudes. In R. C. Friedman, R. N. Richart & R. L. Vande Wiele (Eds.), *Sex differences in behavior.* New York: Wiley.

Kohn, M. L., Naoi, A., Schoenbach, C., Schooler, C., & Slomczynski, K. M. (1990). Position in the class structure and psychological functioning in the United States, Japan, and Poland. *American Journal of Sociology, 95,* 964–1008.

Kohn, M. L., & Schooler, C. (1983). *Work and personality: An inquiry into the impact of social stratification.* Norwood, NJ: Ablex.

Kohnstamm, G. A., Bates, J. E., & Rothbart, M. K. (Eds.). (1989). *Temperament in childhood.* New York: Wiley.

Kolata, G. (1990, June 13). Program helped underweight babies, study shows. *New York Times,* A10.

Kolata, G. (1991a, February 28). Alzheimer's disease: Dangers and trials of denial. *New York Times,* B7.

Kolata, G. (1991b, September 30). Parents of tiny infants find care choices are not theirs. *New York Times,* A1, A11.

Kolata, G. (1992a, October 30). Baby's growth rate not fast enough? Just wait. *New York Times,* A8.

Kolata, G. (1992b, April 29). Ethicists debating a new definition of death. *New York Times,* p. B7.

Kolata, G. (1992c, September 1). Linguists debate study classifying language as innate human skill. *New York Times,* B6.

Kolata, G. (1992d, November 22). The burdens of being overweight. *New York Times,* 1, 18.

Kolata, G. (1993a, May 3). Family aid to elderly is very strong, study shows. *New York Times,* A16 L.

Kolata, G. (1993b, May 3). Strong family aid to elderly is found. *New York Times*, A7.

Kolata, G. (1994a, February 25). Theory on aging is tested, adding 30% to flies' lives. *New York Times*, A8.

Kolata, G. (1994b, July 27). Wrong drugs are given to 1 in 4 of elderly. *New York Times*, B7.

Kolata, G. (1995a, February 28). Man's world, woman's world? Brain studies point to differences. *New York Times*, B5, B8.

Kolata, G. (1995b, February 16). Men and women use brain differently, study discovers. *New York Times*, A1, A8.

Kolata, G. (1997a, April 24). A record and big questions as woman gives birth at 63. *New York Times*, A1.

Kolata, G. (1997b). Childbirth at 63 says what about life? *New York Times* (April 27), A1.

Kolata, G. (1998, February 25). Price of donor eggs soars setting a debate on ethics. *New York Times*, A1, 16.

Kolata, G. B. (1979). Scientists attack report that obstetrical medications endanger children. *Science, 204*, 391–392.

Kolbert, E. (1991, October 11). Sexual harassment at work is pervasive, survey suggests. *New York Times*, A11, A18.

Kolbert, E. (1994, December 14). Television gets closer look as a factor in real violence. *New York Times*, A1, A13.

Koonin, L. M., Smith, J. C., Ramick, M., & Strauss, L. T. (1998). Abortion surveillance: United States 1995. *Morbidity and Mortality Weekly Reports (July 3), 47* (No. SS-2).

Koretz, G. (1992, November 9). Just how welcome is the job market to college grads? *Business Week*, 22.

Korner, A. F., Brown, B. W., Jr., Reade, E. P., Stevenson, D. K., Fernbach, S. A., & Thom, V. A. (1988). State behavior of preterm infants as a function of development, individual and sex differences. *Infant Behavior and Development, 11*, 111–124.

Kornhaber, A., & Woodward, K. L. (1981). *Grandparents/grandchild: The vital connection.* New York: Anchor Press.

Kramer, L., & Houston, D. (1998). Supporting families as they adopt children with special needs. *Family Relations. 47*, 423–432.

Kramer, L., & Radey, C. (1997). Improving sibling relationships among young children: Social skills training model. *Family Relations. 46*, 237–246.

Krampe, R. T. (1994). *Maintaining excellence: Cognitive-motor performance in pianists differing in age and skill level.* Berlin, Germany: Edition Sigma.

Krashen, S. D. (1996). *Under attack: The case against bilingual education.* Culver City, CA: Language Education Associates.

Krause, N., & Goldenhar, L. M. (1992). Acculturation and psychological distress in three groups of elderly Hispanics. *Journal of Gerontology: Social Sciences, 47*, S279–S288.

Kraut, R., Patterson, M., Lundmark, V., Kusler, S., Mukopadhyay, T., & Scherlis, W. (1998). Internet paradox: A social technology that reduces social involvement and psychological well-being? *American Psychologist, 53*, 1017–1031.

Krueger, J. (1992). On the overestimation of between-group differences. *European Review of Social Psychology, 3*, 31–56.

Krueger, J., & Heckhausen, J. (1993). Personality development across the adult life span: Subjective conceptions vs. cross-sectional contrasts. *Journal of Gerontology: Psychological Sciences, 48*, P100–P108.

Krymkowski, D. H. (1991). The process of status attainment among men in Poland, the U.S., and West Germany. *American Sociological Review, 56*, 46–59.

Kübler-Ross, E. (1969). *On death and dying: What the dying have to teach doctors nurses, clergy and their own families.* New York: Macmillan.

Kübler-Ross, E. (1981). *Living with dying.* New York: Macmillan.

Kuczynski, A. (1998, April 12). Anti-aging potion or poison? *New York Times*, 9, 1.

Kuczynski, L., & Kochanska, G. (1995). Function and content of maternal demands: Developmental significance of early demands for competent action. *Child Development, 66*, 616–628.

Kuhl, P. K., Williams, K. A., Lacerda, F., Stevens, K. N., & Lindblom, B. (1992). Linguistic experience alters phonetic perception in infants by 6 months of age. *Science, 255*, 606–608.

Kumar, V., & Suryanarayne. (1989). Problems of the aged in the rural sector. In R. N. Pati & B. Jena (Eds.), *Aged in India: Socio-demographic dimensions.* New Delhi: Ashish.

Kurdek, L. A. (1991). The relations between reported well-being and divorce history, availability of a proximate adult, and gender. *Journal of Marriage and the Family, 53*, 71–78.

Kurdek, L. A. (1993). The allocation of household labor in gay, lesbian, and heterosexual married couples. *Journal of Social Issues, 49*, 127–139.

Kurdek, L. A. (1994a). Areas of conflict for gay, lesbian, and heterosexual couples: What couples argue about influences

relationship satisfaction. *Journal of Marriage and the Family, 56*, 923–934.

Kurdek, L. A. (1994b). Remarriages and stepfamilies are not inherently problematic. In A. Booth & J. Dunn (Eds.), *Stepfamilies: Who benefits? Who does not?* Hillsdale, NJ: Erlbaum.

Kurdek, L. A. (1995). Developmental changes in relationship quality in gay and lesbian cohabiting couples. *Developmental Psychology, 31*, 86–94.

Kurdek, L. A. (1998). Relationship outcomes and their predictors: Longitudinal evidence from heterosexual married, gay cohabiting, and lesbian cohabiting couples. *Journal of Marriage and the Family, 60*, 553–568.

Kurdek, L. A., Fine, M. A., & Sinclair, R. J. (1995). School adjustment in sixth graders: Parenting transitions, family climate, and peer norm effects. *Child Development, 66*, 430–445.

Kutner, L. (1990, February 8). It isn't unusual when the father-to-be wakes up feeling sick. *New York Times*, B8.

Kutner, L. (1993, September 16). Being clumsy. *New York Times*, B4.

L

Labouvie-Vief, G. (1986). Modes of knowledge and the organization of development. In M. L. Commons, L. Kohlberg, F. A. Richards, & J. Sinnot (Eds.), *Models and methods in the study of adult and adolescent thought. Vol. 3: Beyond formal operations.* New York: Praeger.

Labouvie-Vief, G., DeVoe, M., & Bulka, D. (1989). Speaking about feelings: Conceptions of emotion across the life span. *Psychology and Aging, 4*, 425–437.

Lachman, M. E., & James, J. B. (1997). Charting the course of midlife development: An overview. In M. E. Lachman & J. B. James (Eds.), *Multiple paths of midlife development* (pp. 1–17). Chicago: University of Chicago Press.

Lachs, M. S., Williams, C. S., O'Brien, S., Hurst, L., Horwitz, L. (1997). Risk factors for reported elder abuse and neglect: A nine-year observational cohort study. *Gerontologist, 37*, 469–474.

Lachs, M. S., Williams, C. S., O'Brien, S., Pillemer, K. A., & Charlson, M. E. (1998). The mortality of elder mistreatment. *Journal of the American Medical Association, 280*, 428–432.

Lagattuta, K., Wellman, H., & Flavell, J. (1997). Preschoolers' understanding of the link between thinking and feeling. *Child Development, 68*, 1081–1104.

Lagercrantz, H., & Slotkin, T. A. (1986, April). The "stress" of being born. *Scientific American, 254,* 100–107.

LaGreca, A. M., & Wasserstein, S. (1995). What do children worry about? Worries and their relation to anxiety. *Child Development, 66,* 671–686.

Lam, D. H., & Power, M. J. (1991). Social support in a general practice elderly sample. *International Journal of Geriatric Psychiatry, 6,* 89–93.

Lamaze, F. (1958). *Painless childbirth: Psychoprophylactic method.* London: Burke.

Lamb, M. E. (1977). Father-infant and mother-infant interaction in the first year of life. *Child Development, 48,* 167–181.

Lamb, M. E. (Ed.). (1981). *The role of the father in child development.* New York: Wiley.

Lamb, M. E., & Bornstein, M. H. (1987). *Development in infancy: An introduction* (2nd ed.). New York: Random House.

Lamb, M. E., Frodi, A. M., Hwang, C., Frodi, M., & Steinberg, J. (1982). Mother- and father-infant interaction involving play and holding in traditional and nontraditional Swedish families. *Developmental Psychology, 18,* 215–221.

Lamb, M. E., Sternberg, K. J., & Prodromidis, M. (1992). Nonmaternal care and the security of infant-mother attachment: A reanalysis of the data. *Infant Behavior and Development, 15,* 71–83.

Lamborn, S. D., Dornbusch, S. M., & Steinberg, L. (1996). Ethnicity and community context as moderators of the relations between family decision making and adolescent adjustment. *Child Development, 67,* 283–301.

Lampl, M., Cameron, N., Veldhuis, J. D., & Johnson, M. L. (1995). Patterns of human growth. *Science, 268,* 445–447.

Landrigan, P. J. (1997). Illness in Gulf War veterans: Causes and consequences. *Journal of the American Medical Association, 277,* 259–261.

Landry, S. H., Chapieski, M. L., Richardson, M. A., Palmer, J. P., & Hall, S. (1990). The social competence of children born prematurely: Effects of medical complications and parent behaviors. *Child Development, 61,* 1605–1615.

Landsberger, B. H. (1985). *Long-term care for the elderly: A comparative view of layers of care.* New York: St. Martin's Press.

Lang, A. M., & Brody, E. M. (1983). Characteristics of middle-aged daughters and help to their elderly mothers. *Journal of Marriage and the Family, 45,* 193–202.

Lang, O. (1946). *Chinese family and society.* New Haven: Yale University Press.

Lang, S. S., & Patt, R. B. (1994). *You don't have to suffer: A complete guide to relieving cancer pain for patients and their families.* New York: Oxford University Press.

Langer, E. J., & Rodin, J. (1976). The effects of choice and enhanced personal responsibility for the aged: A field experiment in an institutional setting. *Journal of Personality and Social Psychology, 34,* 191–198.

Langer, S. K. (1982). *Philosophy in a new key: A study in the symbolism of reason, rite, and art.* Cambridge, MA: Harvard University Press.

Langlois, J. H., Ritter, J. M., Roggman, L. A., & Vaughn, L. S. (1991). Facial diversity and infant preferences for attractive faces. *Developmental Psychology, 27,* 79–84.

Langmeier, J., & Matějček, Z. (1974). *Psychological deprivation in childhood.* New York: Halsted Press.

Langway, L. (1982, November 1). Growing old, feeling young. *Newsweek,* 56–65.

Larson, J. H., Wilson, S. M., & Beley, R. (1994). The impact of job insecurity on marital and family relationships. *Family Relations, 43,* 138–143.

Larson, R. W. (1983). Adolescents' daily experience with family and friends: Contrasting opportunity systems. *Journal of Marriage and the Family, 45,* 739–750.

Larson, R., & Richards, M. H. (1991). Daily companionship in late childhood and early adolescence: Changing developmental contexts. *Child Development, 62,* 284–300.

Lasky, R. E., Klein, R. E., Yarbrough, C., Engle, P. L., Lechtig, A., & Martorell, R. (1981). The relationship between physical growth and infant behavior development in rural Guatemala. *Child Development, 52,* 219–226.

Laslett, B. (1973). The family as a public and private institution: An historical perspective. *Journal of Marriage and the Family, 35,* 480–492.

Latten, J. J. (1989). Life course and satisfaction: Equal for everyone? *Social Indicators Research, 21,* 599–610.

Lau, R. R., Quadrel, M. J., & Hartman, K. A. (1990). Development and change of young adults' preventive health beliefs and behavior: Influence from parents and peers. *Journal of Health and Social Behavior, 31,* 240–259.

Lauer, J. C., & Lauer, R. H. (1985, June). Marriages made to last. *Psychology Today, 19,* 22–26.

Laumann, E. O., Gagnon, J. H., Michael, R. T., & Michaels, S. (1994). *The social organization of sexuality: Sexual practices in the United States.* Chicago: University of Chicago Press.

Lavoie, J. C., & Looft, W. R. (1973). Parental antecedents of resistance-to-temptation behavior in adolescent males. *Merrill-Palmer Quarterly, 19,* 107–116.

Leach, P. (1998). *Your baby and child from birth to age five.* New York: Knopf.

Learning to see the inevitable signs of aging eyes. (1998). Better Vision Institute. Retrieved from the World Wide Web: http://www.visionsite.org/press/ageeyes.htm

Leary, W. E. (1995, April 27). Study says youth contributes to the risk of problem births. *New York Times,* A12.

Leboyer, F. (1975). *Birth without violence.* New York: Knopf.

Lecanuet, J. P., Granier-Deferre, C., & Busnel, M. C. (1995). Human fetal auditory perception. In: J. P. Lecanuet et al. (Eds.), *Fetal development.* Hillsdale, NJ: Erlbaum.

Lecca, P. J., Quervalu, I., Nunes, J. V., & Gonzales, H. F. (Eds). (1998). *Cultural competency in health, social, and human services: Directions for the twenty-first century.* New York: Garland.

Lee, J. A. (1987). What can homosexual aging studies contribute to theories of aging? *Journal of Homosexuality, 13,* 43–71.

Lee, K., & Chen, L. (1996, September). The development of metacognitive knowledge of basic motor skill: Walking. *Journal of Genetic Psychology,* 361–365.

Lehmann, B. (1996). *The development of posttraumatic stress disorder (PTSD) in a sample of child witness to mother assault.* Poster presentation at the First Annual Conference on Children Exposed to Family Violence, Austin, TX.

Leman, H. C. (1953). *Age and achievement.* Princeton, NJ: Princeton University Press.

Lemonick, M. D. (1997, December 1). The new revolution in making babies. *Time, 150,* 40–46.

Lenneberg, E. H. (1967). *Biological foundations of language.* New York: Wiley.

Lenneberg, E. H. (1969). On explaining language. *Science, 164,* 635–643.

Lennon, M. C. (1982). The psychological consequences of menopause: The importance of timing of a life stage event. *Journal of Health and Social Behavior, 23,* 353–366.

Leonard, M. (1998, February 15). Making a mark on the culture: Body piercing, tattoos, and scarification push the cutting edge, *Boston Globe,* C1.

Lepper, M. R., & Greene, D. (1975). Turning play into work: Effects of adult surveillance and extrinsic rewards on children's intrinsic motivation. *Journal of Personality and Social Psychology, 31,* 479–486.

Lerman, R. I., & Ooms, T. J. (Eds.). (1993). *Young unwed fathers: Changing roles and emerging policies.* Philadelphia: Temple.

Lerner, R. M., Castellino, D. R., Terry, P. A., Villarruel, F. A., & McKinney, M. H. (1995). A developmental contextual perspective on parenting. In M. H. Bornstein (Ed.), *Handbook of parenting* (Vol. 2). Hillsdale, NJ: Erlbaum.

Lessing, E. E., Zagorin, S. W., & Nelson, D. (1970). WISC subtest and IQ score correlates of father absence. *Journal of Genetic Psychology, 117,* 181–195.

Lester, B. (1995). Prenatal cocaine exposure and child outcome: What do we know? In M. Lewis (Ed.), *Mothers, babies, and cocaine.* Hillsdale, NJ: Erlbaum.

Lester, B. (1997). Database of studies on prenatal cocaine exposure and child outcome. *Journal of Drug Issues, 27,* 487–499.

Lester, B. M., & Dreher, M. (1989). Effects of marijuana use during pregnancy on newborn cry. *Child Development, 60,* 765–771.

Lester, B. M., Kotelchuck, M., Spelke, E., Sellers, M. J., & Klein, R. E. (1974). Separation protest in Guatemalan infants: Cross-cultural and cognitive findings. *Developmental Psychology, 10,* 79–85.

Levin, I., & Druyan, S. (1993). When sociocognitive transaction among peers fails: The case of misconceptions in science. *Child Development, 64,* 157.

Levine, L. J. (1995). Young children's understanding of the causes of anger and sadness. *Child Development, 66,* 697–709.

LeVine, R. A. (1970). Cross-cultural study in child psychology. In P. H. Mussen (Ed.), *Carmichael's manual of child psychology* (3rd ed.). New York: Wiley.

Levine, S. (1998). *Sexuality in mid-life.* New York: Plenum Press.

Levinson, D. J. (1986). A conception of adult development. *American Psychologist, 41,* 3–13.

Levinson, D. J. (1996). *The seasons of a woman's life.* New York: Knopf.

Levinson, D. J., Darrow, C. M., Klein, E. B., Levinson, M. H., & McKee, B. (1978). *The seasons of a man's life.* New York: Knopf.

Levinson, H. (1964, March 9). Money aside, why spend life working? *National Observer,* 20.

Levy, D. M. (1943). *Maternal overprotection.* New York: Columbia University Press.

Levy-Shiff, R. (1994). Individual and contextual correlates of marital change across the transition to parenthood. *Developmental Psychology, 30,* 591–601.

Lewin, E. (1993). *Lesbian mothers: Accounts of gender in American culture.* Ithaca, NY: Cornell University Press.

Lewin, R. (1981). Is longevity a positive selection? *Science, 211,* 373.

Lewin, T. (1990, April 22). Too much retirement time? A move is afoot to change it. *New York Times,* 1, 36.

Lewin, T. (1991, January 17). Women found to be frequent victims of assault by intimates. *New York Times,* A12.

Lewin, T. (1992a, October 5). Rise in single parenthood is reshaping U.S. *New York Times,* A1, A16.

Lewin, T. (1992b, December 10). Sex partners on increase among teen-age girls. *New York Times,* A14.

Lewin, T. (1995a, May 11). Study says more women earn half their household income. *New York Times,* A13.

Lewin, T. (1995b, May 26). Income gap between sexes found to widen in retirement. *New York Times,* A15.

Lewis, C. C. (1988). Cooperation and control in Japanese nursery schools. In G. Handel (Ed.), *Childhood socialization* (pp. 125–142). New York: Aldine de Gruyter.

Lewis, J. S. (1985, April 3). Fathers-to-be show signs of pregnancy. *New York Times,* 13.

Lewis, M. (1995). Developmental change in infants' responses to stress. *Child Development, 66,* 657–670.

Lewis, M. (1998). The development and structure of emotions. In M. Mascolo (Ed.), *What develops in emotional development? Emotions, personality, and psychotherapy.* New York: Plenum Press.

Lewis, M., & Starr, M. D. (1979). Developmental continuity. In J. D. Osofsky (Ed.), *Handbook of infant development.* New York: Wiley.

Lewis, M., Young, G., Brooks, J., & Michalson, L. (1975). The beginning of friendship. In M. Lewis & L. A. Rosenblum (Eds.), *Friendship and peer relations.* New York: Wiley.

Lewis, M. D. (1993). Early socioemotional predictors of cognitive competency at 4 years. *Developmental Psychology, 29,* 1036–1045.

Lewis, M. M. (1936/1951). *Infant speech: A study of the beginnings of language.* London: Routledge & Kegan Paul.

Liaw, F., & Brooks-Gunn, J. (1993). Patterns of low-birth-weight children's cognitive development. *Developmental Psychology, 29,* 1024–1035.

Lickona, T. (1976). Research on Piaget's theory of moral development. In T.

Lickona (Ed.), *Moral development and behavior theory, research, and social issues.* New York: Holt, Rinehart & Winston.

Lieberman, M. A., & Coplan, A. S. (1970). Distance from death as a variable in the study of aging. *Developmental Psychology, 2,* 71–84.

Lieberson, S. (1992). Einstein, Renoir, and Greeley: Some thoughts above evidence in sociology. *American Sociological Review, 57,* 1–15.

Liebow, E. (1967). *Tally's corner: A study of negro streetcorner men.* Boston: Little, Brown.

Liebowitz, M. R. (1983). *The chemistry of love.* Boston: Little, Brown.

Lifton, B. J. (1994). *Journey of the adopted self: A quest for wholeness.* New York: Basic Books.

Lili, C. (1999, January 11–17). AIDS: Crisis and counter-measures. *Beijing Review, 42,* 22–23.

Lillo-Martin, D. (1997). In support of the language acquisition device. In M. Marschark and P. Siple (Eds.), *Relations of language and thought: The view from sign language and deaf children.* New York: Oxford University Press.

Lingren, H. G. (1996). *The sandwich generation: A cluttered nest.* NebGuide, Cooperative Extension, University of Nebraska–Lincoln. Retrieved November 5, 1998, from the World Wide Web: http://www.ianr.unl.pubs/family/g1117.htm

Link, B. G., Lennon, M. C., & Dohrenwend, B. P. (1993). Socioeconomic status and depression: The role of occupations involving direction, control, and planning. *American Journal of Sociology, 98,* 1351–1387.

Linn, S., Reznick, J. S., Kagan, J., & Hans, S. (1982). Salience of visual patterns in the human infant. *Developmental Psychology, 18,* 651–657.

Linton, R. (1936). *The study of man.* New York: Appleton-Century-Crofts.

Lipset, S. M. (1989, May 24). Why youth revolt. *New York Times,* 27.

Livesley, W. J., & Bromley, D. B. (1973). *Person perception in childhood and adolescence.* New York: Wiley.

Livingston, M. M., Burley, K., & Springer, T. P. (1996). The importance of being feminine: Gender, sex role, occupational and marital role commitment, and their relationship to anticipated work-family conflict. *Journal of Social Behavior and Personality, 11,* 179–192.

Lobel, T. E., & Menashri, J. (1993). Relations of conceptions of gender-role transgressions and gender constancy to gender-typed toy preferences. *Developmental Psychology, 29,* 150–155.

Lock, M. (1998). Deconstructing the change: Female maturation in Japan and North America. In R. A. Shweder (Ed.), *Welcome to middle age! And other cultural fictions* (pp. 45–74). Chicago: University of Chicago Press.

Lopata, H. Z. (1981). Widowhood and husband satisfaction. *Journal of Marriage and the Family, 43,* 439–450.

Lord, M. G. (1994, October 25). What that survey didn't say. *New York Times,* A17.

Lorenz, K. Z. (1935). Imprinting. In R. C. Birney & R. C. Teevan (Eds.), *Instinct.* London: Van Nostrand.

Lou, H. C., Hansen, D. Nordentoft, M., Pryds, O., Jensen, F., Nim, J., & Hemmingsen, R. (1994). Prenatal stressors of human life affect fetal brain development. *Developmental Medicine of Child Neurology, 37,* 185.

Love, D. O., & Torrence, W. D. (1989). The impact of worker age on unemployment and earnings after plant closings. *Journal of Gerontology, 44,* S190–195.

Love, S. M., & Lindsey, K. (1997). *Dr. Susan Love's hormone book: Making informed choices about menopause.* New York: Random House.

Lovecky, D. V. (1994). Exceptionally different children: Different minds. *Roeper Review, 17,* 116–120.

Lowrie, M. (1998, January 12). 19 European nations sign ban on human cloning. *CNN World News.* Retrieved July 9, 1998, from the World Wide Web: http://www.cnn.com/world/9801/12/cloning.ban/index.html

Lozoff, B., Wolf, A. W., & Davis, N. S. (1984). Cosleeping in urban families with young children in the United States. *Pediatrics, 74,* 171–182.

Luhmann, N. (1986). *Love as passion: The codification of intimacy.* Cambridge, MA: Harvard University Press.

Luker, K. (1996). *Dubious conceptions: The politics of teenage pregnancy.* Cambridge, MA: Harvard University Press.

Lumpkin, C. K., Jr., McClung, J. K., Pereira-Smith, O. M., & Smith, J. R. (1986). Existence of high abundance antiproliferative mRNA's in senescent human diploid fibroblasts. *Science, 232,* 393–395.

Luster, T., & McAdoo, H. P. (1994). Factors related to the achievement and adjustment of young African American children. *Child Development, 65,* 1080–1094.

Luster, T., & Okagaki, L. (Eds.). (1993). *Parenting: An ecological perspective.* Hillsdale, NJ: Erlbaum.

Luster, T., & Small, S. A. (1994a). Adolescent sexual activity: An ecological, risk-factor approach. *Journal of Marriage and the Family, 56,* 181–192.

Luster, T., & Small, S. A. (1994b). Factors associated with sexual risk-taking behaviors among adolescents. *Journal of Marriage and the Family, 56,* 622–632.

Luthar, S. S. (1995). Social competence in the school setting: Prospective cross-domain associations among inner-city teens. *Child Development, 66,* 416–429.

Lykken, D. T., McGue, M., Tellegen, A., & Bouchard, T. J., Jr. (1992). Emergencies: Genetic traits that may not run in families. *American Psychologist, 47,* 1565–1577.

Lyman, K. A. (1993). *Day in, day out with Alzheimer's: Stress in caregiving relationships.* Philadelphia: Temple University Press.

Lynch, E. W., & Hansen, M. J. (1992). *Developing cross-cultural competence: A guide for working with young children and their families.* Baltimore: Paul H. Brookes.

Lyon, T. D., & Flavell, J. H. (1994). Young children's understanding of "remember" and "forget." *Child Development, 65,* 1357–1371.

Lyons-Ruth, K., Alpern, L., & Repacholi, B. (1993). Disorganized infant attachment classification and maternal psychosocial problems as predictors of hostile-aggressive behavior in the preschool classroom. *Child Development, 64,* 572–585.

Lytton, H. (1979). Disciplinary encounters between young boys and their mothers and fathers: Is there a contingency system? *Developmental Psychology, 15,* 256–268.

Lytton, H., & Romney, D. M. (1991). Parents' differential socialization of boys and girls: A meta-analysis. *Psychological Bulletin, 109,* 267–296.

M

Macaulay, A. C., Paradis, G., Potvin, L., Cross, E. J., Saad-Haddad, C., et al. (1997). The Kahnawake Schools diabetes prevention project: Intervention, evaluation, and baseline results of a diabetes primary prevention program with a native community in Canada. *Preventive Medicine, 26,* 779–790.

Maccoby, E. E. (1961). The taking of adult roles in middle childhood. *Journal of Abnormal and Social Psychology, 63,* 493–503.

Maccoby, E. E. (1980). *Social development: Psychological growth and the parent-child relationship.* New York: Harcourt Brace Jovanovich.

Maccoby, E. E. (1983). Social-emotional development and responses to stressors. In N. Garmezy & M. Rutter (Eds.), *Stress, coping, and development in children.* New York: McGraw-Hill.

Maccoby, E. E. (1988). Gender as a social category. *Developmental Psychology, 24,* 755–765.

Maccoby, E. E. (1990). Gender and relationships: A developmental account. *American Psychologist, 45,* 513–520.

Maccoby, E. E. (1991). Different reproductive strategies in males and females. *Child Development, 62,* 676–681.

Maccoby, E. E. (1992). The role of parents in the socialization of children: An historical overview. *Developmental Psychology, 28,* 1006–1017.

Maccoby, E. E., & Jacklin, C. N. (1974). *The psychology of sex differences.* Stanford, CA: Stanford University Press.

Maccoby, E. E., & Jacklin, C. N. (1987). Gender segregation in childhood. In H. W. Reese (Ed.), *Advances in child development and behavior* (Vol. 20). New York: Academic Press.

Maccoby, E. E., & Maccoby, N. (1954). The interview: A tool of social science. In G. Lindzey (Ed.), *Handbook of social psychology.* Reading, MA: Addison-Wesley.

Maccoby, E. E., & Masters, J. C. (1970). Attachment and dependency. In P. H. Mussen (Ed.), *Carmichael's manual of child psychology* (3rd ed.). New York: Wiley.

MacDonald, K. (1992). Warmth as a developmental construct: An evolutionary analysis. *Child Development, 63,* 753–773.

Machan, D. (1997, September 8). A more tolerant generation. *Forbes,* 46–47.

Macmillan, M. (1991). *Freud evaluated: The completed arc.* Amsterdam: North-Holland.

Macsween, M. (1993). *Anorexic bodies: A feminist and sociological perspective on anorexia nervosa.* London: Routledge.

Madden, D. J. (1990). Adult age differences in the time course of visual attention. *Journal of Gerontology, 45,* P9–P16.

Madden, N. A., & Slavin, R. F. (1983). Mainstreaming students with mild handicaps: Academic and social outcomes. *Review of Educational Research, 53,* 519–569.

Makarenko, A. S. (1967). *The collective family: A handbook for Russian parents.* New York: Doubleday.

Malcolm, A. H. (1990, June 9). Giving death a hand: Rending issue. *New York Times,* 6A.

Malinak, D. P., Hoyt, M. F., & Patterson, V. (1979). Adults' reactions to the death of a parent: A preliminary study. *American Journal of Psychiatry, 136,* 1152–1156.

Mallett, P., Apostolidis, T., & Paty, B. (1997). The development of gender schemata about heterosexual and homosexual others. *Journal of General Psychology, 124,* 91–104.

Manegold, C. S. (1994, August 18). Students' skills lag behind expectations. *New York Times,* A6.

Mangelsdorf, S. C. (1992). Developmental changes in infant-stranger interaction. *Infant Behavior and Development, 15,* 191–208.

Manning, A. (1994, October 21). Trouble follows armed students. *USA Today,* D1.

Manning, W. D. (1993). Marriage and cohabitation following premarital conception. *Journal of Marriage and the Family, 55,* 839–850.

Manning, W. D. (1995). Cohabitation, marriage, and entry into motherhood. *Journal of Marriage and the Family, 57,* 191–200.

Manton, K. G., Stallard, E., & Liu, K. (1993). Forecasts of active life expectancy: Policy and fiscal implications. *Journals of Gerontology, 48* (Special Issue), 11–26.

Mantyla, T. (1994). Remembering to remember: Adult age differences in prospective memory. *Journal of Gerontology: Psychological Sciences, 49,* P276–P282.

March of Dimes. (1998). *Ten leading causes of infant mortality: United States, 1995.* March of Dimes. Infant Health Statistics. Retrieved July 27, 1998, from the World Wide Web: http://www.modimes.org/stats/ten.htm

Marcia, J. E. (1966). Development and validation of ego identity status. *Journal of Personality and Social Psychology, 3,* 551–558.

Marcia, J. E. (1991). Identity and self-development. In R. M. Lerner, A. C. Peterson, & J. Brooks-Gunn (Eds.), *Encyclopedia of adolescence* (Vol. 1). New York: Garland.

Marini, Z., & Case, R. (1994). The development of abstract reasoning about the physical and social world. *Child Development, 65,* 147–159.

Marks, N. (1998, April). Women and HIV: Treatment strategies for women examined at physicians' forum. *Positive Living Newsletter.* Retrieved June 11, 1998, from the World Wide Web: http://www.apla.org/apla/9804/womenandhiv.html

Marks, N., & Lambert, J. D. (1998). Marital status continuity and change among young and midlife adults. *Journal of Family Issues, 19,* 652–686.

Marks, N. F. (1996). Flying solo at midlife: Gender, marital status, and psychological well-being. *Journal of Marriage and the Family, 58,* 917–932.

Markus, H. (1977). Self-schemata and processing information about the self. *Journal of Personality and Social Psychology, 35,* 63–78.

Martin, D., & Lyon, P. (1992). The older lesbian. In B. Berzon and R. Leighton (Eds.), *Positively gay* (pp. 111–120). Berkeley: Celestial Arts.

Martin, L. G. (1989). *The graying of Japan.* Washington, DC: Population Reference Bureau.

Martin, N. K., & Dixon, P. N. (1986). Adolescent suicide: Myths, recognition, and evaluation. *School Counselor, 33,* 265–271.

Martin, T. (1998). Of statistics, single mothers, and the politics of language. *Father Magazine.* Retrieved July 30, 1998, from the World Wide Web: http://www.fathermag.com/htmlmodules/Jan97/xTrev1.html

Martin, T. C., & Bumpass, L. L. (1989). Recent trends in marital disruption. *Demography, 26,* 37–51.

Marx, J. L. (1984, December 21). The riddle of development. *Science,* 1406–1408.

Marx, J. L. (1988). Sexual responses are "almost" all in the brain. *Science, 241,* 903–904.

Marx, J. L. (1992). Familial Alzheimer's linked to chromosome 14 gene. *Science, 258,* 550.

Marx, J. L. (1993). Alzheimer's pathology begins to yield its secrets. *Science, 259,* 457–458.

Maslow, A. H. (1968). *Toward a psychology of being* (2nd ed.). New York: Van Nostrand.

Maslow, A. H. (1970). *Motivation and personality* (2nd ed.). New York: Harper & Row.

Mastekaasa, A. (1992). Marriage and psychological well-being: Some evidence on selection into marriage. *Journal of Marriage and the Family, 54,* 901–911.

Masters, W. H., & Johnson, V. E. (1966). *Human sexual response.* Boston: Little, Brown.

Matras, J. (1975). *Social inequality, stratification, and mobility.* Englewood Cliffs, NJ: Prentice Hall.

Matsueda, R. L. (1992). Reflected appraisals, parental labeling, and delinquency: Specifying a symbolic interactionist theory. *American Journal of Sociology, 97,* 1577–1611.

Matthews, D. (1998). *The faith factor: Proof of the healing power of prayer.* New York: Penguin Putnam.

Maurer, D. M., & Maurer, C. E. (1976, October). Newborn babies see better than you think. *Psychology Today, 10,* 85–88.

Maurer, D. M., & Salapatek, P. (1976). Developmental changes in the scanning of faces by young infants. *Child Development, 47,* 523–527.

Maxwell, E. (1998). Exceptionally gifted children. *Gifted Child Quarterly, 39,* 245.

Mayer, R. E. (1996). Learners and information processors: Legacies and limitations of educational psychology's second metaphor. *Educational Psychologist, 31,* 151–161.

McAdams, D. P., & de St. Aubin, E. (1992). A theory of generativity and its assessment through self-report, behavioral acts, and narrative themes in autobiography. *Journal of Personality and Social Psychology, 62,* 1003–1015.

McAdams, D. P., de St. Aubin, E., & Logan, R. L. (1993). Generativity among young, midlife, and older adults. *Psychology and Aging, 8,* 221–230.

McAdoo, M. (1995, March 20). For your child, public or private school? *Investor's Business Daily,* A1, A2.

McAlvaney, D. S. (1996, August). Germ warfare against Americans: Part I—What is Gulf War Illness (GWI)? *McAlvaney Intelligence Advisor.* Retrieved July 31, 1998, from the World Wide Web: http://www.all-natural.com/part-1.html

McAuley, E., Lox, C., & Duncan, T. E. (1993). Long-term maintenance of exercise, self-efficacy, and physiological change in older adults. *Journal of Gerontology: Psychological Sciences, 48,* P218–P224.

McCandless, B. R. (1970). *Adolescents: Behavior and development.* New York: Holt, Rinehart & Winston.

McClelland, D. C., Constantian, C. A., Regalado, D., and Stone, C. (1978, June). Making it to maturity. *Psychology Today, 12,* 42.

McCowage, G. B. (1995, July 6). Transplantation of cord-blood cells. *New England Journal of Medicine, 333,* 67.

McCoy, E. (1982, May 6). Children of single parents. *New York Times,* 19, 21.

McCrae, R. R. & Costa, P. T., Jr. (1990). *Personality in adulthood.* New York: Guilford Press.

McDonald, K. A. (1992, March 4). New studies reveal aging process can be slowed. *Chronicle of Higher Education,* A9, A10.

McDonald, K. A. (1994, September 14). Biology and behavior. *Chronicle of Higher Education,* A10.

McDowd, J. M., & Filion, D. L. (1992). Aging, selective attention, and inhibitory processes: A psychophysiological approach. *Psychology and Aging, 7,* 65–71.

McGee, J., & Wells, K. (1982). Gender typing and androgyny in later life: New directions for theory and research. *Human Development, 25,* 116–139.

McGhee, P. E., & Chapman, A. J. (1980). *Children's humour.* New York: John Wiley & Sons.

McGraw, M. B. (1935). *Growth: A study of Johnny and Jimmy.* New York: Appleton-Century.

McGuire, S., Neiderhiser, J. M., Reiss, D., Hetherington, E. M., & Plomin, R. (1994). Genetic and environmental influences on perceptions of self-worth and competence in adolescence: A study of twins, full siblings, and step-siblings. *Child Development, 65,* 785–799.

McHale, S. M., Crouter, A. C., McGuire, S. A., & Updegraff, K. A. (1995). Congruence between mothers' and fathers' differential treatment of siblings: Links with family relations and children's well-being. *Child Development, 66,* 116–128.

McKenna, J. (1996). Sudden infant death syndrome in cross cultural perspective: Is infant-parent co-sleeping protective. *Annual Review of Anthropology, 25,* 201–216.

McKenna, M. A. J. (1998, March 20). Black teen suicide rate skyrockets: CDC says health experts cite family breakdown, violence, economics. *Atlanta Constitution,* A1.

McKenzie, B. E., Skouteris, H., Day, R. H., Hartman, B., & Yonas, A. (1993). Effective action by infants to contact objects by reaching and leaning. *Child Development, 64,* 415–429.

McKenzie, J. K. (1993). Adoption of children with special needs. *Future of Children: Adoption, 3,* 62–75.

McKeown, R. E., Garrison, C. Z., Cuffe, S. P., Waller, J., et al. (1998). Incidence and predictors of suicidal behaviors in a longitudinal sample of young adolescents. *Journal of the American Academy of Child and Adolescent Psychiatry, 37,* 612–619.

McLannan, S., & Sandefur, G. (1994). *Growing up with a single parent: What hurts, what helps.* Cambridge, MA: Harvard University Press.

McMaster, J., Corme, K., & Pitts, M. (1997). Menstrual and premenstrual experiences of women in a developing country. *Health Care for Women International, 18,* 533–541.

Mead, G. H. (1934). *Mind, self, and other.* Chicago: University of Chicago Press.

Medawar, P. B. (1977, February 3). Unnatural science. *New York Review of Books, 24,* 13–18.

Mederer, H. J. (1993). Division of labor in two-earner homes: Task accomplishment versus household management as critical variables in perceptions about family work. *Journal of Marriage and the Family, 55,* 133–145.

MedicineNet. (1997). *Fetal alcohol syndrome.* Retrieved July 8, 1998, from the World Wide Web: http://www.medicinenet.com/mainmenu/encyclop/article/art_a/alcohol.htm

MedLib Utah. (1998). Prenatal diagnosis. *Internet Pathology Library.* Retrieved July 10, 1998, from the World Wide Web: http://www-medlib.med.utah.edu/WebPath/tutorial/prenatal/prenatal.html

Meer, J. (1986, June 20). The reason of age. *Psychology Today,* 60–64.

Meier, B. (1987, February 5). Companies wrestle with threats to workers' reproductive health. *Wall Street Journal,* 21.

Meier, D. (1998). A national survey of physician assisted suicide and euthanasia in the United States. *New England Journal of Medicine, 338,* 1193–1201.

Menaghan, E. G., & Parcel, T. L. (1990). Parental employment and family life: Research in the 1980s. *Journal of Marriage and the Family, 52,* 1079–1098.

Menten, T. (1991). *Gentle closings: How to say goodbye to someone you love,* Philadelphia: Running Press.

Meredith, D. (1985, June). Dad and the kids. *Psychology Today, 19,* 62–67.

Meredith, H. V. (1973). Somatological development. In B. B. Wolman (Ed.), *Handbook of general psychology.* Englewood Cliffs, NJ: Prentice Hall.

Merline, J. (1994, May 13). Black hole of social security. *Investor's Business Daily,* 1, 2.

Merrill, A. M. (1996). *Report on intercountry adoption.* Boulder, CO: International Concerns for Children.

Merrill, S. S., Seeman, T. E., Kasl, S. V., & Berkman, L. F. (1997). Gender differences in the comparison of self-reported disability and performance measures. *Journals of Gerontology,* Series A, Biological Sciences and Medical Sciences, 52, M19–M26.

Merz, B. (1992, October). Why we get old. *Harvard Health Letter* (Special Supp.), 9–12.

Meschke, L. L., & Silbereisen, R. K. (1997). The influence of puberty, family processes, and leisure activities on the timing of first sexual experience. *Journal of Adolescence, 20,* 403–418.

Messick, S. (1976). Personality consistencies in cognition and creativity. In S. Messick et al. (Eds.), *Individuality in learning* (pp. 4–22). San Francisco: Jossey-Bass.

Messick, S. (1984). The nature of cognitive styles: Problems and promises in educational practice. *Educational Psychologist, 19,* 59–74.

Metzler, C. W., Noell, J., & Biglan, A. (1992). The validation of a construct of high-risk sexual behavior in heterosexual adolescents. *Journal of Adolescent Research, 7,* 233–249.

Meyer, J. W., Ramirez, F. O., & Soysal, Y. N. (1992). World expansion of mass education, 1870–1980. *Sociology of Education, 65,* 128–149.

Michael, R. (1994). *Sex in America: A definitive survey.* Boston: Little, Brown.

Midanik, L. T., Soghikian, K., Ransom, L. J., & Tekawa, I. S. (1995). The effect of retirement on mental health and health behaviors: The Kaiser Permanente Retirement Study. *Journal of Gerontology: Social Sciences, 508,* S59–S61.

Middlekoop, H. A., Smilde-van den Doel, D. A., Neven, A. K., Kamphuisen, H. A., & Springer, C. P. (1996). Subjective sleep characteristics of 1,485 males and females aged 50–93: Effects of sex and age, and factors related to self-evaluated quality of sleep. Journal of Gerontology and Biological Science Medical *Science, 51,* M108–M115.

Midlife passages. (1998a). *Causes of death or impaired health.* Retrieved October 12, 1998, from the World Wide Web: http://www.midlife-passages.com/newpage2.htm

Midlife passages. (1998b). *What is menopause?* Retrieved May 14, 1998, from the World Wide Web: http://www.midlife-passages.com/page33.html

Midlife passages. (1999). *Causes of illness and death.* Retrieved March 25, 1999, from the World Wide Web: http://www.midlife-passages.com/newpage2.htm

Miller, B., & McFall, S. (1991). The effect of caregiver's burden on change in frail older persons' use of formal helpers. Journal of *Health and Social Behavior, 32,* 165–179.

Miller, B. C., & Moore, K. A. (1990). Adolescent sexual behavior, pregnancy, and parenting: Research through the 1980s. *Journal of Marriage and the Family, 52,* 1025–1044.

Miller, J. B. (1991). Relations between young adults and their parents. *Journal of Adolescence, 14,* 179–194.

Miller, M. W. (1985, January 17). Study says birth defects more frequent in areas polluted by technology firms. *Wall Street Journal,* 6.

Miller, M. W. (1994, January 14). Survey sketches new portrait of the mentally ill. *Wall Street Journal*, B1.

Miller, R. A. (1989). The cell biology of aging: Immunological models. *Journal of Gerontology, 44*, B4–8.

Milne, L. (1924). *The home of an eastern clan.* Oxford: Clarendon Press.

Milner, J., Robertson, K., Rogers, D. (1990). Childhood history of abuse and adult child abuse potential, *Journal of Family Violence, 5*, 15–34.

Mintz, S., & Kellogg, S. (1988). *Domestic revolutions: A social history of American family life.* New York: Free Press.

Mirowsky, J. (1995). Age and the sense of control. *Social Psychology Quarterly, 58*, 31–43.

Mirowsky, J. (1996). Age and the gender gap in depression. *Journal of Health and Social Behavior, 37*, 362–380.

Mirowsky, J., & Ross, C. E. (1989). *Social causes of psychological distress.* New York: Aldine de Gruyter.

Mirror images: Depressed mothers, depressed newborns. (1998). *APA Public Communications.* Washington, DC: American Psychological Association. Retrieved July 31, 1998, from the World Wide Web: http://www.apa.org/releases/mom.html

Mischel, W. (1969). Continuity and change in personality. *American Psychologist, 24*, 1012–1018.

Mischel, W. (1973). Toward a cognitive social learning reconceptualization of personality. *Psychological Review, 80*, 252–283.

Mischel, W. (1977). On the future of personality measurement. *American Psychologist, 32*, 246–254.

Mischel, W. (1985). *Diagnosticity of situations.* Paper presented at the October meeting of the Society for Experimental Social Psychology, Evanston, IL.

Mitchell, V., & Helson, R. (1990). Women's prime of life. *Psychology of Women Quarterly, 14*, 451–470.

Mobley, J. A., McKeown, R. E., & Jackson, K. L., Francisco, S., Parham, J. S., & Brenner, E. R. (1998). Risk factors for congenital syphilis in infants of women with syphilis in South Carolina. *American Journal of Public Health, 88*, 597–602.

Modell, J. (1989). *Into one's own: From youth to adulthood in the United States, 1920–1975.* Berkeley: University of California Press.

Moen, P., Dempster-McClain, D., & Williams, R. M., Jr. (1992). Successful aging: A life-course perspective on women's multiple roles and health. *American Journal of Sociology, 97*, 1612–1638.

Moffitt, T. E., Caspi, A., Belsky, J., & Silva, P. A. (1992). Childhood experience and the onset of menarche: A test of a sociobiological model. *Child Development, 63*, 47–58.

Mogul, K. M. (1979). Women in midlife: Decisions, rewards, and conflicts related to work and careers. *American Journal of Psychiatry, 136*, 1139–1143.

Money, J., & Tucker, P. (1975). *Sexual signatures: On being a man or a woman.* Boston: Little, Brown.

Montagu, A. (1964). *Life before birth.* New York: New American Library.

Montagu, A. (1978). *Touching: The human significance of the skin.* New York: Harper & Row.

Montgomery, R. J. V., & Kosloski, K. (1994). A longitudinal analysis of nursing home placement for dependent elders cared for by spouses vs adult children. *Journal of Gerontology: Social Sciences, 49*, S62–S74.

Moody, R. (1976). *Life after life.* New York: Bantam Books.

Moore, C., Bryant, D., & Furrow, D. (1989). Mental terms and the development of certainty. *Child Development, 60*, 167–171.

Moore, D. (1983, January 30). America's neglected elderly. *New York Times Magazine*, 30–35.

Moore, S. M., Rosenthal, D. A., Boldero, J. (1993). Predicting AIDS-preventive behaviour among adolescents. In D. J. Terry et al. (Eds.), *The theory of reasoned action: Its application to AIDS-preventive behavior. (International Series in Experimental Social Psychology, 28)* (pp. 65–80). Oxford, England: Pergamon Press.

Mor, V., Branco, K., Fleishman, J., Hawes, C., Phillips, C., Morris, J., & Fries, B. (1995). The structure of social engagement among nursing home residents. *Journal of Gerontology: Psychological Sciences, 50B*, P1–P8.

Morabia, A., & Costanza, M. C. (1998). International variability in ages at menarche, first livebirth, and menopause. *American Journal of Epidemiology, 148*, 195–205.

Morelli, G., Rogoff, B., Oppenheim, D., Goldsmith, D. (1992). Cultural variation in infants' sleeping arrangements, *Developmental Psychology, 28*, 604–613.

Morgan, G. A., & Ricciuti, H. N. (1969). Infants' responses to strangers during the first year. In B. M. Foss (Ed.), *Determinants of infant behavior* (Vol. 4). New York: Wiley.

Morin, C. M., Stone, J., Trinkle, D., Mercer, J., & Remsberg, S. (1993). Dysfunctional beliefs and attitudes about sleep among older adults with and without insomnia complaints. *Psychology and Aging, 8*, 463–467.

Morokoff, P. J. (1985). Effects of sex guilt, repression, sexual "arousability," and sexual experience on female sexual arousal during erotica and fantasy. *Journal of Personality and Social Psychology, 49*, 177–187.

Morris, M. R. (1998). Elder abuse: What the law requires. *RN, 61*, 52–54.

Morse, M. (1992). *Transformed by the light.* New York: Villard.

Morse, N. C., & Weiss, R. S. (1955). The function and meaning of work and the job. *American Sociological Review, 20*, 191–198.

Mosher, W. D., & Bachrach, C. A. (1996). Understanding U.S. infertility: Continuity and change in the National Survey of Family Growth. *Family Planning Perspectives, 28*, 4–12.

Moskowitz, B. A. (1978, November). The acquisition of language. *Scientific American, 239*, 92–108.

Moss, M., & Moss, S. (1983–1984). The impact of parental death on middle aged children. *Omega: Journal of Death and Dying, 14*, 65–75.

Most want long lives but fear dependence. (1991, November 18). *New York Times*, C10.

Mott, F. L. (1994). Sons, daughters, and fathers' absence: Differentials in father-leaving probabilities and in home environments. *Journal of Family Issues, 15*, 97–128.

MOW International Research Team. (1987). *The meaning of working.* London: Academic Press.

Mowsesian, R. (1987). *Golden goals, rusted realities.* New York: New Horizon/Macmillan.

Moyer, S. & Oliveri, C. (1996). Strengthening families and communities by sharing life stories. Ohio State University Extension Factsheet, Family and Consumer Sciences, Columbus, Ohio. Retrieved September 23, 1998, from the World Wide Web: http://www.ag.ohio-state.edu/~ohioline/hyg-fact/5000/5227.html

Moynihan, D. P. (1985). *Family and nation.* Cambridge, MA: Harvard University Press.

Mueller, E. C., & Cooper, C. R. (Eds.). (1986). *Process and outcome in peer relationships.* Orlando, FL: Academic Press.

Mueller, R. F. (1983). Evaluation of a protocol for post-mortem examination of stillbirths. *New England Journal of Medicine, 309,* 586–590.

Muller, R. T., Hunter, J. E., & Stollack, G. (1995). The intergenerational transmission of corporal punishment: A comparison of social learning and temperament models. *Child Abuse and Neglect, 19,* 1323–1335.

Murdock, G. P. (1934). *Our primitive contemporaries.* New York: Macmillan.

Murdock, G. P. (1935). Comparative data on the division of labor by sex. *Social Forces, 15,* 551–553.

Murdock, G. P. (1949). *Social structure.* New York: Macmillan.

Murdock, G. P. (1957). Anthropology as a comparative science. *Behavioral Science, 2,* 249–254.

Murphy, S. A., Das Gupta, A., Cain, K. C., Johnson, L. C., et al. (1999). Changes in parents' mental distress after the violent death of an adolescent or young adult child: A longitudinal prospective analysis. *Death Studies, 23,* 129–159.

Mussen, P. H., & Jones, M. C. (1957). Self-conceptions, motivations, and interpersonal attitudes of late- and early-maturing boys. *Child Development, 28,* 243–256.

Myers, D. G., & Diener, I. (1995). Who is happy? *Psychological Science, 6,* 10–19.

N

Nadel, S. F. (1951). *The foundations of social anthropology.* New York: Free Press.

Nathanson, C. A. (1991). *Dangerous passage: The social control of sexuality in women's adolescence.* Philadelphia: Temple University Press.

National Adoption Information Clearinghouse. (1996). *Adoption statistics for 1996.* Retrieved December 9, 1998, from the World Wide Web: http://www.naicinfo.com/stats.htm

National Cancer Institute. (1998a). *National Cancer Institute initiatives applicable to prostate cancer research.* Retrieved November 30, 1998, from the World Wide Web: http://www.nci.nih.gov/prostate.html

National Cancer Institute. (1998b). *Prevention of skin cancer.* Retrieved February 10, 1999, from the World Wide Web: http://cancernet.nci.nih.gov/clinpdq/screening/prevention_of_skin_cancer_patient.html

National Cancer Institute. (1998c). *Screening for prostate cancer.* Retrieved June 12, 1998, from the World Wide Web: http://cancernet.nci.nih.gov/clinpdq/screening_for_prostate_cancer_physician.html

National Center for Health Statistics. (1998). *Multiple births.* March of Dimes Perinatal Data Center. Retrieved July 27, 1998, from the World Wide Web: http://www.modimes.org/info/info8.htm

National Center for Learning Disabilities. (1997). *Information about learning disabilities.* Retrieved December 16, 1998, from the World Wide Web: http://www.ncld.org/info_ld.html

National Commission on Excellence in Education. (1983). *A nation at risk: The imperative for educational reform.* Washington, DC: U.S. Department of Education.

National Committee to Prevent Child Abuse. (1996a, December). *Child sexual abuse, 19.* Retrieved August 17, 1998, from the World Wide Web: http://www.childabuse.org/fs19.html

National Committee to Prevent Child Abuse. (1996b, December). *Prevention of child abuse and neglect fatalities, 9.* Retrieved August 17, 1998, from the World Wide Web: http://www.childabuse.org/fs9.html

National Committee to Prevent Child Abuse. (1998). *Child abuse and neglect statistics.* Retrieved August 17, 1998, from the World Wide Web: http://www.childabuse.org/facts97

National Council on Aging. (1998). *Healthy sexuality and vital aging.* Retrieved September 28, 1998, from the World Wide Web: http://www.ncoa.org/news/archives/sexsurvey.htm

National Down Syndrome Society. (1998). *Parent and professional information.* Retrieved July 10, 1998, from the World Wide Web: http://www.ndss.org/information/general_info.html#Down

National Drug Strategy Monograph No. 25. (1998). *DRC online library of drug policy.* Retrieved February 7, 1999, from the World Wide Web: http://www.druglibrary.org

National Human Genome Research Institute. (1998a). *From maps to medicine: About the Human Genome Research Project.* Retrieved July 9, 1998, from the World Wide Web: http://www.nhgri.nih.gov/policy_and_publications/maps_to-medicine/about.html

National Human Genome Research Institute. (1998b). *How to conquer a genetic disease.* Retrieved July 9, 1998, from the World Wide Web: http://www.nhgri.nih.gov/policy_and_publications/maps_to_medicine/how.html#key

National Institute of Allergy and Infectious Disease. (1997). *Pediatric AIDS fact sheet.* National Institutes of Health. Retrieved July 29, 1998, from the World Wide Web: http://www.intellihealth.com

National Institute of Arthritis and Musculoskeletal and Skin Diseases. (1998). *Rheumatoid Arthritis.* Retrieved June 10, 1998, from the World Wide Web: http://www.nih.gov/niams/healthinfo/rahandout_breaks.html

National Institute of Child Health and Human Development. (1999). *NICHD child care study investigators to report on child care quality: Higher quality care related to less problem behavior.* Retrieved February 13, 1999, from the World Wide Web: http://www.nih.gov/nichd/html/news/daycar99.htm

National Institute of Neurological Disorders and Stroke. (1998a). *Brain basics: Preventing stroke.* Retrieved June 10, 1998, from the World Wide Web: http://www.ninds.nih.gov/healinfo/disorder/stroke/strokepr.htm

National Institute of Neurological Disorders and Stroke. (1998b). *Parkinson's disease: Hope through research.* Retrieved June 10, 1998, from the World Wide Web: http://www.ninds.nih.gov/healinfo/disorder/parksinso/pdhtr.htm

National Institute on Aging. (1994). *HIV, AIDS, and older adults.* Retrieved June 10, 1998, from the World Wide Web: http://www.nih.gov/nia/health/pubpub/aids.htm

National Institute on Aging. (1996a). *Osteoporosis: The silent bone thinner. Age Page.* Retrieved June 10, 1998, from the World Wide Web: http://www.nih.gov/nia/health/pubpub/osteo.htm

National Institute on Aging. (1996b). *Skin care and aging.* Retrieved June 10, 1998 from the World Wide Web: http://www.nih.gov/nia/health/pubpub/skin.htm

National Institute on Aging. (1998). *Aging and alcohol abuse.* Retrieved June 10, 1998, from the World Wide Web: http://silk.nih.gov/silks/niaaa1/publication/agepage.htm

National Institutes of Health. (1996). *How to prevent high blood pressure.* National Heart, Lung, and Blood Institute. Retrieved June 10, 1998, from the World Wide Web: http://www.nih.gov/health/htp-hbp/index.htm

National Institutes of Health. (1997). *AIDS and aging: Behavioral sciences prevention research.* NIH Guide, 26. Retrieved June 20, 1997, from the World Wide Web: http://www.nih.gov/grants/guide/pa-files/PA-97-069.html

National Institutes of Health. (1998a). *HIV/AIDS vaccine research.* Retrieved June 10, 1998, from the World Wide Web: http://www.nih.gov/news/pr/may98/od-15a.htm

National Institutes of Health. (1998b). *How rheumatoid arthritis develops and progresses.* Retrieved February 10, 1999, from the World Wide Web: http://www.nih.gov/niams/healthinfo/rahandout/how.html

National Institutes of Health. (1999). *How you can lower your cholesterol level.* Retrieved February 10, 1999, from the World Wide Web: http://rover.nhlbi.nih.gov/chd/lifestyles.htm

National Institute of Mental Health. (1993). *Learning disabilities.* Retrieved February 13, 1999, from the World Wide Web: http://www.nimh.gov/publicat/learndis.htm

National Library of Medicine. (1998) *Cesarean section—A brief history.* Retrieved July 27, 1998, from the World Wide Web: http://www.nlm.hih.gov.exhibition/cesarean/cesarean_4.html

National Organization on Fetal Alcohol Syndrome. (1998). *Fetal alcohol syndrome fact sheet.* Retrieved July 31, 1998, from the World Wide Web: http://www.nofas.org/what.htm

National Public Radio. (1998, January 6). Human cloning plans. *All Things Considered.* Retrieved July 9, 1998, from the World Wide Web: http://www.npr.org/news/health/980106.cloning.html

National Research Council. (1987). *Risking the future: Adolescent sexuality, pregnancy, and childbearing.* Washington, DC: National Academy Press.

National Research Council. (1996). *Understanding child abuse and neglect.* Retrieved July 30, 1998, from the World Wide Web: http://ww2.nas.edu/whatsnew/256E.htm

National Research Council. (1998). *Political debate interferes with research on educating children with limited English proficiency.* Retrieved August 10, 1998, from the World Wide Web: http://www2.nas.edu/whatsnew/2652.html

National Right-to-Life Committee. (1998). *Is abortion safe? Psychological consequences.* Retrieved July 15, 1998, from the World Wide Web: http://www.nrlc.org/abortion/asmf14.html

National survey of family growth. (1999). National Center for Health Statistics. Ann Arbor, MI: Inter-University Consortium for Political and Social Research.

Nazario, S. L. (1989, December 6). Mormon rules aid long life, study discloses. *Wall Street Journal,* B4.

Needham, A., & Baillargeon, R. (1998). Effects of prior experience on 4.5-month-old infants' object segregation. *Infant Behavior and Development, 21,* 1–24.

Needleman, H. L, Schell, A., Bellinger, M. A., Leviton, A., & Allred, E. N. (1996). The long-term effects of exposure to low does of lead in childhood: An 11-year follow-up report. *New England Journal of Medicine, 322,* 83–88.

Neisser, U. (1967). *Cognitive psychology.* New York: Appleton-Century-Crofts.

Neisser, U. (1991). Two perceptually given aspects of the self and their development. *Psychological Review, 11,* 197–209.

Neisser, U. G., Boodoo, T. J., Bouchard, A. W., Boykin, N., Brody, S., et al. (1996). Intelligence: Knowns and unknowns. *American Psychologist, 51,* 77–101.

Nelson, B. (1982, December 7). Early memory: Why is it so elusive? *New York Times,* 17.

Nelson, B. (1983, April 2). Despair among jobless is on rise, studies find. *New York Times,* 8.

Nelson, K. (1972). The relation of form recognition to concept development. *Child Development, 43,* 67–74.

Nelson, K. (1973). Structure and strategy in learning to talk. *Monographs of the Society for Research in Child Development, 38* (No. 149).

Nelson, K., Rescorla, L., Gruendel, J., & Benedict, H. (1978). Early lexicons: What do they mean? *Child Development, 49,* 960–968.

Nelson, N. M., Enkin, M. W., Saigal, S., Bennett, K. J., Milner, R., & Sackett, D. L. (1980). A randomized clinical trial of the Leboyer approach to childbirth. *New England Journal of Medicine, 302,* 655–660.

Neonatology on the Web. (1998). *Computers in the NICU: Lessons from the past, prospects for the future.* Retrieved July 21, 1998, from the World Wide Web: http://www.neonatology.org/cin/survey/cin.10.html

Neugarten, B. L. (1968). The awareness of middle age. In B. L. Neugarten (Ed.), *Middle age and aging.* Chicago: University of Chicago Press.

Neugarten, B. L. (1973). Personality change in later life: A developmental perspective. In C. Eisdorfer & W. P. Lawton (Eds.), *The psychology of adult development and aging.* Washington, DC: American Psychological Association.

Neugarten, B. L. (1982a). Age or need? *National Forum: Phi Kappa Phi Journal, 42,* 25–27.

Neugarten, B. L. (1982b). The aging society. *National Forum: Phi Kappa Phi Journal, 42,* 3.

Neugarten, B. L. & Datan, N. (1974). *The middle years.* In S. Arieti (Ed.), The foundations of psychiatry (pp. 592–608). New York: Basic Books.

Neugarten, B. L., Havighurst, R. J., & Tobin, S. S. (1968). Personality and patterns of aging. In B. L. Neugarten (Ed.), *Middle age and aging.* Chicago: University of Chicago Press.

Neugarten, B. L., Moore, J. W., & Lowe, J. C. (1965). Age norms, age constraints and adult socialization. *American Journal of Sociology, 70,* 710–717.

Neugarten, B. L., & Neugarten, D. A. (1987, May). The changing meanings of age. *Psychology Today, 21,* 29–33.

Neuharth, A. (1997, August 8). What did she prove by living to be 122? *USA Today.* A15:1.

Newcomb, A. F., & Bagwell, C. L. (1995). Children's friendship relations: A meta-analytic review. *Psychological Bulletin, 117,* 306–347.

Newcomb, A. F., Bukowski, W. M., & Pattee, L. (1993). Children's peer relations: A meta-analytic review of popular, rejected, neglected, controversial, and average sociometric status. *Psychological Bulletin, 113,* 99–128.

Newcombe, N., & Fox, N. A. (1994). Infantile amnesia: Through a glass darkly. *Child Development, 65,* 31–40.

Newcombe, N., & Huttenlocher, J. (1992). Children's ability to solve perspective-taking problems. *Developmental Psychology, 28,* 635–643.

Newman, J. L., Roberts, L. R., & Syre, C. R. (1993). Concepts of family among children and adolescents: Effect of cognitive level, gender, and family structure. *Developmental Psychology, 29,* 951–962.

Newman, S. J., Struyk, R., Wright, P., & Rice, M. (1990). Overwhelming odds: Caregiving and the risk of institutionalization. *Journal of Gerontology, 45,* S173–183.

Newport, E. L. (1990). Maturational constraints on language learning. *Cognitive Science, 14,* 11–28.

Nickel. C. (1997, November). Repetitive prostate massage works. *Urology Times.* Retrieved June 12, 1998, from the World Wide Web: http://www.prostate.org/nickelarticle97.html

Nilsson, L., & Hamberger, L. (1990). *A child is born.* New York: Dell.

Nock, S. L. (1995). Commitment and dependency in marriage. *Journal of Marriage and the Family, 57,* 503–514.

Nolen-Hoeksema, S., & Girgus, J. S. (1994). The emergence of gender differences in depression during adolescence. *Psychological Bulletin, 115,* 424–443.

Nolen-Hoeksema, S., Morrow, J., & Frederickson, B. L. (1993). Response styles and the duration of episodes of depressed mood. *Journal of Abnormal Psychology, 102,* 20–28.

Nora, J. J., & Nora, A. H. (1975). A syndrome of multiple congenital anomalies associated with teratogenic exposure. *Archives of Environmental Health, 30,* 17–21.

Norris, F. H., & Murrell, S. A. (1990). Social support, life events, and stress as modifiers of adjustment to bereavement by older adults. *Psychology and Aging, 5,* 429–436.

North, P. (1998). Hospice Care Ring. Retrieved October 31, 1998, from the World Wide Web: http://www.cp-tel.net/pamnorth/hosring.htm

Novak, J., & Gowin, D. (1989). *Learning to learn.* Cambridge: Cambridge University Press.

Nugent, J. K. (1991). Cultural and psychological influences on the father's role in infant development. *Journal of Marriage and the Family, 53,* 475–485.

Nugent, J. K., Lester, B. M., & Brazelton, T. B. (Eds.). (1991). *The cultural context of infancy. Vol. 2: Multicultural and interdisciplinary approaches to parent-infant relations.* Norwood, NJ: Ablex.

Nuland, S. B. (1994). *How we die: Reflections on life's final chapter.* New York: Alfred A. Knopf.

Nursing homes: When a loved one needs care. (1995). *Consumer Reports, 60,* 518–528.

Nutrition: Five servings daily of fruits, veggies may keep doctor away. (1996, September 19). *Detroit News,* F2E.

O

Oakes, L. M. (1994). Development of infants' use of continuity cues in their perception of causality. *Developmental Psychology, 30,* 869–879.

Oaks, J., & Ezell, G. (1993). *Dying and death: Coping, caring, understanding (2nd ed).* Scottsdale, AZ: Gorsuch Scarisbrick.

O'Connor, P. (1992). *Friendships between women: A critical review.* New York: Guilford Press.

O'Connor, R. D. (1969). Modification of social withdrawal through symbolic modeling. *Journal of Applied Behavior Analysis, 2,* 15–22.

Offer, D. (1969). *The psychological world of the teenager: A study of normal adolescent boys.* New York: Basic Books.

Offer, D., & Offer, J. B. (1975). *From teenage to young manhood.* New York: Basic Books.

Offer, D., Ostrov, E., & Howard, K. I. (1981). *The adolescent: A psychological self-portrait.* New York: Basic Books.

Offer, D., Ostrov, E., Howard, K. I., & Atkinson, R. (1988). *The teenage world: Adolescents' self-image in ten countries.* New York: Plenum.

Office of Juvenile Justice and Delinquency Prevention. (1994). [News release, July 17]. Washington, DC: Justice Department.

O'Hara, M. W. (1995). *Postpartum depression: Causes and consequences.* New York: Springer Verlag.

Oleckno, W. A., & Blacconiere, M. J. (1991). Relationship of religiosity to wellness and other health-related outcomes. *Psychological Reports, 68,* 819–826.

Olshansky, S. J. (1992). Estimating the upper limits to human longevity. *Population Today, 20,* 6–8.

Olson, S. L., Bates, J. E., & Bayles, K. (1984). Mother-infant interaction and the development of individual differences in children's cognitive competence. *Developmental Psychology, 20,* 166–179.

O'Maria, N. S., & Santiago, L. (1998). PTSD in children: Move in the rhythm of the child. *Journal of Interpersonal Violence, 13,* 421.

O'Neill, M. (1995, November 11). So it may be true after all: Eating pasta makes you fat. *New York Times,* A8.

Opinion Roundup. (1980, December/January). Work in the 70's. *Public Opinion, 3,* 36.

Oppenheim, D., Sagi, A., & Lamb, M. E. (1988). Infant-adult attachments on the kibbutz and their relation to socioemotional development 4 years later. *Developmental Psychology, 24,* 427–433.

Oppenheimer, J. R. (1955). *The open mind.* New York: Simon & Schuster.

Orbuch, T. L., & Custer, L. (1995). The social context of married women's work and its impact on black husbands and white husbands. *Journal of Marriage and the Family, 57,* 333–345.

O'Regan, K. M., & Quigley, J. M. (1996). Teenage employment and the spatial isolation of minority and poverty households. *Journal of Human Resources, 31,* 692.

Osterweis, M., Solomon, F., & Green, M. (Eds.). (1984). *Bereavement: Reactions, consequences, and care.* Washington, DC: National Academy Press.

Otten, A. L. (1993a, March 17). If retirees hold off, urge to work passes. *Wall Street Journal,* B1.

Otten, A. L. (1993b, February 3). Violent criminals attack older people the least. *Wall Street Journal,* B1.

Otten, A. L. (1994, October 7). Study finds young are often disconnected. *Wall Street Journal,* B1.

Overend, T. J., Cunningham, D. A., Kramer, J. F., Lefcoe, M. S., & Paterson, D. H. (1992). Knee extensor and knee flexor strength: Cross-sectional area ratios in young and elderly men. *Journal of Gerontology: Medical Sciences, 47,* M204–M210.

Owen, J. D. (1995). *Why our kids don't study: An economist's perspective.* Baltimore: Johns Hopkins University Press.

Owen, P. R. (1998) Fears of Hispanic and Anglo children: Real-world fears in the 1990s. *Hispanic Journal of Behavioral Sciences. 20,* 483–491.

Oxman, T. E., Freeman, D. H., & Manheimer, D. E. (1995). Lack of social participation or religious strength and comfort as risk factors for death after cardiac surgery in the elderly. *Psychosomatic Medicine, 57,* 5–15.

P

Paden, S. L., & Buehler, C. (1995). Coping with the dual-income lifestyle. *Journal of Marriage and the Family, 57,* 101–110.

Padian, N. S., Shiboski, S. C., Glass, S. O., & Vittinghoff, E. (1997). Heterosexual transmission of human immunodeficiency virus (HIV) in Northern California: Results from a ten-year study. *American Journal of Epidemiology, 146,* 350–357.

Paikoff, R. L., & Brooks-Gunn, J. (1991). Do parent-child relationships change during puberty? *Psychological Bulletin, 110,* 47–66.

Painter, K. (1992a, August 12). Over 60 and still in the mood for love. *USA Today,* D1.

Painter, K. (1992b, April 7). Study: Most know sexual orientation by 18. *USA Today,* 8D.

Painter, K. (1994a, April 14). Hormone drugs for menopause gain popularity. *USA Today,* D1.

Painter, K. (1994b, June 20). More mothers breast-feed their babies. *USA Today,* D1.

Painter, K. (1995, May 11). Preteens want to be close to their parents. *USA Today,* D1.

Palca, J. (1991). Fetal brain signals time for birth. *Science, 253,* 1360.

Palca, J. (1996). A new national bioethics commission-maybe. *Hastings Center Report, 26,* 5.

Palmore, E., & Cleveland, W. (1976). Aging, terminal decline, and terminal drop. *Journal of Gerontology, 31,* 76–81.

Paone, D., & Chavkin, W. (1993). From the private family domain to the public health forum: Sexual abuse, women and risk for HIV infection. *SIECUS Report, 21,* 13.

Papadatos, D., & Papadatos, C. (Eds.). (1991). *Children and death.* New York: Hemisphere.

Papousek, M., Papousek, H., & Symmes, D. (1991). The meanings of melodies in motherese in tone and stress languages. *Infant Behavior and Development, 14,* 415–440.

Parcel, T. L., & Menaghan, E. G. (1994a). Early parental work, family social capital, and early childhood outcomes. *American Journal of Sociology, 99,* 972–1009.

Parcel, T. L., & Menaghan, E. G. (1994b). *Parents' jobs and children's lives.* New York: Aldine de Gruyter.

Parish, W. L., Hao, L., & Hogan, D. P. (1991). Family support networks, welfare, and work among young mothers. *Journal of Marriage & the Family, 53,* 203–215.

Park, K. A., Lay, K. L., & Ramsay, L. (1993). Individual differences and developmental changes in preschoolers' friendships. *Developmental Psychology, 29,* 264–270.

Parke, R. D. (1974). Rules, roles, and resistance to deviation: Recent advances in punishment, discipline, and self-control. In A. D. Pick (Ed.), *Minnesota symposia on child psychology* (Vol. 8). Minneapolis: University of Minnesota Press.

Parke, R. D. (1979). Perspectives on father-infant interaction. In J. D. Osofsky (Ed.), *Handbook of infant development.* New York: Wiley.

Parke, R. D. (1995). Fathers and families. In M. H. Bornstein (Ed.), *Handbook of parenting* (Vol. 3). Hillsdale, NJ: Erlbaum.

Parke, R. D. (1996). *Fatherhood.* Cambridge, MA: Harvard University Press.

Parke, R. D. (1998). Paternal involvement in infancy: The role of maternal and paternal attitudes. *Journal of Family Psychology, 12,* 268–288.

Parke, R. D., & Deur, J. L. (1972). Schedule of punishment and inhibition of aggression in children. *Developmental Psychology, 7,* 266–269.

Parke, R. D., & Kellam, S. G. (Eds.). (1994). *Exploring family relationships with other social contexts.* Hillsdale, NJ: Erlbaum.

Parker, S. (1987, June 25). Mom's troubles affect the kids. *USA Today,* D1.

Parkes, C.M., Laungani, P., & Young, B. (1996). *Death and bereavement across cultures.* London: Routledge.

Parks, C. A. (1998). Lesbian parenthood: A review of the literature. *American Journal of Orthopsychiatry, 68,* 376–389.

Parnes, H. S., & Sommers, D. G. (1994). Shunning retirement: Work experiences of men in their seventies and early eighties. *Journal of Gerontology: Social Sciences, 49,* S117–S124

Parpal, M., & Maccoby, E. E. (1985). Maternal responsiveness and subsequent child compliance. *Child Development, 56,* 1326–1334.

Parsons, T. (1955). Family structure and the socialization of the child. In T. Parsons & R. Bales (Eds.), *Family, socialization and interaction process.* New York: Free Press.

Pascual-Leone, J. (1988). Organismic processes for neo-Piagetian theories: A dialectical causal count of cognitive development. In A. Demetriou (Ed.), *The neo-Piagetian theories of cognitive development: Toward an integration.* Amsterdam: North-Holland (Elsevier).

Pasley, K., & Ihinger-Tallman, M. (Eds.). (1994). *Stepparenting: Issues in theory, research, and practice.* Westport, CT: Greenwood Press.

Patterson, C. J. (1992). Children of lesbian and gay parents. *Child Development, 63,* 1025–1042.

Patterson, G. R., Littman, R. A., & Bricker, W. (1967). Assertive behavior in children: A step toward a theory of aggression. *Monographs of the Society for Research in Child Development, 32* (5).

Payne, B. B., & McFadden, S. H. (1994). From loneliness to solitude: Religious and spiritual journeys in late life. In L. E. Thomas & S. A. Eisenhandler (Eds.), *Aging and the religious dimension* (pp. 13–28). Westport, CT: Auburn House.

Pearl, D. (1982). *Television and behavior: 10 years of scientific research.* Washington, DC: Government Printing Office.

Peck, R. C. (1968). Psychological developments in the second half of life. In B. L. Neugarten (Ed.), *Middle age and aging.* Chicago: University of Chicago Press.

Perdue, C. W., Dovidio, J. F., Gurtman, M. B., & Tyler, R. B. (1990). Us and them: Social categorization and the process of intergroup bias. *Journal of Personality and Social Psychology, 59,* 475–486.

Perinatal transmission: A mother's story. (1998, June 30). 12th World AIDS Conference. Retrieved July 1, 1998, from the World Wide Web: http://www.AIDS98.ch

Perkins, D. F., Luster, T., Villarruel, F. A., & Small, S. (1998). An ecological, risk-factor examination of adolescents' sexual activity in three ethnic groups. *Journal of Marriage and the Family, 60,* 660–673.

Perlez, J. (1990, January 15). Puberty rite for girls is bitter issue across Africa. *New York Times,* 4.

Perlmutter, M., & Myers, N. A. (1976). Recognition memory in preschool children. *Developmental Psychology, 12,* 271–272.

Perls, T. (1995, January). The oldest old. *Scientific American,* 70–75.

Perner, J., Ruffman, T., & Leekam, S. R. (1994). Theory of mind is contagious: You catch it from your sibs. *Child Development, 65,* 1228–1238.

Perris, E. E., Myers, N. A., & Clifton, R. K. (1990). Long-term memory for a single infancy experience. *Child Development, 61,* 1796–1807.

Perry, C. L., Bishop, D. B., Taylor, G., Murray, D. M., et al. (1998). Changing fruit and vegetable consumption among children: The 5-a-day Power Plus program in St. Paul, Minnesota. *American Journal of Public Health, 88,* 603–609.

Perry, W. G., Jr. (1968). *Forms of intellectual and ethical development in the college years: A scheme.* New York: Holt, Rinehart & Winston.

Perry, W. G., Jr. (1981). Cognitive and ethical growth: The making of meaning. In A. W. Chickering & Associates (Eds.), *The modern American college: Responding to the new realities of diverse students and a changing society* (pp. 76–116). San Francisco: Jossey-Bass.

Perske, R., & Perske, M. (1981). *Hope for the families.* Nashville, TN: Abingdon Press.

Pescitelli, D. (1998). Women's identity development: Out of the inner space and into new territory. *Psybernetika, 3* (Spring). Retrieved August 9, 1998 from the World Wide Web: http://www.sfu.ca/~wwwpsyb/98spring/pescitel.htm

Petersen, A. C. (1983). Menarche: Meaning of measures and measures of meaning. In S. Golub (Ed.), *Menarche: The transition from girl to woman.* Lexington, MA: Heath.

Petersen, S., & Rafuls, S. E. (1998). Receiving the scepter: The generational transition and impact of parent death on adults. *Death Studies, 22,* 493–524.

Peterson, B. E., & Klohnen, E. C. (1995). Realization of generativity in two samples of women at midlife. *Psychology and Aging, 10,* 20–29.

Peterson, C. C. (1996). The ticking of the social clock: Adults' beliefs about the timing of transition events. *International Journal of aging and Human Development, 42,* 189–203.

Peterson, K. S. (1993, April 23). Divorce needn't leave midlife women adrift. *USA Today,* 7D.

Peterson, K. S. (1998, June 19). Teen drug use declines: Increased awareness of parents credited for the turnaround. *USA Today,* 6A.

Peterson, L., & Brown, D. (1994). Integrating child injury and abuse-neglect research: Common histories, etiologies, and solutions. *Psychological Bulletin, 116,* 293–315.

Peterson, R. R. (1996). A re-evaluation of the economic consequences of divorce.

American Sociological Review, 61, 528–536.

Peth-Pierce, R. (1997, September 16). New sudden infant death syndrome (SIDS) fact sheet, AAP statement on bed sharing now available. National Institute on Child Health and Human Development, NIH. Retrieved August 5, 1998, from the World Wide Web: http://www.nih.gov/news/pr/sept97/nichd-16.htm

Petitto, L. A., & Marentette, P. F. (1991). Babbling in the manual mode: Evidence for the ontogeny of language. *Science, 251,* 1493–1496.

Petraitis, J., Flay, B. R., & Miller, T. Q. (1995). Reviewing theories of adolescent substance use: Organizing pieces in the puzzle. *Psychological Bulletin, 117,* 67–86.

Pezdek, K. (1983). Memory for items and their spatial locations by young and elderly adults. *Developmental Psychology, 19,* 895–900.

Philip, N., Neal, P., & Mistry, N. (1998). *The illustrated book of fairy tales.* London: D. K. Publishing.

Physical fitness in midlife lowers risk of late-life osteoarthritis. (1993). *Geriatrics, 48* (9), 27.

Piaget, J. (1932). *The moral judgment of the child* (M. Gaban, Trans.). London: Kegan Paul, Trench, & Trubner.

Piaget, J. (1952). *The origins of intelligence in children* (M. Cook, Trans.). New York: International Universities Press.

Piaget, J. (1962). *Play, dreams and imitation in childhood.* New York: Norton.

Piaget, J. (1963). *The child's conception of the world.* Patterson: Littlefield.

Piaget, J. (1965a). *The child's conception of number.* New York: Norton. (Original work published 1941)

Piaget, J. (1965b). *The child's conception of the world.* New York: Littlefield Adams & Co.

Piaget, J. (1967). *Six psychological studies.* New York: Random House.

Piaget, J. (1970, May). Conversations. *Psychology Today, 3,* 25–32.

Piaget, J. (1981; 1976). *The moral judgment of the child.* New York: Free Press.

Picard, A. (1993, July 1). Women unprepared for labour pain, psychologist says. *Globe and Mail,* A4.

Pick, H. L., Jr. (1992). Eleanor J. Gibson: Learning to perceive and perceiving to learn. *Developmental Psychology, 28,* 787–794.

Pickens, J., & Field, T. (1993). Facial expressivity in infants of depressed mothers. *Developmental Psychology, 29,* 986–988.

Pienta, A. M., Burr, J. A., & Mutchler, J. E. (1994). Women's labor force participation in later life: The effects of early work and family experiences. *Journal of Gerontology: Social Sciences, 49,* S231–S239.

Pillemer, K., & Finkelhor, D. (1988). The prevalence of elder abuse: A random sample survey. *Gerontologist, 28,* 51–57.

Pillemer, K., & Suitor, J. J. (1998, Spring). Baby boom families: Relations with aging parents. *Generations, 22,* 65–69.

Pincus, T., & Callahan, L. F. (1994). Associations of low formal education level and poor health status: Behavioral, in addition to demographic and medical explanations? *Journal of Clinical Epidemiology, 47,* 355.

Pines, M. (1979, June). Good Samaritans at age two? *Psychology Today, 13,* 66–77.

Pines, M. (1983, November). Can a rock walk? *Psychology Today, 17,* 46–54.

Pinker, S. (1991). Rules of language. *Science, 253,* 530–535.

Pinker, S. (1994). *The language instinct: How the mind creates language.* New York: William Morrow.

Pinkowish, M. D. (1996). Who really needs cholesterol-lowering drugs? *Patient Care, 30,* 92–108.

Pipher, M. (1994). *Reviving Ophelia: Saving the selves of adolescent girls.* New York: Putnam.

Pipp, S., Easterbrooks, M. A., & Harmon, R. J. (1992). The relation between attachment and knowledge of self and mother in one- to three-year-old infants. *Child Development, 63,* 738–750.

Pipp-Siegel, S., & Foltz, C. (1997). Toddlers' acquisition of self/other knowledge: Ecological and interpersonal aspects of self and other. *Child Development. 68,* 69–79.

Pirog-Good, M. A. (1993). Child support guidelines and the economic well-being of children in the United States. *Family Relations, 42,* 453–462.

Planned Parenthood of Central and Northern Arizona. (1998). *Methods of birth control.* Retrieved October 26, 1998, from the World Web: http://www.ppcna.org/bcmethds.html

Plomin, R. (1994). *Genetics and experience.* Thousand Oaks, CA: Sage.

Plomin, R., & Daniels, D. (1987). Why are children in the same family so different from one another? *Behavioral and Brain Sciences, 10,* 1–16.

Plomin, R., DeFries, J. C., & Fulker, D. W. (1988). *Nature and nurture during infancy and childhood.* New York: Cambridge University Press.

Plomin, R., & McClearn, G. E. (Eds.). (1993). *Nature-nurture and psychology.* Washington, DC: American Psychological Association.

Plumb, J. H. (1972). *Children.* London: Penguin Books.

Plutchik, R., Botsis, A., Weiner, M., & Kennedy., G. (1996). Clinical measurement of suicidality and coping in late life. In G. J. Kennedy (Ed.), *Suicide and depression in late life.* New York: John Wiley.

Polakow, V. (1993). *Lives on the edge: Single mothers and their children in the other America.* Chicago: University of Chicago Press.

Poll: Discipline OK, abuse isn't. (1994, April 8). *USA Today,* 8A.

Poll: Few would choose leisure over pay. (1995, April 10). *USA Today,* 2A.

Pollitt, K. (1993, July 22). Bothered and bewildered. *New York Times,* A19.

Polymeropoulos, M. H., Higgins, J. J., Golbe, L. I. (1997). Mapping of a gene for Parkinson's disease to chromosome 4q21–q23. *Science, 274,* 1197–1199.

Popkin, B. M., & Udry, R. (1998, April). Adolescent obesity increases significantly in second and third generation U.S. immigrants. *Journal of Nutrition, 128,* 701–706.

Porter, R. H., Makin, J. W., Davis, L. B., & Christensen, K. M. (1992). Breast-fed infants respond to olfactory clues from their own mother and unfamiliar lactating females. *Infant Behavior and Development, 15,* 85–93.

Portman, M. V. (1895). Notes on the Andamanese. Anthropological Institute of Great Britain and Ireland, 25, 361–371.

Posner, J. K., & Vandell, D. L. (1994). Low-income children's after school care: Are there beneficial effects of after school programs? *Child Development, 65,* 440–456.

Poulin-Dubois, D., Serbin, L. A., Kenyon, B., & Derbyshire, A. (1994). Infants' intermodal knowledge about gender. *Developmental Psychology, 30,* 436–442.

Powell, D. H., & Whitla, D. K. (1994). *Profiles in cognitive aging.* Cambridge, MA: Harvard University Press.

Power, T. G., & Parke, R. D. (1983). Patterns of mother and father play with their 8-month-old infant: A multiple analyses approach. *Infant Behavior and Development, 6,* 453–459.

Powers, E. A., & Bultena, G. L. (1976). Sex differences in intimate friendships of old age. *Journal of Marriage and the Family, 38,* 739–747.

Powlishta, K. K., Serbin, L. A., Doyle, A. B., & White, D. R. (1994). Gender, ethnic, and body type biases: The generality of prejudice in childhood. *Developmental Psychology, 30,* 526–536.

Prager, L. O. (1998). Details emerge on Oregon's first assisted suicides. *American Medical News, 41*(33), 9–10.

Pratt, C. C., Walker, A. J., & Wood, B. L. (1992). Bereavement among former caregivers to elderly mothers. *Family Relations, 41,* 278–283.

Pratt, M. W., Diessner, R., Hunsberger, B., Pancer, S. M., & Savoy, K. (1991). Four pathways in the analysis of adult development and aging: Comparing analyses of reasoning about personal-life dilemmas. *Psychology and Aging, 4,* 666–675.

Pratt, M. W., Kerig, P., Cowan, P. A., & Cowan, C. P. (1988). Mothers and fathers teaching 3-year-olds: Authoritative parenting and adult scaffolding of young children's learning. *Developmental Psychology, 24,* 832–839.

Pregnancy and HIV—What women and doctors need to know. (1998). Health Care Financing Administration. Retrieved July 29, 1998, from the World Wide Web: http://www.thebody.com/hcta/medicaid/pregnancy4.html

Price, J. L., Davis, P. B., Morris, J. C., & White, D. L. (1991). The distribution of tangles, plaques, and related immunohistochemical markers in healthy aging and Alzheimer's disease. *Neurobiology of Aging, 12,* 295–312.

Princeton Survey Research Associates, Inc. (1994). *11th prevention index survey.* Emmaus, PA: Prevention Index.

Pritchard, J. A., & MacDonald, P. C. (1976). *Williams Obstetrics, 15 ed.* New York: Appleton-Century-Crofts.

Profile of Older Americans. (1998). Administration on Aging. Retrieved January 14, 1999 from the World Wide Web: http://www.aoa.dhhs.gov/aoa/stats/profile/

Provenzo, E. G., Jr. (1991). *Video kids: Making sense of Nintendo.* Cambridge, MA: Harvard University Press.

Pruchno, R. A., Peters, N. D., Kleban, M. H., & Burant, C. J. (1994). Attachment among adult children and their institutionalized parents. *Journal of Gerontology: Social Sciences, 49,* S209–S218.

Pruett, K. D. (1987). *The nurturing father.* New York: Warner Books.

Puddifoot, J. E., & Johnson, M. P. (1997). The legitimacy of grieving: The partner's experience at miscarriage. *Social Science Medicine, 45,* 837–845.

Pugliesi, K. (1995). Work and well-being: Gender differences in the psychological consequences of employment. *Journal of Health and Social Behavior, 36,* 57–71.

Pulkkinen, L. (1982). Self-control and continuity from childhood to adolescence. In P. Baltes & O. G. Brim (Eds.), *Life-span development and behavior* (Vol. 4). New York: Academic Press.

Pulkkinen, L., & Ronka, A. (1994). Personal control over development, identity formation, and future orientation as components of life orientation: A developmental approach. *Developmental Psychology, 30,* 260–271.

Putallaz, M., & Sheppard, B. H. (1990). Social status and children's orientations to limited resources. *Child Development, 61,* 2022–2027.

Q

Quam, J. K, & Whitford, G. S. (1992). Adaptation and age-related expectations of older gay and lesbian adults. *Gerontologist, 32,* 367–374.

Quill, T. (1993). *Death and dignity: Making choices and taking charge.* New York: W. W. Norton.

R

Radcliffe-Brown, A. R. (1940). On joking relationships. *Africa, 13,* 195–210.

Radetsky, P. (1994). Stopping premature births before it's too late. *Science, 266,* 1486–1488.

Rafferty, C. (1984, April 23). Study of gifted from childhood to old age. *New York Times,* 18.

Ramsey, P. G. (1995). Changing social dynamics in early childhood classrooms. *Child Development, 66,* 764–773.

Raphael, R. (1988). *The men from the boys: Rites of passage in male America.* Lincoln: University of Nebraska Press.

Rausch, M. L. (1977). Paradox, levels, and junctures in person-situation systems. In D. Magnusson & N. S. Endler (Eds.), *Personality at the crossroads.* Hillsdale, NJ: Erlbaum.

Rauscher, F. (1996). *The power of music. Early Childhood News.* Retrieved December 10, 1998, from the World Wide Web: http://www.earlychildhood.com/music.htm

Ravoira, L. W., & Cherry, A. L., Jr. (1992). *Social bonds and teen pregnancy.* Westport, CT: Praeger.

Raymond, C. (1991a, January 23). Cross-cultural study of sounds adults direct to infants shows that "baby talk" can be serious communication. *Chronicle of Higher Education,* A5, A7.

Raymond, C. (1991b, January 23). Pioneering research challenges accepted notions concerning the cognitive abilities of infants. *Chronicle of Higher Education,* A5–A7.

Reber, A. S. (1993). *Implicit learning and tacit knowledge: An essay on the cognitive unconscious.* New York: Oxford University Press.

Reed, E. S. (1988). *James J. Gibson and the psychology of perception.* New Haven, CT: Yale University.

Reeder, H. M. (1996). The subjective experience of love through adult life. *International Journal of Aging and Human Development, 43,* 325–340.

Rees, D. (1997). *Death and bereavement: The psychological, religious, and cultural interfaces.* London: Whurr.

Reich, J. W., & Zautra, A. J. (1991). Experimental and measurement approaches to internal control in at-risk older adults. *Journal of Social Issues, 47,* 143–158.

Reichman, J. (1996). *I'm too young to get old: Health care for women after forty.* New York: Random House.

Reid, J. D. (1995). Development in late life: Older lesbian and gay life. In A. R. D'Augelli & C. J. Patterson (Eds.), *Lesbian, gay, and bisexual identities over the lifespan: Psychological perspectives* (pp. 215–240). New York: Oxford University Press.

Reigner, V. (1994). *Assisted living housing for the elderly: Design innovations from the United States and Europe.* New York: Reinhold.

Renshaw, P. D., & Brown, P. J. (1993). Loneliness in middle childhood: Concurrent and longitudinal predictors. *Child Development, 64,* 1271–1284.

Rescorla, L., (1976). *Concept formation in word learning.* Unpublished doctoral dissertation, Yale University.

Restak, R. (1984). *The brain.* New York: Bantam.

Retherford, R. D., & Sewell, W. H. (1992). Four erroneous assertions regarding the accuracy of the confluence model. *American Sociological Review, 57,* 136–137.

Reynolds, A. J., & Temple, J. A. (1998). Extended early childhood intervention and school achievement: Age thirteen findings from the Chicago Longitudinal Study. *Child Development. 69,* 231–246.

Reynolds, C. F., III, Monk, T. H., Hoch, C. C., Jennings, J. R., Buysse, D. J., Houck, P. R., Jarrett, D. B., & Kupfer, D. J. (1991). Electroencephalographic sleep in the healthy "old old": A comparison with the "young old" in visually scored and automated measures. *Journal of Gerontology, 46,* 39–46.

Reznick, J. S., & Goldfield, B. A. (1992). Rapid change in lexical development in comprehension and production. *Developmental Psychology, 28,* 406–413.

Rheingold, H. L. (1968). Infancy. In D. Sills (Ed.), *International encyclopedia of the social sciences.* New York: Macmillan.

Rheingold, H. L. (1969a). The effect of a strange environment on the behavior of infants. In B. M. Foss (Ed.), *Determinants of infant behavior* (Vol. 4). New York: Wiley.

Rheingold, H. L. (1969b). The social and socializing infant. In D. A. Goslin (Ed.), *Handbook of socialization theory and research.* Chicago: Rand McNally.

Rheingold, H. L. (1985). Development as the acquisition of familiarity. *Annual Review of Psychology, 36,* 105–130.

Rheingold, H. L., & Adams, J. L. (1980). The significance of speech to newborns. *Developmental Psychology, 16,* 397–403.

Rheingold, H. L., & Eckerman, C. O. (1973). Fear of the stranger: A critical examination. In H. W. Reese (Ed.), *Advances in child development and behavior* (Vol. 8). New York: Academic Press.

Rheingold, H. L., Hay, D. F., & West, M. J. (1976). Sharing in the second year of life. *Child Development, 47,* 1148–1158.

Ribble, M. A. (1943). *The rights of infants: Early psychological needs and their satisfaction.* New York: Columbia University Press.

Richards, L. N. (1989). The precarious survival and hard-won satisfactions of white single-parent families. *Family Relations, 38,* 396–403.

Richardson, D., Tyra, J., & McCray, A. (1992). Attenuation of the cutaneous vasoconstrictor response to cold in elderly men. *Journal of Gerontology: Medical Sciences, 47,* M211–M214.

Rickman, M. D., & Davidson, R. J. (1994). Personality and behavior in parents of temperamentally inhibited and uninhibited children. *Developmental Psychology, 30,* 346–354.

Ricks, D. (1979). Making sense of experience to make sensible sounds. In M. Bulowa (Ed.), *Before speech* (pp. 245–268). Cambridge, MA: Cambridge University Press.

Rieser, J., Yonas, A., & Wikner, K. (1976). Radial localization of odors by human newborns. *Child Development, 47,* 856–859.

Riger, S. (1992). Epistemological debates, feminist voices. *American Psychologist, 47,* 730–740.

Rikli, R., & Busch, S. (1986). Motor performance of women as a function of age and physical activity level. *Journal of Gerontology, 41,* 645–649.

Riley, J., & Staimer, M. (1993, October 25). Young (old) as you feel. *USA Today,* D1.

Riley, J. W., Jr. (1983). Dying and the meanings of death: Sociological inquiries. *Annual Review of Sociology, 9,* 191–216.

Rimer, S. (1998, December 23). For aged, dating game is numbers game. *New York Times,* Late Edition, A1.

Rindfuss, R. R. (1991). The young adult years: Diversity, structural change, and fertility. *Demography, 28,* 493–512.

Risman, B. J. (1986). Can men "mother"? Life as a single father. *Family Relations, 35,* 95–102.

Ritchie, K. (1995, April). Marketing to Generation X. *American Demographics,* 34–39.

Rivers, W. H. R. (1906). *The Todas.* New York: Macmillan.

Roach, M. (1983, January 16). Another name for madness. *New York Times Magazine,* 22–31.

Roazen, P. (1990). *Encountering Freud: The politics and histories of psychoanalysis.* New Brunswick, NJ: Transaction.

Robbins, W. J., Brody, S., Hogan, A. G., Jackson, C. M., Greene, C. W. (1928). *Growth.* New Haven, CT: Yale University Press.

Robb-Nicholson, C. (1998). Some tests you may not need. *Harvard Women's Health Watch.* Cambridge, MA: Harvard Medical School.

Roberto, K. A., & Scott, J. P. (1986). Friendships of older men and women: Exchange patterns and satisfaction. *Psychology and Aging, 1,* 103–109.

Roberts, K. (1988). Retrieval of a basic-level category in prelinguistic infants. *Developmental Psychology, 24,* 21–27.

Roberts, P., & Newton, P. M. (1987). Levinsonian studies of women's adult development. *Psychology and Aging, 2,* 154–163.

Roberts, R. K. N., Wasik, B. H., Casto, G., & Ramey, C. T. (1991). Family support in the home: Programs, policy, and social change. *American Psychologist, 46,* 131–137.

Robinson, E. H., Rotter, J. C., Fey, M. A. & Robinson, S. L. (1991). Children's fears: Toward a preventive model. *School Counselor, 38,* 187–202.

Robinson, P. (1993). *Freud and his critics.* Berkeley, CA: University of California Press.

Robinson, R. (1998). Single moms: Raising successful kids. *Parenting, 12,* 171.

Rochat, P., & Striano, T. (1998). Primary action in early ontogeny. *Human Development, 41,* 112–115.

Rodin, J., & Langer, E. J. (1977). Long-term effects of a control-relevant intervention with the institutionalized aged. *Journal of Personality and Social Psychology, 35,* 897–902.

Rodriguez, B. (1998). "It let's the sad out": Using children's art to express emotions. *Early Childhood News.* Retrieved December 12, 1998, from the World Wide Web: http://www.earlychildhoodnews. com/sad.htm

Rogers, A. (1998, December 7). The brain: Thinking differently. *Newsweek,* 60.

Rogers, C. R. (1970). *On becoming a person: A therapist's view of psychotherapy.* Boston: Houghton Mifflin/Sentry.

Rogers, R. G. (1995). Marriage, sex, and mortality. *Journal of Marriage and the Family, 57,* 515–526.

Roggman, L. A., Langlois, J. H., Hubbs-Tait, L., & Rieser-Danner, L. A. (1994). Infant day-care, attachment, and the "file drawer problem." *Child Development, 65,* 1429–1443.

Rohner, R. P., & Pettengill, S. M. (1985). Perceived parental acceptance-rejection and parental control among Korean adolescents. *Child Development, 56,* 524–528.

Rokach, A. (1998). The relation of cultural background to the causes of loneliness. *Journal of Social and Clinical Psychology, 17,* 75–88.

Rolfes, S. B., & DeBruyne, L. (1997). *Life span nutrition: Conception through life.* St. Paul: West.

Romanoff, B. D., & Terenzio, M. (1998). Rituals and the grieving process. *Death Studies, 22,* 697–711.

Rosales-Ruiz, J., & Baer, D. M. (1997). Behavioral cusps: A developmental and pragmatic concept for behavior analysis. *Journal of Applied Behavior Analysis, 30,* 533–544.

Rosen, W. D., Adamson, L. B., & Bakeman, R. (1992). An experimental investigation of infant social referencing: Mothers' messages and gender differences. *Developmental Psychology, 28,* 1172–1178.

Rosenberg, E. B. (1992). *The adoption life cycle: The children and their families through the years.* New York: Free Press.

Rosenblatt, P. C., & Skoogberg, E. H. (1974). Birth order in cross-cultural perspective. *Developmental Psychology, 10,* 48–54.

Rosenbloom, C. A., & Whittington, F. J. (1993). The effects of bereavement on eating behaviors and nutrient intakes in elderly widowed persons. *Journal of Gerontology: Social Sciences, 48,* S223–S229.

Rosenfeld, A., & Stark, E. (1987, May). The prime of our lives. *Psychology Today, 21,* 62–70.

Rosenstein, K. D. & Oster, H. (1988). Differential facial responses to four basic tastes in newborns. *Child Development, 59,* 1555–1568.

Rosenthal, E. (1992, August 18). Troubled marriage? Sibling relations may be at fault. *New York Times,* B5, B8.

Rosenthal, E. (1994, May 26). Panel tells Albany to resist legalizing assisted suicide. *New York Times,* A1, A11.

Rosenthal, T. L., & Zimmerman, B. J. (1978). *Social learning and cognition.* New York: Academic Press.

Rosick, C. H. (1989). The impact of religious orientation in conjugal bereavement among older adults. *International Journal of Aging and Human Development, 28,* 251–260.

Rosinski, R. R., Pellegrino, J. W., & Siegel, A. W. (1977). Developmental changes in the semantic processing of pictures and words. *Journal of Experimental Child Psychology, 23,* 282–291.

Ross, A. O. (1992). *The sense of self: Research and theory.* New York: Springer.

Ross, H. S., Filyer, R. E., Lollis, S. P., Perlman, M., & Martin, J. L. (1994). Administering justice in the family. *Journal of Family Psychology, 8,* 254–273.

Rossi, A. S. (1968). Transition to parenthood. *Journal of Marriage and the Family, 30,* 26–39.

Rossi, A. S. (1977). A biosocial perspective on parenting. *Daedalus, 106,* 1–31.

Rossi, S., & Wittrock, M. C. (1971). Developmental shifts in verbal recall between mental ages two and five. *Child Development, 42,* 333–338.

Rotheram-Borus, M. J., Rosario, M., Van Rossem, R., Reid, H., & Gillis, R. (1995). Prevalence, course and predictors of multiple problem behaviors among gay and bisexual male adolescents. *Developmental Psychology, 31,* 75–85.

Roush, W. (1995). Arguing over why Johnny can't read. *Science, 267,* 1896–1898.

Rovee-Collier, C. (1987). Learning and memory in infancy. In J. D. Osofsky (Ed.), *Handbook of infant development* (2nd ed.). New York: Wiley.

Rowe, D. C. (1994). *The limits of family influence: Genes, experience, and behavior.* New York: Guilford Press.

Rowe, J. W., & Kahn, R. L. (1987). Human aging: Usual and successful. *Science, 237,* 143–149.

Rubenstein, C. (1989, October 8). The baby bomb: Research reveals the astonishingly stressful social and emotional consequences of parenthood. *New York Times Magazine,* 34–41.

Rubenstein, E. (1997, March 10). Right data. *National Review, 49,* 14.

Rubin, J. Z., Provenzano, F. J., & Luria, Z. (1974). The eye of the beholder: Parents' views on sex of newborns. *American Journal of Orthopsychiatry, 43,* 720–731.

Rubin, K. H., Stewart, S. L., & Chen, X. (1995). Parents of aggressive and withdrawn children. In M. H. Bornstein (Ed.), *Handbook of parenting* (Vol. 1). Hillsdale, NJ: Erlbaum.

Rubin, R. M., & Riney, B. J. (1993). *Working wives and dual-earner families.* Westport, CT: Praeger.

Ruble, D. N., & Brooks-Gunn, J. (1982). The experience of menarche. *Child Development, 53,* 1557–1566.

Ruble, D. N., Brooks-Gunn, J., Fleming, A. S., Fitzmaurice, G., Stangor, C., & Deutsch, F. (1990). Transition to motherhood and the self: Measurement, stability, and change. *Journal of Personality and Social Psychology, 58,* 450–463.

Ruch, W. (1998). *The sense of humor: Explorations of a personality characteristic.* New York: Aldine de Gruyter.

Rudman, D., Feller, A. G., Cohn, L., Shetty, K. R., Rudman, I. W., & Draper, W. W. (1991). Effects of human growth hormone on body composition in elderly men. *Hormone Resources, 36,* 73–81.

Rutter, D. R., & Kurkin, K. (1987). Turn-taking in mother-infant interaction: An examination of vocalizations and gaze. *Developmental Psychology, 23,* 54–61.

Rutter, M. (1974). *The qualities of mothering.* New York: Jason Aronson.

Rutter, M. (1983). School effects on pupil progress: Research findings and policy implications. *Child Development, 54,* 1–29.

Ryff, C. (1982). Successful aging: A developmental approach. *Gerontologist, 22,* 209–214.

Ryff, C. D., Lee, Y. H., Essex, M. J., & Schmutte, P. S. (1994). My children and me: Midlife evaluations of grown children and of self. *Psychology and Aging, 9,* 195–205.

S

Saarni, C. (1979). Children's understanding of display rules for expressive behavior. *Developmental Psychology, 15,* 424–429.

Sachs, J. (1987). Preschool boys' and girls' language use in pretend play. In S. U. Phillips, S. Steele, & C. Tanz (Eds.), *Language, gender and sex in comparative perspective.* Cambridge: Cambridge University Press.

Sacker, I., & Zimmer, M. (1987). *Dying to be thin.* New York: Warner Books.

Sadker, M., & Sadker, D. (1994). *Failing at fairness.* New York: Scribner's.

Sadler, W. A. (1978). Dimensions in the problem of loneliness: A phenomenological approach in social psychology. *Journal of Phenomenological Psychology, 9,* 157–187.

Salapatek, P. (1975). Pattern perception in early infancy. In L. B. Cohen & P. Salapatek (Eds.), *Infant perception: From sensation to cognition.* New York: Academic Press.

Salapatek, P., & Kessen, W. (1966). Visual scanning of triangles by the human newborn. *Journal of Experimental Child Psychology, 3,* 155–167.

Salthouse, T. (1991). *Theoretical perspectives on cognitive aging.* Hillsdale, NJ: Erlbaum.

Salthouse, T. A. (1992a). *Mechanisms of age-cognition relations in adulthood.* Hillsdale, NJ: Erlbaum.

Salthouse, T. A. (1992b). Why do adult differences increase with task complexity? *Developmental Psychology, 28,* 905–918.

Salthouse, T. A., & Babcock, R. L. (1991). Decomposing adult age differences in working memory. *Developmental Psychology, 27,* 763.

Saltz, R. (1973). Effects of part-time "mothering" on IQ and SQ of young institutionalized children. *Child Development, 44,* 166–170.

Sameroff, A. J. (1968). The components of sucking in the human newborn. *Journal of Experimental Child Psychology, 6,* 607–623.

Sameroff, A. J., & Cavanagh, P. J. (1979). Learning in infancy: A developmental perspective. In J. D. Osofsky (Ed.), *Handbook of infant development.* New York: Wiley.

Sampson, R. J., & Laub, J. H. (1990). Crime and deviance over the life course: The salience of adult social bonds. *American Sociological Review, 55,* 609–627.

Samuels, S. C. (1997). Midlife crisis: Helping patients cope with stress, anxiety, and depression. *Geriatrics, 52,* 55–63.

Sanders-Phillips, K., Strauss, M. E., & Gutberlet, R. L. (1988). The effect of obstetric medication on newborn infant feeding behavior. *Infant Behavior and Development, 11,* 251–263.

Sanson, A., & Rothbart, M. K. (1995). Child temperament and parenting. In M. H. Bornstein (Ed.), *Handbook of parenting* (Vol. 4). Hillsdale, NJ: Erlbaum.

Santrock, J. W. (1972). Relation of type and onset of father absence to cognitive development. *Child Development, 43,* 455–469.

Sarason, S. B. (1977). *Work, aging, and social change: Professionals and the one life-one career imperative.* New York: Basic Books.

Savage-Rumbaugh, E. S., Murphy, J., Sevcik, R. A., Brakke, K. E., Williams, S. L., and Rumbaugh, D. M. (1993). Language comprehension in ape and child. *Monographs of the Society for Research in Child Development, 58* (3–4, Serial No. 233).

Scafidi, F. A., Field, T. M., Schanberg, S. M., Bauer, C. R., Tucci, K., Roberts, J., Morrow, C., & Kuhn, C. M. (1990). Massage stimulates growth in preterm infants: A replication. *Infant Behavior and Development, 13,* 167–188.

Scarf, M., & Grajek, S. (1982). Similarities and differences among siblings. In M. E. Lamb & B. Sutton-Smith (Eds.), *Sibling relationships.* Hillsdale, NJ: Erlbaum.

Scarr, S. (1985a). An author's frame of mind [Review of Frames of Mind by H. Gardner]. *New Ideas in Psychology, 3,* 95–100.

Scarr, S. (1985b). Constructing psychology. *American Psychologist, 40,* 499–512.

Scarr, S. (1993). Biological and cultural diversity: The legacy of Darwin for development. *Child Development, 64,* 1333–1353.

Scarr, S., & McCartney, K. (1983). How people make their own environments: A theory of genotype-environment effects. *Child Development, 54,* 424–435.

Schachter, F. F., Fuches, M. L., Bijur, P., & Stone, R. K. (1989). Co-sleeping and sleep problems in Hispanic-American urban young children. *Pediatrics, 84,* 522–530.

Schacter, D. L., & Tulving, E. (Eds.) (1994). *Memory systems.* Cambridge, MA: MIT Press.

Schaefer, E. S. (1959). A circumplex model for maternal behavior. *Journal of Abnormal and Social Psychology, 59,* 232.

Schaffer, H. R. (1971). *The growth of sociability.* Baltimore: Penguin Books.

Schaffer, H. R. (1996). *Social development.* Cambridge, MA: Blackwell.

Schaffer, H. R., & Emerson, P. E. (1964). The development of social attachments in infancy. *Monographs of the Society for Research in Child Development, 29* (3).

Schaie, K. W. (1989). Perceptual speed in adulthood: Cross-sectional and longitudinal studies. *Psychology and Aging, 4,* 443–453.

Schaie, K. W. (1994). The course of adult intellectual development. *American Psychologist, 49,* 304–313.

Schaie, K. W. (1995). Brain-astics: Mind exercises to keep you sharp. *New Choices for Retirement Living, 35,* 22–24.

Schaie, K. W. (1996). *Intellectual development in adulthood: The Seattle Longitudinal Study.* New York: Cambridge University Press.

Schaie, K. W., Campbell, R. T., & Rawlings, S. C. (Eds.). (1988). *Methodological issues in aging research.* New York: Springer.

Schaie, K. W., & Willis, S. L. (1986). Can decline in adult intellectual functioning be reversed? *Developmental Psychology, 22,* 223–232.

Schaie, K. W., & Willis, S. L. (1993). Age difference patterns of psychometric intelligence in adulthood: Generalizability within and across ability domains. *Psychology and Aging, 8,* 44–55.

Schaie, K. W., Willis, S. L., & O'Hanlon, A. M. (1994). Perceived intellectual performance change over seven years. *Journal of Gerontology: Psychological Sciences, 49,* P108–P118.

Scharlach, A. E. (1987). Relieving feelings of strain among women with elderly mothers. *Psychology and Aging, 2,* 9–13.

Scheck, A. (1994, October 28). The anti-aging effects of physical fitness. *Investor's Business Daily, 1,* 2.

Scherer, K. R. (1979). Nonlinguistic vocal indicators of emotion and psychopathology. In C. E. Izard (Ed.), *Emotions in personality and psychopathology* (pp. 495–529). New York: Plenum.

Schlegel, A., & Barry, H., III. (1991). *Adolescence: An anthropological inquiry.* New York: Free Press.

Schlesinger, B. (1998). Separating together: How divorce transforms families. *Family Relations, 47,* 308.

Schmeck, H. M., Jr. (1982, January 26). Mysterious thymus gland may hold the key to aging. *New York Times,* 17, 18.

Schmeck, H. M., Jr. (1983, March 22). U.S. panel calls for patients' right to end life. *New York Times,* 1, 18.

Schnaiberg, A., & Goldenberg, S. (1975). Closing the circle: The impact of children on parental status. *Journal of Marriage and the Family, 37,* 937–953.

Schneider, B. H. (1993). *Children's social competence in context: The contributions of family, school, and culture.* Oxford: Pergamon Press.

Schneider, W., & Pressley, M. (1989). *Memory development between 2 and 20.* New York: Springer-Verlag.

Schor, J. B. (1992). *The overworked American: The unexpected decline of leisure.* New York: Basic Books.

Schor, J. B. (1993, August 29). All work and no play: It doesn't pay. *New York Times,* F9.

Schrader, D. (1988). *Exploring metacognition: A description of levels of metacognition and their relation to moral judgment.* Unpublished Dissertation: *Dissertation Abstracts.*

Schrof, J. M. (1993, October 25). Tarnished trophies. *U.S. News & World Report,* 52–59.

Schrof, J. M. (1994, November 28). Brain Power. *U.S. News & World Report,* 89–97.

Schulenberg, J., Wadsworth, K. N., & O'Malley, P. M., et al. (1997). Adolescent risk factors for binge drinking during the transition to young adulthood: Variable- and pattern-centered approaches to change. In G. A. Marlatt and G. R. VandenBox (Eds.), *Addictive behaviors: Readings on etiology, prevention, and treatment* (pp. 129–165). Washington, DC: American Psychological Association.

Schulman, S. (1986, February). Facing the invisible handicap. *Psychology Today, 20,* 58–64.

Schulz, R., & Heckhausen, J. (1996). A life span model of successful aging. *American Psychologist, 51,* 702–714.

Schulz, R., Heckhausen, J., & Locher, J. L. (1991). Adult development, control, and adaptive functioning. *Journal of Social Issues, 47,* 177–196.

Schumm, W. R., & Bugaighis, M. A. (1986). Marital quality over the marital career: Alternative explanations. *Journal of Marriage and the Family, 48,* 165–168.

Schwanenflugel, P. J., Fabricius, W. V., & Alexander, J. (1994). Developing theories of mind: Understanding concepts and relations between mental activities. *Child Development, 65,* 1546–1563.

Schweinhart, L., Barnes, H., & Weikert, D. (1993). *Significant benefits: The High/Scope Perry Preschool Study through age 27.* Ypsilanti, MI: High/Scope Press.

Schweinhart, L. J., & Wikart, D. P. (1986). What do we know so far? A review of the Head Start Synthesis Project. *Young Children, 41,* 45–55

Scott, J. (1994, June 4). Another legacy of Onassis: Facing death on own terms. *New York Times,* 1, 8.

Scrivo, K. L. (1998, March 20). Drinking on campus: Can colleges get it under control? *CQ Researcher, 8,* 241–264.

Search Institute. (1994). *Growing up adopted: A portrait of adolescents and their families.* Minneapolis: Search Institute.

Sears, R. R. (1963). Dependency motivation. In M. Jones (Ed.), *Nebraska symposium on motivation.* Lincoln: University of Nebraska Press.

Sears, R. R. (1972). Attachment, dependency, and frustration. In J. L. Gewirtz (Ed.),

Attachment and dependency. Washington, DC: Winston.

Sears, R. R. (1977). Sources of life satisfactions of the Terman gifted men. *American Psychologist, 32,* 119–128.

Sears, R. R., Maccoby, E. E., & Levin, H. (1957). *Patterns of child rearing.* New York: Harper & Row.

Sears, W., & Sears, M. (1994). *The birth book.* Boston: Little, Brown.

Sebald, H. (1977). *Adolescence: A social psychological analysis* (2nd ed.). Englewood Cliffs, NJ: Prentice Hall.

Sebald, H. (1984). *Adolescence: A social psychological analysis.* Englewood Cliffs, NJ: Prentice Hall.

Sebald, H. (1986). Adolescents' shifting orientation toward parents and peers. *Journal of Marriage and the Family, 48,* 5–13.

Seeman, T. E., & Adler, N. (1998). Older Americans: Who will they be? *National Forum: Phi Kappa Phi Journal, 78,* 22–25.

Seeman, T. E., Charpentier, P. A., Berkman, L. F., Tinetti, M. E., Guralnik, J. M., Albert, M., Blazer, K. D., & Rowe, J. W. (1994). Predicting changes in physical performance in a high-functioning elderly cohort: MacArthur studies of successful aging. *Journal of Gerontology: Medical Sciences, 49,* M97–M108.

Segal, M. (1997). *Women and AIDS. FDA Consumer.* Retrieved June 10, 1998, from the World Wide Web: http://www.fda.gov/opacom/catalog/womaids.html

Segal, N. L. (1993). Twin, sibling, and adoption methods: Tests of evolutionary hypotheses. *American Psychologist, 48,* 943–956.

Seger, C. A. (1994). Implicit learning. *Psychological Bulletin, 115,* 163–196.

Seidman, E., Allen, L., Aber, J. L., Mitchell, C., & Feinman, J. (1994). The impact of school transitions in early adolescence on the self-system and perceived social context of poor urban youth. *Child Development, 65,* 507–522.

Seifer, R., & Sameroff, A. (1987). Multiple determinants of risk and vulnerability. In E. J. Anthony & B. J. Cohler (Eds.), *The invulnerable child.* New York: Guilford.

Seitz, V., Rosenbaum, L. K., & Apfel, N. H. (1985). Effects of family support intervention: A ten-year follow-up. *Child Development, 56,* 376–391.

Seldin, T. (1996). *Every parent's question: Is Montessori worth it? Communication builds trust and confidence.* The Montesssori Foundation. Retrieved from the World Wide Web: http://www.montessori.org/library/ismontworthit.htm

Self, P. A., Horowitz, F. D., & Paden, L. Y. (1972). Olfaction in newborn infants. *Developmental Psychology, 7,* 349–363.

Seligmann, J. (1992, December 14). It's not like Mr. Mom. *Newsweek,* 70–73.

Selman, R. L. (1980). *The youth of interpersonal understanding: Developmental and clinical analyses.* New York: Academic Press.

Selye, H. (1956). *The stress of life.* New York: McGraw-Hill.

Senate Special Committee [Canada] on Euthanasia and Assisted Suicide. (1997). Retrieved November 29, 1997, from the World Wide Web: http://www.rights.org/deathnet/senate.html

Seppa, N. (1997, June). Children's TV remains steeped in violence. *APA Monitor, 28,* 36.

Seppa, N. (1998, September 19). Antiviral suppresses genital herpes. *Science News, 154,* 188.

Setterlund, M. B., & Niedenthal, P. M. (1993). "Who am I? Why am I here?": Self-esteem, self-clarity, and prototype matching. *Journal of Personality and Social Psychology, 65,* 769–780.

Settersten, R. A. Jr. (1997). The salience of age in the life course. *Human Development. 40,* 257–281.

Severy, L., Thapa, S., Askew, I., & Glor, J. (1993). Menstrual experiences and beliefs. *Women and Health, 20,* 1–20.

Sewell, W. H. (1981). Notes on educational, occupational, and economic achievement in American society. *Phi Delta Kappan, 62,* 322–325.

Sewell, W. H., & Mussen, P. H. (1952). The effects of feeding, weaning, and scheduling procedures on childhood adjustment and the formation of oral symptoms. *Child Development, 23,* 185–191.

Shahidullah, S., & Hepper, P. G. (1992). Hearing in the fetus: Prenatal detection of deafness. *International Journal of Prenatal and Perinatal Studies, 4,* 235–240.

Shahidullah, S., Scott, D., & Hepper, P. (1993, July–September). Newborn and fetal response to maternal voice, *Journal of Reproductive and Infant Psychology, 11,* 147–153.

Shapiro, L. (1990, May 28). Guns and dolls. *Newsweek,* 56–65.

Sheehy, G. (1976). *Passages.* New York: Dutton.

Sheehy, G. (1992). *The silent passage: Menopause.* New York: Random House.

Sheehy, G. (1995). *New passages: Mapping your life across time.* New York: Random House.

Sheehy, G. (1998). *Understanding men's passages: Discovering the new map of men's lives.* New York: Random House.

Sheff, D. (1993). *How Nintendo zapped an American industry, captured your dollars, and enslaved your children.* New York: Random House.

Sheler, J. L. (1997, March 31). Heaven in the age of reason. *U.S. News & World Report, 122,* 65–66.

Shellenbarger, S. (1991, September 26). Work and family. *Wall Street Journal,* B1.

Shellenbarger, S. (1994a, July 20). It's hard to do day care right—and survive. *Wall Street Journal,* B1.

Shellenbarger, S. (1994b, June 29). Work and family. *Wall Street Journal,* B1.

Shellenbarger, S. (1997, October 15). Work-life issues are starting to plague teenagers with jobs. *Wall Street Journal,* B1.

Sherman, A. (1997). *Poverty matters: The cost of child poverty in America.* Washington, DC: Children's Defense Fund.

Ship, J. A., Pearson, J. D., Cruise, L. J., Brant, L. J., & Metter, E. J. (1996). Longitudinal changes in smell identification. *Journals of Gerontology: Medical Sciences, 51A,* 86–91.

Ship, J. A., & Weiffenbach, J. M. (1993). Age, gender, medical treatment, and medication effects on smell identification. *Journal of Gerontology: Medical Sciences, 48,* M26–M32.

Shireman, J. F. (1995). Adoptions by single parents. In M. Hanson et al. (Eds.), *Single parent families: Diversity, myth, and realities.* New York: Haworth Press.

Shneidman. E. (1989). The indian summer of life: A preliminary study of septuagenarians. *American Psychologist, 44,* 864–694.

Shonkoff, J. P., Hauser-Cram, P., Krauss, M. W., & Upshur, C. C. (1992). Development of infants with disabilities and their families. *Monographs of the Society for Research in Child Development, 57* (6, Serial No. 230).

Shore, C. (1986). Combinatorial play, conceptual development, and early multiword speech. *Developmental Psychology, 22,* 184–190.

Shorter, E. (1975). *The making of the modern family.* New York: Basic Books.

Shuchman, M., & Wilkes, M. S. (1990, October 7). Dramatic progress against depression. *New York Times Magazine, Part 2,* 12, 30ff.

Shute, N. (1997, August 18–25). Why do we age? *U.S. News & World Report, 123,* 55–57.

Siegel, R. K. (1981, January). Accounting for "afterlife" experiences. *Psychology Today, 15,* 65–75.

Siegler, I. C., & Botwinick, J. (1979). A long-term longitudinal study of intellectual ability of older adults: The matter of selective subject attrition. *Journal of Gerontology, 34,* 242–245.

Sigelman, C., Maddock, A., Epstein, J., & Carpenter, W. (1993). Age differences in understandings of disease causality: AIDS, colds, and cancer. *Child Development, 64,* 272–284.

Silberstein, L. E., & Jefferies, L. C. (1996, July 18). Placental-blood banking—A new frontier in transfusion medicine. *New England Journal of Medicine, 335,* 199–201.

Silverman, P. (1983, November 14). Coping with grief—it can't be rushed. *U.S. News & World Report,* 65–68.

Silverman, W. K., LaGreca, A. M., & Wasserstein, S. (1995). What do children worry about? Worries and their relation to anxiety. *Child Development, 66,* 671–686.

Silverstein, M., Parrott, T. M., & Bengtson, V. L. (1995). Factors that predispose middle-aged sons and daughters to provide social support to older parents. *Journal of Marriage and the Family, 57,* 465–475.

Simon, S. (1991, July 15). Joint custody loses favor for increasing children's feeling of being torn apart. *Wall Street Journal,* B1, B2.

Simons, R. L., Beaman, J., Conger, R. D., & Chao, W. (1993). Stress, support, and antisocial behavior trait as determinants of emotional well-being and parenting practices among single mothers. *Journal of Marriage and the Family, 55,* 385–398.

Simons, R. L., Whitbeck, L. B., Beaman, J., & Conger, R. D. (1994). The impact of mothers' parenting, involvement by nonresidential fathers, and parental conflict on the adjustment of adolescent children. *Journal of Marriage and the Family, 56,* 356–374.

Simons, R. L., Whitbeck, L. B., Conger, R. D., & Chyi-In, W. (1991). Intergenerational transmission of harsh parenting. *Developmental Psychology, 27,* 159–171.

Simonton, D. K. (1988). Age and outstanding achievement: What do we know after a century of research? *Psychological Bulletin, 104,* 251–267.

Simonton, D. K. (1991). Emergence and realization of genius: The lives and works of 120 classical composers. *Journal of Personality and Social Psychology, 61,* 829–840.

Sinaki, M. (1996). Effect of physical activity on bone mass. *Current Opinions in Rheumatology, 8,* 376–383.

Singh, S., Forrest, J. D., & Torres, A. (1990). *Prenatal care in the United States: A state and county inventory.* New York: Alan Guttmacher Institute.

Sinkkonen, J., Anttila, R., & Siimes, M.A. (1998). Pubertal maturation and changes in self-image in early adolescent Finnish boys. *Journal of Youth & Adolescence, 27,* 209–218.

Skinner, B. F. (1957). *Verbal behavior.* New York: Appleton-Century-Crofts.

Slaby, R. G. (1994, January 5). Combating television violence. *Chronicle of Higher Education,* B1, B2.

Slater, A., Mattock, A., Brown, E., & Bremner, J. G. (1991). Form perception at birth: Cohen and Younger (1984) revisited. *Journal of Experimental Child Psychology, 51,* 395–406.

Slater, S. (1995). *The lesbian family life cycle.* New York: Free Press.

Slipp, S. (1993). *The Freudian mystique: Freud, women, and feminism.* New York: New York University Press.

Slobin, D. I. (1972, July). They learn the same way all around the world. *Psychology Today, 6,* 71–82.

Sluckin, W., Herbert, M., & Sluckin. (1983). *Maternal bonding.* Oxford, England: Blackwell.

Sluder, L. C., Kinnison, L. R., & Cates, D. (1997). Prenatal drug exposure: Meeting the challenge. *Childhood Education, 73,* 66–69.

Slusser, W., & Powers, N. G. (1997). Breastfeeding update: Immunology, nutrition, and advocacy. *Pediatrics in Review, 18,* 111–114.

Small, M. (1998). *Our babies, ourselves: How biology and culture shape the way we parent.* New York: Doubleday/Anchor.

Smart, L. S. (1992). The marital helping relationship following pregnancy loss and infant death. *Journal of Family Issues, 18,* 81–91.

Smetana, J. G. (1986). Preschool children's conceptions of sex-role transgressions. *Child Development, 57,* 862–871.

Smetana, J. G. (1995). Parenting styles and conceptions of parental authority during adolescence. *Child Development, 66,* 299–316.

Smetana, J. G., & Asquith, P. (1994). Adolescents' and parents' conceptions of parental authority and personal autonomy. *Child Development, 65,* 1147–1162.

Smiley, P. A., & Dweck, C. S. (1994). Individual differences in achievement goals among young children. *Child Development, 65,* 1723–1743.

Smith, A. D., Park, D. C., Cherry, K., & Berkovsky, K. (1990). Age differences in memory for concrete and abstract pictures. *Journal of Gerontology, 45,* P205–209.

Smith, G. A. (1998). Injuries to children in the United States related to trampolines. *Pediatrics, 101,* 406–412.

Smith, J., & Baltes, P. B. (1997). Profiles of psychological functioning in the old and oldest old. *Psychology and Aging, 12,* 458–472.

Smith, P. K. (Ed.). (1991). *The psychology of grandparenthood: An international perspective.* New York: Routledge, Chapman & Hall.

Smith, R. M., & Smith, C. W. (1981). Child rearing and single-parent fathers. *Family Relations, 30,* 411–417.

Smith, R. P. (1957). *Where did you go? Out. What did you do? Nothing.* New York: W. W. Norton.

Smith, S. (1983). *Ideas of the great psychologists.* Cambridge, MA: Harper & Row.

Smolowe, J. (1990, November 5). To grandma's house we go. *Time,* 86–90.

Snarey, J. (1993). *How fathers care for the next generation: A four decade study.* Cambridge, MA: Harvard University Press.

Snider, M., & Hasson, J. (1993, May 19). Halt urged to "futile" health care. *USA Today,* 1A.

Snow, C. E. (1977). The development of conversation between mothers and babies. *Journal of Child Language, 4,* 1–22.

Soja, N. N. (1994). Young children's concept of color and its relation to the acquisition of color words. *Child Development, 65,* 918–937.

Soken, N. H., & Pick, A. D. (1992). Intermodal perception of happy and angry expressive behaviors by seven-month-old infants. *Child Development, 63,* 787–795.

Sokolov, J. L. (1993). A local contingency analysis of the fine-tuning hypothesis. *Developmental Psychology, 29,* 1008–1023.

Soldz, S. (1988). The construction of meaning: Kegan, Piaget and Psychoanalysis. *Journal of Contemporary Psychology, 18,* 46–59.

Solomon, R. (1995). *Among ourselves: AIDS seen more frequently among older women.* Retrieved June 11, 1998, from the World Wide Web: http://www.apla.org/apla/9511/women.html

Sommer, K., Whitman, T. L., Borkowski, J. G., Schellenbach, C., Maxwell, S., & Keogh, D. (1993). Cognitive readiness and adolescent parenting. *Developmental Psychology, 29,* 389–398.

Sommers, C. H. (1994). *School girls: Young women, self-esteem and the confidence gap.* New York: Doubleday.

Sorce, J. F. (1979). The role of physiognomy in the development of racial awareness. *Journal of Genetic Psychology, 134,* 33–41.

Sorensen, E. S. (1993). *Children's stress and coping: A family perspective.* New York: Guilford Press.

Sorensen, R. C. (1973). *Adolescent sexuality in contemporary America.* New York: World.

South, S. J., & Spitze, G. (1994). Housework in marital and nonmarital households. *American Sociological Review, 59,* 327–347.

Spangler, G., & Grossmann, K. E. (1993). Biobehavioral organization in securely and insecurely attached infants. *Child Development, 64,* 1439–1450.

Sparks, P., & Durkin, K. (1987). Moral reasoning and political orientation: The context sensitivity of individual rights and democratic principles. *Journal of Personality and Social Psychology, 52,* 931–936.

Spearman, C. (1904). "General intelligence" objectively determined and measured. *American Journal of Psychology, 15,* 201–293.

Spearman, C. (1927). *The abilities of man.* New York: Macmillan.

Speicher, B. (1994). Family patterns of moral judgment during adolescence and early adulthood. *Developmental Psychology, 30,* 624–632.

Spelke, E. S., von Hofsten, C., & Kestenbaum, R. (1989). Object perception in infancy: Interaction of spatial and kinetic information for object boundaries. *Developmental Psychology, 25,* 185–196.

Spelman, E. V. (1988). *Inessential woman: Problems of exclusion in feminist thought.* Boston, MA: Beacon Press.

Spence, S. H., & McCathie, H. (1993). The stability of fears in children: A two-year prospective study: A research note. *Journal of Child Psychology and Psychiatry, 34,* 579–585.

Spencer, B., & Gillen, F. J. (1927). *The Arunta* (Vol. 1). London: Macmillan.

Spencer, M. B., & Markstrom-Adams, C. (1990). Identity processes among racial and ethnic minority children in America. *Child Development, 61,* 290–310.

Sperling, D. (1990, July 5). Summer cools off sperm. *USA Today,* 1A.

Sperry, R. W. (1993). The impact and promise of the cognitive revolution. *American Psychologist, 48,* 878–885.

Spiller & Reeves Research. (1996). *Fatigue survey results: Talk of hope at AIDS conference gives added importance to controlling AIDS-related fatigue.* Raritan, NJ: Ortho Biotech.

Spilton, D., & Lee, L. C. (1977). Some determinants of effective communication in four-year-olds. *Child Development, 48,* 968–977.

Spinillo, A. G., & Bryant, P. (1991). Children's proportional judgments: The importance of "half." *Child Development, 62,* 427–440.

Spiro, M. E. (1947). *Ifaluk: A South Sea culture.* Unpublished manuscripts, Coordinated Investigation of Micronesian Anthropology, Pacific Science Board, National Research Council, Washington, DC.

Spitz, R. A. (1946). Hospitalism: A follow-up report. *Psychoanalytic Study of the Child, 2,* 113–117.

Spitz, R. A. (1957). *No and yes: On the genesis of human communication.* Madison, CT: International Universities Press.

Spotlight on the baby milk industry. (1998). McSpotlight Organization. Retrieved July 10, 1998, from the World Wide Web: http://www.mcspotlight.org/beyond/nestle.html

Sprecher, S., & Chandak, R. (1992). Attitude about arranged marriages and dating among men and women from India. *Free Inquiry in Creative Sociology, 20,* 1–11.

Stack, S. (1990). New micro-level data on the impact of divorce on suicide,1959–1980: A test of two theories. *Journal of Marriage and the Family, 52,* 119–127.

The standards: What teachers should know. (1998). National Board for Professional Teaching Standards. Retrieved February 7, 1999, from the World Wide Web: http://www.nbpts.org

Stanley, B., & Sieber, J. E. (Eds.). (1992). *Social research on children and adolescents: Ethical issues.* Newbury Park, CA: Sage.

Stattin, H., & Magnusson, D. (1990). *Pubertal maturation in female development.* Hillsdale, NJ: Erlbaum.

Staudinger, U. M., Smith, J., & Baltes, P. B. (1992). Wisdom-related knowledge in a life review task: Age differences and the role of professional specialization. *Psychology and Aging, 7,* 271–281.

Steele, B. G., & Pollock, C. B. (1968). A psychiatric study of parents who abuse infants and small children. In R. E. Helfer & C. H. Kempe (Eds.), *The battered child.* Chicago: University of Chicago Press.

Steelman, L. C., & Powell, B. (1989). Acquiring capital for college: The constraints of family configuration.

American Sociological Review, 54, 844–855.

Stein, B. E., & Meredith, M. A. (1993). *The merging of the senses.* Cambridge, MA: MIT Press.

Stein, N. L., & Jewett, J. L. (1986). A conceptual analysis of the meaning of negative emotions: Implications for a theory of development. In C. E. Izard & P. B. Read (Eds.), *Measuring emotions in infants and children.* New York: Cambridge University Press.

Stein, S. P., Holzman, S., Karasu, T. B., & Charles, E. S. (1978). Mid-adult development and psychopathology. *American Journal of Psychiatry, 135,* 676–681.

Steinberg, L. (1988). Reciprocal relation between parent-child distance and pubertal maturation. *Developmental Psychology, 24,* 122–128.

Steinberg, L. (1995, April 16). Raising child difficult task in the '90s. *Dispatch* [Columbus, Ohio], 81.

Steinberg, L., Elmen, J. D., & Mounts, N. S. (1989). Authoritative parenting, psychosocial maturity, and academic success among adolescents. *Child Development, 60,* 1424–1436.

Steinberg, L., Lamborn, S. D., Darling, N., Mounts, N. S., & Dornbusch, S. M. (1994). Over-time changes in adjustment and competence among adolescents from authoritative, authoritarian, indulgent, and neglectful families. *Child Development, 65,* 754–770.

Steinberg, L., & Levine, A. (1997). *You and your adolescent.* New York: HarperCollins.

Steinfels, P. (1992, April 3). Bishops warn against withdrawing life supports. *New York Times,* A7.

Steinfels, P. (1993, February 14). Help for the helping hands in death. *New York Times,* E1.

Steinhauer, J. (1995, July 6). Living together without marriage or apologies. *New York Times,* A9.

Stenberg, C., Campos, J. J., & Emde, R. N. (1983). The facial expression of anger in seven-month-old infants. *Child Development, 54,* 178–184.

Stephan, C. W., & Langlois, J. H. (1984). Baby beautiful: Adult attributions of infant competence as a function of infant attractiveness. *Child Development, 55,* 576–585.

Stephens, M. A. P., & Franks, M. M. (1995). Spillover between daughters' roles as caregiver and wife: Interference or enhancement? *Journal of Gerontology: Psychological Sciences, 50B,* P9–P17.

Stephenson, J. (1985). *Death, grief, and mourning: Individual and social realities.* New York: Free Press.

Steptoe, A., Wardle, J., Fuller, R., Holte, A., Justo, J., Sanderman, R., & Wichstrom, L. (1997). Leisure-time physical exercise: Prevalence, attitudinal correlates, and behavioral correlates among young Europeans from 21 countries. *Preventive Medicine, 26,* 845–854.

Stern, D., & Eichorn, D. (Eds.). (1989). *Adolescence and work: Influences of social structure, labor markets, and culture.* Hillsdale, NJ: Erlbaum.

Stern, D. N. (1985). *The interpersonal world of the infant.* New York: Basic Books.

Stern, G. (1991, September 16). Young women insist on career equality, forcing the men in their lives to adjust. *Wall Street Journal,* B1, B3.

Sternberg, R. J. (1984). *Beyond IQ: A triarchic theory of human intelligence.* New York: Cambridge University Press.

Sternberg, R. J. (1986a, March/April). Inside intelligence. *American Scientist, 74,* 137–143.

Sternberg, R. J. (1986b). *Intelligence applied.* San Diego: Harcourt Brace Jovanovich.

Sternberg, R. J. (1988a). *The nature of creativity.* Cambridge, England: Cambridge University Press.

Sternberg, R. J. (1988b). *The triangle of love: Intimacy, passion, commitment.* New York: Basic Books.

Sternberg, R. J. (1990). *Metaphors of mind: Conceptions of the nature of intelligence.* Cambridge, England: Cambridge University Press.

Sternberg, R. J. (1997). Educating intelligence. In: R. Sternberg, (Ed.), *Intelligence, heredity, and environment.* Cambridge: University of Cambridge.

Sternberg, R. J. (1998). Principles of teaching for successful intelligence. *Educational Psychologist, 33,* 65–72.

Sternberg, R. J., & Downing, C. J. (1982). The development of higher-order reasoning in adolescence. *Child Development, 53,* 209–221.

Sternberg, R. J., & Grigorenko, E. L. (1997, July). Are cognitive styles still in style? *American Psychologist, 52,* 700–712.

Sternberg, R. J., & Hojjat, M. (Eds.). (1997). *Satisfaction in close relationships.* New York: Guilford Press.

Steuer, J., LaRue, A., Blum, J. E., & Jarvik, L. F. (1981). "Critical loss" in the eighth and ninth decades. *Journal of Gerontology, 36,* 211–213.

Stevens, C. (1996, November 21). Study: Poverty, death rates linked. *Detroit News,* A4.

Stevens, J. C., Cruz, L. A., Marks, L. E., & Lakatos, S. (1998). A multimodal assessment of sensory thresholds in aging. *Journals of Gerontology: Psychological Sciences and Social Sciences, 53,* 263–272.

Stevenson, H. W., Chen, C., & Lee, S. Y. (1993). Mathematics achievement of Chinese, Japanese, and American children: Ten years later. *Science, 259,* 53–58.

Stevenson-Hinde, J., & Shouldice, A. (1995). Maternal interactions and self-reports related to attachment classifications at 4.5 years. *Child Development, 66,* 583–596.

Stewart, D. E., & Robinson, G. E. (1997). *A clinician's guide to menopause.* Washington: American Psychiatric Press.

Stewart, R. B., Jr. (1990). *The second child: Family transitions and adjustments.* Newbury Park, CA: Sage.

Stipek, D., Recchia, S., & McClintic, S. (1992). Self-evaluation in young children. *Monographs of the Society for Research in Child Development, 57* (1, Serial No. 226).

Stipp, D. (1990, September 13). Alzheimer's is a group of disorders, not a single disease, researchers say. *Wall Street Journal,* B4.

Stocker, C., & Dunn, J. (1994). Sibling relationships in childhood and adolescence. In J. C. DeFries and R. Plomin (Eds.), *Nature and nurture during middle childhood* (pp. 214–232). Oxford, England: Blackwell.

Stone, R. (1994). Environmental estrogens stir debate. *Science, 265,* 308–310.

Stoneman, B. (1998). Beyond rocking the ages. *Demographics, 20,* 44–49.

Storfer, M. D. (1990). *Intelligence and giftedness: The contributions of heredity and early environment.* San Francisco: Jossey-Bass.

Straus, M. A., Gelles, R. J., & Steinmetz, S. K. (1980). *Behind closed doors: Violence in the American family.* Garden City, N.Y.: Doubleday.

Strauss, W., & Howe, N. (1992). *Generations: The history of America's future.* New York: Morrow.

Strauss, W., & Howe, N. (1998). *The fourth turning: An American prophecy.* New York: Broadway Books.

Streissguth, A. P., & Kanter, J. (1997). *The challenge of fetal alcohol syndrome: Overcoming secondary disabilities.* Seattle: University of Washington Press.

Stunkard, A. S. (1990). The body mass index of twins who have been reared apart. *New England Journal of Medicine, 322,* 1483–1487.

Sue, S., & Okazaki, S. (1990). Asian-American educational achievements: A

phenomenon in search of an explanation. *American Psychologist, 45,* 913–920.

Suedfeld, P., & Bluck, S. (1993). Changes in integrative complexity accompanying significant life events: Historical evidence. *Journal of Personality and Social Psychology, 64,* 124–130.

Sugarman, S. (1987). *Piaget's construction of the child's reality.* New York: Cambridge University.

Sugisawa, H., Liang, J., & Liu, X. (1994). Social networks, social support, and mortality among older people in Japan. *Journal of Gerontology: Social Sciences, 49,* S3–S13.

Suh, E., & Diener, E. (1998). Events and subjective well-being: Only recent events matter, *Journal of Personality and Social Psychology, 70,* 1091–1102.

Suicide among black youths: United States 1980–1995. (1998). *Morbidity and Mortality Weekly Reports, 47,* 193–195.

Suitor, J. J. (1991). Marital quality and satisfaction with the division of household labor across the family life cycle. *Journal of Marriage and the Family, 53,* 221–230.

Sullivan, H. S. (1947). *Conceptions of modern psychiatry.* Washington, DC: William A. White Psychiatric Foundation.

Sullivan, H. S. (1953). *The interpersonal theory of psychiatry.* New York: Norton.

Sullivan, M. (1998, February 24). Study finds U.S. high school seniors lag behind global peers in math and science. Boston College. Retrieved December 10, 1998, from the World Wide Web: http://www. bc.edu/

Sullivan, O. (1997). The division of housework among "remarried" couples. *Journal of Family Issues, 18,* 205–223.

Svanborg, A. (1993). A medical-social intervention in a 70-year-old Swedish population: Is it possible to postpone functional decline in aging? *Journals of Gerontology, 48* (Special Issue), 84–88.

Swan, S. H., Elkin, E. P., & Fenster, L. (1997). Have sperm densities declined? A reanalysis of global trend data. *Environmental Health Perspective, 105,* 1228–1232.

Swensen, C. H., & Trahaug, G. (1985). Commitment and the long-term relationship. *Journal of Marriage and the Family, 47,* 939–945.

Swim, J. K., Aikin, K. J., Hall, W. S., & Hunter, B. A. (1995). Sexism and racism: Old-fashioned and modern prejudices. *Journal of Personality and Social Psychology, 68,* 199–214.

Szinovacz, M. E. (1998). Grandparents today: A demographic profile. *Gerontologist, 38,* 37–52.

T

Tajani, E., & Ianniruberto, A. (1990). The uncovering of fetal competence. In M. Papani, A. Pasquinelli, & E. A. Gidoni, (Eds.), *Development handicap and rehabilitation: Practice and theory* (pp. 3–8). Amsterdam: Elsevier Science.

Takanishi, R. (1993). The opportunities of adolescence: Research, interventions, and policy. *American Psychologist, 48,* 85–87.

Tannen, D. (1994). *Talking from 9 to 5.* New York: William Morrow.

Tanner, J. M. (1970). Physical growth. In P. H. Mussen (Ed.), *Carmichael's manual of child psychology* (3rd ed.). New York: Wiley.

Tanner, J. M. (1971, Fall). Twelve to sixteen: Early adolescence. *Daedalus, 100,* 4.

Tanner, J. M. (1972). Sequence, tempo, and individual variation in growth and development of boys and girls aged twelve to sixteen. In J. Kagan & R. Coles (Eds.), *Twelve to sixteen: Early adolescence.* New York: Norton.

Tanner, J. M. (1973, September). Growing up. *Scientific American, 229,* 34–43.

Tapia, J. (1998). The schooling of Puerto Ricans: Philadelphia's most impoverished community. *Anthropology and Education Quarterly, 29,* 297–323.

Tappan, M. B. (1997). Interpretive psychology: Stories, circles, and understanding lived experiences. *Journal of Social Issues, 53,* 645–656.

Task Force on Pediatric AIDS. (1989). Pediatric AIDS and human immunodeficiency virus infection. *American Psychologist, 44,* 258–264.

Tasker, F. L., & Golombok, S. (1997). *Growing up in a lesbian family: Effects on child development.* New York: Guilford Press.

Tate, D. C., Reppucci, N. D., & Mulvey, E. P. (1995). Violent juvenile delinquents: Treatment effectiveness and implications for future action. *American Psychologist, 50,* 777–781.

Taylor, J. M., Gilligan, C., & Sullivan, A. M. (1995). *Between voice and silence: Women and girls, race and relationship.* Cambridge, MA: Harvard University Press.

Taylor, M., Cartwright, B. S., & Carlson, S. M. (1993). A developmental investigation of children's imaginary companions. *Developmental Psychology, 29,* 276–285.

Taylor, M., & Gelman, S. A. (1989). Incorporating new words into the lexicon: Preliminary evidence for language hierarchies in two-year-old children. *Child Development, 60,* 625–636.

Teen Age Drug Use on the Rise, New Government Survey Reports. (1998, August 22). *New York Times,* A3, 11.

Temple, M., & Polk, K. (1986). A dynamic analysis of educational attainment. *Sociology of Education, 59,* 79–84.

Terkel, S. (1974). *Working: People talk about what they do all day and how they feel about what they do.* New York: Ballantine.

Terman, L. M., & Merrill, M. A. (1937). *Measuring intelligence.* Boston: Houghton Mifflin.

Termine, N. T., & Izard, C. E. (1988). Infants' responses to their mothers' expressions of joy and sadness. *Developmental Psychology, 24,* 223–229.

Terry, D. J. (1994). Determinants of coping: The role of stable and situational factors. *Journal of Personality and Social Psychology, 66,* 895–910.

Teti, D. M., & Ablard, K. E. (1989). Security of attachment and infant-sibling relationships: A laboratory study. *Child Development, 60,* 1519–1528.

Thelen, E. (1981). Rhythmical behavior in infancy: An ethological perspective. *Developmental Psychology, 17,* 237–257.

Thelen, E. (1986). Treadmill-elicited stepping in seven-month-old infants. *Child Development, 57,* 1498–1506.

Thelen, E. (1995). Motor development: A new synthesis. *American Psychologist, 50,* 79–95.

Thoits, P. A. (1986). Multiple identities: Examining gender and marital status differences in distress. *American Sociological Review, 51,* 259–272.

Thomas, A., & Chess, S. (1987). Roundtable: What is temperament? *Child Development, 58,* 505–529.

Thomas, A., Chess, S., & Birch, H. G. (1970, August). The origin of personality. *Scientific American, 223,* 102–109.

Thomas, A., Chess, S., Birch, H. G., Hertzig, M. E., & Korn, S. (1963). *Behavioral individuality in early childhood.* New York: New York University Press.

Thomas, C. (1996, August 31). Olajuwon says his marriage was arranged. *Houston Chronicle,* 1.

Thomas, J. L. (1986). Gender differences in satisfaction with grandparenting. *Psychology and Aging, 1,* 215–219.

Thompson, G. (1998, December 14). With obesity in children rising, more get adult type of diabetes. *New York Times,* A1.

Thompson, R. A. (1990). Vulnerability in research: A developmental perspective on research risk. *Child Development, 61,* 1–16.

Thomson, E., & Colella, U. (1992). Cohabitation and marital stability: Quality or commitment? *Journal of Marriage and the Family, 54,* 259–267.

Thorne, B. (1993). *Gender play: Girls and boys in school.* New Brunswick, NJ: Rutgers University Press.

Thornton, A., Young-DeMarco, L., & Goldscheider, F. (1993). Leaving the parental nest: The experience of a young white cohort in the 1980s. *Journal of Marriage and the Family, 55,* 216–229.

Tiedje, L. B., Wortman, C. B., Downey, G., Emmons, C., Biernat, M., & Lang, E. (1990). Women with multiple roles: Role-compatibility perceptions, satisfaction, and mental health. *Journal of Marriage and the Family, 52,* 63–72.

Tikoo, M. (1996). An exploratory study of differences in developmental concerns of middle-aged men and women in India. *Psychological Reports, 78,* 883–887.

Times Mirror Center for the People and the Press. (1990, June 12). [Press release].

Tobin-Richards, M., Boxer, A. M., & Petersen, A. C. (1983). The psychological significance of pubertal change: Sex differences in perceptions of self during early adolescence. In J. Brooks-Gunn & A. C. Petersen (Eds.), *Girls at puberty.* New York: Plenum.

Toda, S., & Fogel, A. (1993). Infant response to the still-face situation at 3 and 6 months. *Developmental Psychology, 29,* 532–538.

Tomasello, M. (1992). *First verbs: A case study of early grammatical development.* New York: Cambridge University Press.

Tomasello, M. (1995). *Joint attention as social cognition.* In C. Moore & P. Dunham (Eds.), Joint attention: Its origins and role in development. Hillsdale, NJ: Erlbaum.

Tomison, A. M. (1996, Winter). Intergenerational transmission of maltreatment. Melbourne, Australia: National Child Protection Clearinghouse. *Issues in child abuse prevention.*

Tomkins, S. S. (1986). Script theory. In J. Aronoff, R. A. Zucker, & A. I. Rabin (Eds.), *Structuring personality.* Orlando, FL: Academic Press.

Toro, P. A., Passero-Rabideau, J. M., Bellavia, C. W., Daeschler, C. V., Wall, D.D., Thomas, D. M., & Smith, S. J. (1997). Evaluating an intervention for homeless persons: Results of a field experiment. *Journal of Consulting and Clinical Psychology, 65,* 476–484.

Toro, P. A., Rickett, E. J., Wall, D. D., & Salem, D. A. (1991). Homelessness in the United States: An ecological perspective. *American Psychologist, 46,* 1208–1218.

Torrey, E. F. (1992). *Freudian fraud: The malignant effect of Freud's theory on American thought and culture.* New York: HarperCollins.

Trachtenberg, S., & Viken, R. J. (1994). Aggressive boys in the classroom: Biased attributions or shared perceptions? *Child Development, 65,* 829–835.

Travis, N. (1993). New piece in Alzheimer's puzzle. *Science, 261,* 828–829.

Treaster, J. B. (1994, July 21). Study finds more drug use but less concern about it. *New York Times,* A8.

Trevarthen, C. (1977). Descriptive analysis of infant communicative behavior. In H. R. Schaffer (Ed.), *Studies in mother-infant interaction.* London: Academic Press.

Trickett, P. K., & Susman, E. J. (1988). Parental perceptions of child-rearing practices in physically abusive and nonabusive families. *Developmental Psychology, 24,* 270–276.

Troike, M. (1992). Achieving coherence in multilingual interaction. *Discourse Processes, 15,* 183–206.

Tronick, E. Z., Morelli, G. A., & Ivey, P. K. (1992). The Efé forager infant and toddler's pattern of social relationships. *Developmental Psychology, 28,* 568–577.

Trotter, R. J. (1987, May). You've come a long way, baby. *Psychology Today, 21,* 34–45.

Tsitouras, P. D., Martin, C. E., & Harman, S. M. (1982). Relationship of serum testosterone to sexual activity in healthy elderly men. *Journal of Gerontology, 37,* 288–293.

Tudge, J. R. H., & Winterhoff, P. A. (1993). Vygotsky, Piaget, and Bandura: Perspectives on the relations between the social world and cognitive development. *Human Development, 36,* 61–81.

Tulving, E. (1968). Theoretical issues in free recall. In T. R. Dixon & D. L. Horton (Eds.), *Verbal behavior and general behavior theory.* Englewood Cliffs, NJ: Prentice Hall.

Turkle, S. (1987, April 5). Hero of the life cycle. *New York Times Book Review,* 36–37.

Turner, A. (1996). A comparison of self-initiated coping behaviors in premature and low-birthweight infants—Toddlers with and without prenatal cocaine exposure, *Dissertation Abstracts, 57(3-A),* 1009.

Turner, P. J. (1991). Relations between attachment, gender and behavior with peers in preschool. *Child Development, 62,* 1475–1488.

Turner, R. J., Wheaton, B., & Lloyd, D. A. (1995). The epidemiology of social stress. *American Sociological Review, 60,* 104–125.

U

U.N. finds teen-age girls at high risk of AIDS. (1993, July 30). *New York Times,* A5.

USA Today/CNN/Gallup. (1995, April 10). Few would choose leisure over pay. *USA Today,* 2A.

U. S. Bureau of the Census. (1994). *Household and family characteristics: March 1993.* Current Population Reports, Series P20–477. Washington, DC: U.S. Government Printing Office.

U. S. Bureau of the Census. (1997a). *Children with single parents—How they fare.* Washington, DC: U.S. Department of Commerce. Retrieved July 31, 1998, from the World Wide Web: http://www.census.gov/prod/www/abs/msp23194.html

U. S. Bureau of the Census. (1997b, March). *Profile of older Americans: Marital status and living arrangements.* Current Population Reports, PPL-90. Retrieved January 14, 1999, from the World Wide Web: http://www.aoa.dhhs.gov/aoa/stats/profile

U. S. Bureau of the Census. (1998a). *Census bureau facts for features.* United States Department of Commerce. Retrieved January 12, 1999, from the World Wide Web: http://www.census.gov/press-release/ff98-03.html

U. S. Bureau of the Census. (1998b). *Number and average monthly benefit in current-payments by type of benefit, sex, and age, June 1998.* Retrieved January 14, 1999, from the World Wide Web: ftp://ftp.ssa.gov/pub/statistics/1b3

U. S. Bureau of the Census. (1998c). *Persons over 100 years of age.* Retrieved February 12, 1999, from the World Wide Web: http://www.census.gov/prod/1/pop/p25-1130/p251130b.pdf

U. S. Bureau of the Census. (1998, October). Marital status and living arrangements: March 1998 (Update). *Current Population Reports, P20-514.* Retrieved February 22, 1999, from the World Wide Web: http://www.census.gov

U. S. Bureau of the Census. (1999). *International data base.* Retrieved January 18, 1999, from the World Wide Web: http://www.census.gov/ipc/www/idbnew.html

U. S. Bureau of Labor Statistics. (1998). *Employment status of the civilian population by sex and age.* Retrieved February 10, 1999, from the World Wide Web: http://stats.bls.gov/news.release/empsit.t01.htm

U. S. Department of Commerce. (1997). *Population profile of the United States.* Retrieved August 5, 1998, from the World Wide Web: http://www.census.gov/prod/www.bs/msp23194.html

U. S. Department of Commerce. (1998). *Cohabitation.* Current Population Reports. (October). Washington, DC: U.S. Government Printing Office.

U. S. Department of Education. (1994, September 13). [News release]. Washington, DC.

U. S. Department of Education. (1996). Eighteenth annual report to Congress on the implementation of the individuals with disabilities education act. *Digest of Education Statistics.* Office of Special Education and Rehabilitative Services. Retrieved August 10, 1998, from the World Wide Web: http://www.ed.gov/offices/OSERS

U. S. Department of Education. (1997). *The condition of education.* Office of Educational Research and Improvement, National Center for Education Statistics. Retrieved August 10, 1998, from the World Wide Web: http://www.ed.gov/NCES/pubs/ce/index.html

U. S. Department of Education. (1998). *Improving opportunities: Strategies of education for Hispanic and limited English proficient students, a response to the Hispanic dropout report.* Office of Bilingual and Minority Languages Affairs (OBEMLA). Retrieved December 10, 1998, from the World Wide Web: http://www.ed.gov/offices/OBEMLA

U. S. Department of Health and Human Services. (1997a). *Americans less likely to use nursing home care today.* Retrieved October 19, 1998, from the World Wide Web: http://www.cdc.gov/nchswww/releases/97news/97news/nurshome.htm

U. S. Department of Health and Human Services. (1997b). *Preventing teenage pregnancy: Fact sheet.* Retrieved February 10, 1999, from the World Wide Web: http://www.dhhs.gov/news/press/1997pres/970911F.html

U. S. Department of Health and Human Services. (1997c). Fertility, family planning, and women's health: New data from the 1995 National Survey of Family Growth. *Vitality and Health Statistics.* Centers for Disease Control and National Center for Health Statistics. Washington, DC: USDHHS.

U. S. Department of Health and Human Services. (1997d). *Trends in the well-being of America's children:* Office of the Assistant Secretary for Planning and Evaluation. Retrieved August 15, 1998, from the World Wide Web: http://aspe.os.dhhs.gov/hsp/97trends/intro-web.htm

U. S. Department of Health and Human Services. (1998). *Profile of older Americans: 1998.* Administration on Aging. Retrieved January 14, 1999, from the World Wide Web: http://www.aoa. dhhs.gov/aoa/stats/profile

U. S. Department of Justice. (1996). *Uniform Crime Reports: 1995 Preliminary Annual Release.* Washington, DC: Federal Bureau of Investigation.

Uchitelle, L. (1991, April 23). Why older men keep on working. *New York Times,* C2.

Udry, J. R. (1971). *The social context of marriage.* Philadelphia: Lippincott.

Udry, J. R. (1988). Biological predispositions and social control in adolescent sexual behavior. *American Sociological Review, 53,* 709–722.

Udry, J. R., & Talbert, L. M. (1988). Sex hormone effects on personality at puberty. *Journal of Personality and Social Psychology, 54,* 291–295.

Uhlenberg, P. R. (1980). Death and the family. *Journal of Family History, 5,* 313–320.

Umberson, D., & Chen, M. D. (1994). Effects of a parent's death on adult children: Relationship salience and reaction to loss. *American Sociological Review, 59,* 152–168.

Upchurch, D. M., & McCarthy, J. (1990). The timing of a first birth and high school completion. *American Sociological Review, 55,* 224–234.

Uzgiris, I. C., & Raeff, C. (1995). Play in parent-child interactions. In M. H. Bornstein (Ed.), *Handbook of parenting* (Vol. 4). Hillsdale, NJ: Erlbaum.

V

Vaillant, C. O., & Vaillant, G. E. (1993). Is the U-curve of marital satisfaction an illusion? A 40-year study of marriage. *Journal of Marriage and the Family, 55,* 230–239.

Vaillant, G. E., & Milofsky, E. (1980). Natural history of male psychological health. IX: Empirical evidence for Erikson's model of the life cycle. *American Journal of Psychiatry, 37,* 1348–1359.

Van Collie, S.-C. (1998). Moving up through mentoring. *Workforce, 77,* 36–42.

Van IJzendoorn, M. H. (1990). Developments in cross-cultural research on attachment: Some methodological notes. *Human Development, 33,* 3–9.

Van IJzendoorn, M. H., & Kroonenberg, P. M. (1988). Cross-cultural patterns of attachment: A meta-analysis of the strange situation. *Child Development, 59,* 147–156.

Van Mechelen, W., Twisk, J., Molendijk, A., Blom, B., Snel, J., & Kemper, H. C. (1996). Subject-related risk factors for sports injuries: A one-year prospective study in young adults. *Medicine and Science in Sports and Exercise,* 1171–1178.

Vandell, D. L., & Corasaniti, M. A. (1988). The relation between third graders' after school care and social, academic, and emotional functioning. *Child Development, 59,* 868–875.

Vandell, D. L., & Ramanan, J. (1991). Children of the National Longitudinal Survey of Youth: Choices in after-school care and child development. *Developmental Psychology, 27,* 637–643.

Vanden Boom, D. C. (1997). Sensitivity and attachment: Next steps for developmentalists. *Child Development, 68,* 592–594.

VandenBos, G. R. (1998). Life-span developmental perspectives on aging: An introductory overview. In I. H. Nordhus et al., (Eds.), *Clinical geropsychology.* Washington, DC: American Psychological Association.

Vander Zanden, J. W. (1987). *Social psychology* (4th ed.). New York: Random House.

Vander Zanden, J. W. (1990). *The social experience* (2nd ed.). New York: McGraw-Hill.

Vander Zanden, J. W., & Pace, A. (1984). *Educational psychology* (2nd ed.). New York: Random House.

Vandewater, E. A., & Lansford, J. E. (1998). Influences of family structure and parental conflict on children's well-being. *Family Relations, 47,* 323–330.

Vandewater, E. A., & Stewart, A. J. (1997). Women's career commitment patterns and personality development. In M. E. Lachman & J. B. James (Eds.), *Multiple paths of midlife development* (pp. 375–410). Chicago: University of Chicago Press.

Vasquez, J. A. (1998). Distinctive traits of Hispanic students. *Prevention Researcher, 5,* 1.

Vatz, R. E. (1994, July 27). Attention deficit delirium. *Wall Street Journal,* A14.

Vaughan, D. (1986). *Uncoupling: Turning points in intimate relationships.* New York: Oxford University Press.

Vaughn, B. E., Block, J. H., & Block, J. (1988). Parental agreement on child rearing during early childhood and the psychological characteristics of adolescents. *Child Development, 59,* 1020–1033.

Venkatraman, M. M. (1995). A cross-cultural study of the subjective well-being of married elderly persons in the United States and India. *Journal of Gerontology: Social Sciences, 50B,* S35–S44.

Ventura, S. J. (1995). *Births to unmarried mothers: United States, 1980–1992.* NCHS Series 21, No. 53. Washington, DC: U. S. Department of Health and Human Services.

Ventura, S. J., Martin, J. A., Mathews, T. J., & Clarke, S. C. (1994). Advance report of final natality statistics. *Monthly Vital Statistics Report, 44.* Hyattsville, MD: National Center for Health Statistics.

Verba, M. (1994). The beginnings of collaboration in peer interaction. *Human Development, 37,* 125–139.

Verbrugge, L. M. (1989). The twain meet: Empirical explanations of sex differences in health and mortality. *Journal of Health and Social Behavior, 30,* 282–304.

Verhaeghen, P., Marcoen, A., & Goossens, L. (1993). Facts and fiction about memory aging: A quantitative integration of research findings. *Journal of Gerontology: Psychological Sciences, 48,* P157–P171.

Victims of elder abuse more likely to die. (1998b). *American Medical News, 41* (30), 23.

Villa, R. F., & Jaime, A. (1993). La fe de la gente. In M. Sotomayer & A. Garcia (Eds.), *Elderly Latinos: Issues and solutions for the 21st century* (pp. 129–142). Washington, DC: National Hispanic Council on Aging.

Visible Embryo Project. (1998). *The visible embryo.* University of California at San Francisco. Retrieved from the World Wide Web: http://visembryo.ucsf.edu/week1/ week1.html

Vitale, B. M. (1986). *Free flight: Celebrating your right brain.* Rolling Hills, CA: Jalmar Press.

Vitality for life: Psychological research for productive aging. (1993). Divison 20: Adult Development and Aging, American Psychological Association. Retrieved September 15, 1998, from: http://www. iog.wayne.edu/APADIV20/vili1.htm

Vitez, M. (1999, February). Through the ages. *Life, 22,* 86–87.

Vobejda, B. (1998a, July 1). Multiple births keep growing in number: Fertility treatments, older mothers cited. *Washington Post,* A08.

Vobejda, B. (1998b, October 15). Teen pregnancy rate has fallen 14% in '90s, to a 23-year low. *Washington Post,* A06.

Volling, B. L., & Belsky, J. (1992). The contribution of mother-child and father-child relationships to the quality of sibling interaction: A longitudinal study. *Child Development, 63,* 1209–1222.

Vollmer, S. (1999). To eat or not to eat: The question of anorexia nervosa. *Family Practice Recertification, 21,* 37–64.

von Hofsten, C. (1982). Eye-hand coordination in the newborn. *Developmental Psychology, 18,* 450–461.

von Hofsten, C. (1993). Prospective control: A basic aspect of action development. *Human Development, 36,* 253–270.

Vuchinich, S., Bank, L., & Patterson, G. R. (1992). Parenting, peers, and the stability of antisocial behavior in preadolescent boys. *Developmental Psychology, 28,* 510–521.

Vygotsky, L. S. (1962). *Thought and language.* Cambridge, MA: MIT Press.

Vygotsky, L. S. (1978). *Mind in society.* Cambridge, MA: Harvard University.

Vygotsky, L. S. (1987). Thinking and speech. In R. W. Rieber & A. S. Carton (Eds.), *The collected works of Vygotsky, L. S. (Vol. 1): Problems of General Psychology.* New York: Plenum.

W

Waber, D. P., Mann, M. B., Merola, J., & Moylan, P. M. (1985). Physical maturation rate and cognitive performance in early adolescence: A longitudinal examination. *Developmental Psychology, 21,* 666–681.

Wadden, T. A., & Van Itallie, T. B. (Eds.). (1992). *Treatment of the seriously obese patient.* New York: Guilford Press.

Waite, L. J., Haggstrom, G. W., & Kanouse, D. E. (1985). The consequences of parenthood for the marital stability of young adults. *American Sociological Review, 50,* 850–857.

Waldman, S., & Caplan, L. (1994, March 21). The politics of adoption. *Newsweek,* 64–65.

Walker, E., Downey, G., & Bergman, A. (1989). The effects of parental psychopathology and maltreatment on child behavior: A test of the diathesis-stress model. *Child Development, 60,* 15–24.

Walker, L. J., de Vries, B., & Bichard, S. L. (1984). The hierarchical nature of stages of moral development. *Developmental Psychology, 20,* 960–966.

Wallace, D. B., Franklin, M. B., & Keegan, R. T. (1994). The observing eye: A century of baby diaries. *Human Development, 37,* 1–29.

Wallerstein, J. (1987). Children of divorce. *American Journal of Orthopsychiatry, 57,* 199–211.

Wallerstein, J. S., & Kelly, J. B. (1980). *Surviving the breakup: How children actually cope with divorce.* New York: Basic Books.

Wallis, C. (1994, July 18). Life on overdrive. *Time,* 42–50.

Wallman, K. (1998). *America's children: Key national Indicators of well-being.* Interagency Forum on Child and Family Statistics. Retrieved July 22, 1998, from the World Wide Web: http://www. ChildStats.gov/ac1998/HIGHLITE.HTM

Walsh, T., & Devlin, M. J. (1998a). Eating disorders: Progress and problems. *Science, 280,* 1387–1390.

Walsh, T., & Devlin, M. J. (1998b). The pharmacologic treatment of eating disorders. *Psychiatric Clinics of North America, 15,* 149–160.

Walters, J. (1991, March 11). Hospice care lets patients die in comfort of home. *USA Today,* 5D.

Walters, R. H., Leat, M., & Mezei, L. (1963). Inhibition and disinhibition of responses through empathetic learning. *Canadian Journal of Psychology, 17,* 235–243.

Walton, G. E., Bower, N. J. A., & Bower, T. G. R. (1992). Recognition of familiar faces by newborns. *Infant Behavior and Development, 15,* 265–269.

Wang, C. T. (1998). *Current trends in child abuse reporting and fatalities: The results of the 1997 annual fifty state survey.* The Center for Child Abuse Prevention Research. Working Paper No. 808. Retrieved August 17, 1998, from the World Wide Web: http://www.childabuse. org/50data97

Wareham, K. A., Lyon, M. F., Glenister, P. H., & Williams, E. D. (1987). Age related reactivation of an X-linked gene. *Nature, 327,* 725–727.

Warner, H. R., & Price, A. R. (1989). Involvement of DNA repair in cancer and aging. *Journal of Gerontology, 44,* 45–54.

Warning Signs. (1999). Youth Crisis Stabilization Program, Community Health and Counseling Services, Bangor, Maine. Retrieved January 27, 1999, from the World Wide Web: http://www.emh. org/suiwarn.htm

Wasik, B. H., Ramey, C. T., Bryant, D. M., & Sparling, J. J. (1990). A longitudinal study of two early intervention strategies: Project CARE. *Child Development, 61,* 1682–1696.

Waterman, A. S. (1993). Two conceptions of happiness: Contrasts of personal expressiveness (eudaimonia) and hedonic enjoyment. *Journal of personality and social psychology, 64,* 678–691.

Waters, E., Matas, L., & Sroufe, L. A. (1975). Infants' reactions to an approaching stranger: Description, validation, and functional significance of wariness. *Child Development, 46,* 348–356.

Waters, J., Roberts, A. R., & Morgan, K. (1997). High risk pregnancies: Teenagers, poverty, and drug abuse. *Journal of Drug Issues, 27,* 541–562.

Watts, W. D., & Ellis, A. M. (1993). Sexual abuse and drinking and drug use: Implications for prevention. *Journal of Drug Education, 23,* 183–200.

Weatherley, D. (1964). Self-perceived rate of physical maturation and personality in late adolescence. *Child Development, 35,* 1197–1210.

Weber, J., Barrett, A., Mandel, M. & Laderman, J. (1998, May 11). The new era of lifestyle drugs. *Business Week,* 92–98.

Webster-Stratton, C. (1989). The relationship of marital support, conflict, and divorce to parent perceptions, behaviors, and childhood conduct problems. *Journal of Marriage and the Family, 51,* 417–430.

Wechsler, D. (1975). Intelligence defined and undefined. *American Psychologist, 30,* 135–139.

Weigel, C., Wertlieb, D., & Feldstein, M. (1989). Percepts of control, competence, and contingency as influences on the stress-behavior symptom relation in school-age children. *Journal of Personality and Social Psychology, 56,* 456–464.

Weihe, V. R. (1991). *Perilous rivalry: When siblings become abusive.* New York: Lexington Books.

Weinberg, M. K., & Tronick, E. Z. (1994). Beyond the face: An empirical study of infant affective configurations of facial, vocal, gestural, and regulatory behaviors. *Child Development, 65,* 1503–1515.

Weinberg, M. S., Williams, C. J., & Pryor, D. W. (1994). *Dual attraction: Understanding bisexuality.* New York: Oxford University Press.

Weiner, B. (1993). On sin versus sickness: A theory of perceived responsibility and social motivation. *American Psychologist, 48,* 957–965.

Weinraub, M., Clemens, L. P., Sockloff, A., Ethridge, T., Gracely, E., & Myers, B. (1984). The development of sex role stereotypes in the third year. *Child Development, 55,* 1493–1503.

Weinstein, S. (1997, March). New attitudes toward menopause. *FDA Consumer,* 26–29.

Weiss, B. (1988). *Many lives, many masters.* New York: Simon & Schuster.

Weiss, B., Dodge, K. A., Bates, J. E., & Pettit, G. S. (1992). Some consequences of early harsh discipline: Child aggression and a maladaptive social information processing style. *Child Development, 63,* 1321–1335.

Weiss, M. J., Zelazo, P. R., & Swain, I. U. (1988). Newborn response to auditory stimulus discrepancy. *Child Development, 59,* 1530–1541.

Weiss, R. S. (1984). The impact of marital dissolution on income and consumption in single-parent households. *Journal of Marriage and the Family, 46,* 115–127.

Weiss, R. S. (1990). *Staying the course: The emotional and social lives of men who do well at work.* New York: Free Press.

Weisz, J. R., Roghbaum, F. M., & Blackburn, T. C. (1984). Standing out and standing in: The psychology of control in America and Japan. *American Psychologist, 39,* 955–969.

Weitzman, L. J. (1990, February 27). Women and children suffer most in divorce. *USA Today,* 9A.

Welford, A. T. (1977). Motor performance. In J. E. Birren & K. W. Schaie (Eds.), *Handbook of the psychology of aging.* New York: Van Nostrand.

Wellman, H. M. (1977). The early development of intentional memory behavior. *Human Development, 20,* 86–101.

Wellman, H. M. (1990). *The child's theory of mind.* Cambridge, MA: MIT Press.

Wellman, H. M., Ritter, K., & Flavell, J. H. (1975). Deliberate memory behavior in the delayed reactions of very young children. *Developmental Psychology, 11,* 780–787.

Welsch, M. C., Pennington, B. F., Ozonoff, S., Rouse, B., & McCabe, E. R. B. (1990). Neuropsychology of early-treated phenylketonuria: Specific executive function deficits. *Child Development, 61,* 1697–1713.

Wentzel, K. R., & Erdley, C. A. (1993). Strategies for making friends: Relations to social behavior and peer acceptance in early adolescence. *Developmental Psychology, 29,* 819–826.

Wenz-Gross, M., & Siperstein, G. N. (1998). Students with learning problems at risk in middle school: Stress, social support, and adjustment. *Exceptional Children, 65,* 91–100.

Werker, J. F., & Stager, C. L. (1997, July 24). Infants listen for more phonetic detail in speech perception than in word-learning tasks. *Nature, 388,* 381–382.

Werner, E. E. (1989). High-risk children in young adulthood: A longitudinal study from birth to 32 years. *American Journal of Orthopsychiatry, 59,* 72–81.

Werner, E. E. (1990). Protective factors and individual resilience. In S. J. Meisel & J. Shonkoff (Eds.), *Handbook of early childhood intervention.* New York: Cambridge University Press.

Werner, L. A., Marean, G. C., Halpin, C. F., Spetner, N. B., & Gillenwater, J. M. (1992). Infant auditory temporal acuity: Gap detection. *Child Development, 63,* 260–272.

West, J. (1995). Child care and early education program participation of infants, toddlers, and preschoolers: Statistics in brief. National Center for Education Statistics: Washington, DC.

Westervelt, K., & Vandenberg, B. (1997). Parental divorce and intimate relationships of young adults. *Psychological Reports, 80,* 923–926.

Westie, F. R. (1964). Race and ethnic relations. In R. E. L. Faris (Ed.), *Handbook of modern sociology.* Chicago: Rand McNally.

Wexler, K., & Culicover, P. W. (1981). *Formal principles of language acquisition.* Cambridge, MA: MIT Press.

Whisman, M. A., & Kwon, P. (1993). Life stress and dysphoria: The role of self-esteem and hopelessness. *Journal of Personality and Social Psychology, 65,* 1054–1060.

Whitbeck, L. B., Hoyt, D. R., Simons, R. L., Conger, R. D., Elder, G. H., Jr., Lorenz, F. O., & Huck, S. (1992). Intergenerational continuity of parental rejection and depressed affect. *Journal of Personality and Social Psychology, 63,* 1036–1045.

Whitbeck, L. B., Simons, R. L., & Kao, M. Y. (1994). The effects of divorced mothers' dating behaviors and sexual attitudes on the sexual attitudes and behaviors of their adolescent children. *Journal of Marriage and the Family, 56,* 615–621.

White, B. L. (1969). Child development research: An edifice without a foundation. *Merrill-Palmer Quarterly, 15,* 49–79.

White, B. L. (1973). Discussions and conclusions. In B. L. White & J. C. Watts (Eds.), *Experience and environment.* Englewood Cliffs, NJ: Prentice Hall.

White, B. L. (1975). *The first three years of life.* Englewood Cliffs, NJ: Prentice Hall.

White, B. L., & Watts, J. C. (Eds.). (1973). *Experience and environment.* Englewood Cliffs, NJ: Prentice Hall.

White, L. (1994). Growing up with single parents and stepparents: Long-term effects on family solidarity. *Journal of Marriage and the Family, 56,* 935–948.

White, L. A. (1949). *The science of culture: A study of man and civilization.* New York: Farrar, Straus.

White, L. K., & Booth, A. (1985). The quality and stability of remarriages: The role of stepchildren. *American Sociological Review, 50,* 689–698.

White, N., & Cunningham, W. R. (1988). Is terminal drop pervasive or specific? *Journal of Gerontology, 43,* P141–P144.

Whiting, B. B., & Edwards, C. P. (1988). *Children of different worlds: The formation of social behavior.* Cambridge, MA: Harvard University Press.

Whorf, B. L. (1956). *Language, thought, and reality.* Cambridge, MA: MIT Press.

Whyte, M. K. (1990). *Dating, mating, and marriage.* New York: Aldine de Gruyter.

Wideman, M. V., & Singer, J. E. (1984). The role of psychological mechanisms in preparation for childbirth. *American Psychologist, 39,* 1357–1371.

Widom, C. S. (1989a). The cycle of violence. *Science, 244,* 160–166.

Widom, C. S. (1989b). Does violence beget violence? A critical examination of the literature. *Psychological Bulletin, 106,* 3–28.

Wiebe, D. J. (1991). Hardiness and stress moderation: A test of proposed mechanisms. *Journal of Personality and Social Psychology, 60,* 89–99.

Wiederman, M. W. (1997a). Extramarital sex: Prevalence and correlates in a national survey. *Journal of Sex Research, 34,* 167–174.

Wiederman, M. W. (1997b). How do earlier rates of EMS compare to more recent findings? *Journal of Sex Research, 34,* 40–45.

Wierson, M., Long, P., & Forehand, R. (1993). Toward a new understanding of early menarche, *Adolescence, 28,* 913–924.

Wiley, D., & Bortz, W. M. (1996). Sexuality and aging—Usual and successful. *Journals of Gerontology: Series A, Biological Sciences and Medical Sciences, 51,* M142–M146.

Wiley, D. C., James, G., Jordan-Belver, C., Furney, S., Calsbeek, F., Benjamin, J., & Kathcart, T. (1996). Assessing the health behaviors of Texas college students. *College Health, 44,* 167–172.

Willems, E. P., & Alexander, J. L. (1982). The naturalistic perspective in research. In B. B. Wolman (Ed.), *Handbook of developmental psychology.* Englewood Cliffs, NJ: Prentice Hall.

Willett, J. B., Singer, J. D., & Martin, N. C. (1998). The design and analysis of longitudinal studies of development and psychopathology in context: Statistical models and methodological recommendations. *Development and Psychopathology, 10,* 395–426.

Willi, J. (1997). The significance of romantic love for marriage. *Family Process, 36,* 171–182.

Williams, C. (1995). *AIDS in post communist Russia and its successor states.* Brookfield, VT: Ashgate.

Williams, C. C. (1998). *Reasons to grow old: Elders: Explorers without maps. Aging and Spirituality.* Retrieved January 15, 1999, from the World Wide Web: http://www. asaging.org/networks/forsa/a&s102.html

Williams, D. A., LoLordo, V. M., & Overmier, J. B. (1992). A reevaluation of Rescorla's early dictums about Pavlovian conditioned inhibition. *Psychological Bulletin, 111,* 275–290.

Williams, L. (1992, February 6). Girl's self-image is mother of the woman. *New York Times,* A1, A12.

Williams, R. (1990). *A Protestant legacy: Attitudes to death and illness among older Aberdonians.* New York: Oxford University Press.

Willis, S. L., & Nesselroade, C. S. (1990). Longterm effects of fluid ability training in old-old age. *Developmental Psychology, 26,* 905–910.

Wilson, B. J., & Gottman, J. M. (1995). Marital interaction and parenting. In M. H. Bornstein (Ed.), *Handbook of parenting* (Vol. 4). Hillsdale, NJ: Erlbaum.

Wilson, J. Q. (1993). *The moral sense.* New York: Free Press.

Wilson, M., & Daly, M. (1997). Life expectancy, economic inequality, homicide, and reproductive timing in Chicago neighborhoods. *British Medical Journal, 314,* 1271–1274.

Wink, P., & Helson, R. (1993). Personality change in women and their partners. *Journal of Personality and Social Psychology, 65,* 597–605.

Winsborough, H. H., Bumpass, L. L., & Aquilino, W. S. (1991). *The death of parents and the transition to old age* (Working Paper 39). Madison: University of Wisconsin, Center for Demography and Ecology.

Winship, C. (1992). Race, poverty, and the American occupational structure. *Contemporary Sociology, 21,* 639–643.

Winslow, R. K. (1990, January 4). Nursing homes get more sick patients due to U.S. policy. *Wall Street Journal,* B3.

Winslow, R. K. (1999, January 6). Medication and psychotherapy help elderly fight depression. *Wall Street Journal,* B4.

Wintre, M. G., & Vallance, D. D. (1994). A developmental sequence in the comprehension of emotions: Intensity, multiple emotions, and valence. *Developmental Psychology, 30,* 509–514.

Wisner, K. L., & Wheeler, S. B. (1994). Prevention of recurrent postpartum onset major depression. *Hospital and Community Psychiatry, 45,* 1191–1196.

Witkin, H. A. (1964). Origins of cognitive style. In C. Sheerer (Ed.), *Cognition: Theory, research, promise.* New York: Harper & Row.

Witkin, H. A. (1975). Some implications of research on cognitive style for problems of education. In J. M. Whitehead (Ed.), *Personality and learning.* London: Hodder & Stoughton.

Wolf, A. M., Gortmaker, S. L., Cheung, L, Gray, H. M., et al. (1993). Activity, inactivity, and obesity: Racial, ethnic, and age differences among schoolgirls. *American Journal of Public Health, 83,* 1625.

Wolf, A. W., Lozoff, B., Latz, S., & Paudette, R. (1996). Parental theories in the management of young children's sleeping Japan, Italy, and the United States. In S. Harkness and C. M. Super (Eds.), *Parents' cultural belief systems* (pp. 364–384). New York: Guilford Press.

Wolf, R. (1998). Domestic elder abuse and neglect. In I. H. Nordhus et al., (Eds.), *Clinical geropsychology* (pp. 161–165). Washington: American Psychological Association.

Wolf, S., & Bruhn, J. G. (1993). *The power of clan: The influence of human relationships on heart disease.* New Brunswick, NJ: Transaction.

Wolff, M. S., Berkowitz, G. S., Forman, J. Leleiko, N., Lason, S., Godbold, J., Kabat, G., & Kase, N. (1998–2001). *Environmental exposures related to early puberty.* New York: Mount Sinai School of Medicine.

Wolff, P. H. (1966). The causes, controls, and organizations of behavior in the neonate. *Psychological Issues, 5,* 1–105.

Wolinsky, F. D., Callahan, C. M., Fitzgerald, J. F., & Johnson, R. J. (1992). The risk of nursing home placement and subsequent death among older adults. *Journal of Gerontology: Social Sciences, 47,* S173–S182.

Wolpe, D. J. (1993). *Teaching your children about God: A modern Jewish approach.* New York: Holt.

Wong, M. M., & Csikszentmihalyi, M. (1991). Affiliation motivation and daily experience: Some issues on gender differences. *Journal of Personality and Social Psychology, 60,* 154–164.

Woodward, K. L. (1994). Erik Erikson: Teaching others how to see. *America, 171,* 6–8.

Woollacott, M. H. (1993). Age-related changes in posture and movement. *Journals of Gerontology, 48* (Special issue), 56–60.

Woolley, J. D., & Wellman, H. M. (1993). Origin and truth: Young children's understanding of imaginary mental representations. *Child Development, 64,* 1–17.

Worthman, C. M. (1986). Developmental dysynchrony as normative experience: Kikuyu adolescents. In J. B. Lancaster & B. A. Hamberg (Eds.), *School-age pregnancy and parenthood: Biosocial dimensions.* Hawthorne, NY: Aldine De Gruyter.

Wright, H. (1967). *Recording and analyzing child behavior.* New York: Harper & Row.

Wright, H., & Barker, R. (1950). *Methods in psychological ecology.* Lawrence: University of Kansas, Department of Psychology.

Wright, J. C., Huston, A. C., Reitz, A. L., & Piemyat, S. (1994). Young children's perceptions of television reality: Determinants and developmental differences. *Developmental Psychology, 30,* 229–239.

Wright, J. D., & Hamilton, R. F. (1978). Work satisfaction and age: Some evidence of the "job change" hypothesis. *Social Forces, 56,* 1140–1158.

Wright, R. (1994). *The moral animal: Evolutionary psychology and everyday life.* New York: Pantheon.

Wu, L. L., & Martinson, B. C. (1993). Family structure and the risk of a premarital birth. *American Sociological Review, 58,* 210–232.

Wu, Z. (1995). The stability of cohabitation relationships: The role of children. *Journal of Marriage and the Family, 57,* 231–236.

Wu, Z., & Balakrishnan, T. R. (1994). Cohabitation after marital disruption in Canada. *Journal of Marriage and the Family, 56,* 723–734.

Wu, Z., & Penning, M. J. (1997). Marital instability after midlife. *Journal of Family Issues, 18,* 459–478.

Wulfert, E., & Biglan, A. (1994). A contextual approach to research on AIDS prevention. *Behavior Analyst, 17,* 353–363.

Wyatt, G. E., & Powell, G. J. (Eds.), (1988). *Lasting effects of child sexual abuse.* Newbury Park, CA: Sage.

Wyshak, G., & Frisch, R. E. (1982). Evidence for a secular trend in age of menarche. *New England Journal of Medicine, 306,* 1033–1035.

Y

Yamada, J. E. (1991). *Laura: A case for the modularity of language.* Cambridge, MA: MIT Press.

Yankelovich, D. (1981, April). New rules in American life. *Psychology Today, 15,* 35–91.

Yao, E. L. (1985). Adjustment needs of Asian immigrant children. *Elementary School Guidance and Counseling, 19*, 222–227.

Yarrow, L. J., MacTurk, R. H., Vietze, P. M., McCarthy, M. E., Klein, R. P., & McQuiston, S. (1984). Developmental course of parental stimulation and its relationship to mastery motivation during infancy. *Developmental Psychology, 20*, 492–503.

Yates, D. J., & Bremner, J. G. (1988). Conditions for Piagetian Stage IV search errors in a task using transparent occluders. *Infant Behavior and Development, 11*, 411–417.

Yoder, J. D., & Kahn, A. S. (1993). Working toward an inclusive psychology of women. *American Psychologist, 48*, 846–850.

Yonas, A., Granrud, C. E., & Pettersen, L. (1985). Infants' sensitivity to relative size information at distance. *Developmental Psychology, 21*, 161–167.

Yoon, K. (1992). New perspective on intrasentential code switching. *Applied Linguistics, 13*, 433–449.

Yoshikawa, K. (1995). A review of the effects of early childhood and family support programs on children's social and delinquency outcomes. *The future of children: Long-term outcomes of early childhood programs, 5*. Retrieved February 12, 1999, from the World Wide Web: http://www.futureofchildren.org/1to/index.htm

Young, K. T. (1990). American conceptions of infant development from 1955 to 1984: What the experts are telling parents. *Child Development, 61*, 17–28.

Younger, B. (1992). Developmental change in infant categorization: The perception of correlations among facial features. *Child Development, 63*, 1526–1535.

Youniss, J., & Smollar, J. (1985). *Adolescent relations with mothers, fathers, and friends.* Chicago: University of Chicago Press.

Z

Zacharias, L., Rand, W. M., & Wurtman, R. J. (1976). A prospective study of sexual development and growth in American girls: The statistics of menarche. *Obstetrical and Gynecological Survey, 31*, 325–337.

Zachary, G. P. (1995, February 9). Parents' gifts to adult children studied. *Wall Street Journal*, A2.

Zachary, G. P., & Ortega, B. (1993, March 10). Workplace revolution boosts productivity at cost of job security. *Wall Street Journal*, A1, A8.

Zachry, W. (1978). Ordinality and interdependence of representation and language development in infancy. *Child Development, 49*, 681–687.

Zahn-Waxler, C. (1990). The ABCs of morality: Affect, behavior, and cognition. *Contemporary Psychology, 35*, 25–26.

Zajonc, R. B. (1976). Family configuration and intelligence. *Science, 192*, 227–236.

Zajonc, R. B. (1986, February). Mining new gold from old research. *Psychology Today, 20*, 47–51.

Zajonc, R. B., Markus, G. B., Berbaum, M. L., Bargh, J. A., & Moreland, R. L. (1991). One justified criticism plus three flawed analyses equals two unwarranted conclusions: A reply to Retherford and Sewell. *American Sociological Review, 56*, 159–165.

Zamanian, K., Thackery, M., Starrett, R. A., Brown, L. G., Lassman, D. K., & Banchard, A. (1992). Acculturation and depression in Mexican American elderly. In T. L. Brink (Ed.), *Hispanic aged mental health* (pp. 109–121). New York: Haworth Press.

Zarit, S. H., Dolan, M., & Leitsch, S. (1998). Interventions in nursing homes and other alternative living settings. In I. H. Nordhus et al. (Eds.). *Clinical geropsychology* (pp. 329–343). Washington: American Psychological Association.

Zarit, S. H., Johansson, L., & Jarrott, S. E. (1998). Family caregiving: Stresses, social programs, and clinical interventions. In I. H. Nordhus et al. (Eds.), *Clinical geropsychology* (pp. 345–357). Washington, DC: American Psychological Association.

Zarling, C. L., Hirsch, B. J., & Landry, S. (1988). Maternal social networks and mother-infant interactions in full-term and very low birthweight, preterm infants. *Child Development, 59*, 178–185.

Zautra, A. J., Reich, J. W., & Guarnaccia, C. A. (1990). Some everyday life consequences of disability and bereavement for older adults. *Journal of Personality and Social Psychology, 59*, 550–561.

Zayas, L. H., Rojas, M., & Malgady, R. (1998). Alcohol and drug use, and depression among Hispanic men in early adulthood. *American Journal of Community Psychology, 26*, 425–438.

Zebrowitz, L. A., Olson, K., & Hoffman, K. (1993). Stability of babyfaceness and attractiveness across the life span. *Journal of Personality and Social Psychology, 64*, 453–466.

Zeller, S. (1998). Fetal abuse laws gain favor. *National Journal, 30*, 1758.

Zeman, J., Penza, S., Shipman, K., & Young, G. (1997). Preschoolers as functionalists: The impact of social context on emotion regulation. *Child Study Journal, 27*, 41–67.

Zeman, N. (1990, Summer). The new rules of courtship. *Newsweek Special Issue*, 24–27.

Zhang, S. Y. (1995). Chinese parents' influence on academic performance. New York State Association for Bilingual Education, 10, 46–53.

Zigler, E. (1970). The environmental mystique: Training the intellect versus development of the child. *Childhood Education, 46*, 402–412.

Zigler, E. (1994, Foreword). In M. Hyson, *The emotional development of young children: Building an emotion-centered curriculum.* New York: Teachers College Press.

Zigler, E., & Styfco, S. J. (1994). Head Start: Criticisms in a constructive context. *American Psychologist, 49*, 127–132.

Zill, N., & Robinson, J. (1995). The generation X difference. *American Demographics, 17*, 24–29.

Zimmer, M. H., & Zimmer, M. (1998). Socioeconomic determinants of smoking behavior during pregnancy. *Social Science Journal, 35*, 133–142.

Zirkel, S. (1992). Developing independence in a life transition: Investing the self in the concerns of the day. *Journal of Personality and Social Psychology, 62*, 506–521.

Zoba, W. M. (1997, February 3). The class of "00." *Christianity Today*, 18–24.

Zsembik, B. A., & Singer, A. (1990). The problem of defining retirement among minorities: The Mexican Americans. *Gerontologist, 30*, 749–757.

Zube, M. (1982). Changing behavior and outlook of aging men and women: Implications for marriage in the middle and later years. *Family Relations, 31*, 147–156.

Zucker, A. (1998). Treatment of the terminally ill. *Death Studies, 22*, 784–786.

Photographs

Part Openers

Part One: Bob Daemmrich/Stock Boston; Part Two: Lennart Nilsson, from BEHOLD MAN, Little Brown, 1974; Part Three: Michael Newman/Photo Edit; Part Four: Joel Gordon; Part Five: Renee Lynn/Photo Researchers; Part Six: Alexi Hay; Part Seven: James Lemass/The Picture Cube; Part Eight: Peter Beck/Stock Market; Part Nine: © Leo Westenberger/LIFE MAGAZINE; Part Ten: Bryan Kelsen/Gamma Liaison.

Chapter One

Opener: Brook Kraft/Sygma; p. 5: Wellesley Public Librarry; p. 7: Bill Aron/Photo Edit; p. 12 left: Tom Miner/The Image Works; p. 12 right: A. Lichtenstein/The Image Works; p. 13: Robert Trippett/Saba; p. 15: DPA/ The Image Works; p. 29: KenHeyman.

Chapter Two

Opener: David Wells/The Image Works; p. 38: Keystone/The Image Works; p. 42: Wellesley Public Library; p. 43: Newsweek; p. 47: Joe McNally/Wheeler Pictures; p. 50: Bill Anderson/Monkmeyer; p. 53: Marcel Schuman Company; p. 54: Thomas McAvoy/Time Life Pictures; p. 55: C.S. Perkins/Magnum; p. 62: Michael Newman/ Photo Edit; p. 63: T.K. Wanstal/The Image Works.

Chapter Three

Opener: Russel D. Curtis/Photo Researchers; p. 70: Lennart Nilsson, from HOW WAS I BORN, Dell, 1993.; p. 73: Lennart Nilsson, from A CHILD IS BORN, Dell, 1990; p. 83: Courtesy Tom Crandell; p. 87: Mouseworks; p. 88: Lennart Nilsson, from A CHILD IS BORN, Dell, 1990; p. 89: Lennart Nilsson, from A CHILD IS BORN, Dell, 1990.

Chapter Four

Opener: Elizabeth Crews/The Image Works; p. 105: H. Gaus/The Image Works; p. 114: Lennart Nilsson, from A CHILD IS BORN, Dell, 1990.; p. 115: Ken Kobre; p. 121 left: Elizabeth Crews/The Image Works; p. 121 right: Laura Dwight; p. 123: M. Algaze/The Image Works; p. 125: Ross Kinne/Comstock; p. 128: David J. Sams/Stock Boston; p. 129: Myrleen Ferguson/Photo Edit.

Chapter Five

Opener: Courtesy Tom Crandell; p. 141: Walter Salinger, courtesy Professor Anthony De Casper; p. 145: Laura Dwight; p. 147: Custom Medical Stock Photo, Inc.; p. 148: Elizabeth Crews/The Image Works; p. 149: Michael Heron/Woodfin Camp; p. 150: Myrleen Ferguson/Photo Edit; p. 153: Michael Nichols/Magnum; p. 155: Marques Chenet/Woodfin Camp and Associates; p. 157: Michael Newman/Photo Edit.

Chapter Six

Opener: Robert Brenner/Photo Edit; p. 166: Bob Daemmrich/Stock Boston; p. 168: Mahaux/The Image Bank; p. 169: Peter Hendrie/The Image Bank; p. 171: Paul Fusco/Magnum; p. 174: Carroll Izard; p. 178: Vic Bider/Photo Edit; p. 180: Keren Su/Stock Boston; p. 183: Carol Palmer/The Picture Cube; p. 188: D. Wells/The Image Works.

Chapter Seven

Opener: Conklin/Monkmeyer; p. 195: Tom Crandell; p. 197: Mary Chind; p. 199: Conklin/Monkmeyer; p. 203: John Eastcott/ Yva Momatuk; p. 204: Culver Pictures; p. 207: Paul Fusco/Magnum; p. 210: Courtesy Tom Crandell; p. 215: Julie O'Neil/The Picture Cube.

Chapter Eight

Opener: David R. Fazier; p. 255: Ellen Senise/The Image Works; p. 277: Charles Harbutt/Actuality; p. 232: John Dewaele/ Stock Boston; p. 235: Roy Kirby/Stock Boston; p. 245: Steve McCurry/Magnum; p. 247: Camilla Smith/Rainbow; p. 249: Eastcott/The Image Works; p. 253: Cynthia R. Benjamin/Picture Group; p. 255: Lee White/Natural Light Productions.

Chapter Nine

Opener: Jose L. Pelaez/The Stock Market; p. 267: Bob Daemmrich/The Image Works; p. 268: Patrick Roddy; p. 279: Ethan Hoffman; p. 283: William E. Sauro/The New York Times; p. 284: Adam Rogers, "Thinking Differently," Newsweek, 12/7/1998, p. 60; p. 285: Bob Daemmrich/Stock Boston; p. 291: Ken Graves/Jeroboam; p. 292: Alan D. Carey/Photo Researchers.

Chapter Ten

Opener: Elizabeth Crews; p. 298: David Young Wolff/Photo Edit; p. 299: Bob Daemmrich/The Image Works; p. 306: Steve Skjold; p. 310: Esbin Anderson/The Image Works; p. 311: Paul Fusco/Magnum; p. 313: Richard Hutchings; p. 316: Ariel Skelly/the Stock Market; p. 317: Patrick Reddy

Chapter Eleven

Opener: Tom and Dee Ann McCarthy; p. 332: Smith/Monkmeyer; p. 334: Wrangham/Anthro-Photo; p. 335: Jack Elness/Comstock; p. 342: Oscar Buriel/Photo Researchers; p. 347: Culver Pictures; p. 351: Charles Gupton/Stock Boston; p. 353: Bob Daemmrich Photo, Inc.

Chapter Twelve

Opener: Bob Daemmrich/Stock Boston; p. 366: Will and Deni McIntyre/Photo Researchers; p. 368: Susan Lapides/Design Conception; p. 372: Douglas Mason/ Woodfin Camp and Associates; p. 380: Michael A. Schwarz/The Image Works; p. 382: Drew Crawford/The Image Works; p. 385: Shackman/Monkmeyer; p. 386: Mark Reinstein/The Image Works

Chapter Thirteen

Opener: Frank Siteman/Stock Boston; p. 394: Costa Manos/Magnum; p. 400: L.D. Gordon/The Image Bank; p. 405: Stephen Simpson/FPG; p. 411, p. 414, p. 416: Steve Skjold Photo CD.

Chapter Fourteen

Opener: Tom Grill/Comstock; p. 433: M. Granitsas/The Image Works; p. 436: Thomas L. Kelly; p. 441: Michael Heron/Woodfin Camp and Associates; p. 442: J. Howard/ Stock Boston; p. 448: Jonathan Nourok/ Photo Edit.

Chapter Fifteen

Opener: Keith Lampher/Gamma Liaison; p. 458: Courtesy of the National Eye Institute, National Institute of Health; p. 464: Inge Morath/Magnum; p. 467: Alan Carey/The Image Works; p. 468: James D. Wilson/ Woodfin Camp and Associates; p. 471: Rob Crandall/Stock Boston; p. 478: Michael A. Keller/Stock Boston; p. 483: Joseph Sohm/Photo Researchers.

Chapter Sixteen

Opener: Michael Krasowitz/FPG; p. 497: Caroline Pallat; p. 503: Steve Skjold; p. 505: Anita Bartsch/Design Conceptions; p. 508: David Schaefer/Jeroboam; p. 511: Jeffrey Dunn/Stock Boston.

Chapter Seventeen

Opener: Roger Dollarhide/Monkmeyer; p. 531: John Running/Stock Boston; p. 533: David W. Hamilton/The Image Bank; p. 535: Kent Reno/Jeroboam; p. 542: Tom Crandell; p. 545: Susan Lapides/Design conceptions; p. 547: Bill Aron/Photo Edit.

Chapter Eighteen

Opener: Farrell Grehan/Photo Researchers; p. 556: Lynn McLaren/Photo Researchers; p. 559: Jeff Greenberg/The Picture Cube; p. 561: Tom Crandell; p. 567: Robert V. Eckert Jr/EKMC Nephenthe; p. 570: Don Mason/Stock Market; p. 571: Joseph Brignolo/The Image Bank; p. 572: Tom Crandell; p. 580: Vince Streano/The Stock Market.

Chapter Nineteen

Opener: Frank Siteman/The Picture Cube; p. 591: The Granger Collection; p. 596: Gramphix/Monkmeyer; p. 600: Irvene De Vore/Anthro-Photo; p. 602: Barbara Kirk/the Stock Market; p. 605/top: AP Photo by Denis Poroy; p. 605/bottom: Hector R. Acebes/Photo Researchers; p. 611: Mark M. Walker/The Picture Cube.

Name Index

Subject Index

Note: **Boldface** page numbers indicate defined key terms.